Amorphous Oxide Semiconductors

Wiley–SID Series in Display Technology

Series Editor: Dr. Ian Sage
Advisory Board: Paul Drzaic, Ioannis (John) Kymissis, Ray Ma, Ian Underwood, Michael Wittek, and Qun (Frank) Yan

---

Flexible Flat Panel Displays, Second Edition
Darran R. Cairns, Dirk J. Broer, and Gregory P. Crawford

Liquid Crystal Displays – Addressing Schemes and Electro-Optical Effects, Third Edition
Ernst Lueder, Peter Knoll, and Seung Hee Lee

Amorphous Oxide Semiconductors: IGZO and Related Materials for Display and Memory
Hideo Hosono and Hideya Kumomi (Eds.)

Introduction to Flat Panel Displays, Second Edition
Jiun-Haw Lee, I-Chun Cheng, Hong Hua, and Shin-Tson Wu

Flat Panel Display Manufacturing
Jun Souk, Shinji Morozumi, Fang-Chen Luo, and Ion Bita

Physics and Technology of Crystalline Oxide Semiconductor CAAC-IGZO: Application to Displays
Shunpei Yamazaki and Tetsuo Tsutsui

OLED Displays: Fundamentals and Applications, Second Edition
Takatoshi Tsujimura

Physics and Technology of Crystalline Oxide Semiconductor CAAC-IGZO: Fundamentals
Noboru Kimizuka and Shunpei Yamazaki

Physics and Technology of Crystalline Oxide Semiconductor CAAC-IGZO: Application to LSI
Shunpei Yamazaki and Masahiro Fujita

Interactive Displays: Natural Human-Interface Techniques
Achintya K. Bhowmik

Addressing Techniques of Liquid Crystal Displays
Temkar N. Ruckmongathan

Modeling and Optimization of LCD Optical Performance
Dmitry A. Yakovlev, Vladimir G. Chigrinov, and Hoi-Sing Kwok

Fundamentals of Liquid Crystal Devices, Second Edition
Deng-Ke Yang and Shin-Tson Wu

3D Displays
Ernst Lueder

Illumination, Color and Imaging: Evaluation and Optimization of Visual Displays
P. Bodrogi and T. Q. Khan

Liquid Crystal Displays: Fundamental Physics and Technology
Robert H. Chen

Transflective Liquid Crystal Displays
Zhibing Ge and Shin-Tson Wu

LCD Backlights
Shunsuke Kobayashi, Shigeo Mikoshiba, and Sungkyoo Lim (Eds.)

Mobile Displays: Technology and Applications
Achintya K. Bhowmik, Zili Li, and Philip Bos (Eds.)

Photoalignment of Liquid Crystalline Materials: Physics and Applications
Vladimir G. Chigrinov, Vladimir M. Kozenkov, and Hoi-Sing Kwok

Projection Displays, Second Edition
Mathew S. Brennesholtz and Edward H. Stupp

Introduction to Microdisplays
David Armitage, Ian Underwood, and Shin-Tson Wu

Polarization Engineering for LCD Projection
Michael G. Robinson, Jianmin Chen, and Gary D. Sharp

Digital Image Display: Algorithms and Implementation
Gheorghe Berbecel

Colour Engineering: Achieving Device Independent Colour
Phil Green and Lindsay MacDonald (Eds.)

Display Interfaces: Fundamentals and Standards
Robert L. Myers

Reflective Liquid Crystal Displays
Shin-Tson Wu and Deng-Ke Yang

Display Systems: Design and Applications
Lindsay W. MacDonald and Anthony C. Lowe (Eds.)

Amorphous Oxide Semiconductors

IGZO and Related Materials for Display and Memory

Edited by

Hideo Hosono and Hideya Kumomi
Tokyo Institute of Technology, Japan

Registered Offices
John Wiley & Sons, Inc., 111 River Street, Hoboken, NJ 07030, USA
John Wiley & Sons Ltd, The Atrium, Southern Gate, Chichester, West Sussex, PO19 8SQ, UK

Editorial Office
The Atrium, Southern Gate, Chichester, West Sussex, PO19 8SQ, UK

For details of our global editorial offices, customer services, and more information about Wiley products visit us at www.wiley.com.

Wiley also publishes its books in a variety of electronic formats and by print-on-demand. Some content that appears in standard print versions of this book may not be available in other formats.

Library of Congress Cataloging-in-Publication Data

Names: Hosono, Hideo, editor. | Kumomi, Hideya, editor.
Title: Amorphous oxide semiconductors : IGZO and related materials for display and memory / edited by Hideo Hosono and Hideya Kumomi, Tokyo Institute of Technology, Japan.
Other titles: Wiley SID series in display technology
Description: Hoboken, NJ : Wiley, 2022. | Series: Wiley-SID Series in Display Technology | Includes bibliographical references and index.
Identifiers: LCCN 2021052142 (print) | LCCN 2021052143 (ebook) | ISBN 9781119715573 (cloth) | ISBN 9781119715610 (adobe pdf) | ISBN 9781119715658 (epub)
Subjects: LCSH: Thin film transistors. | Indium gallium zinc oxide. | Amorphous semiconductors.
Classification: LCC TK7871.96.T45 A46 2022 (print) | LCC TK7871.96.T45 (ebook) | DDC 621.3815/28–dc23/eng/20211227
LC record available at https://lccn.loc.gov/2021052142
LC ebook record available at https://lccn.loc.gov/2021052143

Cover Design: Wiley
Cover Image: Courtesy of Hideya Kumomi

Set in 9.5/12.5pt STIXTwoText by Straive, Chennai, India
Printed and bound by CPI Group (UK) Ltd, Croydon, CR0 4YY

C9781119715573_110522

Contents

Preface

The Fermi level can be controlled in semiconductors via impurity doping or by applying an electric field. Amorphous materials have distinct advantages over their crystalline counterparts in terms of processability and homogeneity (i.e., ease of fabrication of large-sized homogeneous thin films at low temperatures). Amorphous materials with good controllability of their Fermi level would be highly beneficial for large-area electronics, optoelectronics, and flexible applications. These advantages are the major driving force for researching amorphous semiconductors. However, controlling the Fermi level in amorphous semiconductors is impossible because of high-density carrier traps arising from structural randomness. An exception is hydrogenated amorphous silicon (a-Si:H), which is widely used in solar cells and thin-film transistors (TFTs) for liquid crystal displays (LCDs). However, band conduction has not been attained in a-Si:H. Thus, the mobility of a-Si:H remains at $\sim$1 cm^2/Vs, which is less than that of polycrystalline Si by two orders of magnitude.

Transparent amorphous oxide semiconductors (TAOSs) are a novel class of amorphous semiconductors characterized by their ionic bonding nature. In addition to high optical transparency to visible light, their mobility is greater than that of a-Si:H by one order of magnitude and can be fabricated using conventional direct current (DC) sputtering at low temperatures. Presently, TFTs with a channel layer of In-Ga-Zn-O (IGZO) are used to drive pixels of high-definition and energy-saving LCDs in smartphones, tablets, PC monitors, and large-sized organic light-emitting diode (OLED) TVs, and they are being studied for several more applications in X-ray imagers and memory devices.

Although materials science and device physics of TAOS have rapidly advanced in the last decade, the understanding of science and technology is incomplete due to the short research history and the difference in chemical bonding between oxides and covalent-type semiconductors. Consequently, conventional processing used for a-Si:H has caused serious degradation of the resulting devices, such as plasma treatment involving hydrogen. TAOS is the only semiconducting material that can be fabricated by heating the precursor in the ambient atmosphere. Solution-derived processing has been extensively studied for using this unique feature for flexible electronics.

This monograph provides a current understanding of amorphous oxide semiconductors with high mobility and their application to electronics, especially TFTs for displays. The book presents introductory fundamentals and discussions on TFTs and processing, circuits and device simulations, applications to displays and memory devices, and new materials. The authors of each chapter are experts with distinct research achievements in their subject area, and they describe state-of-the-art information along with some fundamental prerequisites for understanding. For further study, some review articles and books are listed in the references.

Notwithstanding that the research history of TAOS and their TFT applications is rather short, they now lead the backplane of advanced displays, with applications to memory devices and imagers soon to begin. Furthermore, application to flexible electronics is expected to employ low-temperature processability. Unfortunately, no monograph on TAOSs and TAOS-TFTs has been published to date. I intend to fill the gap between rapid research progress in this field and the demand for relevant semiconductors and devices by researchers, engineers, and students. The background of this book was the 77th Fujihara Seminar held in Hakone, Japan, in October 2019. We appreciate the financial support from the Fujihara Foundation of Science and the contributors for presenting wonderful talks and active discussions.

The planning of this book was first solicited by Dr. Ian Underwood, the publication committee chair of the Society for Information Display (SID). I appreciate his guidance and patience. I acknowledge my colleagues at Tokyo Tech for organizing the seminar and editing this book. Special thanks to Professors Hideya Kumomi, Junghwan Kim, and Keisuke Ide. Finally, I dedicate this monograph to Dr. Kazunobu Tanaka, a pioneer in amorphous semiconductors. When I was a PhD candidate, I was impressed by his enthusiastic talk on photostructural change in amorphous chalcogenide. I am happy to be presented with the opportunity to publish this book on a novel class of amorphous semiconductors with excellent colleagues.

June 2021

Hideo Hosono
Tokyo Institute of Technology

Series Editor's Foreword

Within the flat-panel display industry, most technological change has occurred in an incremental and evolutionary manner, where developments in materials, processes, and precision have gradually improved products to provide the outstanding performance and capability we see today. Against this background, there have been a small number of truly revolutionary innovations; the adoption of in-plane inductor–capacitor (LC) switching modes and the introduction of organic light-emitting devices (OLEDs) provide examples in which a complete redesign of the display was needed, which in turn provided outstanding advantages in panel performance.

Among these revolutionary changes, the use of transparent oxide semiconductors in place of amorphous or polycrystalline silicon in active-matrix (AM) backplanes has been one of the most dramatic. Although oxide semiconductors have been known and studied for many decades, their use in thin-film transistor (TFT) channels was regarded as impractical before the advances in understanding and materials design that were achieved toward the end of the twentieth century. Once the underlying science was established, the pace at which these novel materials have been introduced into a wide range of commercial products is exceptional. The lead editor of this volume, Professor Hosono, may be regarded as the father of this renaissance in oxide semiconductor technology, due to his decisive contributions to the basic science, technology, and exploitation of new oxide materials. Now, oxide TFTs offer many advantages; they can be fabricated at low temperature and without costly laser annealing, while their high mobility provides routes to higher-resolution displays, faster frame rates, and higher optical power efficiency. The wide bandgap of oxide systems promotes extremely low leakage currents and allows fabrication of transparent TFT arrays. The impact of oxide backplanes on the user's experience of modern displays is also profound, including a leap forward in image quality combined in many cases with reduced power drain.

The development of this new semiconductor class has brought many difficulties, some of which have been overcome while others continue to provide challenges and opportunities to the community. Among the outstanding issues, the difficulty of obtaining high performance in p-type oxide channels is of greater importance, as oxide devices are applied to more complex circuitry for memory, logic, and processing tasks outside the display driver array. Nevertheless, the advantages of oxide devices make such "system-on-panel" integration highly attractive. In the present volume, the editors, Professors Hosono and Hideya Kumomi, have brought together a comprehensive and authoritative collection of contributions from leading scientists in the field, which cover all the important topics touching advanced oxide semiconductors—materials design, electronic properties, characterization and modeling, device design, performance and stability, systems integration, new applications, and the challenges of applying oxide components to new-generation devices on plastic and other flexible substrates. The chapters go beyond the well-established applications of oxide systems to critically examine remaining challenges, such as prospects for oxide-based complementary metal-oxide semiconductors (CMOSs) and alternative circuit architectures as well as applications of oxide to other large-area electronic applications.

The trend to use oxide backplanes in displays is set to accelerate and broaden into new application areas that will place more and more stringent demands on AM components and their peripherals. Oxide TFTs will also find increased use in nondisplay devices. As these trends develop, this book will provide a ready and invaluable source of reference for all those studying, applying, and exploiting oxide semiconductors.

Malvern, UK
2022

Ian Sage

About the Editors

Hideo Hosono is an honorary professor and the founding director of the Materials Research Center for Element Strategy (MCES) of Tokyo Institute of Technology, and a distinguished fellow of the National Institute of Materials Science (NIMS). He is well known as a pioneer of transparent oxide semiconductors, iron-based superconductors, and inorganic electride materials. His proposal of a materials design concept for transparent amorphous oxide semiconductors with high electron mobility in 1995 and his demonstration of the first thin-film transistor (TFT) of crystalline (2003) and amorphous indium–gallium–zinc oxide (IGZO) are milestones in this field. He is a recipient of the Jan Raychman Prize (Society for Information Display [SID]), the von Hippel Prize (Materials Research Society [MRS]), the James C. McGroddy Prize (American Physical Society [APS]), and the Japan Prize, and he is a fellow of the SID and the Royal Society of London. He was awarded a PhD by Tokyo Metropolitan University in 1982.

Hideya Kumomi is a specially appointed professor of the Materials Research Center for Element Strategy (MCS) of Tokyo Institute of Technology, and a program manager of the Tokodai Institute for Element Strategy. He had been in industry for a quarter of a century working on silicon-based semiconductor technologies, and then switched to oxide semiconductor technologies for thin-film transistors. Since joining academia, he has managed a national project called the Element Strategy Initiative in Japan. He received his PhD from Waseda University in 1996.

List of Contributors

Katsumi Abe
Silvaco Japan
Yokohama, Japan

Toshiaki Arai
JOLED
Kyoto, Japan

S. D. Baranovskii
Department of Physics and
Material Sciences Center
Philipps-University
Marburg, Germany
and
Department of Chemistry
University of Cologne
Cologne, Germany

Juan Paolo Soria Bermundo
Graduate School of Science and Technology
Nara Institute of Science and Technology
Ikoma, Japan

Bishal Bhattarai
Department of Physics
Missouri University of Science and Technology
Rolla, Missouri, USA

D. Bruce Buchholz
Department of Materials Science and Engineering
Northwestern University
Evanston, Illinois, USA

Florian De Roose
Sense and Actuate Technologies Department
imec
Heverlee, Belgium

Antonio Facchetti
Department of Chemistry and
Materials Research Center
Northwestern University
and
Evanston, Illinois, USA
Flexterra
Skokie, Illinois, USA

Mami N. Fujii
Graduate School of Science and Technology
Nara Institute of Science and Technology
Ikoma, Japan

F. Gebhard
Department of Physics and
Material Sciences Center
Philipps-University
Marburg, Germany

Paul Heremans
Sense and Actuate Technologies Department
imec
Heverlee, Belgium
and
Department of Electrical Engineering (ESAT)
KU Leuven
Leuven, Belgium

Hideo Hosono
Laboratory for Materials and Structures and
Materials Research Center for Element Strategy
Tokyo Institute of Technology
Yokohama, Japan
and
National Institute for Materials Research
Tsukuba, Japan

Wei Huang
Joint International Research Laboratory of
Information Display and Visualization
Key Laboratory of MEMS of Ministry of Education
School of Electronic Science and Engineering
Southeast University
Nanjing, China

Keisuke Ide
Laboratory for Materials and Structures
Tokyo Institute of Technology
Yokohama, Japan

Kenichi Iga
Laboratory for Future Interdisciplinary Research of
Science and Technology (FIRST)
Tokyo Institute of Technology
Yokohama, Japan

Keiji Ikeda
KIOXIA
Tokyo, Japan

Hye-Won Jang
Department of Advanced Materials Engineering for
Information and Electronics
Kyung Hee University
Seoul, Republic of Korea

Jin Jang
Advanced Display Research Center (ADRC)
Department of Information Display
Kyung Hee University
Seoul, Republic of Korea

Hyun-Jun Jeong
Division of Material Science and Engineering
Hanyang University
Seoul, Republic of Korea

Toshio Kamiya
Laboratory for Materials and Structures and
Materials Research Center for Element Strategy
Tokyo Institute of Technology
Yokohama, Japan

Hyeong-Rae Kim
Department of Advanced Materials Engineering for
Information and Electronics
Kyung Hee University
Seoul, Republic of Korea

Hyo-Eun Kim
Department of Advanced Materials Engineering for
Information and Electronics
Kyung Hee University
Seoul, Republic of Korea

Hyun Jae Kim
School of Electrical and Electronic Engineering
Yonsei University
Seoul, Republic of Korea

Junghwan Kim
Materials Research Center for Element Strategy
Tokyo Institute of Technology
Yokohama, Japan

Mutsumi Kimura
Ryukoku University
Kyoto, Japan
and
Nara Institute of Science and Technology
Ikoma, Japan

Masaharu Kobayashi
d.lab Systems Design Lab
School of Engineering
The University of Tokyo
Tokyo, Japan

Hideya Kumomi
Materials Research Center for Element Strategy
Tokyo Institute of Technology
Yokohama, Japan

Sol-Mi Kwak
Department of Advanced Materials Engineering for Information and Electronics
Kyung Hee University
Seoul, Republic of Korea

Suhui Lee
Advanced Display Research Center (ADRC)
Department of Information Display
Kyung Hee University
Seoul, Republic of Korea

Jiapeng Li
Department of Electronic and Computer Engineering and
State Key Laboratory of Advanced Displays and Optoelectronics Technologies
The Hong Kong University of Science and Technology
Hong Kong, China

Ao Liu
Department of Chemical Engineering
Pohang University of Science and Technology
Pohang, Republic of Korea

Tobin J. Marks
Department of Chemistry and
Materials Research Center
Northwestern University
Evanston, Illinois, USA

Julia E. Medvedeva
Department of Physics
Missouri University of Science and Technology
Rolla, Missouri, USA

K. Meerholz
Department of Chemistry
University of Cologne
Cologne, Germany

Kris Myny
Sense and Actuate Technologies Department
imec
Heverlee, Belgium
and
Department of Electrical Engineering (ESAT)
KU Leuven
Leuven, Belgium

Arokia Nathan
Darwin College
University of Cambridge
Cambridge, UK

A. V. Nenashev
Institute of Semiconductor Physics
Novosibirsk, Russia
and
Department of Physics
Novosibirsk State University
Novosibirsk, Russia

Yong-Young Noh
Department of Chemical Engineering
Pohang University of Science and Technology
Pohang, Republic of Korea

Kenji Nomura
Department of Electrical and Computer Engineering
University of California, San Diego
La Jolla, California, USA

Nikolaos Papadopoulos
Sense and Actuate Technologies Department
imec
Heverlee, Belgium

Jin-Seong Park
Division of Material Science and Engineering
Hanyang University
Seoul, Republic of Korea

Zhang Qun
Department of Materials Science
Fudan University
Shanghai, China

John Robertson
Engineering Department
Cambridge University
Cambridge, UK

Nobuyoshi Saito
KIOXIA
Tokyo, Japan

Robert A. Street
Palo Alto Research Center
Palo Alto, California, USA

Denis Striakhilev
Ignis Innovation
Waterloo, Ontario, Canada

Takanori Takahashi
Graduate School of Science and Technology
Nara Institute of Science and Technology
Ikoma, Japan

Shuenn-Jiun Tang
Ignis Innovation
Waterloo, Ontario, Canada

Yukiharu Uraoka
Graduate School of Science and Technology
Nara Institute of Science and Technology
Ikoma, Japan

John F. Wager
School of Electrical Engineering and Computer Science (EECS)
Oregon State University
Corvallis, Oregon, USA

Binghao Wang
Joint International Research Laboratory of Information Display and Visualization
Key Laboratory of MEMS of Ministry of Education
School of Electronic Science and Engineering
Southeast University
Nanjing, China
and
Department of Chemistry and
Materials Research Center
Northwestern University
Evanston, Illinois, USA

Man Wong
Department of Electronic and Computer Engineering and
State Key Laboratory of Advanced Displays and Optoelectronics Technologies
The Hong Kong University of Science and Technology
Hong Kong, China

Zhihe Xia
Department of Electronic and Computer Engineering and
State Key Laboratory of Advanced Displays and Optoelectronics Technologies
The Hong Kong University of Science and Technology
Hong Kong, China

Ji-Hee Yang
Department of Advanced Materials Engineering for Information and Electronics
Kyung Hee University
Seoul, Republic of Korea

Sung-Min Yoon
Department of Advanced Materials Engineering for Information and Electronics
Kyung Hee University
Seoul, Republic of Korea

Zhaofu Zhang
Engineering Department
Cambridge University
Cambridge, UK

Huihui Zhu
Department of Chemical Engineering
Pohang University of Science and Technology
Pohang, Republic of Korea

Part I

Introduction

1.1

Transparent Amorphous Oxide Semiconductors for Display Applications

Materials, Features, Progress, and Prospects

Hideo Hosono

Material Research Center for Element Strategy, Tokyo Institute of Technology, Tokyo, Japan
National Institute for Materials Research, Tsukuba, Japan

1.1.1 Introduction to Amorphous Semiconductors as Thin-Film Transistor (TFT) Channels

Condensed matter is classified into two categories, crystalline and amorphous materials, depending on whether a unit cell exists or not. This discrimination is performed by X-ray diffraction. Amorphous materials have several distinct advantages over crystalline materials. Large-sized and homogeneous (grain-boundary-free) thin films can be easily fabricated at low temperatures. In addition, material properties may be tuned by varying the chemical composition because there is no limitation to compound formations.

The essence of semiconductors is controllability of the Fermi level (E_F) by intentional operations such as impurity doping and biasing. If the advantage of amorphous materials and the essence of semiconductors could be merged, the resulting amorphous semiconductor should be an ideal semiconductor for giant microelectronics represented by flat-panel displays, as illustrated in Figure 1.1.1.

However, the amorphous semiconductors reported so far are far from this ideal owing to the high concentration of defects (chemical disorders) and tail states (localized states induced by structural randomness). Figure 1.1.2 shows a schematic drawing of electronic states. Since these defects and tail states work as charge-trapping sites, control of the Fermi level is generally impossible in amorphous semiconductors. As a consequence, charge transport is restricted to hopping among localized state like amorphous chalcogenides and semiconducting oxide glasses based on V_2O_5. The requirements for semiconductor thin-film transistor (TFT) channels are rather severe compared with the conventional semiconducting nature. Although these amorphous semiconductors do not work as TFT channels due to their high localized state density, amorphous hydrogenated Si (a-Si:H) is the first amorphous semiconductor in which E_F is controllable by biasing. This is the primary reason why a-Si:H has attracted much attention. A large reduction of dangling bonds giving a midgap level by passivation with hydrogen makes it possible to shift E_F to band edges, but E_F cannot exceed mobility edges. Thus, the mobility of a-Si:H TFTs remains 0.5–1 $cm^2/(Vs)$, which is lower by two orders of magnitude than that of polycrystalline Si-TFTs.

Transparent amorphous oxide semiconductors (TAOSs) based on post transition metal (PTM) oxides are the first category of amorphous semiconductors in which E_F is controllable beyond the mobility edge in the conduction band. As a result, TAOS-TFTs exhibit large mobility, $>10\ cm^2/(Vs)$, which is comparable to that in the corresponding polycrystalline thin films. What is the origin of such a favorable property of TAOSs? This is one focus of this chapter.

Amorphous Oxide Semiconductors: IGZO and Related Materials for Display and Memory, First Edition.
Edited by Hideo Hosono and Hideya Kumomi.

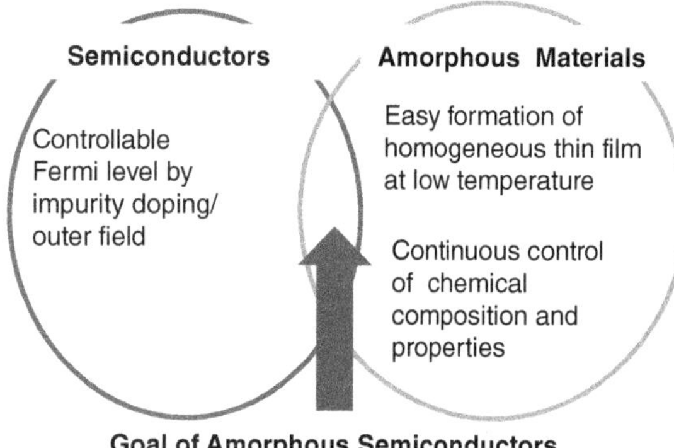

Figure 1.1.1 Ideal amorphous semiconductors.

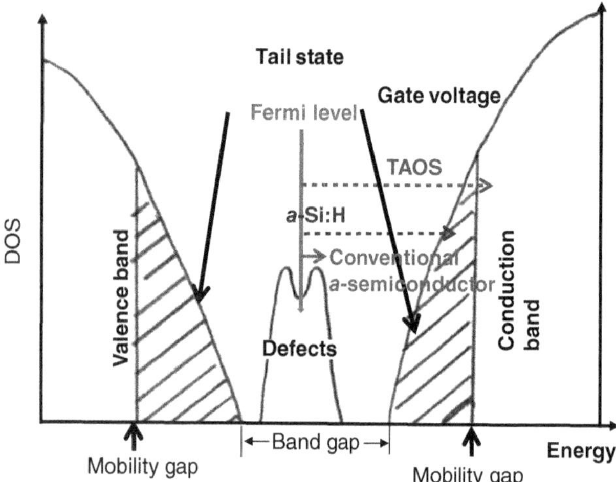

Figure 1.1.2 Electronic structure of amorphous semiconductors and material-dependent Fermi level controllability by biasing gate voltage.

1.1.2 Historical Overview

The history of oxide semiconductors is rather long. Transition metal–based oxides were traditional semiconductors, and so many papers on this topic have been published to date. However, as far as the author knows, neither of them works as the channel layer of TFTs. Transition metal cations have an open-shell structure in their *d* orbitals and give visible absorption originating from a *d*-*d* transition. Since these vacant *d* levels give large densities of states (DOSs) in the gap, it is hard to shift the E_F to significantly exceed these DOSs. This is the reason why transition metal–oxide semiconductors do not work as TFTs. As described in Section 1.1.1, requirements for semiconductors in TFT channels are much more severe than those for *p*-*n* junction formation, because the E_F is needed to shift to valence band maximum (VBM) or conduction band maximum (CBM) by gate voltage. A representative example is Cu_2O, which is well known as a *p*-type semiconductor with high Hall mobility (~100 cm^2/Vs), but its TFT has not operated well (even now) since the first attempt by William Shockley in 1949.

Figure 1.1.3 summarizes the history of oxide TFTs and their relevant TFT technology. The first TFT device structure was proposed in 1926 by Julius Lilienfeld as a patent. Oxide semiconductor TFTs have a long history comparable to that of Si metal-oxide semiconductor field-effect transistors (MOSFETs). In the 1960s, the field effects

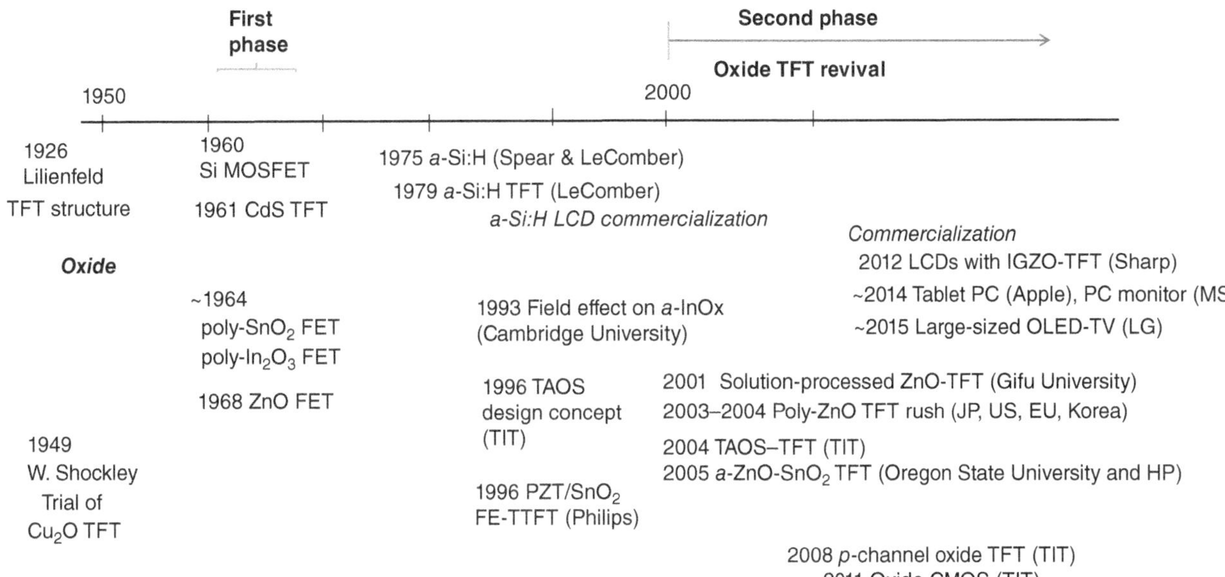

Figure 1.1.3 History of oxide TFTs and relevant technology.

on current modulation in the thin films SnO_2, In_2O_3, and ZnO, representative transparent oxide conductors, were reported, but papers on these oxide TFTs almost disappeared from open domains until circa 2000. Research on ZnO-TFTs was revisited extensively by many groups. Among them is a noteworthy paper in 2001 by Ohya et al., who reported on ZnO-TFTs that were prepared by solution process (i.e., drop-coating of $Zn(CH_3COO)_2$ solution and subsequent heating in air) [1]. This is the first report on oxide TFTs fabricated by nonvacuum processes. Since oxide semiconductors are chemically stable in an ambient atmosphere at elevated temperature, unlike conventional semiconductors, this approach utilizing this intrinsic nature of material became a milestone in the fabrication of solution-derived oxide TFTs, which is now an active subject. Many papers on ZnO-TFTs deposited by sputtering or pulsed-laser deposition were reported [2], but serious issues—such as poor reproducibility and large hysteresis, arising mainly from the complex grain-boundary nature—were pointed out.

A design concept and several examples of TAOSs with large electron mobility were proposed in 1996 [3]. As for TFTs with amorphous oxides, in 1993 Adkins et al. examined field modulation on amorphous InO_x and reported a current on/off ratio of 2–3 [4]. Such a small on/off ratio comes from high carrier concentration. Suppression and carrier concentrations and stabilization of a low carrier state in a-InO_x are still challenging even now. Good-performance AOS-TFTs were reported for IGZO [5] in 2004 and $ZnSnO_x$ [6] in 2005. Since then, a variety of AOS-TFTs have been reported to date [7].

In contrast to *n*-channel AOS-TFTs, the progress in *p*-channel has been much slower, and no satisfactory devices have been realized to date. Although a series of *p*-type transparent oxide semiconductors have been reported since 1997 [8], none of them works as a good TFT channel like Cu_2O. The formation of high-density surface defects arising from oxidation of Cu^+ is likely responsible for this. *P*-channel oxide TFTs were first realized in SnO [9] in 2008, and all complementary metal-oxide semiconductors (CMOSs) were reported [10] in 2011 utilizing the ambipolar nature of SnO. The performance of SnO-TFTs is still insufficient for practical application [11]. As for amorphous *p*-channel TFTs, very few have been reported [12] as far as the author knows. An amorphous oxide *p*-*n* junction using *p*-$ZnRhO_x$/*n*-IGZO exhibiting clear rectifying characteristics was reported [13] in 2003, but amorphous $ZnRhO_x$ did not work as a *p*-channel TFT like a-vanadium-based oxides. High midgap state density arising from a *d* orbital would hinder the smooth E_F shift by gating.

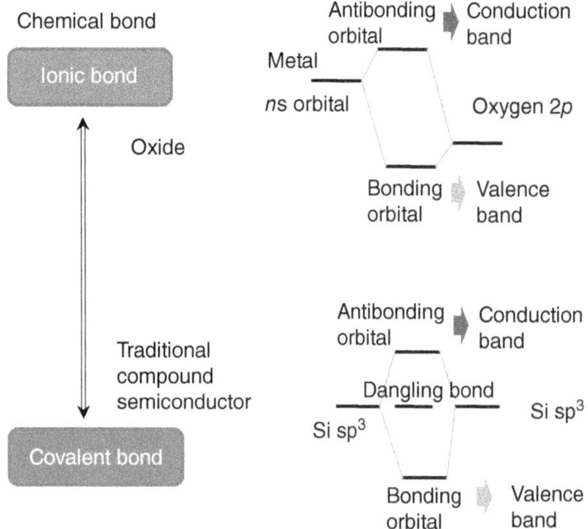

Figure 1.1.4 Comparison in chemical bonding and band formation between oxide and Si.

1.1.3 Oxide and Silicon

Oxide semiconductors made of typical metals are, in general, *n*-types except for several materials. This is a natural consequence of the chemical-bonding nature of oxides. Figure 1.1.4 shows schematic energy diagrams of ionic oxides and silicon. In ionic oxides, the nature of the CBM, which works as an electron pathway, totally differs from that of the VBM, which works as a hole pathway. The CBM in ionic oxides is primarily composed of unoccupied *s* orbitals of cations, and the contribution of oxygen 2*p* orbitals is limited. The spatial spread of this unoccupied *s* orbital is so large that direct overlap between the *s* orbitals of the neighboring cations is possible in PTM oxides; therefore, an effective mass of electrons is small in these oxides. In fact, some ionic oxides satisfy such situations with large electron mobilities of up to $\sim 100\,\text{cm}^2(\text{Vs})^{-1}$ and are called transparent conductive oxides (TCOs) represented by In_2O_3, SnO_2, ZnO, and Ga_2O_3. *N*-type material can be realized with a proper choice of metal cation, but *p*-type material is difficult because oxygen 2*p* orbitals are generally localized at the VBM.

1.1.4 Transparent Amorphous Oxide Semiconductors

1.1.4.1 Electronic Structures

What happens if these TCO materials become an amorphous state? In an amorphous state, structural disorder concentrates on an energetically weak structural unit. In most amorphous materials, structural disorder appears prominently as the bond-angle distribution. When the bond angle has a large distribution, how is the effective mass (i.e., the transfer rate between neighboring cation *s* orbitals or overlap integrals) modified for carrier electrons? We considered the two cases: (i) covalent semiconductors and (ii) ionic semiconductors. In the former case, the magnitude of the overlap between the unoccupied orbitals of the neighboring atoms is very sensitive to the variation in bond angle. As a consequence, rather deep localized states would be created at somewhat high concentrations; thereby, the drift mobility would be largely degraded.

On the other hand, the magnitude of the overlap in the latter case is critically different depending on the choice of metal cations; when the spatial spread of the *s* orbital is larger than the inter-cation distance, the magnitude should be insensitive to the bond-angle distribution because the *s* orbitals are isotropic in shape. As a consequence, we may anticipate that these ionic amorphous materials have large electron mobility comparable to that

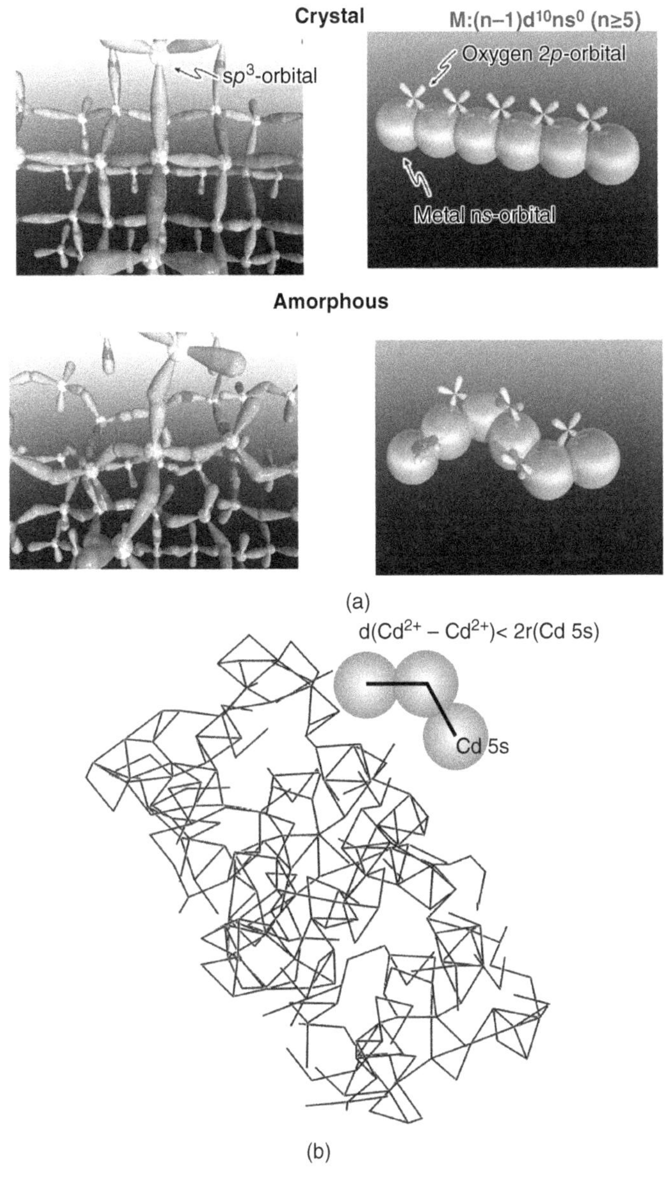

Figure 1.1.5 (a) Comparison in orbital constitution at the conduction band minimum (CBM) between covalent-type semiconductors and post transition metal oxide semiconductors, and (b) a percolated conduction network at the conduction band bottom of amorphous $2CdO{\cdot}GeO_2$.

in the corresponding crystalline phase. In the case that the spatial spread of the metal *s* orbital is small, such a favorable situation cannot be expected [3, 12]. The spatial spread of the *s* orbital of a metal cation is primarily determined by the principal quantum number (n) and is modified by the charge state of the cation, as discussed for the crystalline TCOs. Thus, candidates for high-mobility TAOSs are found in oxides of PTM cations with an electronic configuration of $(n-1)d^{10}ns^0$, where $n \geq 5$ [3, 12] (for crystalline oxide semiconductors, this requirement is relaxed to $n \geq 4$, as exemplified by ZnO with the $(3d)^{10}(4s)^0$ configuration). Figure 1.1.5a shows the difference in orbitals between Si and a PTM oxide and between crystalline and amorphous states. The drastic reduction of the electron mobility in the amorphous state from c-Si may be understood intuitively from that figure, whereas mobility in c-PTM oxides is reserved even in the amorphous state. In a sense, the situation of the CBM in PTM oxides is similar to that in amorphous metal alloys (conductivity is slightly lower than that in crystalline alloys), because metal orbitals dominantly constitute the electron pathways. This simple idea is demonstrated quantitatively by observing the DOS by inverse photoelectron spectroscopy and analyzing the computed DOS on atomic positions determined by a combination of X-ray radial distribution function with reverse Monte Carlo simulation. Figure 1.1.5b illustrates the connectivity of Cd 5*s* orbitals at the CBM of amorphous $2CdO\text{-}GeO_2$ as an example of TAOS. Here, Cd^{2+} with $(4d)^{10}(5s)^0$ meets the requirement for a PTM cation. Here, two Cd^{2+} ions were connected by a line for visualization of orbital overlap when a 2×Cd 5*s* orbital radius (Slater) is larger than the interatomic separation. It is clearly observed that the lines are 3D-connected throughout the sample, forming a percolated electron pathway [12].

1.1.4.2 Materials

Due to the requirements discussed in this chapter for high-mobility oxides, major TAOS materials need to contain In^{3+} or Sn^{4+}, each of which have a $(4d)^{10}(5s)^0$ configuration, and its amount is required to be beyond the percolation threshold of its 5*s* orbitals at the CBM. There is a distinct difference between In and Sn in oxide materials. The valence state of In is stable at +3, while Sn takes two charge states, +4 and +2, depending on its environment. Since the filled 5*s* states are located at above VBM, Sn^{2+} with a $5s^2$ configuration does not work, unlike Sn^{4+}. TAOS materials with many different compositions have been reported to date, and the materials that work as excellent TFT channels are almost restricted to In- and/or Sn-containing systems.

Here, a-In-Ga-Zn-O (a-IGZO) [3, 14] is taken as a representative TAOS material. In a-IGZO, the In^{3+} ions contribute to a large electron mobility, and thus are called "mobility enhancers." Ga with larger ionic strength (ionic charge/ionic radius) forms a stronger chemical bond with oxygen and suppresses the formation of oxygen deficiency and the generation of conduction electrons, and thus it is called a "stabilizer" or "suppressor." The role of Zn taking tetrahedral coordination is not clear but is expected to stabilize amorphous structure; this is known in glass science as a "network former." Therefore, increasing the In content increases the electron mobility but also increases residual electron density and tends to cause negative threshold voltage (V_{th}) in TFTs. Addition of a suppressor like Ga reduces the electron density (N_e) if the same deposition condition is employed, but it decreases the electron mobility. The decrease in mobility is caused by two factors: (i) reduction of the In content and (ii) reduction of electron mobility due to the small N_e, as will be explained further in this chapter. Pure In_2O_3 and ZnO do not form stable amorphous structures, even if deposited at room temperature (RT) without substrate heating. Mixing of two or more metal cations is thus necessary to stabilize the amorphous structure. TAOS materials for TFT channels are made from a combination of mobility enhancers and suppressors. The first material is a-IGZO with the nominal atomic ratio of In:Ga:Zn = 1:1:1 (called "111" composition), and also another composition of In:Ga:Zn = 2:2:1 ("221" composition) has been examined. Oregon State University (OSU) and the HP group [6] proposed Zn-Sn-O (ZTO) and Zn-In-O (ZIO) TFTs just after the first report of a-IGZO TFTs, which exhibited high $\mu_{TFT} > 50\ cm^2/(Vs)$ by annealing at 600 °C. A variety of combinations have been reported for TAOS-TFTs. TAOS materials containing In^{3+} and Sn^{4+} were also reported as ITZO (In-Sn-Zn-O). The advantages of ITZO TFTs are high ($\mu_{TFT} > 30\ cm^2(Vs)^{-1}$), and they have good robustness and selectivity against wet etching for backchannel etched TFTs. In contrast, ITZO requires high oxygen partial pressure during sputtering to suppress the electron density so as to fit to normally off TFTs, but the high P_{O2} condition deteriorates the deposition rate seriously. It is proposed that the addition of water to the sputtering atmosphere improves this issue.

TFT mobility and stability are in a trade-off relationship in many cases. A typical example is a-IZO (ZnO content 10%) material. The amorphous structure of In_2O_3 is much stabilized by the incorporation of ZnO, and μ_{Hall} does not degrade. Conventionally sputtered thin films of this TAOS material are used as amorphous TCOs (transparent metal), not semiconductors. When IZO is sputtered in highly oxidizing conditions, the thin films with low enough

N_e to use as a TFT channel layer can be fabricated. TFTs based on such a thin film exhibit high μ_{FE}, such as 50 cm^2/(Vs), and the off-current continuously increases with time and eventually loses device performance.

1.1.4.3 Characteristic Carrier Transport Properties

TAOS has several common and unique properties that are not seen in conventional amorphous semiconductors [14]. First are their large electron mobilities > 10 cm^2(Vs)$^{-1}$, which are higher by 1–2 orders of magnitude than those in a-Si:H. Second is that a degenerate state can be realized. This is totally different from the other amorphous semiconductors. For instance, c-Si is easily changed to the degenerate state by impurity doping (~10^{16} cm^{-3}), but no such state is attained in a-Si:H. That is, carrier conduction takes place by hopping through localized tail states in conventional amorphous semiconductors. This is the reason why mobility in the amorphous state is so small compared with that in the crystalline state. On the other hand, in TAOS, the E_F can exceed the mobility gap easily by carrier doping, leading to band conduction. It is considered that this striking difference originates from that in the chemical-bonding nature between the materials (i.e., strong ionic bonding with spherical potential is very favorable to forming a shallow tail state with a small DOS).

Hall effect measurements, which are a standard method in crystalline semiconductors, cannot be used for amorphous semiconductors. The reason is that the mean free path is so short, being comparable to or smaller than interatomic separation. On the contrary, TAOS materials give distinct Hall voltages, and the evaluated N_e and mobility of carriers are reliable because the mean free path is several nanometers, which is much larger than the interatomic separation. Carrier transport properties of a-IGZO are shown in Figure 1.1.6 as an example of TAOS. It is noted that mobility (μ_{Hall}) largely depends on N_e due to the presence of potential barriers arising from structural disorder. μ_{Hall} increases with increasing N_e and finally exceeds 10 cm^2(Vs)$^{-1}$ if N_e exceeds ~10^{18} cm^{-3}. The activation energy of mobility is continuously decreased with N_e, and eventually the degenerate state is realized at $N_e = 10^{19}$–10^{20}cm^{-3}. This μ_{Hall} versus N_e behavior is explained by the percolation conduction model, in which

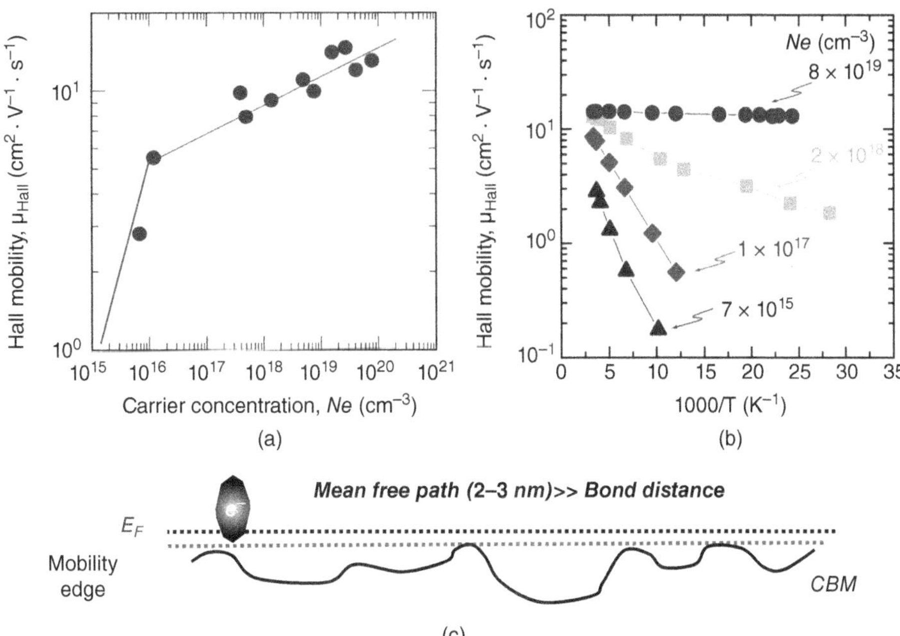

Figure 1.1.6 Electron transport (Hall mobility) in a-IGZO thin films and a schematic potential barrier at the CBM.

electron transport is controlled by distributed potential barriers above CBM (Figure 1.1.6c) [15–17]. If the TAOS is very defective or with very low N_e, hopping conduction would be dominant.

1.1.4.4 Electronic States

TFTs are devices in which source–drain current is modulated by applying a voltage to the gate insulator over several to ten orders of magnitude, as shown in Figure 1.1.7. Thus, both the density of in-gap state and the tail state are called subgap states hereafter; *n*-channel semiconductors are critical for TFT applications because these states work as carrier traps. The dominant factor of μ_{FE} is partly different from that of μ_{Hall} because μ_{FE} is expressed roughly by $\mu_{Hall}\,(N_{ind} - N_{trap}) / N_{ind}$, where N_{ind} is the total electron density induced by the gate voltage and N_{trap} is the density of the induced electrons trapped by subgap defects and tail states. Therefore, low N_{trap} is important to obtain high μ_{FE}, and the low N_{trap} is confirmed by TFT analyses, coefficient of variation (C-V) analyses, and so on, as shown in Figure 1.1.8. These studies have revealed that these localized state densities in TAOSs are

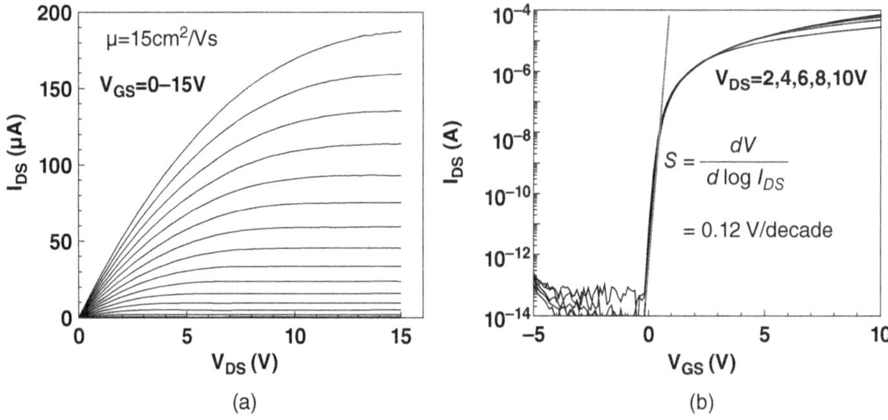

Figure 1.1.7 Typical a-IGZO-TFT characteristics. The thin film was deposited by sputtering using an IGZO (In:Ga:Zn=1:1:1) target. The device size is 30 μm (W), 50 μm (L), and 40 nm (thick). Gate insulator: thermal SiO_2 (150 nm) on Si. Contact: Ti/Au. The device was post-annealed at 300 °C for 1 h in ambient atmosphere before contact formation.

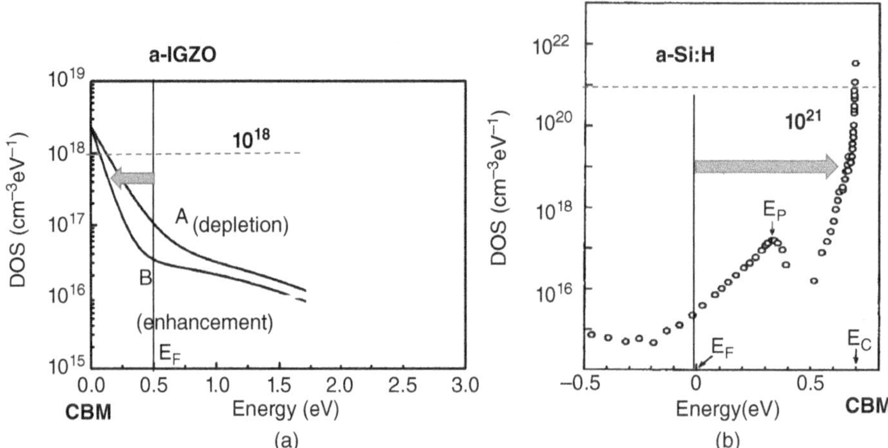

Figure 1.1.8 Tail state density in a-IGZO and a-Si:H. The former was evaluated from a simulation of TFT characteristics. The latter was taken from Ref. [40]. An arrow indicates the Fermi level shift to induce band conduction at the CBM.

2–3 orders of magnitude smaller than that of a-Si:H where E_F is close to CBM. This striking difference is the main reason why the band conduction can be induced in TAOS but not in a-Si:H by biasing gate voltage. (Bob Street [18] successfully explained why E_F cannot exceed the mobility gap.) The low N_{trap} is explained also by the electronic structure specific to the high ionicity of TAOS; that is, since Coulombic potential is independent on angle, the energy level is very insensitive to the variation in bond angle, which is the dominant randomness in amorphous materials, compared with the potential in a covalent bond. Thus, tail state density in TAOS is much lower than that in a-Si:H.

Figure 1.1.9 illustrates these electronic structures of a-IGZO [14]. Noteworthy is the presence of a large DOS (10^{19-20}cm^{-3}) above the VBM. This DOS was first observed [19] for thin films by hard X-ray photoemission spectroscopy (HAX-PES) using a synchrotron radiation facility (6–8 keV). HAX-PES has two advantages over conventional lab-PES; the first is to be bulk sensitive originating from the large escape depth of the photoelectron. This makes it possible to get reliable information from as-prepared thin films without any surface treatment such as sputtering. The other is large ionization cross-section for *s*-state electrons, which are associated with oxygen vacancy, hydrogen anions, and low valence cations with ns^2 electron configurations such as In^+ and Ga^+. Since this DOS is located above the VBM and far below the E_F, an *n*-channel TFT operation does not suffer from their influence, fortunately. However, this DOS plays a critical role in energy saving of the devices and TFT degradation under the dominant operation mode (negative bias under illumination stress [NBIS]), which is called NBIS instability. The E_F cannot push down under negative gate voltage due to the presence of this large DOS (i.e., the *p*-channel does not open when negative bias is applied). This feature makes a contrast with a-Si:H TFTs, which show ambipolar operation (i.e., the drain current is increased when negative bias is further applied to the gate). Since TFTs for LCDs dominantly stay in an off-state, the unipolarity of TAOS-TFTs leads to energy saving. This is the reason why low off-currents of IGZO-TFTs are much lower than those of a-Si:H and low-temperature processed polysilicon (LTPS)-TFTs.

This large DOS above the VBM works as the source of NBIS instability, which is induced by subgap light illumination under negative bias. The electron in the DOS is excited to the CBM by subgap illumination. The electron

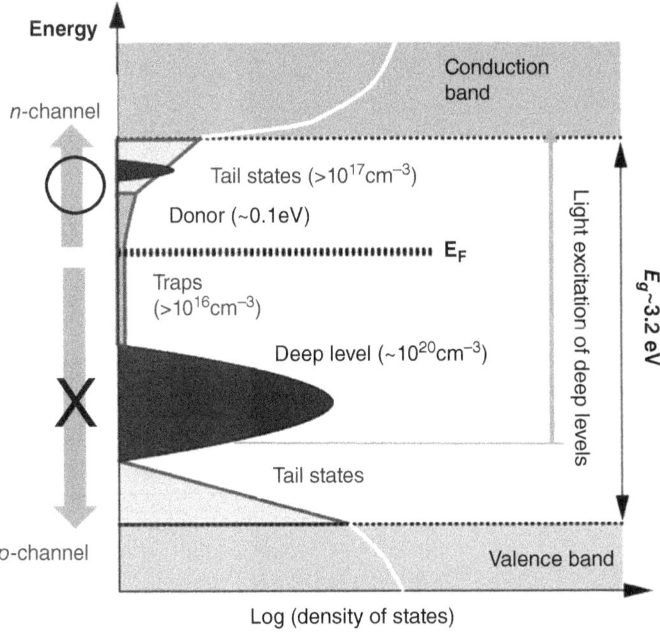

Figure 1.1.9 An experimentally clarified electronic structure of a-IGZO and TFT operation.

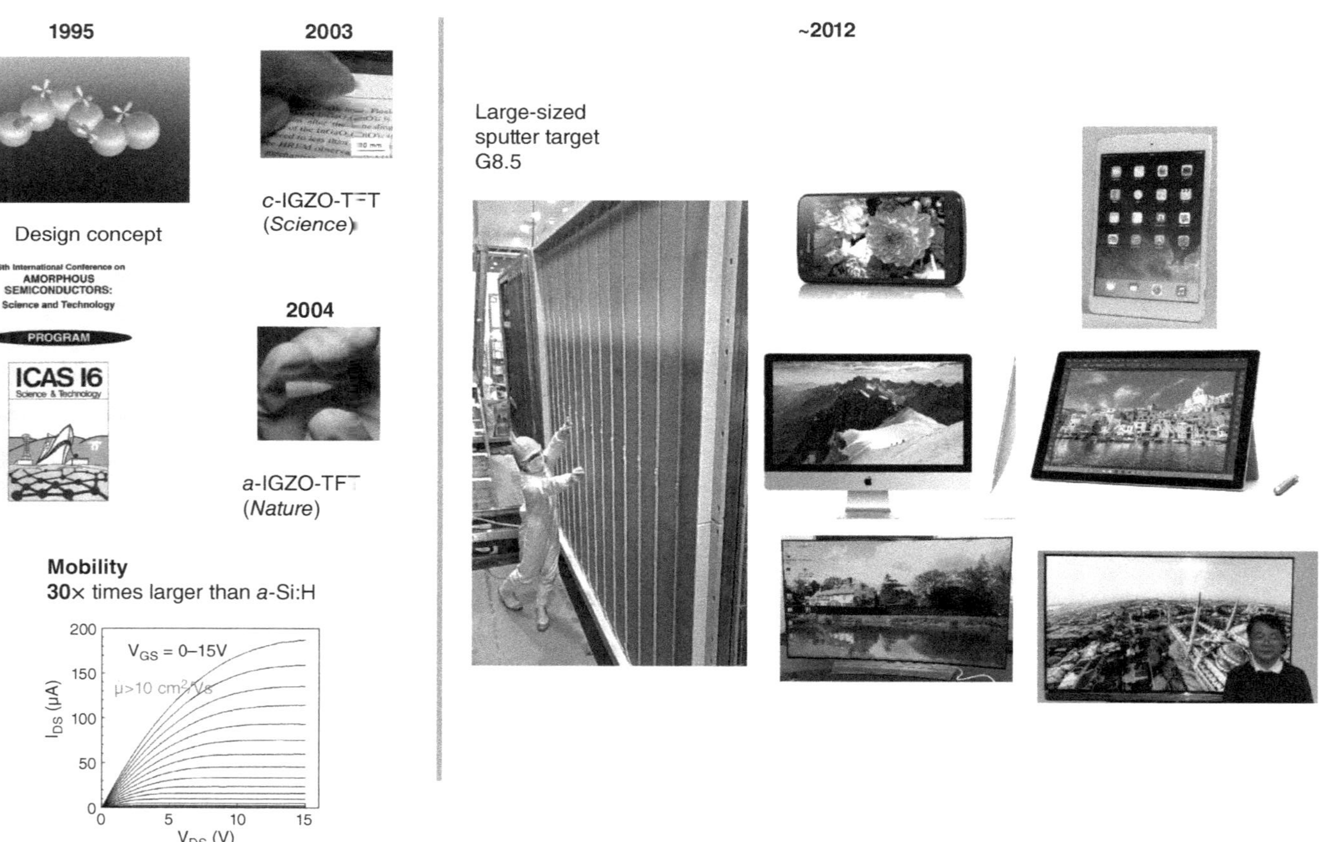

Figure 1.1.10 Progress in IGZO-TFTs and their display applications.

excited to the CBM is easily diffused away due to the high electron mobility at the CBM, while the remaining hole drifts to the interface between the channel and the gate insulator and is trapped there to form a fixed positive charge. As a result, the threshold voltage shifts to the negative voltage side. NBIS instability [20] was first reported in polycrystalline ZnO-TFTs and was found in almost all of the *n*-type oxide TFTs, irrespective of whether they were crystalline or amorphous. This suppression is an important issue for practical application of oxide-TFTs for displays because the threshold voltage shifts result in output current changes, in particular for OLEDs, which are driven by current.

Figure 1.1.10 summarizes the progress of IGZO-TFTs and their application to the backplanes of flat-panel displays in smartphones, tablet PCs, PC monitors, and OLED-TVs period. The first reported IGZO-TFTs were based on epitaxial thin films, and their mobility was ~80 $cm^2/(Vs)$ [21], which is comparable to that of poly-Si TFTs. Amorphous IGZO-TFTs fabricated on plastic substrates were published in 2004, and the mobility was ~10 $cm^2/(Vs)$ [5]. It is obvious from Section 1.1.4.2 that the ratio of Ga to In controls the mobility and stability. A higher In fraction enhances mobility while reducing the stability. A key reason why IGZO-TFTs are widely applied to displays is that since this composition forms a stable crystalline phase, large-sized and dense ceramics can be obtained easily for sputtering targets [22]. Large-sized OLED-TVs are driven by a-IGZO-TFTs. Two major features of a-IGZO-TFTs, high mobility and excellent homogeneity over a large size, are fully utilized in these products.

1.1.5 *P*-Type Oxide Semiconductors for Display Applications

N-Type transparent oxide semiconductors can be designed by selecting metal cations with spatially spread *s* orbitals that constitute the CBM and a crystalline structure with a smaller separation between metal cations, as illustrated in Figure 1.1.11a. However, no guidelines for designing *p*-type semiconductors were presented until 1997 [8]. Needless to say, *p-n* junctions are the origin of various semiconductor functions; therefore, high-quality *p*-type transparent semiconductors are essential for not only transparent oxide electronics but also all-solid dye-sensitized solar cells. For wide-gap oxides, the VBM, which serves as the conduction path of holes, is mainly composed of oxygen 2*p* orbitals, and the contribution of the orbitals of metal cations is generally small. Therefore, the VBM is little dispersed (i.e., the effective mass of holes is large), and the energy level is deep to dope holes. This is why *p*-type transparent oxide semiconductors are difficult to realize. Resolving this problem is a key strategy in realizing *p*-type transparent oxide semiconductors. Three previously proposed approaches are given in this section.

1.1.5.1 Oxides of Transition Metal Cations with an Electronic Configuration of $(n-1)d^{10}ns^0$ ($n = 4$ or 5)

Although transition metal cations have orbitals with energy levels close to those of oxygen 2*p* orbitals, most of them absorb visible light owing to a *d-d* transition [23]. Therefore, oxides of cations with closed-shell *d* orbitals, such as Cu^+ and Ag^+, are considered to have the potential to exhibit *p*-type conductivity. As shown in Figure 1.1.11b, for such oxides, the antibonding orbital component of the bond composed of metal *d* orbitals and oxygen 2*p* orbitals constitutes the VBM, and the holes doped into the VBM are delocalized to realize *p*-type conductivity. A typical example is delafossite $CuMO_2$ (M = Al^{3+}, Ga^{3+}, and In^{3+}) with dumbbell-type O–Cu–O bonds as the building block. Although several *p*-type transparent semiconductors have been found to date, no good TFT operation based on Cu^+ has been reported to date like the Cu_2O case by Shockley. The high concentration of hole traps at the surface associated with oxidation of Cu^+ would be the most plausible cause of these results.

1.1.5.2 Oxides of Metal Cations with an Electronic Configuration of ns^2

Cations with an electronic configuration of ns^2 have lone pairs similar to those of anions [9]. For the oxides of such metal cations, the VBM is mostly occupied by *s* orbitals, as shown in Figure 1.1.11c. Lone pairs occupy the

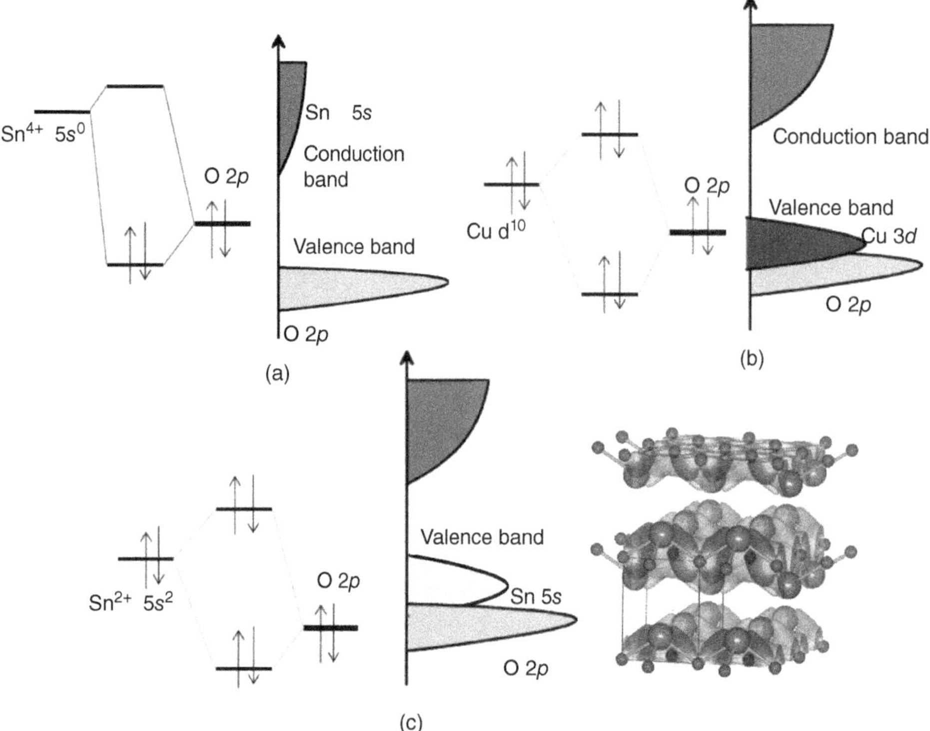

Figure 1.1.11 Schematic energy diagram: (a) *n*-type SnO_2, (b) *p*-type Cu_2O, and (c) *p*-type SnO. The right is an orbital drawing of the VBM. Sn 5*s* electrons with large spread occupy the VBM.

spatially dispersed *s* orbitals, which overlap with the *s* orbitals of adjacent cations via the oxygen, forming a largely dispersed band above the oxygen 2*p* band. Oxides of Sn^{2+} with an electronic configuration of $5s^2$ are typical and exhibit *p*-type conductivity in Hall effect measurements. In addition, the first-ever oxide TFT that can operate as a *p*-channel TFT was realized using SnO for the active layers [9].

The realization of CMOS based on an oxide semiconductor was a long-standing issue in oxide electronics. This objective was first attained in 2011 using SnO [9]. Although much improvement has been reported [11, 24], the CMOS performance is still insufficient for applications. Low-temperature processing as well as improvement of mobility are strongly required for display applications.

It has been reported that chalocogenides and oxides of Pb^{2+} and Bi^{3+}, both with an electronic configuration of $6s^2$, cannot be used to enhance hole transport properties because the energy level of 6*s* electrons is much deeper than that of the VBM. Although oxides of Sb^{3+} with an electronic configuration of $5s^2$, similar to that of Sn^{2+}, are expected to exhibit *p*-type conductivity, no examples of such oxides have been reported to date.

1.1.5.3 Oxides of Metal Cations with an Electronic Configuration of nd^6

For oxides of Rh^{3+} and Ir^{3+}, both with an electronic configuration of $4d^6$ or $5d^6$, these cations stabilize in the low-spin-state octahedral configuration, where electrons occupy the three orbitals d_{xy}, d_{yz}, and d_{zx} [25]. 4*d* and 5*d* orbitals are spatially spread, and the state in which two electrons occupy each of the three *d* orbitals is similar to that in Section 1.1.5.2, that is, a pseudo *s* orbital with a large spread. When doped with holes, these oxides are expected to exhibit *p*-type conductivity. A typical example is $ZnRh_2O_4$, which has a normal spinel structure [25]. Similar to the case of ZnO, the Zn^{2+} ions in this material are coordinated in tetrahedra, which are not continuously connected and do not exhibit *n*-type conductivity. $ZnRh_2O_4$ is the only oxide that is known to exhibit *p*-type conductivity even in the amorphous state, and it has been reported to form *p-n* diodes on plastic substrates [26] when combined with TAOS. Subsequently, *p*-type conductivity was reported for $ZnCo_2O_4$ as an extension of this series to the $3d^6$ system.

These materials work as *p*-type semiconductors for the *p-n*-junction but not as *p*-channel material in TFTs because of the high concentrations of vacant *d* levels, which serve as hole trapping in the band gap.

1.1.6 Novel Amorphous Oxide Semiconductors

Recently, two novel-type AOSs toward display applications were reported based on rather different design concepts: an amorphous electride (a-C12A7:e^-) and amorphous ZnO-SiO_2 (a-ZSO).

Electrides are materials in which electrons serve as anions [27]. This conceptionally novel material was first synthesized in organic crystals. Although these attracted attention as an exotic material, almost no information had been obtained on the physical properties because they are extremely sensitive to heat, O_2, and H_2O.

In 2003, an inorganic electride material derived from $12CaO{\cdot}7Al_2O_3$ crystal (C12A7) was reported [28]. C12A7 is a wide-gap insulator composed of densely packed, subnanometer-sized cages with a positive charge, and it is thermally stable with a melting point of 1415 °C. The unit cell includes two molecules and 12 cages that have a free inner space of ~0.4 nm in diameter and can be represented as $[Ca_{24}Al_{28}O_{64}]^{4+} + 2O^{2-}$. The former denotes the cage framework, and the latter is called “free oxygen ions” that compensate for the positive charge of the framework. These oxygen ions are loosely bound to the cages because the cage diameter is ~50% larger than the O^{2-} size (0.28 nm). They succeeded in extracting this free O^{2-} ion, regarded as the counter anion to the giant framework cation, by a chemical reduction method and injected electrons instead. Since the resulting material (hereafter, C12A7:e^-) with the composition of $[Ca_{24}Al_{28}O_{64}]^{4+}$.(4$e^-$) may be regarded as electride, this material became the first RT-stable electride.

Amorphous C12A7:e^- thin films were successfully fabricated by conventional sputtering of a C12A7:e^- ceramic target in O_2-free atmosphere, and the resulting thin films contain a high anionic electron concentration comparable to that in crystalline C12A7:e^- [29]. Anionic electrons exist in subnanometer-sized space coordinated by 2 Al^{3+} or a Ca^{2+} and an Al^{3+}. The work function of a-C12A7:e^- is ~3.0 eV, which is larger than that of c-C12A7:e^- (2.4 eV) but still rather lower than that of conventional transparent oxide semiconductors (comparable to metal Ca), and it is chemically inert and optically transparent as shown in Figure 1.1.12. There are two types of OLEDs with different stacking, normal type (cathode top) and inverted type (cathode bottom). An inverted type is better than a normal one, as the device structure emitting light can radiate upward through a cathode of transparent electrode (indium tin oxide [ITO]). However, the fabrication of an inverted structure was practically hard due to the lack of an appropriate electron injection layer (EIL) because a combination of LiF and Al does not work for this structure (deposition sequence Al followed by LiF). An inverted structure of OLEDs with comparable performance to a normal structure was realized by using a-C12A7:e^- with a low work function and high optical transparency as the EIL.

Second is a-ZSO, ZnO-SiO_2. A ZSO thin film has a low work function (≈3.5 eV), which is lower by ~1 eV than ZnO and also enables relatively higher mobility (0.3–1.0 $cm^2/(Vs)$) than those of the *n*-type organic semiconductors [30]. In addition to high visible transparency and good chemical stability, a-ZSO can form ohmic contact with various metals irrespective of the work function of partner metals. These properties almost completely meet the requirements to be electron injection/transport layers in OLEDs and perovskite LEDs. The effectiveness of this semiconductor was demonstrated by the realization of high-performance perovskite LEDs [31], as shown in Figure 1.1.13, and efficient tandem OLEDs [32]. These exceptional properties originate from a characteristic nanostructure composed of ZnO nanocrystals separated by thin amorphous zinc silicate layers with larger ionization potential and smaller electron affinity than ZnO. This nanostructure makes it possible to confine wave function in ZnO nanocrystals, which in turn widens the band gap due to the quantum size effect. Since E_F is

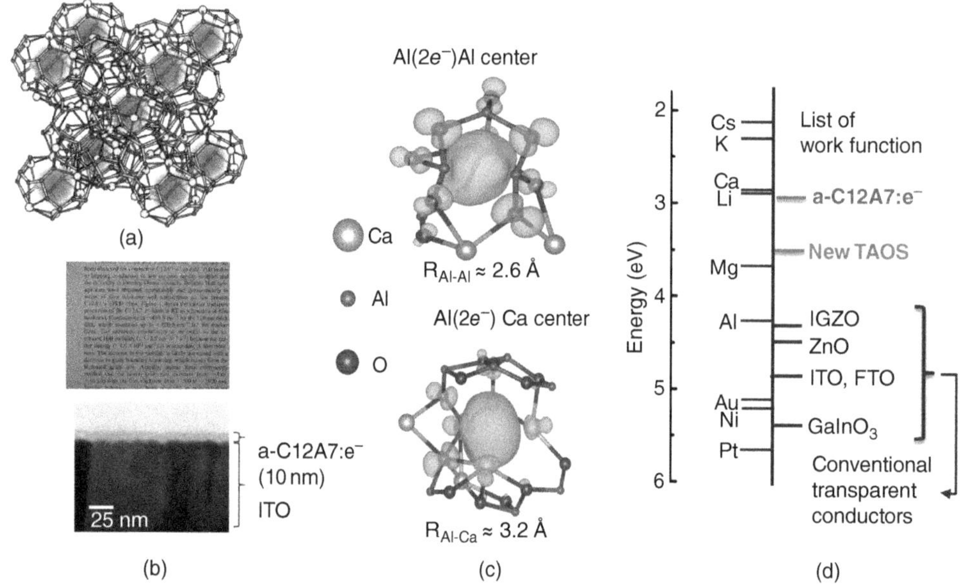

Figure 1.1.12 C12A7 electride. (a) c-C12A7:e$^-$, (b) photos of sputter-deposited a-C12A7:e$^-$, (c) two anionic electron sites in a-C12A7:e$^-$, and (d) the work function of C12A7:e$^-$ and relevant materials.

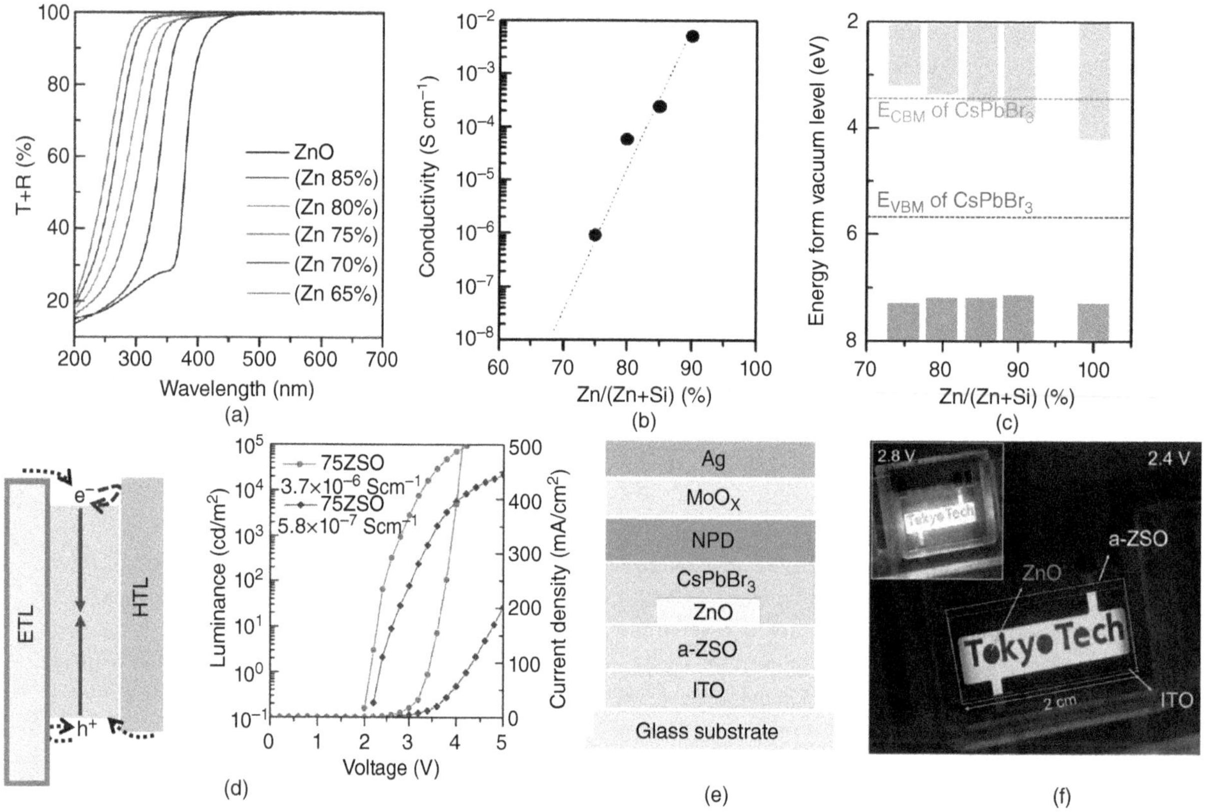

Figure 1.1.13 Amorphous ZnO-SiO$_2$. (a) Optical absorption of 200-nm-thick thin films, (b) conductivity at RT, and (c) variation in the band-edge position with composition. The band-edge positions of green emitting layer CsPbBr$_3$ are also indicated. (d–f) EL device structure and EL characteristics. It is visually evident that a-ZSO works much better than ZnO. Note that the emission intensity goes up to 50,000 cd/cm^2 at 5 V.

located at the CBM, this wide gap widening results in lower work function. ZnO nanocrystals in the material are conducting, due probably to the incorporation of Si to the Zn site, and they are not continuously connected by a thin insulating zinc silicate layer. Thus, mobility observed by hopping conduction through localized levels in the insulating zinc silicate layer is amply large compared with that of organic semiconductors used as the electron transport layer. This X-ray AOS is a kind of nanocomposite material utilizing the optical confinement effect of ZnO nanocrystals and electrically connected conducting ZnO nanocrystals connected by hopping conduction via a thin insulating ZnO-SiO_2 layer.

1.1.7 Summary and Outlook

This chapter outlines the background, fundamental understanding, and recent progress of AOSs for display applications in comparison with crystalline oxide semiconductors and amorphous silicon. An emphasis was placed on the electron transport and electronic structure of TAOSs represented by a-IGZO. Throughout this chapter, the author described design concepts based on a simple consideration of chemical bonding. The tremendous success of Si-based semiconductors has created splendid science and technology. Oxide semiconductors have a longer history than silicon semiconductors, but this had not led to practical device applications. IGZO-TFTs would be the first visible device application of oxide semiconductors. High-mobility TFTs using amorphous semiconductors are based on the intrinsic nature of *p*-block metal cation-based transparent oxides. Crystalline Si cannot meet the requirements for display technology, which needs large-sized thin films. Transparent amorphous oxide semiconductors have huge potential as backplane transistors for high-precision and energy-saving large-sized LCD panels. Application to electron transport/injection layers, which is hard to overcome with the existing technology, may be expected by developing new oxide semiconductors. Oxide is only one semiconductor material that is stable in air at high temperature. Thus, fabricating oxide semiconductors derived from solution is a process allowed for only oxide material. Although the resulting devices are still insufficient in stability and performance, further technical progress would lead to new applications. Review articles [23, 24] are helpful to compensate for the lack of this discussion in the present chapter.

Finally, the author would like to raise two technical challenges in transparent amorphous semiconductors:

1. *High-mobility p-channel TFTs*: Oxides are unfavorable as *p*-type semiconductors. The deep energy level and highly localized nature of O 2*p* dominating the VBM make it hard to have high-mobility *p*-type conduction. Recently, transparent amorphous semiconductors with high mobility (~10 cm^2/Vs) were reported in the Cu-Sn-I system [33]. As shown in Figure 1.1.14, filled I 5*p* orbitals constituting the VBM with Cu 3*d* orbitals may be regarded as the vacant *s* orbital of PTM cations in *n*-type TAOSs. The hole concentration in these materials is too high to operate as TFTs with enough current on/off ratio. Reduction of carrier concentration without degrading mobility is the current issue. If this issue could be overcome, good-performance transparent CMOS devices are possible with a combination with *n*-TAOS TFTs.
2. *High-mobility and high-stability TAOS-TFTs*: Amorphous IGZO-TFTs are now widely applied to flat-panel displays, replacing a-Si:H-TFTs. However, higher mobility TAOS-TFTs are demanded for circuit applications. It is well known that there is a phenomenological observation between mobility and stability (to various stresses) in oxide TFTs. Higher mobility TFTs are obtained in In-rich and/or In-Sn-based TAOSs such as InOx and In-Sn-Zn-O, but each of these TFTs is sensitive to stress (voltage, light, heat, and their combinations). Elucidation of the origin for this trade-off should give an effective clue to resolve the issue.

Materials science of TAOSs has greatly advanced in both fundamentals and device applications in the last two decades. A strong demand for new applications will be the engine to facilitate breadth in this area. There are several excellent review articles and monographs [34–39] on this subject, including those already cited in this chapter. The author encourages the reader to check these reviews if necessary. Very recently, a model from the trade-off described above was proposed and high mobility-highly stable *a*-ITZO TFTs were reported [41].

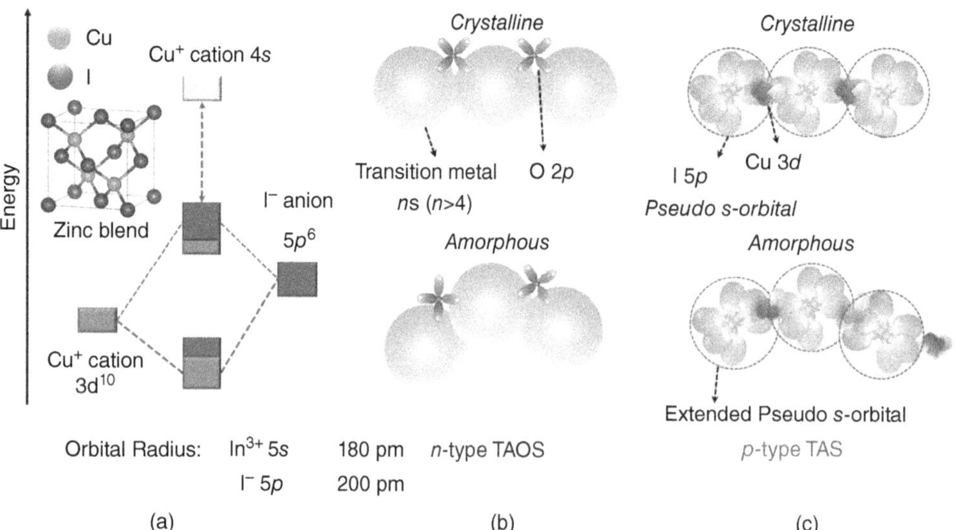

Figure 1.1.14 Amorphous Cu-Sn-I. (a) Crystal structure of γ-CuI zinc blends and schematic band structure. (b) Schematic orbital drawing of the CBM in crystalline and amorphous transparent conductive oxides (TCOs) based on a PTM cation with an electronic configuration $(n-1)d^{10}ns0$, where $n > 4$. (c) Schematic orbital drawing of the VBM in CuI. Three I 5*p* orbitals with a large spatial spread may be regarded as a pseudo *s* orbital similar to the 5*s* orbital with a large spread and a spherical shape.

References

1 Ohya, Y., Niwa, T., Ban, T., and Takahashi, Y. (2001). Thin film transistor of ZnO fabricated by chemical solution deposition. *Japanese Journal of Applied Physics* 40 (1R): 297–298.

2 Özgür, Ü., Hofstetter, D., and Morkoc, H. (2010). ZnO devices and applications: a review of current status and future prospects. *Proceedings of the IEEE* 98 (7): 1255–1268.

3 Hosono, H., Kikuchi, N., Ueda, N., and Kawazoe, H. (1996). Working hypothesis to explore novel wide band gap electrically conducting amorphous oxides and examples. *Journal of Non-Crystalline Solids* 198: 165–169.

4 Adkins, C.I., Hussain, T., and Ahmad, N. (1993). Field-effect measurements of carrier mobilities in transparent conducting films of amorphous indium oxide. *Journal of Physics: Condensed Matter* 5 (36): 6647.

5 Nomura, K., Ohta, H., Takagi, A. et al. (2004). Room-temperature fabrication of transparent flexible thin-film transistors using amorphous oxide semiconductors. *Nature* 432 (7016): 488.

6 Chiang, H.Q., Wager, J.F., Hoffman, R.L. et al. (2005). High mobility transparent thin-film transistors with amorphous zinc tin oxide channel layer. *Applied Physics Letters* 86 (1): 013503.

7 Wager, J.F., Keszler, D.A., and Presley, R.E. (2008). *Devices in Transparent Electronics*, 83–151. Boston: Springer MA. This monograph portrayed research on oxide electronics at the initial stage of oxide TFTs (up to 2006) along with the fundamentals of TFT physics.

8 Kawazoe, H., Yasukawa, M., Hyodo, H. et al. (1997). P-type electrical conduction in transparent thin films of $CuAlO_2$. *Nature* 389 (6654): 939–942.

9 Ogo, Y., Hiramatsu, H., Nomura, K. et al. (2008). *p*-channel thin-film transistor using *p*-type oxide semiconductor SnO. *Applied Physics Letters* 93 (3): 032113.

10 Nomura, K., Kamiya, T., and Hosono, H. (2011). Ambipolar oxide thin-film transistor. *Advanced Materials* 23 (30): 3431–3434.

11 Wang, Z., Nayak, P.K., Caraveo-Frescas, J.A., and Alshareef, H.N. (2016). Recent developments in *p*-type oxide semiconductor materials and devices. *Advanced Materials* 28 (20): 3831–3892.

12 Narushima, S., Orita, M., Hirano, M., and Hosono, H. (2002). Electronic structure and transport properties in the transparent amorphous oxide semiconductor 2 CdO· GeO_2. *Physical Review B* 66 (3): 035203.

13 Narushima, S., Mizoguchi, H., Shimizu, K. et al. (2003). A p-type amorphous oxide semiconductor and room temperature fabrication of amorphous oxide p-n heterojunction diodes. *Advanced Materials* 15 (17): 1409–1413.

14 Kamiya, T., Nomura, K., and Hosono, H. (2010). Present status of amorphous In–Ga–Zn–O thin-film transistors. *Science and Technology of Advanced Materials* 11: 044305-1–044305-23.

15 Kamiya, T., Nomura, K., and Hosono, H. (2009). Electronic structures above mobility edges in crystalline and amorphous In-Ga-Zn-O: percolation conduction examined by analytical model. *Journal of Display Technology* 5 (12): 462–467.

16 Lee, S., Ghaffarzadeh, K., Nathan, A. et al. (2011). Trap-limited and percolation conduction mechanisms in amorphous oxide semiconductor thin film transistors. *Applied Physics Letters* 98 (20): 203508.

17 Nenashev, A.V., Oelerich, J.O., Greiner, S.H.M. et al. (2019). Percolation description of charge transport in amorphous oxide semiconductors. *Physical Review B* 100 (12): 125202.

18 Street, R.A. (1982). Doping and the Fermi energy in amorphous silicon. *Physical Review Letters* 49 (16): 1187.

19 Nomura, K., Kamiya, T., Ikenaga, E. et al. (2011). Depth analysis of subgap electronic states in amorphous oxide semiconductor, a-In-Ga-Zn-O, studied by hard x-ray photoelectron spectroscopy. *Journal of Applied Physics* 109 (7): 073726.

20 Jeong, J.K. (2013). Photo-bias instability of metal oxide thin film transistors for advanced active matrix displays. *Journal of Materials Research* 28 (16): 2071.

21 Nomura, K., Ohta, H., Ueda, K. et al. (2003). Thin-film transistor fabricated in single-crystalline transparent oxide semiconductor. *Science* 300 (5623): 1269–1272.

22 Hosono, H. (2018). How we made the IGZO transistor. *Nature Electronics* 1 (7): 428–428.

23 Hosono, H. (2007). Recent progress in transparent oxide semiconductors: materials and device application. *Thin Solid Films* 515 (15): 6000–6014.

24 Fortunato, E., Barquinha, P., and Martins, R. (2012). Oxide semiconductor thin-film transistors: a review of recent advances. *Advanced Materials* 24 (22): 2945–2986.

25 Mizoguchi, H., Hirano, M., Fujitsu, S. et al. (2002). $ZnRh_2O_4$: a p-type semiconducting oxide with a valence band composed of a low spin state of Rh^{3+} in a $4d^6$ configuration. *Applied Physics Letters* 80 (7): 1207–1209.

26 Narushima, S., Mizoguchi, H., Shimizu, K.I. et al. (2003). A *p*-type amorphous oxide semiconductor and room temperature fabrication of amorphous oxide *p–n* heterojunction diodes. *Advanced Materials* 15 (17): 1409–1413.

27 Hosono, H. and Kitano, M. (2021). Advances in materials and applications of inorganic electrides. *Chemical Reviews* 121 (5): 3121–3185.

28 Matsuishi, S., Toda, Y., Miyakawa, M. et al. (2003). High-density electron anions in a nanoporous single crystal: $[Ca_{24}Al_{28}O_{64}]\ 4^{+}(4e^{-})$. *Science* 301 (5633): 626–629.

29 Hosono, H., Kim, J., Toda, Y. et al. (2017). Transparent amorphous oxide semiconductors for organic electronics: application to inverted OLEDs. *Proceedings of the National Academy of Sciences* 114 (2): 233–238.

30 Nakamura, N., Kim, J., and Hosono, H. (2018). Material design of transparent oxide semiconductors for organic electronics: why do zinc silicate thin films have exceptional properties? *Advanced Electronic Materials* 4 (2): 1700352.

31 Sim, K., Jun, T., Bang, J. et al. (2019). Performance boosting strategy for perovskite light-emitting diodes. *Applied Physics Reviews* 6 (3): 031402.

32 Yang, H., Kim, J., Yamamoto, K., and Hosono, H. (2017). Efficient charge generation layer for tandem OLEDs: bi-layered MoO_3/ZnO-based oxide semiconductor. *Organic Electronics* 46: 133–138.

33 Jun, T., Kim, J., Sasase, M., and Hosono, H. (2018). Material design of *p*-type transparent amorphous semiconductor, Cu–Sn–I. *Advanced Materials* 30 (12): 1706573.

34 Park, J.S., Maeng, W.J., Kim, H.S., and Park, J.S. (2012). Review of recent developments in amorphous oxide semiconductor thin-film transistor devices. *Thin Solid Films* 520 (6): 1679–1693.

35 Kwon, J.Y., Lee, D.J., and Kim, K.B. (2011). Transparent amorphous oxide semiconductor thin film transistor. *Electronic Materials Letters* 7 (1): 1–11.

36 Nathan, A., Lee, S., Jeon, S., and Robertson, J. (2014). Amorphous oxide semiconductor TFTs for displays and imaging. *Journal of Display Technology* 10 (11): 917–927.

37 Yu, X., Marks, T.J., and Facchetti, A. (2016). Metal oxides for optoelectronic applications. *Nature Materials* 15 (4): 383–396.

38 Barquinha, P., Martins, R., Pereira, L., and Fortunato, E. (2012). *Transparent Oxide Electronics: From Materials to Devices*. Chichester, UK: John Wiley and Sons.

39 Facchetti, A. and Marks, T. (ed.) (2010). *Transparent Electronics: From Synthesis to Applications*. Chichester, UK: John Wiley and Sons.

40 Street, R.A. (ed.) (2000). *Technology and Applications of Amorphous Silicon*. Berlin: Springer-Verlag.

41 Shiah, Y.-S., Sim, K., Shi, Y. et al. (2021). Mobility-stability trade-off in oxide thin-film transistors. *Nature Electronics* 4: 800–807.

1.2

Transparent Amorphous Oxide Semiconductors

What's Unique for Device Applications?

Hideya Kumomi

Tokyo Institute of Technology, Tokyo, Japan

1.2.1 Introduction

After a long incubation of half a century, the first product of an active-matrix liquid-crystal display (AM-LCD) driven by backplane circuits composed of oxide semiconductor thin-film transistors (TFTs) [1] was commercialized in 2012. Although the manufacturer did not disclose this detail, a small number of the high-resolution (264 ppi) 9.7-inch-diagonal AM-LCD panels on their tablet PC devices (Apple iPad 3) [2] was driven by backplane circuits driven by AOSs, amorphous In-Ga-Zn-O (a-IGZO) TFTs. After that, the a-IGZO-TFTs were successively commercialized in the AM-LCD screens of smartphones, PC monitors, laptop PCs, and active-matrix organic light-emitting diode (AM-OLED) displays for smartwatches and large-area (up to 88 inches diagonally) and high-resolution (8k×4k) television (TV) screens. Furthermore, even flexible and rollable AM-OLED TVs were launched in October 2020, and a-IGZO-TFT-based TVs dominate the AM-OLED TV market. Nowadays, AOS-TFTs have become indispensable component devices for driving active-matrix flat-panel displays (AM-FPDs), along with conventional hydrogenated amorphous Si (a-Si:H) TFTs and low-temperature polycrystalline Si (LTPS) TFTs.

It is noteworthy that, even after a half-century incubation of oxide TFTs, a-IGZO-TFTs had been commercialized by rapid research and development in a short period of eight years from their first demonstration. It is generally difficult for such a novel and exotic oxide material composed of ternary metal cations to be thrust into an existing industry of established Si-based technologies and to replace a part of them with AOS-TFTs due to both technical and business issues. There are some reasons for the success of a-IGZO-TFTs, and these are discussed and elucidated in this chapter based on the technical requirements of electronics industries, along with a historical review of oxide semiconductor–based TFTs and their uniqueness.

1.2.2 Technical Issues and Requirements of TFTs for AM-FPDs

1.2.2.1 Field-Effect Mobility

AM-LCDs at their dawn can be sufficiently driven by a-Si:H TFTs that have a small field-effect mobility of μ_{FE} ~0.1 $cm^2V^{-1}s^{-1}$ because of their low resolutions (i.e., large pixel size), small panel dimensions, and slow frame rates. The μ_{eff} had been enhanced up to ~1 $cm^2V^{-1}s^{-1}$ by improving a-Si:H materials, device structures, and fabrication processes to meet the requirements of increasing these specifications and creating greater color depth. However, much higher mobility was required from the beginning of the twenty-first century, when screens larger than 50 inches diagonally and with a higher frame rate than 60 Hz for smooth rendering of movies were demanded by markets. Total capacitance, C, of parasitic capacitance of switching TFTs (T_{SW}) is mainly caused by source–drain

Amorphous Oxide Semiconductors: IGZO and Related Materials for Display and Memory, First Edition.
Edited by Hideo Hosono and Hideya Kumomi.

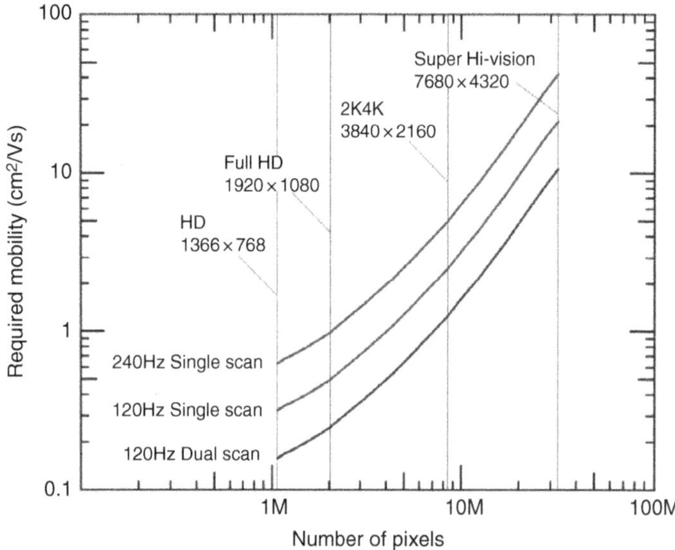

Figure 1.2.1 Dependence of the least mobilities of switching TFTs to drive a 50-inch-diagonal AM-LCD model with a 5-μm-wide Cu gate-line wiring, calculated by Y. Matsueda in 2010 [3]. *Source*: Kamiya, T., Nomura, K., & Hosono, H. (2010). Present status of amorphous In–Ga–Zn–O thin-film transistors. *Science and Technology of Advanced Materials* **11**(4): 044305. doi:10.1088/1468-6996/11/4/044305.

and gate electrodes overlapping, and the storage capacitors in all pixels and so on are parallelly hung on gate (scan)-line metal wirings. Signal transmission from the starting to the ending pixels on a single gate-line must be accompanied by delays with a time constant of RC, where R represents a finite resistance of gate-line wiring determined by the wiring metal materials and their length and cross-sectional areas. This gate-line delay cannot exceed the inverse of the frame rate (i.e., the interval to scan the next gate-line). It is necessary to more rapidly charge the pixel capacitance by enhancing the current driving performance of T_{SW} in order to reduce the gate-line delay. Figure 1.2.1 is a historical plot showing dependence of the least mobilities of T_{SW} to drive a 50-inch-diagonal AM-LCD model with 5-μm-wide Cu gate-line wiring, calculated by Yojiro Matsueda in 2010 [3]. It is estimated here that at least a mobility of ~10 $cm^2V^{-1}s^{-1}$ is required to drive 8k×4k panels by a single scan at 120 Hz, which cannot be achieved by the latest high-end *a*-Si:H TFTs. LTPS-TFTs exhibit high mobilities of ~50–100 $cm^2V^{-1}s^{-1}$ even when their polycrystalline grain size is far smaller than the TFT channel dimensions to suppress the spatial variety of the TFT properties caused by grain-boundary effects. However, LTPS-TFTs still have difficulties with both short- and long-range uniformity.

In cases of current-driving light-emitting diode (LED) devices like AM-OLEDs, the requirements for driving TFTs (T_{DR}) are added to those for T_{SW}. Various estimations suggest that a mobility of ~10 $cm^2V^{-1}s^{-1}$ is required for T_{DR} to provide sufficient current to the directly connected LED component for sufficient visual brightness independent of FPD resolution and size. This cannot be achieved by *a*-Si:H TFTs, and thus AM-OLEDs had been produced only for the small-sized panels of smartphones using LTPS-TFT backplanes. Active-matrix electrochromic (EC) displays also happen to require TFTs with a mobility of ~10 $cm^2V^{-1}s^{-1}$ to rapidly charge the EC cells, although they have not been commercialized so far.

Monolithic implementation of peripheral circuits (composed of TFTs like shift-registers for scanning gate-lines and demultiplexers for dividing data signals into RGB subpixels) becomes necessary if the aim is super-high resolution in small panels when it becomes difficult to implement Si-based integrated circuit (IC) chips on the peripherals of AM-FPDs and connect them to the backplane circuits. Especially in mobile devices, the footprint of TFT-based peripheral circuits should be as small as possible to realize narrow-border frame configurations, and a TFT mobility of ~10–50 $cm^2V^{-1}s^{-1}$ is required to reduce the channel width of the TFTs.

1.2.2.2 Off-State Leakage Current and On/Off Current Ratio

The T_{SW} of AM-FPDs stays in the off-state for most of a frame time after completing the scanning of the gate-line that the T_{SW} is hung on, until the scanning of the gate-line restarts for the next frame. For example, in 8k×4k panels, the T_{SW} remains off for 3,839/3,840 of the time under a relatively high depletion gate bias. Leakage current of the T_{SW} in the off-state should be suppressed to be lower than the reduction of charges in the storage capacitor in one pixel, so as to suppress a change of tone within one bit of color depth. LTPS-TFTs usually exhibit high off-state leakage current and also suffer from inversion operation at a high depletion gate bias; therefore, researchers have attempted to solve this problem by improving their device structures and circuit designs. However, some recent mobile devices try to suppress their power consumption by reducing their frame rate lower than 30 Hz and their power consumption in their peripheral circuits. The lowest frame rate achieved by the LTPS-TFT backplane is only 10 Hz, while AOS-TFTs can easily realize lower than 1 Hz owing to their extremely low off-state leakage current, as described later in this chapter. This feature of low off-state current in AOS-TFTs is actually applied to some mobile products such as smartphones and smartwatches.

Both high mobility and low off-state leakage current lead to enhancement of the on/off current ratio of TFTs. The on/off current ratio should be as large as needed to express necessary color depth if the imaging components are driven by analog control of bias or current. The current standard color depth of eight bits for each primary color has been achieved by conventional Si-based TFTs. However, when higher color depth or a wider color gamut is required, TFTs with a higher on/off current ratio become necessary.

1.2.2.3 Stability and Reliability

TFTs driving AM-FPDs must be stable and reliable against electrical (and thermal) stresses caused by their operation, in addition to thermal, mechanical, and chemical stresses from the environment in electrical devices. The T_{SW} and T_{DR} mainly suffer gate-voltage bias stress and source-to-drain current stress, respectively, and they deteriorate with shifts of TFT characteristics such as threshold voltage, on/off current, and subthreshold swing. TFTs in peripheral circuits are also exposed to various stresses. Some of these shifts can be canceled by special compensation circuits, although the compensation range is limited. Especially in a pixel circuit of AM-OLEDs, characteristics of the OLED component also shift with light-emitting operation, and it is necessary to compensate the characteristic shifts of both TFTs and OLEDs by the compensation circuit composed of the same TFTs. It is very difficult to drive AM-OLEDs with a backplane circuit based on unstable *a*-Si:H TFTs, in addition to the difficulty in driving OLEDs with their small mobilities. Therefore, early products of AM-OLED displays were driven by backplane circuits composed of stable LTPS-TFTs, which limit mother-glass sizes (smaller than generation 6 [Gen. 6]) in mass production and thus final panel sizes up to several inches in diagonal.

1.2.2.4 Uniformity

Characteristics of TFTs driving AM-FPDs should be uniform over their panel dimension to guarantee uniform brightness, colors, and tone rendering. Spatial variations of characteristics of LTPS-TFTs are relatively large due to the in-plane inhomogeneity of polycrystalline grain boundaries, where carrier transfer is hindered by high potential barriers, and the short-range uniformity among neighboring TFTs is not good. Neither is the long-range uniformity between distant TFTs over panel dimensions because of the inhomogeneous formation process with scanning a line-shaped and short-pulsed excimer-laser annealing (ELA) for the melting and recrystallization process of *a*-Si thin films. On the other hand, *a*-Si:H TFTs show good uniformity in both short and long ranges owing to their amorphous nature, with no grain boundaries or film formation by plasma-enhanced chemical-vapor deposition (PECVD). Therefore, middle- and large-area ($\gtrsim$10 inches in diagonal) AM-FPDs have been driven by *a*-Si:H TFT–based backplanes.

1.2.2.5 Large-Area Devices by Large-Area Mother-Glass Substrates

In addition to the uniformity issues in LTPS-TFTs that have been discussed so far in this chapter, there is another reason for difficulty in enlarging mother glasses beyond Gen. 6, and thus in enlarging the final panel size over several inches in diagonal in mass production. In the fabrication processes of LTPS-TFTs, *a*-Si:H thin films are formed over glass substrates by PECVD first, and then the high-density hydrogens in the films have to be removed (dehydrogenation) before the subsequent ELA process to avoid explosion of the films by hydrogen vaporization. The dehydrogenation requires thermal annealing for several hours at a temperature higher than ~450 °C. The high-temperature annealing demands special and expensive glass substrates with very small thermal deformation, but these are not available for mother glasses larger than Gen. 6. Furthermore, it is also difficult to expand the length of line-shaped excimer-laser beams in ELA beyond the dimension of the Gen. 6 mother glasses. Because the cost of an AM-FPD panel is most strongly determined by the number of panels cropped from one mother glass, AM-FPD panel products based on LTPS-TFTs are limited to small-sized (a few to several inches in diagonal) ones, such as those for smartphones.

Mother-glass substrates larger than Gen. 6 (up to >10) had been available to *a*-Si:H TFTs, because the highest process temperature is ~350 °C in PECVD of *a*-Si:H thin films and gate-insulator films, and there is no limitation to substrate sizes in PECVD. Large-sized AM-FPDs had been supplied exclusively with backplanes based on *a*-Si:H TFTs. As mentioned in this chapter, however, there are many AM-FPDs that cannot be driven by *a*-Si:H TFTs, and these had not been commercialized.

1.2.2.6 Low-Temperature Fabrication and Flexibility

A far lower fabrication-processing temperature than ~350 °C is indispensable if substrates are not heat-resistant. Most of the substrates based on polymer plastic sheets for flexible devices cannot endure thermal stresses at temperatures higher than ~200 °C. Some emerging semiconductor materials, such as organic molecules, carbon nanotubes, graphene, transition metal dichalcogenide, halide perovskite, and so on, could be fabricated at low temperatures down to room temperature, but none of them can meet the other requirements mentioned here.

Characteristics of TFTs on flexible devices also have to be stable under fixed or repeated mechanical stresses of bending. Curved or foldable AM-OLED panels driven by LTPS-TFTs are commercialized on smartphones, but further flexible products with rollable panels have not been launched until recently.

1.2.3 History, Features, Uniqueness, Development, and Applications of AOS-TFTs

1.2.3.1 History

In 1939, William Shockley attempted a demonstration of transistors using Cu_2O [4], which is regarded as the earliest challenge to oxide semiconductor–based active devices. Unfortunately, the Cu_2O transistor never operated, so he changed the semiconductor material to germanium (Cu_2O TFTs still do not operate, even in 2021). The first TFT was demonstrated in 1962 using a chalcogenide semiconductor, CdS [5]. The first TFTs using oxide semiconductors, SnO_2 and In_2O_3, appeared in 1964 [6], which was two years earlier than the appearance of polycrystalline Si–based field-effect transistors (FETs) in 1966 [7]. FETs based on single-crystalline ZnO followed these pioneers as a demonstration of oxide semiconductor–based transistors in 1968 [8]. However, further followers have not appeared for more than three decades since then. This was probably due to a lack of killer applications of TFTs during that era. On the other hand, *a*-Si:H thin films emerged in the 1970s [9], and many researchers rushed into them because of their potential applications in backplane circuits for AM-FPDs and in photosensitive and photovoltaic devices, and because of scientific interest in their local and electronic structures. Since the first AM-LCD with an *a*-Si:H TFT backplane was proposed in 1981 [10], Si-based TFTs including polycrystalline Si TFTs dominated the commercialization of AM-FPDs for decades until quite recently. Thus, oxide TFTs had fallen into oblivion for a long time, while oxide semiconductors such as In-Sn-O have played an important role to date as conductors in applications with transparent electrodes.

What re-ignited interest in oxide semiconductors was a working hypothesis, proposed in 1996, about candidate materials for high-mobility amorphous oxides [11]. The hypothesis predicts that oxides of post transition metals with an electronic configuration of $(n-1)d^{10}s^0$ could exhibit high electron mobility, even in amorphous phases. The actual material used in the operation of high-performance TFTs was demonstrated several years later. Meanwhile, oxide semiconductor–based TFTs were revived by aiming at AM-FPD applications via a demonstration of solution-processed polycrystalline ZnO-TFTs in 2001 [12], which was followed by sputter-deposited ZnO-TFTs by many researchers. However, ZnO-TFTs have not been commercialized so far because they cannot escape serious technical issues caused by polycrystalline grain boundaries, and a superior alternative appeared later. In 2003, Nomura *et al.* demonstrated that a multicomponent oxide TFT showed a high field-effect mobility of μ_{FE} ~80 $cm^2V^{-1}s^{-1}$, using *c*-axis-oriented crystalline $InGaO_3(ZnO)_5$ (IGZO) thin films formed by reactive solid-phase epitaxy at high temperature [13]. During exploration of the candidates predicted by the working hypothesis [12], with an insight into electronic configurations and the role of Ga cations, this material system was discovered to be a high-mobility oxide semiconductor. Actually, the first AOS-TFTs were demonstrated in 2004 with μ_{FE} ~9 $cm^2V^{-1}s^{-1}$, using amorphous In-Ga-Zn-O (*a*-IGZO) thin films deposited on plastic substrates at room temperature [14].

AOS-TFTs have attracted keen attention, mainly because their channel mobilities are ~10 times higher than those of conventional *a*-Si:H TFTs. After a few years since the first demonstration of *a*-IGZO-TFTs, an explosive increase in their research and development started, with numerous publications [15]. Finally, in 2012, after remaining in obscurity for over half a century since the first demonstration, the first commercial product based on AOS-TFTs was launched in the worldwide market in a small number of mobile electronic devices. Since then, *a*-IGZO-TFTs have been mainstreamers of oxide semiconductors commercialized in electronic device products due to their useful nature.

1.2.3.2 Features and Uniqueness

As mentioned in Section 1.2.3.1, the choice of complicated quaternary compounds like IGZO—rather than simple binaries like ZnO, SnO_2, and In_2O_3—is based on a working hypothesis for high-mobility amorphous oxides [11]. In these binary compounds, it is difficult to control oxygen deficiencies and to reduce intrinsic carrier density to suppress the off-current and instability of their TFTs. Furthermore, the choice of quaternary compound is based on insight into the coordination number of Ga in IGZO, where Ga^{3+} does not replace Zn^{2+} taking fivefold coordination to suppress excess carrier generation, while Ga^{3+} doped into ZnO replaces Zn^{2+} taking fourfold coordination to generate carriers. Thus, *a*-IGZO was selected as a model compound of candidates for channel materials of high-performance TFTs.

AOS thin films and their TFTs like *a*-IGZO have the following features and uniqueness:

1. Most AOS materials are *n*-type semiconductors.
2. Carrier mobility increases with carrier density via a percolation transport mechanism, unlike conventional crystalline semiconductors with covalent bonding.

3. Films are transparent to the light in a range of visible wavelengths due to wide band gaps ($E_g \gtrsim 3$ eV).
4. It is possible to form the channel layers over noncrystalline substrates at low temperature by sputtering deposition, molecular–organic chemical vapor deposition, atomic layer deposition, and solution process coating.
5. Various conventional gate-insulator materials are available with low densities of interfacial trap states, which warrant sufficiently high breakdown voltages for various applications.
6. It is easy to form good ohmic contact to various source–drain electrodes.
7. Uniformity in both short and long ranges is excellent because the films are free from crystalline grain boundaries.
8. Operation mechanisms are relatively simple so that TFT models for device and circuit simulations are simple. See Chapter 16 for some examples.
9. The features in (4)–(8) enable us to adopt methods and facilities for device and circuit designs and manufacturing, which have been used for Si-based TFTs.
10. A lower tail-state density beneath the conduction band minimum than *a*-Si:H TFTs by a few orders of magnitude makes large band vending possible to push up the Fermi level into the conduction band by field effect at low voltages of a forward gate bias. This leads to operations with band-like transport exhibiting over 10 times higher channel mobilities and low threshold voltages.
11. A low-subgap deep-level density of states (DOS) above the midgap level enables fast switching with a small subthreshold swing (~0.1 $V \cdot dec^{-1}$).
12. A large occupied DOS below the midgap level and above the valence band maximum hinders band vending to pull down the Fermi level into the valence band under a backward gate bias. It also prohibits an inversion operation, which leads to very low off-current.
13. The feature of (12) also suppresses a kink effect in the output characteristics of TFTs.
14. Hot-carrier effect and short-channel effect are also small, and source-to-drain breakdown voltages are high.
15. Stability and reliability under electric or mechanical stresses are much better than those of *a*-Si:H TFTs and comparable to those of LTPS-TFTs.

Furthermore, especially for *a*-IGZO:

16. TFT performance is good around metal cation compositions of crystalline single phases such as In:Ga:Zn:O=1:1:1:4 or 2:2:1:7, and it slowly changes with composition around these single phases.
17. Large-area, stable, polycrystalline sputtering targets with single-phase crystalline compositions are available due to (16).

The features of (10) and (11) just meet the technical requirements for field-effect mobility (see Section 1.2.2.1). The features of (10) and (12) satisfy the requirements for off-state leakage current and on/off current ratio (Section 1.2.2.2). The features of (13), (14), and (15) fulfill the conditions for stability and reliability (Section 1.2.2.3). The feature of (7) meets the needs for uniformity (Section 1.2.2.4). The features of (9) (i.e., (4)–(8)), (16), and (17) satisfy the requirement for large-area devices by large-area mother-glass substrates (Section 1.2.2.5). The features of (4) and (15) fulfill the requirement for low-temperature fabrication and flexibility (Section 1.2.2.6). These are the reasons that AOS-TFTs have been actively researched, developed, and commercialized.

While the features and uniqueness of AOS thin films and their TFTs give great advantages to their applications, there are still two major issues remaining to be solved. The first one is instability under light illumination, even in a range of visible wavelength with energy smaller than E_g despite transparency. The illumination of light shorter than half of E_g excites the deep occupied states below the midgap (featured in (11)) and generates electron–hole pairs to cause persistent photoconductivity and shift threshold voltage, V_{TH}, to the negative direction. Furthermore, when the TFT is stressed by a negative gate bias simultaneously with an illumination stress, V_{TH} largely shifts to the negative direction. This phenomenon is called instability under negative-bias-illumination stress (NBIS) and explained by the following mechanism: The positive charge carriers like holes are drifted toward the interface to the gate insulator through a band bending by the negative gate bias and finally trapped by the interfacial states, and as a result, the effective gate-bias voltage is shifted in the positive direction and V_{TH} moves in the negative direction. The NBIS instability is a serious problem, especially for a switching TFT in a pixel circuit of an AM-LCD, which is biased by a large negative gate voltage for most of a frame time and always illuminated with the bright LCD backlight. The instability issues under illumination have been so far solved in commercial products by introducing some light-shielding structures into TFT device layers. Essential solutions to these problems have been proposed, such as adopting much wider E_g AOS materials [16], but they have not satisfied all of the other properties shown in the features of (1)–(17) that are required for AM-FPD applications.

The second one is a much higher μ_{FE} than $\sim 10\,cm^2V^{-1}s^{-1}$. As mentioned in Section 1.2.2.1, dimensions of peripheral circuits in AM-FPDs depend on the width of channels of implemented TFTs. Higher μ_{FE} reduces the channel width and the width of the peripheral circuits to realize narrower borders of AM-FPDs. It is estimated that $\mu_{FE} \gtrsim 30\,cm^2V^{-1}s^{-1}$ is necessary to design the border width narrower than a few millimeters. Many challenges have been made, such as In-richer compositions in *a*-IGZO or amorphous In-Sn-Zn-O (ITZO), and some of them have started to be adopted in commercial products.

1.2.3.3 Applications

After their revival in the twenty-first century, active research of oxide semiconductor–based TFTs aiming at potential applications started with intense motivation due to their advantages, as described in Section 1.2.3.2. Figure 1.2.2 shows the emergence, evolution, and transition of application targets in the first decade, starting with the first demonstration of *a*-IGZO-TFTs in 2004.

Initially, early challengers attempted to apply AOS-TFTs for AM-FPD backplane circuits of electric papers using electrophoretic display (EPD) devices on flexible polymer substrates, especially for low-temperature fabrication

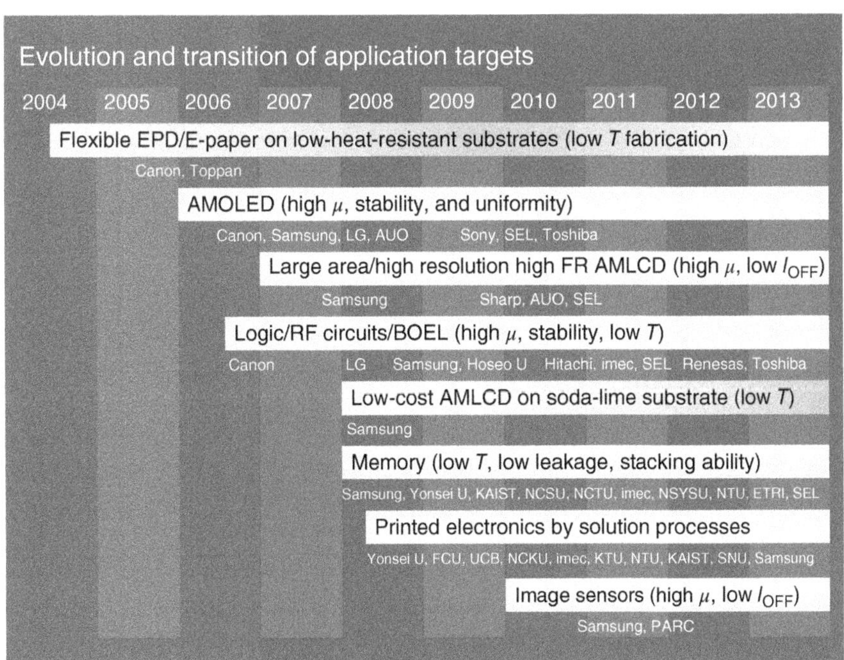

Figure 1.2.2 Emergence, evolution, and transition of application targets in the first decade of the twenty-first century, starting with the first demonstration of *a*-IGZO-TFTs in 2004.

processes. However, no commercial products have been launched because high-mobility TFTs are not always necessary for EPDs that are not able to show movies and full-color pictures due to their slow response and low contrast, and the market for electric papers has not grown.

Applications to AM-OLED backplanes also started from the early days, taking advantage of high mobility, high stability, and high uniformity. As mentioned in Section 1.2.2, it is difficult to drive and fabricate large-sized AM-OLED panels by *a*-Si:H and LTPS-TFTs, respectively, and new, more competent TFTs have been long desired. AOS-TFTs only recently met the demands of large-sized AM-OLEDs, and commercial TV products appeared in 2015. Applications to advanced AM-LCDs with higher pixel resolutions, frame rate, and color depth followed those of AM-OLEDs, utilizing various advantages of AOS-TFTs, and the first commercial product was launched in 2012. The current leading research is introduced in Chapters 15 and 18.

There were some challenges to apply AOS-TFTs to low-cost AM-LCDs fabricated at low temperatures on low-cost soda-lime glass substrates that cannot suffer high-temperature processes, instead of expensive and heat-resistant alkaline-free glass substrates. These challenges have failed to date, because it was found that post-annealing at ≥300 °C was necessary to reduce defect densities of AOS thin films for TFT backplanes of AM-FPDs.

Low-power-consumption AM-FPDs, which can reduce frame rates when displayed pictures are immobile, also have been proposed based on the extremely low off-leakage current of AOS-TFTs. The lower the frame rate is, the lower the power consumption in the peripheral circuits is. The low frame rate driving by AOS-TFTs has been commercialized in mobile devices such as smartphones and smartwatches.

There have been many challenges except applications to AM-FPDs. The high mobility and low-temperature fabrication of AOS-TFTs facilitate their applications to logic and radiofrequency circuits on flexible polymer substrates for wireless communications. The state-of-the art research is introduced in detail in Chapter 17. High breakdown voltage and low-temperature fabrication encourage engineers and scientists engaged in Si-based ultra-large-scale integration (ULSI) circuits to apply AOS-TFTs to the peripheral back-end-of-line circuits of ULSI. The leading research is described in Chapter 20. Volatile and nonvolatile memory devices comprise one of the most promising candidates for AOS-TFT applications, and they have been investigated utilizing low leakage current and low-temperature fabrications. The details are described in Chapters 19, 20, and 21. Neuromorphic applications based on AOS thin-film devices have also attracted attention, and they are described in Chapter 22. Application of AOS-TFT-based active-matrix backplanes to X-ray image sensor arrays is described in Chapter 23. These post-AM-FPD applications are expected to be commercialized soon, taking advantage of accumulated technologies for materials, designs, and fabrications developed in AM-FPD applications.

1.2.3.4 Development and Products of AM-FPDs

Through active research on applications (mentioned in Section 1.2.3.3), many prototype AM-FPDs have been demonstrated using backplanes composed of oxide semiconductor TFTs. Three small panels initially appeared in the same year of 2006: (i) Ito *et al.* demonstrated a color EPD driven by transparent *a*-IGZO-TFTs formed over color filter arrays on a polymer substrate at room temperature [17], (ii) Hirao *et al.* demonstrated a transparent AM-LCD panel driven by ZnO-TFTs [18], and (iii) Park *et al.* demonstrated a transparent AM-OLED panel driven by ZnO-TFTs [19]. These pioneering works were then followed by worldwide activities in academies and industries to demonstrate more sophisticated prototype panels mainly using *a*-IGZO-TFTs with enhancement of panel size, resolution, color rendering, and flexibility.

During this prototyping period, the remaining technical problems of stability, device structures, and mass production had been solved. Finally, in 2012, the first commercial product was launched into market in a small number of shipping tablet PC products (Apple iPad 3) with high-resolution (264 ppi) 9.7-inch-diagonal AM-LCDs for their screen panels driven by the *a*-IGZO-TFT-based backplane circuits. Since then, high-resolution AM-LCDs driven by *a*-IGZO-TFTs have been adopted by some smartphones, PC monitors, laptop PCs, and tablet PCs. In 2015, LG launched the first AM-OLED product, a 55-inch-diagonal full-high-definition (FHD; 2k) TV with a curved

Figure 1.2.3 Market-leading product of a 65-inch-diagonal 4k AM-OLED TV, donated by the manufacturer to the inventor of *a*-IGZO-TFTs.

panel driven by *a*-IGZO-TFTs. The AM-OLED TV products have evolved to a larger size (up to 88 inches diagonal) and a higher resolution (up to ultra-high-definition [UHD]; 8k), and they are more flexible with rollable screens through 2021 (Figure 1.2.3). Hybrid backplanes composed of both LTPS- and *a*-IGZO-TFTs (called "LTPO-TFTs") have also been proposed to partially use AOS-TFTs for frame-rate reduction, and they have been adopted in AM-OLED panels of smartwatch products (since the Apple Watch Series 4 in 2018), where power-consumption savings are critical due to the watches' small batteries and the panels display motionless pictures very often. The LTPO-TFT-based AM-OLED panels also will soon be adopted by smartphones to boost the highest frame rate over 120 Hz and enhance user experiences in motion pictures, gaming, and artificial/virtual-reality content by suppressing response delays.

It is expected that AOS-TFTs will drive emerging and promising AM-FPDs such as halide perovskite LED (PeLED) and micro-LED (μLED) displays. Requirements for backplane TFTs from PeLED displays are almost the same as and rather simpler than AM-OLEDs. Current AOS-TFTs will readily meet the needs of high-performance PeLED displays when they are developed to acquire sufficient stability. On the other hand, μLED displays are composed of arrays of discrete GaN-based LEDs that have too steep a turn-on threshold to control by an analog driving scheme with continuous variation of injected current, and they have to be controlled by digital driving of pulse-width modulation (PWM). The PWM requires very abrupt changes in currents and very high mobilities for driving transistors. Therefore, much higher mobility AOS-TFTs than the current *a*-IGZO are required.

1.2.4 Summary

After incubation for over a half century, oxide semiconductor TFTs were commercialized by research and development of AOS-TFTs for only eight years from the emergence of *a*-IGZO-TFTs. Such relatively quick commercialization is due to not only the excellent properties of AOSs and their fortunate and timely fit with the technical requirements of applications, but also low barriers to entry into AOS-based technologies and commercialization, which are protected by fundamental intellectual properties licensed nonexclusively to everyone. AOSs have become one of the platform materials for semiconductor devices today.

There are still remaining issues in AOSs, as briefly mentioned, but it is expected that, after solving them, AOSs will become much more important in AM-FPD technologies and be used in other applications.

References

1 Kamiya, T., Nomura, K., and Hosono, H. (2010). Present status of amorphous In–Ga–Zn–O thin-film transistors. *Science and Technology of Advanced Materials* 11: 044305-1–044305-23.

2 https://support.apple.com/en-us/HT201471#ipad

3 Matsueda, Y. (2010). Required characteristics of TFTs for next generation flat panel display backplanes. *Digest of International Thin-Film Transistor Conference* 2010: S6.

4 Shockley, W. (1976). The path to the conception of the junction transistor. *Transactions on Electron Devices* ED-23: 597.

5 Weimer, P.K. (1962). The TFT: a new thin-film transistor. *Proceedings of the IRE* 7: 1462–1469.

6 Klasens, H.A. and Koelmans, H. (1964). A tin oxide field-effect transistor. *Solid-State Electronics* 7: 701–702.

7 Fa, C.H. and Jew, T.T. (1966). The poly-silicon insulated-gate field-effect transistor. *IEEE Transactions on Electron Devices* ED-13: 290–291.

8 Boesen, G.F. and Jacob, J.E. (1968). ZnO field-effect transistor. *Proceedings of the IEEE* (November): 2094–2095.

9 Spears, W.E. and LeComber, P.G. (1972). Investigation of the localised state distribution in amorphous Si films. *Journal of Non-Crystalline Solids* 8–10: 727–738.

10 Snell, A.J., Mackenzie, K.D., Spears, W.E. et al. (1981). Application of amorphous silicon field effect transistors in addressable liquid crystal display panels. *Applied Physics* 24: 357–362.

11 Hosono, H., Yasukawa, M., and Kawazoe, H. (1996). Working hypothesis to explore novel wide band gap electrically conducting amorphous oxides and examples. *Journal of Non-Crystalline Solids* 198–200: 165–169.

12 Ohya, Y., Niwa, T., Man, T., and Takahashi, Y. (2001). Thin film transistor of ZnO fabricated by chemical solution deposition. *Japanese Journal of Applied Physics* 40: 297–298.

13 Nomura, K., Ohta, H., Ueda, K. et al. (2003). Thin-film transistor fabricated in single-crystalline transparent oxide semiconductor. *Science* 300: 1269–1272.

14 Nomura, K., Ohta, H., Takagi, A. et al. (2004). Room-temperature fabrication of transparent flexible thin-film transistors using amorphous oxide semiconductors. *Nature* 432: 488–492.

15 Fortunato, E., Barquinha, P., and Martins, R. (2012). Oxide semiconductor thin-film transistors: a review of recent advances. *Advanced Materials* 24: 2945–2986.

16 Kim, J., Sekiya, T., Miyokawa, N. et al. (2017). Conversion of an ultra-wide bandgap amorphous oxide insulator to a semiconductor. *NPG Asia Materials* 9: e359-1–e359-7.

17 Ito, M., Kon, M., Ishizaki, M. et al. (2006). A novel display structure for color electronic paper driven with fully transparent amorphous oxide TFT array. *Proceedings of the IDW/AD '06* EP1-4L: 585–586.

18 Hirao, T., Furuta, M., Furuta, H. et al. (2006). Distinguished paper: high mobility top-gate zinc oxide thin-film transistors (ZnO-TFTs) for active-matrix liquid crystal displays. *SID Symposium Digest of Technical Papers* 37 (4.1): 18–20.

19 Park, S.-H.K., Hwang, C.-S., Lee, J.-I. et al. (2006). Transparent ZnO thin film transistor array for the application of transparent AM-OLED display. *SID Symposium Digest of Technical Papers* 37 (4.3): 25–28.

Part II

Fundamentals

2

Electronic Structure and Structural Randomness

Julia E. Medvedeva[1,†], Bishal Bhattarai[1], and D. Bruce Buchholz[2]

[1] *Department of Physics, Missouri University of Science and Technology, Rolla, Missouri, USA*
[2] *Department of Materials Science and Engineering, Northwestern University, Evanston, Illinois, USA*

2.1 Introduction

Structural disorder has been known to suppress carrier concentration and carrier mobility in common covalent semiconductors, such as silicon, by orders of magnitude. The lack of periodicity and bond irregularities in the amorphous phases with strong directional bonding reduces the orbital overlap of the neighboring atoms and give rise to the formation of localized defect states near the band edges that may cause electron trapping, carrier scattering, and subgap absorption. In striking contrast to the covalent semiconductors, oxides of posttransition metals with ionic bonding are known to remain transparent in the visible range and exhibit not only crystalline-like electron mobility upon amorphization, but also 1–2 orders of magnitude higher carrier concentration in the disordered phases as compared to their crystalline counterparts [1–10].

The weak ionic bonding between posttransition metals and oxygen atoms makes the structural description of the amorphous oxide semiconductors (AOSs) challenging. In marked contrast to the Si- or Ge-based semiconductors or SiO_2-based glasses (composed of the main-group metal oxides) that have strong covalent bonding responsible for distinct symmetry-defined nearest-neighbor polyhedra, the weak ionic metal–oxygen (M-O) bonds in AOSs allow for large bond length and angle deviations resulting in strong distortions in the M-O polyhedra and, therefore, in significant disorder within the short-range (nearest-neighbor) structure. These deviations must be carefully quantified in order to determine their role in the electronic properties of AOSs. For this, theoretical and experimental characterization that involves statistical averaging, although instructive in predicting some macroscopic properties such as optical band gap and electron effective mass using established solid-state theories, will miss important information hidden in the distribution and its tails. It might seem reasonable to assume that the structural outliers are key to understanding carrier generation and carrier scattering in AOSs; the larger the structural deviation from an average or a crystalline value, the more localized the defect should be. However, there is a second important consequence of the weak bonding in AOSs that must be taken into account in order to understand the resulting electronic properties: An undercoordinated atom or a strongly distorted polyhedron may trigger significant bond reconfiguration not only in the immediate vicinity, but also throughout the weakly bound disordered network via a ripple effect. As a result, the AOS's susceptibility to long-range structural rearrangement is beneficial for carrier transport: It may not only improve hybridization of different constituents (e.g., in a multi-cation oxide) in the conduction states, but also *reduce* the electron scattering via extensive structural reconfiguration near defects, dopants, or impurities [11–13]. The latter is not possible in crystalline oxides where the translational periodicity of the well-defined lattice limits structural relaxation around a defect to its nearest or next-nearest neighbors.

† In loving memory of my father, Evgeny Medvedev, who gave me boundless support and encouragement.

Amorphous Oxide Semiconductors: IGZO and Related Materials for Display and Memory, First Edition.
Edited by Hideo Hosono and Hideya Kumomi.

Therefore, structural outliers in AOSs should be considered along with their environment—a disordered, random structure—that requires a systematic description in order to gain insights into carrier generation and mobility in AOSs.

Traditionally, the structural characterization of disordered oxides focuses on the M-O and M-M distributions; however, it is critical to understand how disorder affects the oxygen environment. Owing to the directional nature of the O--*p*-orbitals, the distortions in the O-M polyhedra are expected to have a more pronounced effect on the electronic properties of AOSs as compared to the symmetry-indifferent spherical *s*-orbitals of the posttransition metals [14]. Indeed, the differences in the orbital sensitivity to disorder are manifested in the characteristic asymmetry of the electronic localization of the tail states near the band edges in AOSs: The top of the valence band formed from the O--*p*-states exhibits strongly localized states, whereas the *s*-*p*-hybridized conduction states are generally delocalized in stoichiometric or nearly stoichiometric amorphous oxides and may feature shallow or weakly localized defects associated with metal undercoordination, even in highly sub-stoichiometric oxides [11]. The localized tail states near the top of the valence band arise from nonbonding O--*p*-orbitals for the oxygen atoms that are undercoordinated or in a highly distorted nearest-neighbor (metal) environment. These localized O--*p*-states were shown to contribute to the visible-range absorption in stoichiometric amorphous In_2O_3 [11] and play a key role in H defect formation and H mobility in amorphous In-Ga-O doped with hydrogen [12, 13]. While the most advanced X-ray scattering techniques for short-range structural characterization are not sensitive enough to probe the coordination environment of lightweight oxygen atoms, *ab initio* modeling combined with X-ray photoelectron spectroscopy (XPS) O1*s* or with oxygen nuclear magnetic resonance (NMR) measurements should provide valuable insights into oxygen coordination distribution and its role in the electrical and optical properties of AOSs.

Many important aspects of AOSs have been addressed theoretically. The first molecular dynamics (MD) simulations of amorphous indium oxide appeared in 2009 [15], followed by models of electron transport in multi-cation AOSs [16–21], density functional theory (DFT) calculations of defect formation [22–29], and statistical descriptions of amorphous networks [30–32]. Despite the tremendous progress, the structural randomness that leads to an intricate interplay between bond distortions, coordination morphology, and electron (de)localization in AOSs is far from being understood.

In this work, computationally intensive *ab initio* MD simulations, comprehensive structural analysis, and hybrid density-functional calculations are employed to accurately describe the peculiarities in short- and medium-range structures of amorphous In_2O_3, SnO_2, ZnO, and Ga_2O_3 and their role in the resulting electronic properties. In addition to carefully considering the statistical distributions of the nearest- and next-nearest-neighbor structural characteristics for both metal and oxygen atoms, we examine the individual M and O features and compare the results to the corresponding values in the crystalline oxides. To establish rigorous structure–property relationships, we calculate a so-called effective coordination number (ECN) for every atom based on a weighted average distance in the given polyhedron [33]. This approach provides a significant improvement of the structural description of the ionic AOSs as compared to the typical cutoff-based calculations that neglect bond distribution in individual polyhedra and hence may under- or overestimate the coordination for a large fraction of atoms. Similarly, we avoid the traditional electronic structure tools, such as atom-resolved density-of-states calculations that also rely on a fixed cutoff radius neglecting nonspherical charge density distribution near low-coordinated atoms or those in highly distorted environments. Instead, we employ Bader charge analysis that assigns charge values based on the carefully calculated gradients in the charge density distribution around each atom [34, 35]. Relating the structural peculiarities of each individual atom to its Bader charge contribution in the valence and conduction band helps us establish the microscopic origins of the electron (de)localization in the disordered oxide materials—and develop a framework for an accurate description of complex AOSs' behavior. In addition to the four binary oxides that are common constituents in the AOS phase space, we extend our analysis to multicomponent In-Ga-O (IGO) and In-Ga-Zn-O (IGZO) to highlight how metal composition affects the intricate structure–property relationships of AOSs with tunable properties.

2.2 Brief Description of Methods and Approaches

2.2.1 Computational Approach

All amorphous oxide structures were obtained using an *ab initio* MD liquid-quench approach, as implemented in the Vienna Ab Initio Simulation Package (VASP) [36, 37]. The calculations are based on DFT with periodic boundary conditions and employ Perdew–Burke–Ernzerhof (PBE) exchange–correlation functionals [38] within the projector augmented-wave method [39, 40]. An initial stoichiometric oxide structure with specific density was melted at 3000 K to eliminate any crystalline memory and randomize the composition. The melting step was followed by quench cycles with a specific cooling rate and an equilibration MD step at room temperature to stabilize the amorphous structure. Next, each atomic configuration was fully relaxed using DFT-PBE at 0 K. For optimization, the cutoff energy of 500 eV and the 4×4×4 Γ-centered k-point mesh were used; the atomic positions were relaxed until the Hellmann–Feynman force on each atom was below 0.01 eV/Å. The electronic and optical properties of amorphous In-based oxides were calculated using the hybrid Heyd–Scuseria–Ernzerhof (HSE06) approach [41] with a mixing parameter of 0.25 and a screening parameter μ of 0.2 $Å^{-1}$.

To obtain adequate statistical distributions in the structural and, consequently, the electronic properties, 10–20 separate MD liquid-quench realizations with the same parameters and conditions (density, composition, stoichiometry, initial temperature, quench rate, equilibration, and relaxation) were obtained for each system. The density of an amorphous structure is an important factor and must be carefully determined. A set of independent MD liquid-quench simulations were performed for 4–5 different density values for each composition, resulting in over 300 MD realizations performed for this work. Upon room-temperature equilibration, the DFT total energy was calculated as an average over the final 500 MD steps to remove thermal fluctuations and plotted as a function of density for each oxide.

To validate the MD simulated structures, we calculate the extended X-ray absorption fine-structure (EXAFS) spectra for all amorphous models [42, 43] to directly compare the results with available experimental EXAFS measurements that capture the changes in the M-O bond lengths as well as in the second-shell M-M characteristics responsible for the medium-range structure at different densities.

Structural randomness in amorphous oxides is then quantified by analyzing the characteristics of individual atoms. For this, the effective average distance (l_{av}) for the M-O (O-M) first shell for each M-O (O-M) polyhedron was calculated using a weighted average, where each M-O (O-M) bond distance is compared to the shortest M-O (O-M) distance in each given polyhedron [34, 35]. Next, the ECN for individual M and O atoms was calculated as a sum of the first-shell distances weighted with respect to the effective average length obtained for each polyhedron. In addition, we calculate the distortion of each M-O (O-M) polyhedron σ^2, characterized by the standard deviation of the individual M-O bond lengths from the effective average M-O bond length for the given polyhedron. As mentioned in Section 2.1, this approach provides a significant improvement for the structural analysis of the ionic AOSs as compared to the typical approach that uses a single cutoff distance for all polyhedra, thus neglecting bond distribution in individual polyhedra, and hence may under- or overestimate the coordination of a large fraction of atoms in AOSs.

In addition, the total and partial vibrational density-of-states (VDOS) calculations were performed for the optimized structures to provide crucial information about local bonding and the dynamical stability of a model. The vibrational inverse participation ratio (VIPR) was calculated from the normalized displacement vectors to determine the localization of different vibrational modes. Finally, optical absorption was derived from the frequency-dependent dielectric function calculated within independent particle approximation using the electronic transitions of the hybrid functional (HSE06) solution. The atomic structures and charge densities were plotted using VESTA software [44].

2.2.2 Experimental Approach

Amorphous binary oxide thin films were grown by pulsed-laser deposition (PLD) from dense hot-pressed indium oxide, zinc oxide, tin oxide, or gallium oxide targets (25 mm diameter). PLD was accomplished with a 248 nm KrF excimer laser with 25 ns pulse duration and operated at 2 Hz. The 200 mJ/pulse beam was focused onto a 1 mm × 3 mm spot size. The target was rotated at 5 rpm about its axis to prevent localized heating. The target–substrate separation was fixed at 10 cm. Both the amorphous and crystalline films were grown at O_2 ambient pressure of 7.5 mTorr. The silica substrates were attached to the substrate holder with silver paint and grown at a specific deposition temperature to ensure the films are amorphous. Films grown above 25 °C were attached to a resistively heated substrate holder; films grown below 25 °C were attached to a liquid nitrogen–cooled substrate holder. The amorphous In_2O_3, SnO_2, and Ga_2O_3 films were grown at −25 °C. The amorphous ZnO film was grown at −100 °C. All crystalline films were grown at +600 °C.

X-ray absorption spectroscopy (XAS) was performed at the 5-BMD (bending-moment diagram) beam line of DND-CAT (the DuPont–Northwestern–Dow Collaborative Access Team) at the Advanced Photon Source (APS) of Argonne National Laboratory (Argonne, IL). Metal k α fluorescence emissions from the metal-oxide thin films were measured using a four-element Si-drifted detector (SII) with the incident X-ray angle θ at about 45° with respect to the sample surface. The data were Fourier transformed with a Hanning window over multiple k ranges, where one-shell and three-shell fits were examined. The k ranges of the EXAFS data used in the analyses were k = 2.5 to 11.972 Å^{-1} (ZnO), 2.0 to 15.5 Å^{-1} (In_2O_3), 1.0 to 14.0 Å^{-1} (Ga_2O_3), and 2.3 to 15.8 Å^{-1} (SnO_2), with a k-weight of 3.

2.3 The Structure and Properties of Crystalline and Amorphous In_2O_3

First, the total pair-correlation functions for crystalline (bixbyite) and amorphous stoichiometric In_2O_3 with different densities are shown in Figure 2.1a. Both crystalline and amorphous results were obtained from the MD simulations at 300 K (equilibration with 3,000 MD steps or 6 ps) to include possible temperature fluctuations. The results for amorphous structures represent an average over at least 10 separate MD realizations to provide better statistics. Structural disorder slightly shifts the first-shell peak toward longer In-O distances; in addition, the In-O distance distribution widens to include a small fraction of the In-O bonds with longer distances, between 2.3 Å and 2.6 Å (Figure 2.1a). This behavior is common to other semiconductors such as Si. More striking changes occur in the O-O and In-In distance distributions, with both the O-O peak at 2.8 Å and the In-In peak at 3.4 Å—as well as the longer range peaks—suppressed in the amorphous case. The disorder-induced changes in the O-O distribution set the ionic AOSs apart from the covalent silica-based glasses, where the O-O peak is preserved, signifying that the oxygen environment around the main-group metals (with well-defined O-O distances in the highly symmetric Si-O polyhedra) is maintained upon amorphization. Because the strength of the M-O bonding determines the rigidity of the local polyhedral structure, indium oxide, having the weakest bonding among the posttransition metal oxides considered in this work, features the largest average distortions in the first-shell M-O polyhedra (Table 2.1). Therefore, despite seemingly unchanged first-shell In-O distribution upon amorphization, the short-range (nearest-neighbor) disorder should be expected from the distortions associated with the loss of symmetry and significant bond deviations within individual In-O polyhedra. In the medium range, at ~3.2 Å and above (Figure 2.1a), the broader In-In distance distribution suggests that sharing of the neighboring In-O polyhedra changes significantly upon amorphization, as indeed can be seen from the structural comparison given in Figure 2.1c and 2.1d. The changes in In-O polyhedra sharing will be quantified in Section 2.7.

To determine the optimal density of amorphous In_2O_3, at least 10 independent MD liquid-quench simulations were performed for five different density values. Upon equilibration of each configuration at 300 K for 6 ps, the DFT total energy was calculated as an average over the final (stable) 500 MD steps to remove thermal fluctuations.

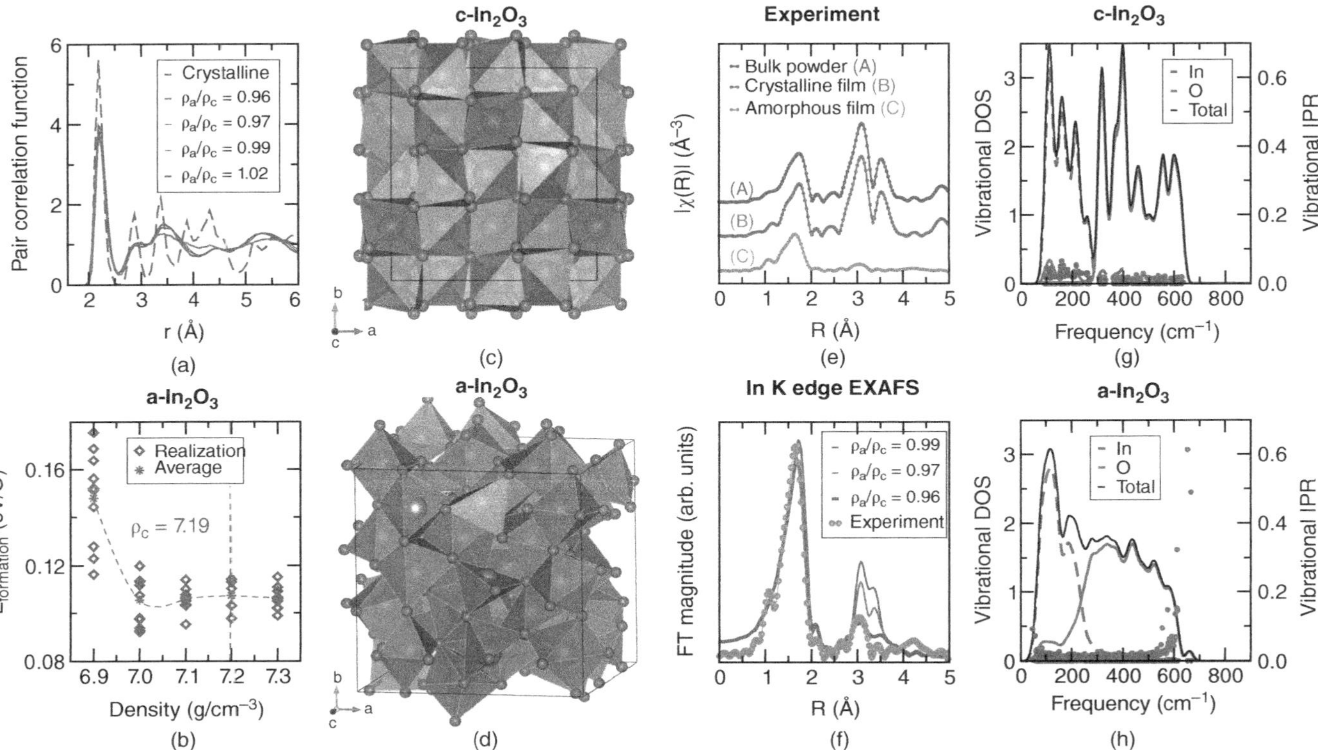

Figure 2.1 General structural properties of In_2O_3. (a) Total pair-correlation function for crystalline (bixbyite) and amorphous In_2O_3 with different densities. Both crystalline (dashed line) and amorphous (solid lines) results were obtained from MD simulations at 300 K (equilibration with 3,000 MD steps = 6 ps). The results for amorphous structures represent an average over at least 10 separate MD realizations at each density. (b) The energy–density curve for amorphous In_2O_3 calculated as $E_{formation} = E_{TOT}(\text{amorphous}) - E_{TOT}(\text{crystalline})$. Upon equilibration of each configuration at 300 K for 6 ps, the DFT total energy was calculated as an average over the final (stable) 500 MD steps to remove thermal fluctuations in each of the 10 realizations (diamond); the average over the realizations is represented by a star symbol. (c,d) Crystalline and amorphous atomic structures with In-O polyhedra highlighted. In and O atoms are represented by large and small spheres, respectively. (e) Experimental extended X-ray absorption fine structure (EXAFS) for an In K edge in bulk powder, crystalline film, and amorphous film samples of In_2O_3. (f) Calculated EXAFS for an In K edge in amorphous In_2O_3 models with different densities (solid line) as compared to experimental spectra for amorphous film (line with circle symbols). (g,h) Total and partial vibrational density of states (VDOS) and vibrational inverse participation ratio (VIPR) calculated for crystalline and the most stable amorphous In_2O_3 structures. In (O) states are represented by dashed (solid) thick lines, and the total VDOS is a thin solid line.

Table 2.1 First-shell effective coordination number (ECN), effective average distance, and distortion for metal and oxygen atoms for crystalline and amorphous binary oxides at different densities.

		Metal–Oxygen			Oxygen–Metal		
	Density, g/cm^3	ECN	l_{av},Å	σ^2, ×10^4 Å^2	ECN	l_{av}, Å	σ^2, ×10^4 Å^2
c-In$_2$O$_3$	**7.20**	**5.84**	**2.17**	**33**	**3.84**	**2.17**	**50**
		5.73	**2.17**	**57**			
a-In$_2$O$_3$	7.30	5.38	2.19	124	3.65	2.20	126
	7.20	5.38	2.20	121	3.64	2.20	126
	7.10	5.35	2.20	127	3.62	2.20	132
	*7.00**	*5.32*	*2.21*	*132*	*3.60*	*2.21*	*135*
	6.90	5.14	2.18	141	3.50	2.19	136
c-SnO$_2$	**6.95**	**5.90**	**2.05**	**19**	**2.95**	**2.05**	**19**
a-SnO$_2$	6.81	5.50	2.09	119	2.80	2.09	97
	*6.48**	*5.41*	*2.09*	*117*	*2.76*	*2.09*	*98*
	6.17	5.26	2.09	116	2.70	2.09	91
	5.88	5.14	2.09	118	2.72	2.10	91
c-ZnO	**5.61**	**3.91**	**1.98**	**25**	**3.91**	**1.98**	**24**
a-ZnO	5.78	3.80	2.00	104	3.80	2.00	106
	5.50	3.70	2.00	99	3.70	2.00	96
	*5.23**	*3.64*	*2.00*	*96*	*3.65*	*2.00*	*106*
	4.98	3.53	1.99	102	3.50	2.00	108
c-Ga$_2$O$_3$	**6.44**	**3.93**	**1.84**	**16**	**3.46**	**1.96**	**139**
		5.59	**1.98**	**75**	**2.89**	**1.87**	**38**
					2.83	**1.90**	**55**
a-Ga$_2$O$_3$	6.12	4.67	1.93	123	3.19	1.94	115
	5.80	4.50	1.92	115	3.10	1.94	110
	5.50	4.34	1.92	97	2.94	1.92	96
	*5.23**	*4.26*	*1.92*	*87*	*2.85*	*1.92*	*90*
	4.97	4.16	1.91	79	2.77	1.91	87

Note: The values represent an average over all metal or oxygen atoms in the supercell and over 10 separate MD realizations at the given density. Bixbyite, rutile, wurtzite, and monoclinic β-phase structures are considered for crystalline In_2O_3, SnO_2, ZnO, and Ga_2O_3, respectively.

The results, shown in Figure 2.1b, suggest that (i) the formation energy of different MD realizations varies within at least 0.02 eV/atom, and the variation is largest for the lowest density case, most likely due to differences in morphology; (ii) several MD realizations (or bigger cells) are required to predict the optimal density; and (iii) there is a substantial overlap between the total energy for the configurations with different densities, giving an extremely shallow minimum in the energy–density plot. The latter signifies that amorphous indium oxide samples with density values within a wide range (7.00–7.30 g/cm^3) could be grown. Different deposition techniques and pre- or postdeposition conditions are likely to affect the density of amorphous In_2O_3 oxide. Moreover, we speculate that the density of amorphous In_2O_3 may change long after the sample was deposited, making the structural characterization challenging. In the discussion here, amorphous indium oxide structures with a density of 7.00 g/cm^3 are used; this optimal density is slightly lower than the crystalline density of 7.19 g/cm^3.

To validate the theoretical amorphous models, the EXAFS is calculated for the In K edge for amorphous In_2O_3 models with different densities. From experimental EXAFS (Figure 2.1e), the peak that corresponds to the indium first shell (In-O) is maintained, whereas the second-shell (In-In) peak is significantly suppressed upon amorphization, in accordance with our pair-correlation function results (Figure 2.1a). Overall, an excellent agreement between the theoretical and experimental EXAFS is obtained (Figure 2.1f). Moreover, the results further corroborate the crucial effect that density has on the In-In distances and coordination (Figure 2.1f).

Next, total and partial VDOSs were calculated for crystalline and the most stable amorphous In_2O_3 structures to verify the dynamical stability of the latter structure and to identify disorder-induced changes. In addition, the VIPR was calculated from the normalized displacement vectors to determine the localization of different vibrational modes. The results, shown in Figure 2.1e and 2.1f, demonstrate that disorder not only makes the VDOS featureless by smearing and suppressing the well-defined peaks within the entire range of frequencies, but also leads to a notable overlap between the In and O VDOSs. The overlap occurs primarily due to a shift of the oxygen VDOS toward the lower frequencies. Similarly, the In VDOS develops low-frequency modes that do not exist in bixbyite In_2O_3. The VDOS shifts toward lower frequencies for both oxygen and indium are characteristic of amorphous phases. At the same time, several oxygen atoms that oscillate at normal modes with high frequencies of around 550–650 cm^{-1} (with the largest VIPR value of 0.6) and several In atoms that oscillate with low frequencies of 50 cm^{-1} (the largest VIPR value of 0.1) appear in the amorphous phase, whereas the remaining frequencies show low VIPR values, comparable to those in the crystalline mode and suggesting an evenly distributed vibration among different atoms within most of the spectrum.

Our comprehensive structural analysis begins with a comparison of the ECN distributions for the first-shell In-O and O-In polyhedra for the crystalline and amorphous indium oxide with perfect stoichiometry (i.e., for the oxygen-defect-free structures; Figure 2.2a). To account for room-temperature atomic fluctuations, the ECN is calculated for each individual In and O atom as a time average obtained from MD simulations at 300 K for 3,000 steps (6 ps). For amorphous phases, we analyzed 10 independent MD realizations at the optimal density; Figure 2.2 combines the results for the 10 realizations.

In bixbyite In_2O_3, there are two non-equivalent In sites, 8b and 24d; the former represents a perfect octahedron with six oxygen neighbors located at a distance of 2.17 Å and an ECN of 6.00, whereas the latter is a distorted octahedra with two O neighbors at 2.13 Å, two at 2.20 Å, and two at 2.23 Å, making the average In-O distance to be 2.18 Å and ECN = 5.91 for the 24d site. At room temperature, the atomic vibrations reduce the ECN of the two In types to 5.84 and 5.73, respectively, and the effective average In-O distance is slightly reduced to 2.17 Å for the 24d site (Table 2.1). Importantly, the temperature effects make the two types of In atoms indistinguishable, representing both with a single-peaked broad distribution of ECN (Figure 2.2a). Oxygen atoms in crystalline In_2O_3 are four-coordinated with In neighbors. At room temperature, the average effective coordination of oxygen atoms is 3.84.

ECN distributions for crystalline and amorphous In_2O_3 are shown in Figure 2.2a. Clearly, both indium and oxygen distributions become nearly two times wider upon amorphization. The number of "fully" coordinated In and O atoms is suppressed; moreover, secondary peaks at ECN(In) = 5 and ECN(O) = 3 become visible. Thus, despite the similarities in the nearest-neighbor indium–oxygen distance distributions in the crystalline and amorphous phases (Figure 2.1a), disorder leads to significant bond deviations in individual polyhedra, reducing the coordination numbers for a large fraction of In-O and O-In. The average ECN values for different densities are given in Table 2.1.

As can be expected, the angle distributions for both O-In-O and In-O-In also widen and become more uniform in the amorphous In_2O_3 (Figure 2.2b). As mentioned in Section 2.1, the distortions in the O-In polyhedra are expected to have a more pronounced effect on the electronic properties of AOSs due to the directional nature of the O--*p*-orbitals as compared to the symmetry-indifferent spherical In-*s*-orbitals [14]. It should be noted here that the In-O-In angle also can be viewed as a measure of the mutual alignment of neighboring InO polyhedra. In bixbyite In_2O_3, the peak at about In-O-In = 128° corresponds to the corner-shared InO_6 polyhedra; this peak

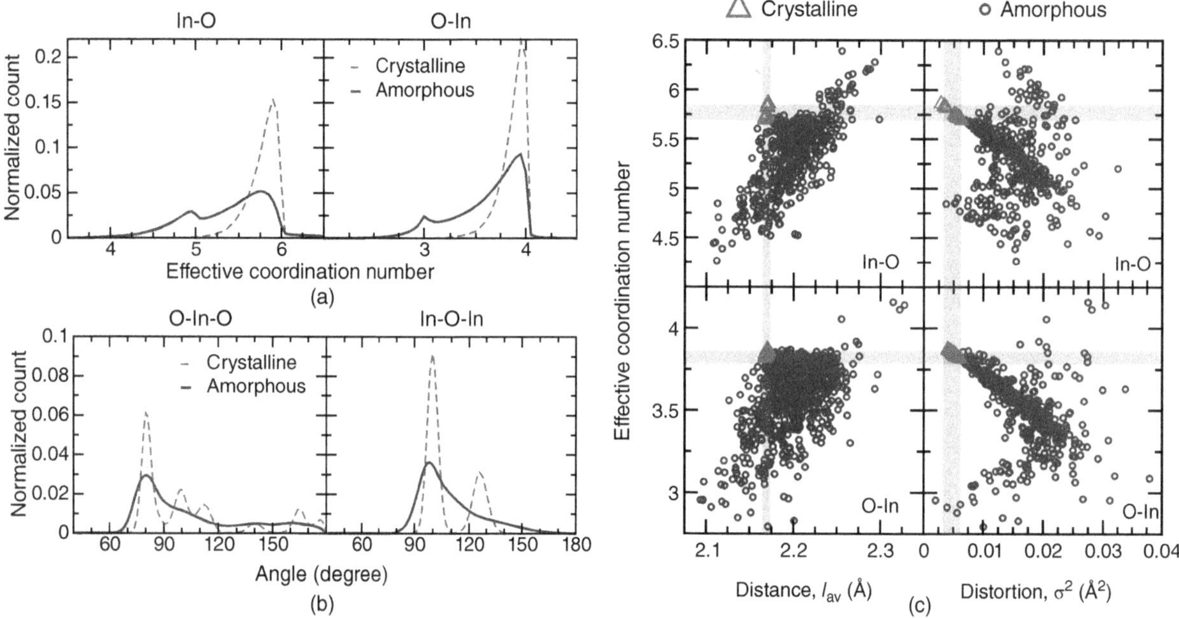

Figure 2.2 Structural randomness: The properties of individual In-O and O-In polyhedra in In_2O_3. (a) Distribution of the effective coordination number of In and O in crystalline (dashed line) and amorphous (solid line) phases. (b) Distribution of the O-In-O and In-O-In angles in crystalline (dashed line) and amorphous (solid line) phases. The results for amorphous structures represent an average over at least 10 separate MD realizations at optimal density. (c) Effective coordination number of individual In and O as a function of the effective average distance or distortion for the given atom in crystalline (triangles and shaded region) and 10 amorphous (circles) phases. All results are calculated based on the MD simulations at 300 K (equilibration with 3,000 MD steps = 6 ps).

is strongly suppressed in the amorphous phase, suggesting a significant disorder in the medium-range structure even though the density of the amorphous structure is only 3% lower than that in the crystalline oxide.

The remarkable difference between crystalline and amorphous In_2O_3 can be best visualized when the time-averaged ECN values for the individual In and O atoms are plotted as a function of their time-averaged effective distance and distortion (Figure 2.2c). Strikingly, despite the similarity of the In-O distance distributions for crystalline and amorphous oxides (Figure 2.1a), only four In atoms out of 540 within the 10 MD realizations have structural values that are similar to those in bixbyite In_2O_3 (i.e., $5.70 < \text{ECN} < 6.00$ and $\sigma^2 < 0.0060$), although all four have the effective average In-O distance of 2.20 Å, which is above the corresponding average crystalline distance of 2.17 Å. Importantly, the amorphous oxide features a notable fraction of In atoms that are undercoordinated (ECN < 5.0) and have low distortions ($\sigma^2 < 0.01$ Å^2) as well as those that are overcoordinated (ECN > 6.0) (Figure 2.2c). Similar observations can be done for oxygen atoms. For the majority of atoms, however, there are two clear correlations: (i) the lower the ECN value, the larger the distortions in the polyhedral are; and (ii) the reduction in ECN value is generally associated with shorter average bond length, as should be expected for an ionic material. While the spherical symmetry of the In *s*-states makes the In atoms indifferent to the exact direction in which the oxygen neighbors are located (the distribution of the O-In-O angles is wide and uniform; Figure 2.1b), the overlap between the spherical *s*-orbital of In and the *p*-orbitals of O atoms may be affected by changes in the In-O distances. This situation will be discussed in the electronic localization section of this chapter (Section 2.7).

Thus, disorder leads to significant bond deviations and reduced effective coordination numbers for individual In and O atoms, even though the density of amorphous oxide is lower by less than 3% as compared to that in the bixbyite In_2O_3 and the amorphous structures are perfectly stoichiometric. This result highlights the importance of investigating ECN and distortion of individual atoms compared to merely examining bond distance averages: Distance distributions do not capture the degree or prevalence of the M-O polyhedra distortions that ultimately determine the metal coordination and quantify structure randomness.

The analyzed structural values averaged over 10 MD realizations, and over 54 In or 81 O atoms in each realization for the amorphous In_2O_3 at each density are compared to those for the crystalline oxide in Table 2.1. As expected, the effective coordination of In and O decreases for lower density, whereas the distortions increase. In agreement with the pair-correlation function shown in Figure 2.1a, the averaged first-shell distances are nearly insensitive to the density changes.

Now, the effect of the broad distributions in the main structural characteristics of In and O atoms in amorphous In_2O_3 on the electronic properties should be determined. For this, charge density distributions in the highest valence band and the lowest conduction band are calculated for the fully relaxed structures using an accurate hybrid functional approach. Figure 2.3a–2.3d compares the results for the crystalline and amorphous oxide. The uniform charge density distributions observed for both the valence and conduction bands of bixbyite In_2O_3 are not maintained upon amorphization. However, the changes in charge density distribution in the valence and conduction bands are markedly different. The conduction band (Figure 2.3a and 2.3b) features various amounts of charge accumulation around different In atoms, with a small fraction having a negligible charge density in their vicinity. Although the conduction charge distribution in amorphous In_2O_3 is clearly not as uniform as it is in the crystalline oxide, a large enough fraction of the In atoms contribute to the conduction states to create conductivity paths with significant overlap between the neighboring atoms' nonspherical density, resulting in interstitial or interatomic bridgelike charge distributions. In contrast, only a few oxygen atoms contribute to the top of the valence band, suggesting a strong electron localization (Figure 2.3c and 2.3d). As mentioned in Section 2.1, this characteristic asymmetry of the electronic localization of the states near the band edges in AOSs stems from the different sensitivity of the directional O-*-p*-orbitals and spherically symmetric In-*-s*-orbitals to disorder.

To relate the specific structural characteristics of individual O or In atoms with their valence or conduction charge, we performed Bader charge analysis and plotted the individual ECN, effective distance (l_{av}), and distortion (σ^2) as a function of Bader charge calculated for the given atom. As expected from the charge distribution plot (Figure 2.3d), the strong electron localization (with Bader contributions as large as 40% of the total charge in the cell; Figure 2.3e) is found for several O atoms. For the In atoms, the largest Bader charge contribution to the conduction band is significantly smaller: only 6% of the total charge in the cell. Overall, the much broader valence Bader charge distribution for the oxygen atoms is asymmetric and shifted toward a lower Bader charge from the expected uniform charge distribution value (shown as a vertical line in Figure 2.3e). This suggests that most of the oxygen atoms give negligible contributions at the top of the valence band. In contrast, the conduction Bader charge distribution is two times narrower for the In atoms and is centered at the expected uniform charge value (Figure 2.3f). Therefore, the majority of the In atoms contribute to the conduction states at a similar level as if the charge is evenly distributed throughout the cell. Most strikingly, no clear correlation is found between the Bader charge and the coordination, distances, or distortion of O or In atoms (Figure 2.3e and 2.3f, respectively). In other words, both fully coordinated and undercoordinated atoms, those with shorter or longer effective distances than the overall average, and those with equivalent or highly uneven bonds in the polyhedra all have a similar probability for a large or a negligible Bader contribution. Therefore, distribution outliers do not necessarily represent electronic "defects" and do not necessarily contribute to the electronic tail states in amorphous In_2O_3. This result implies that the above structural parameters alone do not serve as a predictor for electron (de)localization. As will be shown in this chapter, two or all three of the introduced structural parameters (ECN, l_{av}, and σ^2) should be used in combination and also be compared to the structural features of the surrounding atoms in order to predict the electronic properties.

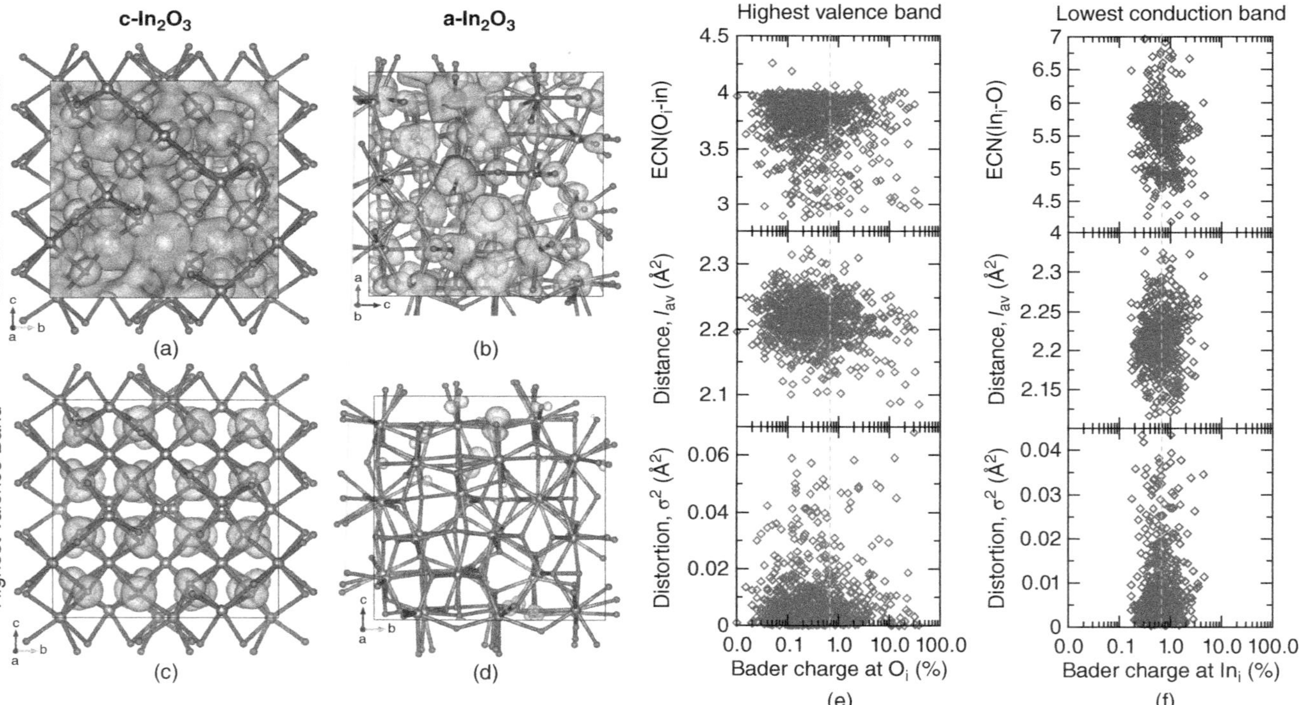

Figure 2.3 Calculated charge density distribution in the (a,b) lowest conduction band and (c,d) highest valence band in (a,c) crystalline and (b,d) amorphous In_2O_3. (e,f) First-shell effective coordination number (ECN), effective average distance, and distortion for individual O and In atoms as a function of the Bader charge contribution to the corresponding band from the given atom. The results are for 10 separate MD realizations at optimal density. The vertical lines represent the calculated charge value for a uniform charge density distribution (i.e., if all O or In atoms in the supercell contribute equally). The electronic properties are calculated for the fully relaxed structures using an accurate hybrid functional approach.

2.4 The Structure and Properties of Crystalline and Amorphous SnO_2

The total pair-correlation functions for crystalline (rutile) and amorphous stoichiometric SnO_2 with different densities are shown in Figure 2.4a. Similar to the In_2O_3 case (Figure 2.1), the nearest-neighbor Sn-O distances are maintained upon amorphization: Structural disorder slightly shifts the first-shell peak toward longer Sn-O distances and widens the Sn-O distance distribution to include a small fraction of the Sn-O bonds with longer distances (Figure 2.4a). Also, the O-O distribution (peaks at 2.6 Å, 2.9 Å, and 3.2 Å in crystalline SnO_2) and the Sn-Sn distribution (peaks at 3.2 Å and 3.7 Å)—as well as the longer range peaks—are nearly completely suppressed in the amorphous case, especially in low-density cases. Hence, similar to the In_2O_3 case, amorphous SnO_2 exhibits strong short-range (nearest-neighbor) disorder and significant changes in the medium-range structure (greater than ~3.2 Å). Strong disorder in the SnO polyhedra sharing can be seen from the structural comparison given in Figure 2.4c and 2.4d.

The optimal density of amorphous SnO_2 was determined based on the energy–volume curve with at least 10 independent MD liquid-quench simulations performed for each density value (Figure 2.4b). The results suggest that (i) the formation energy of different MD realizations varies within 0.03–0.04 eV/atom, most likely due to differences in morphology; (ii) several MD realizations (or bigger cells) are required to predict the optimal density; and (iii) there is a substantial overlap between the total energy for the configurations with different densities, giving a shallow minimum in the energy–density plot. The latter signifies that amorphous tin oxide samples with density values within the considered range (5.9–6.8 g/cm^3) could be grown. In the discussion here, the amorphous tin oxide structures with a density of 6.48 g/cm^3 are used; this optimal density is slightly lower than the crystalline density of 6.95 g/cm^3.

Similar to In_2O_3, disorder strongly suppresses the Sn second-shell (Sn-Sn) peak, whereas the tin first-shell (Sn-O) peak is less affected, as can be seen from experimental EXAFS (Figure 2.4e). This agrees well with our pair-correlation function results (Figure 2.4a). An excellent agreement between theoretical and experimental EXAFS is obtained for the Sn K edge for amorphous SnO_2 (Figure 2.4f), validating the theoretical MD models. In marked contrast to In_2O_3, density in amorphous SnO_2 has a small effect on the Sn-Sn distances and coordination (cf. Figure 2.1f and 2.4f). At the same time, the Sn-O coordination appears to be more sensitive to the amorphous oxide density as compared to that of In-O (Table 2.1).

Next, the total and partial VDOS and the VIPR were calculated for crystalline and the most stable amorphous SnO_2 structures (Figures 2.4e and 2.4f). In rutile SnO_2, the oxygen VDOS is split into low- and high-frequency bands with a pronounced gap between them. The former corresponds to a combination of Sn and O modes with comparable VDOS except for the lowest frequency range, governed by Sn atoms (Figure 2.4e). Similar to In_2O_3, disorder makes the VDOS featureless and shifts both the oxygen and tin VDOS toward the lower frequencies. High-frequency modes (around 650–850 cm^{-1}) associated with oscillations of a few oxygen atoms, with the largest VIPR value of around 0.4, as well as low-frequency modes (50 cm^{-1}) of a few In atoms with the largest VIPR value of less than 0.1 appear in the amorphous phase. However, in the intermediate-frequency range, the VIPR values are comparable to or even lower than those in the crystalline case, suggesting an evenly distributed vibration among different atoms within most of the spectrum.

A comparison of the ECN distributions for the first-shell Sn-O and O-Sn polyhedra for the crystalline and amorphous tin oxide with perfect stoichiometry (i.e., the oxygen-defect-free structures) is given in Figure 2.5a. In rutile SnO_2, there is only one equivalent type of Sn and O sites. Tin atoms have six nearly equidistant Sn-O bonds with a bond length of 2.05 Å, so that even at room temperature, when the atomic fluctuations are included, the symmetry of the Sn-O polyhedral remains close to an octahedron and the average effective coordination of Sn atoms is 5.90, which is higher than that in bixbyite In_2O_3, 5.84 and 5.73 (Table 2.1). Indeed, the polyhedral distortions are smaller in crystalline SnO_2, $\sigma^2 = 0.0019$ Å^2, as compared to 0.0033 Å^2 and 0.0057 Å^2 for two non-equivalent types of In atoms in crystalline In_2O_3. This may originate from a higher structural symmetry of SnO_2 and/or stronger Sn-O bonds as compared to the In-O bonds. All oxygen atoms are three-coordinated with Sn atoms, with an average room-temperature effective coordination of 2.95.

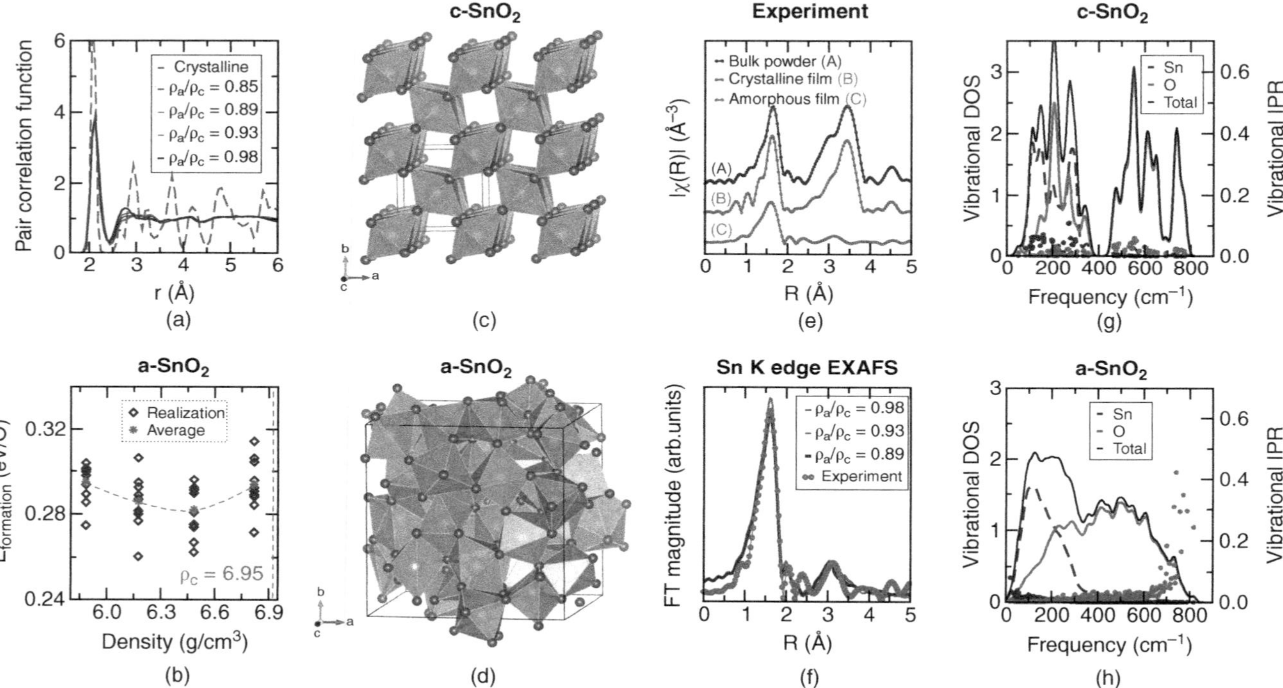

Figure 2.4 General structural properties of SnO_2. (a) Total pair-correlation function for crystalline (rutile) and amorphous SnO_2 with different densities. Both crystalline (dashed line) and amorphous (solid lines) results were obtained from MD simulations at 300 K (equilibration with 3,000 MD steps = 6 ps). The results for amorphous structures represent an average over at least 10 separate MD realizations at each density. (b) The energy–density curve for amorphous SnO_2 calculated as $E_{formation} = E_{TOT}(\text{amorphous}) - E_{TOT}(\text{crystalline})$. Upon equilibration of each configuration at 300 K for 6 ps, the DFT total energy was calculated as an average over the final (stable) 500 MD steps to remove thermal fluctuations in each of the 10 realizations (diamond); the average over the realizations is represented by a star symbol. (c,d) Crystalline and amorphous atomic structures with Sn-O polyhedra highlighted. Sn and O atoms are represented by large and small spheres, respectively. (e) Experimental extended X-ray absorption fine structure (EXAFS) for a Sn K edge in bulk powder, crystalline film, and amorphous film samples of SnO_2. (f) Calculated EXAFS for a Sn K edge in amorphous SnO_2 models with different densities (solid line) as compared to experimental spectra for amorphous film (line with circle symbols). (g,h) Total and partial vibrational density of states (VDOS) and vibrational inverse participation ratio (VIPR) calculated for crystalline and the most stable amorphous SnO_2 structures. Sn (O) states are represented by a dashed (solid) thick line, and the total VDOS is a thin solid line.

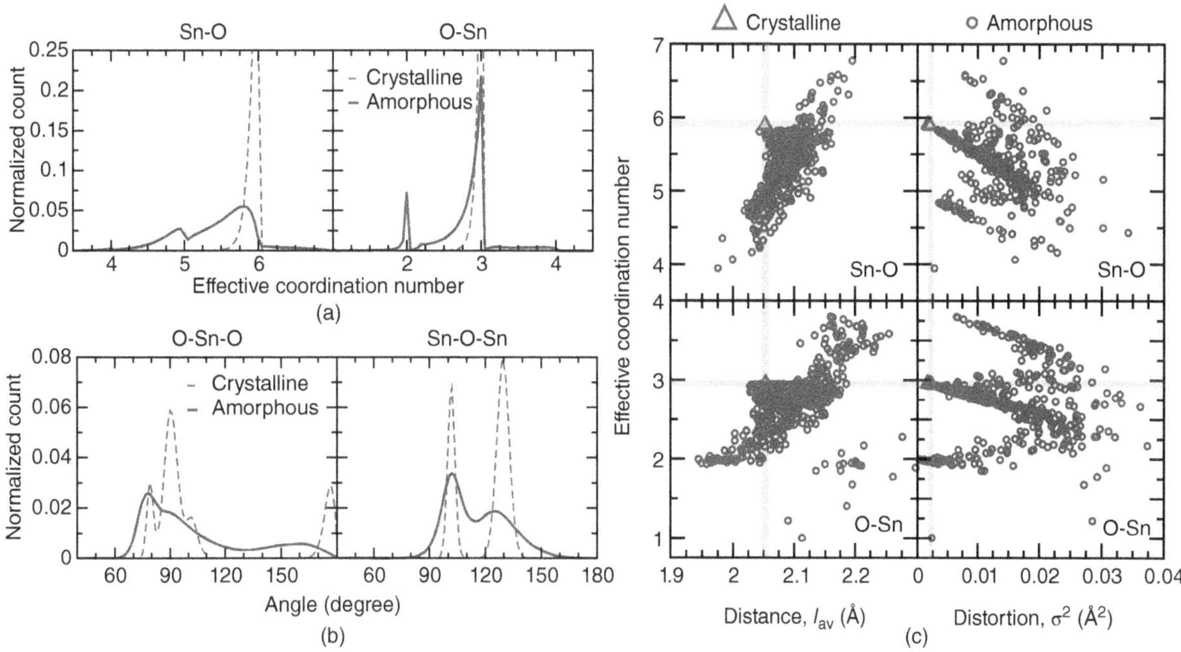

Figure 2.5 Structural randomness: The properties of individual Sn-O and O-Sn polyhedra in SnO_2. (a) Distribution of the effective coordination number of Sn and O in crystalline (dashed line) and amorphous (solid line) phases. (b) Distribution of the O-Sn-O and Sn-O-Sn angles in crystalline (dashed line) and amorphous (solid line) phases. The results for amorphous structures represent an average over at least 10 separate MD realizations at optimal density. (c) Effective coordination number of individual Sn and O as a function of effective average distance or distortion for the given atom in crystalline (triangles and shaded region) and 10 amorphous (circles) phases. All results are calculated based on the MD simulations at 300 K (equilibration with 3,000 MD steps = 6 ps).

Interestingly, the average ECN(Sn) values remain to be higher in amorphous SnO_2 than those in amorphous In_2O_3, although they become similar at lower densities (Table 2.1). Accordingly, the polyhedral distortion remains lower in amorphous SnO_2 than in amorphous In_2O_3 (Table 2.1). Given the lack of ordering in the amorphous phases, the difference stems solely from the weaker In-O bonds as compared to the Sn-O bond strength. The calculated ECN distributions for crystalline and amorphous SnO_2 are shown in Figure 2.5a. Similar to the In_2O_3 case, both tin and oxygen distributions become nearly two times wider upon amorphization. The number of "fully" coordinated Sn atoms is strongly suppressed, although for the O atoms the peak at ECN(O) = 3 value is high. At the same time, secondary peaks at ECN(Sn) = 5 and ECN(O) = 2 become visible. Thus, despite the similarities in the nearest-neighbor tin–oxygen distance distributions in the crystalline and amorphous phases (Figure 2.3a), disorder leads to significant bond deviations in individual polyhedra, reducing the coordination numbers for a large fraction of Sn-O and a smaller fraction of O-Sn. Both Sn and O coordination distributions feature a scarce but persistent number of highly coordinated atoms, above 6 and above 3, respectively. Such coordination values were not part of the crystalline phase.

The angle distributions for both O-Sn-O and Sn-O-Sn widen and become more uniform in the amorphous SnO_2, although the double-peak structure persists for Sn-O-Sn (Figure 2.5b), as expected from the directional nature of the O--*p*-orbitals. The Sn-O-Sn angle represents mutual alignment of neighboring SnO polyhedra. In rutile SnO_2, the majority (80%) of all Sn-Sn neighbors are corner-shared, which corresponds to Sn-O-Sn = 130°. Upon amorphization, the number of corner-shared Sn-Sn pairs decreases to 70%, but the associated Sn-Sn corner-shared distance distribution widens—in accord with the Sn-O-Sn peak at 130°. Yet, the double-peak structure for the Sn-O-Sn angle distribution suggests that the medium-range structure in amorphous SnO_2 is slightly less disordered than it is in amorphous In_2O_3, where the In-O-In angle distribution is single-peaked and long-tailed and the In-In corner-shared distance distribution is significantly broader, as will be shown in Section 2.7.

Disorder leads to significant bond deviations and reduced effective coordination numbers for individual Sn and O atoms, even though the amorphous structures are perfectly stoichiometric. The time-averaged ECN values for the individual In and O atoms, plotted as a function of their time-averaged effective distance and distortion for both crystalline and amorphous SnO_2, are shown in Figure 2.5c. Similar to the In_2O_3 case, only a few Sn and O atoms in the amorphous SnO_2 resemble the characteristics of those in the crystalline phase. Also, the amorphous oxide features a considerable fraction of O atoms that are overcoordinated (ECN > 3.0) and have notably longer effective O-Sn distances. Despite the outliers, the majority of oxygen atoms favor being three-coordinated with tin, even though their effective Sn-O distances vary within a wide range, from 2.02 Å to 2.15 Å. Independent of O coordination, a reduction in ECN(O) is associated with stronger distortion. The majority of tin atoms, similar to the In atoms, follow the expected trend: The lower the ECN value, the shorter the effective average bond length. The analyzed structural values, averaged over 10 MD realizations and over 46 In or 92 O atoms in each realization for the amorphous SnO_2 at each density, are compared to those for the crystalline oxide and to In_2O_3 cases in Table 2.1. As expected, the effective coordination of Sn and O decreases for lower density, although the O coordination seems to saturate at the lowest density. At the same time, the Sn-O polyhedral distortions remain nearly unchanged with density—in contrast to In_2O_3, where the In-O distortions increase when the density decreases. In agreement with the pair-correlation function shown in Figure 2.3a, the averaged first-shell distances are insensitive to the density changes.

Next, the effect of the broad distributions in the main structural characteristics of Sn and O atoms in amorphous SnO_2 on the electronic properties is determined. Figure 2.6a–2.6d compares the charge density distributions in the highest valence band and the lowest conduction band in the fully relaxed crystalline and amorphous structures. The disorder-induced changes are similar to those in In_2O_3 (cf. Figure 2.3a–2.3d): The conduction band charge is not uniform, yet a large enough fraction of the Sn atoms contribute to the conduction states to create conductivity channels with neighbor contributions that are alike in the charge value (Figure 2.6b). A small fraction of Sn atoms have a negligible charge density in their vicinity, and conversely, only a few oxygen atoms contribute to the top of the valence band (Figure 2.6c and 2.6d). The calculated Bader charge contributions from the O atoms to the highest valence band reveal even stronger asymmetry in the distribution as compared to the amorphous In_2O_3 (cf. Figure 2.3e and 2.6e): A significant number of oxygen atoms have zero or negligible contributions, while the Bader charge on several oxygen atoms is as large as 80% of the total charge in the cell. Therefore, the oxygen Bader charge distribution is wider in amorphous SnO_2 than in amorphous In_2O_3, and the electron localization on some nonbonding O-*p*-states is stronger. Similar to In_2O_3, however, we do not find a correlation between the ECN, effective distance (l_{av}), or distortion (σ^2) of oxygen atoms and the corresponding valence Bader charge contribution. For the Sn atoms, the Bader charge is slightly wider and slightly asymmetric with respect to the expected uniform charge value than that of In (cf. Figures 2.3f and 2.6f); yet the majority of Sn atoms contribute to the conduction states at a similar level, as if the charge is evenly distributed throughout the cell.

2.5 The Structure and Properties of Crystalline and Amorphous ZnO

The total pair-correlation function for crystalline (wurtzite) and amorphous stoichiometric ZnO with different densities is shown in Figure 2.7a. Similar to the In_2O_3 and SnO_2 cases (Figures 2.1 and 2.4), the nearest-neighbor Zn-O distances are maintained upon amorphization: Structural disorder only slightly shifts the first-shell peak toward longer Zn-O distances and widens the Zn-O distance distribution to include a fraction of the Zn-O bonds with longer distances (Figure 2.7a). In wurtzite ZnO, the O-O distribution and the Zn-Zn distribution coincide, resulting in a wide peak at 3.23 Å. Upon amorphization, the peak is significantly but not completely suppressed.

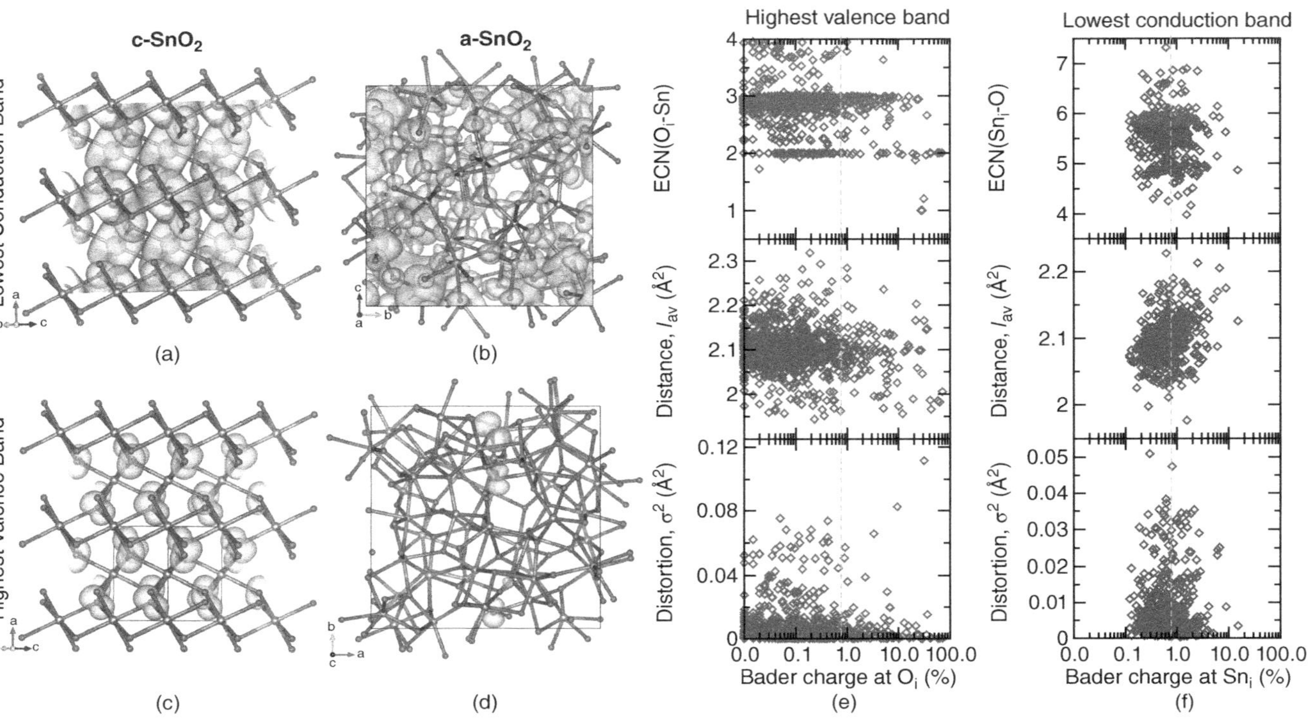

Figure 2.6 Calculated charge density distribution in the (a,b) lowest conduction band and (c,d) highest valence band in (a,c) crystalline and (b,d) amorphous SnO_2. (e,f) First-shell effective coordination number (ECN), effective average distance, and distortion for individual O and Sn atoms as a function of the Bader charge contribution to the corresponding band from the given atom. The results are for 10 separate MD realizations at optimal density. The vertical lines represent the calculated charge value for a uniform charge density distribution (i.e., if all O or Sn atoms in the supercell contribute equally). The electronic properties are calculated for the fully relaxed structures using an accurate hybrid functional approach.

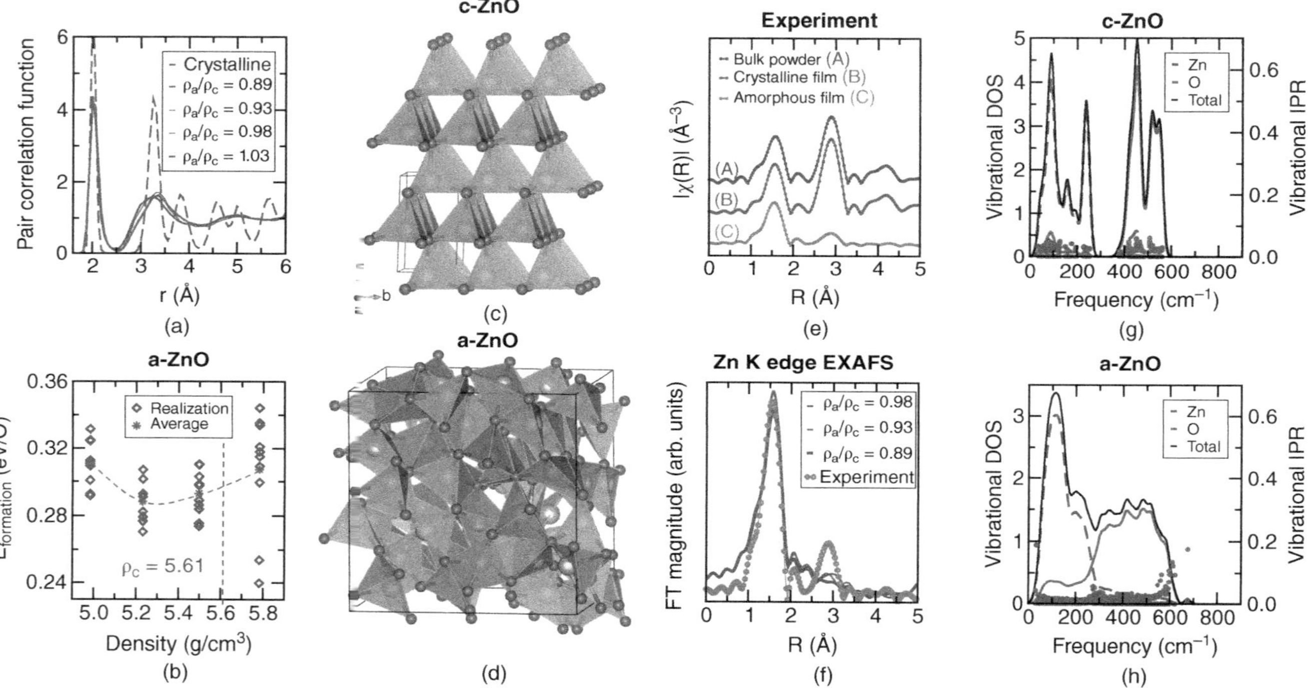

Figure 2.7 General structural properties of ZnO. (a) Total pair-correlation function for crystalline (wurtzite) and amorphous ZnO with different densities. Both crystalline (dashed line) and amorphous (solid lines) results were obtained from MD simulations at 300 K (equilibration with 3,000 MD steps = 6 ps). The results for amorphous structures represent an average over at least 10 separate MD realizations at each density. (b) The energy–density curve for amorphous ZnO calculated as $E_{formation} = E_{TOT}$(amorphous) − E_{TOT}(crystalline). Upon equilibration of each configuration at 300 K for 6 ps, the DFT total energy was calculated as an average over the final (stable) 500 MD steps to remove thermal fluctuations in each of the 10 realizations (diamond); the average over the realizations is represented by a star symbol. (c,d) Crystalline and amorphous atomic structures with Zn-O polyhedra highlighted. Zn and O atoms are represented by large and small spheres, respectively. (e) Experimental extended X-ray absorption fine structure (EXAFS) for a Zn K edge in bulk powder, crystalline film, and amorphous film samples of ZnO. (f) Calculated EXAFS for a Zn K edge in amorphous ZnO models with different densities (solid line) as compared to experimental spectra for amorphous film (line with circle symbols). (g,h) Total and partial vibrational density of states (VDOS) and vibrational inverse participation ratio (VIPR) calculated for crystalline and the most stable amorphous ZnO structures. Zn (O) states are represented by a dashed (solid) thick line, and the total VDOS is a thin solid line.

Hence, similar to the In_2O_3 and SnO_2 cases, amorphous ZnO is expected to feature short-range (nearest-neighbor) disorder and significant changes in the medium range. Strong disorder in the ZnO polyhedra sharing can be seen in Figure 2.7c and 2.7d.

The optimal density of amorphous ZnO was determined based on the energy–volume curve with at least 10 independent MD liquid-quench simulations performed for each density value (Figure 2.7b). The formation energy of different MD realizations varies within 0.04 eV/atom, and there is a substantial overlap between the total energy for the configurations with different densities, giving a shallow minimum in the energy–density plot. It must be noted that at high density values, several realizations resulted in nearly crystalline, well-ordered phases, suggesting that the amorphous zinc oxide could be challenging to grow (i.e., may require fast quench rates, very low deposition temperatures, or an additional constituent in order to stabilize the amorphous phase). In the discussion here, amorphous zinc oxide structures with a density of 5.23 g/cm^3 are used; this optimal density is slightly lower than the crystalline density of 5.61 g/cm^3.

Similar to In_2O_3 and SnO_2, disorder strongly suppresses the Zn second-shell (Zn-Zn) peak, whereas the zinc first-shell (Zn-O) peak is less affected, as can be seen from experimental EXAFS (Figure 2.7e). This agrees well with our pair-correlation function results (Figure 2.7a). An excellent agreement between theoretical and experimental EXAFS is obtained for the Zn K edge for the first shell in amorphous ZnO (Figure 2.7f). The second-shell Zn-Zn peak is completely suppressed in all theoretical models, showing no dependence on the oxide density. The pronounced Zn-Zn peak in experimental EXAFS may signify that the amorphous ZnO sample has a considerable fraction of crystalline nano-inclusions.

The total and partial VDOS and the VIPR, calculated for crystalline and the most stable amorphous ZnO structures, are shown in Figure 2.7e and 2.7f. Similar to rutile SnO_2, the VDOS is split into low- and high-frequency bands with a pronounced gap between them; however, there are only small contributions of the opposite atom type in each band. Disorder makes the VDOS featureless and shifts oxygen but not zinc VDOSs toward the lower frequencies, with a pronounced VDOS(O) tail stretching into the low-frequency Zn vibrational states. High-frequency modes (around 550–650 cm^{-1}) associated with oscillations of a few oxygen atoms, with the largest VIPR value of less than 0.2 and a single low-frequency mode (50 cm^{-1}) for zinc with a similar VIPR value, appear in the amorphous phase. With the exception of the outer band edges and also the band overlap region (around 340 cm^{-1}), the VIPR values are comparable to or even lower than those in the crystalline ZnO case, suggesting an evenly distributed vibration among different atoms within most of the spectrum.

A comparison of ECN distributions for the first-shell Zn-O and O-Zn polyhedra for crystalline and amorphous zinc oxide with perfect stoichiometry (i.e., for the oxygen-defect-free structures) is given in Figure 2.8a. In wurtzite ZnO, Zn and O sites are equivalent by symmetry. Zinc and oxygen atoms have four nearly equidistant Zn-O bonds with a bond length of 1.98 Å. At room temperature, when the atomic fluctuations are included, the symmetry of the Zn-O polyhedral remains close to that of a tetrahedron, and the average effective coordination of both Zn and O atoms is 3.91 (Table 2.1). Upon amorphization, both zinc and oxygen distributions become two times wider, similar to In_2O_3 and SnO_2 cases. Despite the lack of structural order, zinc and oxygen ECN distributions are nearly indistinguishable, with the number of four-coordinated atoms suppressed and a secondary peak at ECN = 3 appearing. Thus, similar to In_2O_3 and SnO_2 cases, the seemingly insignificant broadening of the first-shell distance distribution toward longer nearest-neighbor distances in amorphous ZnO leads to significant bond deviations in individual polyhedra, reducing the coordination numbers for a large fraction of Zn and O atoms. We note here that the average Zn and O effective coordination numbers decrease with decreasing density and begin to deviate from each other slightly at the lowest density value considered (Table 2.1). Compared to both amorphous In_2O_3 and SnO_2, the polyhedral distortions in amorphous ZnO are smaller, as expected from stronger Zn-O bonds as compared to the In-O and Sn-O bonds.

The angle distributions for both O-Zn-O and Zn-O-Zn widen but become less uniform in the amorphous ZnO, due to the presence of a shoulder at the 90° angle—as compared to wurtzite ZnO (Figure 2.8b). However, the 110° angle, characteristic for hexagonal lattice, prevails. In wurtzite ZnO, all Zn-Zn neighbors are corner-shared; upon amorphization, the number of corner-shared Zn-Zn pairs slightly decreases to 90% due to the appearance of edge-shared polyhedra, as will be discussed in Section 2.7.

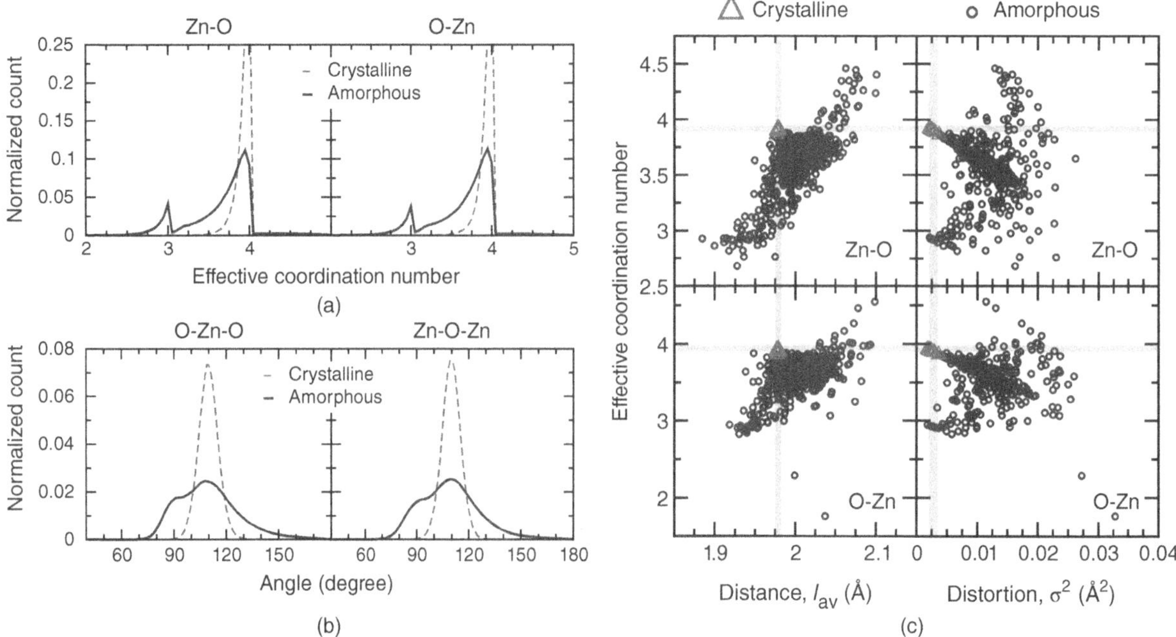

Figure 2.8 Structural randomness: The properties of individual Zn-O and O-Zn polyhedra in ZnO. (a) Distribution of the effective coordination number of Zn and O in crystalline (dashed line) and amorphous (solid line) phases. (b) Distribution of the O-Zn-O and Zn-O-Zn angles in crystalline (dashed line) and amorphous (solid line) phases. The results for amorphous structures represent an average over at least 10 separate MD realizations at optimal density. (c) Effective coordination number of individual Zn and O as a function of effective average distance or distortion for the given atom in crystalline (triangles and shaded region) and 10 amorphous (circles) phases. All results are calculated based on the MD simulations at 300 K (equilibration with 3,000 MD steps = 6 ps).

Similar to In_2O_3 and SnO_2 cases, disorder leads to significant bond distortions and reduced effective coordination numbers for individual Zn and O atoms, even though the amorphous ZnO structures are perfectly stoichiometric. The time-averaged ECN values for the individual Zn and O atoms, plotted as a function of their time-averaged effective distance and distortion for both crystalline and amorphous ZnO, are shown in Figure 2.8c. Despite the nearly identical ECN distributions for Zn and O (Figure 2.8a), the individual structural plots (Figure 2.8c) reveal several important differences between them: There is a smaller number of overcoordinated (ECN $>$ 4) O than that of Zn, and two-coordinated oxygen atoms were found. The former result can be again explained based on the orbital differences: The directional O--*p*-orbitals favor fourfold or threefold symmetry, whereas the spherical *s*-orbitals may add a fifth neighbor and maintain the charge balance by increasing the distances to the remaining O neighbors. The same trend is maintained for the undercoordinated Zn atoms that generally have shorter effective bond lengths (Figure 2.8c).

Figure 2.9a–2.9d compares the charge density distributions in the highest valence band and the lowest conduction band in the fully relaxed crystalline and amorphous ZnO structures. The disorder-induced changes are similar to those in In_2O_3 and SnO_2 (cf. Figures 2.3a–2.3d and 2.6a–2.6d): The conduction band charge is not uniform, yet a significant fraction of the Zn atoms contribute to the conduction states to create a conductivity path through the supercell. A fraction of Zn atoms have a negligible charge in their vicinity, and conversely, only a few oxygen atoms contribute to the top of the valence band (Figure 2.9c and 2.9d). The calculated Bader charge contributions from the O atoms to the highest valence band reveal strong asymmetry in the distribution: A significant number of oxygen atoms have negligible contributions, while several oxygen atoms have a Bader charge as large as 50% of the total charge in the cell, which is a weaker localization than that in SnO_2 and comparable to that in In_2O_3.

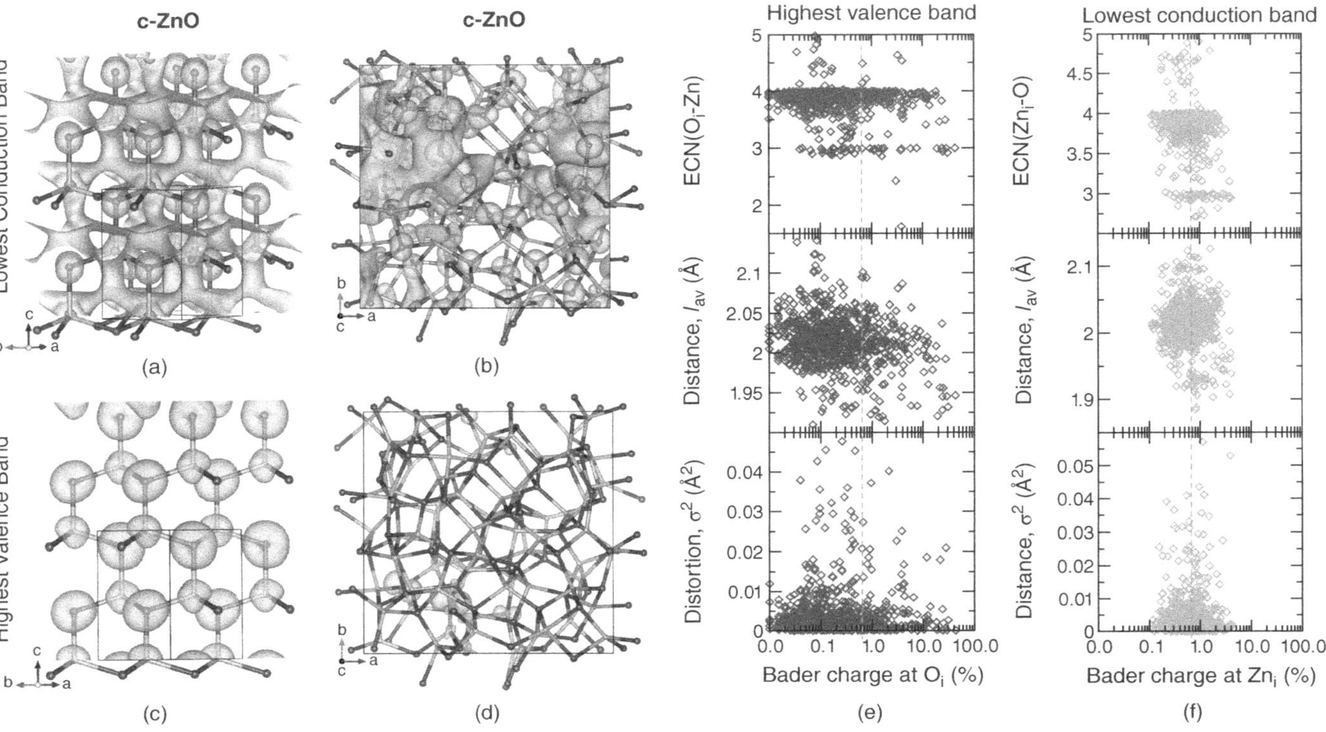

Figure 2.9 Calculated charge density distribution in the (a,b) lowest conduction band and (c,d) highest valence band in (a,c) crystalline and (b,d) amorphous ZnO. (e,f) First-shell effective coordination number (ECN), effective average distance, and distortion for individual O and Zn atoms as a function of the Bader charge contribution to the corresponding band from the given atom. The results are for 10 separate MD realizations at optimal density. The vertical lines represent the calculated charge value for a uniform charge density distribution (i.e., if all O or Zn atoms in the supercell contribute equally). The electronic properties are calculated for the fully relaxed structures using an accurate hybrid functional approach.

Similar to indium and tin oxides, we do not find a correlation between the ECN, effective distance (l_{av}), or distortion (σ^2) of oxygen atoms and the corresponding valence Bader charge contribution. For the Zn atoms, the Bader charge is slightly wider and more asymmetric with respect to the expected uniform charge value than that of In (cf. Figures 2.3f and 2.9f); yet the majority of Zn atoms contribute to the conduction states at a similar level as if the charge is evenly distributed throughout the cell. Therefore, although only a few Zn and O atoms in the amorphous ZnO resemble the same characteristics as those in the crystalline phase, the *s-p* hybridization is maintained and the conduction band remains delocalized.

2.6 The Structure and Properties of Crystalline and Amorphous Ga_2O_3

The total pair-correlation functions for the crystalline (the ground-state monoclinic β-phase) and amorphous stoichiometric Ga_2O_3 with different densities are shown in Figure 2.10a. Similar to the three oxides considered earlier in this chapter (cf. Figures 2.1, 2.4, and 2.7), the nearest-neighbor Ga-O distances are generally maintained upon amorphization. However, the first-shell peak in amorphous Ga_2O_3 becomes higher than that in the crystalline oxide and maintains its width, and the longer Ga-O distances in the distribution are almost negligible (Figure 2.10a). The former result signifies that the coordination of Ga is likely to be more uniform than it is in the monoclinic β-phase Ga_2O_3 with two non-equivalent types of Ga atoms, half being four-coordinated and the other half six-coordinated. The absence of longer Ga-O distances in the distribution is consistent with the strongest Ga-O bonding among the four oxides considered. Given the low symmetry of the monoclinic gallium oxide, the O-O and Ga-Ga distributions coincide, resulting in a wide double-peak structure spread out from about 2.6 Å to 3.8 Å. Upon amorphization, this medium-range distribution becomes featureless. Hence, similar to the other oxides, and despite nearly identical first-shell distance distribution between the crystalline and amorphous Ga_2O_3, the disordered phase is expected to feature short-range (nearest-neighbor) disorder and significant changes in the way how the Ga-O polyhedra connect with each other to form a network, as can be seen in Figure 2.10c and 2.10d.

The optimal density of amorphous Ga_2O_3 was determined based on the energy–volume curve with at least 10 independent MD liquid-quench simulations performed for each density value (Figure 2.10b). The formation energy of different MD realizations varies within at least 0.04 eV/atom, and there is a significant overlap between the total energy for the configurations with different densities, giving a shallow minimum in the energy–density plot. Despite the large deviations in the total energy of different realizations with the same density, an average of 10 independent MD realizations provides a smooth curve and appears to be enough to predict an optimal density; clearly, one should avoid using only a few realizations to make conclusions. Most strikingly, the optimal density in amorphous Ga_2O_3 (5.23 g/cm^3) is found to be as much as 19% lower than that in the monoclinic gallium oxide. This is significantly larger than the reduction of density in amorphous SnO_2 and ZnO (about 7%) and In_2O_3 (almost 3%). To understand this result, first, we note that the monoclinic Ga_2O_3 is the most loosely packed structure among the four binary oxides considered in this work. The low-symmetry β-phase has primarily corner-shared GaO_4 and GaO_6 polyhedra that form ordered chains with large structural voids between them. As will be shown here, in amorphous Ga_2O_3, the majority of GaO polyhedra become undercoordinated and remain to be corner-shared; therefore, the density is strongly reduced. This is in contrast to amorphous ZnO and SnO_2, both of which have (i) more edge-shared metal–metal connections and (ii) longer than typical metal–oxygen bonds as compared to the crystalline phase. These factors allow for better polyhedra packing in disordered ZnO and SnO_2; therefore, both exhibit a moderate density reduction upon amorphization. Finally, bixbyite In_2O_3 not only is the most closely packed structure with half of the In-O polyhedra being edge-shared, but also has the weakest metal–oxygen bonding among the four oxides considered. The latter allows for In overcoordination (ECN > 6 implies that an additional, seventh oxygen atom is in the first shell of such an In atom, albeit at a longer than average distance) and smaller In-O-In angles that help maintain good polyhedral packing.

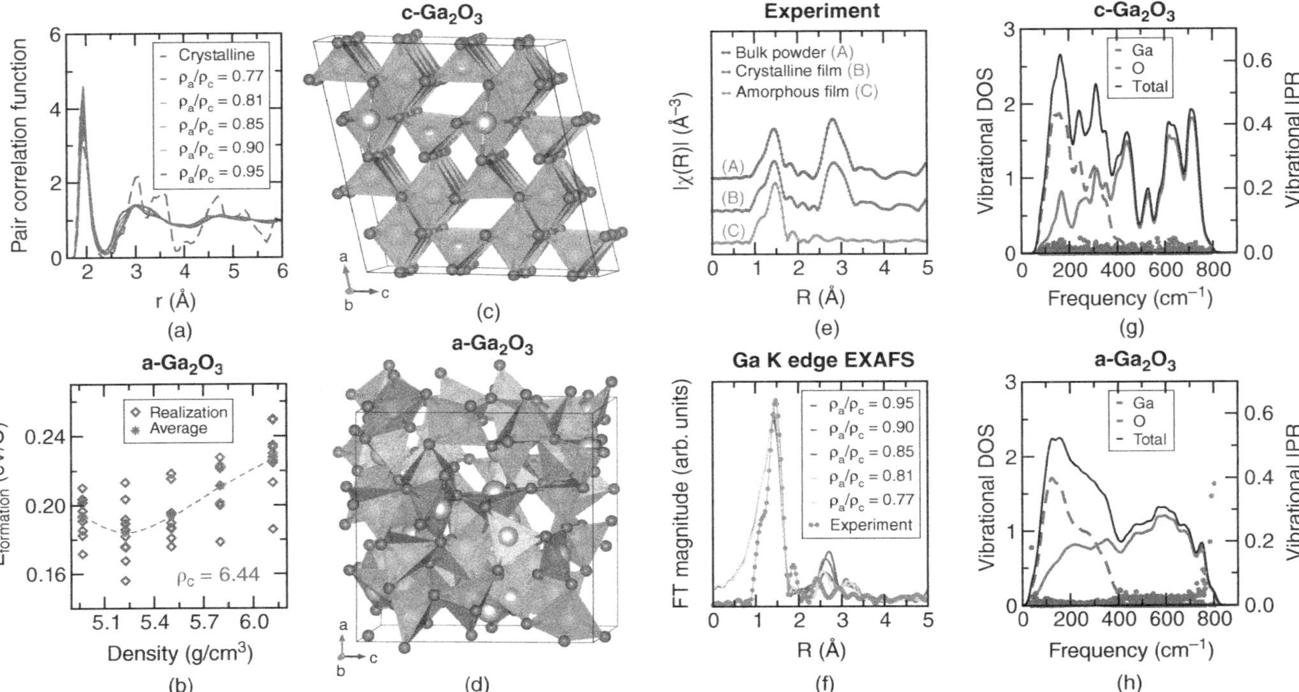

Figure 2.10 General structural properties of Ga_2O_3. (a) Total pair-correlation function for crystalline (monoclinic β-phase) and amorphous Ga_2O_3 with different densities. Both crystalline (dashed line) and amorphous (solid lines) results were obtained from MD simulations at 300 K (equilibration with 3,000 MD steps = 6 ps). The results for amorphous structures represent an average over at least 10 separate MD realizations at each density. (b) The energy–density curve for amorphous Ga_2O_3 calculated as $E_{formation} = E_{TOT}(\text{amorphous}) - E_{TOT}(\text{crystalline})$. Upon equilibration of each configuration at 300 K for 6 ps, the DFT total energy was calculated as an average over the final (stable) 500 MD steps to remove thermal fluctuations in each of the 10 realizations (diamond); the average over the realizations is represented by a star symbol. (c,d) Crystalline and amorphous atomic structures with Ga-O polyhedra highlighted. Ga and O atoms are represented by large and small spheres, respectively. (e) Experimental extended X-ray absorption fine structure (EXAFS) for a Ga K edge in bulk powder, crystalline film, and amorphous film samples of Ga_2O_3. (f) Calculated EXAFS for a Ga K edge in amorphous Ga_2O_3 models with different densities (solid line) as compared to experimental spectra for amorphous film (line with circle symbols). (g,h) Total and partial vibrational density of states (VDOS) and vibrational inverse participation ratio (VIPR) calculated for crystalline and the most stable amorphous Ga_2O_3 structures. Ga (O) states are represented by a dashed (solid) thick line, and the total VDOS is a thin solid line.

The experimental EXAFS measurements for the Ga K edge (Figure 2.10e) suggest that disorder completely suppresses the Ga second-shell (Ga-Ga) peak, whereas the first-shell (Ga-O) peak becomes slightly taller. Both results are in excellent agreement with our calculated pair-correlation functions for amorphous Ga_2O_3 at different densities (Figure 2.10a). As with amorphous In_2O_3 (Figure 2.1f), density has a strong effect on the Ga-Ga distances and coordination, although the calculated EXAFS shows little change in the intensity of the Ga-Ga peak below a density of $0.85\rho_c$ (Figure 2.10f). At the same time, we find that the Ga-O peak becomes taller as the density decreases. While the Ga-O peak position agrees well for the theoretical and experimental EXAFS (Figure 2.10f), the short-distance shoulder observed experimentally in both crystalline and amorphous films but not in a bulk power sample (Figure 2.10e) is not well pronounced in the calculations. We believe that the experimental Ga_2O_3 may possess even lower density than the values studied theoretically ($0.77\rho_c$ and higher).

The low symmetry of monoclinic Ga_2O_3 is also seen from the significant overlap between the Ga and O VDOS. Upon amorphization, the overlap is nearly unchanged, although the VDOS becomes featureless and low-frequency Ga modes appear at about 50 cm^{-1} (Figure 2.10e and 2.10f). High-frequency modes (around 750–800 cm^{-1}), associated with oscillations of a few oxygen atoms with the largest VIPR value of about 0.4 and a single low-frequency mode (50 cm^{-1}) for gallium, appear in the amorphous phase. Within the frequency range of about 80–350 cm^{-1} (i.e., where the Ga and O VDOS overlap), the VIPR values are comparable to or even lower than those in the crystalline Ga_2O_3 case, suggesting an evenly distributed vibration among different atoms.

A comparison of the ECN distributions for the first-shell Ga-O and O-Ga polyhedra for crystalline and amorphous gallium oxide with perfect stoichiometry (i.e., for the oxygen-defect-free structures) is given in Figure 2.11a.

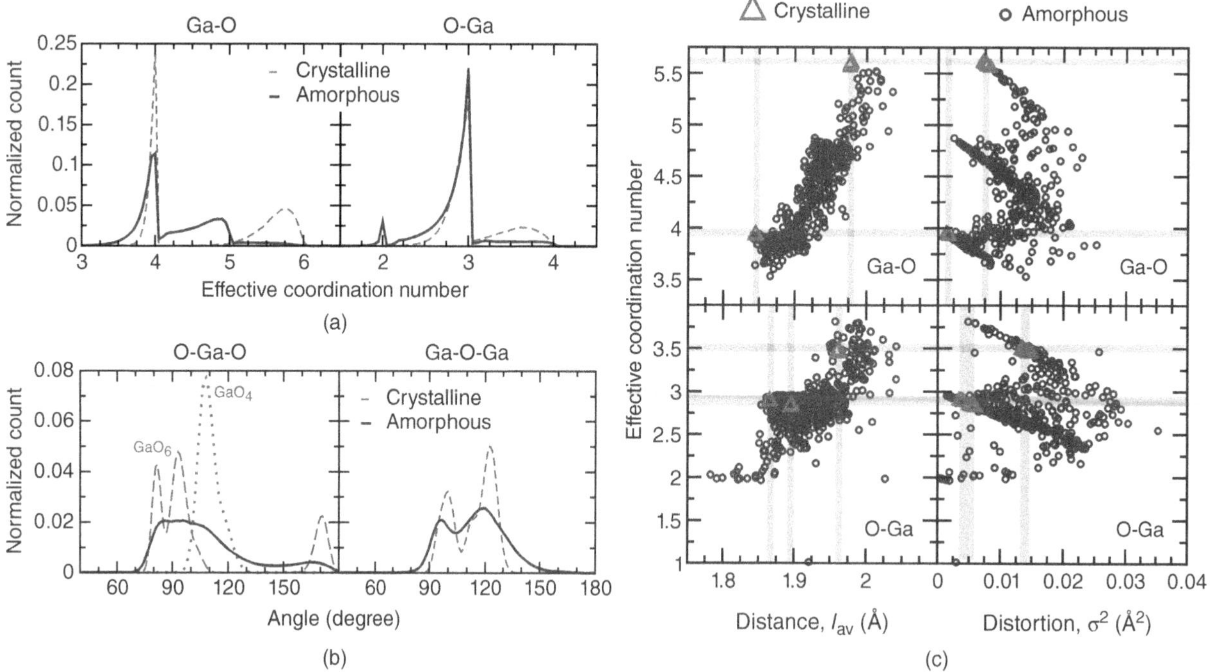

Figure 2.11 Structural randomness: The properties of individual Ga-O and O-Ga polyhedra in Ga_2O_3. (a) Distribution of the effective coordination number of Ga and O in crystalline (dashed line) and amorphous (solid line) phases. (b) Distribution of the O-Ga-O and Ga-O-Ga angles in crystalline (dashed line) and amorphous (solid line) phases. The results for amorphous structures represent an average over at least 10 separate MD realizations at optimal density. (c) Effective coordination number of individual Ga and O as a function of effective average distance or distortion for the given atom in crystalline (triangles and shaded region) and 10 amorphous (circles) phases. All results are calculated based on the MD simulations at 300 K (equilibration with 3,000 MD steps = 6 ps).

In monoclinic Ga_2O_3, there are two Ga and three O non-equivalent sites with different effective coordination and distances. Half of the Ga atoms are four-coordinated (ECN = 3.93 at 300K) with an average Ga-O distance of 1.84 Å, and the other half are six-coordinated (ECN = 5.59) with a distance of 1.98 Å (Table 2.1). The oxygen atoms are three- or four-coordinated (ECN = 2.83, 2.89, and 3.46 at room temperature). Upon amorphization, the coordination distributions for both gallium and oxygen do not become broader, as they do for In_2O_3, SnO_2, and ZnO cases; yet both change as follows. For Ga, the number of atoms with ECN > 5 is nearly completely suppressed, and the distribution is shifted to the effective coordination ranging between four and five (Figure 2.11a). The distribution for the four-coordinated Ga is suppressed and slightly broadened toward lower values. Similarly, the amount of four-coordinated oxygen atoms, ECN(O) = 3–4, becomes negligible, whereas a small fraction of two-coordinated O atoms appears in the distribution (Figure 2.11a). The amount of three-coordinated O atoms increases in amorphous Ga_2O_3, and the corresponding distribution develops a tail that stretches toward ECN = 2. Thus, similar to the other oxides, the seemingly insignificant changes in the first-shell distance distribution (Figure 2.10a) lead to a remarkable coordination transformation in amorphous Ga_2O_3, reducing the ECN values for both Ga and O atoms. We note here that the average Ga and O ECNs decrease with decreasing density (Table 2.1). Interestingly, the polyhedral distortions in amorphous Ga_2O_3 with higher density are comparable to those in amorphous In_2O_3, whereas at lower density the distortions become the smallest among all four oxides considered. This should be expected from the strongest metal–oxygen bonding in Ga_2O_3 as compared to the other three oxides.

The angle distributions for both O-Ga-O and Ga-O-Ga slightly widen and become uniform in the amorphous Ga_2O_3 (Figure 2.11b). Although the amount of six-coordinated Ga atoms is negligible, the characteristic O-Ga-O angles (70°, 95°, and 170°) are still present in the five-coordinated Ga-O polyhedra. The Ga-O-Ga angle distribution preserves the double-peak structure, consistent with negligible changes in the fraction of edge- and corner-shared Ga-Ga pairs, as discussed in section 2.7.

To further understand the coordination transformations in perfectly stoichiometric Ga_2O_3, the time-averaged ECN values for individual Ga and O atoms are plotted as a function of their time-averaged effective distance and distortion for both crystalline and amorphous oxide (Figure 2.11c). No single occurrence of octahedrally coordinated Ga, with an ECN and l_{av} that are the same as in crystalline monoclinic oxide, is found among 10 MD realizations (540 Ga atoms) of the amorphous phase. A characteristic dependence for the two structural parameters is well-defined: The lower the ECN(Ga), the shorter the effective distance for the given polyhedra. Distortion slightly increases with decreasing the coordination within each of the ECN groups. Amorphous Ga_2O_3, is the only case where the oxygen structural distributions include the values for the O atoms in the crystalline β-phase inside the distribution (Figure 2.11c), rather than on its edges (cf. Figures 2.2c, 2.5c, and 2.8c). Yet, despite the similarity of the ECN(O) distributions in the crystalline and amorphous Ga_2O_3 (Figure 2.11a), the majority of O atoms (independent of the coordination) have nearest-neighbor distances and O-Ga polyhedra distortions that deviate significantly from those in the monoclinic gallium oxide.

Figure 2.12a–2.12d compares the charge density distributions in the highest valence band and the lowest conduction band in the fully relaxed crystalline and amorphous Ga_2O_3 structures. Similar to In_2O_3, SnO_2, and ZnO cases (cf. Figures 2.3a–2.3d, 2.6a–2.6d, and 2.9a–2.9d), only a few oxygen atoms contribute at the top of the valence band, and the calculated Bader charge distribution for the O atoms (Figure 2.12e) features a strong asymmetry, with the majority of oxygen atoms giving negligible contributions. The conduction band charge in amorphous Ga_2O_3 is not uniform, and a significant fraction of the Ga atoms have small or negligible contributions to the conduction states, so that several of the 10 structures do not exhibit a fully connected three-dimensional conductivity path throughout the supercell (Figure 2.12b). Indeed, the individual Bader charge distribution for Ga atoms is more asymmetric with respect to the expected uniform charge value than for the other three metals (cf. Figures 2.3f, 2.6f, and 2.9f). Similar to indium, tin, and zinc oxides, we do not find a clear correlation between the ECN, effective distance (l_{av}), or distortion (σ^2) of oxygen atoms and the corresponding valence Bader charge contribution.

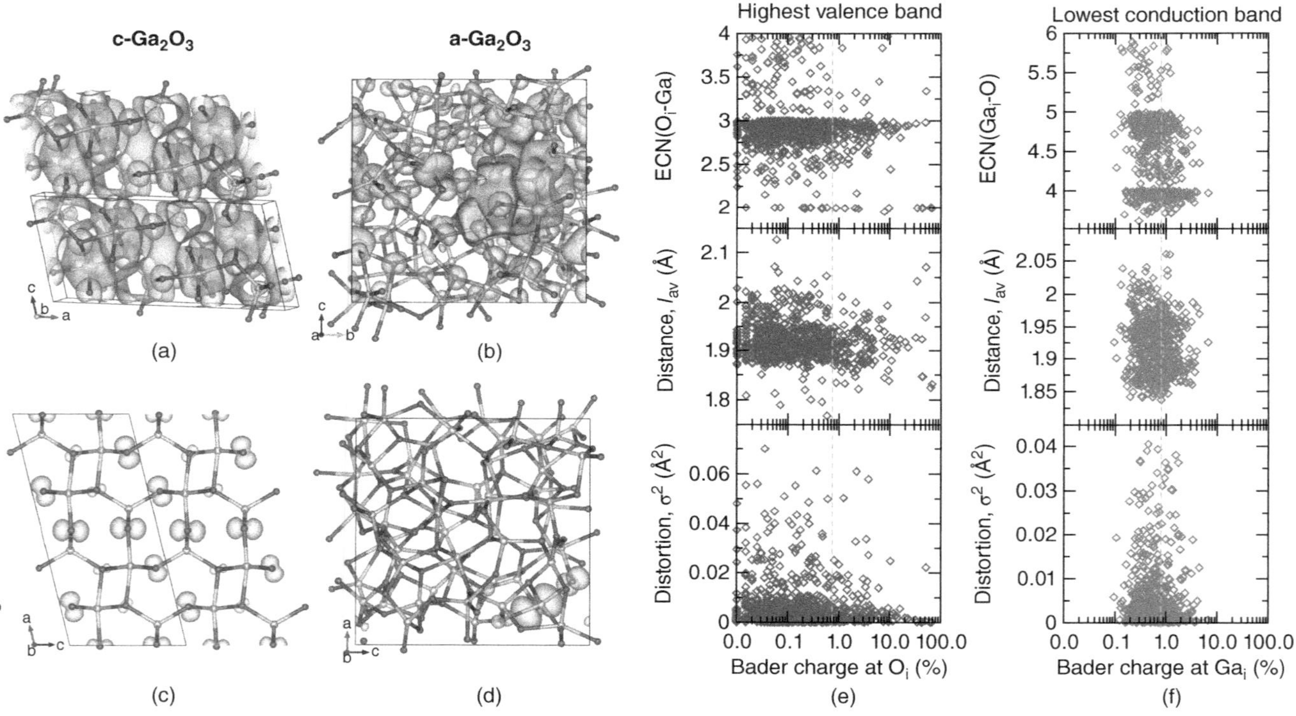

Figure 2.12 Calculated charge density distribution in the (a,b) lowest conduction band and (c,d) highest valence band in (a,c) crystalline and (b,d) amorphous Ga_2O_3. (e,f) First-shell effective coordination number (ECN), effective average distance, and distortion for individual O and Ga atoms as a function of the Bader charge contribution to the corresponding band from the given atom. The results are for 10 separate MD realizations at optimal density. The vertical lines represent the calculated charge value for a uniform charge density distribution (i.e., if all O or Ga atoms in the supercell contribute equally). The electronic properties are calculated for the fully relaxed structures using an accurate hybrid functional approach.

2.7 Role of Morphology in Structure–Property Relationships

In previous sections of this chapter, it was shown that the charge localization in the conduction band (represented by the Bader charge contributions from individual atoms) does not directly correlate with the effective coordination, effective distance, or polyhedral distortion of the given atom in amorphous oxides (Figures 2.3f, 2.6f, 2.9f, and 2.12f). However, further analysis of representative groups of distribution "outliers" suggests that a combination of the above structural parameters (e.g., the actual vs. expected effective distance for a given coordination of an atom, or the asymmetry of polyhedral distortions vs. the conventional coordination geometry) provides valuable insight into the complex structure–property relationships in AOSs.

Among the four binary oxides, a tin atom in amorphous SnO_2 has the largest conduction charge contribution (15% of the total charge for the state). Figure 2.13a shows the oxygen environment around this Sn atom along with the conduction charge density distribution in the vicinity. The metal atom has a square pyramidal coordination, which is not found in crystalline stoichiometric SnO_2, where all Sn atoms are octahedrally coordinated. Clearly, the square pyramidal O arrangement resembles an oxygen vacancy in a sub-stoichiometric crystalline oxide (i.e., a missing apical oxygen atom at the top vertex of an octahedron). However, we recall that the amorphous structures investigated in this work are perfectly stoichiometric (i.e., no oxygen defects have been introduced). Therefore, disorder plays a key role in stabilizing the square pyramidal arrangement—along with the significant charge accumulation around the metal and especially in the void where a sixth oxygen is expected for an octahedral coordination (Figure 2.13a). In the case shown, the calculated Bader charge on the Sn atom is as much as 21 times greater than the expected charge value for a uniform charge distribution (0.7% per atom in the supercell). It should be noted here that the distortion parameter is small for this Sn atom, $\sigma^2 = 0.0034$ Å^2, because the Sn-O distances in the square pyramid are nearly identical (for comparison, the average distortion for all Sn atoms in this SnO_2 structure is 0.0087 Å^2).

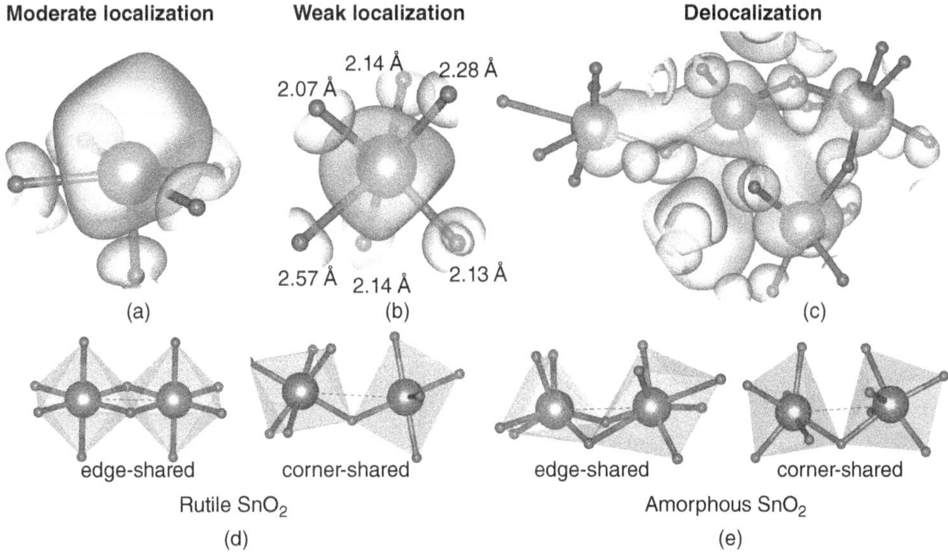

Figure 2.13 Charge density distribution near representative structural defects in the conduction band of amorphous stoichiometric SnO_2. (a) Moderate electron localization caused by a square pyramidal coordination of Sn with the largest conduction charge contribution. (b) Weak electron localization around a Sn atom with longer than expected Sn-O distances. (c) Electron delocalization due to sharing defects, resulting in direct overlap of *s-s* orbitals of neighboring Sn atoms. (d,e) Examples of edge- and corner-shared Sn pairs in rutile and amorphous SnO_2.

The asymmetric distortions in the M-O polyhedra, where an M-O bond is clearly missing so that the central metal atom becomes "exposed" on one side of the polyhedra with no oxygen atom available to bond with in order to achieve a more uniform coordination geometry (cf. Figure 2.13a), are rare in *stoichiometric* AOSs. It has been shown that the lack of lattice periodicity and the ionic nature of AOSs facilitate an extended structural relaxation near defects or impurities [11, 13], and this bond reconfiguration helps either (i) shift a metal atom to the center of mass of the available oxygen atoms to even out the negative charges around the metal atom, or (ii) attract an additional oxygen atom from a nearby polyhedron to be shared by the two metal atoms. In other words, the observed bond disorder in stoichiometric AOSs (namely, the broad distance distributions in the individual M-O polyhedra [Figures 2.2c, 2.5c, 2.8c, and 2.11c] and the accompanying diversity of coordination transformations e.g., from octahedral or trigonal prismatic to square pyramidal to trigonal bipyramidal to tetrahedral, with variable degrees of polyhedral distortions) is a result of the ionic nature of the AOSs that drive the structural reconfigurations toward achieving a uniform distribution of the positive and negative charges throughout the structure. When the structure does not allow for bond rearrangement, for example, due to undercoordination of neighboring metal atoms that have shorter (stronger) M-O bonds and thus cannot share their oxygen atoms, or due to lack of oxygen in non-stoichiometric oxides, the formation of polyhedra with severe asymmetry will lead to charge imbalance and strong electron localization (as typically occur in crystalline ionic oxides or in covalent materials with stronger bonding). Hence, the degree of localization is determined by the ability of the surrounding structure to adjust to a specific defect and reduce charge imbalance.

A more common structural defect that causes electron localization in stoichiometric AOSs is shown in Figure 2.13b. This Sn atom with an effective coordination of 5.0 has a notably longer effective Sn-O distance (2.14 Å) than is expected (2.06 Å) for this ECN value. (We recall here that ECN(Sn) has a distinct slope when plotted as a function of the Sn-O distance, with distances being shorter for smaller coordination numbers; Figure 2.5c.) The Sn atom with longer than expected Sn-O distances, shown in Figure 2.13b, has a conduction band Bader charge contribution (2% of the total charge) that is more than twice the value expected for a uniform charge distribution in the cell (0.7% per atom in the cell). This corresponds to a weak localization associated with a rather small charge imbalance. All four stoichiometric binary oxides possess a notable fraction of such metal atoms with longer than expected distances for a given ECN (Figures 2.2c, 2.5c, 2.8c, and 2.11c), and these atoms are found to have a Bader charge that is larger than the expected value for a uniform distribution (Figures 2.3f, 2.6f, 2.9f, and 2.12f). These "outliers" are spread randomly in the structure; they do not show a tendency to pair up or cluster with each other. It must be noted that these longer than expected distances resemble a three-dimensional strain for the metal atom and are more likely to occur in low-density amorphous materials.

The highly delocalized nature of the conduction band in AOSs is associated with multi-atom formations in which the mutual orientation of undercoordinated and/or highly distorted M-O polyhedra enables direct overlap of the neighboring metals' *s*-orbitals (Figure 2.13c). The charge density extends into the interstitial region connecting two or more metal atoms, resulting in the formation of tree-like branches or chains woven into the rest of the network. Therefore, medium-range structural characteristics (i.e., how the differently coordinated and/or distorted polyhedra are connected with each other to form an amorphous network) must be included in consideration in addition to the individual structural parameters and Bader charges. For the four connected Sn atoms shown in Figure 2.13c, the Bader charge accumulation is 1.6, 1.8, 2.3, and 3.3 times greater than the value for the uniform charge distribution; the corresponding effective coordination numbers (5.8, 4.9, 5.8, and 5.3, respectively), effective average distances (2.12 Å, 2.06 Å, 2.12 Å, and 2.11 Å, respectively), and polyhedral distortions (0.0032 $Å^2$, 0.0017 $Å^2$, 0.0041 $Å^2$, and 0.0169 $Å^2$, respectively) for these Sn atoms do not serve as predictors of the enhanced conduction charge density (cf. Figure 2.12f). Analyzing the individual Sn-O distances that connect the four Sn atoms shown in Figure 2.13c, we find that the majority of them are notably or significantly longer (the average of 10 Sn-O distances is 2.17 Å, ranging from 2.05 Å to 2.39 Å) than the average Sn-O distance (2.09 Å) in amorphous SnO_2 at the same density (Table 2.1). What is more important, however, is the position of the oxygen atom(s) shared by each metal pair with respect to the line connecting these metal atoms, because it ultimately determines the possibility for direct *s-s* orbital overlap. In the SnO_2 case shown in Figure 2.13c, two fully coordinated Sn atoms (ECN = 5.8) that share one oxygen atom with each other are located at a shorter Sn-Sn distance (3.55 Å) and smaller Sn-O-Sn angle (109°) as compared to a corner-shared Sn-Sn pair in crystalline SnO_2 (3.71 Å with a Sn-O-Sn angle of 129°). The combination of a shorter metal–metal distance accompanied by a smaller M-O-M angle originates from elongated M-O distances and signifies that the shared oxygen atom moves farther away from the line connecting the two metal atoms (Figure 2.13d and 2.13e). Similarly, two edge-shared Sn atoms in Figure 2.13c are found at a longer distance from each other (3.50 Å) than the typical Sn-Sn pair that shares two oxygen atoms in crystalline oxide (3.19 Å). In this case, the edge-shared polyhedra tilt with respect to each other, making the angle between two OMO planes to be equal to 130° so that both oxygen atoms are away from the line that connects the metal atoms (Figure 2.13e). In contrast, two metal atoms and their shared oxygen atoms are co-planar in rutile SnO_2 (Figure 2.13d). Thus, sharing defects, arising due to the presence of longer than usual M-O distances, may favor direct *s-s* orbital overlap and are quantified by the individual metal–metal distance combined with the M-O-M angle in the case of corner-shared M-M pairs and by the angle between two O-M-O planes in case of edge-shared pairs.

To understand the effect of disorder on the medium-range structure in AOSs, the M-M distance distributions for the edge- and corner-shared M-O polyhedra are compared for the crystalline and amorphous binary oxides at room temperature (Figure 2.14). In bixbyite In_2O_3, each In atom shares two oxygen atoms with six In neighbors (with an In-In distance of 3.4 Å) and also shares one oxygen atom with another six In atoms (with an In-In distance of 3.7 Å) (i.e., there are equal amounts of edge- and corner-shared pairs for each metal atom). In rutile SnO_2, each Sn atom has two Sn neighbors that are edge-shared (Sn-Sn distance: 3.2 Å) and eight Sn neighbors that are corner-shared (Sn-Sn distance: 3.7 Å). In wurtzite ZnO, all Zn atoms have 12 corner-shared Zn neighbors at a

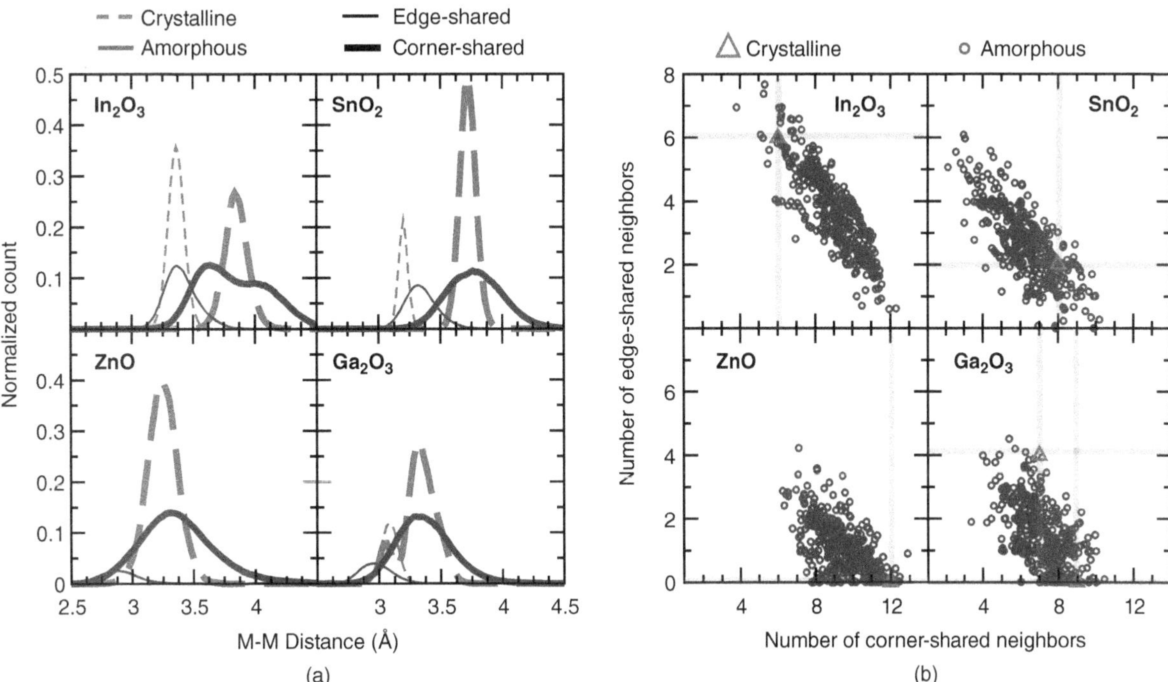

Figure 2.14 Medium-range structure in In_2O_3, SnO_2, ZnO, and Ga_2O_3. (a) Metal–metal distance distribution for edge-shared (thin line) and corner-shared (thick line) metal pairs in crystalline (dashed line) and amorphous (solid line) phases. The results for amorphous structures represent an average over 10 separate MD realizations at the corresponding optimal density. (b) The number of edge-shared neighbors as a function of the number of corner-shared neighbors for individual In atoms in crystalline (triangles and shaded regions) and 10 amorphous (circles) phases. All results are calculated based on the MD simulations at 300 K (equilibration with 3,000 MD steps = 6 ps).

Zn-Zn distance of 3.3 Å. In monoclinic Ga_2O_3, the four-coordinated Ga atom has nine corner-shared neighbors (Ga-Ga distances: 3.1–3.4 Å), while the six-coordinated Ga atom has four edge-shared (Ga-Ga distance: 3.1 Å) and seven corner-shared atoms (Ga-Ga distances: 3.2–3.4 Å). While the room temperature broadens the M-M distance distributions, only in β-phase Ga_2O_3 do the edge- and corner-shared distributions overlap (Figure 2.14a), owing to the low symmetry of the crystal structure.

Disorder has a pronounced effect on M-O polyhedra sharing, as should be expected from the wide bond distance distributions and strong distortions in the M-O polyhedra (Figures 2.2c, 2.5c, 2.8c, and 2.11c). Both edge- and corner-shared distributions broaden and overlap significantly in amorphous oxides, including ZnO where a small fraction of edge-shared pairs appear (Figure 2.14a). Except for the edge-shared distribution in amorphous In_2O_3 and corner-shared distribution in amorphous Ga_2O_3, all other distribution peaks shift from the corresponding crystalline values. The considerable fractions of the longer or shorter than expected M-M distances for the given sharing type (edge- or corner-shared) imply that the sharing defects shown in Figure 2.13e are likely to occur in all four amorphous oxides, yet there are important differences in the morphology of the four oxides. Figure 2.14b plots the number of the edge-shared neighbors as a function of the number of the corner-shared neighbors for every individual M atom in amorphous oxide structures. While the general trend—the lower the number of edge-shared pairs, the higher the number of corner-shared pairs—can be observed, the effect of disorder is different among the oxides: In In_2O_3, the number of edge-shared pairs is suppressed upon amorphization, whereas ZnO and especially SnO_2 feature an increased number of edge-shared polyhedra in the amorphous phases as compared to the crystalline oxides. A reduction in the number of M-O connections (e.g., when edge-shared pairs become corner-shared) and elongation of the M-M distances, both associated with longer M-O bond lengths, are expected in low-density amorphous oxides. However, all four oxides also feature outliers that have either higher than the expected number of edge- or corner-shared neighbors or an increased number of the total shared neighbors (edge- and corner-shared numbers combined; Figure 2.14b). The wide distributions in the numbers of edge-versus-corner neighbors imply that the morphology of the amorphous oxides is complex and the medium-range structural characteristics must be considered when deriving the comprehensive structure–property relationships in AOSs.

All four amorphous oxides show some degree of interatomic charge density chain-like formations in the conduction band (Figures 2.3c, 2.6c, 2.9c, and 2.12c); however, the chains may not extend throughout the entire unit cell. To quantify the differences between the oxides, we compare the width of the single *s*-like conduction band. The bandwidth reflects the degree of hybridization of the conduction states and also determines the electron velocity once the material is doped. Hence, it is instructive to correlate the conduction band widths with the charge density distribution in each amorphous oxide structure and also with its structural morphology. First, the conduction band energy dispersions calculated along the [111] direction (i.e., along the diagonal of the cubic supercell) for both crystalline and amorphous In_2O_3 are shown in Figure 2.15a. The curvature at the bottom of the conduction band is similar in the crystalline and amorphous indium oxide, resulting in effective electron mass of 0.22 m_e and 0.17 m_e, respectively [11]. However, the width of the conduction band is reduced from 2.8 eV in crystalline In_2O_3 to at least 2.21 eV and at most 1.95 eV in the amorphous cases. The range of the bandwidth values in the amorphous In_2O_3 (Figure 2.15b) is represented by 10 different structures obtained via MD liquid-quench simulations at the same density, cooling rate, and stoichiometry, and is a result of morphology differences in these 10 realizations. Importantly, the variable bandwidth values translate into different electron velocities, calculated using a two-electron rigid-band shift to be equal to 1.7×10^6 m/s for crystalline In_2O_3 and 5.6×10^5 m/s or 1.7×10^5 m/s for the two amorphous cases with the widest and the narrowest conduction bands, respectively. Since the conductivity is proportional to the square of the electron velocity, the variation in the bandwidths of these two amorphous configurations is expected to result in up to a 10-time difference in their conductivity, provided the carrier concentration, scattering mechanisms, and relaxation times (that also determine the conductivity) remain the same for the two structures.

The conduction bandwidths for the four amorphous binary oxides, calculated for 10 different realizations in each case, are compared in Figure 2.15b. On average, the conduction bandwidth in indium oxide is wider by at least

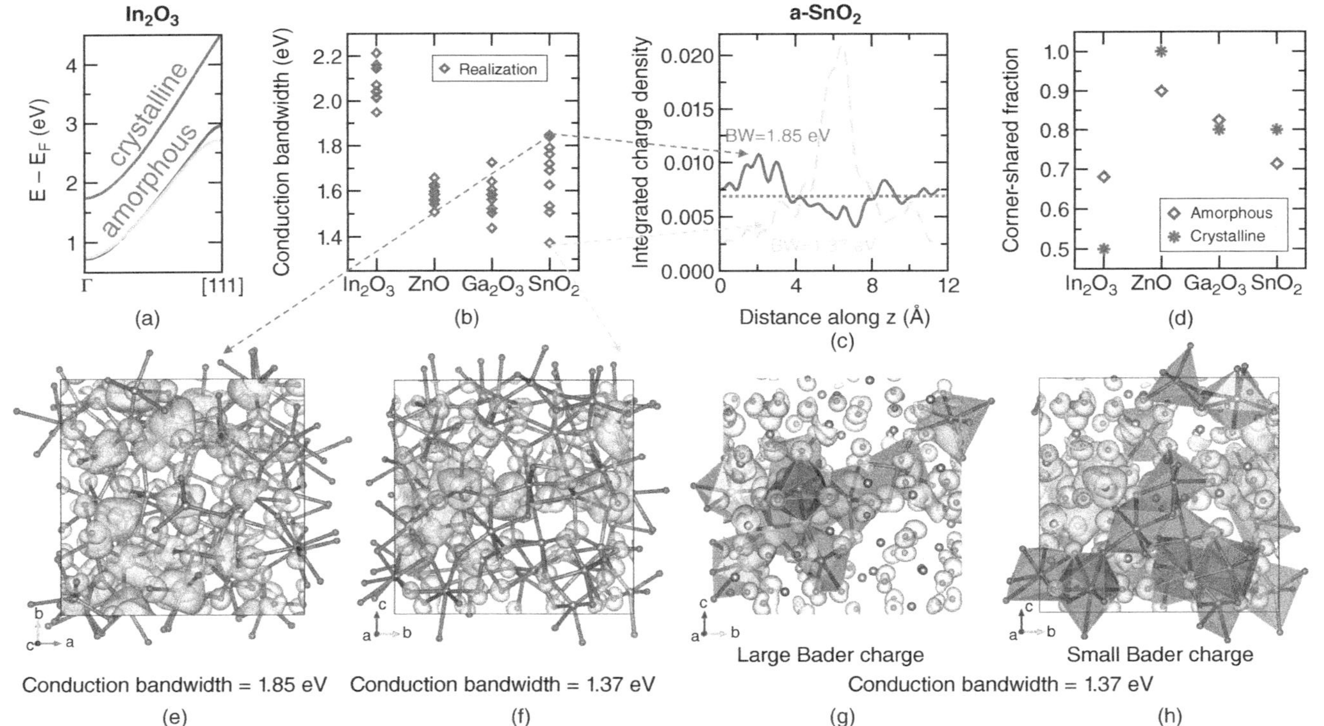

Figure 2.15 Electronic properties of amorphous stoichiometric binary oxides. (a) The lowest conduction-band-like state calculated along the [111] direction in crystalline and two amorphous phases of In_2O_3. (b) Conduction bandwidths calculated along the [111] direction in 10 amorphous phases of binary oxides with optimal density. (c) Integrated charge density projected to the *z*-axis in two amorphous SnO_2 structures with the widest (solid line) and the narrowest (dashed line) conduction-band-like states. The vertical dotted line represents a uniform charge density distribution. (d) Fraction of corner-shared metal–metal pairs calculated out of the total shared M-M pairs in crystalline and amorphous oxides. (e) Charge density distribution in the supercell of the amorphous SnO_2 structure with the widest conduction-band-like state. (f) Charge density distribution in the supercell of the amorphous SnO_2 structure with the narrowest conduction-band-like state. (g,h) Charge density distribution in the supercell of the amorphous SnO_2 structure with the narrowest conduction-band-like state with the Sn-O polyhedra highlighted for the atoms with the largest and smallest Bader charge contributions.

0.5 eV than it is in SnO_2, ZnO, or Ga_2O_3; this may explain the best performance of In-based AOSs. The distribution of the bandwidth values among different realizations for the same metal oxide (prepared at the same density, oxygen stoichiometry, and quench rate) provides further important insights into the complex structure–property relationships in AOSs. Specifically, the conduction bandwidths vary the least in ZnO (from 1.51 eV to 1.66 eV), with the range increasing in In_2O_3 (from 1.95 eV to 2.21 eV), followed by Ga_2O_3 (from 1.44 eV to 1.72 eV), and then SnO_2 (from 1.37 eV to 1.85 eV). This signifies that as-grown tin oxide samples, and to a lesser degree indium and gallium oxides, may have complex morphology, leading to a broad range of properties that are sensitive to deposition conditions and postdeposition treatment. To highlight the effect of morphology on the electronic properties, the charge density distributions in the amorphous SnO_2 structures with the widest and the narrowest conduction bands are shown in Figure 2.15e and 2.15f; the integrated charge densities are compared for the two cases in Figure 2.15c. Clearly, the widest conduction band corresponds to the most uniform charge distribution within the supercell; vice versa, the charge density is uneven in the structure with the narrowest bandwidth, with significant accumulation in some areas and very low charge in the other parts of the supercell. Figure 2.15g and 2.15h show the Sn-O polyhedra with the largest and the lowest Bader charge contributions, respectively, and illustrate how these polyhedra are connected with each other within the supercell. Based on the peculiarities of the structural distributions for the four amorphous binary oxides, the following conclusions can be made:

1. In amorphous SnO_2, the ECN(Sn) distribution is the widest among the four metal oxides (ΔECN(Sn) = 2.8; Figure 2.5); in addition, about a third of the Sn-O polyhedra are edge-shared. Both factors contribute to the morphological richness in the disordered tin oxide. The presence of shorter (longer) than usual corner(edge)-shared Sn-Sn distances is beneficial for the formation of the conduction channels. At the same time, a large fraction of the Sn-O polyhedra possesses an increased number of edge-shared neighbors (from two in rutile SnO_2 to six in amorphous tin oxide; Figure 2.2b), suggesting a tendency for clustering. As a result, while some SnO_2 structures may have a conduction bandwidth comparable to that in In_2O_3, others show the narrowest bandwidth, and hence the lowest electron velocity, among the four binary oxides (Figure 2.15b). Consequently, amorphous SnO_2 is expected to exhibit sample-to-sample stability and reliability issues as well as limited electron transport due to clustering.
2. Amorphous ZnO shows the smallest variance in the conduction bandwidth values for different realizations (Figure 2.15b) that is in accord with the narrowest ECN distribution (Figure 2.8), and also with the finding that 90% of the neighboring polyhedra remain to be corner-shared upon amorphization (Figure 2.15d). The most uniform coordination and polyhedral-sharing morphologies among the four oxides imply that amorphous ZnO-based materials are expected to provide similar electronic properties from sample to sample. At the same time, only a negligible fraction of metal atoms exhibits shorter than expected Zn-Zn corner-shared distances (Figure 2.14a), which suggests that the *s*-*s* overlap is likely to be limited to a few nearest neighbors (Figure 2.9b), restricting the carrier mobility.
3. As expected for the strongest metal–oxygen bond, the Ga-O polyhedral distortions in amorphous Ga_2O_3 are the lowest among the four oxides and comparable to those in the monoclinic phase (Table 2.1). Moreover, the Ga-O-Ga angle, Ga-Ga distance, and edge-versus-corner distributions resemble those in crystalline oxide (Figures 2.11b, 2.14a, 2.14b, and 2.15d), so that disorder in this oxide is mostly driven by the ECN transition from six- to five- and four-coordinated Ga (Figure 2.11). The coordination morphology determines the bandwidth variations in amorphous Ga_2O_3 and may result in charge clustering (Figure 2.12c). A direct *s*-*s* overlap is unlikely due to shorter than usual edge-shared Ga-Ga distances (clustering) and no shorter than expected Ga-Ga corner-shared pairs, leading to the smallest probability of interstitial charge density in the stoichiometric amorphous Ga_2O_3 among the four oxides considered. Nevertheless, the extrapolated experimental bulk electron mobility of β-phase Ga_2O_3 is high, 300 cm^2/Vs for carrier concentrations of 10^{15}–10^{16} cm^{-3}, making the wide-bandgap material an excellent candidate for power electronics in ultrahigh-voltage-switching applications [45]. Moreover, semiconducting amorphous Ga_2O_x thin films with Hall mobility of 8 cm^2/Vs and carrier concentration of 10^{14} cm^{-3} have been successfully fabricated when tuning the oxygen partial pressure [46].

4. The best conductivity is expected in amorphous In_2O_3 and in In-rich AOSs, as indeed observed for the In-X-O materials [47]. Apart from the low formation energy for oxygen defects [11], the origin of the good transport properties in a-In_2O_3 is in the broad ECN(In) distribution and the widest In-In distance distribution among the four oxides. Specifically, the largest fraction of shorter than usual corner-shared In-In distances (Figure 2.14a) facilitates the formation of the conduction channels. At the same time, the coordination and sharing morphologies may lead to sample-to-sample inconsistencies in the observed transport properties, and additional preparation conditions or postdeposition treatment might be required to achieve percolation via the conduction channels and hence best carrier mobility.

Disorder introduces significant optical absorption within the visible range of undoped, perfectly stoichiometric, amorphous In_2O_3, SnO_2, and ZnO, but not Ga_2O_3 (Figure 2.16a). The onset of the absorption from as low as 1.5 eV (3.0 eV in the case of Ga_2O_3) and its slowly rising behavior (as opposed to a sharper absorption onset typical for transparent conducting oxides) is attributed to (i) the non-uniform charge density distributions due to variable contributions in the conduction band (Figures 2.3b, 2.6b, 2.9b, and 2.12b); (ii) the valence tail states associated with the strongly localized nonbonding O--p-states (Figures 2.3d, 2.3e, 2.6d, 2.6e, 2.9d, 2.9e, 2.12d, and 2.12e); and (iii) the weaker metal–oxygen interaction associated with the slightly elongated average M-O distances (Table 2.1). We must stress, however, that the insulating behavior may be difficult to achieve in AOSs because, typically, the amorphous oxide samples are sub-stoichiometric, hydrogenated, and/or possess oxygen-related defects that dope the material n-type by shifting the Fermi level into the conduction band. In crystalline oxides, the O_2 defect is known as the Frenkel defect that forms when an oxygen atom moves away from its lattice site, leaving an oxygen vacancy behind, to combine with another oxygen. Depending on the degree of interaction with the surrounding periodic lattice, the O defect may have different charge states and also may release one or two electron(s) that become free carriers. Among the four amorphous oxides investigated, In_2O_3 and SnO_2 show the strongest tendency for O_2 formation with 100% and 90% of the MD realizations, respectively, having the peroxide defect; whereas only

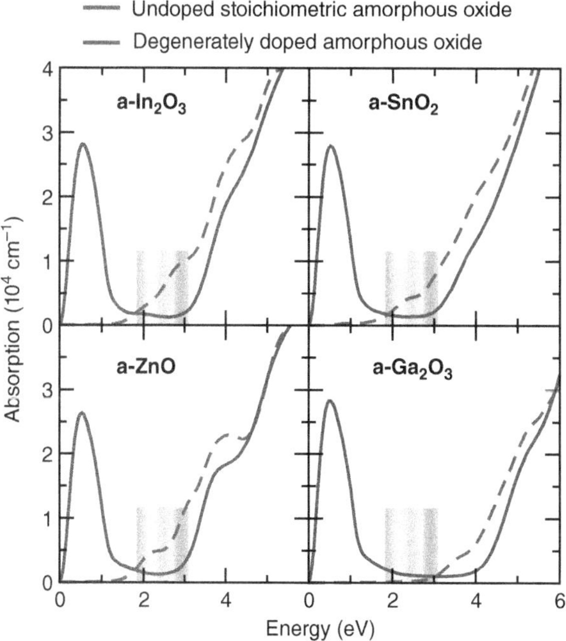

Figure 2.16 Optical absorption in amorphous In_2O_3, SnO_2, ZnO, and Ga_2O_3 in undoped stoichiometric (dashed line) and degenerately doped (solid line) oxide. In the latter case, the additional electron is due to the formation of O_2 defect, resulting in the Fermi-level shift to the conduction band and broadening the optical gap.

8% and 2% of the ZnO and Ga_2O_3 configurations, respectively, have the O_2 formed. In all cases, the O-O distance in the O_2 defect is about 1.5 Å, which is typical for O_2^{2-} peroxide in a lattice with both oxygen atoms in the defect bonded to neighboring metal atoms. We find that in all cases with an O_2 defect, the Fermi level shifts up by about 1.5 eV, 1.3 eV, 1.2 eV, and 1.2 eV for amorphous In_2O_3, SnO_2, ZnO, and Ga_2O_3, respectively, as calculated with respect to the conduction band minimum located at the Γ point. As a result of the pronounced Fermi-level shift, the optical window widens and the stoichiometric materials become nearly transparent within the visible range because of a low absorption associated with the intraband transitions (i.e., within the half-occupied conduction band), characteristic of non-stoichiometric oxides.

2.8 The Role of Composition in Structure–Property Relationships: IGO and IGZO

The structural characteristics of the amorphous binary oxides discussed in this chapter serve as a solid foundation for understanding the complex properties of multi-cation AOSs. In this work, we use amorphous $(In_{0.6}Ga_{0.4})_2O_3$ and $InGaZnO_4$ (with In:Ga:Zn=1:1:1) to illustrate (i) changes in the structural and electronic properties that occur with decreasing In content, and (ii) how the properties are affected by inherent structural preferences of the additional metal oxides.

First, the optimal density of amorphous multi-cation oxides was determined via energy–density curves with at least 10 independent MD liquid-quench simulations performed for each density (Figure 2.17a and 2.17d). Similar to the amorphous binary oxides (Figures 2.1a, 2.4a, 2.7a, and 2.10a), in multi-cation AOSs (i) the formation energy of different MD realizations varies within 2–3 eV/cell, (ii) a single MD realization at each density is not enough to predict optimal density, and (iii) there is a substantial overlap between the total energy for the configurations with different densities, giving a shallow minimum in the energy–density plots, especially in the case of $InGaZnO_4$ (Figure 2.17d). The results imply that amorphous oxide samples are likely to have a wide range of densities, sensitive to the deposition technique and pre- or postdeposition conditions. Moreover, the density may fluctuate from sample to sample, making the structural characterization challenging. In the discussion here, the amorphous $(In_{0.6}Ga_{0.4})_2O_3$ and $InGaZnO_4$ structures with a density of 6.07 g/cm^3 and 5.71 g/cm^3 are used unless density dependence is discussed.

To validate the theoretical amorphous models, the EXAFS is calculated for each metal K edge for the amorphous models with different densities. The simulated structures for both multicomponent oxides show excellent agreement between the calculated and experimental EXAFS for In, Ga, and Zn (Figure 2.17b and 2.17e). In amorphous $InGaZnO_4$, a notable dependence on the density is found for the first-shell In and Ga peaks. Specifically, the In peak decreases and Ga peak increases with density decreasing, making the optimal density structures provide the best agreement with experimental EXAFS [12, 48]. Next, the total and partial VDOS were calculated for the most stable amorphous multi-cation structures to verify their dynamical stability. In addition, the VIPR was calculated from the normalized displacement vectors to determine the localization of different vibrational modes. The results, shown in Figure 2.17c and 2.17f, demonstrate that, similar to binary oxides, the disordered structures exhibit featureless VDOS within the entire range of frequencies and a notable overlap between the metal and O VDOS. In amorphous $(In_{0.6}Ga_{0.4})_2O_3$, although several oxygen atoms that oscillate at normal modes with high frequencies around 550–750 cm^{-1} have a VIPR value of 0.2, it is significantly lower as compared to VIPR(O) in binary In_2O_3 (0.6) (Figure 2.1h) or Ga_2O_3 (0.4) (Figure 2.10h). This suggests that the multi-cation structure may help suppress the localization of specific vibrational modes by broadening the frequency range and the number of atoms with weakly localized modes. In $InGaZnO_4$, a few oxygen atoms that oscillate at normal modes with high frequencies around 700–750 cm^{-1} have a VIPR value of 0.6, which is similar to that of amorphous In_2O_3 (Figure 2.1h). In both amorphous $(In_{0.6}Ga_{0.4})_2O_3$ and $InGaZnO_4$, frequencies below 500 cm^{-1} show low VIPR values, suggesting an evenly distributed vibration among different atoms within most of the spectrum.

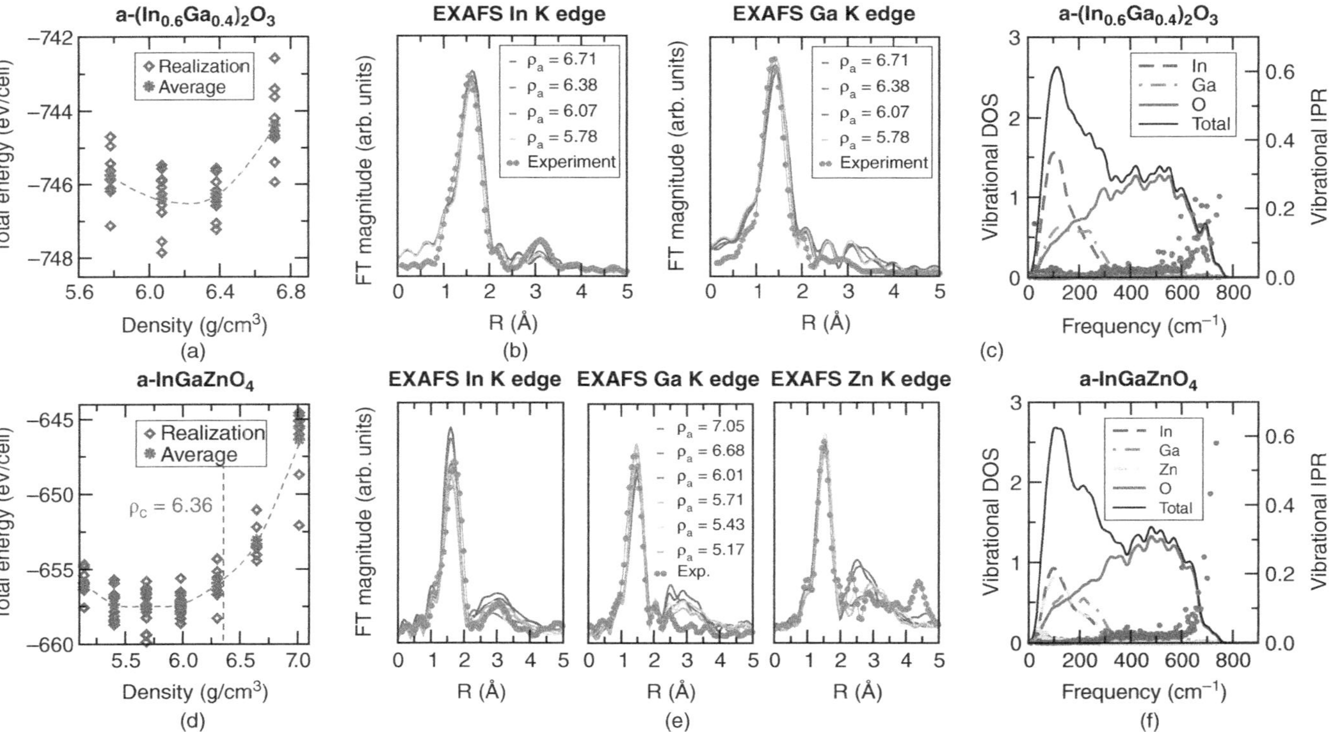

Figure 2.17 General structural properties of amorphous (a,b,c) $(In_{0.6}Ga_{0.4})_2O_3$ and (d,e,f) $InGaZnO_4$. (a,d) The energy–density curves calculated upon equilibration of each configuration at 300 K for 6 ps. The DFT total energy was calculated as an average over the final (stable) 500 MD steps to remove thermal fluctuations in each of the 10 realizations (diamond); the average over the realizations is represented by a star symbol. (b,e) Experimental and theoretical extended X-ray absorption fine structure (EXAFS) for metal K edges in amorphous models with different densities (solid line) as compared to experimental spectra (line with circle symbols). (c,f) Total and partial vibrational density of states (VDOS) and vibrational inverse participation ratio (VIPR) calculated for the most stable amorphous structures. The In, Ga, Zn, and O states are represented by dashed, dashed–dotted, dashed–dashed–dotted, and solid thick lines, and the total VDOS is the thin solid line. *Source*: (b,e) Based on Ref. [12], Huang, W., et al. (2020). Experimental and theoretical evidence for hydrogen doping in polymer solution-processed indium gallium oxide. *Proceedings of the National Academy of Sciences* **117** (31): 18231–18239, doi:10.1073/pnas.2007897117; and Ref. [48], Jia, J., et al. (2018). Evolution of defect structures and deep subgap states during annealing of amorphous In-Ga-Zn oxide for thin-film transistors. *Physical Review Applied* **9** (1): 014018, doi:10.1103/physrevapplied.9.014018.

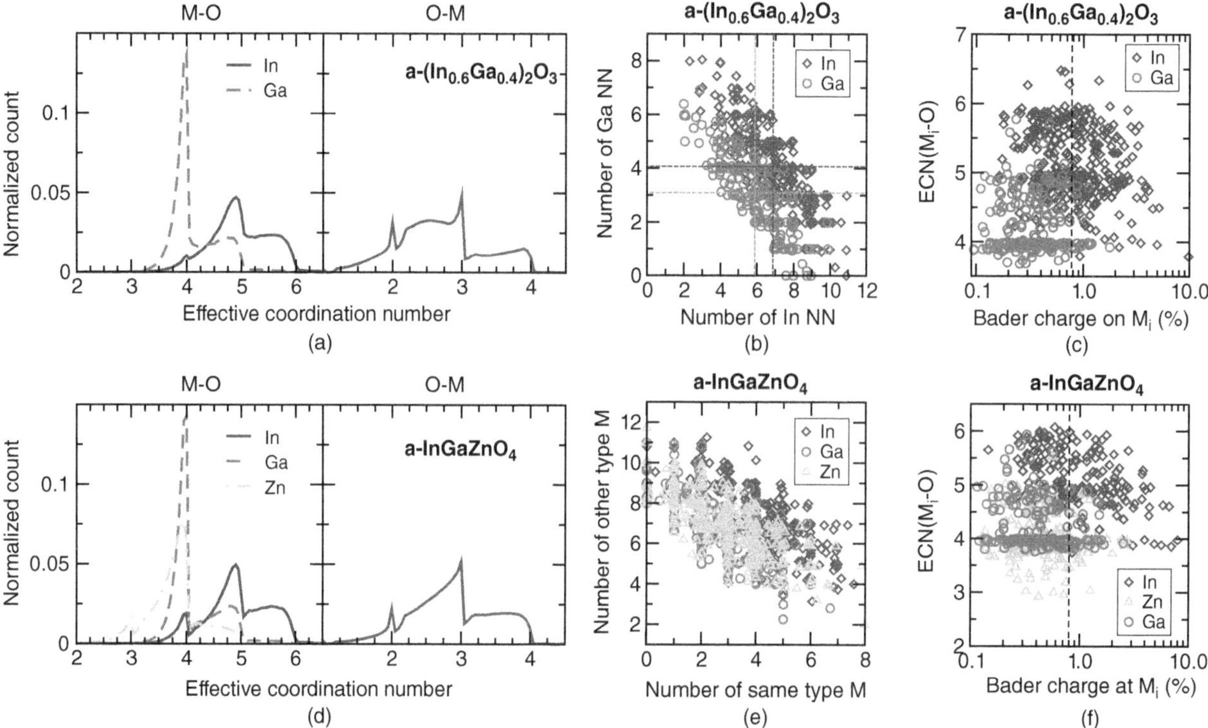

Figure 2.18 Structural randomness: The properties of individual M-O and O-M polyhedra in amorphous (a,b,c) $(In_{0.6}Ga_{0.4})_2O_3$ and (d,e,f) $InGaZnO_4$. (a,d) Distribution of the effective coordination numbers of In (solid line), Ga (dashed line), Zn (dashed–dotted line), and O (solid line). The results represent an average over 10 separate MD realizations at optimal density and are calculated based on the MD simulations at 300 K (equilibration with 3,000 MD steps = 6 ps). (b,e) The number of same M-type neighbors as a function of the number of other M-type neighbors for each individual In (diamond), Ga (circle), or Zn (triangle) atom. Only shared metal neighbors (face-, edge-, or corner-shared) are considered. The results are calculated based on the MD simulations at 300 K (equilibration with 3,000 MD steps = 6 ps) for 10 separate MD realizations at optimal density. (c,f) First-shell effective coordination number (ECN) for individual In (diamond), Ga (circle), or Zn (triangle) atoms as a function of the calculated Bader charge contribution to the conduction band from the given atom. The results are for 10 separate MD realizations at optimal density. The vertical lines represent the calculated charge value for a uniform charge density distribution (i.e., if all M atoms in the supercell contribute equally). The electronic properties are calculated for the fully relaxed structures using an accurate hybrid functional approach.

A comparison of the ECN distributions for the first-shell M-O and O-M polyhedra for the amorphous $(In_{0.6}Ga_{0.4})_2O_3$ and $InGaZnO_4$ structures with optimal density and perfect stoichiometry (i.e., for the oxygen-defect-free structures) is given in Figure 2.18a and 2.18d. The ECN(Ga) distributions in the amorphous $(In_{0.6}Ga_{0.4})_2O_3$ and $InGaZnO_4$ structures are nearly identical and also similar to that in binary Ga_2O_3 (Figure 2.11a), with the exception that the number of four(five)-coordinated Ga atoms increases (decreases) in both multi-cation oxides. The ECN(Zn) in amorphous $InGaZnO_4$ reveals that the number of four-coordinated Zn atoms is slightly reduced, the three-coordinated Zn atoms are nearly completely suppressed, and the small fraction of five-coordinated Zn appears. The presence of five-coordinated Zn atoms is in accord with the Zn environment in a crystalline $InGaZnO_4$ structure. The coordination of In atoms in the amorphous $(In_{0.6}Ga_{0.4})_2O_3$ and $InGaZnO_4$ structures shows the largest deviation from that in binary amorphous oxide: In the latter, only about one-third of In atoms have an effective coordination below five, whereas half of In atoms have an ECN < 5. Moreover, the number of four-coordinated In atoms (ECN < 4) increases in amorphous $InGaZnO_4$ as compared to the amorphous $(In_{0.6}Ga_{0.4})_2O_3$ structure (i.e., it increases with decreasing In content in multi-cation oxide;

Table 2.2 First-shell effective coordination number (ECN), effective average distance, and distortion for metal and oxygen atoms for amorphous $(In_{0.6}Ga_{0.4})_2O_3$ and crystalline and amorphous $InGaZnO_4$ at different densities.

		Metal-Oxygen			Oxygen-Metal		
	Density, g/cm^3	ECN	l_{av}, Å	σ^2, ×10^4 Å^2	ECN	l_{av}, Å	σ^2, ×10^4 Å^2
a-$(In_{0.6}Ga_{0.4})_2O_3$	6.71	In: 5.27 Ga: 4.55	2.18 1.93	159 105	O: 2.92	2.03	245
	6.38	In: 5.09 Ga: 4.35	2.18 1.92	167 89	O: 2.80	2.03	250
	*6.07**	*In: 4.95* *Ga: 4.11*	*2.18* *1.91*	*162* *83*	*O: 2.66*	*2.02*	*261*
	5.78	In: 4.77 Ga: 4.06	2.17 1.90	157 67	O: 2.59	2.01	250
c-InGaZnO$_4$	**6.36**	**In: 5.78** **Ga: 4.21** **Zn: 4.35**	**2.18** **1.89** **2.00**	**46** **136** **140**	**O: 3.22**	**2.00**	**194**
a-$InGaZnO_4$	7.05	In: 5.49 Ga: 4.81 Zn: 4.22	2.15 1.93 2.00	155 102 159	O: 3.34	2.00	208
	6.68	In: 5.38 Ga: 4.66 Zn: 4.09	2.16 1.93 2.01	142 96 155	O: 3.23	2.01	207
	6.01	In: 5.02 Ga: 4.27 Zn: 3.83	2.17 1.92 2.01	155 88 140	O: 2.95	2.00	217
	*5.71**	*In: 4.91* *Ga: 4.13* *Zn: 3.77*	*2.17* *1.91* *2.01*	*146* *75* *141*	*O: 2.86*	*2.00*	*220*
	5.43	In: 4.75 Ga: 3.99 Zn: 3.62	2.17 1.90 2.00	161 59 128	O: 2.76	1.99	212
	5.17	In: 4.58 Ga: 3.96 Zn: 3.53	2.16 1.90 2.00	147 57 140	O: 2.70	1.99	211

Note: The values represent an average over all metal or oxygen atoms in the supercell and over 10 separate MD realizations at the given density.

Figure 2.18a and 2.18d). The trend in the ECN changes with the smallest, moderate, and largest deviations from the effective coordination of binary oxides among Ga, Zn, and In, respectively, is expected from the corresponding metal–oxygen bond strengths. The stronger the first-shell bonding is, the more likely it is that the metal atom will maintain its coordination preference in a multi-cation amorphous oxide [10, 47], and vice versa. Density-dependent average ECN values for the amorphous multicomponent and binary oxides are compared in Tables 2.1 and 2.2. In both amorphous $(In_{0.6}Ga_{0.4})_2O_3$ and $InGaZnO_4$, the coordination of oxygen is significantly reduced as compared to that in binary constituent oxides.

Figure 2.18b and 2.18e show the number of shared neighbors (face-, edge-, or corner-shared) of the same metal type as a function of the number of the shared neighbors of the opposite metal type for every individual M atom in the amorphous oxide structures. In general, the results suggest that there is no segregation of the different metal atoms. A very small fraction of metal atoms (In or Ga) has only In neighbors, and all M atoms have at least two In neighbors in the amorphous $(In_{0.6}Ga_{0.4})_2O_3$. In $InGaZnO_4$, all metal atoms have at least two shared neighbors of an opposite metal type, and a few metal atoms (In, Ga, or Zn) are isolated (i.e., have no shared neighbors of their own type). Consistent with the predominantly corner-sharing preference of Ga and Zn in the corresponding binary oxides (cf. Figures 2.14a, 2.14b, and 2.15d), both have a lower number of shared metal atoms in $(In_{0.6}Ga_{0.4})_2O_3$ and $InGaZnO_4$ as compared to In, which has the largest fraction of edge-shared pairs in amorphous In_2O_3 (about 30%). Despite the well-mixed nature of the multi-cation amorphous structures, the wide distributions in the numbers of shared neighbor types represent the structural randomness at the medium range and imply complex morphology that should be taken into account, especially when doping, defect formation, and/or hydrogenation are studied in AOSs.

Despite the rather uniform spatial distributions of different metal atoms in a-$(In_{0.6}Ga_{0.4})_2O_3$ and a-$InGaZnO_4$, the conduction band is governed primarily by indium in both oxides, as signified by the calculated Bader charge contributions from individual metal atoms (Figure 2.18c and 2.18f). Only a small fraction of Ga and/or Zn have a Bader charge above the average value, which represents a uniform charge distribution. Although the largest Bader contribution is from four-coordinated In (Figure 2.18c and 2.18f), there is no clear correlation between the ECN of any metal atoms and the corresponding conduction Bader charge contribution, similar to the amorphous binary oxides (Figures 2.3f, 2.9f, and 2.12f).

The conduction bandwidths for the two amorphous multi-cation oxides, calculated along the [111] direction for 10 different MD realizations in each case, are compared in Figure 2.19a. On average over the realizations, the conduction bandwidth is wider in a-$(In_{0.6}Ga_{0.4})_2O_3$ (1.66 eV) than in a-$InGaZnO_4$ (1.57 eV), as should be expected from the larger In content in the former case. The distribution of bandwidth values among different realizations for the same composition (obtained at the same density, oxygen stoichiometry, and quench rate) illustrates the effect of morphology on the resulting electronic properties. Interestingly, the bandwidth distribution in a-$(In_{0.6}Ga_{0.4})_2O_3$ resembles the one in binary a-Ga_2O_3 (Figure 2.18b), with only a few configurations having larger bandwidths than that in the binary gallium oxide. The bandwidth distribution shrinks with the addition of Zn in a-$InGaZnO_4$—in agreement with the results for binary a-ZnO (Figure 2.18b). The charge density distributions in the amorphous $(In_{0.6}Ga_{0.4})_2O_3$ structures with the widest and the narrowest conduction bands are shown in Figure 2.19b: The widest conduction band corresponds to a charge accumulation along an In-O polyhedral chain running through the supercell; vice versa, the charge density is clustered in the structure with the narrowest bandwidth, with significant accumulation at an In-O cluster centered in the middle of the supercell. As discussed in Section 2.7, good transport properties in In-based AOSs originate from the widest In-In distance distribution among the four oxides, specifically from the largest fraction of shorter than usual corner-shared In-In distances (Figure 2.14a), which facilitates the formation of conduction channels. Figure 2.19c compares the In-In distance distribution in amorphous In_2O_3, $(In_{0.6}Ga_{0.4})_2O_3$, and $InGaZnO_4$. It is seen that the In-In distributions in the multi-cation oxides are nearly identical and feature a reduced number of edge-shared In-In pairs and a single peak in the corner-shared In-In distribution as compared to the binary indium oxide. A smaller number of the short-distant corner-shared In-In pairs that enable s-s orbital overlap (Figure 2.13) explains the reduction in conduction bandwidths in the multi-cation oxides, as compared to a-In_2O_3. At the same time, it has been shown that the spatial distribution of Ga in amorphous In-Ga-O plays an important role in promoting chain formation of undercoordinated In atoms and hence maintaining good transport properties when Ga content increases from 20 at.% to 40 at.% [49].

Hence, compositional morphology and the ability of amorphous structures to support the formation of extended conduction channels, associated with the structural chains of the In atoms having large (above-average) Bader contributions, determine good mobility in multi-cation AOSs [13, 47]. At the same time, rich morphology may lead to sample-to-sample inconsistencies in the observed transport properties, and additional preparation conditions or postdeposition treatment might be required to achieve a uniform composition morphology that ensures percolation via the conduction channels and thus the best conductivity.

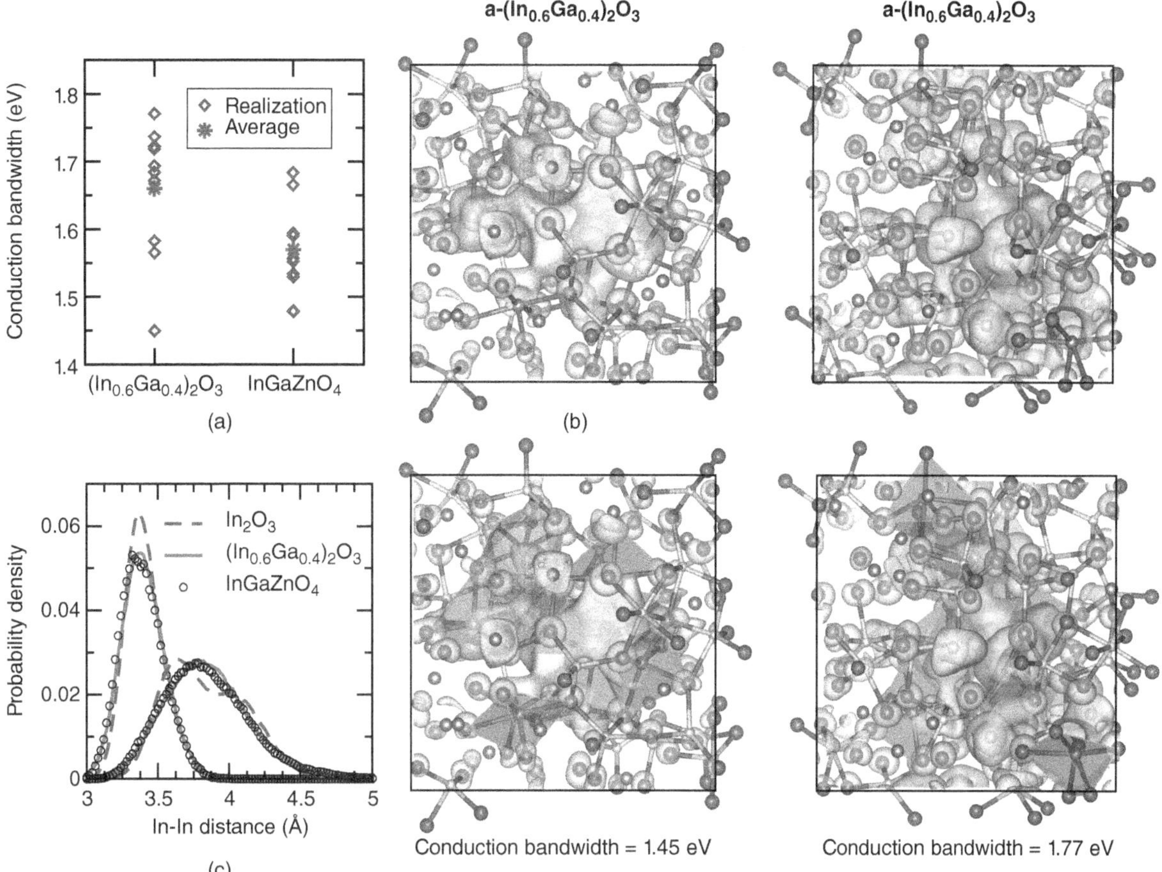

Figure 2.19 Electronic properties of amorphous stoichiometric $(In_{0.6}Ga_{0.4})_2O_3$ and $InGaZnO_4$. (a) Conduction bandwidths calculated along the [111] direction in 10 amorphous phases with optimal density. A star symbol represents an average value. (b) Charge density distribution in the supercell of the amorphous $(In_{0.6}Ga_{0.4})_2O_3$ structure with the narrowest (1.45 eV) and widest (1.77 eV) conduction-band-like states. Below, the In-O polyhedra are highlighted for the In atoms with the largest Bader charge contributions. (c) In-In distance distribution for edge- and corner-shared pairs in amorphous In_2O_3 (dashed line), $(In_{0.6}Ga_{0.4})_2O_3$ (solid line), and $InGaZnO_4$ (circles). The results for amorphous structures represent an average over 10 separate MD realizations at the corresponding optimal density.

We note here that In-free AOSs follow a similar trend: The conductivity is governed by the metal atoms with the weakest bond strength, for example by Sn in amorphous Zn-Sn-O [50].

2.9 Conclusions

Structure randomness in amorphous oxides of posttransition metals originates from weak metal–oxygen bonding, ionic in nature, which is responsible for significant deviations in the bond lengths and bond angles for the majority of metal and oxygen atoms. The nearest-neighbor disorder, hidden in the average distance distributions, becomes apparent from the calculated effective coordination, effective average distances, and polyhedral distortions of individual M and O atoms, only a few of which have structural features that closely match those in the crystalline oxide phases. This implies that distance distributions in AOSs (obtained from pair-correlation functions or EXAFS) do not capture the degree of distortion in *individual* M-O polyhedra, which ultimately determines metal coordination, or prevalence of the distorted M-O polyhedra, which expands the coordination distribution.

Broad distributions in coordination, bond angles, and polyhedral distortions give rise to an intricate morphology in AOSs that manifests itself in sharing disorder (i.e., structural randomness in the way that M-O polyhedra are connected with each other). Sharing disorder may promote clustering (e.g., a large number of edge-shared neighbors in a predominantly corner-shared network of amorphous ZnO and SnO_2) but also enables direct *s-s* orbital overlap via a reduced M-M distance for corner-shared pairs. The latter is key to good mobility in AOSs, provided that the structure favors the formation of extended chains of such metal pairs with distorted sharing, leading to percolation-like conduction paths.

Importantly, the majority of structural "outliers" in the stoichiometric oxides do not cause strong electron localization due to the ability of the disordered structure to rearrange its morphology via an extended bond reconfiguration that minimizes charge imbalance and promotes better hybridization. Therefore, defect-, impurity- or composition-induced transformations in morphology play important roles in the complex structure–property relationships in AOSs and must be carefully studied.

Finally, structural randomness in AOSs with disorder in the local (nearest-neighbor), medium, and long ranges makes a theoretical description of electron transport challenging. Adhering to the Drude model, the electron mobility in an AOS can be represented with the following equation:

$$\frac{1}{\mu} = \frac{m^*}{e}\left(\frac{1}{\tau_{defects}} + \frac{1}{\tau_{composition}} + \frac{1}{\tau_{morphology}} + \frac{1}{\tau_{vibrations}} + \frac{1}{\tau_{nano-crystallinity}}\right)$$

where the contributions to the overall relaxation time are due to (i) structural defects or impurities, (ii) spatial distribution and clustering of various cations, (iii) coordination and sharing morphology, (iv) thermal vibrations, and (v) size and density of nanocrystalline inclusions. The results of this work show that the above factors are interdependent, yet understanding their interplay provides keys to the ways the structural randomness can be tuned to attain percolation that would ensure reduced carrier scattering due to each of these factors.

References

1 Nomura, K., Ohta, H., Takagi, A. et al. (2004). Room-temperature fabrication of transparent flexible thin-film transistors using amorphous oxide semiconductors. *Nature* 432 (7016): 488–492.

2 Hosono, H. (2006). Ionic amorphous oxide semiconductors: material design, carrier transport, and device application. *Journal of Non-Crystalline Solids* 352 (9): 851–858.

3 Fortunato, E., Ginley, D., Hosono, H., and Paine, D.C. (2007). Transparent conducting oxides for photovoltaics. *MRS Bulletin* 32 (March): 242–247.

4 Kamiya, T., Nomura, K., and Hosono, H. (2009). Origins of high mobility and low operation voltage of amorphous oxide TFTs: electronic structure, electron transport, defects and doping. *Journal of Display Technology* 5 (7): 273–288.

5 Park, J.C., Kim, S., Kim, S. et al. (2010). Highly stable transparent amorphous oxide semiconductor thin-film transistors having double-stacked active layers. *Advanced Materials* 22 (48): 5512–5516.

6 Kamiya, T., Nomura, K., and Hosono, H. (2010). Present status of amorphous In-Ga-Zn-O thin-film transistors. *Science and Technology of Advanced Materials* 11 (4): 044305.

7 Park, J.S., Maeng, W.-J., Kim, H.-S., and Park, J.-S. (2012). Review of recent developments in amorphous oxide semiconductor thin-film transistor devices. *Thin Solid Films* 520 (6): 1679–1693.

8 Nathan, A., Lee, S., Jeon, S., and Robertson, J. (2014). Amorphous oxide semiconductor TFTs for displays and imaging. *Journal of Display Technology* 10 (11): 917–927.

9 Yu, X., Marks, T.J., and Facchetti, A. (2016). Metal oxides for optoelectronic applications. *Nature Materials* 15 (4): 383–396.

10 Medvedeva, J.E., Buchholz, D.B., and Chang, R.P.H. (2017). Recent advances in understanding the structure and properties of amorphous oxide semiconductors. *Advanced Electronic Materials* 3 (9): 1700082.

11 Medvedeva, J.E., Zhuravlev, I.A., Burris, C. et al. (2020). Origin of high carrier concentration in amorphous wide-bandgap oxides: role of disorder in defect formation and electron localization in In_2O_{3-x}. *Journal of Applied Physics* 127 (17): 175701.

12 Huang, W., Chien, P.-H., McMillen, K. et al. (2020). Experimental and theoretical evidence for hydrogen doping in polymer solution-processed indium gallium oxide. *Proceedings of the National Academy of Sciences* 117 (31): 18231–18239.

13 Medvedeva, J.E., and Bhattarai, B. (2020). Hydrogen doping in wide-bandgap amorphous In–Ga–O semiconductors. *Journal of Materials Chemistry C* 8 (43): 15436–15449.

14 Medvedeva, J.E. (2010). Combining optical transparency with electrical conductivity: challenges and prospects. In: *Transparent Electronics: From Synthesis to Applications* (ed. A. Facchetti and T. Marks), 1–29. Chichester, UK: John Wiley and Sons.

15 Rosen, J. and Warschkow, O. (2009). Electronic structure of amorphous indium oxide transparent conductors. *Physical Review B* 80: 115215.

16 Takagi, A., Nomura, K., Ohta, H. et al. (2005). Carrier transport and electronic structure in amorphous oxide semiconductor, a-InGaZnO4. *Thin Solid Films* 486 (1–2): 38–41.

17 Cho, D.-Y., Song, J., Na, K.D. et al. (2009). Local structure and conduction mechanism in amorphous In–Ga–Zn–O films. *Applied Physics Letters* 94 (11): 112112.

18 Kimura, M., Kamiya, T., Nakanishi, T. et al. (2010). Intrinsic carrier mobility in amorphous In–Ga–Zn–O thin-film transistors determined by combined field-effect technique. *Applied Physics Letters* 96 (26): 262105.

19 Germs, W.C., Adriaans, W.H., Tripathi, A.K. et al. (2012). Charge transport in amorphous InGaZnO thin-film transistors. *Physical Review B* 86 (15): 155319.

20 Lee, S., Ghaffarzadeh, K., Nathan, A. et al. (2011). Trap-limited and percolation conduction mechanisms in amorphous oxide semiconductor thin film transistors. *Applied Physics Letters* 98 (20): 203508.

21 Smith, J., Zeng, L., Khanal, R. et al. (2015). Cation size effects on the electronic and structural properties of solution-processed In-X-O thin films. *Advanced Electronic Materials* 1 (7): 1500146.

22 Godo, H., Kawae, D., Yoshitomi, S. et al. (2010). Temperature dependence of transistor characteristics and electronic structure for amorphous In-Ga-Zn-Oxide thin film transistor. *Japanese Journal of Applied Physics* 49 (3 Part 2): 03CB04.

23 Kamiya, T., Nomura, K., and Hosono, H. (2010). Subgap states, doping and defect formation energies in amorphous oxide semiconductor a-InGaZnO 4 studied by density functional theory. *Physica Status Solidi (A) Applications and Materials Science* 207 (7): 1698–1703.

24 Noh, H.K., Chang, K.J., Ryu, B., and Lee, W.J. (2011). Electronic structure of oxygen-vacancy defects in amorphous In-Ga-Zn-O semiconductors. *Physical Review B* 84 (11): 115205.

25 Kim, M., Kang, I.J., and Park, C.H. (2012). First-principle study of electronic structure of Sn-doped amorphous In2O3 and the role of O-deficiency. *Current Applied Physics* 12: S25–S28.

26 Deng, H.X., Wei, S.H., Li, S.S. et al. (2013). Electronic origin of the conductivity imbalance between covalent and ionic amorphous semiconductors. *Physical Review B* 87 (12): 125203.

27 Sallis, S., Butler, K.T., Quackenbush, N.F. et al. (2014). Origin of deep subgap states in amorphous indium gallium zinc oxide: chemically disordered coordination of oxygen. *Applied Physics Letters* 104 (23): 232108.

28 Han, W.H., Oh, Y.J., Chang, K.J., and Park, J.S. (2015). Electronic structure of oxygen interstitial defects in amorphous In-Ga-Zn-O semiconductors and implications for device behavior. *Physical Review Applied* 3 (4): 044008.

29 Körner, W., Urban, D.F., and Elsässer, C. (2015). Generic origin of subgap states in transparent amorphous semiconductor oxides illustrated for the cases of In-Zn-O and In-Sn-O. *Physica Status Solidi (A) Applications and Materials Science* 212 (7): 1476–1481.

30 Walsh, A., Da Silva, J.L.F., and Wei, S.H. (2009). Interplay between order and disorder in the high performance of amorphous transparent conducting oxides. *Chemistry of Materials* 21 (21): 5119–5124.

31 Nishio, K., Miyazaki, T., and Nakamura, H. (2013). Universal medium-range order of amorphous metal oxides. *Physical Review Letters* 111 (15): 155502.

32 Zawadzki, P.P., Perkins, J., and Lany, S. (2014). Modeling amorphous thin films: kinetically limited minimization. *Physical Review B* 90 (9): 094203.

33 Hoppe, R., Voigt, S., Glaum, H. et al. (1989). A new route to charge distributions in ionic solids. *Journal of the Less Common Metals* 156 (1): 105–122.

34 Tang, W., Sanville, E., and Henkelman, G. (2009). A grid-based Bader analysis algorithm without lattice bias. *Journal of Physics: Condensed Matter* 21 (8): 084204.

35 Yu, M., and Trinkle, D.R. (2011). Accurate and efficient algorithm for Bader charge integration. *Journal of Chemical Physics* 134 (6): 064111.

36 Kresse, G., and Hafner, J. (1994). Ab initio molecular-dynamics simulation of the liquid-metal–amorphous-semiconductor transition in germanium. *Physical Review B* 49 (20): 14251–14269.

37 Kresse, G., and Furthmüller, J. (1996). Efficient iterative schemes for ab initio total-energy calculations using a plane-wave basis set. *Physical Review B* 54 (16): 11169–11186.

38 Perdew, J.P., Burke, K., and Ernzerhof, M. (1996). Generalized gradient approximation made simple. *Physical Review Letters* 77 (18): 3865–3868.

39 Blöchl, P.E. (1994). Projector augmented-wave method. *Physical Review B* 50 (24): 17953–17979.

40 Kresse, G., and Joubert, D. (1999). From ultrasoft pseudopotentials to the projector augmented-wave method. *Physical Review B* 59 (3): 1758–1775.

41 Heyd, J., Peralta, J.E., Scuseria, G.E., and Martin, R.L. (2005). Energy band gaps and lattice parameters evaluated with the Heyd-Scuseria-Ernzerhof screened hybrid functional. *Journal of Chemical Physics* 123 (17): 174101.

42 Rehr, J.J., Kas, J.J., Prange, M.P. et al. (2009). Ab initio theory and calculations of X-ray spectra. *Comptes Rendus Physique* 10 (6): 548–559.

43 Newville, M. (2001). IFEFFIT: interactive XAFS analysis and FEFF fitting. *Journal of Synchrotron Radiation* 8 (2): 322–324.

44 Momma, K., and Izumi, F. (2011). VESTA 3 for three-dimensional visualization of crystal, volumetric and morphology data. *Journal of Applied Crystallography* 44 (6): 1272–1276.

45 Higashiwaki, M., Sasaki, K., Kuramata, A. et al. (2014). Development of gallium oxide power devices. *Physica Status Solidi (A) Applications and Materials Science* 211 (1): 21–26.

46 Kim, J., Sekiya, T., Miyokawa, N. et al. (2017). Conversion of an ultra-wide bandgap amorphous oxide insulator to a semiconductor. *NPG Asia Materials* 9 (3): e359–e359.

47 Khanal, R., Buchholz, D.B., Chang, R.P.H., and Medvedeva, J.E. (2015). Composition-dependent structural and transport properties of amorphous transparent conducting oxides. *Physical Review B* 91 (20): 205203.

48 Jia, J., Suko, A., Shigesato, Y. et al. (2018). Evolution of defect structures and deep subgap states during annealing of amorphous In-Ga-Zn oxide for thin-film transistors. *Physical Review Applied* 9 (1): 014018.

49 Moffitt, S.L., Zhu, Q., Ma, Q. et al. (2017). Probing the unique role of gallium in amorphous oxide semiconductors through structure-property relationships. *Advanced Electronic Materials* 3 (10): 1700189.

50 Husein, S., Medvedeva, J.E., Perkins, J.D., and Bertoni, M.I. (2020). The role of cation coordination in the electrical and optical properties of amorphous transparent conducting oxides. *Chemistry of Materials* 32 (15): 6444–6455.

3

Electronic Structure of Transparent Amorphous Oxide Semiconductors

John Robertson and Zhaofu Zhang

Engineering Department, Cambridge University, Cambridge, UK

3.1 Introduction

Around 1990, the preeminent large-area semiconductor was hydrogenated amorphous silicon (*a*-Si:H). It could be doped in both polarities, and it could be deposited over large areas at low cost with low surface roughness and without grain boundaries. However, it had two drawbacks: Its carrier mobility of ~0.5 cm^2/V.s was too low, and it suffered from a bias stress instability to charges and an illumination instability (the Staebler–Wronski effect). It was often assumed that by 2000, the next stage of development would be to use nanocrystalline Si, produced by laser crystallization or direct deposition, to greatly increase mobility, retain bipolar doping, and remove the origin of instabilities. However, nanocrystalline Si would still have grain boundaries and be less suitable for depositing reliable gate dielectrics for thin-film transistors (TFTs).

This did not occur. Instead, amorphous oxide semiconductors (AOSs) such as In-Ga-Zn oxide (IGZO) were developed with a much higher electron mobility of ~10 cm^2/V.s; ease of low-temperature, large-area deposition [1–3]; and no grain boundaries, and these films rapidly displaced *a*-Si:H as the dominant semiconductor for display applications [3–5].

It is first necessary to give a description of the electronic structure of AOSs and explain why they have such a high electron mobility compared to *a*-Si:H. The post-transition metal oxides such as ZnO, In_2O_3, and SnO_2 differ considerably from the alkaline–earth oxides like SrO. The divalent oxides like ZnO have electron affinities of ~4.0 to 4.5 eV, not 2.0 eV like SrO [3], and are stable in contact with water. The band structure of a typical member SnO_2 is shown in Figure 3.1 [6]. It has a band gap of around 3.6 V in the near-ultraviolet (near-UV) range. Its bonding is quite ionic, 60%, higher than SiO_2 that still has a covalent network. In contrast, by symmetry, the conduction band minimum (CBM) of SnO_2 is much more polar than the average state, and it consists almost entirely of cation (Sn) s states [7].

3.2 Mobility

The field-effect mobility μ_{FE} is given by the free electron mobility μ_0 times the fraction of the induced charges that remain in extended (free) states rather than in trapped states:

$$\mu_{FE} = \frac{n_{free}}{n_{free} + n_{trapped}} \mu_0 \tag{3.1}$$

Using SnO_2 as the example, the electron conduction occurs between overlapping cation s states rather than between the antibonding sp^3 orbitals of Si, as compared in Figure 3.2. The cation s states are spherically symmetric,

Amorphous Oxide Semiconductors: IGZO and Related Materials for Display and Memory, First Edition.
Edited by Hideo Hosono and Hideya Kumomi.

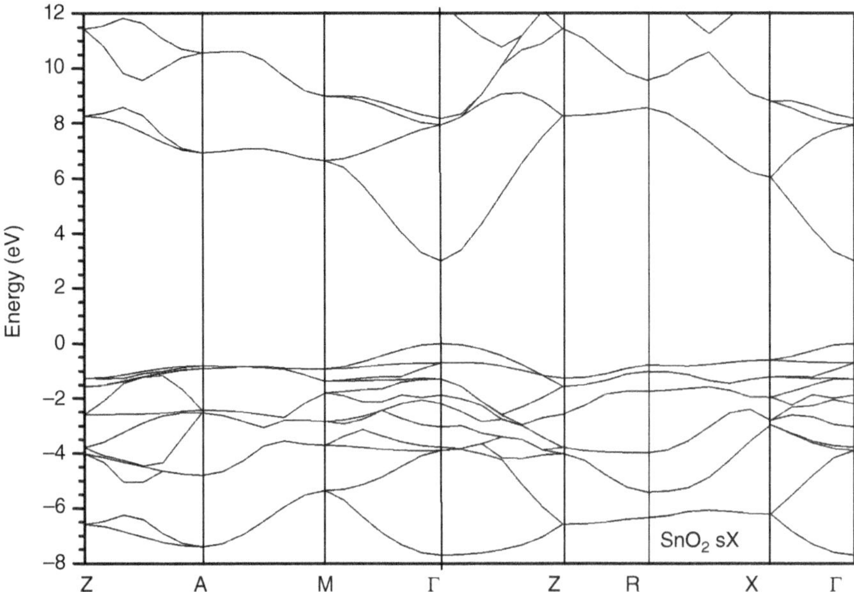

Figure 3.1 Band structure of SnO_2 in rutile structure. Source: Based on Robertson, J., Xiong, K., and Clark, S. J. (2006). Band structure of functional oxides by screened exchange. *Physica Status Solidi (B): Basic Research* **243**: 2054–2070.

and so their overlap with adjacent s states is less sensitive to angular disorder (Figure 3.2b) [2, 8]. This contrasts with the directional antibonding sp^3 states, whose interaction varies greatly with the disorder, as seen in Figure 3.2a. This leads to a much lower electron mobility for the disordered sp^3 states.

However, the argument is deeper than this original explanation. Let us compare the effect of ring disorder on the orbital phases at the CBM state, in crystalline and amorphous Si networks and in the oxides. The diamond lattice of c-Si consists of sixfold rings, whereas the *a*-Si network also has fivefold and sevenfold rings. The phases of the antibonding orbitals around the sixfold rings for the Γ_2' state of c-Si are shown in Figure 3.3a. Despite the sign reversal along each antibond, the signs match up at each vertex, so there is constructive interference around a sixfold ring. Compare this to a fivefold ring, as found in *a*-Si. Now the sign reversal causes a destructive interference at the final (circled) vertex (Figure 3.3a).

However, c-SnO_2 contains two Sn sublattices in its rutile structure. The phases of Sn s orbitals are both positive for each sublattice at the CBM (the oxygen phases are negative), so there is constructive interference between the Sn s states. In the amorphous lattice, there can be even or odd rings of cations, but the phases of all Sn s orbitals are still in-phase for the CBM, giving constructive interference for both even- and odd-membered rings (Figure 3.3b). Similar arguments extend to the oxide alloys.

In *a*-Si:H, these destructive interference effects cause a Hall effect sign anomaly, following the Friedman model of the Hall effect [9, 10]. In contrast, the constructive interference around both even- and odd-membered rings in the oxides means that there is no Hall effect sign anomaly, and there is no decline in the mobility as E_F crosses the mobility edge, as is seen experimentally [11, 12].

3.3 Density of States

This constructive interference has a great effect on the conduction band (CB) tail of localized states for IGZO compared to *a*-Si:H. The density of states (DOS) at the top of the CB tail in *a*-Si:H is on the order of 10^{21} cm^{-3} eV^{-1}, and its slope parameter is ~25 meV [13, 14]. This means that the field effect mobility is only a fraction of the

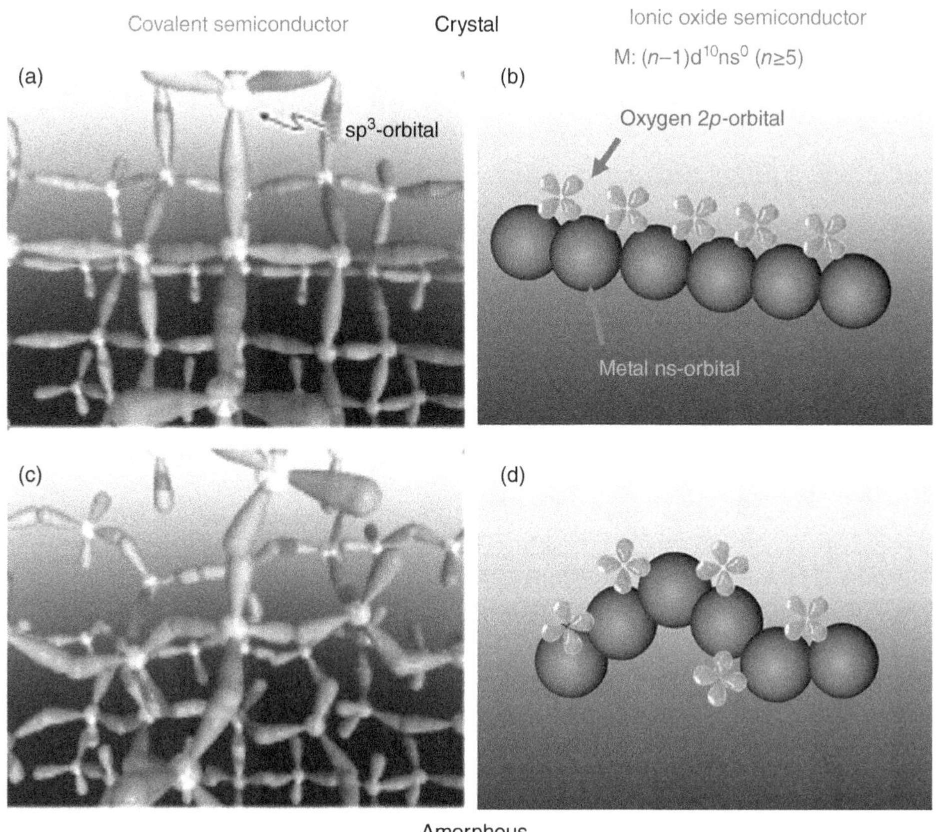

Figure 3.2 A network of (a,c) sp^3 antibonding orbitals (left), (b,d) metal s states (right), for the crystalline (top) and amorphous phases (bottom), compared. Source: Hosono, H. (2006). Ionic amorphous oxide semiconductors: material design, carrier transport, and device application. *Journal of Non-Crystalline Solids* **352** (9–20): 851–858. doi:10.1016/j.jnoncrysol.2006.01.073.

extended state mobility, because the majority of charges induced by the TFT gate voltage remain trapped in tail states by Equation (3.1).

The equivalent values for *a*-IGZO are given in Table 3.1 from various measurement techniques, and are on the order of 10^{18} $cm^{-3}eV^{-1}$ and 20 meV, respectively [11, 15–20]. Thus, the reduced disorder means that the density of tail states for *a*-IGZO is at least 10^3 times lower than that of the *a*-Si:H tail. Thus, the much lower tail DOS of *a*-IGZO means that few of the induced charges are trapped in tail states, and its electron mobility remains close to its crystalline value by Equation (3.1), unlike in *a*-Si:H.

Thus, the DOS across the mobility gap of IGZO is shown schematically in Figure 3.4 and compared to that of *a*-Si:H [15]. There is the much lower DOS at the top of the CB tail at 0 eV in IGZO, a wide region of low DOS in the upper gap, then a higher DOS for 1 eV above the valence band maximum (VBM). The CB has a low effective mass m* from its calculated DOS versus energy of ~0.3 m_e. As the band gap is ~3.0 eV, and IGZO is strongly *n*-type, the intrinsic hole current of IGZO is extremely small.

The small tail state DOS of IGZO also means that the subthreshold swing (S) of their TFTs is lower. S is defined as [3]:

$$S = \frac{dV_{GS}}{d.\log I_{DS}} = \ln 10 * 8\frac{kT}{e}\left(1 + \frac{eD_{sg}}{C_g}\right) \tag{3.2}$$

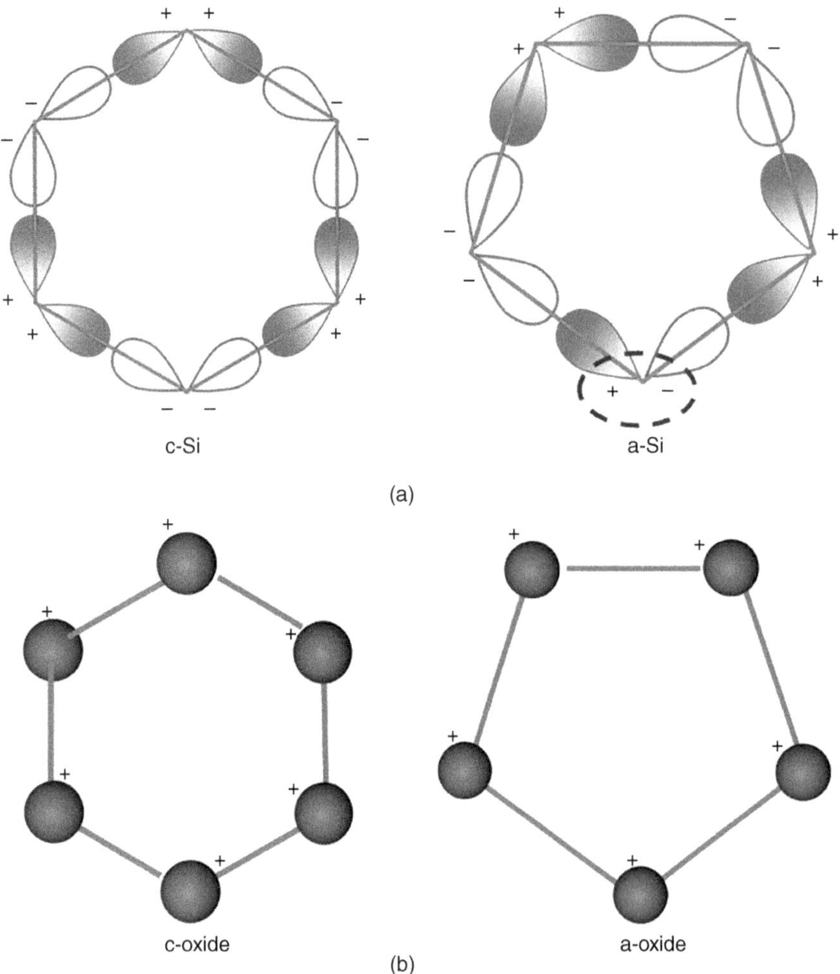

Figure 3.3 Orbital phases for (a) an antibonding conduction band state for a sixfold ring and a fivefold ring in *a*-Si, and (b) a metal s-like state around a sixfold and fivefold ring in an ionic metal oxide. Note the destructive interference at the lowest vertex of a five-membered ring in *a*-Si (circled).

The fundamental lower limit of S at 25 °C is ~60 mV/decade. The subthreshold swings of *a*-Si:H TFTs are typically many times larger. IGZO with its lower tail DOS allows a much lower S and thus lower operating voltage.

The next key point about *a*-IGZO is its response to shifting the Fermi energy [8]. In *a*-Si:H, if the Fermi energy is moved above the CB mobility edge E_C in a TFT, by gate bias or by bulk doping of the bulk, the network will try to rearrange itself by bond switching, which would cause the occupied state above E_C to become a filled defect state (D^-) below E_C, and shift E_F below E_C again [21]. This is a fundamental property first noted by Street [21]. It is a network instability, which could be a bias stress instability in TFTs or a photo-induced instability (Staebler–Wronski effect). The reaction can be expressed as follows (and see Figure 3.5):

$$Si_4^{\ 0} + P_3^{\ 0} \rightarrow Si_3^{\ -} + P_4^{\ +} \tag{3.3}$$

where subscripts are the coordination numbers. The reaction is one of a negative "effective correlation energy" (U) in that the half-filled tetrahedral P dopant site is transformed into an ionized dopant site P^+ and a trivalent defect site D^-.

Table 3.1 Conduction band tail slope (Δ) and density of states at the top of a conduction band tail (N), compared for *a*-IGZO (from various methods) and *a*-Si:H

Method	Author and Reference	Δ (meV)	N ($cm^{-3}.eV^{-1}$)
Modulated PC	Erslev et al. [16]	11	10^{18}
DOS model	Hsieh et al. [19]	80	2×10^{8}
Hall effect	Kamiya et al. [11]		3×10^{18}
CV DOS model	Park et al. [17]	22	1.7×10^{18}
TFT mobility	Lee et al. [18]	35	4×10^{18}
a-Si:H	Powell and Deane [13]	30	10^{22}
	Tiedje et al. [14]	22	

Source: Adapted from Refs. [13–20].

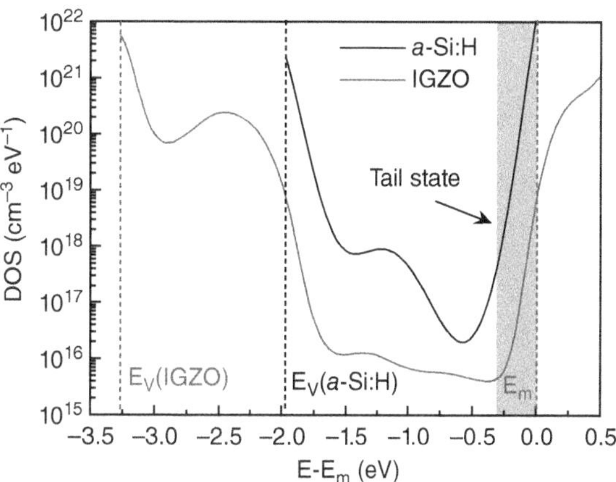

Figure 3.4 Schematic density of states (DOS) of *a*-IGZO and *a*-Si:H compared [15]. Note the 1000-fold reduction of DOS at the conduction-band edge of IGZO compared to *a*-Si:H. Source: Based on Robertson, J. (2012). Properties and doping limits of amorphous oxide semiconductors. *Journal of Non-Crystalline Solids* **358** (17): 2437–2442. doi:10.1016/j.jnoncrysol.2011.12.012.

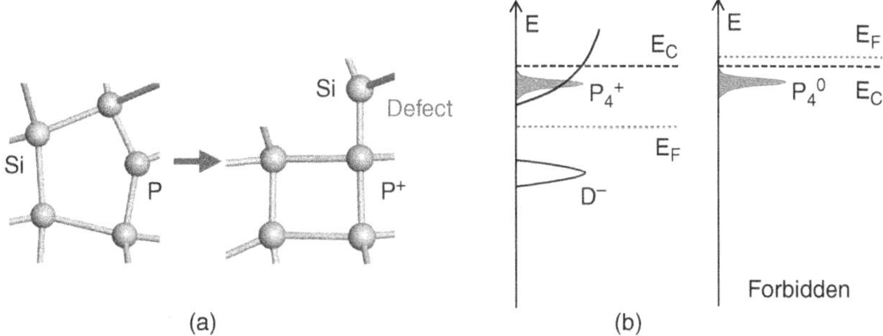

Figure 3.5 (a) Equilibrium-defining doping by P in *a*-Si: The ground state (nondoping) is a trivalent P, and the doping state is an ionized tetrahedral site. (b) If the network tries to form an un-ionized dopant site, which leaves E_F above the conduction-band mobility edge, giving an unpaired spin state P_4^0 (right), this is forbidden. The unpaired spin state P_4^0 decays into paired states P_3^0 or P_4^+/D^-.

This instability does not occur in the amorphous oxide semiconductors. This is because the CB does not consist of directional covalent bonds, so it is not possible to create a distortion that locally creates a gap in the DOS at E_F. The network instead consists of a relatively close-packed array of spheres (the cation s orbitals). The CB of the oxide is largely parabolic in both phases, so the conductance increases with electron density [13, 14].

3.4 Band Structures of *n*-Type Semiconductors

The band structure of SnO_2 was shown in Figure 3.1. Its CBM is due to Sn s states. The valence band (VB) top is due to O 2p lone pair states lying normal to the bonding plane of trivalent O sites. The highest state is arranged so that it gives a direct but optically forbidden transition to the CBM [7]. The first allowed transition is from −0.6 eV. A second deeper group of valence states down to −8 eV forms the Sn-O *p*-bonding states.

In_2O_3 has the cubic bixbyite structure with 80 atoms per unit cell. Its bonding is also quite ionic like SnO_2. The character of its states is similar to that of SnO_2; the CBM consists of In s states, the upper VB consists of O 2p lone pair states, and the deeper states are In-O p bonding states. Its band gap is calculated to be 2.6 eV using the SX approximation. The uppermost O 2p states of the VBM also give a direct forbidden optical band gap down to −0.8 eV [22]. This causes optical transitions from the upper ~0.6 eV of VB states to the CB to be forbidden, giving an observed optical gap of 3.4 eV. This means that the apparent optical band gap is larger than the electrical band gap, which had led to the incorrect idea that In_2O_3 had an indirect gap away from Γ. The gap of In_2O_3 is slightly less than that of ZnO, and the In s orbitals are slightly wider than those of the Zn s states, so the In s states make the greater contribution to electron conduction in the IGZO alloy system [23, 24].

The third AOS member, ZnO, has a polar covalent lattice of tetrahedral Zn and O sites with electron-pair bonds [25, 26]. Its gap is 3.4 eV wide experimentally, and it has well-studied excitons. The Zn 3d states at −7 eV form shallow core levels (Figure 3.6) [26]. Some of these have the same symmetry as the O 2p VBM states. This leads to the Zn 3d states causing a narrower than usual band gap. The energy of the Zn 3d states can be used to grade the quality of a ZnO calculation. Ga_2O_3 has a wider band gap (~4.7 eV) than ZnO, and its bonding is polar covalent. However, its lattice does not consist of 2-electron bonds. It also has CBM states consisting of cation s orbitals.

A simple density functional method like the generalized gradient approximation (GGA) grossly underestimates the band gap of these oxides. The GGA gap is typically 0.9 eV for each oxide. GW gives correct values but is unnecessarily expensive. Hybrid functional methods that mix a fraction of Hartree–Fock (HF) exchange give a good

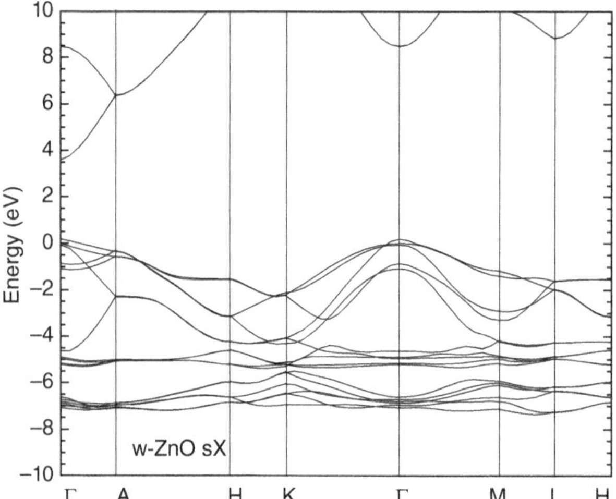

Figure 3.6 Band structure of wurzite ZnO. Source: Robertson, J. (2008). Disorder and instability processes in amorphous conducting oxides. *Physica Status Solidi (B): Basic Research* **245** (6): 1026–1032. doi:10.1002/pssb.200743458.

Table 3.2 Heat of formation of binary oxides and IGZO per O atom, crystalline, and amorphous, and their mass densities

Heat of Formation per O Atom (eV)	Crystalline	Amorphous
ZnO	−3.87 (5.44 g/cm^3)	−3.28 (4.62 g/cm^3)
Ga_2O_3	−4.21	
In_2O_3	−3.92	
IGZO	−3.91 (6.4 g/cm^3)	−3.60 (6.04 g/cm^3)

description of the gap and at lower cost [27, 28]. The SX method, as used here, gives a good description of the bands. The HSE method with its degree of HF content increased to $\alpha = 37\%$ also gives the correct band gap of 3.4 eV for ZnO [23].

A lower cost method to obtain reasonable band structures is to use GGA but add a Hubbard U potential to the Zn 3d states [29]. The U downshifts the Zn 3d states, so they give less repulsion to the VBM O 2p states, thereby widening the gap. However, this U method can need unreasonably large U values to fit the experimental ZnO gap. It is better to apply U to both the Zn 3d states and O 2p states, which repels both Zn 3d and O 2p down. Values of 6 eV for U(d) and U(p) are less extreme, fit the gap, and still have a low GGA-like computational cost [30].

We now discuss the alloys. Crystalline $InGaZnO_4$ was previously known as a semiconducting oxide. It has a fully bonded, layered lattice with a large z lattice constant [27, 28]. The lattice consists of Ga-O_5 bipyramid units forming one layer and Zn-O_4 tetrahedra and In-O_6 units forming intervening layers. Walsh et al. [24] have found that the network of *a*-$InGaZnO_4$ consists of a random packing of cation-centered polyhedra, with the same GaO_5 bipyramids, ZnO_4 tetrahedra, and InO_6 octahedral units. The charge transfer from each cation to oxygens is quite similar to that in the crystalline phase. The mass density of $InGaZnO_4$ falls from 6.4 g/cm^3 for the crystal to 6.04 g/cm^3 for the amorphous phase [30], but the density reduction is greater for pure ZnO. The CBM has a mixed s character on all the cations, with In s being the crucial one.

Table 3.2 gives the heat of formation per O atom for each oxide, and also its value for their amorphous phases. It is interesting that it costs 0.59 eV per O to make ZnO amorphous, a large value due to its poor glass-forming ability (Figure 3.7a). For comparison, IGZO takes only 0.31 eV per O atom to make it amorphous, consistent with its much easier glass-forming ability from the mixed cation effect (Figure 3.7b).

The role of the various components of IGZO is as follows [2, 31]. The In s states provide the main high-mobility electron-conducting path [27, 28]. ZnO is added to In_2O_3 to make it easier to disorder by the mixed cation effect. The function of the third metal oxide in IGZO can be compared by its effect on the mobility, off-current, Urbach slope, and stability. It was found that the electron mobility in amorphous In-metal-zinc alloys follows the effective mass of the parent binary oxide [31]. However, this is only one factor. Ga_2O_3 has a wider gap than In_2O_3, but its electron affinity is similar to that of In_2O_3 (Figure 3.16a), so that it does not add any extra disorder potential for electrons, as seen in the partial density of states (PDOS) (Figure 3.7d).

A major effect of adding Ga_2O_3 is to increase the cost of oxygen deficiency. Each oxygen site is bonded to one or more Ga sites, so the Ga-O bonds increase the formation energy of that O vacancy [30], and this causes *a*-IGZO transistors to have a smaller off-current [3].

Some oxides such as HfO_2, Y_2O_3, SiO_2, and TiO_2 have larger heats of formation [32], which translates into higher oxygen vacancy formation energies, so they would decrease their device off-currents. But HfO_2, Y_2O_3, and SiO_2 have higher energy CBMs (as seen in Figure 3.7f), so these metals would reduce the electron mobility and would be expected to have larger CB tailing. Several alternatives such as TiO_2 cause an increased CB tailing experimentally [31] but have similar electron affinities to Ga_2O_3.

Some oxides like SnO_2 have multiple valences (SnO) and are known to give rise to extra CB tail states and trapping instability. Finally, some oxides like ZnO can bond hydrogen atom pairs at oxygen vacancies, which may be the cause of illumination stress instabilities (see Section 3.5).

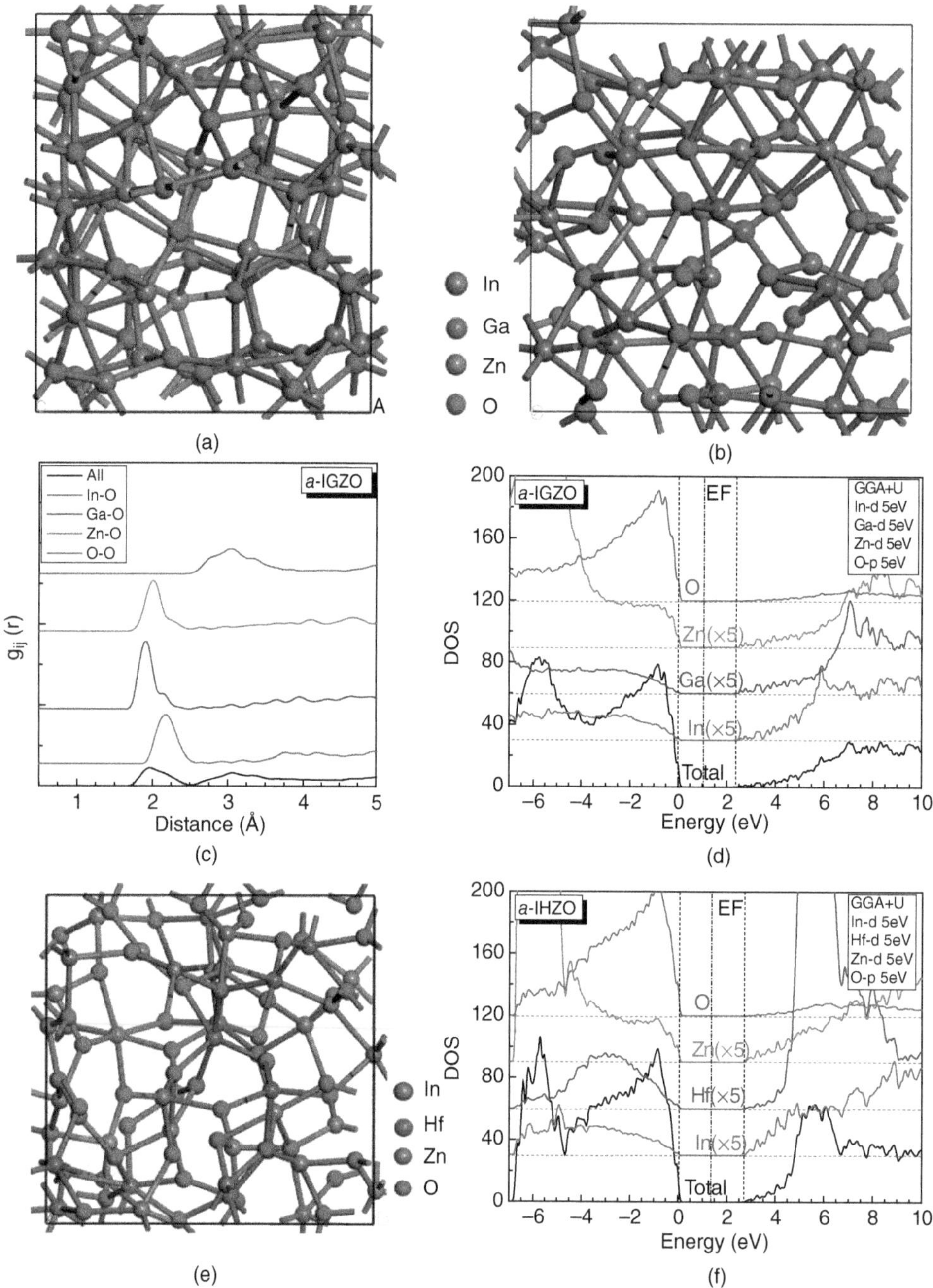

Figure 3.7 (a) An *a*-ZnO network. (b) An *a*-IGZO network. (c) Partial radial distribution function of an *a*-IGZO network. (d) Partial density of states (PDOS) of an *a*-IGZO network by the GGA+U method, showing that the conduction bands of In, Zn, and Ga start at similar energies. (e) An *a*-$InHfZnO_4$ network. (f) PDOS of *a*-$InHfZnO_4$ showing that the Hf conduction states start at ~2 eV above the CBM, adding significant electron disorder to its conduction band. Y is similar.

Table 3.3 Formation energy of the O or cation vacancy in *c*-IGZO, in its neutral state, for O-rich conditions. The O vacancy has the lowest cost.

	Neutral Vacancy Formation Energy (eV)
O	2.0
In	6.1
Zn	4.0
Ga	5.1

Turning to defects, Table 3.3 shows that there is a large difference in defect formation energies for the neutral vacancy, between oxygen and metal vacancies, in the O-rich limit [33–35]. It is estimated that the cost of the neutral Ga vacancy is 5.1 eV for an oxygen chemical potential of 0 eV. The Zn vacancy has the lowest cost of metal vacancies, but it is still higher than for the O vacancy. The behavior of the various point defects of IGZO is compared extensively [33, 34]. These values depend on the O chemical potential and Fermi energy, as discussed later in this chapter. They show that the most prevalent, lowest cost defect is the oxygen vacancy, and that metal site vacancies have much higher energy cost. This makes IGZO typical of other oxides like HfO_2 or $SrTiO_3$, where the oxygen vacancy is the defect of most concern. However, the vacancy is not easy to detect explicitly because it rarely occurs in its paramagnetic form.

So far, we have considered various cation alloys. There is an interesting case of alloying the anion side, as in *a*-ZnON. For the original transparent conducting oxides and SnO_2, it was usual to dope with Sb on the Sn site. An alternative is to dope F on the O site. As a high electron mobility is desired, it is preferable to dope with F than with Sb, because F causes disorder in the VB, so it should not affect electron mobility. In the same way, N-doped *a*-ZnO would generate electrons, not disorder the CB edge, and it gives anion disorder to stabilize the amorphous phase. Kim et al. [36] shows that *a*-ZnON has higher electron mobility than IGZO. The nitrogen adds VB states that reduce the band gap. However, while this improves the stability because it covers up deep traps causing the illumination-induced negative-bias stress instability (NBIS), the reduction of band gap is too much to make this alloy transparent.

3.5 Instabilities

Some of the weaknesses of *a*-Si:H are its instabilities, the light-induced reduction of photoconversion efficiency called the Staebler–Wronski effect, and the bias stress instability of TFTs [13]. These arise from the occupation of states above the CB mobility edge Ec, in a variant of the Street [21] reconstruction. The bias–stress instability is a particular problem in organic light-emitting diode (OLED) driver transistors, where they have a large duty cycle.

Oxide semiconductor devices suffer from instabilities that can be attributed to various factors, such as gate insulator layer trapping or interaction with ambient oxygen or water molecules. Nevertheless, they do not suffer from the bias–stress instability mechanism of *a*-Si:H. However, they do suffer from a specific illumination-induced NBIS [37, 38]. This instability resembles a persistent photoconductivity. NBIS also correlates with the filled gap states lying up to 1 eV above the VB edge, seen by photoemission [39]. There have been various mechanisms proposed to account for it, such as oxygen vacancies [40], oxygen interstitials [41], and hydrogen-related defects [42], but each mechanism has its flaws [30, 43].

1. The oxygen vacancy is a common defect in oxide, and the instability is worse in O-deficient films. The O vacancy has a negative-U property [44] that creates an energy barrier to carrier recombination, which explains

the persistent photoconductivity. However, the filled vacancy states lie too high in energy [31] above the VB compared to their energy seen by photoemission.

2. Oxygen interstitials also possess a negative-U behavior [41]. Their gap states lie at the correct energy in the gap, as seen by photoemission. However, O interstitials are not an oxygen excess defect, which is found experimentally.
3. A third model is that a hydrogen-related defect undergoes a deep-to-shallow transition and becomes donor-like [42]. However, this defect also does not correspond to O deficiency, and there is also not so much evidence for it.

We propose a model corresponding to two hydrogens trapped at a neutral oxygen vacancy [43]. Following work on hydrogen in crystalline ZnO, Du and Biswas [45] noted that two hydrogens can be trapped at an oxygen vacancy. This vacancy consists of four Zn sp^3 hybrids pointing into the vacancy. This has two configurations, a 2+ charge state where the hybrids are empty, so the two hydrogens just make an H_2 molecule (Figure 3.9) [43]. The neutral configuration is interesting. Here each hybrid has ½e, and pairs of them bond to each hydrogen, making two electrons total for each Zn-H-Zn unit. Due to bond polarity, this makes each hydrogen an H^- anion [45]. This configuration has a filled state 1 eV above the ZnO VBM (Figure 3.9b), corresponding to the states seen in photoemission [42]. Subgap light excites electrons from these states into the CB (Figure 3.8).

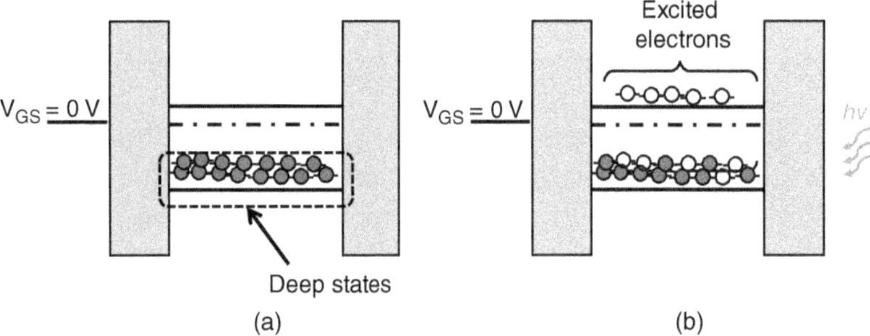

Figure 3.8 Schematic of persistent photoconductivity by illumination (negative-bias stress instability [NBIS]) by photoexcitation of electrons from a deep subgap state near the valence band edge.

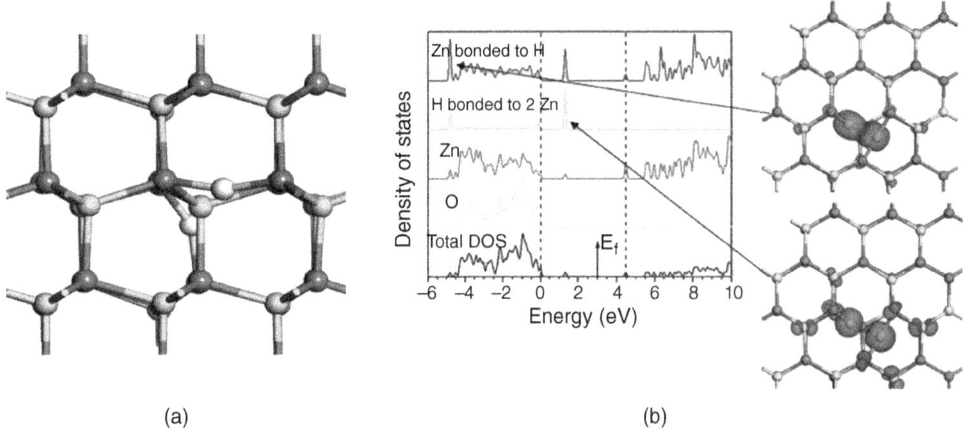

Figure 3.9 (a) Atomic configuration of two hydrogens at neutral O vacancy in *c*-ZnO, making three-center Zn-H-Zn bonds. (b) Wavefunctions for filled states at +1 eV and −5 eV. 0 eV is the VBM of bulk ZnO. Source: Li, H., Guo, Y., & Robertson, J. (2017). Hydrogen and the light-induced bias instability mechanism in amorphous oxide semiconductors. *Scientific Reports* **7** (1). Licensed under CC BY 4.0.

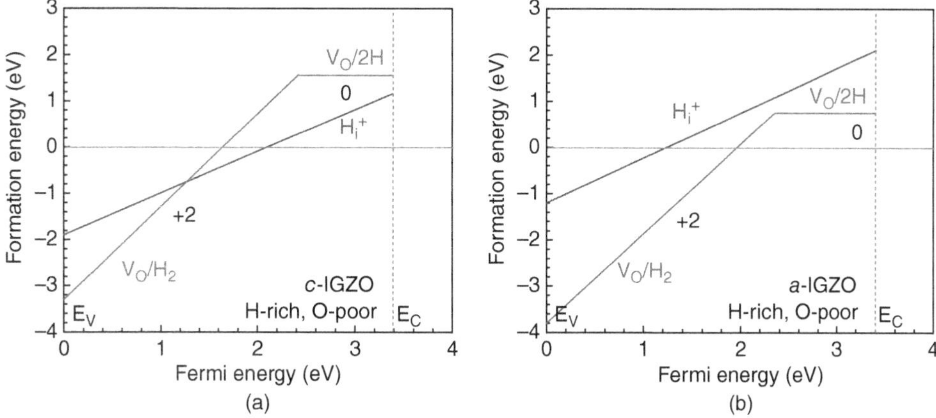

Figure 3.10 Defect formation energy versus charge state for two hydrogens trapped at O vacancy in IGZO: (a) *c*-IGZO, giving low H_2 concentration; and (b) *a*-IGZO, giving high H_2 concentration and a negative U state. (A simplified version of the diagram previously given in Ref. [43].) Source: Based on Li, H., Guo, Y., & Robertson, J. (2017). Hydrogen and the light-induced bias instability mechanism in amorphous oxide semiconductors. *Scientific Reports* **7** (1), 16858.

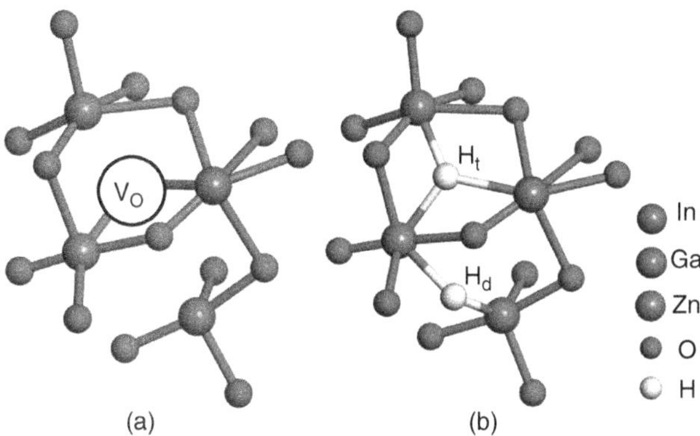

Figure 3.11 (a) An *a*-IGZO network with an oxygen vacancy. (b) An anion hydrogen pair in *a*-IGZO network, seen by FTIR [47], a generalization of the ZnO case. Source: Modified from Bang, J., Matsuishi, S., & Hosono, H. (2017). Hydrogen anion and subgap states in amorphous In–Ga–Zn–O thin films for TFT applications. *Applied Physics Letters* **110** (23): 232105.

In the crystal phase, this defect is stable in all three charge states (+2, +1, and 0), as in Figure 3.10a. In the amorphous phase, the lower mass density stabilizes the neutral state, so that only the +2 and 0 states are stable, a negative-U situation (Figure 3.10b) [43]. The instability of the +1 state means that there is now an energy barrier to recombination of the electrons from the CB back to the deep states, giving the persistent photoconductivity.

The energetics of this defect in *a*-IGZO are consistent with a large density of gap states, a large density of H, and a smaller density of photoexcited electrons (Table 3.4) [46]. Bang et al. [47] have observed the three-centered Zn-H-Zn type bonds by Fourier transform infrared spectroscopy (FTIR) (Figure 3.11), and hydrogen evolution. Interestingly, this configuration is as thermally stable as the more common Zn-O-H species. The H^- anion has been observed as the diamagnetic state in the muon replica [47]. Thus, this is a more complete model for the NBIS process. Ide et al. [48] have suggested sputtering at low-H background pressure, but this may not be economic.

Table 3.4 Approximate density of hydrogen by secondary ion mass spectrometry (SIMS), subgap states seen by X-ray photoelectron spectroscopy (XPS), and excited electrons in the conduction band, under illumination, in negative-bias stress instability (NBIS).

Type	Density (cm^{-3})
Hydrogen	10^{21}
Sub gap states	10^{20}–10^{21}
Excited electrons	10^{18}

p-Type Oxide Semiconductors

IGZO is a good oxide semiconductor, but it is only unipolar. It would be useful to have *p*-type equivalents, not just for *p*–*n* junctions but also to allow more flexible OLED structure design, where the reactivity and compatibility of different electrode materials may dictate certain layer orderings in the manufacturing process.

The band-edge DOS of IGZO shows that m_e* is low (~0.3), while the mass of holes is quite high. The large m_h* is a general problem for oxide semiconductors due to the low k-space curvature of O 2p states, which make up the upper VBs.

A possible solution to this problem, from Kawazoe [49], was to use compounds that hybridize O p states with states of similar energy such as cation Cu 3d states. This interaction creates antibonding combinations of Cu d and O p states, with lower m* values. However, the present oxide choices are disappointing [50, 51].

The simplest case is for Cu_2O, which contains a three-dimensional network of O-Cu-O linear bonding units [52]. However, its three-dimensional network of these units increases the upper VB width too much, and reduces the band gap too much to 2.1 eV. Its band structure and defect properties are described in Ref. [52].

SnO has a VB consisting of disorder-insensitive Sn s states, and it does make a *p*-type oxide of reasonable mobility. However, as Sn^{2+} is a subvalence of Sn^{4+}, it is quite sensitive to process conditions.

It is useful instead to include a third element like Al to make a layered structure, which removes some of the Cu-O backbone, to widen the gap and make it transparent in the visible. Prime examples of this are defossalites [53], $CuAlO_2$, $CuGaO_2$, and $CuInO_2$, and also $CuCrO_2$ and $CuYO_2$. These are layered compounds with octahedral Al sites and linear O-Cu-O sites. The band structure of $CuAlO_2$ by the SX function [53] is shown in Figure 3.12.

An interesting point is how the optical transitions that close the gap are optically forbidden, which widens the effective (optical) gap [54]. It makes the chemical trends of band gap along the series from Al to In unusual. In the case of $CuCrO_2$, the Cr is magnetic and complicates the electronic structure. The related compound $CuYO_2$ occurs in a different polytype than $CuAlO_2$, and its gap is larger.

Another *p*-type oxide with linear O-Cu-O units is $SrCu_2O_2$ with a tetragonal structure [55–57]. It has a direct optical band gap. It is more anisotropic and has a low hole mass in some directions. This has a ~3.3 eV direct optical gap at Γ and mobility of 0.5 cm^2/V.s. However, it is crystalline (Figure 3.13).

The spinel structure compounds $ZnCo_2O_4$, $ZnRh_2O_4$, and $ZnIr_2O_4$ are a third type of *p*-type metal oxide [58–60]. In this case, the gap occurs between the crystal-field split *d*-states of the transition metal (TM) B-site cation. This compound is *p*-type due to its hole mass, and also because of its defect chemistry that favors *p*-type conduction.

The band-edge states are dominated by the TM d states at both the VB top and the lower CBs, as seen in the calculated PDOS in Figure 3.14. The VBM and CBM band-edge states at Γ and X are most important. The crystal field splitting for Co is the least, which leads to a direct forbidden intra-atomic transition at the X point. However, the field splitting increases for the heavier Ir, and the CBM at Γ due to Zn s states goes lower, as shown schematically in Figure 3.14d, so the gap becomes less direct.

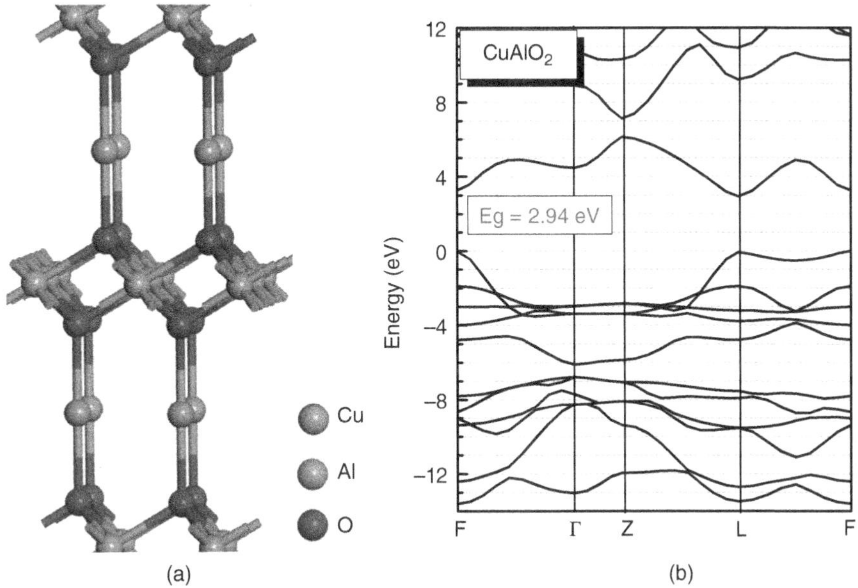

Figure 3.12 Crystal structure of $CuAlO_2$ and its band structure for the 3R phase.

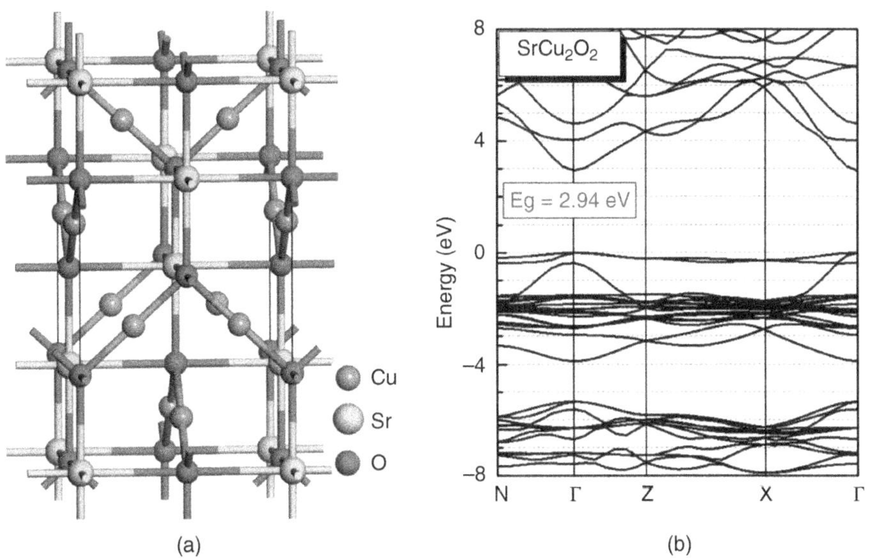

Figure 3.13 Crystal structure of $SrCu_2O_2$ and its band structure by screened exchange (SX).

It was found that $ZnRh_2O_4$ can be made structurally amorphous [58], and it has a hole mobility of 0.5 cm^2V.s. However, the expense of Rh makes this an unfavorable option.

The layered compounds like LaCuOSe are a further variant [61]. These consist of alternative layers of Cu-Se and La-O units; the band gap is the Cu-Se layer. The function of S is to raise the upper VB energy, which also reduces the hole mass. However, many of these examples are unsuitable as amorphous *p*-type oxides. Compounds with

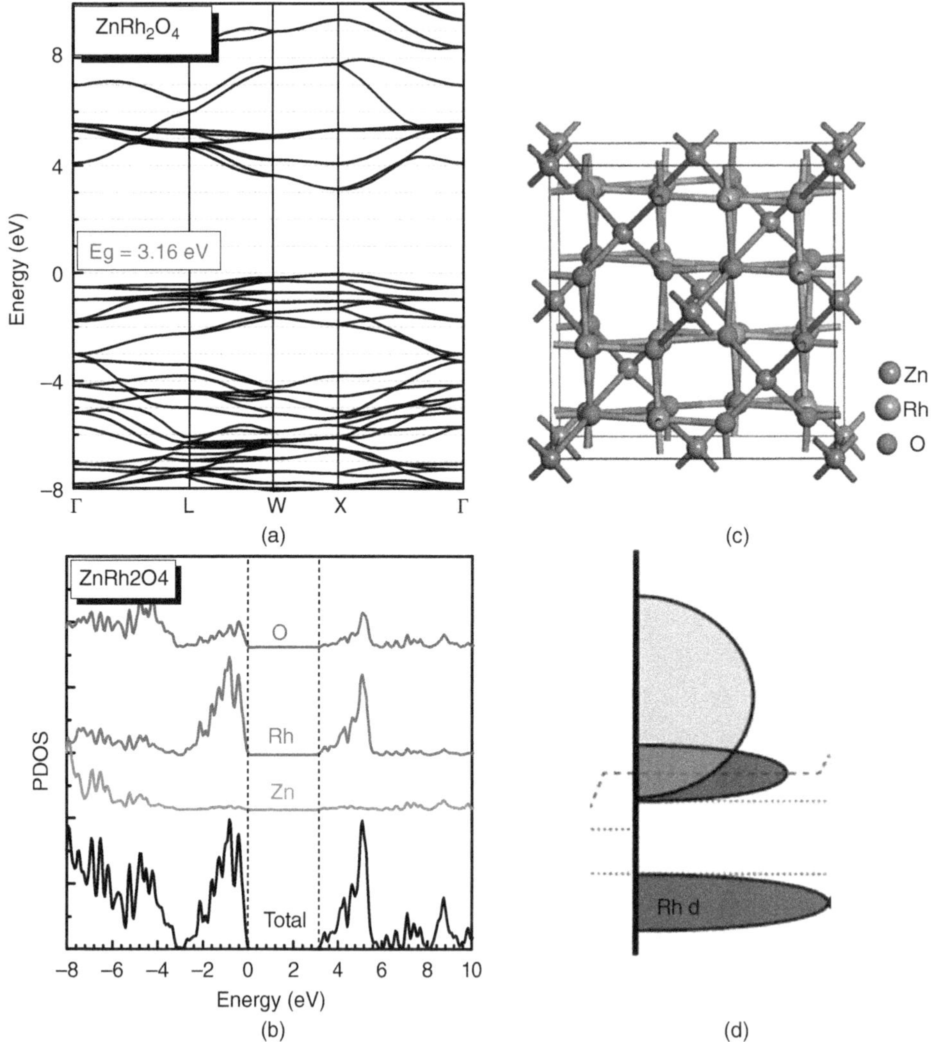

Figure 3.14 The (a) band structure, (b), density of states (DOS), and (c) spinel crystal structure of $ZnRh_2O_4$, and (d) schematic DOS.

open or layered structures like the defossalites, $SrCu_2O_2$, and LaCuSeO will not retain their structural units in an amorphous phase. This leaves $ZnRh_2O_4$ as one of the few possible amorphous *p*-type semiconductors. Materials simulation has yet to find suitable alternatives [50].

3.6 Doping Limits and Finding Effective Oxide Semiconductors

It is useful to give more general conditions for oxides to follow if they are to make good *n*- or *p*-type semiconductors. This is described by the "doping limits" idea, as first given empirically by Walukiewicz [62]. Three factors limit the ability to dope a semiconductor: the dopant solubility, the dopant level depth, and the absence of compensation by a native defect of the opposite polarity, the third being the most fundamental. Thus, a donor is effective if an intrinsic acceptor does not form spontaneously to compensate it. This can be calculated in terms of the formation

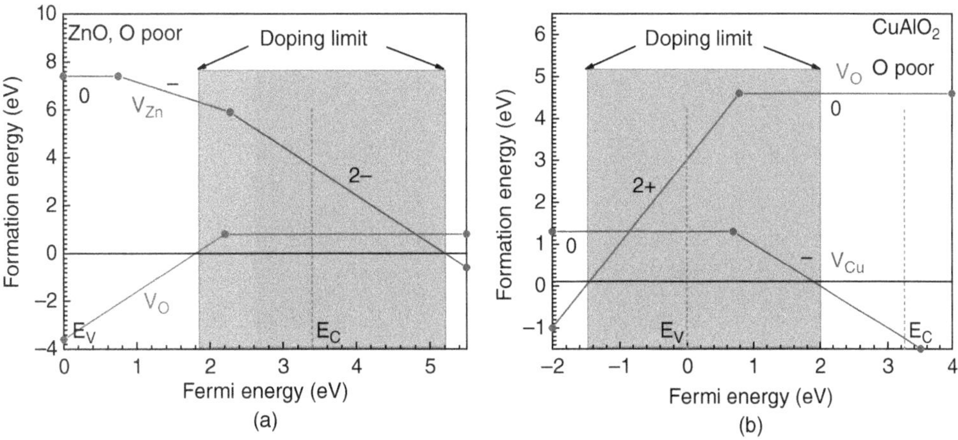

Figure 3.15 How the doping-limit energies arise, by reference to the cases of (a) *n*-type ZnO and (b) *p*-type $CuAlO_2$. The shaded zone indicates the zone where the Fermi energy can lie without compensation by intrinsic defects. Outside this energy zone, any dopant will cause an intrinsic defect to be created to compensate for any doping effect.

energy of the lowest-cost intrinsic defect, as a function of E_F and the oxygen or metal chemical potential [63, 64]. It does not require separate calculations for each dopant. It turns out that this gives relatively universal criteria in terms of the ionization potential and electron affinity of the oxide host.

It is found that oxygen interstitials and metal vacancies, being acceptors, have an increasingly negative formation energy (form spontaneously) in their −2 charge state and compensate donors when E_F rises above some doping-limit energy in the CB (Figure 3.15a). However, metal interstitials or oxygen vacancies, being donors, have an increasingly negative formation energy in their +2 charge state and compensate acceptors for E_F below some doping-limit energy in the VB (Figure 3.15b). The Fermi energy where doping is possible is shown shaded in Figure 3.15a for the case of ZnO, favoring *n*-type doping. A different example is shown in Figure 3.15b for $CuAlO_2$, with similar defects but favoring *p*-type doping.

Thus, the CBM should lie below the electron doping-limit energy for a good *n*-type oxide, and the VBM should lie above the hole doping-limit energy for a good *p*-type oxide. We now place these energies for each compound on a common energy scale, using some reference level. If we use the vacuum level as a reference, then we obtain Figure 3.16a.

It turns out that the doping-limit energies for many compounds lie at relatively constant energies when referring to the vacuum level as a reference (Figure 3.16a), so the good *n*-type and *p*-type oxides can be seen on this universal plot in terms of their electron affinities and ionization potentials. Good *n*-type oxides require an electron affinity well below the vacuum level, and good *p*-type oxides require an ionization potential not too deep. These energies have been measured by photoemission by Hosono [65], Greiner [66], and others, or have been calculated by supercell methods as done here. It means there is an easy means to scan for effective oxide semiconductors.

Interestingly, oxides that satisfy the doping-limits rule also tend to have satisfactory effective band masses. For example, oxides with large ionization potentials also tend to have large effective masses and are unsuitable as *p*-type semiconductors.

The fundamental reason why the vacuum-level reference occurs is tenuous; there are more reasons why the doping-limit energies should lie at roughly constant energies with respect to the charge neutrality level (CNL) of each compound, as in Figure 3.16b. The links between the two scales are accounted for by the alignment of band energies [67].

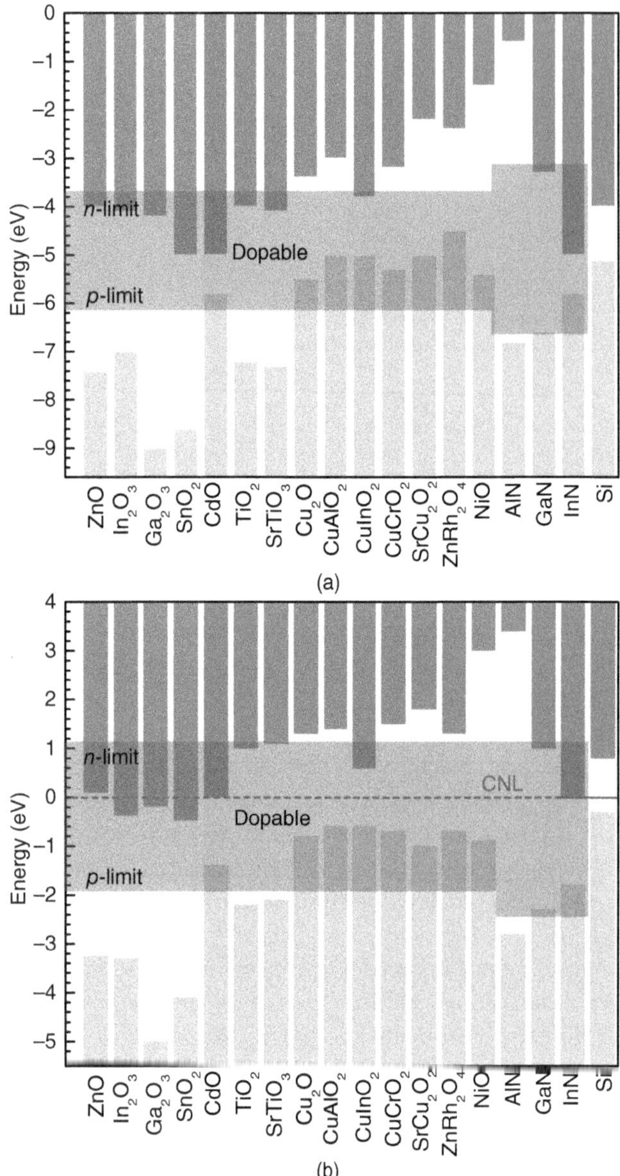

Figure 3.16 (a) Doping energy limits for oxides expressed with reference to vacuum level. Oxides with a band edge in the shaded region are dopable in that polarity. (b) Doping energy limits for oxides referred to their charge neutrality levels (CNLs). Oxides with a CNL near one of the band edges make that oxide dopable with that polarity.

3.7 OLED Electrodes

Designing stable OLEDs has been difficult, as it is necessary to have two electrodes with a significant work function difference to inject electrons and holes. The electron injector was previously a low-work-function metal like Ca, and it was prone to reaction with water or atmosphere. Thus, encapsulation was critical and unsuccessful. Since the discovery that the anode MoO_3 has a work function of 6.5 eV and not 5.3 eV [68], the cathode did not need to be so electronegative or such a reactive metal. It was noticed that the ZnO/SiO_2 phase diagram shows an immiscibility, so suitable heat treatment creates ZnO nanoparticles embedded in a SiO_2 matrix where quantum confinement can raise the ZnO CBM and shift the electron affinity to useful values [69]. This can allow the redesign or inversion of the layer structure in a combined OLED/AOS stack to simplify manufacturing.

3.8 Summary

This chapter has summarized the key aspects of the electronic structure and design of amorphous oxide semiconductors. It notes the advantages of *a*-InGaZnO4 and how its orbitals are more suitable for electron conduction than those of *a*-Si, and it explains at a deeper level how this arises. It discusses how the M element of IMZO might be chosen. It explains a theory of instabilities in IGZO that is reasonably consistent with experimentatiosn. It explains how the search for a *p*-type oxide remains difficult. It explains the background theory on how to select *n*- and *p*-type oxides in terms of their electron affinity and ionization potential.

References

1 Nomura, K., Ohta, H., Takagi, A. et al. (2004). Room-temperature fabrication of transparent flexible thin-film transistors using amorphous oxide semiconductors. *Nature* 432: 488–492.

2 Hosono, H. (2006). Ionic amorphous oxide semiconductors, materials design, carrier transport, device application. *Journal of Non-Crystalline Solids* 352: 851–858.

3 Kamiya, T. and Hosono, H. (2010). Materials characteristics and applications of transparent amorphous oxide semiconductors. *NPG Asia Materials* 2: 15–22.

4 Fortunato, E., Barquinha, P., and Martins, R. (2012). Oxide semiconductor thin film transistors: a review of recent advances. *Advanced Materials* 24: 2945–2986.

5 Park, J.S., Maeng, W.-J., Kim, H.-S., and Park, J.-S. (2012). Review of recent developments in amorphous oxide semiconductor thin film transistor devices. *Thin Solid Films* 520: 1679–1693.

6 Robertson, J., Xiong, K., and Clark, S.J. (2006). Band structure of functional oxides by screened exchange. *Physica Status Solidi (B): Basic Research* 243: 2054–2070.

7 Robertson, J. (1979). Electronic structure of SnO2, GeO2, PbO2, TeO2 and MgF2. *Journal of Physics C: Solid State Physics* 12: 4767–4776.

8 Robertson, J. (2008). Disorder and instability processes in amorphous conducting oxides. *Physica Status Solidi (B): Basic Research* 245: 1026–1032.

9 Friedman, L. (1978). Hall effect of amorphous semiconductors in random phase approximation. *Journal of Non-Crystalline Solids* 6: 329–341.

10 Friedman, L. (1971). Hall effect in ordered and disordered systems. *Philosophical Magazine Part B* 38: 467.

11 Kamiya, T., Nomura, K., and Hosono, H. (2010). Origin of Hall effect voltage and positive slope in mobility donor density relation in disordered oxide semiconductors. *Applied Physics Letters* 96: 122103.

12 Kamiya, T., Nomura, K., and Hosono, H. (2009). Electronic structures above mobility edges in crystalline and amorphous In-Ga-Zn-O percolation conduction examined by analytical model. *Journal of Display Technology* 5: 462–467.

13 Powell, M.J. and Deane, S.C. (1993). Improved defect pool model for charged defects in amorphous silicon. *Physical Review B* 48: 10815–10827.

14 Tiedje, T., Cebulka, J.M., Morel, D.L., and Abeles, B. (1981). Evidence for exponential band tails in amorphous silicon hydride. *Physical Review Letters* 46: 1425–1428.

15 Robertson, J. (2012). Properties and doping limits of amorphous oxide semiconductors. *Journal of Non-Crystalline Solids* 358: 2437–2442.

16 Erslev, P.T., Sundholm, E.S., Presley, R.E. et al. (2009). Mapping out the distribution of electronic states in the mobility gap of amorphous zinc tin oxide. *Applied Physics Letters* 95: 192115.

17 Park, J.-H., Jeon, K., Lee, S. et al. (2010). Self-consistent way of extracting density of states in amorphous InGaZnO thin film transistors. *Journal of the Electrochemical Society* 157: H272–H277.

18 Lee, S. (2008). Subgap states in transparent amorphous oxide semiconductors. *Applied Physics Letters* 98: 202508.

19 Hsieh, H.H., Kamiya, T., Nomura, K. et al. (2008). Modelling of amorphous InGaZnO4 thin film transistors and their subgap density of states. *Applied Physics Letters* 92: 133503.

20 Kimura, M., Kamiya, T., Nakanishi, T. et al. (2010). Intrinsic carrier mobility in amorphous InGaZnO thin film transistors determined by combined field effect technique. *Applied Physics Letters* 96: 262105.

21 Street, R.A. (1982). Doping and the Fermi energy in amorphous silicon. *Physical Review Letters* 49: 1187–1190.

22 Walsh, A., DaSilva, J.L.F., and Wei, S.H. (2008). Nature of the band gap of In2O3. *Physical Review Letters* 100: 167402.

23 Nomura, K., Kamiya, T., Ohta, H., and Hosono, H. (2007). Local coordination number and electronic structure of the large electron mobility amorphous oxide semiconductor InGaZnO. *Physical Review B* 75: 035212.

24 Walsh, A., DaSilva, J.L.F., and Wei, S.H. (2009). Interplay between order and disorder ion the high performance of amorphous transparent conducting oxides. *Chemistry of Materials* 21: 5119–5125.

25 Oba, F., Togo, A., Tanaka, I. et al. (2008). Defect energetic in ZnO, a hybrid Hartree-Fock density functional study. *Physical Review B* 77: 245202.

26 Clark, S.J., Robertson, J., Lany, S., and Zunger, A. (2010). Intrinsic defects in ZnO calculated by screened exchange and hybrid density functionals. *Physical Review B* 81: 115311.

27 Heyd, J., Scuseria, G.E., and Ernzerhof, M. (2003). Hybrid functional based on a screened Coulomb potential. *Journal of Chemical Physics* 118: 8207.

28 Clark, S.J. and Robertson, J. (2010). Screened exchange density functional applied to solids. *Physical Review B* 82: 085208.

29 Janotti, A., Segev, D., and van de Walle, C.G. (2006). Effects of cation d states on the structural and electronic properties of III-nitride and II oxide wide band gap semiconductors. *Physical Review B* 74: 045202.

30 Li, H., Guo, Y., and Robertson, J. (2018). Oxygen vacancies and hydrogen In amorphous InGaZnO and ZnO. *Physical Review Materials* 2: 074601.

31 Kumomi, H., Yaginuma, S., Omura, H. et al. (2009). Materials, devices, and circuits of transparent amorphous oxide semiconductors. *Journal of Display Technology* 5: 531.

32 Robertson, J., Sharia, O., and Demkov, A.A. (2007). Fermi level pinning by defects in HfO2 metal gate stacks. *Applied Physics Letters* 91: 132912.

33 Kamiya, T., Nomura, K., and Hosono, H. (2010). Subgap states, doping, and defect formation energies in amorphous oxide semiconductor InGaZnO4 studied by density function theory. *Physica Status Solidi (A) Applications and Materials Science* 207: 1698–1703.

34 Murat, A., Adler, A.U., Mason, T.O., and Medvedeva, J.E. (2013). Electronic structure of InGaZn oxide. *Journal of the American Chemical Society* 135: 5685.

35 Noh, H.K., Chang, K.J., Ryu, B., and Lee, J.W. (2011). Electronic structure of oxygen vacancy defects in amorphous InGaGaZnO semiconductors. *Physical Review B* 84: 115205.

36 Kim, H.S., Jeon, S.H., Park, J.S. et al. (2013). Anion control as a strategy to achieve high mobility and high stability oxide thin film transistors. *Scientific Reports* 13: 1459.

37 Ghaffarzadeh, K., Nathan, A., Robertson, J. et al. (2010). Instability in threshold voltage and subthreshold behavior in Hf–In–Zn–O thin film transistors induced by bias-and light-stress. *Applied Physics Letters* 97: 113504.

38 Ghaffarzadeh, K., Nathan, A., Robertson, J. et al. (2010). Persistent photoconductivity in HfInZnO thin film transistors. *Applied Physics Letters* 97: 143510.

39 Nomura, K., Kamiya, T., Yanagi, H. et al. (2008). Hydrogen passivation of electron traps in amorphous InGaZnO thin film transistors. *Applied Physics Letters* 92: 202117.

40 Ryu, B., Noh, H.-K., Choi, E.-A., and Chang, K.J. (2010). O-vacancy as the origin of the negative bias illumination stress instability in amorphous InGaZnO thin film transistors. *Applied Physics Letters* 97: 022108.

41 Robertson, J. and Guo, Y. (2014). Light induced instability mechanism in amorphous InGaZn oxide semiconductors. *Applied Physics Letters* 104: 162102.

42 Nahm, H.H. (2014). Bistability of hydrogen in ZnO, origin of doping limit and persistent photoconductivity. *Scientific Reports* 4: 4124.

43 Li, H., Guo, Y., and Robertson, J. (2017). Hydrogen and the light-induced bias instability mechanism in amorphous oxide semiconductors. *Scientific Reports* 7: 16858.

44 Janotti, A. and van de Walle, C.G. (2005). Oxygen vacancies in ZnO. *Applied Physics Letters* 87: 122102.

45 Du, M.H. and Biswas, K. (2011). Anion and hidden hydrogen in ZnO. *Physical Review Letters* 106: 115502.

46 Nomura, K., Kamiya, T., and Hosono, H. (2013). Effects of diffusion of hydrogen and oxygen on electronic properties of amorphous oxide semiconductor in GaZnO. *ECS Journal of Solid State Science and Technology* 2: P5.

47 Bang, J., Matsuishi, S., and Hosono, H. (2017). Hydrogen anion and subgap states on amorphous InGaZnO thin films of TFT applications. *Applied Physics Letters* 110: 232105.

48 Ide, K., Ishikawa, K., Tang, H. et al. (2019). Effects of base pressure on growth and optoelectronic properties of amorphous InGaZnO ultralow optimum oxygen supply. *Physica Status Solidi (A): Applications and Materials Science* 216: 1700832.

49 Kawazoe, H., Yasukawa, M., Hyodo, H. et al. (1997). P-type electrical conduction in transparent thin films of CuAlO2. *Nature* 389: 030.

50 Hautier, G., Miglio, A., Ceder, G. et al. (2013). Identification and design principles of low effective mass p-type transparent conducting oxides. *Nature Communications* 4: 2292.

51 Zhang, K.H.L., Xi, K., Blamire, M.G., and Egdell, R.G. (2016). P-type transparent conducting oxides. *Journal of Physics: Condensed Matter* 28: 383002.

52 Scanlon, D.O. and Watson, G.W. (2010). Conductivity limits in CuAlO2 from screened hybrid density functional theory. *Journal of Physical Chemistry Letters* 1: 3195.

53 Gillen, R. and Robertson, J. (2011). Band structure calculations of CuAlO2, CuGaO2, CuInO2 and CuCrO2. *Physical Review B* 84: 035125.

54 Nie, X., Wei, S.H., and Zhang, S.B. (2002). Bipolar doping and band-gap anomalies in delafossite transparent conductive oxides. *Physical Review Letters* 88: 066405.

55 Ohta, H., Orita, M., Hirano, M. et al. (2002). Electronic structure and optical properties of SrCu2O2. *Journal of Applied Physics* 91: 3074.

56 Godinho, K.G., Carey, J.J., Morgan, B.J. et al. (2010). Understanding conductivity in SrCu2O2. *Journal of Materials Chemistry* 20: 1086.

57 Robertson, J., Peacock, P.W., Towler, M.D., and Needs, R.J. (2002). Electronic structure of p-type transparent conducting oxides. *Thin Solid Films* 411: 96.

58 Narushima, S., Mizoguchi, H., Shimizu, K. et al. (2003). A p-type amorphous oxide semiconductor and room temperature fabrication of amorphous oxide p-n diodes. *Advanced Materials* 15: 1409.

59 Scanlon, D.O. and Watson, G.W. (2011). Band gap anomalies of the ZnCo2O4 spinels. *Physical Chemistry Chemical Physics* 13: 9667–9675.

60 Wahila, M.J., Lebens-Higgins, Z.W., Jackson, A.J., and Scanlon, D.O. (2019). Band edge evolution of transparent ZnM2O4 spinels. *Physical Review B* 100: 085126.

61 Hiramatsu, H., Ueda, K., Ohta, H. et al. (2003). Degenerate p-type conductivity in wide-gap LaCuOS epitaxial films. *Applied Physics Letters* 82: 1048–1050.

62 Walukiewicz, W. (2001). Intrinsic limits to doping of wide gap semiconductors. *Physica B* 302: 123–134.

63 Robertson, J. and Clark, S.J. (2011). Limits to doping in oxides. *Physical Review B* 83: 075205.

64 Zhang, S.B., Wei, S.H., and Zunger, A. (1998). Phenomenological model for prediction of doping limits on II–VI and I–III_VI2 compounds. *Journal of Applied Physics* 83: 3192.

65 Hosono, H. (2013). Exploring electro-active functionality in transparent oxide materials. *Japanese Journal of Applied Physics* 52: 090001.

66 Greiner, M.T., Helander, M.G., Tang, W.M., and Wang, Z.B. (2011). Universal band alignment of molecules on metal oxides. *Nature Materials* 11: 76–81.

67 Guo, Y., Li, H., Clark, S.J., and Robertson, J. (2019). Band offset models of three-dimensional bonded semiconductors and insulators. *Journal of Physical Chemistry C* 123: 5562–5570.

68 Kroger, M., Hamwi, S., Meyer, J. et al. (2009). Role of deep lying electronic states of MoO3 in enhancement of hole injection in organic thin films. *Applied Physics Letters* 95: 123301.

69 Hosono, H., Kim, J., Toda, Y. et al. (2017). Transparent amorphous oxide semiconductors for organic electronics: application to inverted OLEDs. *Proceedings of the National Academy of Sciences* 114: 233–238.

4

Defects and Relevant Properties

Toshio Kamiya[1,2], *Kenji Nomura*[3], *Keisuke Ide*[1], *and Hideo Hosono*[1,2]

[1] *Laboratory for Materials and Structures, Tokyo Institute of Technology, Tokyo, Japan*
[2] *Material Research Center for Element Strategy, Tokyo Institute of Technology, Tokyo, Japan*
[3] *Department of Electrical and Computer Engineering, University of California, San Diego, California, USA*

4.1 Introduction

It is well known that operation characteristics and long-term stability are affected a lot by defects in the band gap (subgap defects) for semiconductor devices, including thin-film transistors (TFTs). For amorphous oxide semiconductors (AOSs), we should note that "amorphous" materials are recognized to have defects everywhere in the structure, but only "electronic defects" are important to understand and control semiconductor properties and characteristics as long as their band gaps are with appropriate E_g values. Therefore, we will focus mainly on electronic defects while referring to structural defects if they affect TFT characteristics and stability.

For review articles, see Ref. [1] for a recent comprehensive review about electronic defects. Other review articles (Refs. [2–4]) would also help to understand atomic and electronic structures of AOSs.

4.2 Typical Deposition Condition

First we would see a typical deposition condition for a representative AOS, amorphous In-Ga-Zn-O (*a*-IGZO), deposited by our small-size radiofrequency (RF) magnetron sputtering (taget size 3 inch ϕ) using a polycrystalline $InGaZnO_4$ target. As will be seen in this chapter, oxygen partial pressure (P_{O2}) is important to suppress formation of oxygen deficiency and also to suppress inclusion of excess or weakly bonded (wb) oxygen, so P_{O2} should be optimized carefully. Total pressure (P_{tot}) and substrate–target distance (D) are strongly correlated because, as known by Paschen's law [5], the pressure × distance (PD) product determines the mean free paths (MFPs) of deposition precursors and the number of collisions by the equations in Figure 4.1. So we can see that deposition precursors collide with Ar more than four times at 0.55 Pa for D = 55 mm with ~13 mm MFP, and with O_2 ~0.1 times at P_{O2} = 0.017 Pa = 0.55 Pa × 3% oxygen flow rate ratio ($R_{O2} = [P_{O2}] / P_{tot}$). It suggests that most deposition precursors, which include oxide cations, do not react in vapor phase, and too many vapor-phase oxidation reactions would not be good to get optimum *a*-IGZO films. We will also see that too high P_{tot} produces low-density films and includes much excess/wb oxygen, which would be understood from the MFP. Deposition precursors get kinetic energies from the sputtering of the target, but lose them by collision with the vapor-phase particles. That means that higher P_{tot} results in more collisions, and deposition precursors lose more kinetic energy and cannot migrate at growing surface to form a stable structure (structural relaxation). Therefore, P_{tot} should be tuned as well. As for sputtering RF power (P_{RF}), usually low P_{RF} is preferred for crystalline thin films in order to reduce ion bombardment damage

Amorphous Oxide Semiconductors: IGZO and Related Materials for Display and Memory, First Edition.
Edited by Hideo Hosono and Hideya Kumomi.

Sputtering target	**: 3" ϕ polycrystalline $InGaZnO_4$**
Base pressure	**: 10^{-4} Pa**
Substrate–target distance (D)	**: 55 mm**
Total pressure (P_{tot})	**: 0.55 Pa**
MFP = 12 mm, $N(D, P) = 4.4$	
Ar: O_2 ratio (R_{O2})	**: 19.4: 06 sccm (R_{O2} = 2–3 %)**
$\Gamma(300K) = 1.3 \times 10^{18}$ (Ar), 4.4×10^{16} (O_2) $cm^{-2}s^{-1}$	
cf. # of O ions per surface area: 1.3×10^{15} cm^{-2} (d(a-$InGaZnO_4$) = 6.0 gcm^{-3})	
Sticking probability of O: ~ 20% for growth rate of 0.5 Å/s	
	without sputtering energy and plasma potential
RF power (P_{RF})	**: 70 W/3" ϕ = 1.5 W/cm²**
Substrate temperature (T_s)	**: No intentional heating**
(surface T raised to ~40°C for a-IGZO / ~140°C for Y_2O_3 [Yabuta et al., APL **89**, 112123 (2006)])	

Mean free path (MFP): $\lambda = \frac{k_B T}{\sqrt{2}\sigma P} = \frac{K}{P[\mathrm{Pa}]}$ [mm] (σ: Collision cross section)

Number of collisions: $N(D,P) = D/\lambda = \frac{\sqrt{2}\sigma}{k_B T}\boxed{PD}$

	H_2	N_2	O_2	Ar	Kr
σ[nm²]	0.231	0.430	0.396	0.402	0.523
K at 300K	12.7	6.81	7.40	7.29	5.60

Surface bombardment flux:

$$\Gamma = \frac{n v_a}{4} = n\sqrt{\frac{k_B T}{2\pi M}} = \frac{P}{\sqrt{2\pi M k_B T}} = K_\Gamma \frac{P[\mathrm{Pa}]}{\sqrt{M}} \quad [(\mathrm{cm}^2\mathrm{s})^{-1}] \quad K_\Gamma(300\mathrm{K}) = 1.5 \times 10^{19}$$

Figure 4.1 Typical deposition condition for a-In-Ga-Zn-O thin films by small-size RF magnetron sputtering and related parameters.

to the resulting film and to reduce defects. However, for AOSs, higher P_{RF} is generally preferred because AOS films are often deposited at low temperatures including room temperature (RT) and cannot get enough energy for structural relaxation from thermal energy. Therefore, kinetic energies of deposition presursors should be the sole source of energy to form stable structures. A typical example will be seen for *a*-Ga-O, where high-power laser ablation and high-rate deposition reduce defect density and produce semiconducting films [6]. Such films deposited by high P_{RF} may include ion bombardment damage, but regardless AOS-TFTs need postdeposition thermal annealing, typically at 300–400 °C, and that damage would be improved in the annealing process [7].

4.3 Overview of Electronic Defects in AOSs

Hereafter, we will review electronic defects that affect the optimum deposition condition explained in Section 4.2 and affect TFT characteristics and stability. Figure 4.2 illustrates an updated version of a schematic electronic structure around the band gap of *a*-IGZO [3].

For conventional covalent semiconductors such as Si, subgap defects allegedly deteriorate device performance and stability, and compensation doping (where, e.g., native donors are compensated by intentional acceptor doping) makes them critically worse. However, as seen as "near-VBM states" (near–valence band maximum states) in Figure 4.2, AOSs have very deep defects just above the VBM at densities even higher than 10^{20} cm^{-3}. Such defect structure can be recognized to capture free electrons and cause the intentional charge compensation, but the bad effects of the near-VBM states are usually not visible in static TFT characteristics at RT because their electronic levels are >2.0 eV below the conduction band minimum (CBM), which is far larger than the band gap of Si (1.1 eV). On the other hand, the near-VBM states work as hole traps, and therefore the Fermi level (E_F) is pinned in those states under a negative gate bias (V_{GS}) to TFT. Negative V_{GS} reduces free electron density, but the E_F pining

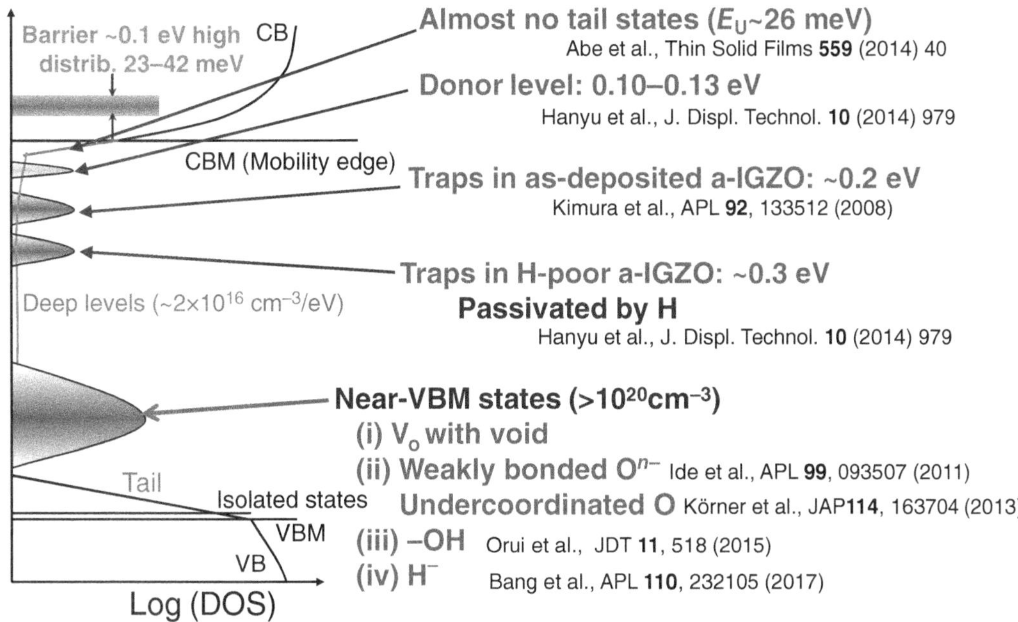

Figure 4.2 Subgap defects in *a*-IGZO known to date.

stops increasing free hole density, so the near-VBM states help to realize a very low source–drain current (I_{DS}) of AOS-TFTs in off states. However, we believe that reducing the near-VBM states would also be important to produce more stable AOS-TFTs, but there has not yet been solid evidence for that. As will be seen later in this chapter, near-VBM states are increased by very low P_{O2} conditions and postdeposition hydrogen doping; therefore, their major origins are thought to be oxygen vacancy (V_O) ("vacancy" cannot be defined well in amorphous structures, but we use V_O as a somewhat large free space coordinated by cations), hydrogen (–OH and/or H^-), and wb-O. Actually, near-VBM states are reduced by ultrahigh vacuum (UHV) sputtering [8].

Shallow defects near the CBM are more important for TFT operations, particularly near on states. In a representative amorphous semiconductor, hydrogenated amorphous Si (*a*-Si:H) defects can be divided into two groups, (i) tail states and (ii) localized electron traps. For most amorphous materials like *a*-Si:H and *a*-SiO_2, tail states are expressed by a Tauc relation with an exponential function, and localized states are expressed with Gaussian functions in many device simulations. For *a*-IGZO, a major part of a TFT transfer curve can be reproduced with a single acceptor-like Gaussian-type state, but the addition of an acceptor-like exponential state is necessary to fully reproduce it (Ref. [9]; representative parameters are given therein). Those results also show that the subgap defect density in *a*-IGZO is 2–3 orders of magnitude smaller than that of *a*-Si:H, in particular near CBM. Figure 4.3 compares those of various *a*-IGZO TFTs (unannealed and annealed, and depletion-type and enhancement-type) obtained by the device simulation [9] and capacitance–voltage (C-V) method [10]. We can see that as-deposited ("unannealed") *a*-IGZO has an extra shallow trap around 0.2 eV from the CBM. "Depletion-type TFT" means that the threshold voltage (V_{th}) is below $V_{GS} = 0\,V$ and includes more shallow donor states and residual free electrons; therefore, its subgap DOS tends to be larger than the enhancement-type *a*-IGZO. In actual TFTs, $V_{th} > 0$ is preferred to use for a simple circuit to control flat-panel displays, and this result also suggests that tuning the fabrication condition while focusing on these subgap states is better for producing enhancement-type TFTs.

We also need to point out that some hydrogens are preferred to reduce subgap defects at ~0.3 eV below the CBM [11]. Those hydrogens are characterized by the effusion temperature around 400 °C, so that *a*-IGZO TFTs tend to be deteriorated by too-high temperature annealing, for example at >400 °C.

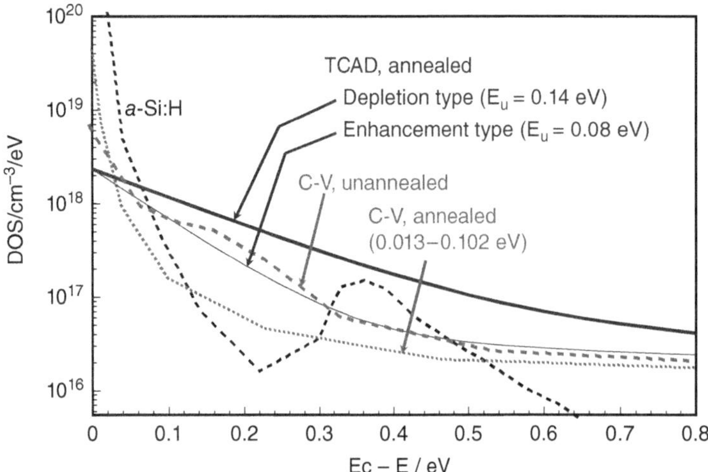

Figure 4.3 Density of states of subgap states in various *a*-IGZO obtained by device simulation (TCAD) and C-V methods.

4.4 Origins of Electron Donors

Donor levels (E_D) are analyzed from the temperature dependence of the Hall effect. It shows that crystalline $InGaZnO_4$ has a donor level at $E_D(N_D)$ [eV] = $0.22 - 2.3\times10^{-7}$ N_D (N_D is the donor density in cm^{-3}; it corresponds to $E_D = 0.15$ eV for $N_D = 3.3\times10^{16}$ cm^{-3}) below the CBM, while *a*-IGZO has shallower donor levels: E_D = 0.11 eV for $N_D = 9.0\times10^{15}$ cm^{-3} [12].

Here we discuss the origins of the donor states. Usually good oxide semiconductors are native *n*-type semiconductors, as known in representative crystalline transparent conductive oxides (TCOs) such as ZnO, SnO_2, In_2O_3, and $InGaO_3(ZnO)_m$ (*m* is an integer). It was thought that V_O was the origin of the native donors in crystalline TCOs; but recent first-principles calculations revealed that, although the formation energy of V_O is low, V_O forms deep donor states (charge transfer levels) and cannot be the source of the native donors. It is because V_O in the crystalline TOSs has a free space that can trap two electrons to form a neutral oxygen vacancy $V_O{}^0$. Therefore, impurity hydrogen is considered as a possible origin, which would be $-O^{2-}H^+$ or H^- at V_O. On the other hand, the local atomic configuration of ions and defects is flexible in AOSs. If an AOS has an oxygen deficiency, it can generate two free electrons to the conduction band, but these electrons can be trapped by the V_O if it retains a large free space similar to that of crystalline TOSs. However, such large V_O space is not necessarily retained in AOSs because structural relaxation during film deposition and annealing can diminish the free space. First-principles calculations suggest that oxygen deficiency works as a deep trap if V_O retains a large free space while as a shallow donor if such free space is diminished [13].

In crystalline oxide semiconductors, substitution doping with aliovalent ions produces shallow donor or acceptor states, such as Ga^{3+}@Zn^{2+} for ZnO:Ga (GZO), Sn^{4+}@In^{3+} for In_2O_3:Sn (indium tin oxide [ITO]), and F^-@O^{2-} for SnO_2:F (F-doped tin oxide [FTO]). However, such substitution doping does not work for AOSs because of the flexibility of their amorphous structure and chemical composition. That is, if we substitute one Zn^{2+} with Ga^{3+} in *a*-$(InGaZnO_4)_{100}$, it may generate a free e^- but an extra 1/2 O^{2-} may be incorporated to trap the e^- released from the Ga^{3+} substitution.

This relation can be extended to a general simple counting rule to speculate charge carrier doping (see Ref. [14, sec. XIII]), as explained here. Remember that oxides of typical metals have closed-shell cations and oxide ions, and the unoccupied states of the cations form conduction bands and the fully occupied O^{2-} 2p states form valence bands, thus producing finite band gaps. Therefore, one may think that a finite band gap is retained without generating extra free charges in the conduction or valence band if the summation of the formal charges of the constituent ions keeps neutral. It in turn means that free electrons or holes should be generated if the total formal charge is not zero. Note that these are not limited to crystalline materials but also applied to amorphous ones. For example, when making an oxygen deficiency in $In_{200}O_{300}$, it should generate two electrons and can be an *n*-type semiconductor, as speculated by the reaction equation $In_{200}O_{299} = In^{3+}{}_{200}O^{2-}{}_{299} + 2e^-$ if the released $2e^-$ stay in the conduction band. Similarly, if we substitute one Zn^{2+} with Ga^{3+} in *a*-$(InGaZnO_4)_{100}$, it can generate a free e^- by $In^{3+}{}_{100}Ga^{3+}{}_{101}Zn^{2+}{}_{99}O^{2-}{}_{400} + e^-$.

We should remind the reader that we usually optimize the deposition condition of AOSs so that we can obtain good semiconductor films, that is, low-carrier-density ones, and therefore the substitution of Ga^{3+} for Zn^{2+} and the released e^- will be compensated by excess $1/2O^{2-}$ incorporated by a higher P_{O2} deposition condition or by postdeposition annealing with O_2 (the resultant chemical composition is expressed as $In_{100}Ga_{101}Zn_{99}O_{400.5}$, which satisfies the charge neutrality with the formal ion charges and therefore does not produce a free electron).

Similarly, hydrogen doping to stoichiometric $In_{200}O_{300}$ produces a free electron by $In_{200}O_{300} + H^+ = In^{3+}{}_{200}O^{2-}{}_{300}H^+ + e^-$ if H is ionized to H^+. On the other hand, if H substitutes an O^{2-}, it is ionized to $H^-@O^{2-}$ and produces a free electron by $In_{200}O_{300} - O^{2-} + H^- = In^{3+}{}_{200}O^{2-}{}_{299}H^- + e^-$. However, as will be seen later in this chapter, AOS films deposited by conventional sputtering apparatus, or even by a UHV pulsed laser deposition one, include high-density impurity hydrogen at densities $>10^{20}$ cm^{-3}. But we use the hydrogenated *a*-IGZO films for TFTs in commercially available flat-panel displays, due to the same reason for the charge compensation mechanism explained earlier in this section for aliovalent ion doping. Incorporating impurity hydrogen would produce $-O^{2-}H^+$ and $H^-@O^{2-}$ donors, as shown by the above equations and will be explained later in the chapter; the released electrons would be compensated by incorporating excess oxygen during film deposition or postdeposition annealing with O_2.

Therefore, we need to recognize that an effective electron doping route would be limited to the following for AOSs: (i) Produce oxygen deficiencies with a low P_{O2} deposition condition or annealing in reducing atmosphere, and (ii) postdeposition hydrogen and cation doping.

Here, we discuss why the H^- state can be stable in ionic materials. Remember that electrostatic potential is different at a different site in ionic material (Madelung potential [MP]), positive MP is formed at a cationic site that is surrounded by anions (an electropositive site), and vice versa. First-principles calculations clarify that an additional H can take different sites, a representative one is a neighbor to O^{2-} and forms –OH, and another is at an electropositive site surrounded by cations (e.g., V_O). The former forms negative MP and pushes down the energy levels in H, giving H^- 1s level down to inside the band gap and stabilizing H^-, as seen in the center two diagrams in Figure 4.4. If the H site becomes more electronegative (like close to O^{2-}), the energy levels in H^- go up, and finally H^- 1s states surpass the CBM, releasing one 1s electron to conduction and then deionizing to H^0.

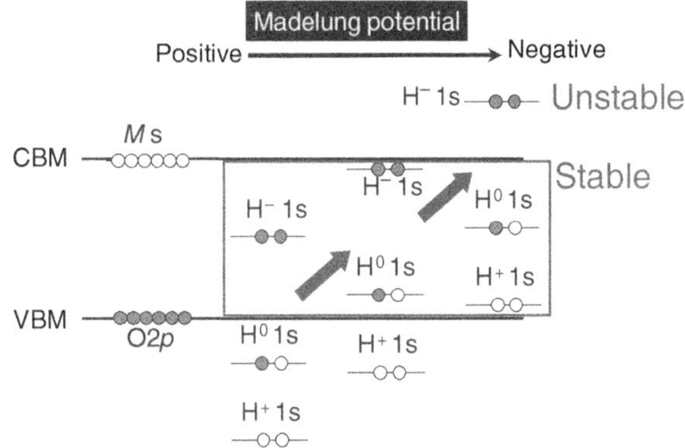

Figure 4.4 Schematic energy-level diagram to explain stabilization of H^- and H^+ in ionic materials.

4.5 Oxygen- and Hydrogen-Related Defects and Near-VBM States

As explained in this chapter, an oxygen deficiency can form a deep electron trap or a shallow donor depending on the local structure. If a large free space remains at a vacancy site, such V_O forms a deep level in the band gap and traps electrons, as seen in the pseudo band structure and the electron density map in Figure 4.5 [13]. If such free space is diminished by additional structural relaxation, E_F goes up by keeping the fundamental band structure of stoichiometric *a*-IGZO, showing this oxygen deficiency works as a shallow donor.

Such deep trap states in the V_O model would explain the near-VBM states that are observed by hard X-ray photoemission spectroscopy (HAXPES) and optical absorption spectra (Figure 4.6) [15]. It shows the *a*-IGZO films have high-density electron traps just above the VBM, peaking at ~2.5 eV from the CBM at densities $\gg 10^{20}$ cm^{-3}. The near-VBM states produce subgap optical absorption at photon energies below the band gap, and they cause photoresponse and light-induced instability in AOS-TFTs against visible light illumination.

On the other hand, although the band structures in Figure 4.5 are obtained by density functional theory (DFT) calculations using generalized gradient approximation (GGA) functionals, which generally underestimate band gaps, more sophisticated self-interaction correction (SIC) and hybrid functional calculations suggest that V_O states are not so deep and are even located close to the mid–band gap [16]. However, experimentally deconvoluted near-VBM states indicate their origins would be V_O and H (Figure 4.7) [1]. Comparing the STD HQ (STD means a conventional sputtering apparatus with a base pressure of ~10^{-4} Pa; HQ means high-quality films deposited with the optimum deposition condition; impurity hydrogen density [H] is $\gg 10^{20}$ cm^{-3}) shows high-density near-VBM states, as seen in Figure 4.6, and the near-VBM states are increased by postdeposition H-plasma treatment. While those are reduced by using a UHV sputtering system with a base pressure of 10^{-7} Pa and an optimum R_{O2} = 1%, which has [H] at the order of 10^{19} cm^{-3}, reducing R_{O2} to 0 during deposition increases the near-VBM states.

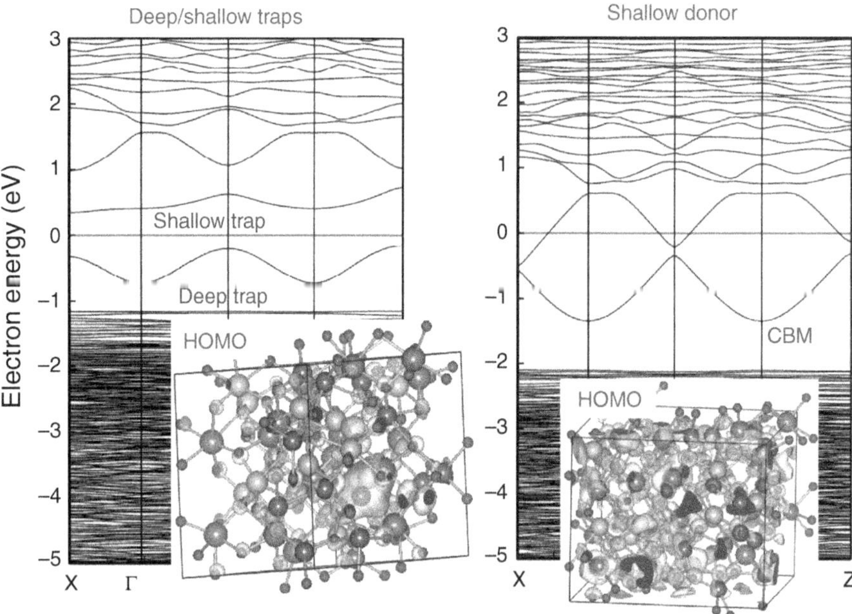

Figure 4.5 Pseudo band structures and electron density maps (insets) of oxygen-deficient *a*-IGZO models. (Left) Model retains a V_O site with free space, and (right) model diminishes free space by additional structural relaxation. Source: Modified from Kamiya, T., Nomura, K., & Hosono, H. (2009). Electronic structure of the amorphous oxide semiconductor a-$InGaZnO_{4-x}$: Tauc–Lorentz optical model and origins of subgap states. *Physica Status Solidi (A) Applications and Materials Science* **206** (5): 860–867. doi:10.1002/pssa.200881303.

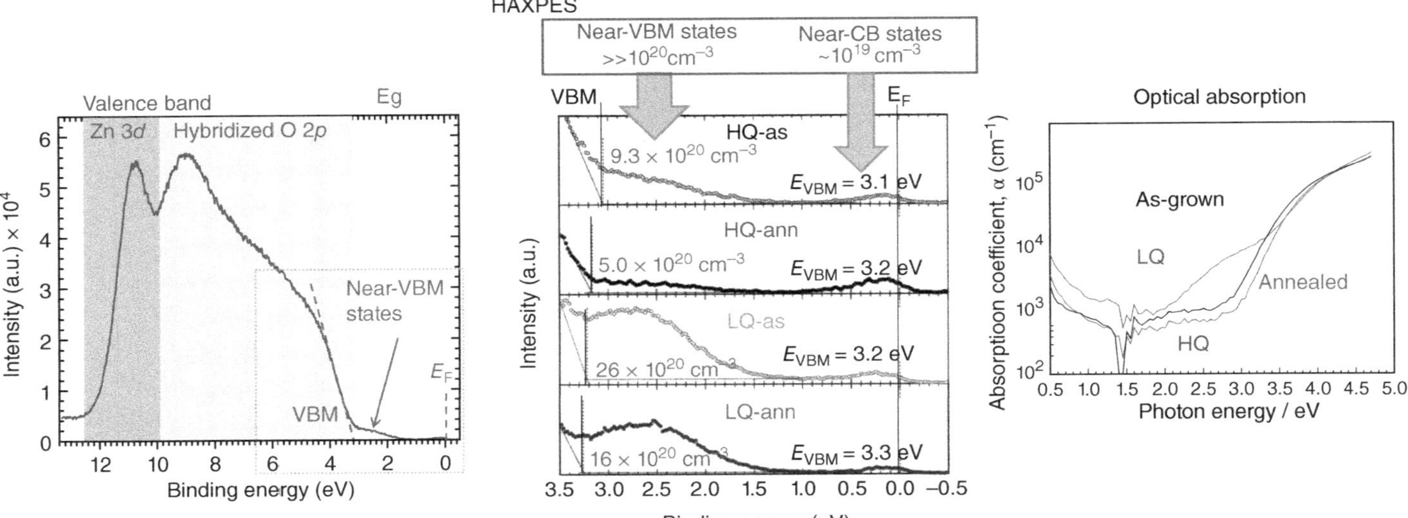

Figure 4.6 Near-VBM states observed by HAXPES and optical absorption spectra. Source: Based on Nomura, K., Kamiya, T., Yanagi, H., Ikenaga, E., Yang, K., Kobayashi, K., Hirano, M., and Hosono, H. (2008). Subgap states in transparent amorphous oxide semiconductor, In–Ga–Zn–O, observed by bulk sensitive x-ray photoelectron spectroscopy. *Applied Physics Letters* **93**: 202117-1–3.

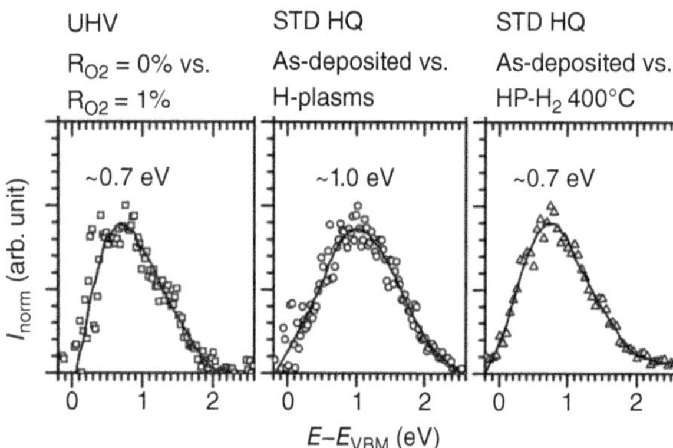

Figure 4.7 Different near-VBM states deconvoluted by different *a*-IGZO films.

High-pressure (HP) H_2 at 400 °C shows the *a*-IGZO films postannealed with HP H_2 atmosphere at 400 °C. HAX-PES O 1s spectra analysis shows that this spectrum does not come from H or OH, and it is thought that the high-temperature H_2 annealing does not incorporate H but reduces the film and generates V_O. This result suggests that near-VBM states include multiple origins and different shape peaks. V_O would be attributed to the one peaked at ~0.7 eV, while the H-related one to that at ~1.0 eV. The latter energy level can be attributed to –OH but also to H^-.

The existence of H^- in *a*-IGZO is confirmed experimentally, showing that conventional sputtered *a*-IGZO films include –OH and H^- at the ratio of ~2:1 [17]. The structures of hydrogens in *a*-IGZO are further studied using muon spin rotation/relaxation (μSR) spectroscopy [18]. Here, muon (μ) works as a probe for hydrogen, and μSR is very sensitive compared with H-NMR (nuclear magnetic resonance). It shows that H in crystalline $InGaZnO_4$ is located at a bond-center site between Zn and O, similar to the case of ZnO. This structure is observed also in STD *a*-IGZO films with [H] ~ 10^{20} cm^{-3}. On the other hand, H-saturated *a*-IGZO (i.e., H-doped after deposition of STD *a*-IGZO) suggests ~20% of Mu forms a defect complex of H^-–Mu at 0.2 pm distance, which is similar to the structure obtained by DFT in Ref. [17]. This result suggests that H in very-high-density hydrogenated *a*-IGZO is deactivated as a donor, as it corresponds to $[2H^-]@O^{2-}$, being consistent with the experimental electron doping limit ~10^{20} cm^{-3} [1, fig. 27b].

Here we like to note that near-VBM states cause an unexpectedly strong charge compensation in oxygen deficiency films. Usually, oxygen-deficient crystalline TCOs (i.e., deposited with lower R_{O2}) exhibit higher electron density and electrical conductivity. However, if we use pulsed laser deposition with an off-optimized deposition condition (i.e., low-power laser ablation and low film density) without O_2 supply, very-high-density near-VBM states are formed, and its resistivity jumps up to >10^4 Ωcm (Figure 4.8). It can be understood as the full self-compensation effect illustrated therein.

Lastly, we like to emphasize that structure and chemical composition of AOS are very flexible, and unusual structures in crystals can be easily formed. We have seen that oxygen deficiency can be a deep electron trap and also a shallow donor, depending on their local structures, and can even compensate free electrons to produce an insulating film at low R_{O2} or vacuum ($R_{O2} = 0$) deposition conditions. Further, AOS has more different oxygen-related defects, weakly bonded (wb)-O, which is first found in 300 °C O_3-annnealed *a*-IGZO films [19] and then confirmed for high-R_{O2} sputtered *a*-IGZO film TFTs. Though oxygen is the major constituent of AOSs, overly strong oxidation conditions (such as high-temperature O_3 annealing and high P_{O2} deposition) incorporate excess oxygens and form wb-O. In the case of 300 °C O_3-annealed *a*-IGZO, it is speculated that a neutral oxygen atom is incorporated to a charge-neutral AOS framework, which produces a TFT with a poor initial transfer curve, like in the lower left panel of Figure 4.9. Once high positive V_{GS} is applied, E_F rises and exceeds the $O_{wb}{}^0$ 2p level, transferring a free electron to ionize it to $O_{wb}{}^-$, as illustrated in the right schematic energy structure, giving rise to the large change

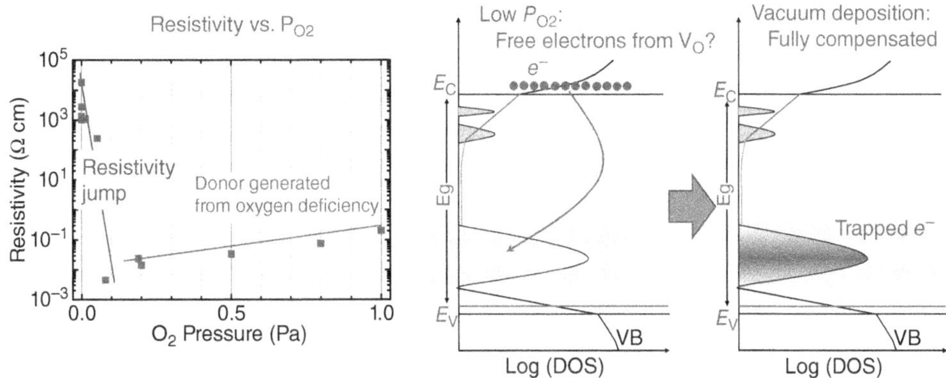

Figure 4.8 Self-compensation of electrons generated from V_O.

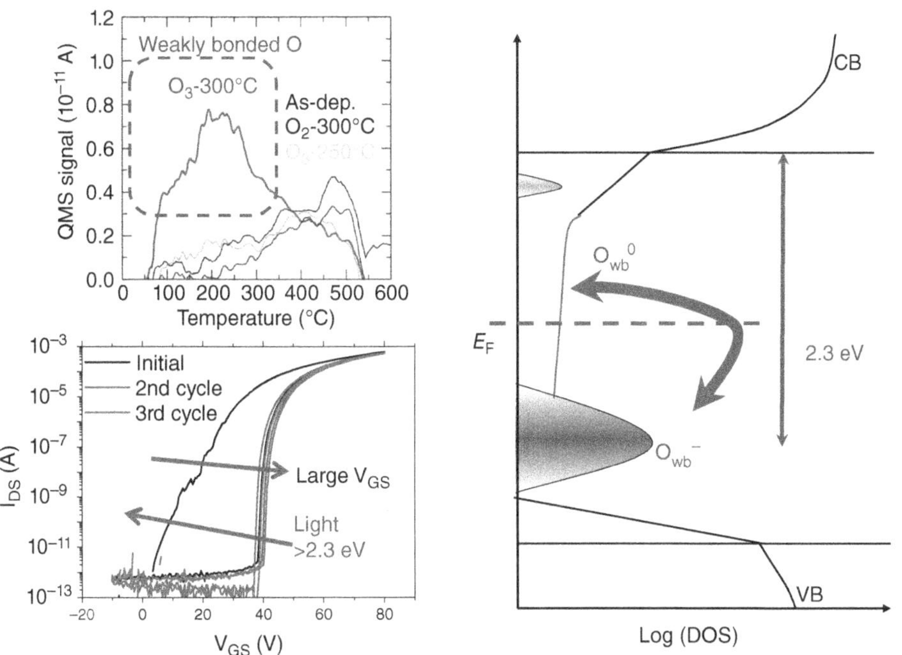

Figure 4.9 Effects of weakly bonded oxygen. (Upper left) Thermal desorption spectra, (lower left) change of TFT transfer curves for cyclic measurements and light illumination, and (right) a schematic electronic structure model to explain the bistability of 300 °C O_3-annealed *a*-IGZO TFTs.

in the transfer curve to those with very high V_{th} (> 40 V, a high-V_{th} state). Light illumination recovers the high-V_{th} state to the initial low-V_{th} state if the photon energy is greater than 2.3 eV. It clarifies that this O_{wb}^- level is located at ~2.3 eV below the CBM, and O_{wb} forms the bistable structures with the negative-U characteristic.

Wb-O also would be important to understand the temperature instability of AOS-TFTs. As seen in the thermal desorption spectra (TDS) in the upper-left panel, wb-O desorps in the very-low-temperature region, even starting from 80 °C. Desorption of charged O_{wb}^- leaves a free electron and causes an increase in electrical conductivity and negative shift of V_{th} in AOS-TFTs.

4.6 Summary

We summarized doping routes and several origins of electronic defects in AOSs. Here we should recognize that impurity hydrogen and stoichimetory of oxygen, and their local structures, are important to understand and to control the electrical properties of AOSs and their devices. In particular, hydrogen can be incorporated anywhere to any thin films unless the entire production system is closed in UHV with oxygen purifiers. Impurity hydrogen has double-sided aspects; some can deactivate electron traps and improve TFT performance, but too-high-density hydrogen causes extra instability. Oxygen deficiency can be an electron trap and a shallow donor. We should not believe oxygen deficiency always increases electrical conductivity in AOSs as in crystalline oxide semiconductors. Overly strong oxidation conditions introduce wb-O and electron traps, producing high-resistivity AOS films and not-working TFTs. These hydrogen and wb-O can desorb at rather low temperatures ($\ll$200 °C), which would cause temperature-induced instability. V_O, H, and wb-O would form near-VBM states and cause light–illumination instability against visible light.

Due to page limitations, we did not refer to microstructural defects and structure relaxation in this book. Please refer to Ref. [1] for details.

References

1 Ide, K., Nomura, K., Hosono, H., and Kamia, T. (2019). Electronic defects in amorphous oxide semiconductors: a review. *Physica Status Solidi (A) Applications and Materials Science* 216: 1800372-1–1800372-28.

2 Kamiya, T. and Hosono, H. (2012). Amorphous In-Ga-Zn-O thin film transistors: fabrication and properties. In: *Handbook of Zinc Oxide and Related Materials*, 2nde. Boca Raton, FL: Taylor & Francis.

3 Kamiya, T., Nomura, K., and Hosono, H. (2010). Present status of amorphous InGaZnO thin-film transistors. *Science and Technology of Advanced Materials* 11: 044305.

4 Kamiya, T., Nomura, K., and Hosono, H. (2009). Origins of high mobility and low operation voltage of amorphous oxide TFTs: electronic structure, electron transport, defects and doping. *IEEE Journal of Display Technology* 5: 273.

5 Paschen, F. (1889). Ueber die zum Funkenübergang in Luft, Wasserstoff und Kohlensäure bei verschiedenen Drucken erforderliche Potentialdifferenz. *Annalen der Physik* 273: 69–96.

6 Kim, J., Sekiya, T., Miyokawa, N. et al. (2017). Conversion of an ultra-wide bandgap amorphous oxide insulator to a semiconductor. *NPG Asia Materials* 9 (3): e359.

7 Ide, K., Kikuchi, M., Ota, M. et al. (2017). Effects of working pressure and annealing on bulk density and nanopore structures in amorphous In–Ga–Zn–O thin-film transistors. *Japanese Journal of Applied Physics* 56: 03Bb03-1–03Bb03-5.

8 Tang, H., Ishikawa, K., Ide, K. et al. (2015). Effects of residual hydrogen in sputtering atmosphere on structures and properties of amorphous In-Ga-Zn-O thin films. *Journal of Applied Physics* 118: 205703-1–205703-6.

9 Hsieh, H.-H., Kamiya, T., Nomura, K. et al. (2008). Modeling of amorphous $InGaZnO_4$ thin film transistors and their subgap density of states. *Applied Physics Letters* 92: 133503-1–133503-3.

10 Kimura, M., Nakanishi, T., Nomura, K. et al. (2008). Trap densities in amorphous-$InGaZnO_4$ thin-film transistors. *Applied Physics Letters* 92: 133512-1–133512-3.

11 Hanyu, Y., Domen, K., Nomura, K. et al. (2013). Hydrogen passivation of electron trap in amorphous In-Ga-Zn-O thin-film transistors. *Applied Physics Letters* 103: 202114-1–202114-3.

12 Kamiya, T., Nomura, K., and Hosono, H. (2009). Electronic structures above mobility edges in crystalline and amorphous In-Ga-Zn-O: percolation conduction examined by analytical model. *Journal of Display Technology* 5: 462–467.

13 Kamiya, T., Nomura, K., and Hosono, H. (2009). Electronic structure of the amorphous oxide semiconductor a-$InGaZnO_{4-x}$: Tauc–Lorentz optical model and origins of subgap states. *Physica Status Solidi (A) Applications and Materials Science* 206: 860–867.

14 Kamiya, T., Nomura, K., and Hosono, H. (2009). Origins of high mobility and low operation voltage of amorphous oxide TFTs: electronic structure, electron transport, defects and doping. *Journal of Display Technology* 5: 273–288.

15 Nomura, K., Kamiya, T., Yanagi, H. et al. (2008). Subgap states in transparent amorphous oxide semiconductor, In–Ga–Zn–O, observed by bulk sensitive x-ray photoelectron spectroscopy. *Applied Physics Letters* 93: 202117-1–202117-3.

16 Körner, W., Urban, D.F., and Elsässer, C. (2013). Origin of subgap states in amorphous In-Ga-Zn-O. *Journal of Applied Physics* 114: 163704.

17 Bang, J., Matsuishi, S., and Hosono, H. (2017). Hydrogen anion and subgap states in amorphous In–Ga–Zn–O thin films for TFT applications. *Applied Physics Letters* 110: 232105-1–232105-5.

18 Kojima, K.M., Hiraishi, M., Okabe, H. et al. (2019). Electronic structure of interstitial hydrogen in In-Ga-Zn-O semiconductor simulated by muon. *Applied Physics Letters* 115: 122104-1–122104-5.

19 Ide, K., Kikuchi, Y., Nomura, K. et al. (2011). Effects of excess oxygen on operation characteristics of amorphous In-Ga-Zn-O thin-film transistors. *Applied Physics Letters* 99: 093507-1–093507-3.

5

Amorphous Semiconductor Mobility Physics and TFT Modeling

John F. Wager

School of Electrical Engineering and Computer Science (EECS), Oregon State University, Corvallis, Oregon, USA

5.1 Amorphous Semiconductor Mobility: An Introduction

Electron or hole transport in a semiconductor (or insulator) is invariably discussed in terms of drift mobility (μ_{drift}) [1]. μ_{drift} is a proportionality constant relating carrier (electron or hole) velocity to the applied electric field (i.e., $v = \mu_{drift}\,\xi$). This linear relationship between v and ξ is valid only at a low electric field, and thus is a defining equation of low-field transport. In contrast, a nonlinear relationship exists between v and ξ for the case of high-field transport. Drift mobility is typically evaluated using [1]:

$$\mu_{drift} = \frac{q < \tau_m >}{m^*} \tag{5.1}$$

where q is electronic charge, $<\tau_m>$ is average momentum relaxation time, and m^* is carrier effective mass.

Equation (5.1) accurately discloses the mobility physics of carrier transport in most crystalline semiconductors. For example, consider intrinsic GaAs with a room-temperature electron mobility of ~8000 $cm^2V^{-1}s^{-1}$ and an effective mass of 0.063 m_0 [2]. Using Equation (5.1), this corresponds to an average momentum relaxation time for intrinsic GaAs of $<\tau_m> \approx 290$ fs and a mean free path $\lambda_{mfp} \approx 140$ nm, where $\lambda_{mfp} = v_{th}<\tau_m>$ and v_{th} is the thermal velocity, given by $v_{th} = (3k_BT/m^*)^{1/2}$, where k_B is Boltzmann's constant and T is temperature. A 140 nm mean free path is very large compared to interatomic distance, corresponding to approximately 560 interatomic distances between scattering events (taking 0.25 nm as a typical interatomic distance). Moreover, the intercollisional drift length, $L_{drift} = \mu_{drift}\,\xi <\tau_m>$, is found to be 22.9 nm (~92 atomic distances) for a very small electric field of $\xi = 1$ $kVcm^{-1}$. (Note that mean free path λ_{mfp} is established by the thermal velocity, while intercollisional drift length L_{drift} pertains to the drift velocity.) Thus, the physical picture that emerges from a quantitative explication of Equation (5.1) is that mobility arises as a consequence of acceleration of an electron due to a gain in energy from the applied electric field, with occasional scattering of this heated electron after it has drifted a relatively large distance (compared to that of an interatomic spacing).

Amorphous semiconductor (or insulator) transport physics differs dramatically from that of the crystalline material transport case just described. As an extreme example, consider hole transport in amorphous SiO_2 [3]. The room-temperature intrinsic hole mobility in SiO_2 is experimentally found to be ~10^{-4} $cm^2V^{-1}s^{-1}$ [4]. Again, using Equation (5.1) and assuming $m^* \approx 10\,m_0$ for holes in SiO_2 [5, 6] lead to $<\tau_m> \approx 0.1$ attosecond (as) and $\lambda_{MFP} \approx 20$ attometer (am) for hole transport in SiO_2. This calculation says that a hole moves only about one billionth of one atomic distance before it experiences a scattering event. As crazy as this sounds, the situation is even more bizarre when SiO_2 is prepared in certain ways. For example, the room temperature hole mobility for a wet oxide of SiO_2 is measured to be ~10^{-9} $cm^2V^{-1}s^{-1}$ [7], that is, five orders of magnitude less than that of the intrinsic (best case) SiO_2 hole mobility discussed here. What's going on?

Amorphous Oxide Semiconductors: IGZO and Related Materials for Display and Memory, First Edition.
Edited by Hideo Hosono and Hideya Kumomi.

Here's the problem. Equation (5.1) is not appropriate for assessing drift mobility in an amorphous solid. Rather, hole drift mobility in an amorphous semiconductor (or insulator) is better described by [8–13]

$$\mu_{drift} = \frac{p}{p + p_T}\mu_0 \tag{5.2}$$

where p is free hole concentration, p_T is trapped hole concentration, and μ_0 is the trap-free (maximum) drift mobility. Note that $\mu_{drift} = \mu_0$ when there is negligible hole trapping (i.e., when $p \gg p_T$). Rationalizing the physics underlying the extraordinarily small hole mobility measured in SiO_2 is straightforward using Equation (5.2); that is, if $\mu_0 \approx 1\,cm^2V^{-1}s^{-1}$, then $p/p_T \approx 10^{-4}$ for the intrinsic hole mobility of $\sim 10^{-4}\,cm^2V^{-1}s^{-1}$, whereas $p/p_T \approx 10^{-9}$ for the wet oxide hole mobility of $\sim 10^{-9}\,cm^2V^{-1}s^{-1}$. Moreover, hole traps in intrinsic SiO_2 are identified as valence band tail states, whereas the hole trap concentration in wet oxide SiO_2 is strongly enhanced due to the presence of a very high density of oxygen vacancies [3].

In summary, electron or hole transport in a crystalline semiconductor is dominated by the physics of scattering, as described by Equation (5.1), whereas electron or hole transport in an amorphous semiconductor (or insulator) is established (primarily) by the physics of trapping, as described (for holes) by Equation (5.2). In the remainder of this chapter, an amorphous semiconductor mobility physics framework is developed and is then employed for device physics modeling of the mobility in a thin-film transistor (TFT). Although mobility physics elucidation is the central theme of this contribution, successful pursuit of this theme requires taking a bit of a detour in order to discuss the important supporting topics of mobility extraction, fitting, and model validation. In Section 5.2, consider the physics underlying diffusive carrier transport.

5.2 Diffusive Mobility

Analogous to the hole drift mobility relationship specified in Equation (5.2), the electron drift mobility for an amorphous semiconductor (or insulator) is given by

$$\mu_{drift} = \frac{n}{n + n_T}\mu_0 \tag{5.3}$$

where n is free electron concentration, n_T is trapped electron concentration, and μ_0 is the (maximum) drift mobility when negligible trapping occurs (i.e., when $n \gg n_T$). In order for Equations (5.2) and (5.3) to be useful, the physics underlying μ_0 needs to be elucidated. Also, a quantitative expression must be developed for μ_0.

To clarify the physics foundational to μ_0, assume (for the sake of argument) that $\mu_0 = 6\,cm^2V^{-1}s^{-1}$ for an amorphous semiconductor with an effective mass equal to $0.34\,m_0$; that is, the effective mass normally assumed for electrons in amorphous indium–gallium–zinc oxide (a-IGZO) [14]. Evaluating this value of $\mu_0 = 6\,cm^2V^{-1}s^{-1}$ within the scattering physics framework of Equation (5.1), it is found that $<\tau_m> = 1.2$ fs and $\lambda_{mfp} = 0.23$ nm. Thus, the mean free path is calculated to be less than that of the interatomic spacing (assuming an interatomic distance of 0.25 nm). This result is inconsistent with a basic tenet of the physics underlying transport in a noncrystalline solid (i.e., "the principle of Ioffe and Regel (1960) that the mean free path cannot be less than the distance between atoms") [9, 15]. Even if the mean free path is a bit larger than the interatomic distance, the scattering physics inherent in the use of Equation (5.1) would appear to require a mean free path of at least ~10 atomic distances (~2.5 nm) before two discreet scattering events are distinguishable from one another (i.e., are adequately separated by an appreciable intercollisional drift distance).

Figure 5.1 attempts to pictorially illustrate essential differences between electron transport in a crystalline versus an amorphous material (ignoring trapping), assuming the scattering physics perspective underlying Equation (5.1). The crystalline transport case is indicated in Figure 5.1a, where a drift mobility of $100\,cm^2V^{-1}s^{-1}$ is assumed. Since $m^* = 0.34\,m_0$ (this is the same effective mass as normally assumed for electrons in a-IGZO [14]), Equation (5.1) can be used to calculate the average momentum relaxation time $<\tau_m> = 19$ fs, leading

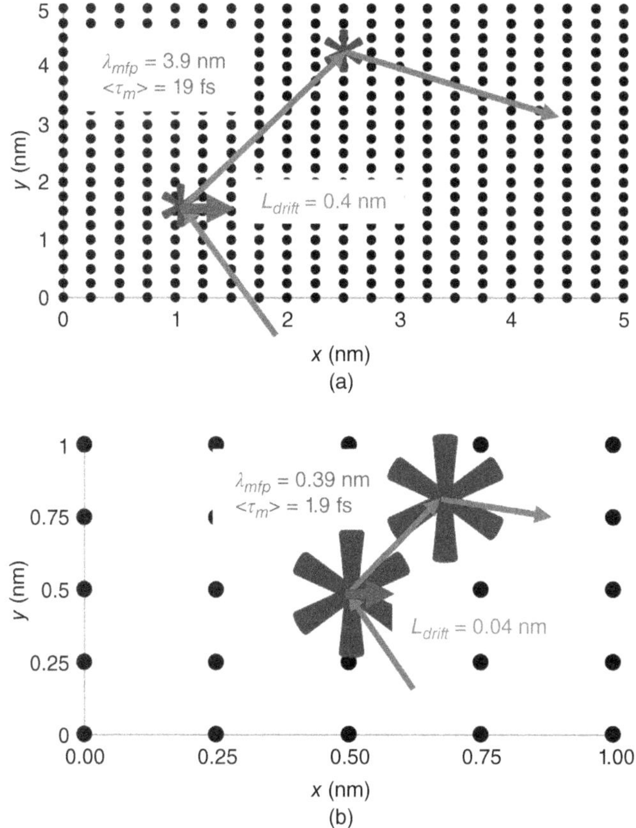

Figure 5.1 A two-dimensional array of atoms separated by an interatomic spacing of 0.25 nm is used to illustrate the scattering physics (asterisks represent scattering events) corresponding to a drift mobility of (a) 100 $cm^2V^{-1}s^{-1}$ (exemplary of crystalline transport) and (b) 10 $cm^2V^{-1}s^{-1}$ (exemplary of amorphous transport) when $m^* = 0.34\,m_0$. λ_{mfp}: Mean free path; $<\tau_m>$: average momentum relaxation time; and L_{drift}: intercollisional drift length for an electric field of magnitude 100 kV/cm that is directed in the negative x direction.

to a mean free path λ_{mfp} = 3.9 nm. Thus, thermal motion between scattering events (indicated as asterisks in Figure 5.1) occurs over a distance corresponding to almost 16 interatomic spacings in this crystalline material. Moreover, the intercollisional drift length is found to be 0.4 nm for an electric field of $\xi = 100$ kVcm^{-1} (100 times larger than the electric field assumed previously for GaAs). In contrast, for the amorphous transport case shown in Figure 5.1b with a drift mobility of 10 $cm^2V^{-1}s^{-1}$, it is found that $<\tau_m>$ = 1.9 fs, λ_{mfp} = 0.39 nm, and L_{drift} = 0.04 nm, all of which indicate that collisions are essentially continuous in an amorphous material. This point is underscored by the almost overlapping nature of the asterisks representing collisional events, as shown in Figure 5.1b. In summary, electron transport in a crystalline material is characterized by distinct collisional events that are separated from one another by a distance of (typically) tens or hundreds of atoms, whereas electron transport in an amorphous material consists of almost continuous scattering that is more similar to Brownian motion than it is to accelerated transit between collisions.

When the mean free path is very small (i.e., a few atomic distances or less), scattering events are essentially indistinguishable from one another such that carrier transport is basically a diffusive process. In fact, "It would be more accurate to describe the carriers as almost continuously under the influence of the scattering centers so that their motion would be more like Brownian motion than wave propagation" [16]. Mott and Davis considered this type of above-the-mobility-edge Brownian-motion-like carrier transport, and they derived a diffusive mobility expression that is now used to describe μ_0 [9]:

$$\mu_0 = \frac{q\hbar}{6m^* k_B T} \tag{5.4}$$

where q is electronic charge, $\hbar$ is reduced Planck constant, m^* is effective mass, k_B is Boltzmann's constant, and T is temperature. Equation (5.4) is a key result in terms of describing amorphous semiconductor (or insulator) carrier transport. Thus, it is worth deriving it, especially since prior published derivations tend to be a bit obscure [9, 10].

The derivation begins with use of the Einstein equation in order to relate mobility to diffusivity:

$$\mu_0 = \frac{D}{k_B T/q} \tag{5.5}$$

where D is diffusivity. Diffusivity is envisaged as a three-dimensional random walk, such that [17]

$$D = \frac{1}{6}\nu_{EL}\langle r^2\rangle \tag{5.6}$$

where ν_{EL} is the electronic jump frequency and $\langle r^2\rangle$ is the mean-square electronic jump distance. The electronic jump frequency, ν_{EL}, is obtained by first invoking Heisenberg's Uncertainty Principle:

$$\Delta k \Delta x \geq \frac{1}{2} \tag{5.7}$$

where Δk is the uncertainty in wave vector and Δx is the uncertainty in distance. Taking $\Delta k \approx k$ and $\Delta x \approx \sqrt{\langle r^2\rangle}$ [11] leads to $k \geq 1/2\sqrt{\langle r^2\rangle}$. Choosing $k = 2\sqrt{\pi/\langle r^2\rangle}$ clearly satisfies Heisenberg's Uncertainty Principle (Equation [5.7]), substituting this choice for k into the dispersion relation $E = \hbar^2 k^2/2m^*$, and recognizing that $E = h\nu_{EL}$ leads to $\nu_{EL} \approx \hbar/(m^*\langle r^2\rangle)$. Finally, substituting this expression for ν_{EL} in Equation (5.6) and then substituting the resulting version of Equation (5.6) in Equation (5.5) lead to our desired result of Equation (5.4), completing the derivation.

Returning to Equation (5.4), note that at a given temperature the diffusive mobility, μ_0, depends only on the effective mass, m^*. Figure 5.2 illustrates this diffusive mobility dependence on effective mass over a range of m^* = 0.1–1 m_0. As argued in Refs. [12, 13], it is hard to imagine any amorphous semiconductor (or insulator) having an effective mass of less than 0.1 m_0, thereby establishing a theoretical upper limit on the drift mobility of approximately 70 $cm^2V^{-1}s^{-1}$ for an amorphous semiconductor (or insulator). It should be noted that this physically based upper bound on mobility is premised on the Brownian-motion, diffusive-transport-physics perspective upon which Equation (5.4) is derived.

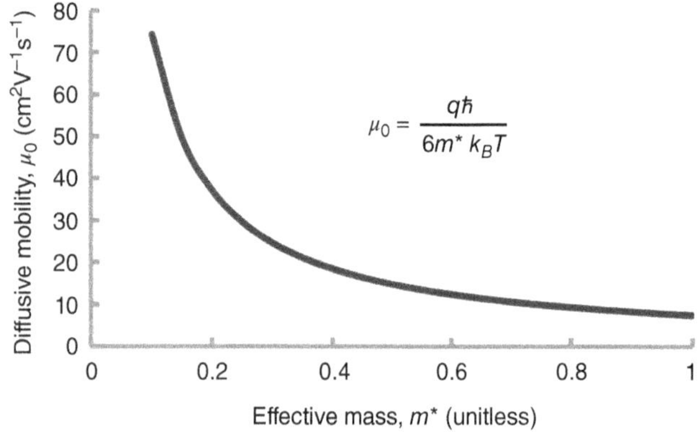

Figure 5.2 Room temperature diffusive mobility versus effective mass.

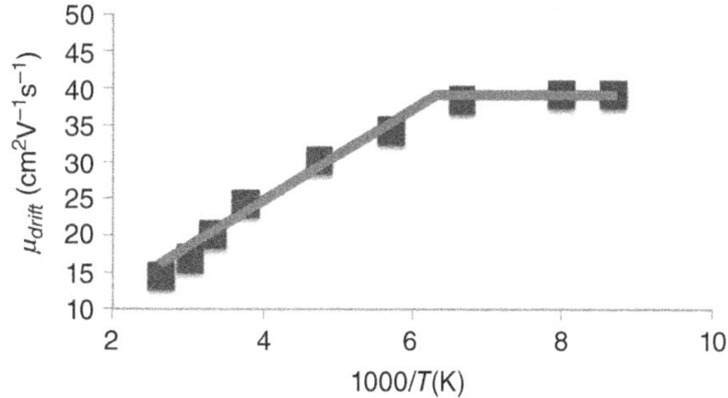

Figure 5.3 Drift mobility versus inverse temperature for electron transport in SiO_2 [3]. Shaded data points are experimental values from Ref. [18]. The striped pattern fits to the data are obtained using Equation (5.4) with $m^* = 0.36\,m_0$ (sloped line) and $W_{TA} = 13$ meV substituted for k_BT (horizontal line). *Source*: Wager, J. F. (2017). Low-field transport in SiO_2. *Journal of Non-Crystalline Solids* **459**: 111–115. Reproduced with permission of Elsevier.

Before leaving this topic of diffusive mobility, one more wrinkle needs to be addressed. Figure 5.3 shows an experimental trend in which electron drift mobility is plotted versus inverse temperature for SiO_2, as measured via transient photoconductivity induced by short (3 ns) X-ray pulse excitation [3, 18]. The sloped portion of the curve fit is generated using Equation (5.4), yielding a best fit when $m^* = 0.36\,m_0$, which is close to that reported for the tunneling effective mass ($0.30 \pm 0.02\,m_0$) [19] and is within the range of values used in the literature to model electron transport in SiO_2 (0.3–1.4 m_0) [20–22]. The horizontal portion of the curve shown in Figure 5.3 is unexpected based on Equation (5.4) and on our prior discussion of diffusive transport. However, Equation (5.4) can still be used to accurately fit the horizontal portion of this curve if k_BT is replaced by a constant, temperature-independent energy term, which turns out to be 13 meV based on a best fit to the data. This 13 meV energy for SiO_2 is identical to that typically employed for the conduction band tail Urbach energy, W_{TA}, used in the assessment of electron transport in *a*-IGZO TFTs [12–14, 23, 24]. Thus, based on this experimental trend, it appears that Equation (5.4) needs to be modified, leading to a refined formulation of the electron diffusive mobility in which

$$\mu_0 = minimum\left[\frac{q\hbar}{6m^*k_BT}, \frac{q\hbar}{6m^*W_{TA}}\right] \tag{5.8}$$

where W_{TA} is the conduction band tail characteristic (Urbach) energy, in which tail states (T) are acceptors (A). A similar expression pertains to the hole diffusive mobility except that W_{TA} is replaced by W_{TD}, the valence band tail characteristic (Urbach) energy, in which tail states (T) are donors (D).

If (naively) Equation (5.8) is equated to Equation (5.1), then it is evident that

$$\langle\tau_m\rangle = minimum\left[\frac{\hbar}{6k_BT}, \frac{\hbar}{6W_{TA}}\right] \tag{5.9}$$

which can be interpreted as an effective average scattering time for diffusive transport. Moreover, once again invoking (another form of) Heisenberg's Uncertainty Principle,

$$\Delta E\Delta t \geq \frac{\hbar}{2} \tag{5.10}$$

and assuming that the uncertainty in time $\Delta t = \langle\tau_m\rangle$ leads to

$$\Delta E = maximum[3k_BT, 3W_{TA}] \tag{5.11}$$

This is an interesting result. It implies that the uncertainty in time associated with the extraordinarily short time scale of diffusive transport "scattering," Δt, leads to an uncertainty in energy state occupation (Equation [5.11]) such that electronic states approximately $\frac{3}{2}k_BT$ or $\frac{3}{2}W_{TA}$ above (or below) the conduction band mobility edge are populated (via uncertainty broadening), and thus these upper populated states are able to contribute to diffusive mobility transport.

This completes the discussion of diffusive mobility. To actually calculate the drift mobility for an amorphous semiconductor (or insulator) using Equations (5.2) or (5.3), it is necessary to specify p and p_T (for holes) or n and n_T (for electrons). To accomplish this task, an appropriate density-of-states model must be developed for the amorphous material of interest. This density-of-states modeling exercise is undertaken in Section 5.3.

5.3 Density of States

The amorphous semiconductor (or insulator) intrinsic density-of-states model employed herein is shown in Figure 5.4. The valence band mobility edge, E_{VME}, separates extended (delocalized) hole states of density, $g_V(E)$, from an exponential distribution of localized donor-like band tail states of density, $g_{TD}(E)$, specified by a peak density, N_{TD}, and a characteristic (Urbach) energy, W_{TD}. Similarly, the conduction band mobility edge, E_{CME}, separates extended electron states of density, $g_C(E)$, from an exponential distribution of localized acceptor-like band tail states of density, $g_{TA}(E)$, specified by a peak density, N_{TA}, and a characteristic (Urbach) energy, W_{TA}. As a physical constraint to the model, the density of states and its first derivative with respect to energy are assumed to be continuous across both mobility edges.

Subsequent discussion in this section focuses exclusively on localized and extended states related to the conduction band. Similar considerations apply for the analogous case of localized and extended states related to the valence band.

Requiring the conduction band density of states and its energy derivative to be continuous across the mobility edge leads to an explicit expression for the peak density of conduction band tail states given by [25]

$$N_{TA} = \frac{1}{2\pi^2}\left(\frac{2m^*}{\hbar^2}\right)^{3/2}\sqrt{\frac{W_{TA}}{2}} = 4.9\times10^{21}\sqrt{W_{TA}}\left(\frac{m^*}{m_0}\right)^{3/2} \quad (cm^{-3}eV^{-1}) \tag{5.12}$$

and for the total density of conduction band tail states equal to

$$n_{TOTAL} = W_{TA}N_{TA} = 4.9\times10^{21}\left(W_{TA}\frac{m^*}{m_0}\right)^{3/2} \quad (cm^{-3}) \tag{5.13}$$

Thus, these two important quantities N_{TA} and n_{TOTAL} are known once the Urbach energy and effective mass have been specified. Moreover, requiring continuity of the conduction band density of states and its energy derivative

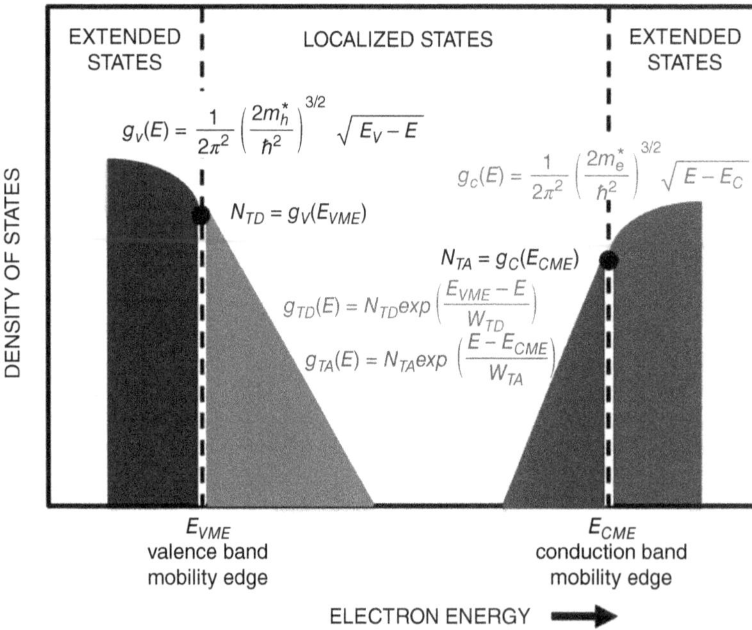

Figure 5.4 Amorphous semiconductor (or insulator) density-of-states model illustrating localized and extended state densities for both valence and conduction bands. *Source*: Reproduced from Wager, J. F. (2017). Real- and reciprocal-space attributes of band tail states, *AIP Advances* **7**: 125321. Reproduced with permission of AIP Publishing.

across the mobility edge leads to the condition that [25]

$$E_{CME} - E_C = \frac{W_{TA}}{2} \tag{5.14}$$

Equation (5.14) reveals that the conduction band minimum, E_C (not shown in Figure 5.4), is positioned below the conduction band mobility edge by an energy equal to half of the Urbach energy.

Knowing the total density of conduction band tail states, n_{TOTAL}, as given in Equation (5.13), establishes the intrinsic disorder density of the cation sublattice for an amorphous semiconductor such as *a*-IGZO. Thus, the average real space distance between conduction band tail states is given by [25]

$$\langle x_{TA} \rangle = 2\left[\frac{3}{4\pi n_{TOTAL}}\right]^{1/3} (nm) \tag{5.15}$$

which indicates that $<x_{TA}>$ is equal to twice the Wigner–Seitz radius, r_s [26]. Thus, the extent of intrinsic conduction band disorder may be quantified in terms of either volume using Equation (5.13) or distance using Equation (5.15).

Now that the amorphous semiconductor intrinsic density-of-states model has been specified, the free electron density (n) and trapped electron density (n_T), as required for assessment of the drift mobility according to Equation (5.3), may be evaluated as

$$n = \int_{E_C}^{\infty} g_C(E) f_{FD}(E, F_n) dE \ (cm^{-3}) \tag{5.16}$$

and

$$n_T = \int_{E_{VME}}^{E_{CME}} g_{TA}(E) f_{FD}(E, F_n) dE \ (cm^{-3}) \tag{5.17}$$

where f_{FD} is the Fermi–Dirac occupancy function and F_n is the electron quasi-Fermi level. Note that only band tail states contribute to the trap density specified by Equation (5.17), since an *intrinsic* amorphous semiconductor model is assumed in Figure 5.4. For an *extrinsic* amorphous semiconductor, trap assessment would also require accounting for trap states associated with the existence of defects and/or impurities.

5.4 TFT Mobility Considerations

Now that it is possible to calculate n (via Equation [5.16]), n_T (via Equation [5.17]), and μ_0 (via Equation [5.8]), it appears as if drift mobility, μ_{drift}, can be assessed using Equation (5.3). In fact, evaluation of Equations (5.16) and (5.17) is only possible if the quasi-Fermi level, F_n, is specified. Thus, calculation of a theory-based drift mobility boils down to finding $\mu_{drift}(F_n)$. Doing this is problematic, however, since experimental mobility is obtained as a consequence of TFT assessment, leading to some type of $\mu_{experimental}(V_G)$ relationship, where V_G is the applied gate voltage used in a TFT measurement. Since independent variables differ, $\mu_{drift}(F_n)$ cannot be directly compared to $\mu_{experimental}(V_G)$.

This problem can be resolved by providing a connecting link between independent variables. Connecting independent variables is accomplished using

$$n^{2/3} + n_T^{\ 2/3} = \frac{C_G(V_G - V_{ON})}{q} \ (cm^{-2}) \tag{5.18}$$

where C_G is gate capacitance density and V_{ON} is turn-on voltage. Equation (5.18) is premised on the use of a charge sheet approximation [2, 27, 28], in which it is assumed that the gate voltage applied in excess of the turn-on voltage (i.e., the overvoltage) induces an accumulation layer charge sheet at the semiconductor–gate insulator interface in which some of the induced charge is trapped. Since n and n_T both implicitly depend on F_n, independent variables

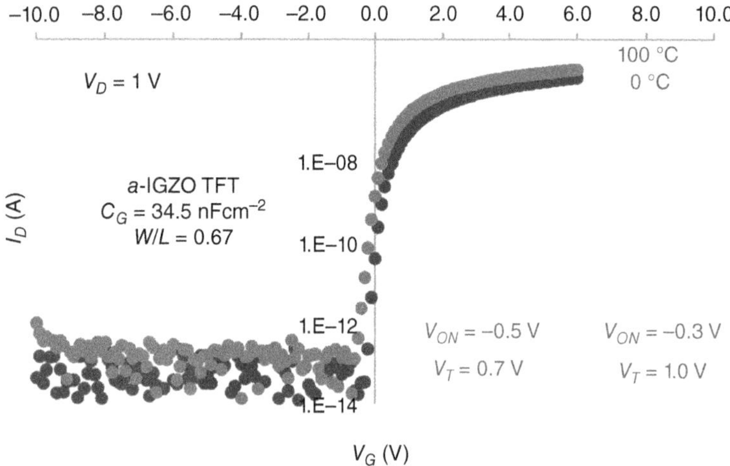

Figure 5.5 Drain current–gate voltage (I_D-V_G) transfer curves for an a-IGZO TFT measured at two temperatures, 0 and 100 °C. These I_D-V_G transfer curves constitute the raw data used extensively in Sections 5.5 and 5.6.

F_n and V_G are indeed linked in Equation (5.18). Note that the charge sheet perspective used in Equation (5.18) differs somewhat from the charge density formulation employed in previous mobility physics formulations [12, 13].

An important implication of using the charge sheet approximation is that the Equation (5.3) drift mobility relationship needs to be modified to

$$\mu_{driftTFT} = \frac{n^{2/3}}{n^{2/3} + n_T{}^{2/3}} \mu_0 \tag{5.19}$$

where the notation $\mu_{driftTFT}$ is a reminder that it pertains to a TFT-extracted drift mobility, while Equations (5.2) and (5.3) are applicable to a bulk drift mobility.

As mentioned in this chapter, experimental mobility may be expressed as $\mu_{experimental}(V_G)$ as obtained from a TFT measurement. More specifically, $\mu_{experimental}(V_G)$ is usually extracted from a drain current–gate voltage (I_D-V_G) transfer curve in which the drain voltage is held constant at a relatively low value. Figure 5.5 shows I_D-V_G transfer curves for an a-IGZO TFT measured at two temperatures, 0 and 100 °C. These two curves, and a 20 °C curve (not shown), constitute the raw data used extensively in Sections 5.5 and 5.6.

Now that a connecting link between the independent variables F_n and V_G has been formulated, it is possible to undertake a physics-based mobility simulation and to compare simulated trends to experimental results. However, this topic must be deferred to Section 5.6; to set the stage, the important topics of TFT mobility extraction, fitting, and model validation are explored in Section 5.5.

5.5 TFT Mobility Extraction, Fitting, and Model Validation

Figure 5.6 displays two experimental mobility trends at two different temperatures, 0 and 100 °C, as extracted from the measured I_D-V_G transfer curves of Figure 5.5. As shown in the equation included in Figure 5.6a, average mobility, $\mu_{AVG}(V_G)$, is obtained from an assessment of the drain conductance as a function of gate voltage, that is, $G_D(V_G) = I_D(V_G)/V_D$, in which the drain voltage, V_D, is held constant at a very low voltage so that it is assured that the drain current, I_D, is operating in a linear manner (with respect to V_D). It is important to understand that $\mu_{AVG}(V_G)$ refers to an average mobility of *all* of the carriers that are induced into the channel by the applied gate overvoltage [29, 30]. In contrast, incremental mobility, $\mu_{INC}(V_G)$, as shown in Figure 5.6b, is evaluated using the

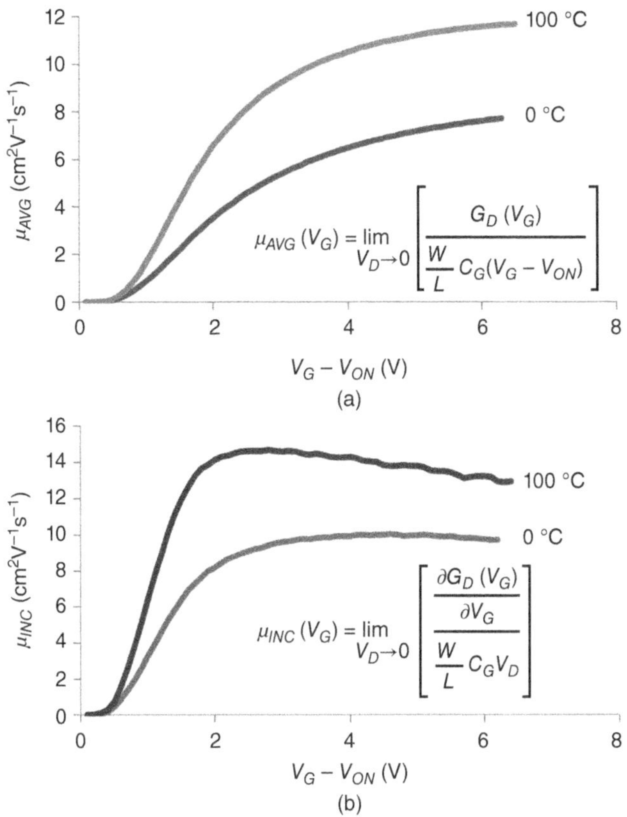

Figure 5.6 (a) Average and (b) incremental mobility as a function of gate overvoltage as measured for an *a*-IGZO TFT at two temperatures, 0 and 100 °C.

differential drain conductance as a function of gate voltage, $\frac{\partial G_D(V_G)}{\partial V_G}$, and is reflective of the *latest carriers induced* into the channel by the applied gate overvoltage [29, 30].

In Figure 5.6a, μ_{AVG} increases monotonically with respect to overvoltage. In general, mobility (both μ_{AVG} and μ_{INC}) tends to increase with overvoltage due to trap filling, as clarified by the drift mobility relationship given in Equation (5.3). At a small overvoltage, almost all of the gate-voltage-induced charge goes into filling traps; thus, the mobility is very small. As overvoltage increases, more of the traps in the channel are filled, so that a larger fraction of the induced charge goes into populating delocalized states, thereby enhancing mobility. μ_{INC} increases to a higher value than μ_{AVG} since the latest carriers induced into the channel experience less trapping; μ_{AVG} is weighed down by all of the trapping that occurred at lower overvoltages.

In Figure 5.6b, μ_{INC} increases and then decreases with respect to overvoltage. This decrease in μ_{INC} with increasing overvoltage is termed mobility degradation. It is a consequence of an increase in the intensity of interfacial roughness scattering as electrons accumulate closer and closer to the insulator–semiconductor interface with increasing overvoltage and experience more intense interfacial scattering [30, 31]. Mobility degradation is more commonly witnessed with μ_{INC} rather than μ_{AVG}. Once again, this trend is ascribed to the fact that μ_{INC} pertains to the *latest carriers induced* into the channel by the applied gate overvoltage.

Which mobility is most relevant for assessing the quality of a TFT? TFT direct current (DC) operation is established by μ_{AVG}, so this mobility is clearly the relevant figure of merit for establishing TFT performance. μ_{INC} is primarily useful for elucidating device physics aspects of TFT operation (e.g., differential trapping or interfacial scattering), as discussed in this chapter.

The elephant-in-the-room aspect of Figure 5.6, not yet discussed, is the dramatic temperature dependence of the mobility. Both μ_{AVG} and μ_{INC} significantly increase with increasing temperature. This trend argues against scattering as its source, since the scattering rate is known to increase with increasing temperature, thus resulting in a decrease in mobility with increasing temperature. Rather, this type of temperature-dependent behavior suggests that a thermally activated process is operative, as explored in Section 5.6.

Mobility estimation can be tricky, as evident from the disparity of the three trends shown in Figure 5.7. The μ_{AVG} equation included as an inset to Figure 5.6a is almost identical to the defining relationship for effective mobility, μ_{EFF}, except that the turn-on voltage, V_{ON}, is replaced by the threshold voltage, V_T [30, 31]. (A few points of common confusion should be clarified here. In the literature, the definition of μ_{AVG} or μ_{EFF} sometimes involves the large-signal drain conductance, G_D, and sometimes involves $\frac{\partial I_D}{\partial V_D}$ or g_D, the small-signal drain conductance. In the limit $V_D \to 0$ in which linear TFT device behavior prevails, $G_D \equiv \frac{I_D}{V_D} = \frac{\partial I_D}{\partial V_D} \equiv g_D$ so that these definitions are equivalent. Similarly, the definition of μ_{INC} or field-effect mobility, μ_{FE}, sometimes involves $\frac{\partial G_D}{\partial V_G}$ and sometimes involves either $\frac{1}{V_D}\frac{\partial I_D}{\partial V_G}$ or $\frac{g_m}{V_D}$, where g_m is the small-signal transconductance. In the limit $V_D \to 0$ in which linear TFT device behavior prevails, $\frac{\partial G_D}{\partial V_G} = \frac{1}{V_D}\frac{\partial I_D}{\partial V_G} = \frac{g_m}{V_D}$ so that these definitions are equivalent.)

μ_{AVG} and μ_{EFF} are plotted for two different temperatures in Figure 5.7. The advantage of using μ_{AVG} is that it is possible to specify mobility for all overvoltages above the turn-on voltage, so that, for example, trap-filling trends can be revealed. However, using μ_{AVG} leads to a lower estimate of maximum mobility. In contrast, the advantage

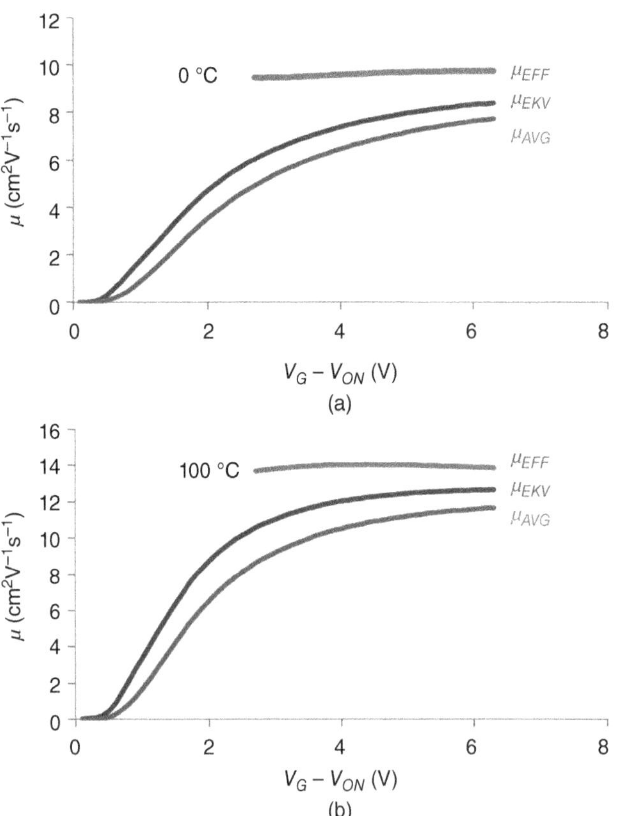

Figure 5.7 Effective, EKV, and average mobility versus gate overvoltage for an a-IGZO TFT at two temperatures, (a) 0 °C and (b) 100 °C.

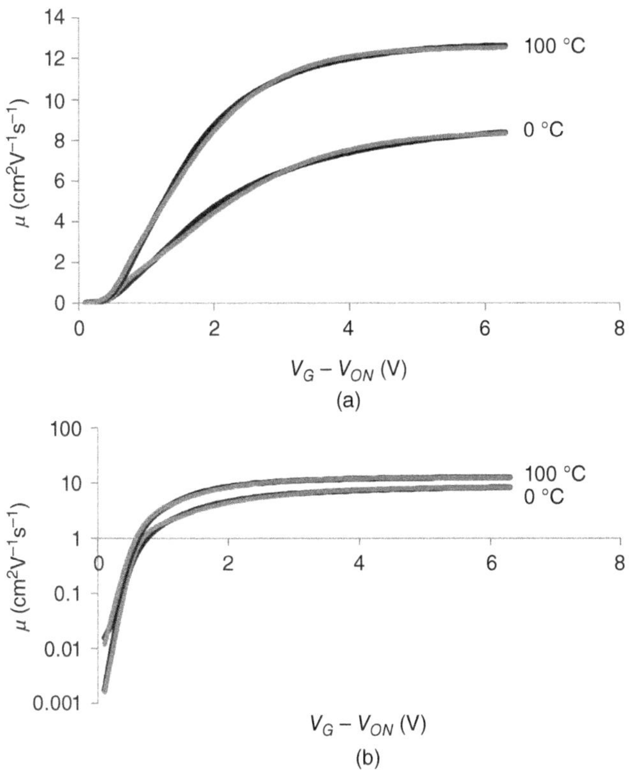

Figure 5.8 (a) Linear and (b) logarithmic plots of EKV mobility (black curves) and five-parameter μ_{GZ+SUB} fits to the EKV mobility (shaded curves) versus overvoltage at 0 and 100 °C for an *a*-IGZO TFT. Fit parameters are as follows: 0 °C: μ_{0GZ} = 8.5 cm^{2}V^{-1}s^{-1}, k_{GZ} = 3.5, V_{GZ} = 1.19 V, $\mu_{MIN} = 4 \times 10^{-4}$ cm^{2}V^{-1}s^{-1}, n_{SUB} = 1.5; and 100 °C: μ_{0GZ} = 12.6 cm^{2}V^{-1}s^{-1}, k_{GZ} = 4.0, V_{GZ} = 0.87 V, $\mu_{MIN} = 4 \times 10^{-3}$ cm^{2}V^{-1}s^{-1}, n_{SUB} = 1.9.

of using μ_{EFF} is that it provides a larger and a simple, relatively constant estimate of mobility. However, using μ_{EFF} is problematic since it is singular when $V_G = V_T$, making it impossible to estimate mobility just above, at, or below threshold. Note that the 100 °C μ_{EFF} curve in Figure 5.8 decreases slightly with increasing overvoltage; this trend is a singularity artifact, rather than a real physical trend, underscoring the problematic nature of using μ_{EFF} to characterize mobility. Given all of these considerations, which mobility is correct, μ_{AVG} or μ_{EFF}? Neither!

Instead of extracting mobility using μ_{AVG} or μ_{EFF}, a new approach is proposed herein that is based on the Enz, Krummenacher, and Vittoz (EKV) compact model [27]. From an EKV modeling perspective, the drain current of a TFT can be expressed as

$$I_{D,EKV} = 2\left(\frac{k_BT}{q}\right)^2 \mu_n C_G \frac{W}{L}\left\{\left[ln\left(1 + Exp\left(\frac{q(V_G - V_{ON})}{2k_BT}\right)\right)\right]^2 - \left[ln\left(1 + Exp\left(\frac{q(V_G - V_{ON} - V_D)}{2k_BT}\right)\right)\right]^2\right\} \quad (5.20)$$

If the EKV model electron mobility from Equation (5.20) is set as equal to one (i.e., if $\mu_n = 1\,\text{cm}^2\text{V}^{-1}\text{s}^{-1}$), then a new type of EKV mobility can be defined as

$$\mu_{EKV}(V_G, V_D) = \frac{I_{D,MEASURED}(V_G, V_D)}{I_{D,EKV,\mu=1}(V_G, V_D)} \quad (5.21)$$

where $I_{D,MEASURED}(V_G, V_D)$ refers to a *measured* drain current versus gate voltage transfer curve in which $V_D \to 0$ is a constant and $I_{D,EKV,\,\mu=1}(V_G, V_D)$ refers to the use of Equation (5.20), but with $\mu_n = 1\,\text{cm}^2\text{V}^{-1}\text{s}^{-1}$.

This EKV mobility is included in Figure 5.7. It is clear from Figure 5.7 that μ_{EKV} is intermediate to μ_{EFF} and μ_{AVG}. Like μ_{AVG}, it is defined over the full domain of overvoltage beyond V_{ON}. In fact, μ_{EKV} is unique since it is well defined and continuous for all three regimes of TFT behavior (i.e., subthreshold, linear, and saturation), as shown by the black curves of Figure 5.8. As defined by Equation (5.21), μ_{EKV} accounts for any deviation between measured and simulated I_D data. Thus, μ_{EKV} relies on the accuracy of the interpolating function used in Equation (5.20) to account for basic electrostatic considerations embedded within $I_D(V_G,V_D)$ that give rise to subthreshold, linear, or saturation TFT behavior; and then everything not properly accounted for by the interpolating function is lumped into μ_{EKV}. This perspective makes the physical viability of μ_{EKV} appear to be questionable (i.e., perhaps μ_{EKV} is correcting for non-idealities in the interpolating function that are not related to mobility). As discussed further here, it appears that above-threshold μ_{EKV} is an accurate estimator of mobility, while below-threshold μ_{EKV} corrects for non-ideality in the subthreshold swing. A unique strength of μ_{EKV} is that model validation is automatically assured in the sense that the estimated data sets $\mu_{EKV}(V_G,V_D)$ and $I_{D,EKV,\mu=1}(V_G,V_D)$ can be used to accurately reconstruct the measured data set $I_{D,MEASURED}(V_G,V_D)$ in a trivial manner as a consequence of having Equation (5.20) as the defining equation for μ_{EKV}.

An important aspect of μ_{EKV} mobility extraction is revealed by the shaded curves included in Figure 5.8. Each shaded curve is a five-parameter fit to its corresponding extracted EKV mobility (black curve). The fit is almost perfect. It is accomplished using a three-parameter Gompertz function to fit the above-turn-on portion of the curve:

$$\mu_{GZ} = \mu_{0GZ} e^{-k_{GZ} e^{\frac{-(V_G - V_{ON})}{V_{GZ}}}} \tag{5.22}$$

where μ_{GZ0} is a mobility prefactor that controls the maximum value of the fit, k_{GZ} is an exponential prefactor term that accounts for shifting of the fit curve along the V_G–V_{ON} axis, and V_{GZ} is a characteristic voltage that establishes how much the fit curve is stretched along the V_G–V_{ON} axis. The subthreshold portion of the μ_{EKV} curve is fit using a two-parameter EKV-inspired expression:

$$\mu_{SUB} = \mu_{MIN} \left[ln \left(1 + Exp \left(\frac{q(V_G - V_{ON})}{2n_{SUB} k_B T} \right) \right) \right]^2 \tag{5.23}$$

where μ_{MIN} is a mobility prefactor equal to the smallest value of μ_{GZ} in the extracted data set and n_{SUB} is a subthreshold ideality factor that controls the mobility subthreshold slope. Equation (5.23) reveals that the subthreshold portion of μ_{EKV} corrects for subthreshold swing non-ideality [24, 30]. The overall fit to $\mu_{EKV}(V_G)$ is denoted $\mu_{GZ+SUB}(V_G)$ and is obtained by adding Gompertz and subthreshold mobility fits in a reciprocal manner:

$$\mu_{GZ+SUB}{}^{-1} = \mu_{SUB}{}^{-1} + \mu_{GZ}{}^{-1} \tag{5.24}$$

When mobility degradation is encountered (e.g., as witnessed for the incremental mobility plots of Figure 5.6), the Gompertz fitting function must be modified to include a mobility degradation factor, θ [30, 31], resulting in

$$\mu_{GZ+\theta} = \frac{\mu_{0GZ} e^{-k_{GZ} e^{\frac{-(V_G - V_{ON})}{V_{GZ}}}}}{1 - \theta(V_G - V_{ON})} \tag{5.25}$$

Figure 5.9 illustrates incremental mobility trends for two curves exhibiting mobility degradation. Although the quality of these $\mu_{GZ+\theta}$ fits (curves) appears to be excellent, as evident from the large adjusted R^2's reported (as obtained using NonlinearModelFit in Mathematica), this Gompertz fit significantly overestimates the magnitude of the subthreshold mobility. This is easily fixed using a μ_{SUB} subthreshold correction (Equation [5.23]), leading to a six-parameter fit (not shown) given by $\mu_{GZ+\theta+SUB}{}^{-1} = \mu_{SUB}{}^{-1} + \mu_{GZ+\theta}{}^{-1}$. Such a correction is entirely unwarranted, however, if attention is restricted to above-threshold TFT operation. Notice that the mobility degradation factor increases by a factor of 59 when going from 0 to 100 °C in the fits shown in Figure 5.9.

A key aspect of mobility estimation (which is, in fact, almost always ignored) is model validation. This simply means that once a mobility has been estimated, it should then be plugged back into its appropriate device

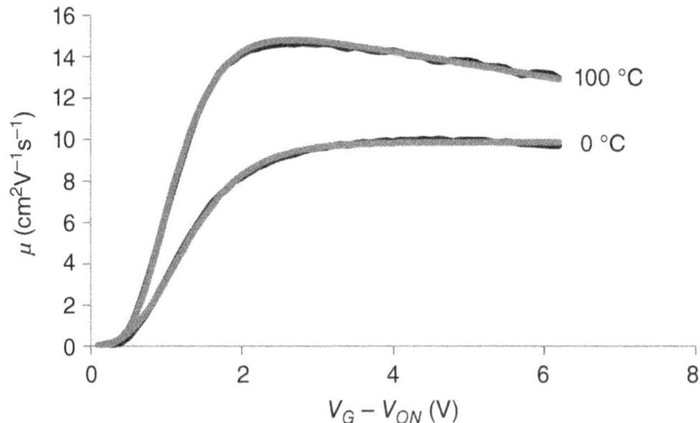

Figure 5.9 Incremental mobility (black curves) and four-parameter $\mu_{GZ+\theta}$ Gompertz fits (shaded curves) versus overvoltage at 0 and 100 °C for an *a*-IGZO TFT. Fit parameters are as follows: 0 °C: $\mu_{0GZ} = 9.93\ \text{cm}^2\text{V}^{-1}\text{s}^{-1}$, $k_{GZ} = 6.93$, $V_{GZ} = 0.55$ V, $\theta = 0.00076\ \text{V}^{-1}$, adj. $R^2 = 0.99987$; and 100 °C: $\mu_{0GZ} = 16.68\ \text{cm}^2\text{V}^{-1}\text{s}^{-1}$, $k_{GZ} = 7.81$, $V_{GZ} = 0.42$ V, $\theta = 0.045\ \text{V}^{-1}$, adj. $R^2 = 0.999965$.

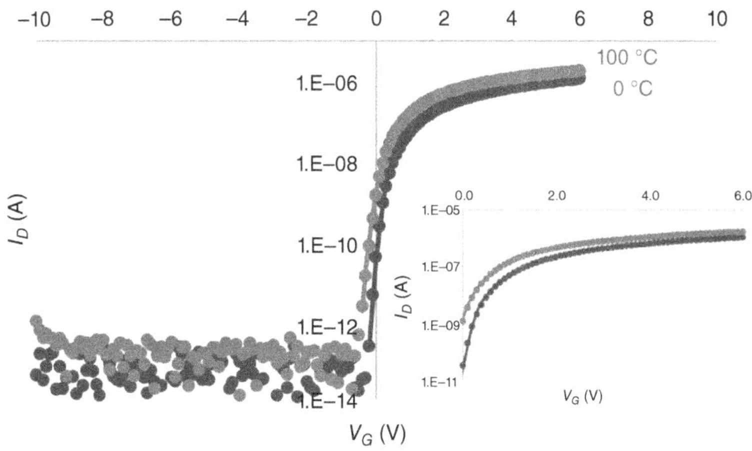

Figure 5.10 Model validation drain current versus gate voltage transfer curves at 0 and 100 °C for an *a*-IGZO TFT. Raw data (dots) are compared to the simulation (solid lines) based on a five-parameter μ_{GZ+SUB} fit to the EKV mobility (model parameters are specified in Figure 5.8's caption) and the use of Equation (5.20) with μ_{GZ+SUB} replacing μ_n.

physics I_D equation to see if it realistically reproduces the raw $I_D(V_G,V_D)$ data used for its estimation. An exquisite example of model validation is given in Figure 5.10, in which μ_{GZ+SUB} (the shaded curve of Figure 5.8) replaces μ_n in Equation (5.20), leading to an almost perfect reproduction of the raw data I_D-V_G transfer curve. Very few mobility estimates are capable of demonstrating the degree of model validation witnessed in Figure 5.10.

A simple but effective mobility validation "reality check" is to calculate $I_D = W/L\ \mu_{ESTIMATED}\ C_G\ (V_G - V_T)\ V_D$ (linear) or $I_D = W/2L\ \mu_{ESTIMATED}\ C_G\ (V_G - V_T)^2$ (saturation) for the maximum value of I_D in an I_D-V_G transfer curve and to then compare this calculated I_D to its corresponding raw-data equivalent. Employing this procedure suggests that the maximum mobility of a plasma-enhanced atomic layer deposition (PEALD) *a*-IGZO TFT that is reported to be ~70 $\text{cm}^2\text{V}^{-1}\text{s}^{-1}$ is actually closer to ~50 $\text{cm}^2\text{V}^{-1}\text{s}^{-1}$ [32]. This type of mobility overestimation is very common in the existent TFT literature, and yet it is avoidable if the simple reality check just proposed is employed and if a careful mobility assessment strategy is followed [33].

5.6 Physics-Based TFT Mobility Modeling

A primary objective of this chapter is to compare a physics-based, simulated mobility (μ_{SIM}) to a measured mobility; this is accomplished for three temperatures in Figure 5.11 using μ_{EKV} as the measured mobility estimator. Table 5.1 summarizes the device physics equations employed, as well as model, measured device, physical operating, and structure parameters used in the simulation. Additionally, actual parameter simulation values are listed at the bottom of Table 5.1. Each model parameter density is specified as both a volume density (cm^{-3}) and (equivalently) a sheet density (cm^{-2}); for most readers, volume density is a more intuitively appealing quantity, but sheet density is more in keeping with the charge sheet approximation philosophy upon which this device physics modeling approach is based. Physics-based mobility simulations are accomplished using Mathematica.

Three mobility curves are shown in Figure 5.11a, all corresponding to a temperature of 20 °C. The (shaded) curve labeled μ_{SIM1} has a final slope similar to that of the experimental (black) μ_{EKV} curve, but the simulated trend is much more abrupt at its onset and significantly overestimates the measured mobility curve across the entire range of overvoltage. The μ_{SIM1} simulation is generated using an intrinsic trap model involving only conduction band tail states (see the shaded curve of Figure 5.12). The shaded μ_{SIM1} curve shown in Figure 5.11a

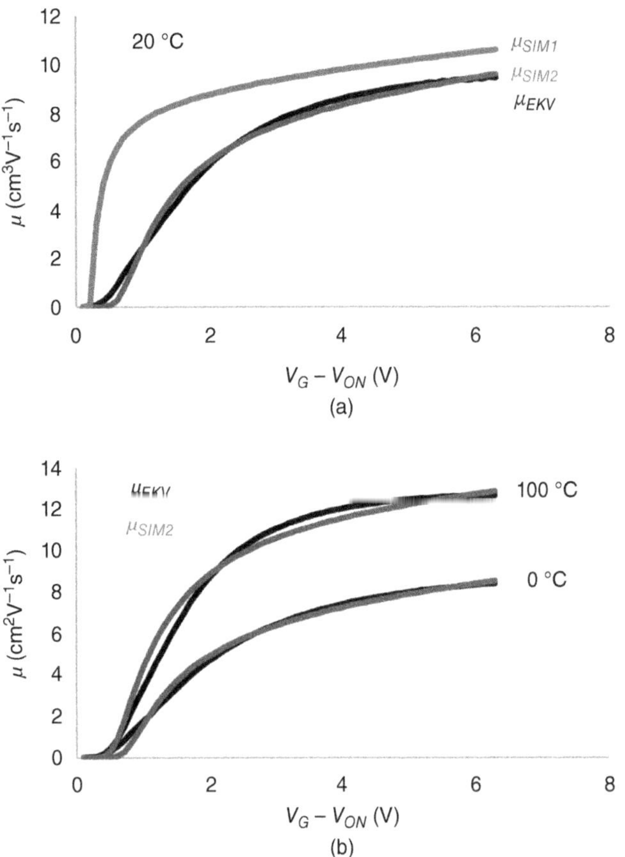

Figure 5.11 Mobility versus overvoltage for an *a*-IGZO TFT at a temperature of (a) 20 °C and (b) 0 and 100 °C. Black curves are EKV-extracted (measured) mobilities, while shaded curves are device physics simulated drift mobility curves based on modeling traps as intrinsic (involving only band tail states; μSIM1) or extrinsic (involving an exponential distribution of trap states and a shallow donor, as well as band tail states; μSIM2), respectively.

Table 5.1 *a*-IGZO TFT physics-based electron drift mobility simulation summary

Model Equations

$$\text{(i)}\ \mu_{driftTFT} = \frac{n^{\frac{2}{3}}}{n^{\frac{2}{3}} + n_T^{\frac{2}{3}}}\mu_0\ (\text{cm}^2\text{V}^{-1}\text{s}^{-1})$$

$$\text{(ii)}\ \mu_0 = \frac{q\hbar}{6m^* k_B T} e^{\frac{q(E_{SD} - E_{ECME})}{k_B}\left(\frac{1}{293} - \frac{1}{T}\right)}\ (\text{cm}^2\text{V}^{-1}\text{s}^{-1})$$

$$\text{(iii)}\ n = \int_{E_C}^{\infty} g_C(E) f_{FD}(E, F_n)\, dE\ (\text{cm}^{-3})$$

$$\text{(iv)}\ n_T = n_{Tmin} + \int_{E_{VME}}^{E_{CME}} [\underline{g_{ET}(E)} + \underline{g_{SD}(E)} + g_{TA}(E)] f_{FD}(E, F_n)\, dE\ (\text{cm}^{-3})$$

$$\text{(v)}\ g_C(E) = \frac{1}{2\pi^2}\left(\frac{2m^*}{\hbar^2}\right)^{3/2}\sqrt{E - E_C}\ (\text{cm}^{-3}\text{eV}^{-1})$$

$$\text{(vi)}\ \underline{g_{ET}(E) = N_{ET} e^{\frac{E - E_{CME}}{W_{ET}}}\ (\text{cm}^{-3}\text{eV}^{-1})}$$

$$\text{(vii)}\ \underline{g_{SD}(E) = N_{SD} e^{\left(\frac{E - E_{SD}}{W_{SD}}\right)^2}\ (\text{cm}^{-3}\text{eV}^{-1})}$$

$$\text{(viii)}\ g_{TA}(E) = N_{TA} e^{\frac{E - E_{CME}}{W_{TA}}}\ (\text{cm}^{-3}\text{eV}^{-1})$$

$$\text{(ix)}\ f_{FD}(E, F_n) = \frac{1}{1 + e^{\frac{E - F_n}{k_B T}}}\ (\text{unitless})$$

$$\text{(x)}\ E_{CME} - E_C = \frac{W_{TA}}{2}\ (\text{eV})$$

$$\text{(xi)}\ N_{TA} = \frac{1}{2\pi^2}\left(\frac{2m^*}{\hbar^2}\right)^{3/2}\sqrt{\frac{W_{TA}}{2}}\ (\text{cm}^{-3}\text{eV}^{-1})$$

$$\text{(xii)}\ n^{2/3} + n_T^{2/3} = \left[\frac{C_G(V_G - V_{ON})}{q}\right]\ (\text{cm}^{-2})$$

Model Parameters: m^*, W_{TA}, n_{Tmin}, $\underline{N_{ET}}$, $\underline{W_{ET}}$, $\underline{N_{SD}}$, $\underline{E_{SD}}$, $\underline{W_{SD}}$

Measured Device Parameters: V_{ON}

Physical Operating Parameters: T

Structure Parameters: $C_G = 34.5$ nFcm^{-1}

Model Parameters: $m^* = 0.34\,\text{m}_0$; $W_{TA} = 23$ meV; $n_{Tmin} = 10^{16}$ cm^{-3}

Exponential Distribution of Traps: $N_{ET} = 1.0 \times 10^{18}$ cm^{-3}eV^{-1}; $W_{ET} = 190$ meV

Shallow Donor: $N_{SD} = 1.0 \times 10^{19}$ cm^{-3}eV^{-1}; $E_{SD} - E_{CME} = -45$ meV; $W_{SD} = 5$ meV

Volume Density: $n_{TOTAL} = 3.4 \times 10^{18}$ cm^{-3}; $\underline{n_{ET} = 1.9 \times 10^{17}\ \text{cm}^{-3}}$; $\underline{n_{SD} = 4.5 \times 10^{17}\ \text{cm}^{-3}}$

Sheet Density: $n_{TOTAL} = 2.3 \times 10^{12}$ cm^{-2}; $\underline{n_{ET} = 3.3 \times 10^{11}\ \text{cm}^{-2}}$; $\underline{n_{SD} = 5.9 \times 10^{11}\ \text{cm}^{-2}}$

Gray indicates that this simulation parameter is invariant (not adjustable). Underscored text indicates that these quantities pertain to extrinsic traps associated with defects and/or impurities.

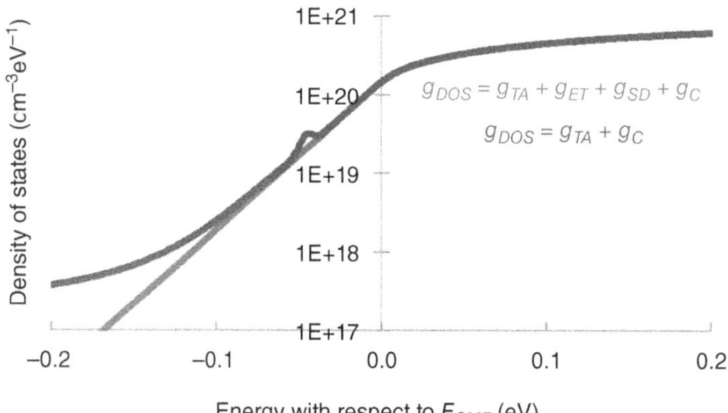

Figure 5.12 Density-of-states model for *a*-IGZO. The intrinsic model (shaded curve) considers only conduction band tail states (g_{TA}) and conduction band delocalized states (g_C), while the extrinsic model (dark curve) also includes an exponential distribution of traps (g_{ET}) and a small bump associated with a shallow donor (g_{SD}).

assumes a conduction band tail state Urbach energy of W_{TA} = 23 meV, leading to a bending of the curve near 8 $cm^2V^{-1}s^{-1}$. If a more commonly reported Urbach energy of W_{TA} = 13 meV is used in the simulation, this bend occurs near 13 $cm^2V^{-1}s^{-1}$, and the postbend slope is slightly increased compared to that of the W_{TA} = 23 meV simulation, although both curves exhibit the same basic shape. No matter how much model parameters are adjusted, this intrinsic trap model is incapable of closely fitting the experimental (black) μ_{EKV} trend, especially for gate overvoltages of less than about 3 V.

The (shaded) simulated curve labeled μ_{SIM2} (Figure 5.11a) very closely resembles the μ_{EKV} measured mobility trend. This agreement between simulation and measurement is achieved by (i) using W_{TA} = 23 meV rather than the normally reported value of W_{TA} = 13 meV as the conduction band tail Urbach energy to properly fit the upper portion of the measured curve (i.e., for gate overvoltages greater than about 2–3 V); (ii) including an additional exponential distribution of traps of density $n_{ET} = 1.9 \times 10^{17}$ cm^{-3} and characteristic energy W_{ET} = 190 meV to account for the non-abrupt onset of the measured curve (i.e., for overvoltages less than about 2–3 V); and (iii) adding a Gaussian shallow donor with energy $E_{SD} - E_{CME} = -45$ meV, density $n_{SD} = 4.5 \times 10^{17}$ cm^{-3}, and width W_{SD} = 5 meV in order to account for the temperature dependence of the measured curves as witnessed in Figure 5.11a and 5.11b at temperatures of 0, 20, and 100 °C.

To a large extent, optimizing the device physics model to obtain a best fit to a measured μ_{EKV} curve involves properly accounting for trap and conduction band charging, as will be discussed here. The total sheet density induced into an accumulation layer at a given gate voltage is given by $C_G (V_G - V_{ON}) / q = 2.16 \times 10^{11} (V_G - V_{ON})$ cm^{-2} when a 100-nm-thick SiO_2 (C_G = 34.5 $nFcm^{-2}$) gate insulator is used. When the quasi-Fermi level is positioned far from E_{CME}, as it is at a low overvoltage, most of the induced sheet density goes into filling trap states far removed from E_{CME}. Thus, the upward transition in μ_{EKV} between a gate voltage of about 0.5 to 2 V (see Figure 5.11a) corresponds to a total induced sheet density of 3.2×10^{11} cm^{-2} or (equivalently) a volume density 1.8×10^{17} cm^{-3}. This volume density is almost identical to the total exponential trap density of $n_{ET} = 1.9 \times 10^{17}$ cm^{-3}. Therefore, the upward transition in μ_{EKV} between a gate voltage of about 0.5 to 2 V (Figure 5.11a) is ascribed to charging of the exponential distribution of traps characterized by W_{ET} = 190 meV. The slope of the (black) μ_{EKV} curve (Figure 5.11a) decreases beyond a gate overvoltage of ~3 V and is relatively constant for gate overvoltages greater than ~4–5 V. This upper portion of the μ_{EKV} curve with a relatively constant slope is distinguished by simultaneous charging of both conduction band tail states as well as conduction band delocalized states. Although a 10 V overvoltage would be sufficient to fully charge a total conduction band tail state density of $n_{TOTAL} = 3.4 \times 10^{18}$ cm^{-3} if these states were located remote from E_{CME}, a much larger overvoltage than this is required to completely fill them since band tail states and conduction band states are charged simultaneously.

Since the density of the shallow donors ($n_{SD} = 4.7 \times 10^{17}$ cm^{-3}) included in the μ_{SIM2} simulation is almost an order of magnitude less than that of the total density of conduction band tail states ($n_{TOTAL} = 3.4 \times 10^{18}$ cm^{-3}), the shallow donor contributes very little to the shape of the (shaded) μ_{SIM2} simulation curve due to charging of this shallow donor state. However, it is important to include this shallow donor state in the simulation since it accounts for the temperature dependence of the mobility. This is accomplished by interpreting the shallow donor ionization energy as a diffusive mobility activation energy (see Equation [ii] of Table 5.1). Thermal activation of μ_0 is not surprising since μ_0 is derived from D (see Equation [5.5]), and it is well known that diffusivity is a thermally activated process [17]. However, the mathematical formulation employed in Equation (ii) may be confusing to some readers. Typically, a thermally activated process is expected to involve a prefactor that is equal to the 0 K intercept, and an qE_A/k_BT exponential term (e.g., $\mu_0 = \mu_{00} e^{\frac{qE_A}{k_BT}}$). However, our original equation for μ_0, Equation (5.8), depends inversely on temperature, so that the 0 K intercept of the μ_0 prefactor is singular and therefore not defined. Thus, Equation (ii) is formulated in such a manner that the prefactor corresponds to its room temperature (20 °C) value of μ_0, rather than its 0 K intercept. Returning to Figure 5.11, notice how well the shallow donor ionization energy $E_{SD} - E_{CME} = -45$ meV allows the simulated (shaded) μ_{SIM2} curves to account for the temperature dependence of the (black) measured μ_{EKV} curves.

The minimum trap density, n_{Tmin}, is an important simulation parameter included in Table 5.1, but as yet undiscussed. It is not a model parameter in the normal sense, since it is always held constant at $n_{Tmin} = 10^{16}$ cm^{-3} for all simulations. Physically, n_{Tmin} accounts for the fact that a finite density of traps on the order of 10^{16} cm^{-3} (or more) always exists across the subgap region of any real semiconductor (or insulator). If this minimum trap density is not accounted for in simulation, the simulated mobility for an intrinsic trap model (involving only band tail states) tends to a constant value below threshold, and does not tend to zero, as expected. This is a simulation artifact that is corrected for by inclusion of n_{Tmin}. In earlier physics-based mobility models, this simulation artifact was addressed by introducing a diffusion mobility and then adding drift and diffusion mobility in a reciprocal fashion [12, 13]. Work undertaken during the completion of this chapter revealed this diffusion-mobility correction strategy to be unreliable for quantitative comparisons of simulated and measured trends.

In summary, the physics-based model summarized in Table 5.1 is capable of accurately fitting the set of temperature-dependent mobility experimental curves shown in Figure 5.11 using the shaded density-of-states distribution given in Figure 5.12. Only two model parameters, m^* and W_{TA}, are required to account for *intrinsic* properties of *a*-IGZO associated with conduction band tail states and conduction band extended states. In contrast, modeling *extrinsic* effects necessitates the use of five model parameters (i.e., N_{ET} and W_{ET} for the secondary exponential trap distribution and N_{SD}, E_{SD}, and W_{SD} for the shallow donor).

It is not entirely clear which defects and/or impurities give rise to the extrinsic density-of-states features modeled in Figure 5.12; oxygen vacancies and hydrogen appear to be the most likely candidates. The simulated shallow donor density of $n_{SD} = 4.7 \times 10^{17}$ cm^{-3} is very close to the 5×10^{17} cm^{-3} estimate of the oxygen vacancy shallow donor concentration inferred from a charge balance assessment of *a*-IGZO based on measurement of the subgap density of states by ultrabroadband photoconduction [34]. This suggests that shallow oxygen vacancies may be the origin of the shallow donor. Atomic identification of the exponential distribution of trap states used in the simulation is more difficult. Many attempts were made to fit the initial upward transition in μ_{EKV} (Figure 5.11a) with one, two, or three Gaussian traps, but obtaining a best fit was found to require the use of an exponential distribution of traps. It is possible that this exponential distribution of traps arises from the superposition of a large number of oxygen vacancy states or complexes, but the precise atomic identity of this exponential distribution of traps remains elusive at this time.

5.7 Conclusions

Mobility modeling can be accomplished in various ways.

Analytical physics-based modeling of the drift mobility of an amorphous semiconductor as it manifests itself in a TFT is rather complicated, as evident from the 12 equations required for its description, which are collected in Table 5.1. This model is capable of accurately duplicating measured TFT device operation, as revealed in the temperature-dependent mobility comparisons shown in Figure 5.11. Accurate simulation requires the inclusion of *extrinsic* trap states associated with impurities and/or defects (requiring the use of five model parameters), as well as accounting for *intrinsic* conduction band tail and extended states (requiring the use of only two model parameters).

An alternative approach to mobility modeling is development of a compact model in which a minimal set of equations and model parameters are specified in order to provide a description of general TFT device behavior, including current–voltage and mobility trends. In this regard, the EKV modeling approach in conjunction with Gompertz function fitting of the EKV-extracted mobility appears to be a promising new avenue for undertaking TFT modeling. The fact that the compact model needs only one measured device parameter (V_{ON}), while it requires (at least) three mobility-fitting parameters (μ_{0GZ}, k_{GZ}, V_{GZ}, and also θ [when mobility degradation occurs]), underscores that accurate TFT device simulation depends primarily on the careful modeling of mobility.

Most TFT modeling endeavors rely on introducing mobility within the context of linear-regime TFT device operation. This is a mistake. The key advantage of the EKV modeling approach employed herein is that it accounts for all three regimes of TFT device operation (i.e., subthreshold, linear, and saturation), in such a way that the EKV mobility is well-defined and continuous across all three regimes of operation.

References

1 Lundstrom, M.S. (2000). *Fundamentals of Carrier Transport*, 2nde. Cambridge: Cambridge University Press.
2 Sze, S.M. and Ng, K.K. (2007). *Physics of Semiconductor Devices*, 3rde. Hoboken, NJ: Wiley.
3 Wager, J.F. (2017). Low-field transport in SiO_2. *Journal of Non-Crystalline Solids* 459: 111–115.
4 Hughes, R.C. (1975). Hole transport in MOS oxides. *IEEE Transactions on Nuclear Science* 22: 2227–2233.
5 Hughes, R.C. (1978). High field electronic properties of SiO_2. *Solid State Electronics* 21: 251–258.
6 Schneider, P.M. and Fowler, W.B. (1976). Band structure and optical properties of silicon dioxide. *Physical Review Letters* 36: 425–428.
7 Hughes, R.C. (1975). Hole mobility and transport in thin SiO_2 films. *Applied Physics Letters* 231: 436–438.
8 Tickle, A.C. (1969). *Thin-Film Transistors*. New York: Wiley.
9 Mott, N.F. and Davis, E.A. (1979). *Electronic Processes in Non-Crystalline Materials*, 2nde. Oxford: Clarendon Press.
10 Elliot, S.R. (1984). *Physics of Amorphous Materials*. London: Longman Group.
11 Street, R.A. (1991). *Hydrogenated Amorphous Silicon*. Cambridge: Cambridge University Press.
12 Stewart, K.A., Yeh, B.-S., and Wager, J.F. (2016). Amorphous semiconductor mobility limits. *Journal of Non-Crystalline Solids* 432B: 196–199.
13 Stewart, K.A. and Wager, J.F. (2016). Thin-film transistor mobility limits considerations. *Journal of the Society for Information Display* 24: 386–393.
14 Kamiya, T., Nomura, K., and Hosono, H. (2010). Present status of amorphous In-Ga-Zn-O thin-film transistors. *Science and Technology of Advanced Materials* 11: 044305.
15 Ioffe, A.F. and Regel, A.R. (1960). Non-crystalline, amorphous, and liquid electronic semiconductors. *Progress in Semiconductors* 4: 237–291.
16 Cohen, M.H. (1970). Review of the theory of amorphous semiconductors. *Journal of Non-Crystalline Solids (B)* 4: 391–409.
17 Balluffi, R.W., Allen, S.M., and Carter, W.C. (2005). *Kinetics of Materials*. New York: Wiley.
18 Hughes, R.C. (1973). Charge-carrier transport phenomena in amorphous SiO_2: direct measurement of the drift mobility and lifetime. *Physical Review Letters* 30: 1333–1336.
19 Brar, B., Wilk, G.D., and Seabaugh, A.C. (1996). Direct extraction of the electron tunneling effective mass in ultrathin SiO_2. *Applied Physics Letters* 69: 2728–2730.
20 Ferry, D.K. (1979). Electron transport and breakdown in SiO_2. *Journal of Applied Physics* 50: 1422–1427.
21 Fischetti, M.V., DiMaria, D.J., Brorson, S.D. et al. (1985). Theory of high-field transport in silicon dioxide. *Physical Review B* 31: 8124–8142.
22 Gritsenko, V.A., Ivanov, R.M., and Morkov, Y.N. (1995). Electronic structure of amorphous SiO_2: experiment and numerical simulation. *Journal of Experimental and Theoretical Physics* 81: 1208–1216.
23 Fung, T.-C., Chuang, C.-S., Chen, C. et al. (2009). Two-dimensional numerical simulation of radio frequency sputter amorphous In-Ga-Zn-O thin-film transistors. *Journal of Applied Physics* 106: 084511.
24 Wager, J.F. and Yeh, B. (2013). Oxide thin-film transistors: device physics. In: *Semiconductors and Semimetals*, vol. 88 (ed. B.G. Svensson, S.J. Pearton and C. Jagadish), 283–315. San Diego, CA: Academic Press.
25 Wager, J.F. (2017). Real- and reciprocal-space attributes of band tail states. *AIP Advances* 7: 125321.
26 Girifalco, L.A. (2000). *Statistical Mechanics of Solids*. Oxford: Oxford University Press.

27 Enz, C.C. and Vittoz, E.A. (2006). *Charge-Based MOS Transistor Modeling*. Chichester, UK: Wiley.

28 Tsividis, Y. and McAndrew, C. (2011). *Operation and Modeling of the MOS Transistor*, 3rde. Oxford: Oxford University Press.

29 Hoffman, R.L. (2004). ZnO-channel thin-film transistors: channel mobility. *Journal of Applied Physics* 95: 5813–5819.

30 Hong, D., Yerubandi, G., Chiang, H.Q. et al. (2008). Electrical modeling of thin-film transistors. *Critical Reviews in Solid State and Materials Science* 33: 101–132.

31 Schroder, D.K. (2006). *Semiconductor Material and Device Characterization*, 3rde. Hoboken, NJ: Wiley.

32 Sheng, J., Hong, T.H., Lee, H.-M. et al. (2019). Amorphous IGZO TFT with high mobility of $\sim 70\,cm^2V^{-1}s^{-1}$ via vertical dimension control using PEALD. *ACS Applied Materials & Interfaces* 43: 40300–40309.

33 Wager, J.F. (2010). Transfer-curve assessment of oxide thin-film transistors. *Journal of the Society for Information Display* 18 (2010): 749–752.

34 Vogt, K.T., Malmberg, C.E., Buchanan, J.C. et al. (2020). Ultrabroadband density of states of amorphous In-Ga-Zn-O. *Physical Review Research* 2 (2020): 033358.

6

Percolation Description of Charge Transport in Amorphous Oxide Semiconductors: Band Conduction Dominated by Disorder

A. V. Nenashev[1,2], *F. Gebhard*[3], *K. Meerholz*[4], *and S. D. Baranovskii*[3,4]

[1] *Institute of Semiconductor Physics, Novosibirsk, Russia*
[2] *Department of Physics, Novosibirsk State University, Novosibirsk, Russia*
[3] *Department of Physics and Material Sciences Center, Philipps-University, Marburg, Germany*
[4] *Department of Chemistry, University of Cologne, Cologne, Germany*

6.1 Introduction

While amorphous oxide semiconductors (AOSs) such as InGaZnO (IGZO) have been in the focus of intensive experimental and theoretical research for at least a couple of decades [1], the charge transport mechanism in such systems is still a matter of controversial debates. Interest in AOSs is mainly caused by applications of such materials in modern electronic devices, for instance, in thin-film transistors for transparent and flexible flat-panel displays [1, 2]. Although charge transport plays a decisive role for such applications, there is no agreement among researchers on the transport mechanism of charge carriers in IGZO materials, albeit the consensus that disorder potential should be taken into account by all means, because the magnitude of the carrier mobility $\mu \sim 10\ \text{cm}^2/\text{Vs}$ measured in IGZO systems is too low for the conventional band transport, which takes place in perfectly ordered crystals. For materials with strong effects of disorder, hopping transport is often discussed, which occurs by tunneling of charge carriers between localized states created by a disorder potential. However, hopping transport can be probably excluded for IGZO systems. First, the values of $\mu \sim 10\ \text{cm}^2/\text{Vs}$ are too high for hopping transport. The latter mechanism usually accounts for mobility values far below $\mu \sim 10^{-2}\ \text{cm}^2/\text{Vs}$, for instance, in organic semiconductors. In order to fit the high mobilities into the theory of hopping transport, one has to assume unrealistically large values (>4 nm) for the localization length of carriers in the localized states created by the disorder potential [3]. Second, the hopping transport mechanism is not compatible with the well-developed Hall effect observed in IGZO. The latter observation rather points at traditional band transport [3, 4].

Therefore, it looks reasonable to assume that the basic transport mechanism in IGZO materials is band transport decisively affected by a disorder potential. Remarkably, already in the pioneering works [1, 4–7], Hosono and collaborators proposed this charge transport mechanism as the most probable one for IGZO materials. Furthermore, they pointed at the necessity to take into account that charge transport in a system with strong disorder potential is essentially a percolation process. However, most theoretical studies so far neglected the percolation nature of the phenomenon. Referring to recent theoretical studies [8, 9], we describe in this chapter a theory for charge transport in IGZO materials based essentially on percolation arguments. In order to illustrate the drastic difference between the percolation approach on one hand, and various averaging procedures suggested in the literature on the other hand, we start in Section 6.2 with a random-barrier model (RBM), used by Kamiya et al. [4]. Being perhaps an oversimplified model, the RBM allows analytical solutions that help to explain basic theoretical concepts. In Section 6.3, we describe a more realistic scenario, the so-called random band-edge model (RBEM) suggested

Amorphous Oxide Semiconductors: IGZO and Related Materials for Display and Memory, First Edition.
Edited by Hideo Hosono and Hideya Kumomi.

for transport in IGZO by Fishchuk et al. [10]. In Section 6.4, a percolation description of charge transport in the RBEM is given. In Section 6.5, the results are compared with those given by the effective medium approximation (EMA). The latter approach is often used for description of charge transport in disordered materials. Percolation theory appears superior as compared to EMA. In Section 6.6, the results of the percolation theory in the framework of the RBEM are compared with experimental data. Conclusions are gathered in Section 6.7.

Although this chapter is dedicated to the description of charge transport in AOSs, particularly in IGZO compounds, the basic theoretical concepts addressed here should be applicable to the description of charge transport in a much broader class of materials. These are materials with charge carrier mobilities of the order $\mu \sim$1–10 $cm^2/(Vs)$, which are too low for perfectly ordered systems while too high for strongly disordered materials with hopping transport. Among organic materials, the highest mobility values are usually obtained for single crystals of rod- or disk-like molecules such as copperphtalocyanine [11] (1 $cm^2/(Vs)$), tetracene [12] (2.4 $cm^2/(Vs)$), pentacene, or rubrene [13, 14] (40 $cm^2/(Vs)$). Also in the essentially disordered organic systems, mobility values of the order $\mu \sim 1$ $cm^2/(Vs)$ have been reported. For instance, $\mu \sim 0.4$ $cm^2/(Vs)$ has been recently measured for electrons in the fungi-derived pigment xylindein [15]. With respect to inorganic systems, the widely studied and used amorphous silicon demonstrates mobility values [16] up to $\mu \sim 0.45$ $cm^2/(Vs)$.

Moreover, the theoretical description of charge transport suggested in this chapter can be applicable to such a broad class of disordered materials as semiconducting mixed crystals. In accord with the localization landscape theory [17–20], the potential relief caused by disorder potential, which affects the carrier transport in mixed crystals, can be perfectly mimicked by the RBEM as described in Section 6.3. Therefore, percolation theory presented in Section 6.4 is suitable to describe charge transport in such materials.

The content of this chapter is essentially based on recent publications [8, 9], where more details on the topic can be found.

6.2 Band Transport via Extended States in the Random-Barrier Model (RBM)

In this section, we highlight the drastic difference between the percolation approach to the description of charge transport in disordered systems and the approaches based on various averaging procedures. For this purpose we consider the RBM addressed already in the pioneering studies dedicated to charge transport in IGZO [1, 4–7]. In the RBM it is assumed that charge carriers can move above the band edge E_m, where their motion is affected by disorder in the form of random potential barriers with a Gaussian distribution of heights,

$$G_B(V) = \frac{1}{\delta_\phi \sqrt{2\pi}} \exp\left(-\frac{(V-\phi_0)^2}{2{\delta_\phi}^2}\right), \tag{6.1}$$

where ϕ_0 is the average height of the barriers, and δ_ϕ is the standard deviation in the distribution of the barrier heights. This model is sketched in Figure 6.1.

Charge transport in the RBM is often described in the framework of the Drude approach, which is based on the average relaxation time $\langle \tau \rangle$ for free carriers in the states above the band edge E_m. In this approximation, the carrier

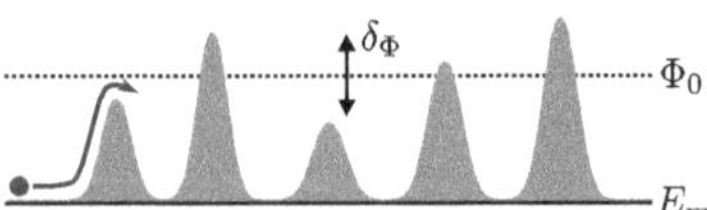

Figure 6.1 Random barrier model for band transport above the band edge E_m affected by random potential barriers. *Source:* Reproduced with permission from Ref. [8], Baranovskii, S. D., et al. (2019). Percolation description of charge transport in the random barrier model applied to amorphous oxide semiconductors. *EPL (Europhysics Letters)* **127** (5): 57004. Copyright 2019 by the Institute of Physics.

mobility is determined as $\mu = e\langle\tau\rangle/m$, where m is the effective mass. For the band transport in the absence of a disorder potential, the Drude approach leads to the expression [3, 4, 21, 22]

$$\mu = -\frac{e}{m \cdot n}\int_{E_m}^{\infty} \tau(E)v_z(E)\frac{\partial f_e(E)}{\partial v_z}D_m(E)\mathrm{d}E, \tag{6.2}$$

where n is the concentration of carriers, $\tau(E)$ is the momentum relaxation time at the electron energy E, $v_z(E)$ is the electron velocity along the transport direction (z axis), $D_m(E)$ is the density of states above the band edge E_m, and $f_e(E)$ is the Fermi function.

In the presence of substantial disorder in AOSs, the Drude approach is then modified heuristically by the introduction of the weight function [4]

$$\varrho(E) = \int_{E}^{\infty} G_B(\varepsilon)\mathrm{d}\varepsilon, \tag{6.3}$$

which was termed "transmission probability" [4]. The expression for the charge carrier mobility in AOSs thus attains the form

$$\mu = -\frac{e}{m \cdot n}\int_{E_m}^{\infty} \tau(E)v_z(E)\varrho(E)\frac{\partial f_e(E)}{\partial v_z}D_m(E)\mathrm{d}E. \tag{6.4}$$

The introduction of the weight function $\varrho(E)$ into Equation (6.4) is sometimes interpreted [4] as taking into account the percolation arguments suggested by Adler et al. [23]. However, percolation has little to do with Equations (6.3) and (6.4), as is readily seen from the fact that the percolation threshold does not appear in the above equations. Sometimes percolation is misinterpreted by mixing up this term with the averaging of hopping rates [5, 24].

Let us consider this issue in more detail. The rate of the carrier activation over the barrier with height V is equal to

$$\nu(V) = \nu_0 \exp\left(-\frac{eV}{kT}\right), \tag{6.5}$$

where e is the elementary charge, and ν_0 is the attempt-to-escape frequency. Averaging the activation rates given by Equation (6.5) over the distribution of barriers given by Equation (6.1) yields the average rate $\langle\nu\rangle$

$$\langle\nu\rangle = \nu_0 \exp\left[-\frac{e\phi_0}{kT} + \frac{(e\delta_\phi)^2}{2(kT)^2}\right], \tag{6.6}$$

which is dominated by barriers with the heights close to

$$V_\nu = \phi_0 - \frac{e\delta_\phi^2}{kT}. \tag{6.7}$$

Sometimes, the band mobility μ_0 in the presence of disorder is replaced [24] by

$$\mu_{\langle\nu\rangle} = \mu_0 \exp\left[-\frac{e\phi_0}{kT} + \frac{(e\delta_\phi)^2}{2(kT)^2}\right], \tag{6.8}$$

and the extra factor $\exp\left[-\frac{e\phi_0}{kT} + \frac{(e\delta_\phi)^2}{2(kT)^2}\right]$ is called [24] a "percolation term."

In Sections 6.2.1 and 6.2.2, the above results will be compared with those of percolation theory. In Section 6.2.1, the rate averaging will be analyzed using transparent electrotechnical analogies. In Section 6.2.2, the recipe for the description of charge transport via a system of random barriers based on percolation theory will be formulated. The description based on percolation theory contrasts that based on the averaging of transition rates.

6.2.1 Deficiencies of the Rate-Averaging Approach: Electrotechnical Analogy

The irrelevance of the average rate $\langle\nu\rangle$ to characterize charge transport in disordered systems with a broad distribution of microscopic transition rates, as given by Equation (6.5), had been recognized already in the early

1970s, when the theoretical approach to the description of charge transport based on percolation theory was developed [25–27]. This issue became a matter of monographs [28], edited books [29], and topical reviews [30, 31]. Nevertheless, this particularly approach is often used for charge transport in AOSs. Therefore, it is instructive to analyze this approach once again. The deficiencies of the rate averaging in the form of Equations (6.6) and (6.8) are mostly transparent for one-dimensional (1D) transport. In 1D, carriers move via thermal activation over the barriers, being forced to climb over the barrier tops without an option to avoid the highest barriers. On the contrary, in three dimensions (3D), charge carriers can avoid the activation to the tops of the barriers percolating through valleys in the potential relief. The percolation nature of charge transport in the RBM neglected in Equations (6.6) and (6.8) and the corresponding theory will be discussed in Section 6.2.2. Before addressing the percolation approach, let us analyze in more detail the rate averaging, which yields Equations (6.6) and (6.8), and show that this method can hardly be considered as reliable even in the case of 1D transport, when percolation is not possible.

In the RBM illustrated in Figure 6.1, it is assumed that charge carriers can rapidly move through valleys at the mobility edge E_m between the barriers and that this fast movement is interrupted by slow activation of carriers over the barriers. In such an incoherent process, charge transport can be described using a transparent electrotechnical analogy [28]. Each potential barrier can be viewed as a resistance, whose magnitude $R(V)$ is proportional to the time $\tau(V) = \nu^{-1}(V)$ necessary to overcome the potential barrier. Using Equation (6.5), one obtains

$$R(V) = R_0 \exp\left(\frac{eV}{kT}\right), \tag{6.9}$$

where the factor R_0 does not depend on the barrier height V. The scheme of the RBM illustrated in Figure 6.1 can be then considered as a scheme of resistances connected in series.

The activation rates $\nu(V)$ given by Equation (6.5) appear in this picture analogous to the conductances, that is, to the inverse resistances $R^{-1}(V)$. Averaging the activation rates is then analogous to the calculation of the resistivity of a series of resistances by averaging the conductances. This is surely not correct, particularly in the case when the resistances of single participants in a series have an exponentially broad distribution of magnitudes. This broad distribution is governed by the exponential dependence of the resistances on the barrier heights given by Equation (6.9). In 1D, when percolation is not possible, the resistivity of a series of resistances is determined by the average resistance, that is, by the average time for an activation over the barriers

$$\langle \tau \rangle = \nu_0^{-1} \exp\left[\frac{e\phi_0}{kT} + \frac{(e\delta_\phi)^2}{2(kT)^2}\right]. \tag{6.10}$$

The average time $\langle \tau \rangle$ is dominated by barriers with the heights close to

$$V_\nu = \phi_0 + \frac{e\delta_\psi^2}{kT}. \tag{6.11}$$

The mobility of charge carriers $\mu_{\langle \tau \rangle}$ is then proportional to the inverse of the average activation time $\langle \tau(V) \rangle^{-1}$,

$$\mu_{\langle \tau \rangle} = \mu_0 \exp\left[-\frac{e\phi_0}{kT} - \frac{(e\delta_\phi)^2}{2(kT)^2}\right]. \tag{6.12}$$

It is worth emphasizing once again that the rate averaging, which yields Equations (6.6)–(6.8), is equivalent to the calculation of the resistivity of a series of resistances by averaging the inverse resistances, that is, the conductances. Due to the exponentially broad distribution of conductances described by the exponential distribution of rates given by Equation (6.5), the average value is dominated by the very large conductances, that is, by exponentially small resistances. It is apparent that such small resistances cannot be responsible for the resistivity of a system of resistances connected in series. Ascribing the dominant role to such small resistances leads to overestimating the mobility by the factor $\exp\left[\frac{(e\delta_\phi)^2}{(kT)^2}\right]$. This is apparent from a comparison between Equation (6.8) obtained by the rate averaging and the correct result given by Equation (6.12) obtained by the averaging of times for carrier activation over the potential barriers. The error caused by the rate averaging is decisive when $e\delta_\phi \gg kT$.

Another comment is necessary with respect to the averaging procedure that leads to Equations (6.6)–(6.8). This approach was borrowed [5] from the study of the barrier inhomogeneities at Schottky contacts by Werner and Güttler [32]. In the latter case, the potential barriers are acting parallel to each other along the area of a Schottky contact. The electrotechnical analogy is a system of resistances connected parallel to each other, where each barrier can be represented by an effective resistance determined by Equation (6.9). In order to calculate the resistivity of such a system, it is appropriate to average the inverse resistances, that is, conductances represented by rates of carrier activation over the barriers. Therefore, the average activation rate given by Equation (6.6) is responsible for carrier injection through inhomogeneous Schottky contacts [32]. This average rate has, however, nothing to do with charge transport through a series of potential barriers illustrated in Figure 6.1. For a series of barriers, not the average activation rate, but rather the average activation time given by Equation (6.10), is the characteristic quantity responsible for charge transport. Hence, the carrier mobility is to be described by Equation (6.12) and not by Equation (6.8).

So far, we considered a one-dimensional model, in which charge carriers move via activation over the barrier tops. In Section 6.2.2, we will consider a 3D case, in which carriers can move avoiding the tops of the highest barriers. Percolation arguments will play a decisive role in the description of charge transport in such a case.

6.2.2 Percolation Approach to Charge Transport in the RBM

In three dimensions, charge carriers are not obliged to climb the tops of the highest energy barriers. They are able to avoid such high barriers by percolating aside the tops. The essence of the percolation approach is the idea that transport is determined not by the average rates, or by the average times, but rather by the rates and times of those transitions that are most difficult among the ones still relevant for long-range transport. Conduction in disordered systems is in fact a percolation process, in which the slowest transitions still needed to provide a connected path through the system determine the charge transport.

In order to formulate a percolation criterion for charge transport in 3D via a system of random barriers, as applied to AOSs, let us modify slightly the model illustrated in Figure 6.1. Let the volume of the material be occupied by cubic cells of two distinct types. One type of cells, called "valleys," does not contain potential barriers for charge carriers. Such cells provide a uniform energy level equal to the position of the mobility edge E_m in the absence of barriers. The other type of cells, called "barriers," provides potential barriers with the distribution of heights V described by Equation (6.1). Let the volume fraction of valleys be ξ and that of barriers $1 - \xi$. Let us assume for simplicity that the volume of a single valley is equal to the volume of a single barrier and that valleys and barriers occupy cells on a 3D lattice grid. Then the percolation problem in the RBM can be mapped onto the site percolation problem on the corresponding lattice grid [28]. In such a problem, the current can flow via sites arranged on a regular lattice grid. Some sites are blocked and they prevent transport to neighboring sites, while other sites are unblocked and they transfer current to the neighboring sites. Let the fraction of unblocked sites be x. The site percolation problem provides a solution x_c for the minimal fraction of unblocked sites, which allow the current flow through the system over macroscopic distances. The value x_c, called the "percolation threshold," depends on the particular structure of the lattice grid.

According to the percolation approach to charge transport in the RBM, one has to find the "percolation level," that is, the minimal height of the potential barriers V_c, which the carriers still have to overcome in order to enable transport over macroscopic distances. This level is related to x_c as follows:

$$x_c = \xi + (1 - \xi) \int_{-\infty}^{V_c} G_B(V)dV. \tag{6.13}$$

Then the carrier mobility μ_{perc} is determined as

$$\mu_{perc} = \mu_0 \exp\left[-\frac{eV_c}{kT}\right]. \tag{6.14}$$

In Equation (6.14), it is assumed that the prefactor μ_0 for charge transport at the percolation level V_c is equal to the band mobility at the level E_m.

It is convenient for calculations to introduce a variable $\widetilde{V}_c \equiv V_c - \phi_0$, which describes the position of the percolation level V_c relative to the middle of the barrier distribution ϕ_0. For the Gaussian distribution of barrier heights given by Equation (6.1), the percolation level $\widetilde{V}_c$ is related to the percolation threshold of the site percolation problem x_c via the equation

$$x_c = \xi + (1 - \xi) \int_{-\infty}^{\widetilde{V}_c/\delta_\phi} \frac{1}{\sqrt{2\pi}} \exp\left(-\frac{t^2}{2}\right) dt. \tag{6.15}$$

Equation (6.15) yields the percolation level $\widetilde{V}_c$ as a function of the volume fraction of valleys ξ. If this fraction ξ is larger than the percolation threshold x_c, no thermal activation is necessary for charge transport, and carriers can move to macroscopic distances via spatially connected valleys at the energy level E_m. If $\xi < x_c$, percolation via valleys is interrupted by barriers, and thermal activation to the percolation level $V_c = \widetilde{V}_c + \phi_0$, where $\widetilde{V}_c$ is determined by Equation (6.15), becomes necessary. This yields the carrier mobility

$$\mu_{perc} = \mu_0 \exp\left[-\frac{e\phi_0}{kT} - \frac{e\widetilde{V}_c}{kT}\right]. \tag{6.16}$$

In order to assess the importance of the percolation arguments, one should compare the results for carrier mobility given by Equation (6.16) with those given by Equation (6.8). The value of the percolation threshold x_c is necessary for calculations of μ_{perc} via Equations (6.15) and (6.16). Let us assume for simplicity that valleys and barriers form a simple cubic lattice grid. For such a case, one should use in Equation (6.15) the value [33–36] $x_c \simeq 0.312$. At $\xi \geq 0.312$, barriers can be completely avoided by current flow at the level E_m. In order to bring barriers into play, ξ should be smaller than x_c. The values $\xi = 0.1$ and $\xi = 0.2$ will be used in the calculations. Since both Equations (6.16) and (6.8) contain the factor $\tilde{\mu} \equiv \mu_0 \exp[-e\phi_0/(kT)]$, it would be convenient to compare the ratios $\mu_{perc}/\tilde{\mu}$ and $\mu_{\langle v\rangle}/\tilde{\mu}$. In Figure 6.2, the results of the percolation theory $\mu_{perc}/\tilde{\mu}$ expressed by Equations (6.15) and (6.16) are plotted as functions of $e\delta_\phi/kT$ along with the result of the rate averaging [5, 24] $\mu_{\langle v\rangle}/\tilde{\mu}$ expressed by Equation (6.8). At low temperatures, as compared to the width of the barrier distribution, $kT \ll e\delta_\phi$, the results of the percolation theory (depicted in Figure 6.2 by the solid line for $\xi = 0.1$ and by the dotted line for $\xi = 0.2$) differ by many orders of magnitude from the results given by the rate averaging depicted in Figure 6.2 by the dashed

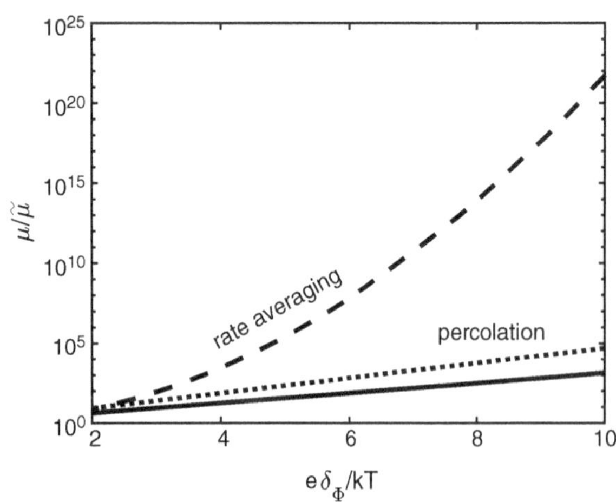

Figure 6.2 Results of the percolation theory (solid line for $\xi = 0.1$ and dotted line for $\xi = 0.2$) compared to the results of the rate averaging (dashed line). Plotted are μ_{perc} given by Equations (6.15) and (6.16) and $\mu_{\langle v\rangle}$ given by Equation (6.8) normalized by $\tilde{\mu} = \mu_0 \exp[-e\phi_0/(kT)]$. *Source:* Reproduced with permission from Ref. [8], Baranovskii, S. D., et al. (2019). Percolation description of charge transport in the random barrier model applied to amorphous oxide semiconductors. *EPL* (*Europhysics Letters*) **127** (5): 57004. Copyright 2019 by the Institute of Physics.

line. At high temperatures, when kT is comparable with the width $e\delta_\phi$ of the barrier distribution, the distribution of the barrier heights does not play any role, and the carrier mobility is equal to the value

$$\mu_{perc} = \mu_0 \exp\left[-\frac{e\phi_0}{kT}\right], \tag{6.17}$$

determined solely by the average barrier height $e\phi_0$. In such a case, taking the distribution of barrier heights into account is not necessary.

The drastic difference between the results of the percolation theory expressed by Equations (6.15) and (6.16) as compared to those of the rate averaging expressed by Equation (6.8) proves the importance of the percolation treatment of charge transport in the framework of the RBM that is widely used for electrical conduction in AOSs. However, the RBM itself is probably not the best model for charge transport in AOSs. If disorder creates potential barriers above the band edge, as sketched in Figure 6.1, it will also create potential wells below the band edge. The statistical distribution of these wells must be taken into account as well, which makes E_m a regional, random quantity. A model that takes this effect into account has been suggested by Adler et al. [23] and recently exploited by Fishchuk et al. [10]. We will call this model a random band-edge model. The rest of our chapter is dedicated to the description of charge transport in the framework of the RBEM.

6.3 Random Band-Edge Model (RBEM) for Charge Transport in AOSs

The RBEM assumes that the position of the band edge E_m varies in space due to disorder potential. The spatial fluctuations of E_m are assumed to be Gaussian with the distribution function [10]

$$G(E_m) = \frac{1}{\delta\sqrt{2\pi}} \exp\left[-\frac{1}{2}\left(\frac{E_m}{\delta}\right)^2\right], \tag{6.18}$$

where δ is the standard deviation, and the position of the band edge E_m is counted from the position of the band edge without disorder potential. The model is illustrated in Figure 6.3, where the local position of the band edge E_m corresponds to the upper border between the gray and white areas.

Fishchuk et al. [10] applied the EMA to study charge transport theoretically. In contrast, we use percolation arguments to develop a theoretical description of charge transport in the random band-edge model. At $kT \ll \delta$, where kT is the thermal energy and δ is the scale of disorder in Equation (6.18), the results of the percolation theory

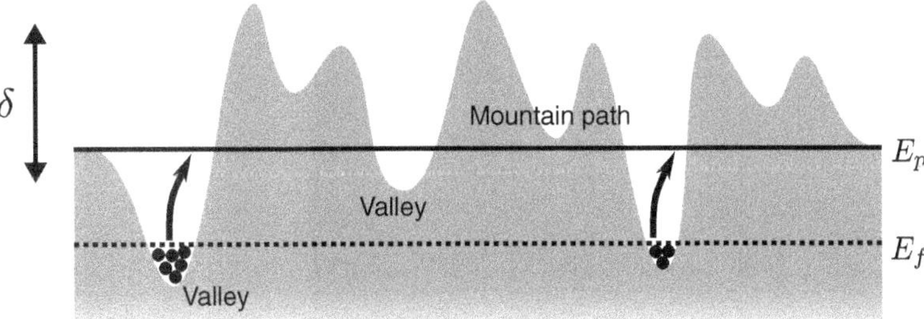

Figure 6.3 Schematic representation of the spatial fluctuations of the band edge E_m in the random band-edge model. The local position of E_m corresponds to the upper edge between the gray and the white areas. The carrier motion is due to activation from the Fermi level E_f toward the percolation level E_p. *Source:* Reproduced with permission from Ref. [9], Nenashev, A. V., et al. (2019). Percolation description of charge transport in amorphous oxide semiconductors. *Physical Review B* **100**(12). Copyright 2019 by the American Physical Society.

are reliable and they substantially differ from those of the EMA, as shown in Section 6.5. Therefore, the percolation theory seems superior to the EMA for the description of charge transport in the framework of the RBEM.

In Section 6.4, we will show how to calculate the carrier mobility using percolation theory. In Section 6.5, we will compare the percolation approach with the EMA. In Section 6.6, the ability of the theory to account for experimental data in IGZO materials will be addressed. It will be shown that percolation theory is capable of accounting for the dependencies of the charge carrier mobility on temperature and on the concentration of charge carriers in IGZO materials. Moreover, a comparison between the results of percolation theory with experimental data reveals the characteristic parameter δ of the band-edge disorder in Equation (6.18) and the conduction-electron mobility μ_0.

Following Fishchuk et al. [10], we also include localized states with energies below E_m, whose density of states is assumed to be exponential:

$$g(E) = N_m \exp\left(\frac{E}{E_0}\right), \tag{6.19}$$

where E is the energy of the trap counted from the band edge E_m, E_0 is the energy scale, and N_m is the density of localized states at the band edge E_m. The value of E_0 depends on the material, and it is usually estimated in inorganic amorphous semiconductors to lie below 0.1 eV [29, 37–40].

Again following Fishchuk et al. [10], we assume that the delocalized states with energies above E_m are characterized by the density of states

$$g(E - E_m) = g_c\sqrt{E - E_m + \Delta E}, \tag{6.20}$$

where the value $g_c = 1.4 \cdot 10^{21}\ \text{cm}^{-3}\text{eV}^{-3/2}$ has been reported for a-IGZO thin films [4]. Equations (6.20) and (6.19) can be combined to the *regional* density of states

$$\begin{aligned} g(E - E_m) = {} & \Theta(E - E_m)\, g_c \sqrt{E - E_m + (N_m/g_c)^2} \\ & + \left[1 - \Theta(E - E_m)\right]\, N_m \exp\left(\frac{E - E_m}{E_0}\right), \end{aligned} \tag{6.21}$$

where $\Theta(x)$ is the Heaviside step function. The first term on the right-hand side describes the density of delocalized states above E_m, and the second term describes the density of localized states below E_m. The energy shift ΔE in Equation (6.20) guarantees the continuity of the density of states at $E = E_m$ when we choose $\Delta E = (N_m/g_c)^2$. The admixture of localized states will not play a significant role further on.

For a given value of E_m, one can find a corresponding *regional* electron density,

$$n_{\text{region}}(E_m) = \int_{-\infty}^{+\infty} g(E - E_m)\, f(E)\, dE, \tag{6.22}$$

where $f(E)$ is the Fermi function,

$$f(E) = \left[\exp\left(\frac{E - E_f}{kT}\right) + 1\right]^{-1}, \tag{6.23}$$

and E_f is the Fermi level. The *total* electron density n, averaged over the regional positions of the mobility edge E_m, is

$$n = \int_{-\infty}^{+\infty} G(E_m)\, n_{\text{region}}(E_m)\, dE_m. \tag{6.24}$$

For given temperature T and electron concentration n, the position of the Fermi level E_f can be found from the system of Equations (6.21)–(6.24).

The AOS material is assumed to be a medium with a smoothly varying regional conductivity σ_{region}, which is a product of the elementary charge e, the conduction-band electron mobility μ_0, and the local concentration of mobile electrons (with energies above E_m),

$$\sigma_{\text{region}}(E_m) = e\mu_0 \int_{E_m}^{+\infty} g(E - E_m)\, f(E)\, \mathrm{d}E. \tag{6.25}$$

This assumption is justified when the electron mean-free path is small compared to the spatial scale for variations of E_m. The coefficient μ_0 is the intrinsic (band) charge-carrier mobility in extended states determined by the electron effective mass and the average scattering time. It may depend on temperature and also on the electron concentration in the case of a degenerate system. However, one cannot study the dependence of μ_0 on T and n without a deep understanding of the electron-scattering mechanisms in the given material. Therefore, μ_0 is considered as a constant for the sake of simplicity. Here, we will focus on the exponential factor in the temperature dependence of the drift mobility leaving the pre-exponential factor containing μ_0 as a fitting parameter.

The global (macroscopic) conductivity σ is to be found by some "averaging" of the regional values $\sigma_{\text{region}}(E_m)$, taking into account the Gaussian distribution $G(E_m)$ of the mobility edge E_m, that is, $\sigma = \langle \sigma_{\text{region}} \rangle$. When the global σ is found, one can calculate the (measured) mobility μ as

$$\mu = \frac{\sigma}{en}. \tag{6.26}$$

In the case of an exponentially broad scatter of regional conductivities, a proper choice of the averaging procedure is crucial for a correct determination of the global conductivity σ. We will consider three methods of "averaging": The first one is based on the EMA, expressed by Equation (6.42) (see Section 6.5); the other two procedures are based on percolation theory expressed by Equations (6.30) and (6.36) (see Section 6.4).

For the experimentally accessible regions of the (n, T) phase diagram of IGZO materials, Fishchuk et al. [10] have shown that the conductivity $\sigma(n, T)$ and the mobility $\mu(n, T)$ depend on the carrier-concentration n and the temperature T mostly through the variations of the conduction-band edge E_m, and are not limited by the localized states. Therefore, localized states might be disregarded in IGZO films.

6.4 Percolation Theory for Charge Transport in the RBEM

6.4.1 From Regional to Global Conductivities in Continuum Percolation Theory

The random band-edge model belongs to the class of "continuum percolation problems" [28]. The transport is determined by charge carriers with energies above the percolation level E_p, which is defined as the minimal energy that allows a transport path via connected regions with E_m not exceeding E_p.

Let $p(E)$ denote the volume fraction of regions where the mobility edge E_m is below E,

$$p(E) = \int_{-\infty}^{E} G(E_m)\, \mathrm{d}E_m. \tag{6.27}$$

The quantity

$$\vartheta_c = p(E_p) \tag{6.28}$$

plays the role of the dimensionless percolation threshold determined as the minimal volume fraction of the conducting material that enables electrical connection throughout the infinitely large sample. Numerical studies for Gaussian energy distributions with various spatial correlation properties yield the value $\vartheta_c = 0.17 \pm 0.01$ for the three-dimensional continuum percolation problem [28]. Using the value $\vartheta_c = 0.17$ in Equation (6.28), one obtains the value of the percolation level E_p,

$$E_p = -0.95\,\delta. \tag{6.29}$$

It remains to determine the macroscopic conductivity σ from the regional conductivities $\sigma_{\text{region}}(E_m)$.

According to the percolation approach, only electrons with energies above the threshold E_p determine the macroscopic conductivity. An electron cannot travel through the whole sample without "climbing" to the regions where $E_m > E_p$, since regions with $E_m < E_p$ do not provide a connected path from one side of the sample to another one. Therefore, each conducting path goes not only through places with $E_m < E_p$ ("valleys"), but also through short regions with $E_m > E_p$ ("mountain passes"). Only the latter regions are important for overall conduction, because they are the most resistive parts of the conductive path and they are connected in series with less resistive "valleys." The higher the band edge E_m, the lower the local electron concentration, and consequently the larger the local resistivity. For this reason, when calculating the conductivity or mobility, only places with band edge E_m above the threshold E_p are decisive for charge transport. These places possess the highest resistances among those that belong to the percolation path. Regions with $E_m < E_p$ can also belong to the percolation path, but they represent shortcuts with small resistances, and, therefore, such regions do not contribute essentially to the global resistance of the transport path.

The simplest recipe to calculate σ on the basis of percolation theory is to average the regional conductivities over the regions where $E_m > E_p$,

$$\sigma = \frac{1}{1 - \vartheta_c} \int_{E_p}^{+\infty} \sigma_{\text{region}}(E_m)\, G(E_m)\, \mathrm{d}E_m, \tag{6.30}$$

where $G(E_m)$ is the Gaussian distribution of the local mobility edges E_m. The mobility $\mu = \sigma/en$ that corresponds to Equation (6.30) can be expressed as

$$\mu = \mu_0 \frac{n_{\text{mob}}}{n}, \tag{6.31}$$

where n_{mob} is the average concentration of *mobile* electrons in the regions with $E_m > E_p$, and n is the total electron concentration. It is easy to recognize that Equation (6.30) gives the correct value of the conductivity $\sigma = e\mu_0 n$ in the absence of disorder, $\delta = 0$.

Equation (6.30) also gives the correct value in the opposite limit of very pronounced disorder, $kT \ll \delta$, for a non-degenerate occupation of states above E_p, when the regional conductivities $\sigma_{\text{region}}(E_m)$ have an exponentially broad distribution of values. In this case, the Fermi function can be approximated as $f(E) = \exp\left[(E_f - E)/kT\right]$, and Equation (6.25) yields the exponential dependence

$$\sigma_{\text{region}}(E) = e\mu_0 N_c \exp\left(\frac{E_f - E}{kT}\right), \tag{6.32}$$

where N_c is the effective density of states in the conduction band,

$$N_c = \int_0^{+\infty} g(E) \exp(-E/kT)\, \mathrm{d}E. \tag{6.33}$$

Inserting Equation (6.32) into Equation (6.30) gives the asymptotic expression for the carrier mobility,

$$\mu = \frac{\sigma}{en} = \frac{\mu_0 N_c}{n(1 - \vartheta_c)} \int_{E_p}^{+\infty} \exp\left(\frac{E_f - E}{kT}\right) G(E)\, \mathrm{d}E. \tag{6.34}$$

At low temperatures, $kT \ll \delta$, the main contribution to the integral comes from the vicinity of the percolation level E_p. Therefore, one can approximately replace $G(E)$ by $G(E_p)$ and bring the constant factor $G(E_p)$ out of the integral. The remaining integral is easy to calculate, and the carrier mobility given by Equation (6.31) assumes its asymptotic form,

$$\mu \approx \mu_0 \frac{N_c}{n} \frac{G(E_p)\, kT}{1 - \vartheta_c} \exp\left(\frac{E_f - E_p}{kT}\right). \tag{6.35}$$

The exponential term in Equation (6.35) shows that the charge transport is dominated by thermal activation of electrons to the percolation level E_p, as schematically depicted in Figure 6.3. More sophisticated considerations

[28, 41] lead to a marginal correction of the pre-exponential factor in this equation. In fact, the pre-exponential factor should contain $(kT)^{\nu}$ instead of kT, where $\nu \simeq 0.88$ is the critical exponent for the correlation length of the percolation cluster [35, 42, 43]. In Section 6.4.2, we will use Equation (6.35) and ignore this marginal correction.

6.4.2 Averaging Procedure by Adler et al.

In several theoretical studies of charge transport in AOSs, the percolation approach suggested by Adler et al. [23] has been invoked [4, 44]. Adler et al. [23] considered a system with a Gaussian distribution of the regional band edges as given by Equation (6.18). They suggested that the global conductivity σ can be obtained as

$$\sigma_{\mathrm{A}} = \frac{1}{kT} \int_{E_p}^{+\infty} \sigma(E) f(E) \left[1 - f(E)\right] \, \mathrm{d}E. \tag{6.36}$$

Here, $\sigma(E)$ is the contribution to the conductivity of carriers with energy E,

$$\sigma(E) = \tilde{B}\left[p(E) - \vartheta_c\right]^2, \tag{6.37}$$

where $\tilde{B}$ is some unspecified constant.

In order to compare with our approach in Section 6.4.1, let us analyze the conductivity σ_{A} given by Equation (6.36) when the Fermi level is far below the percolation level, $E_f < E_p$ and $kT \ll E_p - E_f$. In such a case, one can use the Boltzmann approximation $f(E) \approx \exp\left[(E_f - E)/kT\right]$ and $1 - f(E) \approx 1$ for the Fermi functions in Equations (6.36) and (6.37) in the non-degenerate case, which leads to

$$\sigma_{\mathrm{A}} = \frac{\tilde{B}}{kT} \int_{E_p}^{+\infty} \exp\left(\frac{E_f - E}{kT}\right) \left[p(E) - \vartheta_c\right]^2 \, \mathrm{d}E. \tag{6.38}$$

At low temperatures, $kT \ll \delta$, the major contribution to this integral comes from the region $E \approx E_p$, providing $G(E) \approx G(E_p)$. Using Equations (6.27) and (6.28), the factor $\left[p(E) - \vartheta_c\right]$ can be simplified to

$$p(E) - \vartheta_c = \int_{E_p}^{E} G(E') \, \mathrm{d}E' \approx (E - E_p) G(E_p). \tag{6.39}$$

Concomitantly, Equation (6.38) simplifies to

$$\sigma_{\mathrm{A}} \approx \frac{\tilde{B}}{kT} \left[G(E_p)\right]^2 \int_{E_p}^{+\infty} \exp\left(\frac{E_f - E}{kT}\right) (E - E_p)^2 \, \mathrm{d}E. \tag{6.40}$$

The integral can be calculated easily. Using its value in Equation (6.40), we obtain the asymptotic expression for the mobility,

$$\mu_{\mathrm{A}} \approx \frac{2\tilde{B}[G(E_p)]^2 (kT)^2}{en} \exp\left(\frac{E_f - E_p}{kT}\right). \tag{6.41}$$

A comparison of this expression with the result of the percolation theory in Equation (6.35) shows that the exponential term is correctly reproduced by Equation (6.41). However, besides the unknown coefficient $\tilde{B}$, Equation (6.41) displays an incorrect temperature dependence of the pre-exponential factor due to the assumption of a quadratic energy dependence of the regional conductivity above the threshold; see Equation (6.37).

The (global) band-edge E_m in the random-barrier model, as described in Section 6.2, and the percolation threshold E_p are unrelated conceptually. However, one could raise the question of whether it is possible to interpret the energy level E_m in the model sketched in Figure 6.1 as the percolation level E_p in Figure 6.3, so that the RBM could be viewed as a part of the RBEM for energies above the percolation level, $E > E_p$. Unfortunately, this is not the case. First, the RBM in Figure 6.1 cannot contain a recipe on how to calculate the percolation level E_p. Second, since E_m is a regional feature determined by the distribution, as shown in Equation (6.18), one cannot consider the value E in Equations (6.2) and (6.4) as if E_m were uniform for the whole system. Therefore, an approach based on replacing E_m in Figure 6.1 with the percolation level E_p would not make sense.

In Section 6.6, we will compare the predictions of the percolation theory expressed by Equation (6.30) with experimental data obtained by several experimental groups for the dependences of the carrier mobility $\mu(n, T)$ on the carrier concentration n and on temperature T in IGZO materials [3–5, 10]. In Section 6.5, we stay with the theory and compare the percolation approach that was described in this section with the EMA used by Fishchuk et al. [10].

6.5 Comparison between Percolation Theory and EMA

The EMA and percolation theory are often considered as complementary to each other in their ability to account for charge transport in disordered systems. Percolation theory is considered to be valid for strongly disordered systems [28], while in systems with a weak disorder the EMA is often applied [10]. In fact, percolation theory gives reliable results not only for the case of strong disorder, $kT \ll \delta$, but also for the opposite case of $\delta \to 0$, as discussed in Section 6.4 in the context of Equation (6.30). Therefore, it is instructive to estimate the difference between the results of percolation theory and those of the EMA in the case of strong disorder.

In the EMA used by Fishchuk et al. [10], the conductivity σ was determined from its regional values $\sigma_{\text{region}}(E_m)$ via the equation

$$\left\langle \frac{\sigma_{\text{region}} - \sigma}{\sigma_{\text{region}} + (d-1)\sigma} \right\rangle = 0, \tag{6.42}$$

where d is the spatial dimension, and the angular brackets mean the averaging over the density distribution function $G(E_m)$,

$$\langle \mathcal{A} \rangle \equiv \int_{-\infty}^{+\infty} G(E_m)\mathcal{A}(E_m)\, \mathrm{d}E_m. \tag{6.43}$$

In order to calculate the carrier mobility $\mu(n, T)$ in the framework of the EMA, one should calculate the Fermi level E_f from Equations (6.21)–(6.24), and then determine the dependence of the regional conductivity $\sigma_{\text{region}}(E_m)$ on the regional mobility edge E_m, which then leads to the global conductivity σ via Equation (6.42) and to the carrier mobility μ via Equation (6.26).

Let us rewrite the averaging condition shown in Equation (6.42) in the following equivalent form in three dimensions, $d = 3$:

$$\left\langle \frac{\sigma_{\text{region}}}{\sigma_{\text{region}} + 2\sigma} \right\rangle = \frac{1}{3}. \tag{6.44}$$

In the limit of low temperature and low electron concentration, the dependence $\sigma_{\text{region}}(E_m)$ of the regional conductivity on the regional conduction band edge E_m is very steep, that is, for almost all values of E_m we have either $\sigma_{\text{region}}(E_m) \gg 2\sigma$ or $\sigma_{\text{region}}(E_m) \ll 2\sigma$. In the former case, the expression inside the angular brackets in Equation (6.44) is close to unity; in the latter case, it is close to zero. Therefore,

$$\frac{\sigma_{\text{region}}(E_m)}{\sigma_{\text{region}}(E_m) + 2\sigma} \approx \begin{cases} 1 & \text{if } E_m < E^*, \\ 0 & \text{if } E_m > E^*, \end{cases} \tag{6.45}$$

where E^* is the value of E_m that separates these two limits,

$$\sigma_{\text{region}}(E^*) = 2\sigma. \tag{6.46}$$

Inserting Equation (6.45) into Equation (6.44), and using the rule of averaging shown in Equation (6.43), one can evaluate Equation (6.44) as

$$\int_{-\infty}^{E^*} G(E_m)\, \mathrm{d}E_m = \frac{1}{3}. \tag{6.47}$$

This equation defines the energy E^*.

In the case of Gaussian distribution function $G(E_m)$, as shown in Equation (6.18), the solution is

$$E^* \approx -0.43\ \delta. \tag{6.48}$$

With this value for E^*, one can find the macroscopic conductivity σ from Equation (6.46), where $\sigma_{\text{region}}(E^*)$ is to be calculated from Equation (6.25). Inserting Equation (6.32) into Equation (6.46) provides the following asymptotic expression for the mobility $\mu = \sigma/en$:

$$\mu \approx \mu_0 \frac{N_c}{2n} \exp\left(\frac{E_f - E^*}{kT}\right). \tag{6.49}$$

The expression in Equation (6.49) is valid when $E_f < E^*$, $kT \ll E^* - E_f$, and $kT \ll \delta$.

According to Equation (6.49), one can interpret the transport in the low-temperature and low-concentration case as thermal activation of electrons to the energy level E^*. This result is to be compared with that of the percolation theory given by Equation (6.35). Even if we ignore the differences in the pre-exponential factors, we can conclude that the results given by Equations (6.49) and (6.35) differ by an exponential factor $\propto \exp(-0.52\ \delta/kT)$, which is essential for strong disorder, $\delta \gg kT$.

However, this result is specific to the considered case, and it does not imply that the EMA always leads to an exponentially large error in the case of strong disorder. As has been shown in several studies [45, 46], one can achieve a better description of the conductivity within the EMA by replacing the spatial dimension $d = 3$ with the inverse of the percolation threshold $1/\vartheta_c \approx 6$. However, conceptual improvements of the EMA are beyond the scope of our presentation. In Section 6.6, we turn to the comparison between the results of the percolation theory described in Section 6.4 and some experimental data.

6.6 Comparison with Experimental Data

In this section, we show that the percolation approach developed in Section 6.4 and applied to the random band-edge model presented in Section 6.3 is able to reproduce experimental data on charge transport in InGaZnO materials. The main theoretical result to be compared with experimental data is Equation (6.30) for the conductivity $\sigma(n, T)$, as discussed in Section 6.4. The regional conductivity $\sigma_{\text{region}}(E_m)$ in this equation is given by Equation (6.25), where the regional density of states $g(E - E_m)$ is taken in the form of Equation (6.21) with $g_c = 10^{21}\ \text{cm}^{-3}(\text{eV})^{-3/2}$. Following Ref. [10], we take into account the distribution $G(E_m)$ of the regional positions of the band edge E_m in the form of Equation (6.18), and neglect for simplicity the presence of localized states with energies below E_m by setting $N_m = 0$. We address experimental data for the temperature dependencies of conductivity $\sigma(T)$ and mobility $\mu(T)$ at different concentrations of charge carriers n. The carrier concentration is changed experimentally, either by varying the doping level [4] or by varying the gate voltage in the field-effect transistors [3, 10, 47].

In their pioneering works [4, 5], Kamiya et al. investigated two series of n-type IGZO films: crystalline (c-IGZO) and amorphous (a-IGZO) films. Samples of c-IGZO are crystalline materials, but they contain inherent disorder due to the statistical distribution of Ga and Zn ions. Therefore, such materials are to be considered as disordered materials with respect to charge transport [4]. In each series of samples, the conductivity $\sigma(T)$ was measured, varying the carrier concentrations n between $n < 10^{16}\ \text{cm}^{-3}$ and $n \sim 10^{20}\ \text{cm}^{-3}$ [4].

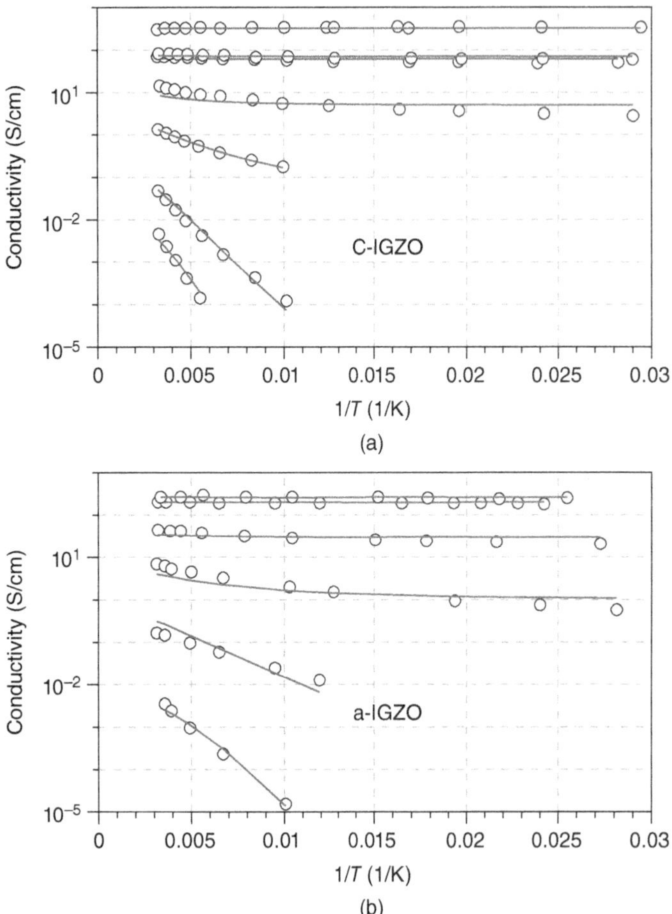

Figure 6.4 Temperature dependence of the conductivity $\sigma(T)$ in c-IGZO and a-IGZO samples with different carrier concentrations n. Circles: Experimental data [4]; solid lines: fit to Equation (6.30). The values of fitting parameters are specified in the text. *Source:* Reproduced with permission from Ref. [9], Nenashev, A. V., et al. (2019). Percolation description of charge transport in amorphous oxide semiconductors. *Physical Review B* **100**(12). Copyright 2019 by the American Physical Society.

Experimental data on the temperature dependencies of the conductivity $\sigma(T)$ are shown by circles in Figure 6.4a for c-IGZO and in Figure 6.4b for a-IGZO. These data are copied from figure 1b and figure 1d of Ref. [4], respectively. Theoretical results given by Equation (6.30) are shown in Figure 6.4 by solid lines. These results are obtained by adjusting the band-edge disorder parameter δ in Equation (6.18) and the conduction-band mobility μ_0, keeping these parameters fixed for *each group of samples*. Since the values of the carrier concentration n in different samples were not exactly specified [4], we use n as an adjustable parameter. Values for n in the range between $n = 10^{15}$ cm^{-3} and $n = 6 \cdot 10^{19}$ cm^{-3} give the best fits, in good agreement with experimental estimates [4]. The parameters δ, μ_0, and n are considered as independent of temperature. The values of δ and μ_0 that provide the best fits to the experimental data are $\delta = 0.057$ eV, $\mu_0 = 39$ cm^2/(Vs) for c-IGZO, and $\delta = 0.036$ eV, $\mu_0 = 47$ cm^2/(Vs) for a-IGZO. It is not obvious why amorphous samples should possess a slightly higher carrier mobility as compared to the crystalline ones. One should, however, take into account that c-IGZO and a-IGZO films were created by different deposition techniques: a reactive solid-phase epitaxy method for c-IGZO, and pulsed laser deposition for a-IGZO. One can expect different factors resulting in energy disorder and electron scattering (non-stoichiometry, dangling bonds, etc.) due to different deposition techniques.

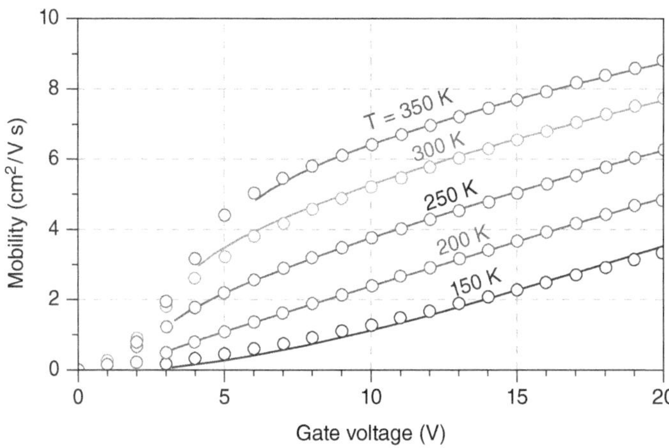

Figure 6.5 Dependence of the carrier mobility μ on the gate voltage V_g at different temperatures. Circles: Experimental data [3]; solid lines: fit $\mu(n, T) = \sigma(n, T)/(en)$, where $\sigma(n, T)$ is given by Equation (6.30). The values of fitting parameters are specified in the text. *Source:* Reproduced with permission from Ref. [9], Nenashev, A. V., et al. (2019). Percolation description of charge transport in amorphous oxide semiconductors. *Physical Review B* **100**(12). Copyright 2019 by the American Physical Society.

Another set of experimental data to be compared with theoretical predictions is related to the carrier mobility $\mu(n, T)$ measured in thin-film transistors with IGZO channels [3, 10]. In Figure 6.5, experimental data for the dependencies of μ on the gate voltage V_g at different temperatures are reproduced from Ref. [3], as depicted by circles. Solid lines are the theoretical results for $\mu(n, T) = \sigma(n, T)/(en)$, where $\sigma(n, T)$ is obtained from Equation (6.30). The carrier concentration n is assumed to be linearly dependent on the gate voltage V_g,

$$n = \lambda[V_g - V^*(T)], \tag{6.50}$$

where the proportionality constant λ serves as a fitting parameter. It depends on the relative capacitance between the gate and the channel, and on the thickness of the electron accumulation layer. The value $\lambda = 9.11 \cdot 10^{16}\ \text{cm}^{-3}\text{V}^{-1}$ gives the best fits. Following Ref. [3], the experimental gate voltage is counted from its threshold value. The flat-band voltage V^* is also treated as an adjustable parameter. Following Ref. [10], V^* is considered temperature-dependent. This dependence could be caused by the temperature dependence of the density of surface charges at the interface IGZO–SiO_2. The values $V^* = 2.69$ V, 2.32 V, 2.79 V, 3.61 V, and 5.57 V are used for $T = 150$ K, 200 K, 250 K, 300 K, and 350 K, respectively. The best fits to the experimental data in Figure 6.5 are achieved by choosing the band-edge disorder parameter $\delta = 0.063$ eV in Equation (6.18) and the conduction-band mobility $\mu_0 = 30\ \text{cm}^2/(\text{Vs})$. The lower bound on the curves in Figure 6.5 is 0.5 Volts above V^*, so it is different at different temperatures. A noticeable conduction below V^* at $T = 250$ K and at higher T could be due to electrons in the bulk of IGZO films.

Experimental data on the carrier mobility in a-IGZO thin-film transistors, analogous to those in Figure 6.5, were obtained by Fishchuk et al. [10], who converted the data into $\mu(T)$ at different gate voltages V_g. The data are shown by circles in Figure 6.6. Solid lines are fits to Equation (6.30). The temperature dependence of V^* in Equation (6.50) is taken from Ref. [10], $V^*(T) = -1.61\ \text{V} + 109\ \text{V}/(T/\text{K})$. The value $\lambda = 1.32 \cdot 10^{17}\ \text{cm}^{-3}\text{V}^{-1}$ gives the best fit. The best agreement with experimental data in Figure 6.6 is achieved by choosing the band-edge disorder parameter $\delta = 0.05$ eV in Equation (6.18) and the conduction-band mobility $\mu_0 = 36\ \text{cm}^2/(\text{Vs})$. These values are close to the values $\delta = 0.04$ eV and $\mu_0 = 22\ \text{cm}^2/(\text{Vs})$ obtained by Fishchuk et al. [10] from a comparison of their experimental data and their theory based on the EMA. This shows that there is not much difference between the results of percolation theory and those of the EMA for the range of parameters δ, kT, n, and μ_0 relevant to the experimental situation studied in Ref. [10].

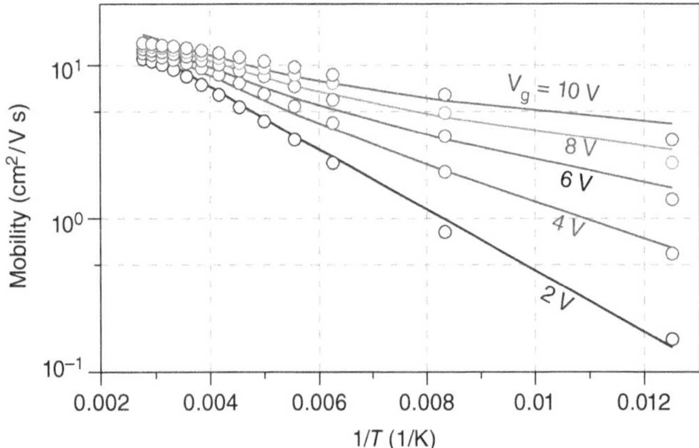

Figure 6.6 Dependence of the carrier mobility $\mu(V_g, T)$ on temperature at different gate voltages V_g. Circles: Experimental data [10]; solid lines: fit of $\mu(n, T) = \sigma(n, T)/(en)$, where $\sigma(n, T)$ is given by Equation (6.30). The values of fitting parameters are specified in the text. *Source:* Reproduced with permission from Ref. [9], Nenashev, A. V., et al. (2019). Percolation description of charge transport in amorphous oxide semiconductors. *Physical Review B* **100**(12). Copyright 2019 by the American Physical Society.

From the comparison between theory and experiment given here, one can conclude that percolation theory reliably reproduces experimental data in IGZO materials over a broad range of temperatures and charge carrier densities, revealing the band-edge disorder parameter δ in Equation (6.18) in the range of 36 meV $< \delta <$ 63 meV and the conduction-band mobility μ_0 in the range of 30 cm^2/(Vs) $< \mu_0 <$ 47 cm^2/(Vs). These values for δ and μ_0 do not look unreasonable, manifesting a success of the described theory.

6.7 Discussion and Conclusions

Theory presented in this chapter proves that charge transport in disordered amorphous oxide semiconductors (AOSs) can be described in full analogy with the textbook description of charge transport in traditional crystalline semiconductors (TCSs), with only one modification related to the role of the band edge E_c. Below we briefly compare the textbook description of charge transport in TCSs with the approach formulated in this chapter for description of charge transport in disordered AOSs, such as IGZO materials. For certainty, we consider an n-type material with electrons as the majority of charge carriers. Transport of holes in p-type materials can be described in a similar way.

6.7.1 Textbook Description of Charge Transport in Traditional Crystalline Semiconductors (TCSs)

As usual, electrical conductivity is the product $\sigma = e \cdot \mu \cdot n$ between the electron charge e, the electron mobility μ, and the concentration of electrons n possessing mobility μ and charge e. In TCSs, μ is equal to the mobility μ_0 of free electrons above the band edge E_c. The concentration of conducting electrons n in thermal equilibrium at temperature T is equal to $n = N_c \cdot \exp[-(E_c - E_f)/kT]$, where k is the Boltzmann constant, N_c is the effective concentration of conducting states with energies above E_c, and E_f denotes the Fermi energy. In the intrinsic TCSs, Fermi energy is close to the middle of the band gap E_g. Concomitantly, the energy distance $E_c - E_f$ between the band edge E_c and the Fermi energy E_f is close to $E_g/2$. In doped TCSs, Fermi energy is related to the energy depth of the donor levels with respect to the band edge E_c. As long as the Fermi energy E_f is situated below the band edge E_c, the material is a semiconductor, so that its electrical conductivity exponentially decreases with decreasing temperature T. If, however, the doping is so high that the large electron concentration brings the Fermi energy E_f

above the band edge E_c, the electrical conduction of TCSs attains a metallic character. In the latter case, electrical conductivity usually increases with decreasing T, because the scattering of electrons due to interactions with phonons becomes weaker at low T. Very important for description of charge transport in TCSs is the fact that the energy position of the band edge in TCSs is independent of spatial coordinates and possesses a constant value E_c.

6.7.2 Results of This Chapter for Charge Transport in Amorphous Oxide Semiconductors (AOSs)

In AOSs, such as IGZO materials, the band-edge position depends on spatial coordinates, and it cannot be characterized by some universal value E_c. This happens because the local position of the band edge is subjected to spatial fluctuations caused by disorder potential created by various sources of disorder, for instance by fluctuations in the spatial concentrations of the chemically different constituents. In such a case, the position of the band edge fluctuates in space, creating a disorder-potential landscape as illustrated in Figure 6.3. This landscape consists of potential hills and potential valleys. In this chapter, Gaussian fluctuations of the band edge E_c are considered. The fluctuations are described by Equation (6.18), where the quantity E_m is counted from the position of the band edge without disorder potential taken as a reference energy level. The fluctuations are characterized by the energy amplitude δ.

According to rich experimental literature, the concentration of electrons n in IGZO materials desired for device applications is so high that electrons fill the deep valleys in the band-edge potential landscape depicted in Figure 6.3. In such a case, the Fermi energy E_f is situated above the bottoms of the deepest valleys in this potential landscape. Although the Fermi energy is situated above the local band edges in such valleys, the conduction of AOSs does not have a metallic character, because electrons at the Fermi level are spatially localized in the deep valleys. In order to contribute to the electrical conduction, such electrons have to be thermally activated toward the percolation level E_p. This percolation level E_p is determined as the lowest energy level that allows electrons to move through the system without climbing to higher energies than E_p. For the Gaussian distribution of local mobility edges characterized by the amplitude δ, as described by Equation (6.18), the percolation level is related to δ via Equation (6.29): $E_p = -0.95\delta$. This percolation level E_p plays for disordered AOSs a role similar to the role of the band edge E_c in the case of ordered TSCs, described in Section 6.7.1.

The main result for charge carrier mobility μ in this chapter is Equation (6.35), which looks very similar to the standard equation for carrier mobility in the textbooks for TSCs, $\sigma = e \cdot \mu_0 \cdot N_c \cdot \exp[-(E_c - E_f)/kT]$, though with one significant modification. The role of the band edge E_c in the latter equation derived for the case of no disorder is played in Equation (6.35) for a disordered system by the percolation level E_p. The exponential term in Equation (6.35) shows that the charge transport is dominated by thermal activation of electrons from the Fermi level E_f to the percolation level E_p, as schematically depicted in Figure 6.3.

Acknowledgments

The authors are indebted to Professor A. Kadashchuk and Professor H. Hosono for bringing their attention to the problem of charge transport in AOSs. Financial support of the Deutsche Forschungsgemeinschaft (TG 2591 and GRK 1782) and that of the key profile area "Quantum Matter and Materials" (QM2) at the University of Cologne are gratefully acknowledged.

References

1 Nomura, K., Ohta, H., Takagi, A., Kamiya, T., Hirano, M., and Hosono, H. (2004). Room-temperature fabrication of transparent flexible thin film transistors using amorphous oxide semiconductors. *Nature* 432: 488.
2 Hosono, H. (2018). How we made the IGZO transistor. *Nature Electronics* 1: 428.

3 Germs, W. C., Adriaans, W. H., Tripathi, A. K., Roelofs, W. S. C., Cobb, B., Janssen, R. A. J., Gelinck, G. H., and Kemerink, M. (2012). Charge transport in amorphous InGaZnO thin-film transistors. *Physical Review B* 86: 155319.

4 Kamiya, T., Nomura, K., and Hosono, H. (2010). Origin of definite Hall voltage and positive slope in mobility-donor density relation in disordered oxide semiconductors. *Applied Physics Letters* 96: 122103.

5 Kamiya, T., Nomura, K., and Hosono, H. (2009). Electronic structures above mobility edges in crystalline and amorphous In-Ga-Zn-O: percolation conduction examined by analytical model. *Journal of Display Technology* 5: 462.

6 Takagi, A., Nomura, K., Ohta, H., Yanagi, H., Kamiya, T., Hirano, M., and Hosono, H. (2005). Carrier transport and electronic structure in amorphous oxide semiconductor, a-$InGaZnO_4$. *Thin Solid Films* 486 (1): 38.

7 Kimura, M., Kamiya, T., Nakanishi, T., Nomura, K., and Hosono, H. (2010). Intrinsic carrier mobility in amorphous InGaZnO thin-film transistors determined by combined field-effect technique. *Applied Physics Letters* 96 (26): 262105.

8 Baranovskii, S. D., Nenashev, A. V., Oelerich, J. O., Greiner, S. H. M., Dvurechenskii, A. V., and Gebhard, F. (2019). Percolation description of charge transport in the random barrier model applied to amorphous oxide semiconductors. *EPL (Europhysics Letters)* 127: 57004.

9 Nenashev, A. V., Oelerich, J. O., Greiner, S. H. M., Dvurechenskii, A. V., Gebhard, F., and Baranovskii, S. D. (2019). Percolation description of charge transport in amorphous oxide semiconductors. *Physical Review B* 100: 125202.

10 Fishchuk, I. I., Kadashchuk, A., Bhoolokam, A., de Jamblinne de Meux, A., Pourtois, G., Gavrilyuk, M. M., and Genoe, J. (2016). Interplay between hopping and band transport in high-mobility disordered semiconductors at large carrier concentrations: the case of the amorphous oxide InGaZnO. *Physical Review B* 93: 195204.

11 Zeis, R., Siegrist, T., and Kloc, C. (2005). Single-crystal field-effect transistors based on copper phthalocyanine. *Applied Physics Letters* 86 (2): 022103.

12 Reese, C., Chung, W.-J., Ling, M.-m., Roberts, M., and Bao, Z. (2006). High-performance microscale single-crystal transistors by lithography on an elastomer dielectric. *Applied Physics Letters* 89 (20): 202108.

13 Jurchescu, O. D., Popinciuc, M., van Wees, B. J., and Palstra, T. T. M. (2007). Interface-controlled, high-mobility organic transistors. *Advanced Materials* 19 (5): 688–692.

14 Takeya, J., Yamagishi, M., Tominari, Y., Hirahara, R., Nakazawa, Y., Nishikawa, T., and Ogawa, S. (2007). Very high-mobility organic single-crystal transistors with in-crystal conduction channels. *Applied Physics Letters* 90 (10): 102120.

15 Giesbers, G., Van Schenck, J., Quinn, A., Van Court, R., Vega Gutierrez, S. M., Robinson, S. C., and Ostroverkhova, O. (2019). Xylindein: naturally produced fungal compound for sustainable (opto)electronics. *ACS Omega* 4 (8): 13309–13318.

16 Gleskova, H., and Wagner, S. (2001). Electron mobility in amorphous silicon thin-film transistors under compressive strain. *Applied Physics Letters* 79 (20): 3347–3349.

17 Arnold, D. N., David, G., Jerison, D., Mayboroda, S., and Filoche, M. (2016). Effective confining potential of quantum states in disordered media. *Physical Review Letters* 116: 056602.

18 Li, C.-K., Piccardo, M., Lu, L.-S., Mayboroda, S., Martinelli, L., Peretti, J., and Wu, Y.-R. (2017). Localization landscape theory of disorder in semiconductors. III. Application to carrier transport and recombination in light emitting diodes. *Physical Review B* 95: 144206.

19 Piccardo, M., Li, C.-K., Wu, Y.-R., Speck, J. S., Bonef, B., Farrell, R. M., and Weisbuch, C. (2017). Localization landscape theory of disorder in semiconductors. II. Urbach tails of disordered quantum well layers. *Physical Review B* 95: 144205.

20 Filoche, M., Piccardo, M., Wu, Y.-R., Li, C.-K., Weisbuch, C., and Mayboroda, S. (2017). Localization landscape theory of disorder in semiconductors. I. Theory and modeling. *Physical Review B* 95: 144204.

21 Bube, R. H. (1974). *Electronic Properties of Crystalline Solids: An Introduction to Fundamentals*. Academic Press, New York.

22 Sze, S. M. (1981). *Physics of Semiconductor Devices*, 2nd ed. John Wiley & Sons, New York.

23 Adler, D., Flora, L. P., and Sentuna, S. D. (1973). Electrical conductivity in disordered systems. *Solid State Communications* 12: 9.

24 Lee, S., Ghaffarzadeh, K., Nathan, A., Robertson, J., Jeon, S., Kim, C., Song, I.-H., and Chung, U.-I. (2011). Trap-limited and percolation conduction mechanisms in amorphous oxide semiconductor thin film transistors. *Applied Physics Letters* 98 (20): 203508.

25 Shklovskii, B. I., and Efros, A. L. (1971). Impurity band and conductivity of compensated semiconductors. *Soviet Physics: Journal of Experimental and Theoretical Physics* 33: 468.

26 Ambegaokar, V., Halperin, B. I., and Langer, J. S. (1971). Hopping conductivity in disordered systems. *Physical Review B* 4: 2612.

27 Pollak, M. (1972). A percolation treatment of DC hopping conduction. *Journal of Non-Crystalline Solids* 11: 1.

28 Shklovskii, B. I., and Efros, A. L. (1984). *Electronic Properties of Doped Semiconductors*. Springer, Berlin.

29 Baranovski, S. (Ed.) (2006). *Charge Transport in Disordered Solids with Applications in Electronics*. John Wiley & Sons, Chichester, UK.

30 Baranovskii, S. D. (2014). Theoretical description of charge transport in disordered organic semiconductors. *Physica Status Solidi (B): Basic Research* 251: 487.

31 Nenashev, A. V., Oelerich, J. O., and Baranovskii, S. D. (2015). Theoretical tools for the description of charge transport in disordered organic semiconductors. *Journal of Physics: Condensed Matter* 27: 093201.

32 Werner, J. H., and Güttler, H. H. (1991). Barrier inhomogeneities at Schottky contacts. *Journal of Applied Physics* 69 (3): 1522.

33 Sykes, M. F., and Essam, J. W. (1964). Critical percolation probabilities by series methods. *Physical Review* 133: A310.

34 Škvor, J., and Nezbeda, I. (2009). Percolation threshold parameters of fluids. *Physical Review E* 79: 041141.

35 Wang, J., Zhou, Z., Zhang, W., Garoni, T. M., and Deng, Y. (2013). Bond and site percolation in three dimensions. *Physical Review E* 87: 052107.

36 Xu, X., Wang, J., Lv, J.-P., and Deng, Y. (2014). Simultaneous analysis of three-dimensional percolation models. *Frontiers of Physics* 9: 113.

37 Mott, N. F., and Davis, E. A. (1979). *Electronic Processes in Non-Crystalline Materials*, 2nd ed. Clarendon Press, Oxford.

38 Overhof, H., and Thomas, P. (1989). *Electronic Transport in Hydrogenated Amorphous Semiconductors*. Springer, Heidelberg, Germany.

39 Street, R. A. (1991). *Hydrogenated Amorphous Silicon*. Cambridge Solid State Science Series. Cambridge University Press, Cambridge.

40 Semeniuk, O., Juska, G., Oelerich, J. O., Jandieri, K., Baranovskii, S. D., and Reznik, A. (2017). Transport of electrons in lead oxide studied by CELIV technique. *Journal of Physics D: Applied Physics* 50: 035103.

41 Nenashev, A. V., Jansson, F., Oelerich, J. O., Huemmer, D., Dvurechenskii, A. V., Gebhard, F., and Baranovskii, S. D. (2013). Advanced percolation solution for hopping conductivity. *Physical Review B* 87: 235204.

42 Hu, H., Blöte, H. W. J., Ziff, R. M., and Deng, Y. (2014). Short-range correlations in percolation at criticality. *Physical Review E* 90: 042106.

43 Koza, Z., and Poła, J. (2016). From discrete to continuous percolation in dimensions 3 to 7. *Journal of Statistical Mechanics: Theory and Experiment* **2016**: 103206.

44 Germs, W. C., van der Holst, J. J. M., van Mensfoort, S. L. M., Bobbert, P. A., and Coehoorn, R. (2011). Modeling of the transient mobility in disordered organic semiconductors with a Gaussian density of states. *Physical Review B* 84: 165210.

45 Nakamura, M. (1984). Conductivity for the site-percolation problem by an improved effective-medium theory. *Physical Review B* 29: 3691–3693.

46 Chen, Y., and Schuh, C. A. (2006). Diffusion on grain boundary networks: percolation theory and effective medium approximations. *Acta Materialia* 54: 4709.

47 Lee, S., and Nathan, A. (2012). Localized tail state distribution in amorphous oxide transistors deduced from low temperature measurements. *Applied Physics Letters* 101 (11): 113502.

7

State and Role of Hydrogen in Amorphous Oxide Semiconductors

Hideo Hosono[1,2] *and Toshio Kamiya*[1,3]

[1] *Materials Research Center for Element Strategy, Tokyo Institute of Technology, Tokyo, Japan*
[2] *National Institute for Materials Research, Tsukuba, Japan*
[3] *Laboratory for Materials and Structures, Tokyo Institute of Technology, Tokyo, Japan*

7.1 Introduction

Hydrogen is one of the most common impurities in semiconductors. The critical roles of hydrogen on electrical properties have dual faces, favorable and unfavorable roles. A representative example of the former is hydrogen in amorphous silicon (*a*-Si:H), where silicon dangling bonds are passivated effectively in *a*-Si:H by forming Si-H bonds, making it possible to control *p*/*n* conduction by impurity doping even in an amorphous material [1]. The latter example is seen in GaN, as impurity hydrogen incorporated during thin-film deposition by chemical vapor deposition. The incorporated hydrogen works as charge compensators for struggling to develop *p*-type GaN [2]. It was reported that amorphous oxide semiconductor (AOS) thin films contain high-density impurity hydrogen at 10^{20-21} cm^{-3}, which were considered to be mainly in the form of hydroxyl (OH) groups in the early 2010s [3]. Since thermal annealing around 300–400 °C is required to stabilize AOS-TFTs (thin-film transistors), it is natural to suggest a close relation between the stabilization/desorption of these hydrogen impurities and the TFT electrical properties such as mobility and subthreshold swing. In 2017, Bang et al. [4] found hydrogen directly bonding to metal cations as a hidden form of impurity hydrogens, and their concentration is comparable to that of OH groups. Figure 7.1 summarizes three representative charge states and associated chemical structures of hydrogen in solids. In ionic solids such as oxides, the neutral state of hydrogen (hydrogen atom, H^0) is unstable and is stabilized in terms of charged states, H^+ and H^-. In fact, while neutral hydrogen with an unpaired electron is easily detected by electron paramagnetic resonance (EPR), there has been no paper reporting the detection of H^0 in AOS materials until now.

It is of interest to note that an electron loosely bound to positively charged hydrogen was detected by electron-nuclear double-resonance spectroscopy in single-crystalline ZnO at 4K [5]. This state may be regarded as an excited state of H^0 (explained as a hydrogen-like shallow donor state in semiconductor textbooks), but this bound electron is converted to carrier around room temperature. The chemical state of hydrogen is thus restricted to H^+ (OH group) and H^- (M-H) in AOS materials. In this chapter, the chemical states and origin of impurity hydrogen and their influence on structural and electrical properties of AOSs are briefly summarized.

7.2 Concentration and Chemical States

Figure 7.2 shows the depth concentration of H measured by secondary-ion mass spectroscopy (SIMS) in amorphous In–Ga–Zn-O (*a*-IGZO) thin films sputter-deposited in the conventional vacuum chamber (base pressure

Amorphous Oxide Semiconductors: IGZO and Related Materials for Display and Memory, First Edition.
Edited by Hideo Hosono and Hideya Kumomi.

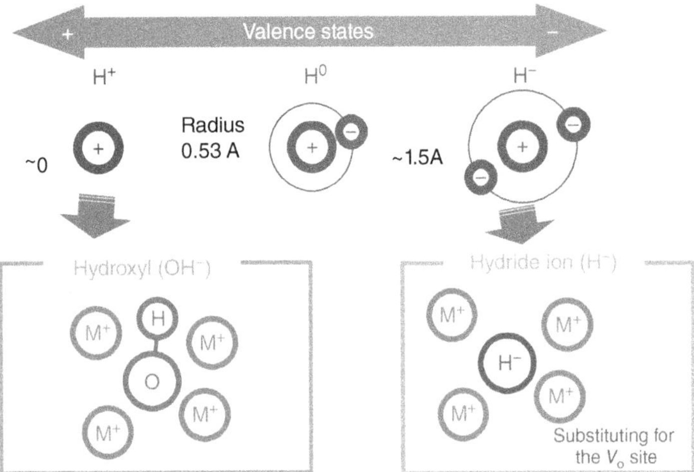

Figure 7.1 Three charge states of hydrogen in solids. Neutral H^0 has been reported to be unstable compared to H^+ and H^- in AOS materials.

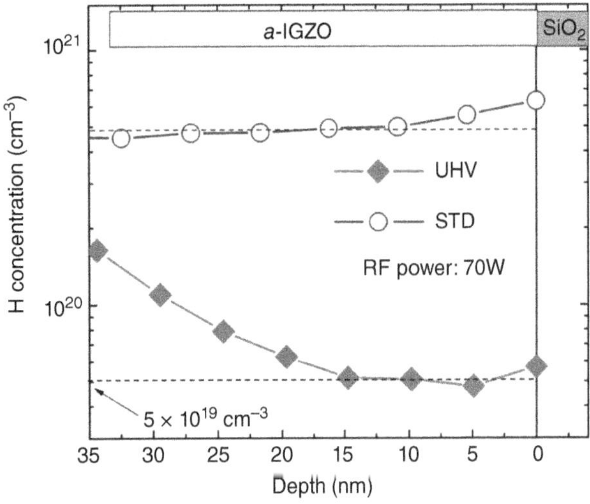

Figure 7.2 SIMS depth concentration of H in *a*-IGZO thin films deposited on an SiO_2/Si substrate. Sputtering was performed at the same conditions except back pressures (UHV: ~10^{-7} Pa; and STD: ~10^{-4} Pa). The conventionally adopted condition corresponds to the STD.

~10^{-4} Pa, denoted as the standard condition or "STD"). The H-concentrations in the as-deposited thin films go up to ~5×10^{20} cm^{-3} throughout the sample. The reduction of the concentration is restricted to ~5×10^{19} cm^{-3}, even in the thin films deposited in ultrahigh vacuum (UHV) chamber at this deposition condition (~10^{-7} Pa). Figure 7.3 shows thermal desorption spectra (TDS) of STD and UHV as-deposited thin films [6]. The desorbed gases are H_2, O_2, and H_2O, and each of them starts to desorb from ~100 °C; the desorption distinctly increases up to >350 °C and reaches a maximum at ~450 °C for H_2, ~500 °C for O_2, and ~450 °C for H_2O. It is noted that desorption of Zn is distinct compared with In and Ga, and its TDS profile is similar to that of H_2O, suggesting that thermal dehydration accompanies evaporation of Zn. A similar reaction process would occur in STD sputtering, and thus resultant STD-deposited *a*-IGZO thin films are Zn-poor, and heating to >300 °C further reduces Zn concentration. Since the main residual gas species in high-vacuum chambers evacuated with turbo molecular pumps are

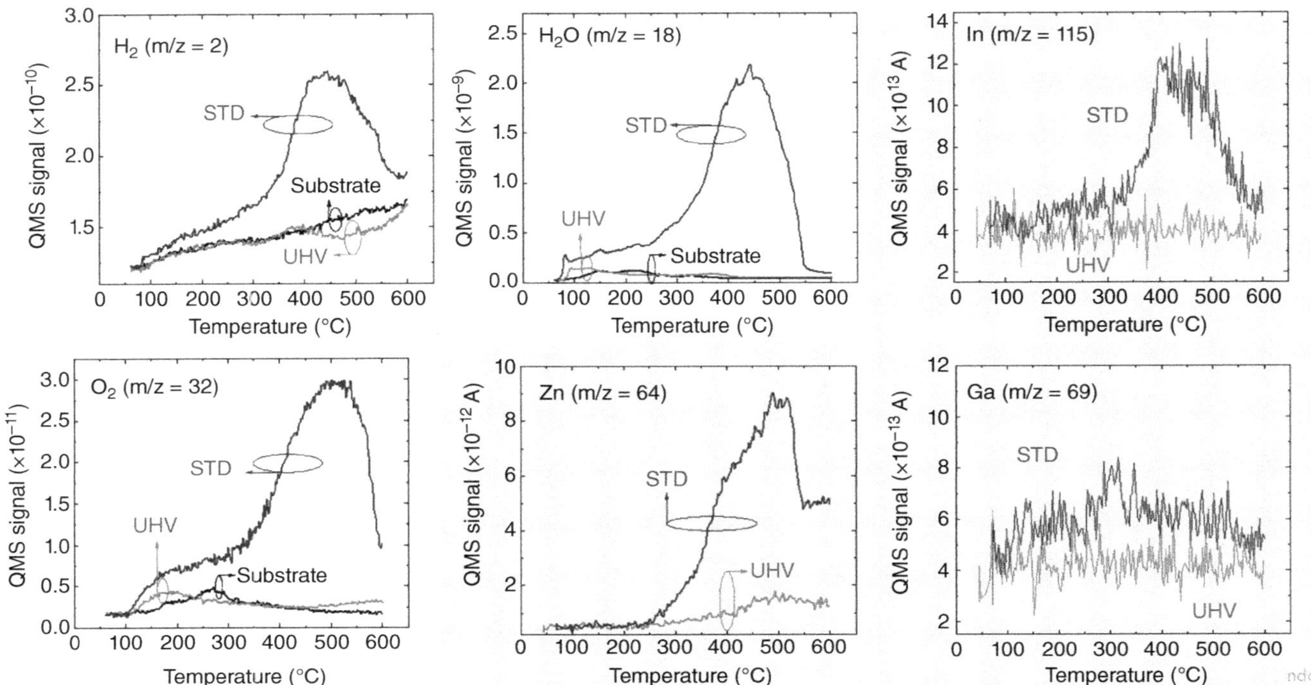

Figure 7.3 Comparisons in thermal desorption spectra between UHV- and STD-deposited *a*-IGZO thin films. The desorbed amounts of H_2, H_2O, O_2, and Zn from STD samples are distinctly larger than those from UHV samples.

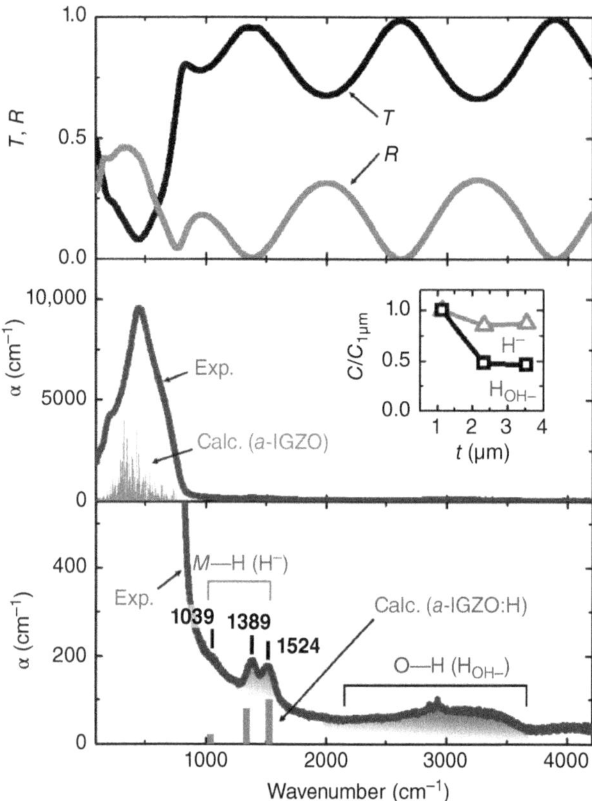

Figure 7.4 Infrared spectra of ~1-μm-thick self-standing *a*-IGZO thin film. T: Transmission; R: reflection; α: absorption coefficient.

H_2 and H_2O, these results may be understood by considering that AOS has a strong affinity to hydrogen species in the sputter-deposition conditions. This idea suggests that a rather high-temperature state (valence of metal ion is reduced) is generated by collision of positively charged gas ions with the target and gas plasma with extremely high electronic temperature in the sputtering process, and thereby high density of oxygen vacancy is created at the deposited state. It is likely to occur that these oxygen vacancies react readily with H_2 and H_2O as the residual gases to give rise to the formation of OH and M-H.

Chemical states of incorporated H-species in *a*-IGZO thin films obtained by the STD sputtering conditions (base pressure ~10^{-4} Pa) were revealed by infrared absorption spectroscopy. Sample thin films used were deposited on NaCl single-crystal substrates, and then the substrate was dissolved out to obtain self-standing thin films for measurements in transmission mode [4]. As is shown in Figure 7.4, there are two H-associated bands, a broad absorption around ~3000 cm^{-1}, and three sharp bands located in the range of ~1000–1500 cm^{-1}. The former is attributed to strong hydrogen-bonding OH. We assigned the latter three bands to In-H, Ga-H, and Zn-H bonds, referring to the previously reported data on respective metal hydrides [4]. This assignment is substantiated by DFT (density functional theory) calculations on the model structure generated by molecular dynamics (MD) simulation (see Section 7.3). Figure 7.5a shows variations of infrared absorption spectra of *a*-IGZO thin films with thermal heating. The network vibrations remain unchanged except splitting of the broad peak by crystallization at >700 °C. There is a conspicuous difference in thermal stability between the OH and the M-H: Upon heating, the OH intensity is continuously reduced and reaches a very weak level at 600 °C, while the reduction of the M-H intensity is much smaller compared with that of the OH and almost half intensity remains even after heating up

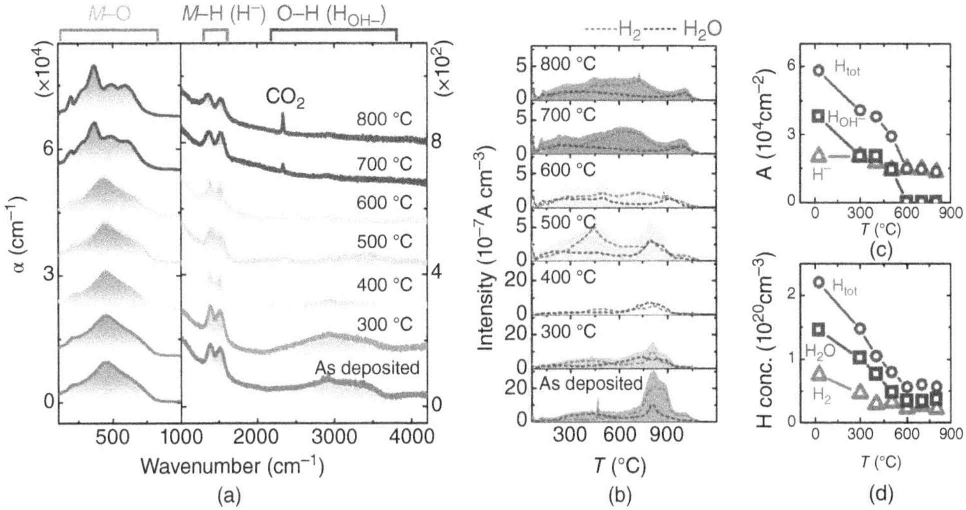

Figure 7.5 Thermal stability of OH and H. (a) Infrared and (b) TDS spectra of self-standing *a*-IGZO thin films postannealed under O_2 flow. (c) Integrated peak areas (A) for M–H and O–H vibrational modes. (d) Hydrogen concentrations desorbed as H_2 and H_2O upon heating. Data were obtained from TDS spectra.

at 800 °C, where crystallization appears complete. TDS measurements (Figure 7.5b) indicates H_2O is the major desorbed component at low temperatures (up to ~500 °C), and H_2 desorption is dominant at higher temperatures. This means desorbed H_2O and H_2 come mainly from OH and M-H, respectively. Figure 7.5b, 7.5c, and 7.5d summarize the variation in the desorbed total hydrogens and each hydrogen species with temperature. The concentration of H_2 is ~1/2 of that of H_2O, indicating the concentration ratio of M-H to OH is ~1:2. Since the total hydrogen impurities in *a*-IGZO thin films sputter-deposited in the standard condition are $\sim 5\times10^{20}cm^{-3}$ and there is no EPR signal of H^0, the M-H concentration is estimated to be $\sim 1.7\times10^{20}cm^{-3}$.

What are the precursors for these OH and M-H? These thin films are fabricated by extremely rapid quenching of a high-temperature state (a metal-reducing condition) created by the collision of high-kinetic-energy deposition precursors to substrate. The presence of strained bonds is commonly observed for unannealed thin films deposited by sputtering or chemical vapor deposition (CVD). A representative example is *a*-SiO_2 thin films by CVD. The peak position of the infrared absorption band by the Si-O-Si bond in *a*-SiO_2 by CVD is red shifted compared with thermally oxidized *a*-SiO_2, and this shift is caused by a decrease in the Si-O-Si bond angle [7, 8].

Here, we assume a simple precursor model by referring to that in *a*-SiO_2 in which hydrogen impurities exist in the form of Si-OH and Si-H. In *a*-SiO_2 thin films fabricated by CVD, SiOH is formed by reactions with strained Si-O-Si bonds with H_2O molecules following the reaction [9]:

$$\equiv \mathrm{Si-O-Si} \equiv + \mathrm{H_2O} = 2 \equiv \mathrm{Si-OH} \tag{7.1}$$

where ≡Si indicate Si coordinated with three oxygens.

On the other hand, the Si-H bond is created by the reactions of H_2 with ≡Si-Si ≡ [10], which may be regarded as relaxed "oxygen vacancy" [11], following Equation (7.2):

$$\equiv \mathrm{Si-Si} \equiv + \mathrm{H_2} = 2 \equiv \mathrm{Si-H} \tag{7.2}$$

If this analogy is valid to AOSs, the precursor of OH would be strained M-O-M, OH is created through the reaction with H_2O, and M-H is created by reactions of an oxygen vacancy coordinated by M cations with H_2. When the numbers of residual H_2 and H_2O in the sputtering chamber are higher than that of the precursor, the numbers of the precursor strained M-O-M bond and the oxygen vacancy are evaluated to be $\sim 3\times10^{20}$ cm^{-3} and $\sim 1\times10^{20}$ cm^{-3}, respectively.

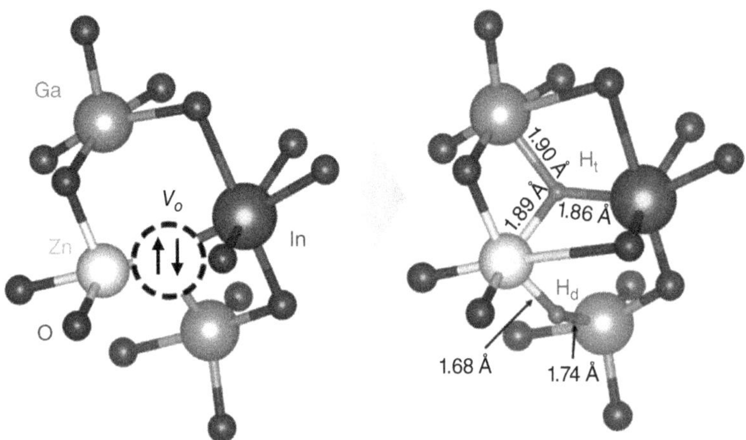

Figure 7.6 MD-simulated oxygen vacancy model [13] before the structural relaxation as an initial structure for hydrogen containing *a*-IGZO. The local structure of the hydrogen incorporated *a*-IGZO. Ht and Hd indicate the triply and doubly coordinated hydrogen atoms, respectively. This reaction of neutral oxygen vacancy with H_2 occurs readily if H_2 is present near neutral oxygen vacancy, V_o. Source: Bang, J., Matsuishi, S., & Hosono, H. (2017). Hydrogen anion and subgap states in amorphous In–Ga–Zn–O thin films for TFT applications. *Applied Physics Letters* **110** (23): 232105.

The formation model for M-H by reaction of H_2 with neutral oxygen vacancy (NOV, trapping two electrons at the oxygen vacancy) was justified by a computation. Bang et al. [4] examined what happens if NOV and H_2 coexist in an *a*-IGZO structure model [12] by DFT calculation. Local structure around the NOV is generated by a classical MD simulation on an oxygen-deficient *a*-IGZO model [13], and an H_2 molecule is placed nearby the NOV. The H_2 reacts easily with the NOV to split to a pair of hydrogens doubly and triply coordinated with metal cations, as shown in Figure 7.6, demonstrating the validity of the oxygen vacancy precursor model for the M-H bond. The validity of the strained M-O-M is discussed in Section 7.3.

Li et al. [14] reported that the M-H-M bridges may cause negative-bias illumination-stress (NBIS) instability in AOS-TFTs.

7.3 Carrier Generation and Hydrogen

The roles of hydrogen in AOS materials are not simple, and its effect on carrier generation is especially complex, as discussed in several publications to date. In this section, the relation between hydrogen and carrier is described by separating low temperature and high temperature.

7.3.1 Carrier Generation by H Injection at Low Temperatures

Passivation of dangling bonds in *a*-Si:H makes it possible to control the Fermi level by impurity doping like crystalline Si, although efficiency is much lower. At the early stage of AOS-TFTs, H-containing-plasma processing was applied to deposition of gate insulators, but the AOS became so conducting that the TFT could not switch off due to the high carrier generation. This result is easily predicted by remembering the early research on AOSs by proton implantation at room temperature (RT) [15]. On the other hand, no carrier generation is observed by exposing *a*-SiO_2 and *a*-Ga_2O_x to hydrogen-containing plasma. This conspicuous difference in carrier generation by hydrogen may be understood in terms of the concept of the universal charge transition level of hydrogen in compound semiconductors proposed by Van de Walle [16]. According to this rule, the transition energy, $E(H^+/H^-)$ from H^- to H^+, is commonly present around 4.5 eV below the vacuum level irrespective of materials including aqueous

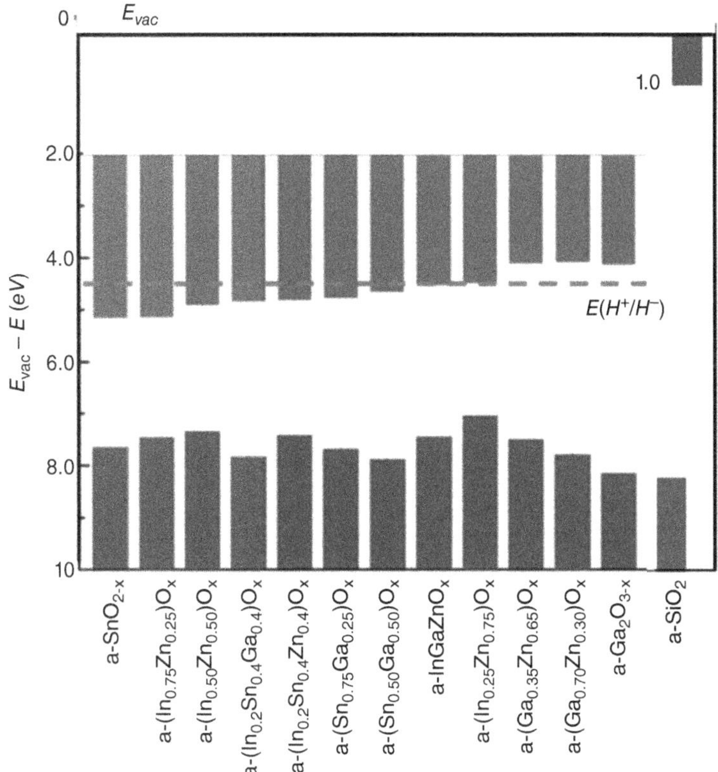

Figure 7.7 Energy levels of VBM and CBM in various AOS materials and a-SiO_2 along with a hydrogen charge transmission level (4.5 eV).

solutions. This level is pretty close to the chemical potential (−4.44 eV) of electrons under the standard hydrogen electrode condition. Figure 7.7 summarizes the band-edge energy of various AOS materials and a-SiO_2 along with the universal level of $E(H^+/H^-)$. The energy levels of the conduction band minimum (CBM), E_{CBM}, of a-SnO_2, a-In-Sn-O, In-Zn-O, and a-In-Ga-Zn-O (IGZO), are lower than $E(H^+/H^-)$, while those of a-Ga-O and a-Ga-Zn-O are higher than $E(H^+/H^-)$. When E_{CBM} is lower than $E(H^+/H^-)$, H is ionized to $H^+ + e^-$, and the electron and H^+ are stabilized by transferring e^- to the CBM and forming OH through attaching to oxygen ions. In contrast, H^0 is not ionized in the AOS materials, as H^0 has higher energy states than H^+ and H^-, and thus is higher than $E(H^+/H^-)$. The experimental data agree with this prediction. Conventional wide-gap amorphous oxide insulators represented by a-SiO_2 have much smaller electron affinity (higher E_{CBM}). Thus, no carrier generation occurs in these materials, even by exposing H-plasma or implantation.

7.3.2 Carrier Generation and Annihilation by Thermal Treatment

As described in Section 7.1, a-IGZO thin films sputter-deposited at standard conditions contain impurity hydrogen at $\sim 5\times10^{20}$ cm^{-3} but the carrier density remains $\sim 10^{16} cm^{-3}$, which is workable as the TFT channel [17]. However, when the as-deposited thin films are heated in an N_2 atmosphere to ~400 °C, the conductivity drastically increases from 10^{-4} $S{\cdot}cm^{-1}$ at RT to $\sim 10^3$ $S{\cdot}cm^{-1}$ while the high conductivity remains constant ($\sim 10^2$ $S{\cdot}cm^{-1}$) during the cooling process [17], as shown in Figure 7.8. This result indicates the thin films after heating in N_2 to >400 °C are not applicable to TFT channels. The carrier concentration in the as-deposited thin film before and after heating

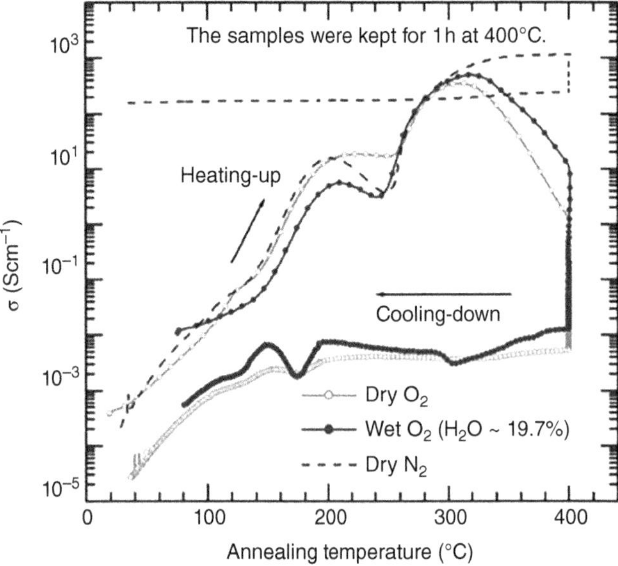

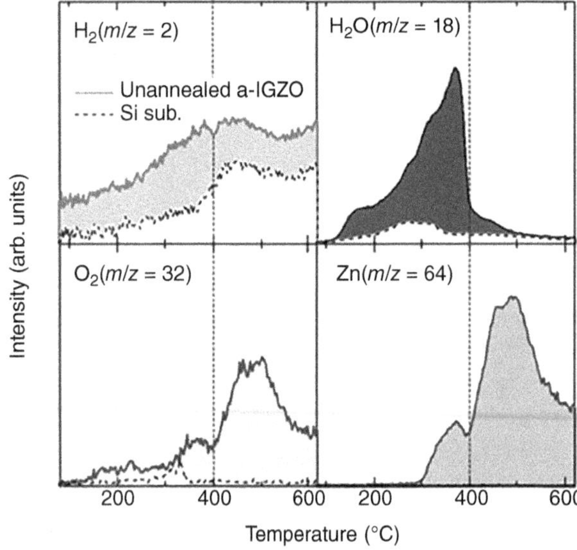

Figure 7.8 (Top) Change in in situ conductivity of *a*-IGZO (sputter-deposited in standard conditions) during thermal annealing in dry O_2, wet O_2, and dry N_2 atmospheres. (Bottom) TDS of as-deposited, dry and wet O_2-annealed *a*-IGZO thin films. The dotted curve shows the blank data on a Si/SiO_2 substrate.

is estimated to be $\sim 10^{16}$ cm^{-3} and $\sim 10^{20}$ cm^{-3}, respectively. The activation energies estimated from the slopes of the Arrhenius plots of Figure 7.8 at 300–400 K are 0.35–0.6 eV, which implies that a high-energy reaction process, such as desorption of H_2O and O_2, would dominate the activation energies and the generation of high-density carriers. The TDS data in Figure 7.8 indicate that the major desorption species is H_2O molecule and its temperature profile is similar to that of conductivity. This finding suggests the following reactions occur during the heating process in N_2:

$$M - OH + H - M = M - V_O - M + H_2O \quad (3)$$

$$M - OH + HO - M = M - O - M + H_2O \quad (4)$$

where V_O is a shallow oxygen vacancy that serves as a donor if the V_O size becomes small by relaxation, as will be described later in the chapter.

It is worth noting a striking difference in conductivity during the cooling process between O_2 and N_2 atmospheres. Although increasing behaviors of conductivity during the heating process up to ~350 °C are similar, the conductivity is drastically reduced by further heating to ~450 °C in O_2. This cause is tentatively attributed to the annihilation of the shallow V_O by the reaction with O_2 or H_2O in the atmosphere at these temperatures, where these molecules may diffuse to the thin films [18].

7.4 Energy Levels and Electrical Properties

Figure 7.9 shows energy levels of hydrogen impurities and relevant oxygen defects in *a*-IGZO thin films. The energy level of an M-OH bond is ~2 eV below the valance band maximum (VBM), as evidenced by oxygen X-ray photoelectron spectroscopy (O-XPS) data [19]. Thus, the OH does not participate directly in the photo effects or chemical reactions with H_2, O_2, or H_2O. On the other hand, M-H bonding gives the electronic state just above the VBM. This location is below the deep V_O [4, 18]. Since the concentrations of M-H and deep V_O are rather high ($>10^{20}$ cm^{-3}), both levels are experimentally observed by hard X-ray emission spectroscopy [20], which has high sensitivity to *s*-electrons and bulk state and gives sub-bandgap optical absorptions at >2.3 eV (~460 nm). There are at least two types of oxygen vacancies, deep and shallow V_O. Shallow V_O is created by thermal dehydration of as-deposited AOSs and works as a donor, while some V_O can trap two electrons and work as a deep electron trap, as described in Section 7.3. These V_Os are differentiated by the void size, as schematically shown in Figure 7.9. The deep trap has a large void space similar to an O^{2-} size, while a small V_O creates a shallow level. In the as-deposited thin films, electrons are predominantly trapped in the deep V_O without generating carriers. However, dehydration

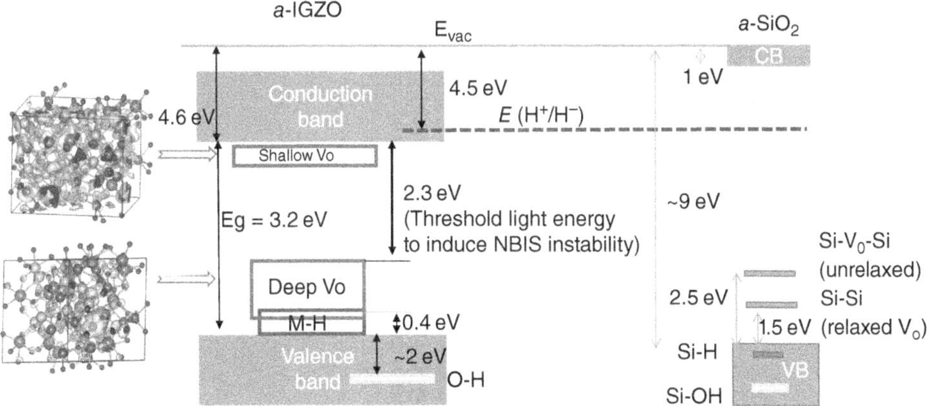

Figure 7.9 Energy level of hydrogen-related species in *a*-IGZO. Note there are two oxygen vacancies with different energy levels. Shallow and deep oxygen vacancies are characterized by small and large voids, respectively [13]. For comparison, corresponding energy levels of *a*-SiO_2 are drawn using data in literature [20, 21]. Source: Adapted from Kamiya, T., Nomura, K., Hirano, M., & Hosono, H. (2008). Electronic structure of oxygen deficient amorphous oxide semiconductor a-$InGaZnO_{4-x}$: optical analyses and first-principle calculations. *Physica Status Solidi (c)* **5** (9): 3098–3100; O'Reilly, E. P., & Robertson, J. (1983). Theory of defects in vitreous silicon dioxide. *Physical Review B* **27** (6): 3780–3795; and Pacchioni, G., & Ieraño, G. (1998). Ab initio theory of optical transitions of point defects in SiO_2. *Physical Review B* **57** (2): 818–832.

from M-OH and M-H by heating to >400 °C in N_2 atmosphere creates high-density shallow V_O, which works as a donor. This is the reason why high-density carriers are generated upon this heating.

It is of interest to compare V_O between AOSs and *a*-SiO_2. Both have two types of V_O with distinctly different energy levels, but in the latter, there is no shallow V_O and the deep V_O is largely relaxed to Si-Si direct bonding [20, 21]. Such differences would come from differences in the location of the CBM and ionic nature of M-O bonding.

Although the deep V_O is related to the NBIS instability as the electron source by photoexcitation with sub-gap photons, it has no direct influence on *n*-channel TFT operation. However, the situation is totally different in *p*-channel TFT operation. The deep V_O locating above the VBM works as a hole trap when gate voltage is negatively biased. Recently, Lee et al. [22] reported that the *p*-channel SnO-TFT performance was improved by heating as-deposited SnO thin films at 360 °C in NH_3 atmosphere. This treatment may be understood as conversion of the deep V_O to electronically inactive Sn-H, since NH_3 is easily dissociated to H and NH_2 even at low temperatures.

7.5 Incorporation and Conversion of H Impurities

Figure 7.10 summarizes the features of the various *a*-IGZO thin films that were deposited at different conditions and annealed under different atmospheres. Hydrogen impurities are incorporated at surprisingly high concentrations (>10^{20} cm^{-3}) for AOS thin films sputter-deposited at STD (base pressure ~10^{-4} Pa). This concentration can be reduced by an order of magnitude by deposition in higher vacuum conditions (UHV, ~10^{-7} Pa). The chemical forms of impurity hydrogen are OH and M-H, and each of them has comparable concentrations. Carrier concentrations in the former STD thin films are ~10^{16} cm^{-3}, which works as the TFT channel, but the latter UHV thin films are not suitable due to too-low carrier concentrations to apply TFTs. Orui et al. [23] proposed that this carrier reduction in the latter was ascribed to the presence of excess oxygen, which compensates the shallow V_O based on an observation of distinct reduction of the deep V_O giving the subgap absorption above the VBM. The UHV chamber contains less residual H_2 concentration, and thereby the conventionally applied partial pressure of O_2 gas appropriate to fabricate AOS thin films for TFTs becomes excess because of less consumption of O_2 due to less formation

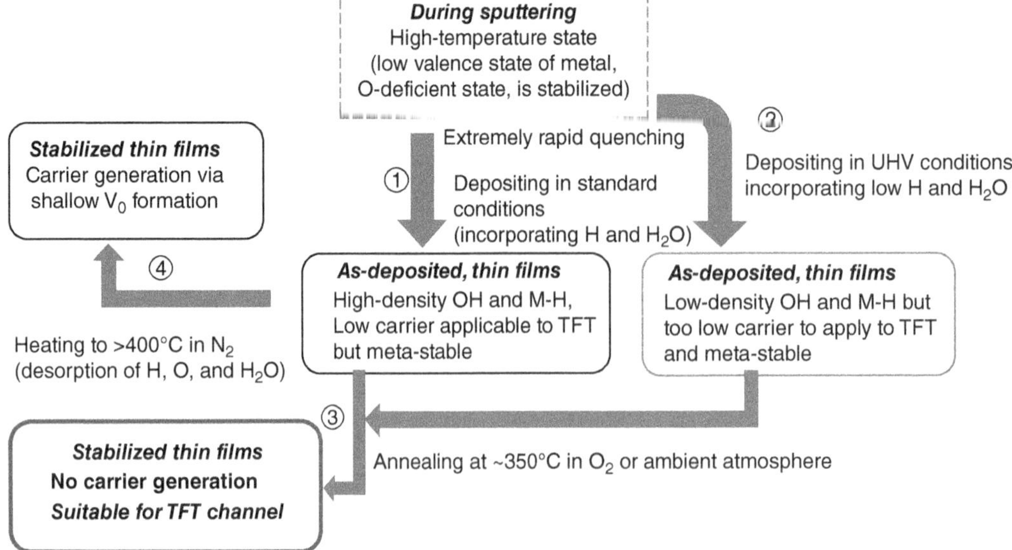

Figure 7.10 Incorporation and conversion of hydrogen-related species in AOS thin films.

of H_2O through the reaction with residual H_2. These results may be understood by considering the sputtering process. The gas plasma state and collision of sputtering ions with an oxide ceramic target create a high-temperature state, which makes the lower valence state of metal oxygen favorable (i.e., oxygen-deficient states are stabilized). Since sputter deposition is an extremely rapid quenching process (quenching from the high-temperature gas and high-energy precursor states to the stabilized states at the growing surface), the resulting thin films have highly strained M-O-M and oxygen vacancies. Hydrogen and water are the main residual gas species in the standard vacuum chamber. Water molecules react with the strained M-O-M bonds to form OH, while H_2 does this with the oxygen vacancy to give M-H bonds.

When the as-deposited STD thin films are heated to >400 °C in an N_2 atmosphere, the carrier concentration increases to $\sim 10^{20}$ cm^{-3}, which is too high to be applied to a TFT channel. This carrier generation originates from the shallow oxygen vacancy, which is created by desorption of H_2O and H_2 from OH and M-H following Equation (7.3). When the as-deposited thin films are annealed in an O_2 atmosphere around ~350 °C, carrier generation is suppressed and structurally stabilized thin films are obtained. Since a wet O_2 atmosphere promotes the oxygen diffusion and oxidation and the structural relaxation during annealing better than dry O_2 annealing, annealing at an ambient atmosphere (note that it includes H_2O) works well to produce good *a*-IGZO thin films for TFT channels [17, 18, 24]. However, we should note that wet O_2 annealing incorporates extra H into the *a*-IGZO films; therefore, it is not conformably concluded that wet O_2 annealing is better than dry O_2 annealing in terms of TFT stability [18].

7.6 Concluding Remarks

Roles of impurity hydrogen in semiconductors have been revealed over the past three decades, as represented by its roles as a hole compensator in GaN and electron donor in ZnO. However, hydrogen impurity in AOS thin films is rather different from the above examples in several aspects. First is its very high density of $\sim 5\times10^{20}$ cm^{-3}, which corresponds to ~1% of all the constituents. Second is that the concentration of hydride ion H^- taking the M-H is comparable to that of OH. The presence of such a large concentration of hydride ion was surprising to researchers in oxide semiconductors. The major driving forces of incorporating high-density H impurities in AOS thin films are the high affinity of oxygen to hydrogen and the structural flexibility of AOSs arising from the presence of free volume. Recent studies have revealed that hydride ions exist in bulk oxides and play important roles. For instance, the substitution of O^{2-} with H^- enabled the realization of heavy electron doping to 6×10^{21} cm^{-3}, which was impossible by F-doping to O^{2-} in iron-based superconductors, LnFeAsO (Ln: lanthanoid) [25]. This success is based on utilization of a reaction between the oxygen vacancy and H_2 at high pressure. This marked difference is tentatively attributed to that in host materials (i.e., crystal and amorphous materials). In amorphous materials, the effect of a dopant ion is locally compensated by a flexible amorphous structure to form a relaxed state. During the thin-film deposition process, oxygen vacancies easily react with H_2, the major residual gas, in the chamber at the thin-film surface. Since this reaction occurs near the top surface during deposition, the resulting H^- ions would be relaxed to form two stable M-H bonds without producing carriers. This situation rather changes for the reaction of H in postdeposited thin films with low E_{CBM} than $E(H^+/H^-)$ at low temperatures. Since the amorphous structure is rigid at low temperatures, H^0 injected to AOS ionizes to generate carrier e^- and OH, as described in Section 7.3.1. What is quite surprising is the high-density carrier generation up to $\sim 10^{20}$ cm^{-3} upon heating the as-deposited AOS thin films in an O_2/H_2O-free atmosphere. This carrier generation is attributed to the formation of an oxygen vacancy, which works as a shallow electron donor by dehydration reaction between M-H and MOH. These findings tell us that the major part of electron donors in AOS is passivated with hydrogen species.

In contrast to crystalline ZnO [5], and maybe in contrast to other transparent conductive oxides too, there has been no report that hydrogen incorporated in as-deposited AOS thin films forms a shallow donor as far as we know. This fact would be understood in terms of the formation of stable M-H bonds that give deep energy levels just above the VBM, not just below the CBM. Since the majority of oxygen vacancy in AOSs gives deep levels above the VBM, carrier concentrations in the as-deposited AOSs are retained at a relatively low ($\sim 10^{16}$ cm^{-3}) level, which is applicable to a TFT channel. This result would be beneficial for application to flexible electronics in which heat treatments at >150 °C are not allowed for conventional plastic substrates.

High concentrations of H impurities taking the form of M-OH and M-H appear to be specific to the sputter- or CVD-deposited AOS thin films. Chemical reactions of residual gases of H_2 and H_2O in the deposition chamber with the strained M-O bonds and oxygen vacancy V_O created during deposition are the primary origins of the M-OH and M-H bonds. Postannealing treatments to stabilize the as-deposited thin films often lead to high-density carrier generation via desorption of H_2 and H_2O through the reaction between M-H and M-H or M-OH. Careful annealing at an appropriate atmosphere and temperatures, as described in Figure 7.10, is required for TFT applications.

Acknowledgments

The authors thank Drs. J. Kim and K. Ide of Tokyo Tech for discussions. This work was supported in part by the MEXT Element Strategy Initiative to Form Core Research Center (No. JPMXP0112101001).

References

1 Spear, W.E. and Le Comber, P.G. (1975). Substitutional doping of amorphous silicon. *Solid State Communications* 17 (9): 1193–1196.
2 Nakamura, S., Mukai, T., Senoh, M., and Iwasa, N. (1992). Thermal annealing effects on p-type Mg-doped GaN films. *Japanese Journal of Applied Physics* 31 (2B): L139.
3 Kamiya, T. and Hosono, H. (2013). Roles of hydrogen in amorphous oxide semiconductor. *ECS Transactions* 54 (1): 103.
4 Bang, J., Matsuishi, S., and Hosono, H. (2017). Hydrogen anion and subgap states in amorphous In–Ga–Zn–O thin films for TFT applications. *Applied Physics Letters* 110 (23): 232105.
5 Hofmann, D.M., Hofstaetter, A., Leiter, F. et al. (2002). Hydrogen: a relevant shallow donor in zinc oxide. *Physical Review Letters* 88 (4): 045504.
6 Miyase, T., Watanabe, K., Sakaguchi, I. et al. (2014). Roles of hydrogen in amorphous oxide semiconductor In-Ga-Zn-O: comparison of conventional and ultra-high-vacuum sputtering. *ECS Journal of Solid State Science and Technology* 3 (9): Q3085.
7 Devine, R.A.B. (1996). Structural nature of the Si/SiO_2 interface through infrared spectroscopy. *Applied Physics Letters* 68 (22): 3108–3110.
8 Sameshima, T. and Satoh, M. (1997). Improvement of SiO_2 properties by heating treatment in high pressure H_2O vapor. *Japanese Journal of Applied Physics* 36: L687–L689.
9 Ramkumar, K. and Saxena, A.N. (1992). Stress in SiO_2 films deposited by plasma and ozone tetraethylorthosilicate chemical vapor deposition processes. *Journal of the Electrochemical Society* 139 (5): 1437.
10 Hosono, H., Abe, Y., Imagawa, H. et al. (1991). Experimental evidence for the Si-Si bond model of the 7.6-eV band in SiO_2 glass. *Physical Review B* 44 (21): 12043.
11 Imai, H., Arai, K., Imagawa, H. et al. (1988). Two types of oxygen-deficient centers in synthetic silica glass. *Physical Review B* 38 (17): 12772.
12 Nomura, K., Kamiya, T., Ohta, H. et al. (2007). Local coordination structure and electronic structure of the large electron mobility amorphous oxide semiconductor In-Ga-Zn-O: experiment and ab initio calculations. *Physical Review B* 75 (3): 035212.

13 Kamiya, T., Nomura, K., Hirano, M., and Hosono, H. (2008). Electronic structure of oxygen deficient amorphous oxide semiconductor *a*-$InGaZnO_{4-x}$: optical analyses and first-principle calculations. *Physica Status Solidi C* 5 (9): 3098–3100.

14 Li, H., Guo, Y., and Robertson, J. (2017). Hydrogen and the light-induced bias instability mechanism in amorphous oxide semiconductors. *Scientific Reports* 7 (1): 16858.

15 Hosono, H., Kikuchi, N., Ueda, N., and Kawazoe, H. (1995). Amorphous transparent electroconductor 2CdO·GeO_2: conversion of amorphous insulating cadmium germanate by ion implantation. *Applied Physics Letters* 67 (18): 2663–2665.

16 Van de Walle, C.G. and Neugebauer, J. (2003). Universal alignment of hydrogen levels in semiconductors, insulators and solutions. *Nature* 423: 626–628.

17 Nomura, K., Kamiya, T., Ohta, H. et al. (2008). Defect passivation and homogenization of amorphous oxide thin-film transistor by wet O_2 annealing. *Applied Physics Letters* 93 (19): 192107.

18 Ide, K., Nomura, K., Hosono, H., and Kamiya, T. (2019). Electronic defects in amorphous oxide semiconductors: a review. *Physica Status Solidi A* 216 (5): 1800372.

19 Nomura, K., Kamiya, T., Ikenaga, E. et al. (2011). Depth analysis of subgap electronic states in amorphous oxide semiconductor, a-In-Ga-Zn-O, studied by hard x-ray photoelectron spectroscopy. *Journal of Applied Physics* 109 (7): 073726.

20 O'Reilly, E.P. and Robertson, J. (1983). Theory of defects in vitreous silicon dioxide. *Physical Review B* 27 (6): 3780.

21 Pacchioni, G. and Ieraño, G. (1998). Ab initio theory of optical transitions of point defects in SiO_2. *Physical Review B* 57 (2): 818.

22 Lee, A.W., Le, D., Matsuzaki, K., and Nomura, K. (2020). Hydrogen-defect termination in SnO for p-channel TFTs. *ACS Applied Electronic Materials* 2 (4): 1162–1168.

23 Orui, T., Herms, J., Hanyu, Y. et al. (2015). Charge compensation by excess oxygen in amorphous In–Ga–Zn–O films deposited by pulsed laser deposition. *Journal of Display Technology* 11 (6): 518–522.

24 Watanabe, K., Lee, D.H., Sakaguchi, I. et al. (2013). Surface reactivity and oxygen migration in amorphous indium-gallium-zinc oxide films annealed in humid atmosphere. *Applied Physics Letters* 103 (20): 201904.

25 Iimura, S., Matsuishi, S., Sato, H. et al. (2012). Two-dome structure in electron-doped iron arsenide superconductors. *Nature Communications* 3: 943.

Part III

Processing

8

Low-Temperature Thin-Film Combustion Synthesis of Metal-Oxide Semiconductors: Science and Technology

Binghao Wang[1,2], *Wei Huang*[1], *Antonio Facchetti*[2,3], *and Tobin J. Marks*[2]

[1]*Joint International Research Laboratory of Information Display and Visualization, Key Laboratory of MEMS of Ministry of Education, School of Electronic Science and Engineering, Southeast University, Nanjing, China*
[2]*Department of Chemistry and the Materials Research Center, Northwestern University, Evanston, Illinois, USA*
[3]*Flexterra, Skokie, Illinois, USA*

8.1 Introduction

Amorphous metal-oxide (a-MO) semiconductors are an exceptionally promising class of electronic materials that have made impressive progress since the first 2004 demonstration of their utility as the semiconductor layer in thin-film transistors (TFTs) [1–5]. Due to their outstanding electrical and optical properties, remarkable large-area uniformity, and mechanical flexibility, a-MO semiconductors challenge amorphous and polycrystalline silicon not only for conventional display and energy applications, but also in new fields such as flexible/wearable electronics [6–9]. Importantly, unlike silicon, which can only be deposited by vacuum-based growth technologies or using highly flammable silane precursors [10], a-MO films can be deposited from inert metal salt solutions under ambient conditions by low-cost, high-throughput coating and printing processes [4, 11]. Since this chapter primarily focuses on the use of metal oxides (MOs) in TFTs, we first briefly describe the device structure and operation. Figure 8.1a and 8.1b shows the schematic structure of a bottom-gate/top-contact TFT and a photo of the first flexible indium–gallium–zinc oxide (IGZO) TFT reported in 2004. The output and transfer plots for an n-channel TFT are shown in Figure 8.1c and 8.1d. Specifically, the source–drain current (I_D) begins to increase when the gate voltage (V_G) surpasses the turn-on voltage (V_{ON}), which creates mobile charge carriers (here, electrons) in the semiconductor adjacent to the interface with the dielectric layer. Other important TFT performance metrics extracted from these plots are the field-effect mobility (μ), threshold voltage (V_T), current on/off ratio (I_{on}/I_{off}), and subthreshold swing (SS).

TFTs fabricated using solution-processed MO films typically require postdeposition annealing of the MO precursor(s) films at elevated temperatures >400 °C to ensure high-quality films and stable charge transport characteristics [12]. Thus, MO fabrication on, and TFT integration with, inexpensive and typically temperature-sensitive polymer substrates remain challenging. Furthermore, reducing MO processing temperature and time is of paramount importance for implementing in fabrication (FAB) processing lines for commercializing an MO TFT technology [13]. Consequently, new processing methodologies addressing these challenges are being developed, such as combustion synthesis, high-pressure annealing (HPA), microwave-assisted annealing, and photonic annealing, and have been widely used for various types of MO films (Figure 8.2) [14]. However, to date, no commonly accepted standard method of (post)processing MO films on flexible substrates has been identified [15]. Thus, despite significant progress, the microwave-assisted annealing, HPA, and photonic annealing require costly external setups during postannealing. To this end, combustion synthesis is a promising solution to minimize the manufacturing costs and overall production throughput [16, 17].

Amorphous Oxide Semiconductors: IGZO and Related Materials for Display and Memory, First Edition.
Edited by Hideo Hosono and Hideya Kumomi.

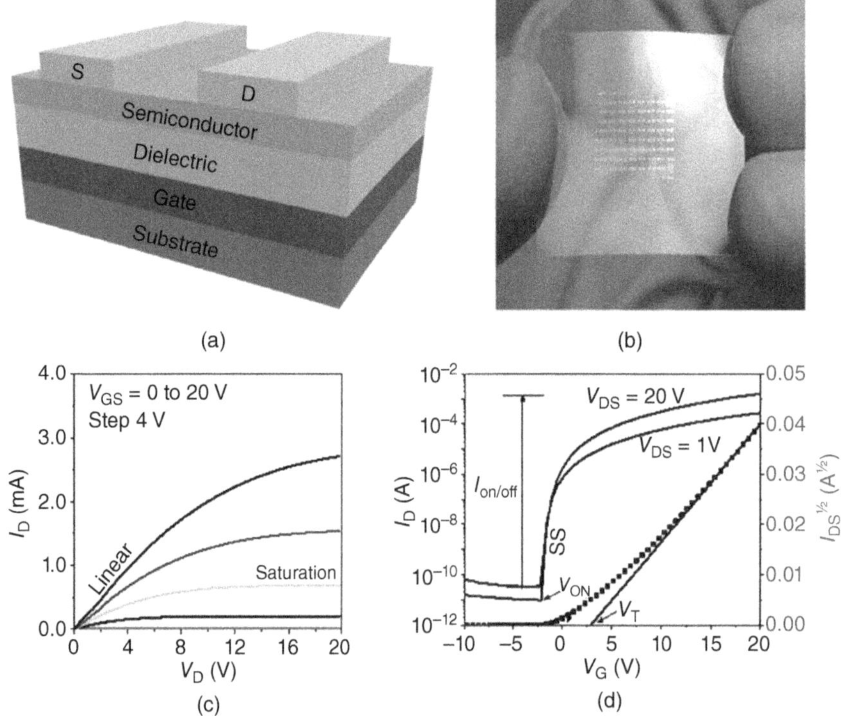

Figure 8.1 (a) Schematic structure of a bottom-gate/top-contact TFT. (b) A photo of the first flexible IGZO TFTs on a polyethylene terephthalate substrate. Examples of (c) output characteristics and (d) transfer characteristics of an *n*-channel MO TFT. *Source*: (a,c,d) Based on Ref. [5], Nomura, K., et al. (2004). Room-temperature fabrication of transparent flexible thin-film transistors using amorphous oxide semiconductors. *Nature* **432**: 488–492. (b) From Ref. [5], fig. 03, p. 490, with permission of *Springer Nature*. https://doi.org/10.1038/nature03090.

In Section 8.2, we discuss methodologies to solution-process metal-oxide semiconducting films at low temperature, except for combustion synthesis. In the sections that follow, we then focus on thin-film combustion synthesis of MO films and recent studies addressing both *n*-type and *p*-type MO TFTs. Finally, we conclude with a summary, and overview the major challenges and opportunities in this field.

8.2 Low-Temperature Solution-Processing Methodologies

In this section, we summarize the pioneering studies addressing MO solution methodologies that effectively reduced the processing temperatures, highlighting the strengths and weaknesses of each method.

8.2.1 Alkoxide Precursors

In 2010, the Sirringhaus group reported organic–inorganic metal alkoxides of the metal ions used for the synthesis of indium–zinc oxide (IZO) and IGZO films [18]. Specifically, the indium alkoxide cluster $[In_5(\mu^5\text{-}O)(\mu^3\text{-}OPr^i)_4(\mu^2\text{-}OPr^i)_4(OPr^i)_5]$, zinc-*bis*-methoxyethoxide, $[Zn(OC_2H_4OCH_3)_2]$, and gallium-*tris*-isopropoxide $[Ga(OCH(CH_3)_2)_3]$ were used as In, Zn, and Ga sources, respectively (Figure 8.3a). These metal alkoxides

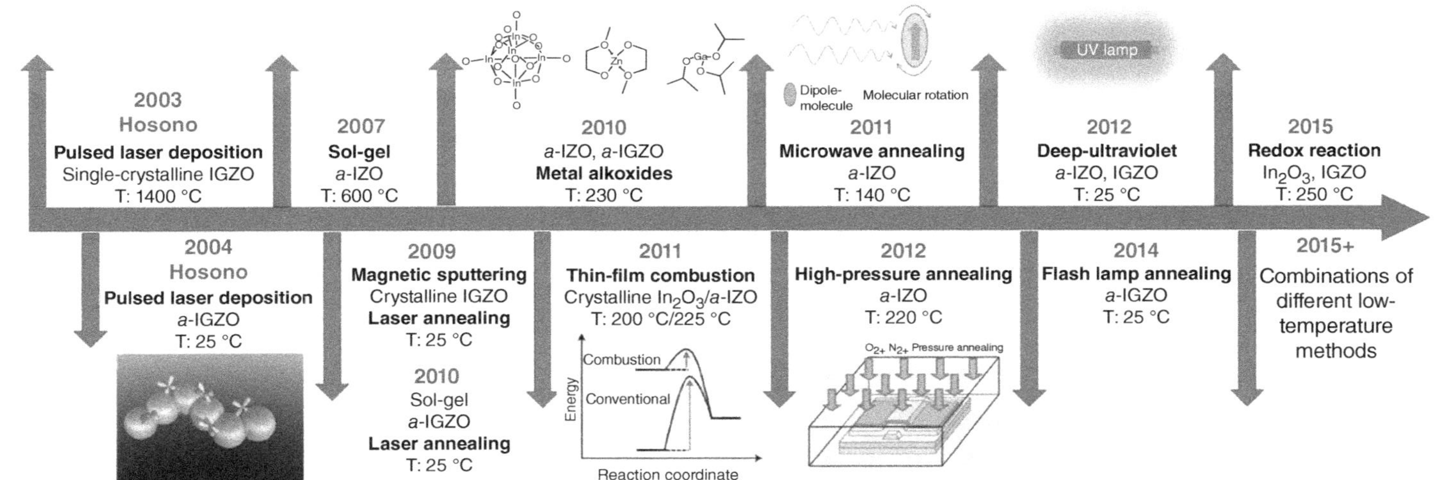

Figure 8.2 Development timeline of metal-oxide semiconductors composed of post-transition metal cations. *Source*: Reproduced with permission from Ref. [5], Nomura, K., et al. (2004). Room-temperature fabrication of transparent flexible thin-film transistors using amorphous oxide semiconductors. *Nature* **432**(7016): 488–492. Copyright 2004 Nature Publishing Group. Reproduced with permission from Ref. [18], Banger, K. K., et al. (2010). Low-temperature, high-performance solution-processed metal oxide thin-film transistors formed by a "sol–gel on chip" process. *Nature Materials* **10**(1): 45–50. Reproduced with permission from Ref. [19], Kim, M.-G., et al. (2011). Low-temperature fabrication of high-performance metal oxide thin-film electronics via combustion processing. *Nature Materials* **10**(5): 382–388. doi:10.1038/nmat3011. Reproduced with permission from Ref. [20], Song, K., Young Koo, C., Jun, T., Lee, D., Jeong, Y., & Moon, J. (2011). Low-temperature soluble InZnO thin film transistors by microwave annealing. *Journal of Crystal Growth* **326**(1): 23–27. Copyright 2011 Elsevier B.V. Reproduced with permission from Ref. [21], Rim, Y. S., et al. (2012). Simultaneous modification of pyrolysis and densification for low-temperature solution-processed flexible oxide thin-film transistors. *Journal of Materials Chemistry* **22**(25): 12491. Copyright 2012 The Royal Society of Chemistry. Reproduced with permission from Ref. [15], Yarali, E., et al. (2019). Recent progress in photonic processing of metal-oxide transistors. *Advanced Functional Materials* 1906022. Copyright 2019 Wiley-VCH Verlag GmbH & Co. KGaA, Weinheim.

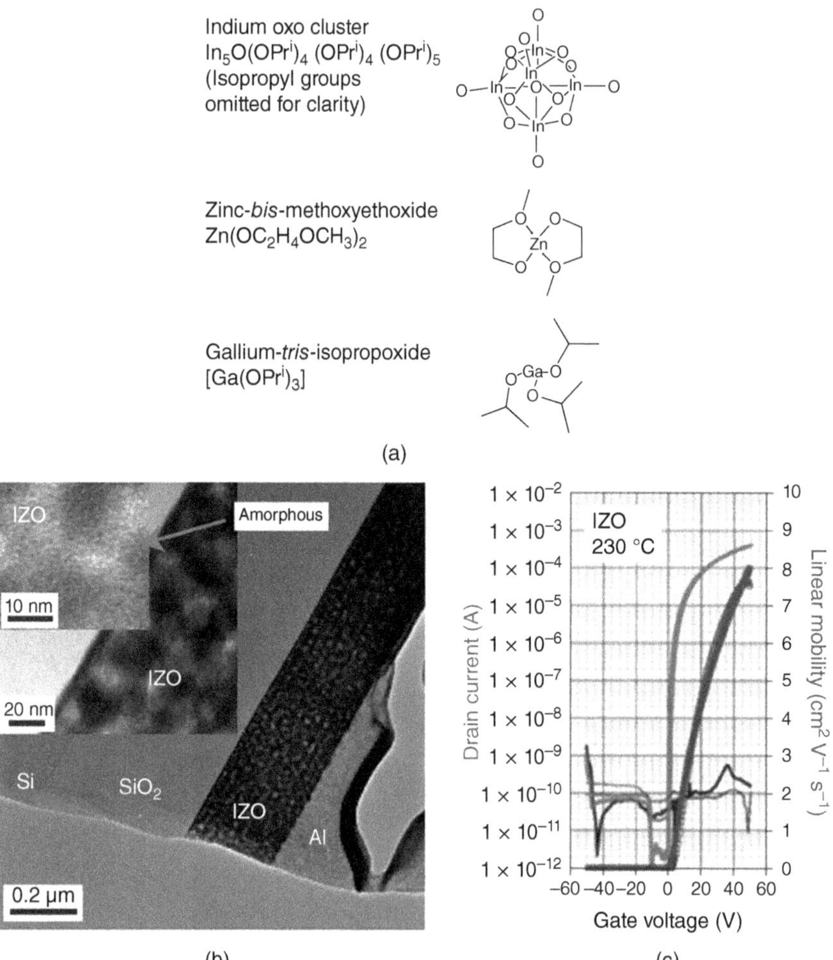

Figure 8.3 (a) Chemical structures of the metal alkoxide precursors. (b) Cross-sectional TEM image of an IZO film fabricated on a SiO_2/Si substrate. (c) Transfer curve of a TFT based on 230 °C–annealed IZO film. *Source*: (a,c) Reproduced with permission from Ref. [18], Banger, K. K., et al. (2010). Low-temperature high-performance solution-processed metal oxide thin-film transistors formed by a "sol–gel on chip" process. *Nature Materials* **10**(1): 45–50. (b) From Ref. [18], fig. 02, p. 47, with permission of *Springer Nature*. https://doi.org/10.1038/nmat2914.

enabled in-situ hydrolysis of the coated precursor films into metal-oxide films at temperatures as low as 200 °C. The selected-area electron diffraction of the resulting IZO films showed formation of an amorphous phase (Figure 8.3b). The IZO films had a smooth surface topography with a root mean square (rms) surface roughness of 0.15 nm. As shown in Figure 8.3c, the IZO (70/30 = at.% In/Zn) TFTs fabricated at 230 °C on a 100 nm-thick SiO_2 dielectric (C_i = 34 nF cm^{-2}) showed a maximum field-effect mobility (μ) of ~8 cm^2 V^{-1} s^{-1}, a high on/off current ratio (I_{on}:I_{off}) of ~10^8, and good bias-stress stability (ΔV_T = 3.4 V after 14 h bias stress). TFTs based on IGZO films (5% Ga) exhibited a μ of ~4.0 and 6.1 cm^2 V^{-1} s^{-1} when fabricated at 275 °C and 300 °C, respectively. This approach showed for the first time that high-performance solution-processed MO TFTs can function well also if processed at relatively low temperatures. However, no additional studies have utilized this approach, possibly because of the labor-intensive and costly syntheses of the metal alkoxide precursors.

8.2.2 Microwave-Assisted Annealing

The Moon group reported the microwave-assisted annealing of solution-processed zinc oxide (ZnO) and IZO TFTs in 2011 (Figure 8.4a) [20, 22]. Microwave radiation can produce dense MO films via efficient and large volumetric energy realized by the conversion of vibrational energy into thermal activation energy [23]. For the ZnO film fabrication, the authors first prepared an aqueous solution of zinc hydroxide (0.1 M) and ammonium hydroxide as the metal precursor solution. After spin-coating, the corresponding films were annealed at 140, 220, and 320 °C for 30 min in a microwave oven. Atomic force microscopy (AFM) images showed that the average grain size of the microwave-annealed ZnO films was larger (~23 nm) than that of the hotplate samples annealed at the same processing temperatures (~12 nm). The resulting ZnO TFTs on 100-nm-thick SiO_2/Si substrates fabricated at 140 °C and 220 °C exhibited an average μ of 1.7 $cm^2\ V^{-1}\ s^{-1}$ and 2.8 $cm^2\ V^{-1}\ s^{-1}$, respectively. Regarding IZO (55/45 = at.% In/Zn) film fabrication, β-diketone was added to the metal salt precursor solution (indium nitrate and zinc nitrate) for stabilization and providing additional combustion heat during the microwave-assisted annealing. Cross-sectional transmission electron microscopy (TEM) images of the 13-nm-thick IZO films microwave-annealed at 150 °C indicate the formation of a quasi-amorphous phase with 3–4 nm size nanocrystals. As shown in Figure 8.4b and 8.4c, the microwave-annealed IZO TFTs begin to function after 150 °C processing (μ = 0.1 $cm^2\ V^{-1}\ s^{-1}$), whereas the hotplate-annealed devices processed at 150 °C were inactive. The IZO TFTs showed enhanced device characteristics after 250 °C microwave-assisted annealing with a μ of 1.0 $cm^2\ V^{-1}\ s^{-1}$, I_{on}/I_{off} of 10^6, and *SS* of 0.8 V dec^{-1}. Since this demonstration, other groups reported the microwave-assisted annealing of several other MO systems, including IGZO and ZnSnO (ZTO), achieving μ of 3.8 $cm^2\ V^{-1}\ s^{-1}$ (300 °C) and 1.8 $cm^2\ V^{-1}\ s^{-1}$ (350 °C), respectively [24–28]. Nevertheless, microwave-assisted annealing may be difficult to incorporate into roll-to-roll industrial settings.

8.2.3 High-Pressure Annealing

The HPA process for MO TFTs was reported in 2011 for optimizing sputtered IGZO, yielding far more negative-bias illumination stress (NBIS)-stable devices [29, 30]. The next year, the Kim group applied HPA to solution-processed flexible MO TFTs, annealing at low temperatures [21]. The authors claimed that HPA influenced the thermodynamics of the decomposition reaction, resulting in more densified MO films (Figure 8.5a). Several types of MOs, including IZO (75/25 = at.% In/Zn), IGZO (67/11/22 = at.% In/Ga/Zn), and ZTO (67/33 = at.% Zn/Sn), were investigated under different pressures (0.1 MPa and 1.0 MPa) and gas atmospheres (O_2, N_2, and air). Figure 8.5b and 8.5c shows that the thickness of IZO film annealed under 1 MPa of O_2 was about 23 nm, thinner than that (28 nm) of IZO film annealed in air. As shown in Figure 8.5d, the IZO TFTs (200-nm-thick SiO_2 gate dielectric) processed under optimized conditions (1.0 MPa O_2, 220 °C) have a μ of 2.43 $cm^2\ V^{-1}\ s^{-1}$, *SS* of 0.56 V dec^{-1}, and I_{on}/I_{off} = 10^6, respectively, while the IGZO and ZTO TFTs processed under similar conditions exhibited a μ of 1.81 $cm^2\ V^{-1}\ s^{-1}$ and 0.85 $cm^2\ V^{-1}\ s^{-1}$, respectively. In later work, the optimized HPA condition (1.0 MPa O_2) was used for IGZO (62.5/12.5/25 = at.% In/Ga/Zn) and IGO (75/25 = at.% In/Ga) TFTs, which exhibited a μ of 3.3 and 2.2 $cm^2\ V^{-1}\ s^{-1}$, respectively, after 350 °C/1 h annealing [31, 32]. Similar to microwave annealing, the HPA requires a closed high-pressure environment, making integration into roll-to-roll processing daunting.

8.2.4 Photonic Annealing

The direct interaction of light with MO precursor films promotes photochemically-induced precursor decomposition and MO film synthesis/formation without the need of a high thermal budget (Figure 8.6) [15]. To this end, light–matter interactions offer precise energy delivery close to the surface and control over the physicochemical processes, leading to temperature gradients throughout the entire structure, which allows the substrate

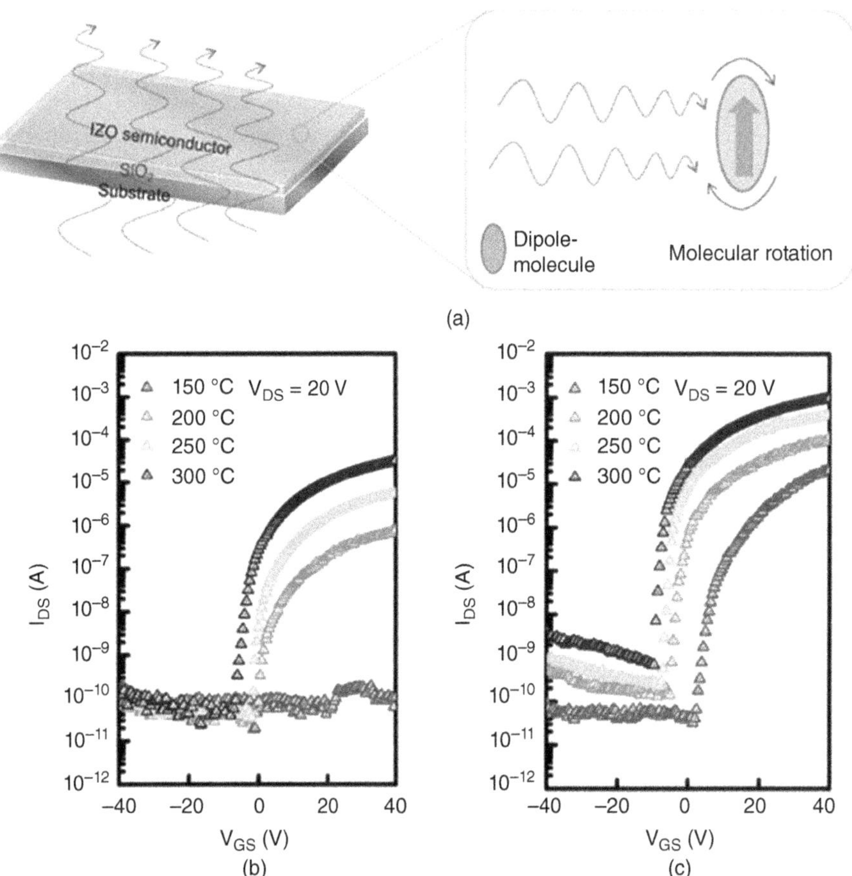

Figure 8.4 (a) Schematic of the device structure and the microwave annealing process. Transfer curves of (b) hot plate–annealed and (c) microwave-annealed IZO TFTs fabricated on SiO_2/ Si substrates at different temperatures. *Source:* (a–c) Reproduced with permission from Ref. [20], Song, K., et al. (2011), Low-temperature soluble InZnO thin film transistors by microwave annealing. *Journal of Crystal Growth* **326**: 23–27. Copyright 2011 Elsevier B.V.

material to remain intact and at significantly lower temperatures than the upper MO precursor film undergoing chemical transformation. Here we report three examples of these processes for producing TFT metal-oxide films.

8.2.4.1 Laser Annealing

Laser annealing is a simple and versatile methodology enabling freedom of patterning design and fast processing while offering compatibility with large-scale manufacturing on inexpensive flexible substrates [15]. In most cases, the wavelength of the laser light is in the ultraviolet (UV) region, ensuring the highest possible energy absorption by most MO precursors. The monochromatic intense light beam induces localized heating over an area of micrometers and a time of a few nanoseconds.

In 2009 Nakata et al. reported a series of pioneering studies demonstrating excimer laser annealing on sputtered IGZO (33.3/33.3/33.3 = at.% In/Ga/Zn) films. A XeCl excimer laser with a wavelength of 308 nm and a beam width of 400 μm was used. The 20-nm-thick IGZO films were sputter-grown under a gas mixture of argon and oxygen at a gas pressure of 0.5 Pa and a radiofrequency (RF) power density of 2 W cm^{-2}, then laser irradiation was carried out at room temperature in ambient air. The bottom-gate/top-contact TFTs based on IGZO films irradiated

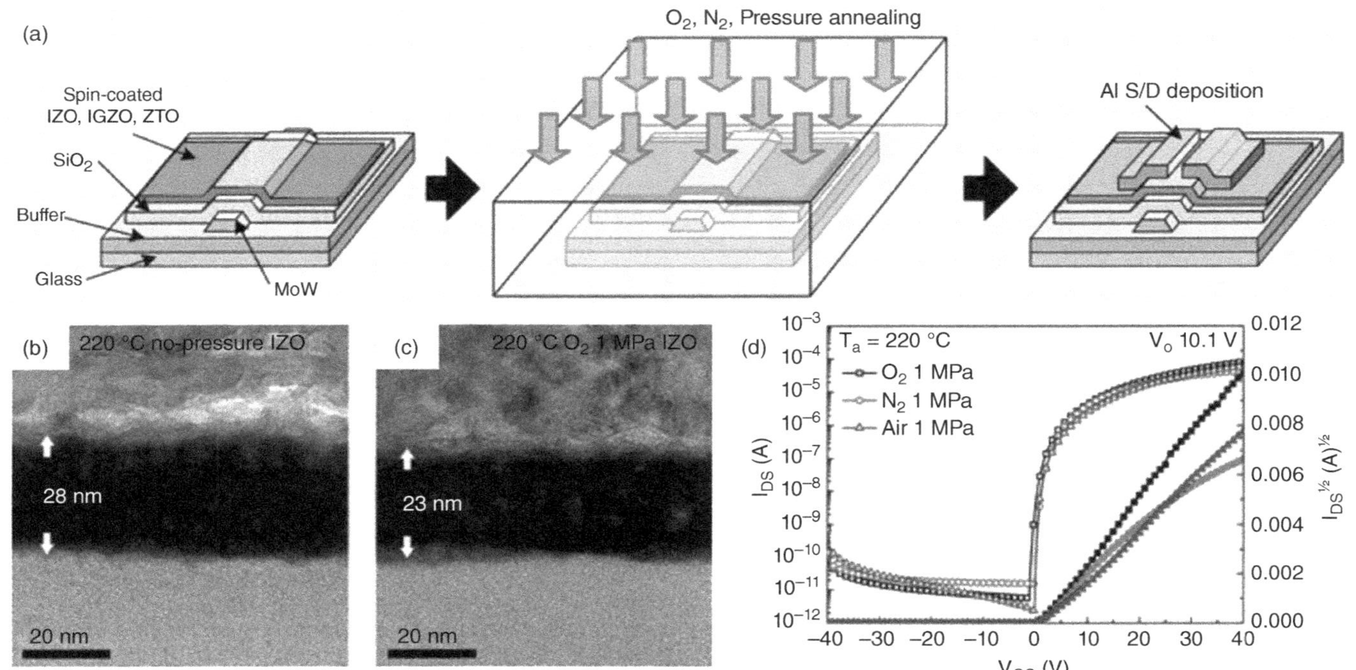

Figure 8.5 (a) HPA (high-pressure annealing) for solution-processed MO TFTs. Cross-sectional TEM image of an IZO thin film annealed at 220 °C (b) in air and (c) in 1 MPa O_2. (d) Transfer curves of TFTs based on IZO films annealed at different HPA atmosphere. *Source*: (a,d) Reproduced with permission from Ref. [20], Song, K., et al. (2011), Low-temperature soluble InZnO thin film transistors by microwave annealing. *Journal of Crystal Growth* **326**, 23–27. Copyright 2011 Elsevier B.V. (b) From Ref. [20], fig. 02, p. 12792, with permission of Royal Society of Chemistry. https://doi.org/10.1039/C2JM16846D.

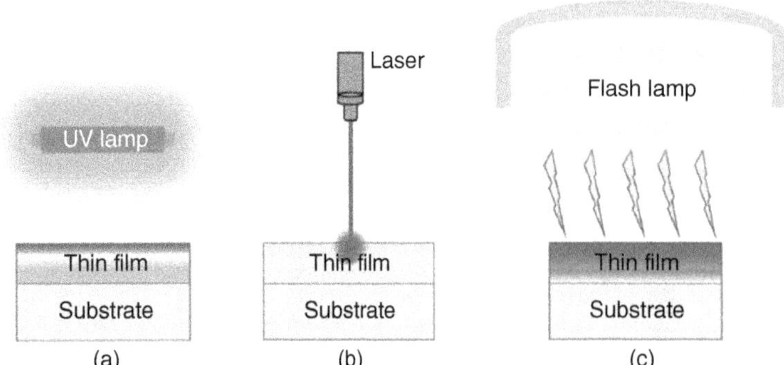

Figure 8.6 Schematic drawings representing: (a) photochemical reaction induced by continuous exposure to UV light, (b) laser annealing, and (c) flash lamp annealing. *Source*: (a–c) Reproduced with permission from Ref. [15], Yarali, E., et al. (2019). Recent progress in photonic processing of metal-oxide transistors. *Advanced Functional Materials* 1906022. Copyright 2019 Wiley-VCH Verlag GmbH & Co. KGaA, Weinheim.

with a total energy density of 130 mJ cm^{-2} exhibited a high μ of 14 $cm^2\ V^{-1}\ s^{-1}$ using 100-nm-thick SiO_2 as the gate dielectric [33, 34]. A year later, Yang et al. applied laser annealing to spin-coated IGZO (33.3/33.3/33.3 = at.% In/Ga/Zn) films [35]. The authors irradiated the spin-coated precursor films using Nd:YAG laser pulses with a pulse duration of 5 ns ($\lambda = 355$ nm) at a repetition rate of 50 Hz. Each pulse provided an energy density of 15 mJ cm^{-2}, and the irradiation was carried out at room temperature in air. The TFTs made from IGZO films irradiated at a laser dosage of 740 mJ exhibited superior performance with a μ of 7.65 $cm^2\ V^{-1}\ s^{-1}$, the highest value reported up to then for solution-processed MO TFTs fabricated using a solution process methodology. However, lasers have inherent limitations in the size of the operating area, limited by their beam size in addition to their high cost.

8.2.4.2 Deep-Ultraviolet Illumination

The most frequently used type of deep-ultraviolet (DUV) lamp is the low-pressure mercury-based lamp, which features strong and clearly defined emission peaks at 253.7 nm and 184.9 nm. In the case of solution-processed MO films, DUV photochemical reactions can either be carried out by photoinduced chemical conversion of the MO precursor to a stable metal-oxide phase or used as an auxiliary tool in combination with thermal annealing, which could thus be carried out at lower temperatures. The Salleo group was the first to report the use of UV irradiation with a high-pressure mercury lamp for densifying ZrO_2 dielectric films [36]. In 2012, Park's group reported for the first time MO semiconductor films and corresponding MO TFTs via DUV irradiation [37]. Indium nitrate, gallium nitrate, and zinc acetate dissolved in 2-methoxyethanol were used as In, Ga, and Zn sources, respectively. Film fabrication involved spin-coating of the sol–gel precursors followed by DUV irradiation under inert N_2 atmosphere. Figure 8.7a and 8.7b shows that the as-spun films (25–35 nm thick) before DUV irradiation contain a significant amount of residual organic components, as confirmed by a high carbon content in the film as determined by X-ray photoelectron spectroscopy (XPS). The high-energy DUV photons induce photochemical cleavage of alkoxy groups and facilitate M–O–M network formation. Note that although the DUV irradiation was conducted at room temperature, the process could increase the sample temperature up to 150 °C due to irradiation heating. The DUV-derived IGZO TFTs on a 200-nm-thick SiO_2/Si substrate showed comparable device performance ($\mu = 2.64\ cm^2\ V^{-1}\ s^{-1}$) to those fabricated by thermal annealing at 350 °C. Figure 8.7c shows the saturation mobility distribution of photo-annealed and 350 °C–annealed oxide TFTs with a channel length and width of 10 and 100 μm, respectively, and with a 35-nm-thick Al_2O_3 as the gate dielectric (capacitance = 138 nF cm^{-2}) on glass substrates. The photo-annealed TFTs exhibited a μ of 8.8 $cm^2\ V^{-1}\ s^{-1}$ for IGZO, 4.4 $cm^2\ V^{-1}\ s^{-1}$ for IZO, and 11.3 $cm^2\ V^{-1}\ s^{-1}$ for In_2O_3. Compared with the TFTs annealed at 350 °C, the photo-annealed devices

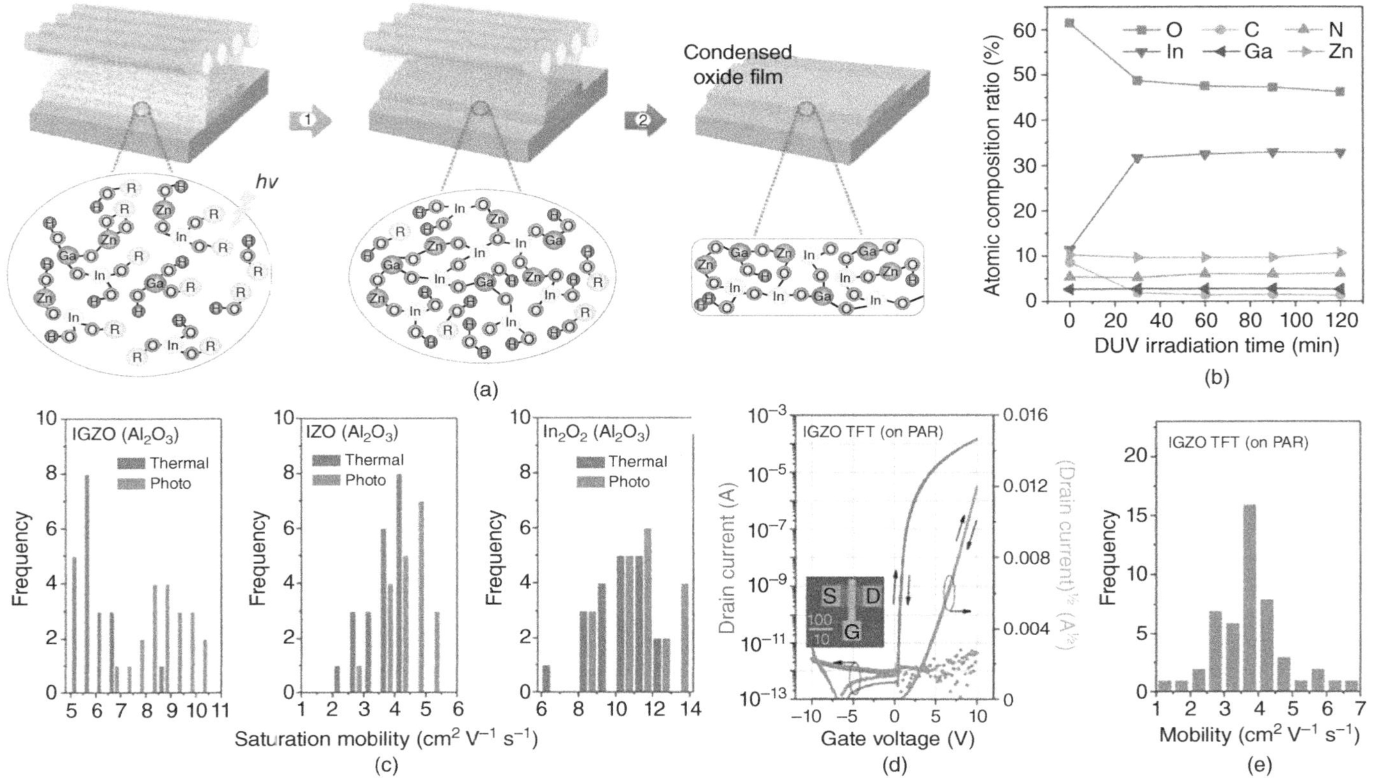

Figure 8.7 (a) Schemes showing the condensation mechanism of metal-oxide precursors by DUV irradiation. (b) Atomic composition ratios of IGZO thin films as a function of DUV irradiation time. (c) Saturation mobility distribution of photo-annealed and 350 °C–annealed IGZO, IZO, and In_2O_3 TFTs. (d) Transfer characteristics and (e) distribution of saturation mobilities of photo-annealed flexible IGZO TFTs. *Source*: (a–e) Reproduced with permission from Ref. [37], Kim, Y.-H., et al. (2012). Flexible metal-oxide devices made by room-temperature photochemical activation of sol–gel films. *Nature* **489**(7414): 128–132. Doi:10.1038/nature11434.

exhibit comparable or enhanced charge transport characteristics. Figure 8.7d and 8.7e show typical device characteristics for photo-annealed IGZO TFTs on polyarylate substrates. The devices exhibited a μ of 3.77 $cm^2\ V^{-1}\ s^{-1}$ with a narrow distribution, high I_{on}/I_{off} of 10^8, V_T of 2.7 V, and *SS* of 95.8 mV dec^{-1}. Although DUV irradiation enables MO film processing at a low temperature of <150 °C, the process accelerates the degradation of the plastic substrates and should be carried out under inert atmosphere to prevent reactive ozone formation [15]. However, since this demonstration, the UV (wavelength <400 nm) and far-UV (wavelength <200 nm) irradiation, along with DUV (wavelength <300 nm), have been widely investigated for various metal-oxide semiconductors, metal-oxide dielectrics, and direct patterning, to cite a few [38–42].

8.2.4.3 Flash Lamp Annealing

Flash lamp annealing (FLA) relies on the use of a flash lamp (e.g., xenon) that emits photons over a broad spectrum, from the UV to the near-infrared range (i.e., 200–1100 nm), to decompose MO precursors. During flashing, photons with different wavelengths interact with the absorptive layer, and they are capable of delivering thermal energy within a time range of microseconds to milliseconds [15]. As a result, high temperatures can be reached momentarily without subjecting the substrate to high thermal loads, due to the mismatch between the thermal transport properties of the substrate and the absorber layer. FLA has been utilized for the ultrafast fabrication of shallow Si junctions and has recently been adapted to MO film growth [43–45].

The use of FLA to anneal sputter-deposited IGZO films on Si/SiO_2 substrates was first reported by the Rack group [46]. As compared to IGZO (33.3/33.3/33.3 = at.% In/Ga/Zn) TFTs post-thermally annealed at 250 °C for 1 h in air (μ = 6.7 $cm^2\ V^{-1}\ s^{-1}$), the FLA-derived TFTs exhibited a higher μ of 7.8 $cm^2\ V^{-1}\ s^{-1}$ and enhanced bias-stress stability. FLA of solution-based IGZO films for TFTs was first reported by the Song group in 2014 [43]. Here, intense pulsed white light was used and offered the advantages of room-temperature processing and high processing speed in the order of milliseconds under ambient conditions. The TFTs based on spin-coated IGZO (50/25/25 = at.% In/Ga/Zn) films annealed by FLA irradiation exhibited improved electrical performance with a μ of 2.67 $cm^2\ V^{-1}\ s^{-1}$ and $I_{on}/I_{off} = 10^8$, while the μ of IGZO TFTs thermally annealed at 500 °C/1 h was only 0.3 $cm^2\ V^{-1}\ s^{-1}$. The unique combination of short processing time and unmatched scalability makes FLA a promising technique for solution-processed large-area MO electronics. However, the high-energy density radiation pulses instantaneously generate temperatures >1000 °C on the film surface, creating extremely large vertical temperature gradients. These temperature non-uniformities can yield non-uniform films, large thermal stresses, and permanently damaged underlying plastic substrates [47, 48].

8.2.5 Redox Reactions

Yang and Rim's groups reported a redox reaction for low-temperature MO processing in 2015 [49]. Chloride-based metal precursors have been widely used to fabricate MO films from solution, but they usually require post-deposition annealing ~500 °C to decompose into high-purity oxide films [50]. In this work, the authors used a combination of perchloric acid ($HClO_4$) and chloride-based metal precursors. $HClO_4$ is a strong oxidizer that does not form complexes with metal ions in solution; however, some metal perchlorates as solids form explosive mixtures with organics; accordingly, $HClO_4$ can effectively eliminate the Cl residues without introducing any new impurities via the redox reaction during thermal annealing (Figure 8.8a). Using this approach, In_2O_3 TFTs exhibited a high μ of ~8.8 $cm^2\ V^{-1}\ s^{-1}$ and an I_{on}/I_{off} of ~10^7 for a processing temperature of 250 °C for 4 h (Figure 8.8b and 8.8c). In particular, the μ increased by >20-fold compared to that (0.44 $cm^2\ V^{-1}\ s^{-1}$) of In_2O_3 TFTs based on an $InCl_3$ precursor. However, for IGZO (90/5/5 = at.% In/Ga/Zn) TFTs processed at 250 °C/4 h, the μ increased from 0.10 to 1.30 $cm^2\ V^{-1}\ s^{-1}$ via addition of $HClO_4$ and exhibited greater negative/positive bias-stress stability. The generality of this method was also demonstrated for both a high-*k* HfO_2 dielectric and amorphous ZTO semiconductor [51]. Specifically, the ZTO (50/50 = at.% Zn/Sn) precursor solution with a concentration of 0.2 M was prepared by dissolving $SnCl_2{\cdot}2H_2O$, $Zn(NO_3)_2{\cdot}H_2O$, and $HClO_4$ in 2-methoxyethanol. Next, the

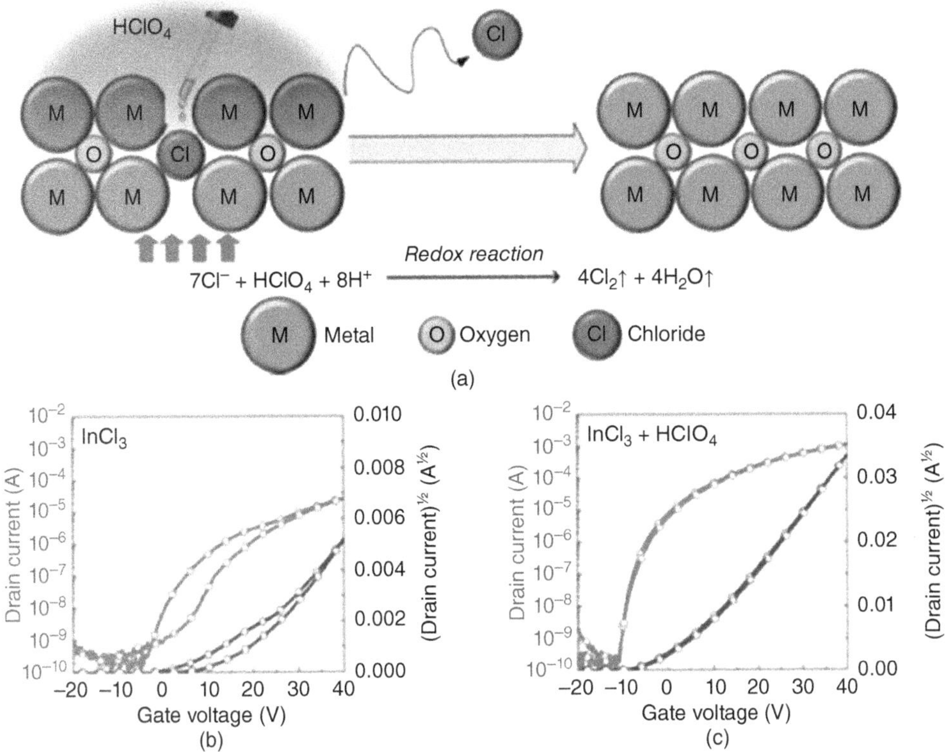

Figure 8.8 (a) Illustration of metal-oxide film growth using a redox reaction. Transfer characteristics of In_2O_3 TFTs. (b) $InCl_3$ precursor annealed at 250 °C. (c) Optimized $InCl_3$ + $HClO_4$ precursor annealed at 250 °C. *Source*: (a–c) Reproduced with permission from Ref. [49], Chen, H., et al. (2015). Low-impurity high-performance solution-processed metal oxide semiconductors via a facile redox reaction. *Chemistry of Materials* **27**(13): 4713–4718. Copyright 2015 American Chemical Society.

spin-coated ZTO precursor films were thermally annealed at 300–400 °C for 2 h. The resulting ZTO TFTs on a 300-nm-thick SiO_2/Si substrate annealed at 350 °C/2h exhibited the best electrical performance, including a high μ of 8.8 $cm^2\ V^{-1}\ s^{-1}$, a large I_{on}/I_{off} of ~10^8, and a small *SS* of 680 mV dec^{-1}.

Since some metal nitrates, such as those of Sn(IV) and Hf(IV), are not stable, the use of redox reactions may facilitate the decomposition of the chloride-based metal precursors into MO films upon milder heating. However, the processing temperature and/or processing time remain larger compared with those based on stable metal nitrate sources. For large-scale manufacture, any process using metal salts + organics with $HClO_4$ should avoid any conditions that could lead to explosions.

8.3 Combustion Synthesis for MO TFTs

Solution combustion synthesis is a process involving propagation of self-sustained exothermic reactions in aqueous or sol-gel media. Such media have the fuels and oxidizers that can react self-propagating when ignited and convert the precursors into solid products by self-generated heat. This process enables the synthesis of a variety of materials, including oxides, metals, alloys, and sulfides [16]. The solution combustion synthesis was first invented in the mid-1980s for the low-temperature thermal decomposition of metal hydrazinecarboxylate hydrates into metal-oxide powders [52]. Epifani et al. first reported the combustion synthesis of indium oxide (In_2O_3) semiconductor films in 2003 [53]. The process consisted of dissolving the appropriate amount of $In(NO_3)_3{\cdot}5H_2O$ and acetylacetone (AcAcH) with an In/AcAcH molar ratio of 2, followed by the addition of ammonium hydroxide (30 wt.% solution in water) with a NH_3/In molar ratio = 2. The films were prepared by spin coating 0.3 M solutions at 2000 rpm onto glass or alumina substrates. The 500 °C–annealed In_2O_3 films were used as O_3 gas sensors with a detection limit down to ppb level. In 2011, Kim et al. at Northwestern University implemented the first use of combustion synthesis for producing ultrathin and dense metal oxides at temperature as low as 200 °C for TFT applications [19]. Since then, numerous groups, including the Marks group, have used this methodology for many film growth processes [54–57]. In this section, we summarize these results by dividing them into *n*-type MO TFTs and *p*-type MO TFTs.

8.3.1 *n*-Type MO TFTs

In pioneering studies of the Northwestern group, AcAcH and urea were chosen as the fuels and metal nitrates as oxidizers for thin-film combustion growth of several MO films (Figure 8.9a) [19]. A critical factor for efficient combustion synthesis of TFT MO films is to achieve not only very pure MO compositions but also smooth and ultradense films, which favor charge transport. In typical combustion syntheses, films and powders are produced with great porosities due to the raid heating during the reaction and large gas evolution in a short time. To avoid porosity, low precursor concentrations (0.05 M) are used that afford ultrathin films incapable of trapping gas, thus preventing nanopore formation during the combustion process. Therefore, to produce films of thickness usable as a TFT channel (>10 nm, preferably ~50 nm), the spin-coating and combustion processes are repeated several times (in Northwestern work, typically four times) to achieve the desired film thickness.

The Northwestern studies have highlighted how the combustion composition, and those based on sol–gel processes, decompose upon thermal annealing (Figure 8.9a). Thus, thermogravimetric analysis (TGA) and differential thermal analysis (DTA) were conducted on dried conventional and combustion precursor compositions (Figure 8.9b). Unlike conventional systems, which exhibit broad endotherms for oxide lattice formation and exotherms for organic impurity removal, the combustion systems exhibit a single, intense exotherm in the DTA that corresponds exactly to the abrupt mass loss in the TGA and is sufficient to drive the reaction rapidly to completion. Both grazing incidence angle X-ray diffraction (XRD) and XPS of In_2O_3 precursors demonstrate combustion precursor conversion to the desired crystalline oxides and metal–oxygen–metal lattice, at far lower temperatures than for conventional precursors.

The TFTs used in this study utilized different substrates, gate contacts, gate insulators, and 30-nm-thick Al source and drain electrodes thermally evaporated on the MO films. As shown in Figure 8.9c, the combustion-derived MO TFTs exhibit much higher carrier mobility, and combustion-derived ITO films exhibited higher conductivity, especially those processed at lower temperature (<250 °C), than the conventional sol–gel counterparts. Specifically, the In_2O_3 TFTs on doped Si(gate)/300-nm-thick SiO_2 dielectric (C_i = 10.5 nF cm^{-2}) exhibited a μ of ~0.8 cm^2 V^{-1} s^{-1} and 3.4 cm^2 V^{-1} s^{-1} at 200 °C and 250 °C processing temperatures, respectively, while the IZO TFTs exhibited a μ of ~0.3 cm^2 V^{-1} s^{-1} and ~0.9 cm^2 V^{-1} s^{-1} at 225 °C and 250 °C processing temperatures, respectively. Furthermore, this approach was implemented for flexible In_2O_3 TFTs fabricated on an AryLite polyester substrate, a 30-nm-thick Al gate, and a ~40-nm-thick Al_2O_3 dielectric, achieving a μ of 6 cm^2 V^{-1} s^{-1} for thermal annealing at 200 °C (Figures 8.9d–8.9f).

In a significant advance, a two-component mixture of metal salts whose ligands function as both a fuel and oxidizer was developed by the Cho group in a process called self-combustion (Figure 8.10a) [55]. Zinc acetylacetonate (fuel) and indium nitrate (oxidizer) were dissolved in 2-methoxyethanol to form a 0.1 M precursor solution. After >6 h stirring, the IZO precursor was spin-coated on 300-nm-thick SiO_2/Si substrates. These films were prebaked at 100 °C for 10 min to remove the residual solvent and organic materials, followed by 250–350 °C/1 h annealing in ambient. Top-contact source and drain electrodes (100-nm-thick Al) were thermally evaporated through a metal

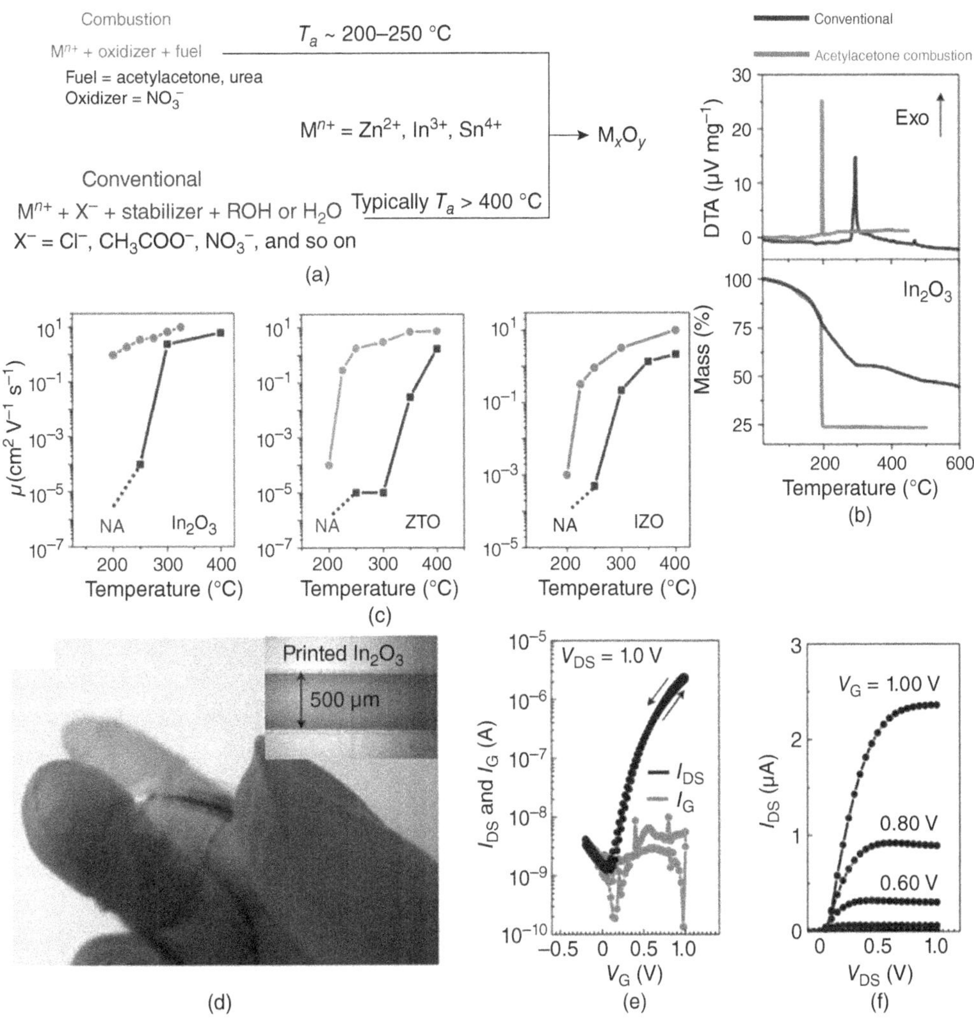

Figure 8.9 (a) Depiction of the approaches for combustion and conventional sol-gel film growth. (b) DTA and TGA of dried In_2O_3 precursors. (c) μ versus annealing temperature for In_2O_3 precursors, ZTO precursors, and IZO precursors. (d) Optical image of a flexible combustion-processed In_2O_3 device on an AryLite substrate. (e) Transfer characteristics and (f) output characteristics of flexible In_2O_3 TFTs. *Source*: (a–c,e,f) Reproduced with permission from Ref. [19], Kim, M. G., et al. (2011). Low-temperature fabrication of high-performance metal oxide thin-film electronics via combustion processing. *Nature Materials* **10**(5): 382–388. (d) From Ref. [19], fig. 05, p. 387, with permisision of *Springer Nature*. https://doi.org/10.1038/nmat3011.

shadow mask to complete the device. As shown in Figure 8.10b, the thermal analysis of the fuel and oxidizer precursors confirmed the generation of heat at ~200 °C. The IZO (50/50 = at.% In/Zn) TFTs began to function after 250 °C annealing with an average μ of 1.2 cm^2 V^{-1} s^{-1}, then it increased to 3.7 cm^2 V^{-1} s^{-1} and 8.7 cm^2 V^{-1} s^{-1}, respectively, after 300 °C and 350 °C annealing. In comparison, the 350 °C-derived IZO TFTs based on indium/zinc acetylacetonates and indium/zinc nitrates precursors showed an average μ of 0.7 cm^2 V^{-1} s^{-1} and 6.2 cm^2 V^{-1} s^{-1}, respectively (Figure 8.10c).

In 2013, Kim's group combined UV irradiation and combustion synthesis to achieve photo-patternable high-performance IGZO TFTs [56]. First, a 0.3 M In_2O_3 solution was prepared by dissolving indium nitrates in 2-methoxyethanol, then HNO_3 and AcAcH were added, and the solution was stirred/aged for 24 h. As shown in

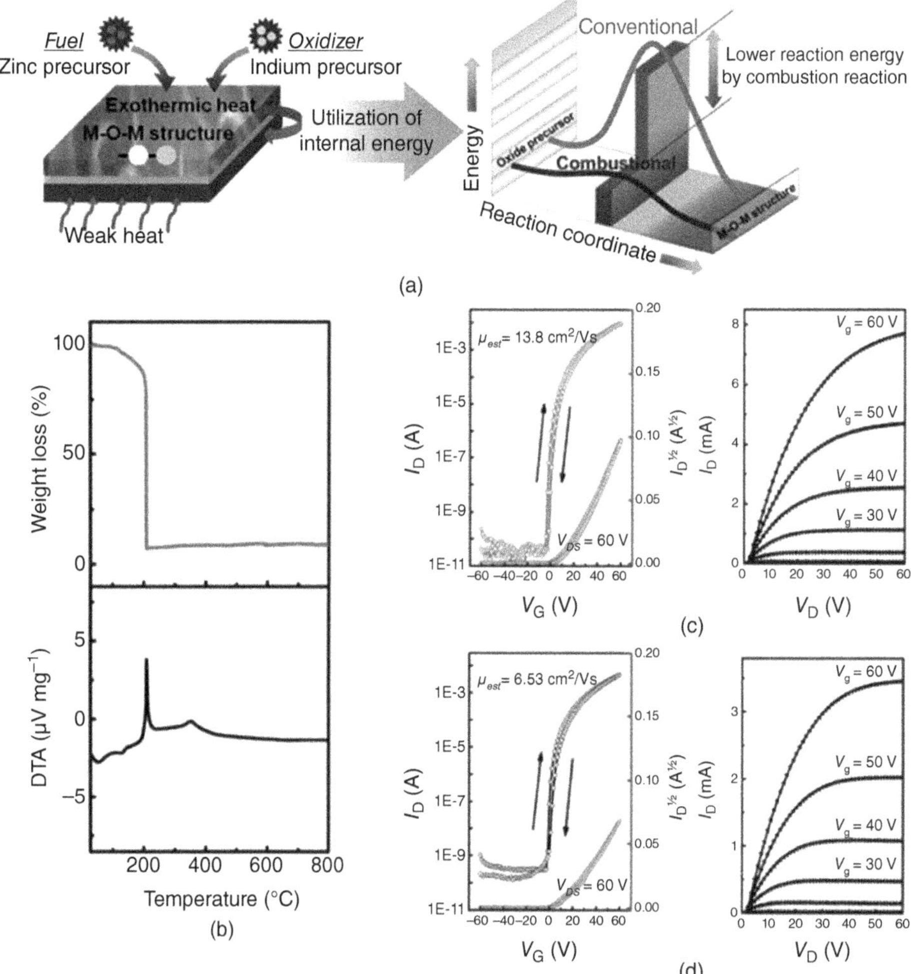

Figure 8.10 (a) A schematic diagram of the metal-oxide film synthesis via self-combustion of metal precursors bearing coordinated fuel and oxidizer ligands. (b) Thermal behavior of IZO precursors prepared from zinc acetylacetonate and indium nitrate. (c) Transfer and output characteristics of IZO TFTs fabricated from zinc acetylacetonate and indium nitrate. (d) Transfer and output characteristics of IZO TFTs fabricated from zinc nitrate and indium nitrate. *Source*: (a–d) Reproduced with permission from Ref. [55], Kang, Y. H., et al. (2014). Two-component solution processing of oxide semiconductors for thin-film transistors via self-combustion reaction. *J. Mater. Chem. C* **2**(21): 4247–4256. Copyright 2014 The Royal Society of Chemistry.

Figure 8.11a, In_2O_3-based TFTs were fabricated on 20-μm-thick polyimide substrates with SiO_x/SiN_x (200/200 nm) as the gate dielectric. After spin-coating and prebaking the In_2O_3 precursor films, they were irradiated using a high-pressure mercury UV lamp through a channel mask. Then, the films were soaked in ethanol for 1 min, and the non-UV-irradiated film regions were removed. The patterned In_2O_3 films were annealed at 250 °C for 1 h in air, followed by depositing Al source and drain electrodes. Figure 8.11b shows the UV-Vis spectra for the In_2O_3 precursors with different additives. The one with HNO_3+AcAc exhibits broader absorption in the UV region (250–400 nm), which is attributed to the AcAcH chelate ring π–π* transition. The highly exothermic redox reaction occurred at 136 °C for In_2O_3 combustion precursors (Figure 8.11c). To demonstrate this combined approach, the authors fabricated the directly photo-patterned In_2O_3 TFTs on flexible polyimide substrates, reporting $\mu = 2.24\,cm^2\,V^{-1}\,s^{-1}$, $SS = 0.45\,V\,dec^{-1}$, and $I_{on}/I_{off} \sim 10^8$.

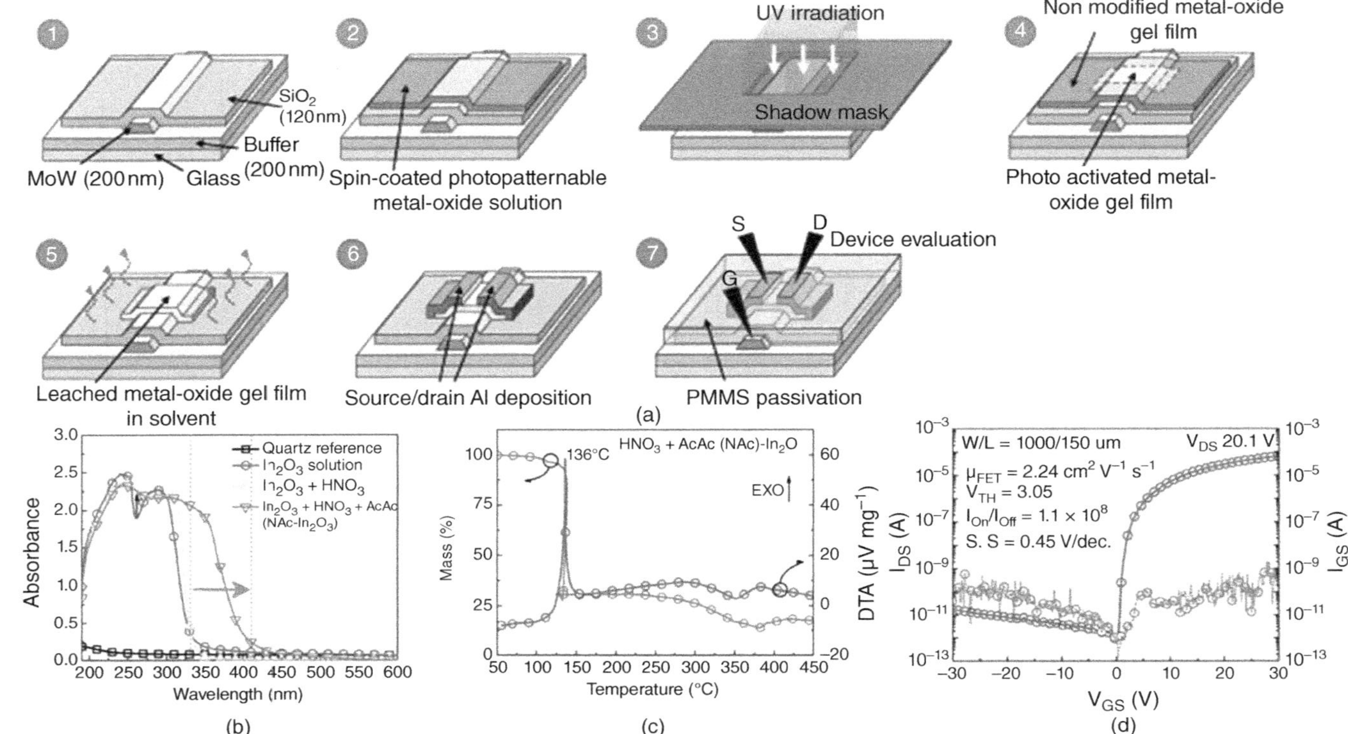

Figure 8.11 (a) Process flow used for the UV-irradiated directly photo-patternable MO films and the fabrication of MO TFTs. (b) UV-vis absorbance spectra In_2O_3 precursors with various additives, such as no additives, HNO_3, and HNO_3 + AcAcH. (c) TGA and DTA scans of the In_2O_3 combustion precursor. (d) Transfer curve of an In_2O_3 TFT fabricated on a flexible substrate. *Source*: (a–d) Reproduced with permission from Ref. [56], Rim, Y. S., Lim, H. S., & Kim, H. J. (2013). Low-temperature metal-oxide thin-film transistors formed by directly photopatternable and combustible solution synthesis. *ACS Applied Materials & Interfaces* **5**(9): 3565–3571. Copyright 2013 American Chemical Society.

In 2015, the Northwestern group reported a technique combining an efficient spray-pyrolysis process and highly exothermic combustion synthesis (spray-CS) that enables fabrication of high-quality nanoscopically thick MO films at low temperature (Figure 8.12a) [58, 59]. Spray-coating techniques transfer small ink droplets onto a substrate using a pneumatic airbrush that is widely utilized in industrial processes for film deposition [60]. The spray-CS solutions were prepared with metal nitrates in 2-methoxyethanol to yield 0.05 or 0.5 M solutions. For 0.05 (or 0.5) M solutions, 55 (or 110) μL NH_4OH and 100 (or 200) μL acetylacetone were added to 10 (or 2) mL precursor solutions and stirred overnight at room temperature. Before spin- or spray-coating, the precursor solutions were combined in the desired molar ratios and stirred for 2 h. Then, the substrates were maintained at 200–350 °C on a hot plate, whereas precursor solutions were loaded into the spray gun and sprayed intermittently (60-s cycles) on the substrates until the desired film thickness (20 or 50 nm) was obtained. The nozzle–substrate distance was maintained at ~20 cm. Owing to the continuous drop-size deposition feature, the spray–combustion synthesis produces high-density, macroscopically continuous, conformal films.

Selected-area energy-filtered nanobeam diffraction images of spray-CS and spin-CS In_2O_3 films processed at 300 °C indicate that the former was more crystalline than the latter, in agreement with the XRD data. Regarding the TFT performance based on 20-nm-thick IGZO (71/7/22 = at.% In/Ga/Zn) films, the 250 °C–processed spray-CS-derived TFTs on 300-nm-thick SiO_2/Si substrates exhibit a μ of 2.0 cm^2 V^{-1} s^{-1} and V_T of 20 V, while those of the spin-CS-derived TFTs were 0.8 cm^2 V^{-1} s^{-1} and 23 V, respectively. Next, the spray-CS growth of thicker, single-layer 50-nm-thick IGZO films was investigated, as this thickness is the minimum required for IGZO TFT–based display manufacturing. Spray-CS significantly reduces device fabrication time from several hours to 30 min with high performance. A maximum μ = 7.6 cm^2 V^{-1} s^{-1} and I_{on}:I_{off} ~ 10^8 are obtained for 300 °C–processed spray-CS IGZO TFTs, ~10^3–10^4× greater than those in the sol–gel and spin-CS devices and approaching those of sputtered IGZO (33.3/33.3/33.3 = at.% In/Ga/Zn) devices (Figure 8.12b and 8.12c). As shown in Figure 8.12d, positron annihilation spectroscopy (PAS) measurements, a methodology to correlate positron lifetime with porosity, demonstrate that spray-CS IGZO (71/7/22 = at.% In/Ga/Zn) sputtered IGZO (33.3/33.3/33.3 = at.% In/Ga/Zn) films are the least porous, with the sub-nanometer porosities estimated as ~6% with a mean pore size (Φ) of 0.4–0.6 nm and a cavity number density (η) of ~9×10^{20} cm^{-3} followed by a spray-CS IGZO (33.3/33.3/33.3 = at% In/Ga/Zn) porosity of ~10% (Φ = 1–2 nm; η = ~2×10^{20} cm^{-3}). The sol–gel and spin-CS film porosities were >15%, with Φ = 2–3 nm and η = ~1×10^{20} cm^{-3}, demonstrating the great advantages of spray-CS for not only reducing processing times but also enhancing IGZO film quality.

To reduce the film-annealing times, Choi et al. combined laser annealing and combustion precursors to fabricate In_2O_3 films as a TFT channel material [61]. Urea was used as the fuel, and indium nitrate was used as the oxidizer. A continuous-wave blue laser with a spot size of 8×2 μm^2 was used for laser annealing the spin-coated In_2O_3 thin films. The radiation time was varied from 0.1 ms to 1 ms to create energy density fluences of 20.6–250 J cm^{-2}. These devices were completed by liftoff of 100-nm-thick tungsten as source–drain electrodes. The authors showed that 0.2 mol% urea and a laser radiation with a 250 J cm^{-2} energy fluence provided sufficient thermal energy to convert the $In(OH)_3$ to polycrystalline In_2O_3 film without furnace annealing. The resulting In_2O_3 TFTs exhibited a μ of 2.8 cm^2 V^{-1} s^{-1} with an *SS* of 0.1 V dec^{-1}.

To avoid photonic annealing, the Northwestern group reported a highly efficient thin-film combustion process by adding a fluorinated cofuel (1,1,1-trifluoro-2,4-pentanedione) to the MO combustion precursors and using a low-temperature prebaking step (120 °C/60 s) [62]. The removal of the extraneous solvent from the coated films prior to combustion onset and enhancement of combustion heat dramatically shortens the postannealing time to as little as 10–60 s for each layer (Figure 8.13a). The scope of this approach was demonstrated for several MO semiconductors (In_2O_3, IZO [70/30 at.% In/Zn], and IGZO [71/7/22 at.% In/Ga/Zn]) and an Al_2O_3 dialectic. The μs of the spin-coated 250 °C/60-s-annealed MO film on 300 nm SiO_2/Si substrates are 2.9, 2.3, and 0.3 cm^2 V^{-1} s^{-1} for In_2O_3, IZO, and IGZO TFTs, respectively. As shown in Figure 8.13b, a blade-coating combustion process is also effective for the expeditious wafer-scale fabrication of IGZO TFTs. The μs of the 300 °C/60-s-derived IGZO TFTs on Al_2O_3 dielectric are as high as 25.2 cm^2 V^{-1} s^{-1}, with negligible threshold-voltage deterioration in a 4000 s

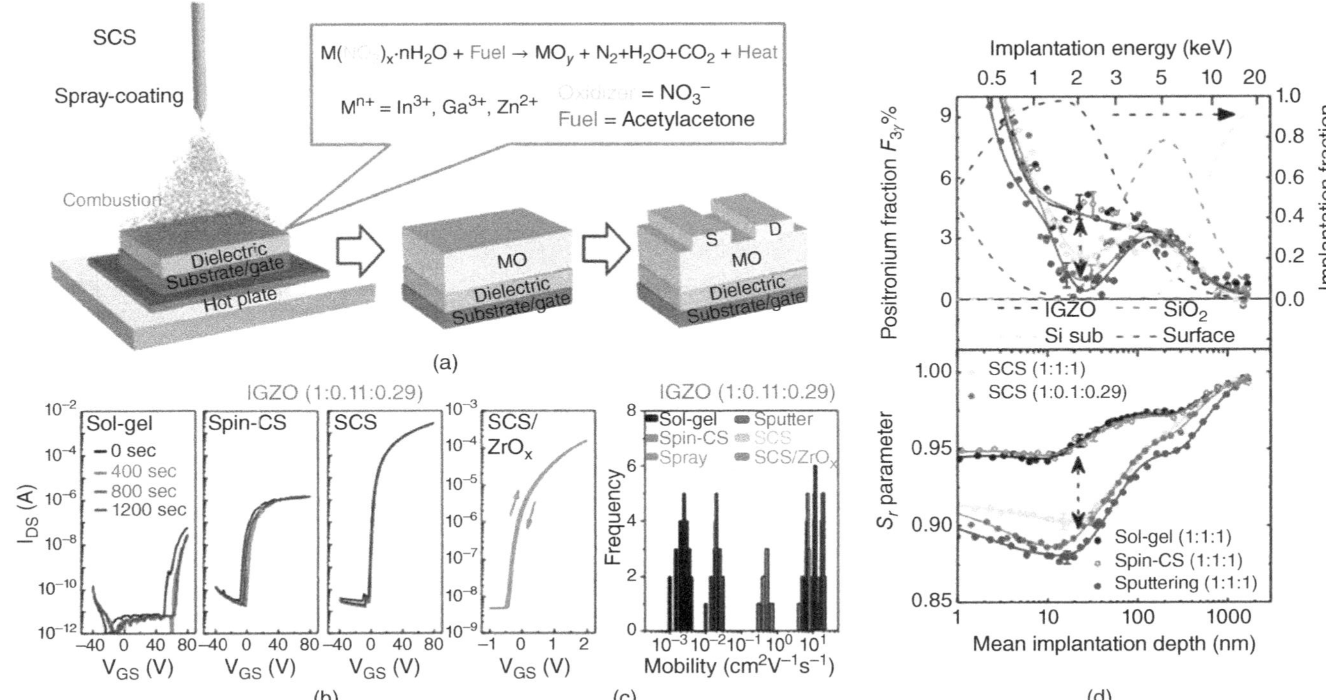

Figure 8.12 (a) Schematic of the spray-CS process used for growing MO films under ambient conditions and the corresponding bottom-gate/top-contact TFT structure. (b) Transfer characteristics bias-stress data, and (c) mobility distribution statistics for 50-nm-thick, single-layer IGZO TFTs fabricated by the indicated deposition methods. (d) Positron annihilation spectroscopy (PAS) S_r and three-γ ortho-Ps annihilations ($F_{3\gamma}$) parameters for IGZO films as a function of the positron mean implantation depth. *Source*: (a–d) Reproduced with permission from Ref. [58], Yu, X., et al. (2015). Spray-combustion synthesis: efficient solution route to high-performance oxide transistors. *Proceedings of the National Academy of Sciences* **112**(11): 3217–3222. Copyright 2013 National Academy of Sciences.

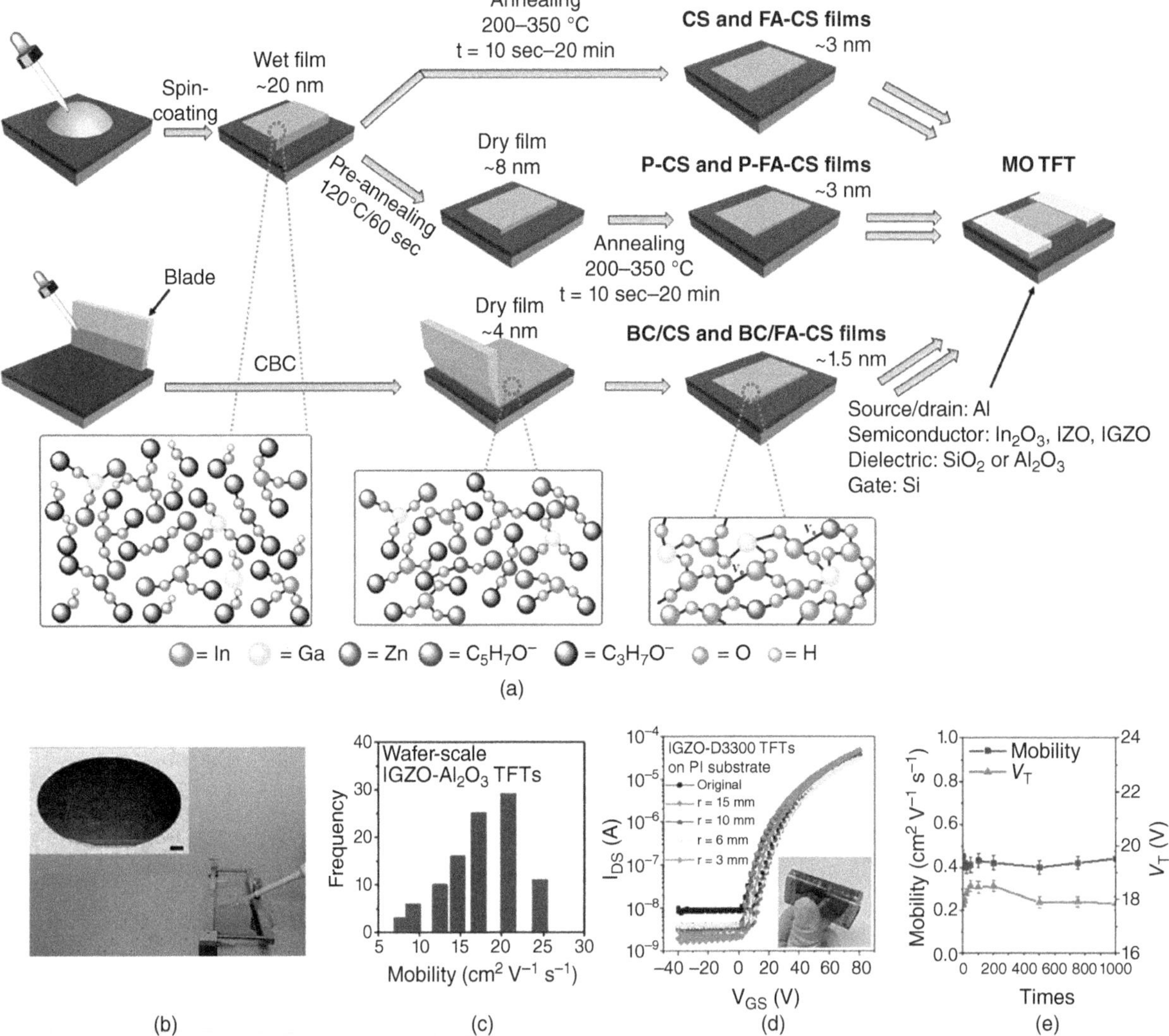

Figure 8.13 (a) Film-fabrication processes with the schematic evolution of MO-precursor coordination chemistry and film thickness. (b) Photo of the blade-coating process on a 4-inch Si wafer; the scale bar is 1 cm. (c) Distribution of saturation mobilities of wafer-scale IGZO TFTs on high-*k* Al_2O_3 dielectric. (d) Transfer curves of top-gate flexible IGZO TFTs as a function of the bending radius. (e) Mobility and threshold voltage stability for bending test cycles with a radius of 6 mm. *Source*: (a,c–e) Reproduced with permission from Ref. [62], Wang, B., et al. (2019). Expeditious, scalable solution growth of metal oxide films by combustion blade coating for flexible electronics. *Proceedings of the National Academy of Sciences* 201901492. Copyright 2019 National Academy of Sciences. (b) From Ref. [62], fig. 05, p. 07, with permission of National Academy of Sciences. https://doi.org/10.1073/pnas.1901492116. (d1) From Ref. [63], fig. 05, p. 07, with permission of National Academy of Sciences. https://doi.org/10.1073/pnas.1901492116.

bias-stress test (Figure 8.13c), while the 300 °C/60-s-derived IGZO TFTs with a top-gated polymer dielectric on flexible polyimide (PI) substrates exhibit high flexibility when bent to a 3-mm radius and good TFT bending stability over 1000 cycles (Figure 8.13d and 8.13e).

8.3.2 *p*-Type MO TFTs

Currently, all commercially available MO semiconductors are *n*-type, indicating the slow development of *p*-type MO semiconductors. Although several *p*-type semiconductors, such as Cu_xO, $CuAlO_2$, SnO, NiO, CuI, and CuSCN, have been reported, TFT performance based on these *p*-type inorganic materials is far lower than that of their *n*-type counterparts.

In 2017, Liu at al. implemented the thin-film combustion method to fabricate Cu-doped NiO (Cu:NiO) films at a low temperature of 150 °C [63]. AcAcH was used as the fuel, and copper nitrate and nickel nitrate as oxidizers. The Cu:NiO combustion precursor solutions with various Cu concentrations (0–10 at.%) were first spin-coated on ZrO_2 dielectrics, followed by 150 °C annealing. The TFTs were completed by thermal-evaporating Ni source and drain electrodes through a shadow mask. The thermal behavior of the Cu:NiO precursors (5 at.% Cu) with two fuels (acetylacetone and urea) was analyzed, and TGA scans are shown in Figure 8.14a. The acetylacetone-assisted precursor exhibits an abrupt mass loss at 130 °C, which coincides well with the exothermic peak observed in

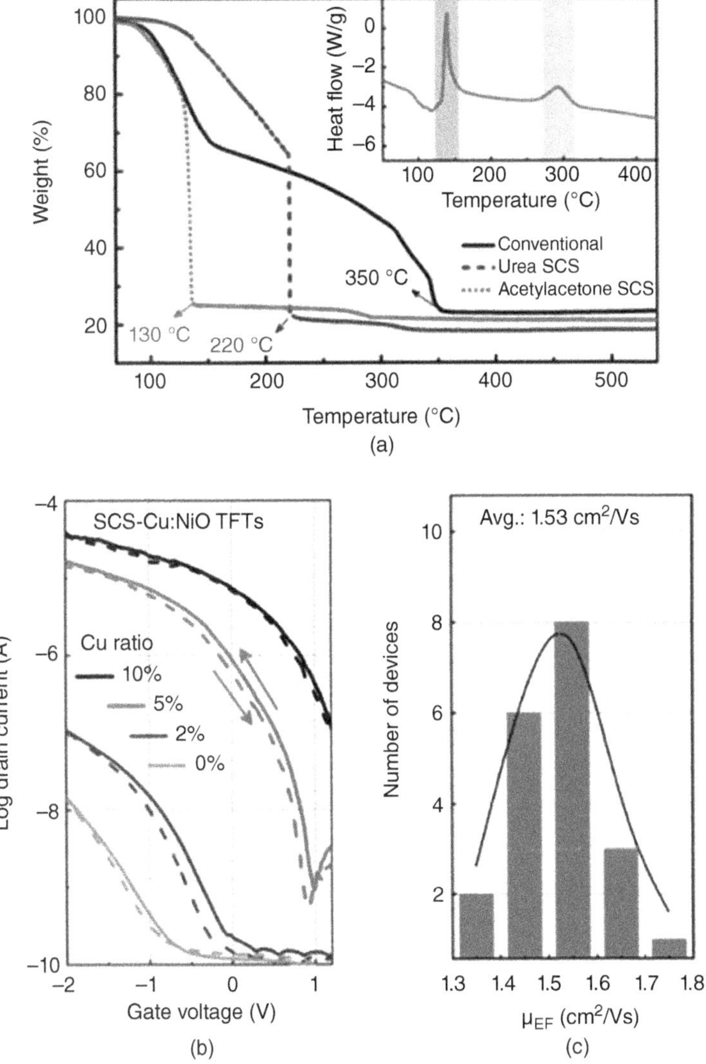

Figure 8.14 (a) TGA scans of the acetylacetone- and urea-based combustion and conventional Cu:NiO (5 at% Cu) precursors. The inset shows the corresponding DSC scan of the acetylacetone-based precursor. (b) Transfer characteristics of Cu:NiO TFTs with various Cu doping concentrations. (c) Histogram of the μ of Cu:NiO TFTs with 5% Cu. *Source*: Reproduced with permission from Ref. [4], Liu, A., et al. (2017). Solution combustion synthesis: low-temperature processing for p-type Cu:NiO thin films for transparent electronics. *Advanced Materials* **29**(34): 1701599. Copyright 2017 Wiley-VCH Verlag GmbH & Co. KGaA, Weinheim.

the DSC scan (inset of Figure 8.14a). As shown in Figure 8.14b and 8.14c, using a high-k ZrO_2 gate dielectric enables TFT operation at voltages <2 V. Optimized Cu:NiO TFTs with 5 mol% Cu show the best performance with a $\mu = 1.5\,cm^2\,V^{-1}\,s^{-1}$, $I_{on}/I_{off} = 10^4$, and an SS value of 0.13 V dec^{-1}.

In the same year, the Pei group reported solution-processed TFTs based on NiO films by thin-film combustion synthesis [64]. AcAcH was used as the fuel, and nickel nitrate as the oxidizer. The authors investigated the effects of precursor aging time (1–15 days) on the thermal behavior of the dried precursors. The DSC scans showed a sharp exothermic peak at ~205 °C and a broad exothermic process at ~350 °C. With increased aging time, the sharp exothermic peak at ~205 °C declined, while the broad exothermic peak intensity of ~350 °C was enhanced, demonstrating an increased degree of chelation between AcAcH and the metal ions. The TGA results showed that the largest total weight loss of 63.3% at 205 °C was obtained for the precursors with 10 days' aging. Regarding the NiO TFT performance for precursors with 10 days' aging, the results showed that 350 °C annealing yielded the best TFT performance with a μ of 0.015 $cm^2\,V^{-1}\,s^{-1}$ and I_{on}/I_{off} of 2. In 2018, Yang et al. reported the combination of thin-film combustion and DUV irradiation to achieve p-type Li:NiO_x films [65]. AcAcH was used as the fuel, and metal nitrates were used as the oxidizers. After spin-coating the precursors on ZrO_2 dielectrics, the wet films were subjected to DUV treatment and 150 °C thermal annealing for 30 min. The TFTs were completed by thermal evaporation of 200-nm-thick Ni electrodes as source and drain electrodes. The optimized Li:NiO_x TFTs with 5% Li exhibited the best performance with a μ of 1.7 $cm^2\,V^{-1}\,s^{-1}$, I_{on}/I_{off} of 10^6, and an SS value of 0.21 V dec^{-1}. Furthermore, counterclockwise hysteresis in the transfer curves was negligible (0.1 V).

8.4 Summary and Perspectives

In this chapter, we have summarized the major advances in low-temperature solution-processed MO films for TFT applications. Particular attention is paid to thin-film combustion technology owing to a reduction in processing temperature and compatibility with large-area, high-throughput manufacturing. The metal nitrate–based precursors are preferred due to their high oxidizing power, good solubility, and low decomposition temperature, while the acetylacetone is the most efficient coordinating fuel.

The broad generality of thin-film combustion methodology has been validated for diverse MO materials from n-type In_2O_3, IZO, and IGZO to p-type Cu_xO and NiO. Depending on the MO charge transport characteristics, processing temperatures from 150 °C to 300 °C have been utilized to achieve high-quality MO films for TFTs as well as other devices such as memories, solar cells, and batteries [66, 67]. More promisingly, recent reports have demonstrated that the annealing times for combustion synthesis can be greatly shortened from ~1.5 h to <5 min, making combustion synthesis truly compatible with roll-to-roll fabrication [62, 68, 69]. Nevertheless, processing temperatures near 250 °C are still required to achieve high-performance ternary or quaternary MO films having strong oxygen getters, bias-stress stabilizers, and metals such as Ga^{3+}, Sc^{3+}, Y^{3+}, and La^{3+} [54]. The combination of combustion synthesis with other low-temperature methods has proven to be an attractive and realistic strategy. Recent reports on UV-assisted combustion synthesis for several MO systems have made important steps toward future roll-to-roll, high-throughput manufacturing of flexible and wearable electronics [61, 65, 70, 71]. More studies will broaden applications even further.

Acknowledgments

We thank the Air Force Office of Scientific Research (AFOSR; grant no. FA9550-18-1-0320), the Northwestern University Materials Research Science and Engineering Center (MRSEC; grant no. NSF DMR-1720139), and Flexterra Inc. for support of this research. We thank our many colleagues and collaborators, whose names are cited in the reference list, for their essential contributions to the work described here.

References

1 Yu, X.G., Marks, T.J., and Facchetti, A. (2016). Metal oxides for optoelectronic applications. *Nature Materials* 15: 383–396.

2 Wang, B., Huang, W., Chi, L. et al. (2018). High-*k* gate dielectrics for emerging flexible and stretchable electronics. *Chemical Reviews* 118: 5690–5754.

3 Fortunato, E., Barquinha, P., and Martins, R. (2012). Oxide semiconductor thin-film transistors: a review of recent advances. *Advanced Materials* 24: 2945–2986.

4 Xu, W., Li, H., Xu, J.B., and Wang, L. (2018). Recent advances of solution-processed metal oxide thin-film transistors. *ACS Applied Materials & Interfaces* 10: 25878–25901.

5 Nomura, K., Ohta, H., Takagi, A. et al. (2004). Room-temperature fabrication of transparent flexible thin-film transistors using amorphous oxide semiconductors. *Nature* 432: 488–492.

6 Gutierrez-Heredia, G., Rodriguez-Lopez, O., Garcia-Sandoval, A., and Voit, W.E. (2017). Highly stable indium-gallium-zinc-oxide thin-film transistors on deformable softening polymer substrates. *Advanced Electronic Materials* 3: 1700221.

7 Wang, B., Thukral, A., Xie, Z. et al. (2020). Flexible and stretchable metal oxide nanofiber networks for multimodal and monolithically integrated wearable electronics. *Nature Communications* 11: 2405.

8 Lee, C.-T., Huang, Y.-Y., Tsai, C.-C. et al. (2015). A novel highly transparent 6-in. AMOLED display consisting of IGZO TFTs. *SID International Symposium Digest of Technical Papers* 46: 872–875.

9 Munzenrieder, N., Cantarella, G., Vogt, C. et al. (2015). Stretchable and conformable oxide thin-film electronics. *Advanced Electronic Materials* 1: 1400038.

10 Shimoda, T., Matsuki, Y., Furusawa, M. et al. (2006). Solution-processed silicon films and transistors. *Nature* 440: 783–786.

11 Kim, S.J., Yoon, S., and Kim, H.J. (2014). Review of solution-processed oxide thin-film transistors. *Japanese Journal of Applied Physics* 53: 02BA02.

12 Wang, B., Zeng, L., Huang, W. et al. (2016). Carbohydrate-assisted combustion synthesis to realize high-performance oxide transistors. *Journal of the American Chemical Society* 138: 7067–7074.

13 Park, J.W., Kang, B.H., and Kim, H.J. (2019). A review of low-temperature solution-processed metal oxide thin-film transistors for flexible electronics. *Advanced Functional Materials* 30: 1904632.

14 Chen, R. and Lan, L. (2019). Solution-processed metal-oxide thin-film transistors: a review of recent developments. *Nanotechnology* 30: 312001.

15 Yarali, E., Koutsiaki, C., Faber, H. et al. (2019). Recent progress in photonic processing of metal-oxide transistors. *Advanced Functional Materials* 30: 1906022.

16 Varma, A., Mukasyan, A.S., Rogachev, A.S., and Manukyan, K.V. (2016). Solution combustion synthesis of nanoscale materials. *Chemical Reviews* 116: 14493–14586.

17 Carlos, E., Martins, R., Fortunato, E., and Branquinho, R. (2020). Solution combustion synthesis: towards a sustainable approach for metal oxides. *Chemistry—A European Journal* 10.1002/chem.202000678.

18 Banger, K.K., Yamashita, Y., Mori, K. et al. (2011). Low-temperature, high-performance solution-processed metal oxide thin-film transistors formed by a "sol–gel on chip" process. *Nature Materials* 10: 45–50.

19 Kim, M.G., Kanatzidis, M.G., Facchetti, A., and Marks, T.J. (2011). Low-temperature fabrication of high-performance metal oxide thin-film electronics via combustion processing. *Nature Materials* 10: 382–388.

20 Song, K., Young Koo, C., Jun, T. et al. (2011). Low-temperature soluble InZnO thin film transistors by microwave annealing. *Journal of Crystal Growth* 326: 23–27.

21 Rim, Y.S., Jeong, W.H., Kim, D.L. et al. (2012). Simultaneous modification of pyrolysis and densification for low-temperature solution-processed flexible oxide thin-film transistors. *Journal of Materials Chemistry* 22: 12491–12497.

22 Jun, T., Song, K., Jeong, Y. et al. (2011). High-performance low-temperature solution-processable Zno thin film transistors by microwave-assisted annealing. *Journal of Materials Chemistry* 21: 1102–1108.

23 Yu, B.S., Jeon, J.Y., Kang, B.C. et al. (2019). Wearable 1 V operating thin-film transistors with solution-processed metal-oxide semiconductor and dielectric films fabricated by deep ultra-violet photo annealing at low temperature. *Scientific Reports* 9: 8416.

24 Teng, L.-F., Liu, P.-T., Lo, Y.-J., and Lee, Y.-J. (2012). Effects of microwave annealing on electrical enhancement of amorphous oxide semiconductor thin film transistor. *Applied Physics Letters* 101: 132901.

25 Park, J., Ha, T.-J., and Cho, W.-J. (2015). Improvement of air stability on solution-processed InZnO thin-film transistors by microwave irradiation and in:Zn composition ratio. *Physica Status Solidi (A): Applications and Materials Science* 212: 1719–1724.

26 Moon, S.-W. and Cho, W.-J. (2015). Improvement of device characteristic on solution-processed InGaZnO thin-film-transistor (TFTs) using microwave irradiation. *Journal of Semiconductor Technology and Science* 15: 249–254.

27 Yoo, Y.B., Park, J.H., Lee, S.J., and Song, K.M. (2012). Low-temperature, solution-processed zinc tin oxide thin-film transistors fabricated by thermal annealing and microwave irradiation. *Japanese Journal of Applied Physics* 51: 040201.

28 Cho, W.-J., Ahn, M.-J., and Jo, K.-W. (2017). Performance enhancement of solution-derived zinc-tin-oxide thin film transistors by low-temperature microwave irradiation. *Physica Status Solidi (A): Applications and Materials Science* 214: 1700350.

29 Ji, K.H., Kim, J.-I., Jung, H.Y. et al. (2011). Effect of high-pressure oxygen annealing on negative bias illumination stress-induced instability of InGaZNo thin film transistors. *Applied Physics Letters* 98: 103509.

30 Shin, H.S., Rim, Y.S., Mo, Y.-G. et al. (2011). Effects of high-pressure H_2O-annealing on amorphous IGZO thin-film transistors. *Physica Status Solidi (A): Applications and Materials Science* 208: 2231–2234.

31 Rim, Y.S. and Kim, H.J. (2014). Densification effects on solution-processed indium-gallium-zinc-oxide films and their thin-film transistors. *Physica Status Solidi (A): Applications and Materials Science* 211: 2195–2198.

32 Rim, Y.S., Choi, H.-W., Kim, K.H., and Kim, H.J. (2016). Effects of structural modification via high-pressure annealing on solution-processed InGaO films and thin-film transistors. *Journal of Physics D: Applied Physics* 49: 075112.

33 Nakata, M., Takechi, K., Yamaguchi, S. et al. (2009). Effects of excimer laser annealing on $InGaZnO_4$ thin-film transistors having different active-layer thicknesses compared with those on polycrystalline silicon. *Japanese Journal of Applied Physics* 48: 115505.

34 Nakata, M., Takechi, K., Azuma, K. et al. (2009). Improvement of $InGaZnO_4$ thin film transistors characteristics utilizing excimer laser annealing. *Applied Physics Express* 2: 021102.

35 Yang, Y.-H., Yang, S.S., and Chou, K.-S. (2010). Characteristic enhancement of solution-processed In–Ga–Zn oxide thin-film transistors by laser annealing. *IEEE Electron Device Letters* 31: 969–971.

36 Park, Y.M., Daniel, J., Heeney, M., and Salleo, A. (2011). Room-temperature fabrication of ultrathin oxide gate dielectrics for low-voltage operation of organic field-effect transistors. *Advanced Materials* 23: 971–974.

37 Kim, Y.H., Heo, J.S., Kim, T.H. et al. (2012). Flexible metal-oxide devices made by room-temperature photochemical activation of sol-gel films. *Nature* 489: 128–132.

38 Park, S., Kim, K.-H., Jo, J.-W. et al. (2015). In-depth studies on rapid photochemical activation of various sol-gel metal oxide films for flexible transparent electronics. *Advanced Functional Materials* 25: 2807–2815.

39 Jo, J.W., Kim, J., Kim, K.T. et al. (2015). Highly stable and imperceptible electronics utilizing photoactivated heterogeneous sol-gel metal-oxide dielectrics and semiconductors. *Advanced Materials* 27: 1182–1188.

40 Wang, K.-H., Zan, H.-W., and Soppera, O. (2018). The zinc-loss effect and mobility enhancement of DUV-patterned sol–gel IGZO thin-film transistors. *Semiconductor Science and Technology* 33: 035003.

41 Leppaniemi, J., Eiroma, K., Majumdar, H., and Alastalo, A. (2017). Far-UV annealed inkjet-printed In_2O_3 semiconductor layers for thin-film transistors on a flexible polyethylene naphthalate substrate. *ACS Applied Materials & Interfaces* 9: 8774–8782.

42 Leppäniemi, J., Ojanperä, K., Kololuoma, T. et al. (2014). Rapid low-temperature processing of metal-oxide thin film transistors with combined far ultraviolet and thermal annealing. *Applied Physics Letters* 105: 113514.

43 Yoo, T.-H., Kwon, S.-J., Kim, H.-S. et al. (2014). Sub-second photo-annealing of solution-processed metal oxide thin-film transistors via irradiation of intensely pulsed white light. *RSC Advances* 4: 19375–19379.

44 Kang, C.-M., Kim, H., Oh, Y.-W. et al. (2016). High-performance, solution-processed indium-oxide TFTs using rapid flash lamp annealing. *IEEE Electron Device Letters* 37: 595–598.

45 Kamiyama, S., Miura, T., and Nara, Y. (2005). Ultrathin HfO_2 films treated by Xenon flash lamp annealing for use as transistor gate dielectric replacements. *Electrochemical and Solid-State Letters* 8: G367.

46 Noh, J.H., Joshi, P.C., Kuruganti, T., and Rack, P.D. (2015). Pulse thermal processing for low thermal budget integration of IGZO thin film transistors. *IEEE Journal of the Electron Devices Society* 3: 297–301.

47 Gebel, T., Rebohle, L., Fendler, R. et al. (2006). Millisecond annealing with flashlamps: tool and process. In. In: *14th IEEE International Conference on Advanced Thermal Processing of Semiconductors*, October, 47–55. https://ieeexplore.ieee.org/xpl/conhome/4200427/proceeding.

48 Acharya, N. and Timans, P.J. (2004). Fundamental issues in millisecond annealing. *Electrochemical Society Proceedings* 1: 11–18.

49 Chen, H., Rim, Y.S., Jiang, C., and Yang, Y. (2015). Low-impurity high-performance solution-processed metal oxide semiconductors via a facile redox reaction. *Chemistry of Materials* 27: 4713–4718.

50 Gao, P., Lan, L., Lin, Z. et al. (2017). Low-temperature, high-mobility, solution-processed metal oxide semiconductors fabricated with oxygen radical assisting perchlorate aqueous precursors. *Chemical Communications* 53: 6436–6439.

51 Liu, A., Guo, Z., Liu, G. et al. (2017). Redox chloride elimination reaction: facile solution route for indium-free, low-voltage, and high-performance transistors. *Advanced Electronic Materials* 3: 1600513.

52 Ravindranathan, P. and Patil, K.C. (1985). Preparation, characterization and thermal analysis of metal hydrazinocarboxylate derivatives. *Proceedings of the Indian Academy of Sciences* 95: 345–356.

53 Epifani, M., Capone, S., Rella, R. et al. (2003). In_2O_3 thin films obtained through a chemical complexation based sol-gel process and their application as gas sensor devices. *Journal of Sol-Gel Science and Technology* 26: 741–744.

54 Hennek, J.W., Smith, J., Yan, A. et al. (2013). Oxygen "getter" effects on microstructure and carrier transport in low temperature combustion-processed a-inxzno (x = Ga, Sc, Y, La) transistors. *Journal of the American Chemical Society* 135: 10729–10741.

55 Kang, Y.H., Jeong, S., Ko, J.M. et al. (2014). Two-component solution processing of oxide semiconductors for thin-film transistors via self-combustion reaction. *Journal of Materials Chemistry C* 2: 4247–4256.

56 Rim, Y.S., Lim, H.S., and Kim, H.J. (2013). Low-temperature metal-oxide thin-film transistors formed by directly photopatternable and combustible solution synthesis. *ACS Applied Materials & Interfaces* 5: 3565–3571.

57 Branquinho, R., Salgueiro, D., Santos, L. et al. (2014). Aqueous combustion synthesis of aluminum oxide thin films and application as gate dielectric in GZTO solution-based tfts. *ACS Applied Materials & Interfaces* 6: 19592–19599.

58 Yu, X., Smith, J., Zhou, N.J. et al. (2015). Spray-combustion synthesis: efficient solution route to high-performance oxide transistors. *Proceedings of the National Academy of the USA* 112: 3217–3222.

59 Wang, B., Yu, X., Guo, P. et al. (2016). Solution-processed all-oxide transparent high-performance transistors fabricated by spray-combustion synthesis. *Advanced Electronic Materials* 2: 1500427.

60 Sondergaard, R.R., Hosel, M., and Krebs, F.C. (2013). Roll-to-roll fabrication of large area functional organic materials. *Journal of Polymer Science Part B: Polymer Physics* 51: 16–34.

61 Choi, J.-W., Han, S.-Y., Nguyen, M.-C. et al. (2017). Low-temperature solution-based In_2O_3 channel formation for thin-film transistors using a visible laser-assisted combustion process. *IEEE Electron Device Letters* 38: 1259–1262.

62 Wang, B., Guo, P., Zeng, L. et al. (2019). Expeditious, scalable solution growth of metal oxide films by combustion blade coating for flexible electronics. *Proceedings of the National Academy of the USA* 116: 9230–9238.

63 Liu, A., Zhu, H., Guo, Z. et al. (2017). Solution combustion synthesis: low-temperature processing for p-type cu:NiO thin films for transparent electronics. *Advanced Materials* 29: 1701599.

64 Li, Y., Liu, C., Wang, G., and Pei, Y. (2017). Investigation of solution combustion-processed nickel oxide p-channel thin film transistors. *Semiconductor Science and Technology* 32: 085004.

65 Yang, J., Wang, B., Zhang, Y. et al. (2018). Low-temperature combustion synthesis and UV treatment processed p-type li:Niox active semiconductors for high-performance electronics. *Journal of Materials Chemistry C* 6: 12584–12591.

66 Zheng, D., Wang, G., Huang, W. et al. (2019). Combustion synthesized zinc oxide electron-transport layers for efficient and stable perovskite solar cells. *Advanced Functional Materials* 29: 1900265.

67 Huang, W., Wang, G., Luo, C. et al. (2019). Controllable growth of $LiMn_2O_4$ by carbohydrate-assisted combustion synthesis for high performance Li-ion batteries. *Nano Energy* 64: 103936.

68 Zhuang, X., Patel, S., Zhang, C. et al. (2020). Frequency-agile low-temperature solution-processed alumina dielectrics for inorganic and organic electronics enhanced by fluoride doping. *Journal of the American Chemical Society* 142: 12440–12452.

69 Sil, A., Avazpour, L., Goldfine, E.A. et al. (2019). Structure–charge transport relationships in fluoride-doped amorphous semiconducting indium oxide: combined experimental and theoretical analysis. *Chemistry of Materials* 32: 805–820.

70 Carlos, E., Dellis, S., Kalfagiannis, N. et al. (2020). Laser induced ultrafast combustion synthesis of solution-based AlO_x for thin film transistors. *Journal of Materials Chemistry C* 8: 6176–6184.

71 Carlos, E., Leppäniemi, J., Sneck, A. et al. (2020). Printed, highly stable metal oxide thin-film transistors with ultra-thin high-*k* oxide dielectric). *Advanced Electronic Materials* 6: 1901071.

9

Solution-Processed Metal-Oxide Thin-Film Transistors for Flexible Electronics

Hyun Jae Kim

School of Electrical and Electronic Engineering, Yonsei University, Seoul, Republic of Korea

9.1 Introduction

In the last couple of decades, flexible electronic systems that can cover a large area on a flexible substrate have received growing attention for broadening classes of applications lying outside those effortlessly addressed with conventional wafer-based electronics, including sensory skin, electronic textiles, and organic light-emitting diodes (OLEDs) in flexible displays [1]. In this type of electronic system, thin-film transistors (TFTs) distributed over a large flexible surface provide switching elements for device operation, assuming the role played by metal-oxide semiconductor field-effect transistors (MOSFETs) in typical wafer-based rigid electronics. Therefore, successful fabrication of TFTs on flexible substrates is critically important in large-area flexible electronics for effective operation in the desired applications. However, three issues still remain in the fabrication of TFTs for flexible electronics: (1) limits in material selection for TFTs, (2) the processing cost of film deposition on flexible substrates, and (3) the low thermal budget of flexible substrates. As an optimal solution for these constraints, solution-processed metal-oxide TFTs will be discussed in detail throughout this chapter.

Selection of appropriate materials for the active layer of TFTs is a primary consideration, because flexible electronics require both excellent device performance and high-level uniformity over a large-area flexible surface. Material candidates for the active layer include amorphous silicon (*a*-Si), low-temperature polycrystalline silicon (LTPS), organic semiconductors, and oxide semiconductors. Large-area uniform deposition has been achieved for *a*-Si; however, its poor electrical performance, with mobility under $1\,cm^2/Vs$, limits its practical use as an active layer. Similarly, the carrier mobilities of organic semiconductors are mostly limited to relatively low values similar to those of *a*-Si-TFTs. On the other hand, LTPS exhibits superior electrical characteristics, with mobility over $100\,cm^2/Vs$; however, there is a limitation to deposition over a large area because the grain boundary distribution can exert a negative impact on individual device characteristics. By contrast, oxide semiconductors can maintain reasonable electrical performance with mobility over $10\,cm^2/Vs$, with uniform deposition over a large area in the amorphous phase, which makes oxide semiconductors highly suited to flexible electronics [2]. Therefore, hereafter, oxide semiconductors will be exclusively discussed as a material for flexible electronics.

Reducing the processing cost of manufacturing TFTs is essential for realization of commercialized flexible electronics. The processing cost is closely related to the material deposition method. For oxide semiconductors, there are two general categories of deposition method: vacuum processes and solution processes. Vacuum processes, such as sputtering or pulsed laser deposition (PLD), are predominantly used compared to solution processes due to their capacity for fabrication of relatively high-performance oxide semiconductor TFTs [3]. However, the large area of flexible substrates compared to conventional wafer or glass substrates poses a significant challenge with respect to adapting vacuum processes to flexible electronics; in particular, there is an astronomical increase in

Amorphous Oxide Semiconductors: IGZO and Related Materials for Display and Memory, First Edition.
Edited by Hideo Hosono and Hideya Kumomi.

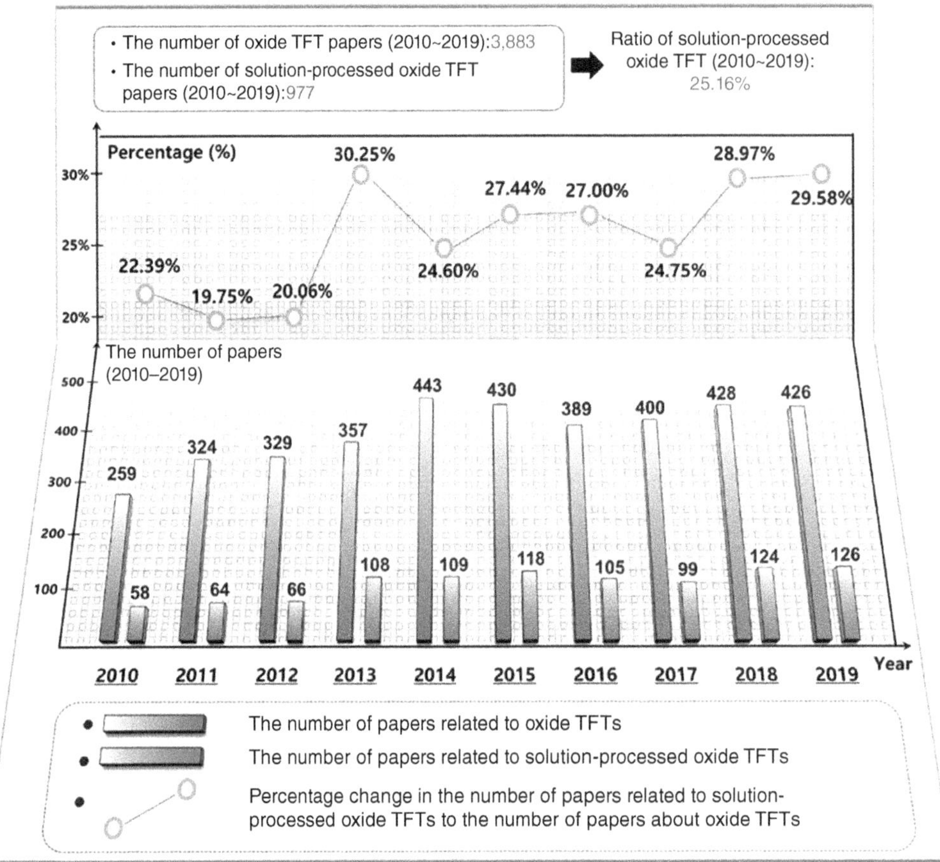

Figure 9.1 The number of research papers on oxide TFTs and solution-processed oxide TFTs published from 2010 to 2019. Ratio of the studies on solution-processed oxide TFTs to the studies on overall oxide-based TFTs. *Source*: Based on Web of Science searches and Google Scholar.

cost associated with the required vacuum chamber. By contrast, solution processing may provide a simple and inexpensive method for manufacturing flexible TFTs. Since solution processes can be combined with roll-to-roll (R2R) technology, they are likely to exhibit far superior efficiency to a vacuum process in terms of both time and cost. Following the introduction of ZnO solution-processed oxide TFTs, and as solution processes are expected to be widely suitable for the fabrication of flexible electronics, the number of solution-processed oxide TFT studies has increased explosively over the past decade. Figure 9.1 shows in detail the number of research papers on oxide TFTs and solution-processed oxide TFTs published from 2010 to 2019 (based on Google Scholar and Web of Science searches). Overall, the number of papers on oxide TFTs and solution-processed oxide TFTs increased steadily to a maximum, and then remained steady without any significant drops, indicating that research and development in this field has entered a stable period. In addition, the ratio of the number of studies on solution-processed oxide TFTs to the overall number of studies on oxide-based TFTs has increased continuously, apart from in 2014 and 2017.

Finally, low-temperature techniques are essential for fabrication of devices on flexible plastic substrates with low processable temperatures of under ~300 °C. Plastic substrates (e.g., polyethylene naphthalate [PEN], polyethersulfone [PES], polycarbonate [PC], polyether ether ketone [PEEK], polyethylene terephthalate [PET], polyimide [PI], and colorless polyimide [CPI]), which are essential for flexible electronics, experience a transition from the

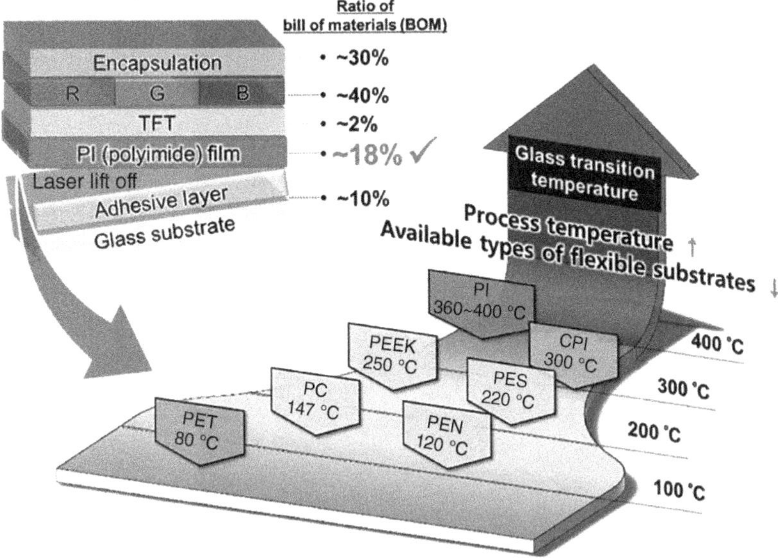

Figure 9.2 Ratio of bill of materials of currently adopted mobile organic light-emitting diode (OLED) display panels, and various types of plastic substrates with their glass transition temperatures.

solid state to a rubbery state, thus losing their capacity as a substrate, when exposed to a temperature higher than the critical temperature; this is known as the glass transition temperature. Various types of plastic substrates with their glass transition temperatures are shown in Figure 9.2 [4]. Several PI substrate types have a relatively high thermal budget with glass transition temperatures in the range of 360–400 °C, and they may be appropriate for use as flexible solution-processed oxide TFT substrates. However, PI tends to be expensive, such that the bill of materials (BOM) may comprise up to 18% of the total cost of a whole mobile OLED display panel, making it unsuitable for large-area flexible display applications that require a low cost. Therefore, reduction of the processing temperature is essential to broaden the available range of low-cost plastic substrates. A number of studies have focused on processing techniques that lower the temperature of solution processes while maintaining performance. The trend of decreasing processing temperatures for solution-processed oxide TFTs over the past decade is illustrated in Figure 9.3. This graph shows journals in the top 25% of the Journal Citation Reports (JCR, by Clarivate Analytics), and indicates that the average process temperature in studies since 2015 is concentrated below the glass transition temperature of PI. Looking at the 10-year trend, this tendency is expected to continue, and solution processes will be applied to more diverse plastic substrates in the future.

The importance of low-temperature solution-processed metal-oxide TFTs has been discussed so far. In Section 9.2, the basics of solution processes and methods for realizing low-temperature processes will be introduced, and finally, flexible electronics applications based on these technologies will be presented.

9.2 Fundamentals of Solution-Processed Metal-Oxide Thin-Film Transistors

9.2.1 Deposition Methods for Solution-Processed Oxide Semiconductors

In this section, various solution deposition methods for oxide thin films are discussed. In Figure 9.4a, a number of solution-based deposition methods are classified as either coating- or printing-based type. Coating-based deposition methods are mainly used for "entire-area deposition," whereas printing-based processes are used for

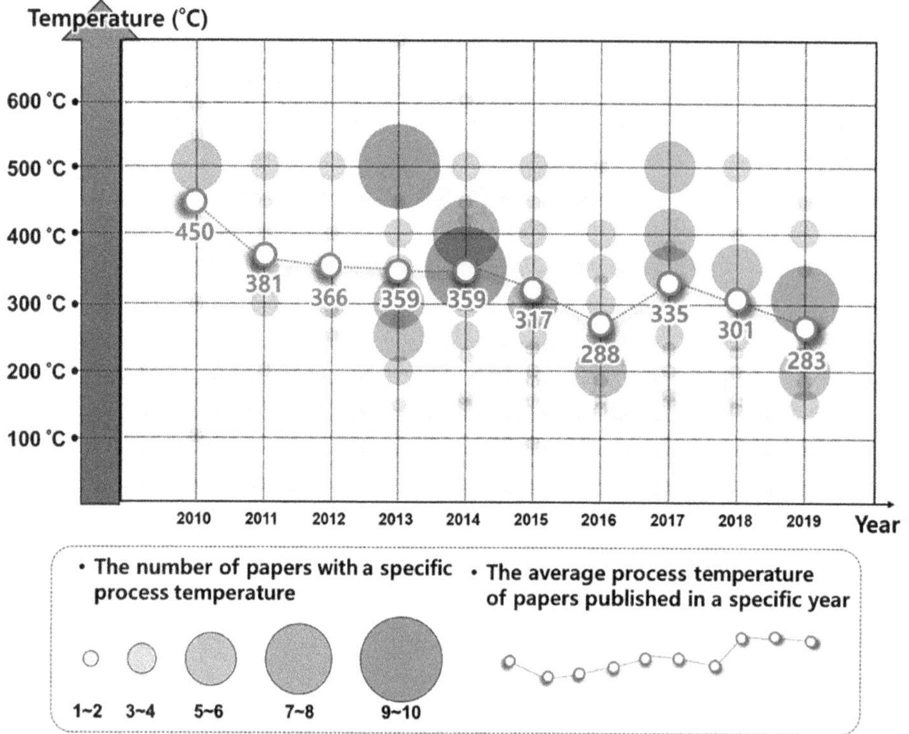

Figure 9.3 A line graph with circles showing trends and specific figures of process temperatures for fabricating solution-processed oxide TFTs.

selective-area deposition. Conventionally, coating-based methods are most widely used because the resulting oxide TFTs exhibit higher electrical performance and reliability than their printed counterparts. However, with more widespread adoption in recent years, printing-based methods have become increasingly essential in a range of applications such as flexible OLED displays, photovoltaics, and integrated circuits, due to their material cost savings, cost-effectiveness, ecofriendly processing, and easy formulation of patterns. Printing-based methods can be categorized into contact (e.g., flexography, gravure, and screen printing) and noncontact (e.g., inkjet printing, spray printing, and electrohydrodynamic [EHD] jet printing) approaches [5]. The choice of solution deposition method is influenced by the envisaged application and by the requirements for deposition temperature and/or the compatibility of the solution chemistry with the substrate.

Figure 9.4b shows the number of papers and authors on specific processing techniques for solution-processed oxide TFTs, which is closely related to preference at the level of the academic institution. The bubble size in the graph indicates the number of cumulative papers multiplied by the number of cumulative authors for each solution deposition method. The graph is divided into three regions (spin coating, inkjet printing, and others) due to the large differences in scale. The bubble sizes for inkjet printing and spin coating have been reduced to 1/7 and 1/225 of their original sizes, respectively, for clarity and readability. As expected, spin coating is the most extensively used technique at the laboratory scale because of its simplicity and low cost, but with comparable performance to fabricated devices. Following spin coating, inkjet printing has the second largest number of both papers and authors; as the most widely applied printing-based method, it is used in a range of electronics. Other noncontact-printing deposition methods, such as spray printing and EHD jet printing, have been utilized recently for fabricating solution-processed oxide TFTs, but their use is still much less widespread than spin coating and

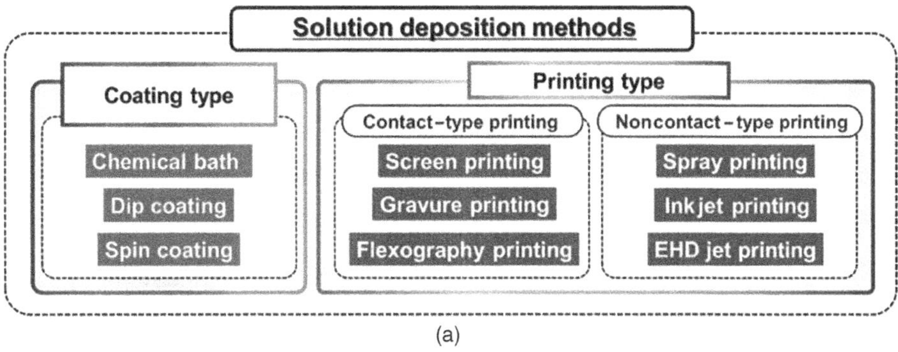

(a)

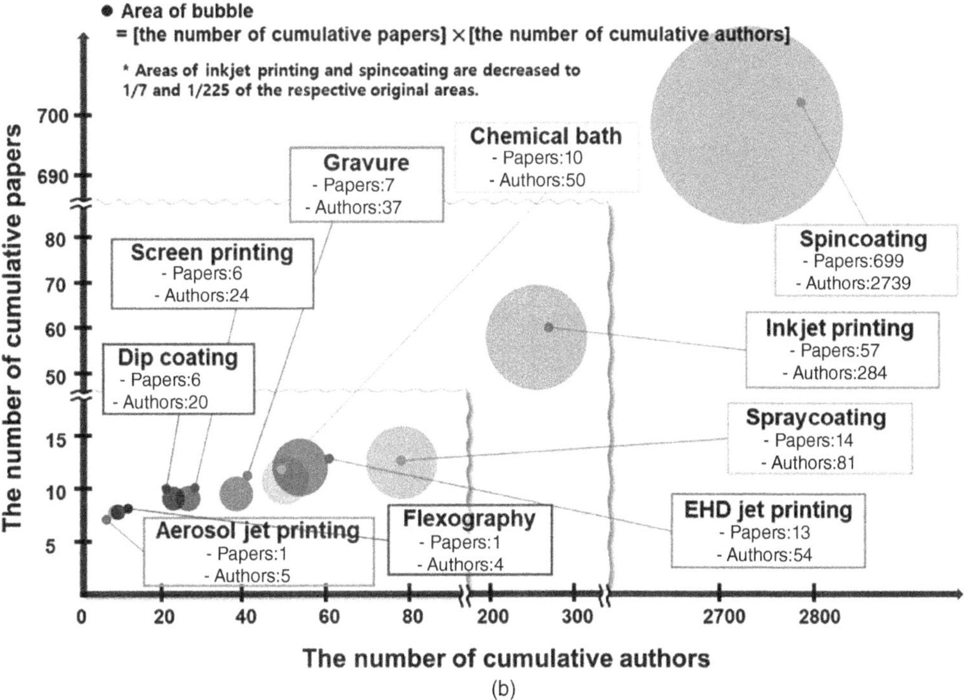

(b)

Figure 9.4 (a) Classification of solution deposition methods for fabricating oxide thin films. (b) The number of cumulative papers and authors on specific deposition methods for solution-processed oxide TFTs over a decade. *Source*: (b) Reproduced with permission from Park, J. W., et al. (2020). A review of low-temperature solution-processed metal oxide thin-film transistors for flexible electronics. *Advanced Functional Materials* **30** (20): 1904632-1–1904632-40. Copyright 2020, Wiley-VCH.

inkjet printing. Furthermore, the number of researchers and papers on gravure printing, screen printing, dip coating, and flexography is low, indicating that they are not yet compatible with oxide TFTs.

Here, nine representative solution deposition methods will be introduced according to their classification in Figure 9.4b, and their deposition principle, influencing parameters, and advantages will be discussed. Although there are other solution deposition methods, such as doctor blading, slot casting, slot die printing, gravure-offset printing, microcontact printing, nano-imprinting, and transfer printing, they have been excluded from this section because they are rarely used for the deposition of oxide semiconductors.

9.2.1.1 Coating-Based Deposition Methods

Coating is used to describe solution deposition methods whereby layers of ink are transferred onto the substrates by pouring, painting, casting, or smearing over the surface. Coating deposition can be divided into two types: chemical growth and liquid coating. In the former type, materials are chemically grown directly on the substrate during deposition; the chemical bath method is representative of this type. In the latter type, liquid coating requires additional processing to evaporate the solvent and remove residues from the desired phase. Dip coating and spin coating are typical liquid-coating methods. Both types of coating methods are described in Figure 9.5.

Chemical Bath The chemical bath is a coating deposition technique for fabrication of thin films, which can be adopted for continuous deposition or large-area batch processing. In the chemical bath process, deposition involves only two steps: nucleation and particle growth by an additional energy source, such as thermal annealing or electrical induction; the method is hence based on forming a solid phase from a solution [6]. The growth of thin films depends strongly on growth conditions, such as the composition ratio, duration of deposition, pH and temperature of the solution, and chemical and topographical nature of the substrate [7]. The major advantage of the chemical bath method is the fabrication of stable, adherent, uniform, and hard films with high reproducibility via a relatively simple process. Additionally, this method does not cause physical damage to the substrate. However, a major drawback of the chemical bath method is wastage of the solution after every deposition.

Dip Coating Dip coating refers to a simple coating method wherein the substrate is immersed in a solution, and a thin film is formed by raising the substrate in the vertical direction at a constant rate. The process of dip coating can be divided into five sequential stages: immersion, start-up, deposition, drainage, and evaporation [8]. First, the substrate is immersed in the solution at a constant speed. After being held in the solution for a certain time, the substrate starts to be pulled up, in a process known as "start-up." As it is pulled up, a thin layer is formed on both sides of the substrate. Drainage and evaporation follow, to remove the liquid and form the thin-film layer. Dip-coated film structure and thickness can be modulated by many factors, including the submersion time, functionality of the initial substrate surface, number of dipping cycles, withdrawal speed, solution concentration and composition, humidity, and temperature. The advantages of the dip-coating technique are that it yields uniform and high-quality films even on complex, bulky shapes with a simple technique.

Spin Coating Spin coating is one of the most popular solution deposition methods at the laboratory scale for solution-process fabrication of oxide thin films. The typical spin-coating process involves application of a solution to a substrate, followed by accelerating substrate to a desired rotational speed. The principle of this method involves the creation of an equilibrium state between the centrifugal force produced by the rotating substrate and the viscous force arising from the viscosity of the solution. The rotational speed of the substrate with overlying

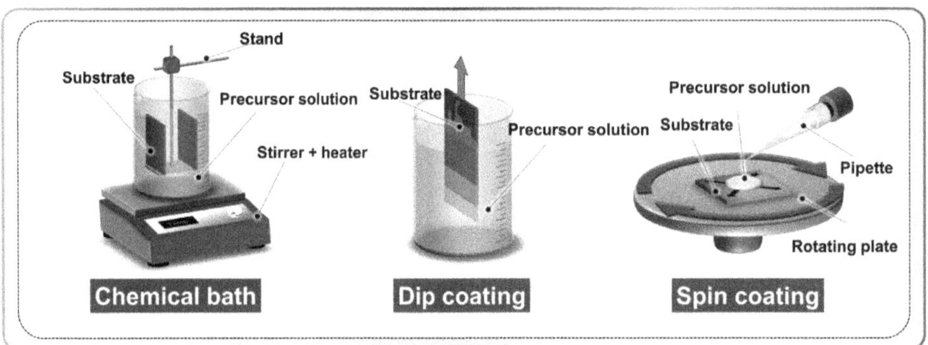

Figure 9.5 Coating-type solution deposition methods.

liquid results in ejecting most of the applied liquid, with only a thin film left on the substrate. The morphology, thickness, and surface topography of the coated film, obtained from a particular solution, are highly reproducible, and they are known to depend on the angular velocity, and on the diffusivity, volatility, viscosity, concentration, and molecular weight of the solutes. By contrast, the dependence on the amount of solution deposited and rate of deposition is relatively small. The resulting film thickness, d, obtained through spin coating can be expressed by the empirical relationship:

$$d = k\omega^{\alpha} \tag{9.1}$$

where ω is the angular velocity, and α and k are empiric constants regarding the physical properties of the solute, solvent, and substrate [9]. The main reasons for the universal application of this method are its advantages of high simplicity, uniformity, reproducibility, and compatibility with diverse substrates. However, the method is an inherently single-substrate process that results in a relatively low throughput compared to other solution deposition methods, such as R2R processes.

9.2.1.2 Printing-Based Deposition Methods

In printing-based deposition methods, layers of ink are transferred from a nozzle or a stamp to a substrate. These methods are revolutionizing the burgeoning field of flexible electronics by giving cost-effective routes to process various materials at temperatures that are compatible with flexible substrates. The advantages of such processes include simplicity, low fabrication costs, reduced material wastage, and simple patterning techniques, which are attractive for cost-effective manufacturing.

Although all printing-based methods operate on a similar principle, in that they transfer the desired ink from the stamp onto the substrate, the ink viscosity and line width of the formed thin film vary depending on the method. The characteristics of various printing-based deposition methods are compared in Figure 9.6 in terms of ink viscosity and line width. In general, contact-based printing methods form wider film line widths and can print higher viscosity ink compared to noncontact-based methods. However, in the case of EHD jet printing, high-viscosity inks comparable to those used in contact-based printing can be printed, while the line width is the smallest among the

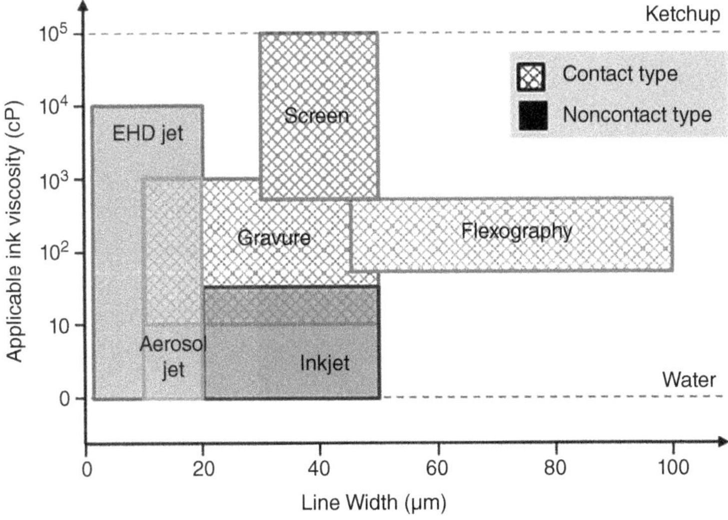

Figure 9.6 Comparison of printing characteristics in terms of ink viscosity and printed line width. *Source*: Reproduced with permission from Park, J. W., et al. (2020). A review of low-temperature solution-processed metal oxide thin-film transistors for flexible electronics. *Advanced Functional Materials* **30** (20): 1904632-1–1904632-40. Copyright 2020, Wiley-VCH.

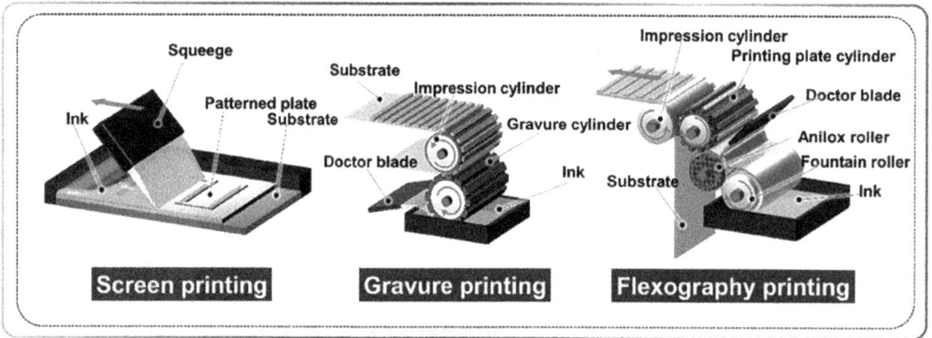

Figure 9.7 Contact printing-type solution deposition methods.

methods. Detailed explanations of each printing method, categorized as contact and noncontact, are provided in the remainder of this subsection.

Contact-Based Printing In contact-based-printing deposition methods, patterned structures with inked surfaces are brought into physical contact with the substrate. These methods include gravure, flexographic, and screen printing, which are shown in Figuree 9.7.

Screen Printing Screen printing is a versatile printing-type deposition method that allows for full two-dimensional patterns on the printed layer, and it is one of the most popular and mature technologies for printed electronics. The screen-printing system has a simple setup comprising a screen, a press bed, a squeegee, and the substrate. In screen printing, ink poured on the screen is squeegeed across the screen, resulting in transfer onto the substrate beneath through the stencil openings. Despite its very simple process, print quality and characteristics can vary considerably according to a range of factors like the solution viscosity, printing speed, geometry and angle of the squeegee, mesh size, snap-off between the screen and substrate, and material of substrate. Screen printing is widely used because the results can be reproduced by repeating only a few steps, and an optimum operating range can be developed quickly.

Gravure Printing Gravure printing transfers functional inks directly by physically contacting engraved structures of the substrate. The tool used for gravure printing is a large cylinder electroplated with metal like copper and engraved with micro cells. The engraved cells are filled with ink from a reservoir beneath the rotating gravure cylinder, where extra ink on the rotating cylinder can also be removed by a doctor blade. After the engraved cells have been filled with ink, the ink is transferred through a capillary action onto a rollable substrate as it is placed between the engraved and impression cylinders. Solution properties such as viscosity, wettability, and the shape of the cell (like the width/depth ratio) play important roles in gravure printing. In particular, low-viscosity inks are often applied to prevent ink bleeding out from the gravure micro cells, to speed up the process and allow emptying of the cells, thus achieving better line resolution. As a result, gravure printing can produce high-quality patterns on a large-area substrate in a cost-effective manner, typically in a R2R process.

Flexography Printing Flexography printing is a similar method to gravure printing, with the exception that the ink is adsorbed on the outer surface of the patterned printing roller, which is typically made of rubber, instead of filling the engraved cells. A flexography printing system consists of two rollers and two cylinders: a fountain roller, an anilox roller, a printing plate cylinder, and an impression cylinder. The fountain roller transfers ink to the anilox roller, and the anilox roller primarily controls the ink quantity to be transferred onto the printing plate roller (and subsequently to the substrate). Then, by contacting with the inked areas in the anilox cylinder, these raised patterns on the printing plate cylinder print on the substrate, running between the printing plate and the impression cylinders. This results in uniform thin layers, sharper edges, and improved pattern reliability. The anilox volume (i.e., the size and frequency of engraved cells) has a significant effect on the printed network tracks and sheet resistance. Flexographically printed films have been reported to be more uniform and slightly less smooth than spin-coated/gravure-printed films. However, this method is susceptible to de-wetting and film instability, which promote defects such as overlapped lines, open lines, and edge waviness.

Noncontact Printing In noncontact printing, the solution is dispensed through nozzles or openings, and structures are defined by the movement of the stage in a pre-programmed pattern. The foremost noncontact-printing methods include spray printing, inkjet printing, and EHD jet printing. Here, noncontact-printing methods with distinct advantages such as simplicity, affordability, speed, and adaptability will be explained in detail.

Spray Printing Spray printing is an attractive noncontact-printing-based deposition method for large-area printing, in which the printing ink is transferred to the desired portion of the substrate through a nozzle, where a fine aerosol is formed. The spray-printing process is conducted in a chamber containing carrier gases, and the combined force of gravity and carrier gas sprays the droplets to the substrate surface. The roughness, thickness, and quality of the sprayed films are affected by the solution viscosity, droplet size, number of coating cycles, and substrate temperature during printing. Spray printing has the advantages of facile material doping, compatibility with large-area substrates, and multicomponent film integration via multiple printing passes. However, the film quality achieved with spray printing has the limitations of the availability of suitable solutions with low viscosity and the relatively high roughness of spray-printed films. Additionally, this process tends to require a long processing time, due to the use of minute droplets to create the films.

Inkjet Printing Inkjet printing is a rapidly emerging method for direct patterning during solution deposition. Various methods for actuation of an inkjet nozzle head have been developed. Among these methods, the most prominent systems are based on thermal and piezoelectric inkjetting mechanisms. Droplets are ejected in response to a pulse generated by either thermal or piezoelectric actuators in the inkjet nozzle head. In the thermal inkjet system, a tiny amount of the solvent is heated and evaporated, where the evaporated gas makes ejection of the ink from the nozzle. In the case of the piezoelectric inkjet, a voltage is applied to a metal nozzle tip constructed from a piezoelectric material, and the nozzle changes its shape (open/closed) to force ink out of the nozzle according to the electrical signals. The diameter, temperature, and type of nozzle directly affect the morphology of the printed ink. For instance, the size of printed droplets and the line width of the ejected ink decrease as the nozzle diameter decreases; however, there are physical limitations to minimize the nozzle size. The viscosity of printable ink and the width of inkjet-printed lines are in the ranges of 1–50 cP and 20–50 μm, respectively. Inkjet printing offers a range of advantages compared to other printing methods, including the selective printing of a small quantity of ink on small-area substrates, the ability to pattern ink without photolithography, ease of alignment when printing multiple layers, and materials savings. However, the throughput of this method is lower than that of roller-based methods, such as gravure and flexography printing. Furthermore, there is a limited range of ink choices, because high-viscosity inks cause agglomeration and nozzle clogging.

EHD Jet Printing EHD jet printing is a recently developed printing technique for noncontact high-resolution patterning that addresses the limitation of inkjet printing. An EHD jet printer system consists of a nozzle, conductive substrate, syringe pump, transition stage, and high-voltage power supplier. The most prominent difference between EHD jet printing and other inkjet printing is the utilization of a conductive substrate and printing nozzle for electric field actuation for the stable jetting characteristics. Thus, the electric field is the driving force in EHD jet printing, unlike thermal or piezoelectric actuation inkjet-printing techniques. In particular, the diameter of the nozzle is usually larger than the other printing nozzles. This large nozzle prevents clogging for easier printing

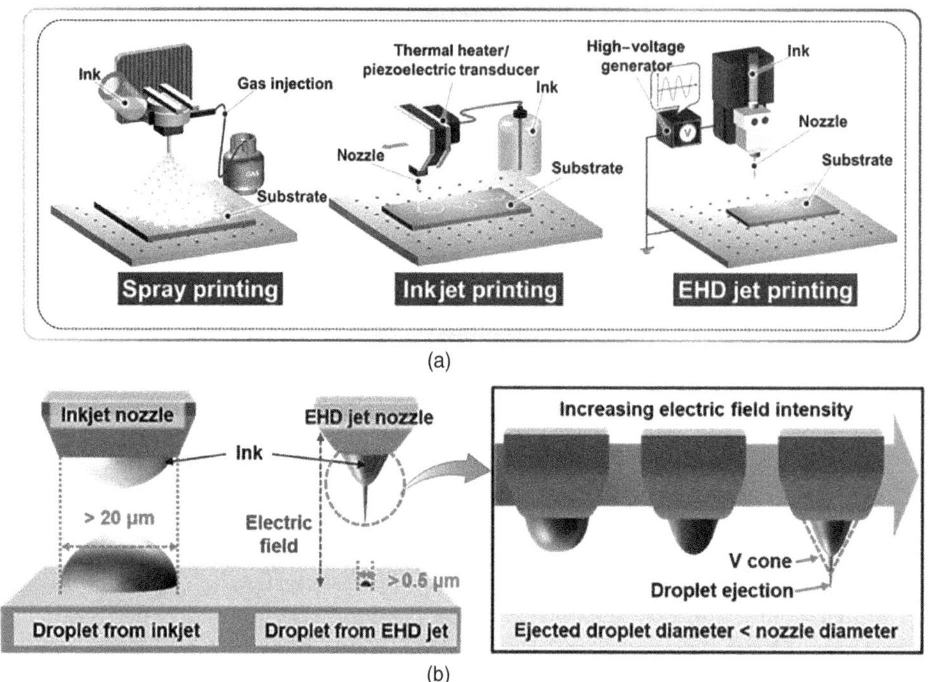

Figure 9.8 (a) Noncontact printing solution deposition methods. (b) Comparison of inkjet and electrohydrodynamic (EHD) jet printing. *Source*: (b) Reproduced with permission from Park, J. W., et al. (2020). A review of low-temperature solution-processed metal oxide thin-film transistors for flexible electronics. *Advanced Functional Materials* **30** (20): 1904632-1–1904632-40. Copyright 2020, Wiley-VCH.

of viscous colloids. The electrical charge also helps to prevent agglomeration of solution particles, because of the Coulombic repulsion in printed droplets charged to the same pole. Although an EHD jet printer has a large nozzle, the droplets generated are even finer than those of the other jet printers, which results from the unique shape of the ink-ejecting part. If voltage is first applied, a small quantity of ink converges at the edge of the nozzle, forming a meniscus. As the electric field gets stronger, Coulombic repulsion from mobile ions gives a tangential force to the ink surface, transforming the meniscus into a "V" shape, which is known as the Taylor cone [10]. If the applied electric potential exceeds a threshold value, the surface tension of the ink is overcome by the Coulombic force, and minute droplets can be ejected from the narrow tip of the V cone, which is described in Figure 9.8b. Even if the general viscosity of the printable ink and the line width of the EHD jet printing range from 1 to 10,000 cP and from 0.5 to 20 μm, respectively, these parameters vary greatly depending on the pulse conditions and the intensity of the applied voltage. As a result, EHD jet printing can provide higher printing resolution (submicrometer scale), the capacity for high-viscosity printable inks, the ability to modulate the ejection form (e.g., spray mode), and control over electric field intensity and the nozzle shape [11].

9.2.2 The Formation Mechanism of Solution-Processed Oxide Semiconductor Films

Solution-processed oxide thin films have been intensively studied through two main approaches based on nanoparticle and sol–gel-based metal precursors. Nanoparticles are common precursors that are converted to a film by thermally/chemically induced structural transformations; however, the method based on the nanoparticles is limited by nonuniform dispersion over large areas and low reliability. Thus, nanoparticle-based channel TFTs exhibit low stability because of the poor gate dielectric–channel interface roughness resulting from the voids between nanoparticles [12].

By contrast, the process using sol–gel-based metal precursors can be adapted to produce multicompositional, highly homogeneous, and higher quality films compared to the nanoparticle-based process [13]. In addition, the films tend to be uniform and highly reliable, even for large-area processing applications, due to improved homogeneity through mixing at the molecular level. However, producing high-quality films through a low-temperature process using the precursor method remains challenging.

Specific steps must be followed to form an oxide thin film using the sol–gel-based metal precursor method. The first step is to select a proper metal precursor for solution preparation; acetates, chlorides, and nitrates are viable options because of their cost-effectiveness and stability. An appropriate solvent and stabilizer are also needed, and chemical reactivity within the components should be considered. Conventional solvents include organic solvents and deionized water. Organic solvents are commonly utilized because of their high solubility; however, the impurities in organic solvents require higher annealing temperatures (above 350 °C) for decomposition of chemical forms of organic compounds. Aqueous solvents enable low-temperature processing, because they are free from organic compounds that act as impurities. In aqueous solvents, stabilizers are added along with precursors such as a homogenizer, chelating agent, wettability improver, and capping agent to effectively form high-quality thin films. The chemical reactivity can also act as a stimulus to accelerate the chemical reaction in the synthesizing solution.

Next, the deposition process is conducted with the prepared solvent, as described in detail in Section 9.2.1. The deposited oxide thin film is dried to evaporate the solvent and annealed for activation of semiconducting properties. Then, the oxide thin film is subjected to three processing steps for film formation, which commonly occur in both organic and aqueous solvents, as shown in Figure 9.9a: (1) decomposition, (2) hydrolysis, and (3) dehydroxylation and condensation. The precursors consist of metal cations and organic anions (i.e., ligands), dissociate into the separate ions while synthesized with the solvent, and thermally decompose at low temperature. The metal cation is surrounded by aquo ligand molecules (H_2O) or organic ligand molecules (ROH). These molecules transform into either a hydroxo ligand (OH^-) or oxo ligand ($O_2{}^-$), forming M-OH or M-O bonds; this process is referred to as a hydrolysis reaction. Finally, dehydroxylation and condensation reactions eliminate the remaining impurities to generate an oxo-bridge, which is a basic metal–oxygen–metal (M-O-M) bond formation. Based on this formation, the metal-oxide framework or lattice becomes denser, ultimately leading to oxide thin-film formation [14].

This film formation mechanism in solution-processed indium–gallium–zinc oxide (IGZO) thin films was reported in 2009, and it will be used here as an example [15]. First, the three stages of the thermogravimetry–differential scanning calorimetry (TG-DSC) curves of the IGZO sol are shown in Figure 9.9b. The first endothermic reaction, with a range of 60–130 °C, demonstrated a large weight loss. In the IGZO sol, zinc acetates ($Zn(OAc)_2{\cdot}2H_2O$), indium nitrates ($In(NO_3)_3{\cdot}xH_2O$), and gallium nitrate ($Ga(NO_3)_3{\cdot}xH_2O$) were decomposed and hydrolyzed to M-OH, such as $Zn(OH)_2$, $In(OH)_3$, and $Ga(OH)_3$. H_2O and 2-methoxyethanol (2-ME) solvents were vaporized at 60–130 °C, as shown by the following reaction:

$$\begin{aligned}&[\text{Decomposition and hydrolysis}]\ 2\text{Zn}(\text{CH}_3COO)_2 \cdot 2H_2O + Ga(NO_3)_3 \cdot 3H_2O\\&\quad + In(NO_3)_3 \cdot 3H_2O \rightarrow 2Zn(OH)_2 + Ga(OH)_3 + 4CH_3COOH(\uparrow) + 6HNO_3(\uparrow)\end{aligned} \tag{9.2}$$

Then, in the range of 190–210 °C, a large exothermic reaction was observed at 196 °C with small weight loss. This peak indicates the formation of the IGZO compound by dehydration and alloy reactions of M-OH. The small weight loss is due to the combustion of a stabilizer material and H_2O produced by the dehydroxylation. The formation reactions of the IGZO compound can be expressed as

$$[\text{Dehydroxylation}]\ 2\text{Zn}(\text{OH})_2 + Ga(OH)_3 + In(OH)_3 \rightarrow InGaZn_2O_5 + 5H_2O(\uparrow) \tag{9.3}$$

A final exothermic reaction at 305–420 °C can be attributed to the crystallization of IGZO. Generally, most oxide thin films, including IGZO, undergo a phase transition from an amorphous phase to a polycrystalline phase when a higher temperature is continuously applied. In Figure 9.9c, the film annealed at 275 °C shows an amorphous

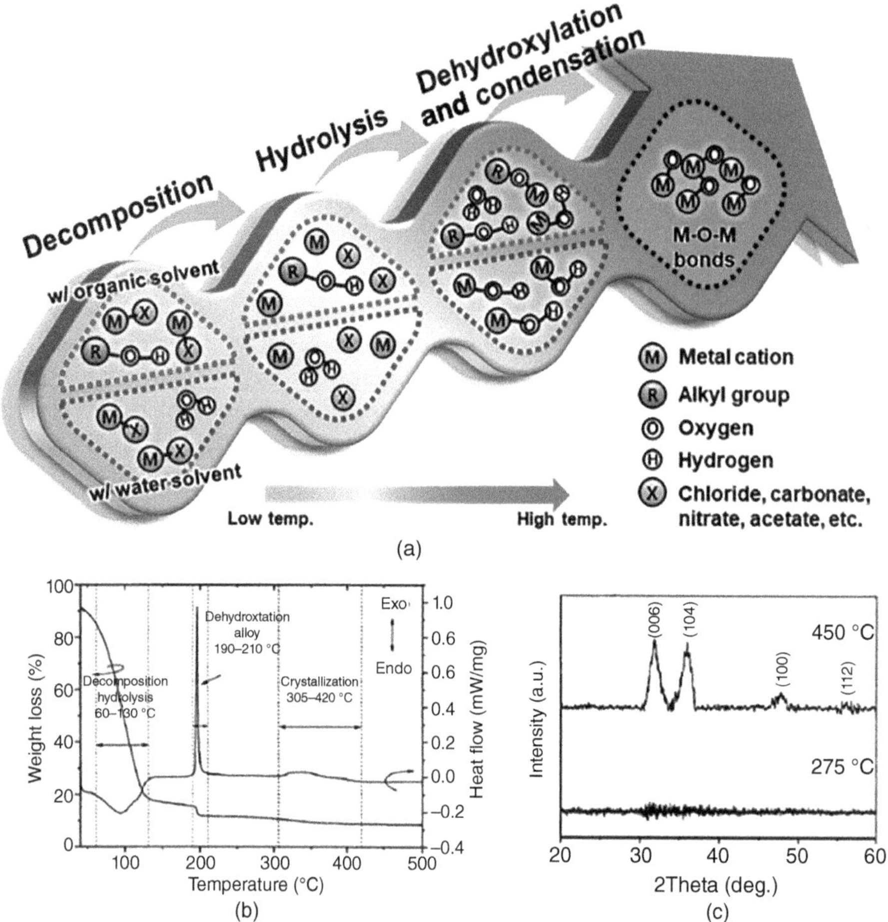

Figure 9.9 (a) Formation mechanism of solution-processed oxide thin film. (b) Steps for forming indium–gallium–zinc oxide (IGZO) thin film from the solution state, as shown in the thermogravimetric-differential scanning calorimetry (TG-DSC) curves. (c) X-ray diffraction (XRD) results of the solution-processed IGZO thin film. *Source*: (b) Reproduced with permission from Kim, G. H., et al. (2008). Formation mechanism of solution-processed nanocrystalline InGaZnO thin film as active channel layer in thin-film transistor. *Journal of the Electrochemical Society* **156** (1): H7–H9. Copyright 2009, The Electrochemical Society.

phase, while the film with an annealing temperature of 450 °C exhibits a polycrystalline phase with (008), (104), (109), and (112) X-ray diffraction (XRD) peaks. These results strongly indicate that the last exothermic reaction corresponds to the crystallization of $InGaZn_2O_5$.

9.3 Low-Temperature Technologies for Active-Layer Engineering of Solution-Processed Oxide TFTs

9.3.1 Overview

As discussed in Section 9.2.2, the formation of solution-processed oxide films normally requires an annealing temperature above 450 °C [16]. This is because a sol–gel solution in general involves a large-volume fragment of chemically bonded organic functional groups originating from solvents, additives, and ligands. As such, production of

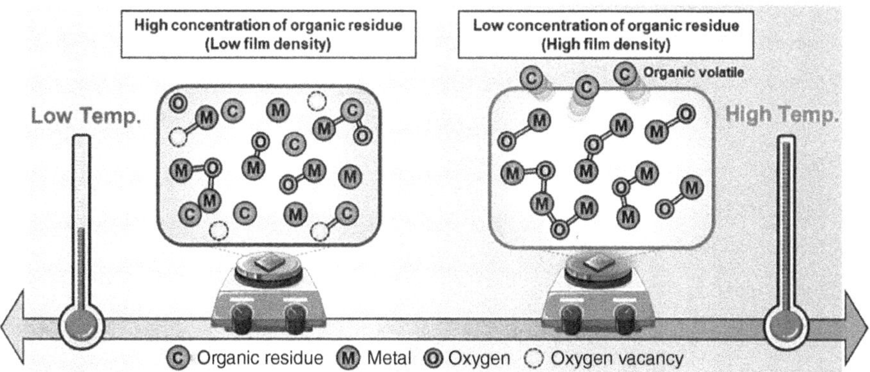

Figure 9.10 Schematic illustration of the thermal annealing process for oxidation and impurity removal according to the annealing temperature for solution-processed oxide semiconductors.

a pure-phase metal-oxide thin film from this solution phase requires high-temperature annealing. In other words, as shown in Figure 9.10, the organic components must be completely decomposed before the metal precursors undergo thermal conversion to form a metal–oxygen–metal (M-O-M) framework, through sufficient energy from the high-temperature annealing. Any remaining organic residues act as charge trap sites and adversely affect the quality of the thin films, which results in large differences between low- and high-temperature-annealed devices. Moreover, many voids and defects that degrade the electrical characteristics are generated in the oxide active layers during the annealing process, because of the evaporation of impurities such as large organic molecules, among other impurities, in the solution. Therefore, interconnected gel films with substantial voids must be compressed to achieve high density through high-temperature annealing; this explains why oxide TFTs, in which the annealing process is carried out below 300 °C, may exhibit inactivity [17]. Hence, achieving the optimal trade-off between performance and annealing temperature for solution-processed metal-oxide TFTs remains a great challenge.

Consequently, because of this high temperature requirement, substrate selection is limited; in particular, most polymer substrates for flexible electronics are excluded. This is because the high-temperature annealing process can cause cracking of these flexible substrates due to the differences in thermal expansion coefficients between the channel layer and substrate. As a result, the PI substrate has become one of the most widely used flexible polymer substrates, as it has a glass transition temperature of around 360 °C (or higher) and thus can withstand a relatively high temperature. However, PI exhibits high etchant reactivity and low transparency due to its pale-yellow color. Solving the high-temperature problem for solution-processed oxide semiconductors may permit the use of flexible polymer substrates that are inexpensive and have superior material properties for full realization of flexible and large-area oxide electronics.

In Sections 9.3.2, 9.3.3, and 9.3.4, approaches to lowering the process temperature for solution-processed oxide TFTs will be discussed. In particular, research on TFT active layers, which are among the most important determinants of electrical performance, will be discussed in detail. The techniques can be classified into three categories: solution modulation, process modulation, and structure modulation.

9.3.2 Solution Modulation

To ensure superior electrical performance of solution-processed metal-oxide thin films as an active layer for oxide TFTs, a high-temperature annealing process is required for well-linked M-O-M networks, a lower number of ligand-related impurities, and a dense film through high thermal energy. To meet these requirements using a low-temperature annealing process, various chemical approaches have been used to carefully design ligands and counterions, to reduce the activation energy and lower the thermal budget. In the remainder of this

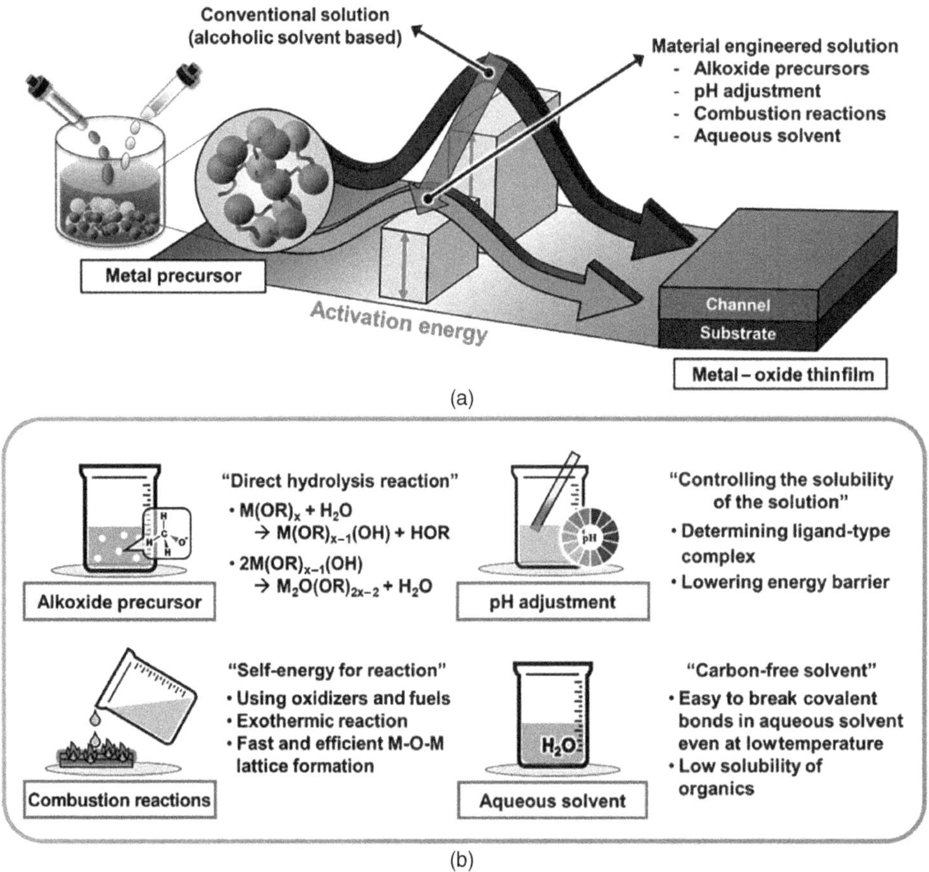

Figure 9.11 (a) Schematic illustration of a mechanism for solution-processed oxide TFTs with a low-temperature process through "solution modulation" and (b) the representative examples.

subsection, four representative solution modulation methods (Figure 9.11) are reviewed, including precursor and solvent-engineering methods for achieving effective chemical reactions during the fabrication process.

9.3.2.1 Alkoxide Precursors

Acetates, chlorides, and nitrates are typically used as precursors for the formation of solution-processed oxide semiconductor thin films, because of their high cost-effectiveness and relatively low impurity concentrations. Metal-oxide thin films using nitrate-based solutions form at lower temperatures compared to acetate- and chloride-based solutions. However, nitrate precursor-based oxide films still require a high-temperature annealing process for high TFT performance [18]. To lower the annealing temperature, a novel method, "sol–gel on chip," based on metal-alkoxide precursors has been reported [16]. These metal-alkoxide precursor-based oxide semiconductors have relatively low hydrogen and carbon impurity concentrations and can thus form metal-oxide films at low temperatures through decomposition processes that require less energy. As reported in the research, this method produced metal-oxide TFTs at only 230 °C through a direct hydrolysis reaction without an intermediate hydrolysis step. As a result, it can be concluded that the use of proper precursors could reduce the impurity residues generated by pyrolysis during the solution process and lower the annealing temperature.

9.3.2.2 pH Adjustment

The pH (hydrogen ion concentration) of the precursor solution is one of the most important determinants of final device performance, because the solubility of the solutions varies depending on the pH. According to previous research, soluble species in oxide precursor solutions at high pH form metal-hydroxy ligands (M-OH), which require only a simple dehydroxylation/dehydration reaction to be converted into metal oxide (M-O) [19]. By contrast, stable species at low pH are hydrated metal cations, such as $M(H_2O)^{z+}$, where "z" is the valence number of the metal. Hydrated metal cations can undergo more complex reactions to form M-O compared to those under the basic conditions, such as anionic decomposition during the annealing process. Therefore, through adjustment of the pH in the solution, it is possible to achieve superior TFT performance (even at a low temperature) by lowering the energy required to form M-O bonds.

To demonstrate this relationship, the research showed improvement of TFT device performance at an annealing temperature of 150 °C, which was achieved by controlling the pH and ligands of solutions in ZnO-TFTs. According to this previous work, a basic Zn precursor has a stable hydroxy ligand, and it has a lower energy barrier for solution-to-solid conversion under basic conditions compared to acid solutions. Hence, controlling the pH in solutions is important to determine the primary ligand-type complex in a solution, and it promotes both superior performance and low-temperature processing of oxide TFTs.

9.3.2.3 Combustion Reactions

Various studies have also reported using additives to control the combustion reactions and thus lower the process temperature, instead of synthesizing a new precursor [20, 21]. When mixed with organic fuels and an oxidizer, the additives cause a rapid exothermic combustion reaction. In detail, the formation reaction of oxide thin films using conventional precursors composed with metal hydroxide or/and alkoxide is endothermic, whereas formation using combustion reactions is exothermic and hence does not require large external energy input during the process. Therefore, the combustion reaction could support a low-temperature process by lowering the energy required for decomposition and oxidation of the precursor via self-energy generated from the additives.

According to previous research, when a generic metal with nitrate as the oxidizer and acetylacetone as the fuel is introduced into the precursor solution, the chemical potential of the oxide precursor increases remarkably compared to when the combustion process is not applied [21]. Therefore, when spin coating and annealing are performed using a precursor solution that has a higher chemical potential, exothermic chemical transformation occurs in the oxide thin film, enabling efficient and fast M-O-M lattice formation even at a low temperature of around 200 °C. Also, since organic impurities are removed by the local oxidizer during the combustion process, high temperatures are not required. A comparison of the required energy for conversion is described in Figure 9.12a and 9.12b.

9.3.2.4 Aqueous Solvent

In general, in solution processes using organic solvents, high-temperature annealing is required to effectively remove impurities, such as carbon residues, chloride, nitrate, and acetate, from the solvent and to promote oxidation and dehydrogenation. To solve this fundamental problem, studies have been conducted to replace the organic solvent with an aqueous solvent, because covalent bonds in aqueous solvents can be broken at low temperatures. According to previous research, the use of aqueous solutions is advantageous because oxide TFTs that can perform as well at high temperatures can be manufactured at low temperatures, using a simple and cost-effective process [22]. As illustrated in Figure 9.12c, thermogravimetric analysis of different precursor solutions (aqueous solution and 2-ME solution) makes out that the thermal decomposition temperature is considerably lower with the aqueous solution. Also, as illustrated in Figure 9.12d, nitrate-based precursors with an aqueous solution could be readily decomposed at low temperatures during solvation to form the hexaaqua metal cation ($[M(H_2O)^6]^{3+}$) [23]. As a result, in the case of the aqueous solvent, most of the weight changes are complete at about 150 °C, and

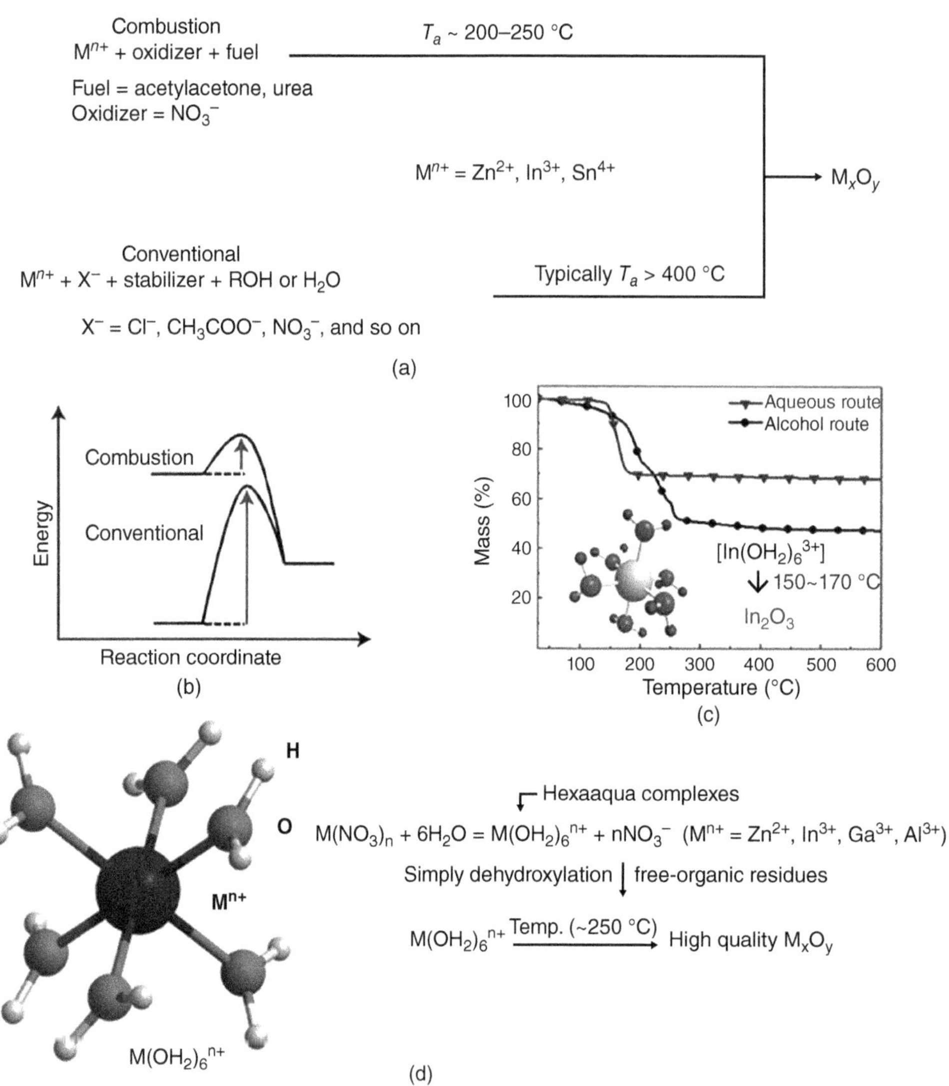

Figure 9.12 (a) Example of combustion reaction for a generic metal (M) with acetylacetone as the fuel and nitrate acting as the oxidizer (b) Energetics of the combustion synthesis-based processes versus conventional processes. (c) Thermogravimetric analysis of both aqueous and 2-methoxyethanol (2-ME)-based solutions for In_2O_3 films. (d) Structure and distinctive processing of hexaaqua metal complexes. *Source*: (a,b) Reprinted with permission from Hennek, J. W., et al. (2013). Oxygen "getter" effects on microstructure and carrier transport in low temperature combustion-processed a-InXZnO (X = Ga, Sc, Y, La) transistors. *Journal of the American Chemical Society* **135** (29): 10729–10741. Copyright 2013, American Chemical Society. (c) Reproduced with permission from Hwang, Y. H., et al. (2013). An "aqueous route" for the fabrication of low-temperature-processable oxide flexible transparent thin-film transistors on plastic substrates. *NPG Asia Materials* **5** (4): e45-1–e45-8. Copyright 2013, Springer Nature. (d) Reprinted with permission from Rim, Y. S., et al. (2015). Hexaaqua metal complexes for low-temperature formation of fully metal oxide thin-film transistors. *Chemistry of Materials* **27** (16): 5808–5812. Copyright 2015, American Chemical Society.

the oxide thin film can be formed at a lower temperature than the decomposition temperature (>230 °C) of the precursor using an organic solvent.

However, the aqueous solvent has low solubility in the organic-based precursor; this makes formation of a thin film difficult due to precipitation, which limits the selectivity of the precursors. Thus, to overcome this limitation, additives (such as a strong oxidizer) that can effectively control any impurities remaining in the thin film during the annealing process have recently been developed, along with aqueous solvents.

9.3.3 Process Modulation

Various research has attempted to reduce the process temperature by substituting other energy sources (e.g., through optical, chemical, and physical processes), as opposed to decreasing the activation energy through solution modulation. Among the various methods for process modulation, the four most representative methods (shown in Figure 9.13) will be discussed in the remainder of this subsection, particularly with respect to lowering the process temperature. Examples of the technologies will also be discussed [24–27].

9.3.3.1 Photoactivation Process

Photoactivation via light energy is a common method for lowering the annealing temperature [24, 28, 29]. This is because photoactivation has a large process window with a range of parameters (e.g., wavelength, intensity, time, and source [lamp or laser]). Through control of these parameters, the photo energy can excite the atoms in the solution and cause the molecules of the material to vibrate, ultimately leading to breakage of the molecular bonds.

As an example of a photoactivation process, replacement of high-temperature thermal energy with high energy of deep ultraviolet (DUV) has been reported [30]. The M-O-M network was formed by inducing photochemical cleavage to alkoxy groups in the solution even at room temperature, using mercury lamp irradiation at 184.9 nm (10%) and 253.7 nm (90%). When the oxide semiconductor was annealed using this method, the surface temperature of the substrate due to the radiant heat energy of the mercury lamp was reported to be 130–150 °C. Therefore, DUV-assisted oxide thin-film formation can cause densification, polycondensation, and impurity decomposition, even at low temperatures.

As an extension of the previous DUV activation approach, a more efficient DUV irradiation–water treatment–DUV irradiation (DWD) method has also been reported (Figure 9.14a) [28]. The adsorption of water on oxygen vacancies and subsequent photochemical reaction during the DWD process result in efficient passivation, more complete formation of M–O–M bonds, and greater removal of defect sites, such as oxygen vacancies and hydroxyl groups (–OH), in amorphous metal-oxide semiconductor thin films. With structural relaxation and removal of charge traps, devices fabricated using the DWD method achieved better electrical performance, with denser and more stable M-O-M networks than conventional DUV-derived devices.

Methods where light energy was not used after deposition of the channel, but instead used in solution conditions to produce oxide TFTs at low temperature, have also been reported [29]. For example, one technology used photocatalytic reaction of TiO_2 (PRT) and demonstrated solution-processed oxide TFTs with increased electrical performance at a lower process temperature. The PRT is induced through mixing the TiO_2 powder with a metal-oxide precursor solution and exposing the solution to UV light (wavelength: 365 nm). Then, the TiO_2 reacts with H_2O on its surface under UV illumination, resulting in the generation of hydroxyl radicals ($OH^{\bullet}$). As shown in Figure 9.14b, the generated hydroxyl radicals oxidize the organic solvent and decompose it into smaller organic molecules. These decomposed organic molecules have lower molar masses and boiling points. Therefore, they not only improve electrical performance by reducing defect sites inside the oxide channel layers during thermal treatment, but also enable solution-processed fabrication at low temperature. Additionally, hydroxyl radicals are strong oxidants; thus, they can combine with oxygen vacancies or chemically induced defects, thereby improving the electrical characteristics of oxide TFTs.

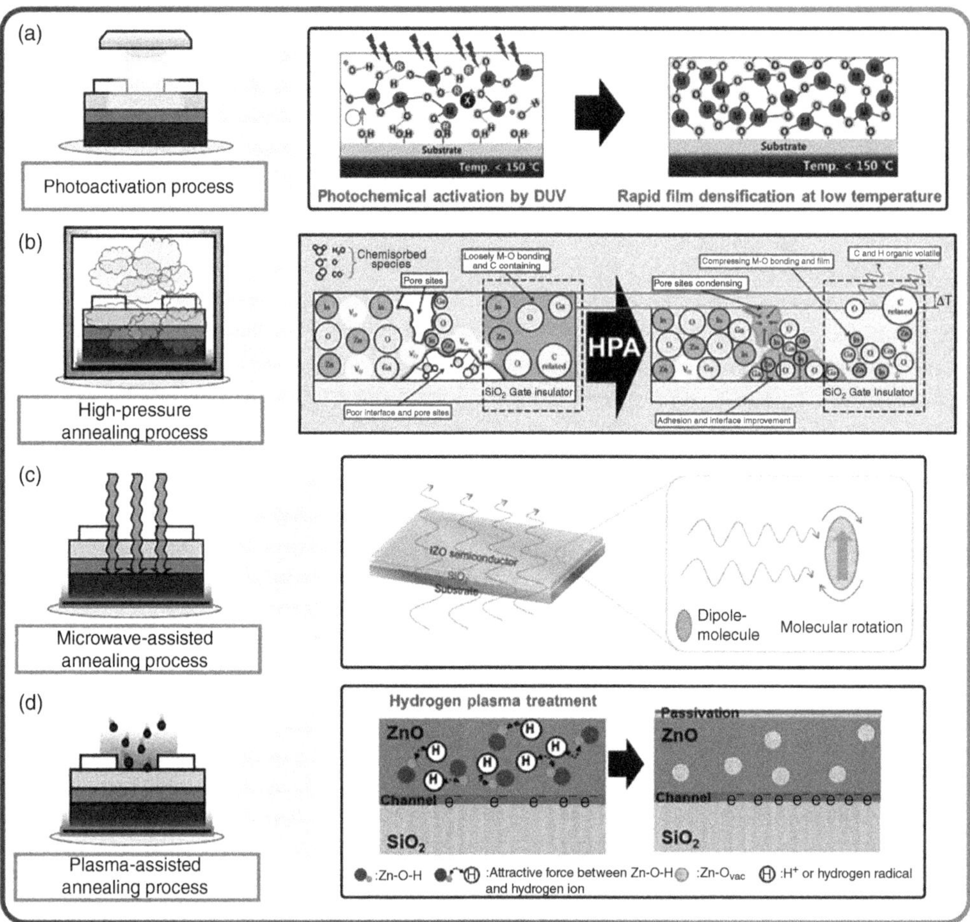

Figure 9.13 Schematic illustration of mechanism for solution-processed oxide TFTs with low-temperature process through "process modulation" and the representative examples: the (a) photoactivation process, (b) high-pressure annealing process, (c) microwave-assisted annealing process, and (d) plasma-assisted annealing process. *Source*: (a) Reprinted with permission from John, R. A., et al. (2016). Low-temperature chemical transformations for high-performance solution-processed oxide transistors. *Chemistry of Materials* **28** (22): 8305–8313. Copyright 2016, American Chemical Society. (b) Reproduced with permission from Rim, Y. S., et al. (2012). Simultaneous modification of pyrolysis and densification for low-temperature solution-processed flexible oxide thin-film transistors. *Journal of Materials Chemistry C* **22** (25): 12491–12497. Copyright 2012, Royal Society of Chemistry. (c) Reproduced with permission from Song, K., et al. (2011). Low-temperature soluble InZnO thin film transistors by microwave annealing. *Journal of Crystal Growth* **326** (1): 23–27. Copyright 2011, Elsevier BV. (d) Reproduced with permission from Jeong, Y., et al. (2014). Effects of hydrogen plasma treatment on the electrical behavior of solution-processed ZnO transistors. *Journal of Applied Physics* **116** (7): 074509-1–074509-7. Copyright 2014, American Institute of Physics.

9.3.3.2 High-Pressure Annealing (HPA) Process

This section introduces a technology whereby the energy from high-temperature annealing is compensated with high pressure, by exploiting the thermodynamics of solution-processed oxide thin-film formation. The thermodynamics can be explained through the mechanism of the Gibbs free energy change. The Gibbs free energy change ($\Delta G(T,P)$) for the formation of a homogeneous oxide film via decomposition of an initial film can be expressed as

$$\Delta G(T,P) = \frac{\left(\frac{1}{6}\right)\pi d^3}{V^C}(\Delta G^{gf\rightarrow of} + E) + \pi d^2\sigma + P\Delta V \tag{9.4}$$

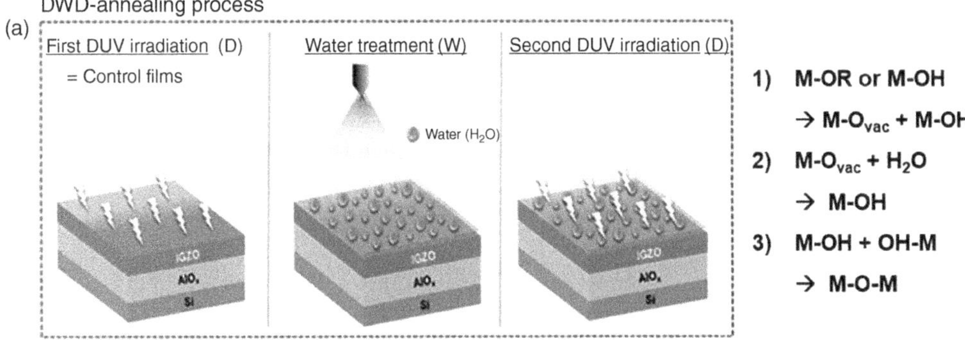

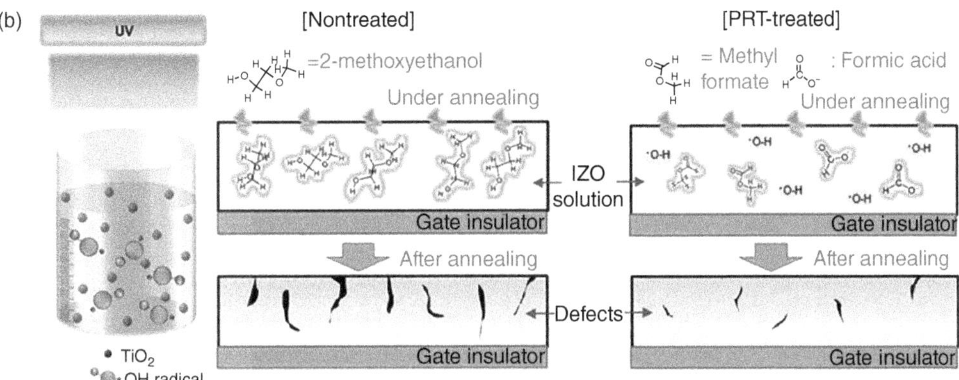

Figure 9.14 (a) Schematic illustrations of a deep ultraviolet (DUV) irradiation–water treatment–DUV irradiation (DWD) process. (b) Mechanism for improving electrical characteristics through a photocatalytic reaction of TiO_2 (PRT). *Source*: (a) Reprinted with permission Heo, J. S., et al. (2016). Water-mediated photochemical treatments for low-temperature passivation of metal-oxide thin-film transistors. *ACS Applied Materials & Interfaces* **8** (16): 10403–10412. Copyright 2016, American Chemical Society. (b) Reprinted with permission from Kang, J. K., et al. (2018). Improvement in electrical characteristics of eco-friendly indium zinc oxide thin-film transistors by photocatalytic reaction. *ACS Applied Materials & Interfaces* **10** (22): 18837–18844. Copyright 2018, American Chemical Society.

where $\Delta G^{gf \rightarrow of}$ is the free energy change of the transformation from the gel films to the oxide films, σ is the free energy for formation of the interface between the gel films and the oxide film matrix, V^C is the molar volume of the oxide films, ΔV is the change in volume during the formation of oxide in the mechanically and chemically modified films, and E is the elastic energy induced by the change in volume during decomposition of the intrinsic components of the films. P is an applied pressure that affects the kinetics of oxide formation through changes in the density and volume of the redundant portion of the oxide films. High pressure applied to the oxide film can result in $\Delta V < 0$, thereby lowering the Gibbs free energy. As a result, high-pressure annealing (HPA) can decompose the precursors and form an oxide lattice at a lower temperature by pressurizing the film. Moreover, HPA has the advantages of easy control of film density control and effective confirmation of the gas effect in the oxide film.

To increase the efficiency of the HPA process described here, a novel method called sequential pressure annealing (SPA) has been reported [31]. The SPA treatment is a sequential process from hydrogen-based HPA (H-PA) to oxygen-based HPA (O-PA), which can construct a vertically graded oxygen deficiency structure (Figure 9.15). H-PA produces more oxygen vacancies in all channel regions via a reduction reaction in weak chemical bonds, and the O-PA treatment selectively forms a region with fewer oxygen vacancies in the backchannel through an oxidation reaction. As a result, gradual vertical oxygen deficiencies can form in the film. Through this technology, the electrical performance and reliability of the device can be improved, even at a process temperature of 200 °C.

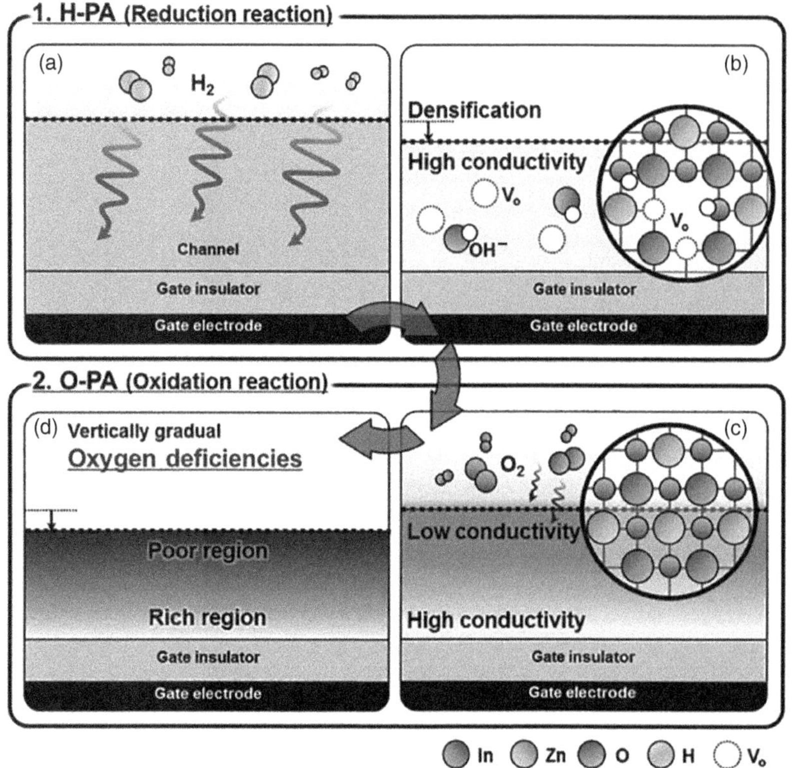

Figure 9.15 Schematic images of the sequential pressure annealing (SPA) treatment mechanism: (a) during hydrogen-based HPA (H-PA) treatment, (b) after H-PA treatment, (c) during oxygen-based HPA (O-PA) treatment, and (d) after O-PA treatment. *Source*: Reproduced with permission from Tak, Y. J., et al. (2016). Modified stoichiometry in homogeneous indium–zinc oxide system as vertically graded oxygen deficiencies by controlling redox reactions. *Advanced Materials Interfaces* **3** (4): 1500606-1–1500606-7. Copyright 2016, Wiley-VCH.

9.3.3.3 Microwave-Assisted Annealing Process

The process technology using microwave is also one of the methods forming oxide semiconductor thin films by converting electromagnetic energy into thermal energy to facilitate low-temperature processing. This is based on the principle of exposing the solution-processed metal-oxide semiconductor to microwave energy to heat the solution quickly and uniformly, thereby accelerating the reaction rate by directly coupling to the molecules within the solution through conduction or polarization. Furthermore, the microwave energy can produce heat selectively, by adjusting the wavelengths to the preferred energy absorption range of the high dielectric/ohmic loss material. Dielectric loss is in accordance with the size of the external electric field applied to molecules of the material and can result in a large heating effect in the material. In previous studies, the characteristics of devices heat-treated at high temperature (>300 °C) using a hot plate, and at low temperature (<150 °C) using microwave radiation, showed similar electrical performance even after a very short sintering time [26].

9.3.3.4 Plasma-Assisted Annealing Process

The plasma-assisted annealing process is one of the most promising methods for enhancing dihydroxylation or dehydration reactions, which could lead to improved electrical performance in solution-processed oxide TFTs even at low temperature. Among the various plasmas, hydrogen and NH_3 plasmas are mainly applied for inducing hydrogen diffusion. Hydrogen exposure has been reported to improve electrical performance parameters, such as carrier mobility. This improvement is caused by the reductions of weakly bound oxygen species, such as $-CO_3$,

–OH, or adsorbed O_2, on the surface of the films. According to previous research, plasma treatment can passivate unstable and weakly bonded groups with hydrogen ions, release H_2O, and create oxygen vacancies even at low temperature (<150 °C) [27]. Moreover, sufficient energy can be generated during the plasma-assisted annealing process to remove any water molecules from the oxide layer.

9.3.4 Structure Modulation

Compared to vacuum-processed oxide TFTs, solution-processed oxide TFTs have lower mobility due to the relatively poor film quality and impurities induced by solvents or precursors. In addition, this difference in performance is more severe in oxide TFTs fabricated via a low-temperature process. Thus, a number of methods have been proposed to improve these inferior characteristics, including the solution and process modulation methods discussed in this chapter, as well as those based on composition modification, dopant addition, and additive addition. However, these methods cannot address an intrinsic limitation of solution-processed AOS films—low film density—which is caused by the numerous pores and pinholes created during solvent volatilization [32]. To overcome this limitation, a number of studies have reported novel structures for solution-processed oxide TFTs, to enhance electrical performance even after low-temperature fabrication. Among these, multistacked active layers such as dual-active layers and multiactive layers (as shown in Figure 9.16) have been widely studied [33].

Multistacked layers can be classified into two types: homojunction and heterojunction active layers. A homojunction is formed by two layers of the same material, whereas a heterojunction is formed from different materials.

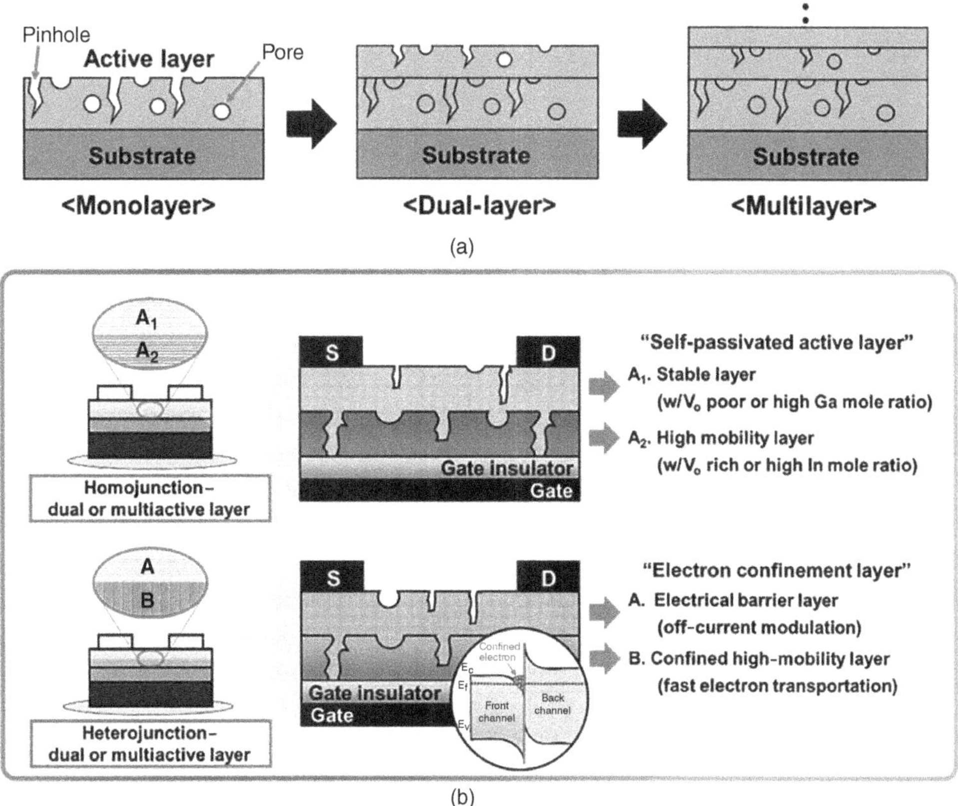

Figure 9.16 (a) Schematic illustration of mechanism for solution-processed oxide TFTs with low-temperature process through "structure modulation" and (b) two representative examples.

Heterojunction active layers have poor interface properties compared to homojunctions; however, they have the advantage of easily controlling the electrical characteristics, chemical stability, and reliability in TFTs. This section will introduce oxide TFTs based on these two types of multistacked layers and discuss in detail why they exhibit superior electrical performance even at low temperatures.

9.3.4.1 Homojunction Dual-Active or Multiactive Layer

A simple method using homojunction active layers has been reported to reduce the process temperature for a quaternary oxide: the vertical diffusion technique (VDT) shown in Figure 9.17 [34]. The VDT is used to deposit two oxide layers (IZO/GaO) successively, and anneal them simultaneously to facilitate effective diffusion between each layer. Generally, conventional IGZO-TFTs do not exhibit proper transfer characteristics at low temperature (<300 °C), as it is difficult to create a metal–oxide–metal framework from four atoms at this low temperature. However, with the VDT, binary and ternary oxide atoms are effectively diffused into the other layer, resulting in the formation of a quaternary oxide film at low temperature, and thus the fabrication of IGZO-TFTs with superior electrical performance with no additional treatment. Consequently, the VDT could enable a significant reduction in processing temperature to below 300 °C, while maintaining the electrical characteristics of IGZO-TFTs.

On the other hand, introducing the dual-active layer could also increase the off current because of the improved film quality of the whole channel. To solve this problem, various studies have reported the inclusion of a homojunction active layer that lowers the conductivity in the backchannel far from the gate and increases the conductivity in the front-channel layer near the gate. For example, research has been conducted using self-passivated dual-active layers, including a backchannel layer with an increased Ga molar ratio compared to the front-channel layer [35]. When the TFTs are exposed to air, this structure can effectively decrease the unexpected off-current in the exposed active layer, because the carriers and defects related to oxygen vacancies are reduced with increased Ga content.

9.3.4.2 Heterojunction Dual- or Multiactive Layer

Structure modulation via heterojunction multistacked channel layers, including electron confinement, has been reported to improve the electrical performance of devices fabricated at low temperature. The key to increasing

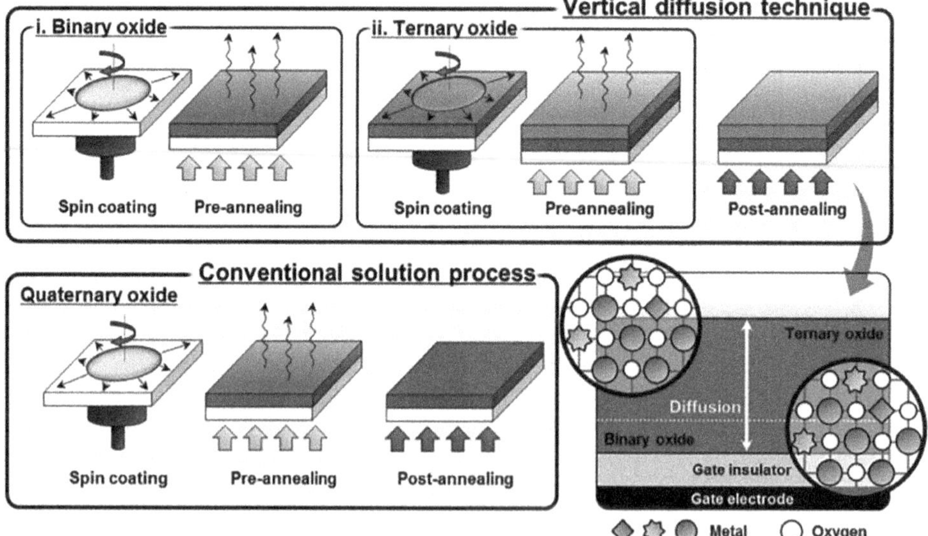

Figure 9.17 Schematic diagram of the vertical diffusion technique (VDT) and conventional processes for quaternary oxide. *Source*: Yoon, S., et al. (2017). A solution-processed quaternary oxide system obtained at low-temperature using a vertical diffusion technique. *Scientific Reports* **7**: 43216-1–43216-7. Licensed under CC BY 4.0.

mobility in heterojunction dual-active-layer oxide TFTs is electron confinement at the interface between the two active layers. The improved carrier mobility in the heterojunction dual-active layers depends on the difference in Fermi energy level between the two active layers, and their electron confinement in the potential well. As the difference between the two Fermi energy levels becomes larger, the concentration of confined electrons in the potential well also increases.

Two representative studies have investigated the fabrication of solution-processed oxide TFTs via electron confinement in a low-temperature process. First, growth of an ultrathin (≤13 nm) low-dimensional ZnO/In_2O_3 heterojunction at 250 °C as a heterojunction dual-active layer, through sequential ultrasonic spray pyrolysis and spin-coating processes, was reported [36]. The TFTs exhibited superior electron mobility because of the sharp ZnO/In_2O_3 heterojunction and 2D free-electron confined layer in its vicinity. Figure 9.18a shows a high-resolution transmission electron microscopy (HR-TEM) image of the ZnO/In_2O_3 heterojunctions; the heterojunctions were continuous, without an intermixing process of materials. The hexagonal wurtzite structure of the nanocrystalline ZnO layer and the body-centered cubic (bcc) structure of the In_2O_3 layer were confirmed by reference to the fast Fourier transform patterns shown in Figure 9.18b. In addition, Figure 9.18c shows the energy band alignment of the In_2O_3/ZnO heterojunction and suggests the existence of an electron confinement layer. This result may be attributed to the generation of a local electron confinement layer due to band alignment and band offset,

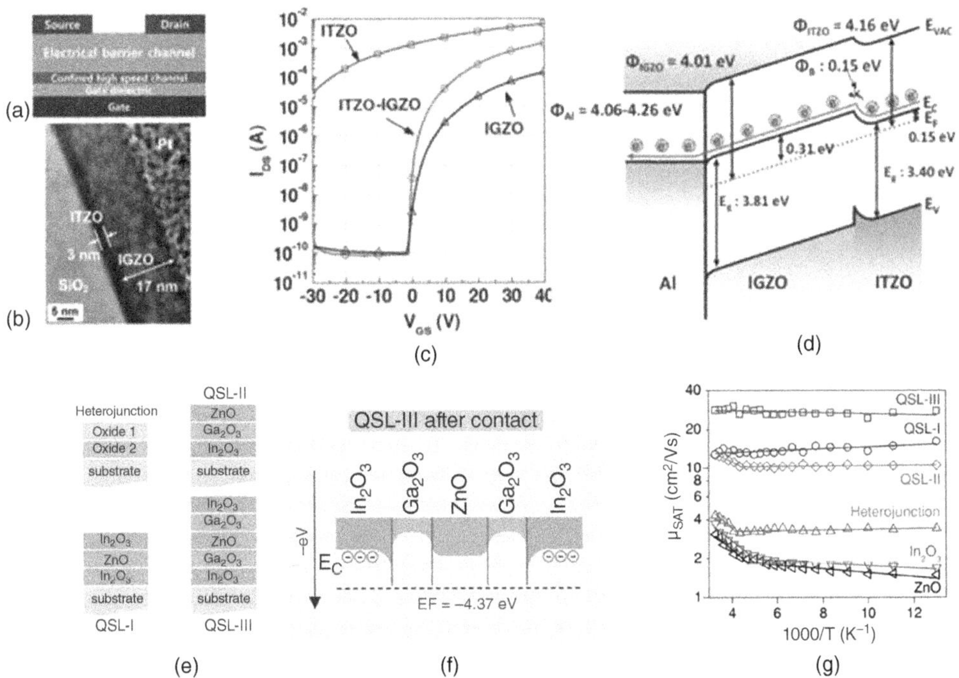

Figure 9.18 a) Device schematic of a heterojunction TFT with dual-stacked channel layers comprised of an electron confined layer and an electrical barrier layer. b) Cross-sectional high-resolution transmission electron microscopy (HR-TEM) images of the IGZO layer confined ultra-thin ITZO film. c) Transfer characteristics of IGZO, ITZO, and ITZO-IGZO TFTs. d) ITZO-IGZO bend structure schematic for the barrier formation on electron path confined between IGZO and ITZO layers. Reproduced with permission [36]. Copyright 2014, Wiley-VCH. d) Schematic of a structure with different heterojunctions and three types of quasi-superlattice (QSL; QSL-I, QSL-II, QSL-III). e) Schematic energy band diagram of the QSL-III after contact. f) Arrhenius plot showing the saturation mobility of the heterojunction and the three types of QSL depending on the temperature. *Source*: Reproduced with permission [37]. Copyright 2015, Wiley-VCH.

resulting from the formation of an interface with fewer defects between the high-crystallinity In_2O_3 layer and the ZnO layer.

Following demonstration of this heterogeneous dual-active layer, much research on heterojunction multi-active layers has been reported. As a typical example, low-dimensional polycrystalline heterojunctions and quasi-superlattices (QSLs) have been reported, with large mobility enhancement even with a low-temperature process [37]. Figure 9.18d shows a schematic diagram of TFTs with various heterojunction multiactive layer structures, which consist of alternating layers of In_2O_3, Ga_2O_3, and ZnO deposited by sequential spin coating of the corresponding precursors in air at relatively low temperatures (180–200 °C). Through time-of-flight secondary ion mass spectrometry (Figure 9.18e), the structure of the QSL-III multiactive layer was indirectly certified by the changes in element signals according to the film thickness. Finally, the Arrhenius plot (Figure 9.18f) shows the temperature dependence of the saturation mobility; In_2O_3-TFTs and ZnO-TFTs with the structure of heterojunction active layers maintained greater mobility, independent of temperature, compared to the TFTs with the structure of a monoactive layer. This result arises from the presence of two-dimensional electron gas (2DEG) systems with electron confinement at the carefully modulated oxide heterointerfaces in the novel structures, giving rise to a new perspective on design principles that can be applied to develop next-generation oxide semiconductors, devices, and circuits.

9.4 Applications of Flexible Electronics with Low-Temperature Solution-Processed Oxide TFTs

The improvements recently achieved in the electrical performance and process technologies of low-temperature solution-processed oxide TFTs, combined with special features such as large-area conformability, low cost, and high compatibility with various fabrication methods, suggest a wide range of possible applications including flexible electronics [38]. Even though research in this area has only recently shown significant advances, several systems have already been developed and brought to at least the prototype stage. This section will describe the progress achieved in flexible electronics based on solution-processed oxide TFTs, covering systems for sensors, displays, and integrated circuits.

9.4.1 Flexible Displays

The motivation for research on oxide TFTs is to meet the requirements for high-quality next-generation active-matrix displays, especially flexible displays. However, there are no commercialized displays that use solution-processed oxide TFTs; only prototypes exist. The first LCD display (a 12.1-inch WXGA TFT-LCD), with solution-processed oxide TFTs annealed at 350 °C, was demonstrated by AU Optronics in 2013 [39]. Although the solution-processed oxide TFTs used in that study showed low saturation mobility, and the display had image defects, the study is noteworthy as it was at the forefront of the development of the flexible displays using low-temperature solution-processed oxide TFTs.

In another example, Chunghwa Picture Tubes developed a 5.5-inch LCD with solution-processed oxide TFTs in 2018 [40]. The solution-processed oxide TFTs were fabricated using slot-die coating technology and showed highly uniform electrical performance on Gen. 4.5 glass substrates. The display has a narrow bezel (<1 mm) and is considered to have potential as a commercialized product.

9.4.2 Flexible Sensors

Flexible sensors have various applications, including in wearable electronics and as electronic skin for robots. Low-cost bendable sensors for sensing pH, X-rays, and temperature based on solution-processed oxide TFTs have been widely demonstrated. These applications can be realized because oxide TFTs have high sensitivity to various external stimuli, such as water, temperature, gas molecules, and light, via their bonding properties. The metal–oxide bonding exhibits ionic bonding, which is relatively vulnerable to redox processes between external molecules and the oxide, compared to covalent bonding or van der Waals bonding. Adsorbed molecules on the surface change the potential barrier to transportation of electrons, allowing variations in electrical characteristics. The high transparency, flexibility, and printability of solution-processed oxide TFTs are likely to be appropriate for next-generation devices, such as Internet-of-things (IoT) and sensor applications.

In particular, biosensing systems based on solution-processed oxide TFTs have been investigated comprehensively due to their high sensitivity, as well as the possibility of fabricating ultrathin, uniform films. This is because the thinness of ultrathin films could facilitate the depletion of carriers in a channel with a thickness within the Debye length (~20 nm). For a recent example of such sensors, flexible biosensors using a solution-processed oxide semiconductor that adhere to the skin have been reported: Ultrathin (3.5 nm) In_2O_3 films were prepared for a biosensor application via a one-step process of spin coating and an aqueous In_2O_3 solution [41]. The resulting biosensor platforms were used to detect pH and glucose. This biosensor has the advantages that it can be attached to an artificial eye, as well as the skin, and then easily removed again. Thus, it was demonstrated that solution-processed oxide TFT–based biosensors have potential as wearable biosensors.

9.4.3 Flexible Integrated Circuits

A high degree of TFT integration is essential to fabricate flexible circuits for future applications. In accordance with this requirement, various vacuum-processed oxide TFTs have been applied to a gate driver circuit, complementary metal-oxide semiconductor (CMOS) circuit, and complicated logic circuit, and for radiofrequency identification (RFID), beyond an inverter, ring oscillator, and operational amplifier. However, since no solution-processed oxide TFT has been reported that satisfies both the low-temperature and high-performance requirements, development is slower than for vacuum-processed devices. To maximize the benefits of the low-cost and simple fabrication process, flexible circuits based on solution-processed oxide TFTs, such as unipolar inverters and ring oscillators, have been demonstrated.

As a representative example, a high-performance and flexible common-source amplifier circuit composed with an EHD jet-printed In_2O_3 TFT array was reported in 2016 [42]. The amplifier exhibited voltage transfer characteristics, with a gain of 16 at a supply voltage of 4 V, a well-defined output signal (output high voltage: 3.9 V; output low voltage: 0.3 V), wide noise margins (noise margin high: 2 V; noise margin low: 0.9 V), and a transconductance of 200 μS at a V_{DD} of 4 V. Additionally, the flexible and stretchable circuits could be realized in a wavy configuration due to the EHD jet-printed In_2O_3 TFT, fabricated at a low temperature of 250 °C.

References

1 Sun, Y. and Rogers, J.A. (2007). Inorganic semiconductors for flexible electronics. *Advanced Materials* 19 (15): 1897–1916.

2 Park, J.W., Kang, B.H., and Kim, H.J. (2020). A review of low-temperature solution-processed metal oxide thin-film transistors for flexible electronics. *Advanced Functional Materials* 30 (20): 1904632-1–1904632-40.

3 Kim, H.J., Park, K., and Kim, H.J. (2020). High-performance vacuum-processed metal oxide thin-film transistors: a review of recent developments. *Journal of the Society for Information Display* 28 (7): 591–622.

4 Tong, S., Sun, J., and Yang, J. (2018). Printed thin-film transistors: research from China. *ACS Applied Materials & Interfaces* 10 (31): 25902–25924.

5 Choi, Y., Kim, G.H., Jeong, W.H. et al. (2010). Characteristics of gravure printed InGaZnO thin films as an active channel layer in thin film transistors. *Thin Solid Films* 518 (22): 6249–6252.

6 Mitzi, D. (2008). *Solution processing of inorganic materials*. Hoboken, NJ: John Wiley & Sons.

7 Patil, M., Sharma, D., Dive, A. et al. (2018). Synthesis and characterization of Cu_2S thin film deposited by chemical bath deposition method. *Procedia Manufacturing* 20: 505–508.

8 Rahaman, M. (2007). *Sintering of ceramics*. Boca Raton, FL: Taylor & Francis.

9 Norrman, K., Ghanbari-Siahkali, A., and Larsen, N.B. (2005). 6 Studies of spin-coated polymer films. *Annual Reports Section "C" (Physical Chemistry)* 101: 174–201.

10 Taylor, G.I. (1964). Disintegration of water drops in an electric field. *Proceedings of the Royal Society A: Mathematical, Physical and Engineering Sciences* 280 (1382): 383–397.

11 Choi, W.S. (2019). Electrical properties of a ZTO thin-film transistor prepared with near-field electrohydrodynamic jet spraying. *Electronic Materials Letters* 15 (2): 171–178.

12 Jeong, S. and Moon, J. (2012). Low-temperature, solution-processed metal oxide thin film transistors. *Journal of Materials Chemistry C* 22 (4): 1243–1250.

13 Garlapati, S.K., Mishra, N., Dehm, S. et al. (2013). Electrolyte-gated, high mobility inorganic oxide transistors from printed metal halides. *ACS Applied Materials & Interfaces* 5 (22): 11498–11502.

14 Park, S., Kim, C.H., Lee, W.J. et al. (2017). Sol-gel metal oxide dielectrics for all-solution-processed electronics. *Materials Science & Engineering R: Reports* 114: 1–22.

15 Kim, G.H., Shin, H.S., Du Ahn, B. et al. (2008). Formation mechanism of solution-processed nanocrystalline InGaZnO thin film as active channel layer in thin-film transistor. *Journal of the Electrochemical Society* 156 (1): H7–H9.

16 Banger, K., Yamashita, Y., Mori, K. et al. (2011). Low-temperature, high-performance solution-processed metal oxide thin-film transistors formed by a "sol–gel on chip" process. *Nature Materials* 10 (1): 45–50.

17 Shao, F. and Wan, Q. (2019). Recent progress on jet printing of oxide-based thin film transistors. *Journal of Physics D: Applied Physics* 52 (14): 143002-1–143002-26.

18 Xu, W., Li, H., Xu, J.B., and Wang, L. (2018). Recent advances of solution-processed metal oxide thin-film transistors. *ACS Applied Materials & Interfaces* 10 (31): 25878–25901.

19 Jun, T., Jung, Y., Song, K., and Moon, J. (2011). Influences of pH and ligand type on the performance of inorganic aqueous precursor-derived ZnO thin film transistors. *ACS Applied Materials & Interfaces* 3 (3): 774–781.

20 Hennek, J.W., Smith, J., Yan, A. et al. (2013). Oxygen "getter" effects on microstructure and carrier transport in low temperature combustion-processed a-InXZnO (X= Ga, Sc, Y, La) transistors. *Journal of the American Chemical Society* 135 (29): 10729–10741.

21 Kim, M.-G., Kanatzidis, M.G., Facchetti, A., and Marks, T.J. (2011). Low-temperature fabrication of high-performance metal oxide thin-film electronics via combustion processing. *Nature Materials* 10 (5): 382–388.

22 Hwang, Y.H., Seo, J.S., Yun, J.M. et al. (2013). An "aqueous route" for the fabrication of low-temperature-processable oxide flexible transparent thin-film transistors on plastic substrates. *NPG Asia Materials* 5 (4): e45-1–e45-8.

23 Rim, Y.S., Chen, H., Song, T.B. et al. (2015). Hexaaqua metal complexes for low-temperature formation of fully metal oxide thin-film transistors. *Chemistry of Materials* 27 (16): 5808–5812.

24 John, R.A., Chien, N.A., Shukla, S. et al. (2016). Low-temperature chemical transformations for high-performance solution-processed oxide transistors. *Chemistry of Materials* 28 (22): 8305–8313.

25 Rim, Y.S., Jeong, W.H., Kim, D.L. et al. (2012). Simultaneous modification of pyrolysis and densification for low-temperature solution-processed flexible oxide thin-film transistors. *Journal of Materials Chemistry C* 22 (25): 12491–12497.

26 Song, K., Koo, C.Y., Jun, T. et al. (2011). Low-temperature soluble InZnO thin film transistors by microwave annealing. *Journal of Crystal Growth* 326 (1): 23–27.

27 Jeong, Y., Pearson, C., Lee, Y.U. et al. (2014). Effects of hydrogen plasma treatment on the electrical behavior of solution-processed ZnO transistors. *Journal of Applied Physics* 116 (7): 074509-1–074509-7.

28 Heo, J.S., Jo, J.W., Kang, J. et al. (2016). Water-mediated photochemical treatments for low-temperature passivation of metal-oxide thin-film transistors. *ACS Applied Materials & Interfaces* 8 (16): 10403–10412.

29 Kang, J.K., Park, S.P., Na, J.W. et al. (2018). Improvement in electrical characteristics of eco-friendly indium zinc oxide thin-film transistors by photocatalytic reaction. *ACS Applied Materials & Interfaces* 10 (22): 18837–18844.

30 Park, S., Kim, K.H., Jo, J.W. et al. (2015). In-depth studies on rapid photochemical activation of various sol–gel metal oxide films for flexible transparent electronics. *Advanced Functional Materials* 25 (19): 2807–2815.

31 Tak, Y.J., Rim, Y.S., Yoon, D.H. et al. (2016). Modified stoichiometry in homogeneous indium–zinc oxide system as vertically graded oxygen deficiencies by controlling redox reactions. *Advanced Materials. Interfaces* 3 (4): 1500606-1–1500606-7.

32 Kim, D.J., Kim, D.L., Rim, Y.S. et al. (2012). Improved electrical performance of an oxide thin-film transistor having multistacked active layers using a solution process. *ACS Applied Materials & Interfaces* 4 (8): 4001–4005.

33 Hong, S., Park, J.W., Kim, H.J. et al. (2016). A review of multi-stacked active-layer structures for solution-processed oxide semiconductor thin-film transistors. *Journal of Information Display* 17 (3): 93–101.

34 Yoon, S., Kim, S.J., Tak, Y.J., and Kim, H.J. (2017). A solution-processed quaternary oxide system obtained at low-temperature using a vertical diffusion technique. *Scientific Reports* 7: 43216-1–43216-7.

35 Kim, D.J., Rim, Y.S., and Kim, H.J. (2013). Enhanced electrical properties of thin-film transistor with self-passivated multistacked active layers. *ACS Applied Materials & Interfaces* 5 (10): 4190–4194.

36 Faber, H., Das, S., Lin, Y.H. et al. (2017). Heterojunction oxide thin-film transistors with unprecedented electron mobility grown from solution. *Science Advances* 3 (3): e1602640-1–e1602640-9.

37 Lin, Y.H., Faber, H., Labram, J.G. et al. (2015). High electron mobility thin-film transistors based on solution-processed semiconducting metal oxide heterojunctions and quasi-superlattices. *Advanced Science (Weinheim)* 2 (7): 1500058-1–1500058-12.

38 Kim, W.G., Tak, Y.J., and Kim, H.J. (2018). Nitrocellulose-based collodion gate insulator for amorphous indium zinc gallium oxide thin-film transistors. *Journal of Information Display* 19 (1): 39–43.

39 Lin, L.Y., Hung, W.Y., Cheng, C.C. et al. (2013). 12.1-inch WXGA TFT-LCD driven by solution-processed metal oxide TFTs. In: *SID Symposium Digest of Technical Papers*. Vancouver, BC, Canada, May 19–24. Blackwell Publishing, Oxford.

40 Cai, S., Han, Z., Wang, F. et al. (2018). Review on flexible photonics/electronics integrated devices and fabrication strategy. *Science China. Information Sciences* 61 (6): 060410-1–060410-27.

41 Rim, Y.S., Bae, S.-H., Chen, H. et al. (2015). Printable ultrathin metal oxide semiconductor-based conformal biosensors. *ACS Nano* 9 (12): 12174–12181.

42 Kim, S.Y., Kim, K., Hwang, Y.H. et al. (2016). High-resolution electrohydrodynamic inkjet printing of stretchable metal oxide semiconductor transistors with high performance. *Nanoscale* 8 (39): 17113–17121.

10

Recent Progress on Amorphous Oxide Semiconductor Thin-Film Transistors Using the Atomic Layer Deposition Technique

Hyun-Jun Jeong and Jin-Seong Park

Division of Material Science and Engineering, Hanyang University, Seoul, Republic of Korea

10.1 Atomic Layer Deposition (ALD) for Amorphous Oxide Semiconductor (AOS) Applications

10.1.1 The ALD Technique

Atomic layer deposition (ALD) is a thin-film deposition technique, which is a special variant of chemical vapor deposition (CVD) to grow atomic-scale films by using self-limiting surface reaction. Based on the research history of ALD, Russian professors V. B. Aleskovskii and S. I. Kolt'sov worked on "molecular layering" in the 1960s [1]. Independently, Suntola and coworkers also worked on this method, which they called "atomic layer epitaxy." Since Suntola and Antson filed the first patent of ALD in 1974, thin-film electroluminescent (TFEL) display was the original motivation for the development of ALD technology in the mid-1970s [1, 2]. As shown in Figure 10.1, the previous TFEL should be operated under high electrical fields (1–2 MV/cm) across an insulator–luminescent–insulator tandem structure, thereby fabricating large-area devices with pinhole-free layers. The thin-film materials deposited by ALD for the TFEL display are ZnS:Mn as the luminescent layer (>1 μm), Al_2O_3 or Al_xTi_xO as the insulators (>200 nm), and Al_2O_3 as the passivation and protective layers. Mostly monochrome TFEL displays commercially provided a very wide viewing angle and full transparency within an exceptionally wide operating temperature range. Thus, ALD easily enables one to make high-quality thin films over a large area with excellent uniformity.

Conventionally, ALD has two basic mechanisms: One is complementary reaction, and the other is self-limited surface reaction. The first means that ALD should consider at least two different molecules for forming the thin film on the redox reactions. The growth surface is alternately exposed to only one of two complementary chemical environments. The other means that the surface reactions will stop and self-saturate until the surface reactive sites are entirely depleted. The unique growth technique may provide "atomic layer" deposition because each reaction is self-limiting on the surface. Thus, the self-limiting nature of the sequences is the basis of ALD. The distinction between ALD and CVD lies in the self-limiting characteristics for precursor adsorption, and alternate and sequential introduction of the precursors and reactants. As seen in Figure 10.2 [3, 4], the ALD process for two precursors is usually used when depositing metal-oxide films. The formation of Al_2O_3 thin film is an example of an ALD process where metal precursor (trimethylaluminum [TMA]) and oxidant source (water [H_2O]) are in the thermal ALD reactor. The step of a single-cycle deposition process is as follows (Figure 10.2): (i) TMA precursor exposure to the substrate; (ii) evacuation or purging of the precursor as well as byproducts from the reactor; (iii) oxygen reactant (water) exposure, typically oxidants or other reagents; and (iv) evacuation or purging of the reactant and byproduct molecules from the reactor. While the number of cycles increases, the Al_2O_3 thin film grows linearly on the substrate with a certain thickness, called "growth per cycle" (GPC). This ALD characteristic may provide outstanding

Amorphous Oxide Semiconductors: IGZO and Related Materials for Display and Memory, First Edition.
Edited by Hideo Hosono and Hideya Kumomi.

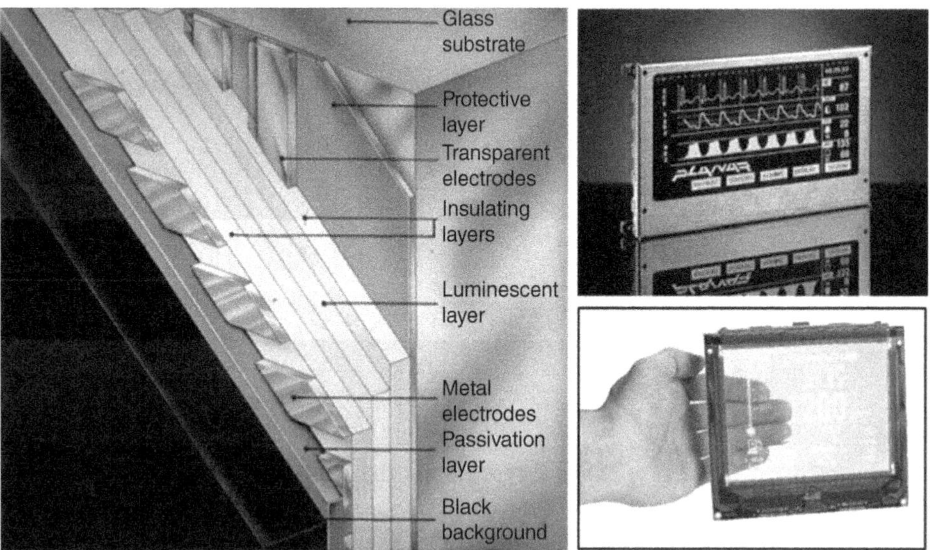

Figure 10.1 Schematics and photos of TFEL displays. The luminescent ZnS:Mn (about 1 μm in thickness), insulating Al_2O_3 or Al_xTi_yO (about 200 nm each), as well as protective and passivating Al_2O_3 layers are produced by ALD. (Lower photo) In the transparent display, the metal electrode is replaced with a transparent electrode, and the black background is left out. The TFEL display is operated by applying about 200 V AC voltage to the electrodes crossing a chosen pixel. *Source*: Copyright LUMINEQ Displays, Beneq Oy.

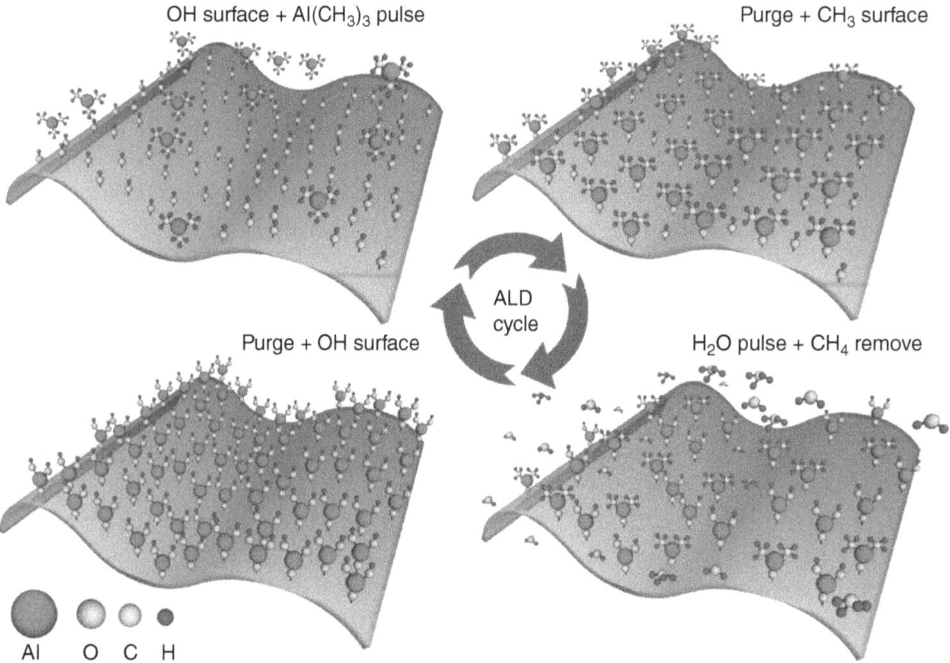

Figure 10.2 Schematic diagram depicting the general growth process of ALD using trimethylaluminum (TMA) as metal precursor and H_2O as reactant.

merits of thin-film growth, including conformality (step coverage) on complex 3D nanostructures, accurate thickness control under the atomic scale, uniform thickness and composition on a large area, and high-quality thin film at low deposition temperature.

10.1.2 Research Motivation for ALD AOS Applications

Recently, amorphous oxide semiconductors (AOSs) have been adopted as an active layer for commercial flat-panel display (FPD) products such as active-matrix liquid crystal displays (AMLCDs) and active-matrix organic light-emitting diodes (AMOLEDs) [5, 6]. Most AOS thin films have been intensively deposited on large-area substrates (over the eighth generation [8G; 2200 mm × 2500 mm]) by physical vapor deposition (PVD) such as the reactive sputtering method. Since 2011, the representative amorphous InGaZnO (*a*-IGZO) has been used as a channel layer in various FPD products (AMLCDs and AMOLEDs) because of its reasonable field-effect mobility (~10 cm^2/V.Sec), low off-current (below 10^{-18} A/um), and excellent on/off ratio. Many researchers have anticipated that its unique properties, such as extremely low off-current, may apply to novel active-matrix devices for the emerging power-saved electronics. However, many researchers have still attempted to improve the following issues for the emerging applications (high-resolution small and medium-sized displays, mixed reality [MR, augment + virtual reality], memory and logic devices, etc.): higher mobility and reliability, process repeatability under nanoscale dimensions, uniformity in atomic thickness and multicomposition, conformality on 3D nanostructure, novel *p*-type oxide semiconductor process and devices, and so on.

As shown in Figure 10.3, conventional sputtering systems for AOS materials are well-matured and established in FPD industries. Commercial *a*-IGZO thin films are currently deposited on the large-area substrate (over 8G) by reactive sputtering with a mixture of Ar and oxygen gas [7]. They have also used the conductive IGZO sputter target, which can be applied for the DC sputtering method. These are the great advantages in mass production in terms of scalability and throughput. However, there may be nonuniformity issues with the large-area substrate because of the limitation of reactive sputtering fundamentals, including the difficulties of an identical composition–stoichiometry ratio (In, Ga, Zn, and oxygen) and thickness (less than 30 nm). Moreover, the limited process parameters such as annealing process and oxygen partial pressure have made it difficult to manipulate the uniformity and to improve the performance of *a*-IGZO thin-film transistors (TFTs). So, many researchers have investigated the relationship and optimization between the annealing condition and oxygen partial pressure [8]. By now, they have reported the critical challenge to improve both mobility and reliability simultaneously with only one single target material. Thus, they have suggested several possible approaches to improve the device performance via controlling the anion–cation combination and designing vertical channel stacks [9, 10]. Unfortunately, it is very hard to use reactive sputtering to control nanoscale thickness and composition and to deposit a conformal nanoscale-thick coating on the complex nanostructure.

Currently, AOS materials have been developed to make a technical breakthrough via the ALD process, because the unique nature of ALD enables one to solve the many concerns discussed here. Unlike the reactive sputtering method, ALD can easily synthesize various AOS material combinations via low deposition temperatures (100~250 °C) as well as vertically manipulate each chemical element in a thin AOS channel layer. Moreover, it may accurately control sub-nano-thickness and reproduce identical AOS films repeatably on a large-area substrate. Very recently, a few groups have reported high mobility and reliable stability on ALD AOS-TFTs [11–14]. Although ALD shows great potential for the emerging applications, it still has fundamental problems for practical mass production. ALD is a very slow and expensive process, which will be a big hurdle for practical mass fabrication. Also, a metal precursor and suitable oxidant should be developed, including key elements such as In, Ga, Zn, and Sn. Most of all, AOS ALD equipment should be considered for a large substate and high throughput. Fortunately, a few leading companies have started to develop both ALD equipment and precursors for AOS applications [15–18].

In this chapter, various AOS materials deposited by ALD will be introduced, from binary oxide semiconductors (ZnO, In_2O_3, and SnO_2) to quaternary oxide semiconductors (*a*-InGaZnO and *a*-InSnZnO). The properties of ALD

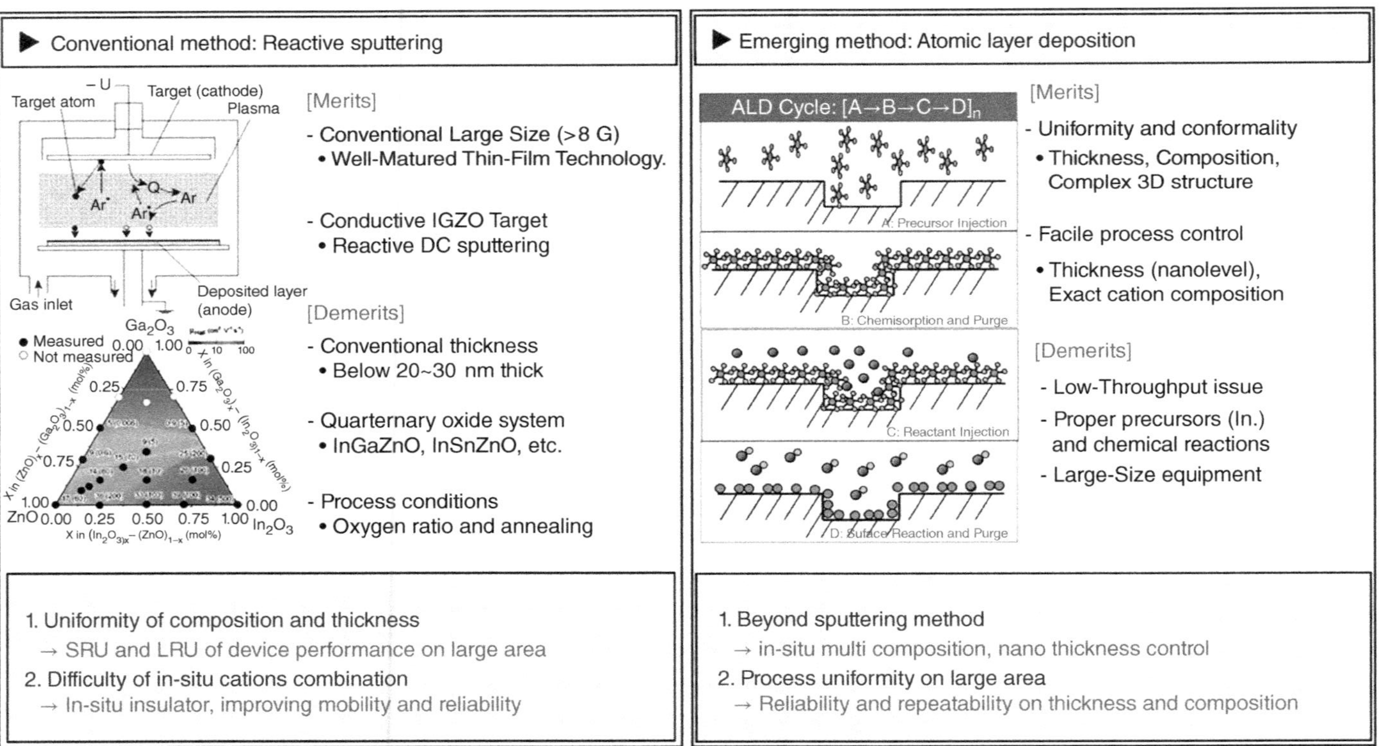

Figure 10.3 The advantages and disadvantages between reactive sputtering and atomic layer deposition, considering the AOS semiconductor materials and process. SRU: Short-range uniformity; LRU: long-range uniformity.

films strongly depend on their chemical molecules (precursor and reactant), the deposition temperature, and the combination of ALD cycle, because of complementary and self-limiting surface reactions. Section 10.2 will cover several recent works to develop AOS materials and the associated TFT performance based on ALD processes, including material combination, improvement of device performances, novel ALD equipment, as well as *p*-type oxide semiconductors.

10.2 AOS-TFTs Based on ALD

10.2.1 Binary Oxide Semiconductor TFTs Based on ALD

10.2.1.1 ZnO-TFTs

Initial research on ALD oxide semiconductors mainly used ZnO. Most ZnO thin films fabricated by ALD are used with diethylzinc (DEZ) as precursor and deionized (DI) water and ozone (O_3) as reactant [19–28]. Table 10.1 summarizes the precursor, reactant, and deposition temperature of ALD ZnO. Most ALD ZnO using DEZ has a high film-growth rate (over 1.0 Å/cycle) with a low deposition temperature (≲200 °C).

Cho et al. reported ALD ZnO using DEZ and water as precursor and reactant, respectively [28]. The ZnO growth temperature was changed from 100 to 200, and carrier concentration and Hall mobility are increased (carrier concentration: 10^{14} to 10^{19} cm^{-3}; Hall mobility: 1.3 to 15 cm^2/Vs). More oxygen vacancies are expected in ZnO thin film grown at higher temperature with high carrier concentration. The transfer characteristics reliability of ALD ZnO-TFTs was optimized when the deposition temperature was over 175 °C; this originates from large grain formation and suppressing grain boundary effects.

Lim et al. reported plasma-enhanced ALD (PEALD) ZnO-TFTs using ultraviolet (UV) exposure treatment [27]. Device performance was optimized when UV exposure time was 30 min (with a threshold voltage [V_{th}] of 1.7 V, field-effect mobility [μ_{FE}] of 0.011 cm^2/Vs, and subthreshold swing [SS] of 1.6 V/decade). When UV light is exposed to ZnO thin film, (i) direct electron–hole pair generation, (ii) oxygen vacancy formation, and (iii) creation of ionized oxygen vacancies (V_o^+ and V_o^{++}) occur. Based on these mechanisms, UV exposure made ZnO thin films conductive, and device parameters changed with UV exposure time.

Table 10.1 Summary of recent reports for thermal ALD and plasma-enhanced ALD for ZnO thin films

Zn Precursor	Reactant	Deposition Temperature (°C)	Reference
DEZ	H_2O	200	[19]
DEZ	H_2O, O_2	150	[20]
DEZ	H_2O	150	[21]
DEZ	NH_3	200	[22]
DEZ	H_2O	130	[23]
DEZ	O_2 plasma	150	[24]
DEZ	N_2O plasma	200	[25]
DEZ	O_2	170	[26]
DEZ	O_2 plasma	200	[27]
DEZ	H_2O	100–200	[28]

ALD, Atomic layer deposition; DEZ, diethylzinc.

10.2.1.2 InOx-TFTs

Indium oxide has higher electrical properties because indium has large metal 5*s* orbitals, which act as an electron path near the conduction band. ALD InO_x has a large direct bandgap (≥3.0 eV) and high electrical properties, and it has been investigated for various optoelectronics applications, such as displays, photovoltaic devices, and optical sensors. Various precursors such as cyclopentadienyl indium (InCp), [1,1,1-trimethyl-N-(trimethylsilyl)silanaminato]-indium (InCA-1), trimethyl indium (TMIn), triethyl indium (TEIn), and (3-dimethylaminopropyl)dimethylindium (DADI) were used in ALD and PEALD InO_x processes. The most representative indium precursors are TMIn and indium acetylacetonate ($In(acac)_3$), and various other indium precursors have been developed to secure a high growth rate and low defect level by making a liquid precursor. Table 10.2 summarizes the characterizations of ALD InO_x films using different precursors and reactants [17, 18, 29–34].

Sheng et al. reported ALD InO_x-TFTs using InCA-1 and H_2O_2 as precursor and reactant, respectively [31]. The growth rate saturates to approximately 0.6 Å/cycle when the precursor and reactant dose time were 0.5 s and 0.4 s, respectively. As the growth temperature was increased from 100 °C to 250 °C, there was an increase in the stable metal–oxygen bond and crystallization in the InO_x thin films, which induced highly conductive properties (carrier concentration: $4.6 \cdot 10^{19}/cm^3 \rightarrow 2.6 \cdot 10^{21}/cm^3$). Figure 10.4a shows the transfer characteristics of InO_x flexible TFTs. The field-effect mobility of InO_x-TFTs was 15 cm^2/Vs with a threshold voltage near 0 V when the deposition temperature was 125 °C.

Ma et al. reported ALD InO_x using InCp and H_2O_2 as precursor and reactant, respectively [29]. Figure 10.4b and 10.4c show the ALD windows and linear growth characteristics of InO_x. The growth rate of ALD InOx using InCp (~1.46 Å/cycle) is relatively higher compared to other precursors. The optical bandgap of InO_x thin films increased from 3.42 eV to 3.75 eV when the growth temperature was increased from 150 °C to 200 °C. In addition, the carbon impurity is diminished 6.9% → 0%) and the atomic ratio of indium and oxygen (In:O) is increased from 1.17 to 1.36. Finally, the transfer characteristics of InO_x-TFTs differed depending on the annealing times (annealing temperature: 300 °C). The InO_x-TFTs with 10 hours of annealing exhibited a μ_{FE} of 7.8 cm^2/Vs, V_{th} of −3.7 V, SS of 0.32 V/dec, and on/off current ratio (I_{on}/I_{off}) of 10^7.

10.2.1.3 SnOx-TFTs

Tin oxide (SnO_2) also has electrical properties and high transparency (≥3.6 eV) in the visible-light range. Tin metal has large metal 5*s* orbitals, which is consistent with indium metal. In addition, SnO_2 has strong chemical

Table 10.2 Summary of recent reports for thermal ALD and plasma-enhanced ALD for InO_x thin films

Indium Precursor	Reactant	Deposition Temperature (°C)	Reference
InCp	H_2O_2	160–200	[29]
InCp	H_2O_2	160	[30]
InCA-1	H_2O_2	125–225	[31]
TMIn	H_2O	200–250	[32]
TEIn	O_3	100–200	[18]
DADI	O_3	150–225	[33]
$(In(CH_3)_3[CH_3OCH_2CH_2NHtBu])$	H_2O	100–250	[17]
$Et_2InN(SiMe_3)_2$	O_2 plasma	200–250	[34]

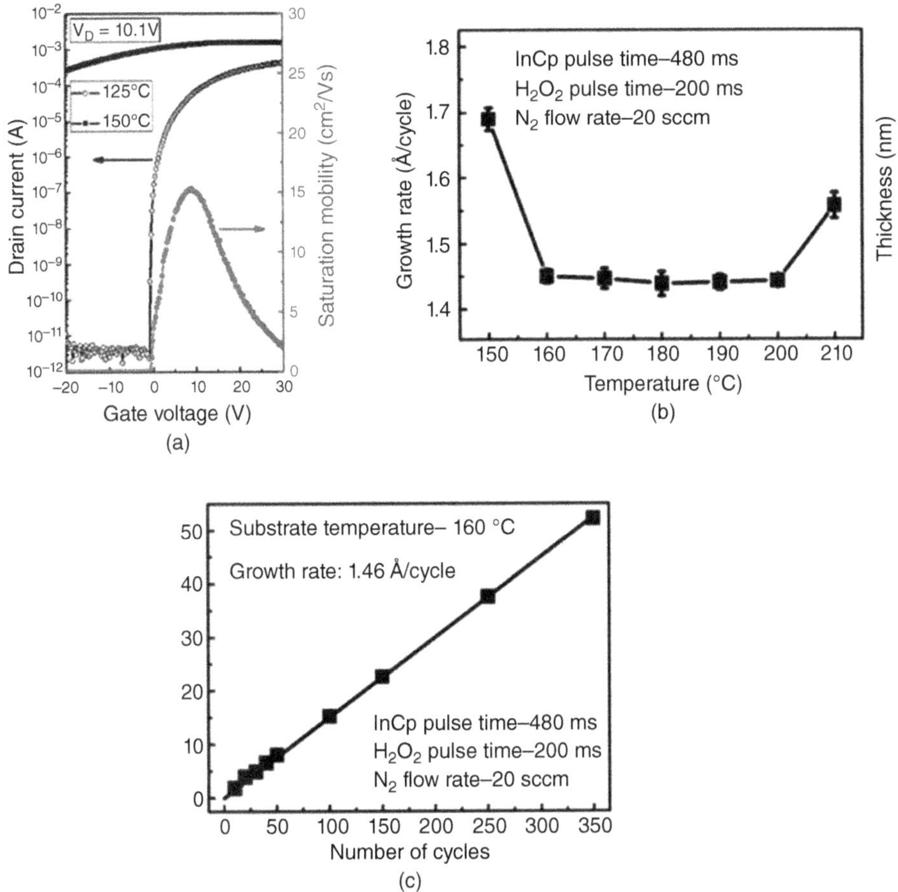

Figure 10.4 (a) Transfer characteristics and saturation mobility of InO_x-TFTs. (b) ALD windows with different deposition temperatures, and (c) dependence of the InO_x thin-film thickness as a function of ALD cycles. *Source*: (a) Sheng, J., et al. (2016). Performance modulation of transparent ALD indium oxide films on flexible substrates: transition between metal-like conductor and high-performance semiconductor states. *Journal of Materials Chemistry C* **4**: 7571. (b,c) Ma, Q., et al. (2018). Atomic-layer-deposition of indium oxide nano-films for thin-film transistors. *Nanoscale Research Letters* **13** (1): 4. Licensed under CC BY 4.0.

stability in acid conditions. SnO_2 thin films are used in gas sensors and battery anodes. Various precursors include $SnCl_4$, SnI_4, tetrakis-dimethyl-amine-tin (TDMASn), $Sn(dmamp)_2$ [dmamp = $OCMe_2CH_2NMe_2$], and N,N′-tert-butyl-1,1-dimethylethylenediamine stannylene(II). Table 10.3 summarizes characterizations of ALD SnO_x films using different precursors and reactants [35–43]. However, only a few papers on SnO_2 thin films reported the TFTs' performance, because SnO_2 thin films have photolithography process issues.

Mai et al. reported SnO_2-TFTs using a new functionalized alkyl precursor, $Sn[(DMP)_4]$ [43]. The ALD window was determined by temperature-dependent growth studies in a temperature range of 60–220 °C. The growth rate was saturated for deposition temperatures between 150 and 220 °C. For substrate temperatures below 150 °C, the growth rates increase at a linear rate to 0.41 Å/cycles at 60 °C. The increasing growth rates of SnO_2 thin films at low deposition temperatures might originate from condensation of precursors. SnO_2 thin films fabricated in the ALD window show no impurities (carbon: 44.1% [60 °C] → 0% [over 150 °C]; nitrogen: 18.3% [60 °C] → 0% [over 150 °C]). In addition, stable Sn-O bonding increased when the deposition temperature was over 150 °C from X-ray photoelectron spectroscopy (XPS) O 1s results. The SnO_2-TFTs were fabricated using 60 °C PEALD SnO_2. As SnO_2

Table 10.3 Summary of recent reports for thermal ALD and plasma-enhanced ALD for SnO_x thin films

Sn Precursor	Reactant	Deposition Temperature (°C)	Reference
$SnCl_4$	O_2 plasma	150–350	[35]
TDMASn	O_3	50–250	[36]
TEMASn	O_2 plasma	50–200	[27]
$Sn(dmamp)_2$	O_2 plasma	150–200	[38]
Dibutyl tin diacetate	O_2	200–400	[39]
SnI_4	O_2	400–750	[40]
N,N′-tert-butyl-1,1-dimethyl-ethylenediamine stannylene(II)	O_3	50–250	[41]
N^2,N^3-di-tert-butyl-butane-2,3-diamidotin(II)	H_2O_2	25–250	[42]
$Sn[(DMP)_4]$	O_2 plasma	60–220	[43]

thickness was increased from 4 to 8 nm, the transfer characteristics changed. As active thickness became thicker, the V_{th} underwent a negative shift and I_{on} increased. The SnO_2 devices with 6 nm active thickness showed the highest field-effect mobility (12 cm^2/Vs). When channel thickness is less than 6 nm, the TFTs' mobility is lower because the surface roughness effect, such as surface scattering, is strong.

10.2.2 Ternary and Quaternary Oxide Semiconductor TFTs Based on ALD

Binary oxide is a crystalline structure easily formed by postfabrication and annealing processes, even at low temperatures. Electrons are scattered and trapped at the grain boundary, which causes degradation of electrical performance and in the reliability of oxide TFTs. Mixing the two different metal oxides, which have different crystal structures, suppresses grain boundary formation. In this section, ternary and quaternary oxides, such as indium–zinc oxide, indium–gallium oxide, zinc–tin oxide, indium–gallium–zinc oxide, and indium–tin–zinc oxide, will be discussed. Interestingly, when two or more precursors are used in the ALD process, the growth properties of oxide thin films might vary according to the precursor structure, reactant type, and ALD process sequence.

10.2.2.1 Indium–Zinc Oxide (IZO) and Indium–Gallium Oxide (IGO)

Sheng et al. reported the ALD IZO-TFTs using InCA-1 and DEZ precursors with different deposition temperatures [11]. Figure 10.5a shows the growth rate of InO_x, ZnO, and IZO as a function of growth temperature. The InO_x thin films show little change in growth rate (0.68 A/cycle) in terms of deposition temperature, while the growth rate of ZnO thin films is increased from 0.7 Å/cycle at 150 °C to 1.7 Å/cycle at 200 °C. However, the growth properties of IZO thin films show a 1.4 Å/cycle at 150 °C and a 1.8 Å/cycle at 200 °C, which are less than the sum of the individual growth rates of thin film at 200 °C (2.4 Å/cycle). InCA-1 ligands ($-N[Si(CH_3)_3]_2$) and $-CH_3$ react with two –OH on the surface and remain one $-CH_3$ ligand (Case I). However, only ($-N[Si(CH_3)_3]_2$) reacts with –OH from the surface and remains two $-CH_3$ ligands (Case II). The proposed surface reactions (Cases I and II) are illustrated in Figure 10.5b. After finishing a half reaction, the number of sites that can react with DEZ varies depending on whether it is Case I or Case II. The number of –OH ligands that react with DEZ precursor is 4

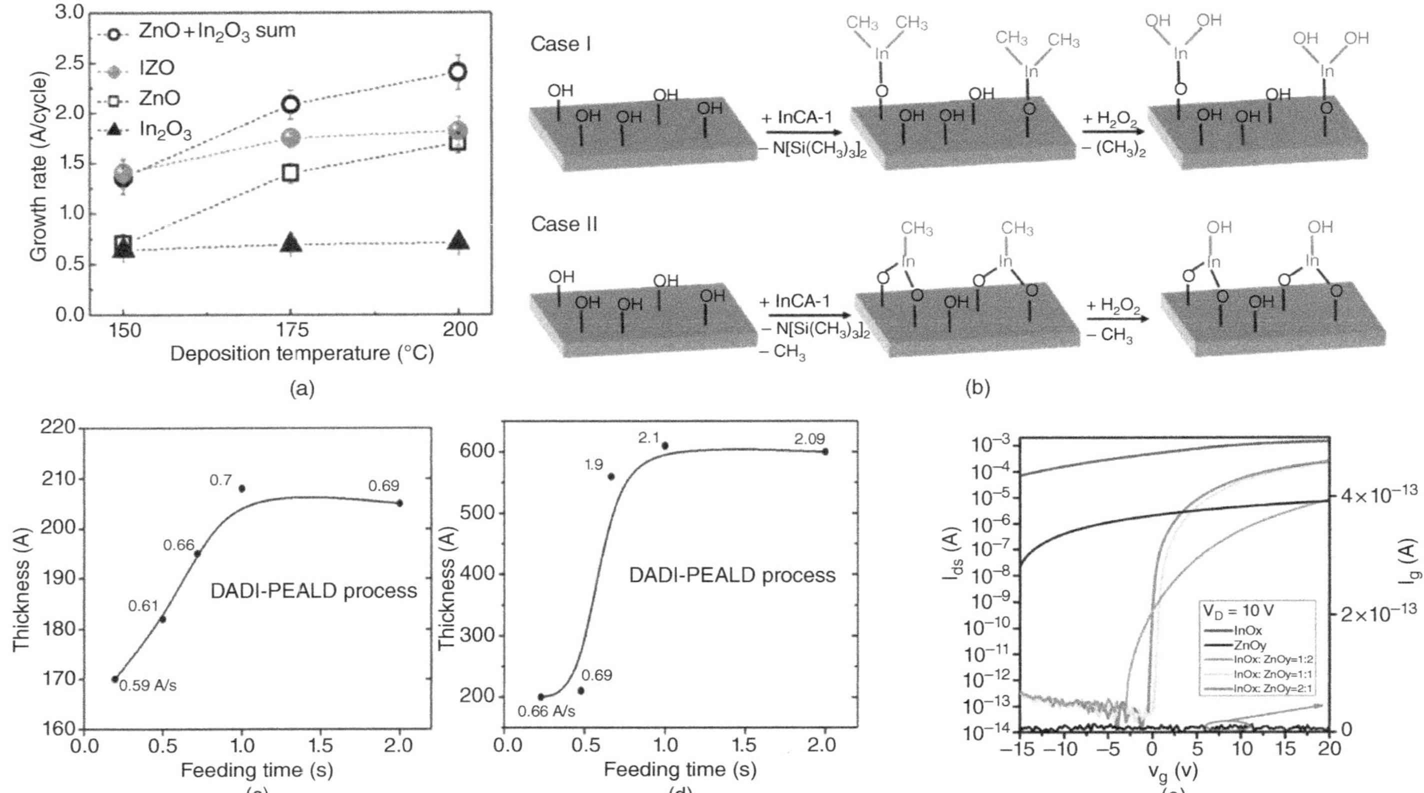

Figure 10.5 (a) The growth rate of IZO, ZnO, and InO_x thin films with different deposition temperatures. (b) The proposed growth mechanisms of IZO thin films. (c,d) The saturation curves of (c) InO_x and (d) ZnO with precursor dosage time. (e) The transfer curves of IZO-TFTs with different ALD sequences. *Source*: (a,b) Sheng, J., et al. (2016). Flexible and high-performance amorphous indium zinc oxide thin film transistor using low-temperature atomic layer deposition. *ACS Applied Materials & Interfaces* **8**: 33821–33828. (e) Lee, J.-M., et al. (2019). All-oxide thin-film transistors with channels of mixed InOx-ZnOy formed by plasma-enhanced atomic layer deposition process. *Journal of Vacuum Science & Technology A* **37**: 060910.

in Case I. However, it reduces to 2 in Case II. To confirm the metal composition of IZO thin films with different deposition temperatures, atomic emission spectroscopy (AES) analysis was conducted. The indium and zinc metal composition are 0.66:1 at a 150 °C deposition temperature. However, the ratio is decreased to 0.57:1 when the deposition temperature is 200 °C. When the deposition temperature is low, Case 1's reaction is dominant. However, when the deposition temperature is increased, Case 2 is dominant. As a result, the actual growth rate of IZO thin film at the high temperature is lower than the summation between single InO_x/ZnO thin films, and the indium ratio of the thin film might be decreased as the deposition temperature increases. The electrical properties of IZO-TFTs using ALD were also evaluated at different deposition temperatures. 200 °C IZO-TFTs show the best performance (μ_{FE}: 42.1 cm^2/Vs; SS: 0.29 V/decade) and stability (ΔV_{th}: 1.0 V at V_{GS}: 20 V; temperature: 60 for 1 hour), which originate from lower deep-level defect states (corresponding to better stability) and larger shallow defect states (corresponding to higher electrical performance).

Lee et al. reported PEALD IZO-TFTs with different In:Zn subcycle ratios (2:1, 1:1, and 1:2) [44]. The growth rates of single InO_x and ZnO thin films are saturated at 0.7 Å/cycle and 2.1 Å/cycle, respectively (Figure 10.5c and 10.5d). DADI and DEZ are both liquid precursors, giving uniform transport in the PEALD facility and stable oxide growth rates. Amorphous IZO films were observed regardless of subcycle ratio, which might be attributed to zinc ion blockings that hinder the indium occupancy of the octahedral site. In addition, the oxygen deficiency site in IZO thin films is decreased as the indium subcycle is increased. Figure 10.5e shows the transfer curves and device parameters of ALD IZO-TFTs with different In:Zn subcycles. TFTs achieved their best performance (μ_{FE}: 30.3 cm^2/Vs; and SS: 0.14 V/decade) when the In:Zn subcycle was 2:1.

Sheng et al. fabricated IGO thin films using InCA-1 and trimethyl gallium (TMGa) as precursor and H_2O_2 as reactant with different ALD sequences [12], consisting of (i) TMGa–H_2O_2 (GaO), (ii) InCA-1–H_2O_2 (InO), (iii) InCA-1–TMGa–H_2O_2 (In-Ga), (iv) TMGa–InCA-1–H_2O_2 (Ga-In), and (v) InCA-1–H_2O_2–TMGa–H_2O_2 (InO-GaO). Table 10.4 shows the growth rate and atomic concentration of thin films under different ALD supercycles. GaO_x deposition using only TMGa and H_2O_2 could not be deposited; however, gallium was successfully deposited when the indium subcycle was accompanied. This phenomenon was induced by the half-reaction difference between TMGa on the SiOx-OH substrate and TMGa on the InO_x-OH substrate. Figure 10.6a and 10.6b show the density functional theory (DFT) calculated energy profiles when TMGa reacts at SiO_x-OH and InO_x-OH surfaces. When TMGa physisorpted at SiO_x-OH surfaces, the half reactions hardly occur because the barrier energy of the chemisorption state is higher than that of the initial states. However, when TMGa physisorpted at InO_x-OH surfaces, the reaction is spontaneous and irreversible because of the low barrier energy of chemisorption. Figure 10.6c shows the schematic illustrations of the top-gate IGO-TFTs and transfer curves of IGO-TFTs with three different IGO ALD sequences (1:1, 2:1, and 3:1). As indium subcycles increase from 1 to 3, field-effect mobility is increased (0.17 cm^2/Vs → 9.45 cm^2/Vs), and SS is improved (0.42 V/decade → 0.26 V/decade).

Table 10.4 The atomic composition of IGO thin films grown via ALD with various sequences

Sequence	Supercycle	Atom % C	Atom % Ga	Atom % In	Atom % O
GaO	(TMGa)-H_2O_2				
InO	(InCA-1)-H_2O_2	0.5	0	42.8	56.7
In-Ga	(InCA-1)-(TMGa)-H_2O_2	0.4	3.8	39.6	55.9
Ga-In	(TMGa)-(InCA-1)-H_2O_2	0.7	28.3	14.9	56.4
InO-GaO	(InCA-1)-H_2O_2-(TMGa)-H_2O_2	0.6	20.6	22.7	56.2

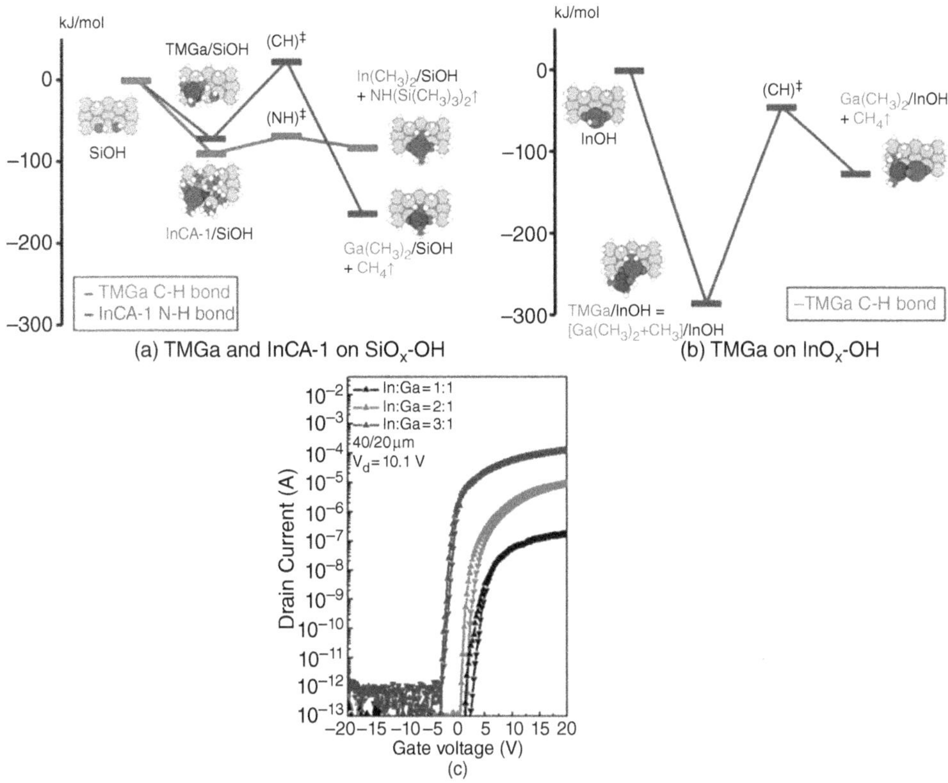

Figure 10.6 The DFT-calculated energy profiles of chemisorption and optimized structures of TMGa and InCA-1 on (a) SiO_x-OH and (b) InO_x-OH surfaces. (c) Transfer curves of IGO-TFTs with different In:Ga subcycle ratios. *Source*: Sheng, J., et al. (2017). Atomic layer deposition of an indium gallium oxide thin film for thin-film transistor applications. *ACS Applied Materials & Interfaces* **9**: 23934–23940.

10.2.2.2 Zinc–Tin Oxide (ZTO)

Ahn et al. reported the ALD ZTO-TFTs using TDMASn and DEZ as Sn and Zn precursor, respectively [45]. The Sn and Zn subcycle was 1:1. After fabricating TFTs, the postannealing temperature was increased from 300 °C to 500 °C. SnO_2-TFTs have conducting properties regardless of their postannealing temperature. In addition, ZnO-TFTs have low electrical properties despite a high annealing temperature (500 °C). However, ZTO-TFTs show good transfer characteristics (V_{th}: 0.34 V; μ_{FE}: 13.2 cm^2/Vs; and SS: 0.15 V/decade) with a high I_{on}/I_{off} ratio of over 10^7. In addition, the device reliability of ZTO-TFTs under bias stress at 400 °C annealing ($V_{GS} = -20$ V; $V_{DS} =$ 10.1 V; stress time = 3,000 sec) shows that the threshold voltage shift is nearly unchanged (ΔV_{th}: −0.2 V). This phenomenon originates from the decrease in oxygen-related defects as the annealing temperature is increased. To define the electrical performance and reliability change of ZTO-TFTs, the subgap states near the conduction band are analyzed with different ZTO annealing temperatures. The deconvoluted Gaussian band-edge states (D1 and D2), which indicate the shallow band-edge state (D1) and the deep band-edge state (D2), decrease when the annealing temperature is increased.

10.2.2.3 Indium–Gallium–Zinc Oxide (IGZO)

Yoon et al. reported ALD IGZO-TFTs with different ALD process temperatures using a newly synthesized indium–gallium single precursor, DEZ, and ozone (O_3) as In, Ga, and Zn precursor and reactant, respectively [46]. The ALD sequence consists of In-Ga-O and Zn-O. The concentration of zinc is relatively high (~60%) compared to indium and gallium (Figure 10.7a). This might be the original precursor structure. In and Ga

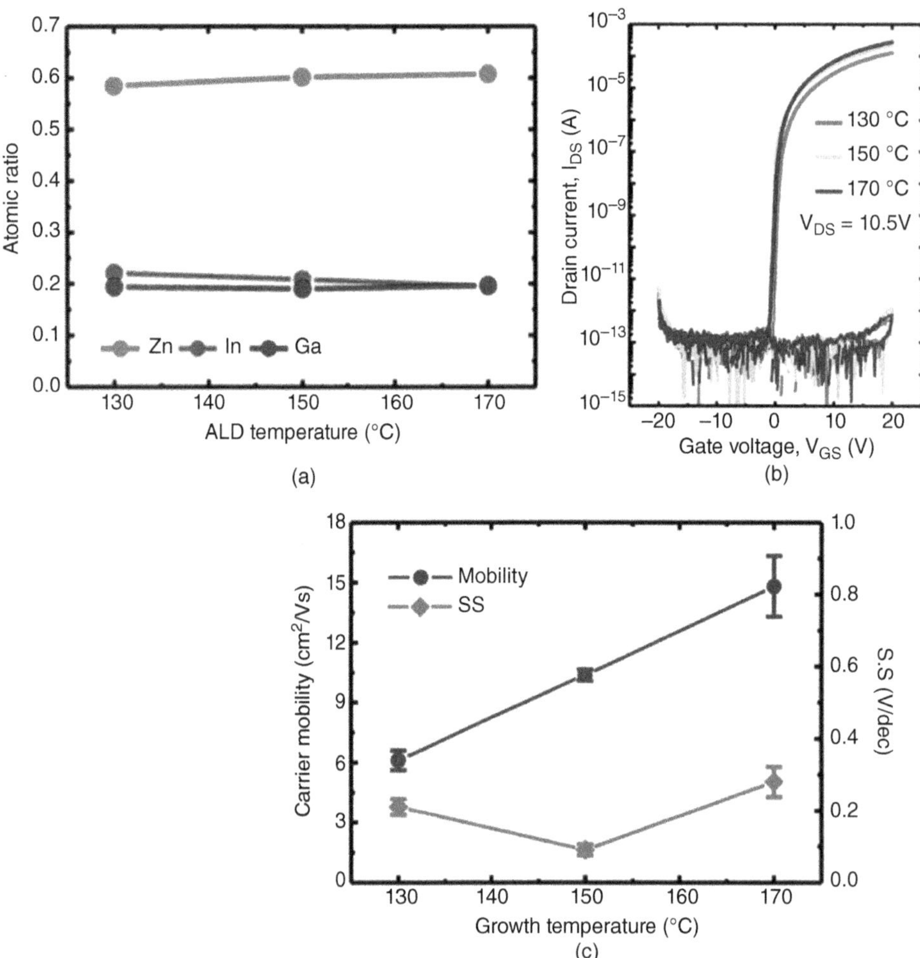

Figure 10.7 (a) The metal compositions of IGZO thin films as a function of ALD temperature. (b) Transfer curves and (c) device parameters of IGZO-TFTs with different deposition temperatures. *Source*: Yoon, S.-M., et al. (2017). Effects of deposition temperature on the device characteristics of oxide thin-film transistors using In–Ga–Zn–O active channels prepared by atomic-layer deposition. *ACS Applied Materials & Interfaces* **9**: 22676–22684.

are composed of single precursors, and the DEZ precursor has a high deposition rate. In addition, the atomic concentration of IGZO thin films does not change significantly regardless of process temperature. However, XPS results show that oxygen-related defects in IGZO thin films increased as deposition temperature is increased (37.1% → 58.6%). In addition, the optical bandgap is decreased from 3.81 eV to 3.21 eV. Also, the top-gate structure of IGZO-TFTs was fabricated using ALD IGZO thin films. Figure 10.7b and 10.7c show the transfer characteristics of IGZO-TFTs with different deposition temperatures and device parameters, such as carrier mobility and SS. Mobility is increased from 6.1 cm^2/Vs to 14.8 cm^2/Vs with increasing deposition temperature, which originates from an oxygen-bonding status change. Device reliability, positive bias stress (PBS, $V_{GS} = +20$ V), and negative bias illumination stress (NBIS, $V_{GS} = -20$ V, 0.1 mW/cm^2) are also measured as functions of deposition temperature. All the devices have excellent PBS reliability ($\Delta V_{th} < 1$ V). However, the NBIS reliability of ALD IGZO-TFTs deposited at 130 °C and 170 °C experienced ΔV_{th}'s of −10.7 and −16.7 V, respectively. These originated from the difference in oxygen-related defects.

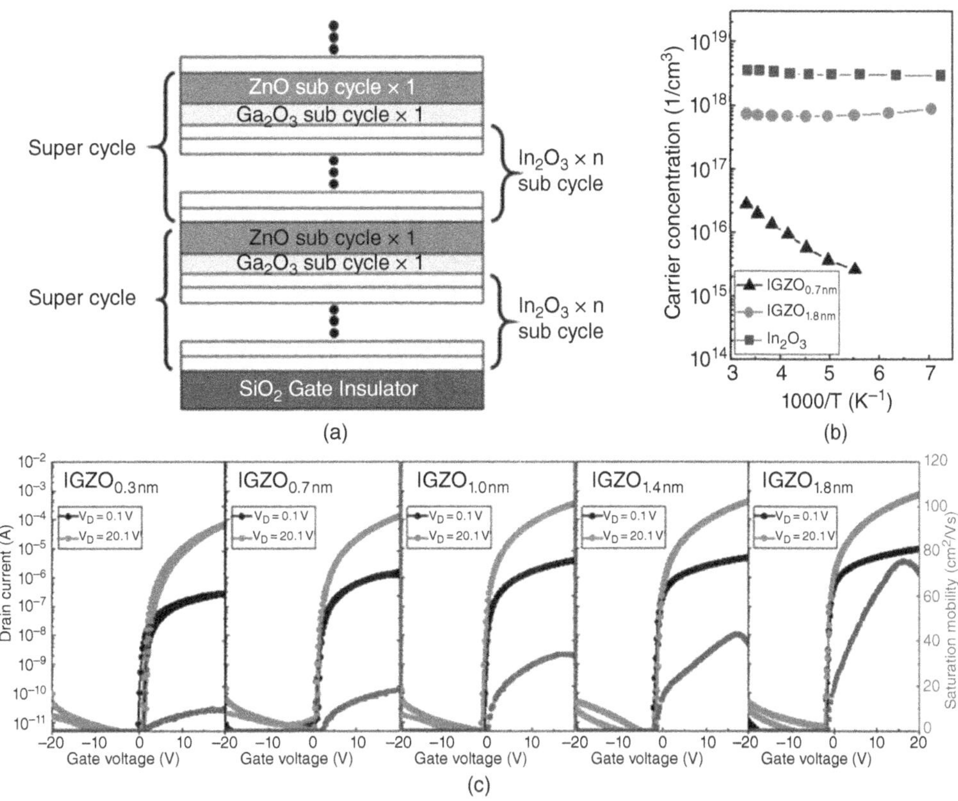

Figure 10.8 (a) Illustrations of PEALD IGZO ALD sequences. (b) Carrier concentrations of three different oxide thin films as a function of temperature. (c) Transfer curves and saturation mobility of IGZO-TFTs with different IGZO conditions. *Source*: Sheng, J., et al. (2019). Amorphous IGZO TFT with high mobility of ~70 $cm^2/(Vs)$ via vertical dimension control using PEALD. *ACS Applied Materials & Interfaces* **11**: 40300–40309.

Sheng et al. also reported the heterostructure of PEALD IGZO-TFTs using DADI, TMGa, DEZ, and oxygen plasma as metal precursors and reactant, respectively [14]. The PEALD IGZO supercycle consists of [(InO_x × *n* cycles) – (GaO_x × 1 cycle) – (ZnO_x × 1 cycle)], where *n* increased from 4 to 20 (Figure 10.8a). Table 10.5 summarizes the elemental compositions of PEALD IGZO with different InO_x thickness. As InO_x thickness is increased, the difference between observed oxygen (Obsd O%) and calculated oxygen (Calcd O%) is increased, which means oxygen vacancies in the PEALD IGZO films are increased. The relationship between carrier concentration and measurement temperature is shown in Figure 10.8b. When the carrier concentration is greater than 10^{17} cm^{-3}, the Fermi level exceeds the potential barriers. Thus, the carrier transport is no longer affected by the potential barrier and exhibits temperature independence. Figure 10.8c shows the transfer curves of PEALD IGZO-TFTs with different InO_x thicknesses. When InO_x thickness is 0.3 nm, the field-effect mobility of TFTs is 9.9 cm^2/Vs. However, the field-effect mobility is increased significantly (74.3 cm^2/Vs) at $IGZO_{1.8nm}$. In addition, the device parameter shifts under PBS were measured. In the case of $IGZO_{0.3nm}$-TFTs, the V_{th} shift is over +5 V and the degradation of mobility originates from the large content of oxygen. When the InO_x thickness was thinner, the high oxygen content in the IGZO thin films acted as occupied electron-trapping sites. However, when the thickness of the InO_x sub-layer increases to 1.0 nm, the excess oxygen is complemented by the oxygen deficiencies generated during indium oxide deposition. However, the InO_x thickness is over 1.0 nm, the weak In-O bond is easily dissociated, and excess defects are formed that resulted in SS degradation.

Table 10.5 The elemental ratio of IGZO thin films and the differences between observed oxygen (Obsd O%) and calculated oxygen (Calcd O%) with different InO_x thicknesses

	In%	Zn%	Ga%	O%	Calcd O%	Calcd O%– Obsd O%
$IGZO_{0.3nm}$	11.5	20.3	12.5	55.8	56.3	0.5
$IGZO_{0.7nm}$	16.4	17.6	9.9	56.1	57.1	1.0
$IGZO_{1.0nm}$	19.0	16.2	8.9	55.9	58.1	2.2
$IGZO_{1.4nm}$	22.1	14.0	8.2	55.7	59.5	3.8
$IGZO_{1.8nm}$	22.9	13.6	7.9	55.6	59.8	4.2

10.2.2.4 Indium–Tin–Zinc Oxide (ITZO)

Baek et al. published research on ALD ITZO-TFTs using $Me_2In(EDPA)$, $[Sn(dmamp)_2]$, and DEZ as indium, tin, and zinc precursor, respectively. They changed ALD supercycles and evaluated TFTs to confirm a positive turn-on voltage (V_{on}). A positive V_{on} of ITZO-TFTs is observed when the Zn composition is over 60 at.%. Within this region, further ITZO composition modulation was conducted to optimize the device parameters. There are two different series of compositions: (i) the variation of Zn composition at a fixed In/Sn concentration (42 and 58 at.%), and (ii) the variation of In/Sn composition at a fixed Zn concentration (70 at.%). Figure 10.9a and 10.9b show the transfer curves and μ_{FE} of ITZO-TFTs with different Zn contents (Figure 10.9a) and In/Sn contents (Figure 10.9b). As the Zn concentration is increased in the ITZO thin film, V_{on} is positive shift and positive V_{on} was confirmed at the Zn content of 71 at.%. In addition, when the In/Sn ratio of the ITZO thin film was 10/20 at.%, the best mobility characteristic (22 cm^2/Vs) was observed. Consequently, an ITZO thin film of In:Zn:Sn = 10:70:20 at.% was used for the following vertical TFT fabrication process. Finally, the Beak group fabricated vertical TFTs using an Al_2O_3 gate insulator and ALD ITZO-TFTs. A high on/off current ratio of 10^8 is obtained with a reasonably high device performance of μ_{FE} of 10 cm^2/Vs and SS of 0.24 V/decade.

Sheng et al. also reported ALD ITZO-TFTs using InCA-1, TDMASn, and DEZ as indium, tin, and zinc precursor, respectively. The Sn subcycle is changed from 0 to 2 (IZO, IZTO [111], and IZTO [112]). As the Sn subcycle is changed to 1 and 2, the Sn concentration in the IZTO thin film is also changed to 9 at.% and 15.8 at.%, respectively. The transfer curves of ITZO (111) and the band alignment of IZTO-TFTs with different Sn subcycles are shown in Figure 10.9c and 10.9d. The best electrical performance of TFTs is obtained (μ_{FE} = 27.8 cm^2/Vs) when the Sn subcycle is 1. However, band diagrams of ITZO-TFTs show that the barrier energy of free electrons is 0.63 eV and 0.53 eV when the Sn subcycle is 1 and 2, respectively. Conventionally, the electrical performance of TFTs improves when the Fermi level is close to the conduction band. However, the electrical performance of ALD ITZO-TFTs is not matched in previous results. To investigate the origin of this discrepancy, XPS and spectroscopy ellipsometry (SE) analysis were conducted. As the Sn cycle increased, the oxygen-related defect of XPS O 1s and the deep-level defect are decreased. The device reliability of ITZO-TFTs is also improved (V_{th} shift: −2.2 V → 0.7 V) with increasing Sn subcycles.

10.3 Challenging Issues of AOS Applications Using ALD

10.3.1 *p*-Type Oxide Semiconductors

Most oxide semiconductors are usually *n*-type semiconductors because the formation energy of metal vacancy, which generates excess holes, is higher than that of oxygen vacancy [47]. In addition, the defect sites near the valence band form mostly electron donor states, in which it is also difficult to generate holes. The detailed *p*-type

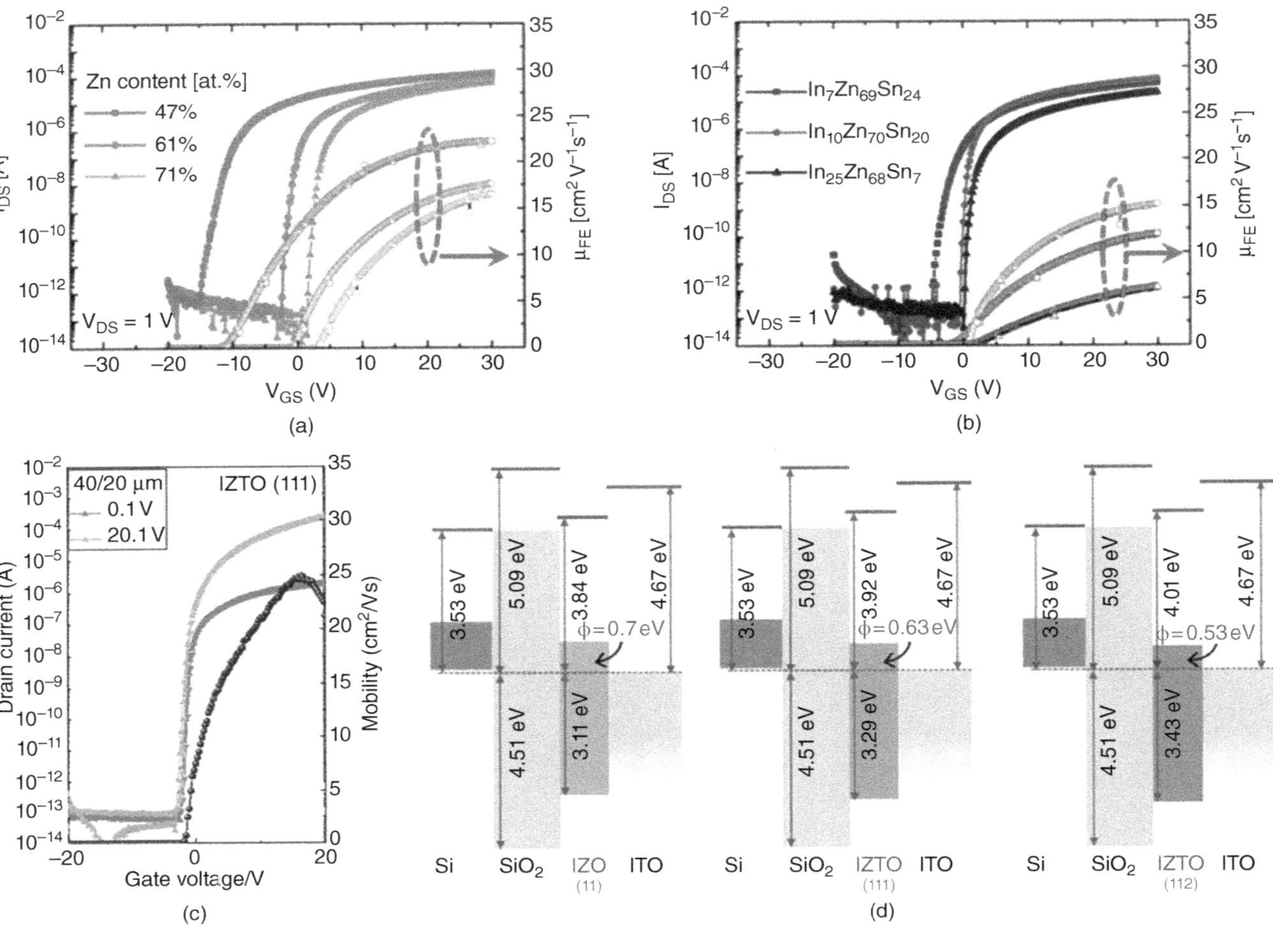

Figure 10.9 The transfer curves of ITZO-TFTs with different (a) Zn content and (b) In/Zn ratio. (c) The transfer curves of ITZO (111)-TFTs and (b) band alignment of ITZO-TFTs with different ALD sequences. *Source*: (a,b) Baek, I.-H., et al. (2019). High-performance thin film transistors of quaternary indium-zinc-tin oxide films grown by atomic layer deposition. *ACS Applied Materials & Interfaces* **11** (16). doi:10.1021/acsami.9b03331. (c,d) Sheng, J., et al. (2019). Design of InZnSnO semiconductor alloys synthesized by supercycle atomic layer deposition and their rollable applications. *ACS Applied Materials & Interfaces* **11**: 12683–12692.

Table 10.6 Summary of recent reports for thermal ALD and plasma-enhanced ALD for *p*-type thin films (SnO and Cu_xO)

p-Types	Precursor	Reactant	Deposition Temperature (°C)	Reference
SnO	$Sn(dmamp)_2$	H_2O	210	[48]
SnO	Bis[bis(trimethysilyl) amino]tin(II)	H_2O	100–250	[49]
SnO	$Sn(dmamp)_2$	H_2O	210	[50]
SnO	N,N′-tert-butyl-1,1-dimethylethylenediamine stannylene(II)	H_2O	60–180	[41]
Cu_xO	Cu(I)(hfac)(TMVS)	H_2O	225	[51]
Cu_xO	ffacCu(I)(DMB)	H_2/O_2 plasma	110	[52]
Cu_xO	(hfac)Cu-(I)(DMB)	O_3	100	[53]
Cu_xO	$Cu(dmamp)_2$	H_2O	120–240	[54]

formation mechanisms and conduction mechanisms of *p*-type oxide semiconductors are discussed in Part VII (Chapter 24, "Toward the Development of High-Performance *p*-Channel Oxide-TFTs and All-Oxide Complementary Circuits"). Tin monoxide and copper oxide are representative *p*-type oxide materials, and this chapter will discuss *p*-type oxide semiconductors using ALD. Table 10.6 summarizes recent research on *p*-type oxide semiconductors and TFTs using the ALD process [41, 48–54].

10.3.1.1 Tin Monoxide (SnO)

Lee et al. reported *p*-type SnO and *n*-type SnO_2 thin films using the same precursor, N,N′-tert-butyl-1,1-dimethylethylenediamine stannylene(II) [41]. When ozone was used as reactant, *n*-type SnO_2 was formed, but when water was used as reactant, *p*-type SnO was formed. The growth rate of SnO_x thin film using ozone reactant is saturated at ~3 Å/cycles, while water reactant is saturated at ~0.5 Å/cycles. In addition, the ALD window of ozone and water is 80–250 °C, whereas the water process exhibits an unusual growth behavior. Figure 10.10a and 10.10b show the XRD analysis using (a) water and (b) ozone reactant. In addition, band alignment using XPS, SE, and UPS also confirms that the thin film using water as reactant has *p*-type characteristics and the thin film using ozone as reactant has *n*-type characteristics. DFT calculations confirmed (see Figure 10.10c and 10.10d) the growth behavior difference when using water or ozone reactant. A highly reactive ozone molecule directly attacks the Sn and N ions of a Sn precursor. The dissociative binding of ozone thermodynamically drives the formation of an SnO_2-like local structural motif, which can be spontaneously desorbed from a Sn precursor. However, when water reactant meets the Sn precursor, water forms the Sn-OH group and $Sn(OH)_2$ made 2(SnO) with second Sn precursors. Finally, a diode was fabricated using ALD SnO and SnO_2 thin films, and the rectifying ratio was about 15.

Kim's group reported SnO-TFTs with different deposition temperatures (150 °C ~210 °C) using $Sn(dmamp)_2$ and H_2O as precursor and reactant, respectively [50]. Also, they changed the SnO thickness from 6.4 nm to 14 nm. The switching characteristics of SnO-TFTs were secured when the deposition temperature was 210 °C ($I_{on}/I_{off} \geq 10^3$). The crystallinity of SnO is also formed over 180 °C of deposition temperature. In addition, they changed the SnO thickness under different conditions, (i) before and (ii) after passivation, and measured transfer and output characteristics. Compared to unpassivated SnO-TFTs, passivated SnO-TFTs show negative V_{th} shift and low I_{off}. In addition, the field-effect mobility and V_{th} of SnO-TFTs increase up to a SnO thin-film thickness of ~9 nm and

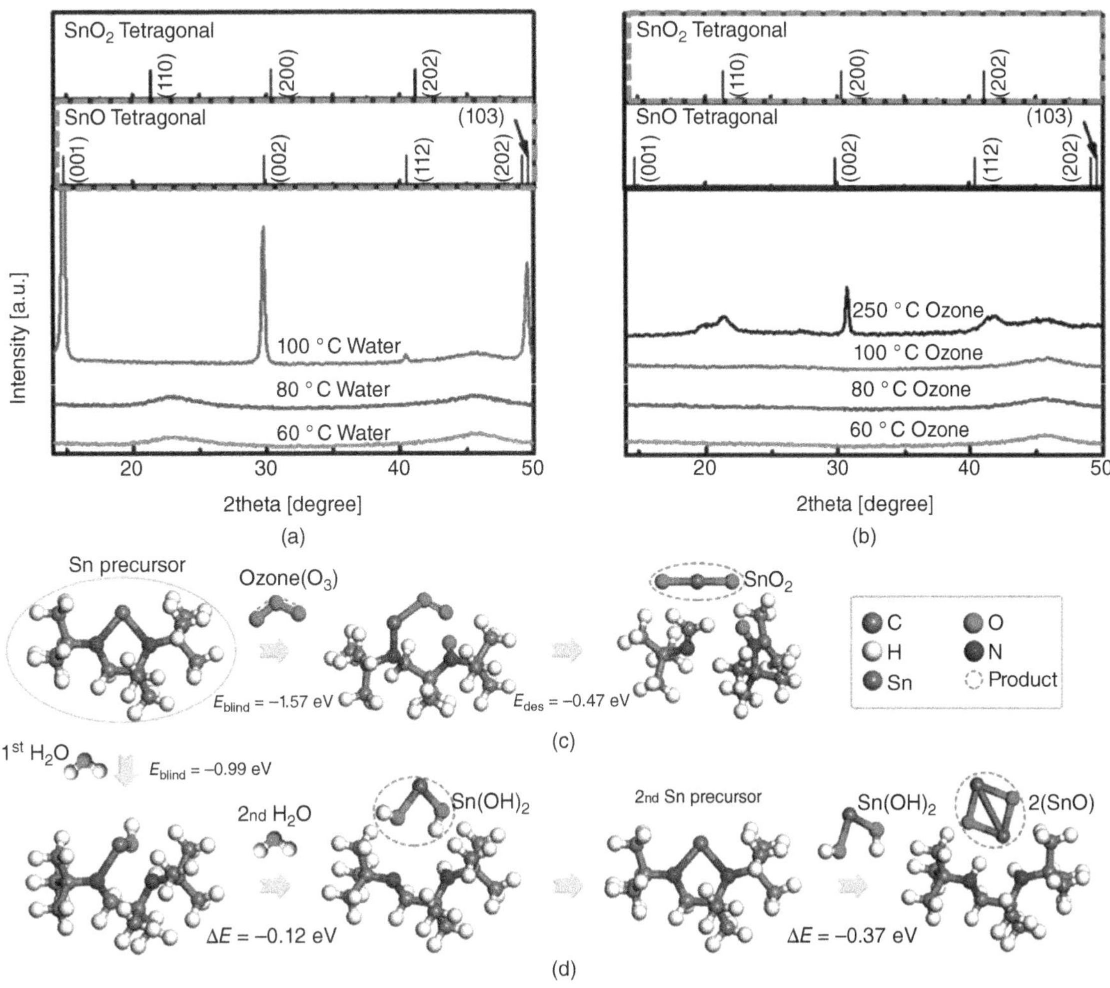

Figure 10.10 XRD patterns of SnO_x thin films with different uses of reactant: (a) water and (b) ozone. The DFT-studied formation mechanism of initial (c) SnO_2 and (d) SnO units. *Source*: (a–c) Lee, J.-H., et al. (2018). Selective SnOx atomic layer deposition driven by oxygen reactants. *ACS Applied Materials & Interfaces* **10**: 33335–33342.

saturated over 9 nm. Finally, they optimized the device performance, with a high I_{on}/I_{off} ratio of 2×10^6, SS of 1.8 V/decade, and μ_{FE} of ~1 cm^2/Vs within a thermal limit of 250 °C.

10.3.1.2 Copper Oxide (Cu_xO)

Maeng et al. reported ALD CuO_x thin film at a low deposition temperature (100 °C) using hexafluoroacetylacetonateCu(I)(3,3-dimethyl-1-butene) ((hfac)Cu(I)(DMB)) and ozone (O_3) as the copper precursor and oxidant, respectively [53]. The saturated growth rate of CuOx thin film is ~0.31 A/cycle when precursor and reactant dose time are 1 sec and 5 sec, respectively, as shown in Figure 10.11a and 10.11b. Also, ALD CuO_x-TFTs are fabricated with different annealing temperatures (Figure 10.11c). The CuO_x devices show the best device parameters, such as field-effect mobility (5.64 cm^2/Vs) and high I_{on}/I_{off} ratio ($\geq10^5$), when the annealing temperature is 300 °C. However, it was confirmed that the electrical characteristics of the device were poor in subsequent heat treatment processes other than 300 °C. To confirm this origin, XPS analysis was performed of CuO_x thin films with different annealing temperatures, as shown in Figure 10.11d. As the annealing temperature

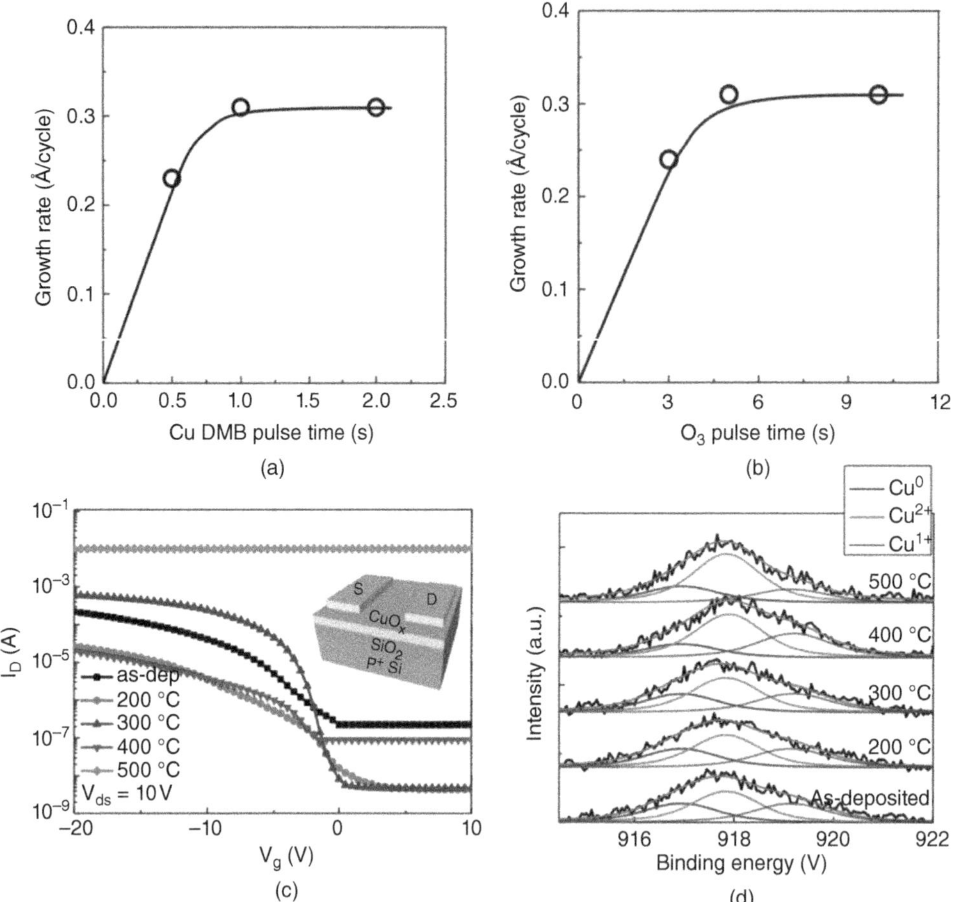

Figure 10.11 The saturation curves of (a) Cu precursor dosage time and (b) ozone dosage time. (c) Transfer curve of CuO_x-TFTs as a function of annealing temperature. (d) XPS Cu LMM peaks with different annealing temperatures. *Source*: Maeng, W., et al. (2016). Atomic layer deposited p-type copper oxide thin films and the associated thin film transistor properties. *Ceramics International* **42**: 5517–5522.

is increased to 300 °C, metallic Cu (Cu^{+0}) peaks in XPS Cu LMM are decreased and the Cu^{2+} peak is increased. However, when the annealing temperature is increased up to 500 °C, the Cu^{2+} peak is decreased, while the Cu^{+} peak is increased. Consequently, the optimum ratio of Cu_2O and CuO maximized the electrical performances of CuO_x-TFTs.

10.3.2 Enhancing Device Performance: Mobility and Stability

As mentioned in this chapter, the ALD process can precisely control even the nanometer scale of film composition and structure compared to other deposition methods. Recently, TFTs have been reported with (i) a gradient composition of an oxide semiconductor and (ii) a two-dimensional electron gas (2DEG) oxide semiconductor using a heterojunction of oxide thin films via the ALD process and its improved electrical properties.

10.3.2.1 Composition Gradient Oxide Semiconductors

Ahn et al. reported aluminum–zinc oxide (AZO) semiconductors using trimethyl aluminum (TMA) and DEZ as Al and Zn precursor, respectively [10]. In addition, they deposited AZO semiconductors using a step-composition

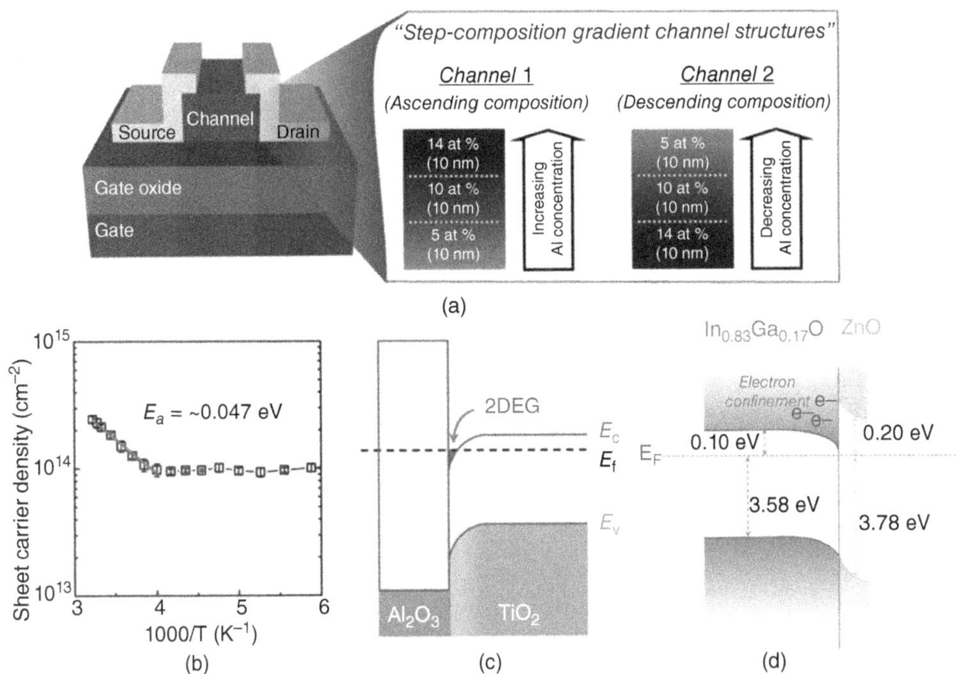

Figure 10.12 (a) Schematic illustration of the oxide TFTs using AZO channels with an Al step composition gradient. (b) Carrier density of the Al_2O_3/TiO_2 heterostructure as a function of temperature. (c) Energy band diagram of the ZnO/0.83$Ga_{0.17}$O heterointerface. *Source*: (a) Ahn, C. H., et al. (2014). Design of step composition gradient thin film transistor channel layers grown by atomic layer deposition. *Applied Physics Letters* **105**: 223513. (b) Seok, T. J., et al. (2018). Field-effect device using quasi-two-dimensional electron gas in mass-producible atomic-layer-deposited Al_2O_3/TiO_2 ultrathin (<10 nm) film heterostructures. *ACS Nano* **12**: 10403–10409. (c) Seul, H. J., et al. (2020). Atomic layer deposition process-enabled carrier mobility boosting in field-effect transistors through a nanoscale ZnO/IGO heterojunction. *ACS Applied Materials & Interfaces* **12**: 33887–33898.

gradient, which deposited the three different aluminum concentrations (5, 10, and 14 at.%) of AZO thin films sequentially. Channel 1 is ascending step composition (5 → 10 → 14 at.% Al composition), and channel 2 is descending step composition (14 →10 → 5 at.% Al composition), as shown in Figure 10.12a. The electrical performance of channel 1 TFTs (μ_{FE}: 0.6 cm^2/Vs) is relatively superior to that of channel 2 TFTs (μ_{FE}: 0.05 cm^2/Vs). To investigate the difference in mobility in terms of channel stack, the band alignment of channels 1 and 2 is observed. As the Al concentration is increased in AZO thin film, the band offset shifts toward the upper level. In the case of channel 1, electrons easily accumulate to the gate insulator. However, channel 2 has relatively more difficulty when accumulating electrons. In addition, the device reliability under PBS is different at 2 hours. The ΔV_{th} of channel 1 TFTs is 1.4 V, whereas that of channel 2 TFTs is 4.2 V. This difference originates from the backchannel condition. When the backchannel is an AZO thin film with a high Al concentration, the reaction between the backchannel and ambient oxygen is relatively small compared to that of AZO thin films with a low Al concentration. Consequently, the device reliability of TFTs is different.

10.3.2.2 Two-Dimensional Electron Gas (2DEG) Oxide Semiconductors

Seok et al. reported quasi-2DEG using Al_2O_3/TiO_2 thin films deposited by ALD [55]. The carrier density of the Al_2O_3/TiO_2 heterostructure as a function of temperature was plotted in Figure 10.12b. When the temperature is high, activation of electrons in the Al_2O_3/TiO_2 heterostructure is 0.047 eV, which is comparable with those of donor levels in typical extrinsic (As- or B-doped) Si (0.03–0.05 eV) and slightly higher than those from single-crystalline STO or TiO_2 (0.01–0.025 eV). This means the generated electrons are confined in the quantum well built by

Table 10.7 Summary of electrical parameters (μ_{FE}, SS, V_{TH}, and ON/OFF) of the IGO-TFTs and ZnO/IGO-TFTs with different cation ratios

Sample	μ_{FE} (cm^2/Vs)	SS (V/dec)	V_{TH} (V)	$I_{ON/OFF}$
$In_{0.65}Ga_{0.35}O$	24.5 ± 0.72	0.28 ± 0.05	-0.27 ± 0.58	5×10^7
$In_{0.75}Ga_{0.25}O$	34.7 ± 1.51	0.43 ± 0.02	-1.34 ± 0.52	2×10^7
$In_{0.83}Ga_{0.17}O$	41.7 ± 1.43	0.44 ± 0.07	-1.55 ± 0.71	1×10^7
$ZnO/In_{0.65}Ga_{0.35}O$	24.8 ± 0.83	0.35 ± 0.07	-1.23 ± 0.36	3×10^7
$ZnO/In_{0.75}Ga_{0.25}O$	44.6 ± 0.86	0.38 ± 0.11	-1.00 ± 0.57	9×10^7
$ZnO/In_{0.83}Ga_{0.17}O$	63.2 ± 0.26	0.26 ± 0.03	-0.84 ± 0.85	9×10^7

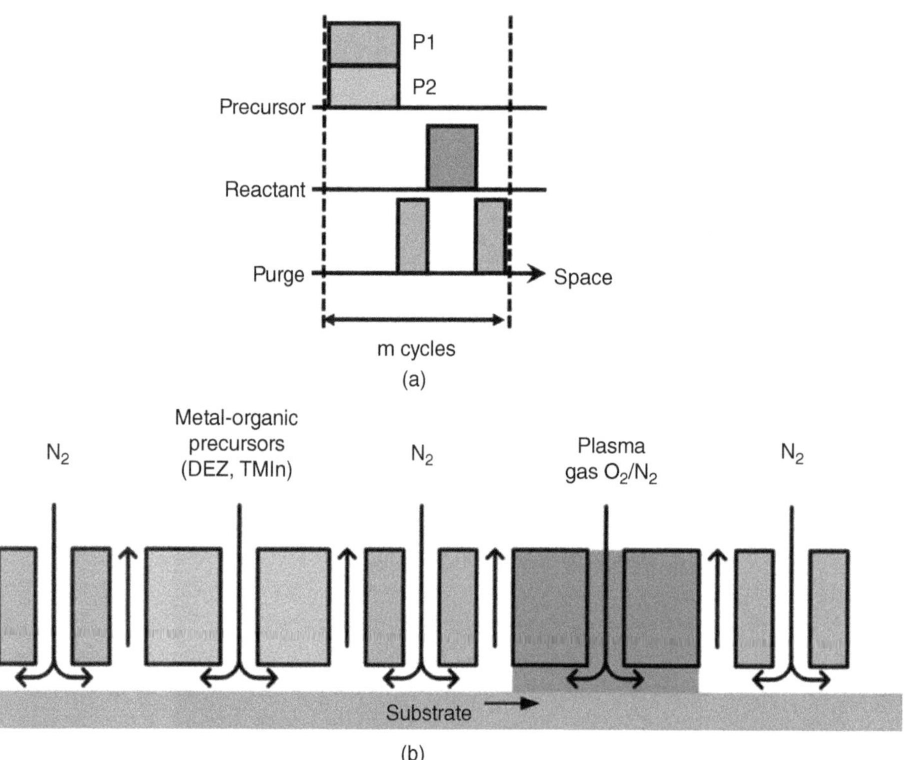

Figure 10.13 (a) Schematic illustration of SALD cycles. (b) Schematic illustration of SALD concept. *Source*: (b) Illiberi, A., et al. (2018). Atmospheric plasma-enhanced spatial-ALD of InZnO for high mobility thin film transistors. *Journal of Vacuum Science & Technology A* **36**: 04F401.

Al_2O_3/TiO_2 interfaces (Figure 10.12c). However, when the temperature is low, electrons might move via hopping conduction rather than quantum well. Therefore, Seok et al. called Al_2O_3/TiO_2 structures "quasi-2DEG" [55]. The transfer curves of quasi-2DEG transistors were measured at high temperature (h-FET) and low temperature (l-FET). h-FET shows superior device characteristics (I_{on}: 12 μA/μm; SS: ≤100 mV/decade) compared to l-FET, which is originated from high sheet carrier density.

Seul at al. also reported the IGO/ZnO heterostructure TFTs using the ALD deposition process [56]. They controlled three different IGO compositions ($In_{0.65}Ga_{0.35}O$, $In_{0.75}Zn_{0.25}O$, and $In_{0.83}Ga_{0.17}O$, each 10 nm) and added

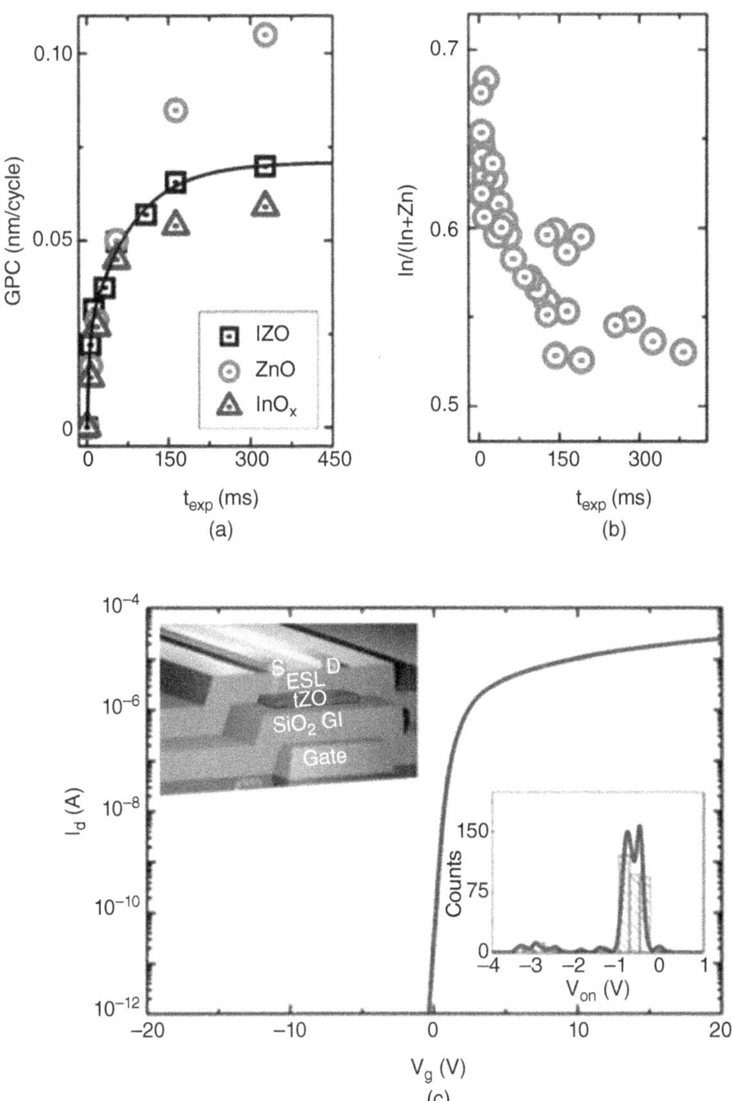

Figure 10.14 (a) Growth rate and (b) In/(In+Zn) ratio change of SALD IZO, ZnO, and InO_x thin films as a function of exposure time. (c) Device structure and transfer curves of SALD-IZO-TFTs. *Source*: Illiberi, A., et al. (2018). Atmospheric plasma-enhanced spatial-ALD of InZnO for high mobility thin film transistors. *Journal of Vacuum Science & Technology A* **36**: 04F401. (c) Illiberi et al. (2018), Figure 8 [p. 04F401-6], with permission of American Vacuum Society.

an additional ZnO layer (3 nm). The transfer curves of the IGO single-layer and IGO/ZnO heterostructure TFTs were measured, and Table 10.7 summarizes the device parameters of TFTs. Compared to single IGO ($In_{1-x}Ga_xO$, x = 0.25 and 0.17) active-layer TFTs, heterojunction active-layer TFTs show high field-effect mobility (34.7cm^2/Vs and 41.7cm^2/Vs → 44.6 cm^2/Vs and 63.2 cm^2/Vs). Moreover, the SS of heterojunction TFTs is also improved (to 0.38 V/decade and 0.26 V/decade) as compared to single IGO-TFTs (0.43 V/decade and 0.44 V/decade). The optical bandgap of 3 nm ZnO and $In_{0.83}Ga_{0.17}O$ thin films is 3.98 eV and 3.68 eV, respectively. In addition, the valence band measured by the UPS of $In_{0.83}Ga_{0.17}O$ and $ZnO/In_{0.83}Ga_{0.17}O$ is also measured. In the case of $In_{0.83}Ga_{0.17}O$, the valence band is not changed with the etching time. In contrast, the valence band of $ZnO/In_{0.83}Ga_{0.17}O$ thin film is changed from 3.78 to 3.58 eV as the etching section moved from the ZnO region to the $In_{0.83}Ga_{0.17}O$ region. Based on the optical bandgap and valence band, the band alignment of $ZnO/In_{0.83}Ga_{0.17}O$ is illustrated in Figure 10.12d. The formation of the 2DEG near $In_{0.83}Ga_{0.17}O$ boosted mobility up to 63.2 cm^2/Vs.

10.3.2.3 Spatial and Atmospheric ALD for Oxide Semiconductors

The ALD and PEALD oxide semiconductors introduced in this chapter were fabricated by vacuum- and time-divided based ALD processes. However, time-divided ALD has a slow deposition rate, which strongly relies on purge time. In addition, ALD with a slow deposition rate (10^{-2}~10^{-1} nm/sec) [3, 11–14, 16–18] cannot increase throughput, making it difficult to apply to mass production. To overcome this drawback, spatial ALD (S-ALD), in which the dosage of the precursors occurs in different zones of the reactor and a moving substrate (or reactor) is sequentially exposed to each of these zones, is introduced. SALD has the advantage of a high growth rate because it reduces the purge process compared to conventional ALD (Figure 10.13a). Figure 10.13b illustrates the representative of SALD. The growth characteristics of oxide thin films fabricated by SALD can be controlled by (i) exposure time (moving speed of substrate or precursor/reactant inlet), (ii) precursor dosage ratio (precursor partial pressure), (iii) precursor/reactant flow rate, and (iv) deposition temperature. Here we discuss the oxide semiconductor and its TFT research using SALD.

Illiberi et al. reported InZnO-TFTs using TMIn and DEZ as indium and zinc precursor, respectively, via SALD [15]. Figure 10.14a shows the growth rate of InOx, ZnO, and IZO thin films as a function of exposure time. Deposition temperature was 400 °C, and precursor and the flows from the DEZ (0.01 SLM) and TMI (0.15 SLM) bubblers were mixed and injected by argon (0.75 SLM). The growth rate of the IZO thin film is saturated at 0.7 A/cycle when the exposure time is over 150 ms. Also, the In/(In+Zn) ratio as a function of exposure time was plotted at Figure 10.14b. The In/(In+Zn) ratio decreases from ~0.7 to ~0.5 when the exposure time is increased from 5 to 420 ms, and precursor flows are kept constant at 0.150 SLM for TMIn and 0.010 SLM for DEZ. Etch-stopper layer (ESL) TFTs are fabricated using SALD IZO (In/(In+Zn) ~0.6) active layers. The postannealing process is conducted at 450 °C for 1 hour. Illustrations of ESL-TFTs, the transfer curve, and Von variations of IZO-SALD-TFTs are shown in Figure 10.14c. The maximum linear mobility achieved amounted to 32 cm^2/Vs and a subthreshold swing of 250 mV/decade. Finally, a 19-stage ring oscillator circuit was fabricated using IZO-SALD-TFTs. At V_{dd} = 20 V, the maximum oscillation frequency is approximately 110 kHz, corresponding to a propagation delay of less than 240 ns per stage.

References

1 Puurunen, R.L. (2014). A short history of atomic layer deposition: Tuomon Suntola's atomic layer epitaxy. *Chemical Vapor Deposition* 20: 332–344.

2 Ritala, M. and Niinisto, J. (2009). Industrial applications of atomic layer deposition. *ECS Transactions* 25 (8): 641–652.

3 Sheng, J., Lee, J., Choi, W. et al. (2018). Review article: atomic layer deposition for oxide semiconductor thin film transistors. *Advances in Research and Development JVST A* 36 (6): 060801.

4 Oviroh, P.O., Akbarzadeh, R., Pang, D. et al. (2019). New development of atomic layer deposition: process, methods, and applications. *Science and Technology of Advanced Materials* 20 (1): 465–496.

5 Park, J., Maeng, W., Kim, H., and Park, J.-S. (2012). Review of recent developments in amorphous oxide semiconductor thin film transistor devices. *Thin Solid Films* 520 (6): 1679–1693.

6 Park, J.-S., Kim, H., and Kim, I. (2014). Overview of electroceramics materials for oxide semiconductor thin film transistors. *Journal of Electroceramics* 32: 117–140.

7 Lee, K.-H., Ok, K.-C., Kim, H., and Park, J.-S. (2014). The influence of oxygen partial pressure on the performance and stability of Ge-doped InGaO thin film transistors. *Ceramics International* 40: 3215–3220.

8 Chiu, C.J., Pei, Z.W., Chang, S.T. et al. (2011). Effect of oxygen partial pressure on electrical characteristics of amorphous indium gallium zinc oxide thin-film transistors fabricated by thermal annealing. *Vacuum* 86: 246–249.

9 Park, J., Jeong, H.-J., Lee, H.-M. et al. (2017). The resonant interaction between anions or vacancies in ZnON semiconductors and their effects on thin film device properties. *Scientific Reports* 7: 2111.

10 Ahn, C.H., Kim, S.H., Yun, M.G., and Cho, H.K. (2014). Design of step composition gradient thin film transistor channel layers grown by atomic layer deposition. *Applied Physics Letters* 105: 223513.

11 Sheng, J., Lee, H.-J., Oh, S., and Park, J.-S. (2016). Flexible and high-performance amorphous indium zinc oxide thin film transistor using low-temperature atomic layer deposition. *ACS Applied Materials & Interfaces* 8: 33821–33828.

12 Sheng, J., Park, E.J., Shong, B., and Park, J.-S. (2017). Atomic layer deposition of an indium gallium oxide thin film for thin-film transistor applications. *ACS Applied Materials & Interfaces* 9: 23934–23940.

13 Sheng, J., Hong, T., Kang, D. et al. (2019). Design of InZnSnO semiconductor alloys synthesized by supercycle atomic layer deposition and their rollable applications. *ACS Applied Materials & Interfaces* 11: 12683–12692.

14 Sheng, J., Hong, T., Lee, H.-M. et al. (2019). Amorphous IGZO TFT with high mobility of $\sim$70 $cm^2/(Vs)$ via vertical dimension control using PEALD. *ACS Applied Materials & Interfaces* 11: 40300–40309.

15 Illiberi, A., Katsouras, I., Gazibegovic, S. et al. (2018). Atmospheric plasma-enhanced spatial-ALD of InZnO for high mobility thin film transistors. *Journal of Vacuum Science & Technology A* 36: 04F401.

16 Sheng, J., Park, J., Choi, D.-W. et al. (2016). A study on the electrical properties of atomic layer deposition grown InO_x on flexible substrates with respect to N_2O plasma treatment and the associated thin-film transistor behavior under repetitive mechanical stress. *ACS Applied Materials & Interfaces* 8: 31136–31143.

17 Lee, J.-H., Sheng, J., An, H. et al. (2020). Metastable rhombohedral phase transition of semiconducting indium oxide controlled by thermal atomic layer deposition. *Chemistry of Materials* 32: 7397–7403.

18 Maeng, W.J., Choi, D.-W., Park, J., and Park, J.-S. (2015). Indium oxide thin film prepared by low temperature atomic layer deposition using liquid precursors and ozone oxidant. *Journal of Alloys and Compounds* 649: 216–221.

19 Levy, D.H., Nelson, S.F., and Freeman, D. (2009). Oxide electronics by spatial atomic layer deposition. *Journal of Display Technology* 5 (12): 484–494.

20 Nelson, S.F., Ellinger, C.R., and Tutt, L.W. (2014). Design considerations for ZnO transistors made using spatial ALD. *ECS Transactions* 64 (9): 73–83.

21 Hwang, C.S., Park, S.-H.K., Oh, H. et al. (2014). Vertical channel ZnO thin-film transistors using an atomic layer deposition method. *IEEE Electron Device Letters* 35 (3): 360–362.

22 Nelson, S.F., Ellinger, C.R., and Levy, D.H. (2015). Improving yield and performance in ZnO thin-film transistors made using selective area deposition. *ACS Applied Materials & Interfaces* 7: 2754–2759.

23 Ma, A.M., Shen, M., Afshar, A. et al. (2016). Interfacial contact effects in top gated zinc oxide thin film transistors grown by atomic layer deposition. *IEEE Transactions on Electron Device* 63 (9): 3540–3546.

24 Soga, I., Komuro, A., and Tsuboi, O. (2012). Rectifying characteristics of thin film self-switching devices with ZnO deposited by atomic layer deposition. *Electronics Letters* 48 (15): 914–916.

25 Sun, K.G., Choi, K., and Jackson, T.N. (2016). Low-power double-gate ZnO TFT active rectifier. *IEEE Electron Device Letters* 37 (4): 426–428.

26 Park, H.K., Yang, B.S., Kim, M.S. et al. (2015). Effects of in-situ molecular oxygen exposure on the modulation of electrical properties of zinc oxide thin films grown by atomic layer deposition. *Physica Status Solidi A* 212 (2): 323–328.

27 Lim, S.J., Kim, J.-M., Kim, D. et al. (2010). The effects of UV exposure on plasma-enhanced atomic layer deposition ZnO thin film transistor. *Electrochemical and Solid-State Letters* 13 (5): H151–H154.

28 Cho, S.W., Ahn, C.H., Yun, M.G. et al. (2014). Effects of growth temperature on performance and stability of zinc oxide thin film transistors fabricated by thermal atomic layer deposition. *Thin Solid Films* 562: 597–602.

29 Ma, Q., Zheng, H.-M., Shao, Y. et al. (2018). Atomic-layer-deposition of indium oxide nano-films for thin-film transistors. *Nanoscale Research Letters* 13 (1): 4.

30 Ma, Q., Shao, Y., Wang, Y.-P. et al. (2018). Rapid improvement in thin film transistors with atomic-layer-deposited InO_x channels via O_2 plasma treatment. *IEEE Electron Device Letters* 39 (11): 1672–1676.

31 Sheng, J., Choi, D.-W., Lee, S.-H. et al. (2016). Performance modulation of transparent ALD indium oxide films on flexible substrates: transition between metal-like conductor and high-performance semiconductor states. *Journal of Materials Chemistry C* 4: 7571.

32 Lee, D.-J., Kwon, J.-Y., Lee, J.I., and Kim, K.-B. (2011). Self-limiting film growth of transparent conducting In_2O_3 by atomic layer deposition using trimethylindium and water vapor. *The Journal of Physical Chemistry C* 115: 15384–15389.

33 Maeng, W.J., Choi, D.-W., Park, J., and Park, J.-S. (2015). Atomic layer deposition of highly conductive indium oxide using a liquid precursor and water oxidant. *Ceramics International* 41: 10782–10787.

34 Yeom, H.-I., Ko, J.B., Mun, G., and Ko Park, S.-H. (2016). High mobility polycrystalline indium oxide thin-film transistors by means of plasma-enhanced atomic layer deposition. *Journal of Materials Chemistry C* 4: 6873–6880.

35 Lee, D.-K., Wan, Z., Bae, J.-S. et al. (2016). Plasma-enhanced atomic layer deposition of SnO2 thin films using $SnCl_4$ and O_2 plasma. *Materials Letters* 166: 163–166.

36 Choi, D.-W. and Park, J.-S. (2014). Highly conductive SnO_2 thin films deposited by atomic layer deposition using tetrakis-dimethyl-amine-tin precursor and ozone reactant. *Surface & Coatings Technology* 259: 238–243.

37 Choi, W.-S. (2010). Effects of seed layer and thermal treatment on atomic layer deposition-grown tin oxide. *Transactions on Electrical and Electronic Materials* 11 (5): 222–225.

38 Lee, B.K., Jung, E., Kim, S.H. et al. (2012). Physical/chemical properties of tin oxide thin film transistors prepared using plasma-enhanced atomic layer deposition. *Materials Research Bulletin* 47: 3052–3055.

39 Choi, G., Satyanarayana, L., and Park, J.-S. (2006). Effect of process parameters on surface morphology and characterization of PE-ALD SnO_2 thin films for gas sensing. *Applied Surface Science* 252: 7878–7883.

40 Sundqvist, J., Lu, J., Ottosson, M., and Harsta, A. (2006). Growth of SnO_2 thin films by atomic layer deposition and chemical vapour deposition: a comparative study. *Thin Solid Films* 514: 63–68.

41 Lee, J.-H., Yoo, M., Kang, D. et al. (2018). Selective SnOx atomic layer deposition driven by oxygen reactants. *ACS Applied Materials & Interfaces* 10: 33335–33342.

42 Heo, J., Hock, A.S., and Gordon, R.G. (2010). Low temperature atomic layer deposition of tin oxide. *Chemistry of Materials* 22: 4964–4973.

43 Mai, L., Zanders, D., Subasi, E. et al. (2019). Low-temperature plasma-enhanced atomic layer deposition of tin(IV) oxide from a functionalized alkyl precursor: fabrication and evaluation of SnO_2-based thin-film transistor devices. *ACS Applied Materials & Interfaces* 11: 3169–3180.

44 Lee, J.-M., Lee, H.-J., Pi, J.-E. et al. (2019). All-oxide thin-film transistors with channels of mixed InOx-ZnOy formed by plasma-enhanced atomic layer deposition process. *Journal of Vacuum Science & Technology A* 37: 060910.

45 Ahn, B.D., Choi, D.-W., Choi, C., and Park, J.-S. (2014). The effect of the annealing temperature on the transition from conductor to semiconductor behavior in zinc tin oxide deposited atomic layer deposition. *Applied Physics Letters* 105: 092103.

46 Yoon, S.-M., Seong, N.-J., Choi, K. et al. (2017). Effects of deposition temperature on the device characteristics of oxide thin-film transistors using In–Ga–Zn–O active channels prepared by atomic-layer deposition. *ACS Applied Materials & Interfaces* 9: 22676–22684.

47 Wang, Z., Nayak, P.K., Caraveo-Frescas, J.A., and Alshareef, H.N. (2016). Recent developments in p-type oxide semiconductor materials and devices. *Advanced Materials* 28: 3831–3892.

48 Han, J.H., Chung, Y.J., Park, B.K. et al. (2014). Growth of p-type tin(II) monoxide thin films by atomic layer deposition from bis(1-dimethylamino-2-methyl-2propoxy)tin and H_2O. *Chemistry of Materials* 26: 6088–6091.

49 Tupala, J., Kemell, M., Mattinen, M. et al. (2017). Atomic layer deposition of tin oxide thin films from bis[bis(trimethylsilyl)amino]tin(II) with ozone and water. *Journal of Vacuum Science & Technology A* 35: 041506.

50 Kim, S.H., Baek, I.-H., Kim, D.H. et al. (2017). Fabrication of high-performance p-type thin film transistors using atomic-layer-deposited SnO films. *Journal of Materials Chemistry C* 5: 3139–3145.

51 Muñoz-Rojas, D., Jordan, M., Yeoh, C. et al. (2012). Growth of $\sim 5\,cm^2V^{-1}s^{-1}$ mobility, p-type Copper(I) oxide (Cu_2O) films by fast atmospheric atomic layer deposition (AALD) at 225°C and below. *AIP Advances* 2: 042179.

52 Kwon, J.-D., Kwon, S.-H., Jung, T.-H. et al. (2013). Controlled growth and properties of p-type cuprous oxide films by plasma-enhanced atomic layer deposition at low temperature. *Applied Surface Science* 285P: 373–379.

53 Maeng, W., Lee, S.-H., Kwon, J.-D. et al. (2016). Atomic layer deposited p-type copper oxide thin films and the associated thin film transistor properties. *Ceramics International* 42: 5517–5522.

54 Kim, H., Lee, M.Y., Kim, S.-H. et al. (2015). Highly-conformal p-type copper(I) oxide (Cu_2O) thin films by atomic layer deposition using a fluorine-free amino-alkoxide precursor. *Applied Surface Science* 349: 673–682.

55 Seok, T.J., Liu, Y., Jung, H.J. et al. (2018). Field-effect device using quasi-two dimensional electron gas in mass-producible atomic-layer-deposited Al_2O_3/TiO_2 ultrathin (<10 nm) film heterostructures. *ACS Nano* 12: 10403–10409.

56 Seul, H.J., Kim, M.J., Yang, H.J. et al. (2020). Atomic layer deposition process-enabled carrier mobility boosting in field-effect transistors through a nanoscale ZnO/IGO heterojunction. *ACS Applied Materials & Interfaces* 12: 33887–33898.

Part IV

Thin-Film Transistors

11

Control of Carrier Concentrations in AOSs and Application to Bulk-Accumulation TFTs

Suhui Lee and Jin Jang

Advanced Display Research Center (ADRC), Department of Information Display, Kyung Hee University, Seoul, Republic of Korea

11.1 Introduction

With the demand for consumer electronic products, display technology needs to satisfy the requirements of low power consumption, high resolution, and high frame rate. These can be possible with high-drain-current, thin-film transistors (TFTs) with excellent stability that are applied to small- and large-sized displays [1]. To satisfy those requirements for display devices such as organic light-emitting diodes (OLEDs) and micro-LEDs, low-temperature polycrystalline silicon (LTPS) and amorphous oxide semiconductor (AOS) TFTs are currently being used. The TFT backplane can be realized with LTPS and AOS [2–4] because of the high mobility and excellent bias stability of the TFTs made of these materials. LTPS TFTs are being used for small-sized OLED displays, mainly smartphone displays, and oxide TFTs are for OLED TV displays. Note that amorphous-silicon (*a*-Si:H) TFTs have an inherent issue of threshold-voltage (V_{TH}) shift during pixel operation, so they cannot be used for OLED TFT backplanes [5].

LTPS TFT backplanes have been widely used to drive smartphone display, due to their high hole mobility (>100 cm^2/Vs) and excellent bias stability. However, there are some issues such as nonuniformity in LTPS material due to excimer laser annealing (ELA), robustness under high mechanical strain, and high manufacturing cost [5, 6]. The crystalline grain can be easily broken, and defects could be generated when applying high mechanical strain to the LTPS TFT.

AOSs are of increasing interest due to their advantages, such as relatively high mobility, a simple manufacturing process, and transparency in the visible region and low-temperature process [7]. This can lead to applying this material to highly robust and low-cost manufacturing of OLED displays and flexible TFT electronics [8–11].

Amorphous indium–gallium–zinc oxide (*a*-IGZO), the most popular material, has been studied intensively by many industry and academic groups [12–14]. Its carrier concentration can be controlled to $\sim 10^{16}$ cm^{-3}, which is good to achieve accumulation-mode TFTs.

The carrier concentration and field-effect mobility are very important parameters of AOS TFTs for display applications. There are many reports on the control of channel carrier concentration by changing the Ar/O_2 ratio during reactive sputtering [15], and H_2 partial pressure during sputtering [16, 17]. UV/O_3 exposure [18, 19], thermal annealing [20], F plasma exposure [21, 22], and He/Ar treatment [23] on the films can change the carrier concentration.

It is known that the generation of positively charged oxygen vacancies ($V_O^{+/++}$) leads to increased carrier concentration. The generation of positively charged $V_O^{+/++}$ in *a*-IGZO increases its conductivity, but the conductivity is unstable after high-temperature annealing (>300 °C). Therefore, doping by O vacancy generation is not stable upon high-temperature annealing, for example higher than 300 °C.

Amorphous Oxide Semiconductors: IGZO and Related Materials for Display and Memory, First Edition.
Edited by Hideo Hosono and Hideya Kumomi.

It is well-known that percolation transport takes place in IGZO. The potential barriers for the electron transport decrease with increasing the Fermi level (E_F) toward the conduction band edge (Ec). The shift of the E_F toward the E_c is correlated with the carrier concentration; the E_F shifts to the E_c by increasing electron concentration. Another method for controlling the E_F is by using the dual-gate (DG) TFT structures. The carrier concentration in the channel layer can be controlled more efficiently by DG driving [24, 25].

A DG TFT has a similar device structure as a single-gate (SG) TFT, using the bottom as a main gate and the top as an additional gate to control the carrier concentration in the channel.

We proposed the concept of bulk-accumulation (BA) mode TFTs, which can be achieved by DG driving (when the two gates are electrically tied together) of DG *a*-IGZO TFTs. The concept of BA is that the induced charges by top- and bottom-gate potentials are placed in the bulk and at the bottom–top interface regions of the channel. The BA oxide TFT has advantages such as a high drain current (I_{DS}), 3–5 times that of a SG TFT, with turn-on voltage (V_{ON}) near zero volts, and low subthreshold swing (SS). In addition, the BA TFT is more robust under stress compared to the conventional TFT, regarding issues such as bias stress, temperature stress, light illumination stress, and mechanical stress. The increase in the drain currents is partially due to the higher electron mobility with increasing carrier concentration in the channel, which leads to a fast filling-up of the gap states as the E_F moves toward the conduction band edge (Ec).

11.2 Control of Carrier Concentration in *a*-IGZO

The carrier concentration and field-effect mobility are very important parameters in oxide TFTs for display applications. The carrier concentration in *a*-IGZO can be basically controlled by the Ar/O_2 ratio during reactive sputtering. It also depends on substrate temperature, sputtering power, and the composition ratio between In, Ga, Zn, and O.

Hosono et al. reported the effect of oxygen partial pressure in Ar/O_2 during the deposition processes on the carrier concentration in IGZO and *a*-IZO films (shown in Figure 11.1). The carrier concentration in *a*-IGZO could be decreased lower than 10^{13} cm^{-3} by increasing oxygen partial pressure (PO_2), whereas it is higher than 10^{18} cm^{-3} when the PO_2 is very low. This is due to more O inclusion in *a*-IGZO during the sputtering at a high O_2 ratio. The O vacancy concentration decreases with increasing O concentration in *a*-IGZO; this can be done by reactive sputtering of *a*-IGZO in a higher O_2 environment. A similar trend can be seen in Figure 11.1 for IZO.

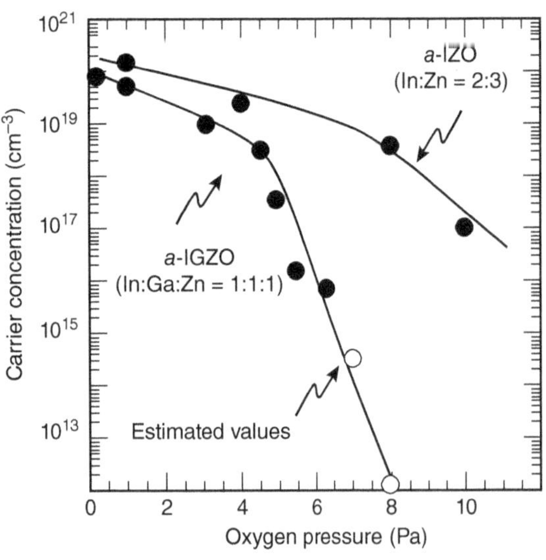

Figure 11.1 The carrier concentration plotted against O_2 pressure during the reactive sputtering for the deposition of *a*-IGZO and *a*-IZO [15, 16]. *Source*: Hosono, H. (2006). Ionic amorphous oxide semiconductors: material design, carrier transport, and device application. *Journal of Non-Crystalline Solids* **352**: 851–858.

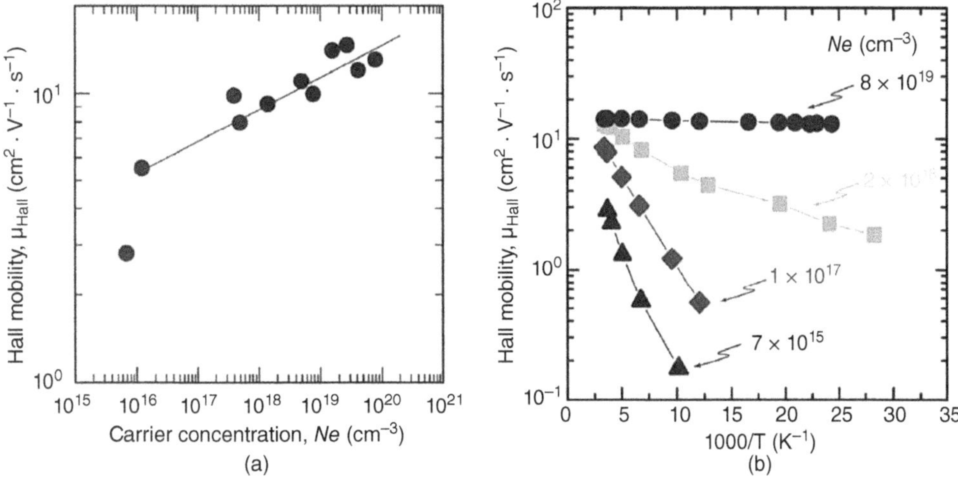

Figure 11.2 Hall mobility of *a*-IGZO plotted as a function of (a) carrier concentration and (b) an inverse absolute temperature [15, 16]. *Source*: Hosono, H. (2006). Ionic amorphous oxide semiconductors: material design, carrier transport, and device application. *Journal of Non-Crystalline Solids* **352**: 851–858.

Figure 11.2a shows the Hall mobility in *a*-IGZO as a function of carrier concentration. Hall mobility increases with carrier concentration, which is well-known in oxide semiconductors because of percolation transport in *a*-IGZO. The potential barriers for the electron transport decrease by shifting the E_F toward the Ec. The shift of the E_F toward the Ec is correlated with the carrier concentration; the E_F shifts to the Ec by increasing electron concentration. Figure 11.2b shows the temperature dependence of Hall mobility, indicating an activated form with inverse absolute temperature. The mobility activation energy obtained from the straight lines decreases with increasing carrier concentration, because it is related to the energy gap between the E_C and E_F.

The structural and electronic properties of *a*-IGZO depend on thermal annealing. It is reported that the amorphous phase keeps until 550 °C thermal annealing. But it changes into the crystalline phase by annealing at >600 °C [17]. It is also reported that Hall mobility increases with increasing annealing temperature, which is due to the shift of the E_F toward the E_C by annealing. The exposure of the IGZO to UV results in the formation of O vacancy, which generates the carriers. Therefore, the resistivity of IGZO decreases with increasing UV exposure time [18, 19].

The role of hydrogen in *a*-IGZO is very complicated, and thus a lot of research is going on. H can be incorporated during the deposition or can be diffused into *a*-IGZO after deposition. H can have bonding with O and metal in *a*-IGZO. The mobility depends on the quantity of incorporated H [20]. The small increase in H after H_2 annealing leads to a minor increase in both oxygen vacancies and unoccupied states under the Ec. Alternatively, a larger increase in the H concentration incorporated after the H-plasma treatment results in a larger increase in the oxygen vacancies and the unoccupied states below Ec, thereby inducing an increase in the carrier concentration and a decrease in mobility. Therefore, control of the incorporated H quantity is an applicable method to modify the electrical properties of *a*-IGZO TFTs.

F incorporation in *a*-IGZO is also a very important topic to control the carrier concentration. The electrical properties of *a*-IGZO films change by exposure to F plasma. The carrier concentration, resistivity, and Hall mobility as a function of F volume concentration are plotted in Figure 11.3b, 11.3c, and 11.3d, respectively. By increasing F in *a*-IGZO, the carrier concentration increases from 1×10^{13} to 6×10^{19} cm^{-3}, and thus its resistivity decreases from 1×10^2 to 3×10^{-3} Ω cm, while the Hall mobility increases from 8 to 22 cm^2/V s. These results indicate that some of the F atoms substitute O, leading to the generation of free electrons because of the difference in valence electrons between O and F [21].

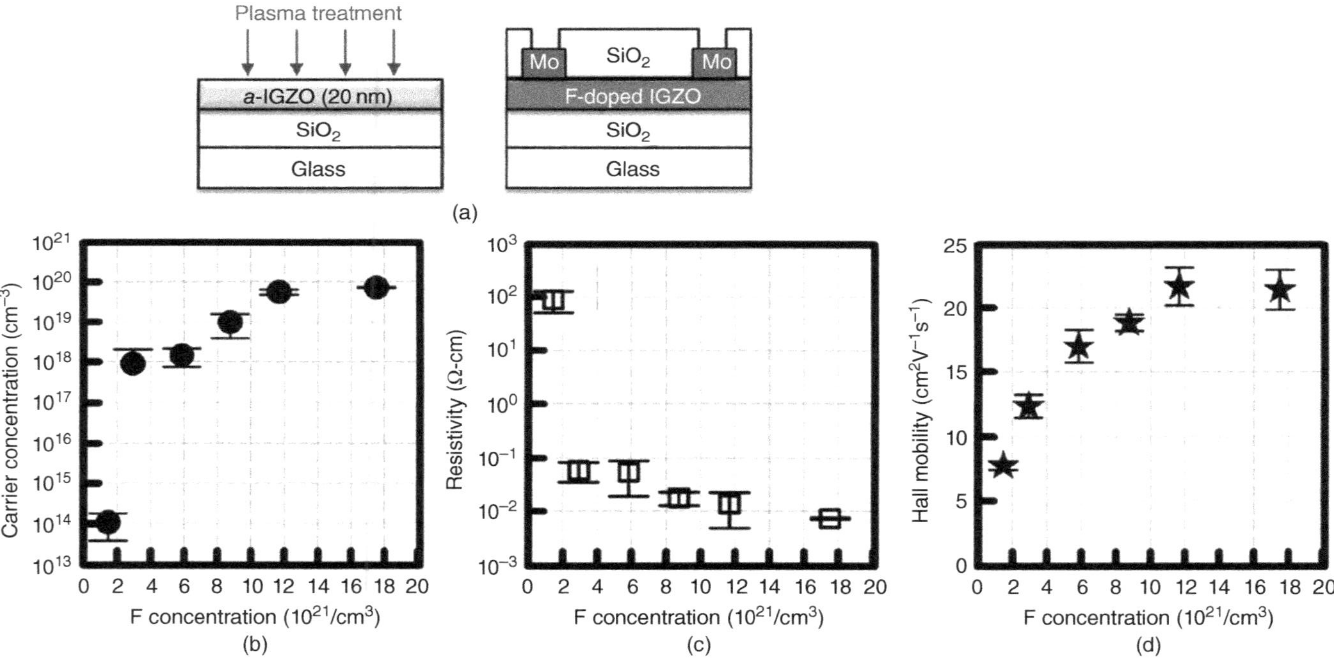

Figure 11.3 Hall effect results for the 20-nm-thick *a*-IGZO film as a function of F concentration (atoms/cm^3) before annealing. (a) Sample structure for the Hall effect measurement; SiO_2/plasma-treated *a*-IGZO/SiO_2 with the Mo contact via holes, (b) carrier concentration, (c) resistivity, and (d) Hall mobility [21]. *Source*: Um, J. G., and Jang, J. (2018). Heavily doped n-type a-IGZO by F plasma treatment and its thermal stability up to 600 °C. *Applied Physics Letters* **112**: 162104.

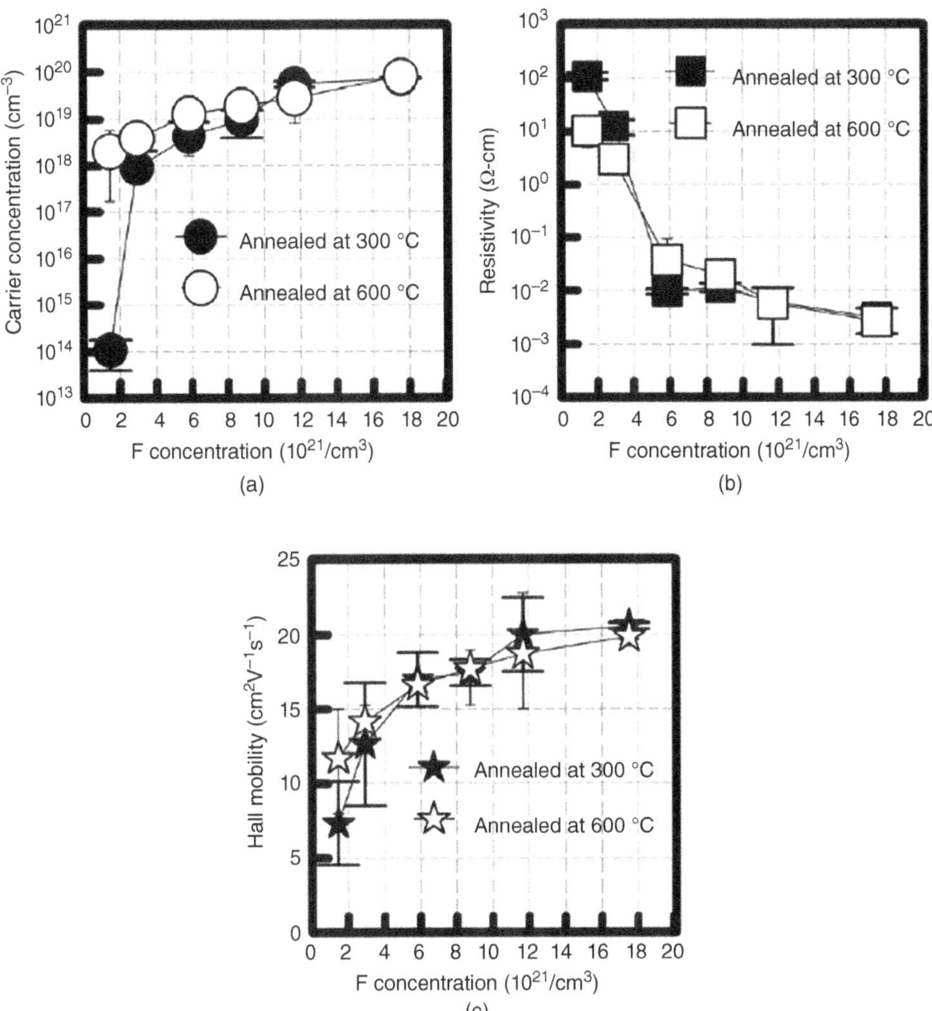

Figure 11.4 Hall effect results for the *a*-IGZO layer annealed at 300 °C or 600 °C, plotted as a function of F concentration (atoms/cm^3). (a) Carrier concentration, (b) resistivity, and (c) Hall mobility. The doping efficiencies, defined as the carrier concentration divided by the F concentration, are 15.7% and 17.9% for the *a*-IGZO annealed at 300 °C and 600 °C, respectively [21]. *Source*: Um, J. G., and Jang, J. (2018). Heavily doped n-type a-IGZO by F plasma treatment and its thermal stability up to 600 °C. *Applied Physics Letters* **112**: 162104.

Figure 11.4 shows the carrier concentration, Hall mobility, and resistivity for the *a*-IGZO annealed at 300 °C or 600 °C plotted as a function of F concentration. Here, the carrier concentration of *a*-IGZO is almost unchanged [$(6.3 - 6.6) \times 10^{19}$ cm^{-3}] as the annealing temperature varies from 300 °C to 600 °C when the F concentration is 17.5×10^{21}/cm^3. Therefore, the heavily F-doped *a*-IGZO exhibits excellent thermal stability.

X-ray photoelectron spectroscopy (XPS) depth profiles, measured for the SiO_2/*a*-IGZO/SiO_2 samples with undoped and F-doped *a*-IGZO films, are shown in Figure 11.5a and 11.5b, respectively. Note that the F 1s peak can be seen only within 10 nm of the top *a*-IGZO surface and that the F diffusion depth is almost unchanged even after annealing at 600 °C.

When the F-doping concentration in *a*-IGZO is 17.5×10^{21} cm^{-3}, the carrier concentration and Hall mobility are 6×10^{19} cm^{-3} and 22 cm^2/Vs, respectively. Therefore, F is a suitable *n*-type dopant in *a*-IGZO and thus can be used for Ohmic contact of oxide devices.

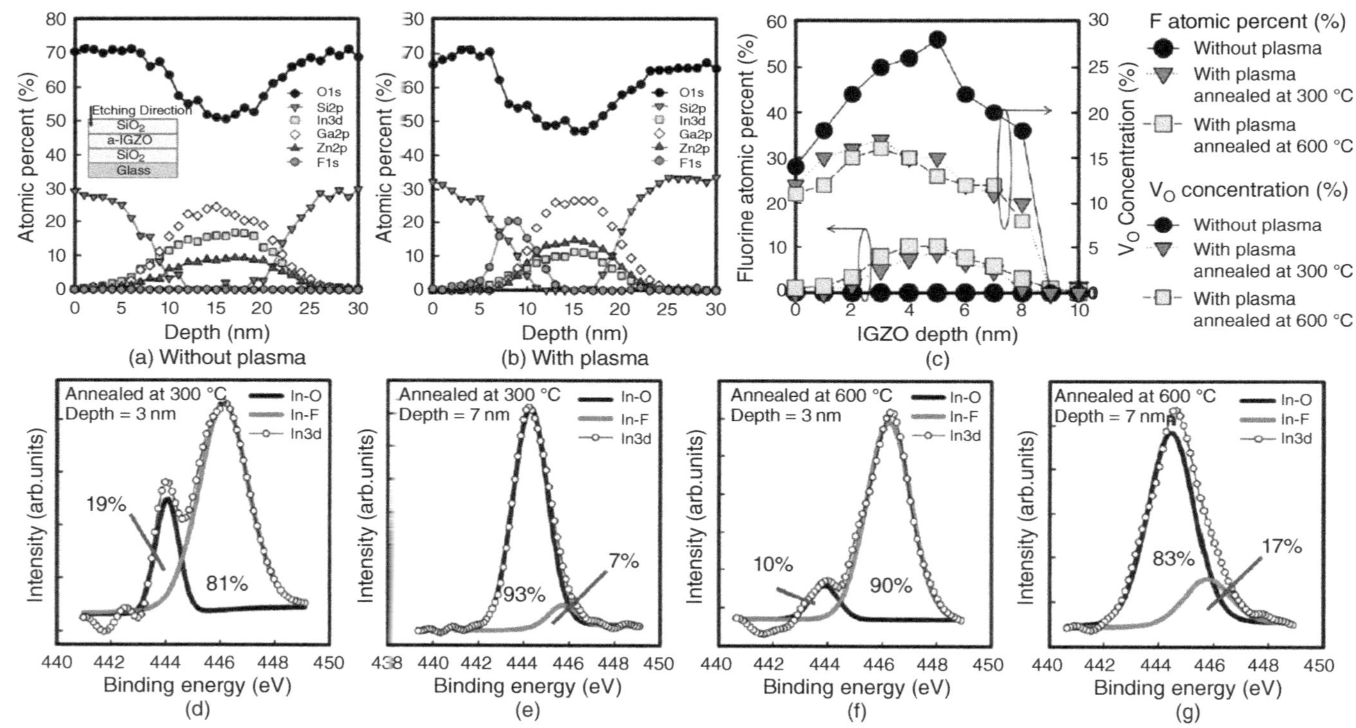

Figure 11.5 XPS analysis for the F-doped *a*-IGZO film. Depth profiles of O1s, Si2p, In3d, Ga2p, Zn2p, and F1s for SiO_2/*a*-IGZO/SiO_2; (a) without plasma treatment and (b) with plasma treatment, both annealed at 300 °C. F exists until a depth of 10 nm from the top surface. Atomic percent of F and concentration of oxygen vacancy (V_O) as a function of *a*-IGZO depth are plotted for the *a*-IGZO without or with plasma treatment. A significant reduction in the V_O concentration can be seen by F incorporation. Deconvolutions of In3d with the plasma-treated *a*-IGZO film at the *a*-IGZO depths of (d) 3 nm and (e) 7 nm at annealed at 300 °C, and (f) 3 nm and (g) 7 nm at annealed at 600 °C, respectively. The bond energies are 444.0 eV for In-O and 446.2 eV for In-F [21]. *Source*: Um, J. G., and Jang, J. (2018). Heavily doped n-type a-IGZO by F plasma treatment and its thermal stability up to 600 °C. *Applied Physics Letters* **112**: 162104.

11.3 Effect of Carrier Concentration on the Performance of *a*-IGZO TFTs with a Dual-Gate Structure

11.3.1 Inverted Staggered TFTs

Figure 11.6a shows a cross-sectional view of etch-stopper (ES) *a*-IGZO TFTs with a DG structure. The transfer curves measured at the bottom-gate (BG) driving are shown in Figure 11.6b and 11.6c, where the top-gate (TG) potential is controlled from positive to negative. The shift of the threshold voltage is apparent. Figure 11.6d shows the plot of the V_{TH} shifts (ΔV_{TH}) as a function of V_{TG}, indicating a straight line. The application of V_{TG} controls the carrier density in the channel layer such that it is a depletion-mode or accumulation-mode operation, with respect to the BG. The shift in the V_{TH} is mainly due to charge accumulation at the active-layer/gate-insulator interface.

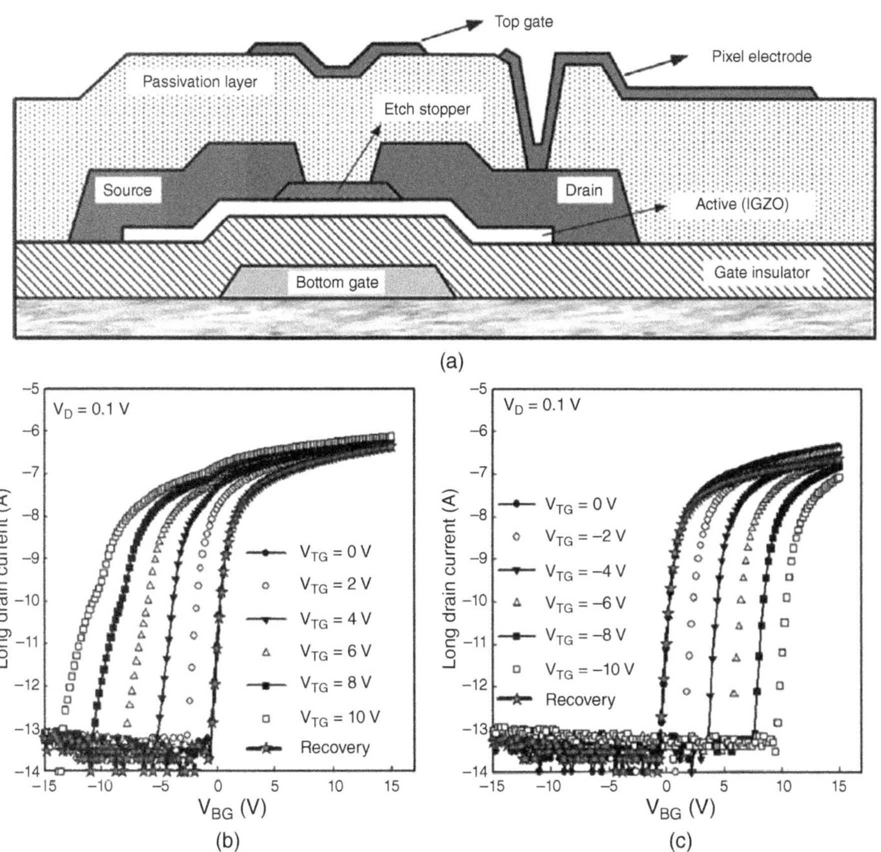

Figure 11.6 Cross-sectional view of an *a*-IGZO TFT with (a) dual-gate structures. Transfer characteristics of a dual-gate TFT when the bottom gate (V_{BG}) is driving from −15 to 15 V while biasing the top gate (V_{TG}) with various (b) positive and (c) negative voltages at a drain voltage (V_D) of 0.1 V. The star symbols indicate the transfer characteristics that were measured soon after the application of the top gate. (d) Plot of the threshold voltage shifts (ΔV_{TH}) as a function of the applied top-gate voltage (V_{TG}). The V_{TH} was calculated from the transfer characteristics in (b,c) as the bottom-gate voltage (V_{BG}) corresponding to a drain current (I_D) of W/L × 10 nA at V_{DS} = 0.1 V [26]. *Source*: Seok, M. J., et al. (2011). A full-swing a-IGZO TFT-based inverter with a top-gate-bias-induced depletion load. *IEEE Electron Device Letters* **32** (8): 1089–1091.

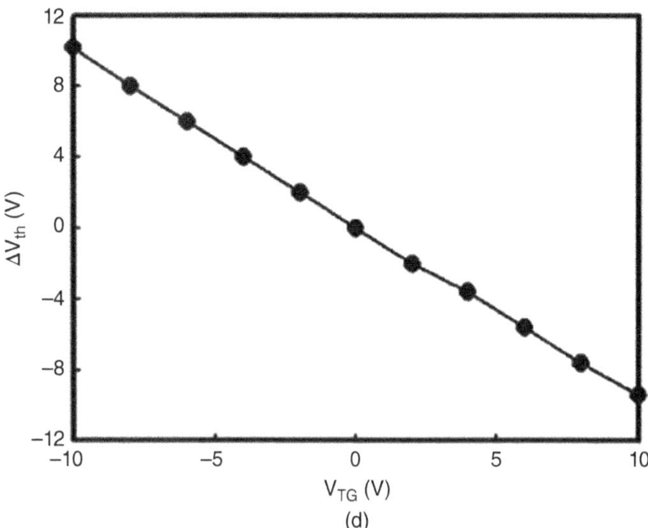

Figure 11.6 (*Continued*)

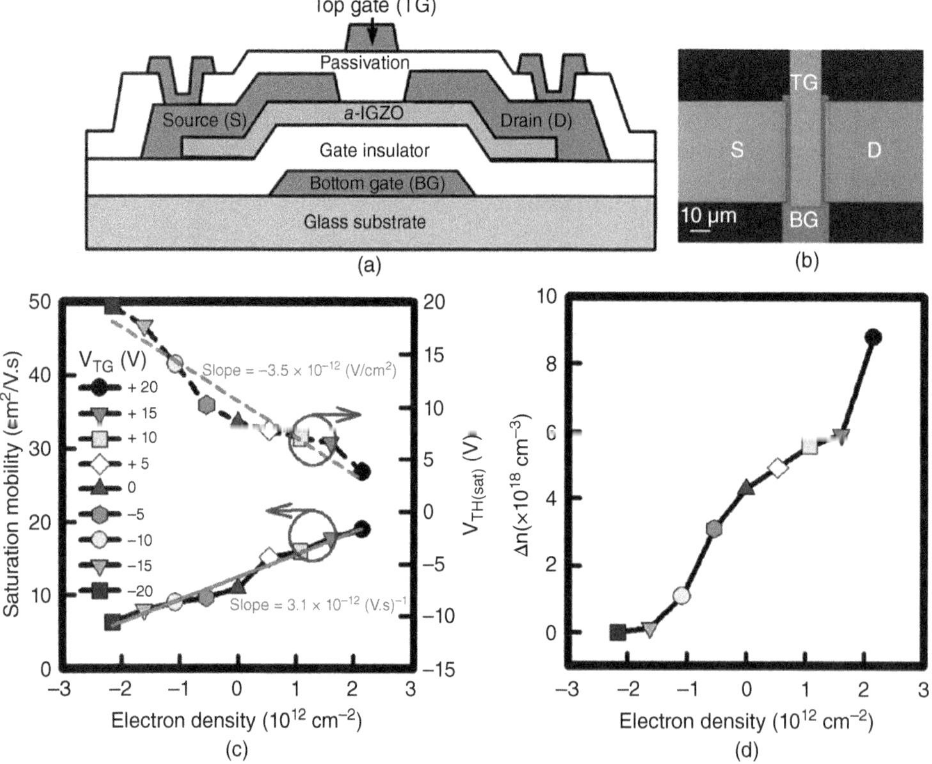

Figure 11.7 Structure and characteristics of dual-gate *a*-IGZO TFTs with an offset top gate. (a) Schematic cross section; (b) optical image of a dual-gate TFT with offsets of 1 μm at the source/drain electrodes. (c) Saturation mobility and threshold voltage extracted from a bottom-gate sweep in the saturation regime versus electron concentration induced by a top gate. The density is obtained with the multiplication of various top-gate biases (from −20 to +20 V) and top-gate insulator capacitance. (d) The increase of electron density (Δn) in the channel plotted against the electron density induced by top-gate potential [27]. *Source*: Chun, M., et al. (2015). Semiconductor to metallic transition in bulk accumulated amorphous indium-gallium-zinc-oxide dual gate thin-film transistor. *AIP Advances* **5** (5): 057165.

The negative shift of the transfer characteristics under the application of the positive V_{TG} is thus believed to be caused by the increase in the concentration of free electrons in the active layer.

A DG TFT has top and bottom gates, and its device structure is similar to that of a SG TFT, using the bottom as the main gate and the top as an additional gate to control the carrier concentration in the channel (see Figure 11.7a and 11.7b). A TG voltage-dependent V_{TH} shift (Figure 11.7c), changing the (c) mobility and (d) carrier concentrations, can be explained by the movement of E_F by applying TG potential. The carrier concentration in the TFT channel increases to ~10^{19} cm^{-3}, as shown in Figure 11.7d.

Temperature-dependent saturation mobility, measured as a variation of BG potential with a fixed TG voltage, indicates that the TFT shows a metallic behavior when a higher TG bias is applied. This is due to the shift of the E_F above the E_C. On the other hand, applying negative TG bias places the E_F in the gap well below the E_C, and the mobility shows a thermally activated process. This is a typical semiconductor property. This trend indicates clearly the transition from semiconducting to metallic behavior by controlling TG voltage.

We designed an inverted staggered DG TFT whose primary gate is the BG and the secondary gate is the TG that covers only a small portion of the channel, as shown in Figure 11.8a and 11.8b. We show that if constant positive bias is applied to the BG, while driving the TG, the transfer shifts to negative gate voltage and the on-current also

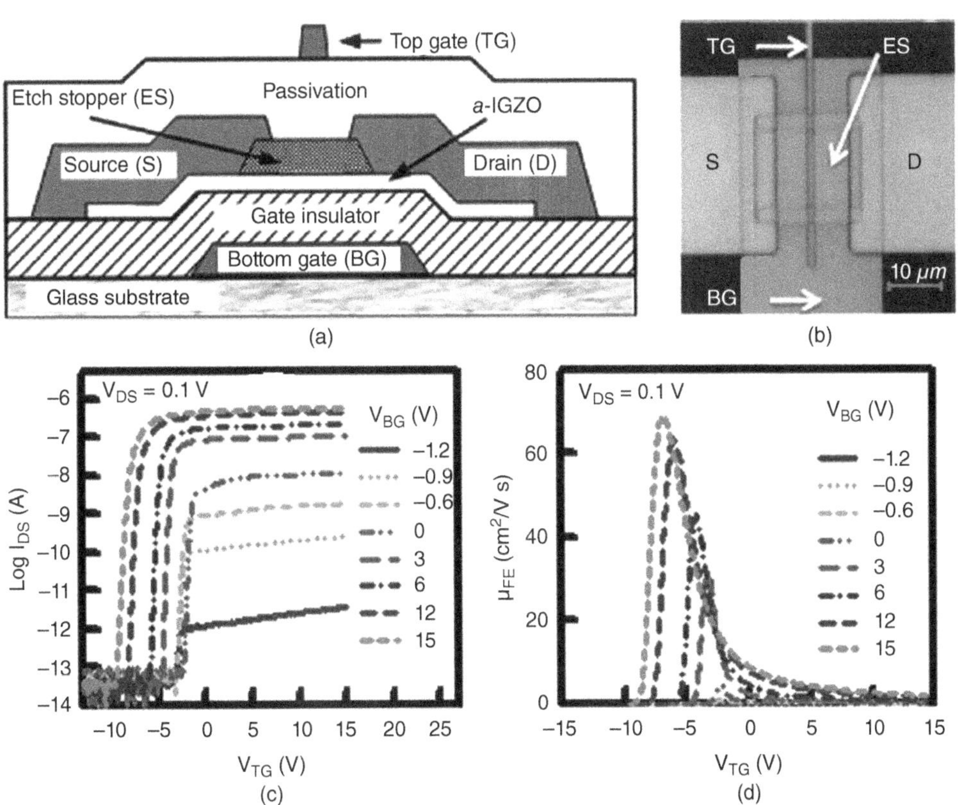

Figure 11.8 (a) Schematic cross-sectional and (b) optical image of the fabricated dual-gate *a*-IGZO TFTs. Channel width (W) and channel length (L) are both 20 μm. The length of the BG is 26 μm, whereas that of the TG is only 2 μm. (c) Transfer curves and (d) μ_{FE} curves obtained by TG driving, while applying constant voltage to the BG. The voltages applied to the bottom gate and top gate are denoted V_{BG} and V_{TG}, respectively. V_{BG} is varied from −1.2 to 15 V [28]. *Source*: Park, S. H., et al. (2013). Increase of mobility in dual gate amorphous-InGaZnO4 thin-film transistors by pseudo-doping. *Applied Physics Letters* **103** (4): 043509.

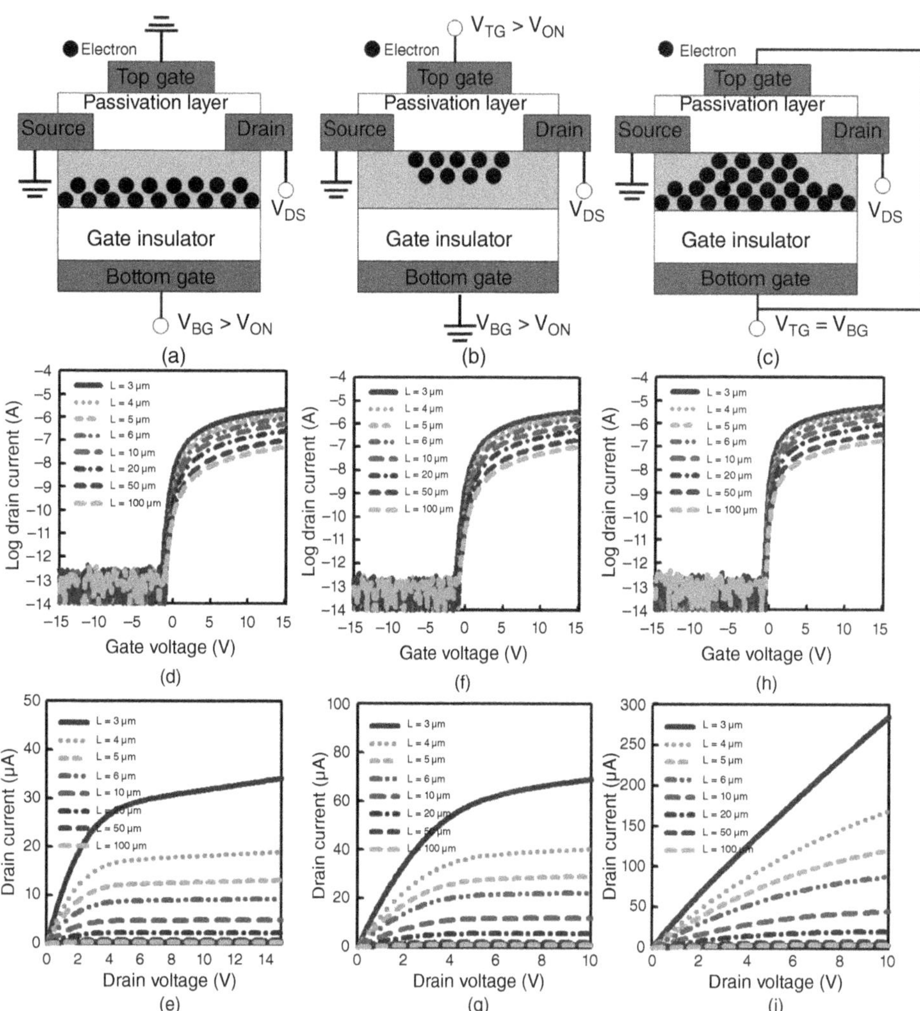

Figure 11.9 Schematic cross section of TFT view for (a) BG driving with grounded top gate (bottom-gate sweep), (b) TG driving with grounded bottom gate (top-gate sweep), and (c) the BG is electrically tied to the TG (dual-gate driving). Evolution of the transfer and output characteristics of *a*-IGZO TFTs with a fixed L of 10 μm and various lengths for (d,e) BG driving, (f,g) TG driving, and (h,i) DG driving, respectively.

increases (see Figure 11.8c), resulting in a dramatic increase in μ_{FE} (see Figure 11.8d). The increase in mobility is explained by high electron concentration in the channel induced by the BG potential.

We proposed the device concept of a BA oxide TFT for high drain currents. BA *a*-IGZO TFTs consist of a DG structure in which the active oxide semiconductor should have a low density of states in the gap along with a low interface state density, and the TG is electrically tied to the BG [24, 25]. The BA oxide TFT gives advantages for driving displays, given that the drain current increases by 3–5 times (shown in Figure 11.9), with V_{ON} near zero volts. In addition, it gives very robust device performance under bias stress, temperature stress, light illumination stress, and mechanical stress [29–34]. This also gives the advantages of fast switching speed and reducing power consumption [35–42].

11.3.2 Coplanar TFTs

The coplanar TFT is an ideal device structure for display applications because of its lowest parasitic capacitance between the gate and source/drain (S/D) electrodes. Note that the LTPS TFTs being used for smartphone active-matrix organic light-emitting diodes (AMOLEDs) and the IGZO TFTs for OLED TVs all have coplanar structure. In particular, coplanar TFTs have more advantages for large-size display because of their lower capacitance. In TG IGZO TFTs, the channel is formed under the gate, but there is a S/D offset region in coplanar *a*-IGZO TFTs that could not be controlled by the TG. Therefore, the resistivity of the offset region should be very low. In the case of LTPS TFTs, the offset region is heavily doped by ion implantation/ion doping. We need n^+ doping in IGZO, which was done by Ar or He plasma treatment. This leads to generating O vacancy, which increases the carrier concentration. But the O vacancy can be changed easily by the following process step and/or thermal treatment.

We reported coplanar *a*-IGZO TFTs with heavily doped n^+ *a*-IGZO S/D regions by using He plasma treatment. The resistivity decreases from 2.98 Ωcm to 2.79×10^{-3} Ωcm by He plasma, but it increases to 7.92×10^{-2} Ωcm after the annealing at 300 °C, as shown in Figure 11.10f.

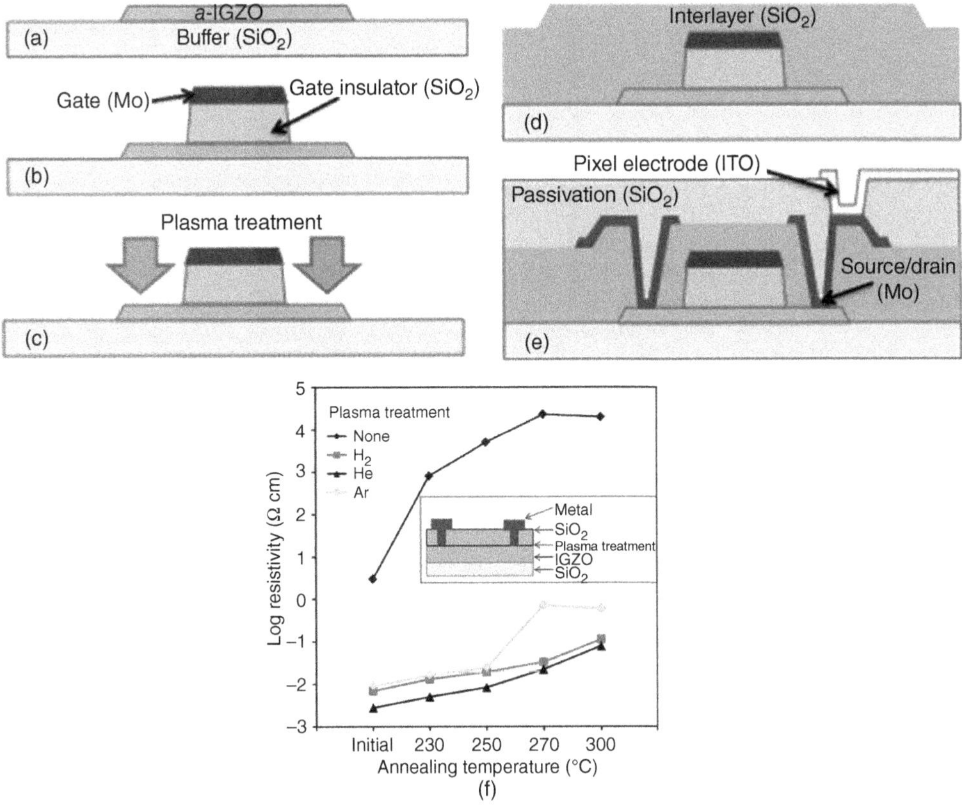

Figure 11.10 Fabrication process of the coplanar *a*-IGZO TFT: (a) *a*-IGZO deposition and patterning, (b) gate insulator and gate metal deposition and patterning using a self-aligned process, (c) plasma treatments for forming n^+ *a*-IGZO, (d) interlayer dielectric deposition, and (e) formation of source/drain and pixel electrodes. (f) Dependence of n^+ IGZO resistivity on annealing temperature [23]. *Source*: Jeong, H. Y., et al. (2014). Coplanar amorphous-indium-gallium-zinc-oxide thin film transistor with He plasma treated heavily doped layer. *Applied Physics Letters* **104** (2): 022115.

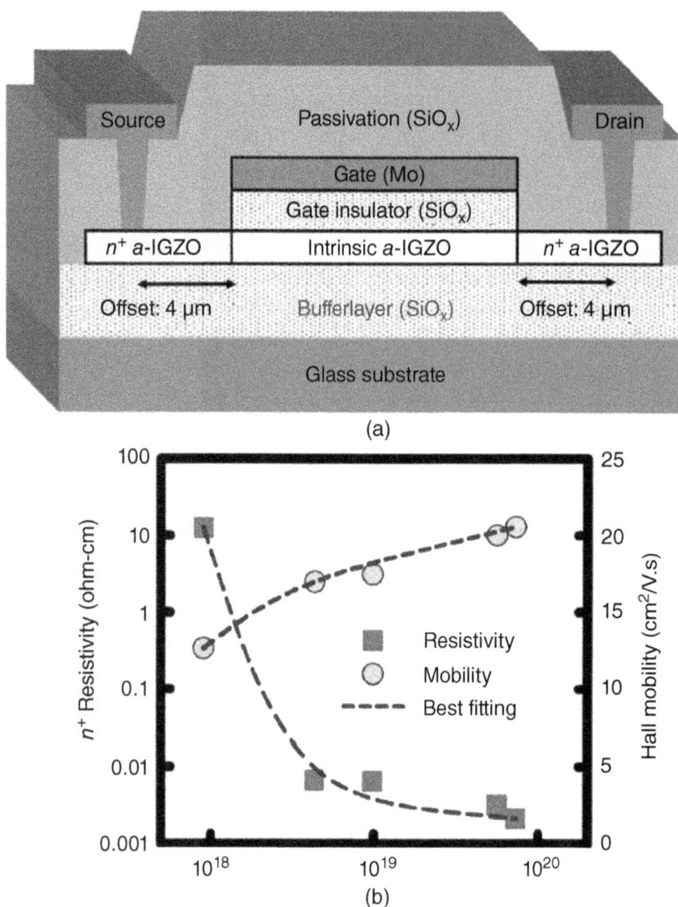

Figure 11.11 (a) Schematic cross-sectional view of a coplanar *a*-IGZO TFT. (b) Resistivity and Hall mobility plotted as a function of carrier concentration in the offset *a*-IGZO region [41]. *Source*: Rahaman, A., et al. (2018). Effect of doping fluorine in offset region on performance of coplanar a-IGZO TFTs. *IEEE Electron Device Letters* **39** (9): 1318–1321.

F plasma treatment on *a*-IGZO decreases its resistivity from 12 Ωcm to 0.0021 Ωcm and increases the Hall mobility from 12 cm^2/V·s to 20 cm^2/V·s, as shown in Figure 11.11. This low-resistive IGZO keeps its low value until 600 °C thermal annealing. This low-resistive IGZO can be used in the offset region of coplanar TFTs, as shown in Figure 11.12. The variation of drain currents shown in the output characteristics of IGZO TFTs could be fitted well by using a technology computer-aided design (TCAD) simulation, as shown in Figure 11.12. The carrier concentration at the S/D offset region varies from 3×10^{17} cm^{-3} to 2×10^{18} cm^{-3} in the TCAD fitting. Based on the experimental results and TCAD device simulation, the optimized carrier concentration is ~10^{19}cm^{-3} by F doping at the S/D offset region.

11.4 High-Drain-Current, Dual-Gate Oxide TFTs

Dual-gate TFTs with various channel lengths are manufactured at Kyung Hee University's Advanced Display Research Center (ADRC) with ES, back-channel-etched (BCE), and coplanar device structures, as shown in

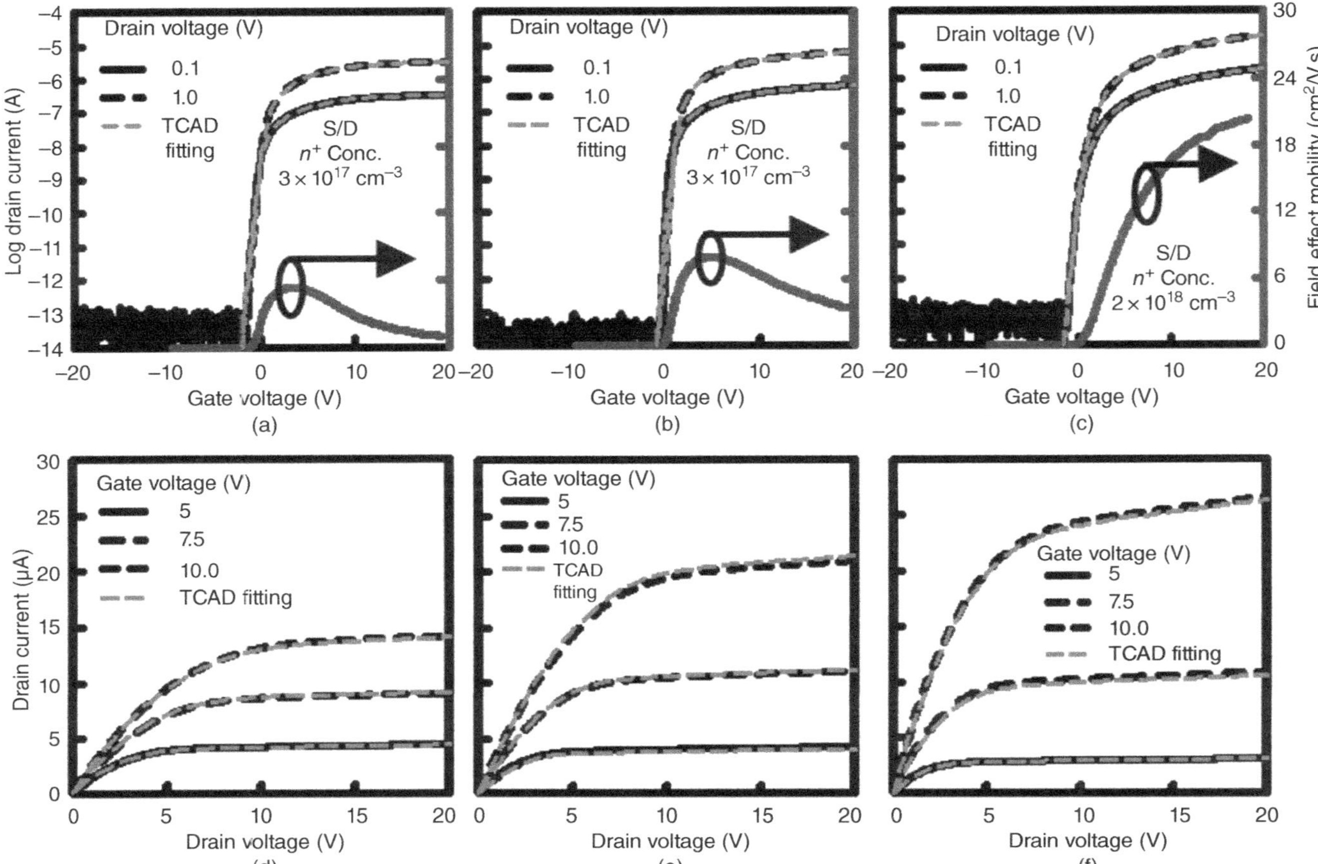

Figure 11.12 (a–c) Transfer and (d–f) output characteristics of coplanar *a*-IGZO TFTs with different n^+ carrier concentration. Red dotted lines are TCAD fittings, and black/blue lines are the measurement data [41]. *Source*: Rahaman, A., et al. (2018). Effect of doping fluorine in offset region on performance of coplanar a-IGZO TFTs. *IEEE Electron Device Letters* **39** (9): 1318–1321.

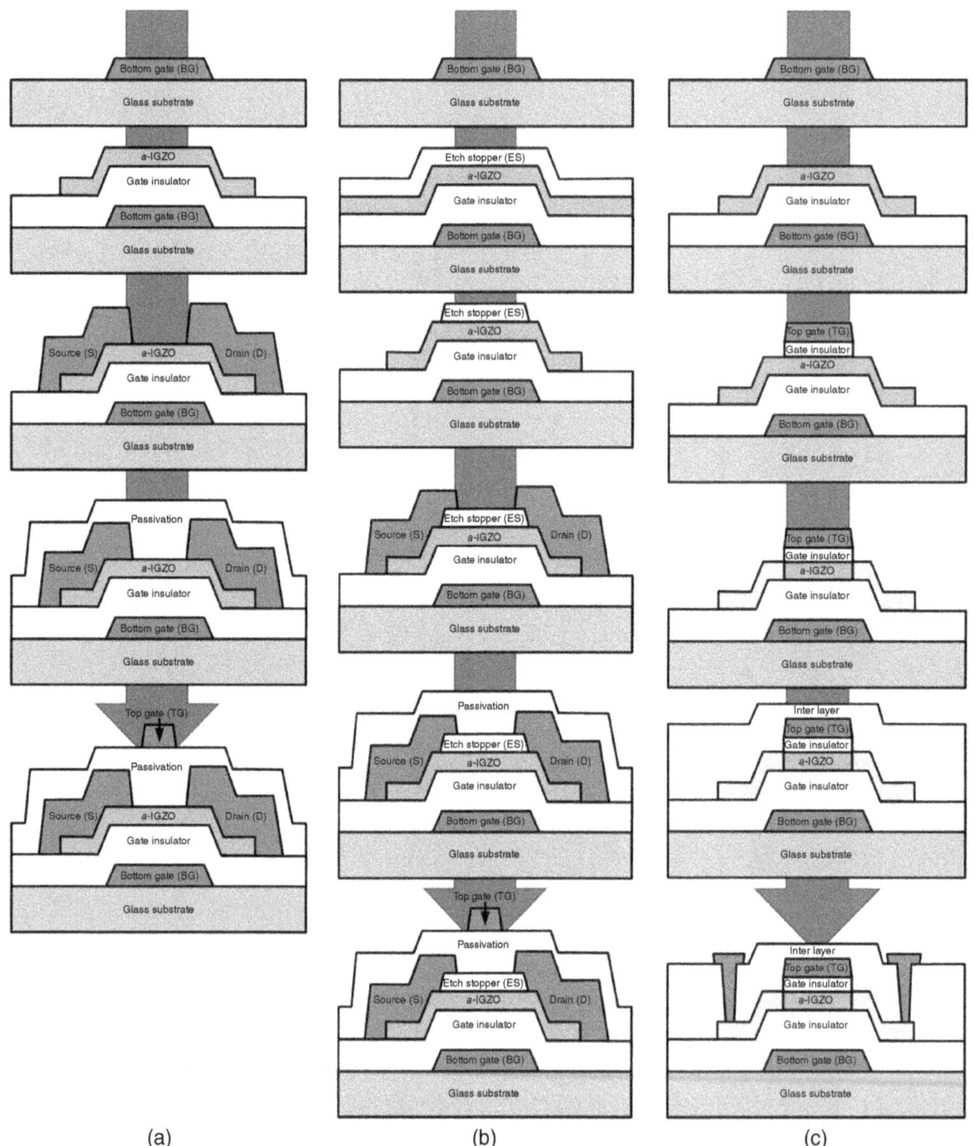

Figure 11.13 Fabrication process flow of oxide TFTs. Schematic cross-sectional views of the fabrication process of dual-gate TFTs with (a) back-channel-etched (BCE), (b) etch-stopper (ES), and (c) coplanar structures, respectively.

Figure 11.13. Figure 11.14 shows the transfer and output characteristics of the DG *a*-IGZO TFTs with BCE, ES, and coplanar structures, and a statistical summary is shown in Table 11.1.

The DG *a*-IGZO TFTs were measured under three gate-driving modes: BG driving (sweeping BG with a TG grounded), TG driving (sweeping TG with a BG grounded), and DG driving (synchronized sweep in which the TG is connected to the BG electrically). The SG driving can be measured with BG or TG driving. Compared to SG-driving TFTs, the benefits of DG driving of BA *a*-IGZO TFTs are higher drain currents (3–6 times higher than those of SG driving; Figure 11.15), turn-on voltage closer to 0 V, a smaller subthreshold voltage swing, and better

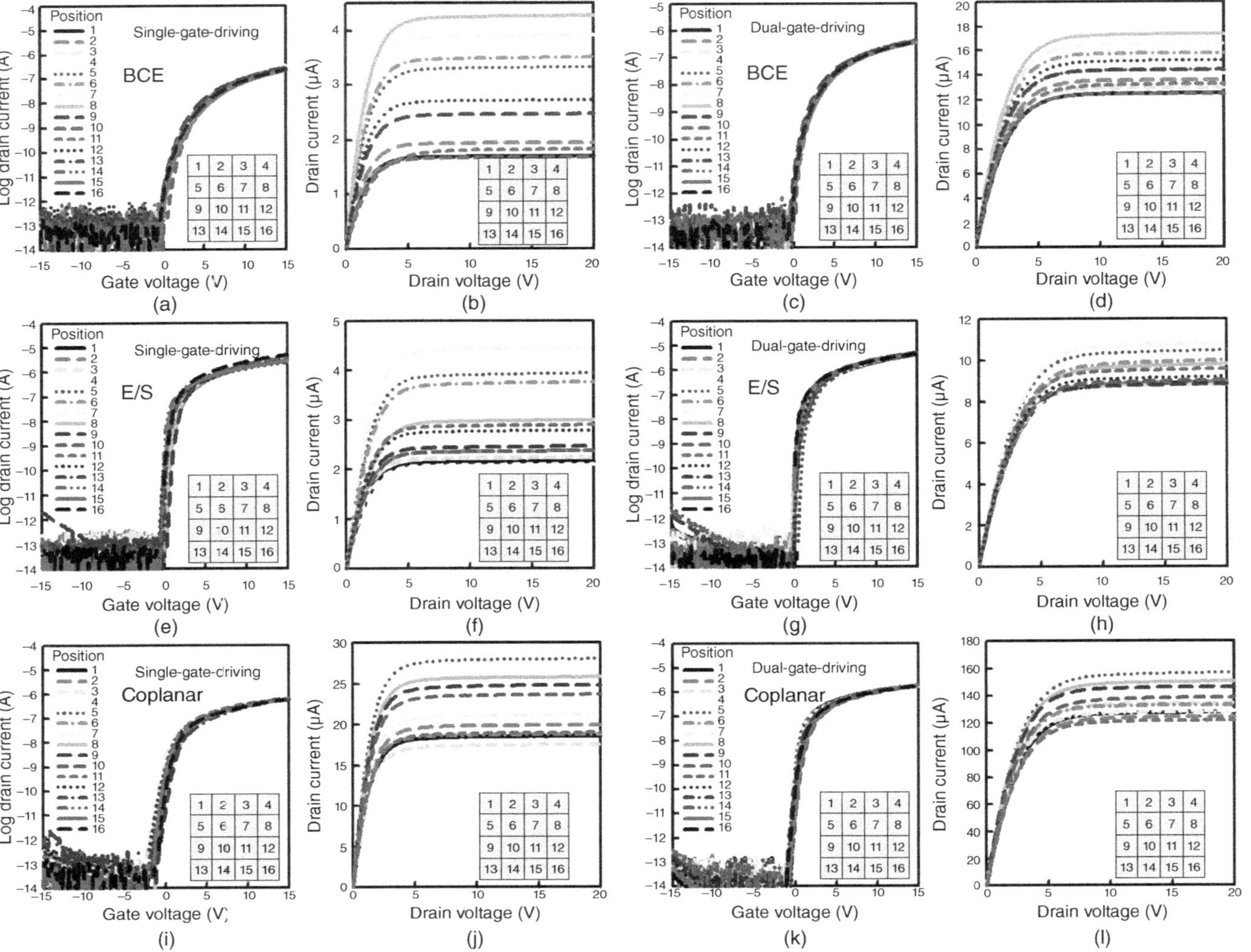

Figure 11.14 Uniformity of transfer and output characteristics for conventional single-gate and dual-gate TFTs. Transfer and output characteristics of BCE TFTs with (a,b) single-gate driving and (c,d) dual-gate driving, ES TFTs with (e,f) single-gate driving and (g,h) dual-gate driving, and coplanar TFTs with (i,j) single-gate driving and (k,l) dual-gate driving, respectively. 16 single-gate and 16 dual-gate *a*-IGZO TFTs on the 15 cm $\times$ 15 cm glass substrate (inset figures show the TFT positions). All TFTs have channel width (W) = 20 μm and length (L) = 15 μm (ADRC data).

Table 11.1 Drain currents and threshold voltages for *a*-IGZO TFTs with BCE, ES, and coplanar structures.

		V_{TH} (V)		I_{DS} (μA)	
TFT Structure		**Mean**	**Standard Deviation**	**Mean**	**Standard Deviation**
BCE (*a*-IGZO)	SG (BG driving)	0.84	0.24	2.61	0.93
	DG (DG driving)	0.63	0.12	14.32	1.56
ES (*a*-IGZO)	SG (BG driving)	0.28	0.32	2.88	0.78
	DG (DG driving)	0.16	0.21	9.44	0.70
Coplanar (*a*-IGZO)	SG (TG driving)	−0.61	−0.10	21.26	3.7
	DG (DG driving)	0.35	0.20	135.34	11.4

V_{TH} and I_{DS} denote the threshold voltage and the on current (V_{GS} = 10 V, V_{DS} = 20 V), respectively. The error bars indicate the standard deviation for 16 devices.

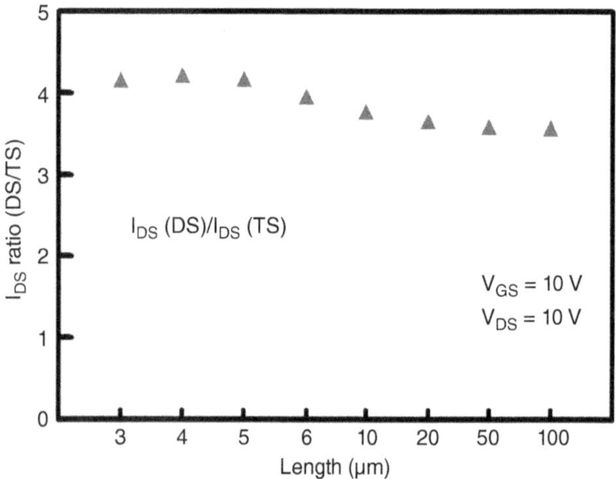

Figure 11.15 The ratio of drain currents of dual-gate driving to those of top-gate driving, or the I_{DS} ratio, is plotted for various channel lengths. Note that the drain currents at |V_{DS} = 10 V| and |V_{GS} = 10 V| are used to calculate the ratio.

device-to-device uniformity. These result from the fast filling-up of gap states as the Fermi level moves toward the conduction band. The requirements for the BA *a*-IGZO TFTs are a low density of states (DOS) in the gap of active-layer AOSs because the accumulation depth from the interface decreases with increasing gap state density, and also a low density of interface states with bottom- and top-gate insulators.

The transfer characteristics of the TFTs as a variation of channel length are shown in Figure 11.16: (a) SG TFT and (b–d) DG TFTs with (b) TG driving, (c) BG driving, and (d) DG driving. The channel resistance (R_{CH}) and external resistance (R_{EXT}) for single- and dual-gate TFTs could be achieved from the total resistances plotted as a function of channel length. The total resistance (R_{TOT}) achieved from the transfer characteristics at low drain voltage V_{DS} = 0.1 V, with various channel lengths, are shown in Figure 11.16e–h. Extracted resistance parameters of single- and dual-gate TFTs with various driving are shown in Table 11.2. The channel resistance of DG *a*-IGZO

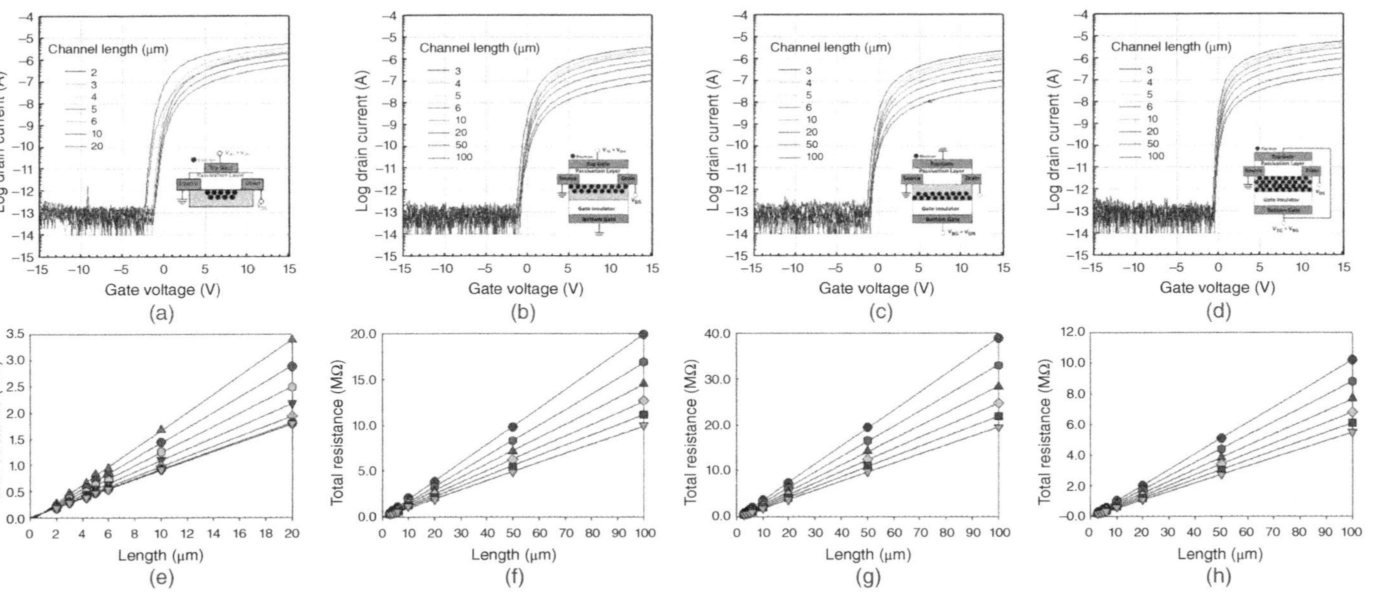

Figure 11.16 Channel length (L) dependency of *a*-IGZO TFTs. Drain current–voltage ($I_{DS}-V_{GS}$) characteristics with $V_{DS} = 0.1$ V for the TFTs with various channel lengths with a fixed channel width (W) for (a) a single-gate TFT and dual-gate TFTs with (b) top-gate driving, (c) bottom-gate driving, and (d) dual-gate driving. Linear regression of the extrinsic total resistance (R_{TOT}) as a function of the design channel length (L) for (e) single-gate TFTs and dual-gate TFTs with (f) TG driving, (g) BG driving, and (h) DG driving.

Table 11.2 Extracted resistance parameters of various TFTs.

TFT Structure	Gate Driving	R_{ch} (kΩ)/μm	R_{ext} (kΩ)	R_{par} (kΩ)	ΔL (μm)
SG	SG	8.87	1.75	9.97	0.99
DG	Top gate	9.95	0.24	9.72	0.99
DG	Bottom gate	19.7	−13.8	8.55	0.99
DG	Dual gate	5.45	1.73	6.22	0.99

The extrinsic total resistance (R_{TOT}), external resistance (R_{ext}), parasitic resistance (R_{par}), and ΔL for the single-gate TFTs, and dual-gate TFTs with TG driving, BG driving, and DG driving, respectively.

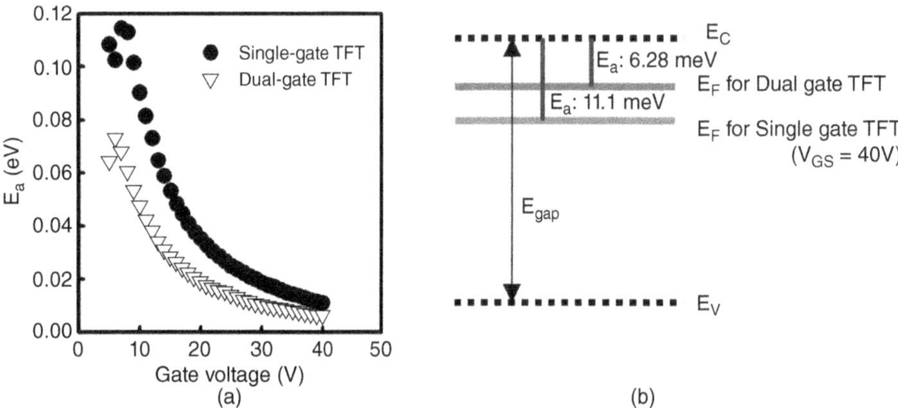

Figure 11.17 (a) The activation energy extracted from the slope in an Arrhenius plot by single-gate driving and dual-gate driving. (b) Energy-level diagram for the effective Fermi level with single-gate driving and dual-gate driving at $|V_{GS} = 40\ V|$.

TFTs with DG driving is much lower than that of *a*-IGZO TFTs with SG driving. This is due to the difference in carrier concentration and carrier mobility in DG-driving TFTs.

The activation energy of the drain currents extracted from the slope in an Arrhenius plot by SG driving and DG driving is shown in Figure 11.17. The Fermi levels for the TFT at $V_{GS} = 40$ V with SG and DG driving are found to be 11.1 meV and 6.28 meV below E_C, respectively. This clarifies that the conduction mechanism is through band transport with traps to the localized states. Figure 11.17b shows the energy-level diagram for the effective Fermi level with SG and dual-sweep driving at $V_{GS} = 40$ V.

To enhance the BA effect, the density of states in the gap (including the interface states) should be as low as possible. Better electrical characteristics are attributed to high gate drive and less carrier scattering at the interfaces. But still there are effects on the interface state density, which means that a requirement for the enhancement of BA TFTs is low density of interface states with bottom- and top-gate insulators. Another important condition is active-layer thickness, because the area density of the gap states increases with semiconductor thickness. The two channels formed at the back and front interfaces during DG driving can strongly overlap if the active layer is very thin and/or the density of states in the gap is low enough. This is achieved by employing a DG structure in which the semiconductor layer is thin (<50 nm) and the TG and BGs are electrically tied together, as shown in Figure 11.18.

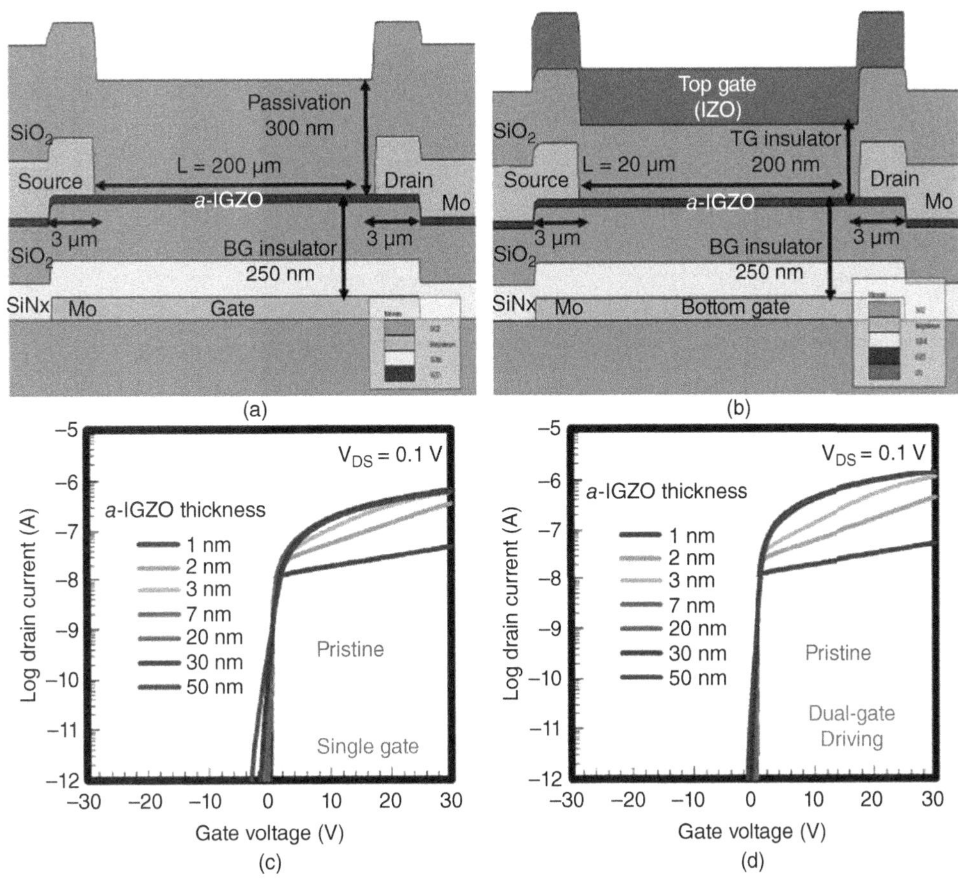

Figure 11.18 Schematic cross section of *a*-IGZO (a) single-gate and (b) dual-gate TFTs. Simulated pristine transfer characteristics of (c) single-gate and (d) dual-gate driving for various active-layer thicknesses [30]. *Source*: Billah, M. M., et al. (2016). Analysis of improved performance under negative bias illumination stress of dual gate driving a-IGZO TFT by TCAD simulation. *IEEE Electron Device Letters* **37** (6): 735–738.

11.5 Stability of Oxide TFTs: PBTS, NBIS, HCTS, Hysteresis, and Mechanical Strain

The results on the stability of DG *a*-IGZO TFTs under negative bias illumination stress (NBIS; Figure 11.19), hysteresis (Figure 11.20) and high current temperature stress (HCTS; Figure 11.20), and positive bias temperature stress (PBTS; Figure 11.21) are shown. Compared to SG-driving TFTs, bulk-accumulation TFTs show much better bias stability. A DG-driving TFT has less ΔV_{TH} (V) shift under stress time as compared with a SG-driving TFT because of higher vertical electric field and higher induced electron concentration, which rapidly shifts the quasi-Fermi level toward the conduction band in *a*-IGZO. Better stability is attributed to a high gate drive.

The *a*-IGZO TFT with DG driving exhibits better performance compared with SG driving under the mechanical strain. A TCAD simulation for the mechanical strain confirms 25% less strain at the center part of the *a*-IGZO channel for DG TFTs than for SG TFTs, and it can explain the experimental data for the strain dependence of

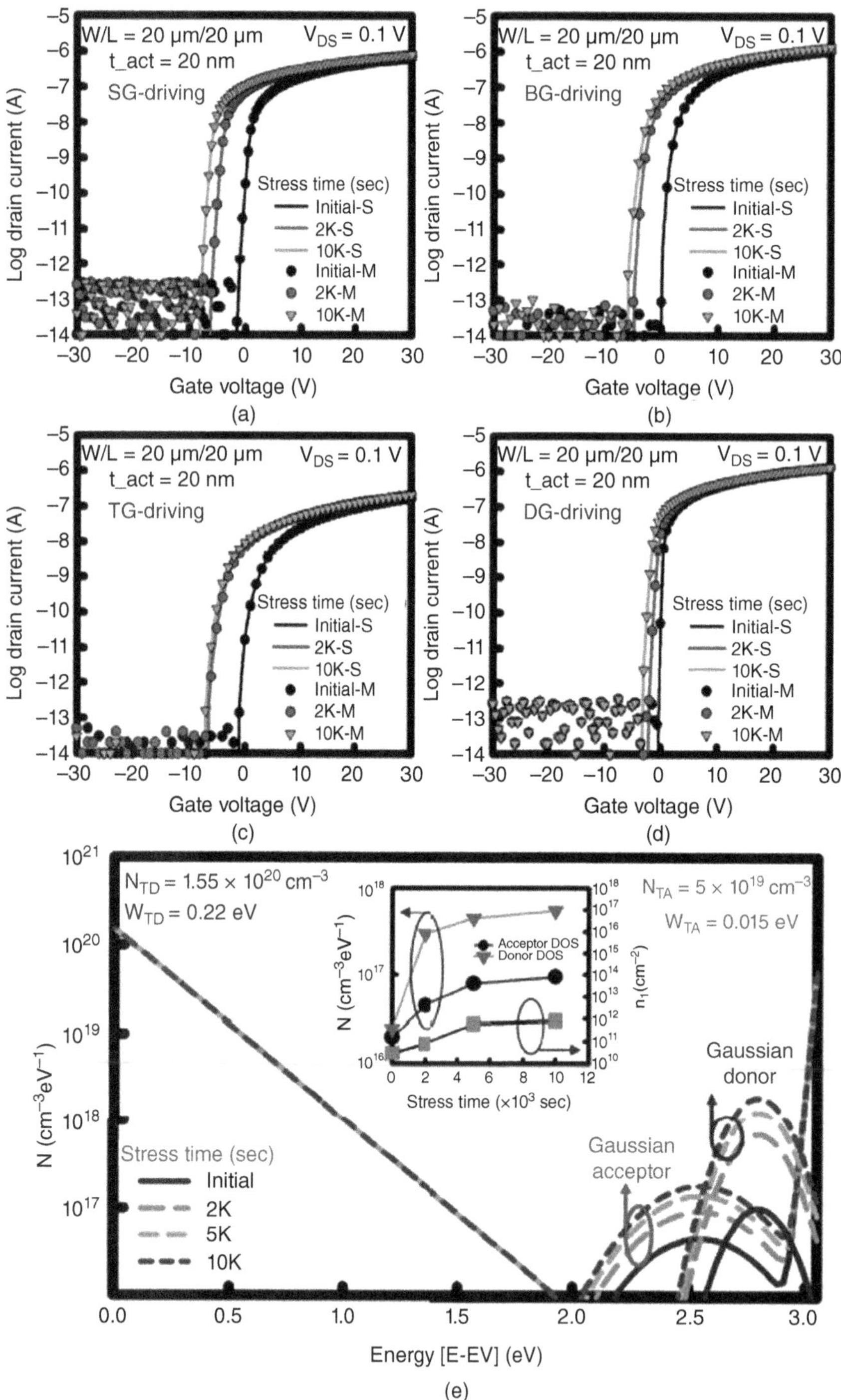

Figure 11.19 Effect of NBIS (negative bias illumination stress) on transfer characteristics of single-gate and dual-gate *a*-IGZO TFTs. (a) A SG (with one gate, only the BG) TFT. (b) A DG TFT driven by the BG, while grounding the TG. (c) A DG TFT driven by the TG, while grounding the BG. (d) A DG TFT with the TG and BG electrically tied together (DG driving). The SG and DG TFTs have *a*-IGZO thickness of 20 nm, and channel length (L) = 20 μm and width (W) = 20 μm. Symbols represent the measurement data (M), and solid lines are TCAD fitting (S). (e) TCAD DOS as a function of NBIS stress time [30]. *Source*: Billah, M. M., et al. (2016). Analysis of improved performance under negative bias illumination stress of dual gate driving a-IGZO TFT by TCAD simulation. *IEEE Electron Device Letters* **37** (6): 735–738.

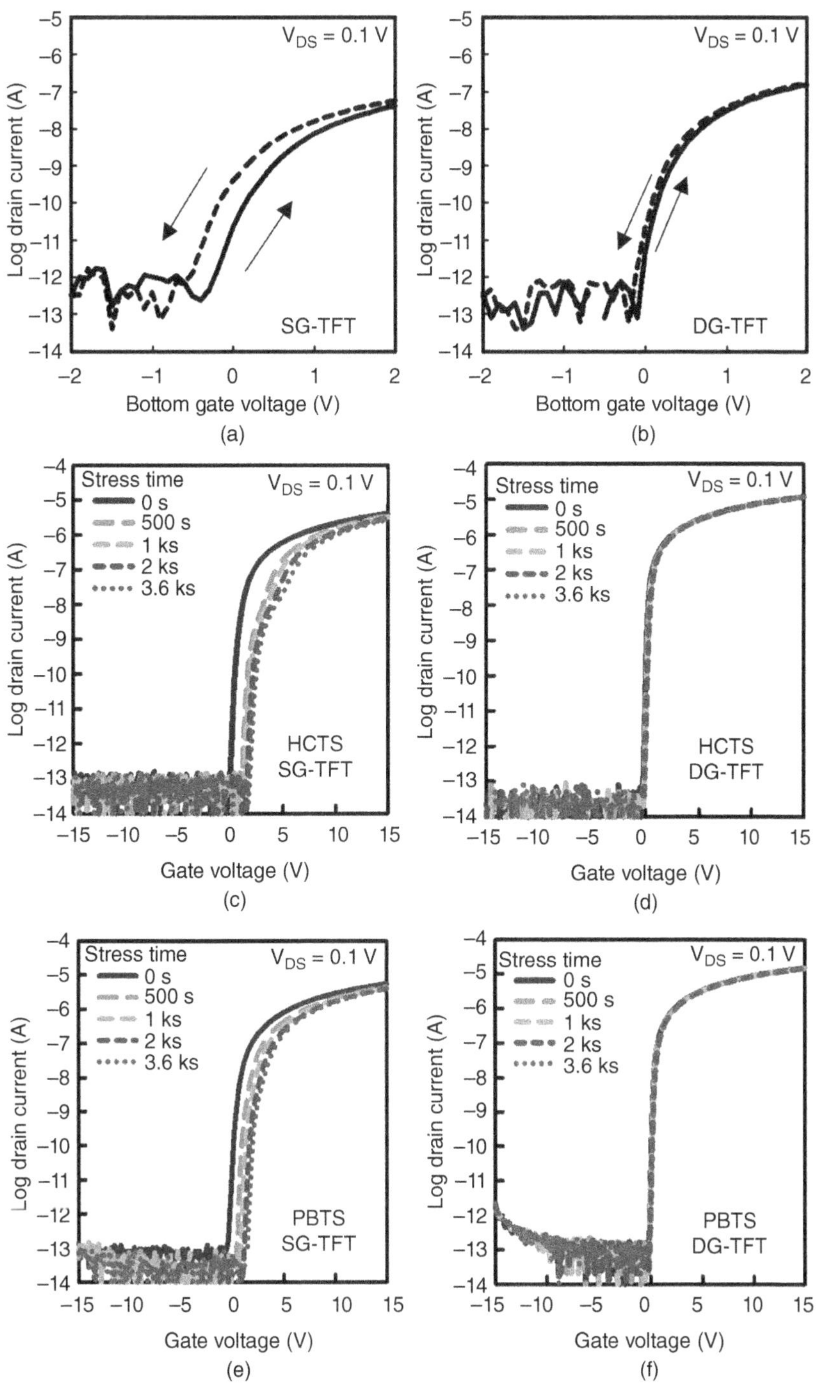

Figure 11.20 The hysteresis, high current temperature stress (HCTS), and positive bias temperature stress (PBTS) effects for the transfer curves of the single-gate and dual-gate *a*-IGZO TFTs. The hysteresis of transfer characteristics for TFTs with (a) single gate and (b) dual gate (gate voltage sweep from −2 V to +2 V to see the changes). The HCTS and PBTS of transfer characteristics for the TFTs with (c,e) single-gate and (d,f) dual-gate/dual-gate driving, respectively [31, 32]. *Source*: Hong, S., et al. (2014). Reduction of negative bias and light instability of a-IGZO TFTs by dual-gate driving. *IEEE Electron Device Letters* **35** (1): 93–95; and Lee, S., et al. (2014). Removal of negative-bias-illumination-stress instability in amorphous-InGaZnO thin-film transistors by top-gate offset structure. *IEEE Electron Device Letters* **35** (9): 930–932.

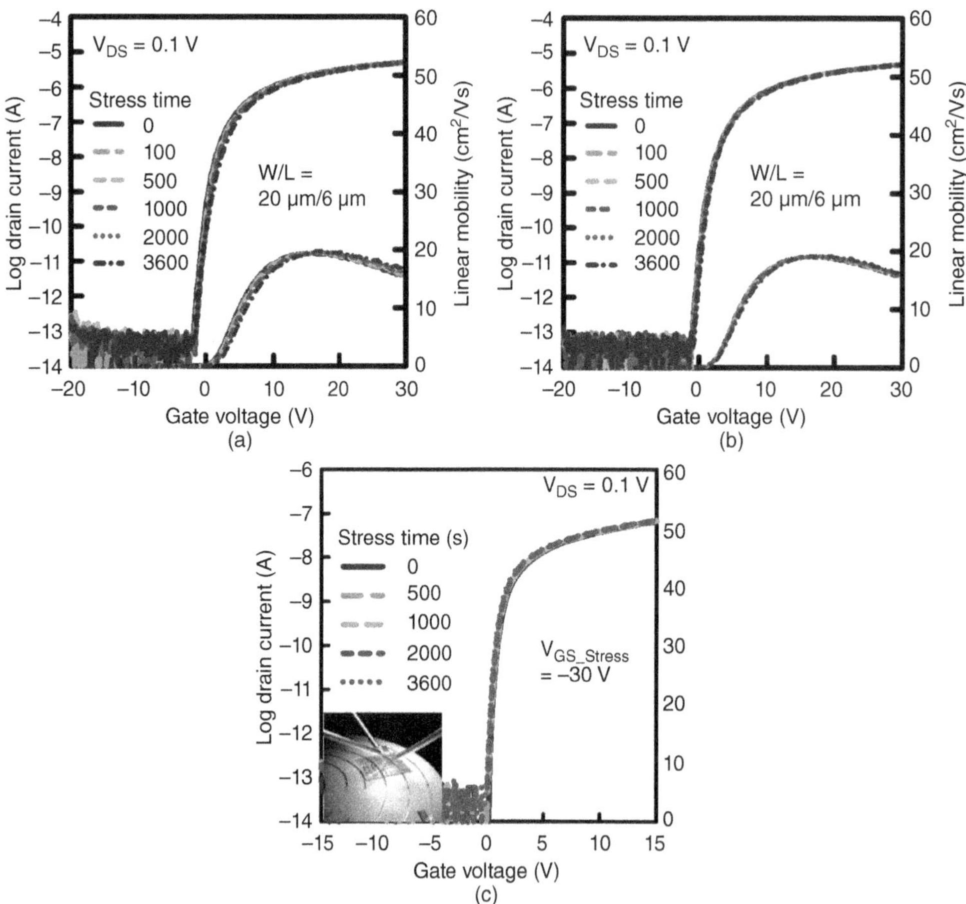

Figure 11.21 Device stability of dual-gate oxide TFTs with dual-gate driving as a variation of PBTS time, and NBIS time, respectively. The stress conditions of PBTS are $V_{GS} = +30$ V at (a) 80 °C and (b) 100 °C for 1 h. The stress condition of (c) NBIS is $V_{GS} = -20$ V under the white light, 10,000nit [31, 32]. *Source*: Hong, S., et al. (2014). Reduction of negative bias and light instability of a-IGZO TFTs by dual-gate driving. *IEEE Electron Device Letters* **35** (1): 93–95; and Lee, S., et al. (2014). Removal of negative-bias-illumination-stress instability in amorphous-InGaZnO thin-film transistors by top-gate offset structure. *IEEE Electron Device Letters* **35** (9): 930–932.

the transfer characteristics, as shown in Figures 11.22 and 11.23. In DG TFT structures, one additional gate can affect the mechanical strain effect. An additional TG metal with higher Young's modulus might be beneficial for the absorption of the applied strain by modulating the neutral plane (NP) position. From TCAD simulations, the immunity to slight variations in carrier concentration under mechanical strain is found to be a result of the high-gate-drive TFT with DG driving, as shown in Figure 11.24.

Note that the bottom and top interfaces are very different because the bottom SiO_2 is usually deposited at 300~420 °C, but the TG interface (SiO_2) is deposited at 150~200 °C. The IGZO can be damaged when the top SiO_2 is deposited at high temperatures of 300 °C to 400 °C. Therefore, the TG driving is significantly lower than the BG driving because the top interface state density is much higher compared to the bottom interface state. Note that the mass density of the gate insulator SiO_2 depends on the substrate temperature; lower mass density is deposited at lower temperature.

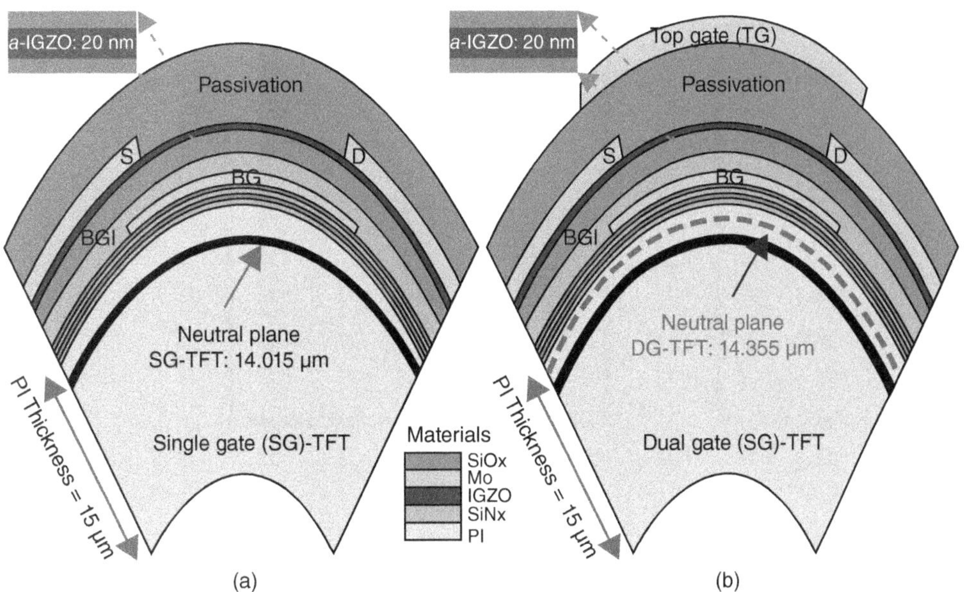

Figure 11.22 Schematic cross section of a flexible *a*-IGZO TFT on a bendable substrate using TCAD simulation. (a) Single-gate and (b) dual-gate *a*-IGZO TFTs with a bending radius, R (mm) [33, 34]. *Source*: Billah, M. M., et al. (2018). Reduced mechanical strain in bendable a-IGZO TFTs under dual gate driving. *IEEE Electron Device Letters* **36** (6): 835–838.

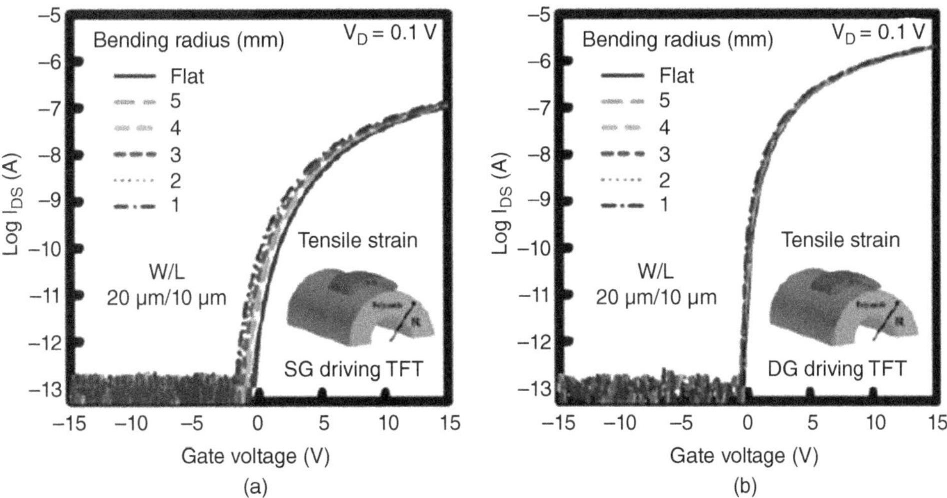

Figure 11.23 Evolution of transfer characteristics of flexible *a*-IGZO TFTs as a function of the bending radius (R) for (a) single-gate TFTs and (b) dual-gate driving TFTs, respectively. (c,d) The plots of transfer characteristics for flexible *a*-IGZO TFTs at pristine condition, and after 100, 500, and 1,000 bending cycles under 1.65% tensile strain (R = 0.48 mm, $\theta = 60°$) for SG and DG driving, respectively [33, 34]. *Source*: Billah, M. M., et al. (2018). Reduced mechanical strain in bendable a-IGZO TFTs under dual gate driving. *IEEE Electron Device Letters* **36** (6): 835–838.

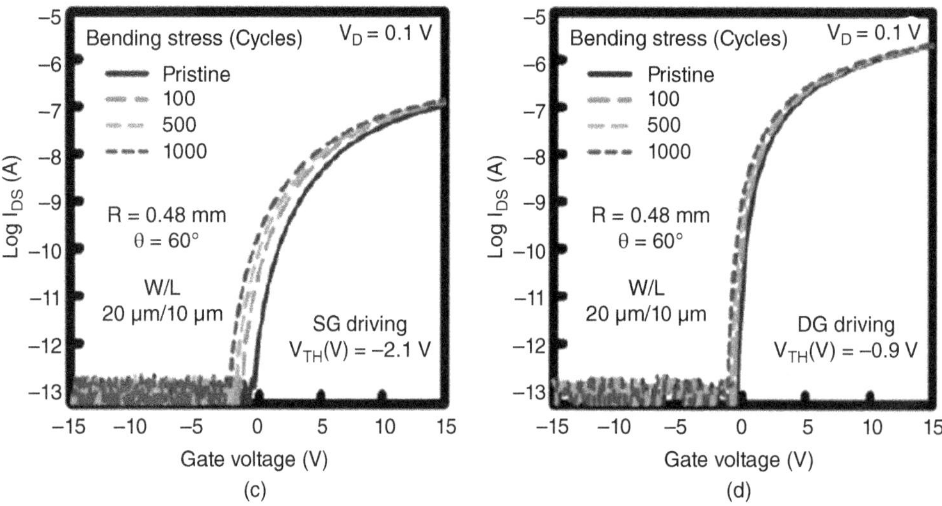

Figure 11.23 (*Continued*)

The excellent performance of split TFTs appears to be due to the improvement of the top interface by metal-F bonding and a subsequent decrease in the interface state density. Compared to conventional TFTs, the drain currents of the TG swing of split TFTs are ~30 times higher, as shown in Figure 11.25. Note that the current obtained by DG driving is much higher than the sum of currents of TG driving and BG driving due to the bulk-accumulation effect.

The quality of the interface may be improved by incorporating F, which can react with the metal ions located at the top interface, probably In and Zn atoms diffused into the top interface from the bulk IGZO. The F atoms

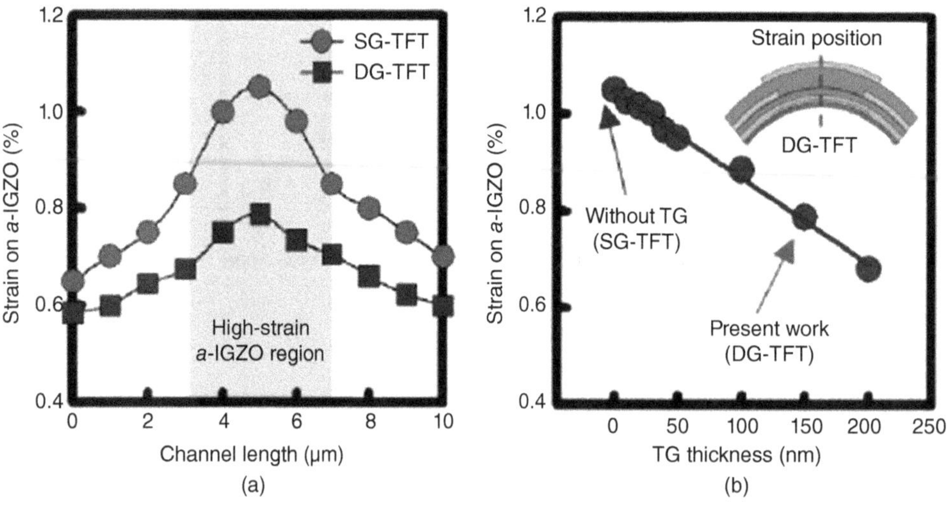

Figure 11.24 (a) Lateral strain distribution of single-gate and dual-gate *a*-IGZO TFTs under a bending radius (R) (corresponds to tensile strain vertical to the *a*-IGZO channel). (b) Influence of TG thickness (nm) on strain (%) in an *a*-IGZO layer associated with DG structure. Electron concentrations at the center of a TFT channel with respect to *a*-IGZO depth from the bottom GI for (d) SG and (d) DG, respectively [33]. *Source*: Billah, M. M., et al. (2018). Reduced mechanical strain in bendable a-IGZO TFTs under dual gate driving. *IEEE Electron Device Letters* **36** (6): 835–838.

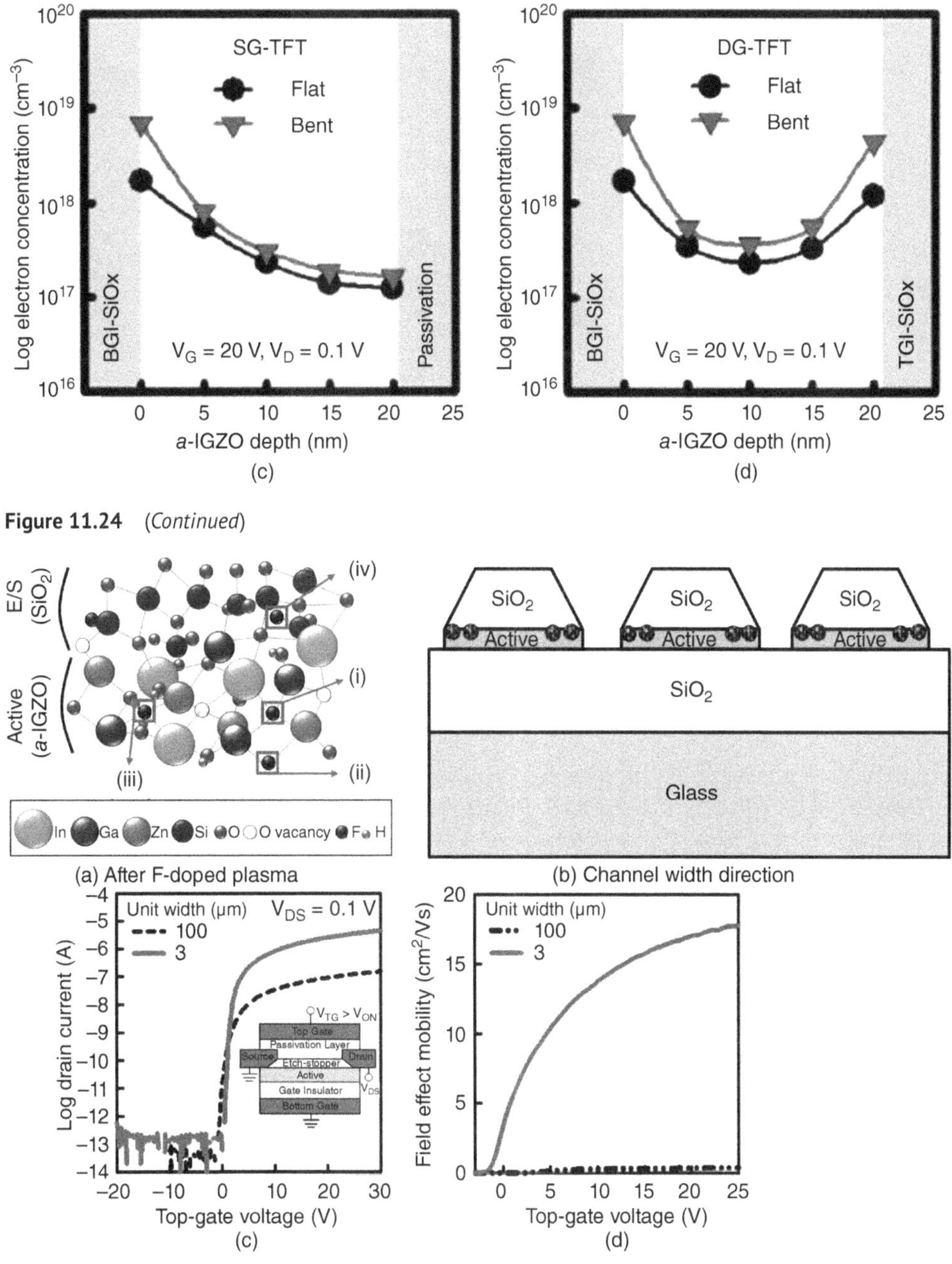

Figure 11.24 (Continued)

Figure 11.25 (a) Schematic of F doping in IGZO network: (i) substitution of oxygen lattice generating free electrons; (ii) occupy the oxygen vacancies reducing the electron trap; (iii) form hydrogen bond and passivation of hydroxyl group; (iv) fluorine ion passivates the defects (Si dangling bond) at the interface. (b) Cross-sectional views along the channel width direction for ES split-active TFTs. (c) Comparison of top-gate driving, transfer characteristics for the TFTs with and without an active layer split. Inset shows the schematic TFT cross-sectional view for TG driving with a grounded bottom gate. (d) Comparison of the mobility in the linear region for the two ES TFTs achieved from top-gate driving with a grounded bottom gate at V_{DS} = 0.1 V [43–46]. *Source*: (a,c,d) Lee, S., et al. (2017). Top interface engineering of flexible oxide thin-film transistors by splitting active layer. *Advanced Functional Materials* **27** (11): 1604921. (b) Adapted from Lee, S., et al. (2017). Top interface engineering of flexible oxide thin-film transistors by splitting active layer. *Advanced Functional Materials* **27** (11): 1604921; Lee, S., et al. (2018). Transparent AMOLED display driven by split oxide TFT backplane. *Journal of the SID* **26** (3): 164–168; Lee, S., et al. (2017). Highly robust oxide thin film transistors with split active semiconductor and source/drain electrodes. *2017 IEEE International Electron Devices Meeting (IEDM)* **2** (8.2.1): 187–190; and Lee, S., et al. (2019). Highly robust oxide TFT with bulk accumulation and source/drain/active layer splitting. *Journal of the SID* **27** (8): 507–513.

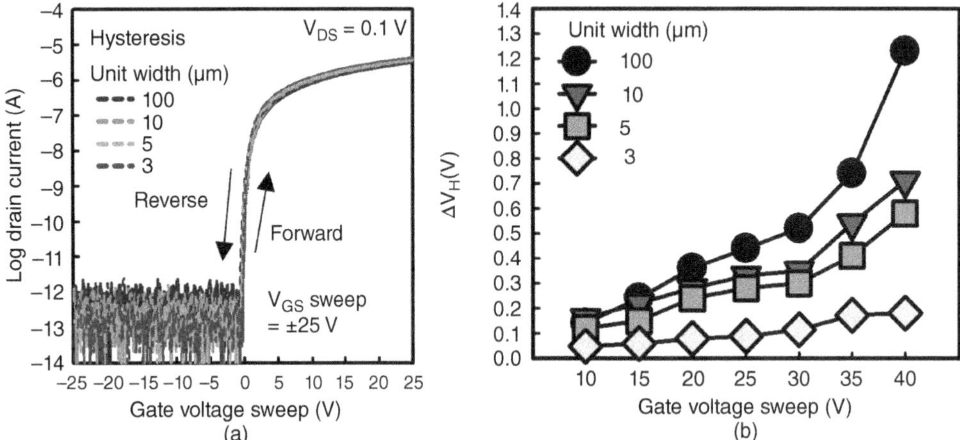

Figure 11.26 Hysteresis test for the analysis of device stability conducted for the *a*-IGZO TFTs with unit widths from 3 to 100 μm. The hysteresis of (a) transfer characteristics for a single-gate TFT from the gate voltage sweeps between the forward sweep (V_{GS} = −25 to 25 V) and reverse sweep (V_{GS} = 25 to −25 V). (b) The hysteresis voltages achieved from various gate voltage sweep ranges for the TFTs with different unit widths from 3 μm to 100 μm. (Gate voltage sweep from −10 V to +10 V and then to −10 V; from −20 V to +20 V and then to −20 V; and up to −40 V.) *Source*: [43] Lee, S., et al. (2017). Top interface engineering of flexible oxide thin-film transistors by splitting active layer. *Advanced Functional Materials* **27** (11): 1604921.

diffused to the top interface from the edge regions passivate the charge-trapping sites arisen from the diffused metal ions (In^+) by forming strong In-F (506 kJ mol^{-1}). Note that the dissociation energy of In-F (5.80 eV) is much stronger than that of In-O (3.58 eV). Also, Zn-F (3.81 eV) dissociation energy is higher than that of Zn-O (2.90 eV). This result indicates that the performance can be improved by F incorporation. This improvement is correlated with bonding of metal (In and Zn) with F at the *a*-IGZO top interface. The hysteresis is also remarkably improved, as shown in Figure 11.26.

11.6 TFT Circuits: Ring Oscillators and Amplifier Circuits

The advantage of DG TFTs with DG driving can be confirmed by comparing the speed of the ring oscillators (ROs) made of SG and DG oxide TFTs. We confirmed the advantage of DG TFTs through a comparison of the TFT circuits made of SG and DG TFTs.

Figure 11.27 shows the RO performance with SG and DG driving. Typical output waveforms of SG-driving and DG-driving ROs are shown in Figure 11.27 for V_{DD} = 20 V. The oscillation frequencies are ~334.1 and 780.6 kHz, respectively, by using SG driving and DG driving. The RO implemented with BA driving of TFTs oscillates at a higher frequency. This is mainly due to two reasons: The drain current under DG driving is higher than that under SG driving, and there is better uniformity under DG driving (BA mode).

The speed of *a*-IGZO TFT-based circuits can be further enhanced by proper circuit design. Figure 11.28 shows a proposed pseudo-CMOS RO. The advantage of the pseudo-CMOS circuits compared to the ratioed inverters can be seen in Ref. [38]. Figure 11.28e shows the V_{DD} frequency dependency, including the comparison between pseudo-CMOS-type ROs using SG and BA TFTs with L = 10 or 6 μm. A RO with an L = 6 μm shows faster speed, but both results for BA TFTs show much higher speed compared to those of a SG TFT-based RO. Therefore, proves that due to the better switching characteristics of BA TFTs, which include higher on-state TFT current and lower switching speed, the switching speed of pseudo-CMOS circuits can also be improved by the implementation of BA TFTs.

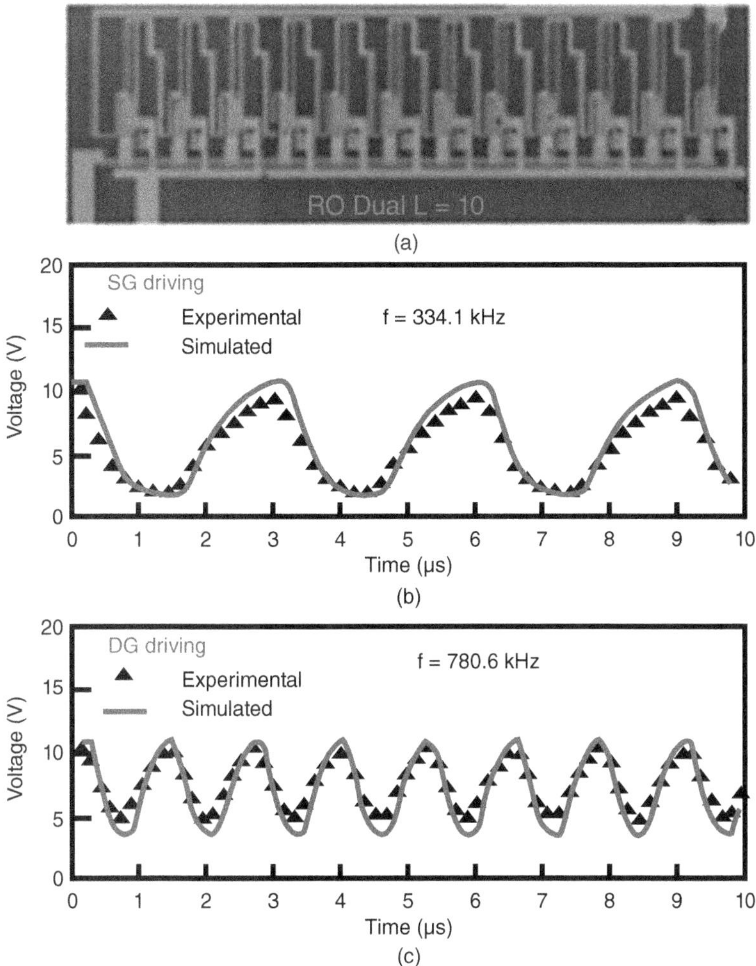

Figure 11.27 Ring oscillator (RO) performance. (a) Optical image of a fabricated DG-driving RO. (b,c) Typical output waveform of a (b) SG-driving RO and (c) DG-driving RO. Supply voltage V_{DD} was 20 V [36, 37]. *Source*: Adapted from Li, X., et al. (2014). Effect of bulk-accumulation on switching speed of dual-gate a-IGZO TFT based circuits. *IEEE Electron Device Letters* **35** (12): 1242–1244, Figure 03, (p 03) / with permission of IEEE; and Lee, S., et al. (2015). Bulk-accumulation oxide thin-film transistor circuits with zero gate-to-drain overlap capacitance for high speed. *IEEE Electron Device Letters* **36** (12): 1329–1331.

We confirmed the advantage of DG TFTs through a comparison of the *a*-IGZO TFT with an operational amplifier (op-amp) composed with the coplanar structures made of SG and DG TFTs. The circuit consisted of 19 TFTs and was designed on a glass substrate in both DG and SG structure for performance evaluation, as shown in Figure 11.29.

Having the yield of a total voltage gain (Av) of 23.5 dB, a cutoff frequency (fc) of 500 kHz, a unit gain frequency (fug) of 2.37 MHz, a gain-bandwidth product (GBWP) of 7500 kHz, a slew rate (up/down) of (2.1/1.2) V/μs, and a phase margin (PM) of 102° at a supply voltage of ±10 V, the fabricated DG TFT op-amp demonstrates good performance among all *a*-IGZO-based literature, as shown in Figure 11.30 [42].

The DG coplanar op-amp stands as the best performer among the *a*-IGZO op-amp circuits. As the coplanar DG structure is promising for high-speed circuits, this op-amp has the potential to keep pace with the demand for faster operation.

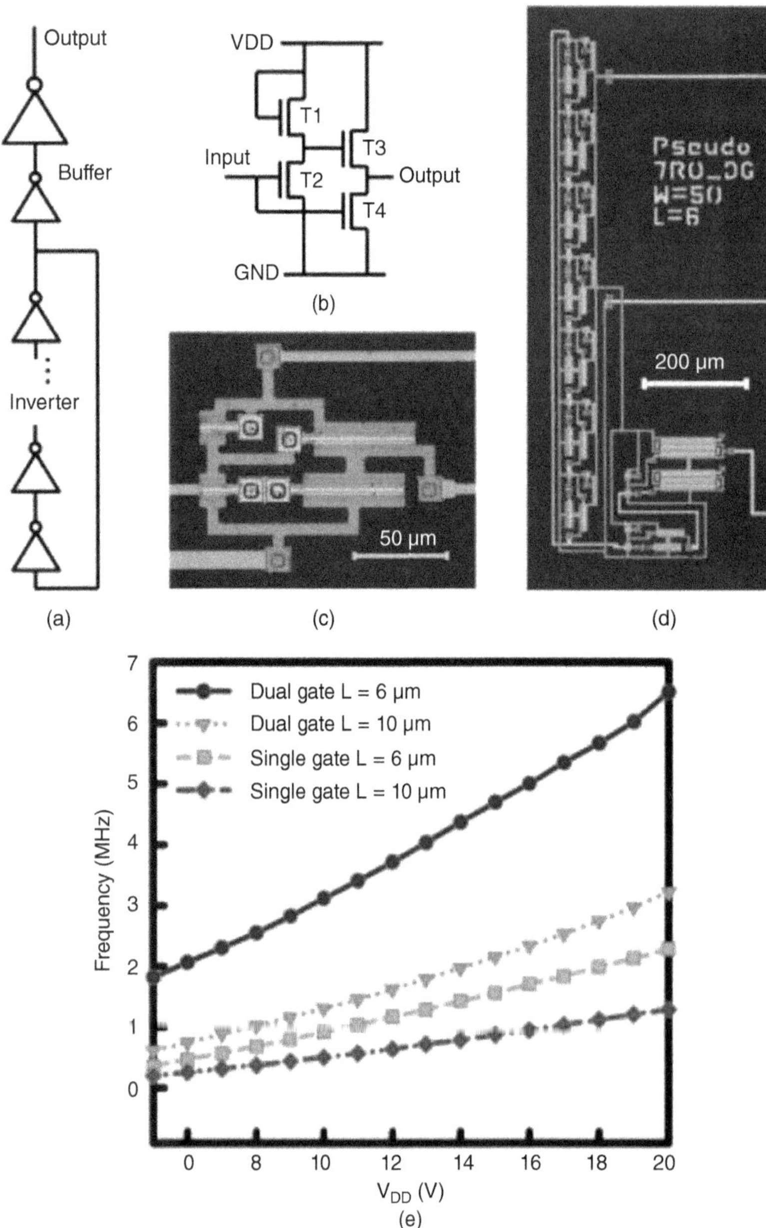

Figure 11.28 Pseudo-CMOS circuit design. (a) Equivalent circuit schematic of a 7-stage ring oscillator with an output buffer. (b) Circuit schematic of a pseudo-CMOS inverter. Optical image of a fabricated (c) one inverter stage and (d) 7-stage ring oscillator. (e) Output frequency dependency on supply voltage V_{DD} for single-gate and dual-gate (bulk-accumulation) pseudo-CMOS ring oscillators, with channel lengths (L) of 6 and 10 μm, respectively [38, 39]. *Source*: Chen, Y., et al. (2015). High-speed pseudo-CMOS circuits using bulk accumulation a-IGZO TFTs. *IEEE Electron Device Letters* **36** (2): 153–155, Figure 02, p. 02 / with permission of IEEE.

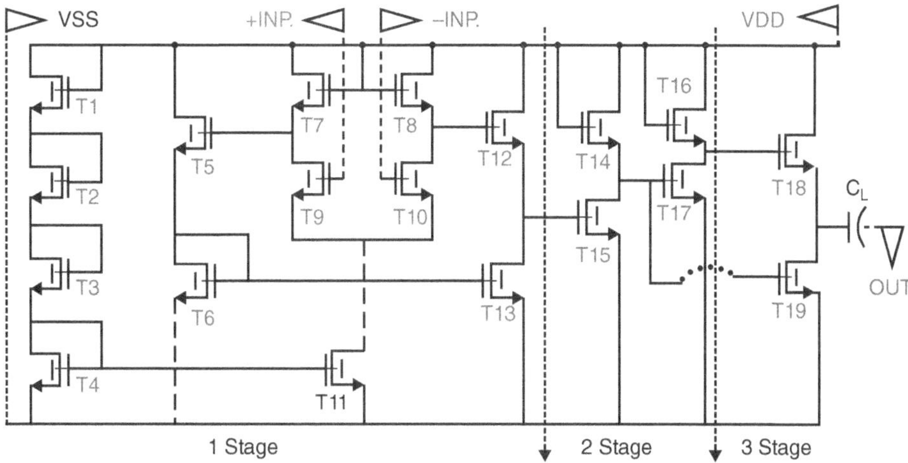

Figure 11.29 The op-amp circuit schematic, which can be subdivided in three stages. The first stage comprises the input terminal, current mirror, and single-ended converter stage. The second stage ensures proper amplification, and the third stage works as an output buffer [42]. *Source*: Rahaman, A., et al. (2019). A high performance operational amplifier using coplanar dual gate a-IGZO TFTs. *Journal of the IEEE Electron Devices Society* **7** (7): 655–661.

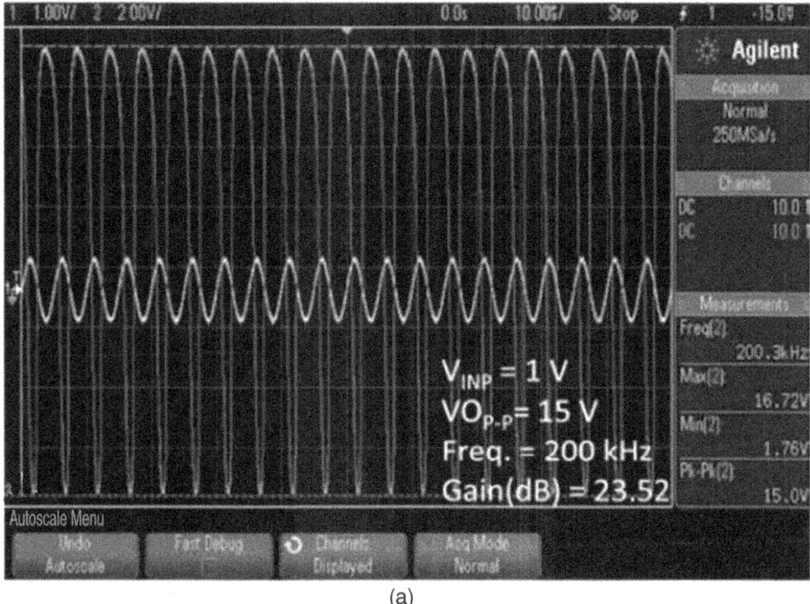

(a)

Figure 11.30 (a) The measured input and output waveforms for the frequency of 200 kHz. Gain and phase response with frequency is plotted in (b). Cutoff (fc) and unit gain frequency (fug) of DG TFT op-amps are 500 kHz and 2.37 MHz, while those for SG are 350 kHz and 1.6 MHz, respectively. It is clear that the DG TFT op-amp is almost 1.5 times broader in the aspect of fc and fug. Phase response is similar for both types with a phase margin (PM) of 102°. Bode plots show the distinguishable superiority of the DG TFT op-amp over the SG TFT op-amp [42]. *Source*: Rahaman, A., et al. (2019). A high performance operational amplifier using coplanar dual gate a-IGZO TFTs. *Journal of the IEEE Electron Devices Society* **7** (7): 655–661, [42]. Figure 04, p. 04 / CC BY 4.0.

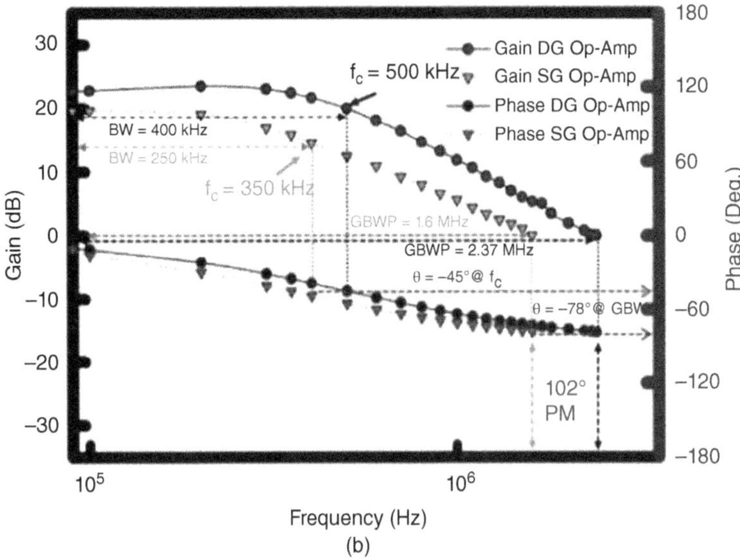

Figure 11.30 *(Continued)*

11.7 Conclusion

Carrier concentration in AOS is a very important factor to decide the field-effect mobility and threshold voltage of the TFT. The field-effect mobility of AOS TFTs increases with increasing carrier concentration because of the reduction of the potential barriers for carrier transport. We report much experimental evidence of the advantages of BA-mode TFTs, which can be possible by connecting top and bottom gates to have high carrier concentration in TFT channels in dual-gate structure. With *a*-IGZO channel layers in dual-gate TFTs, the drain currents can be 3–6 times that of single-gate TFTs. This is confirmed in coplanar, ES, and BCE of inverted staggered *a*-IGZO TFTs. The advantages of BA TFTs could be further confirmed from the experimental results on RO, SR, and amplifier circuits.

References

1 Labram, J.G., Lin, Y., and Anthopoulos, T.D. (2015). Exploring two-dimensional transport phenomena in metal oxide heterointerfaces for next-generation, high-performance, thin-film transistor technologies. *Small* 11 (41): 5472–5482.

2 Wang, Z.L. (2012). Self-powered nanosensors and nanosystems. *Advanced Materials* 24 (2): 280–285.

3 Hosono, H., Yasukawa, M., and Kawazoe, H. (1996). Novel oxide amorphous semiconductors: transparent conducting amorphous oxides. *Journal of Non-Crystalline Solids* 203 (1): 334–344.

4 Gelinck, B.G., Heremans, P., Nomoto, K., and Anthopoulos, T.D. (2010). Organic transistors in optical displays and microelectronic applications. *Advanced Materials* 22 (34): 3778–3798.

5 Lin, C. and Chen, Y. (2007). A novel LTPS-TFT pixel circuit compensating for TFT threshold-voltage shift and OLED degradation for AMOLED. *IEEE Electron Device Letters* 28 (2): 129–131.

6 Zawa, M.O. and Atsumura, M.M. (1998). A novel phase-modulated excimer-laser crystallization method of silicon thin films. *Japanese Journal of Applied Physics* 37: 492–495.

7 Liu, B.J., Buchholz, D.B., Chang, R.P.H. et al. (2010). High-performance flexible transparent thin-film transistors using a hybrid gate dielectric and an amorphous zinc indium tin oxide channel. *Advanced Materials* 22 (21): 2333–2337.

8 Yu, X., Zeng, L., Zhou, N. et al. (2015). Ultra-flexible, "invisible" thin-film transistors enabled by amorphous metal oxide/polymer channel layer blends. *Advanced Materials* 27 (14): 2390–2399.

9 Kim, M., Kanatzidis, M.G., Facchetti, A., and Marks, T.J. (2011). Low-temperature fabrication of high-performance metal oxide thin-film electronics via combustion processing. *Nature Materials* 10: 382–388.

10 Hwang, Y.H., Seo, J., Yun, J.M. et al. (2013). Aqueous route for the fabrication of low-temperature-processable oxide flexible transparent thin-film transistors on plastic substrates. *NPG Asia Materials* 5: e45–e48.

11 Arai, T. and Sasaoka, T. (2011). Emergent oxide TFT technologies for next-generation AMOLED displays. *SID International Symposium Digest of Technical Papers* 42 (1): 710–713.

12 Lin, Y., Faber, H., Labram, J.G. et al. (2015). High electron mobility thin-film transistors based on solution-processed semiconducting metal oxide heterojunctions and quasi-superlattices. *Advanced Science* 2 (7): 1–12.

13 Rim, Y.S., Chen, H., Kou, X. et al. (2014). Boost up mobility of solution-processed metal oxide thin-film transistors via confining structure on electron pathways. *Advanced Materials* 26 (25): 4273–4278.

14 Shih, T., Ting, H., Chen, C. et al. (2013). A 56-inch high mobility metal oxide thin film transistors active-matrix organic light-emitting diode television. In. In: *IEEE Active-Matrix Flatpanel Displays and Devices (AM-FPD), 2014 21st International Workshop on 2013,*, 913–914.

15 Hosono, H. (2006). Ionic amorphous oxide semiconductors: material design, carrier transport, and device application. *Journal of Non-Crystalline Solids* 352: 851–858.

16 Kamiya, T., Nomura, K., and Hosono, H. (2009). Origins of high mobility and low operation voltage of amorphous oxide TFTs: electronic structure, electron transport, defects and doping. *Journal of Display Technology* 5 (7): 273–288.

17 Nomura, K., Kamiya, T., Ohta, H. et al. (2008). Relationship between non-localized tail states and carrier transport in amorphous oxide semiconductor, In–Ga–Zn–O. *Physica Status Solidi (A): Applications and Materials Science* 205 (8): 1910–1914.

18 Liu, P., Chen, T.P., Li, X.D. et al. (2013). Effect of exposure to ultraviolet-activated oxygen on the electrical characteristics of amorphous indium gallium zinc oxide thin film transistors. *ECS Solid State Letters* 2 (4): Q21–Q24.

19 Ramanathan, S., Wilk, G.D., Muller, D.A. et al. (2001). Growth and characterization of ultrathin zirconia dielectrics grown by ultraviolet ozone oxidation. *Applied Physics Letters* 79 (16): 2621–2623.

20 Song, A., Park, H.-W., Chung, K.-B. et al. (2017). Modulation of the electrical properties in amorphous indium-gallium zinc-oxide semiconductor films using hydrogen incorporation. *Applied Physics Letters* 111 (24): 243507.

21 Um, J.G. and Jang, J. (2018). Heavily doped n-type a-IGZO by F plasma treatment and its thermal stability up to 600 °C. *Applied Physics Letters* 112: 162104.

22 Jiang, J., Furuta, M., and Wang, D. (2014). Self-aligned bottom-gate In–Ga–Zn–O thin-film transistor with source/drain regions formed by direct deposition of fluorinated silicon nitride. *IEEE Electron Device Letters* 35 (9): 933–935.

23 Jeong, H.Y., Lee, B.Y., Lee, Y.J. et al. (2014). Coplanar amorphous-indium-gallium-zinc-oxide thin film transistor with He plasma treated heavily doped layer. *Applied Physics Letters* 104 (2): 022115.

24 Mativenga, M., An, S., and Jang, J. (2013). Bulk accumulation a-IGZO TFT for high current and turn-on voltage uniformity. *IEEE Electron Device Letters* 34 (12): 1533–1535.

25 Jang, J. (2016). Bulk-accumulation oxide-TFT backplane technology for flexible and rollable AMOLEDs: part I. *Information Display* 2 (16): 1–7.

26 Seok, M.J., Choi, M.H., Mativenga, M. et al. (2011). A full-swing a-IGZO TFT-based inverter with a top-gate-bias-induced depletion load. *IEEE Electron Device Letters* 32 (8): 1089–1091.

27 Chun, M., Chowdhury, M.D.H., and Jang, J. (2015). Semiconductor to metallic transition in bulk accumulated amorphous indium-gallium-zinc-oxide dual gate thin-film transistor. *AIP Advances* 5 (5): 057165.

28 Park, S.H., Mativenga, M., and Jang, J. (2013). Increase of mobility in dual gate amorphous-InGaZnO4 thin-film transistors by pseudo-doping. *Applied Physics Letters* 103 (4): 043509.

29 Billah, M.M., Hasan, M.M., and Jang, J. (2017). Effect of tensile and compressive bending stress on electrical performance of flexible a-IGZO TFTs. *IEEE Electron Device Letters* 38 (7): 890–893.

30 Billah, M.M., Chowdhury, M.D.H., Mativenga, M. et al. (2016). Analysis of improved performance under negative bias illumination stress of dual gate driving a-IGZO TFT by TCAD simulation. *IEEE Electron Device Letters* 37 (6): 735–738.

31 Hong, S., Lee, S., Mativenga, M., and Jang, J. (2014). Reduction of negative bias and light instability of a-IGZO TFTs by dual-gate driving. *IEEE Electron Device Letters* 35 (1): 93–95.

32 Lee, S., Mativenga, M., and Jang, J. (2014). Removal of negative-bias-illumination-stress instability in amorphous-InGaZnO thin-film transistors by top-gate offset structure. *IEEE Electron Device Letters* 35 (9): 930–932.

33 Billah, M.M., Han, J.-U., Hasan, M.M., and Jang, J. (2018). Reduced mechanical strain in bendable a-IGZO TFTs under dual gate driving. *IEEE Electron Device Letters* 36 (6): 835–838.

34 Li, X., Billah, M.M., Mativenga, M. et al. (2015). Highly robust flexible oxide thin-film transistors by bulk accumulation. *IEEE Electron Device Letters* 36 (8): 811–813.

35 Li, X., Geng, D., Mativenga, M., and Jang, J. (2014). High-speed dual-gate a-IGZO TFT-based circuits with top-gate offset structure. *IEEE Electron Device Letters* 35 (4): 461–463.

36 Li, X., Geng, D., Mativenga, M. et al. (2014). Effect of bulk-accumulation on switching speed of dual-gate a-IGZO TFT based circuits. *IEEE Electron Device Letters* 35 (12): 1242–1244.

37 Lee, S., Li, X., Mativenga, M., and Jang, J. (2015). Bulk-accumulation oxide thin-film transistor circuits with zero gate-to-drain overlap capacitance for high speed. *IEEE Electron Device Letters* 36 (12): 1329–1331.

38 Chen, Y., Geng, D., Mativenga, M. et al. (2015). High-speed pseudo-CMOS circuits using bulk accumulation a-IGZO TFTs. *IEEE Electron Device Letters* 36 (2): 153–155.

39 Chen, Y., Geng, D., Lin, T. et al. (2016). Full-swing clock generating circuits on plastic using a-IGZO dual-gate TFTs with pseudo-CMOS and bootstrapping. *IEEE Electron Device Letters* 37 (7): 882–885.

40 Geng, D., Chen, Y.F., Mativenga, M., and Jang, J. (2015). 30-μm-pitch oxide-TFT-based gate-driver design for small-size, high-resolution, and narrow-bezel displays. *IEEE Electron Device Letters* 36 (8): 805–807.

41 Rahaman, A., Billah, M.M., Um, J.G. et al. (2018). Effect of doping fluorine in offset region on performance of coplanar a-IGZO TFTs. *IEEE Electron Device Letters* 39 (9): 1318–1321.

42 Rahaman, A., Chen, Y., Hasan, M.M., and Jang, J. (2019). A high performance operational amplifier using coplanar dual gate a-IGZO TFTs. *Journal of the IEEE Electron Devices Society* 7 (7): 655–661.

43 Lee, S., Shin, J., and Jang, J. (2017). Top interface engineering of flexible oxide thin-film transistors by splitting active layer. *Advanced Functional Materials* 27 (11): 1604921.

44 Lee, S., Chen, Y., Kim, J. et al. (2018). Transparent AMOLED display driven by split oxide TFT backplane. *Journal of the SID* 26 (3): 164–168.

45 Lee, S., Geng, D., Li, L. et al. (2017). Highly robust oxide thin film transistors with split active semiconductor and source/drain electrodes. *2017 IEEE International Electron Devices Meeting (IEDM)* 2 (8.2.1): 187–190.

46 Lee, S., Chen, Y., Kim, J. et al. (2019). Highly robust oxide TFT with bulk accumulation and source/drain/active layer splitting. *Journal of the SID* 27 (8): 507–513.

12

Elevated-Metal Metal-Oxide Thin-Film Transistors: A Back-Gate Transistor Architecture with Annealing-Induced Source/Drain Regions

Man Wong[1,2], *Zhihe Xia*[1,2], *and Jiapeng Li*[1,2]

[1] *Department of Electronic and Computer Engineering, The Hong Kong University of Science and Technology, Hong Kong, China*
[2] *State Key Laboratory of Advanced Displays and Optoelectronics Technologies, The Hong Kong University of Science and Technology, Hong Kong, China*

12.1 Introduction

A thin-film transistor (TFT) is a field-effect device built on a semiconducting thin film. With a patent issued to Lilienfeld in 1930, its invention predated by almost two decades that of the point-contact transistor by Bardeen and Brattain in 1947 at Bell Telephone Laboratories. Emerging later from the same laboratories were the bipolar junction transistor and metal-oxide semiconductor field-effect transistor (MOSFET), both built on a bulk single-crystal semiconducting substrate. The contrast between the thin-film and bulk-type devices is significant in terms of both performance and technology. Benefiting from the better material quality of the single-crystal substrate and the compatibility of the substrate with high-temperature processing, a bulk-type device exhibits much better electrical attributes. It is the combination of the monolithic integration of transistors and the relentless drive of Moore's law that is responsible for the ubiquitous incorporation of these devices in almost all things electronic. Today, it is hard to imagine life without such electronic gadgets, with the smartphone being a prime example.

Often prepared on a noncrystalline heterogeneous substrate, the semiconducting thin film for a TFT is typically amorphous or polycrystalline – and hence more "defective" than the same semiconductor prepared in its single-crystalline form. The poorer material quality, coupled with a process temperature capped by the lower thermal tolerance of commonly used substrates, gives rise to relatively degraded device characteristics. It is not surprising that TFTs, despite their earlier debut, lost out in competition with bulk-type transistors for computationally intensive applications requiring "fast-switching" devices. A solution seeking a problem, its large-scale commercial deployment would wait for the eventual arrival of a "killer app" that came in the form of an active-matrix liquid crystal display (AM-LCD).

An AM-LCD consists of a regular array of voltage-driven liquid crystal (LC) pixels, and the optical state of each is controlled by an electrical signal regulated using a TFT. The TFTs are dispersed over a "mother glass," on which multiple display units are made. Although the relatively low glass-transition temperature of commonly used amorphous substrate limits the process temperature, and hence also the device performance, the control of a pixel does not demand a fast-switching TFT. Among a range of candidate materials, hydrogenated amorphous silicon (*a*-Si:H) prepared by plasma-enhanced chemical vapor deposition (PECVD) offers an optimal combination of performance and technology that fits the purpose. For higher resolution mobile displays and for displays based on current-driven organic light-emitting diodes (OLEDs), *a*-Si:H has been replaced by low-temperature polycrystalline silicon (LTPS) as the semiconductor for realizing TFTs. As the primary conduits for man–machine interaction, sensor-enabled displays are ubiquitously deployed. The smartphone comes to mind again, and it is hard to imagine one without a spiffy display!

Amorphous Oxide Semiconductors: IGZO and Related Materials for Display and Memory, First Edition.
Edited by Hideo Hosono and Hideya Kumomi.

The characteristics of a TFT depend on both the properties of its semiconducting channel and its architecture. The incumbent *a*-Si:H and LTPS TFT technologies have their respective origins in the thin-film photovoltaic and integrated circuits industries. While they have been serving the industry well since their adaptation by early flat-panel display manufacturers, their limitations are becoming apparent as more stringent requirements are placed on the TFTs to meet the demand for displays with higher specifications. The shortcomings to be overcome for an *a*-Si:H-TFT are its lower carrier mobility and poorer stability as a current-delivering device; for an LTPS TFT, these would be its higher leakage current and poorer uniformity of device characteristics.

Being investigated to address the shortcomings mentioned here, TFTs based on metal-oxide (MO) semiconductors have been actively pursued as alternatives to their silicon-based counterparts. Those realized with suitable MOs could be uniquely suited to the construction of active-matrix flat-panel displays with advanced specifications. Since much has already been written on many facets of MO TFT technologies [1–3], the present report is not an attempt to add to the already comprehensive body of published literature. Instead, it is a focused discourse on a unique annealing behavior of MO semiconductors [4], and the resulting device architecture [5] enabled by this behavior.

12.1.1 Semiconducting Materials for a TFT

12.1.1.1 Amorphous Silicon

Pure amorphous silicon (*a*-Si) is populated by a high density of localized trap states. A large amount of hydrogen (H) is incorporated to passivate these states in *a*-Si:H, thus making it possible to modulate its carrier concentration by doping and electrical bias. Compatible with a large-area substrate up to a tenth-generation mother glass (G10: $\sim 3 \times 3\,\text{m}^2$), PECVD is the preferred film preparation technique. Capable also of in-situ doping, the deposition temperature is typically below 250 °C. A larger ratio of on- to off-current ($I_{on}/I_{off} \sim 10^7$) is possible, because of the relatively larger mobility gap of *a*-Si:H than the bandgap of crystalline silicon. The field-effect mobility μ_{FE} for electrons is on the order of 0.5~1 cm²/Vs, and a couple orders of magnitude lower for holes. This disparity in electron and hole μ_{FE} makes *a*-Si:H unsuitable for implementing circuits with complementary *n*- and *p*-type TFTs; hence, only the *n*-type variety is deployed in display applications. The low μ_{FE} also places a limit on how small a TFT can be scaled to satisfy a given current-drive requirement. Consequently, *a*-Si:H-TFTs are more compatible with the implementation of displays containing relatively larger pixel sizes, rather than those demanding finer pixel dimensions.

12.1.1.2 Low-Temperature Polycrystalline Silicon

Consisting of fine crystalline grains separated by grain boundaries, as is evident in the secondary electron micrograph (SEM) shown in Figure 12.1, LTPS can be prepared by either deposition directly in the polycrystalline form or thermal crystallization from *a*-Si. The latter technique, including solid-phase crystallization, metal-induced crystallization, and excimer-laser crystallization (ELC), is preferred because of the resulting larger average grain size, and hence better average film quality and device characteristics. However, this is often accompanied by increased film roughness and nonuniformity in device characteristics, particularly compared to those obtainable in an *a*-Si:H thin film. Hitherto, ELC is the technique that is industrially deployed but constrained by equipment availability to a maximum size of G8 ($\sim 2.5 \times 2.5\,\text{m}^2$) mother glass. The μ_{FE} of electrons is on the order of $\sim 10^2$ cm²/Vs, and that of holes is comparable but lower. The higher values of μ_{FE} and their rough parity make it possible to implement a range of complementary peripheral circuits and to use a relatively smaller TFT to meet a given current-drive requirement. However, *p*-type TFTs are more commonly deployed within the pixel matrix of a flat-panel display because of their higher stability against hot-carrier stress. Relative to those of an *a*-Si:H-TFT, the I_{off} is higher, and the I_{on}/I_{off} is generally one to two orders of magnitude smaller. Multigate device architecture is often employed to reduce I_{off}, thus resulting in a device footprint larger than that of the conventional single-gate variety. However, the positive attributes outweigh the relative drawbacks and make LTPS TFTs the preferred device

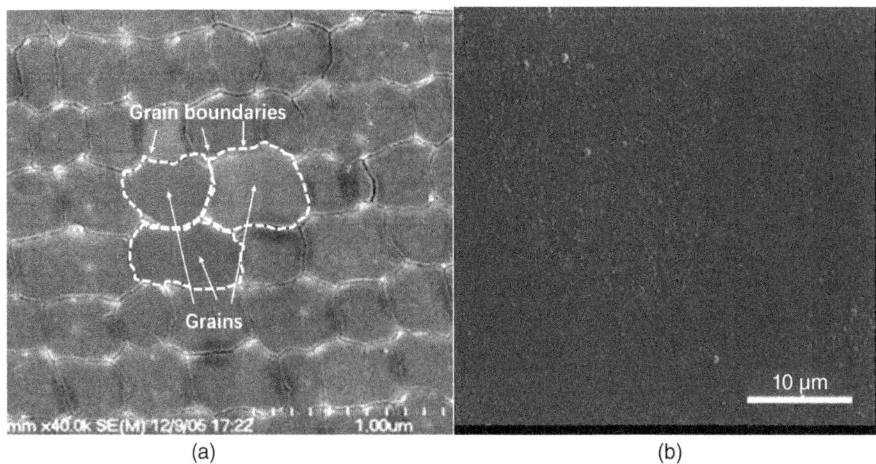

Figure 12.1 Plan-view SEMs of (a) an ELC LTPS film with three of the grains schematically highlighted, and (b) an a-Si:H film. *Source*: (a) Ref. [6], Wu, G. M., et al. (2009). Improved AMOLED with aligned poly-Si thin-film transistors by laser annealing and chemical solution treatments. *Physical Review B: Condensed Matter and Materials Physics* **404** (23–24): 4649–4652, Figure 1, p. 2 / with permission of Elsevier; and Ref. [7], Hamasha, K., et al. (2016). Aluminum-induced crystallization of hydrogenated amorphous silicon thin films with assistance of electric field for solar photovoltaic applications. *Journal of Display Technology* **12** (1): 82–88, Figure 9, p. 5 / with permission of Elsevier.

technology for implementing high-resolution mobile displays with fine pixel dimensions. On substrates capable of handling higher process temperature, such as quartz, high-temperature polycrystalline silicon (HTPS) TFTs have been deployed in niche applications.

12.1.1.3 MO Semiconductors

There is a large family of MO semiconductors, including single-cation binary oxides such as zinc oxide (ZnO), indium oxide (In_2O_3), and tin (IV) oxide (SnO_2); double-cation ternary oxides, such as indium–zinc oxide (IZO), zinc–tin oxide (ZTO), and indium–gallium oxide (IGO); and triple-cation quaternary oxides such as indium–gallium–zinc oxide (IGZO) [8] and indium–gallium–tin oxide (IGTO). Steady progress in the development of MO device technologies has been made, starting from their investigation in the mid-1960s [9] of the use of binary oxides in the construction of TFTs with the realization of polycrystalline ZnO-TFTs at a temperature comparable to that used for the construction of a-Si:H-TFTs while offering a much higher electron $\mu_{FE} \gg 10$ cm^2/Vs. Similar to a-Si:H, most MO semiconductors exhibit unipolar n-type conduction behavior, making it necessary to employ different materials, such as tin (II) oxide (SnO), copper (I) oxide (Cu_2O), or LTPS, for the realization of p-type TFTs [10]. Not for a lack of attention or effort, the development of p-type MO TFTs still lags far behind that of n-type TFTs, and the commercial deployment of circuits based on complementary MO TFTs remains a distant dream.

The attributes cited for MO TFTs being particularly suitable for display applications are their low process temperature, compatible with common glass substrates; typically tens of cm^2/Vs and, between those of a-Si:H and LTPS, a μ_{FE} that fits the purpose; a large mobility gap and strongly localized valence-band tail states, leading to a lower I_{off}; and a large optical bandgap, resulting in higher transparency within the visible spectrum. However, because of device instability issues when subjected to a combination of high electrical bias and short-wavelength illumination, the potential of realizing a higher light transmission implied by the last attribute when using a MO semiconductor as a channel material is overstated and currently unattained.

Significant progress has been made in the commercial deployment of MO TFT technology, particularly since the seminal publication on amorphous IGZO by Hosono et al. [11] in 2004. Hitherto, MO is the only technology

Table 12.1 Comparison of MO and the incumbent silicon-based TFT technologies.

Parameters	*a*-Si:H-TFT	Poly-Si-TFT	MO TFT
Mother glass	>G10	G8 (LTPS)	G8.5
Channel semiconductor	*a*-Si:H	ELC LTPS/SPC HTPS	IGZO
Number of masking steps	4–6	5–9	4–6
μ_{FE} (cm^2/Vs)	<1	30–100	3–50
TFT uniformity	Good	Poor/better	Good
TFT polarity	*n*-type	*n*- and *p*-type	*n*-type
Cost	Low	High	Low
Current-stress stability	Poor	Good	Good
Bias-stress stability	Adequate	Good	Adequate
Circuit integration	Limited	Yes	Yes
Complementary circuit	No	Yes	No
Process temperature (°C)	150–350	250–550	RT–450
Compatible pixel types	LC	LC, OLED	LC, OLED

Source: Adapted from [12], Park, J. S., et al. (2012). Review of recent developments in amorphous oxide semiconductor thin-film transistor devices. *Thin Solid Films* **520** (6): 1679–1693; and [13], Xia, Z. (2020). Semiconductor metal-oxide thin-film transistors for display. In *Semiconducting Metal Oxide Thin-Film Transistors* (Y. Zhou, ed.). IOP Publishing, Bristol, UK.

compatible with the construction of commercially viable large OLED televisions on G8.5 mother glass. Summarized in Table 12.1 is a comparison of the silicon- and MO-based TFT technologies.

12.1.2 TFT Architectures

Although both are field-effect devices, a bulk-type MOSFET and a TFT are constructed in very different architectures. The former generally adopts a top-gate (TG), self-aligned (SA) structure, with the gate electrode located above the channel and the source/drain (S/D) regions self-aligned to the projected edges of the gate electrode normal to the channel current. The structures for the latter are more varied, with an LTPS TFT also commonly adopting the TG SA architecture. Due to technological constraints, the architectures adopted for an *a*-Si:H-TFT generally are not self-aligned – with the resulting overlap between the gate and the S/D electrodes contributing to a parasitic capacitance. There are four architectural flavors for an *a*-Si:H-TFT: the staggered ones (Figure 12.2a and 12.2b) with the gate and the S/D electrodes located on opposite sides of the channel, and the coplanar ones (Figure 12.2c and 12.2d) with them located on the same side. Besides the top-gate varieties (Figure 12.2a and 12.2c), there are also the bottom-gate (BG) structures (Figure 12.2b and 12.2d) with the channels located above the gate electrodes. Unlike a bulk-type MOSFET, a TFT is a floating-body device because of the general absence of a body contact.

The technology applied to the construction of a TFT plays a significant role in the selection of the device architecture, resulting in the BG-staggered structure (Figure 12.2b) being the most commonly deployed on an industrial scale. Since the characteristics of a TFT are highly sensitive to the quality of certain structural interfaces, it is important that the technology used to construct such interfaces be carefully controlled. A typical *a*-Si:H-TFT fabrication process starts with the preparation and patterning (Mask 1) of the gate electrode. This is followed by the sequential PECVD of a hydrogenated silicon nitride (*a*-SiN:H) gate dielectric, an intrinsic *a*-Si:H active layer, and a thin n$^+$ *a*-Si:H contact layer heavily in-situ doped with phosphorus. The sequential and blanket deposition without vacuum breaks is vital for delivering the required quality of two critical interfaces: one between the gate dielectric

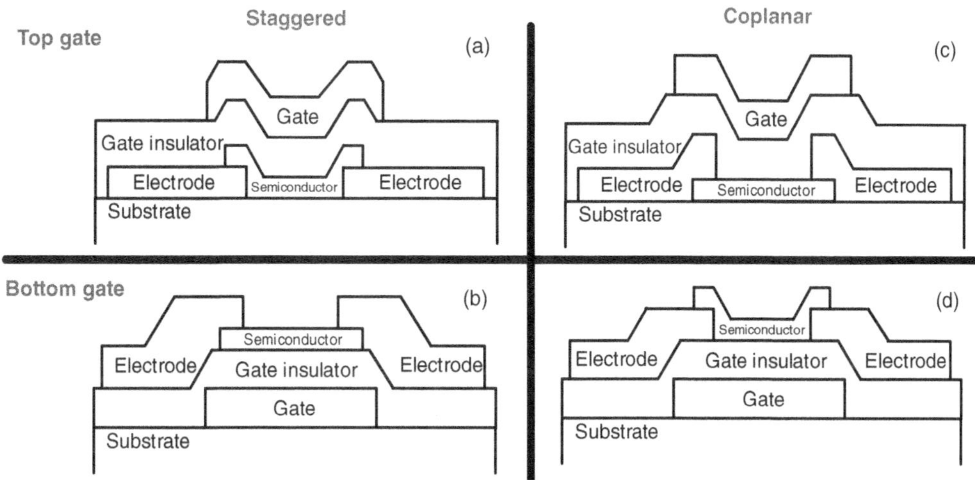

Figure 12.2 Non-self-aligned TFT architectures: (a) top-gate staggered, (b) bottom-gate staggered, (c) top-gate coplanar, and (d) bottom-gate coplanar.

and the active layer, in the vicinity of which the field-induced channel is formed; and the other between the active layer and the contact layer, across which the current flows. A metal electrode layer is subsequently sputtered after the individual active islands for the TFTs are patterned (Mask 2). The stack of metal and contact layers on the islands is subsequently patterned (Mask 3), forming the S/D electrodes. While the metal electrode layer is often wet-etched, the n^+ *a*-Si:H contact layer is plasma-etched. Any finite etch-selectivity between the contact layer and the underlying *a*-Si:H layer would lead to the etching of a certain amount of the latter during the definition of the S/D electrodes, including the contact layer. The portion of the active layer located away from the gate electrode is designated the "backchannel" region; hence, the structure resulting from this fabrication sequence is given the designation of "backchannel-etched" (BCE) architecture (Figure 12.3a). The etching of the backchannel invariably leads to the creation of localized defects. If unmitigated, these defects would lead to degradation of the TFT characteristics, including reduced μ_{FE} and device reliability, and increased I_{off} and threshold voltage V_{th}. If the population of the defects were high enough to cause pinning of the Fermi level, the gate might completely lose field-effect control of the backchannel.

The highest pixel density achievable in a display is constrained by the minimum-geometry TFT that can be realized for a given lithographic resolution limit λ. For a BCE TFT, the minimum channel length L_{min} is limited by the smallest lithographically definable separation between the S/D electrodes; hence, $L_{min} = \lambda$. If the field-effect control of the channel is to be guaranteed, the minimum length of the gate electrode along the channel-length direction would be 3λ if an alignment accuracy of λ were assumed. This would lead to an overlap of λ between the gate and each of the S/D electrodes, contributing to the parasitic capacitance.

One way of preventing the detrimental etching of the active layer during the patterning of the S/D electrodes is the insertion of an etch-stop (ES) layer between the n^+ *a*-Si:H contact layer and the *a*-Si:H active layer in the backchannel region. The triple layer of *a*-SiN:H/*a*-Si:H/n^+ *a*-Si:H in a BCE TFT is replaced by *a*-SiN:H/*a*-Si:H/*a*-SiN:H. Islands of TFTs are next defined (Mask 2) before the top *a*-SiN:H ES layer is patterned (Mask 3) to selectively expose portions of the *a*-Si:H active layer. The contact and metal layers are next prepared, followed by the patterning (Mask 4) of the S/D electrodes. The resulting structure is designated the "ES" architecture (Figure 12.3b).

For a minimum-geometry ES TFT, the separation between the S/D electrodes, the length of the ES (hence also the channel) layer, and the length of the gate electrode are λ, 3λ, and 5λ, respectively. The overlap between the gate and each of the S/D electrodes is 2λ. The improved characteristics of an ES TFT come with a number of

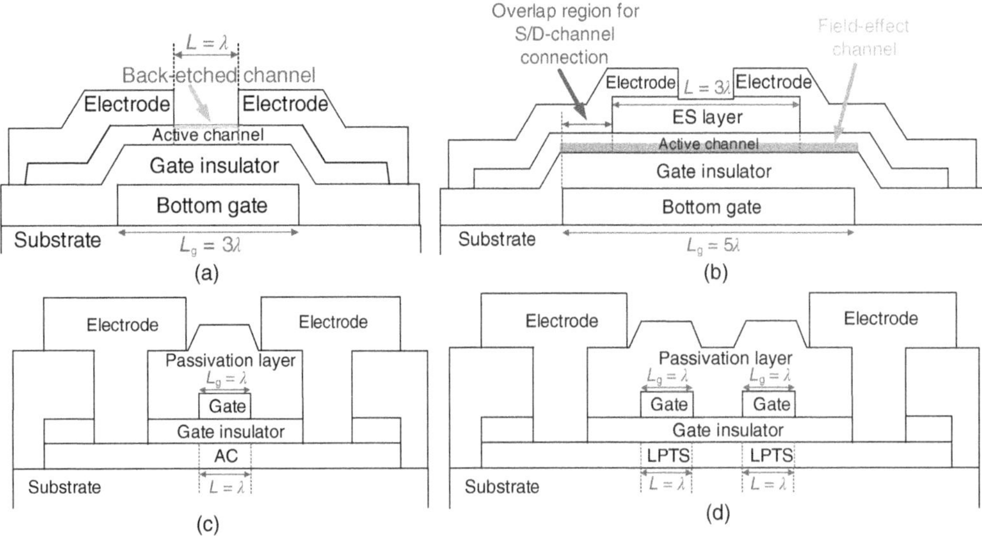

Figure 12.3 Schematic cross-sections of (a) a BCE, (b) an ES, (c) a conventional single-gate TG SA TFT architecture and (d) a double-gate TG SA TFT architecture typically employed to reduce the leakage current of an LTPS TFT. L_g denotes the length of the gate electrode in the direction of the channel current, and L denotes the channel length.

Table 12.2 Comparison of common TFT architectures.

Structures	BCE	ES	TG SA
Channel damage	Yes	No	No
Capacitance	Medium (1λ overlap)	Large (2λ overlap)	Minimal (self-aligned)
TFT footprint	Small ($L_{min} = 1\lambda$)	Large ($L_{min} = 3\lambda$)	Small/medium (single/double gates)
Masking steps (manufacturing cost)	3 (Low)	4 (High)	4 (High)

drawbacks: The requirement of an extra mask-count implies a higher manufacturing cost, the vacuum break before the deposition of the n^+ a-Si:H contact layer could lead to a degraded contact quality, the larger gate and overlap parasitic capacitance lead to a longer propagation delay, and the larger device footprint puts a lower bound on the maximum achievable pixel density. It is for these reasons that the ES architecture is often an option of last resort, when adequate means of mitigating the etch damage of the backchannel of a BCE TFT are not yet available. A comparison of the dominant TFT architectures is given in Table 12.2, including the TG SA varieties (Figure 12.3c and 12.3d).

A new TFT architecture (Figure 12.4) is presently described, in which the S/D metal electrodes are "elevated" above an ES layer inserted into a BCE TFT (Figure 12.3a) while maintaining the length of the gate electrode at 3λ. The edges of the channel are self-aligned to those of the S/D electrodes projected in the direction normal to that of the channel current, thus retaining $L_{min} = 1\lambda$ for a minimum-geometry TFT. The realization of the new architecture requires the provision of conducting S/D regions connecting the field-induced channel to the contacts with the S/D electrodes. Hitherto, it has been difficult to realize this architecture for a silicon-based TFT without a significant increase in process complexity. However, it can be more readily realized for a MO TFT because of a unique annealing behavior of MO semiconductors [4]. The resulting structure is dubbed an elevated-metal

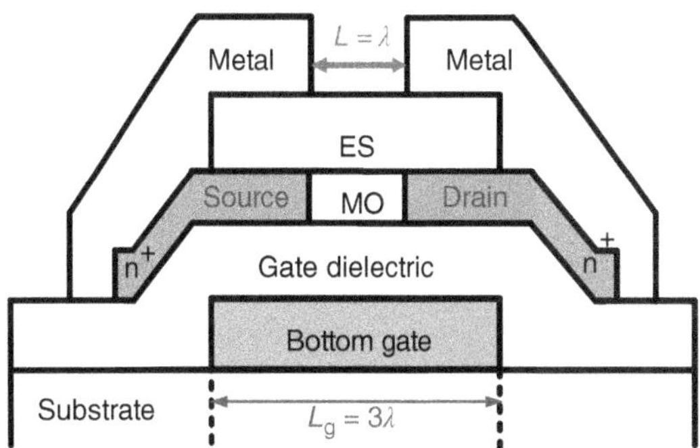

Figure 12.4 Schematic cross-section of a minimum-geometry EMMO TFT.

metal-oxide (EMMO) TFT [14], offering smaller gate-to-S/D overlap and a device footprint similar to those of a BCE TFT (Figure 12.3a), and the protection of the backchannel of an ES TFT (Figure 12.3b).

This chapter is organized as follows: Section 12.2, following this introduction, covers the annealing behavior of MO semiconductors. Section 12.3 describes how this behavior is applied to the realization of an EMMO TFT. Section 12.4 shows how the technology and architecture of EMMO TFTs can be further enhanced. Finally, a brief conclusion is given in Section 12.5.

12.2 Annealing-Induced Generation of Donor Defects

Contributing to both spreading and contact resistance, the S/D resistance of a field-effect transistor is another critical parasitic parameter to be minimized. In a bulk-type MOSFET, this is accomplished by heavily doping the S/D regions. Ion implantation is preferred over thermal diffusion doping for better control of the dopant dose and junction depth. More sophisticated schemes are available, such as the provision of a conductive shunt based on a transition-metal silicide.

For a TFT with a staggered architecture, the path across the intrinsic active layer between the field-induced channel and the contact layer contributes an additional component to the parasitic S/D resistance. The remedy is to re-create the conductive S/D regions of a bulk-type MOSFET, thus reducing the resistivity of the "S/D portions" of the active layer on the two sides of the channel. For TFTs built on a class of MO semiconductors, including IGZO, the formation of such conductive regions has been accomplished by either exposure to an argon (Ar) plasma, leading to the generation of donor defects; or the incorporation of hydrogen (H), a donor impurity, upon exposure to a H-containing plasma or diffusion from a H-containing thin film. Although a low resistivity on the order of a few mΩ·cm has been obtained, the incorporation of H is often thermally unstable and would decrease upon further thermal treatment [15]. For fast-diffusing H with significant migration even at a relatively low temperature of 150 °C [16], its unintended diffusion into the adjacent channel region leads to a negative shift of V_{th} or, if excessive, electrical shorting of the channel.

Based on the effects of thermal annealing on the population of donor defects in a MO semiconductor when subjected to a suitable combination of material boundary condition and annealing atmosphere [17], an alternative and thermally stable technique of controlling the conductivity of a MO semiconductor is now described.

12.2.1 Effects of Annealing on the Resistivity of IGZO

From a target of composition In_2O_3:Ga_2O_3:ZnO = 1:1:1, 100-nm-thick IGZO samples were sputter-deposited on oxidized silicon substrates in an atmosphere of 10% oxygen (O_2) and 90% Ar with no deliberate substrate heating. Samples with different cover configurations were subsequently prepared [18]: (i) Sample B consisted of bare, hence "exposed," IGZO; (ii) Sample O was Sample B covered by a 100-nm-thick PECVD silicon oxide (SiO_x) layer deposited using tetraethylorthosilicate (TEOS) and O_2 as the source gases; (iii) Sample ON was Sample O covered by an additional 200-nm-thick PECVD silicon nitride (SiN_y) layer deposited using silane (SiH_4) and ammonia (NH_3) as the source gases; and (iv) Sample OT was similar to Sample ON, except with the SiN_y layer replaced by 200-nm-thick sputtered titanium (Ti).

These samples were subjected to a series of thermal anneals at different temperatures, for different durations, and in different atmospheres, including nitrogen (N_2), O_2, and Ar. After the cover layers were removed in a reactive-ion etcher running a sulfur hexafluoride (SF_6) chemistry, the resistivity ρ of the exposed IGZO was measured using a four-point probe. The effects of the etch process on the ρ were minimal, as inferred from the unchanged ρ of a simultaneously etched, unannealed Sample B. Not surprisingly, all samples exhibited *n*-type conductivity, verified using Hall effect measurement.

The dependence of the ρ on the annealing atmosphere (Figure 12.5) was investigated by subjecting the samples to 20-min isochronal anneals at 450 °C in alternating atmospheres of O_2 or N_2. A corresponding cyclical change by over 10^4 times in the ρ of Sample B was revealed. While the ρ was high after annealing in O_2, a comparably low $\rho \sim 1$ Ω·cm was obtained on Sample B after annealing in N_2 or Ar. Since Ar is not a dopant in IGZO, it is unlikely that nitrogen doping was responsible for the increased population of charge carriers giving rise to the same low ρ for the sample annealed in N_2. Rather, this cyclical behavior can be attributed to the respective generation or annihilation of shallow donor-like defects [4] within the IGZO thin film upon annealing in N_2 or O_2, and resulting from the exchange of oxygen-carrying species between the IGZO and the atmosphere during the anneal. The generation of such defects in IGZO during an annealing in N_2 has been reported previously [19]. Similar dependence of the ρ of Sample O on the cyclical change in the annealing atmosphere was observed, only with a much higher ρ than that measured on Sample B annealed in N_2. This indicates that the oxygen-permeable SiO_x cover layer either acted like a diffusion barrier that slowed down the transport of the oxygen-carrying species or was itself a finite source or sink of such species.

Unlike the ρ of Samples B and O, that of Sample ON was insensitive to the cyclical change in the annealing atmosphere and remained at a relatively low value of ~10 mΩ·cm. This behavior is consistent with the top SiN_y cover acting like an impermeable seal that prevented the exchange of the oxygen-carrying species between the IGZO and the atmosphere. It is unlikely that the low ρ resulted from H migrated from the SiN_y [20] cover layer, when comparably low ρ was obtained also for Sample OT and in which the H-containing SiN_y was replaced by

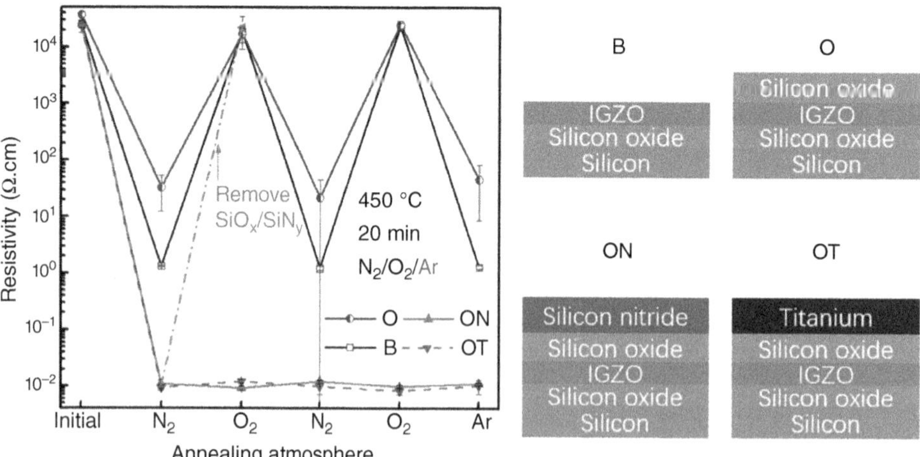

Figure 12.5 The dependence of the ρ on the annealing atmosphere for the IGZO Samples B, O, ON, and OT. *Source*: Ref. [18], Lu, L., and Wong, M. (2015). A bottom-gate indium-gallium-zinc oxide thin-film transistor with an inherent etch-stop and annealing-induced source and drain regions. *IEEE Transactions on Electron Devices* **62** (2): 574–579.

Ti. The ρ of the annealed Sample ON or OT was notably lower than that of Sample B annealed in N_2, indicating a difference in either the mechanisms or the rates of the generation of the donor-like defects among these samples.

For isothermal annealing between 400 °C and 700 °C, the evolution of the ρ of Samples B and ON (Figure 12.6) exhibited a sharp initial drop with the annealing time before reaching a steady-state resistivity ρ_{ss}. The time to reach ρ_{ss} is denoted by t_{ss}. Both ρ_{ss} and t_{ss} increased with decreasing annealing temperature for all samples; both were reduced for Sample ON, indicating more efficient and faster defect generation in the sealed configuration. At a temperature of 500 °C or higher and with a higher defect generation rate overwhelming the defect annihilation rate induced by the oxygen-carrying species, the time evolution of the ρ of Sample O approached that of Sample B. The consistently high ρ for Sample O annealed at 400 °C suggests a corresponding t_{ss} longer than 60 min.

The temperature dependence of the ρ of the samples subjected to 60-min isochronal anneals in N_2 is extracted from Figure 12.6 and redrawn in Figure 12.7. A step-like dependence is revealed, characterized by a low- and

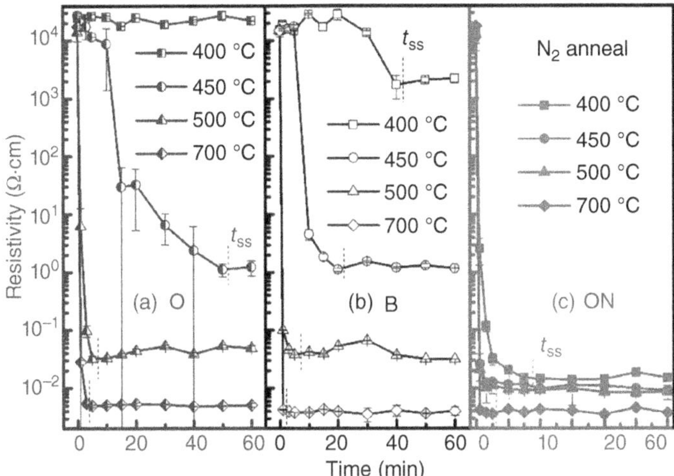

Figure 12.6 The dependence of the ρ on the annealing time for the IGZO Samples O, B, and ON annealed at different temperatures in N_2. *Source*: Ref. [18], Lu, L., and Wong, M. (2015). A bottom-gate indium–gallium–zinc oxide thin-film transistor with an inherent etch-stop and annealing-induced source and drain regions. *IEEE Transactions on Electron Devices* **62** (2): 574–579.

Figure 12.7 The dependence of the ρ on the annealing temperature for the IGZO Samples B, O, and ON annealed in N_2 for 60 min. *Source*: Ref. [18], Lu, L., and Wong, M. (2015). A bottom-gate indium–gallium–zinc oxide thin-film transistor with an inherent etch-stop and annealing-induced source and drain regions. *IEEE Transactions on Electron Devices* **62** (2): 574–579.

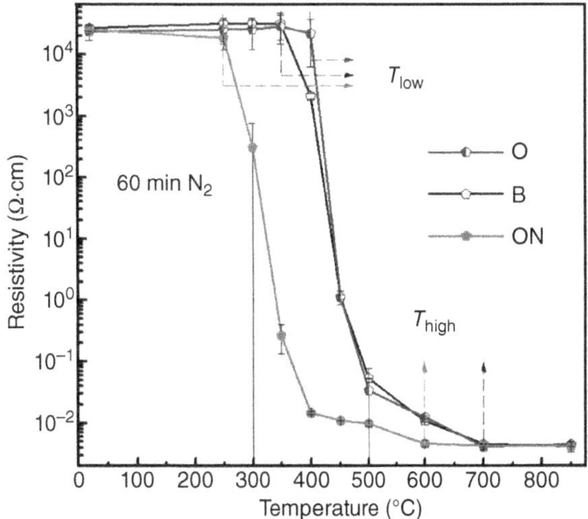

a high-transition temperature denoted respectively by T_{low} and T_{high}. The ρ stayed relatively constant and high for an annealing temperature below T_{low}, and it rapidly dropped to a low and saturated value for an annealing temperature above T_{high}. At below 10 mΩ·cm, the saturated values for all three types of samples were similar.

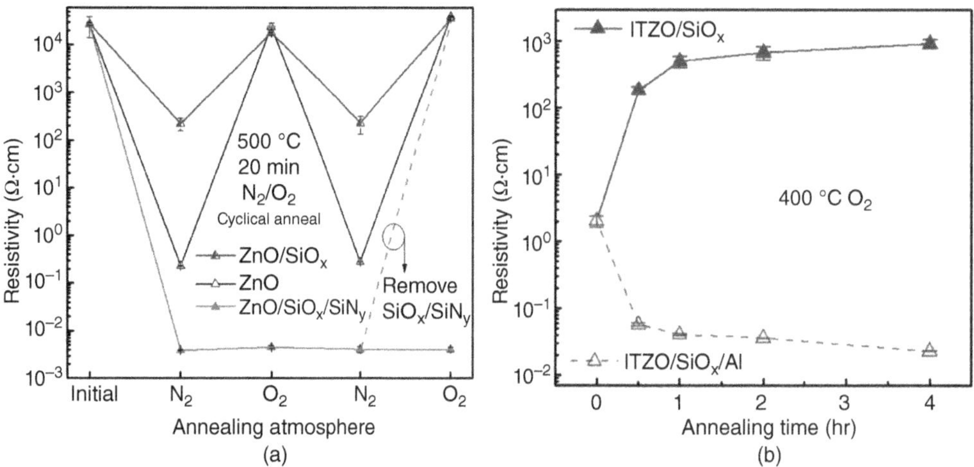

Figure 12.8 Dependence of ρ on (a) the annealing atmosphere for ZnO-based Samples B, O, and ON; and (b) the annealing time for impermeable Al- and permeable oxide-covered ITZO. *Source*: Ref. [4], Lu, L., and Wong, M. (2014). The resistivity of zinc oxide under different annealing configurations and its impact on the leakage characteristics of zinc oxide thin-film transistors. *IEEE Transactions on Electron Devices* **61** (4): 1077–1084; and adapted from Ref. [21], Xia, Z., et al. (2017). Characteristics of elevated-metal metal-oxide thin-film transistors based on indium-tin-zinc oxide. *IEEE Electron Device Letters* **38** (7): 894–897.

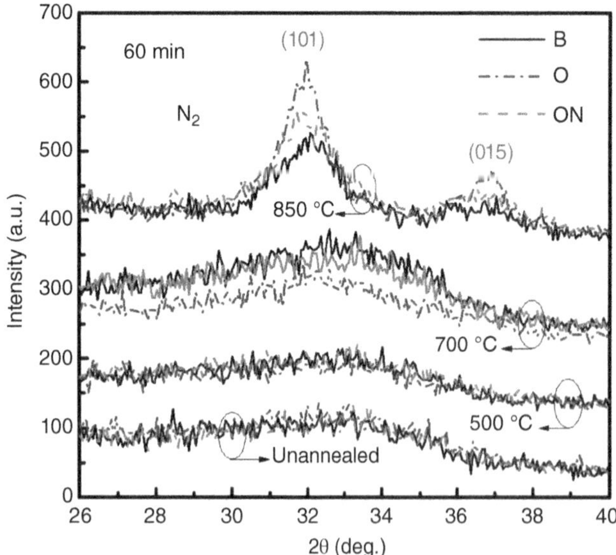

Figure 12.9 The XRD spectra of the IGZO Samples B, O, and ON annealed in N_2 for 60 min from 500 °C to 850 °C. *Source*: Ref. [18], Lu, L., and Wong, M. (2015). A bottom-gate indium–gallium–zinc oxide thin-film transistor with an inherent etch-stop and annealing-induced source and drain regions. *IEEE Transactions on Electron Devices* **62** (2): 574–579.

Besides IGZO, the annealing-induced generation of donor defects has been similarly observed in other MO semiconductors, such as polycrystalline ZnO (Figure 12.8a) [4], amorphous indium–tin–zinc oxide (ITZO) (Figure 12.8b) [21], and indium–aluminum–zinc oxide [5].

12.2.2 Microanalyses of the Thermally Annealed Samples

Crystallization is an unlikely cause of the annealing-induced change in ρ, since none is revealed by the X-ray diffraction (XRD) spectra (Figure 12.9) for all three sample types annealed for 60 min at a temperature as high as ~700 °C.

Taken after the surface contaminants (such as dissociated O_2 or other adsorbed species like $-OH$, $-CO_3$, H_2O, or O_2) have been removed using in-situ Ar ion bombardment, the oxygen 1*s* X-ray photoelectron spectra (XPS) are displayed in Figure 12.10 for an as-deposited IGZO (Sample A) and samples annealed for 1 h in O_2 at 400 °C (Sample O-1), for 4 h in O_2 at 400 °C (Sample O-4), and for an additional 5-min anneal in N_2 at 350 °C following 4 h in O_2 at 400 °C (Sample N). Each spectrum can be resolved into two Gaussian–Lorenz components and labeled as "O_I" and "O_{II}," centered respectively at ~530.8 and ~531.6 eV. The former is associated with the fully coordinated

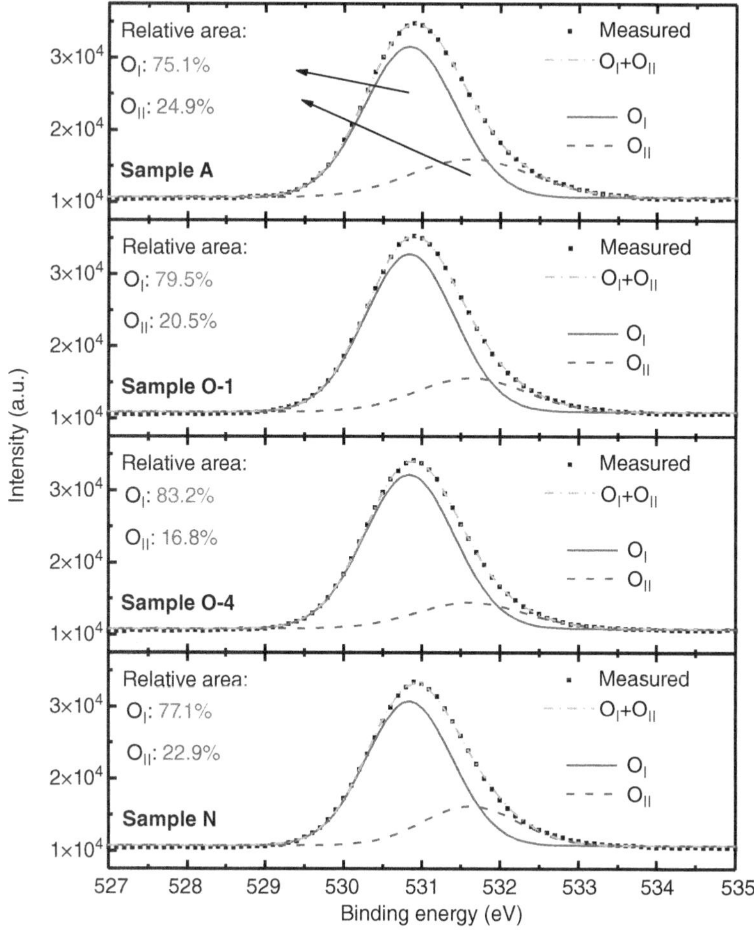

Figure 12.10 The XPS spectra of the O 1*s* state in the IGZO films subjected to different annealing conditions, together with the Gaussian–Lorenz decompositions. *Source*: Ref. [24], Li, J., et al. (2017). An oxidation-last annealing for enhancing the reliability of indium-gallium-zinc oxide thin-film transistors. *Applied Physics Letters* **110** (14): 142102.

O^{2-} ions in stoichiometric IGZO [22], and the latter is attributed to undercoordinated O^- ions or OH^- species in an oxygen-deficient IGZO [23]. Normalized by the total area of the corresponding composite spectrum and ranked in decreasing magnitude, the calculated percentage area of the O_{II} component changed in the order of Samples A, N, O-1, and O-4. It can be concluded that annealing in an oxidizing atmosphere leads to a reduction in the percentage area of O_{II}, and the opposite for annealing in a non-oxidizing atmosphere. A smaller percentage area is correlated with a higher ρ and a smaller population of donor defects.

12.2.3 Lateral Migration of the Annealing-Induced Donor Defects

Controlled placement of the annealing-induced donor defects is important for many applications, including the construction of a TFT. The lateral migration of the annealing-induced donor defects was characterized using a "windowed bridge structure" (Figure 12.11). Fabrication of the structure started with a heavily doped, *n*-type silicon substrate covered with 100-nm-thick thermal oxide. A stack of insulating PECVD films consisting of 75-nm-thick SiO_x on 50-nm-thick SiN_y was deposited. The respective carrier gases were 8 sccm SiH_4/1400 sccm nitrous oxide (N_2O) and 40 sccm SiH_4/40 sccm NH_3; the temperature was 300 °C, and the pressure was 0.9 Torr. A 20-nm-thick IGZO active layer was subsequently deposited at room temperature in a radiofrequency (RF) magnetron-sputtering machine at a process pressure of 3 mTorr in an atmosphere of 10% O_2 and 90% Ar. The composition of the target was In_2O_3:Ga_2O_3:ZnO = 1:1:1, and the base pressure was ~1 μTorr. An active island of width W_{IS} was patterned using 1/1000 molar aqueous hydrofluoric (HF) acid solution, before a 300-nm-thick gas-permeable SiO_x passivation layer and an 80-nm-thick impermeable SiN_y cover layer were sequentially deposited using the same PECVD processes. The SiN_y cover layer was patterned, and a window of length L_A and width $>W_{IS}$ was opened by etching in an inductively coupled plasma (ICP) etcher running a SF_6 chemistry, thus exposing the underlying gas-permeable SiO_x passivation layer in the windows. Contact holes were subsequently

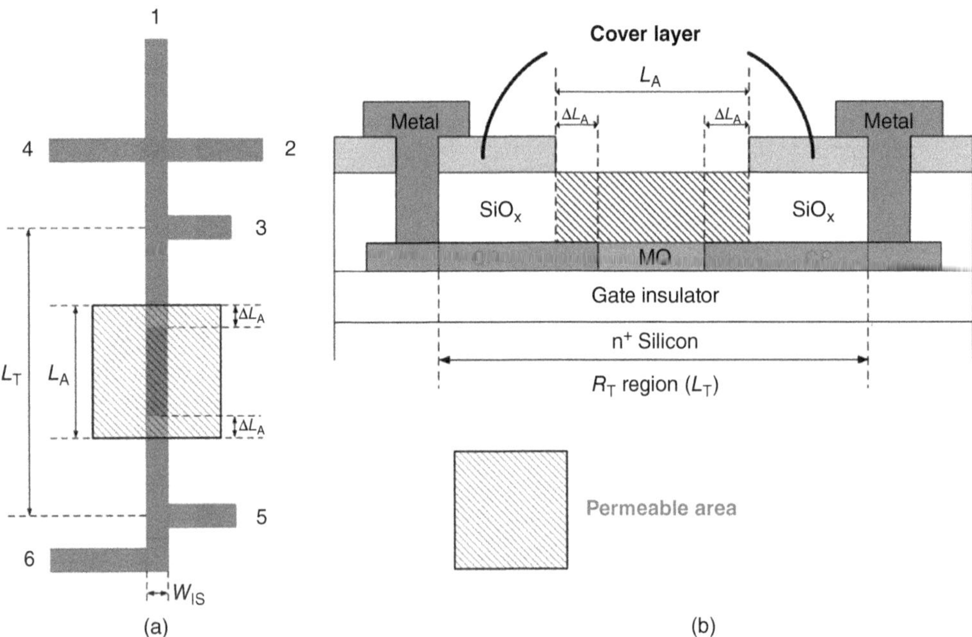

Figure 12.11 The (a) layout and (b) schematic cross-section of a windowed bridge structure. *Source*: Adapted from Ref. [25], Xia, Z., et al. (2018). A bottom-gate metal–oxide thin-film transistor with self-aligned source/drain regions. *IEEE Transactions on Electron Devices* **65** (7): 2820–2826.

opened before the metal electrodes, consisting of sputtered and patterned stacks of 300-nm-thick aluminum (Al) on 50-nm-thick molybdenum (Mo), were defined. Finally, the devices were annealed in a conventional resistively heated horizontal atmospheric pressure furnace at 400 °C for 2 h with a N_2 flow rate of 8 sccm.

Consistent with the formation of annealing-induced donor defects under the condition of non-oxidizing heat treatment, conductive regions (CRs) are formed under the gas-impermeable SiN_y cover layer adjacent to the window. There is an inevitable misalignment ΔL_A between the edge of a CR and that of the cover layer. The total resistance R_T measured between the metal contacts is given by

$$R_T = \frac{L_A + 2\Delta L_A}{W_{IS}}\left(R'_A - R'_{CR}\right) + \frac{L_T}{W_{IS}} R'_{CR} \tag{12.1}$$

where R'_A and R'_{CR} are the sheet resistance of the IGZO active layer in the window and in the adjoining CRs, respectively; and L_T is the separation between the two contact holes, as indicated in Figure 12.11.

Recognizing the similarity between the bridge structure and a bottom-gate TFT, one can modulate the conductance of the IGZO in the window by applying a "gate" bias V_g to the heavily doped silicon substrate. The dependence of R_T on L_A and V_g after an initial 2-h, N_2-anneal at 400 °C is displayed in Figure 12.12a. The relatively constant R_T implies $R'_A \approx R'_{CR}$. This is reasonable, since the IGZO inside the window was also subjected to a non-oxidizing anneal, resulting in the creation of donor defects, the large population of which is relatively insensitive to modulation by V_g.

This population decreases after a subsequent oxidizing anneal, and when it becomes negligible compared to the field-induced carrier population, one obtains

$$R'_A \approx \frac{1}{\mu_{FE} C'_{GI}(V_g - V_{th})} \tag{12.2}$$

where C'_{GI} is the gate dielectric capacitance per unit area, and V_{th} is the linearly extrapolated threshold voltage. It should be noted that $(V_g - V_{th}) \gg V_A$ is assumed in Equation (12.2), where V_A is the bias applied to one of the two metal electrodes. Consequently,

$$R_T \approx \frac{L_A + 2\Delta L_A}{W_{IS}}\left[\frac{1}{\mu_{FE} C'_{GI}(V_g - V_{th})} - R'_{CR}\right] + \frac{L_T}{W_{IS}} R'_{CR} \tag{12.3}$$

The typical dependence of R_T on L_A and V_g after a series of oxidizing thermal anneals with increasing duration is displayed in Figure 12.12c and 12.12d. V_g was varied between 18 and 22 V, and V_A was fixed at 0.5 V. A linear fit to the set of measured data at a given V_g is generated using the least-squares fit technique. For the initial state after the annealing in N_2, the entire IGZO active layer was populated by donor defects (Figure 12.12a) without any junctions separating the "channel" in the window and the adjacent CRs. For each duration of annealing in O_2 at 400 °C, a unique ΔL_A can be extracted from the common intersection of the least-squares fit lines at various values of V_g.

The total resistance R_{CR} of an adjacent CR, obtained from a point of common intersection shown in Figure 12.12, is extracted from a series of TFTs with different lengths of CRs. Figure 12.13 shows the dependence of R_{CR} on the length of CRs, and from the slope of the plot, a ρ of ~7.1 mΩ·cm can be obtained. This ρ is consistent with that shown in Figure 12.5. The intercept of the plot with the R_{CR}-axis, at ~35 kΩ, is taken to be a resistance associated with the 10-μm-wide junction between a CR and the active channel region of an EMMO TFT.

The dependence of ΔL_A on the O_2-annealing time (Figure 12.14) can be roughly divided into two regimes, with ΔL_A changing more quickly for annealing times less than 4 h and changing more gradually for longer annealing times.

In the initial phase of an oxidizing anneal, O_2 entered the thin 20-nm-thick IGZO in the window and annihilated the donor defects. This is a regime characterized by "vertical" oxidation, since the entire surface of the channel is "exposed" to O_2 diffusing through the SiO_x passivation layer. The annihilation in the edge regions is incomplete due to the supply of the donor defects from the SiN_y-covered CRs, resulting in a small but negative

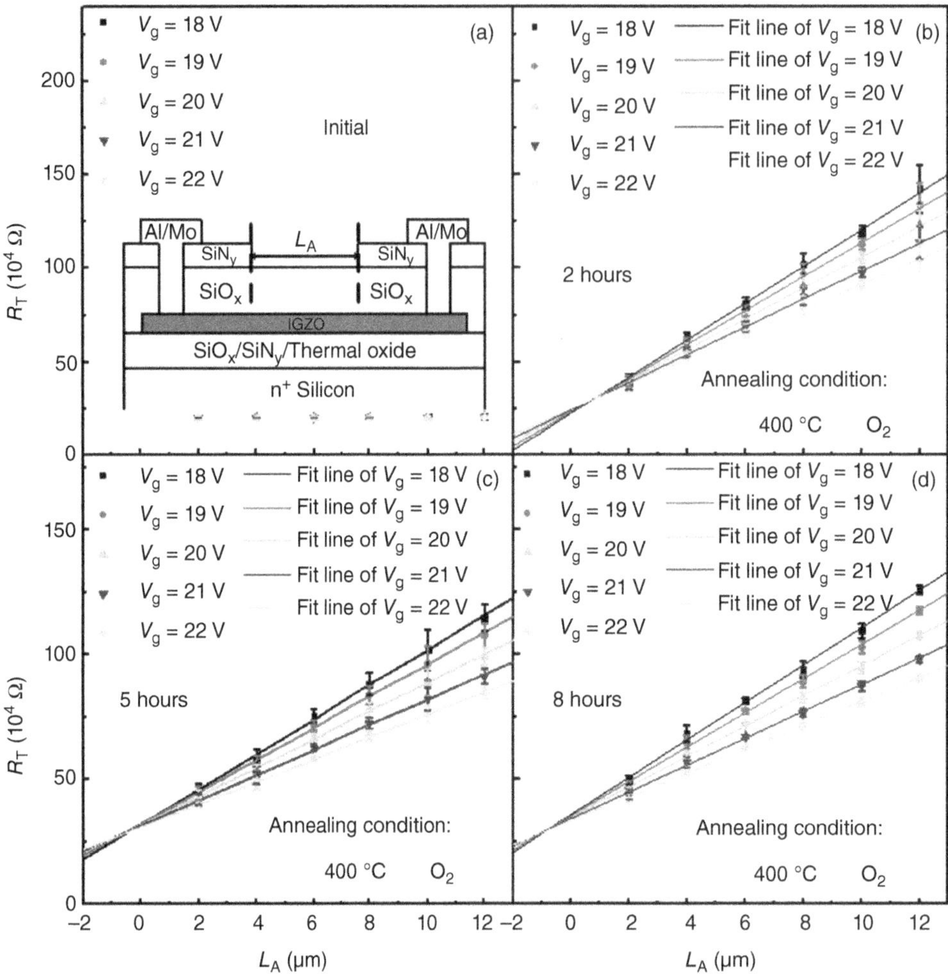

Figure 12.12 R_T at different V_g as a function of drawn L_A, obtained from the bridge structure subjected to oxidizing anneals at 400 °C for different durations. Each corresponding ΔL_A is extracted from the intersection of the lines obtained using the least-squares-fit technique. *Source:* Ref. [25], Xia, Z., et al. (2010). A bottom gate metal–oxide thin-film transistor with self-aligned source/drain regions. *IEEE Transactions on Electron Devices* **65** (7): 2820–2826.

ΔL_A (Inset I). Upon further annealing beyond 3.5 h, the defect-annihilation fronts (hence junctions) were pushed into the SiN_y-covered regions. ΔL_A turned positive (Inset II), and its time-rate of change slowed down appreciably. This is a regime characterized by "lateral" oxidation, and hence annihilation, of the donor defects. In this regime, the edge of a CR is said to be self-aligned to that of the cover layer.

12.3 Elevated-Metal Metal-Oxide (EMMO) TFT Technology

The relevance of the process of annealing-induced generation of donor defects to the realization of an EMMO TFT can be understood with reference to Figure 12.15. A MO active layer is covered by a gas-permeable layer laterally sandwiched between two impermeable layers. Upon thermal annealing in an oxidizing atmosphere, the two adjacent portions of the active layer under the impermeable covers become conductive, making it suitable

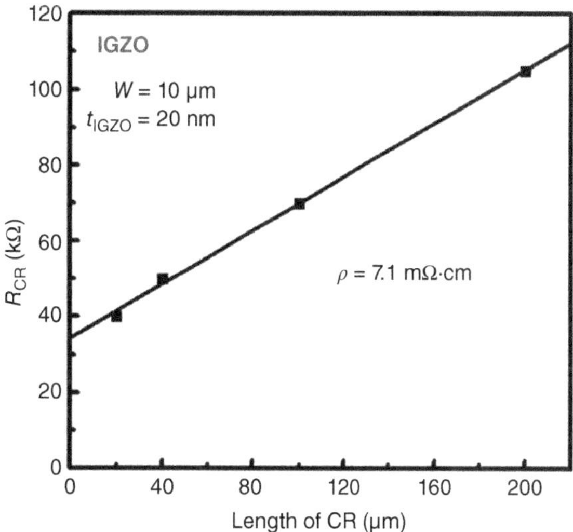

Figure 12.13 Dependence of R_{CR} on the length of CR.

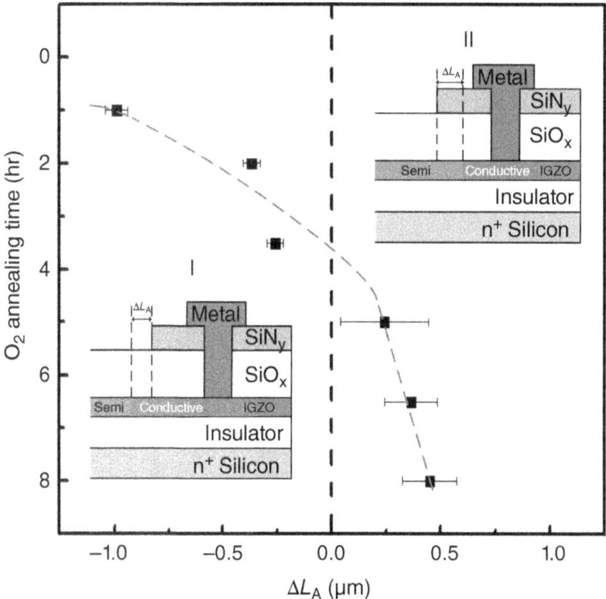

Figure 12.14 Dependence of the ΔL_A on a different O_2-annealing time. *Source*: Ref. [25], Xia, Z., et al. (2018). A bottom-gate metal–oxide thin-film transistor with self-aligned source/drain regions. *IEEE Transactions on Electron Devices* **65** (7): 2820–2826.

for realizing the S/D regions of an EMMO TFT. The middle portion is exposed to the atmosphere through the permeable layer, and its continuous oxidation helps promote its semiconducting properties, making it suitable for implementing the channel region of a TFT.

12.3.1 Technology and Characteristics of IGZO EMMO TFTs

On an oxidized silicon wafer, the construction of an EMMO TFT [14] started with the patterning of sputtered indium–tin oxide (ITO) as the bottom-gate electrode. A gate dielectric of 100-nm-thick SiO_x was next deposited

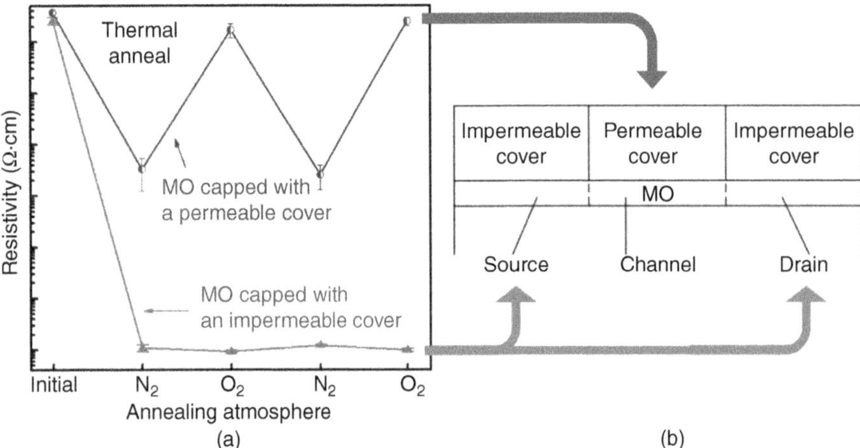

Figure 12.15 Illustration of the technology of implementing an EMMO TFT with annealing-induced S/D regions.

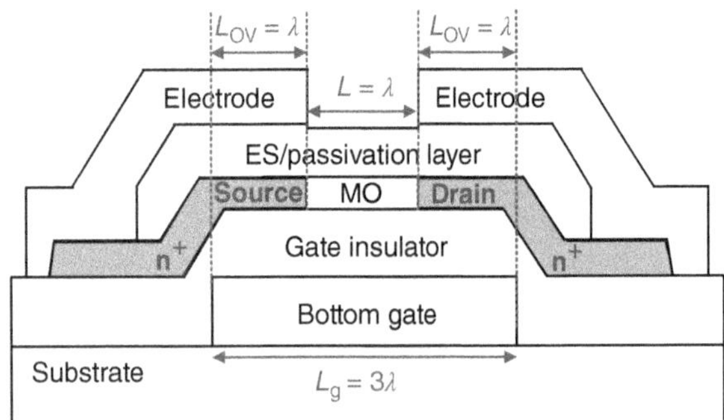

Figure 12.16 Schematic cross-section of a minimum-geometry EMMO TFT.

at 420 °C in a low-pressure chemical vapor deposition reactor using SiH_4 and O_2 as the source gases. In a RF magnetron-sputtering machine with an atmosphere of 10% O_2 and 90% Ar at a total pressure of 3 mTorr and a target composition of In_2O_3:Ga_2O_3:ZnO = 1:1:1, an IGZO active layer was deposited at room temperature. The active island, patterned using a dilute aqueous HF acid solution, was then covered with a passivation layer of 300-nm-thick SiO_x deposited using the same process as that for the gate dielectric. The contact holes were next opened in an ICP etcher running a SF_6 chemistry. With the IGZO intrinsic channel region protected by the SiO_x passivation (thus equivalent to an inherent ES) layer, a sputtered Al-on-Mo bilayer was patterned to form the S/D electrodes with an overlap (L_{OV}) with the gate electrode. Finally, the TFT was annealed at 400 °C in an O_2 atmosphere, leading to the annealing-induced formation of the conductive S/D regions with edges self-aligned to the projected edges of the gas-impermeable S/D electrodes. A schematic cross-section of the TFT is shown in Figure 12.16, similar to that shown in Figure 12.4.

The drain current (I_d) versus V_g transfer characteristics at different drain voltages (V_d) of an EMMO TFT, annealed at 400 °C in O_2 for 2 h, are shown in Figure 12.17. A low I_{off} was obtained, below the ~10^{-14} A detection limit of the semiconductor parameter analyzer used for electrical characterization.For conventional BCE or ES IGZO-TFTs, both the I_{off} and the hysteresis could be degraded by the incorporation of unintentional external impurities such as H during the fabrication process or the generation of donor defects commonly attributed to "oxygen vacancies" (V_O) [19] or undercoordinated cations [23]. The culprit responsible for the degradation could

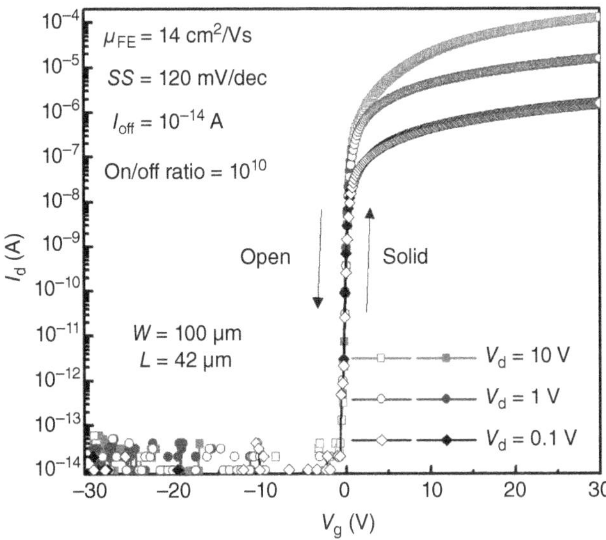

Figure 12.17 The transfer characteristics of an EMMO TFT with a channel length of 42 μm and width of 100 μm. Solid and hollow symbols denote, respectively, forward and reverse V_g sweeps. *Source*: Ref. [14], Lu, L., et al. (2016). Elevated-metal-metal-oxide thin-film transistor: technology and characteristics. *IEEE Electron Device Letters* **37** (6): 728–730.

be the backchannel etch when patterning the S/D electrodes, the plasma bombardment when depositing the ES or passivation layer [26], or other improperly designed thermal processes [4, 27]. In contrast, the oxidizing annealing of the EMMO-TFT performed toward the end of its fabrication process and after the formation of the S/D electrodes helped annihilate the donor defects, thus ensuring a semiconducting active channel region with a high ρ and a low I_{off} [27].

From the transfer characteristic measured at a V_d of 10 V, a steepest pseudo-subthreshold slope SS of ~120 mV/decade, a saturation V_{th} of ~0.4 V extracted from the dependence of $I_d^{0.5}$ on V_g, and a peak μ_{FE} of ~14 cm²/Vs extracted from the maximum $\partial I_d^{0.5}/\partial V_g$ are obtained. The relatively high μ_{FE} originates from the combined effects of the reduced density of defects in a strongly oxidized channel region and the low resistance of the conductive S/D regions. The low I_{off}, combined with the reasonably high μ_{FE}, leads to an I_{on}/I_{off} of better than ~10^{10}. The I_d versus V_d output characteristics (Figure 12.18) were also measured at various V_g. The nicely linear dependence of I_d on V_d shown in the inset of Figure 12.18 is a reflection of the good ohmic contact between the highly conductive S/D regions and the Al–Mo S/D electrodes.

Figure 12.19 shows a demonstration of the effectiveness of the 300-nm-thick SiO_x ES/passivation layer in protecting an EMMO TFT against environmental stress. The measured transfer characteristics hardly changed after storage in 90% humidity at 80 °C for 10 h. The reliability against positive bias stress (PBS) and negative bias stress (NBS) was also investigated, with grounded S/D and respective V_g of +20 and −20 V. The transfer characteristics changed minimally throughout the 30,000-s stress, as is evident from their negligible time evolution exhibited in Figure 12.19a and 12.19b.

Various approaches for enhancing TFT reliability have been proposed: by improving the passivation layer [26], adjusting the metal-cation and oxygen-anion content of the MO [28], modifying the TFT architecture [29], and designing proper process technologies [30]. The PBS/NBS-induced degradation in a MO TFT is commonly attributed to the presence of defects, such as V_O or undercoordinated cations, in its channel. Suppressing the population of such defects is thus an effective method of enhancing the reliability of a TFT, such as by heat

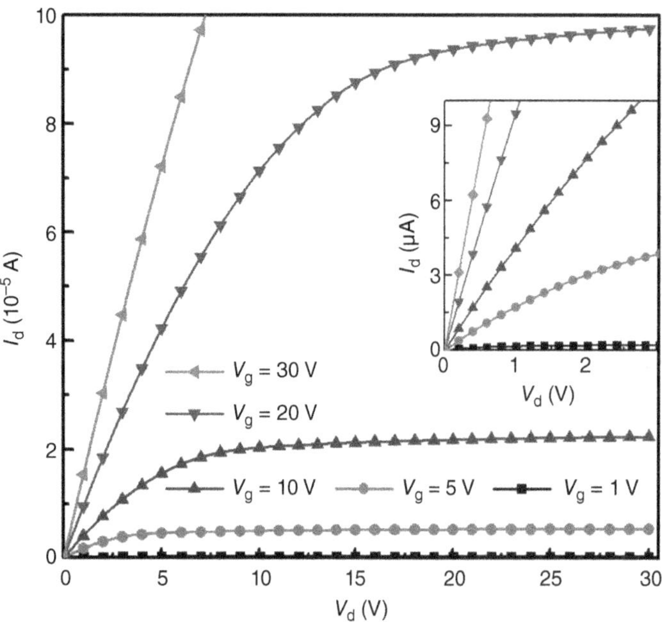

Figure 12.18 The output characteristics at various V_g for an EMMO TFT. (Inset) The output characteristics at low V_d. *Source*: Ref. [14], Lu, L., et al. (2016). Elevated-metal-metal-oxide thin-film transistor: technology and characteristics. *IEEE Electron Device Letters* **37** (6): 728–730.

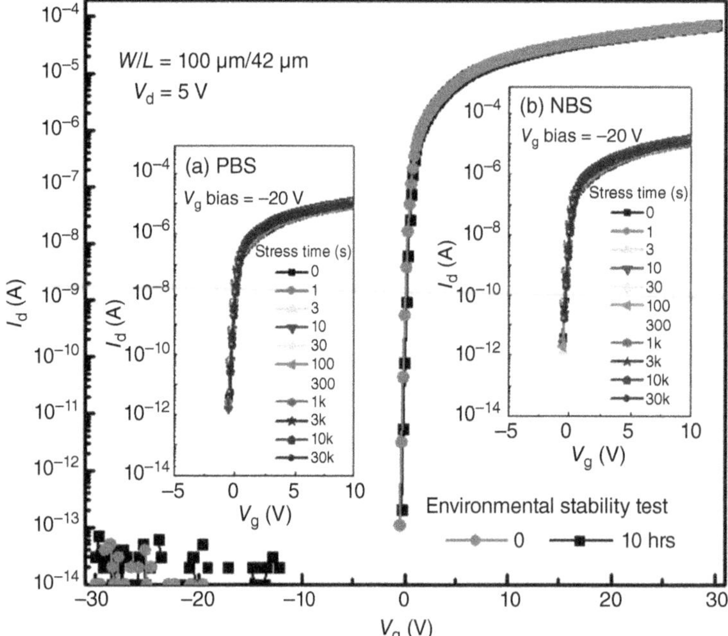

Figure 12.19 Comparison of the transfer characteristics of an EMMO TFT before and after an environmental stress in 90% humidity at 80 °C for 10 h. (Insets) The time evolution of the transfer characteristics of EMMO TFTs under (a) PBS and (b) NBS. *Source*: Ref. [14], Lu, L., et al. (2016). Elevated-metal-metal-oxide thin-film transistor: technology and characteristics. *IEEE Electron Device Letters* **37** (6): 728–730.

treatment in an oxidizing atmosphere of water vapor or air. The good reliability of an EMMO TFT against both environmental and electrical stress is consistent with the oxidizing anneal used for its construction.

12.3.2 Applicability of EMMO Technology to Other MO Materials

TFTs with a higher μ_{FE} are needed to realize advanced displays, such as a minimum of 16 cm^2/Vs for an 8k × 4k active-matrix OLED display driven at 120 Hz [31]. This is higher than the typical ~10 cm^2/Vs offered by an IGZO-TFT. Amorphous ITZO has been investigated as a channel material, because of the replacement of Ga by the more abundantly available Sn and reports of high μ_{FE}~50 cm^2/Vs [32]. Figure 12.20a shows the I_d versus V_g transfer characteristics of an ITZO EMMO TFT subjected to different V_d. Hysteresis is negligible, as reflected by the excellent overlap of the transfer characteristics resulting from both forward (FW) and reverse (RV) V_g sweep at a rate of 1.5 V/s, as indicated by the arrows in the figure. The I_d versus V_d output characteristics, shown in the inset of Figure 12.20, reveal a nice linearity at small V_d and indicate a good ohmic contact with the Al–Mo metal electrode. A $\mu_{FE} \equiv Lg_m/\left(WC'_{GI}V_d\right)$ of 23.2 ± 0.8 cm^2/Vs is extracted from the maximum transconductance g_m at a low V_d of 0.5 V, where W is the width of the channel. With the measurement of I_{off} limited by the noise level of ~2.3×10^{-14} A, a lower bound of ~3×10^{10} is estimated for I_{on}/I_{off}. For a more accurate estimation of I_{off}, a TFT with a large width/length (W/L) of 10^5 μm/10 μm [33] was fabricated, and the corresponding transfer characteristic is shown in Figure 12.20b. From the noise-limited I_{off} of ~8.1×10^{-14} A, a W-normalized $\tilde{I}_{off}$ of no larger than ~8.1×10^{-19} A/μm is obtained. Clearly, the low $\tilde{I}_{off}$ is a consequence of the suppression of the population of the residual donor defects in the channel by the oxidizing anneal.

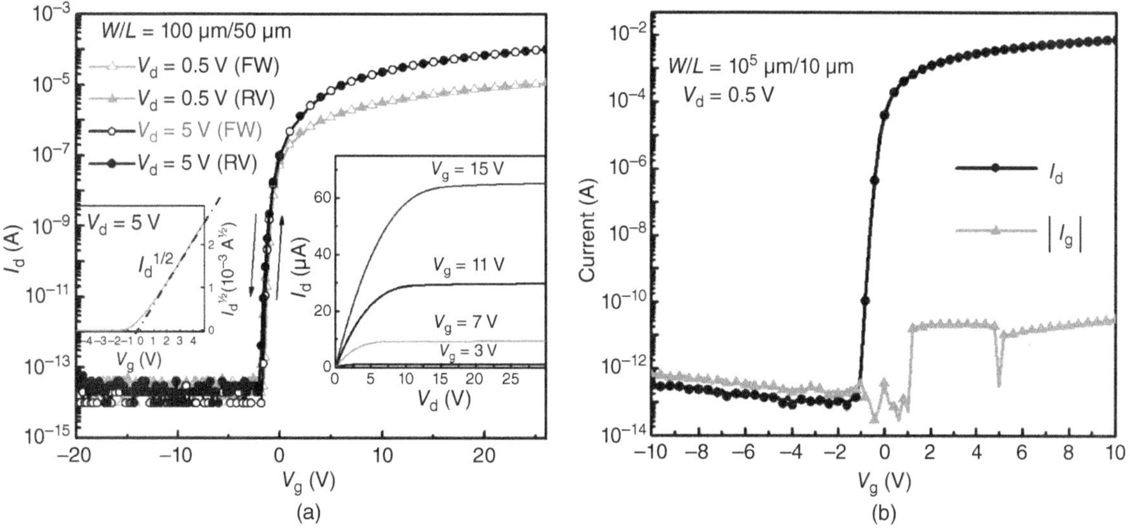

Figure 12.20 (a) The forward (FW) and reverse (RV) transfer characteristics of an EMMO ITZO-TFT at V_d of 0.5 and 5 V. (Left inset) $I_d^{0.5}$ versus V_g of the same TFT; and (right inset) the output characteristics of the same TFT at various V_g. (b) The transfer characteristics of a wide EMMO ITZO-TFT with a W of 105 μm. *Source*: Ref. [21], Xia, Z., et al. (2017). Characteristics of elevated-metal metal-oxide thin-film transistors based on indium-tin-zinc oxide. *IEEE Electron Device Letters* **38** (7): 894–897.

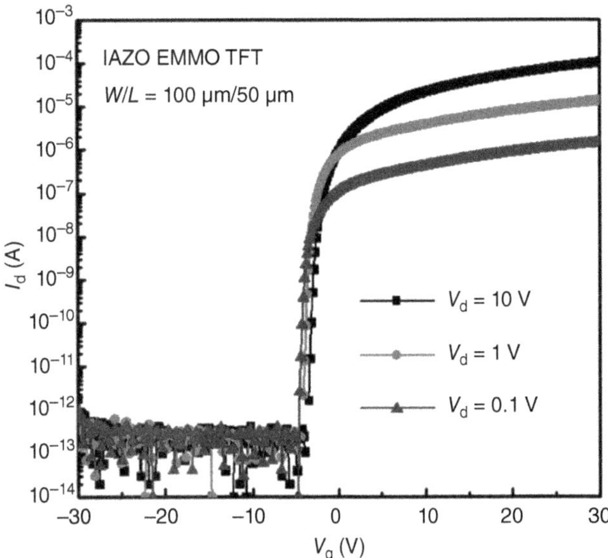

Figure 12.21 The transfer characteristics of an EMMO TFT implemented using IAZO as the active layer. *Source*: Ref. [5], Lu, L., et al. (2017). High-performance and reliable elevated-metal metal-oxide thin-film transistor for high-resolution displays. In *2016 IEEE International Electron Devices Meeting* (IEDM), pp. 32.2.1–32.2.4. IEEE, Piscataway, NJ.

Other than ITZO, EMMO TFTs have also been implemented using indium–aluminum–zinc oxide (IAZO) [5] as the active layer. The representative transfer characteristics of IAZO are shown in Figure 12.21.

12.3.3 Fluorinated EMMO TFTs

The performance of an EMMO TFT is found to improve with a longer oxidizing anneal, up to the longest reported duration of 6 h. The manufacturing efficiency would be enhanced if a comparable or better performance could be obtained with a shorter anneal. This has been achieved with the fluorination of the MO, first demonstrated on ZnO using techniques such as fluorine (F) implantation [34] and treatment in a F-containing plasma [35]. Diffusion from SiN_y:F [36] as a source of F has also been reported. Since beam-line implanters are not readily available in a display foundry, the last two techniques are more compatible with processing on a large-area substrate.

Sample preparation for studying the effects of plasma fluorination started with the deposition of 20-nm-thick IGZO thin film on an oxidized *p*-type silicon substrate. The composition of the IGZO target was In_2O_3:Ga_2O_3:ZnO = 1:1:1. Some of the samples were subsequently treated at 300 °C in a tetrafluoromethane (CF_4) plasma for 10 min to form fluorinated IGZO, hereafter designated as IGZO:F. A 300-nm-thick gas-permeable PECVD SiO_x passivation layer was deposited, before a thermal treatment at 400 °C for 2 h in an O_2 atmosphere was conducted. The presence of F near the surface of the IGZO directly exposed to the plasma and its drive-in during the subsequent thermal treatment were confirmed (Figure 12.22) using secondary ion mass spectrometry (SIMS) [37].

The effects of fluorination followed by oxidation on the population of defects were characterized by comparing the XPS (Figure 12.24) of the IGZO:F samples and the reference IGZO samples. The top 5 nm of the samples, together with any surface contaminants, were removed using in-situ Ar ion bombardment prior to the measurement. Oxygen 1*s* spectra were obtained from unannealed samples of IGZO (Figure 12.23a) and IGZO:F

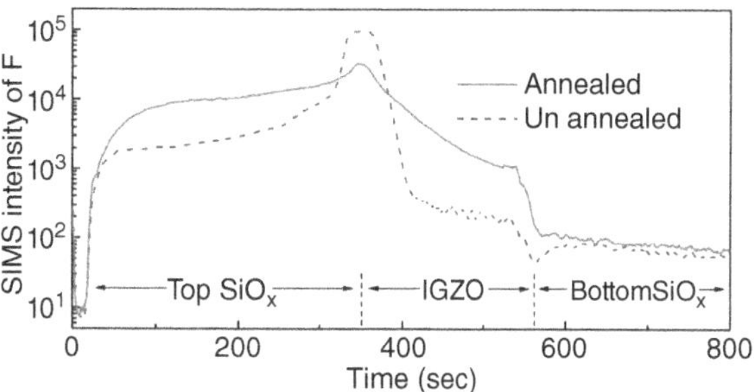

Figure 12.22 SIMS depth profiles of F in IGZO:F before and after O_2 annealing. *Source*: Ref. [37], Feng, Z., et al. (2018). Fluorination-enabled monolithic integration of enhancement- and depletion-mode indium-gallium-zinc oxide TFTs. *IEEE Electron Device Letters* **39** (5): 692–695.

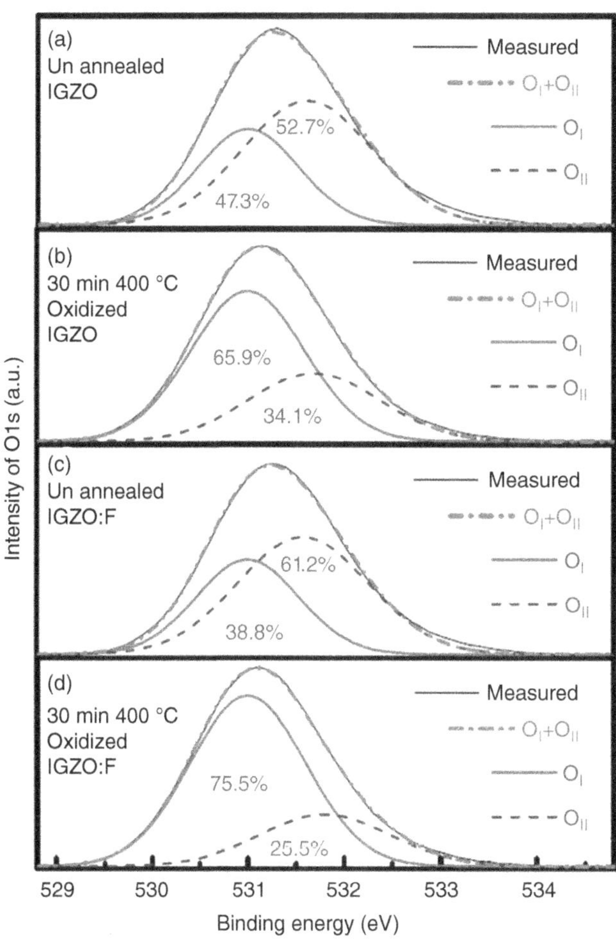

Figure 12.23 The XPS spectra of the O 1*s* state for (a) unannealed IGZO, (b) oxidized IGZO, (c) unannealed IGZO:F, and (d) oxidized IGZO:F, together with the Gaussian–Lorenz decompositions. *Source*: Ref. [35], Lu, L., et al. (2018). A comparative study on fluorination and oxidation of indium–gallium–zinc oxide thin-film transistors. *IEEE Electron Device Letters* **39** (2): 196–199.

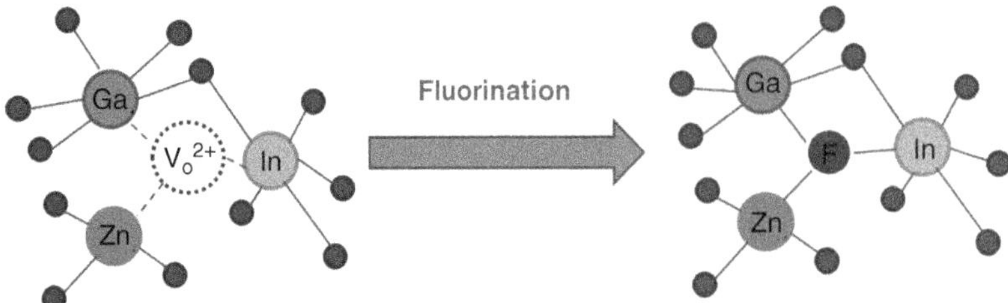

Figure 12.24 Schematic illustration of the passivation of a defect by the incorporation of F.

(Figure 12.23c), and from samples of IGZO (Figure 12.23b) and IGZO:F (Figure 12.23d) annealed at 400 °C for 30 min in O_2.

Each spectrum is resolved into two Gaussian–Lorenz components and again labeled "O_I" and "O_{II}," centered respectively at ~530.8 and ~531.6 eV. Normalized by the total area of the corresponding composite spectrum, the percentage area of O_{II} is calculated and taken as a measure of the population of defects [38]. Probably due to ion bombardment–induced damage, it can be seen that the unannealed IGZO:F, with O_{II} at 61.2%, is more "defective" than the unannealed IGZO, with O_{II} at 52.7%. After the oxidation and F drive-in, O_{II} for IGZO dropped to 34.1%, and that for IGZO:F is the lowest at 25.5%. This is an indication of the effectiveness of F in passivating the defects in a MO semiconductor, as schematically shown in Figure 12.24 [39].

The fabrication of a fluorinated EMMO TFT started with the sputter deposition and patterning of a 100-nm-thick Mo gate electrode (Figure 12.25a) on an oxidized *p*-type silicon substrate. A stack of insulating films consisting of PECVD SiO_x on SiN_y was prepared. A 25-nm-thick IGZO or ITZO active layer was subsequently deposited at room temperature, respectively, by the sputtering of an IGZO target or the co-sputtering of ZnO and ITZO targets in a RF magnetron-sputtering machine at a process pressure of 3 mTorr in a mixed atmosphere of O_2 and Ar. The base pressure was ~1 μTorr. Some of the samples were subsequently treated in a CF_4 plasma (Figure 12.25b) for 10 min to form IGZO:F and ITZO:F. Active islands of MO film were patterned using 1/1000 molar aqueous HF acid solution, before a 300-nm-thick gas-permeable SiO_x ES/passivation layer was deposited using the same PECVD processes. Following a thermal treatment at 400 °C for 2 h in an O_2 atmosphere, the contact areas (Figure 12.25c) were opened before the metal electrodes, consisting of sputtered and patterned stacks of 300-nm-thick Al on 50-nm-thick Mo, were defined. The overlap (Figure 12.25d) between each end of the gate electrode and the corresponding S/D electrode was 4 μm. Finally, conductive S/D regions were formed using a thermal treatment at 400 °C in O_2 for 4 h.

The I_d versus V_g transfer characteristics were measured for EMMO TFTs with channels made of IGZO (Figure 12.26a) and IGZO:F (Figure 12.26b). For $L = 2$ μm, the field-effect modulation of the channel was lost in the former but nicely maintained in the latter. For TFTs with longer L, the SS and V_{on} were enhanced by the fluorination treatment, improving from 315 mV/decade and −1.8 V to 232 mV/decade and −0.1 V, respectively. Different from the mechanism responsible for conventional drain-induced barrier lowering, V_{on} is sensitive to the density of residual charge carriers in the channel, and its variation is attributed to the migration of donor defects into the channel region from the thermally induced S/D regions. The observed L dependence of V_{on} is consistent with more of the defects reaching the middle of a shorter channel. The significantly smaller $|V_{on}|$ of the IGZO:F TFT agrees with a F-induced reduction in the population of donor defects. The respective μ_{FE} is 10.4 and 10.3 cm^2/Vs for IGZO and IGZO:F TFTs, slightly smaller for the fluorinated device and attributed to additional carrier scattering by ionized F.

Positive-bias thermal stress (PBTS) was carried out at 85 °C with V_g set at V_{on} + 20 V. For the nonfluorinated EMMO TFTs, the family of transfer curves shifted in the negative V_g direction (Figure 12.27) with increasing stress time. A shift in V_{on} (ΔV_{on}) of ~ − 2.3 V was obtained after 10^4 s of stress, as shown in Figure 12.27a. Such negative ΔV_{on} possibly resulted from holes transporting to the backchannel–ES interface, thermal emission over

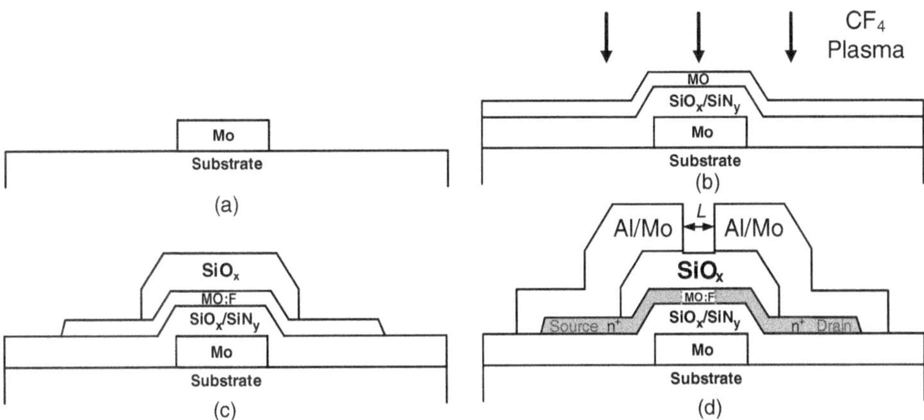

Figure 12.25 The evolution of the schematic cross-sections of a fluorinated EMMO TFT during its construction: (a) Mo gate electrode formation, (b) fluorination in a CF4 plasma, (c) patterning active island and passivation etch-stop layer, and (d) metal electrode definition and S/D formation. *Source*: Ref. [40], Xia, Z., et al. (2018). P-15: the use of fluorination to enhance the performance and the reliability of elevated-metal metal-oxide thin-film transistors. *SID Symposium Digest of Technical Papers* **49** (1): 1235–1238.

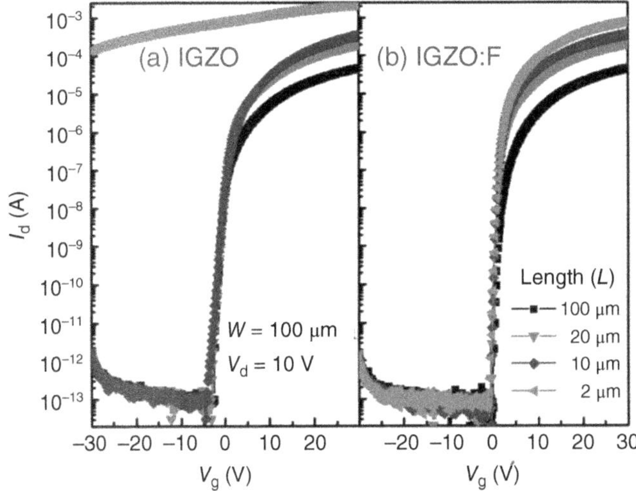

Figure 12.26 The transfer characteristics measured at V_d of 10 V for EMMO TFTs with (a) IGZO and (b) IGZO:F channels thermally oxidized at 400 °C for 2 h. *Source*: Ref. [40], Xia, Z., et al. (2018). P-15: the use of fluorination to enhance the performance and the reliability of elevated-metal metal-oxide thin-film transistors. *SID Symposium Digest of Technical Papers* **49** (1): 1235–1238.

the interfacial barrier, and trapping in the ES layer [41]. In comparison, significantly reduced degradation was observed for the fluorinated EMMO TFTs against PBTS, with ΔV_{on} of ~ -0.2 V measured after 10^4 s of stress, as shown in Figure 12.27b. This weaker dependence of ΔV_{on} on PBTS is a demonstration of the effectiveness of fluorination in further eliminating defects in the channel region of a MO TFT.

The case of ITZO EMMO TFTs is similar. A comparison of the characteristics of nonfluorinated and fluorinated TFTs is given in Table 12.3. The better metrics (more positive V_{on} and better reliability) could be attributed to a more stable F–metal bond than O–metal bond and thus a more effective resistance against the formation of oxygen-related defects in IGZO and ITZO [40].

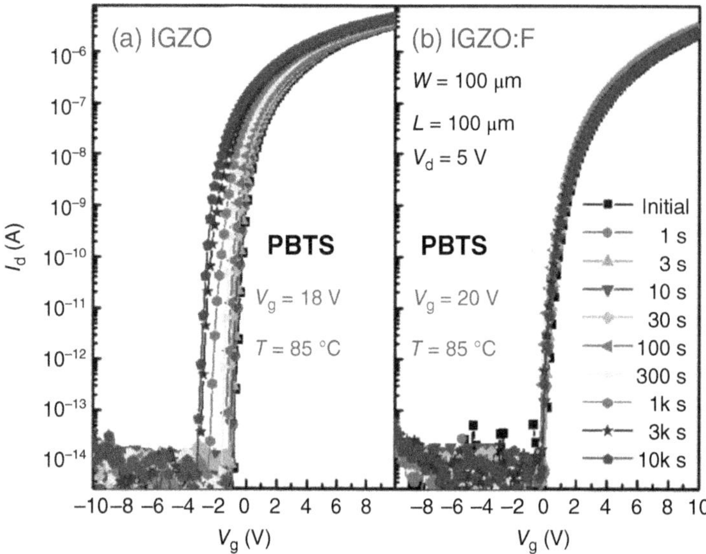

Figure 12.27 Time evolution of the transfer curves for (a) nonfluorinated and (b) fluorinated EMMO IGZO-TFTs (W/L = 100 μm/100 μm) subjected to the same PBTS ($V_g = V_{on}$+20 V at 85 °C). *Source*: Ref. [40], Xia, Z., et al. (2018). P-15: the use of fluorination to enhance the performance and the reliability of elevated-metal metal-oxide thin-film transistors. *SID Symposium Digest of Technical Papers* **49** (1): 1235–1238.

Table 12.3 The electrical attributes of EMMO TFTs with different active layers.

	μ_{FE} (cm²/Vs)	*SS* (mV/decade)	V_{on} (V)	Maximum I_{on}/I_{off}
IGZO	10.4	315	−1.8	>10^{10}
IGZO:F	10.3	232	0.1	>10^{10}
ITZO	19.2	135	−1.5	>10^{10}
ITZO:F	18.6	170	−0.9	>10^{10}

Source: Ref. [40], Xia, Z., et al. (2018). P-15: the use of fluorination to enhance the performance and the reliability of elevated-metal metal-oxide thin-film transistors. *SID Symposium Digest of Technical Papers* 49 (1). 1235–1238.

12.3.4 Resilience of Fluorinated MO against Hydrogen Doping

Because H is an efficient donor impurity in many MO semiconductors, its detrimental effects in the channel region of a TFT could be reduced by making the channel less sensitive to or preventing its incorporation. Aluminum oxide (Al_2O_3) is popularly used as a barrier against gas and moisture intrusion in a variety of devices [42]. Al_2O_3 formed using atomic layer deposition (ALD) has also been adopted [43, 44] to limit the migration of ambient H to the channel region of an IGZO-TFT. However, the H originating from the precursors, such as trimethyaluminum and water, used for the ALD remains a problem. IGZO:F is found to effectively resist the detrimental influence of H [45]. This resilience against H doping has been investigated using PECVD *a*-Si:H to simulate a potentially H-rich processing environment to which a TFT could be exposed. As is evident from the comparison shown in Figure 12.28, IGZO:F-TFTs fluorinated for 10 min in a CF_4 plasma exhibit significantly improved intrinsic resistance against H-induced degradation, including that induced by the H generated during thermal ALD Al_2O_3.

The reduced H content in IGZO:F compared to that in IGZO upon exposure to H is revealed using SIMS (Figure 12.29). The depth profiles of In, Zn, Ga, Si, H, and F were obtained across film stacks consisting of

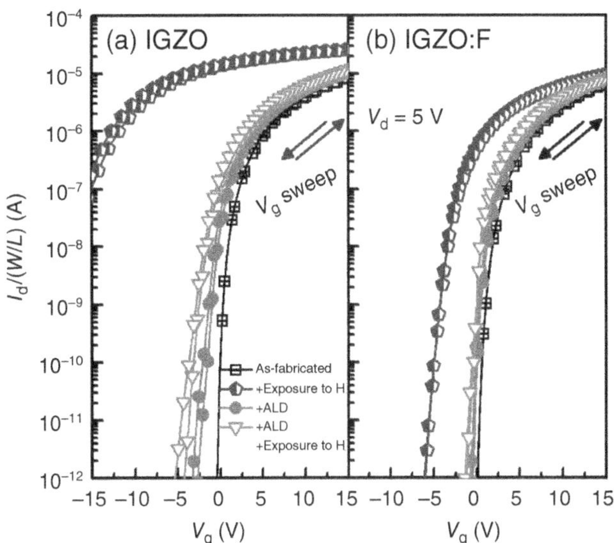

Figure 12.28 The transfer characteristics of EMMO TFTs built with (a) IGZO and (b) IGZO:F TFTs, including as-fabricated; directly subjected to *a*-Si:H deposition (+exposure to H), covered with a 40-nm-thick ALD-Al_2O_3 diffusion barrier (+ALD), and subsequently subjected to *a*-Si:H deposition (+ALD+exposure to H). *Source*: Ref. [45], Wang, S., et al. (2020). Resilience of fluorinated indium-gallium-zinc oxide thin-film transistor against hydrogen-induced degradation. *IEEE Electron Device Letters* **41** (5): 729–732.

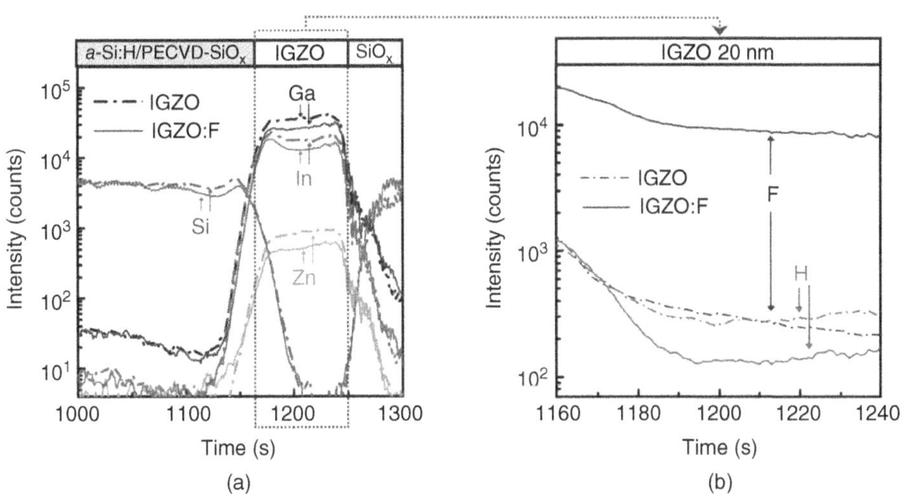

Figure 12.29 SIMS depth profiles of (a) the intensities of In, Zn, Ga, and Si across the stacks of SiO_x/IGZO or IGZO:F after exposure to H (*a*-Si:H deposition); and (b) the intensities of H and F across the IGZO or IGZO:F in the same stacks. *Source*: Ref. [45], Wang, S., et al. (2020). Resilience of fluorinated indium-gallium-zinc oxide thin-film transistor against hydrogen-induced degradation. *IEEE Electron Device Letters* **41** (5): 729–732.

50-nm-thick SiO_x on 20-nm-thick IGZO or IGZO:F, capped with a 150-nm-thick *a*-Si:H layer to mimic the exposure to H. All stacks were annealed at 400 °C for 2 h in O_2 after the deposition of the SiO_x layer. It is not surprising that the profiles of In, Zn, and Ga track each other well, because they are associated with the IGZO or IGZO:F host matrices. Figure 12.29a shows the profiles of In, Zn, Ga, and Si, with the solid and dashed lines, respectively, representing those built with IGZO:F and IGZO. The top and bottom boundaries of IGZO and IGZO:F are taken to be the locations where the intensities of Ga drop to 50% of its peak value. The intensities of H and F across IGZO or IGZO:F are plotted and compared in Figure 12.29b.

It is clear the F content in the stack built with IGZO is about two orders of magnitude lower than that in the stack built with IGZO:F. In fact, the F signal measured in the former could have resulted from H_3O^+ having an ionic mass similar to the atomic mass of F. It is also clear the H content in the stack built with IGZO is higher than that in the stack built with IGZO:F. This is direct evidence of the increased resistance of IGZO:F against the incorporation of H. Despite the mechanism suggested in Figure 12.24, there has been a relative dearth of theoretical studies on how F is incorporated in amorphous IGZO [46]. One could speculate that the population of M:H or M:OH could have been reduced by the presence of F due to the formation of M:F, where M stands for one of the cations in IGZO. Consequently, when used in combination with an ALD Al_2O_3 diffusion barrier, IGZO:F-TFTs are more suitable than IGZO-TFTs for monolithic integration with H-containing devices, such as *a*-Si:H photo-diodes and LTPS TFTs [47, 48].

12.3.5 Technology and Display Resolution Trend

An active-matrix display consists of an array of pixels, each containing at least an "address" TFT (Figure 12.30a) controlling the optical state of its host pixel (Figure 12.30b). Two prototype pixels are considered: Type I consists of an address TFT charging a capacitor of total effective capacitance C_T; Type II consists of an address TFT charging a "driving" TFT and a capacitor. To the address TFT of a Type II pixel and in the context of timing control, the driving TFT appears to contribute an additional capacitive load. Consequently, Type II is conceptually equivalent to Type I, again with an effective C_T obtained by combining the capacitance contributed by the driving TFT and other capacitive loads in the pixel. It is easy to recognize Types I and II as the respective pixels for prototype AM-LCDs and active-matrix OLED displays. The gate electrodes of each row of TFTs are connected to a scan line (Figure 12.31b); the "scan signal" V_{SC} driving the TFTs is provided by a corresponding output of the scan drivers schematically located at the left edge of the display. The TFTs in a column share a data line, and the "data signal" V_{DT} carried by which is provided by a corresponding output of the data drivers schematically located at the top edge of the display. The voltage across C_T is denoted by V_{PX}.

A pixel is typically composed of three subpixels (Figure 12.30a), one each red, green, and blue. There is a delay T_{DL} for signals propagating along the scan and data lines, with distributed finite resistance (R_{PX}) and capacitive loading (C_{PX}) associated with each subpixel. The pixel located at the top-left corner of the display is in the immediate vicinity of the drivers and suffers from a negligible delay, whereas the "farthest" one located at the lower-right corner (Figure 12.30a) suffers from the longest delay. When an address TFT is switched on, C_T is charged. However, it takes a finite amount of charging time T_{CH} for the pixel voltage V_{PX} on C_T to approach V_{DT}. V_{PX} in turn determines the amount of light (hence the gray level) transmitted or emitted by the pixel. For proper driving, V_{DT} must not be changed until the address TFT is adequately switched off. Denoting T_{SC} and T_{DT} (Figure 12.31), respectively, as the propagation delay associated with the scan and data lines reaching the farthest pixel, one has $T_{DL} \equiv T_{SC} + T_{DT}$ and

$$T_{LA} \equiv \underbrace{T_{SC} + T_{DT}}_{T_{DL}} + T_{CH} < T^*_{LA} \equiv \frac{1}{fN_V}, \tag{12.4}$$

where T^*_{LA} and T_{LA} are the respective maximum allowable and the actual line-address times, f is the frame rate, and N_V is the number of scan lines.

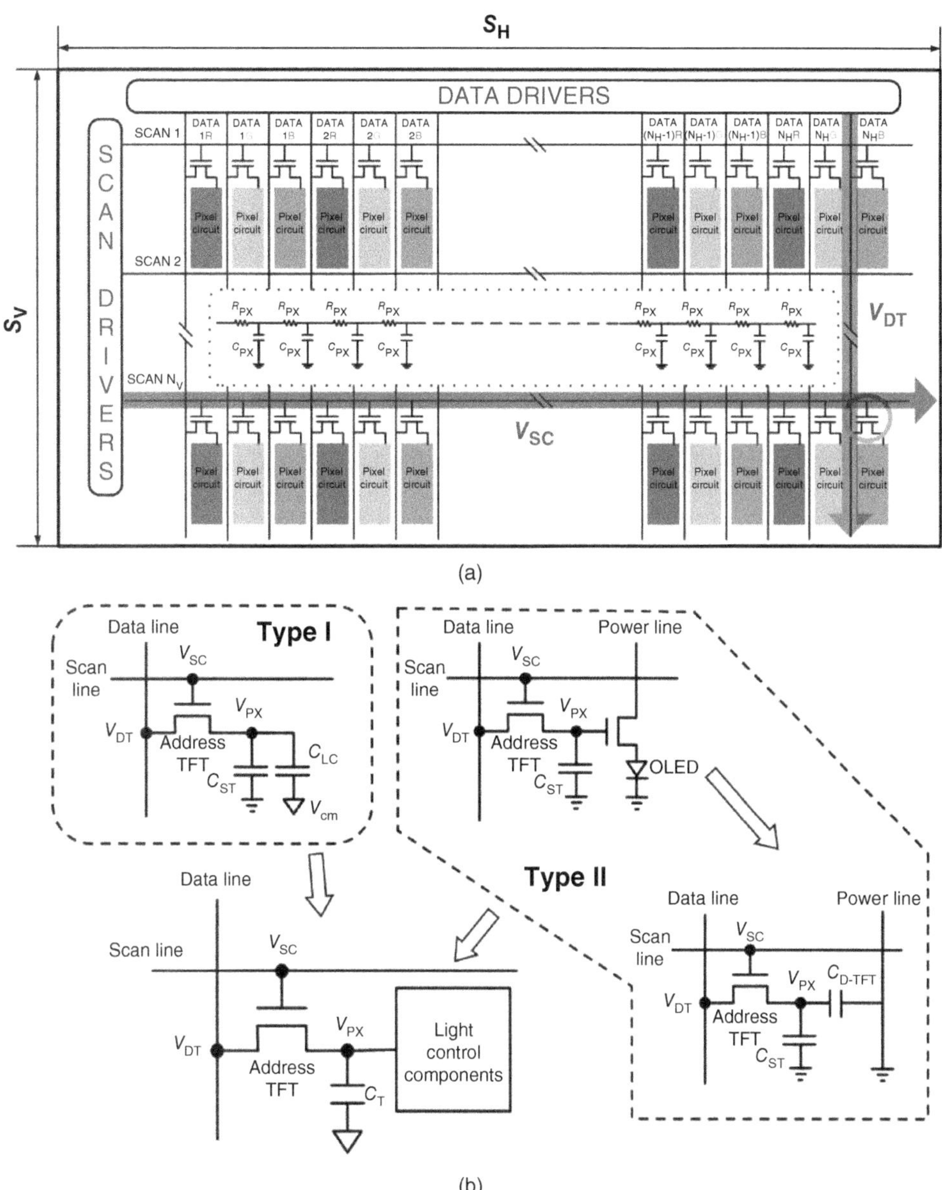

Figure 12.30 (a) A display panel shown schematically as a distributed resistor–capacitor network. (b) Schematic types I and II pixel circuits. *Source*: Ref. [49], Xia, Z., et al. (2020). A timing model for the optimal design of a prototype active-matrix display. *IEEE Transactions on Electron Devices* **67** (8): 3167–3174.

Signal propagation along an interconnect is determined by the product of its resistance (R) and capacitance (C), with a larger RC time-constant corresponding to a longer delay. The two main components making up C are the capacitance associated with the attached address TFTs and that associated with the crossover at the intersection with other interconnects. The TFT-related capacitance can be further decomposed to include an intrinsic gate-channel capacitance and a parasitic gate–S/D overlap capacitance, with respective values sensitive to the architecture of the TFT. When a scan line crosses over a data line, an overlap area of $W_{SC}W_{DT}$ contributes a

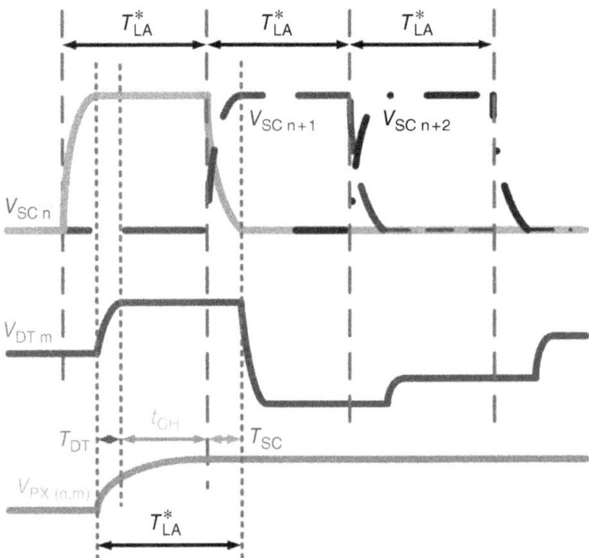

Figure 12.31 Timing diagrams illustrating the charging of a capacitor of a selected pixel in a matrix. *Source*: Ref. [49], Xia, Z., et al. (2020). A timing model for the optimal design of a prototype active-matrix display. *IEEE Transactions on Electron Devices* **67** (8): 3167–3174.

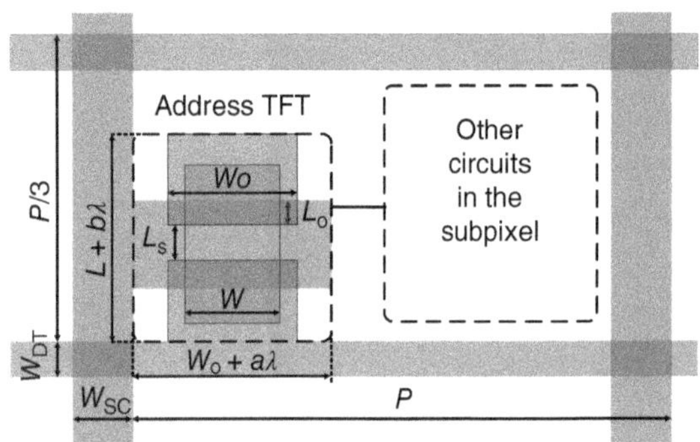

Figure 12.32 Schematic layout of a subpixel. *Source*: Ref. [49], Xia, Z., et al. (2020). A timing model for the optimal design of a prototype active-matrix display. *IEEE Transactions on Electron Devices* **67** (8): 3167–3174.

capacitance of $W_{SC}W_{DT}C'_X$ to both C_{PX-S} and C_{PX-D}, where W_{SC}, W_{DT}, C'_X, C_{PX-S}, and C_{PX-D} are the respective width of the scan line, width of the data line, crossover capacitance per unit area, and effective capacitance per subpixel along the scan and data lines.

The capacitive loading attributed to the address TFT is different for the scan and data lines. For the scan line, C_{PX-S} includes also the capacitance of the active channel, and that arising from the overlap between the gate electrode and both S/D regions. Consequently,

$$C_{PX-S} = \underbrace{W_{SC}W_{DT}C'_X}_{\text{Crossover}} + \underbrace{(2W_oL_o + WL_s)C'_{GI}}_{\text{Overlap and channel capacitance}} \tag{12.5}$$

where L_s (Figure 12.32) is the separation between the S/D electrodes along the channel length direction. Since all but one of the TFTs along the data line are turned off, C_{PX-D} can be approximated by including the overlap capacitance on only one side of each TFT. Consequently,

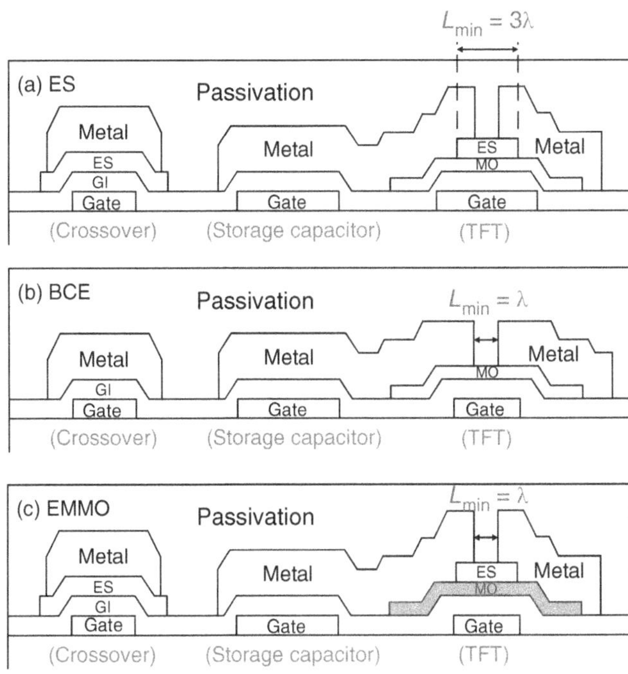

Figure 12.33 The schematic cross-sections of (a) an ES, (b) a BCE and (c) an EMMO TFT, showing also the corresponding crossover capacitors.

$$C_{\text{PX-D}} \approx \underbrace{W_{\text{SC}} W_{\text{DT}} C'_{\text{X}}}_{\text{Crossover}} + \underbrace{W_{\text{o}} L_{\text{o}} C'_{\text{GI}}}_{\text{Overlap capacitance}} \tag{12.6}$$

Figure 12.33 illustrates schematic cross-sections of the ES, BCE and EMMO TFT architectures. The L_{min} of an ES TFT is triple those of a BCE or EMMO TFT, which leads to a ninefold increase in the gate capacitance of a minimum-geometry ES TFT with current-drive ability (i.e., W/L) equal to that of a minimum-geometry BCE or EMMO TFT. However, the C'_{X} offered by an ES TFT technology is smaller, because of the availability of an extra insulating ES layer. It is clear an EMMO TFT technology combines the advantages of both, offering the smaller TFT-related capacitance of a BCE TFT technology and the smaller C'_{X} of an ES TFT technology.

With the number of pixels in the horizontal dimension of a display denoted by N_{H} and the display driven in the quad-drive configuration [49], Figure 12.34 shows the dependence of the maximum N_{H} on the μ_{FE} of the address TFT for different TFT architectures. For any given μ_{FE} and subject to the constraint given in Equation (3.1), the bound on the maximum N_{H} is the lowest for a display realized using the ES technology because of the larger TFT footprint and its related parasitic gate–S/D overlap capacitance. On the other hand, the bound is higher for the EMMO TFT technology than for the BCE TFT technology because of the reduced C'_{X}.

12.4 Enhanced EMMO TFT Technologies

A comparison of the MO TFT technologies based on the BCE, TG SA, and EMMO architectures is given in Table 12.4, providing the context for the desire for enhanced EMMO TFT technologies. Two variations are presently described, one with reduced mask-count and the other with reduced parasitic overlap capacitance.

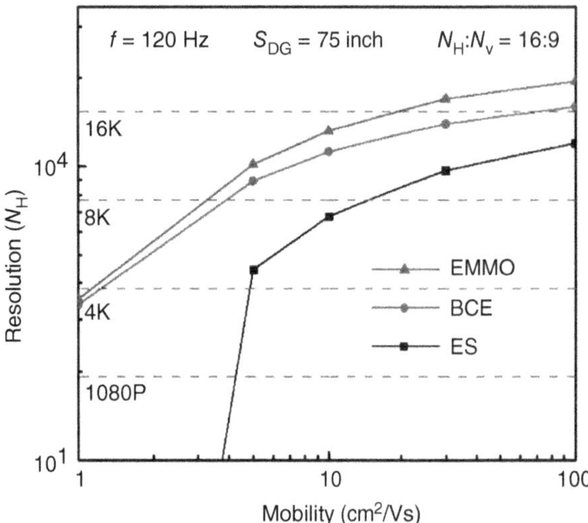

Figure 12.34 The dependence of maximum N_H on μ_{FE} of the address TFT for ES, BCE, and EMMO TFT architectures. S_{DG} is the length of the diagonal of the display; and μ_{FE} of 1, 10, and 100 cm_2/Vs roughly correspond to those of *a*-Si:H, MO, and LTPS, respectively.

Table 12.4 Comparison of MO TFT technologies based on the BCE, TG SA, and EMMO architectures.

Structures	BCE	TG SA	EMMO
Channel damage	Yes	Some	No
Capacitance	Medium (1λ overlap)	Small (self-aligned)	Medium (1λ overlap)
L_{min}	1λ	1λ/2λ (single/double gates)	1λ
Masking steps (manufacturing cost)	3	4	4

12.4.1 3-EMMO TFT Technology

Enabled by the annealing-induced formation of the S/D regions self-aligned to the edges of the impermeable S/D electrodes, the four-mask EMMO (4-EMMO) TFT technology described in Section 12.3 enjoys the combined benefits of the protection of the backchannel by the ES layer and the small footprint of a BCE TFT. However, the advantages are gained at the expense of an extra masking step employed for the definition of the ES layer, and hence a higher manufacturing cost than that of a three-mask BCE technology (Table 12.4). The first enhancement presently described is a 3-EMMO TFT technology: Instead of the active island being conventionally patterned using a separate mask, its postponed patterning is self-aligned to the combined patterns of the ES layer and the S/D electrodes.

The fabrication of a 3-EMMO TFT started with the sputtering and patterning (Mask 1) of 80-nm-thick ITO as the bottom-gate electrode (Figure 12.35a) on a thermally oxidized silicon wafer substrate. A layer of 100-nm-thick PECVD SiO_x as the gate dielectric (Figure 12.35b) was prepared at 300 °C before a 30-nm-thick IGZO active layer (Figure 12.35c) was sputter-deposited from a target with a composition of In_2O_3:Ga_2O_3:ZnO = 1:1:1. The sputtering atmosphere was 10% O_2 and 90% Ar at a total pressure of 3 mTorr. This was immediately followed by the deposition of another 200-nm-thick PECVD SiO_x as the ES layer, and subsequently etched and patterned (Mask 2; Figure 12.35d) in an ICP etcher running a SF_6 chemistry. The sputtered impermeable S/D electrodes

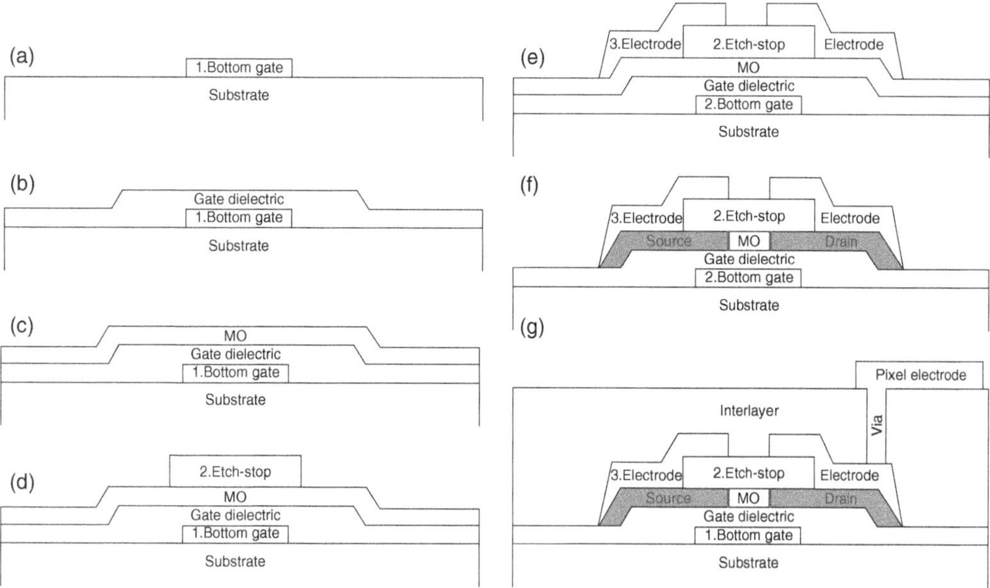

Figure 12.35 Evolution of the schematic device cross-sections during the fabrication of a 3-EMMO TFT. *Source*: (e–g) Ref. [50], Li, J., et al. (2018). Three-mask elevated-metal metal-oxide thin-film transistor with self-aligned definition of the active island. *IEEE Electron Device Letters* **39** (1): 35–38.

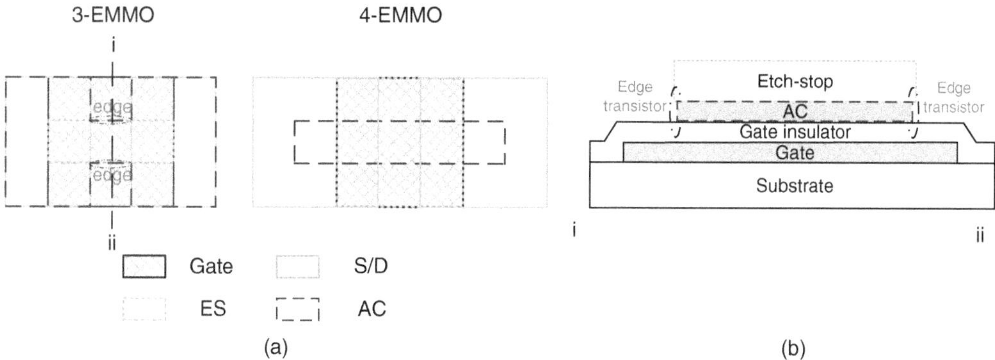

Figure 12.36 (a) Top views of the schematic layouts of 3- and 4-EMMO TFTs. (b) Cross-sectional schematic along the dashed line (i–ii), showing the location of the two edge-transistors of a 3-EMMO TFT. *Source*: Ref. [51], Li, J., et al. (2018). P-1.5: Edge effects of three-mask elevated-metal metal-oxide thin-film transistor and their elimination. *SID Symposium Digest of Technical Papers* **49**: 531–534.

(Figure 12.35e), consisting of a stack of 50-nm-thick Mo under 300-nm-thick Al, were next patterned (Mask 3). Self-aligned to the combined and projected shadows of the ES layer and the S/D electrodes, the active island was defined (Figure 12.35f) by etching the exposed IGZO in dilute aqueous HF acid. The post-TFT backend process consisted of the deposition of a 300-nm-thick SiO_x passivation layer and the subsequent opening of electrical access to the electrodes (Figure 12.35g). A final heat treatment was performed at 400 °C in O_2 for 4 h to thermally "activate" and form the conductive S/D regions [3].

The respective masking levels used for the construction of the 3- and 4-EMMO TFTs are schematically compared in Figure 12.36. Since the need to accommodate lithographic misalignment between the active island (AC) and the ES layer or the S/D electrodes is eliminated, the footprint of a minimum 3-EMMO TFT can be made even smaller than that of an equivalent 4-EMMO TFT. However, parasitic "edge"-transistors are formed because of the self-aligned patterning of the active island [51], making a 3-EMMO TFT more sensitive than a 4-EMMO TFT to variations in process conditions. A range of techniques [51] have been investigated for mitigating the detrimental effects of such edge-transistors.

The I_d versus V_g transfer characteristics of 3-EMMO TFTs measured at various V_d are shown in Figure 12.37a. Negligible hysteresis is observed during the cyclic V_g-sweep. Defined as the V_g at which an exponential increase in I_d is first observed, a turn-on voltage V_{on} of ~ -1 V is obtained. Also extracted are a peak μ_{FE} of ~5 cm²/Vs, an SS of ~300 mV/decade, and an I_{on}/I_{off} of over 10^7. These performance metrics are comparable to those reported for 4-EMMO TFTs [41–43]. The I_d versus V_d output characteristics at various V_g are shown in Figure 12.37b. The nicely linear dependence of I_d on V_d at small V_d reflects the good ohmic contact between the thermally induced S/D regions and the S/D electrodes.

Good consistency in the transfer curves of 3-EMMO TFTs (Figure 12.38a) was obtained across a 100-mm substrate. After a short 30-min activation anneal, the ρ of the S/D regions (Figure 12.38b) and the specific contact resistance with the Mo contact layer (Figure 12.38c) quickly reached respective steady-state values of ~20 mΩ·cm and $\sim 10^{-4}$ Ω·cm².

Figure 12.39a shows the temperature dependence of the transfer characteristics of a single 3-EMMO TFT measured between 25 and 160 °C. The I_{on} at high V_g slightly increases with temperature, consistent with the weakly thermally activated conduction mechanism in IGZO [44]. With its I_{off} and V_{on} hardly changed, the 3-EMMO TFT exhibited a temperature operating range wider than that of silicon-based TFTs. Figure 12.39b shows a comparison of the characteristics taken in the dark and those measured under white-light illumination at a power density of 5 W/m² generated using a halogen lamp. The good overlap of the characteristics is a strong indication of their relative insensitivity to illumination.

3-EMMO TFTs have been subjected to a variety of stress tests, including pure NBS/PBS, bias thermal stress (NBTS/PBTS), and bias stress with illumination (NBIS/PBIS). The S/D electrodes are grounded, and V_g of −20 and

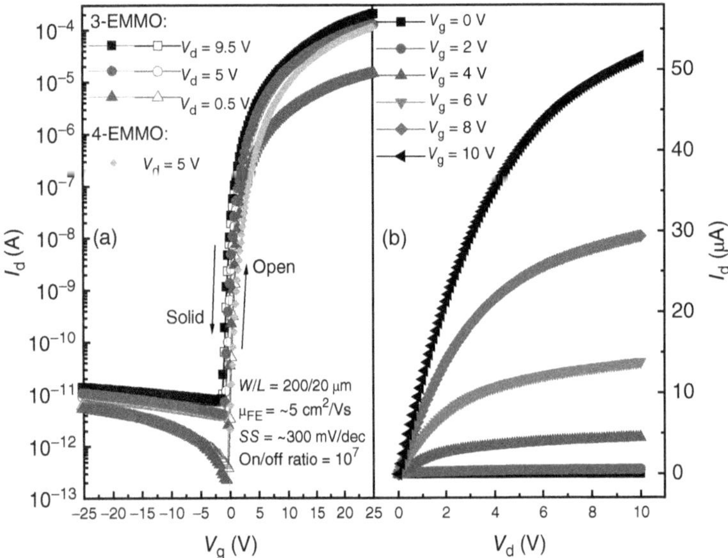

Figure 12.37 The (a) transfer and (b) output characteristics of a 3-EMMO IGZO-TFT. The nice linearity at small V_d is an indication of the good ohmic contact to the S/D electrodes. *Source*: Ref. [50], Li, J., et al. (2018). Three-mask elevated-metal metal-oxide thin-film transistor with self-aligned definition of the active island. *IEEE Electron Device Letters* **39** (1): 35–38.

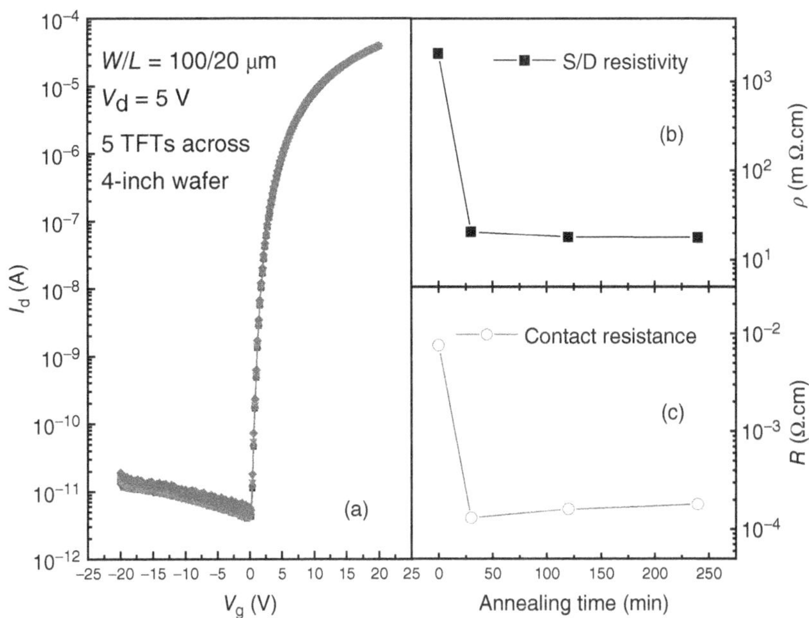

Figure 12.38 (a) Good uniformity of the transfer curves of 3-EMMO IGZO-TFTs across a 100-mm wafer. The dependence on the activation annealing time of the (b) resistivity of and (c) contact resistance with the S/D regions.

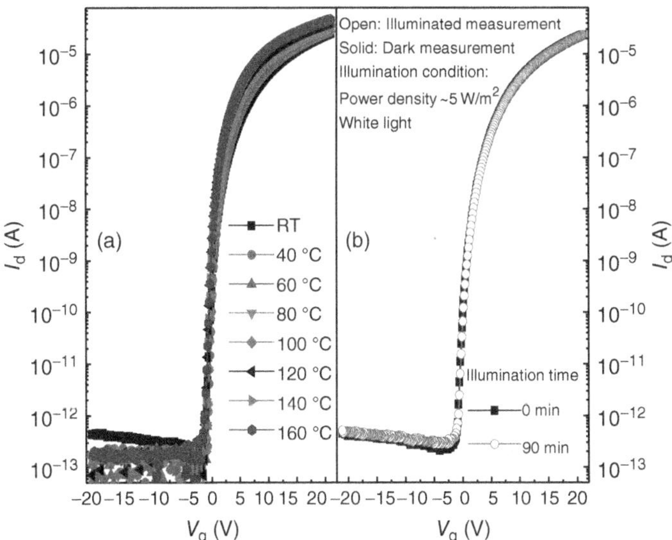

Figure 12.39 (a) The dependence of the transfer characteristics of a 3-EMMO IGZO-TFT on the operating temperature. (b) A comparison of the transfer characteristics of a 3-EMMO IGZO-TFT measured in the dark and under white-light illumination.

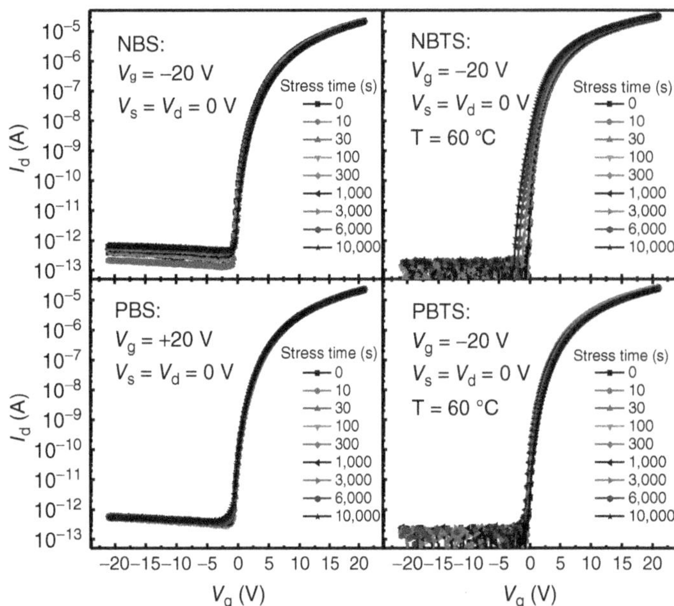

Figure 12.40 The transfer characteristics of a 3-EMMO TFT subjected to NBS, PBS, NBTS, and PBTS. A small amount of degradation is observed after NBTS.

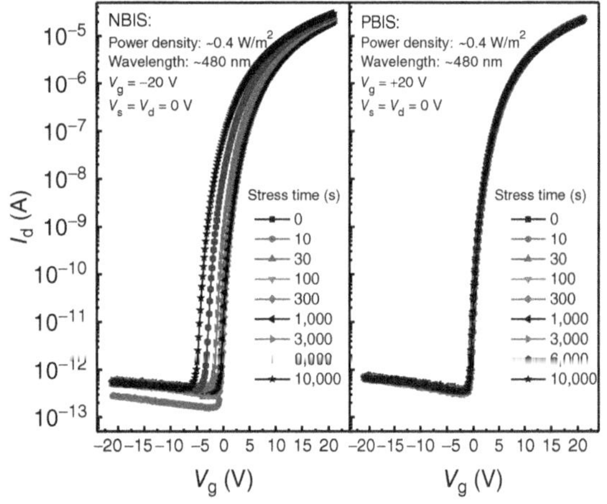

Figure 12.41 The transfer characteristics of a 3-EMMO TFT subjected to (left) NBIS and (right) PBIS.

+20 V are applied for the respective NBS and PBS. Little degradation (Figure 12.40) is observed in the 3-EMMO TFTs subjected to NBS/PBS. This can be attributed to the protection of the inherent ES layer and the effective passivation of the defects in the channel region during the oxidizing anneal [43]. When subjected to stress at 60 °C, the transfer characteristics (Figure 12.40) of the 3-EMMO TFTs are relatively stable under PBTS but shift slightly under NBTS. For the latter, a $\Delta V_{on} \sim -2$ V without *SS* deterioration is observed. Such degradation is often attributed to the creation of dangling bonds and the trapping of holes at the interface [52].

A 3-EMMO TFT suffers from the severest degradation (Figure 12.41) under NBIS. The transfer characteristics were measured in the dark after each illumination session, and the delay between light-off and electrical measurement was less than 1 s. A continuous negative ΔV_{on} up to ~ -6 V is observed without significant change in *SS*. It has been reported that positive charges, such as holes, are photogenerated with the assistance of V_O and

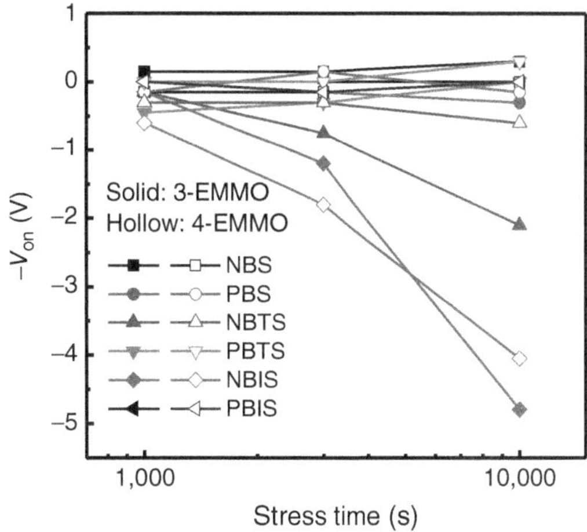

Figure 12.42 A comparison of the reliability characteristics of 3- and 4-EMMO TFTs. *Source*: Ref. [50], Li, J., et al. (2018). Three-mask elevated-metal metal-oxide thin-film transistor with self-aligned definition of the active island. *IEEE Electron Device Letters* **39** (1): 35–38.

drift under the applied electric field to the front channel interface. Some of the holes arriving at the interface are trapped, thus accounting for the negative ΔV_{on}.

The reliability of 3- and 4-EMMO TFTs is compared in Figure 12.42, measured by the stress-induced ΔV_{on} under the aforementioned stress conditions. It can be seen that the reliability of both is largely comparable; the 3-EMMO TFTs exhibit a larger $|\Delta V_{on}|$ after stress durations of 3000 and 10,000 s. This could be attributed to the poorer reliability of the edge-transistors resulting from the self-aligned definition of the active island of a 3-EMMO TFT.

12.4.2 Self-Aligned EMMO TFTs

Unlike the minimal overlap between the S/D regions and the gate electrode of a SA TFT, the corresponding overlap is finite and lithographically constrained in a non-self-aligned EMMO TFT. The increase in the parasitic overlap capacitance increases the delay of signals propagating along the scan and data lines of an active-matrix flat-panel display [49] made of non-self-aligned TFTs (Table 12.4). While a TG architecture is predominantly self-aligned, the low-resistance metal used to implement the TG electrode is typically gas-impermeable. Consequently, any thermal process after the preparation of such an electrode would be equivalent to a non-oxidizing anneal of the channel region. Consequently, the quality of the channel is degraded due to the annealing-induced generation of defects. Despite this potential pitfall, TG SA IGZO-TFTs are deployed for the construction of active-matrix OLED television panels [53] because of a decreased propagation delay due to reduced parasitic capacitance. It would be advantageous if self-alignment can be achieved in a BG TFT. Based on the transparency of the glass substrate on which a display is made, the second enhancement presently described is a BG SA EMMO TFT: exhibiting a minimal overlap capacitance while retaining the ability of annealing the channel in an oxidizing atmosphere.

The fabrication of a BG SA EMMO TFT started with the sputter deposition and patterning of a 120-nm-thick Mo gate electrode on a transparent glass substrate. A stack of PECVD SiO_x/SiN_y gate insulator, IGZO semiconductor, gas-permeable SiO_x passivation layer, and gas-impermeable SiN_y cover layer was deposited and patterned. A negative-tone photoresist was subsequently coated on the cover layer before it was flood-exposed from the back side of the glass substrate (Figure 12.43a), thus capturing the image of the opaque Mo gate electrode. A window was etched through the exposed SiN_y cover layer (Figure 12.43b) to reveal the underlying SiO_x passivation layer. The edges of the window along the channel direction were self-aligned to those of the gate electrode. Contact regions exposing the IGZO active layer on the two sides were opened (Figure 12.43c) before the S/D electrodes consisting of the same stack of Al on Mo were patterned. Finally, conductive S/D regions self-aligned to the edges of the SiN_y cover layer, and thus also to those of the gate electrode, were formed after a heat treatment at 400 °C in O_2 (Figure 12.43d).

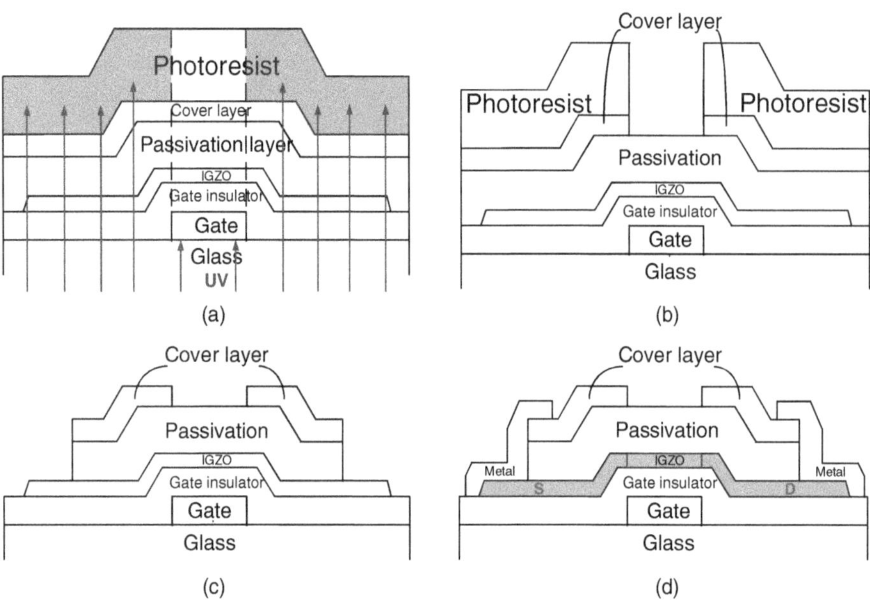

Figure 12.43 Evolution of the schematic device cross-sections of a self-aligned BG EMMO IGZO-TFT realized using backside exposure. *Source*: Ref. [25], Xia, Z., et al. (2018). A bottom-gate metal–oxide thin-film transistor with self-aligned source/drain regions. *IEEE Transactions on Electron Devices* **65** (7): 2820–2826.

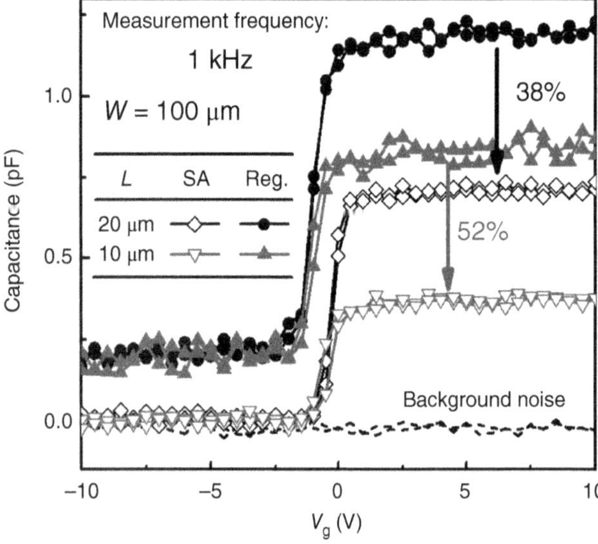

Figure 12.44 Capacitance–voltage characteristics of regular (Reg) and self-aligned (SA) EMMO IGZO-TFTs with two different values of *L*. *Source*: Ref. [25], Xia, Z., et al. (2018). A bottom-gate metal–oxide thin-film transistor with self-aligned source/drain regions. *IEEE Transactions on Electron Devices* **65** (7): 2820–2826.

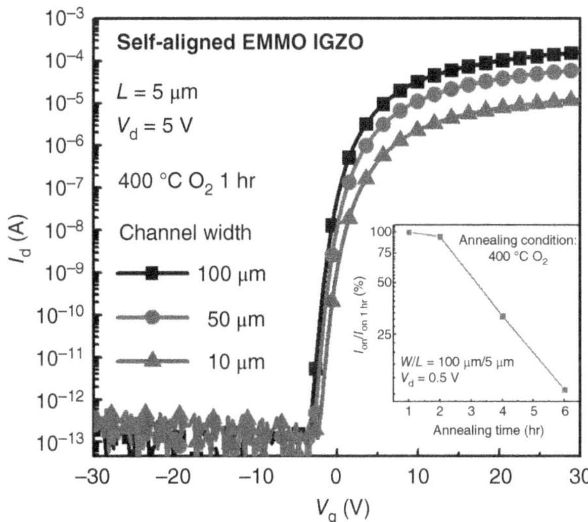

Figure 12.45 Transfer characteristics of self-aligned EMMO IGZO-TFTs at V_d = 5V with different channel widths. (Inset) The dependence of normalized I_{on} on different annealing time. *Source*: Ref. [25], Xia, Z., et al. (2018). A bottom-gate metal-oxide thin-film transistor with self-aligned source/drain regions. *IEEE Transactions on Electron Devices* **65** (7): 2820–2826.

With an overlap of ~4 μm between the gate and the S or D electrode, regular 4-EMMO TFTs with lithographically defined S/D electrodes were also constructed for comparison. The V_g-dependent gate capacitance of the regular and self-aligned TFTs is compared in Figure 12.44 for TFTs with different L. The values at sufficiently negative V_g are insensitive to L and provide estimates of the parasitic gate–S/D capacitance of ~241 and ~30 fF for the regular and the self-aligned TFTs, respectively. Clearly the parasitic capacitance of the latter, at merely ~13% of that of the former, is much smaller.

Figure 12.45 shows the I_d versus V_g transfer characteristics of self-aligned EMMO TFTs with different W at the same L = 5 μm. After 1 h of O_2 annealing at 400 °C, all devices exhibit a relatively low I_{off} of ~10^{-13} A, giving rise to a high I_{on}/I_{off} of at least 10^9. This is consistent with the effectiveness of the oxidizing annealing of the channel in reducing the population of defects in a MO semiconductor.

The inset of Figure 4.11 shows the degradation of the normalized I_{on} of TFTs subjected to different annealing durations. Compared with the I_{on} after 1 h of O_2 annealing at 400 °C, that after 2 h of annealing was slightly reduced. This is related to the continuous migration of the junctions toward the edges defined by the S/D SiN_y cover layer, and hence a slight increase in the effective L. Upon further extending the annealing duration to 4 h and then 6 h, a precipitous drop in the I_{on} was observed. This behavior is consistent with the migration of the junctions (Figure 12.13) to beyond the edges of the gate electrode, thus weakening the connection between the field-induced channel and the annealing-induced S/D regions. In the present demonstration, annealing at 400 °C in O_2 for between 1 and 3 h appears acceptable, reflecting a fairly wide process window.

12.5 Conclusion

An elevated-metal TFT architecture is described that combines some important merits of the conventional BCE and ES TFT structures. However, the realization of such a TFT with a minimum geometry for a given lithographic resolution requires the insertion of conductive S/D regions to bridge the gaps between its field-induced channel and S/D contacts. The resistivity of a class of MO semiconductors has been reported to depend on the material boundary condition and the atmosphere during a thermal anneal. For such a MO thin film with a gas-permeable cover layer annealed in an oxidizing atmosphere, its semiconducting attributes are promoted, and the resulting resistivity is high. For the same film with an impermeable cover layer, the annealing-induced generation of donor defects leads to a resistivity that is low and independent of the nature of the annealing atmosphere. This distinct property of MO semiconductors is applied to the realization of an elevated-metal metal-oxide (EMMO) TFT with annealing-induced S/D regions.

The choice of TFT architecture also affects the capacitive loading that determines the delay associated with a signal propagating along an interconnect in an active-matrix display. A shorter delay associated with a smaller loading is considered desirable. Compared to a minimum ES TFT, a minimum BCE TFT with an equivalent current-drive ability presents a smaller capacitance to the interconnect to which it is connected; but an ES TFT technology makes available a thicker insulation layer, and hence a smaller interconnect crossover capacitance. The EMMO TFT technology presents an added advantage of having a smaller device capacitance similar to that exhibited by a BCE TFT and a smaller interconnect crossover capacitance similar to that implemented using the ES TFT technology.

Further enhancement of the technology and architecture of EMMO TFTs has been described: (i) The device characteristics and reliability are improved with the incorporation of F in the channel region upon its exposure to a plasma-fluorination treatment; (ii) the mask-count is reduced by one and made equal to that required to realize a BCE TFT when the definition of the active island is postponed and made self-aligned to the combined patterns of the etch-stop layer and the S/D electrodes; and (iii) a self-aligned bottom-gate EMMO TFT with minimal parasitic gate-to-S/D overlap capacitance is realized using a maskless flood exposure through a transparent substrate, making the edges of the annealing-induced S/D regions self-aligned to those of an opaque backgate electrode.

Acknowledgments

The contributions of Lei Lu, Sisi Wang, and Zhi Ye are gratefully acknowledged; a portion of their great body of research work has been incorporated in this chapter. The devices were fabricated in the Nanosystem Fabrication Facility, and the microanalyses were carried out in the Materials Characterization and Preparation Facility, both at The Hong Kong University of Science and Technology.

References

1 Kamiya, T., Nomura, K., and Hosono, H. (2010). Present status of amorphous In–Ga–Zn–O thin-film transistors. *Science and Technology of Advanced Materials* 11117 (11): 44305–44323.

2 Kamiya, T. and Hosono, H. (2015). Oxide TFTs. In: *Handbook of Visual Display Technology* (ed. J. Chen, W. Cranton and M. Fihn), 1–28. Berlin: Springer.

3 Ide, K., Nomura, K., Hosono, H., and Kamiya, T. (2019). Electronic defects in amorphous oxide semiconductors: a review. *Physica Status Solidi (A): Applications and Materials Science* 216 (5): 1800372.

4 Lu, L. and Wong, M. (2014). The resistivity of zinc oxide under different annealing configurations and its impact on the leakage characteristics of zinc oxide thin-film transistors. *IEEE Transactions on Electron Devices* 61 (4): 1077–1084.

5 Lu, L., Li, J., Kwok, H.S., and Wong, M. (2017). High-performance and reliable elevated-metal metal-oxide thin-film transistor for high-resolution displays. In: *2016 IEEE International Electron Devices Meeting* (IEDM), 32.2.1–32.2.4. Piscataway, NJ: IEEE.

6 Wu, G.M., Chen, C.N., Feng, W.S., and Lu, H.C. (2009). Improved AMOLED with aligned poly-Si thin-film transistors by laser annealing and chemical solution treatments. *Physical Review B: Condensed Matter and Materials Physics* 404 (23–24): 4649–4652.

7 Hamasha, K., Hamasha, E., Masadeh, G. et al. (2016). Aluminum-induced crystallization of hydrogenated amorphous silicon thin films with assistance of electric field for solar photovoltaic applications. *Journal of Display Technology* 12 (1): 82–88.

8 Hosono, H., Yasukawa, M., and Kawazoe, H. (1996). Novel oxide amorphous semiconductors: transparent conducting amorphous oxides. *Journal of Non-Crystalline Solids* 203: 334–344.

9 Boesen, G.F. and Jacobs, J.E. (1968). ZnO field-effect transistor. *Proceedings of the IEEE* 56 (11): 2094–2095.

10 Fortunato, E., Barquinha, P., and Martins, R. (2012). Oxide semiconductor thin-film transistors: a review of recent advances. *Advanced Materials* 24 (22): 2945–2986.

11 Nomura, K., Ohta, H., Takagi, A. et al. (2004). Room-temperature fabrication of transparent flexible thin-film transistors using amorphous oxide semiconductors. *Nature* 432 (7016): 488–492.

12 Park, J.S., Maeng, W.J., Kim, H.S., and Park, J.S. (2012). Review of recent developments in amorphous oxide semiconductor thin-film transistor devices. *Thin Solid Films* 520 (6): 1679–1693.

13 Xia, Z. (2020). Semiconductor metal-oxide thin-film transistors for display. In: *Semiconducting Metal Oxide Thin-Film Transistors* (ed. Y. Zhou), 7.1–7.25. Bristol, UK: IOP Publishing.

14 Lu, L., Li, J., Feng, Z. et al. (2016). Elevated-metal-metal-oxide thin-film transistor: technology and characteristics. *IEEE Electron Device Letters* 37 (6): 728–730.

15 Kamiya, T. and Hosono, H. (2013). (Invited) roles of hydrogen in amorphous oxide semiconductor. *ECS Transactions* 54 (1): 103–113.

16 Nomura, K., Kamiya, T., and Hosono, H. (2013). Effects of diffusion of hydrogen and oxygen on electrical properties of amorphous oxide semiconductor, In-Ga-Zn-O. *ECS Journal of Solid State Science and Technology* 2 (1): P5–P8.

17 Medvedeva, J.E., Buchholz, D.B., and Chang, R.P.H. (2017). Recent advances in understanding the structure and properties of amorphous oxide semiconductors. *Advanced Electronic Materials (Md.)* 1700082.

18 Lu, L. and Wong, M. (2015). A bottom-gate indium-gallium-zinc oxide thin-film transistor with an inherent etch-stop and annealing-induced source and drain regions. *IEEE Transactions on Electron Devices* 62 (2): 574–579.

19 Kamiya, T., Nomura, K., and Hosono, H. (2009). Origins of high mobility and low operation voltage of amorphous oxide TFTs: electronic structure, electron transport, defects and doping. *Journal of Display Technology* 5 (12): 468–483.

20 Sato, A., Abe, K., Hayashi, R. et al. (2009). Amorphous In-Ga-Zn-O coplanar homojunction thin-film transistor. *Applied Physics Letters* 94: 133502.

21 Xia, Z., Lu, L., Li, J. et al. (2017). Characteristics of elevated-metal metal-oxide thin-film transistors based on indium-tin-zinc oxide. *IEEE Electron Device Letters* 38 (7): 894–897.

22 Bang, S., Lee, S., Park, J. et al. (2009). Investigation of the effects of interface carrier concentration on ZnO thin film transistors fabricated by atomic layer deposition. *Journal of Physics D: Applied Physics* 42: 235102.

23 Dupin, J.-C., Gonbeau, D., Vinatier, P., and Levasseur, A. (2000). Systematic XPS studies of metal oxides, hydroxides and peroxides. *Physical Chemistry Chemical Physics* 2 (6): 1319–1324.

24 Li, J., Lu, L., Feng, Z. et al. (2017). An oxidation-last annealing for enhancing the reliability of indium-gallium-zinc oxide thin-film transistors. *Applied Physics Letters* 110 (14): 142102.

25 Xia, Z., Lu, L., Li, J. et al. (2018). A bottom-gate metal–oxide thin-film transistor with self-aligned source/drain regions. *IEEE Transactions on Electron Devices* 65 (7): 2820–2826.

26 Chowdhury, M.D.H., Mativenga, M., Um, J.G. et al. (2015). Effect of SiO2 and SiO2SiNx passivation on the stability of amorphous indium-gallium zinc-oxide thin-film transistors under high humidity. *IEEE Transactions on Electron Devices* 62 (3): 869–874.

27 Lu, L., Li, J., and Wong, M. (2014). A comparative study on the effects of annealing on the characteristics of zinc oxide thin-film transistors with gate-stacks of different gas-permeability. *IEEE Electron Device Letters* 35 (8): 841–843.

28 Liu, L.C., Chen, J.S., and Jeng, J.S. (2014). Role of oxygen vacancies on the bias illumination stress stability of solution-processed zinc tin oxide thin film transistors. *Applied Physics Letters* 105: 023509.

29 Chen, Y.C., Chang, T.C., Li, H.W. et al. (2012). The suppressed negative bias illumination-induced instability in In-Ga-Zn-O thin film transistors with fringe field structure. *Applied Physics Letters* 101 (22): 5–8.

30 Rim, Y.S., Jeong, W., Ahn, B.D., and Kim, H.J. (2013). Defect reduction in photon-accelerated negative bias instability of InGaZnO thin-film transistors by high-pressure water vapor annealing. *Applied Physics Letters* 102 (14): 143503.

31 Arai, T. and Sasaoka, T. (2011). 49.1: Invited paper: emergent oxide TFT technologies for next-generation AM-OLED displays. *SID Symposium Digest of Technical Papers* 42 (1): 710–713.

32 Song, J.H., Kim, K.S., Mo, Y.G. et al. (2014). Achieving high field-effect mobility exceeding 50 cm^2/Vs in In-Zn-Sn-O thin-film transistors. *IEEE Electron Device Letters* 35 (8): 853–855.

33 Kato, K., Shionoiri, Y., Sekine, Y. et al. (2012). Evaluation of off-state current characteristics of transistor using oxide semiconductor material, indium-gallium-zinc oxide. *Japanese Journal of Applied Physics* 51: 021201.

34 Ye, Z., Wong, M., Ng, M.T., and Luo, J.K. (2013). High stability fluorinated zinc oxide thin film transistor and its application on high precision active-matrix touch panel. In: *2013 IEEE International Electron Devices Meeting*, 27.2.1–27.2.4. Piscataway, NJ: IEEE.

35 Lu, L., Xia, Z., Li, J. et al. (2018). A comparative study on fluorination and oxidation of indium–gallium–zinc oxide thin-film transistors. *IEEE Electron Device Letters* 39 (2): 196–199.

36 Wang, D., Jiang, J., and Furuta, M. (2015). Suppression of positive gate bias temperature stress and negative gate bias illumination stress induced degradations by fluorine-passivated In-Ga-Zn-O thin-film transistors. *ECS Journal of Solid State Science and Technology* 5 (3): Q88–Q91.

37 Feng, Z., Lu, L., Wang, S. et al. (2018). Fluorination-enabled monolithic integration of enhancement- and depletion-mode indium-gallium-zinc oxide TFTs. *IEEE Electron Device Letters* 39 (5): 692–695.

38 Jiang, J., Furuta, M., and Wang, D. (2014). Self-aligned bottom-gate In-Ga-Zn-O thin-film transistor with source/drain regions formed by direct deposition of fluorinated silicon nitride. *IEEE Electron Device Letters* 35 (9): 933–935.

39 Park, Y.C., Um, J.G., Mativenga, M., and Jang, J. (2018). Thermal stability improvement of back channel etched a-IGZO TFTs by using fluorinated organic passivation. *ECS Journal of Solid State Science and Technology* 7 (6): Q123–Q126.

40 Xia, Z., Lu, L., Li, J. et al. (2018). P-15: the use of fluorination to enhance the performance and the reliability of elevated-metal metal-oxide thin-film transistors. *SID Symposium Digest of Technical Papers* 49 (1): 1235–1238.

41 Conley, J.F. (2009). Instabilities in oxide semiconductor transparent thin film transistors. *IEEE Transactions on Device and Materials Reliability* 10 (4): 460–475.

42 George, S.M. (2010). Atomic layer deposition: an overview. *Chemical Reviews* 110 (1): 111–131.

43 Nguyen, T.T.T., Aventurier, B., Renault, O. et al. (2014). Impact of hydrogen diffusion on electrical characteristics of IGZO TFTs passivated by SiO_2 or Al_2O_3. In: *2014 21st International Workshop on Active-Matrix Flatpanel Displays and Devices (AM-FPD)*, 149–152. Piscataway, NJ: IEEE.

44 Park, S.K., Kim, J.W., Ryu, M.-K. et al. (2013). Bilayered etch-stop layer of Al_2O_3/SiO_2 for high-mobility In–Ga–Zn–O thin-film transistors. *Japanese Journal of Applied Physics* 52 (10R): 100209.

45 Wang, S., Shi, R., Li, J. et al. (2020). Resilience of fluorinated indium-gallium-zinc oxide thin-film transistor against hydrogen-induced degradation. *IEEE Electron Device Letters* 41 (5): 729–732.

46 Sil, A., Avazpour, L., Goldfine, E.A. et al. (2020). Structure-charge transport relationships in fluoride-doped amorphous semiconducting indium oxide: combined experimental and theoretical analysis. *Chemistry of Materials* 32 (2): 805–820.

47 Wong, M., Kwok, H.-S., and Ye, Z. (2014). Metal-oxide based thin-film transistors with fluorinated active layer. US Patent No. 8878176B2.

48 Wang, S., Shi, R., Li, J. et al. (2020). 24-2: Distinguished student paper: fluorination for enhancing the resistance of indium-gallium-zinc oxide thin-film transistor against hydrogen-induced degradation. *SID Symposium Digest of Technical Papers* 51 (1): 347–350.

49 Xia, Z., Liu, X., Shi, R. et al. (2020). A timing model for the optimal design of a prototype active-matrix display. *IEEE Transactions on Electron Devices* 67 (8): 3167–3174.

50 Li, J., Lu, L., Xia, Z. et al. (2018). Three-mask elevated-metal metal-oxide thin-film transistor with self-aligned definition of the active island. *IEEE Electron Device Letters* 39 (1): 35–38.

51 Li, J., Lu, L., Xia, Z. et al. (2018). P-1.5: Edge effects of three-mask elevated-metal metal-oxide thin-film transistor and their elimination. *SID Symposium Digest of Technical Papers* 49: 531–534.

52 Lu, L., Zhou, W., Wong, M., and Kwok, H.-S. (2019). Integration of silicon thin-film transistors and metal-oxide thin-film transistors. US Patent No. 10,504,939 B2.

53 Tani, R., Yoon, J.-S., Yun, S.-I. et al. (2015). 64.2: Panel and circuit designs for the world's first 65-inch UHD OLED TV. *SID Symposium Digest of Technical Papers* 46 (1): 950–953.

13

Hot Carrier Effects in Oxide-TFTs

Mami N. Fujii, Takanori Takahashi, Juan Paolo Soria Bermundo, and Yukiharu Uraoka

Graduate School of Science and Technology, Nara Institute of Science and Technology

13.1 Introduction

Research related to oxide semiconductors has been carried out for a long time, but it increased explosively since the report in 2004 from Professor Hosono's group. Now, oxide semiconductor–based devices are attracting attention as driving devices, memories, sensors, and other applications that will facilitate building the next-generation society [1, 2]. On the other hand, compared to amorphous silicon (*a*-Si) [3–5] and low-temperature polycrystalline silicon (LTPS) [6–10], oxide devices suffer from degradation phenomena against electrical [11–24] and optical [25–34] stress. Common degradation phenomena of oxide thin-film transistors (oxide-TFTs) have been reported as a threshold voltage shift when positive and negative voltage is applied, and the threshold voltage shift toward the negative gate voltage direction against light irradiation under a negative bias stress. These phenomena are reported to be due to charge trapping at the semiconductor bulk or at the semiconductor–gate insulator interface [35–38]. Apart from these degradation phenomena, there are reports suggesting the hot carrier generation under a strong drain electric field or high drain current region. Recently, we achieved results that clearly showed its existence. This chapter outlines the proposal of an evaluation method using a photoemission microscope (Hamamatsu Photonics K.K. PHEMOS-200, Figure 13.1) for the purpose of further improving the reliability and performance of oxide-TFTs.

13.2 Analysis of Hot Carrier Effect in IGZO-TFTs

13.2.1 Photoemission from IGZO-TFTs

We observed photoemission in conventional oxide-TFTs with an IGZO channel using a photoemission microscope with a Si-CCD camera and photon-counting camera. This system can separate the light wavelength using bandpass filters of 500, 600, 700, 800, and 900 nm with a full width half maximum (FWHM) of 10 nm. An etching-stopper type IGZO-TFT with field-effect mobility of 11.1 cm^2/Vs was adopted.

Figure 13.2a is the TFT structure used in this measurement; Figure 13.2b (top) is the TFT pattern image; Figure 13.2b (bottom) is a superimposed image of the TFT pattern with an observed photon emission image when constant drain voltage (V_d) = 30 V with gate voltage (V_g) = 12 V; and Figure 13.2c is the normalized photon counts in each photon energy through the bandpass filter. It can be clearly seen that the photon emission near the drain edge in the lower photon energy region that is 1.38–1.77 eV is dominant. These results do not completely dismiss the possibility of electron–hole recombination or transition between the conduction and valence band; however, at least, this photon energy peak has lower energies than the IGZO optical bandgap. Moreover, we can

Amorphous Oxide Semiconductors: IGZO and Related Materials for Display and Memory, First Edition.
Edited by Hideo Hosono and Hideya Kumomi.

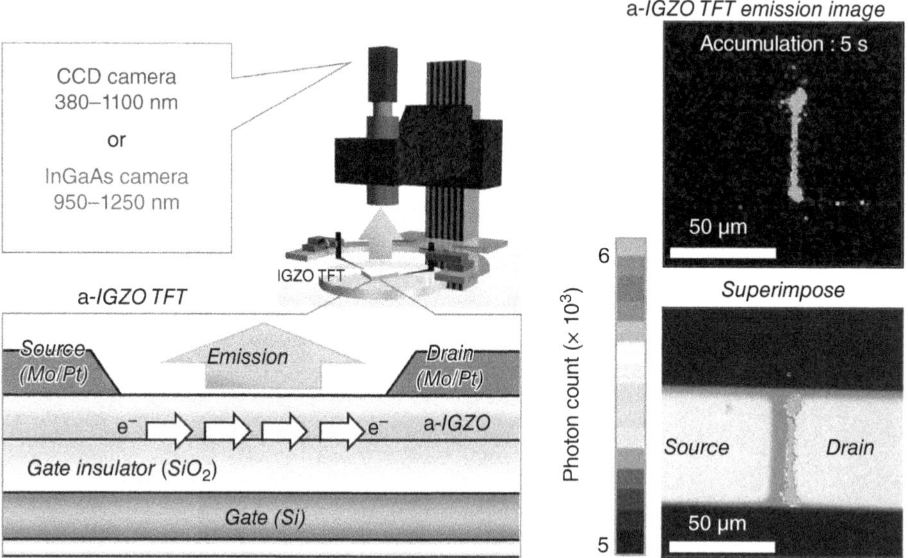

Figure 13.1 Emission microscope (PHEMOS-200) used in this work and the example of a photon emission image. *Source*: Copyright 2018 PhD thesis of Kahori Kise, Nara Institute of Science and Technology.

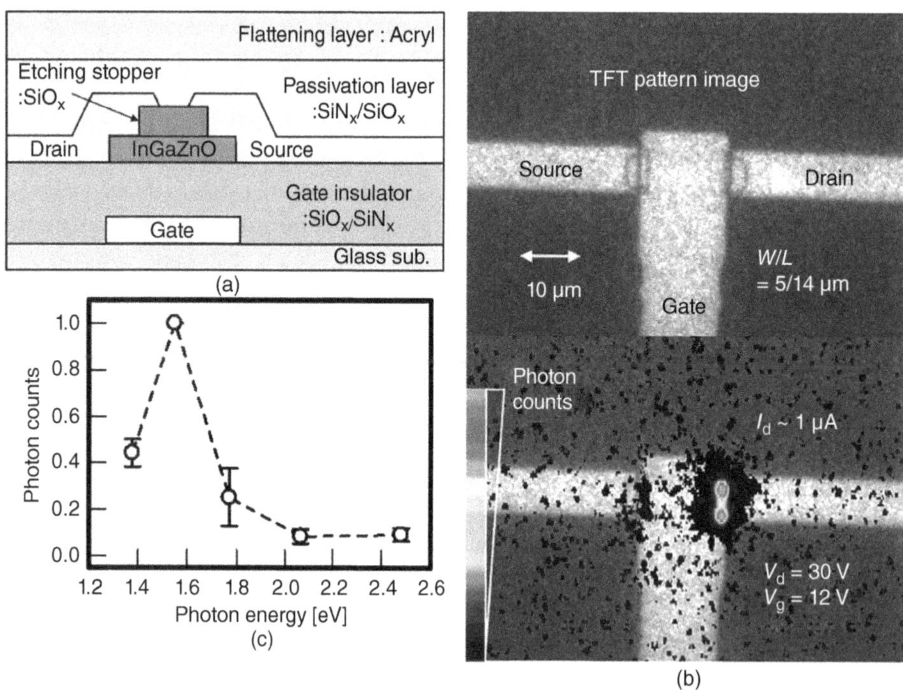

Figure 13.2 (a) Cross-sectional image of the IGZO-TFT. (b) TFT pattern image (upper) and emission image (lower) of the IGZO-TFT obtained from the emission microscope. (c) Normalized photon emission spectrum of the IGZO-TFT obtained through the bandpass filters. *Source*: Takahashi, T., et al. Hot carrier effects in InGaZnO thin-film transistor. Reproducing from permission Appl. Phys. Express 12, 094007 (2019), Copyright (2019) The Japan Society of Applied Physics.

eliminate the notion that the photon emissions are due to leakage currents and breakdown of gate insulator because normal transfer characteristics were obtained after this emission analysis.

From previous investigations of the relationship between the photon counts and the wavelength, it is considered that the photon emission follows the Maxwell–Boltzmann distribution and is due to intraband transition such as hot carrier scattering or bremsstrahlung [39–42]. These previous reports show that photon emission induced by the hot carrier does not have specific energy. For the IGZO-TFT, although there is a difference in photon counts, the photon energies have a wide range from around 1.38 to 2.48 eV, which suggests being caused by hot carriers. Furthermore, the location where the photon emission was observed in the drain edge was where hot carriers are generated, which supports this hypothesis [39, 40, 43, 44].

Next, we show the bias dependence of photon emission intensity of a top-contact/bottom-gate-type IGZO-TFT in Figure 13.3. The photon counts increase monotonously as the V_d increases; however, the photon counts-V_g characteristic shows a peak when the V_g is about ½ of V_d. Moreover, the photon counts strongly depend on the gate length, and for the same bias voltage, the photon counts increased as the gate length became shorter, as shown in Figure 13.4. Therefore, it was found that this photon emission strongly depends on the electric field of the tangential component with respect to the channel, which was generated by the drain voltage. Note that unlike self-heating [35, 45], it does not depend monotonously on the V_g and is not due to the current.

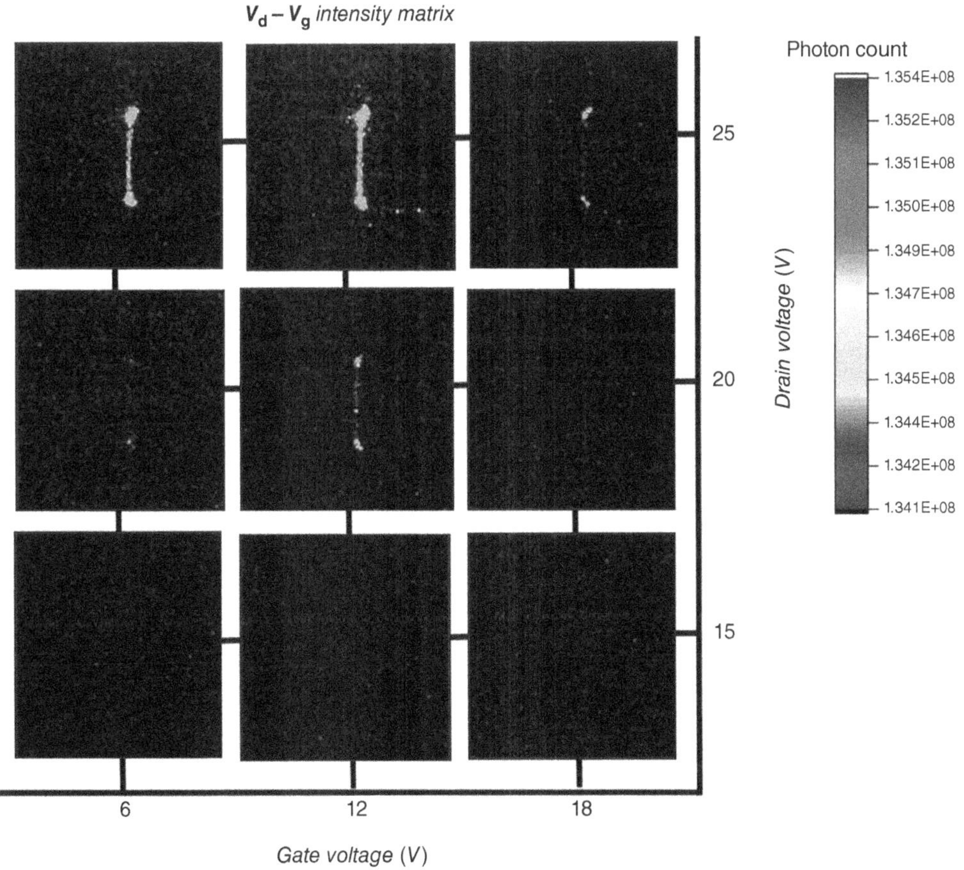

Figure 13.3 V_g and V_d dependence of photon emission. Note that these photon emission images were obtained with a Si-CCD camera with a wavelength sensitivity of 400 to 1100 nm. *Source*: Copyright 2018 PhD thesis of Kahori Kise, Nara Institute of Science and Technology.

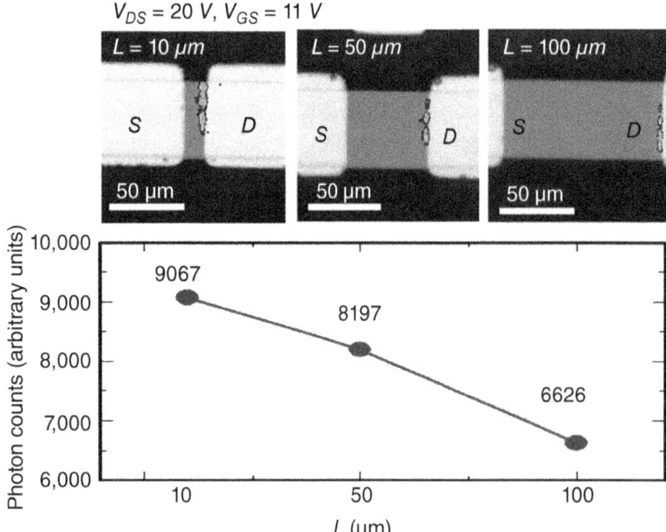

Figure 13.4 The channel length dependence of photon counts. *Source*: Copyright 2018 PhD thesis of Kahori Kise, Nara Institute of Science and Technology.

Based on these results, we presume that the photon emission from our IGZO-TFT was induced by hot carrier effects. However, we consider that further investigations are required to clarify the photon emission mechanism in oxide-TFTs.

13.2.2 Kink Current in Photon Emission Condition

It is reported that the amount of photon emission is closely related to the kink current in Si-based transistors [40]. Therefore, the kink current in the high V_d region of the output characteristics was extracted, as shown in Figure 13.5. The kink current was obtained from the difference between the tangent line drawn at the inflection point of the drain current. As a result, it became clear that the kink current does not depend monotonously on the gate voltage and shows a peak with respect to the gate voltage. Figure 13.6 shows the relationship between the photon counts and the kink current at each V_g. As a result, the emission intensity showed a gate voltage dependence very similar to the kink current. From this relationship, it has been clarified that photon emission is caused by an impact ionization phenomenon that occurs in the high V_d region, that is, a hot carrier phenomenon.

13.2.3 Hot Carrier–Induced Degradation of *a*-IGZO-TFTs

We have checked the electrical degradations due to electron traps related to the hot carrier effect in the etching-stopper type IGZO-TFTs, namely local depletion of the channel at the drain region. In order to analyze this degradation, the TFTs were driven at a high $V_d = 30\,\text{V}$ that corresponds to the photon emission condition of our TFTs, and the accumulation and depletion states of the channel in the drain and source regions were evaluated through the C–V characteristics.

Figure 13.7a shows the transfer characteristics of IGZO-TFTs; see also the C–V characteristics of Figure 13.7b's gate to drain ($C_{G\text{–}D}$) and Figure 13.7c's gate to source ($C_{G\text{–}S}$). These measurements were performed using the following experimental flow as:

I. C–V [Initial]
II. Transfer characteristics with V_d of 30 V, swept twice [1st and 2nd]

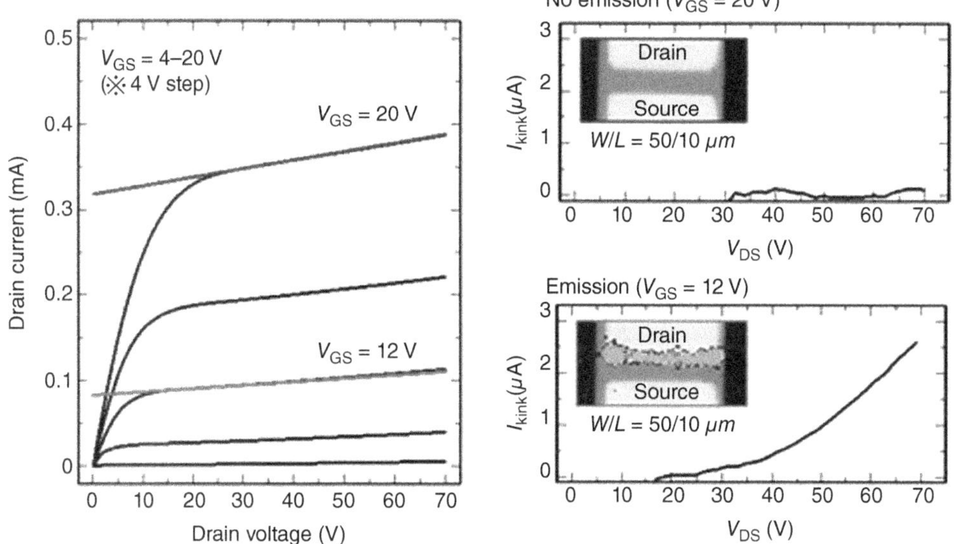

Figure 13.5 Kink current in an IGZO-TFT. The kink current was obtained from the difference between the tangent line drawn at the inflection point of the drain current. This image was obtained with a Si-CCD camera. *Source*: Copyright 2018 PhD thesis of Kahori Kise, Nara Institute of Science and Technology.

Figure 13.6 Photon counts and kink current in an IGZO-TFT. The emission intensity showed a V_g dependence very similar to the kink current. *Source*: Copyright 2018 PhD thesis of Kahori Kise, Nara Institute of Science and Technology.

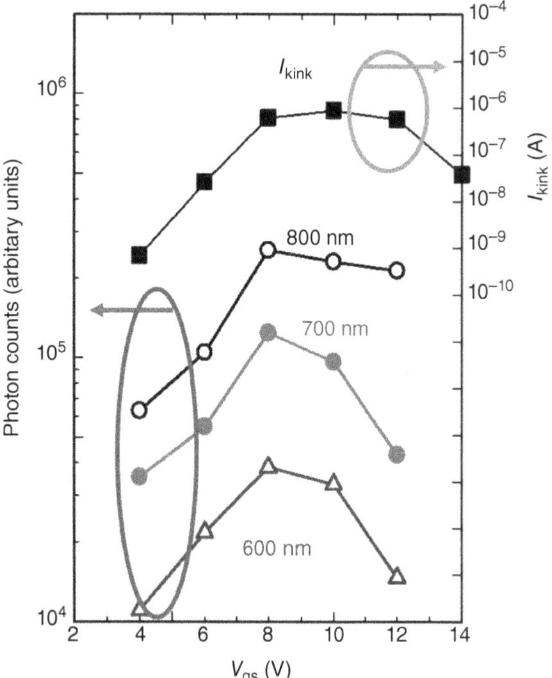

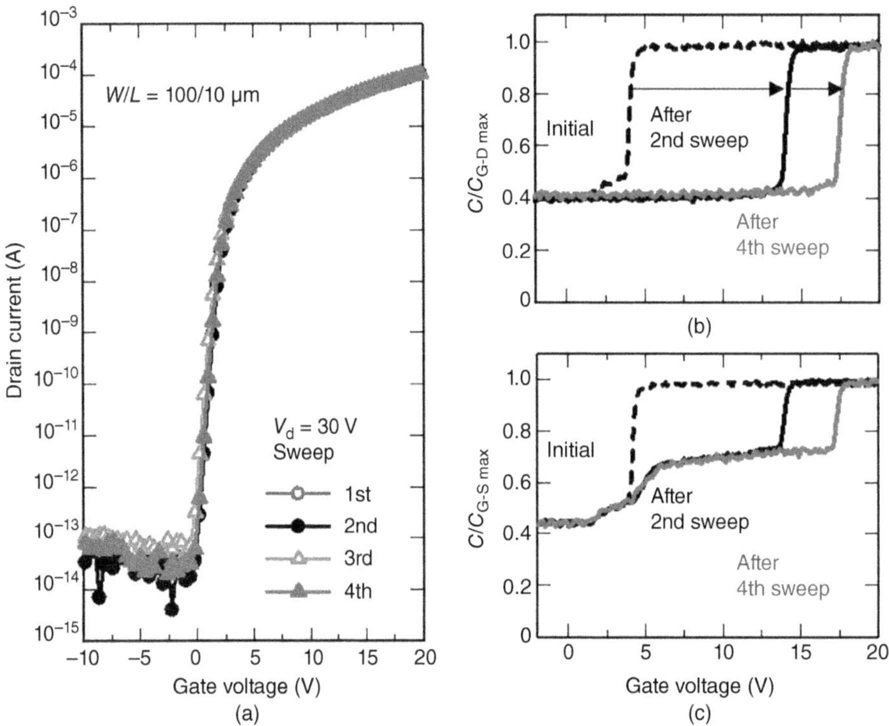

Figure 13.7 (a) Transfer characteristics of the IGZO-TFT driven by V_d of 20 V. C–V characteristic of (b) $C_{G\text{-}D}$ and (c) $C_{G\text{-}S}$. *Source*: Takahashi, T., et al. Hot carrier effects in InGaZnO thin-film transistor. Reproducing from permission Appl. Phys. Express 12, 094007 (2019), Copyright (2019) The Japan Society of Applied Physics.

III. C–V [After 2nd transfer sweep]
IV. Transfer characteristics with V_d of 30 V, swept twice [3rd and 4th]
V. C–V [After 4nd transfer sweep]

In this TFT, the W and L were 100 and 10 μm, respectively, and the typical device parameters were μ_{FE} of 9.6 cm^2 V^{-1} s^{-1}, V_{th} of 1.8 V, and S.S. of 0.34 V/decade. The V_{th}-shift of transfer characteristics during this experimental flow is extremely small, and the TFT did not show clear degradation. In the case of C–V characteristics, the initial state (flow number I) of $C_{G–D}$ and $C_{G–S}$ shows almost the same behaviors, as shown in Figure 13.7b and 13.7c. This is due to channel at the source and the drain regions has same potentials, as explained in Figure 13.8a. However, in Figure 13.7b, the C–V curve of $C_{G–D}$ in after second transfer sweep was parallel shifted in the positive direction from initial state. This behavior indicates that the channel at the drain region was depleted by TFT measurements with a high V_d = 30 V, as explained in Figure 13.8b. In contrast, the C–V characteristics of $C_{G–S}$ shows two-step accumulations that different from the case of $C_{G–D}$.

We believe that the causes of this phenomenon are as follows. Due to the potential barrier formation around the drain region after high V_d is applied, the channel does not accumulate uniformly and local accumulation appears around the source region in the low V_g region, namely the "first accumulation," as explained in Figure 13.8e. Applying higher V_g is necessary to obtain uniform accumulation, namely "second accumulation" in the channel, as explained in Figure 13.8f. This second accumulation at high V_g range corresponds to the capacitance of the gate dielectric due to channel formation. The differences in these C–V characteristics between $C_{G–D}$ and $C_{G–S}$ were induced by the potential barrier at the drain region, and we believed that this is generated by hot carrier effects.

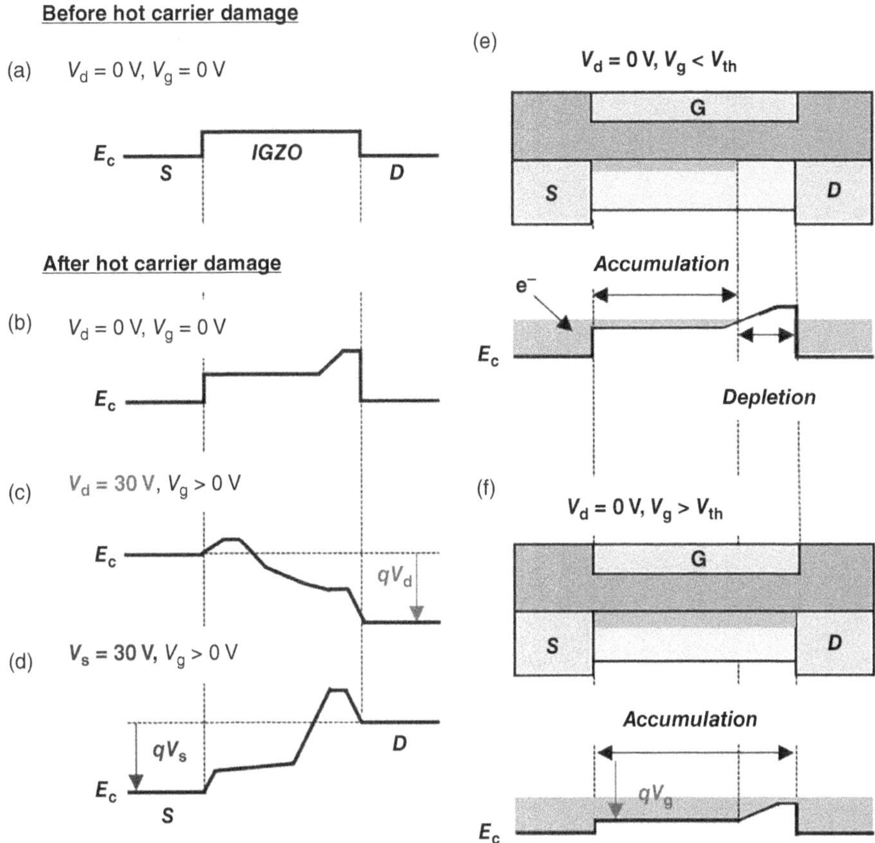

Figure 13.8 Conceptual diagrams of conduction band in the IGZO-TFT (a) before and (b–d) after hot carrier–induced degradation, and (e,f) status of carrier accumulation and depletion in channel. *Source*: Takahashi, T., et al. Hot carrier effects in InGaZnO thin-film transistor. Reproducing from permission Appl. Phys. Express 12, 094007 (2019), Copyright (2019) The Japan Society of Applied Physics.

As an experiment related to this consideration, we sought to confirm the influence of the tangential component of the electric field on the channel from the change of transfer characteristics. In the transfer characteristics, even though there is a potential barrier at the drain region, the V_{th}-shift does not appear as in Figure 13.7a. This is because the carrier accumulation is undisturbed with the potential barrier of the drain region and the barrier is bent downwards enough by qV_d, as illustrated in Figure 13.8c. When the source voltage (V_s) is applied after the formation of a potential barrier in the drain region due to high drain voltage, the potential level in the source region was reduced by qV_s, as shown in Figure 13.8d. However, the potential barrier in the drain region should become higher compared with that of Figure 13.8a. This means that the result of requiring higher V_g to form carrier accumulation causes the V_{th}-shift of the transfer characteristics.

The transfer characteristics of the TFT were also swept while applying the V_s of 30 V in order to understand the effects of the potential barrier at the drain region. This confirmation measurements were performed using the following experimental flow as:

I. Transfer characteristics with V_d of 30 V, swept twice
II. Transfer characteristics with V_s of 30 V, swept twice

Figure 13.9 shows the transfer characteristics of an IGZO-TFT operated at V_d or V_s = 30 V. In the first and second sweeps with V_d = 30 V, the TFT showed no V_{th}-shift. In contrast, in the case of two sweeps with V_s = 30 V after

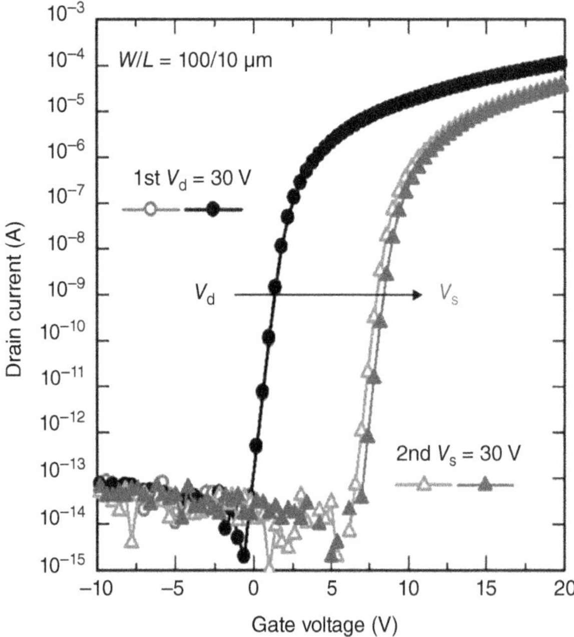

Figure 13.9 Transfer characteristics of the IGZO-TFT driven by V_d or V_s of 30 V. *Source*: Takahashi, T., et al. Hot carrier effects in InGaZnO thin-film transistor. Reproducing from permission Appl. Phys. Express 12, 094007 (2019), Copyright (2019) The Japan Society of Applied Physics.

two sweeps at $V_d = 30$ V, the TFT showed a large positive V_{th}-shift of 7.1 V from the initial state. These results strongly support the existence of the potential barrier at the drain region and proved our hypothesis. Based on the obtained electrical properties, we conclude that local degradation at the drain region was induced by hot carrier effects. Therefore, we consider that the photon emission from the IGZO-TFT was due to hot carrier effects caused by a tangential component electric field on the channel.

13.3 Analysis of the Hot Carrier Effect in High-Mobility Oxide-TFTs

To realize high-resolution displays and high-frequency operation in driver circuits, which are societal requirements in modern times, it is necessary to search for high-mobility oxide semiconductors that exhibit electron mobility far exceeding 10 cm^2/Vs [46]. However, improving the performance of oxide-TFTs may cause new degradation phenomena that have never existed before. As one of them, we are concerned about the effect of hot carriers. Furthermore, investigating the reliability under AC stress is also meaningful because the TFTs in display or circuit applications are driven by pulsed voltage [47, 48]. We believe that understanding such new degradation phenomena and developing evaluation technology are the immediate challenges. We will discuss the effect of pulse voltage applied to the gate and the relation with the hot carrier effect.

13.3.1 Bias Stability under DC Stresses in a High-Mobility IWZO-TFT

Here, we focused on the IWZO film as the high-mobility and stable oxide semiconductor [49]. The fabricated IWZO-TFT achieved mobilities of over 40 cm^2/Vs, even though it has an amorphous phase [50]. These TFTs show excellent bias stability against conventional positive bias stress (PBS) and negative bias stress (NBS), as shown in

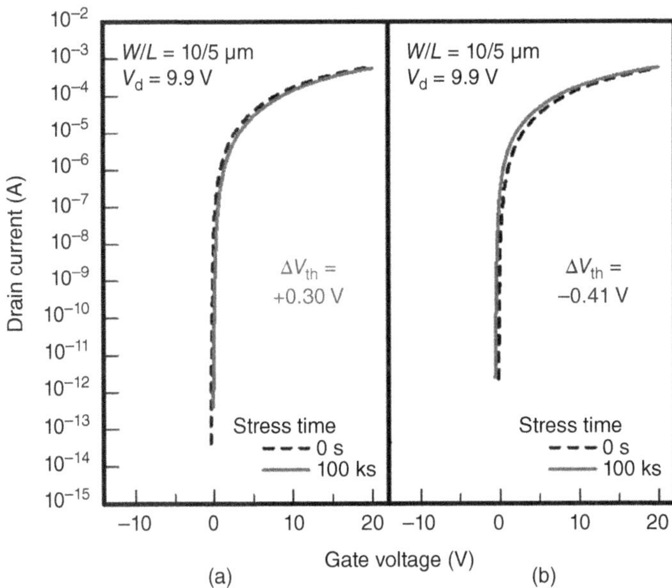

Figure 13.10 Transfer characteristic of the IWZO-TFT against (a) PBS of $V_g = 20\,\text{V}$ and (b) NBS of $V_g = -20\,\text{V}$ for 100 ks. *Source*: Takahashi, T., et al. (2020). Unique degradation under AC stress in high-mobility amorphous In-W-Zn-O thin-film transistors. Reproducing from permission Appl. Phys. Express 13 (5): 054003 (2020), Copyright (2020) The Japan Society of Applied Physics.

Figure 13.10. The stress conditions are $V_g = 20\,\text{V}$ and $V_g = -20\,\text{V}$, respectively, for 100 ks. These TFTs show no significant degradation in threshold voltage, on-current, and subthreshold swing (*SS*).

13.3.2 Analysis of Dynamic Stress in Oxide-TFTs

In a circuit application, there is a possibility of performance degradation during pulse driving. We have already reported the reliability under AC stress of the IGZO-TFT, as shown in Figure 13.11. We observed the unique degradation under AC stress for *SS* values that are not typically observed under DC bias stress in IGZO-TFTs [51]. To further investigate this degradation, the degradation amount was compared by changing the various transition times of gate pulse, as shown in Figure 13.12. From these comparisons, it was obviously dependent on the falling time of pulse voltage. We found that the shorter the transition of falling time is, the more severe the TFT degradations will be. This behavior follows the LTPS degradation mechanism. We concluded that high-energy electrons form a defect level at the interface between IGZO and the gate insulator during the falling edge of the gate pulse. The existence of hot carrier during AC operation was suggested according to these results.

When gate pulse as AC stress was applied to the high-mobility IWZO-TFT, as shown in Figure 13.13, the threshold voltage did not essentially change, and the on-current and *SS* value degraded significantly, which is different from the degradation observed under DC stress [50]. In other reports, the major degradation mode under AC stress in conventional oxide-TFTs is the threshold voltage shift [52–56]. This result indicates that new factors such as hot carrier effects should be considered for degradation in high-mobility oxide-TFTs.

13.3.3 Photon Emission from the IWZO-TFT under Pulse Stress

A photon emission analysis of IWZO-TFTs during pulse driving was performed. Figure 13.14 shows the photon counts through several kinds of bandpass filters and a top-view picture of the IWZO-TFT with photon emission. Significant photon emission around the drain and source edges of the IWZO-TFT was observed. When a strong

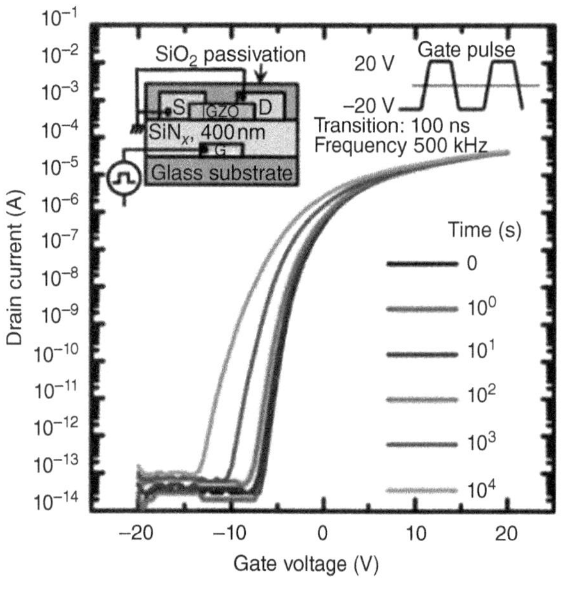

Figure 13.11 Transfer characteristic of the IGZO-TFT under AC (±20 V, 500 kHz) stress. *Source*: Fujii, M., et al. (2011). Unique phenomenon in degradation of amorphous In2O 3-Ga2O3-ZnO thin-film transistors under dynamic stress. Reproducing from permission Appl. Phys. Express 4 (10): 3–5 (2011), Copyright (2011) The Japan Society of Applied Physics.

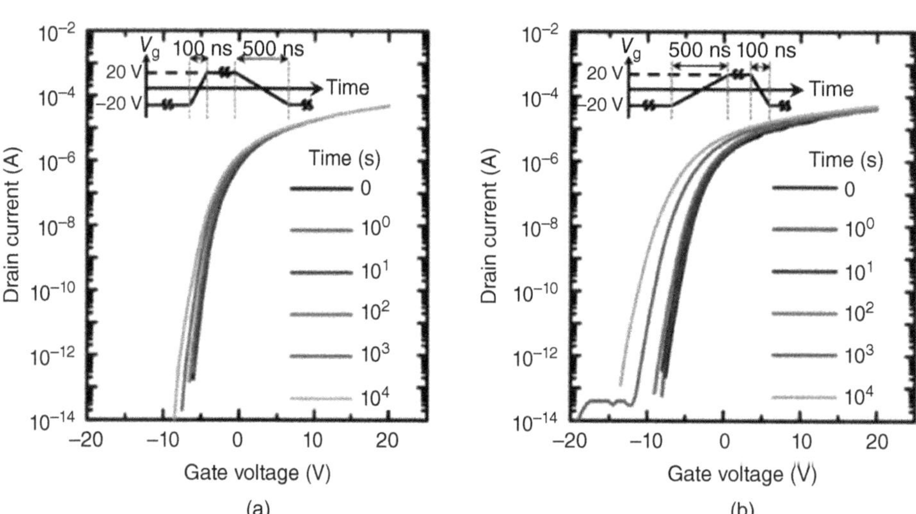

Figure 13.12 Transfer characteristic under AC stress with various transition times. Results under the AC pulse stress, including (a) the sharp rising edge and the slow falling edge, and (b) the slow rising edge and sharp falling edge. *Source*: Fujii, M., et al. (2011). Unique phenomenon in degradation of amorphous In2O 3-Ga2O3-ZnO thin-film transistors under dynamic stress. Reproducing from permission Appl. Phys. Express 4 (10): 3–5 (2011), Copyright (2011) The Japan Society of Applied Physics.

electric field was induced by the gate pulse voltage, the electrons obtained high energy and flowed rapidly to the source and drain electrodes. Here, the photon emission spectrum from the IWZO-TFT is not limited to a specific energy. Photon emission is not monochromatic and has continuous energy below the optical bandgap of IWZO, which is 2.63 eV. The photon counts are considered to be similar in the range of 1.55 to 2.48 eV (500–800 nm) from Figure 13.14. As we mentioned, the photon emission mechanism of the Si-based devices was suggested to be the intraband transition of hot carriers or their scattering, and the energy of these photons covers a wide range [39, 40, 57]. Our results suggest that the dominant photon emission origin is not the transition in the bandgap,

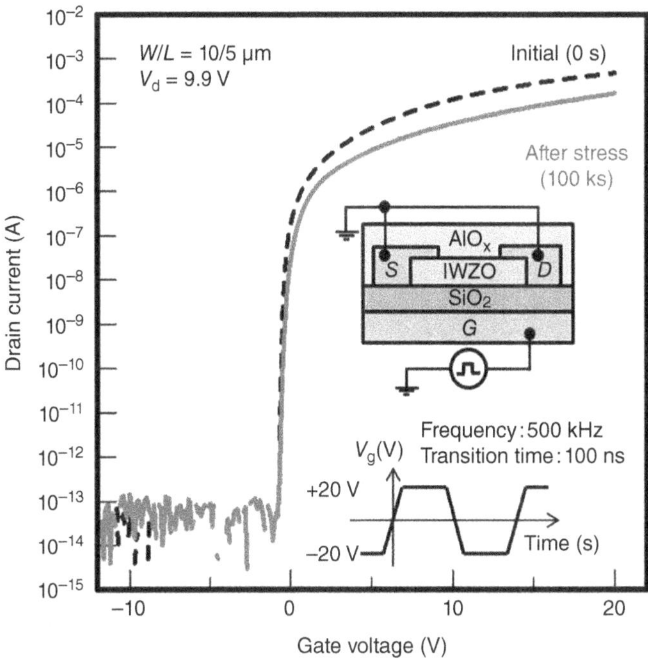

Figure 13.13 Degrading transfer characteristics of the IWZO-TFT under AC stress. Inset: Illustrated cross-sectional view of the IWZO-TFT device and stress conditions; amplitude of the gate pulse, frequency, transition time, and duty ratio: 20 V, 500 kHz, 100 ns, and 50%, respectively. *Source*: Takahashi, T., et al. (2020). Unique degradation under AC stress in high-mobility amorphous In-W-Zn-O thin-film transistors. Reproducing from permission Appl. Phys. Express 13 (5): 054003 (2020), Copyright (2020) The Japan Society of Applied Physics.

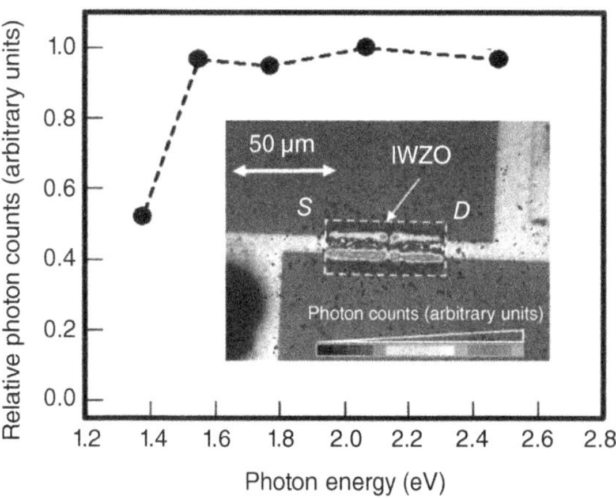

Figure 13.14 Normalized photon counts obtained using bandpass filters and a top-view image of the IWZO-TFT with photon emission around the edges of the source and drain electrodes under AC stress observed by an emission microscope with a photon-counting camera. Images were obtained by integrating 100 images without a bandpass filter. *Source*: Takahashi, T., et al. (2020). Unique degradation under AC stress in high-mobility amorphous In-W-Zn-O thin-film transistors. Reproducing from permission Appl. Phys. Express 13 (5): 054003 (2020), Copyright (2020) The Japan Society of Applied Physics.

and instead is from hot carrier effects. However, further investigations are required to clarify whether the photon emission mechanism in IWZO-TFTs is the same as with IGZO.

Figure 13.15a shows the output characteristics of the IWZO-TFT, including the kink current in the high drain voltage region. The electric field was calculated by the device simulator (ATLAS, Silvaco Inc.). The maximum electric field at the interface between the IWZO channel and gate insulator near the drain electrode region was 0.3 MV/cm, when V_d = 30 V with V_g = 20 V was applied. Furthermore, the calculated electric field at the point of V_g = −20 V during the pulse voltage application was 0.7 MV/cm around the source and drain edges, as shown

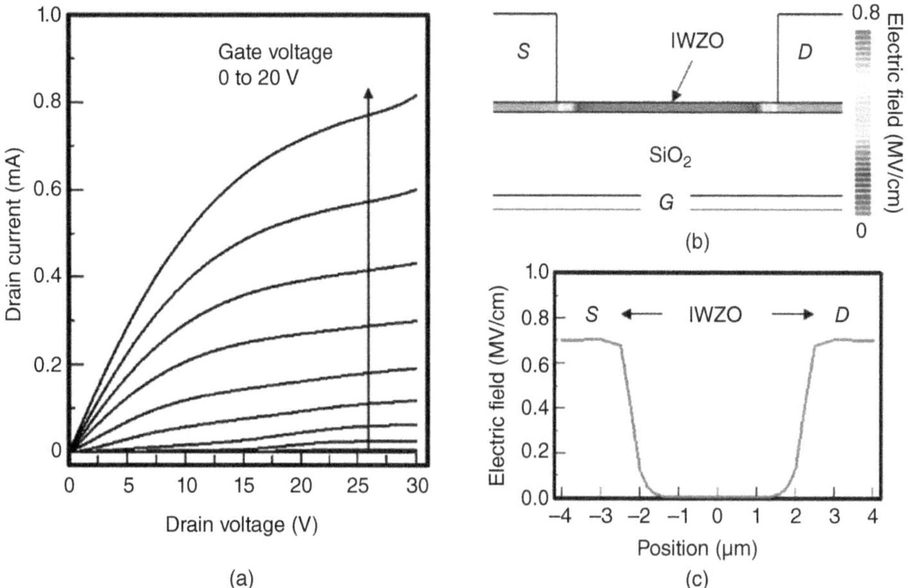

Figure 13.15 (a) Output characteristics of the IWZO-TFT. Simulated (b) distribution of the E-field in the IWZO-TFT and (c) horizontal distribution of the E-field in the IWZO layer when $V_g = -20$ V was applied as the gate pulse. *Source*: Takahashi, T., et al. (2020). Unique degradation under AC stress in high-mobility amorphous In-W-Zn-O thin-film transistors. Reproducing from permission Appl. Phys. Express 13 (5): 054003 (2020), Copyright (2020) The Japan Society of Applied Physics.

in Figure 13.16b and 13.16c. This electric field is greater than the degradation condition due to the hot carrier in LTPS (0.5 MV/cm) and the kink current–generated condition in IWZO-TFTs, as we mentioned here. Degradation of the on-current and *SS* occurred as a result of bond breaking when the LTPS-TFT with mobility of 45 $cm^2 V^{-1} s^{-1}$ is exposed to an electric field of 0.5 MV/cm [47]. The bond dissociation energies of Si-H, Si-Si, and Si-O are 3.04, 3.21, and 8.28 eV, respectively [58]. Thus, the hot electron energy is estimated to be least 3.04 eV in the LTPS TFTs. In the case of our IWZO-TFT, the bond dissociation energies in the components In-O, W-O, and Zn-O are 3.59, 7.46, and <2.59 eV, respectively [58]. We consider that weak metal–oxygen bonds such as Zn-O around the source and drain electrodes' edge were broken by hot carriers when the IWZO-TFT with high μ_{FE} (over 40 $cm^2 V^{-1} s^{-1}$) was driven under gate pulse stress causing the strong electric field over 0.7 mV/cm. Weak metal–oxygen bonds of IGZO have been suggested to break under electrical stress [51, 53, 59, 60]. A large shift of V_{th} accompanied by an increase in *SS* was caused by a localized acceptor-like shallow defect from the conduction band to approximately 1.0 eV [59].

To explain the degradation model of IWZO-TFTs under pulse voltage, we focused on decreasing the on-current. This on-current decrease indicates that the electron transport property of the semiconductor has degraded. Other reports suggested that shallow acceptor-like defects related to Zn generated by bond breaking reduce the electron transport properties in IGZO, and consequently impair the current conduction capability [60]. Generally, the conduction band minimum (CBM) of InOx-based amorphous oxide semiconductors (AOSs) is composed of the 5s orbital of In [61], and the origin of the tail state near the CMB is due to the randomness of the amorphous structure (i.e., a variation in the In-O-metal bonding angle) [62]. Thus, the tail state at the CBM has a large effect on the electron transport properties of AOSs. The device simulations showed that the on-current decrease and the SS value increase were due to the increasing tail-state density [62]. Based on these other reports and our experimental results, we consider that the tail state near the CBM increased as a result of structural degradation of the

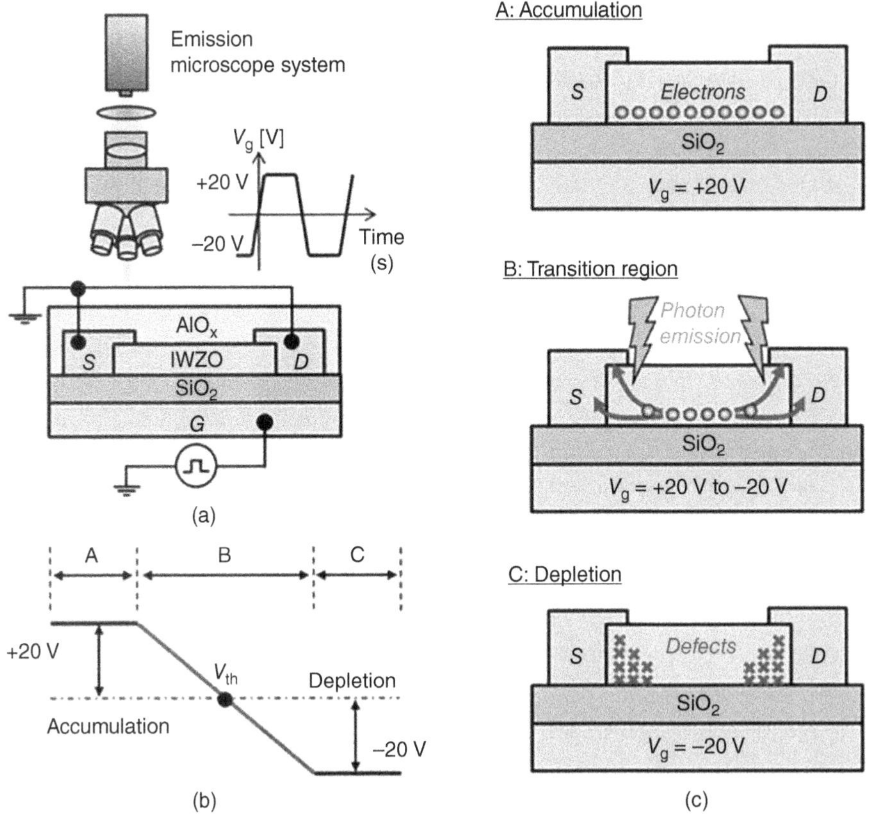

Figure 13.16 Schematic concept of the degradation under AC stress in the IWZO-TFT. (a) Illustration of device structure with stress condition of gate pulse and emission microscope system. (b) Transition region in gate pulse. (c) Illustration of increase of the defects in the IWZO-TFT under AC stress. *Source*: Takahashi, T., et al. (2020). Unique degradation under AC stress in high-mobility amorphous In-W-Zn-O thin-film transistors. Reproducing from permission Appl. Phys. Express 13 (5): 054003 (2020), Copyright (2020) The Japan Society of Applied Physics.

IWZO when weak metal–oxygen bonds such as Zn-O are broken by hot electrons. This degradation occurred at the interface between IWZO and the gate insulator around the source and drain electrodes where a strong electric field and photon emission appeared.

Figure 13.16 is a mechanism for device degradation during pulse application that we propose. When $V_g = 20\,V$ was applied to the gate electrode, electrons flowed to the interface between the IWZO and the gate insulator due to the electric field, as shown in Figure 13.16c region A, which corresponds to channel formation. Accumulated electrons are swept out to the source and drain electrodes when a negative voltage is applied, which corresponds to depletion. However, certain electrons cannot follow rapidly because they are trapped in the IWZO bulk and at the interface between the IWZO and the gate insulator, and these electrons are exposed to a strong electric field of around 0.7 MV/cm. The electrons obtain high energy from the high electric field that appeared in electrode edges and become hot carriers. Accordingly, defects are generated at the interface between the semiconductor and the gate insulator around the electrode edge region, as shown in Figure 13.16c region B, and photon emission was observed. These hot carriers induce structural degradation in the IWZO and a shallow defect density increases, as shown in Figure 13.16c region C. The hot carrier was confirmed from our photon emission analysis, and it revealed that the degradation by pulse stress in IWZO-TFTs was caused by the hot carrier.

13.4 Conclusion

In this chapter, we have discussed the reliability degradation phenomenon in metal oxide thin-film transistors with high performance, which will be increasingly researched in the future. For the first time, we confirmed the hot carrier degradation phenomenon with light emission in thin-film transistors with high mobility and thin-film transistors under pulse stress assumed to be in actual operation. Unlike in conventional degradation mechanisms, the degradation induced by hot carriers caused the ON current to decrease without changing the threshold voltage. In oxide thin-film transistors, which are expected to have wider applications in the future, it is necessary to consider the hot carrier effect to ensure high reliability.

References

1 Ide, K., Nomura, K., Hosono, H., and Kamiya, T. (2019). Electronic defects in amorphous oxide semiconductors: a review. *Physica Status Solidi (A) Applications and Materials* 216 (5): 1–28.

2 Fortunato, E., Barquinha, P., and Martins, R. (2012). Oxide semiconductor thin-film transistors: a review of recent advances. *Advanced Materials* 24 (22): 2945–2986.

3 Spear, W.E. and Le Comber, P.G. (1975). Substitutional doping of amorphous silicon. *Solid State Communications* 17 (9): 1193–1196.

4 Spear, W.E. and Le Comber, P.G. (1976). Electronic properties of substitutionally doped amorphous Si and Ge. *Philosophical Magazine* 33 (6): 935–949.

5 Le Comber, P.G., Spear, W.E., and Ghaith, A. (1979). Amorphous-silicon field-effect device and possible application. *Electronics Letters* 15 (6): 179–181.

6 Sameshima, T., Usui, S., and Sekiya, M. (1986). XeCl excimer laser annealing used in the fabrication of poly-Si TFT'S'. *Electron Device Letters* EDL-7 (5): 276–278.

7 Kuriyama, H., Kiyama, S., Noguchi, S. et al. (1991). Enlargement of poly-Si film grain size by excimer laser annealing and its application to high-performance poly-Si thin film transistor. *Japanese Journal of Applied Physics* 30 (12): 3700–3703.

8 Kirimura, H., Uraoka, Y., Fuyuki, T. et al. (2005). Study of low-temperature crystallization of amorphous Si films obtained using ferritin with Ni nanoparticles. *Applied Physics Letters* 86 (26): 1–3.

9 Uraoka, Y., Hirai, N., Yano, H. et al. (2003). Hot carrier analysis in low-temperature poly-Si thin-film transistors using pico-second time-resolved emission microscope. *IEEE Electron Device Letters* 24 (4): 236–238.

10 Uraoka, Y., Hatayama, T., Fuyuki, T. et al. (2000). Reliability of high-frequency operation of low-temperature polysilicon thin film transistors under dynamic stress. *Japanese Journal of Applied Physics* 39 (12 A): 1209.

11 Uraoka, Y., Bermundo, J.P., Fujii, M.N. et al. (2019). Degradation phenomenon in metal-oxide-semiconductor thin-film transistors and techniques for its reliability evaluation and suppression. *Japanese Journal of Applied Physics* 58 (9): 090502.

12 Cross, R.B.M. and De Souza, M.M. (2006). Investigating the stability of zinc oxide thin film transistors. *Applied Physics Letters* 89: 263513.

13 Hoshino, K., Hong, D., Chiang, H.Q., and Wager, J.F. (2009). Constant-voltage-bias stress testing of a-IGZO thin-film transistors. *IEEE Transactions on Electron Devices* 56 (7): 1365–1370.

14 Tsai, C.T., Chang, T.C., Chen, S.C. et al. (2010). Influence of positive bias stress on N2 O plasma improved InGaZnO thin film transistor. *Applied Physics Letters* 96 (24): 242105.

15 Mativenga, M., Seok, M., and Jang, J. (2011). Gate bias-stress induced hump-effect in transfer characteristics of amorphous-indium-galium-zinc-oxide thin-fim transistors with various channel widths. *Applied Physics Letters* 99 (12): 122107.

16 Chowdhury, M.D.H., Migliorato, P., and Jang, J. (2011). Time-temperature dependence of positive gate bias stress and recovery in amorphous indium-gallium-zinc-oxide thin-film-transistors. *Applied Physics Letters* 98 (15): 153511.

17 Chiang, H.Q., McFarlane, B.R., Hong, D. et al. (2008). Processing effects on the stability of amorphous indium gallium zinc oxide thin-film transistors. *Journal of Non-Crystalline Solids* 354 (19–25): 2826–2830.

18 Suresh, A. and Muth, J.F. (2008). Bias stress stability of indium gallium zinc oxide channel based transparent thin film transistors. *Applied Physics Letters* 92 (3): 033502.

19 Lee, J.M., Cho, I.T., Lee, J.H., and Kwon, H.I. (2008). Bias-stress-induced stretched-exponential time dependence of threshold voltage shift in InGaZnO thin film transistors. *Applied Physics Letters* 93 (9): 2008–2010.

20 Cross, R.B.M., De Souza, M.M., Deane, S.C., and Young, N.D. (2008). A comparison of the performance and stability of ZnO-TFTs with silicon dioxide and nitride as gate insulators. *IEEE Transactions on Electron Devices* 55 (5): 1109–1115.

21 Lopes, M.E., Gomes, H.L., Medeiros, M.C.R. et al. (2009). Gate-bias stress in amorphous oxide semiconductors thin-film transistors. *Applied Physics Letters* 95 (6): 063502.

22 Lee, J.-M., Cho, I.-T., Lee, J.-H. et al. (2009). Comparative study of electrical instabilities in top-gate InGaZnO thin film transistors with Al2O3 and Al2O3/SiNx gate dielectrics. *Applied Physics Letters* 94 (22): 222112.

23 Moon, Y.K., Lee, S., Kim, W.S. et al. (2009). Improvement in the bias stability of amorphous indium gallium zinc oxide thin-film transistors using an O2 plasma-treated insulator. *Applied Physics Letters* 95 (1): 9–11.

24 Ryu, M.K., Yang, S., Park, S.H.K. et al. (2009). Impact of Sn/Zn ratio on the gate bias and temperature-induced instability of Zn-In-Sn-O thin film transistors. *Applied Physics Letters* 95 (17): 173508.

25 Lee, K.H., Jung, J.S., Son, K.S. et al. (2009). The effect of moisture on the photon-enhanced negative bias thermal instability in Ga-In-Zn-O thin film transistors. *Applied Physics Letters* 95 (23): 232106.

26 Gosain, D.P. and Tanaka, T. (2009). Instability of amorphous indium gallium zinc oxide thin film transistors under light illumination. *Japanese Journal of Applied Physics* 48 (3 Part 3): 1–5.

27 Ryu, B., Noh, H.K., Choi, E.A., and Chang, K.J. (2010). O-vacancy as the origin of negative bias illumination stress instability in amorphous In-Ga-Zn-O thin film transistors. *Applied Physics Letters* 97 (2): 022108.

28 Chen, T.C., Chang, T.C., Hsieh, T.Y. et al. (2010). Light-induced instability of an InGaZnO thin film transistor with and without SiOx passivation layer formed by plasma-enhanced-chemical- vapor-deposition. *Applied Physics Letters* 97 (19): 192103.

29 Chowdhury, M.D.H., Migliorato, P., and Jang, J. (2010). Light induced instabilities in amorphous indium-gallium-zinc-oxide thin-film transistors. *Applied Physics Letters* 97 (17): 173506.

30 Oh, H., Yoon, S.M., Ryu, M.K. et al. (2010). Photon-accelerated negative bias instability involving subgap states creation in amorphous In-Ga-Zn-O thin film transistor. *Applied Physics Letters* 97 (18): 183502.

31 Yang, B.S., Huh, M.S., Oh, S. et al. (2011). Role of ZrO2 incorporation in the suppression of negative bias illumination-induced instability in Zn-Sn-O thin film transistors. *Applied Physics Letters* 98 (12): 122110.

32 Park, J.S., Kim, T.S., Son, K.S. et al. (2011). The effect of UV-assisted cleaning on the performance and stability of amorphous oxide semiconductor thin-film transistors under illumination. *Applied Physics Letters* 98 (1): 012107.

33 Wang, C.L., Cheng, H.C., Wu, C.Y. et al. (2012). The recovery mechanism of the light-induced instability of the amorphous InGaZnO thin film transistors. *Applied Physics Letters* 100 (21): 212112.

34 Migliorato, P., Chowdhury, M.D.H., Um, J.G. et al. (2012). Light/negative bias stress instabilities in indium gallium zinc oxide thin film transistors explained by creation of a double donor. *Applied Physics Letters* 101 (12): 123502.

35 Fujii, M., Yano, H., Hatayama, T. et al. (2008). Thermal analysis of degradation in Ga2O3-In 2O3-ZnO thin-film transistors. *Japanese Journal of Applied Physics* 47 (8): 6236–6240.

36 Kimura, M., Nakanishi, T., Nomura, K. et al. (2008). Trap densities in amorphous-InGaZn O4 thin-film transistors. *Applied Physics Letters* 92 (13): 3–5.

37 Wang, D., Hung, M.P., Jiang, J. et al. (2014). Suppression of degradation induced by negative gate bias and illumination stress in amorphous InGaZnO thin-film transistors by applying negative drain bias. *ACS Applied Materials & Interfaces* 6 (8): 5713–5718.

38 Nomura, K., Kamiya, T., and Hosono, H. (2010). Interface and bulk effects for bias–light-illumination instability in amorphous-In–Ga–Zn–O thin-film transistors. *Journal of the Society for Information Display* 18: 789–795.

39 Tam, S. and Hu, C. (1984). Hot-electron-induced photon and photocarrier generation in silicon MOSFETs. *IEEE Transactions on Electron Devices* 31 (9): 1264–1273.

40 Toriumi, A., Yoshimi, M., Iwase, M. et al. (1987). A study of photon emission from n-channel MOSFET's. *IEEE Transactions on Electron Devices* 34 (7): 1501–1508.

41 Lanzoni, M., Sangiorgi, E., Fiegna, C. et al. (1991). Extended (1.1-2.9 eV) hot-carrier-induced photon emission in n-channel Si MOSFET's. *IEEE Electron Device Letters* 12 (6): 341–343.

42 Bigliardi, S. and Manfredi, M. (1993). On the Bremsstrahlung origin of hot-carrier-induced photons in silicon devices. *IEEE Transactions on Electron Devices* 40 (3): 577–582.

43 Farmakis, F.V., Dimitriadis, C.A., Brini, J. et al. (1999). Photon emission and related hot-carrier effects in polycrystalline silicon thin-film transistors. *Journal of Applied Physics* 85 (9): 6917–6919.

44 Uraoka, Y., Hatayama, T., Fuyuki, T. et al. (2001). Hot carrier effects in low-temperature polysilicon thin-film transistors. *Japanese Journal of Applied Physics* 40: 2833–2836.

45 Kise, K., Fujii, M.N., Urakawa, S. et al. (2016). Self-heating induced instability of oxide thin film transistors under dynamic stress. *Applied Physics Letters* 108 (2): 0–4.

46 Fukumoto, E., Arai, T., Morosawa, N. et al. (2011). High-mobility oxide TFT for circuit integration of AMOLEDs. *Journal of the Society for Information Display* 19 (12): 867.

47 Uraoka, Y., Hirai, N., Yano, H. et al. (2004). Hot carrier analysis in low-temperature poly-Si TFTs using picosecond emission microscope. *IEEE Transactions on Electron Devices* 51 (1): 28–35.

48 Toyota, Y., Shiba, T., and Ohkura, M. (2005). Effects of the timing of AC stress on device degradation produced by trap states in low-temperature polycrystalline-silicon TFTs. *IEEE Transactions on Electron Devices* 52 (8): 1766–1771.

49 Kizu, T., Mitoma, N., Miyanaga, M. et al. (2015). Codoping of zinc and tungsten for practical high-performance amorphous indium-based oxide thin film transistors. *Journal of Applied Physics* 118 (12): 125702.

50 Takahashi, T., Fujii, M.N., Miyanaga, R. et al. (2020). Unique degradation under AC stress in high-mobility amorphous In-W-Zn-O thin-film transistors. *Applied Physics Express* 13 (5): 054003.

51 Fujii, M., Ishikawa, Y., Horita, M., and Uraoka, Y. (2011). Unique phenomenon in degradation of amorphous In2O 3-Ga2O3-ZnO thin-film transistors under dynamic stress. *Applied Physics Express* 4 (10): 3–5.

52 Lee, S., Jeon, K., Park, J.H. et al. (2009). Electrical stress-induced instability of amorphous indium-gallium-zinc oxide thin-film transistors under bipolar ac stress. *Applied Physics Letters* 95 (13): 4–6.

53 Kim, D.H., Kong, D., Kim, S. et al. (2011). AC stress-induced degradation of amorphous InGaZnO thin film transistor inverter. *Japanese Journal of Applied Physics* 50 (9 Part 1): 090202.

54 Chen, T.C., Chang, T.C., Hsieh, T.Y. et al. (2011). Investigating the degradation behavior caused by charge trapping effect under DC and AC gate-bias stress for InGaZnO thin film transistor. *Applied Physics Letters* 99: 022104.

55 Wang, H., Wang, M., and Shan, Q. (2015). Dynamic degradation of a-InGaZnO thin-film transistors under pulsed gate voltage stress. *Applied Physics Letters* 106 (13): 2–6.

56 Yang, Y., Zhang, D., Wang, M. et al. (2018). Suppressed degradation of elevated-metal metal-oxide thin-film transistors under bipolar gate pulse stress. *IEEE Electron Device Letters* 39 (5): 707–710.

57 Bude, J., Sano, N., and Yoshii, A. (1992). Hot-carrier luminescence in Si. *Physical Review B* 45 (11): 5848–5856.

58 Luo, Y.R. (2007). *Comprehensive Handbook of Chemical Bond Energies*. Boca Raton, FL: CRC Press.

59 Nomura, K., Kamiya, T., Hirano, M., and Hosono, H. (2009). Origins of threshold voltage shifts in room-temperature deposited and annealed a-In-Ga-Zn-O thin-film transistors. *Applied Physics Letters* 95 (1): 013502.

60 Lee, H.J., Cho, S.H., Abe, K. et al. (2017). Impact of transient currents caused by alternating drain stress in oxide semiconductors. *Scientific Reports* 7 (1): 1–9.

61 Kamiya, T. and Hosono, H. (2010). Material characteristics and applications of transparent amorphous oxide semiconductors. *NPG Asia Materials* 2 (1): 15–22.

62 Hsieh, H.H., Kamiya, T., Nomura, K. et al. (2008). Modeling of amorphous InGaZn O4 thin film transistors and their subgap density of states. *Applied Physics Letters* 92 (13): 2008–2010.

14

Carbon-Related Impurities and NBS Instability in AOS-TFTs

Junghwan Kim[1] *and Hideo Hosono*[1,2,3]

[1] *Materials Research Center for Element Strategy, Tokyo Institute of Technology, Tokyo, Japan*
[2] *Laboratory for Materials and Structures, Tokyo Institute of Technology, Tokyo, Japan*
[3] *National Institute for Materials Science, Tsukuba, Japan*

14.1 Introduction

Amorphous oxide semiconductors (AOSs) represented by amorphous indium–gallium–zinc oxide (*a*-IGZO) have attracted extensive attention in the flat-panel display (FPD) industry due to their large electron mobility, good uniformity over a large area, low-temperature process, and largely tunable mobility by choosing the appropriate cation [1–4]. Nevertheless, currently used AOSs (mobility around 10 cm^2/Vs) for pixel-driving devices have been considered to be improved for meeting future applications such as Super Hi-Vision (SHV) mainly conducted by organic light-emitting diodes (OLEDs). Thin-film transistors (TFTs) with electron mobility higher than 40 cm^2/Vs are a basic requirement [5]. Such a situation stimulates reconsideration of material design concepts for higher electron mobility AOS-TFTs. In terms of high electron mobility, various AOSs with different cation compositions have been reported recently [5–8]. In particular, amorphous In–Sn–Zn–O (ITZO) has drawn a lot of attention. ITZO is an analog of IGZO with Ga^{3+} replaced by Sn^{4+}. Ga^{3+} with high ionic potential Z/r (Z: valence; r: ionic radius) works as a suppressor of oxygen vacancy but does not contribute to conduction because of less participation of its vacant 4*s*-orbital to the conduction band minimum (CBM) [4, 9]. Compared with Ga^{3+}, Sn^{4+} has higher ionic potential, and its 5*s*-orbital largely contributes to the CBM. It is thus reasonable to consider ITZO a good candidate of higher mobility AOS-TFTs. Indeed, electron mobility is much increased from 10 to >40 cm^2/Vs, even at the similar indium content. However, ITZO TFT exhibits, unfortunately, serious bias instability problems [10, 11]. To guarantee the uniform light-emitting characteristic over the whole area, the threshold voltage variation of driving TFTs in pixel circuits should be less than 0.2 V [12]. Therefore, a lot of efforts have been devoted to exploring elucidation of the origin for bias instability. To date, there has been less focus on the fabrication process effect, since there are almost no influences from inevitable photoresist (PR) or developer exposure on conventional AOSs (*a*-IGZO) during a photolithography procedure. In this study, we noticed that NBS in ITZO-TFTs is strongly correlated to PR-derived carbon-related impurities measured by thermal desorption spectroscopy (TDS) of ITZO thin films with various treatments. To further understand the underlying mechanism, we also performed hard X-ray photoemission spectroscopy (HAX-PES) measurement and compared the results with the TDS results. It is well known that the surface conductivity of SnO_2 is sensitive to carbon-related gases such as CO. This characteristic of SnO_2 is applied to gas sensing. The present finding and the accumulated information about SnO_2 led us to an idea that carbon-related impurity would be a major origin of instability in ITZO TFTs. To examine this hypothesis, the device was exposed to pure carbon monoxide (CO) and pure carbon dioxide (CO_2) atmosphere for 1 h. Distinct instability was observed under negative stress for the device exposed to carbon monoxide. Based

Amorphous Oxide Semiconductors: IGZO and Related Materials for Display and Memory, First Edition.
Edited by Hideo Hosono and Hideya Kumomi.

on this idea, we eliminated carbon-related impurities and successfully fabricated bias instability–free ITZO TFTs with high field-effect mobility of ~50 cm^2/Vs.

14.2 Experimental

Bottom-gate/top-contact TFTs were fabricated on heavily doped silicon substrate with 150-nm-thick thermally grown SiO_2. A 20 nm-thick ITZO (20/40/40 = at.% In/Sn/Zn) layer was deposited on the SiO_2 by 150 W radiofrequency (RF)-sputtering under an oxygen partial pressure [O_2 / (Ar+O_2) = 25%]. The film and the substrate were then annealed at 400 °C in air. The S/D (source/drain) electrodes were composed of a 5-nm-thick conductive ITZO layer (deposited under 0% oxygen partial pressure) and a 60-nm-thick gold layer. The S/D layers were patterned to form a TFT channel width and length of 60 μm and 30 μm, respectively. Lastly, UV ozone treatment was conducted. A lift-off process was adopted for all patterning procedures. The electrical properties of each TFT device were measured at source-to-drain voltage of $V_{DS} = -0.1$ V under vacuum by KEYSIGHT B1500A. Voltages for NBS and PBS were set up to be at $V_{th}-20$ V and $V_{th}+20$ V, respectively, where V_{th} is the threshold voltage of the TFT. Forty-nanometer-thick ITZO films deposited at the same deposition condition were prepared for the TDS (ESCO) measurement. A HAX-PES measurement was conducted on 100-nm-thick ITZO films at SPring-8. UV ozone (SKB401Y-01, SUN ENERGY) treatment was conducted for the cleaning device and film surface. The field-effect mobility of the device was calculated by the following equation: $\mu_{Linear} = \frac{L}{WC_{ox}V_{ds}} g_{m_{max}}$, where μ_{Linear} is the linear mobility of the TFT, $g_{m_{max}}$ is the maximum transconductance, C_{ox} is the unit area capacitance of the gate dielectric, V_{ds} is the drain voltage, and W and L are the channel width and length, respectively. The threshold voltage of the device was estimated as a gate voltage, V_g, where the source-to-drain current, I_{ds}, is equal to $\frac{Channel\ Width(W)}{Channel\ Length(L)} \times 10\ nA$.

14.3 Results and Discussion

At the beginning of this study, we compared the bias instability of ITZO TFTs to that of IGZO TFTs. Transfer and output characteristics of an ITZO TFT are shown in Figure 14.1b and 14.1c, respectively. As expected, the ITZO TFT exhibited rather higher mobility of ~50 cm^2/Vs compared to the IGZO TFT (~20 cm^2/Vs). However, a huge difference was found in NBS instabilities between IGZO and ITZO TFTs (see Figure 14.1e and 14.1f). Although the completely same TFT fabrication process was applied to both IGZO and ITZO TFTs, a striking difference was observed: A serious negative threshold voltage (V_{th}) shift of −7.6 V was observed only for ITZO TFTs, while neither IGZO nor ITZO TFTs suffered from positive bias stress (as shown in Figure 14.1d).

There are two possible origins for the negative V_{th} shift. The first is the generation of static electric field at the gate insulator (GI) surface by accumulated/fixed charges, such as hole/proton trapping at the GI. The second is the Fermi-level shift of the active layer itself. Nevertheless, the former origin is unlikely dominant for the following reasons: (i) The amount of the trapped holes at the GI induced by negative bias is extremely small under no light illumination [13], and (ii) the amount of protons (H^+) is not so different between IGZO and ITZO if the same deposition process was conducted. In this respect, we paid more attention to the model of Fermi-level shift in the active layer. To date, several reports have been reported on the Fermi-level shift model originating from bistable states [9, 14]. However, it has remained controversial since direct evidence was not provided.

Consequently, we found that a large number of carbon-related impurities are formed on the ITZO surface after the photolithography process, and PR is responsible for the carbon-related impurities. Figure 14.2a–14.2c shows the TDS and HAX-PES results of ITZO thin films with different treatment (as-deposited, after a photolithography process, and after UV–ozone treatment). From the TDS measurement results, it is evident that unexpected peaks

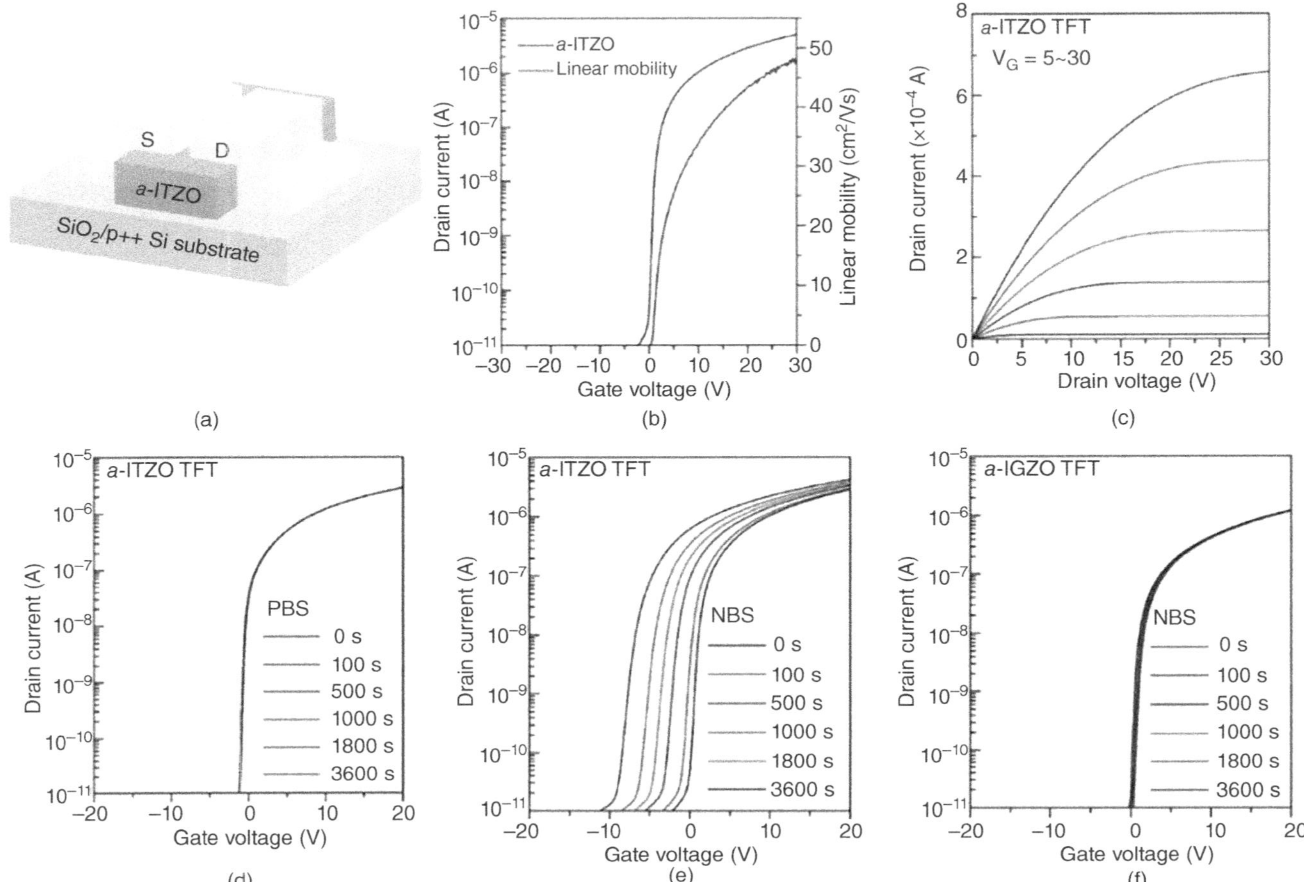

Figure 14.1 TFT performances and gate-bias stress tests of ITZO and *a*-IGZO TFTs. (a) Schematic TFT structure. (b) Transfer characteristic, field-effect mobility, and (c) output characteristic. (d) PBS and (e) NBS instability. (f) NBS instability of IGZO TFTs (negative bias stress: $V_{th}-20$ V; positive bias stress: $V_{th}+20$ V; V_{DS}: 0.1 V).

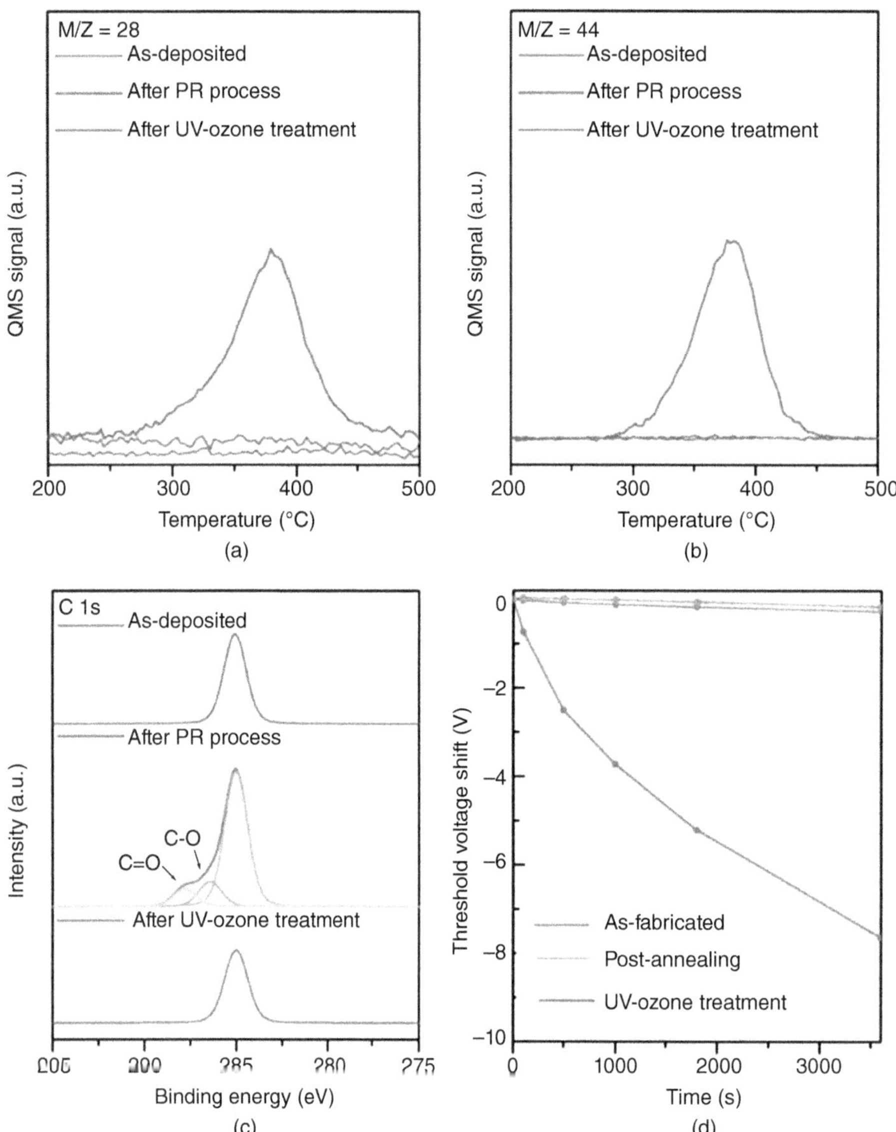

Figure 14.2 Carbon-related impurity and the relevant bias instability of ITZO TFTs. (a) Thermal desorption spectra for mass number 28 (CO). (b) Thermal desorption spectra for mass number 44 (CO_2). (c) Carbon 1 s HAX-PES spectra of as-deposited and posttreated ITZO films. The commonly observed peak at 285 eV is due to well-known surface carbon contamination (C-C bond), and the peaks located at 286.5 and 288 eV are attributed to carbon in a C-O single bond and double bond, respectively. (d) Threshold voltage shift of TFTs with different treatment under NBS (negative bias stress: V_{th}−20 V; V_{DS}: 0.1 V).

(mass number 28 and 44) are detected only from the PR-contaminated ITZO thin film (see Figure 14.2a and 14.2b). Furthermore, it was revealed that mass number 28 and 44 peaks originated from carbon monoxide and carbon dioxide, respectively. This result infers that there may be some chemical reactions between the ITZO and the PR.

As seen in the TDS spectra, elimination of the carbon-related impurities is possible just by post-annealing. However, it should be noted that the desorption temperature of carbon-related impurity is higher than 350 °C. The post-annealing process must be applied after the fabrication of source/drain electrodes, which places restrictions on device structure and electrode materials. To achieve low-temperature removal of the carbon-related impurity, we conducted a room temperature UV–ozone treatment. As seen in Figure 14.2a and 14.2b, the peaks of mass number 28 and 44 were fully removed after the UV–ozone treatment for 5 min.

To reveal the correlation between the carbon-related impurities and the NBS instability, we compared the PR-contaminated ITZO TFT and impurity-free ITZO TFT. To prepare the impurity-free ITZO TFT, UV–ozone treatment was performed at room temperature. The results shown in Figure 14.2d prove that carbon-related impurity significantly affects the NBS stability.

It still remains unclear which specific underlying mechanism can explain this phenomenon. HAX-PES was performed to extract further information on surface-adsorbed carbon-related impurities; it was confirmed that the CO and CO_2 existed only at the surface region by take-off angle-dependent HAX-PES measurement (data not shown here). Figure 14.2c exhibits HAX-PES data of carbon 1 s peaks for uncontaminated, PR-contaminated, and UV–ozone-treated ITZO films. For the PR-contaminated sample, three peaks locating at 285, 286.5, and 288 eV are observed by deconvolution based on Gaussian and Lorentz functions from a carbon 1 s peak. Referring to the data already known, these peaks are assigned to a C-C bond, C-O single bond, and C-O double bond, respectively [15]. The aforementioned C-O bonding from a carbon 1 s peak and the easy formation of carbon-related impurities actually remind us of the gas-sensing nature of tin dioxide. Indium oxide and zinc oxide were conventionally utilized for gas-sensing purposes as well [16–18]. Among them, SnO_2 is best known for its outstanding sensitivity regarding carbon monoxide [18]. This sensing property of post-transition metal oxides is due mainly to charge transfer between the metal oxide itself and surface impurity, and it is described by the Wolkenstein model [19, 20]. In this model, chemisorbed particles may be viewed as surface impurities. Depending on adsorbates' charge states, they can be categorized into weak chemisorption or strong chemisorption. When chemisorbed particles remain electrically neutral, which means that no free carriers are involved in the bonding between adsorbent lattice and the adsorbate, this chemisorption is defined as weak chemisorption. In contrast, when free carriers are captured by chemisorbed particles and directly contribute to the chemisorption bond, this chemisorption is defined as strong chemisorption. Additionally, the interchange between weak chemisorption and strong chemisorption, which corresponds to charge transferring between adsorbent and adsorbate, are effectively governed by the Fermi level at the adsorbent surface. Following this model, conductivity of the adsorbent can be drastically changed by varying the Fermi level at the surface of the adsorbent. In fact, field-effect transistors based on metal oxide (e.g., SnO_2) have already been proposed to control weak chemisorption and strong chemisorption and hence manipulate its sensitivity against gas [21, 22].

To verify our idea, we cut a device chip into several pieces and exposed two of them to pure carbon dioxide and carbon monoxide atmosphere, respectively. After 1 h exposure, the samples were taken out and the NBS test was examined in a vacuum (to exclude the effect of humidity). As a result, it was revealed that a serious negative V_{th} shift occurs only when exposed to carbon monoxide atmosphere (see Figure 14.3b and 14.3c). As for the sample exposed to carbon dioxide (Figure 14.3b), no significant threshold voltage was observed. This result is attributed to less sensitivity of metal oxide against carbon dioxide and is consistent with an already-known report [23]. This finding confirms the validity of our hypothesis. Although specific carbon-related impurity is not clear now, we proposed that carbon-related impurities on the surface are the major cause of negative bias instability in ITZO TFTs. Based on this hypothesis, high-performance TFTs with very high stability were successfully fabricated. Figure 14.4 exhibits threshold voltage stability against the positive bias and the negative bias. Threshold voltage shifts of PBS and NBS were 0.14 V and −0.24 V, respectively. Electrical parameters are summarized in Table 14.1.

14.4 Summary

In this chapter, we clarified carbon-related impurities as the major origin of NBS instability in ITZO-TFTs. A strong correlation between NBS instability and carbon-related species was experimentally found. Easy adsorption of

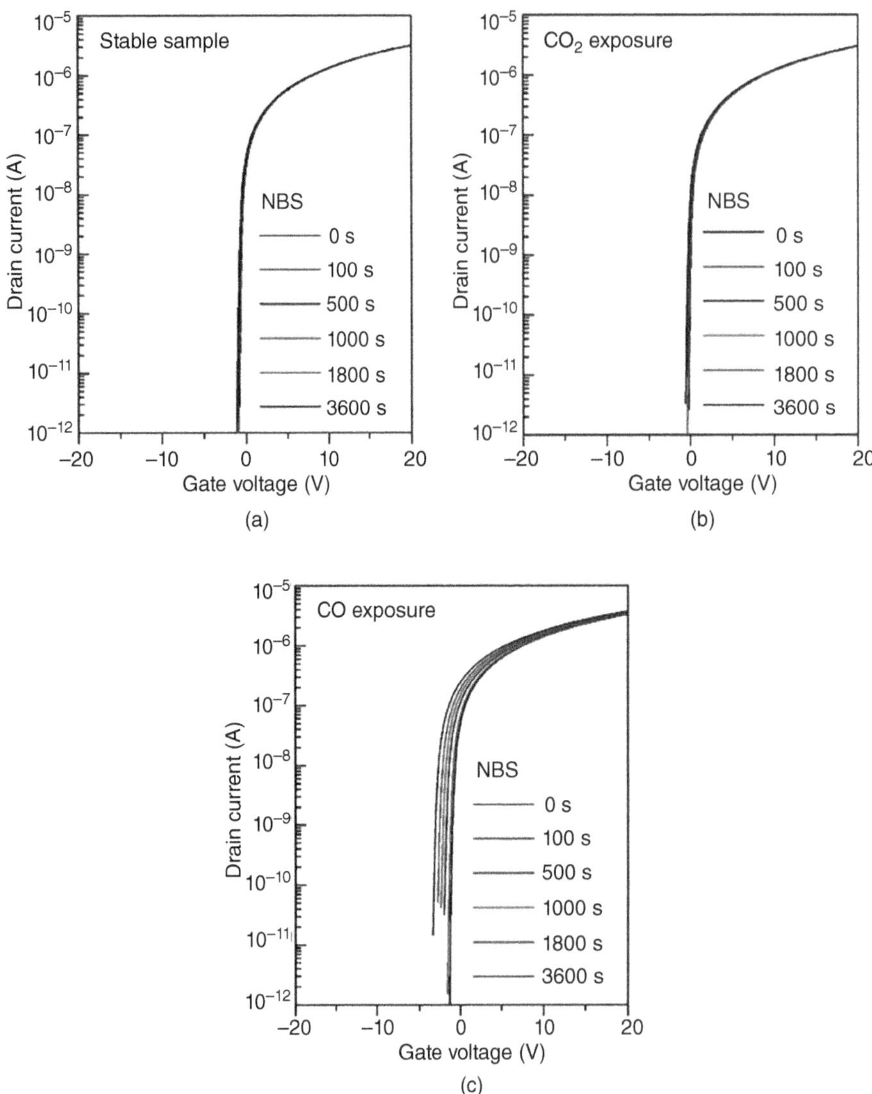

Figure 14.3 NBS instability of ITZO-TFTs (a) without CO exposure, (b) with CO_2 exposure, and (c) with CO exposure (negative bias stress: V_{th}−20 V; V_{DS}: 0.1 V).

metastable carbon species on the ITZO surface is tentatively ascribed to the gas-sensing nature of post-transition metal oxides. Based on understanding of the bias-induced instability mechanism, device instability against the negative bias can be cured by eliminating carbon-related species through simple UV–ozone treatment. Consequently, we successfully realized a bias instability-free ITZO TFT with high field-effect mobility of ~50 cm^2/Vs [24, 25]. Nevertheless, it is noticeable that the carbon-related impurity rarely appears in conventional IGZO TFTs.

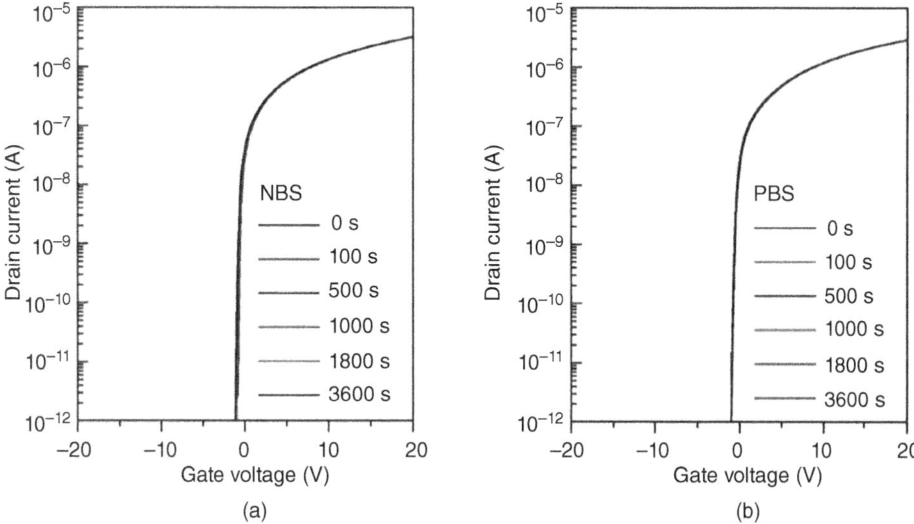

Figure 14.4 (a) NBS stability and (b) PBS stability of ITZO TFTs (negative bias stress: V_{th}−20 V; positive bias stress: V_{th}+20 V; V_{DS}: 0.1 V).

Table 14.1 Device performances of ITZO TFTs with and without PR contamination.

	V_{th} (V)	Mobility (cm^2/Vs)	Subthreshold Swing (V/decade)	I_{on}/I_{off}	NBS (−20 V, 3,600 s)
ITZO	−0.1	49.3	0.10	10^7	−0.24
PR-exposed ITZO	0.4	48.4	0.41	10^7	−7.64

References

1 Park, J., Maeng, W., Kim, H., and Park, J. (2012). Review of recent developments in amorphous oxide semiconductor thin-film transistor devices. *Thin Solid Films* 520: 1679–1693.

2 Ide, K., Nomura, K., Hosono, H., and Kamiya, T. (2019). Electronic defects in amorphous oxide semiconductors: a review. *Physica Status Solidi (A): Applications and Materials Science* 216: 1800372.

3 Kamiya, T. and Hosono, H. (2010). Material characteristics and applications of transparent amorphous oxide semiconductors. *NPG Asia Materials* 2: 15–22.

4 Nomura, K., Ohta, H., Takagi, A. et al. (2004). Room-temperature fabrication of transparent flexible thin-film transistors using amorphous oxide semiconductors. *Nature* 432: 488.

5 Song, J., Kim, K., Mo, Y. et al. (2014). Achieving high field-effect mobility exceeding 50 cm2/Vs in In-Zn-Sn-O thin-film transistors. *IEEE Electron Device Letters* 35: 853–855.

6 Choi, I.M., Kim, M.J., On, N. et al. (2020). Achieving high mobility and excellent stability in amorphous In–Ga–Zn–Sn–O thin-film transistors. *IEEE Transactions on Electron Devices* 67: 1014–1020.

7 Jeong, H., Ok, K., Park, J. et al. (2015). Stability improvement of In-Sn-Ga-O thin-film transistors at low annealing temperatures. *IEEE Electron Device Letters* 36: 1160–1162.

8 Jeon, G., Yang, J., Lee, S. et al. (2021). Abnormal thermal instability of Al-InSnZnO thin-film transistor by hydroxyl-induced oxygen vacancy at SiO_x/active interface. *IEEE Electron Device Letters* 42: 363–366.
9 Hosono, H. (2006). Ionic amorphous oxide semiconductors: material design, carrier transport, and device application. *Journal of Non-Crystalline Solids* 352: 851–858.
10 Kang, Y., Ahn, B.D., Song, J.H. et al. (2015). Hydrogen bistability as the origin of photo-bias-thermal instabilities in amorphous oxide semiconductors. *Advanced Electronic Materials* 1: 140006.
11 Park, J., Rim, Y., Li, C. et al. (2018). Defect-induced instability mechanisms of sputtered amorphous indium tin zinc oxide thin-film transistors. *Journal of Applied Physics* 123: 161568.
12 Jeong, J.K., Chung, H.J., Mo, Y.G., and Kim, H.D. (2008). A new era of oxide thin-film transistors for large-sized AMOLED displays. *Information Display* 24: 20–23.
13 Jeong, J.K. (2013). Photo-bias instability of metal oxide thin film transistors for advanced active matrix displays. *Journal of Materials Research* 28: 2071–2084.
14 Janotti, A. and Van de Walle, C. (2005). Oxygen vacancies in ZnO. *Applied Physics Letters* 87: 122102.
15 Naumkin, A. V., Kraut-Vass, A., Gaarenstroom, S. W., and Powell, C. J. (2012). NIST X-ray Photoelectron Spectroscopy Database. http://dx.doi.org/10.18434/T4T88K
16 Wang, C., Yin, L., Zhang, L. et al. (2010). Metal oxide gas sensors: sensitivity and influencing factors. *Sensors* 10: 2088–2106.
17 Doll, T., Velasco-Velez, J.J., Rosenthal, D. et al. (2013). Direct observation of the electroadsorptive effect on ultrathin films for microsensor and catalytic-surface control. *ChemPhysChem* 14: 2505–2510.
18 Harrison, M. and Willett, P. (1988). The mechanism of operation of tin(iv) oxide carbon monoxide sensors. *Nature* 332: 337–339.
19 Wolkenstein, T. (1963). *The Electron Theory of Catalysis on Semiconductors*. Oxford: Pergamon.
20 Geistlinger, H. (1993). Electron theory of thin-film gas sensors. *Sensors & Actuators, B: Chemical* 17: 47–60.
21 Yang, S., Jiang, C., and Wei, S. (2017). Gas sensing in 2D materials. *Applied Physics Reviews* 4: 021304.
22 Velasco-Velez, J.J., Chaiyboun, A., Wilbertz, C. et al. (2009). CMOS-compatible nanoscale gas-sensor based on field effect. *Physica Status Solidi (A): Applications and Materials Science* 206: 474–483.
23 Hoefer, U., Kühner, G., Schweizer, W. et al. (1994). CO and CO_2 thin-film SnO_2 gas sensors on Si substrates. *Sensors & Actuators, B: Chemical* 22: 115–119.
24 Shiah, Y. S., Sim, K., Shi, Y. et al. (2021). Mobility–stability trade-off in oxide thin-film transistors. *Nature Electronics* 4 (11): 800–807.
25 Shiah, Y. S., Sim, K., Ueda, S. et al. (2021). Unintended Carbon-Related Impurity and Negative Bias Instability in High-Mobility Oxide TFTs. *IEEE Electron Device Letters* 42 (9): 1319–1322.

Part V

TFTs and Circuits

15

Oxide TFTs for Advanced Signal-Processing Architectures

Arokia Nathan[1], Denis Striakhilev[2], and Shuenn-Jiun Tang[2]

[1] *Darwin College, University of Cambridge, Cambridge, UK*
[2] *Ignis Innovation Inc., Waterloo, Ontario, Canada*

15.1 Introduction

The growing maturity of thin-film transistor (TFT) technology coupled with newly emerging materials and processes are enabling integration of circuits and systems. This is driven by the insatiable demand for larger active-matrix pixelated arrays. Here, the TFT is a key building block that provides switching, driving, or sensor interface readout functions [1, 2].

While the flat-panel technology continues to evolve, producing TFTs with improved performance, the implementation of circuits and systems is still somewhat constrained. TFT-based circuits and systems are becoming increasingly important for realization of next-generation device platforms ranging from displays and imaging to energy-autonomous sensing systems that are flexible, disposable, and eventually recyclable [3–8]. This has fueled the quest for new material systems with high field-effect mobility along with the ability to be processed at low temperatures [9, 10].

In addition, transparency of both the active electronics and substrate are desirable attributes, in particular for flexible patch-like electronics that can be seamlessly embedded on objects for realization of smart, immersive ambients [11].

While amorphous-silicon (*a*-Si) and low-temperature polysilicon (LTPS)-TFTs are widely used in pixels and gate drivers in liquid crystal and organic light-emitting diode displays, the optical transparency is poor and the mobility is low in the amorphous phase [12].

In addition, there are cost issues associated with LTPS, particularly when it comes to its scalability to very large areas. Recently, amorphous oxide semiconductors (AOSs), such as amorphous In-Ga-Zn-O (*a*-IGZO), have gained significant traction in view of their large bandgap, which gives rise to high transparency, and relatively high mobility at the range of 15–50 $cm^2/(V{\cdot}s)$. The latter stems from its strong ionic bonding structure, where charge transfer occurs between metal and oxygen atoms [9, 10]. This leads to a higher operating frequency (>1 MHz) compared to *a*-Si and organic TFTs. Transparent electronic systems, which were once viewed as science fiction, can become a reality in the not-so-distant future.

15.1.1 Device–Circuit Interactions

When realizing circuits with devices of relatively poor performance and high non-ideality [13, 14], features that haunt most TFT technology families, it is important to account for device–circuit interactions (DCIs) in the design process. In most cases, the intrinsic performance of TFTs does not meet the requirements of a desired application. As shown in Figure 15.1, if the performance requirement of the desired application is much lower than the

Amorphous Oxide Semiconductors: IGZO and Related Materials for Display and Memory, First Edition.
Edited by Hideo Hosono and Hideya Kumomi.

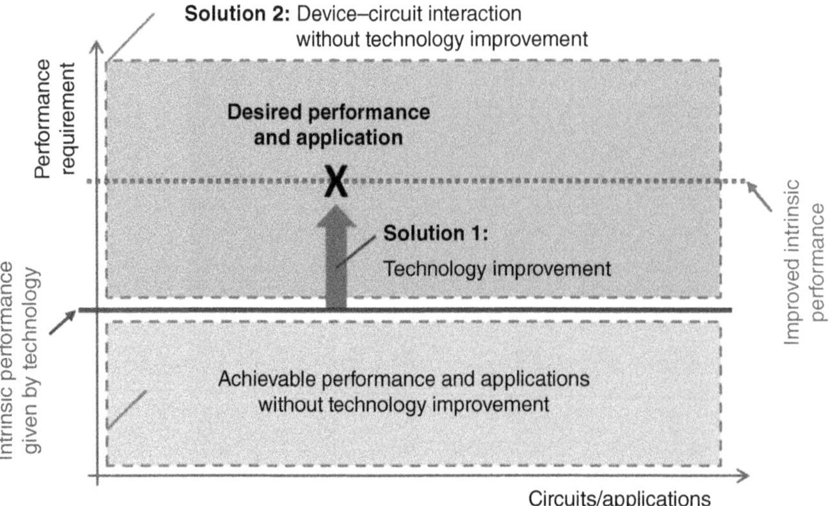

Figure 15.1 Role of device–circuit interaction and performance requirements of circuits and applications. *Source*: Ref. [13], Cheng, X., et al. (2017). Device-circuit interactions and impact on TFT circuit-system design. *IEEE Journal on Emerging and Selected Topics in Circuits and Systems* **7**: 71–80.

maximum achievable intrinsic performance of the TFT, the circuit can be independently designed without considering device non-idealities. For example, when the error in the TFT's output current created by the threshold voltage (V_T) shift is much lower than the required accuracy, the V_T shift problem is not of concern. However, when the performance requirement needs to be higher than the intrinsic performance, the designer should seek a compensation solution based on the device–circuit interaction rather than wait for improvements in the technology.

We will discuss the intrinsic performance regions of TFTs (Figure 15.2), along with compensation methods. Above-threshold operation and application to active-matrix organic light-emitting diode (AMOLED) displays are reviewed in Section 15.2. Subthreshold operation in ultralow-power devices and circuits is discussed in Section 15.3. Another aspect of DCI stems from the material and processing attributes of the TFTs, which usually come with specific and often self-limiting properties. For example, in analog front-end and digital designs,

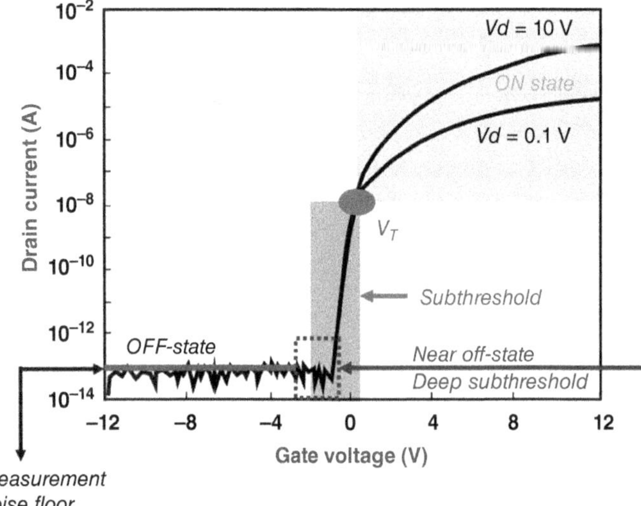

Figure 15.2 Sketch of transfer characteristics of metal-oxide TFTs with threshold voltage (V_T), above-threshold (on-state), subthreshold, and off-state regions as indicated.

alternative circuit architectures are needed to match the properties of the CMOS (see, e.g., Ref. [13, refs. 20–24]). This will be discussed in Section 15.4 along with solutions to deal with, for example, light-induced non-ideality in oxide photo-TFTs and image sensors.

15.2 Above-Threshold TFT Operation and Defect Compensation: AMOLED Displays

15.2.1 AMOLED Display Challenges

A schematic structure of an AMOLED display pixel array is presented in Figure 15.3 with reference to the simplest possible pixel circuit. The amount of light generated by the OLED in each pixel is determined by the current supplied to it, which is controlled by the voltage applied to the gate of the drive transistor. This voltage is set during each pixel programming cycle by turning on the TFT switch and connecting the gate of the drive TFT to the display column (source) driver via the data line (Figure 15.3a). The TFT backplane can be implemented using several TFT technologies. Metal-oxide TFTs are the preferred backplane technology for larger size displays such as TV screens. The metal-oxide TFT has far superior mobility, better current driving capability, higher electrical stability, and lower off-current than *a*-Si-TFTs. At the same time, their advantages over LTPS-TFT technology are the much lower off-current, scalability to large-area substrates, and lower production cost. However, a number of technical issues need to be overcome to design and manufacture AMOLED displays. The most common challenges are summarized in Table 15.1 and briefly described here, following Refs. [15–17].

Process nonuniformity: The active semiconductor layer is a multicomponent film typically deposited by reactive sputtering. It usually exhibits both thickness and composition variations across the substrate area. In addition, dielectric films are usually allowed a few percent variation in thickness. All of the above result in spatial variation in TFT performance parameters and current, leading to brightness variations across the display panel. Pixel-to-pixel variations in TFT characteristics in the fabricated AMOLED panel are evaluated and discussed in Section 15.2.4.

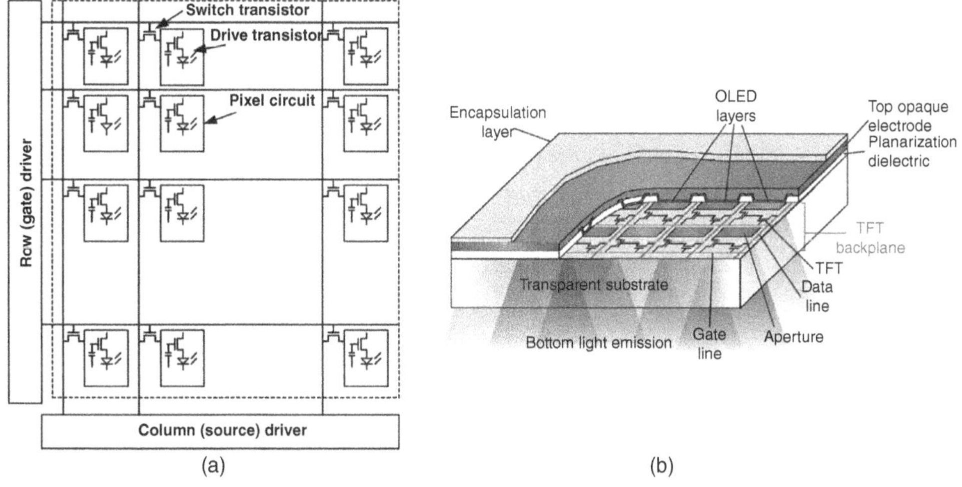

Figure 15.3 Schematic structure of AMOLED display: (a) Simplified schematic display module showing pixel array and row (gate) and column (source) driving circuitry. (b) Simplified illustration of display panel showing bottom emission.
Source: (a) Reproduced with permission from Ref. [2], Chaji, R., and Nathan, A., (2013). *Thin-Film Transistor Circuits and Systems*. Cambridge University Press, Cambridge. (b) Ref. [18], Nathan, A. (2019). Ultra-low power devices for advanced signal processing architectures. Short course presented at the IEEE 65th International Electron Devices Meeting in San Francisco, CA, December 8.

Table 15.1 AMOLED Displays with Oxide TFT Backplanes: Technical Challenges.

Challenge	Illustration	Time to 1JND	Challenge	Illustration	Time to 1JND
Process non uniformity		N/A	OLED aging		Tens of hours
TFT V_T-shift (electrical stress)		Hours	Thermal effects		Seconds
Illumination stress		Hours	TFT hysteresis		Seconds

Note: JND denotes just noticeable difference. *Source*: Ref. [17], Zahirovic, N. (2017). Metal-oxide integration challenges for OLED TV applications. Paper presented at the International Conference on Display Technology in Fuzhou, China, February 19. Copyright 2017 Ignis Innovation.

TFT V_T*-shift by electrical stress*: When a display is powered up, each pixel is emitting light continuously (unless a pixel is turned to display black), and the drive TFTs in the above-threshold gate bias, are subject to continuous electrical stress. This results in a positive V_T-shift and reducing the drive current capability over time. Thus, displaying bright patterns for a prolonged period of time would create burn-in images.

Illumination stress: Unlike LTPS-TFTs, oxide TFTs suffer from negative V_T-shift under certain stress conditions. If the TFT is under negative bias stress while being exposed to high-energy photons, the TFT will have a negative V_T-shift. The source of illumination can be light emitted by neighbouring (sub)pixels and scattered back. For example, if the red or green subpixels are off while the blue subpixel is on, the red or green drive TFTs will have a negative V_T-shift. A negative V_T-shift would increase the drive current capability over time, resulting in brighter areas on the panel.

OLED aging: One factor independent of backplane technology is the OLED efficiency degradation. For example, displaying a brighter pattern over a long enough period of time and then switching to a uniform image would reveal darker burned-in patterns (Table 15.1). Maximizing the aperture ratio (Figure 15.3b) is an important target in panel design, since it helps reduce OLED current density, hence slowing down luminance degradation. While this may be tolerable for small displays, due to the short lifespan requirements, in large-area applications such as TVs, the OLED efficiency degradation needs to be compensated by adjusting the OLED current through the usage cycle of the display.

Thermal effects: TFT characteristics are sensitive to ambient temperature. Increasing the temperature results in increase in the TFT's drain current. The operational temperature may vary because of external factors and can also vary across the panel due to nonuniform power dissipation. The temperature sensitivity of IGZO-TFTs is discussed in Section 15.2.3.

TFT hysteresis: Because of the slow interface traps, the TFT does not react instantly to stepwise changes in its gate voltage. The effect can be visible on a display panel as a residual image from previous frame(s) that would gradually fade over a period of time ranging from several seconds to several tens of seconds. The hysteresis effect in metal-oxide TFT backplanes, can be improved to practically acceptable levels by optimizing the TFT manufacturing process [19].

One common way to address the adverse TFT-related effects is by in-pixel compensation [2, 20]. Basic principles of voltage-programming and current-programming in-pixel compensation schemes were reviewed in Refs. [18] and [21]. In-pixel compensation designs rely on more advanced and complex driving schemes and multitransistor pixel circuits. At the same time, they do not address OLED aging. An alternative approach is external compensation [2, 22], which includes regular pixel-level measurements of the drive TFT and OLED states over the display usage cycle and appropriately adjusting the image data sent to the display. External compensation will be reviewed in greater detail in Section 15.2.5.

15.2.2 Above-Threshold Operation

Typical TFT transfer characteristics are shown in Figure 15.2. In AMOLED displays, the drive TFTs that supply current to subpixel OLEDs spent most of the time biased above-threshold.

According to Ref. [13], the gate voltage (V_{GS}) dependence of drain current, I_{DS}, can be approximated by the following equation:

$$I_{DS} \approx K \times (V_{GS} - V_T)^{(\alpha_p+1)} \tag{15.1}$$

which includes three temperature- and process-dependent parameters, K, V_T, and α_p. Along with its full differential,

$$dI_{DS} = (V_{GS} - V_T)^{(\alpha_p+1)} dK + K(\alpha_p + 1)(V_{GS} - V_T)^{\alpha_p} dV_T + Kln(V_{GS} - V_T)(V_{GS} - V_T)^{(\alpha_p+1)} d\alpha_p \tag{15.2}$$

Equation (15.1) will be used in subsequent sections of this chapter, where we analyze the effects of temperature and process uniformity on IGZO-TFT characteristics.

15.2.3 Temperature Dependence

Following Equations (15.1) and (15.2), the temperature sensitivity of the overall current can be separated into three parts, whereby each part of the function is determined by the temperature sensitivity of K, V_T, or α_p:

$$dI_{DS} = (V_{GS} - V_T)^{(\alpha_p+1)} \frac{\partial K}{\partial T} dT - K(\alpha_p + 1)(V_{GS} - V_T)^{\alpha_p} \frac{\partial V_T}{\partial T} dT + Kln(V_{GS} - V_T)(V_{GS} - V_T)^{(\alpha_p+1)} \frac{\partial \alpha_p}{\partial T} dT \tag{15.3}$$

The contribution of each parameter can then be calculated through extraction of the temperature sensitivity of the three parameters.

Consider the transfer characteristic for an IGZO-TFT, measured every 10 °C from 40 °C to 80 °C, shown in Figure 15.4. As seen, the overall current would increase when temperature increases. To further investigate the degree of influence of the three parameters, their values have been extracted from the measured transfer characteristics according to Equation (15.1). Here, we extract the threshold voltage (V_T) independently from a multi-derivative method [23], and then use it to calibrate the gate voltage as $V_{GS}-V_T$. With this, *I-V* data are plotted on a log-log plot. In this plot, all the data turn into linear behavior, where the intercept on the *y*-axis, $log(I_{DS})$, is $log(K)$, with slope α_p. From this, we get K and α_p independently. The results are shown in Figure 15.5. Through a linear fitting of $log(K)$, $log(V_T)$, and α_p, we get the empirical models $a_p(T)$, $K(T)$, and $V_T(T)$ with the following relations:

$$\alpha_p(T) = \alpha_0 + \frac{E_\alpha}{kT} \tag{15.4}$$

$$K(T) = K_0 exp\left(-\frac{E_K}{kT}\right) \tag{15.5}$$

$$V_T(T) = V_{T_0} exp\left(-\frac{E_{V_T}}{kT}\right) \tag{15.6}$$

where a_0 and E_a, K_0 and E_K, V_{T0}, and E_{VT} are fitting parameters in $a_p(T)$, $K(T)$, and $V_T(T)$ models, respectively.

Combining Equations (15.3)–(15.6), the temperature sensitivity of the drain current can be derived as

$$\frac{dI_{DS}}{dT} = (V_{GS} - V_T)^{(\alpha_p+1)} K_0 \frac{E_k}{kT^2} exp\left(-\frac{E_K}{kT}\right) - K(\alpha_p + 1)(V_{GS} - V_T)^{\alpha_p} V_{T_0} \frac{E_{V_T}}{kT^2} exp\left(-\frac{E_{V_T}}{kT}\right)$$
$$- Kln(V_{GS} - V_T) \times (V_{GS} - V_T)^{\alpha_p+1} \frac{E_\alpha}{kT^2} \quad (15.7)$$

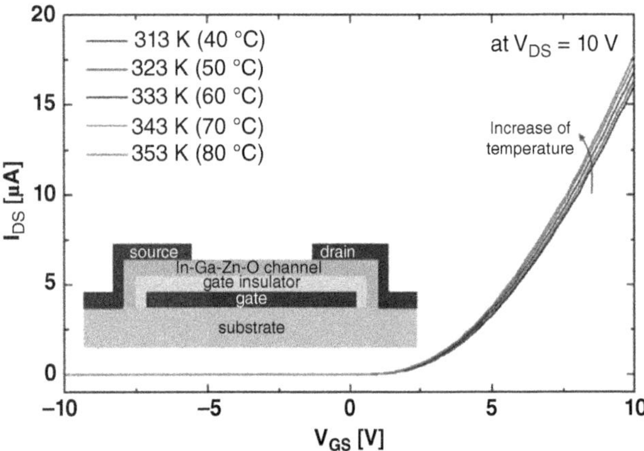

Figure 15.4 *I-V* characteristics of IGZO-TFTs measured at different temperatures. *Source*: Ref. [13], Cheng, X., et al. (2017). Device-circuit interactions and impact on TFT circuit-system design. *IEEE Journal on Emerging and Selected Topics in Circuits and Systems* **7**: 71–80.

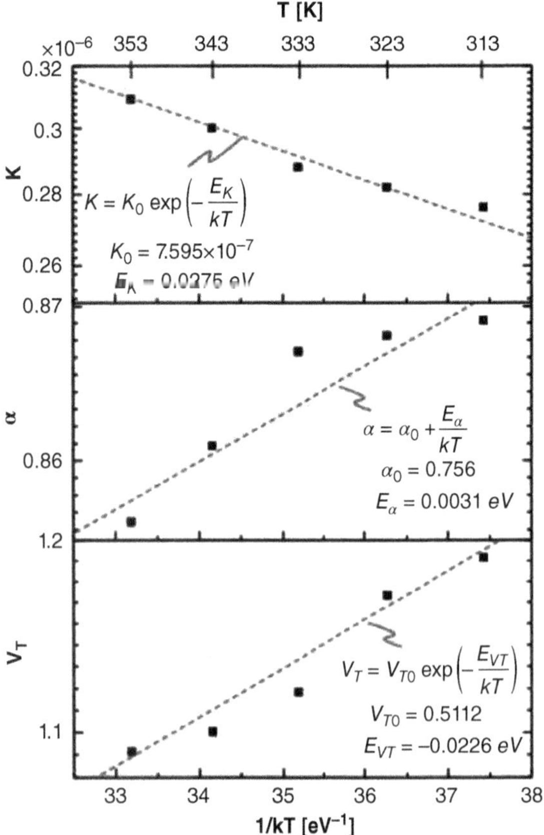

Figure 15.5 Extracted values of K, α_p, and V_T at different temperatures. *Source*: Ref. [13], Cheng, X., et al. (2017). Device-circuit interactions and impact on TFT circuit-system design. *IEEE Journal on Emerging and Selected Topics in Circuits and Systems* **7**: 71–80.

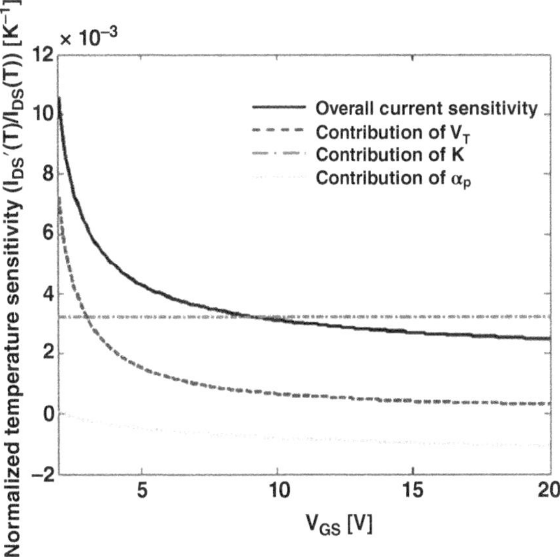

Figure 15.6 Normalized temperature sensitivity of current and contribution of different parameters at 313 K. *Source*: Ref. [13], Cheng, X., et al. (2017). Device-circuit interactions and impact on TFT circuit-system design. *IEEE Journal on Emerging and Selected Topics in Circuits and Systems* **7**: 71–80.

Note that the three terms on the right-hand side of Equation (15.7) define the contribution of K, V_T, and α_p, respectively.

To understand how much the overall current is affected by ambient temperature and the associated contribution of the different parameters, the normalized temperature sensitivity $(dI_{DS}/dT)/I_{DS}$ is plotted in Figure 15.6 as a function of gate bias (V_{GS}). As can be seen, the overall temperature sensitivity drops with increase in V_{GS} due to the fact that the contribution of V_T drops while V_{GS} increases. This tendency starts to saturate when V_{GS} increases to 4 V, when the contribution of K becomes dominant. This analysis suggests that TFTs can be much more sensitive to temperature variations when biased at a voltage near V_T. Although higher stability can be achieved through intentionally biasing the transistor at higher voltage levels, the maximum achievable level will be determined by the temperature sensitivity of K. As the temperature dependence of α_p has a negative contribution to current with respect to temperature and its contribution increases at higher bias, the temperature dependence of K can be compensated by α_p, resulting in decreased sensitivity. However, a higher bias level implies increased power consumption of the circuit. Therefore, an appropriate bias point should be chosen for sufficient stability, albeit with acceptable power consumption.

15.2.4 Effects of Process-Induced Spatial Nonuniformity

Apart from time- or temperature-dependent variations in device parameters, processing-induced spatial variations should be considered (Table 15.1), especially in pixelated arrays or analog circuit applications. These variations would cause pixel nonuniformity in displays or imagers and create errors or undesired behavior in analog circuit design.

Analysis of spatial nonuniformity effects can follow a similar route as with temperature dependence. As different parameters would follow different probability distributions, the overall current will be determined by the randomness of all three parameters, according to Equation (15.1). As variations are usually smaller than the respective mean values, the mismatch in I_{DS} can be expressed as a first-order approximation (15.2), where K, V_T, and α_p are represented by their mean values, while the differentials dK, dV_T, and $d\alpha_p$ are replaced by respective variations. Assuming K, α_p, and V_T are independent variables and that all follow a normal distribution, the variance of I_{DS} can be expressed as [13]

$$\sigma_{I_{DS}}^2 = (V_{GS} - V_T)^{2(\alpha_p+1)}\sigma_K^2 + K^2(\alpha_p + 1)^2(V_{GS} - V_T)^{2\alpha_p}\sigma_{V_T}^2 + K^2[ln(V_{GS} - V_T)]^2 \times (V_{GS} - V_T)^{2(\alpha_p+1)}\sigma_{\alpha_p}^2 \quad (15.8)$$

where σ_K, σ_{ap}, and σ_{VT} are the standard deviations of K, α_p, and V_T, respectively. As expected, the standard deviation of the overall current is determined by the deviation of all three parameters.

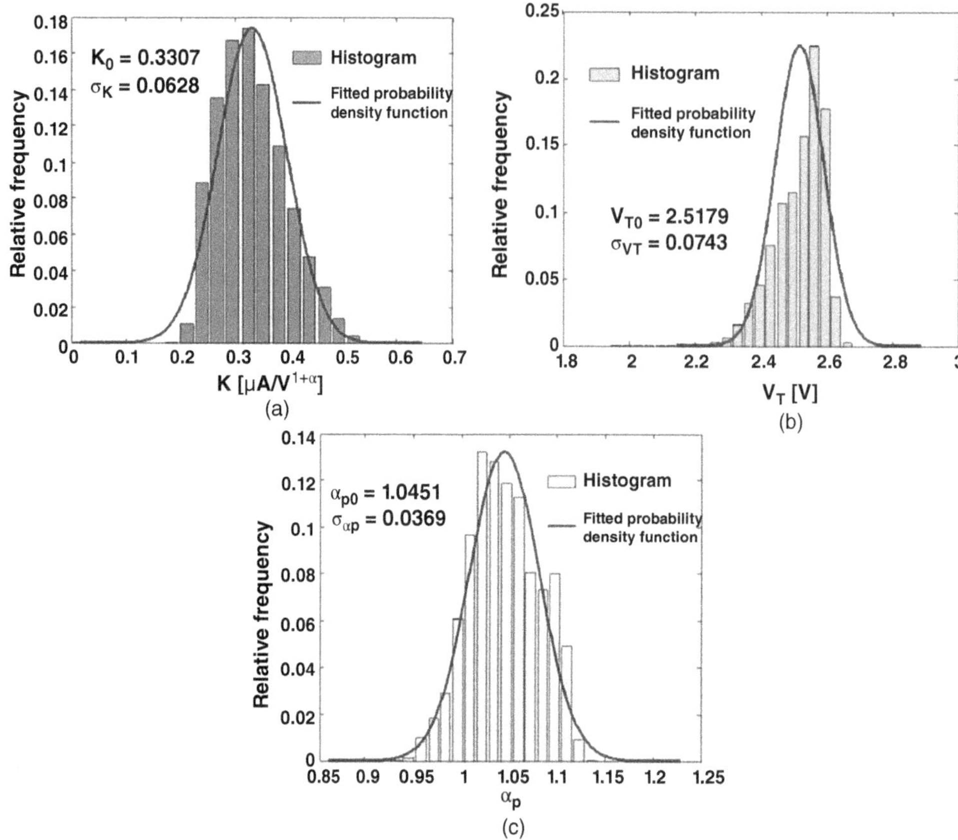

Figure 15.7 Probability distribution and histogram of (a) K, (b) V_T, and (c) α_p. The data were extracted at room temperature from a 1080×1920 RGBW AMOLED panel. *Source*: Ref. [13], Cheng, X., et al. (2017). Device-circuit interactions and impact on TFT circuit-system design. *IEEE Journal on Emerging and Selected Topics in Circuits and Systems* **7**: 71–80.

To analyze the variation sensitivity of the overall current, we need statistical data for all three parameters. Here, we acquired statistical data of the transfer characteristics by measuring a 1080×1920 RGBW OLED display panel (Figure 15.7). By using the pixel circuit described in Refs. [2, 22], which will be reviewed in Section 15.2.3, we could extract the *I-V* characteristics of the drive TFTs within the panel and extract statistical data for the three parameters, K, V_T, and α_p. Here, the TFTs for driving the green OLED pixels are shown. Histogram data were fitted with a normal distribution to obtain the mean and standard deviation values. Their values were then substituted in Equation (15.8) to calculate the overall variance of the drain current and relative contributions of each parameter. The results are plotted as a function of V_{GS} in Figure 15.8. Similar to temperature sensitivity data (Figure 15.6), the effect of spatial current variation is higher when TFTs are biased near V_T. However, unlike the temperature dependence where sensitivity drops with increasing bias, the spatial variation shows a minimum at around $V_{GS} = 6$ V. This is due to the decreasing contribution of V_T and the increasing contribution of α_p. Therefore, analog designers can intentionally choose a bias point close to the minimum point for the output transistor, when designing circuits to reduce the output current sensitivity to process variations.

The analysis of temperature and spatial nonuniformity effects presented here is done using a generic approach, since the current–voltage behavior is estimated by three key parameters: K, V_T, and α_p. This is adaptable to most TFT types because of the similar working principle, albeit with different parameter values. Therefore, the methodologies and derivations presented here are generic and approached empirically for applicability to other material families, including organic thin-film transistors (OTFTs) and related material families.

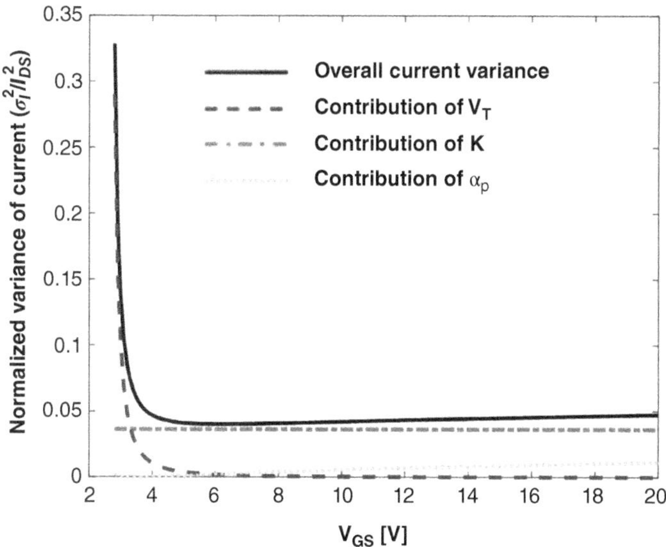

Figure 15.8 Normalized variance of drain current and the contributions of different parameters. *Source*: Calculated from the data in Ref. [13], Cheng, X., et al. (2017). Device-circuit interactions and impact on TFT circuit-system design. *IEEE Journal on Emerging and Selected Topics in Circuits and Systems* **7**: 71–80.

15.2.5 Overview of External Compensation for AMOLED Displays

The pixel structure for defect extraction reported in Refs. [2, 22] is shown in Figure 15.9a. The pixel circuit has a similar structure to the ubiquitous 2T1C circuit—the difference being the addition of a monitor line, readout (RD) control line, and extra TFT (RD switch) to enable access to the pixel OLED and drive TFT for external sensing circuitry. Therefore, the TFT and OLED characteristics can be measured, and their defect and aging status can be extracted. The operation of the pixel circuit is illustrated in Figure 15.9b–15.9d. Programming a pixel (Figure 15.9b) has close similarity with that of the basic 2T1C circuit: The data line voltage is set to V_{DATA} by drive integrated chip (IC) (1), and the monitor line voltage is set to V_{ref} (2); enabling WR and RD switches will let the storage capacitor charge to $V{\sim}V_{DATA}{-}V_{ref}$ (3); WR and RD switches are then disabled, and the voltage V is stored in the capacitor until the next programming cycle. Measurement of the TFT state (Figure 15.9c) requires programming the pixel with forward-biased data (1), setting the monitor line voltage $V_{MONITOR}$ below the turn-on voltage of the OLED (2), and activating the RD switch to direct the TFT current to the external sensing circuit via the monitor line (3). For OLED measurement (Figure 15.9d), the pixel is programmed with V_{DATA} that would ensure that the drive TFT is off (1); monitor line voltage $V_{MONITOR}$ is set to above the OLED turn-on voltage (2); then, activating the RD switch will direct the current into the OLED (3). As the I-V characteristic of the OLED can be a signature of its efficiency [24, 25], the aging of OLED can then be captured from the pre-acquired data for the OLED.

Examples of the aging and defect status of TFTs and OLEDs, as extracted through the monitor lines, are given in Figure 15.10a and 15.10b, respectively. Here, the pixels were measured after the AMOLED panel displayed a checkerboard pattern for a period of time. The areas denoted as "W" were set to display bright white. The drive TFTs and OLEDs in these pixels were therefore affected by aging. The areas denoted as "B" were set to display black. Sharp features that appear randomly across the panel correspond to panel fabrication defects. The patterns in red, yellow, and green express the degree of aging of the pixels. As the center of the displayed white square has a higher temperature due to self-heating of the surrounding pixels, the aging of these pixels will be faster compared to other pixels.

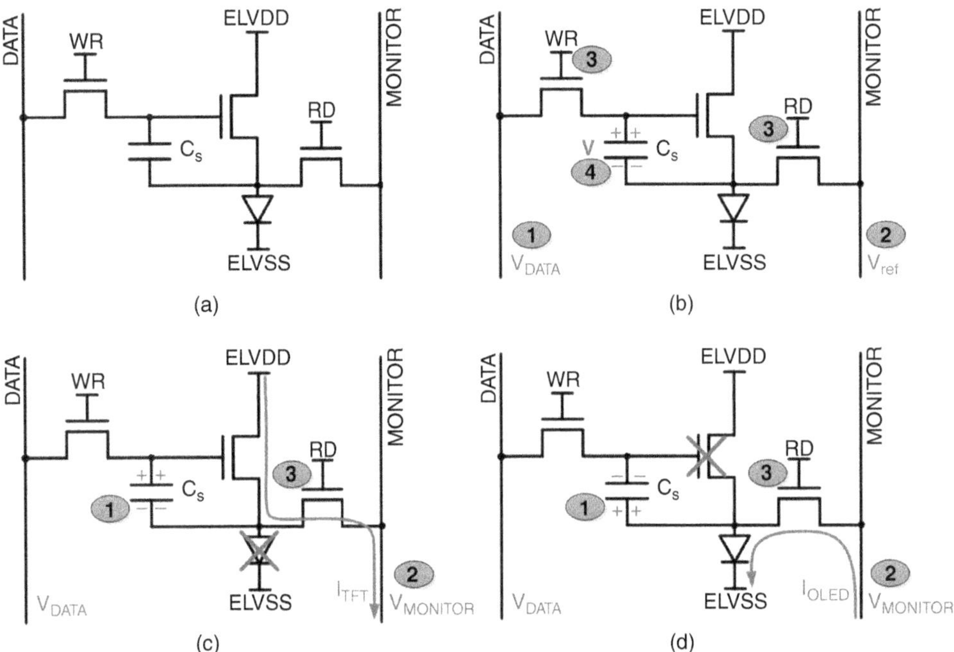

Figure 15.9 AMOLED pixel circuit [2, 22] for off-pixel defect extraction and feedback. (a) Circuit schematic and illustration of circuit operation for (b) pixel programming, (c) TFT measurement, and (d) OLED measurement. *Source*: Ref. [17], Zahirovic, N. (2017). Metal-oxide integration challenges for OLED TV applications. Paper presented at the International Conference on Display Technology in Fuzhou, China, February 19. With permission of Ignis Innovation Inc.

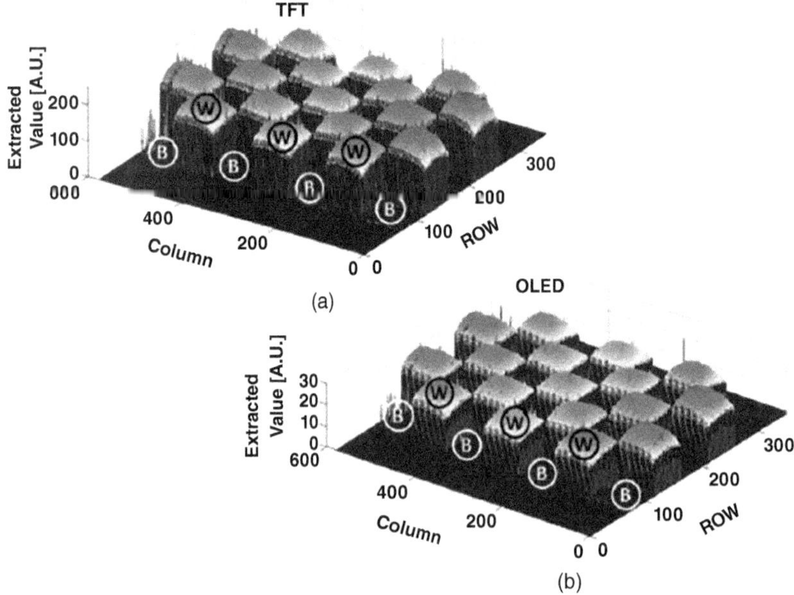

Figure 15.10 Extracted aging parameters for (a) TFTs and (b) OLEDs after continuously displaying a checkerboard pattern (W: displayed with white squares; B: displayed with black squares). *Source*: Ref. [13], Cheng, X., et al. (2017). Device-circuit interactions and impact on TFT circuit-system design. *IEEE Journal on Emerging and Selected Topics in Circuits and Systems* **7**: 71–80.

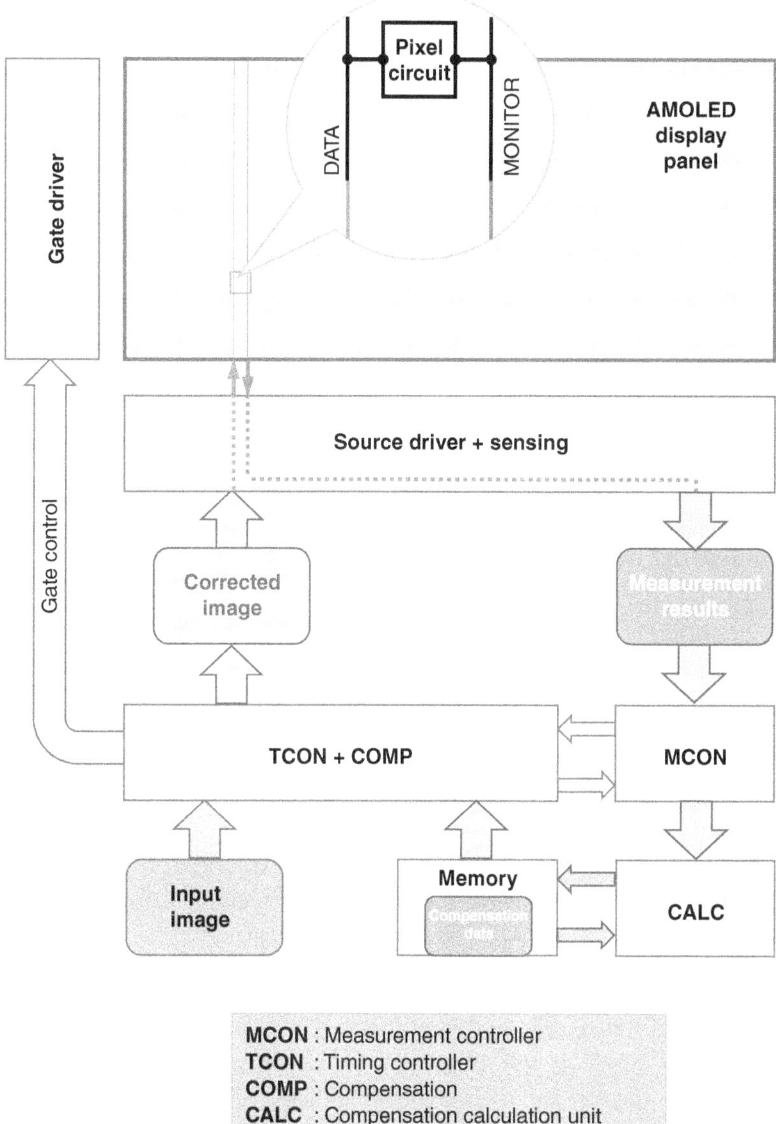

Figure 15.11 Block-diagram and data flow of AMOLED display system that realizes external compensation of TFT and OLED degradation. *Source*: Compiled from Ref. [15], Chaji, R., and Nathan, A. (2014). LTPS vs oxide backplanes for AMOLED displays: system design considerations and compensation techniques. *SID Symposium Digest of Technical Papers* **45** (1): 153–156. Ref. [17], Zahirovic, N. (2017). Metal-oxide integration challenges for OLED TV applications. Paper presented at the International Conference on Display Technology in Fuzhou, China, February 19. Copyright 2017, Ignis Innovation. Ref. [25], Tang, S.-J., and He, J. (2018). Advanced compensation technology for AMOLED automotive displays. In *SID Vehicle Displays and Interfaces Symposium Digest of Technical Papers*, p. 4.6 with permission of Ignis Innovation Inc.

A high-level block diagram and data flow of an AMOLED display system that realizes external compensation for TFT and OLED effects are shown in Figure 15.11. Sensing circuitry implemented as a discrete chip or embedded in a drive IC performs TFT and OLED measurements on display pixels and generates digital measurement data that are applied to calculate the compensation model parameters. The compensation data for the pixels are stored in system memory and updated at each measurement cycle. Compensation correction is extracted from the memory and combined with input image data to generate the corrected image data, which are then loaded into the source driver.

Figure 15.12 An IGZO-TFT AMOLED display showing image nonuniformity due to the panel-manufacturing process: (a) as is (i.e., uncompensated), and (b) compensated. *Source*: Ref. [17], Zahirovic, N. (2017). Metal-oxide integration challenges for OLED TV applications. Paper presented at the International Conference on Display Technology in Fuzhou, China, February 19. With permission of Ignis Innovation Inc.

The display system using this method has the ability to compensate all kinds of defects that can be extracted through the monitor line, including any manufacturing-process-related nonuniformity and aging of TFTs and OLEDs. For example, the images on Figure 15.12a that were displayed on an as-fabricated AMOLED panel by applying uncorrected video data show obvious nonuniformity due to spatial variation in TFT characteristics (as discussed in Section 15.2.4) and due to nonuniformity in OLED deposition. Measuring the TFT and OLED states in each subpixel, as described in this chapter, and subsequently calculating the data correction and finally applying the corrected data to the pixels result in a uniform image, as shown in Figure 15.12b. Figure 15.13 illustrates the effect of compensation of TFT and OLED aging in an IGZO-TFT AMOLED panel. The aged display has dark and bright burn-in patterns (Figure 15.13a) due to TFT bias stress, illumination stress, and OLED aging. These are eliminated almost perfectly (Figure 15.13c) after applying both TFT and OLED compensation algorithms. The general approach described here relies on external compensation of both TFT- and OLED-related non-idealities. It is also possible to use a hybrid compensation scheme [2], which relies on in-pixel compensation for fast and small variations associated with some TFT effects and uses external compensation for OLED aging.

For example, a variant of the hybrid scheme equipped with an advanced OLED aging algorithm was able to maintain a burn-in-free uniform image through more than 1200 h of high-temperature (85 °C) operation, even though the OLED efficiency due to high brightness and high-temperature stress dropped to approximately half of its initial value [25].

15.3 Ultralow-Power TFT Operation in a Deep Subthreshold (Near Off-State) Regime

For deployment of TFTs in mobile devices, such as wearables, low voltage and low power are crucial because the operation of the wearable device is challenged by the limited battery lifetime, even if it is augmented with energy harvesting [26–29]. Existing TFT technologies are unlikely to meet the low-power and high-intrinsic-gain requirements because of the typically high bias current (microamperes) and limited signal amplification (i.e., intrinsic gain of a few tens), which is insufficient to process weak signals at the nanoampere level [2, 30–32]. Moreover, the

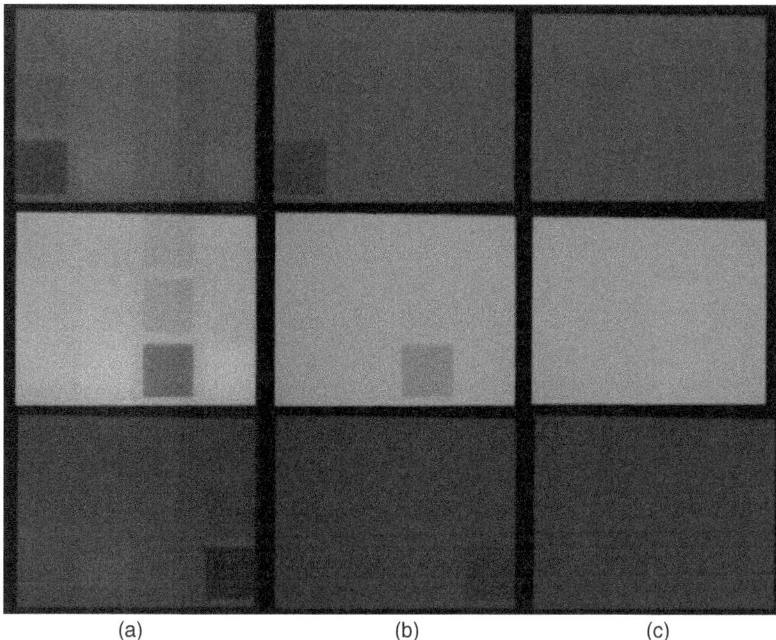

Figure 15.13 An IGZO-TFT AMOLED display after aging showing R, G, and B flat-field images: (a) without compensation, (b) with TFT compensation, and (c) with TFT and OLED compensation. *Source*: Ref. [17], Zahirovic, N. (2017). Metal-oxide integration challenges for OLED TV applications. Paper presented at the International Conference on Display Technology in Fuzhou, China, February 19. With permission of Ignis Innovation Inc.

Table 15.2 Off-Current and Standby Power for TFT Technologies.

TFT Technologies	LTPS	*a*-Si	Organics	IGZO
Bandgap	~1.1 eV	~1.7–1.8 eV	1–2 eV	>3 eV
Off-current	$>10^{-12}$ A	$\sim10^{-12}$ A	$\sim10^{-12}$ A	$<10^{-13}$ A
Minimum stand-by power (for $V_{DD} = 1$ V)	$>10^{-12}$ W	$\sim10^{-12}$ W	$\sim10^{-12}$ W	$<10^{-13}$ W

Source: Adapted from Ref. [4], Lee, S., and Nathan, A. (2016). Subthreshold Schottky-barrier thin-film transistors with ultralow power and high intrinsic gain. *Science* **354**: 302–304.

off-current of other TFT technologies is higher due to their narrower bandgap compared to IGZO, thus producing a higher power consumption in their off-state. Indeed, according to the literature [9, 10, 26–29, 33–36], LTPS, *a*-Si, organics, and oxide (e.g., IGZO) technologies have $>10^{-12}$, $\sim10^{-12}$, $\sim10^{-12}$, and $<10^{-13}$ amperes as their off-current, respectively (Table 15.2). Clearly the IGZO-TFT's off-current is the lowest [36]. For an estimation of power consumption (P_{out}), the theoretically achievable minimum P_{out} would be $I_{OFF}{\cdot}V_{DD}$, where V_{DD} is the supply voltage and I_{OFF} is the off-current.

15.3.1 Schottky Barrier TFTs

The approach to achieve ultralow power is to operate the transistor in the deep subthreshold (sub-T) regime, that is, near the off-state (Figure 15.14). In addition, as discussed in this section, TFTs with Schottky barrier (SB) source–drain contact rather than ohmic contact can be employed to achieve high performance (e.g., high gain) in spite of operating in the deep sub-T regime [4].

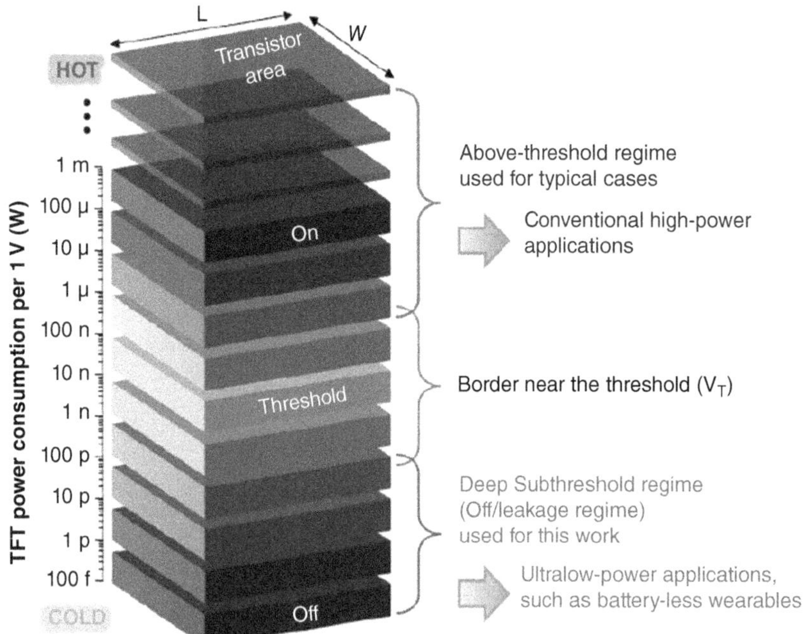

Figure 15.14 TFT power consumption per 1 V at different operating modes from the off- to on-states. *Source*: Ref. [4], Lee, S., and Nathan, A. (2016). Subthreshold Schottky-barrier thin-film transistors with ultralow power and high intrinsic gain. *Science* **354**: 302–304.

Within this regime, the saturation drain current (I_{DS}) of the SB-TFT is independent of drain voltage (V_{DS}), yielding an infinite output resistance (i.e., $r_o = \partial V_{DS}/\partial I_{DS} \rightarrow \infty$). Indeed, the magnitude of the current is scaled geometrically only by the channel width (W), as opposed to the channel width-to-length ratio (W/L), as in conventional transistors (see the inset of Figure 15.15b). The insulated gate provides an effective means of modulating the SB height at the source contact and hence the thermionic emission (TE) and thermionic field emission (TFE) properties. Thus, the emission current into the channel is determined by the reverse saturation current of the Schottky diode at the source, which in turn is modulated by the gate voltage. As a result, the SB-TFT yields a large intrinsic gain that is independent of both geometry and bias. This bias independence of the intrinsic gain and zero input current by virtue of the insulated gate make the SB-TFT capture the best of the bipolar junction transistor (BJT) and metal-oxide semiconductor field-effect transistor (MOSFET) technology families [2, 4, 24, 38].

To form a Schottky contact at the source–drain contact of the IGZO-TFT, we decreased the electron concentration of the IGZO film [4] by using a high oxygen-gas partial pressure relative to argon, that is, $P_{ox}=O_2/(O_2+Ar)$, during the radiofrequency (RF) sputtering process, with subsequent thermal annealing for a more reliable contact (Figure 15.15a). Here, a high P_{ox} serves to compensate oxygen vacancies (V_{ox}), which act as electron donors [4, 23, 25, 40]. Indeed, in the measured output characteristics, more compensated (MC)-IGZO-TFT (at P_{ox} = 15%) provided Schottky characteristics at low V_{DS}, whereas less compensated (LC)-IGZO-TFT (at P_{ox} = 4%) showed the usual ohmic behavior [4]. At higher V_{DS}, both devices show current saturation (Figure 15.15b). More important, the SB-TFT (i.e., MC-IGZO-TFT) has a much flatter output curvature compared to the ohmic device, yielding a much higher r_o. In particular, the output characteristics of the ohmic device had an L dependence [4]. In contrast, the output characteristics of the SB-TFT were almost independent of L (Figure 15.15b). This can be explained with the saturation drain–current (I_{sat}) relation as [41–43]

$$I_{sat} = A_J J_{sat}(V_{GS}) exp\left(-\frac{\varphi_{B0}}{v_{th}}\right)\left(1 - exp\left(-\frac{V_{DS}}{nv_{th}}\right)\right) \tag{15.9}$$

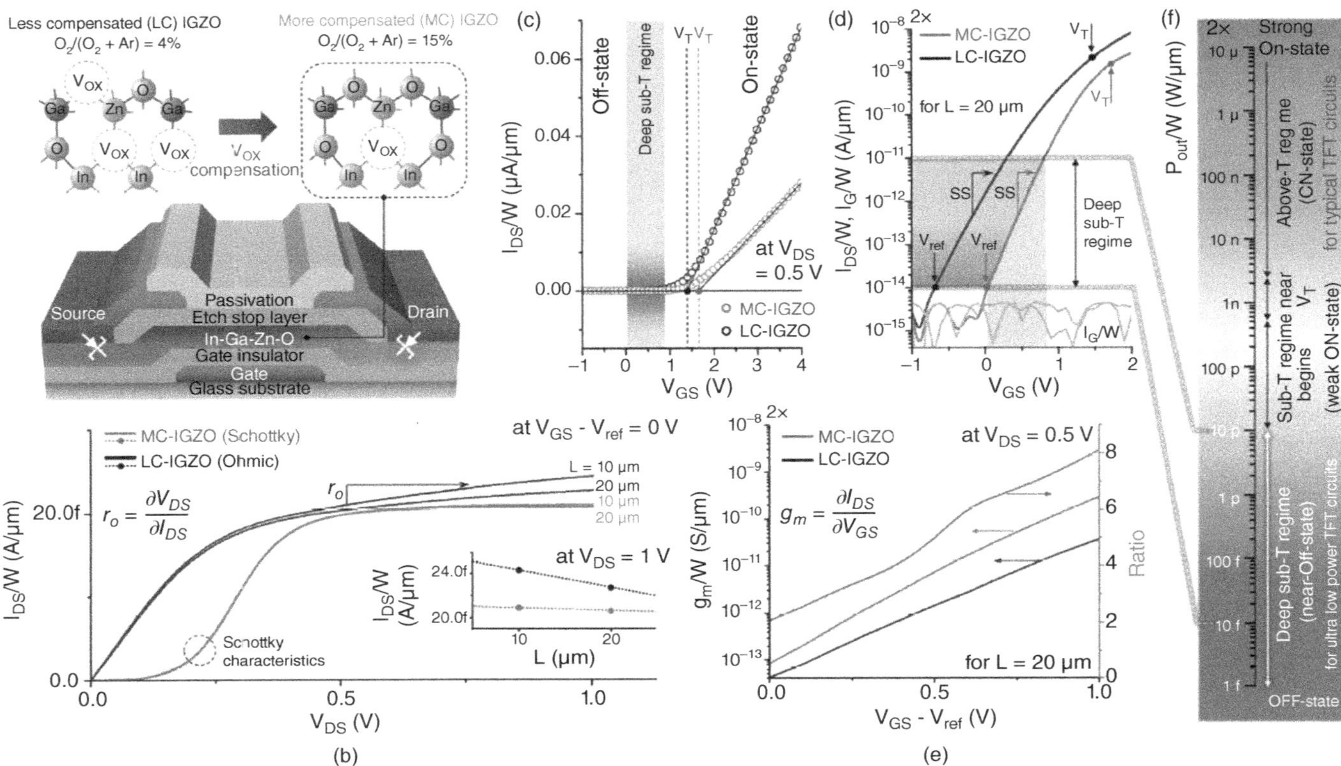

Figure 15.15 Device structure and basic electrical characteristics. (a) Schematic cross-section of the examined device; (inset) illustration of atomic structures for less compensated (LC) and more compensated (MC) IGZO films, respectively. (b) Measured I_{DS}/W versus V_{DS} for MC and LC devices (inset: I_{DS}/W versus L). Measured input characteristics: (c) in linear scale, indicating V_T (the threshold voltage), and (d) in logarithmic scale, indicating V_{ref}, respectively (I_G is the gate leakage current). (e) Measured g_m/W of each device along with the ratio between them. (F) Conceptual color bar of P_{out} normalized with W for a 1 V supply. Here, the sub-T and above-T denote the subthreshold and above-threshold, respectively. *Source*: Ref. [4], Lee, S., and Nathan, A. (2016). Subthreshold Schottky-barrier thin-film transistors with ultralow power and high intrinsic gain. *Science* **354**: 302–304.

where n is the ideality factor (~1.7 of the examined device); A_J is the contact area, where electrons are emitted through TE-TFE rather than the drift-diffusion process; v_{th} is the thermal voltage (i.e., k_BT/q, where k_B is Boltzmann's constant, T the absolute temperature, and q the elementary charge); and J_{sat} is the saturation current density as a function of gate voltage (V_{GS}). I_{sat} is linearly proportional to A_J, and scales with W. In addition, the term $(1 - exp(-V_{DS}/nv_{th}))$ is almost unity in the saturation regime because $V_{DS} \gg nv_{th}$ at 300 K, and thus is independent of V_{DS}. These are consistent with the results in Figure 15.15b. Besides, Figure 15.15c–15.15e shows input characteristics, in which a conceptual color bar of output power consumption (P_{out}) for a 1 V supply, normalized with W, is shown, clearly indicating each operational regime (Figure 15.15f). In particular, as seen in Figure 15.15e, the SB-TFT has a higher transconductance (g_m). This can be explained with its smaller subthreshold slope (SS) of ~0.28 V/decade compared to the ohmic device (Figure 15.15d, Ref. [4]). Because the intrinsic gain (A_i) of a transistor is defined as $g_m \cdot r_o$, the SB-TFT provides a higher A_i associated with its higher r_o and g_m compared to the ohmic device [4].

To theoretically explain the results of the SB-TFT, we describe its operating principle in Figure 15.16. At a given V_{GS}, the nonlinear response of the drain current at $V_{DS} < V_{tran}$ (the transition voltage) suggests a forward-biased Schottky diode at the drain junction (Figure 15.16a). Here, the electron collection is modulated with V_{GS} through an effective SB lowering at the drain side ($\Delta\varphi_{BD}$) (Figure 15.16b). When $V_{DS} > V_{tran}$, I_{DS} becomes firmly saturated because of the reverse-biased Schottky diode at the source (Figure 15.16c). This regime satisfies the condition of $L \gg W_D$, suggesting a negligible image-charge effect from the drain to source, where W_D is the depletion width at the drain (Figure 15.16c, Ref. [4]). Thus, the SB lowering at the source ($\Delta\varphi_{BS}$) is mainly a function of V_{GS} modulating the current density, expressed as

$$J_{sat}(V_{GS}) = J_0 exp\left(-\frac{\Delta\varphi_{BS}(V_{GS})}{v_{th}}\right) \quad (15.10)$$

where J_0 is a reference current density. As seen in Figure 15.16d, the intercept ($\Delta\varphi_0$) ~0.165 V can be considered as an initial SB lowering corresponding to the reference current at $V_{GS} = V_{ref}$ (the reference voltage). An effective SB-lowering (e.g., $\Delta\varphi_{BS}$) approximation is used to account for changes in the SB width (W_S) and, hence, the degree of the quantum mechanical tunneling [4, 43].

15.3.2 Device Characteristics and Small Signal Parameters

Based on the theory discussed along with Figure 15.16, the output characteristics of the SB device for different V_{GS} (ranging from 0 to 1 V, in steps of 0.1 V) were measured and modeled (Figure 15.17a). The results show good agreement with each other. Figure 15.17b shows the transfer characteristics for V_{DS} of 0.5 and 1 V. They are nearly identical, implying current saturation for $V_{DS} > V_{tran}$ ~0.48 V. Also, it shows an exponential dependency on V_{GS}, which can be explained with Equation (15.10), where $\Delta\varphi_{BS}(V_{GS}) = \zeta 0(V_{GS} - V_{ref}) + \Delta\varphi_0$. Here, ζ_0 is a coefficient that describes the sensitivity of barrier lowering to V_{GS}. The retrieved r_o and g_m for $V_{DS} = 1$ V are shown as a function of V_{GS} (Figure 15.17c). Both follow an exponential law with an opposite proportionality on V_{GS}, as described in Equations (15.11) and (15.12), and their product gives a signal amplification factor, that is, intrinsic gain (A_i):

$$r_0 = n\frac{v_{th}}{A_J J_{B0}} exp\left(-\frac{\zeta_0(V_{GS} - V_{ref}) - \Delta\varphi_0}{v_{th}}\right) exp\left(\frac{v_{sat}}{nv_{th}}\right) \quad (15.11)$$

$$g_m = \zeta_0\frac{A_J J_{B0}}{v_{th}} exp\left(\frac{\zeta_0(V_{GS} - V_{ref}) + \Delta\varphi_0}{v_{th}}\right) \quad (15.12)$$

$$A_i = \zeta_0 n exp\left(\frac{v_{sat}}{nv_{th}}\right) \quad (15.13)$$

where $J_{B0} = J_0\, exp(-\varphi_{B0}/v_{th})$ [4]. As seen in Equation (15.13), A_i is not a function of either bias (e.g., V_{GS}, V_{DS}) or geometry (e.g., W and L), but rather is a function of intrinsic parameters (e.g., ζ_0, n, v_{th}, and a saturation voltage v_{sat}). So, it is just a constant, unlike the ohmic device. With Equation (15.13), A_i is calculated as ~450 with the

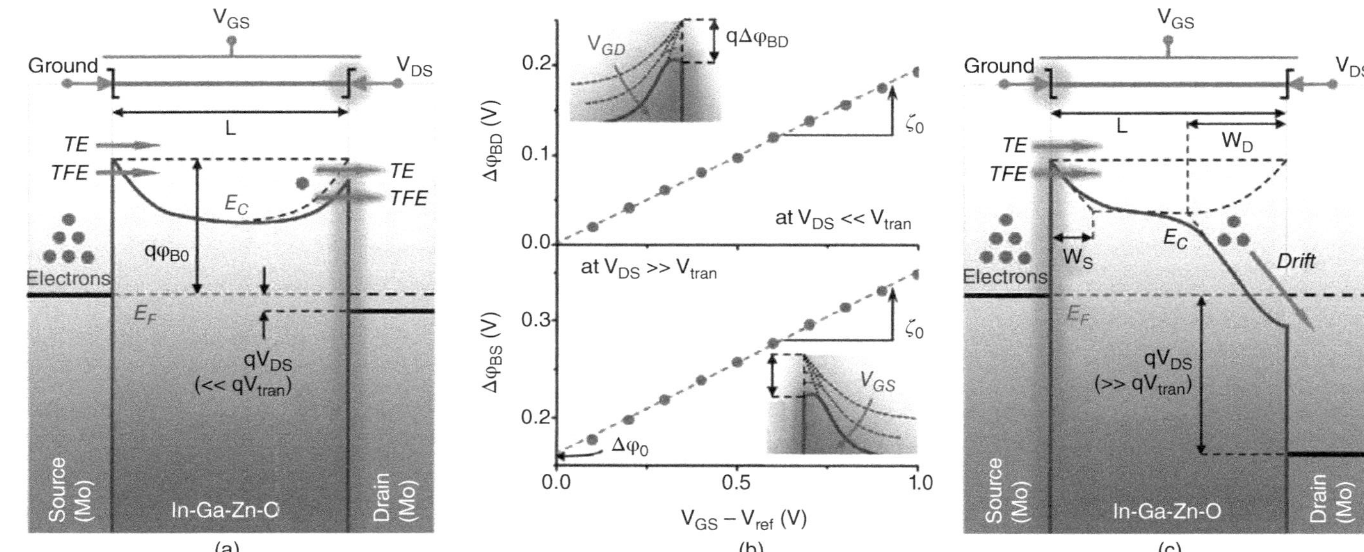

Figure 15.16 Operating principle of a deep-subthreshold Schottky barrier TFT. (a) Band diagram along L when $V_{DS} < V_{tran}$. (b) Retrieved $\Delta\varphi_{\mathrm{BD}}$ at drain contact as a function of V_{GS} calibrated with V_{ref} (i.e., $V_{GS} - V_{ref}$). (c) Band diagram along L when $\mathrm{V_{DS}} > \mathrm{V_{tran}}$. (d) Extracted $\Delta\varphi_{\mathrm{BS}}$ at source contact as a function of $V_{GS} - V_{ref}$. Here, E_C and E_F denote the conduction band minima and Fermi level, respectively. Insets in (a) and (c) are equivalent circuit representations, and those of (b) and (d) are schematic diagrams to describe the bias-dependent SB lowering, respectively. *Source*: Ref. [4], Lee, S., and Nathan, A. (2016). Subthreshold Schottky-barrier thin-film transistors with ultralow power and high intrinsic gain. *Science* **354**: 302–304.

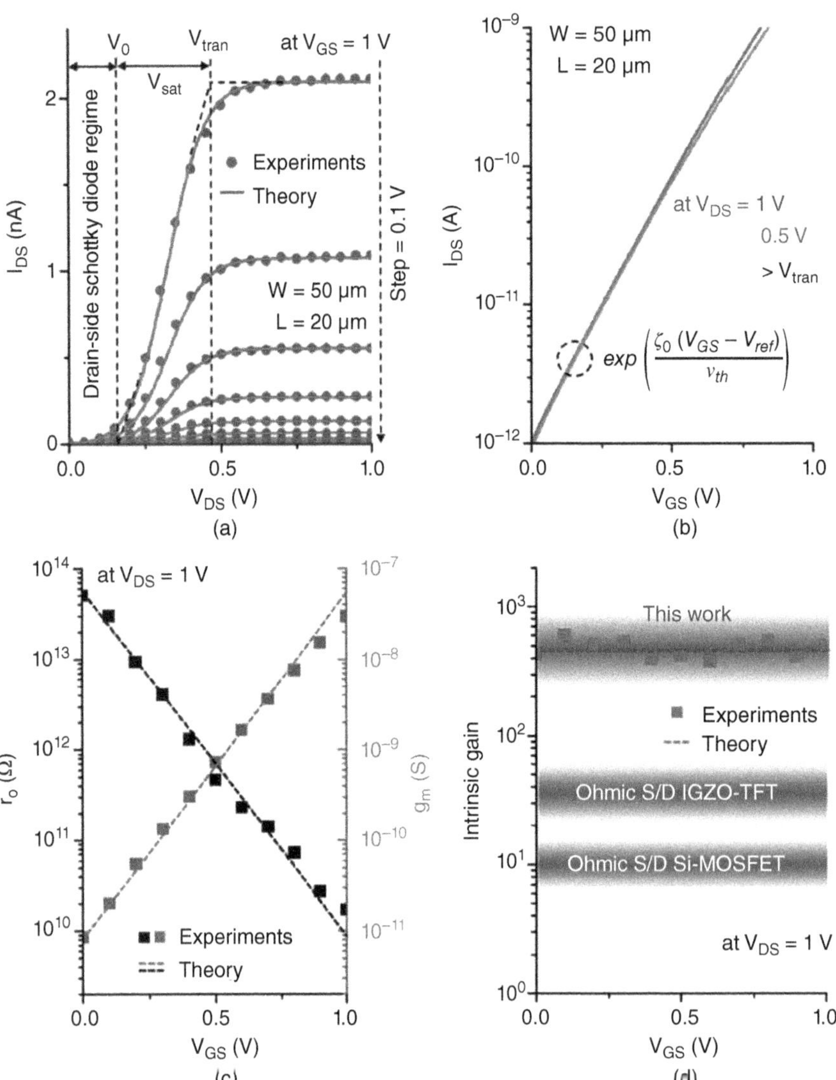

Figure 15.17 Terminal characteristics and small-signal parameters of the SB-TFT in the deep subthreshold regime. (a) Measured I_{DS} versus V_{DS} for a different V_{GS} along with theoretical prediction. Here, V_0 is a threshold at which the source-side Schottky diode starts dominating. (b) Transfer characteristics for V_{DS} = 0.5 and 1 V. (c) Experimental values of r_o and g_m as a function of V_{GS} along with theoretical prediction (dashed lines). (d) Measured A_i as a function of V_{GS}. *Source*: Ref. [4], Lee, S., and Nathan, A. (2016). Subthreshold Schottky-barrier thin-film transistors with ultralow power and high intrinsic gain. *Science* **354**: 302–304.

retrieved values of intrinsic parameters [4], which is consistent with measurements (Figure 15.17). Here, A_i of the SB-TFT is at least an order of magnitude higher compared to the ohmic IGZO-TFT [4] or any typical Si-MOSFET [37, 38]. Because the SB-TFT operates at low voltage/current, it is also electrically stable over time [4].

15.3.3 Common Source Amplifier

The low power and high-gain performance of the SB-TFTs were assessed with a common source amplifier [4], as demonstrated in Figure 15.18. As seen in Figure 15.18a and Figure 15.18b, the TFT-2 (load), whose V_{ref} (and V_T) was shifted to negative because of light stress, was used as a depletion load [4, 44]. Alternatively, a TFT with a

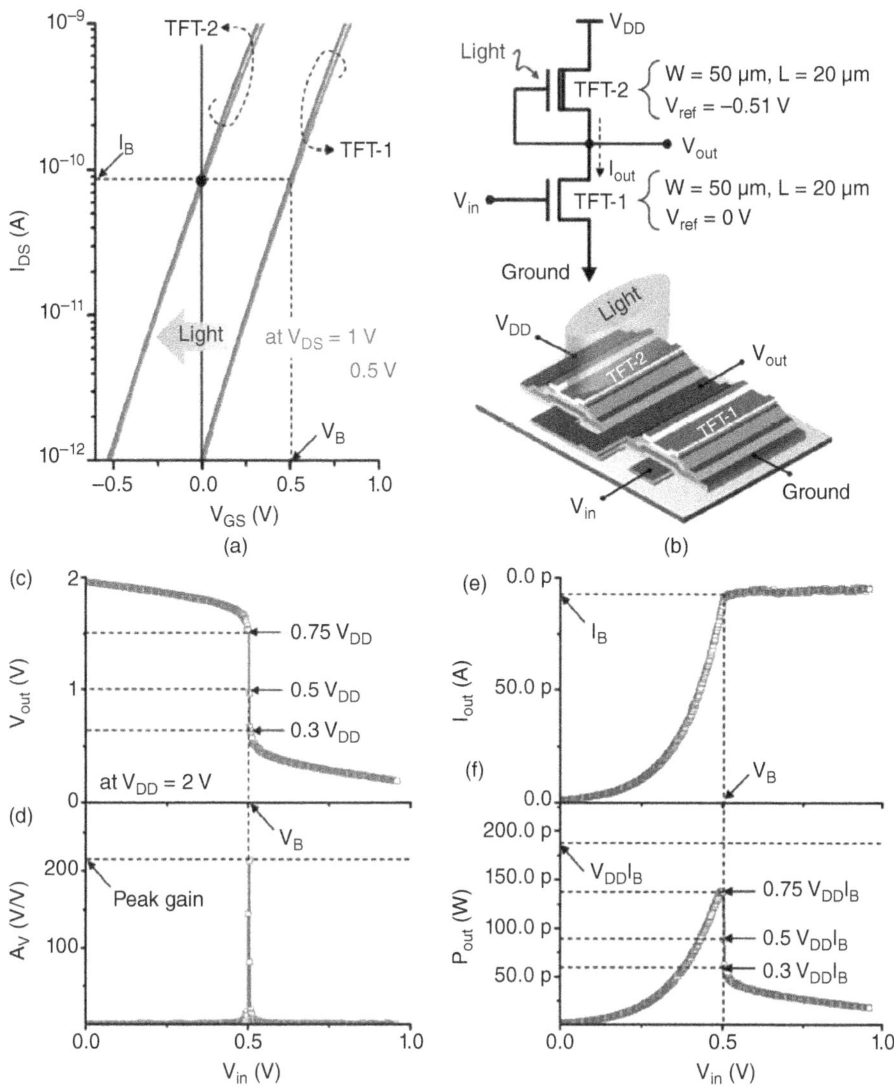

Figure 15.18 Circuit-level demonstrations with the SB-TFTs. (a) Transfer characteristics of TFT-1 (driver) and TFT-2 (depletion load due to light stress), where I_B and V_B are determined as ~90 pA and ~0.5 V, respectively. (b) Common-source circuit and its three-dimensional view. (c) Measured output voltage (V_{out}) as a function of input voltage (V_{in}) while using 2 V supply (V_{DD}), (d) A_V, (e) output current (I_{out}), and (f) P_{out} versus V_{in}, respectively. *Source*: Ref. [4], Lee, S., and Nathan, A. (2016). Subthreshold Schottky-barrier thin-film transistors with ultralow power and high intrinsic gain. *Science* **354**: 302–304.

larger W as a design parameter can also be employed [4]. As shown in Figure 15.18d, the circuit exhibits a high voltage gain (A_V) >220, and its output-power consumption (P_{out}) is very low, <150 pW (Figure 15.18f). Thus, it can even be driven by a nanowatt power source. The deep subthreshold operating SB-TFT is fundamentally an ultralow-power and high-gain device. This opens up possibilities for innovative system design in many applications, including wearables and implantable devices, where low-power, low-voltage analog signal processing is an essential requirement. In addition, the operating principle of the SB-TFT in the deep subthreshold regime (i.e., near-off-state) brings together the best of two transistor families: the bias independence of gain of the BJT and the virtually zero-input current of the MOSFET [4]. Thus, the SB-TFT will bring about a new design paradigm for near-off-state sensor interfaces and analog front-end circuits [4].

15.4 Oxide TFT-Based Image Sensors

Although the AOS is well known for its optical transparency, this, in principle, is very much dependent on the composition of the material, which can be tailored to enhance optical absorption and induce photoconductivity when irradiated with green and shorter wavelength light. This is true as long as there are oxygen vacancy defects (V_O) in the energy gap, which then give rise to subgap absorption. For typical display applications, the AOS is incorporated with additional metal atoms (e.g., Ga) to suppress oxygen defects. Thus, the oxide semiconductor material can be custom-designed for photodetection.

15.4.1 Heterojunction Oxide Photo-TFTs

Optical absorption spectra of IZO and IGZO oxide semiconductors are plotted in Figure 15.19, where IGZO is used interchangeably with GIZO. IZO has much higher absorption of visible photons than IGZO thanks to lower band gap energy (E_g) and a broad subgap absorption tail. In particular, at photon energies smaller than the bandgap energy (E_g) of IZO, absorption is mainly due to ionization of V_O defect states rather than band-to-band excitation, even under visible illumination [46].

High optical absorption of IZO favors its application as a light-sensitive layer in photosensors. At the same time, V_O states located within the bandgap come with a critical drawback, persistent photoconductivity (PPC) [39], especially at shorter wavelengths [49]. PPC causes the semiconductor material to remain conductive for hours/days, even in the absence of light [49]. This increases the response times and limits the frame rates. Two-terminal photosensor devices do not allow for rapid recovery from the effects of PPC [49, 50]. However, as we discuss further in this section, PPC effect can be eliminated in three terminal devices (e.g., photo-TFT) with an appropriate bias scheme [12, 46, 51, 52]. Note that IZO-TFTs with high V_O concentration in the active material exhibit relatively large negative V_T [46, 51]. It is more beneficial to use IZO in a hetero-structure with IGZO [46, 51] or HIZO [52] (Figure 15.20a). Here, IZO functions as a light-absorption layer, while IGZO (or HIZO) allows for adjustment of V_T to more positive values. Moreover, a favorable conduction band offset along with gate-modulated band bending leads to efficient transfer of photogenerated electrons at the hetero-interface to the IGZO (or HIZO) transport layer in which the V_O concentration is far less. This, coupled with localization of holes, retards recombination with the ionized V_O and/or holes, giving rise to extended electron lifetime, τ_n, and hence, high photoconductive gain, G_{ph} [47], and high photocurrent, I_{ph}:

$$I_{ph} = G_{ph} WLq \frac{P_{abs}\lambda}{hc}, \tag{15.14}$$

where P_{abs} is the absorbed optical power, W the channel width, L the channel length, q the electron charge, and λ and c the wavelength and speed of light, respectively. G_{ph} is defined as the number of collected carriers per absorbed photon and can be expressed as [53]

$$G_{ph} = \frac{\tau_n}{\tau_t} \tag{15.15}$$

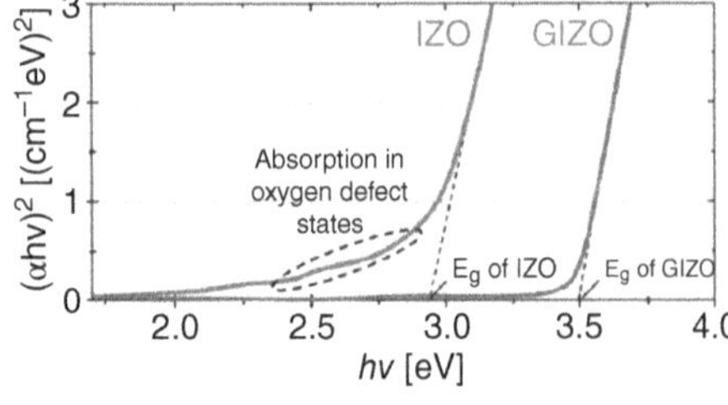

Figure 15.19 Measured optical absorption spectra for IGZO (denoted as GIZO) and IZO, respectively. *Source*: Ref. [12], Lee, S., et al. (2015). Transparent semiconducting oxide technology for touch free interactive flexible displays. *Proceedings of the IEEE* **103**: 644–664. Adapted from Refs. [45–48].

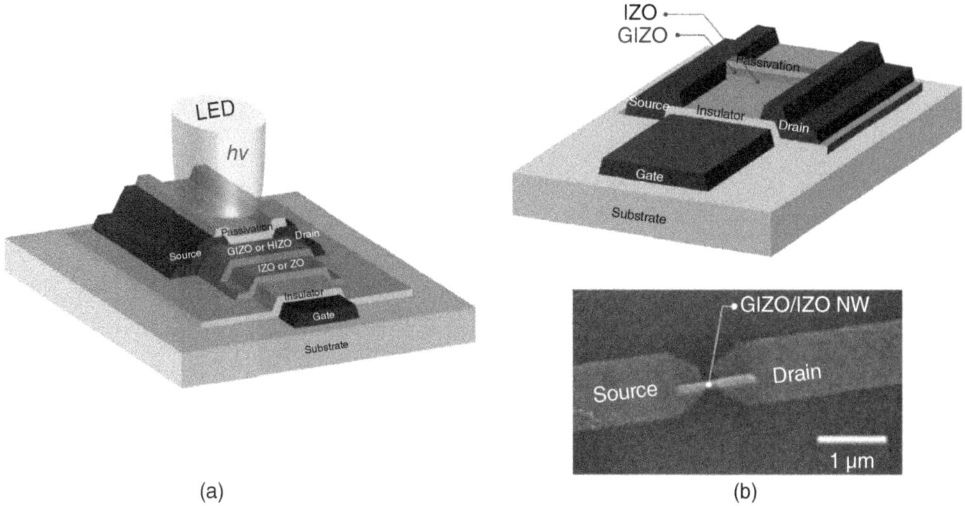

Figure 15.20 (a) Schematic three-dimensional view of a heterojunction oxide photo-TFT. (b) Schematic three-dimensional view of a single NW heterojunction photo-TFT [54] and scanning electron microscopy (SEM) image of a fabricated device. *Source*: Ref. [12], Lee, S., et al. (2015). Transparent semiconducting oxide technology for touch free interactive flexible displays. *Proceedings of the IEEE* **103**: 644–664.

where transit time τ_t can be approximated as

$$\tau_t = \frac{L^2}{\mu_{eff} V_{DS}} \tag{15.16}$$

Indeed, a very high I_{ph} can be achieved in optimal device structures, with carefully optimized absorption and transport layers. Transfer characteristics of bottom-gate photo-TFTs with an IZO/IGZO active layer in the dark and under green-light illumination are shown in Figure 15.21a [51]. The photo-TFT exhibits a high photocurrent to dark-current ratio (I_{photo}/I_{dark}) of ~10^7 at a light power of 260 μW/cm^2. This is four orders of magnitude higher than that of the reference amorphous silicon photo-TFT under the same illumination conditions [51].

Transfer characteristics of photo-TFTs with an IGZO–IZO–IGZO tri-layer of active material in the dark and under 400, 450, 500, 550, 600, and 650 nm illumination are shown in Figure 15.21b [46]. The photo-TFT is sensitive to wavelengths $\lambda < 550$ nm. Here, the increased absorption coefficient observed at shorter wavelengths (Figure 15.19) leads to an increased photocurrent. Again, high (I_{photo}/I_{dark}) up to ~10^7 was observed under light exposure of ~100 W/cm^2 at λ ~400 nm, when the gate-to-source V_{GS} was set to −5 V.

The effect of source–drain contact on the performance of photo-TFTs with nanocrystalline HIZO/IZO heterojunction channels was systematically studied in Ref. [52]. It was found that the photocurrent can be further increased by about an order of magnitude in photo-TFTs with transparent IZO/Au contacts as compared to the reference device with Mo metal contacts. This is because the transparent electrodes facilitate relatively high injection of photogenerated carriers due to a significant lowering of the SB at the source side in comparison to Mo metal electrodes [52].

The basic advantages of a heterojunction channel structure are also maintained in scaled-down single-nanowire (NW) photo-TFTs. The single-NW heterojunction oxide TFT [54] shown in Figure 15.20b can be fabricated using conventional thin-film semiconductor processing technology. All fabrication processes are identical to that of the TFT shown in Figure 15.20a, the only difference being the NW was patterned using e-beam lithography followed by dry etching to realize the small feature (L, W down to 200 nm). Typical characteristics of the single NW photo-TFT are illustrated in Figure 15.22a. The NW oxide photo-TFT with a gate width of 200 nm and gate length of 200 nm demonstrates well-behaved transfer characteristics in the dark, with a high I_{on}/I_{off} modulation ratio of ~ $1{\times}10^{8}$.

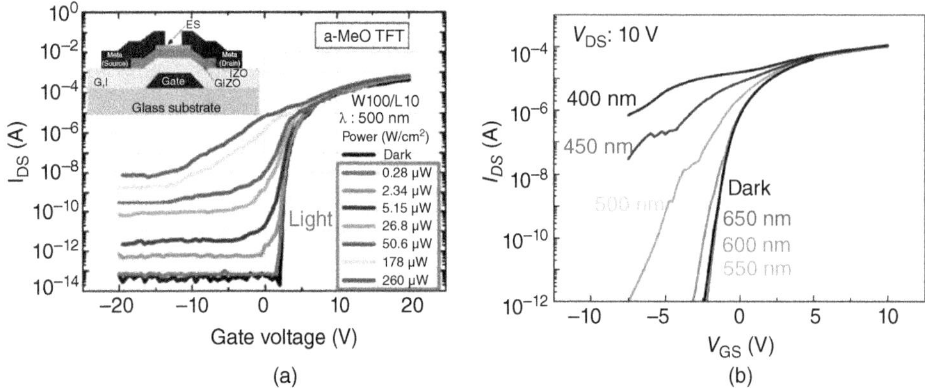

Figure 15.21 Transfer characteristics of metal-oxide photo-TFTs. (a) TFTs with IZO–IGZO active material in the dark and under illumination at a wavelength of 500 nm and light power of 0.28, 2.34, 5.15, 26.8, 50.6, 178, and 260 μW/cm^2. (b) IGZO–IZO–IGZO-TFTs under exposure to 400, 450, 500, 550, 600, and 650 nm light with a power of ~100 μW/cm^2. *Source*: (a) Ref. [51], Ahn, S.-E., et al. (2012). Metal oxide thin film phototransistor for remote touch interactive displays. *Advanced Materials* **24**: 2631–2636. (b) Ref. [46], Jeon, S., et al. (2012). Gated three-terminal device architecture to eliminate persistent photoconductivity in oxide semiconductor photosensor arrays. *Nature Materials* **11**: 301–305.

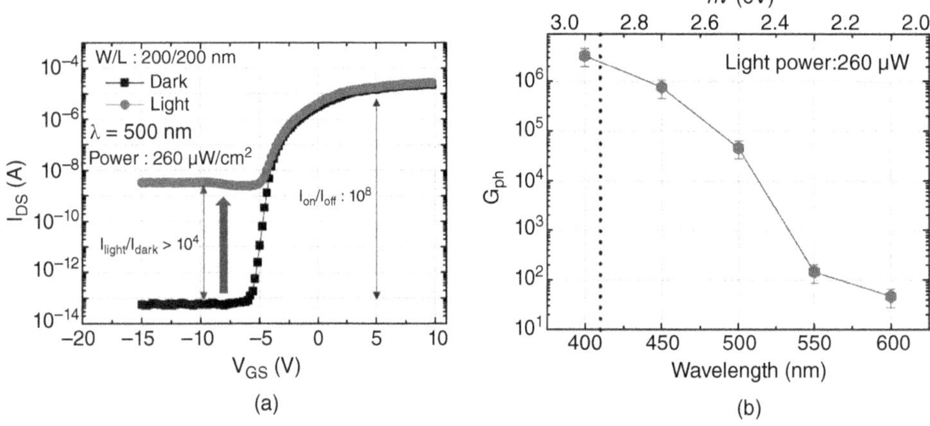

Figure 15.22 (a) Transfer curves of a single NW oxide photo-TFT in the dark and under illumination (500 nm wavelength and 260 μW/cm^{-2} light intensity). (b) Photoconductive gain (G_{ph}) as a function of wavelength and photon energy at a fixed light intensity of 260 μW/cm^{-2}. *Source*: Ref. [54], Ahn, S.-E., et al. (2013). High-performance nanowire oxide photo-thin film transistors. *Advanced Materials* **25**: 5549–5554.

When a single NW photo-TFT was exposed to illumination (with light intensity of 260 μW cm^{-2} and a wavelength of 500 nm), a high photocurrent with a I_{photo}/I_{dark} ratio of 10^4 was registered in the off-current region, even though the light absorption area $W{\times}L$ in a nanoscale device is small. A very high photoconductive gain of up to ~10^5–10^7 is achieved for illumination wavelengths of ~400–480 nm. This can be explained by the increased optical absorption and photocurrent for shorter wavelength light, maintaining a long electron lifetime τ_n [12], while transit time τ_t is expected to be very small due to the short nanoscale L, as implied by Equation (15.16).

15.4.2 Persistent Photocurrent

All of the devices discussed here exhibit a persistent photocurrent (PPC) that is associated with the high concentration of V_O states in the IZO absorption layer. The PPC phenomenon and its compensation by gate voltage pulse

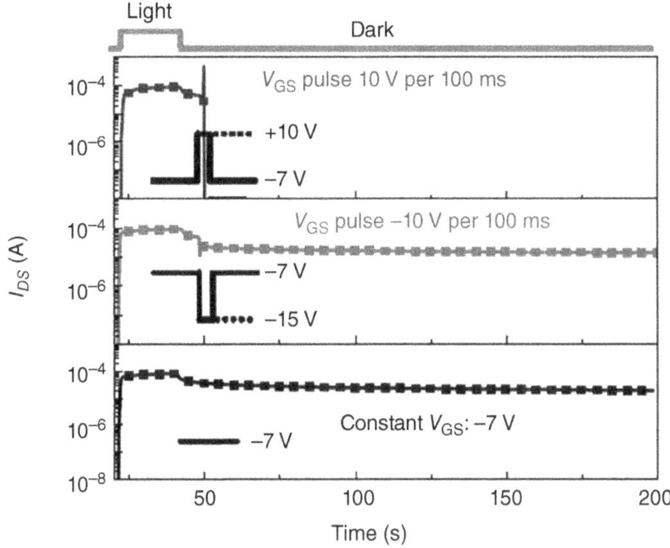

Figure 15.23 Drain–source current as a function of time when TFTs are subject to light and/or gate bias pulses. Even after the illumination is stopped, the photocurrent persists. Only a positive gate pulse is effective in resetting the original dark state, that is, a 10 ns positive gate voltage pulse leads to recovery from PPC. *Source*: Ref. [46], Jeon, S., et al. (2012). Gated three-terminal device architecture to eliminate persistent photoconductivity in oxide semiconductor photosensor arrays. *Nature Materials* **11**: 301–305.

in oxide photo-TFTs is illustrated in Figure 15.23 [46]. As seen, an illumination pulse causes the drain current of the photo-TFT to rapidly increase (due to the I_{ph} contribution). When the TFT gate is held continuously under negative gate bias or if a negative V_{GS} pulse is applied after illumination, switching off the light does not remove this high I_{DS}. Instead, the photocurrent can remain for another day or even more due to PPC [46, 51], apparently rendering these devices unusable as photosensors. However, the PPC can be erased by applying a positive gate voltage pulse. As detailed in Ref. [46], light-induced ionization of oxygen vacancies V_O generates photoelectrons and positively charged defects V_O^{2+}, where the energy level of V_O is located deep in the bandgap while the ionized state is a shallow donor state [40]. After turning off the illumination, the photocurrent decay is controlled by the neutralization rate of charged defects.

$$V_O^{2+} + 2e^- \rightarrow V_O \tag{15.17}$$

The relaxation of chemical bonds surrounding the V_O sites then creates an energy barrier against neutralization of V_O^{2+} sites, thus keeping the material in a state of high conductivity [39]. Furthermore, under negative V_{GS}, electrons become localized near the back interface of the active channel (see the top interface on Figure 15.20a and 15.20b), that is, spatially separated from ionized vacancies V_O^{2+} and holes slowing the (15.17) even more [46].

A positive gate pulse results in electron accumulation at the front channel (bottom interface on Figure 15.20a and Figure 15.20b), thereby accelerating the (15.17) and hence dark current recovery [46]. PPC compensation by positive gate pulse makes the oxide photo-TFT usable in sensor arrays, allowing for practical frame rates.

15.4.3 All-Oxide Photosensor Array

A TFT photosensor circuit [12, 46, 51] that supports PPC compensation using the above compensation/driving scheme is shown in Figure 15.24a. In addition to the photo-TFT sensing element, another TFT functioning as a sampling switch is required to extract the signal through the sensing line to the readout integrated circuit. This switch TFT needs to be shielded from light. When a positive voltage is applied to the switch TFT via the gate line

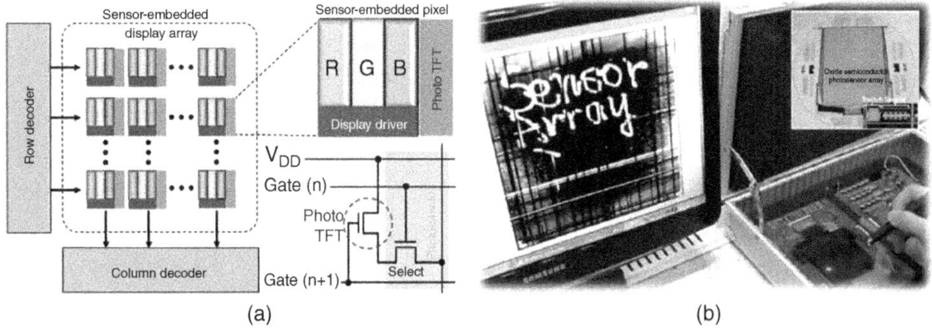

Figure 15.24 (a) Architecture of a sensor-embedded display. (b) Demonstration of touch-free interactive displays with green laser-pointing devices. *Source*: Ref. [12], Lee, S., et al. (2015). Transparent semiconducting oxide technology for touch free interactive flexible displays. *Proceedings of the IEEE* **103**: 644–664. (a) Redrawn and adapted from Refs. [46, 51], Jeon, S., et al. (2012). Gated three-terminal device architecture to eliminate persistent photoconductivity in oxide semiconductor photosensor arrays. *Nature Materials* **11**: 301–305; and Ahn, S.-E., et al. (2012). Metal oxide thin film phototransistor for remote touch interactive displays. *Advanced Materials* **24**: 2631–2636. (b) Redrawn and adapted from Ref. [46], Jeon, S., et al. (2012). Gated three-terminal device architecture to eliminate persistent photoconductivity in oxide semiconductor photosensor arrays. *Nature Materials* **11**: 301–305.

(n), and the other gate lines are not selected and thus negatively biased, the light-sensing signals generated by the photo-TFTs in row n are retrieved by the respective sensing lines. To overcome PPC effects, a positive gate pulse is provided by a high-level select signal used for the next (n+1)th row of pixels, and the photo-TFT is reset so that the next light sensing operation can be performed. Thereafter, the oxide TFT is in a reset state while the negative bias is applied again to the row (n+1) until the next light-sensing operation is carried out. The above sensor architecture was integrated with a display array to realize a touch-free interactive display [51]. Because of the high photosensitivity and speed of the oxide photo-TFT, the sensing pixel can be fit into a relatively small area, and thus the display aperture ratio can be dramatically improved.

Figure 15.24b demonstrates an all-oxide photosensor array prototype based on the bias-assisted PPC recovery scheme (the inset shows the fabricated array panel). Here, the words "Sensor Array" were written using a green laser pointer held 3 cm in front of the panel. The array has a response time of 25 μs, yielding a frame rate of 150 Hz of a fully fabricated oxide-based 256×192 sensor array [12, 46, 55].

References

1 Street, R.A. (2009). Thin film transistors. *Adv. Mater.* 21: 2007–2022.

2 Chaji, R. and Nathan, A. (2013). *Thin-film transistor circuits and systems*. Cambridge: Cambridge University Press.

3 Nathan, A., Ahnood, A., Cole, M.T. et al. (2012). Flexible electronics: the next uniquitous platform. *Proc. IEEE (Centennial Issue)* 100: 1479–1510.

4 Lee, S. and Nathan, A. (2016). Subthreshold Schottky-barrier thin-film transistors with ultralow power and high intrinsic gain. *Science* 354: 302–304.

5 Jiang, C., Choi, H.W., Cheng, X. et al. (2019). Printed subthreshold organic transistors operating at high gain and ultralow power. *Science* 363: 719–723.

6 Park, J.-S., Kim, T.-W., Stryakhilev, D. et al. (2009). Flexible full color organic light-emitting diode display on polyimide plastic substrate driven by amorphous indium gallium zinc oxide thin-film transistors. *Appl. Phys. Lett* 95: 013503 (1-3).

7 Martins, R.F.P., Ahnood, A., Correia, N. et al. (2013). Recyclable, flexible, low-power oxide electronics. *Advanced Functional Materials* 23: 2153–2161.

8 Martins, R., Nathan, A., Barros, R. et al. (2011). Complementary metal oxide semiconductor technology with and on paper. *Adv. Mater.* 23: 4491–4496.

9 Nomura, K., Ohta, H., Takagi, A. et al. (2004). Room-temperature fabrication of transparent flexible thin-film transistors using amorphous oxide semiconductors. *Nature* 432: 488–492.

10 Wager, J.F. (2003). Transparent electronics. *Science* 300: 1245–1246.

11 Jiang, C., Cheng, X., and Nathan, A. (2019). Flexible ultralow power sensor interfaces for E-Skin. *Proceedings of the IEEE* 107: 2084–2105.

12 Lee, S., Jeon, S., Chaji, R., and Nathan, A. (2015). Transparent semiconducting oxide technology for touch free interactive flexible displays. *Proceedings of the IEEE* 103: 644–664.

13 Cheng, X., Lee, S., Chaji, R., and Nathan, A. (2017). Device-circuit interactions and impact on TFT circuit-system design. *IEEE Journal on Emerging and Selected Topics in Circuits and Systems* 7: 71–80.

14 Powell, M.J. (1989). The physics of amorphous-silicon thin-film transistors, IEEE Trans. *Electron Devices* 36: 2753–2763.

15 Chaji, R. and Nathan, A. (2014). *LTPS vs Oxide Backplanes for AMOLED Displays: System Design Considerations and Compensation Techniques*. SID Symp. Dig: 153-156.

16 Nathan, A., Lee, S., Jeon, S., and Chaji, R. (2015). Oxide thin film transistor technology: capturing device-circuit interactions. *IEEE IEDM Tech. Dig* 149–152.

17 Zahirovic, N. (2017), Metal-oxide integration challenges for OLED TV applications, presented at International Conference on Display Technology in Fuzhou, China (19 February, 2017).

18 Nathan, A. (2019). Ultra-low power devices for advanced signal processing architectures. Short course presented at the IEEE 65th International Electron Devices Meeting in San Francisco, CA, December 8.

19 Ide, K., Nomura, K., Hosono, H., and Kamiya, T. (2019). Electronic defects in amorphous oxide semiconductors: a review. *Physica Status Solidi (A): Applications and Materials Science* 216 (1–28): 1800372.

20 Nathan, A., Chaji, G.R., and Ashtiani, S.J. (2005). Driving schemes for a-Si and LTPS AMOLED displays. *Journal of Display Technology* 1: 267–277.

21 Sanford, J.L. and Libsch, F.R. (2003). TFT AMOLED pixel circuits and driving methods. *SID Symposium Digest of Technical Papers* 34: 10–13.

22 Chaji, G.R., Alexander, S., Dionne, J.M. et al. (2010). Stable RGBW AMOLED display with OLED degradation compensation using electrical feedback. In: *2010 IEEE International Solid-State Circuits Conference (ISSCC)*, 118–119.

23 Lee, S. and Nathan, A. (2015). Conduction threshold in accumulation-mode InGaZnO thin film transistors. *Scientific Reports* 6: 22567.

24 Chaji, G.R., Ng, C., Nathan, A. et al. (2007). Electrical compensation of OLED luminance degradation. *IEEE Electron Device Letters* 28 (12): 1108–1110.

25 Tang, S.-J. and He, J. (2018). Advanced compensation technology for AMOLED automotive displays. In: *SID Vehicle Displays and Interfaces Symposium Digest of Technical Papers*, 4.6.

26 Nishide, H. and Oyaizu, K. (2008). Toward flexible batteries. *Science* 319: 737–738.

27 Mitcheson, P.D., Yeatman, E.M., Rao, G.K. et al. (2008). Energy harvesting from human and machine motion for wireless electronic devices. *Proceedings of the IEEE* 96: 1457–1486.

28 Donelan, J.M., Li, Q., Naing, V. et al. (2008). Biomechanical energy harvesting: generating electricity during walking with minimal user effort. *Science* 319: 807–810.

29 Vullers, R.J.M., van Schaijk, R., Doms, I. et al. (2009). Micropower energy harvesting. *Solid-State Electronics* 53: 684–693.

30 Kodaira, T., Hirabayashi, S., Komatsu, Y. et al. (2008). A flexible 2.1-in. active-matrix electrophoretic display with high resolution and a thickness of 100 μm. *Journal of the Society for Information Display* 16: 107–111.

31 Gleskova, H., Wagner, S., Soboyejo, W., and Suo, Z. (2002). Electrical response of amorphous silicon thin-film transistors under mechanical strain. *Journal of Applied Physics* 92: 6224–6229.

32 Sekitani, T., Zschieschang, U., Klauk, H., and Someya, T. (2010). Flexible organic transistors and circuits with extreme bending stability. *Nature Materials* 9: 1015–1022.

33 Nomura, K., Ohta, H., Ueda, K. et al. (2003). Thin-film transistor fabricated in single-crystalline transparent oxide semiconductor. *Science* 300: 1269–1272.

34 Fortunato, E., Barquinha, P., and Martins, R. (2012). Oxide semiconductor thin-film transistors: a review of recent advances. *Advanced Materials* 24: 2945–2986.

35 Tak, Y.J., Ahn, B.D., Park, S.P. et al. (2016). Activation of sputter-processed indium–gallium–zinc oxide films by simultaneous ultraviolet and thermal treatments. *Scientific Reports* 6: 21869.

36 Wager, J.F., Yeh, B., Hoffman, R.L., and Keszler, D.A. (2014). An amorphous oxide semiconductor thin-film transistor route to oxide electronics. *Current Opinion in Solid State & Materials Science* 18: 53–61.

37 Razavi, B. (2001). *Design of Analog CMOS Integrated Circuits*. New York: McGraw-Hill.

38 Sedra, A.S. and Smith, K.C. (2007). *Microelectronic Circuits*. New York: Oxford University Press.

39 Lany, S. and Zunger, A. (2005). Anion vacancies as a source of persistent photoconductivity in II-VI and chalcopyrite semiconductors. *Physical Review B* 72 (1–13): 035215.

40 Janotti, A. and Van de Walle, C.G. (2005). Oxygen vacancies in ZnO. *Applied Physics Letters* 87: 122102.

41 Sze, S.M. (1981). *Physics of Semiconductor Devices*, 2nde. Hoboken, NJ: Wiley.

42 Lee, S., Jeon, S., and Nathan, A. (2013). Modeling sub-threshold current–voltage characteristics in thin film transistors. *Journal of Display Technology* 9: 883–889.

43 Umemoto, Y., Schaff, W.J., Park, H., and Eastman, L.F. (1993). Effect of thermionic-field emission on effective barrier height lowering in In0.52Al0.48As Schottky diodes. *Applied Physics Letters* 62: 1964–1966.

44 Lee, S., Nathan, A., Jeon, S., and Robertson, J. (2015). Oxygen defect-induced metastability in oxide semiconductors probed by gate pulse spectroscopy. *Scientific Reports* 5 (1–10): 14902.

45 Robertson, J. (2012). Properties and doping limits of amorphous oxide semiconductors. *Journal of Non-Crystalline Solids* 358: 2437–2442.

46 Jeon, S., Ahn, S.-E., Song, I. et al. (2012). Gated three-terminal device architecture to eliminate persistent photoconductivity in oxide semiconductor photosensor arrays. *Nature Materials* 11: 301–305.

47 Lee, S., Jeon, S., Robertson, J., and Nathan, A. (2012). How to achieve ultra high photoconductive gain for transparent oxide semiconductor image sensors. In: *IEDM Technical Digest: IEEE International Electron Devices Meeting*, 24.3.1.1–24.3.1.4.

48 Robertson, J. (2008). Disorder and instability processes in amorphous conducting oxides. *Physica Status Solidi (B): Basic Research* 245: 1026–1032.

49 Su, Y.K., Peng, S.M., Ji, L.W. et al. (2010). Ultraviolet ZnO nanorod photosensors. *Langmuir* 26: 603–606.

50 Suehiro, J., Nakagawa, N., Hidaka, S.-I. et al. (2006). Dielectrophoretic fabrication and characterization of a ZnO, nanowire-based UV photosensor. *Nanotechnology* 17: 2567–2573.

51 Ahn, S.-E., Song, I., Jeon, S. et al. (2012). Metal oxide thin film phototransistor for remote touch interactive displays. *Advanced Materials* 24: 2631–2636.

52 Jeon, S., Song, I., Lee, S. et al. (2014). Origin of high photoconductive gain in fully transparent heterojunction nanocrystalline oxide image sensors and interconnects. *Advanced Materials* 26: 7102–7109.

53 Jeon, S., Kim, S.I., Park, S. et al. (2010). Low-frequency noise performance of a bilayer InZnO-InGaZnO thin-film transistor for analog device applications. *IEEE Electron Device Letters* 31: 1128–1130.

54 Ahn, S.-E., Jeon, S., Jeon, Y.W. et al. (2013). High-performance nanowire oxide photo-thin film transistors. *Advanced Materials* 25: 5549–5554.

55 Jeon, S., Park, S., Song, I. et al. (2010). 180 nm gate length amorphous InGaZnO thin film transistor for high density image sensor applications. In: *IEDM Technical Digest: IEEE International Electron Devices Meeting*, 21.3.1–21.3.4.

16

Device Modeling and Simulation of TAOS-TFTs

Katsumi Abe

Silvaco Japan, Yokohama, Japan

16.1 Introduction

Device simulations are often used to aid in understanding of the operation, failure analysis, and design of semiconductor devices. These simulations calculate the device characteristics by numerically solving basic semiconductor equations such as Poisson's equation, carrier continuity equations, and drift-diffusion equations [1, 2]. These equations contain physical parameters (e.g., potential, carrier density, carrier mobility, and generation/recombination rate) specified in device models that reflect specific material properties and device structures. Device modeling of TAOS-TFTs therefore establishes device models that include the physical properties of the TAOS channels to reproduce the TAOS-TFT characteristics.

Most TAOSs are *n*-type semiconductors with wide bandgaps. The conduction band bottom is composed of the large spherical *s*-orbitals of the metal cations, and these orbitals overlap each other. Although TAOSs have amorphous structures, this overlapping of their orbitals leads to high electron mobility [3–5]. Their electron transport properties can be described using a percolation model [6]. In addition, various subgap states also exist within the wide bandgaps in these materials. The subgap states reflect the TAOS's electronic states, and the distribution of the density of the subgap states (DOS) affects the TAOS-TFT characteristics [7]. These TAOS properties differ from those of well-known conventional semiconductors such as amorphous Si (*a*-Si).

In this chapter, we first introduce new mobility and DOS models that reflect the physical properties of TAOS films. Device simulations using these models can reproduce the characteristics of TAOS-TFTs over wide temperature and bias ranges, thus indicating that the proposed models are reasonable. Furthermore, the effects of the TAOS-TFT structures are investigated using device simulations based on the models. Finally, we discuss the reliability of TAOS-TFTs under bias stresses.

16.2 Device Models for TAOS-TFTs

16.2.1 Mobility Model

Hall measurements of TAOS films show that the Hall electron mobility in the films rises with increasing electron concentration, subsequently becoming constant (degeneration conduction) when the electron density exceeds 10^{19} cm^{-3} [8]. The electron transport properties are described using the Boltzmann transport equation with the percolation model, which supposes a distribution of potential barriers on the conduction band bottom [6]. The properties of these films are clearly different to those of conventional semiconductors such as crystalline Si (*c*-Si) and *a*-Si.

Amorphous Oxide Semiconductors: IGZO and Related Materials for Display and Memory, First Edition.
Edited by Hideo Hosono and Hideya Kumomi.

We have previously proposed an empirical mobility model for the electron transport within TAOSs [9]. This model assumes that the Hall mobility is equivalent to the drift mobility and can be represented using a power function of the electron density. This assumption implies that the momentum relaxation time in the Boltzmann equation is not dependent on the energy of the carriers, and the assumption is also used in the percolation model [6]. The electron drift mobility μ_e in this model is given by

$$\mu_e = \mu_0 \left(\frac{n_e}{n_{CR}} \right)^{\gamma/2} \tag{16.1}$$

where n_e is the carrier electron density, μ_0 is the mobility under the degeneration conduction condition, n_{CR} is the critical electron density, and γ is the power index, which is expressed as

$$\gamma = \gamma_0 + \frac{T_\gamma}{T} \tag{16.2}$$

where T is the temperature, and γ_0 and T_γ are the parameters used to express the temperature dependence. When the mobility is used, an expression for the drain current I_D of the TAOS-TFT at a low drain voltage V_D is given approximately by

$$I_D \approx \frac{W}{L} \frac{\mu_0}{1+\gamma} \left(\frac{V_G - V_{TH}}{V_{CR}} \right)^{\gamma} C_I (V_G - V_{TH}) V_D \tag{16.3}$$

where W is the TFT channel width, L is the TFT channel length, V_G is the gate voltage, V_{TH} is the threshold voltage, C_I is the gate capacitance per unit area, and V_{CR} is the critical voltage, which is defined as

$$V_{CR} \equiv \frac{\left(\frac{2q\varepsilon_S n_{CR}}{\beta} \right)^{\frac{1}{2}}}{C_I} \tag{16.4}$$

where q is the elementary charge, ε_S is the permittivity of the TAOS, $\beta \equiv (k_B T)/q$, and k_B is the Boltzmann constant.

The positive V_G during the ON operation forms a nonuniform electron density distribution along the TAOS thickness, called the "channel." The factor $1/(1+\gamma)$ on the right-hand side of Equation (16.3) originates from the electron density distribution. The field-effect mobility $(\partial I_D/\partial V_G)/(WC_I V_D/L)$, which is derived from Equation (16.3), then has the same expression as that for the drift mobility given by Equation (16.1).

Figure 16.1a shows the measured transfer curves of a coplanar homojunction bottom-gate amorphous In-Ga-Zn-O (*a*-IGZO) TFT at $V_D = 0.1$ and 12 V (open circles) [9]. The *a*-IGZO is a typical TAOS, and the *a*-IGZO film was deposited using a standard sputtering method. The mobility model parameters are extracted from the transfer curve at $V_D = 0.1$ V using Equation (16.3). The lines shown in Figure 16.1a were calculated in a device simulator [1] using the mobility model with the extracted parameters, and they reproduce the measured curves. These results verify that the proposed mobility model is reasonable.

Figure 16.1b shows the electron density dependence of the Hall mobility of the *a*-IGZO films. The closed circles represent the measured data from *a*-IGZO films formed under various conditions, and the open circles represent measured data from the *a*-IGZO films formed under the same conditions as those used for the TFT channel and source/drain regions. The line was calculated using the mobility model (Equation [16.1]) and the parameters extracted from the TFT curve, and this line overlaps the closed circles. This overlap indicates that the assumption made in the mobility model (i.e., that the Hall mobility is almost equivalent to the drift mobility) is satisfied. Furthermore, it suggests that the *a*-IGZO has a low DOS near the conduction band bottom, and that it has both a good channel interface and good contacts.

In the mobility model discussed here, the drift mobility is expressed using the power function of the electron density, and the applicable range of this density is wide, extending from 10^{16} to 10^{20} cm^{-3}. The simple percolation transport model cannot reproduce the electron density dependence of the mobility shown in Figure 16.1b because the mobility of that model remains almost constant when the electron density is below the degenerate density

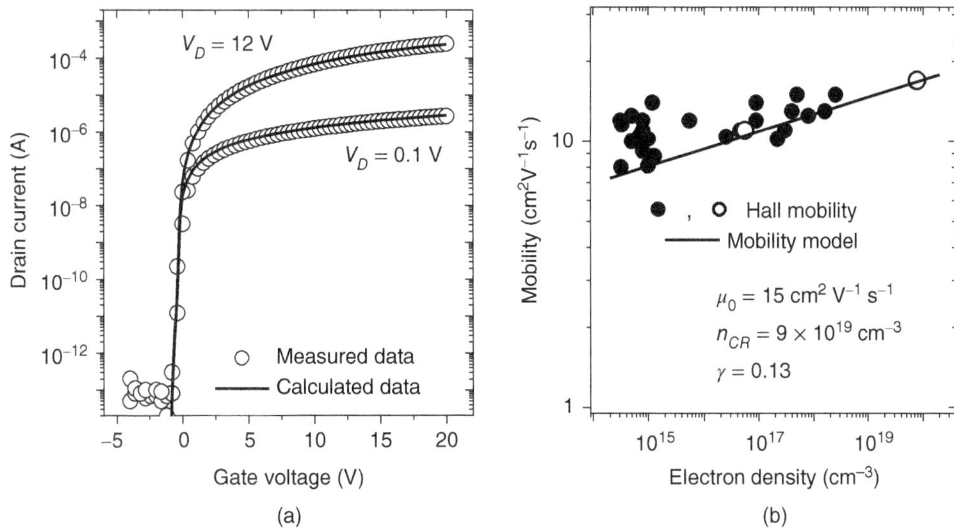

Figure 16.1 (a) Measured (open circles) and calculated (lines) transfer curves of *a*-IGZO-TFT at $V_D = 0.1$ and 12 V, and (b) Hall mobility of *a*-IGZO films with respect to electron density (closed circles: *a*-IGZO films formed under various conditions; open circles: *a*-IGZO films formed under the same conditions as the channel and source/drain of the *a*-IGZO-TFT) and the mobility model (line). *Source*: Abe, K., et al. (2019). Device modeling of amorphous oxide semiconductor TFTs. *Japanese Journal of Applied Physics* **58**: 090505-1–9.

value. A device model using the simple percolation transport model considers the electron density dependence of the mobility during operation at low electron densities (<10^{18} cm^{-3}) to originate from a multiple trapping and release process of the electrons due to subgap states within the TAOS [10]. Unfortunately, it is difficult to explain the electron density dependence of the Hall mobility using the device model. To solve this issue, the percolation transport model is revised by adding potential barriers that are located at long distances that exceed the mean-free path of the electrons and thus do not affect the Hall effect [6]. The mobility model given by Equation (16.1) is considered to have included the potential barrier effect inherently.

16.2.2 Density of Subgap States (DOS) Model

According to a recent review paper [7], the subgap states of a TAOS reflect its electronic states and include both acceptor-like and donor-like states. The acceptor-like states consist of excess oxygen, point defects, unstable structures such as hydrogen-poor systems, undercoordinated oxygen, and negative-ionized hydrogen. The donor-like states consist of oxygen vacancies, hydrogen species (e.g., –OH), and hydrogen-related defects. The subgap states are estimated using optical and electrical methods. The estimated results for standard *a*-IGZO films show that while the DOS near the top of the valence band, denoted by E_V, is approximately 10^{20} cm^{-3}, the DOS near the bottom of the conduction band, denoted by E_C, is low at less than 10^{18} cm^{-3}. These subgap states affect the operation of TAOS-TFTs. For example, the subthreshold operation of a TAOS-TFT is influenced by the subgap states around E_C–0.3 eV. When the TFT operation enters the ON state, the mobility characterizes the operation because of the low DOS value near E_C.

We have proposed a simple method to estimate the energy distribution of the DOS from a single transfer curve at low V_D [9]. The method uses the assumption that the potential is uniform along the channel thickness direction. Under this assumption, the expression for I_D at low V_D is given by

$$I_D \approx qt_S \frac{W}{L} \frac{1 - e^{-\beta V_D}}{\beta} \mu_0 \left(\frac{n_{e0} e^{\beta \phi_0}}{n_{CR}} \right)^{\gamma/2} n_{e0} e^{\beta \phi_0} \tag{16.5}$$

where t_S is the channel thickness, n_{e0} is the electron density at a flat band, ϕ_0 is the potential variation from the flat-band value, and n_{e0} exp $[\beta\phi_0]$ is the electron density at the potential ϕ_0. The factor $\left(1 - e^{-\beta V_D}\right)/\beta$ on the right-hand side of Equation (16.5) is found when both the diffusion and the drift are considered as part of the electron transport [9]. If the drift alone is considered, this factor is replaced with V_D. Additionally, Gauss's law provides the following relation:

$$\phi_0 \approx V_G - V_{FB} + \frac{qt_s n_{tot}}{C_I} \tag{16.6}$$

where V_{FB} is the flat-band voltage, and n_{tot} is the charged particle density per unit volume, which is given by

$$n_{tot} = n_{e0}\left(1 - e^{\beta\phi_0}\right) - \int_{E_{F0}-E_C}^{E_{F0}+q\phi_0-E_C} n_t(E)\, dE \tag{16.7}$$

where E_{F0} is the Fermi energy at the flat band, and $n_t(E)$ is the DOS per unit volume and per unit energy at an energy E. These equations provide the following expression for E at V_G, where

$$E - E_C \approx \frac{k_B T}{q} \ln\left[\frac{1}{N_C}\left(\frac{I_D n_{CR}^{\gamma/2} L}{k_B T t_S W \mu_0 \left(1 - e^{-\beta V_D}\right)}\right)^{2/(2+\gamma)}\right] \tag{16.8}$$

and the following expression for n_t at V_G, where

$$qn_t \approx \frac{C_I}{qt_S}\left\{\left(1 + \frac{\gamma}{2}\right)\frac{\beta S}{\ln 10} - 1\right\} - \beta N_C e^{(E-E_C)/(k_B T)} \tag{16.9}$$

where S is the subthreshold swing given by $dV_G/d\log(I_D)$; $E - E_C = E_{F0} + q\phi_0 - E_C$; $n_{e0}\exp[\beta\phi_0] = N_C \exp[(E - E_C)/(k_B T)]$; and N_C is the effective density of states in the conduction band.

These expressions can be used to easily extract the energy distribution of the DOS from a single transfer curve at low V_D. A detailed derivation of the expressions is given in the literature (in Ref. [9] and its supplement). The method used is based on an assumption of uniform potential along the channel thickness. This assumption is a good approximation when the total charged particle density (the sum of the electron density and the DOS around E_{F0}) is approximately 10^{17} cm^{-3} or less. Because this condition is satisfied during subthreshold operation of a typical TAOS-TFT, a DOS extraction method using Equations (16.8) and (16.9) is applicable.

16.2.3 Self-Heating Model

Similar to other electronic devices, the self-heating effect affects the ON characteristics of TAOS-TFTs [9, 11]. The self-heating effect requires both a self-heating model and an electron transport model with temperature dependence. The self-heating model is described using Joule heat generation and the heat flow equation. The electron transport model uses the mobility model in Section 16.2.1.

The Joule heat generation H has the following form:

$$H = (\vec{j}_n + \vec{j}_p) \cdot \vec{E} \tag{16.10}$$

where $\vec{j}_n$ is the electron current density, $\vec{j}_p$ is the hole current density, and $\vec{E}$ is the electric field. In the TAOS-TFT case, because the carriers are electrons, the hole current density can be neglected, except in the case where V_D is high. When $V_D - V_G$ is high (i.e., ≥30 V), impact ionization can affect the TAOS-TFT properties. The heat flow equation is then given by

$$c\frac{\partial T_L}{\partial t} = \nabla(\kappa \nabla T_L) + H \tag{16.11}$$

where T_L is the lattice temperature, and c and κ are the heat capacitance and thermal conductivity of the material, respectively. In addition, boundary conditions are required to solve the equations in the device simulator. We suppose that the following relationship exists at the boundary:

$$\vec{J_{tot}} \cdot \vec{s} = \alpha(T_L - T_{Ext}) \tag{16.12}$$

where $\vec{J_{tot}} \cdot \vec{s}$ is the inner product of the total energy flux and the unit external normal of the boundary, T_{Ext} is the external temperature, and α is the thermal conductance at the boundary.

To the best of our knowledge, this is the minimum set of device models required to consider the operation of TAOS-TFTs. The applications of these models to the TAOS-TFTs are discussed in Section 16.3.

16.3 Applications

16.3.1 Temperature Dependence

We examine the validity of the device models via the application of these models to a TAOS-TFT with long L [9]. Figure 16.2a shows a diagram of the structure of a coplanar homojunction bottom-gate a-IGZO-TFT. The symbols in Figure 16.2b represent the transfer curves of the TFT with $W/L = 10/40$ μm at $V_D = 0.1$ V measured over a range of temperatures between 253 and 358 K.

These curves can be used to extract the mobility model parameters using Equation (16.3) and to estimate the energy distribution of the DOS via Equations (16.8) and (16.9). Figure 16.2c and 16.2d show the Arrhenius plot of the extracted mobility parameter γ and the estimated DOS distribution near E_C, respectively. Device simulations using the device models with the parameters determined above are then able to calculate the transfer curves. The calculated curves are shown as the lines in Figure 16.2b and reproduce the measured curves. These results demonstrate the validity of the device model. Table 16.1 summarizes the model parameters and the physical properties of the a-IGZO-TFT used in the calculations.

16.3.2 Channel-Length Dependence

The ON current of a TAOS-TFT depends on its channel length L, even if the current is normalized using the geometric effect, W/L. The source of this L dependence is clarified via device simulations using the device models [9].

The closed circles in Figure 16.3a and 16.3b show the measured L dependences of the normalized ON current (I_D at $V_G = 20$ V) at $V_D = 0.1$ and 12 V, respectively, for the coplanar homojunction bottom-gate a-IGZO-TFTs with $W = 10$ μm. The normalized ON current at $V_D = 0.1$ V is independent of L and is constant. In contrast, the normalized ON current at $V_D = 12$ V exhibits an L dependence and increases with decreasing L.

There are several possible mechanisms for the increasing ON current with decreasing L at high V_D, including the channel-length modulation (CLM) effect, the drain-induced barrier-lowering (DIBL) effect, the impact ionization effect, and the self-heating effect. The basic semiconductor equations have already considered both the CLM and DIBL effects. The open triangles in Figure 16.3a and 16.3b show the normalized ON currents calculated in the device simulations using the device models described in Section 16.2. The results at $V_D = 0.1$ V reproduce the measured L dependence, but the calculated results at $V_D = 12$ V show large differences from the measured values when L is less than 10 μm. Furthermore, even when the impact ionization effect is added, the calculated ON current does not change. These results indicate that the three mechanisms are not sufficient to describe the effects of the short L value.

We therefore add the self-heating effect to the device models to calculate the ON current. The self-heating model was described in Section 16.2.3. This model uses the values of the a-IGZO thermal conductivity and the thermal capacitance reported by Yoshikawa et al. [12]. Additionally, the thermal boundary conditions are specified as $\alpha = 20$ W/(cm^{2}K) and $T_{Ext} = 300$ K at each electrode. The open circles in Figure 16.3a and 16.3b represent the

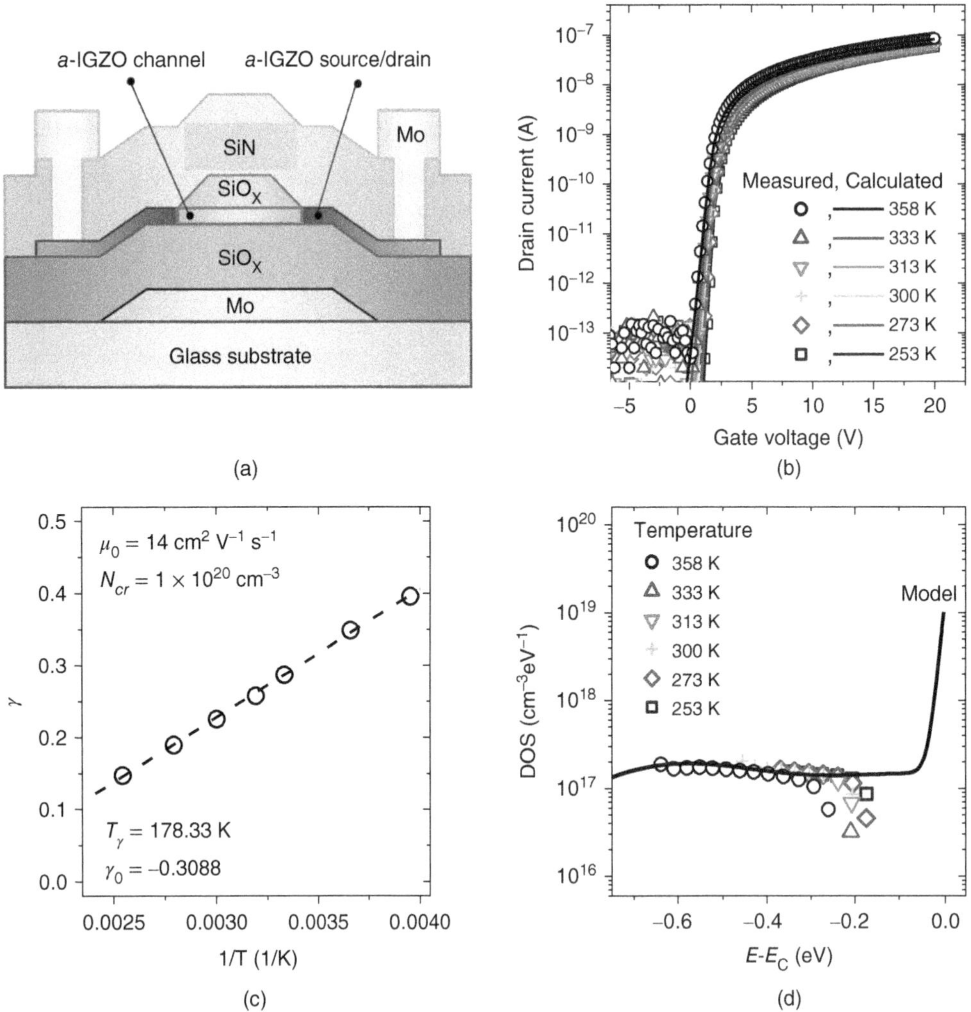

Figure 16.2 (a) Structural diagram of a coplanar homojunction bottom-gate *a*-IGZO-TFT. (b) Measured (symbols) and calculated (lines) transfer curves of *a*-IGZO-TFTs at $V_D = 0.1$ V and $T = 253$ to 358 K. (c) Arrhenius plots of γ (open circles) and mobility model (dotted line), and (d) estimated energy distributions of the DOS (symbols) and DOS model (line). *Source*: Abe, K., et al. (2019). Device modeling of amorphous oxide semiconductor TFTs. *Japanese Journal of Applied Physics* **58**: 090505-1–9.

calculated ON currents. These results indicate that the calculated ON current reproduces the L dependence of the measured current at both $V_D = 0.1$ V and $V_D = 12$ V.

Figure 16.3c shows the calculated temperature distribution within the TFT with $L = 6$ μm at $V_G = 20$ V and $V_D = 12$ V. The temperature is highest near the drain and is around 25 K higher than the external temperature of 300 K. Figure 16.3d shows two curves for the mobility distributions with and without the self-heating effect (the line and the dotted line, respectively) along the channel direction at the center of the thickness. Because the mobility model presented here has a positive temperature dependence, the temperature distribution caused by the self-heating effect leads to both the higher mobility and the peak near the drain. The increased mobility results in high ON current.

Table 16.1 Model parameters and physical properties of an *a*-IGZO-TFT.

Description	Symbol	Values
Acceptor-like tail trap density at $E = E_C$ [(cm^{3}eV)$^{-1}$]	N_{TA}	10^{19}
Slope of acceptor-like tail trap [eV]	E_{TA}	0.008
Peak density of acceptor-like Gaussian trap [(cm^{3}eV)$^{-1}$]	N_{GA}	1.5×10^{17}
Peak energy of acceptor-like Gaussian trap [eV]	E_{GA}	0
Decay energy of acceptor-like Gaussian trap [eV]	W_{GA}	0.64
Donor-like exponential-tail trap density at $E = E_V$ [(cm^{3}eV)$^{-1}$]	N_{TD}	3.0×10^{20}
Slope of donor-like exponential-tail trap [eV]	E_{TD}	0.12
Peak density of donor-like Gaussian trap [(cm^{3}eV)$^{-1}$]	N_{GD}	1.3×10^{17}
Peak energy of donor-like Gaussian trap [eV]	E_{GD}	2.48
Decay energy of donor-like Gaussian trap [eV]	W_{GD}	0.24
Intrinsic mobility under degeneration conduction [cm^2/(Vs)]	μ_0	14
Critical electron density [cm^{-3}]	n_{CR}	10^{20}
γ intrinsic temperature [K]	T_γ	178.4
γ at $1/T = 0$	γ_0	−0.31
Effective conduction band density of states of *a*-IGZO [cm^{-3}]	N_C	5.0×10^{18}
Effective valence band density of states of *a*-IGZO [cm^{-3}]	N_V	4.6×10^{19}
Relative permittivity of *a*-IGZO	ε_S	13
Bandgap of *a*-IGZO [eV]	E_g	3.1
Electron affinity of *a*-IGZO [eV]	χ	4.2
Thickness of *a*-IGZO [nm]	t_S	40
Thickness of SiO_x gate insulator [nm]	t_I	200
SiO_x gate insulator capacitance per unit area [F/cm^2]	C_I	1.86×10^{-8}

The total DOS is given by a sum of four functions, acceptor- and donor-like tail and Gaussian functions, as follows:
Tail functions: N_{TA} exp $[-(E_C - E)/(E_{TA})]$ and N_{TD} exp $[-(E - E_V)/(E_{TD})]$.
Gaussian functions: N_{GA} exp $[-((E_{GA} - E)/W_{GA})^2]$ and N_{GD} exp $[-((E_{GD} - E)/W_{GD})^2]$.

These results indicate that the self-heating effect affects the ON currents of *a*-IGZO-TFTs. Additionally, device simulations using the device models with the self-heating effect are available to calculate the operational behavior of *a*-IGZO-TFTs with short L values of less than 10 μm.

16.3.3 Channel-Width Dependence

The ON current is also dependent on the size of W [13]. The open triangles and circles in Figure 16.4a show the W dependence of the normalized ON-current ratio of the *a*-IGZO-TFT at $V_D = 0.1$ and 12 V, respectively. The normalized ON-current ratio $R_{I_{ON}}$ is defined as the ratio of the ON current (I_D at $V_G = 20$ V) normalized using W/L with respect to that of the TFT with $W = 10$ μm, which is given by

$$R_{I_{ON}} \equiv \frac{I_D(V_G = 20\ V)}{W/L} \bigg/ \frac{I_D(V_G = 20\ V)}{10/L} \tag{16.13}$$

where L is fixed at 10 μm. In the case where $V_D = 0.1$ V, $R_{I_{ON}}$ decreases monotonically with increasing W and reaches saturation at approximately 0.95. In the case where $V_D = 12$ V, $R_{I_{ON}}$ decreases with increasing W when W

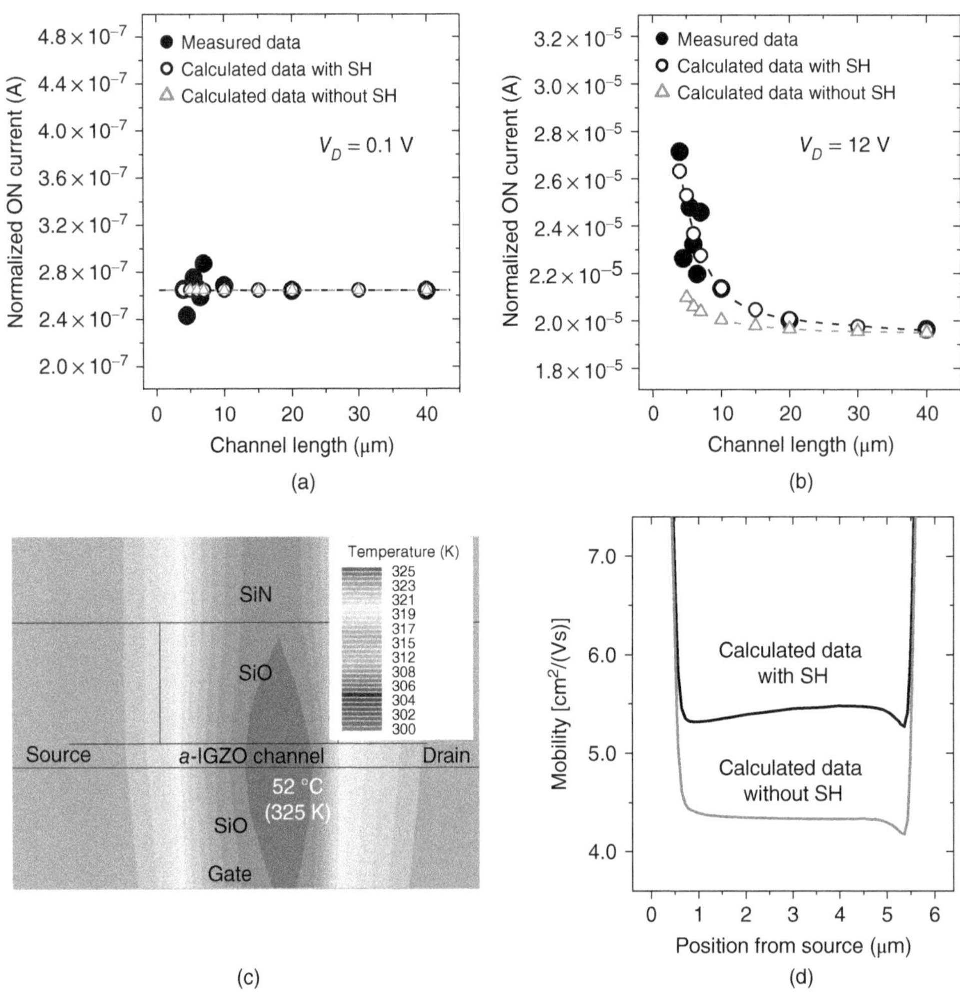

Figure 16.3 (a,b) L dependences of normalized ON currents at $V_D = 0.1$ and 12 V (closed circles: measured data; open triangles: calculated data without self-heating effect; open circles: calculated data with self-heating effect). (c) Calculated temperature distribution of an a-IGZO-TFT with $L = 6$ μm at $V_G = 20$ V and $V_D = 12$ V, and (d) calculated distribution of mobility along the channel at the channel center in the thickness direction of the a-IGZO-TFT with $L = 6$ μm at $V_G = 20$ V and $V_D = 12$ V (dotted line: calculated data without self-heating effect; line: calculated data with self-heating effect). *Source*: Abe, K., et al. (2019). Device modeling of amorphous oxide semiconductor TFTs. *Japanese Journal of Applied Physics* **58**: 090505-1–9.

is 20 μm or less. This trend is similar to that shown at $V_D = 0.1$ V. In contrast, $R_{I_{ON}}$ increases with increasing W when W exceeds 20 μm. The maximum value is approximately 1.05.

A three-dimensional (3D) device simulation is used to investigate the W dependence. Figure 16.4b shows the simulation structures. The upper figure shows the 3D structural diagram of the TFT, while the lower figures show the cross-sectional diagrams along the L and W directions, respectively. Figure 16.4c shows the distributions of the electric field and the electron current density obtained within the cross-section along the W direction when $V_G = 20$ V and $V_D = 12$ V are applied to a TFT with $W = 10$ μm. The high electric field regions near both channel edges along the W direction are formed by the edge-shape effect and generate high electron current-density regions there. Because the high current densities near the edges are independent of W, $R_{I_{ON}}$ decreases with increasing W.

Figure 16.4d shows the W dependence of the maximum temperature within the device at $V_G = 20$ V and $V_D = 12$ V. The maximum temperature rises in tandem with increasing W. As described in the discussion of the

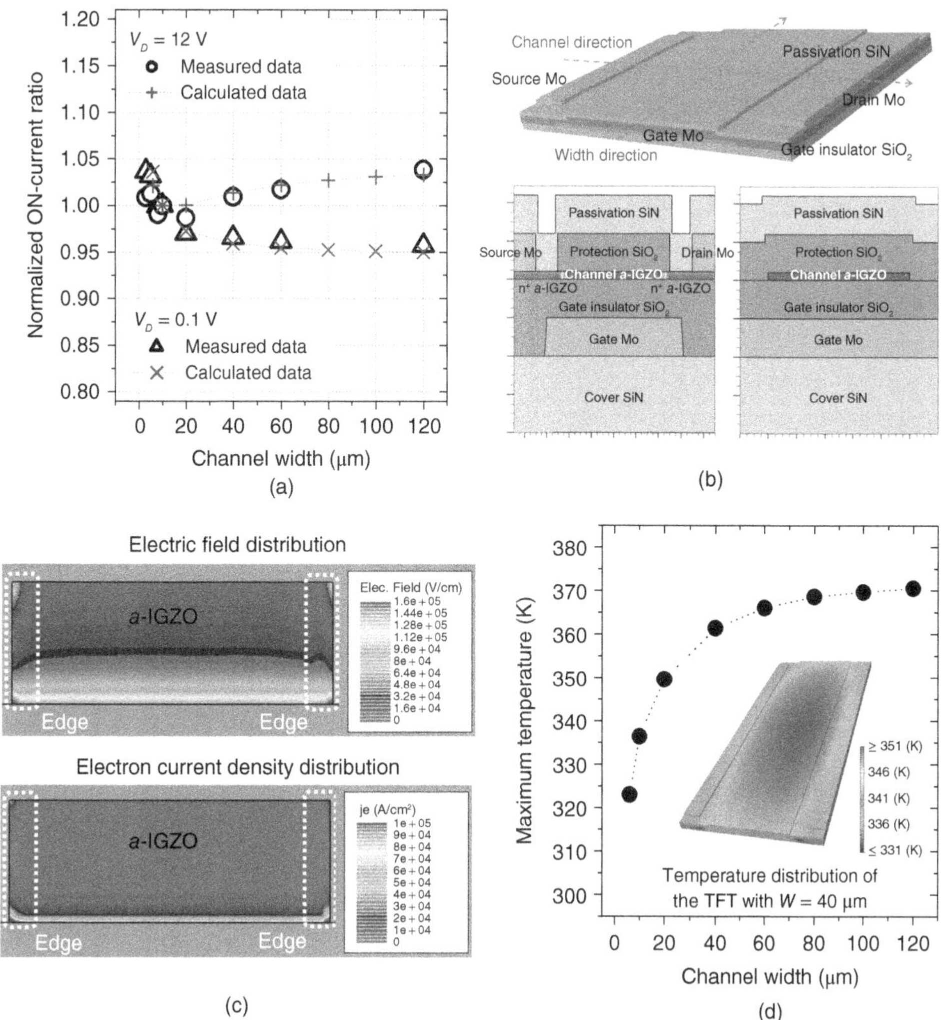

Figure 16.4 (a) W dependence of the normalized ON current ratio at $V_D = 0.1$ and 12 V (open triangles: measured data at $V_D = 0.1$ V; open circles: measured data at $V_D = 12$ V; crosses: calculated data at $V_D = 0.1$ V; and plus signs: calculated data at $V_D = 12$ V); (b) simulation structures: 3D diagram (top figure), cross-sectional diagram along the L direction (lower-left figure), and cross-sectional diagram along the W direction (lower-right figure); (c) calculated distributions of the electric field (upper figure) and electron current density (lower figure) within the cross-section along the W direction at $V_G = 20$ V and $V_D = 12$ V; and (d) calculated W dependence of the maximum temperature within the device at $V_G = 20$ V and $V_D = 12$ V (closed circles); the inset shows the temperature distribution of the TFT with $W = 40$ μm. *Source*: Abe, K., et al. (2019). Simulation study of self-heating and edge effects on oxide-semiconductor TFTs: channel-width dependence. In *Proceedings of the International Display Workshops*, vol. 26, Sapporo, Japan, November 26–29.

channel-length dependence in Section 16.3.2, a temperature rise induces increases in both the mobility and the current. This is the source of the rise in $R_{I_{ON}}$ with increasing W. The inset in Figure 16.4d shows the temperature distribution of the TFT with $W = 10$ μm. The maximum temperature within this device is found near the drain. The temperature distribution is consistent with the corresponding distributions shown in Figure 16.3c and in Ref. [11, fig. 7]. Additionally, the cross and plus symbols added to Figure 16.4a show the $R_{I_{ON}}$ values that were calculated at $V_D = 0.1$ and 12 V, respectively. These distributions reproduce the measured W dependences. These results therefore demonstrate that the W dependence of $R_{I_{ON}}$ can be described using the W edge-shape effect and the self-heating effect.

16.3.4 Dual-Gate Structure

TAOSs generally have optical bandgaps of more than 2.5 eV. Additionally, the ionization potential of TAOSs is greater than 7 eV and is deeper than the work functions of metals. The band properties reduce the thermal generation of electron–hole pairs and hole injection from the electrodes. Therefore, the formation of an inversion layer with holes appears to be almost impossible. The difficulty in formation of an inversion layer is important for TAOS-TFT operation. For example, the comparatively low OFF-current characteristics of TAOS-TFTs are mainly caused by this difficulty.

Figure 16.5a shows a structural diagram of a dual-gate (DG) *a*-IGZO-TFT, which has a gate electrode below the channel and a back-gate electrode above the channel [14]. Because of the difficulty involved in inversion

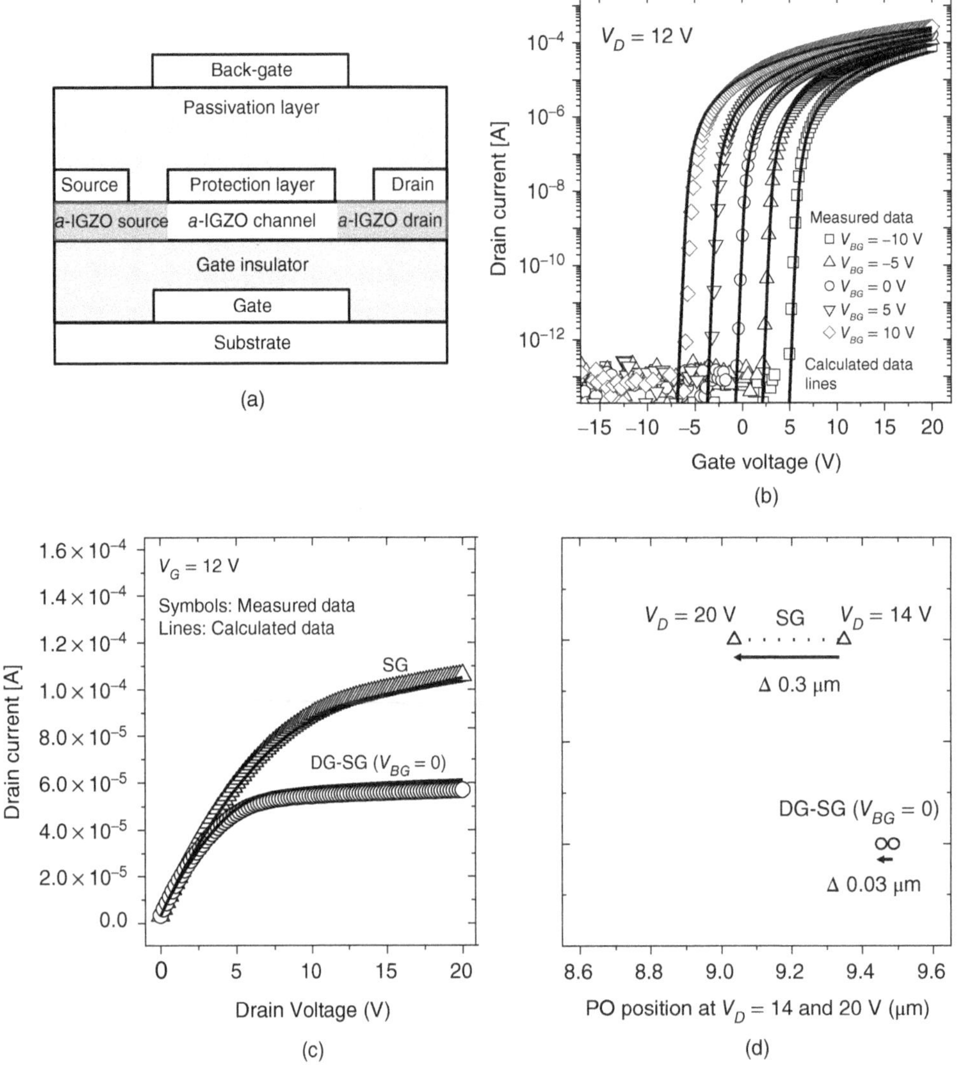

Figure 16.5 (a) Structural diagram of a DG *a*-IGZO-TFT; (b) transfer curves of a ZDG *a*-IGZO-TFT at $V_D = 12$ V and $V_{BG} = -10$, −5, 0, 5, and 10 V (symbols: measured data; lines: calculated data); (c) output curves of SG and DG *a*-IGZO-TFTs at $V_G = 12$ V (symbols: measured data; lines: calculated data); and (d) calculated pinch-off (PO) position of SG and DG *a*-IGZO-TFTs at $V_D = 14$ and 20 V (open triangles: SG *a*-IGZO-TFT; open circles: DG *a*-IGZO-TFT under DG-SG operation conditions).

layer formation, the DG a-IGZO-TFT has electrical properties that are widely controlled using the back-gate voltage V_{BG}. The symbols in Figure 16.5b show transfer curves for the DG a-IGZO-TFT with $W/L = 60/10$ μm at $V_{BG} = 10$, 5, 0, −5, and −10 V. When V_{BG} is positive, the transfer curves will shift negatively, depending on the V_{BG} value. In contrast, when V_{BG} is negative, the transfer curves will shift positively, depending on the V_{BG} value. Figure 16.5b also shows lines that represent the transfer curves calculated via device simulations using the device model described earlier in this chapter. The calculated curves are close to their measured counterparts. This agreement indicates that the device models are appropriate for use in descriptions of the operation of DG a-IGZO-TFTs.

The back-gate effect is also available to suppress the V_D dependence of the saturation I_D. Figure 16.5c shows two measured output curves. The open triangles represent the output curve of the standard single-gate (SG) a-IGZO-TFT with its gate electrode at $V_G = 12$ V. The open circles represent the output curve of the DG TFT during single-gate (DG-SG) operation at $V_G = 12$ V and $V_{BG} = 0$ V. The DG-SG saturation I_D value shows "hard saturation," and the change in I_D with respect to variations in V_D is smaller than that in the SG case. The hard-saturation property is valuable for driving current-load devices such as organic light-emitting diodes.

Figure 16.5c also shows two lines, which are the SG and DG-SG output curves calculated via device simulations using the device models described in Section 16.2. These curves almost overlap the corresponding measured curves. Furthermore, the calculations can determine the pinch-off (PO) position at which the vertical electric field at the interface between the channel and the gate insulator reaches zero [15]. Figure 16.5d shows the PO positions for the SG and DG a-IGZO-TFTs (DG-SG operation) at $V_D = 14$ and 20 V. While the change in the PO positions of the SG TFT is 0.3 μm, the change in the PO positions of the DG TFT (DG-SG operation) is only 0.03 μm because of the back-gate effect. This small change in the PO position is equivalent to a small CLM effect and leads to the hard saturation.

As described here, we can understand the effects of the TAOS-TFT structure on the device electrical characteristics through device simulations using the appropriate device models.

16.4 Reliability

One important issue for TAOS-TFTs is their reliability under bias stress. There are five types of bias stress:

1. Positive gate bias (temperature) stress (PB[T]S) [7, 16, 17];
2. Negative gate bias (temperature) stress (NB[T]S) [7, 16];
3. Negative gate bias illumination (temperature) stress (NBI[T]S) [7, 16, 18];
4. Positive high gate and drain bias (temperature) stress (high-current stress [HCS]) [11, 19]; and
5. Positive high-voltage drain stress (PHVDS) [20, 21].

These bias stresses cause degradation of the electrical characteristics of TAOS-TFTs. This degradation includes a parallel shift and a shape change in the transfer curve. Figure 16.6a and 16.6b show degraded transfer curves with parallel shifts and shape changes, respectively. In this section, the relationships between the degradation of the TAOS-TFTs and the device models are discussed.

The parallel shift in the transfer curve appears when the device is under PBS, NBS, and NBIS. When these stresses are applied, charged particles accumulate at the gate interface (IF) between the gate insulator and the channel. These charged particles are then captured by the IF traps or by the gate insulator near the IF. As a result, the transfer curve shift along the positive/negative V_G direction (i.e., the "positive/negative shift") occurs. Furthermore, because standard TFTs do not have a back-gate electrode, the back-gate potential is a floating potential and changes along with the V_G variation. This behavior suggests that the charged particles accumulate at the back-gate IF. Therefore, even if traps exist near the back-gate IF, the positive/negative shift will still arise. The charged particles are negatively charged electrons for PBS and photoinduced positive charges, such as holes and/or ions, for

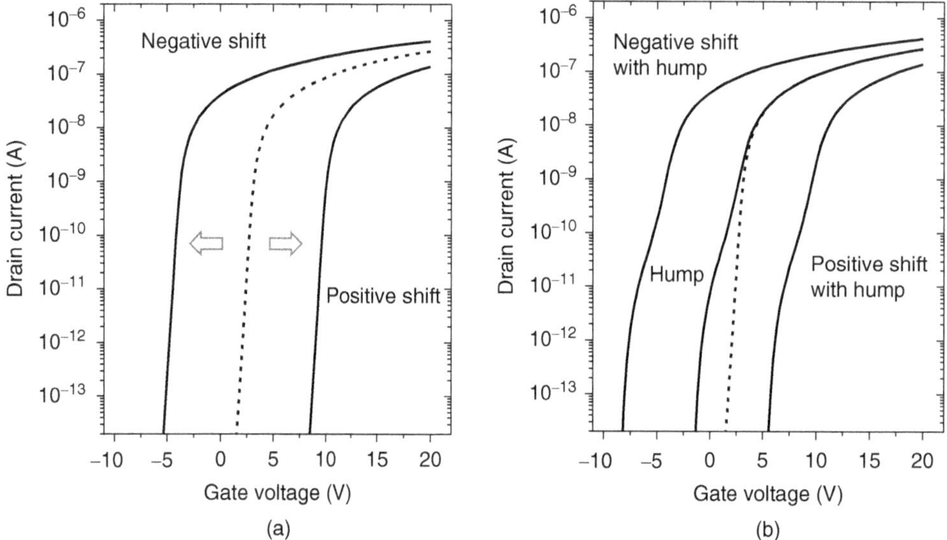

Figure 16.6 (a) Transfer curves with positive/negative shifts, and (b) transfer curves with shape changes.

NBIS. In contrast, the charge accumulation model is not applicable to NBS because of the lack of photoinduced positive charges. The degradation that occurs under NBS requires different processes for introduction of the positive charges, such as injections of positive ions along the cracks in the structure. The device simulations can express the parallel shifts through the addition of traps deeper than 0.6 eV from E_C or the addition of fixed charges to the IFs.

The shape change in the transfer curve can be found under NBIS, HCS, and HVDS. An increase in the value of S and formation of a hump represent the typical shape changes, and these changes sometimes appear in conjunction with the negative shift for NBIS and the positive shifts for HCS and HVDS. Each stress seems to have inherent processes that play important roles in the degradation: the photogeneration of the charged particles (holes or positive ions) under NBIS, the temperature increase due to the self-heating effect under HCS, and the impact ionization near the drain under HVDS.

Two causes of the shape changes are considered. The first is the addition of subgap states in an energy range from E_C−0.6 to E_C−0.2 eV. As discussed in Section 16.2.2, the DOS distribution is related directly to S. When a peak in the energy range above is added to the DOS, the transfer curve contains a hump, as shown in Figure 16.7a. The second cause involves the formation of two channel regions that have different electrical properties. For example, when a positive/negative shift is caused near the edge along the W direction or near the drain edge along the L direction, the difference in the increase in current between the regions with and without the shift forms a hump in the transfer curve, as shown in Figure 16.7b. These causes can be distinguished based on the W dependence of the electrical characteristics or on the changes in the characteristics that arise with exchange of the voltages applied to the source and drain.

The device model presented here can describe the degraded electrical characteristics of TAOS-TFTs after the application of bias stress. However, it is difficult to calculate the transient degradation behavior because the calculations require detailed knowledge of the degradation processes (e.g., the type, formation, transport, and reactions of the charged particles/states) and the construction of models to describe these processes. This will be one of the next challenges in our work. We expect that the models presented here will be developed further and make it possible to perform device simulations to predict TAOS-TFT degradation in future.

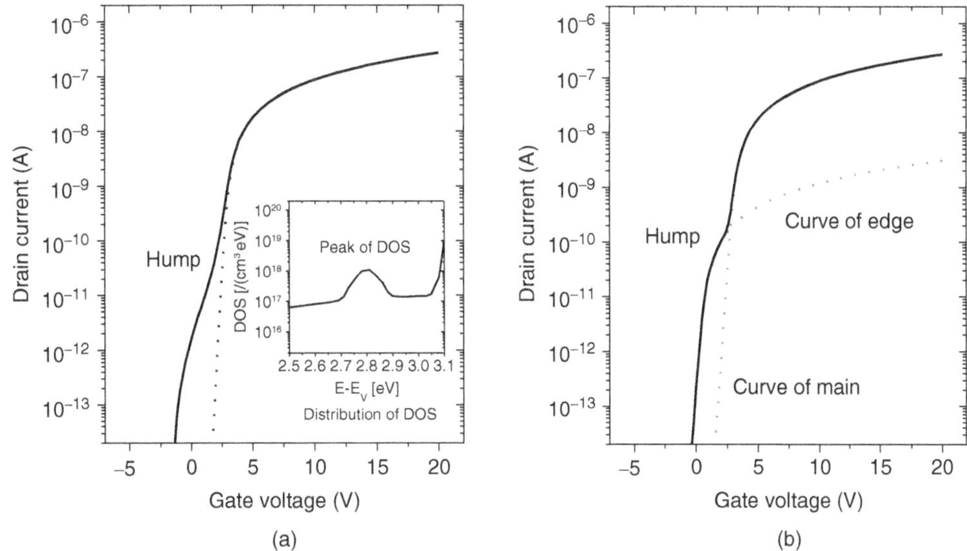

Figure 16.7 (a) Transfer curve with hump caused by the peak of the DOS (see the inset), and (b) transfer curve with hump, which is the sum of the main and edge curves.

16.5 Summary

We have discussed device models for TAOS-TFTs and device simulations performed using these models in this chapter.

Mobility and DOS models that reflect the electron transport properties and the electronic states of TAOSs were proposed. Device simulations using these device models reproduced the electrical characteristics of a TAOS-TFT over wide bias and temperature ranges. This indicated that the proposed device models are reasonable. Furthermore, with the addition of the self-heating effect, these device models can be used to explain the structural dependences of the TAOS-TFT characteristics, such as the L and W size effects and the effects of dual-gate structures. The degradation of the TAOS-TFTs after the application of bias stresses was described via device simulations using the device models. The results demonstrate that the device models and the device simulations using these models are valuable for understanding TAOS operation, device design, and failure analysis.

In addition, the development of a model of the transient properties is one of the issues to be addressed in future work. If device models that include the degradation processes can be developed, it will be possible for the device simulations to predict the device degradation.

Acknowledgments

I would like to express my gratitude to Professor H. Hosono, Professor T. Kamiya, Professor K. Nomura, and many others at the Hosono–Kamiya Laboratory for their guidance and support. I would also like to thank Professor H. Kumomi and other previous associates at Canon for their cooperation. This chapter is based on collaborative work with them.

References

1 Silvaco Inc. (2019). *ATLAS Device Simulation Software User's Manual.*

2 Silvaco Inc. (2019). *Victory Device Simulation Software User's Manual.*

3 Hosono, H., Yasukawa, M., and Kawazoe, H. (1996). Novel oxide amorphous semiconductors: transparent conducting amorphous oxides. *Journal of Non-Crystalline Solids* 203: 334–344.

4 Nomura, K., Ohta, H., Takagi, A. et al. (2004). Room-temperature fabrication of transparent flexible thin-film transistors using amorphous oxide semiconductors. *Nature* 432: 488–492.

5 Kamiya, T., Nomura, K., and Hosono, H. (2010). Present status of amorphous In–Ga–Zn–O thin-film transistors. *Science and Technology of Advanced Materials* 11: 044305-1–044305-23.

6 Kamiya, T., Nomura, K., and Hosono, H. (2010). Origin of definite Hall voltage and positive slope in mobility-donor density relation in disordered oxide semiconductors. *Applied Physics Letters* 96: 122103-1–122103-3.

7 Ide, K., Nomura, K., Hosono, H., and Kamiya, T. (2019). Electronic defects in amorphous oxide semiconductors: a review. *Physica Status Solidi (A): Applications and Materials Science* 216: 1800372-1–1800372-28.

8 Takagi, A., Nomura, K., Ohta, H. et al. (2005). Carrier transport and electronic structure in amorphous oxide semiconductor, a-$InGaZnO_4$. *Thin Solid Films* 486: 38–41.

9 Abe, K., Ota, K., and Kuwagaki, T. (2019). Device modeling of amorphous oxide semiconductor TFTs. *Japanese Journal of Applied Physics* 58: 090505-1–090505-9.

10 Lee, S., Ghaffazadeh, K., Nathan, A. et al. (2011). Trap-limited and percolation conduction mechanisms in amorphous oxide semiconductor thin film transistors. *Applied Physics Letters* 98: 203508-1–203508-3.

11 Fujii, M., Uraoka, Y., Fuyuki, T. et al. (2009). Experimental and theoretical analysis of degradation in Ga_2O_3–In_2O_3–ZnO thin-film transistors. *Japanese Journal of Applied Physics* 48: 04C091-1–6.

12 Yoshikawa, T., Yagi, T., Oka, N. et al. (2013). Thermal conductivity of amorphous indium–gallium–zinc oxide thin films. *Applied Physics Express* 6: 021101-1–021101-3.

13 Abe, K., Ota, K., and Kuwagaki, T. (2019). Simulation study of self-heating and edge effects on oxide-semiconductor TFTs: channel-width dependence. In: *Proceedings of the International Display Workshops*, vol. 26. Japan: Sapporo November 26–29.

14 Abe, K., Takahashi, K., Sato, A. et al. (2012). Amorphous In–Ga–Zn–O dual-gate TFTs: current–voltage characteristics and electrical stress instabilities. *IEEE Transactions on Electron Devices* 59: 1928–1935.

15 Sze, S.M. and Ng, K.K. (2007). *Physics of Semiconductor Devices*, (third edition), Hoboken, NJ: Wiley.

16 Park, J.S., Maeng, W.-J., Kim, H.-S., and Park, J.-S. (2012). Review of recent developments in amorphous oxide semiconductor thin-film transistor devices. *Thin Solid Films* 520: 1679–1693.

17 Chowdhury, M.D.H., Migliorato, P., and Jang, J. (2011). Time-temperature dependence of positive gate bias stress and recovery in amorphous indium-gallium-zinc-oxide thin-film-transistors. *Applied Physics Letters* 98: 153511-1–153511-3.

18 Oh, H., Yoon, S.-M., Ryu, M.K. et al. (2010). Photon-accelerated negative bias instability involving subgap states creation in amorphous In–Ga–Zn–O thin film transistor. *Applied Physics Letters* 97: 183502-1–183502-3.

19 Hsieh, T.-Y., Chang, T.-C., Chen, T.-C. et al. (2012). Systematic investigations on self-heating-effect-induced degradation behavior in a-InGaZnO thin-film transistors. *IEEE Transactions on Electron Devices* 59: 3389–3395.

20 Tsai, M.-Y., Chang, T.-C., Chu, A.-K. et al. (2013). High temperature-induced abnormal suppression of sub-threshold swing and on-current degradations under hot-carrier stress in a-InGaZnO thin film transistors. *Applied Physics Letters* 103: 012101-1–012101-5.

21 Lee, H.-J. and Abe, K. (2020). A study on the effect of pulse rising and falling time on amorphous oxide semiconductor transistors in driver circuits. *IEEE Electron Device Letters* 41: 896–899.

17

Oxide Circuits for Flexible Electronics

Kris Myny[1,2], Nikolaos Papadopoulos[1], Florian De Roose[1], and Paul Heremans[1,2]

[1] *imec, Sense and Actuate Technologies Department, Heverlee, Belgium*
[2] *KU Leuven, ESAT Department, Heverlee, Belgium*

17.1 Introduction

Metal-oxide electronics are present today in numerous devices, mainly for backplane applications such as active-matrix displays. In that application, the metal-oxide thin-film transistors (TFTs) serve as switching devices for typical liquid crystal displays (LCDs) or as current-control elements for organic light-emitting displays (OLEDs). Amorphous metal oxides, mainly based on indium–gallium–zinc oxide (IGZO), are very attractive, thanks to their reasonable charge carrier mobility, in between organics and low-temperature polycrystalline silicon; their extremely low off-current; and their low short-range and long-range variability thanks to their amorphous nature. Most metal-oxide semiconductors are electron-transporting devices and therefore result in *n*-type transistor behavior. In a display application, a unipolar transistor is mostly sufficient, although the *n*-type nature poses extra challenges for driving the anode of normally configured OLEDs. When designing more complex circuits, the unipolar nature poses even more challenges. They will be elaborated in this chapter, and we will investigate technology-aware design options to circumvent or solve them. We will discuss the impact of including an additional gate to the TFT, Moore's law for thin-film electronics, and how technology downscaling may help to meet the specifications of the emerging applications. Product specifications and standards today are determined for Si complementary metal-oxide semiconductors (CMOS), so IGZO circuits need to comply with those standards prior to product integration. By exploiting tailored techniques, we have been able to realize a thin-film metal-oxide-based near-field communication (NFC) tag that complies to the official ISO norm. Besides thin-film NFC tags, we will discuss other recent developments and implementations published in this field, divided into digital and analog electronics subsections. Section 17.3 on digital electronics details communication chips, such as inductive and capacitive radiofrequency identification (RFID) tags and more complex digital chips. We show an innovative tag that communicates directly with the touchscreen of a smartphone, and elaborate on recent developments related to flexible microprocessors, a hardwired machine-learning chip, and a security cryptographic chip-on-flex. In Section 17.4, we analyze different implementations for IGZO-based analog-to-digital converters (ADCs) and their specifications; several opportunities and demonstrators in this field are also discussed, such as analog converters for optical and fingerprint imagers and healthcare patches.

17.2 Technology-Aware Design Considerations

A typical CMOS designer does not need to know every detail about the technology and exact process steps with which the design will be fabricated, as long as a good process design kit (PDK) is available. For flexible thin-film

Amorphous Oxide Semiconductors: IGZO and Related Materials for Display and Memory, First Edition.
Edited by Hideo Hosono and Hideya Kumomi.

oxide technology, this is not so straightforward, considering both the novelty and limitations the devices bring. In addition, the process tends to evolve continuously. Therefore, there is typically a lot more interaction needed between the designer and the technologist compared to a CMOS design flow. This section will discuss the most important technological aspects impacting the performance of state-of-the-art circuits. First, the fundamental structure of the transistor stack and fabrication thereof will be discussed. This is followed by a quick dive into the addition and use of a backgate. Finally, a brief discussion will shed light on the dynamics of scaling in thin-film technologies and will put forward the drivers for a Moore's law for flexible electronics.

17.2.1 Etch-Stop Layer, Backchannel Etch, and Self-Aligned Transistors

Like in all semiconductor circuits, the performance of a flexible oxide circuit depends heavily on the selected technology. Many different stacks incorporating oxide transistors can be found in the literature [1, 2], but the most prominent ones for industrial production are etch-stop layer (ESL) [3, 4], backchannel etch (BCE) [5, 6], and self-aligned (SA) [7, 8] technologies. The selection of any of these stacks will have a large impact on the speed, power, and footprint of circuits. However, the technology selection for each project is very often based on other parameters, like the process availability, yield, and cost of the different fabrication flows. It is therefore essential for a designer to determine the impact of the technology selection on the circuit behavior and the effects on system-level performance.

17.2.1.1 Etch-Stop Layer

Etch-stop layer is the most mature device architecture for thin-film devices. Figure 17.1a1 shows a typical cross-section of such a device. To fabricate an ESL device, the gate is first deposited and patterned, followed by the deposition of the gate dielectric. Next, the semiconductor is deposited and patterned. The direct patterning of a metal layer on top of the channel of a transistor should be avoided because, for example, the etchant might reduce the quality of the semiconductor. Therefore, in the ESL process, an ESL is deposited on top of the semiconductor. The contacts of the transistor can then be etched into this ESL layer, thereby exposing the semiconductor only at the position of the contacts. Next, the source–drain (SD) metal contacts can be deposited and patterned.

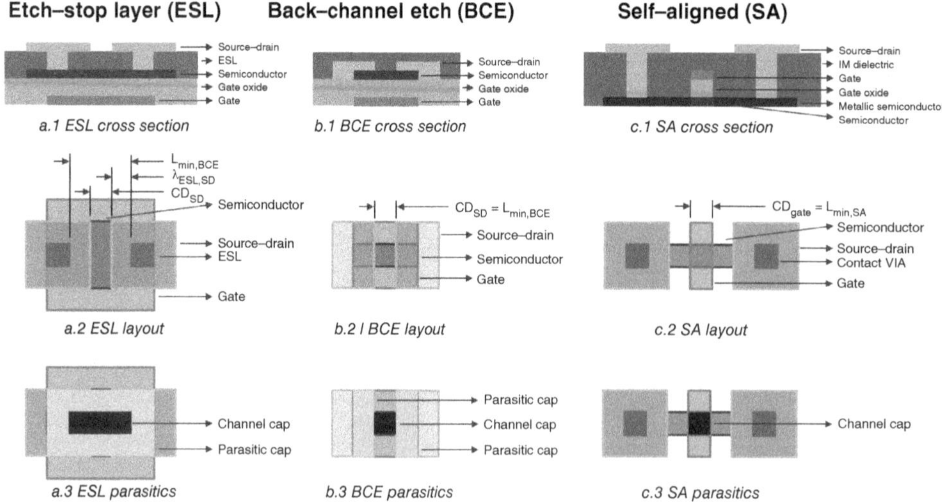

Figure 17.1 Cross-section and layout for three device architectures for thin-film devices: etch-stop layer, back-channel etch, and self-aligned transistors.

Due to the steps involved in the ESL process, some margins must be foreseen in the layout between the transistor contact hole and the SD metal. A typical minimal transistor layout is shown in Figure 17.1a2. The shortest transistor that can be fabricated has a channel length of:

$$L_{min,ESL} = CD_{SD} + 2\lambda_{ESL,SD}$$

where CD_{SD} is the minimal spacing between two pieces of SD metal; and $\lambda_{ESL,\,SD}$ is the worst-case misalignment between the ESL patterning and the SD patterning. For an example technology with all design rules equal to 5 μm, the minimal channel length is 15 μm.

In addition, a lot of parasitic overlap capacitance is present in the layout of the minimal transistor. Figure 17.1a2 also indicates the size of the intrinsic and desired channel capacitance compared to the parasitic channel capacitance. For the transistor in the example shown here, this parasitic area is 225 μm^2, compared to the 75 μm^2 channel area shown in Figure 17.1a3.

The parasitic capacitance and the long channel length are detrimental to the performance of the device. Therefore, it is recommended to limit the use of this device to technologies where the lack of stability of the other architectures is unacceptable, or applications where transistor performance is not a big driver.

17.2.1.2 Backchannel Etch

To overcome the issues of the long channel length and large footprint of ESL, another architecture can be used. The BCE architecture has smaller parasitics and a shorter channel length for the same technology node compared to ESL. As the name suggests, the SD metal etch is here directly on top of the semiconductor. Therefore, processing must be more careful to avoid damage to the channel. Up to the semiconductor patterning, the same processing steps are used as for ESL. However, after that, the SD metal is directly deposited on the semiconductor and etched. In many cases, the channel is then protected by an isolator. The resulting stack can be seen in Figure 17.1b1.

The number of lithographic steps and consequently the number of masks are reduced by one. This lowers the fabrication cost, as the number of lithographic steps is a good estimator of overall fabrication cost. In addition, the layout can be made much more compact, as shown in Figure 17.1b2. The channel length is no longer limited by the alignment accuracy, but only by the SD patterning accuracy: $L_{min,\,BCE} = CD_{SD}$. The BCE structure still suffers from a significant amount of parasitics. For a technology with 5 μm design rules, a minimum of 150 μm^2 parasitic overlap is unavoidable, compared to a 25 μm^2 channel area, as indicated in Figure 17.1b3.

17.2.1.3 Self-Aligned Transistors

The SA process flow tries to mitigate the main problem to achieve high circuit performance: the large amount of parasitic capacitance in the device. To achieve this, the active region of the semiconductor is defined by its overlap with the gate metal. The gate protects the semiconductor from doping by the intermetal dielectric (IMD) during one of the processing steps. This looks very similar to the standard CMOS transistor architecture, where the gate shields the semiconductor from doping during the ion implants that define the contacts.

A typical cross-section of the SA TFT can be found in Figure 17.1c1. Contrary to ESL and BCE, the gate is on top of the semiconductor. The process starts by depositing and patterning the semiconductor. Next, the gate oxide is deposited, followed by the gate metal. Both are then patterned by the same lithography mask. Next, the IMD is deposited and patterned. This material has the property that it will dope all the semiconductor not protected by the gate, turning it into a metal-like conductive material, indicated by the lightly colored region in the cross-section. The last step in the technology is the patterning of the SD metal layer.

Due to its structure, the SA transistor has no intrinsic parasitic capacitance, as shown in Figure 17.1c3. This makes it much more suitable for circuit applications compared to the aforementioned stacks. The channel length of this device is limited by the patterning accuracy of the gate: $L_{min,\,SA} = CD_{gate}$, where CD_{gate} is the minimal track width of the gate metal. To avoid overlapping gate and SD contacts, though, an additional margin must be foreseen between both, resulting in the longest footprint of the three TFT architectures. However, thanks to the

Table 17.1 Comparison between the different TFT stacks, assuming a 5 μm technology

	Etch-Stop Layer (ESL)	Backchannel Etch (BCE)	Self-Aligned (SA)
Number of litho steps (~cost)	5	4	4
L_{min} for 5 μm tech	15 μm	5 μm	5 μm
Minimum TFT area for 5 μm tech	25 μm × 35 μm = 875 μm^2	15 μm × 25 μm 375 μm^2	15 μm × 45 μm 675 μm^2
Minimum C_{gate} area for 5 μm tech	375 μm^2	175 μm^2	25 μm^2
Processing quality demands	Small	Large	Medium

clear separation of contact and gate, the transistor can be quite narrow, resulting in an overall smaller footprint than ESL, but a larger one than BCE, as can be seen in Figure 17.1c2.

17.2.1.4 Comparison

An overall comparison of the different topologies can be found in Table 17.1. In general, ESL is likely to be the most stable of the three architectures. It offers good protection to the semiconductor layer during processing and can be easily implemented in a large-volume fabrication process. However, it has large devices, with long channel lengths and large parasitic capacitance. The most interesting applications for this technology are cases where performance and TFT area are not as important (e.g., LCD displays and X-ray imagers).

BCE is the best alternative for applications with a small footprint. It has short channel lengths but rather large parasitics. Applications could include high-resolution displays and mid-spec, low-cost circuitry. Finally, SA has the highest speed performance, as it promises short channel lengths and no parasitic capacitance. However, the footprint is quite considerable, and processing can be complicated due to the doping requirements. Typical applications are speed-critical circuits and high-sensitivity thin-film image sensors.

17.2.2 Dual-Gate Transistors

Designers in oxide technologies typically have only an *n*-type transistor at their disposal, as either no *p*-type is integrated in the process flow, or the *p*-type devices are significantly lacking in performance [9]. That makes the design of digital circuits challenging, as there are no good pull-ups available. The traditional unipolar diode-load logic gate has quite poor robustness, making it unsuitable for large-scale integration. A solution to this challenge may be the introduction of multi-transistor multistage logic style, such as pseudo-CMOS [3]. The drawback of such solutions is the footprint of the logic gate, and the corresponding integrated circuit. Another proposed solution to this problem is creating a new type of device with two gates: the dual-gate (DG) transistor [10, 11]. The additional gate (backgate [BG]) can be used to influence the behavior of the transistor, most notably by changing the apparent threshold voltage V_T of the device.

17.2.2.1 Stack Architecture

Figure 17.2 shows how the different metal-oxide transistor stacks can be changed to accommodate such an additional gate. For ESL, the backgate can be incorporated with the current layer stack, but with a different layout. In this case, the contacts are moved away from each other, and a strip of SD metal is added in between the contacts, as shown in Figure 17.2a. The main advantage of this technique is the simplicity: No extra processing steps are required. The main disadvantage is that the transistor length now becomes even longer, with a minimum of 25 μm for the reference CD of 5 μm, with a corresponding increase in gate capacitance and transistor area. The backgate oxide is now the ESL layer, so it brings some limitations to optimizing the coupling of the backgate to the channel.

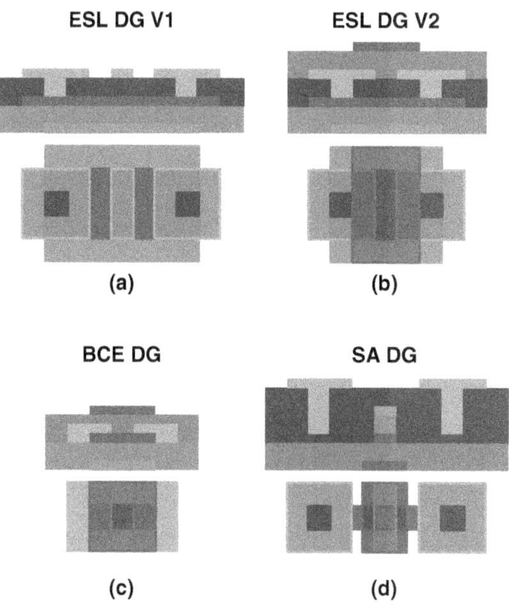

Figure 17.2 Cross-section and layout for four dual-gate device architectures for thin-film devices: ESL with backgate in the SD layer, ESL with backgate in a separate layer, and BCE and SA transistors.

Table 17.2 Comparison of the different transistors with backgate

	ESL DG in SD	ESL DG separate	BCE DG	SA DG
Number of litho steps (~cost)	5	7	6	6
L_{min} for 5 μm tech	25 μm	15 μm	5 μm	5 μm
Minimum TFT area for 5 μm tech	25 μm × 45 μm 1125 μm^2	25 μm × 35 μm 875 μm^2	15 μm × 25 μm 375 μm^2	15 μm × 45 μm 675 μm^2
Minimum C_{gate} area for 5 μm tech	600 μm^2	600 μm^2	350 μm^2	75 μm^2

BCE: Backchannel etch; DG: dual-gate; ESL: etch-stop layer; SA: self-aligned; SD: source–drain.

Alternatively, in ESL, an extra, separate metallization layer can be added on top to serve as a backgate. This adds two mask steps: one for the dielectric (i.e., the via) and one for the metal, as shown in Figure 17.2b. The area of the TFT remains the same as for the single-gate (SG) device, but the capacitance increases, as the extra gate overlaps with the channel and part of the contacts. The backgate coupling is through the ESL layer and the new dielectric, typically resulting in a low coupling level. Also for BCE, an extra metallization layer must be added on top of the stack, as shown in Figure 17.2c. The backgate is now only coupled through the new dielectric, and the capacitive coupling level can be tuned independently of the rest of the stack. The area of the backgate is identical to the gate, so the total capacitance area doubles.

For the self-aligned transistor, the additional gate metallization must be the first layer at the bottom of the stack, as shown in Figure 17.2d. The new backgate cannot be self-aligned to the transistor channel, and therefore some additional margin has to be added, increasing the capacitor area to triple the original amount. The transistor area can remain the same, however.

Table 17.2 compares the implementation of the backgate in the different technologies. The ESL transistor with backgate in the SD layer has mainly technological simplicity as its advantage, requiring no special technology development. It further increases the minimum channel length. For the other devices, similar trade-offs as for the SG devices mentioned in this chapter are valid.

17.2.2.2 Effect of the Backgate

The most basic use of the backgate is electrically connecting it to the gate itself, thereby creating a sort of gate-all-around device. We call this device the pseudo-single-gate (pseudo-SG) transistor. Compared to their SG counterparts, it has a better electrostatic control of the channel, resulting in double the on-current for a device with matched dielectrics and a sharper subthreshold swing. In addition, the device boasts a higher output resistance and faster saturation. Figure 17.3 compares the performance of a SA SG and pseudo-SG device of the same size, where the mentioned improvements can be observed.

However, there are other relevant use cases for the backgate connection. The backgate can shift the apparent V_T of the transistors, that is, the V_T one would observe when measuring the transfer curve (I_D-V_G) at a fixed backgate voltage V_{BG}. This phenomenon can be observed in Figure 17.4 for a self-aligned transistor with matched gate and backgate dielectric (i.e., $C_{ox,gate} = C_{ox,backgate}$). Here, the apparent shift of V_T is equal to $-V_{BG}$. This allows creating transistors with arbitrary threshold voltages by selecting the right fixed V_{BG}, as long as the source voltage V_S is constant. By carefully using the backgate, we can create circuits that significantly outperform circuits without DG transistors. The drawback of this modus operandi of the backgate is a lower observed mobility in the V_T-shifted device: The transfer curve of a transistor with fixed V_{BG} is much lower than the pseudo-SG or even the SG device,

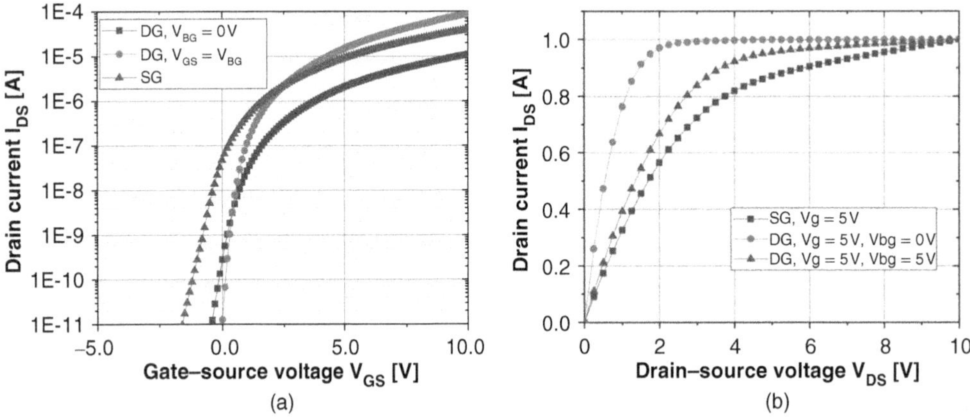

Figure 17.3 Comparison between single-gate (SG), dual-gate (DG), and pseudo-SG self-aligned transistors in (a) transfer characteristic and (b) output characteristic.

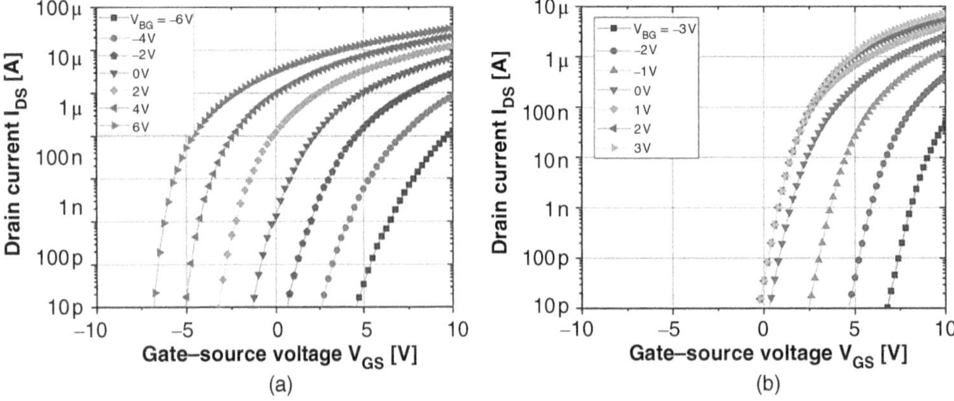

Figure 17.4 Transfer characteristic of (left) a self-aligned dual-gate device with variable backgate voltage and (right) an ESL dual-gate device with a backgate in the SD layer with variable backgate voltage.

as shown in Figure 17.3. Nevertheless, Figure 17.3 also shows that the output characteristics of pseudo-SG and DG are very similar, and much better than those of a single-gate device.

The ratio between the coupling of the frontgate and backgate has to be carefully considered when developing a technology. It is always advisable to keep the coupling of the gate equal to or larger than the coupling of the backgate. Having equal coupling is advantageous when the main goal of the dual-gate stack is the enhanced performance of the pseudo-SG devices, or when the backgate is used as a full-blown, equally powerful, independent gate. However, when only using the backgate to shift the V_T in the pulldown networks, it might be more opportune to reduce the coupling of the backgate to, for example, one-third or one-fourth of the gate coupling. It reduces the additional capacitive load caused by the backgate. In addition, it is now possible to shift the backgate with full-swing negative signals ($-V_{DD}$), without the need for explicit voltage dividers to create smaller backgate voltages used for more moderate V_T shifts.

Dual-gate transistors in ESL technologies do not behave entirely the same as SA or BCE DG transistors due to their intrinsic structure. Figure 17.4 shows the transfer curve of an ESL transistor with a backgate in the SD layer. It is possible to increase the V_T, but not to decrease it beyond a certain limit. Part of the source–drain metal used to create the contact is shielding the channel from the backgate's electrostatic influence. As such, these parts of the channel will maintain their original V_T. When the V_T is lowered below the original $V_{T,}$ the resistance of the regions that are electrostatically controlled by the backgate will decrease further, but the shielded region's resistance remains the same, yielding a lower limit to the overall observed V_T.

Dual-gate transistors have many advantages over their SG counterparts. For display and array applications, the trade-off between the added cost for the extra layers must be weighed against the improved performance of the devices. However, they have proven indispensable for high-performance integrated circuit design in a unipolar environment like oxide technologies.

17.2.3 Moore's Law for TFT Technologies

17.2.3.1 CMOS

We are all very familiar with Moore's law for CMOS technologies, which states that the number of transistors per integrated circuit (IC) doubles every two years [12]. This self-fulfilling prophecy has held true for many decades, as shown in Figure 17.5. It has driven semiconductor manufacturers to scale down their devices to continuously improve the cost–performance–power balance [12]. Indeed, Moore's law is primarily an economic one. When devices are smaller, more of them will fit on a wafer. If the cost to fabricate a single wafer does not increase too dramatically, then the cost per functionality will decrease.

Coincidentally, the performance of a smaller device is also better than that of a larger one, as predicted by the Dennard scaling laws [13]. The speed of circuits is determined by the current driving capability of a transistor, divided by the capacitive load of the next stage. Roughly speaking, the gate capacitance of a device scales down quadratically with its linear dimensions, while the driving current remains the same. As such, CMOS devices became ever faster and more performant.

17.2.3.2 Thin-Film Electronics Historically

In thin-film electronics, the market drivers are very different. For CMOS devices, consumer products become more appealing when they are more performant in a smaller form factor. Thin-film electronics, however, have historically always been used for large-area matrix-based applications, like screens and sensor arrays. The cost of such a product cannot be reduced by shrinking the device size, as the overall active area should stay the same or even increase. The only push for smaller TFT dimensions comes from the desire for higher resolution displays (e.g., by going toward 8k or augmented-reality [AR] displays). The typical circuit used in LCD displays (i.e., a pass gate) does not fundamentally benefit from the advantages of scaling like higher drive currents. Within a given budget, scaling will have to compete with the desire for a larger active area. In the last 30 years, we have observed that

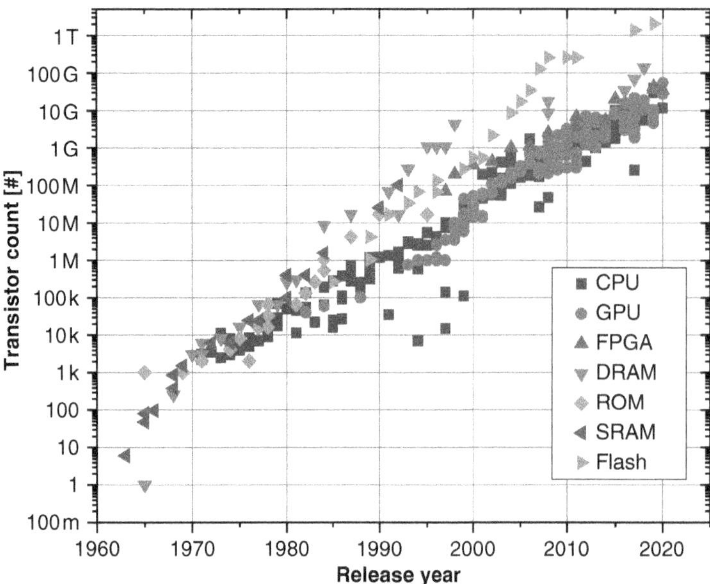

Figure 17.5 Moore's law: illustrated number of transistors per chip over the years.

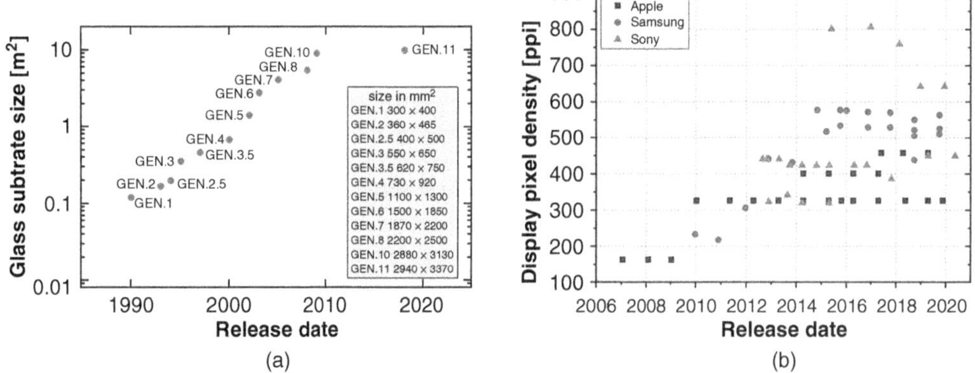

Figure 17.6 (Left) Growth of the substrate sizes for flat-panel display manufacturing over the years, and (right) display resolution for prominent flagship smartphones over the years.

flat-panel display (FPD) manufacturers have pushed for larger and larger substrates, as shown in Figure 17.6. The reasons are both economic and market driven. A larger substrate can handle more displays during the same manufacturing lot. Simultaneously, the market requires ever larger displays for personal use, which requires subsequently a larger substrate. Meanwhile, there have been only limited improvements in display resolutions, and conversely in transistor sizes, as shown in Figure 17.6.

17.2.3.3 New Drivers for Thin-Film Scaling: Circuits

The traditional markets for thin-film electronics, displays, and sensor arrays do not benefit significantly from downscaling, as explained. However, new application fields may shift this balance, such as flexible thin-film integrated circuits (TFICs). For TFICs, the same economical drivers as for CMOS are present. Manufacturers can reduce their production cost for a given functionality by reducing the size of their transistors. In addition, they might even open new markets thanks to the higher performance of their short-channel transistors. Right now, with a 2 μm channel and IGZO, it is already possible to create thin-film chips that can directly communicate with a smartphone through the NFC protocol [14]. Future improvements in the mobility of the oxide semiconductors and reduction in the channel length will allow even faster communication and more complex circuitry, like microprocessors or ADCs having operating speeds within the range of interest for microcontroller applications.

17.2.3.4 L-Scaling

The most obvious way to scale down the transistor is to reduce only the channel length L. This can be achieved by optimizing one process step: For BCE and SA, the critical dimension of the SD and gate, respectively, has to be improved. The extent of scaling is often referred to by the reduction ratio S between the old and new lengths: $L_{new} = L_{old} / S$. If all other parameters are kept constant, while only L is scaled, the drive current of the transistor will increase linearly, while the input capacitance will decrease linearly. This results in a quadratic increase in speed, but also a linear increase in power consumption. Notice that for unipolar oxide circuits, the power consumption scales with the drive current, and not with the gate capacitance due to the dominant leakage in traditional logic topologies. This is different compared to complementary technologies. When scaling L only, the overall area consumption does not scale down significantly, as there is a substantial footprint overhead for the transistor. As a consequence, the power density will scale up linearly. Power density matters for oxide circuits, as they are manufactured on substrates with a low thermal conductivity, like glass or plastic, in contrast to crystalline silicon with a very high thermal conductivity. As such, we suggest including the evaluation of the local power density, which is the power dissipation locally at the channel location, relates to $W \times L$, and is irrespective of the footprint overhead. Since the drive current increases with S and the channel length decreases by a factor of S, the increase of the local power density is quadratic. Peaking power issues like catastrophic breakdown are more related to the local power density, while long-term, averaged power density problems like thermally assisted degradation are related to the global power density. Local power density is a short-term issue (i.e., until the power properly distributes). In general, the advantages of only scaling L are limited to quadratically higher speed, with a linear power increase as a cost.

17.2.3.5 W and L Scaling

If power consumption is a concern, as is mostly the case, then scaling of the transistor width W along with L can be a solution. W scaling only requires the optimization of the patterning accuracy of the semiconductor layer. In this case, the transistor drive current remains constant, while the gate capacitance scales down quadratically. As such, the (static) power consumption stays the same, while the speed increases quadratically too. The transistor footprint overhead will still dominate the overall area, and only very small improvements to the area consumption can be made. The local power density still increases quadratically, but the global power consumption is now also constant. Again, the main advantages of W and L scaling are the higher speed, but no significant gains in area consumption can be made.

17.2.3.6 Overall Lateral Scaling

Finally, all lateral dimensions can be scaled simultaneously, including W, L, via sizes, and alignment accuracy. Electrically, the improvement is identical to the W and L scaling scenario, but now the area consumption can be scaled down quadratically. This also implies that global power density goes up quadratically, like the local one. This scenario is mostly interesting with respect to area consumption, providing the economic stimulus for scaling down.

17.2.3.7 Oxide Thickness and Supply Voltage Scaling

Finally, on top of the scaling in all lateral dimensions, it is also possible to thin down the oxide thickness t_{ox}. Typically, the thickness of the oxides in thin-film technologies is selected to be around 100–200 nm, so they can handle the voltage requirements of around 10–20 V from the display and sensor applications. However, some TFIC

Table 17.3 Change in key parameters under several scaling scenarios for TFICs

	L	W and L	All lateral dimensions	All dimensions and supply
Length	1/S	1/S	1/S	1/S
Width	1	1/S	1/S	1/S
Oxide thickness	1	1	1	1/S
Supply voltage	1	1	1	1/S
Gate capacitance	1/S	$1/S^2$	$1/S^2$	1/S
Drive current	S	1	1	1/S
Speed	S^2	S^2	S^2	1
Area consumption	1	1	$1/S^2$	$1/S^2$
Power consumption	S	1	1	$1/S^2$
Local power density	S^2	S^2	S^2	1
Global power density	S	1	S^2	1

applications do not demand high voltages. If the supply voltages can be scaled down for a specific application, and the gate dielectric is scaled accordingly, further gains in power consumption can be expected. In general, it is advised to scale the lateral and thickness parameters at about an equal rate to avoid short-channel effects.

Scaling down the oxide provides linearly more drive current, while scaling down the supply voltage quadratically decreases it due to the transistor saturation regime, causing an overall linear decrease in drive current. Gate capacitance, including W, L, and t_{ox} scaling, goes down only linearly, so overall the speed remains constant. Power consumption decreases quadratically with both current and voltage going down linearly. Power density stays constant, and, as just shown, the total area consumption goes down quadratically.

17.2.4 Conclusion

In Table 17.3, all discussed options for oxide circuit downscaling are compared. If the application requires a faster circuit, scaling L only will suffice. However, if power cannot increase, both W and L must be scaled down simultaneously. If the circuit area should decrease to reduce chip cost or meet the specifications, then all dimensions should be scaled. Finally, if the speed is sufficiently high, but power consumption needs to be reduced, then scaling including the oxide thickness and supply voltage can be beneficial.

17.3 Digital Electronics

Digital electronics are an important part of any electronic system, for example processing digitized sensor data or detecting and understanding incoming data from a communication block. Digital electronics operate on binary data (i.e., 1's and 0's) or with signals occurring close to supply or ground. In this section, we will elaborate a few complex TFT-based digital realizations.

17.3.1 Communication Chips

In an electronics system, the role of a communication block is to interface with the outer world. The field of TFT technologies focuses mainly on RFID communication chips, as those frequencies are within reach for the

Table 17.4 Four categories of communication devices for thin-film electronics

	Capacitive		Inductive	
	Custom Reader [19, 20]	**Touchscreen [16]**	**Custom Reader [17, 18]**	**NFC Reader [14]**
Base carrier frequency	Self-selected, 13.56 MHz demonstrators	No base carrier	Self-selected, 13.56 MHz demonstrators	13.56 MHz, imposed by reader
Antenna	2 parallel plates	2 parallel plates	Coil antenna	Coil antenna
Power	Capacitive	Battery or photodetector	Inductive	Inductive
Reading distance	Millimeters	Touch	Few millimeters–10 cm (*Note*: ISO15793 allows larger reading distances)	Few millimeters–10 cm
Data rates	Custom protocol, self-selected	Low frequency, depending on touchscreen technology	Custom protocol, self-selected	ISO14443 105.9 kbit/s

technology. In addition, TFT-based RFID tags may be seamlessly embedded into objects, potentially at low cost, enabling the Internet-of-things. In such a system, the tag communicates to a reader device, which is subsequently connected to the cloud. The operating frequency for TFT-based tags is mainly 13.56 MHz.

Table 17.4 details four categories of communication chips for thin-film electronics. The main distinction can be found between capacitive-coupled devices versus inductive-coupled devices; see Figure 17.7. In a fully capacitive system, the power transfer is proportional to the capacitance (i.e., increasing the area of the capacitor increases the power transfer). In contrast, however, a larger distance between both plates reduces the power transfer. For capacitive systems, it is important to align the capacitor plates accurately and maintain a close coupling in the millimeter range. For applications that require a high-security data transfer, this presents a great opportunity. Inductively coupled devices, on the other hand, benefit from a larger power transfer efficiency and can be read out at larger distances, typically in the range of a few centimeters. Inductive antennas for 13.56 MHz are credit-card sized with several windings. Smaller antennas are possible, but the connection has lower efficiency, therefore reducing power transfer and readout distance. In many of the 13.56 MHz wireless systems, capacitive antennas can be made monolithically with the same technology as TFTs. Inductive antennas require a sufficiently low resistance and thicker metals, usually not available in low-cost transistor technologies. The integration step of an inductive antenna with the TFT chip will add to the system cost.

Figure 17.7b shows the basic building blocks of such an RFID transponder chip for both cases. These general blocks need to be developed specifically for each application. Table 17.4 further distinguishes between custom-developed reader hardware and a widespread omnipresent reader. Such a custom reader is ideally for closed-loop systems, whereby a specific protocol is defined with specific hardware and RFID tags that can only be used in this system. In such an example, the design blocks of Figure 17.7 can be kept very simple. An 8-bit or 16-bit RFID tag would be sufficient for many of those applications and employs only a few hundred logic gates, resulting in a high-yield, small-sized transponder chip.

For a full NFC tag, on the other hand, the chip needs to adhere to the ISO14443 standard; therefore, specific developments are required. The main challenges for the implementation of this standard in TFT technologies are

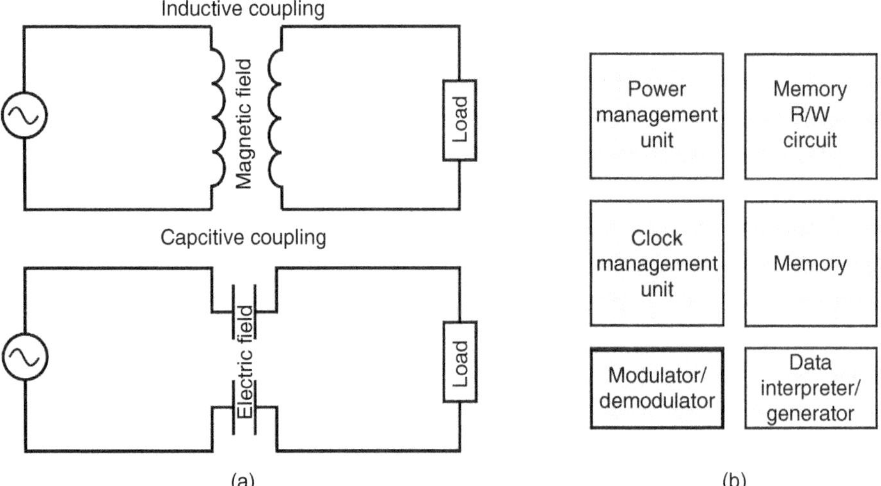

Figure 17.7 (a) Inductive and capacitive power transfer and identification, and (b) basic building blocks of an RFID transponder chip.

the low power budget, the high data rate derived from the 13.56 MHz carrier frequency, and a minimum memory of 128 bits, including a cyclic redundancy check. Section 17.2.3 about scaling showed that downscaling will enable faster electronics. We downscaled the self-aligned IGZO transistors to 1.5 μm channel length. Scaling together with the optimization of the pseudo-CMOS logic gate resulted in 2.4 ns gate delays, well within specifications for the ISO norm. The resulting IC showed the first IGZO-based 13.56 MHz clock division circuit [15]. In addition, several measures have been taken to reduce power consumption at the system level. The clock management unit generates all system clocks from the 13.56 MHz incoming carrier frequency. Only the first stage needs to operate at high speed, the second stage operates at half the carrier frequency, and the third at quarter of the carrier frequency. Therefore, we have optimized the pseudo-CMOS logic by selecting the W and L carefully to be able to operate at the right frequency while minimizing the power consumption. For example, the channel length for the first flip-flop of the frequency divider was 1.5 μm, while the channel length for the seventh flip-flop was increased to 4 μm. In addition, the code generator has been designed using the implementation of the pseudo-CMOS logic gates with the lowest power consumption. Lastly, the cyclic redundancy check was hardwired in order to reduce gate count and power consumption. All these design improvements lead to a fully functional, flexible IGZO-based NFC tag that communicates directly to a smartphone with an embedded NFC reader [15].

The quality of the Internet-of-things service depends on how widespread the readers are. Therefore, we have investigated direct communication to a capacitive touchscreen, as these are more ubiquitous than NFC readers. This device is called a capacitive touchscreen tag or C-touch tag. There is a wide variety of touch screens available (e.g., with respect to resolution, sensitivity, etc.), which are driven by a host of different control chips. This imposes several challenges to the system, as we target a C-touch tag that communicates to all sorts of capacitive touchscreens, independent of resolution and controller. Therefore, we have analyzed a broad range of touchscreen displays [16] and defined a protocol: First, we create a touch event with one of the capacitive electrodes, and subsequently we trigger a swiping action on the touchscreen by applying a pulse sequence to the other electrode. Decoding the swiping sequence in the software of the touchscreen device allows deciphering the transmitted code. In contrast to the NFC tag, the C-touch tag with swiping has a data rate limited to the order of tens of bits per second. Faster data rates may be obtained by increasing the number of electrodes and swiping actions (i.e., enabling multiswipe events). The clock for such small data rates is generated by a slow ring oscillator with long-channel-length devices operating at very low supply voltages. As the capacitive power from the touchscreen is very limited, we have developed two alternatives: integrating a thin-film battery operating at 1.5–3 V or integrating a thin-film photovoltaic cell that captures the light of the touchscreen. In the latter case, the supply voltage of the system was about 0.6 V. In this realization, various system-level blocks were implemented in different logic styles. The power consumption of the tag was about 31 nW at 600 mV supply voltage for a simple 12-bit C-touch tag [16].

17.3.2 Complex Metal-Oxide-Based Digital Chips

This section elaborates on more recent complex metal-oxide-based digital chips, whereby repetitive digital circuits such as gate drivers are out of scope. The selected chips (Table 17.5) are all related to the field of the Internet-of-things and sensor tags. The high-level building blocks are shown in Figure 17.8 and comprise one or multiple sensors to be read out. The sensor signals are subsequently converted to digital signals, and the raw data can be transmitted to the reader via the communication unit. There is an option to first process the data prior to transmitting it outside. In addition, for some applications, secure data transfer is a must-have that requires a dedicated encryption block.

One can consider several options to perform data signal processing on chip, ranging from a general- or specific-purpose microprocessor to a machine-learning-based classifier. A general-purpose microprocessor is a very versatile solution, as it can be utilized for many varying sensor systems. It will require custom programming for executing a specific function. In 2014, we realized a general-purpose microprocessor based on an arithmetic unit, a register bank, and a programmable memory chip. This chip could be programmed by additive printing steps. The technology combined an organic evaporated semiconductor with a solution-processed oxide semiconductor, leading to hybrid complementary logic. In total, the full processor comprised 3,504 TFTs and operated at 6.5 V [21].

Table 17.5 Overview of three selected complex digital chips

	Microprocessor [21]	Odor Classification Chip [22]	Security Chip [23]
Technology	Hybrid organic *p*-type oxide *n*-type	Metal-oxide based	Metal-oxide based
Channel length	5 μm	0.8 μm	5 μm
Number of routable metal layers	2 (gate and source–drain)	4 (including gate and source–drain)	2 (gate and source–drain)
Chip area	225.6 mm^2	5.6 mm^2	331.5 mm^2
Supply voltage	6.5 V	4.5 V	15 V (largest supply)
Logic type	CMOS (2 transistors per inverter)	Resistive load (1 transistor + resistor per inverter)	Pseudo-CMOS (4 transistors per inverter)
Number of devices	3,504 TFTs	2,084 TFTs and 1,048 resistors	4,044 TFTs

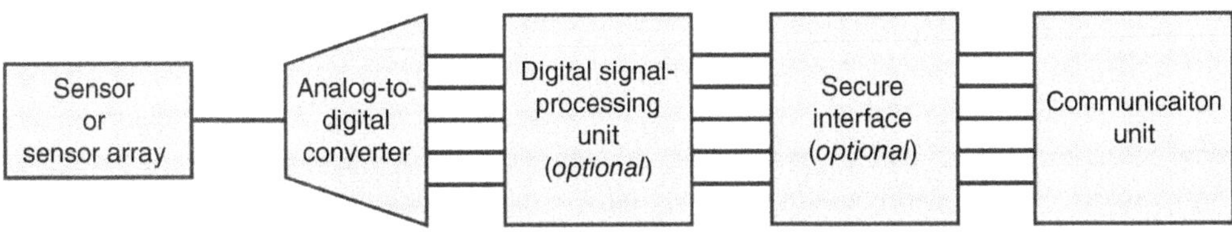

Figure 17.8 High-level building blocks of a sensor tag.

Classification techniques by means of artificial intelligence and machine-learning algorithms may be very attractive to interpret sensor data. Ozer and his coworkers have recently demonstrated a hardwired machine-learning processing engine to classify odor data from a set of gas sensors [22]. A typical artificial-intelligence engine is computationally extremely demanding during training. As this is very memory- and hardware-intensive, this is not viable on the flexible TFT platform. Ozer et al. investigated several machine-learning algorithms based on prediction accuracy versus data quantization bits versus gate count. Afterward, the odor-classification training was performed off-chip. The outcome of the training is subsequently implemented in a TFT-based hardwired machine-learning chip, employing 3,132 TFTs and resistor devices. The chip operated at 4.5 V and was implemented in an 0.8 μm IGZO TFT technology [22]. The resulting machine-learning chip was based only on combinatorial logic except for an 8-bit sensor data buffer. The described methodology may be very efficient for specific classification algorithms and could be used for low-power sensor classification tasks on flexible substrates.

Sensor tags, particularly healthcare patches, may require built-in security for data treatment and data transfer. To transmit encrypted data, a system needs to embed cryptographic core blocks to generate encrypted data that only the targeted reader will understand. Cryptographic cores have a high complexity and typically employ a very large number of transistors, which is challenging in TFT technologies. Mentens et al. have recently demonstrated a lightweight security algorithm to cope with this challenge, reducing the number of transistors to 4,044 TFTs in pseudo-CMOS logic [23]. The selected security algorithm is based on KTANTAN encryption and was implemented with a 32-bit block size and an 80-bit key. In thin-film technologies, the content of hardwired memory blocks could be easily detected by optical microscopy due to their large size. Mentens et al. stored the key in a set of logic gates. Programming was done by an invisible laser treatment that shifts the threshold voltage, thereby eliminating the optical security flaw.

17.4 Analog Electronics

The maturity of metal-oxide large-area technologies (LAEs) on rigid and flexible substrates using IGZO-TFTs in analog applications (such as OLED TVs) and in digital applications (such as RFID) is leading academia and industry to explore the expansion of the technology to other applications. Analog electronics are needed to establish a connection to physical processes such as temperature [24, 25], humidity, pressure, fingerprint sensing [26–28], bio-potential measurements [29–35], and many more. The explored building blocks vary from amplifiers to complex ADCs. For analog electronics, variability is a key factor for achieving good accuracy. In Figure 17.9, the distribution (σ) of the threshold voltage (V_T) is shown for various transistor sizes over a 150 mm round wafer. Larger sized devices have less variability, which can be used to cope with mismatch in analog electronics. In this section, we will discuss key analog building blocks and techniques to build readout circuits for applications such as fingerprint sensing and healthcare patches.

17.4.1 Thin-Film ADC Topologies

Multiple TFT-based implementations of ADC topologies on glass or flexible substrates have been presented in the literature [24–27, 36–42] (Table 17.6). The first ADC implementations have been realized using complementary low-temperature polycrystalline silicon (LTPS) or organic thin-film transistor (*c*-OTFT) technology. The high-performance LTPS technology achieves the fastest ADCs (up to 3 MSps) [37] and the most accurate ones (up to 11.2-bit) [36]. The complementary OTFT successive approximation (SAR) [38] achieved an impressive 6-bit resolution using cascaded inverters as a comparator at 3 V supply, but the footprint is large and the speed very slow. Speed and footprint are the drawbacks of the other three implementations [39, 40, 42]. Nevertheless, the voltage-controlled oscillator (VCO)-based unipolar OTFT ADC achieves 7.7-bit resolution [42], albeit only for quasi-static signals. The main drawbacks of organic technologies are their limited stability and uniformity;

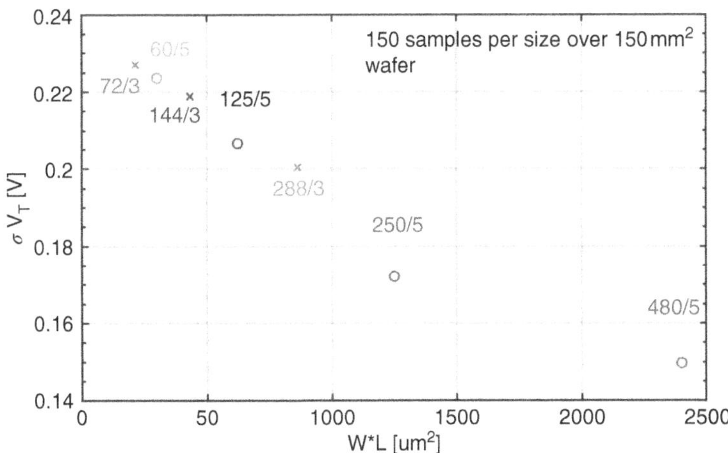

Figure 17.9 Extracted threshold voltage distribution from mapping data of various SADG IGZO TFT W/L sizes. Per size, 150 samples were measured over a 150 mm wafer.

therefore, the feasibility of a robust ADC implementation is questionable. On the other hand, the similarly performant *a*-Si:H TFT technology is more industry-relevant and uniform. A 5-bit flash ADC at 2 kSps was demonstrated, touching the limits of the topology and showcasing the maturity of the technology. The growth of IGZO technology had led to the demonstration of several implementations as well, starting in the ESL architecture [24, 25]. A ΔΣ ADC with 8-bit resolution at 10 Sps or 6.3-bit at 300 Sps was demonstrated, yielding a figure of merit (FoM) of 390 and 39 nJ/cs, respectively. In the same technology, a C-2C SAR ADC was also demonstrated with much lower power dissipation, but at only 26 Sps it yielded a 33% better FoM. Finally, self-aligned dual-gate (SADG) IGZO-TFTs can provide faster responses thanks to negligible overlap parasitics. An SADG implementation using indium–tin–zinc oxide (ITZO) demonstrated 6-bit resolution at 1 kSps at 15 V supply, yielding the state-of-the-art FoM of 8 nJ/cs for unipolar implementations and reaching the performance of the complementary designs (Table 17.6) [27]. This can be attributed to the higher intrinsic charge carrier mobility of the ITZO semiconductor.

In Figure 17.10, the speed-to-resolution performance of all the TFT–ADC implementations are plotted as points. Also, an estimation/extrapolation of the performance is made for the different basic ADC topologies for oxide and LTPS technologies, shown by the squares. Improvement in both speed and resolution is predicted due to the higher mobility and the availability of complementary LTPS technology compared to the unipolar oxide technologies. Although local variability of LTPS technology is worse than the *a*-IGZO, the complementary nature of the LTPS technology is estimated to give an advantage to the latter.

17.4.2 Imager Readout Peripherals

Large-area high-resolution imagers need multiple connections to off-panel electronics to synchronously read out all pixels of a row. A high-speed and stable in-panel charge-sense amplifier (CSA) [27, 28] can interface with the pixels and multiplex the output to the off-panel electronics, thereby reducing the number of connections and lowering the cost of, for example, fingerprint sensors, as shown in Figure 17.11. For an imager compatible with the FBI standard, the pixel density must be at least 500 ppi, corresponding to a 50 μm pixel pitch and row width. IGZO amplifiers [27] are able to achieve the minimum gain and phase margin specifications for the CSA, and the ADC can achieve sufficient resolution, but the readout system lacks in speed, limiting the readout speed to 1 fps for a 1 MP array. LTPS technology provides higher speed and a smaller footprint, as demonstrated in Refs. [43, 44]. Complementary amplifiers achieve gains higher than 50 dB and bandwidths exceeding 200 kHz, in a smaller

Table 17.6 Detailed comparison table of ADC implementations in different technologies

	Complementary				Unipolar					
Technology	LTPS (2009)	LTPS (2010)	*c*-OTFT (2010)	*c*-OTFT (2013)	DG OTFT (2011)	*a*-Si:H-TFT (2012)	DG OTFT (2013)	DG ESL IGZO (2017)	DG ESL IGZO (2018)	SADG IGZO and ITZO (2019)
Architecture	2nd ΔΣ	Flash	SAR	Counting	1st ΔΣ	Flash	VCO-based	ADSM	C2C SAR	C2C SAR
DNL (LSB)	11.2-b	0.25	−0.6	0.24	4.1-b	5-b	7.7-b	6.3-b and 8-b	0.8	0.84
INL (LSB)		0.25	0.6	0.42					0.69	0.9
FoM (nJ/cs)	8.6[b)]	—	<5[a)]	17,000[b)]	3,450[b)]	—	690	39 and 390	26	16 and 8[a)]
Power (mW)	63.3	(13 V)	0.004 (3 V)	0.54	1.5	13.6	0.048	2	0.052	0.212 and 0.55
Speed	1.56k	3 M	10	2	15.6	2 k	167 m	300 and 10	26	133 and 1 kS
Area (mm^2)	26	3.75	700	2,450	260	—	19.4	27.9	27.5	8.88
Integration	Only analog	Decoder	Only analog	Counter, DAC	Only analog	Only analog	+ Metal resistor + logic	PWM out only	+ Offset cancellation + DC biasing	+ Offset cancellation
Authors	Lin et al. [36]	Jamshidi-Roudbari et al. [37]	Xiong et al. [38]	Abdinia et al. [39]	Mariën et al. [40]	Dey et al. [41]	Raiteri et al. [42]	Garripoli et al. [25]	Papadopoulos et al. [24]	Papadopoulos et al. [26, 27]

a) Calculated assumed minimum: eNOB = 5b.
b) Calculated.
ADSM: Asynchronous delta sigma modulation; C2C: chip-to-chip; *c*-OTFT: complementary organic thin-film transistor; DAC: digital-to-analog converter; DC: direct current; DG: dual gate; DNL: differential nonlinearity; ESL: etch-stop layer; FoM, figure of merit; INL: integral nonlinearity; LSB: least significant bit; LTPS: low-temperature polycrystalline silicon; PWM: pulse width modulated; SADG: self-aligned dual-gate; SAR; successive approximation; VCO: voltage-controlled oscillator.

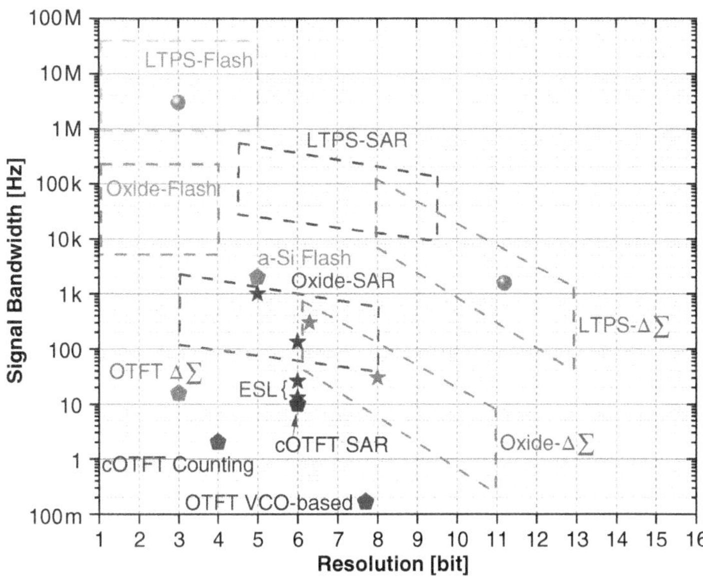

Figure 17.10 Performance comparison of various ADC topologies implemented in all TFT technologies.

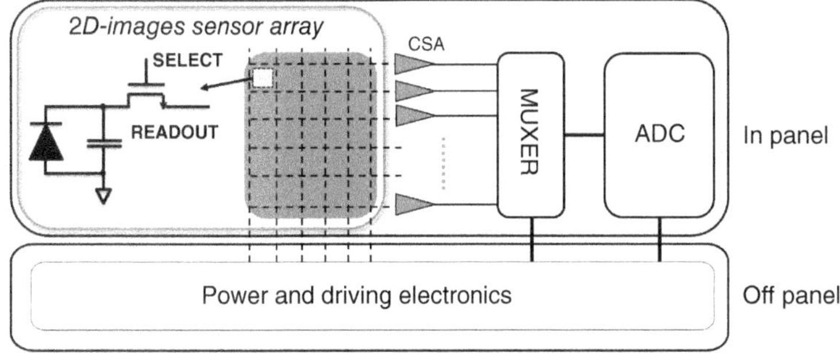

Figure 17.11 Block diagram of an in-panel readout system for imager applications.

footprint than unipolar amplifiers. One drawback of LTPS, though, is the lower output resistance of the transistors due to the kink effect, limiting the minimum L; using LTPS decreases the footprint of the readout electronics, and a higher readout speed (fps) can be achieved. Table 17.6 shows that the sampling speed of a LTPS Flash ADC can reach 3 MSps with a large L = 8 μm technology [37]. Assuming a 6-bit 10 fps readout of a 1 MP imager array and 1:10 multiplexing, a 100 kSps ADC is needed. This is feasible with SAR-ADC, according to our estimations from Figure 17.10. For lower resolutions, a Flash ADC can be used to multiplex more rows and decrease the number of connections by up to 100×. Multiplexing is limited by the leakage in the mux (multiplexer). By integrating TFT-CSA and a TFT-ADC on an imager panel, one can reduce the number of connections to off-panel electronics by one to two orders of magnitude, depending on the resolution.

17.4.3 Healthcare Patches

Due to the small bandwidth of biopotential signals (electrocardiogram [ECG], electroencephalogram [EEG], electromyogram [EMG], etc.) and the benefit of spatial resolution, interest is growing to develop healthcare patches with thin TFTs [29–35]. Ultra-flexible patches have demonstrated the measurement of EMG [29] and ECG signals [30] using OTFT. EEG signals have been measured with active amplifier and multiplexing electronics using *a*-Si:H-TFTs [31]. Organics suffer from mismatch, and therefore a rather complex process was applied to suppress offset in the differential amplifier used. Oxide electronics with lower noise levels [46] and good uniformity and stability demonstrated improved ECG measurements [32–35]. Another critical parameter for monitoring biopotential is the noise performance of the technology and the electronics. Various methods have been employed to improve the overall noise performance, compensating for non-uniformities and 1/f noise, and to achieve high signal integrity and sensitivity. One method is to improve the noise performance of the TFTs by using self-aligned topologies [35], better semiconductors and dielectrics [47], or a second gate. The second method is to design less noisy amplifiers by using the backgate of the TFTs appropriately [32], by using chopping [31] and buffering the input signal [33]. The state of the art achieves an integrated noise of 2.3 μV_{rms} at 100 Hz [31]. This is worse than silicon performance, but the overall state-of-the-art power–efficiency factor is just 1.20×10^5 [35], which is promising. Further optimizations are possible using a combination of the aforementioned techniques.

17.5 Summary

This chapter elaborated on the opportunities and challenges for metal-oxide electronics beyond the active-matrix display product field. The emerging applications range from readout schemes for large-area IGZO-based imaging arrays, to flexible RFID communication chips, to artificial intelligence and readout chips for flexible healthcare patches. Several design techniques have been discussed to overcome the challenges and to bridge the gaps with the application requirements, such as the introduction of an additional gate and downscaling of the thin-film transistor in order to obtain power, speed, and area advantages.

Acknowledgments

We would like to thank our coworkers at imec and TNO/Holst Centre, and M. Zulqarnain and E. Cantatore from Eindhoven University of Technology (TU/e), for their valuable contributions to this work. Part of this work has received funding from the European Research Council (ERC) under the European Union's Horizon 2020 research and innovation program under grant agreement No. 716426 (FLICs project).

References

1 Lin, C., Chien, C., Wu, C. et al. (2012). Top-gate staggered a-IGZO TFTs adopting the bilayer gate insulator for driving AMOLED. *IEEE Transactions on Electron Devices* https://www.doi.org/10.1109/TED.2012.2191409.
2 Cho, D., Yang, S., Shin, J. et al. (2009). Passivation of bottom-gate IGZO thin film transistors. *Journal of the Korean Physical Society* https://www.doi.org/10.3938/jkps.54.531.
3 Wang, H., Chan, F., Cheng, C. et al. (2017). Ultra low voltage 1-V RFID tag implement in a-IGZO TFT technology on plastic. In: *2017 IEEE International Conference on RFID*. https://www.doi.org/10.1109/RFID.2017.7945608.
4 Nag, M., Steudel, S., Bhoolokam, A. et al. (2014). High performance a-IGZO thin-film transistors with mf-PVD SiO_2 as an etch-stop-layer. *Journal of the Society for Information Display* https://www.doi.org/10.1002/jsid.212.
5 Street, R.A. (1991). *Hydrogenated Amorphous Silicon*. Cambridge: Cambridge University Press https://www.doi.org/10.1017/CBO9780511525247.

6 Morita, S., Ochi, M., and Kugimiya, T. (2016). Amorphous oxide semiconductor adopting back-channel-etch type thin-film transistors. *Kobelco Technology Review* 34: 52–58.

7 Nag, M., Obata, K., Fukui, Y. et al. (2014). 20.1: Flexible AMOLED display and gate-driver with self-aligned IGZO TFT on plastic foil. In: *SID Symposium Digest of Technical Papers*. https://www.doi.org/10.1002/j.2168-0159.2014.tb00068.x.

8 Morosawa, N., Ohshima, Y., Morooka, M. et al. (2012). Novel self-aligned top-gate oxide TFT for AMOLED displays. *Journal of the Society for Information Display* https://www.doi.org/10.1889/JSID20.1.47.

9 Rockelé, M., Vasseur, K., Mityashin, A. et al. (2018). Integrated tin monoxide p-channel thin-film transistors for digital circuit applications. *IEEE Transactions on Electron Devices* https://www.doi.org/10.1109/TED.2017.2781618.

10 Nag, M., De Roose, F., Myny, K. et al. (2017). Characteristics improvement of top-gate self-aligned amorphous indium gallium zinc oxide thin-film transistors using a dual-gate control. *Journal of the Society for Information Display* https://www.doi.org/10.1002/jsid.558.

11 Myny, K., Tripathi, A., Van der Steen, J.-L., and Cobb, B. (2015). Flexible thin-film NFC tags. *IEEE Communications Magazine* https://www.doi.org/10.1109/MCOM.2015.7295482.

12 Mack, C. (2011). Fifty years of Moore's law. *IEEE Transactions on Semiconductor Manufacturing* https://www.doi.org/10.1109/TSM.2010.2096437.

13 Dennard, R., Gaensslen, F., Yu, H.-N. et al. (1974). Design of ion-implanted MOSFET's with very small physical dimensions. *IEEE Journal of Solid-State Circuits* https://www.doi.org/10.1109/JSSC.1974.1050511.

14 Myny, K., Lai, Y., Papadopoulos, N. et al. (2017). 15.2 A flexible ISO14443-A compliant 7.5mW 128b metal-oxide NFC barcode tag with direct clock division circuit from 13.56MHz carrier. In: *2017 IEEE International Solid-State Circuits Conference*. https://www.doi.org/10.1109/ISSCC.2017.7870359.

15 Myny, K. (2018). The development of flexible integrated circuits based on thin-film transistors. *Nature Electronics* https://www.doi.org/10.1038/s41928-017-0008-6.

16 Papadopoulos, N., Qiu, W., Ameys, M. et al. (2019). Touchscreen tags based on thin-film electronics for the Internet of Everything. *Nature Electronics* https://www.doi.org/10.1038/s41928-019-0333-z.

17 Ozaki, H., Kawamura, T., Wakana, H. et al. (2011). 20-μW Operation of an a-IGZO TFT-based RFID chip using purely NMOS "active" load logic gates with ultra-low-consumption power. In: *2011 Symposium on VLSI Circuits*. https://www.doi.org/10.1109/RFID.2017.7945608.

18 Yang, B., Oh, J., Kang, H. et al. (2013). A transparent logic circuit for RFID tag in a-IGZO TFT technology. *ETRI Journal* https://www.doi.org/10.4218/etrij.13.1912.0004.

19 Cantatore, E., Geuns, T., Gelinck, G. et al. (2007). A 13.56-MHz RFID system based on organic transponders. *IEEE Journal of Solid-State Circuits* https://www.doi.org/10.1109/JSSC.2006.886556.

20 Papadopoulos, N., Smout, S., Willegems, M. et al. (2018). 2-D smart surface object localization by flexible 160-nW monolithic capacitively coupled 12-b identification tags based on metal–oxide TFTs. *IEEE Transactions on Electron Devices* https://www.doi.org/10.1109/TED.2018.2869666.

21 Myny, K., Smout, S., Rockelé, M. et al. (2014). A thin-film microprocessor with inkjet print-programmable memory. *Scientific Reports* https://www.doi.org/10.1038/srep07398.

22 Liu, F., Dahiya, A., and Dahiya, R. (2020). A flexible chip with embedded intelligence. *Nature Electronics* https://www.doi.org/10.1038/s41928-020-0446-4.

23 Mentens, N., Genoe, J., Vandenabeele, T. et al. (2019). Security on plastics: fake or real? *IACR Transactions on Cryptographic Hardware and Embedded Systems* https://www.doi.org/10.13154/tches.v2019.i4.1-16.

24 Papadopoulos, N., De Roose, F., Van der Steen, J. et al. (2018). Toward temperature tracking with unipolar metal-oxide thin-film SAR C-2C ADC on plastic. *IEEE Journal of Solid-State Circuits* https://www.doi.org/10.1109/JSSC.2018.2831211.

25 Garripoli, C., Van der Steen, J., Smits, E. et al. (2017). 15.3 An a-IGZO asynchronous delta-sigma modulator on foil achieving up to 43dB SNR and 40dB SNDR in 300Hz bandwidth. *2017 IEEE International Solid-State Circuits Conference* https://www.doi.org/10.1109/ISSCC.2017.7870360.

26 Papadopoulos, N., Steudel, S., Kronemeijer, A. et al. (2019). Flexible 16nJ/c.s. 134S/s 6b MIM C-2C ADC using dual gate self-aligned unipolar metal-oxide TFTs. In: *2019 IEEE Custom Integrated Circuits Conference*. https://www.doi.org/10.1109/CICC.2019.8780122.

27 Papadopoulos, N., Steudel, S., Smout, S. et al. (2019). 9-1: Invited paper: metal-oxide readout electronics based on indium-gallium-zinc-oxide and indium-tin-zinc-oxide for in-panel fingerprint detection application. *SID Symposium Digest of Technical Papers* https://www.doi.org/10.1002/sdtp.12863.

28 Papadopoulos, N., Steudel, S., De Roose, F. et al. (2018). In-panel 31.17dB 140kHz 87μW unipolar dual-gate In-Ga-Zn-O charge-sense amplifier for 500dpi sensor array on flexible displays. In: *IEEE 44th European Solid-State Circuits Conference*. https://www.doi.org/10.1109/ESSCIRC.2018.8494260.

29 Fuketa, H., Yoshioka, K., Shinozuka, Y. et al. (2014). 1 μm-thickness ultra-flexible and high electrode-density surface electromyogram measurement sheet with 2 V organic transistors for prosthetic hand control. *IEEE Transactions on Biomedical Circuits and Systems* https://www.doi.org/10.1109/TBCAS.2014.2314135.

30 Sugiyama, M., Uemura, T., Kondo, M. et al. (2019). An ultraflexible organic differential amplifier for recording electrocardiograms. *Nature Electronics* https://www.doi.org/10.1038/s41928-019-0283-5.

31 Moy, T., Huang, L., Rieutort-Louis, W. et al. (2017). An EEG acquisition and biomarker-extraction system using low-noise-amplifier and compressive-sensing circuits based on flexible, thin-film electronics. *IEEE Journal of Solid-State Circuits* https://www.doi.org/10.1109/JSSC.2016.2598295.

32 Garripoli, C., Van der Steen, J., Torricelli, F. et al. (2017). Analogue frontend amplifiers for bio-potential measurements manufactured with a-IGZO TFTs on flexible substrate. *IEEE Journal on Emerging and Selected Topics in Circuits and Systems* https://www.doi.org/10.1109/JETCAS.2016.2616723.

33 Garripoli, C., Abdinia, S., van der Steen, J. et al. (2018). A fully integrated 11.2-mm^2 a-IGZO EMG front-end circuit on flexible substrate achieving up to 41dB SNR and 29MΩ input impedance. *IEEE Solid-State Letters* https://www.doi.org/10.1109/LSSC.2018.2878184.

34 Zulqarnain, M., Stanzione, S., van der Steen, J. et al. (2018). A 52 μW heart-rate measurement interface fabricated on a flexible foil with a-IGZO TFTs. In: *IEEE 44th European Solid State Circuits Conference*. https://www.doi.org/10.1109/ESSCIRC.2018.8494298.

35 Zulqarnain, M., Stanzione, S., Rathinavel, G. et al. (2020). A flexible ECG patch compatible with NFC RF communication. *NPJ Flexible Electronics* https://www.doi.org/10.1038/s41528-020-0077-x.

36 Lin, W., Lin, C., and Liu, S. (2009). A CBSC second order sigma delta modulator in 3μm LTPS-TFT technology. In: *2009 IEEE Asian Solid-State Circuit Conference*. https://www.doi.org/10.1109/ASSCC.2009.5357196.

37 Jamshidi-Roudbari, A., Kuo, P., and Hatalis, M. (2010). A flash analog to digital converter on stainless steel foil substrate. *Solid-State Electronics* https://www.doi.org/10.1016/j.sse.2009.10.020.

38 Xiong, W., Zschieschang, U., Klauk, H., and Murmann, B. (2010). A 3V 6b successive-approximation ADC using complementary organic thin-film transistors on glass. In: *2010 IEEE International Solid-State Circuits Conference*. https://www.doi.org/10.1109/ISSCC.2010.5434017.

39 Abdinia, S., Benwadih, M., Coppard, R. et al. (2013). A 4b ADC manufactured in a fully-printed organic complementary technology including resistors. In: *2013 IEEE International Solid-State Circuits Conference*. https://www.doi.org/10.1109/ISSCC.2013.6487657.

40 Marien, H., Steyaert, M., Van Veenendaal, E., and Heremans, P. (2011). A fully integrated ΔΣ ADC in organic thin-film transistor technology on flexible plastic foil. *IEEE Journal Solid-State Circuits* https://www.doi.org/10.1109/JSSC.2010.2073230.

41 Dey, A. and Allee, D. (2012). Amorphous silicon 5 bit flash analog to digital converter. In: *2012 Custom Integrated Circuits Conference*. https://www.doi.org/10.1109/CICC.2012.6330647.

42 Raiteri, D., van Lieshout, P., van Roermund, A., and Cantatore, E. (2013). An organic VCO-based ADC for quasi-static signals achieving 1LSB INL at 6b resolution. In: *2013 International Solid-State Circuits Conference*. https://www.doi.org/10.1109/ISSCC.2013.6487658.

43 Yang, H., Fluxman, S., Reita, C., and Migliorato, P. (1994). Design, measurement and analysis of CMOS polysilicon TFT operational amplifiers. *IEEE Journal of Solid-State Circuits* https://www.doi.org/10.1109/4.293120.

44 Kim, S., Baytok, S., and Roy, K. (2011). Scaled LTPS TFTs for low-cost low-power applications. In: *2011 12th International Symposium on Quality Electronic Design*. https://www.doi.org/10.1109/ISQED.2011.5770812.

45 Bisht, R. and Mazhari, B. (2006). Impact of kink effect on performance of poly-silicon based TFT differential amplifiers. In: *Proceedings of ASID*, October 8–12. New Delhi, India.

46 Fung, T., Baek, G., and Kanicki, J. (2010). Low frequency noise in long channel amorphous In–Ga–Zn–O thin film transistors. *Journal of Applied Physics* https://www.doi.org/10.1063/1.3490193.

47 Ramírez, H., Chien, Y., Hoffman, L. et al. (2019). Low-frequency noise in InGaZnO thin-film transistors with Al_2O_3 gate dielectric. In: *25th International Conference on Noise and Fluctuations*. https://www.doi.org/10.5075/epfl-ICLAB-ICNF-269292.

Part VI

Display and Memory Applications

18

Oxide TFT Technology for Printed Electronics

Toshiaki Arai

JOLED, Kanagawa, Japan

An oxide thin-film transistor (TFT) is fabricated by a simple process flow and comparatively low-temperature process (around 300 °C). Although the manufacturing infrastructure of oxide TFTs resembles that of amorphous silicon (a-Si)-TFTs, which are used for backplanes in liquid crystal displays (LCD), their electric performance such as high mobility and low off-current is superior to that of a-Si-TFTs. The oxide TFT is used for applications of active-matrix elements for displays and sensors, and for switching elements for integrated circuits (IC), memory, and so on. Contrary to its superior electrical characteristics, its deterioration by synthetic stresses was a serious issue for commercialization [1, 2]; however, effective technologies for high reliability began to be reported in the early 2010s.

In this chapter, a highly reliable oxide-TFT technology securing reliability not inferior to that of conventional low-temperature polysilicon (LTPS) TFTs is introduced [3].

18.1 OLEDs

18.1.1 OLED Displays

Information displays have been widely used for electric devices since the advent of the cathode ray tube (CRT). In the early 1990s, the trend of information displays has dramatically changed from the CRT to the flat-panel display (FPD). Several display modes have been proposed for FPDs, and the active-matrix liquid crystal display (AM-LCD) led the change owing to its image quality and its simple manufacturing methodology. In 2000, the LCD has become the majority in the information display market; however, its image quality, such as color reproduction, contrast, viewing angle, and response time, still has room for improvement.

In 1987, C. W. Tang reported an organic light-emitting diode (OLED) device with high luminous efficiency (1.5 lm/W) and brightness (>1000 cd/m^2) [4]. After this report, many researchers studied OLED devices for use in displays [5–8]. An OLED device has a high color gamut, high contrast ratio, wide viewing angle, and quick response. These characteristics are suitable for application in information displays, and also to overcome the drawbacks of LCDs. Therefore, OLED displays are expected to be the future of information displays.

Figure 18.1 shows the structures of an OLED display and a LCD. A LCD consists of a liquid crystal cell with a TFT substrate, a color filter (CF) substrate, some optical films including two polarizers, and a backlight module. A liquid crystal cell just acts as an optical switch; therefore, a backlight module as a light source and some optical films to realize uniform luminance are necessary. In contrast, an OLED is a self-emitting device and can have a very simple structure. This simple structure contributes to lightweight devices or flexible devices by changing glass substrates to flexible substrates.

Amorphous Oxide Semiconductors: IGZO and Related Materials for Display and Memory, First Edition.
Edited by Hideo Hosono and Hideya Kumomi.

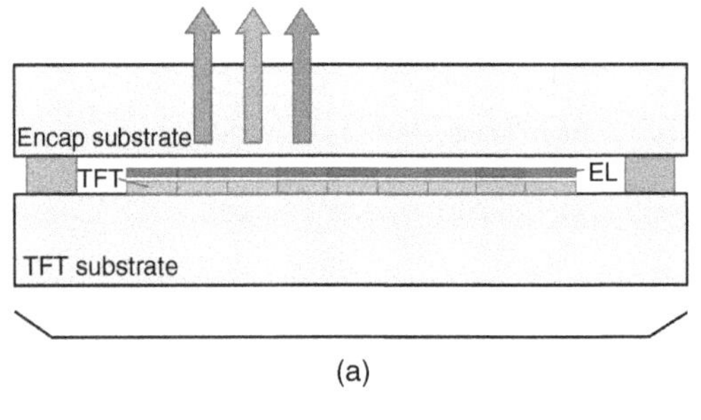

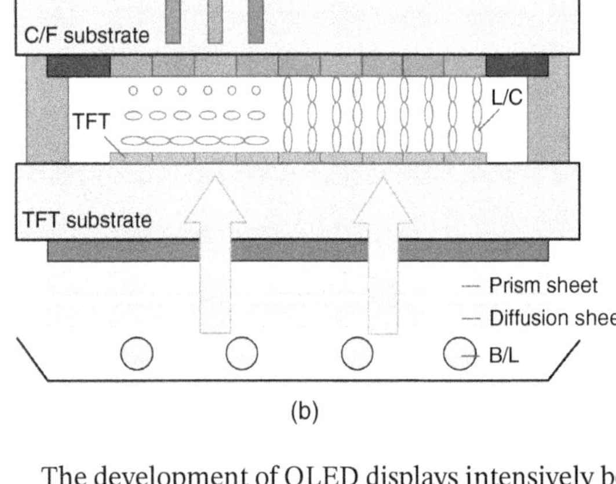

Figure 18.1 Cross-sectional structures of (a) an OLED display and (b) a LCD.

The development of OLED displays intensively began in the 1990s, and small active-matrix OLED (AM-OLED) displays have been commercialized in mobile phones and watches since the beginning of the 2000s. Medium-sized to large AM-OLED displays, including their use in televisions (TVs), also started to be commercialized in the second half of the 2000s.

In recent years, the image quality of OLED displays has improved dramatically, and has come to be preferred by customers as compared with that of LCD displays. The interest of customers has then moved to cost and form factor. For the cost improvement, low-cost materials, low-cost manufacturing processes and equipment, and low-cost and simple device structures have been the focus of study. As for the form factor, low-temperature manufacturing processes have been researched for applying flexible substrates instead of glass substrates.

18.1.2 Organic Light-Emitting Diodes

Electroluminescence (EL) was first reported by A. Bernanose at the Nancy-Université in France in 1953. He applied high alternating voltages to materials such as acridine orange, deposited on cellulose film. Light emission was very dim, and it degraded within an extremely short period of time. In 1987, C. W. Tang reported epoch-making technology and achieved both high luminance and long-term emission [4]. The organic EL materials were formed between the anode and the cathode, and luminescent light was observed through the transparent anode (Figure 18.2). This OLED device achieved both high luminance and long-term emission. That result was mainly achieved by

1. A functionally separated structure,
2. a stable and low work function cathode, and
3. particle-less thin films formed by a vapor deposition method.

After their report, a lot of EL device structures and materials were reported, and the luminance, efficiency, and reliability have been considerably improved; however, the basic EL device structure was not changed. Because the

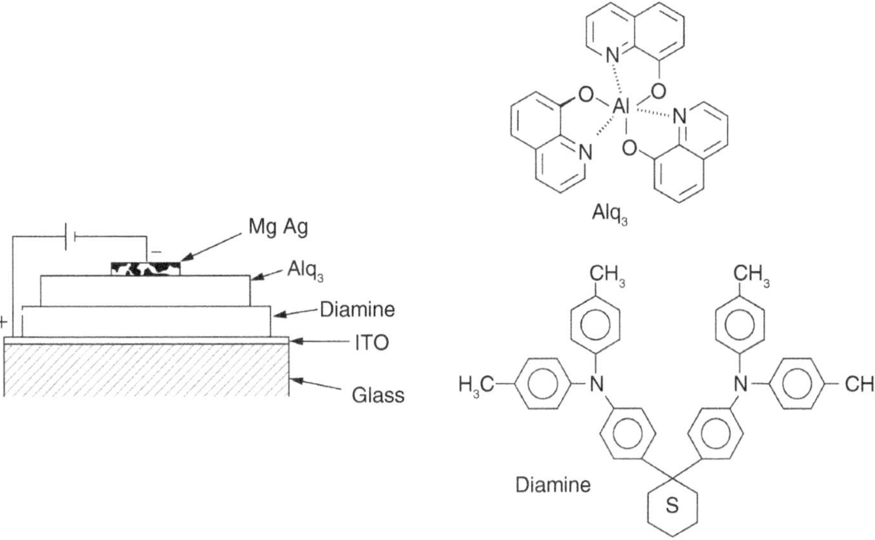

Figure 18.2 OLED device structure of the first report.

EL device can emit light at a wide color range with the three colors red, green, and blue (RGB), OLED devices have been considered for applications in information displays.

For display applications, the OLED device is fabricated on a TFT array. After forming a planarization layer and the bank pattern that separates each pixel, an OLED device is formed on the TFT array. Red (R), green (G), and blue (B) emitting devices are formed side-by-side (SBS), and color image is displayed by controlling the current for each pixel.

Figure 18.3 shows two kinds of device structures for the AM-OLED display with different emitting directions. A bottom emission device (Figure 18.3a) looks easy to fabricate because all of the devices are covered by the cathode metal, which acts as passivation. However, the emitting area needs to be small, because the OLED device is formed beside TFT devices. On the contrary, a top emission device (Figure 18.3b) has difficulty in capping the device with an organic encapsulating layer and a thin transparent cathode. However, the emitting area can be large and further storage capacitances can be freely designed, because the OLED device is formed on the TFT device. The advantages of the top emission device are, for example:

- Wide emission area
- High luminescence
- Long life due to low operating bias
- Free TFT array design
- High compensation ability due to enough circuit space
- High manufacturing yield by enlarging the space between TFT devices (e.g., TFTs, capacitors, and bass lines)
- Long life due to low operating bias (large channel width)
- Uniform image due to uniform TFT current (large channel length)

As a matter of course, the top emission structure is preferable in AM-OLED displays.

18.1.3 Printed OLEDs

For the AM-OLED display, the OLED device is fabricated on the TFT array. Table 18.1 summarizes the differences among three manufacturing methods for OLEDs [9]. A SBS evaporation technology is used for smartphones, and a white OLED + CF technology is used for TVs.

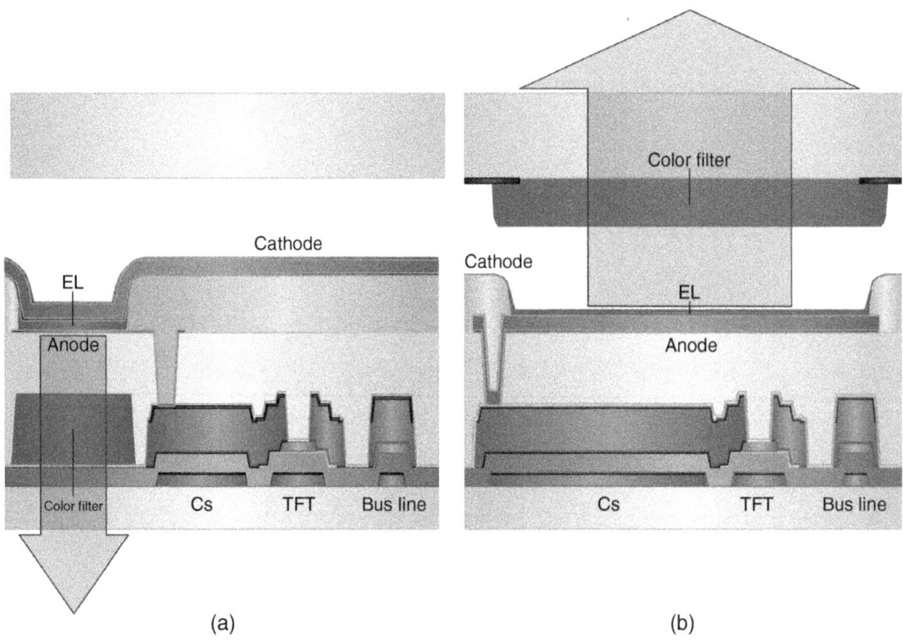

Figure 18.3 Cross-sectional OLED device structures. (a) Bottom emission type. (b) Top emission type. (The arrow indicates the light-emitting direction.)

Table 18.1 OLED manufacturing method.

OLED technology	**Evaporation**		**Printing**
	RGB SBS	**White OLED + CF**	**RGB SBS**
Overview	Fine metal mask (FMM); Evaporation source	Metal mask	Printing head; Substrate
Structure		C/F	Bank; Cathode; EML; Anode; TFT; Substrate
Quality	Good	Fair	Fair → Good
Resolution	Excellent	Excellent	Fair → Good (~350 ppi)
Scalability	Bad	Good	Excellent
Advantages		-Not fine metal mask	-Atmospheric condition -High material usability -Metal maskless
Issues	-Large FMM alignment error (Size limitation/low yield)	-CF necessity (Low light-utilization efficiency) -Long process flow (Massive investment, large FAB)	-Material performance (Low luminous efficiency and short life)

The RGB SBS evaporation technology is suitable for small, high-resolution mobile phone applications. The RGB materials are evaporated and deposited on each pixel of the TFT array through a fine metal mask (FMM). However, the FMM fabrication and precise alignment with the TFT backplane are difficult, especially in the case of a large substrate. Thin metal with a small thermal expansion rate such as "invar" is used as a mask, but it is difficult to make it precisely because of the thermal extension of the metal. Therefore, several narrow-width FMMs are attached to a frame while giving tension to the FMMs. In addition, the FMM is reused in several substrates, so that many particles are generated by repeating evaporation processes several times and are transferred to the FMM, thus causing defects. According to the limitation of the width of FMMs and the defect formation frequency, this technique is used for small-panel manufacturing.

The white OLED + CF technology has been developed to escape the issues posed by FMMs, and it is suitable for application in large TVs. While the issues with the FMM technique are wiped out, large-panel manufacturing is enabled. However, other issues arise. One is long manufacturing steps for white emission. Many process steps are needed for a highly stacked device structure, and a huge production line is needed for manufacturing. Another issue is low light utilization efficiency. As two-thirds of the light is absorbed by the CF, the power consumption becomes large.

The RGB printing method is advantageous in scalable manufacturing with a large substrate under atmospheric conditions, and in highly efficient material utilization without FMMs. Because the printing materials are formed only in the light-emitting area, the material utilization efficiency is predominantly higher than that of conventional evaporation methods. The operation ratio of the printing system is very high, because the RGB printing process of different panel designs can be executed only by changing recipes. Currently, there is room for improving the luminous efficiency and lifetime of the printed OLED. However, its performance nowadays has come to a stage where it is very close to that of evaporation technology.

From the beginning of the 2000s, several high-quality prototypes of printed OLED display have been reported. However, their quality was not high enough for commercialization. For mass production, an inkjet-type printing method has been focused in a non-contact type of printing to avoid particle-related defects. More than ten years have been required for improving the printing quality and the performance of the printing materials.

In 2017, JOLED finally started the production of a printed OLED display (Figure 18.4 and Table 18.2). According to this printing technology, it has become easy to manufacture medium-sized to large OLED displays.

Because the OLED is a self-emitting device, the OLED display does not need a backlight unit and has no viewing-angle issues, unlike LCDs. Therefore, an AM-OLED display is suitable for flexible, foldable displays. In the field of mobile phones, the flexible display has already been put to practical use, and its market is expected to

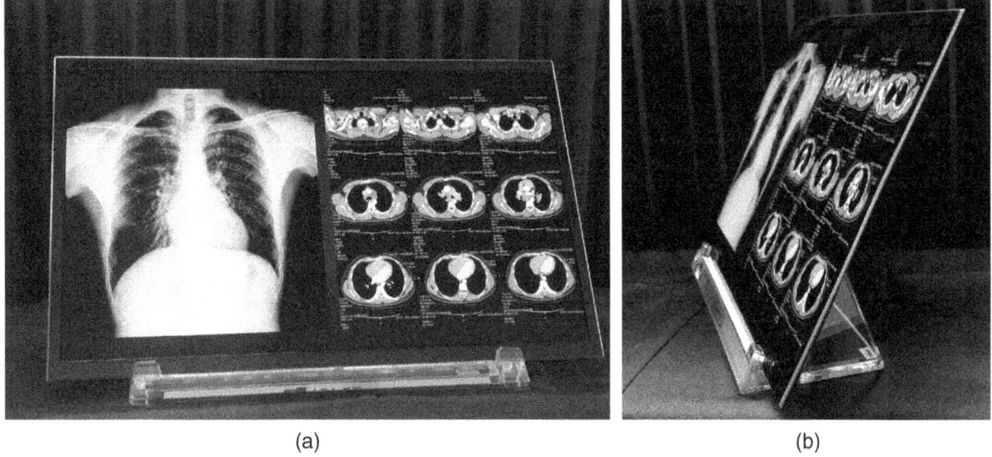

(a) (b)

Figure 18.4 World's first printed OLED display product. *Source*: JOLED, https://www.j-oled.com/eng/.

Table 18.2 Specifications of the world's first printed OLED display.

Panel size	21.6-inch (297 mm) diagonal
Format	4K
Number of pixels	3840 RGB × 2160
Brightness	All white: 140 cd/m^2
	Peak: >350 cd/m^2
Contrast ratio	>1,000,000:1 (Dark)
Number of colors	10-bit RGB
Color saturation	>125% (sRGB)

gradually expand. New applications to the field of medium-sized to large displays are expected, too. Rigid-type large OLED displays have been manufactured only through white OLED + CF technology; however, the printed OLED technology is suitable for large flexible-display manufacturing. Flexible displays are going to change how we use information displays. It is expected that various medium-sized to large applications will be developed in the future by using the printed OLED technology (Figure 18.5).

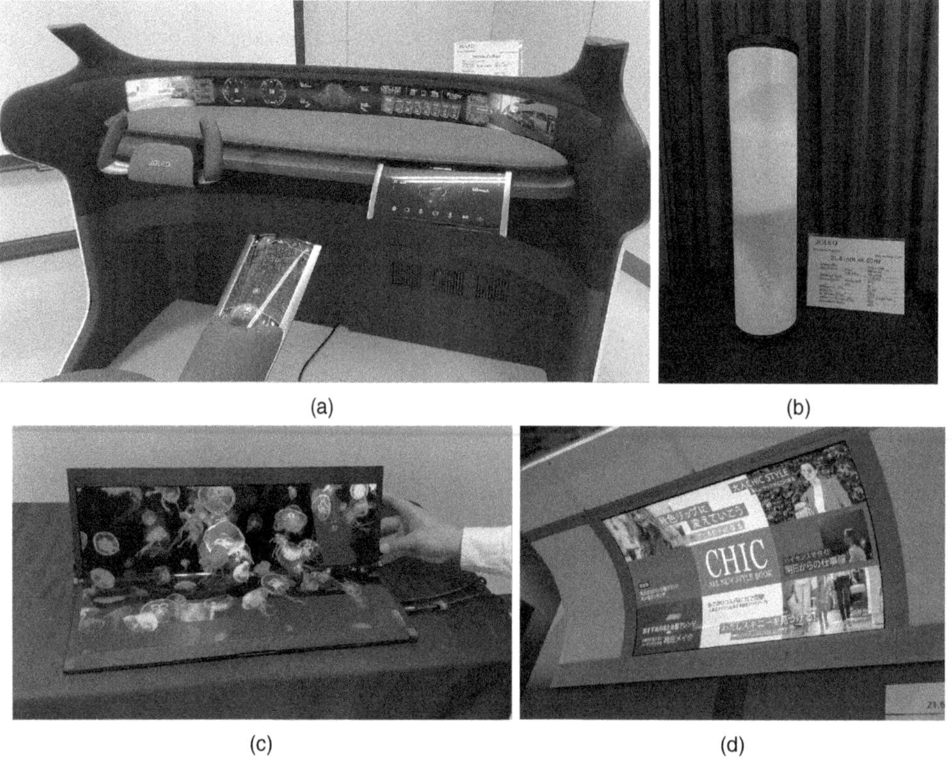

Figure 18.5 Medium-sized flexible OLED display prototypes. (a) 12.3-inch, 14-inch FHD vehicle cockpit; (b) 21.6-inch 4K column display; (c) 21.6-inch 4K foldable screen; and (d) 21.6-inch 4K curved advertisement screen. *Source*: JOLED, https://www.j-oled.com/eng/.

18.2 TFTs for OLED Driving

18.2.1 TFT Candidates

To drive an AM-OLED display, a TFT is used as a switching device. In the case of AM-LCDs, amorphous silicon (*a*-Si) is widely used as a channel material of TFTs because of its high productivity. However, for the AM-OLED display, *a*-Si-TFTs have disadvantages in their field-effect mobility and stability, and they are hard to manipulate. Because the OLED is a current-driven device, high current flow is required for the TFT to realize high luminance. In the case of *a*-Si-TFTs, a high bias is applied to the TFTs to realize high current flow, and the threshold voltage of the TFTs easily shifts due to the bias temperature stress, which is continuously applied during the driving period. Therefore, other TFTs with higher mobility and more stable channel materials have been studied (Table 18.3) [10]. LTPS is a strong candidate, especially in the stability aspect; however, its low productivity owing to its long manufacturing steps is a serious issue. The manufacturing of LTPS TFTs requires many manufacturing apparatuses and materials, so the initial investment and manufacturing costs are much higher than for other TFTs. Manufacturing also requires special apparatuses for the excimer laser annealing (ELA), ion doping, and high-temperature annealing processes in the Si crystallization. These apparatuses have demerits because they not only are expensive, but also have restrictions in substrate sizes.

18.2.2 Pixel Circuits

To drive the AM-OLED display, several TFTs are required for the pixel circuit. Figure 18.6 shows examples of compensation circuits. In 1998–1999, Dawson's team reported a voltage-programmed compensation pixel circuit [11]. Such a circuit is composed of four transistors and two capacitors (4Tr2C), and it compensates the threshold voltage variation of the TFTs in a panel. After the report, many compensation circuits were reported to compensate the threshold voltage variation of TFTs [12–15].

The compensation circuit is more complicated than the pixel circuit of the LCD. From the manufacturing point of view, a simpler circuit is preferable to realize a high-resolution display with high manufacturing yield; however, compensating ability, required bias, driving speed, and so on are often taken into consideration as the decisive factors.

I_{OLED} is given by the following equation [12]:

$$\begin{aligned} I_{OLED} &= \frac{1}{2}k\mu(\text{Vgs-Vth})^2 = \frac{1}{2}k\mu(\text{Vsig-Vofs-}\Delta\text{Vs}(\mu)\text{-Vth})^2 \\ &= \frac{1}{2}k\mu(\text{Vsig-Vofs-}\Delta\text{Vs}(\mu))^2 \\ &\left(\text{k} = \frac{\text{W}}{\text{L}}\text{Cox}\right) \end{aligned} \tag{18.1}$$

Table 18.3 TFT candidates for AM-OLED displays.

Channel material	*a*-Si	μc-Si	LTPS	Oxide
Field effect mobility (cm²/Vs)	0.5	2~5	~150	10~50
Stability	Low	Medium~high	High	Low~high
Uniformity	Good	Good	Fair	Good
Productivity	High	Medium	Low	High
Plate size	Gen. 10 (3.1 × 2.8m)	Gen. 10	Gen. 6	Gen. 10
Mask number (for TFTs only)	5	6	7~10	4~5

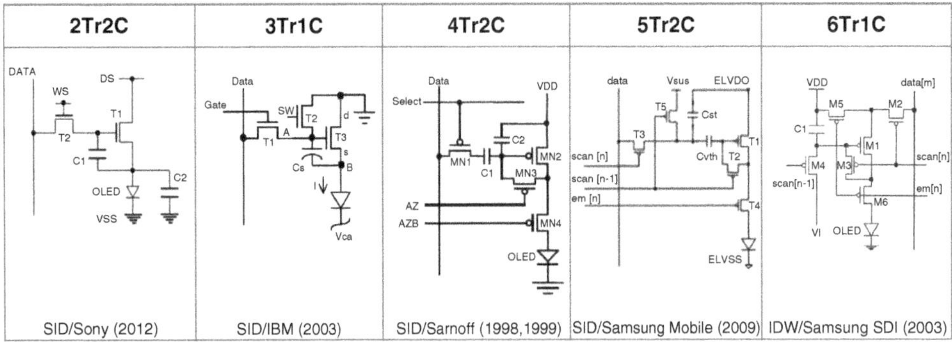

Figure 18.6 Schematics of the compensation circuits where I_{OLED}: current through the OLED; μ: field-effect mobility; Vsig: signal bias; Vofs: offset bias; ΔVs(μ): bias shift according to the mobility; W: channel width of the TFT; L: channel length of the TFT; and Cox: capacitance of the gate insulator.

Equation (18.1) implies that the variations of both Vth and the mobility of T1 can be canceled out. Consequently, the compensation circuit enables realization of uniform pixel-to-pixel luminance.

This operation is to link the gate potential Vg of the drive transistor (T1 in the 2Tr2C circuit in Figure 18.6) with the source potential Vs according to the charge-up operation by the write transistor (T2 in the 2Tr2C circuit in Figure 18.6) into the storage capacitor (C1 in the 2Tr2C circuit in Figure 18.6); thus, it is called a bootstrap operation. The ability of the bootstrap operation is called a bootstrap gain (Gb), and it is expressed by the following equation [16]:

$$\mathrm{Gb} = \mathrm{Cs}/(\mathrm{Cs} + \mathrm{Cw} + \mathrm{Cp}) \tag{18.2}$$

Here, Cs is the capacitance of the storage capacitor C1, Cw is the capacitor between the gate and source nodes of the write scan transistor, and Cp is the parasitic capacitor between the gate node of the drive transistor and the other structurally nearby nodes. The bootstrap gain Gb closer to 1 means a higher gain Gb. That is, the gain Gb closer to 1 means a greater ability to correct the drive current of the drive transistor and the emitting current of the OLED device I_{OLED}.

To realize higher Gb, it looks easy to prepare large Cs; however, a larger area in a pixel and large electron charge are required for operation. It is contradictory for high-resolution or low-power-consumption displays. Therefore, the reduction and uniformization of the parasitic capacitances (Cw and Cp) are effective to realize effective compensation driving.

18.2.3 Oxide TFTs

Transistors for the AM-FPD have been completely dominated by silicon-based semiconductor technology. A microcrystalline silicon TFT is a strong candidate to realize both electrical uniformity for large substrates and stability against bias stress; however, there is a trade-off relationship between the electrical uniformity and mobility. When necessary electrical uniformity is secured for the TFT, there is a problem that the mobility is suppressed by 3~5 $\mathrm{cm}^2/\mathrm{Vs}$. On the other hand, high-resolution or high-frame-rate images increased from the late 2000s, and higher mobility came to be expected for the TFTs of AM-FPDs. In 2004, Hosono's group reported an amorphous metal oxide–based semiconductor [17]. The proposed amorphous oxide semiconductor from the In-Ga-Zn-O system (amorphous IGZO, or *a*-IGZO) for the active channel has high mobility exceeding 10 $\mathrm{cm}^2/\mathrm{Vs}$, and it is formed by a conventional direct current (DC)-sputtering apparatus with a low-temperature deposition condition. That means an oxide TFT can just follow the infrastructure of the amorphous silicon TFT production, and has the potential to realize low-cost manufacturing with large substrates.

The a-IGZO TFT has high mobility exceeding 10 cm^2/Vs and an excellent on/off ratio exceeding 10^8. It has a strong potential to be used for high-resolution, high-frame-rate AM-OLED displays, which require higher mobility than amorphous Si-TFTs and much higher uniformity than LTPS-TFTs. In order to apply an oxide TFT to AM-OLED displays, the most crucial point would be its stability.

Because the OLED is a current-driven device, both the TFT and OLED generate heat with current flow and light emission. Actually, the TFT is placed under a severe bias–temperature stress (BTS) environment. In the case of amorphous Si-TFTs, the device instability has been attributed to two different mechanisms: charge trapping in the gate dielectric and the creation of a metastable dangling bond in the amorphous Si [18–20]. The former charge-trapping mechanism is known to be also applicable to oxide TFTs [21–24]; therefore, the formation of a gate insulator and the treatment of an interface between the gate insulator and oxide semiconductor are important for TFT stability.

Much severe instability due to the adsorption/desorption dynamics of water, oxygen, hydrogen, and so on has been reported [25–28]. In the case of OLED displays, various materials for the OLED device are formed on the TFTs and sealed; therefore, the TFTs are affected by those materials or the environment in which they are formed. Furthermore, the photo illumination also affects the instability. Despite the wide bandgap property, the oxide semiconductor absorbs shortwave-length light, resulting in leakage [29]. The effect of moisture on photon-enhanced negative BTS instability is also reported [30].

To avoid those instabilities owing to various stresses (bias, temperature, humidity, photo irradiation, etc.), annealing or passivating technologies have been proposed [10, 31–35].

18.2.3.1 Bottom-Gate TFTs

As shown in Table 18.3, the oxide TFT attracted attention because of its high mobility and high uniformity; however, there have been several reliability issues, which are those mentioned in Section 8.2.3. Improvement of reliability is a key for wide acceptance.

TFT Structure It is well known that the properties of oxide semiconductors are highly dependent on the oxygen content, since oxygen vacancies provide free carriers. Even in the case of In-Ga-Zn-O material, which is said to be a stable oxide semiconductor, the content of oxygen controls its film properties in applications from conductor to insulator. The oxygen in the oxide film may be absorbed or desorbed even after film deposition by the disclosed conditions such as heat, plasma, or a vacuum. Therefore, a bottom-gate TFT structure is easy to fabricate so as to be as little affected as possible by the manufacturing process after the a-IGZO film deposition (Figure 18.7) [10]. An etching stopper–type TFT has advantages in terms of stability, because the etching stopper acts as not only an etching stopper in the source/drain patterning but also a passivating film in the manufacturing process, an

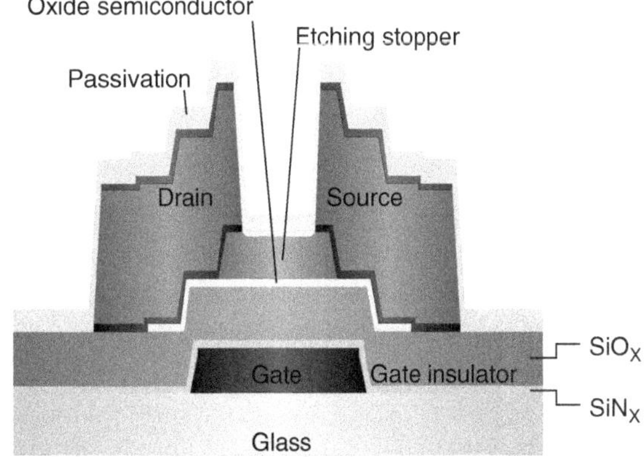

Figure 18.7 Cross-sectional view of a TFT.

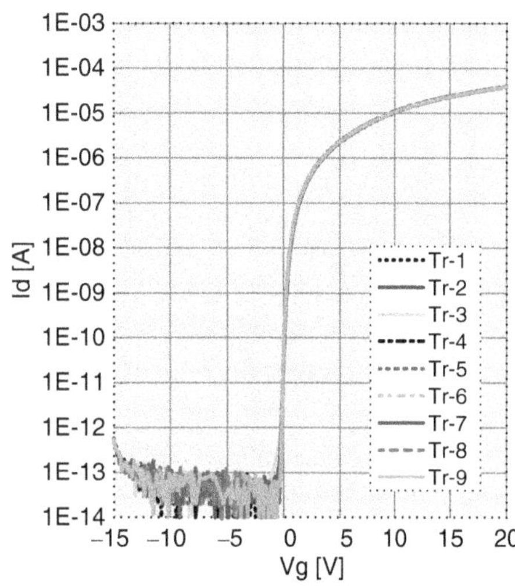

Figure 18.8 Transfer curves of the bottom-gate TFTs.

oxygen stabilizer for the oxide semiconductor, and a precise channel-length promoter. For the gate insulator and an etching stopper, conventional silicon oxide (SiO_x) films were deposited by chemical vapor deposition (CVD), and a very thin silicon nitride (SiN_x) film was inserted on the bottom of the gate insulator to avoid contamination from lower layers.

Figure 18.8 shows the transfer curves of the nine TFTs located in a Gen. 1 substrate (300 mm × 350 mm). The mobility was 11.5 cm^2/Vs, the S factor was 0.27 V/decade, and the Vth was 0.3 V. TFTs have very high uniformity over both short range and wide range. The wide-range uniformity of on-current at Vg = 5 V was smaller than 5%.

Passivation Material for High Reliability The passivating films for oxide semiconductor TFTs are key technologies to improving reliability. A conventional SiN_x passivation deposited by CVD easily degrades TFT performance. It is caused by its poor passivation ability in the face of oxygen or water penetration, and electron donors generated by the excess hydrogen in the SiN_x film or SiH_4 gas during film formation. Some researchers have reported that the Al_2O_3 (alumina) film deposited by atomic layered deposition (ALD) or SiO_x film deposited by radiofrequency (RF) sputter shows good passivation properties [34, 35]. However, for large FPD manufacturing, ALD and RF sputtering have several limitations for application in industry due to their difficulty of deposition over a large substrate. On the other hand, DC sputtering is widely adopted for FPD manufacturing; therefore, a DC- or AC-sputtered Al_2O_3 is chosen as the passivation. Al_2O_3 is widely used as a surface protection material or a gas barrier film, and it is expected that it functions as the passivation material to water and/or hydrogen diffusions, which change the properties of oxide semiconductors.

Figure 18.9 shows the Vth shift after BTS. The bias, temperature, and period of BTS are Vg = 0∼15 V, Vd = 0∼15 V; 50 °C; and 10,000 sec, respectively [10]. The conventional SiN_x passivation with a SiO_x underlayer showed large Vth shifts depending on the stress condition. These large shifts are supposed to originate from the hydrogen diffusion from the SiN_x film. SiO_x passivation showed better reliability than the conventional CVD passivation, but had negative Vth shifts after BTS in the range of Vd larger than Vg. On the other hand, the Al_2O_3 passivation showed superior reliability in all bias ranges, and the Vth shifts after BTS were only 0.2 V at the maximum. The Al_2O_3 is known to have negative fixed charge; however, it does not seem to adversely affect the TFT characteristics and the reliability. This might have been caused by the TFT structure using the Al_2O_3 film as a passivation without having direct contact with the channel layer. When current flow through the driving TFT was set to 2.2 μA with the TFT size of W (width) / L (length) = 20/8 μm, the extrapolated shift is smaller than 1 V even after 100,000 hours, if the square of the shift is in proportion to the square of the stress time (Figure 18.10a). Similarly, when the gate bias and drain bias were set to 15 V, the extrapolated shift is almost 1 V even after 100,000 hours, if the

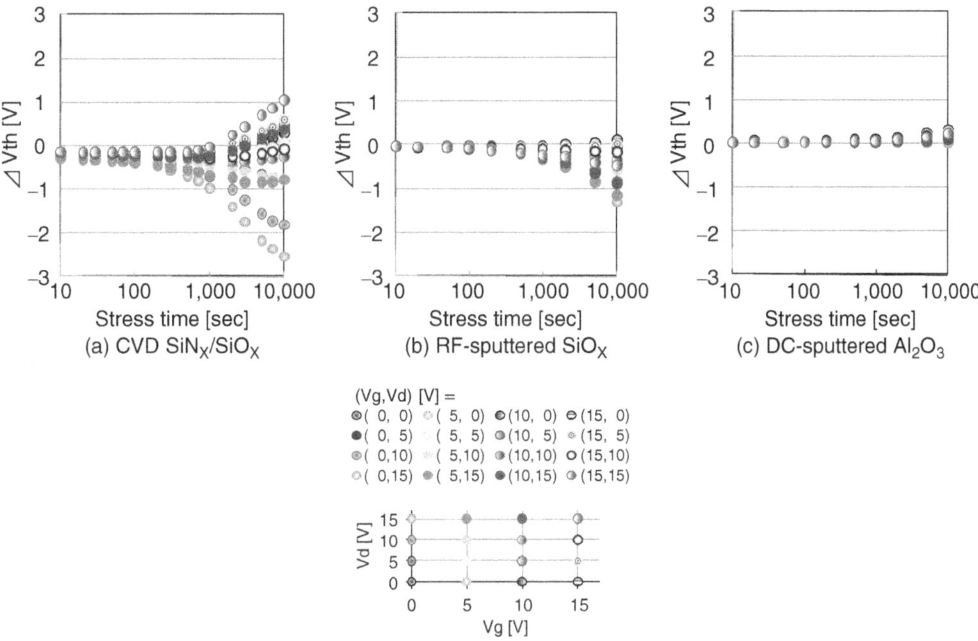

Figure 18.9 Vth shifts for various passivation after BTS. (a) CVD SiN_x/SiO_x; (b) RF-sputtered SiO_x; and (c) DC-sputtered Al_2O_3. (Vg and Vd were set from 0 V to 15 V at 5 V steps, and the stress temperature was 50 °C.)

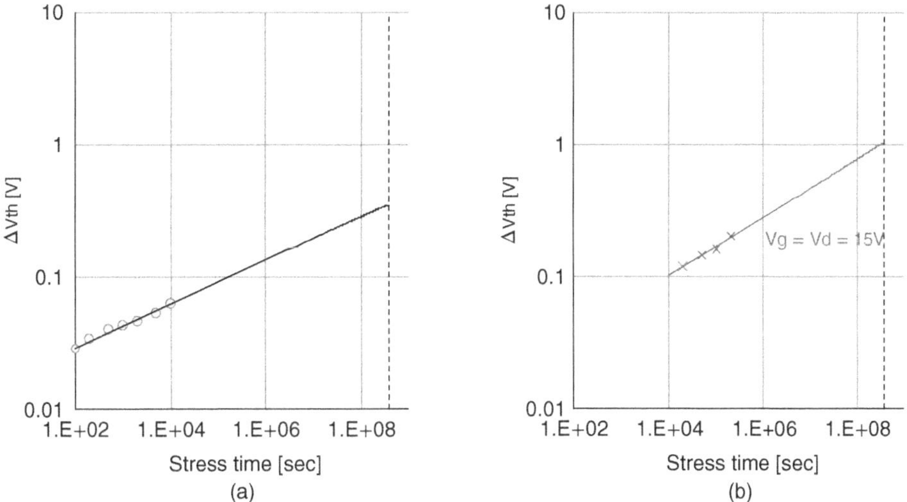

Figure 18.10 Estimated Vth shift after 10 years. (a) Id = 2.2 μA setting case; and (b) Vg = Vd=15 V setting case. (A solid line means the extrapolated Vth shift calculated from data until 10,000 sec.)

square of the shift is in proportion to the square of the stress time (Figure 18.10b). These values are small enough for driving AM-OLED displays, and they indicate a panel lifetime of over 10 years.

11.7-Inch Diagonal AM-OLED Prototype Figure 18.11 shows an 11.7-inch diagonal qHD *a*-IGZO-TFT-driven AM-OLED prototype, and Table 18.4 shows its specifications. Even in DC-mode driving, the brightness uniformity was only 10% in wide distribution.

Figure 18.11 11.7-inch qHD *a*-IGZO TFT–driven AM-OLED display prototype.

Table 18.4 Specifications of the 11.7-inch *a*-IGZO-TFT-driven AM-OLED display prototype.

Panel size	11.7-inch (297 mm) diagonal
Format	qHD
Number of pixels	960 RGB × 540
Brightness	All white: 200 cd/m^2
	Peak: >600 cd/m^2
Contrast ratio	>1,000,000:1 (Dark)
Number of colors	10-bit RGB
Color saturation	>100% (NTSC)

18.2.3.2 Top-Gate TFTs

A self-aligned top-gate TFT structure is a strong candidate to realize small and uniform parasitic capacitance. For the top-gate TFT structure, a coplanar-type source/drain formation method is required. In 2009, Sato et al. reported that the hydrogen diffusion into the IGZO layer through the SiO_x island gave the coplanar-type source/drain (Figure 18.12) [28]. By applying this technique to the top-gate TFT structure, it is possible to form a top-gate TFT. In such cases, impinged hydrogen forms an *n*-channel region. However, this TFT is not so stable

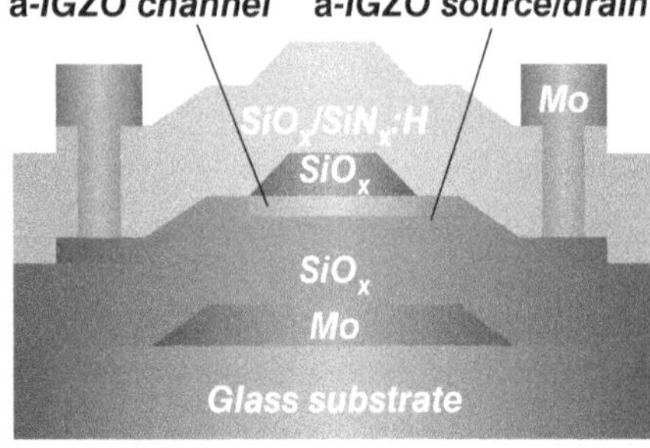

Figure 18.12 Cross-sectional schematic of an *a*-IGZO TFT with a coplanar homojunction structure.

under thermal stress, because excess hydrogen in the film is not stable under the thermal stress and degrades the TFT characteristics.

TFT Structure and Source/Drain Formation In 2011, Morosawa et al. reported an aluminum-reaction method to form a coplanar homojunction [31]. The process flow of a self-aligned top-gate oxide TFT is shown in Figure 18.13. At first, an oxide semiconductor as an active layer was deposited by DC sputtering and was patterned by wet etching. A SiO_2 layer as gate insulator was deposited by plasma-enhanced CVD on the active layer. Stacked layers of Al and Ti as a gate electrode were sequentially sputtered, and gate electrode and gate insulator were continuously dry-etched using a gate pattern as shown in Figure 18.13a. An Al layer as reactive metal was sputtered to make source/drain regions of the active layer, as can be seen in Figure 18.13b. Then, an annealing process was executed at 200 °C in the presence of oxygen atmosphere. As illustrated in Figure 18.13c, source/drain regions and a protection layer of aluminum oxide (Al_2O_3) were fabricated at the same time by thermal diffusion of aluminum into oxide semiconductor and oxidation of aluminum, respectively. After that, an organic insulator layer was coated and patterned. Then, the Al_2O_3 layer was removed to contact with the source/drain electrode by dry etching. Finally, a stacked layer of molybdenum (Mo) and Al as the source/drain electrode was sputtered and then patterned by wet etching.

According to this process, the resistivity of the oxide semiconductor is reduced by the chemical reaction with the deposited metal, and the top-gate TFT was fabricated with high stability.

Table 18.5 shows typical TFT structures for the oxide TFT and those features. Three types of TFT structure have been proposed for the oxide semiconductor. An etching stopper (ES)-type bottom-gate TFT is the most popular

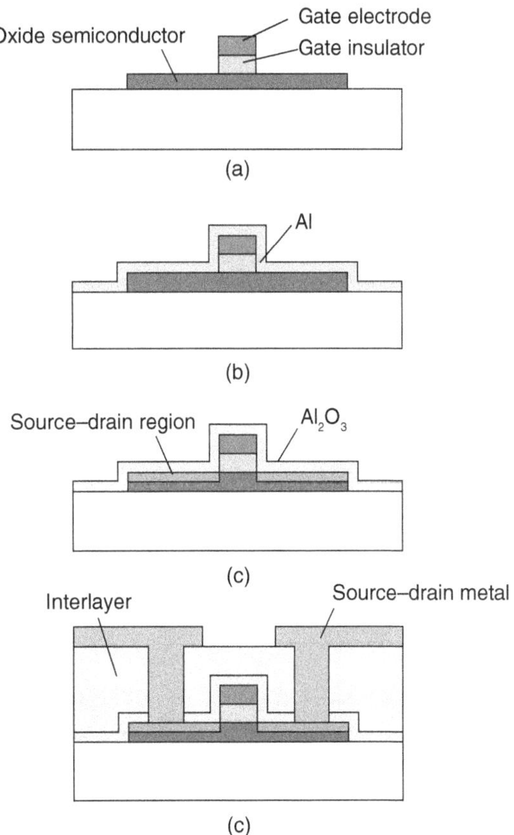

Figure 18.13 Process flow for the fabrication of a self-aligned top-gate oxide TFT.

Table 18.5 TFT structure comparison.

	ES-type Bottom-gate	CE-type Bottom-gate	Self-aligned Top-gate
TFT structure	Lc, Cp	Lc, Cp	Cp, Lc
Mask count	6	5	4
Channel length (Lc)	Long	Short	Short
Parasitic capacitance (Cp)	Large	Medium	Small

structure, because it is relatively easy to fabricate. However, many mask steps are needed in this structure. In addition, its channel length must be longer and parasitic capacitance must be larger than those of other structures due to large overlap regions. A self-aligned top-gate TFT is the ideal structure because it requires fewer mask steps to fabricate. Furthermore, a short channel length and small parasitic capacitance are also possible as a result of the self-aligned top-gate structure and the self-aligned structure, respectively.

TFT Characteristics and Reliability Figure 18.14 shows typical transfer curves of the self-aligned top-gate TFT with a channel width/length of 10/4 μm. It exhibits good transfer characteristics at a drain voltage of 10 V, such as a subthreshold swing of 0.22 V/decade, minimum off-current of 10 fA, threshold voltage (Vth) of −1.5 V, field-effect mobility of 9.8 cm^2/Vs, and on/off ratio on the order of 10^{10}.

The device structure affects reliability [36]. Figure 18.15 shows the cross-sectional view of the top emission AM-OLED device. According to this structure, only 5-Mask is needed for the backplane process of AM-OLED display manufacturing. In case of a top-gate TFT structure, there is no need to deposit a thick gate insulator because the step height of the oxide layer underneath is small. The high gate electric field due to a thin gate insulator provides high current flow; however, it tends to accelerate the threshold voltage shift in the positive bias temperature stress (PBTS) test. On the other hand, in the case of a top emission AM-OLED device, there are several organic layers around the channel region. These organic layers contain impurities such as hydrogen and water. In particular,

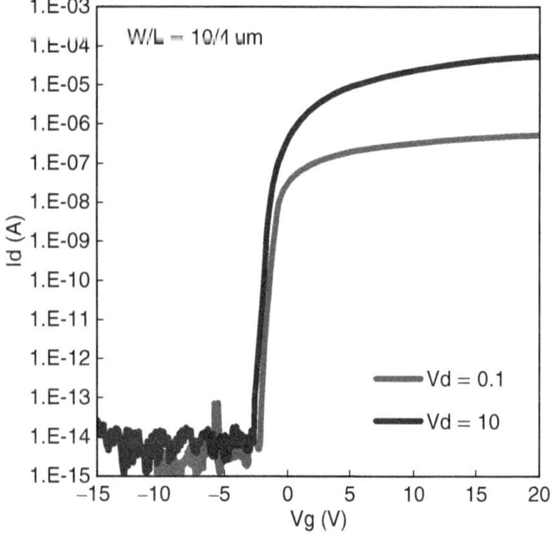

Figure 18.14 Transfer curves of the self-aligned top-gate TFTs.

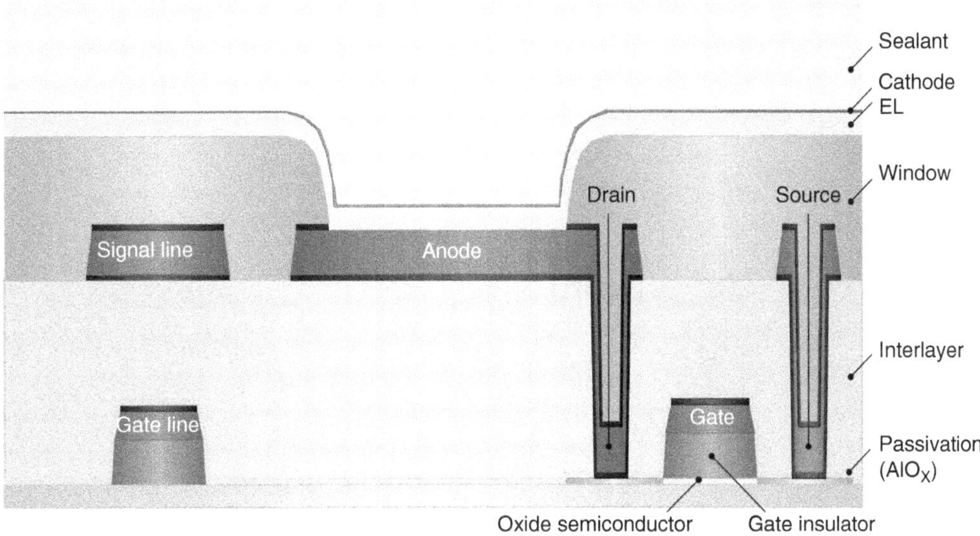

Figure 18.15 Cross-sectional view of a top emission AM-OLED device.

the upper layers tend to contain more impurities because of their lower baking temperature. They diffuse through the device to the channel region and cause the threshold voltage to negatively shift. These kinds of impurities often accelerate the threshold voltage shift in the negative bias temperature stress (NBTS) test. Figure 18.16 shows the differences in diffusion-origin degradation due to the TFT structure. In the case of a top-gate TFT, the gate electrode acts as a passivation. Moreover, there is a thin alumina passivation layer on the oxide semiconductor layer as a by-product of the source/drain formation process. An alumina has an excellent passivating property, even though it has thin thickness; therefore, this top-gate TFT tends to have excellent reliability against NBTS.

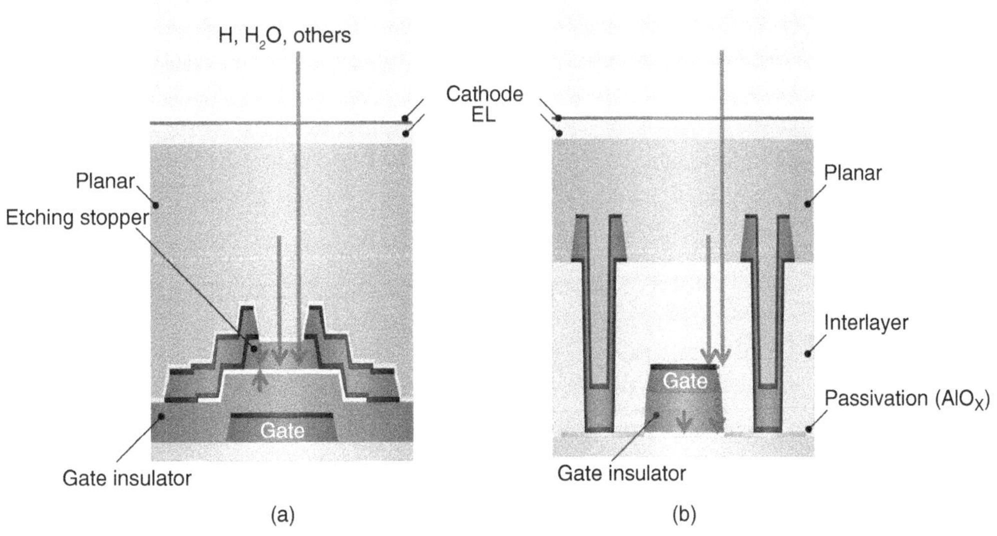

Figure 18.16 Difference in diffusion-originated degradation due to the TFT structure. (a) Etching stopper-type bottom-gate TFT; and (b) self-aligned top-gate TFT.

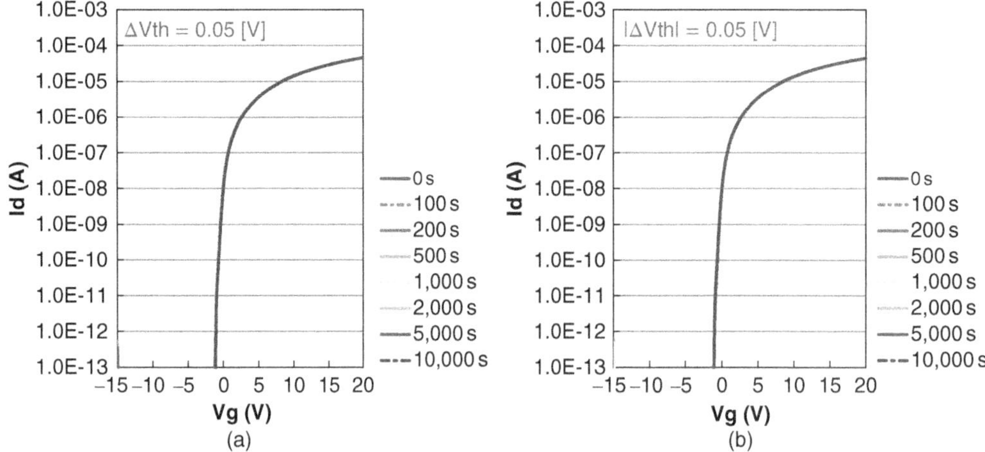

Figure 18.17 Results of PBTS and NBTS tests. (a) PBTS (Vg = +15 V; Vd = +15 V; T = 50 °C). (b) NBTS (Vg = −15 V; Vd = +15 V; T = 50 °C).

Figure 18.17 shows the results of PBTS and NBTS tests. The stress bias voltages were set to Vgs = ±15 V and Vds = 15 V. The stress temperature was set to 50 °C, and the stress period was 10,000 sec. The shifts of threshold voltage after PBTS and NBTS tests became 0.05 V and 0.05 V, respectively. These values are small enough for the OLED driving, and they indicate a panel lifetime of over 10 years.

9.9-Inch qHD AM-OLED Prototypes When the parasitic capacitances are large or non-uniform in this circuit, brightness uniformity deteriorates because the pixel circuit cannot work properly and the variation of the parasitic capacitances worsens the uniformity of the image. Figure 18.18 shows the brightness uniformity at a low gray level of the AM-OLED panel, depending on the TFT structures [31]. On the panel driven by ES-type bottom-gate TFTs, the brightness variation depending on the panel position was detected. On the other hand, the panel driven by self-aligned TFTs shows excellent uniformity. The brightness uniformity of the AM-OLED panel is degraded by the variation of the TFT characteristics of each pixel. As a result, in AM-OLED displays, compensation pixel circuits are used to achieve uniform brightness. On the panel using ES-type bottom-gate TFTs, pixel circuits cannot compensate the variation of the TFT characteristics due to the large parasitic capacitance. However, on the panel with self-aligned TFTs, uniform brightness can be obtained due to the small parasitic capacitances. To realize high frame-rate driving such as 240 Hz or 480 Hz, RC delay is a crucial issue that leads to brightness non uniformity due to signal delay. The above-proposed self-aligned TFTs can drive the AM-OLED panel with a much higher frame rate than the conventional ES-type bottom-gate TFTs. This is because the self-aligned TFTs have lower

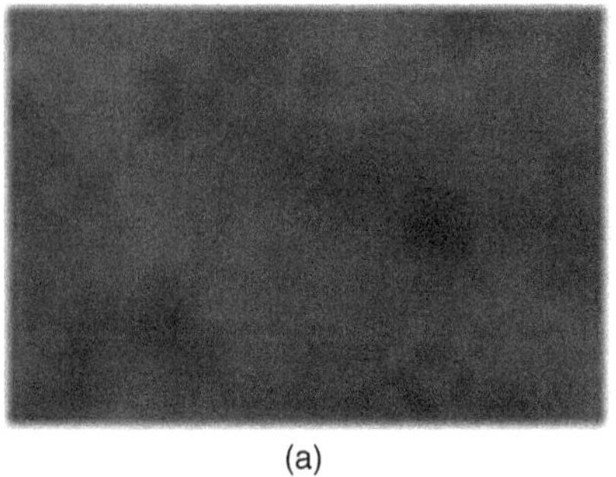

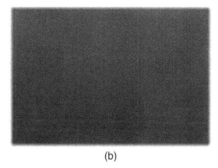

(a)

(b)

Figure 18.18 AM-OLED panels with different TFT structures at low gray levels. (a) AM-OLED panel by ES-type TFT. (b) AM-OLED panel by self-aligned TFT.

interlayer capacitance owing to a thicker organic insulator compared to the ES-type bottom-gate TFTs, as shown in Figure 18.18. In addition, because the organic insulator has small relative permittivity, the interlayer capacitance of the self-aligned TFT further decreases. To this point, the proposed TFT structure has an advantage over the conventional ES-type TFT structure in achieving a higher frame-rate operation.

Figure 18.19 shows cross-sectional views of an ES-type TFT and self-aligned TFT, and Figure 18.20 shows a 9.9-inch-diagonal AM-OLED prototype driven by 5-Mask-processed self-aligned top-gate TFTs using an Al-metal

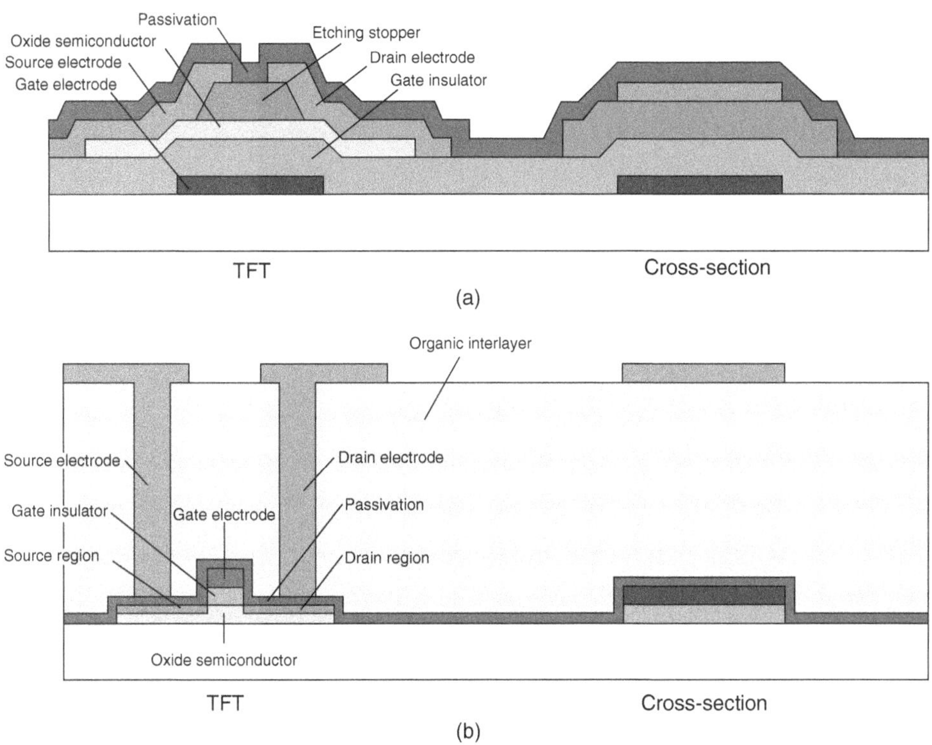

Figure 18.19 Cross-sectional views of an ES-type TFT and a self-aligned TFT.

Figure 18.20 A 9.9-inch qHD AM-OLED display prototype by 5-Mask processed self-aligned top-gate *a*-IGZO TFT.

Table 18.6 Specifications of the 9.9-inch AM-OLED display prototype.

Panel size	9.9-inch diagonal
Format	qHD
Number of pixels	960 (H) × 540 (V)
Resolution	111 ppi
Brightness	All white: 200 cd/m^2
	Peak: >600 cd/m^2
Contrast ratio	>1,000,000:1 (Dark)
Color saturation	96% (NTSC)
Pixel circuit	2Tr1C

reaction method to make source/drain regions (see also Table 18.6). IGZO (In:Ga:Zn = 2:2:1) was used for the active layer of TFTs in this panel.

18.3 Oxide TFT–Driven Printed OLED Displays

One of the most serious problems for the OLED display is its price. Its specification is higher than that of LCD; however, a lower price is desirable. There are two major reasons for the high cost. One is the high price of the OLED-related materials, and the other is the small scale of the OLED display production. The former problem will be settled to some extent by the expansion of OLED display production volume from now on. The latter problem is mainly caused by the maximum substrate side for TFT and OLED device production. The *a*-Si-TFT for the LCD is fabricated with a substrate larger than Gen. 10.5 (3370×2940 mm); however, the LTPS-TFT for OLED displays is fabricated with a substrate smaller than Gen. 6 (1800×1500 mm).

Because the oxide TFT can be manufactured from the same facilities as those for the *a*-Si-TFT, scalable manufacturing of OLED displays is expected in place of LTPS-TFTs. Moreover, the conventional evaporation technology for OLEDs is proceeding, after cutting the substrate to half or one-sixth of Gen. 6 substrate at this moment. According to such nonscalability, the production cost of the OLED display is high at this moment.

The proposed highly reliable oxide TFT technology is expected for low-cost production of OLED displays with the application of OLED printing technology. Both technologies can be performed with a large substrate, and thus a lower cost production is realized (Figure 18.21).

Figure 18.22 shows a cross-sectional view of the printed OLED display. Because the printing process is a wet process and can be executed in atmospheric conditions, the printed layer contains some mobile ions and water. In addition, the neighboring layers such as the planar, bank, and encapsulation layer are also fabricated by a wet process and contain some mobile ions and water. The excess mobile ions and water often diffuse into the OLED and TFT region by the thermal or bias stress, and degrade those performances. To secure device performance, it is effective to prevent their diffusion by the passivation layer or the device structure. As introduced in Section 18.2, the alumina passivation and the top-gate TFT have a superior passivation property, and the TFT has excellent reliability for flexible printed OLED displays.

Table 18.7 and Figure 18.23 show a 12-inch-diagonal printed AM-OLED prototype [37]. According to a highly reliable TFT structure with small and uniform parasitic capacitances, an excellent image quality was achieved. Figure 18.24 shows panel stability against drive stress and storage stress in acceleration tests [9]. As driving stress, 350 cd/m^2 was emitted, and as storage stress, 80 °C was applied to the panel. Very stable driving stability and high temperature storage stability were achieved according to the aluminum-passivated top-gate TFT structure.

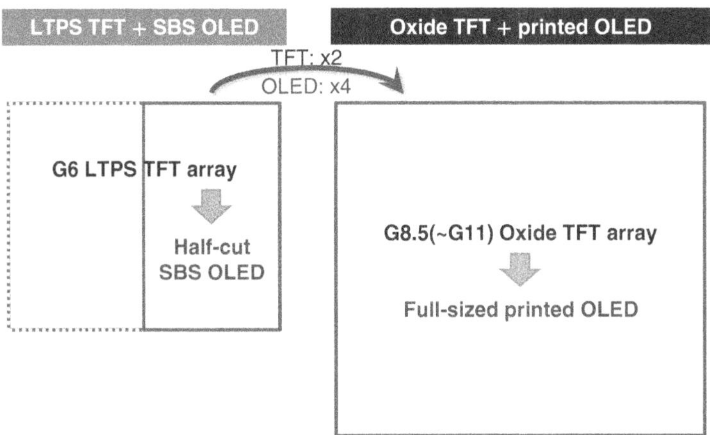

Figure 18.21 Schematics of substrate size for OLED display manufacturing.

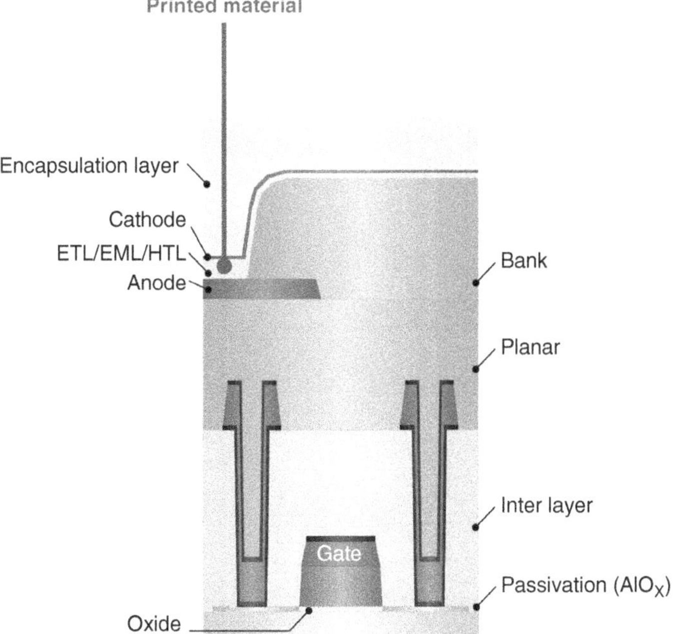

Figure 18.22 Cross-sectional device structure of a printed OLED display.

The oxide TFT is expected for flexible display manufacturing. In general, a poly-imide film is currently used as a substrate of the LTPS-TFT-driven flexible OLED display. The poly-imide film has high thermal durability and is applicable for the substrate of the LTPS-TFT. However, there are two issues; one is high cost, and the other is transparency. It becomes brown to secure durability against high-temperature (>450 °C) processes of LTPS-TFT fabrication. For lower cost and transparent flexible substrate, a lower process temperature is required. The oxide TFT, which has a lower process temperature than 350 °C, is therefore preferable for the flexible display.

Figure 18.25 and Table 18.8 show a prototype of the oxide TFT–driven flexible printed OLED display [38, 39] and its specifications. Even though the OLED printing technology and the oxide TFT technology are both suitable for scalable production, high-resolution flexible displays can be manufactured by these technologies.

Table 18.7 Specification of the 12-inch AM-OLED display prototype.

Panel size	12-inch diagonal
Format	FHD
Resolution	180 ppi
Brightness	Peak: >350 cd/m^2
Contrast ratio	>1,000,000:1 (Dark)
Color saturation	>125% (sRGB)

Figure 18.23 12 -inch oxide-TFT-driven printed OLED display prototype.

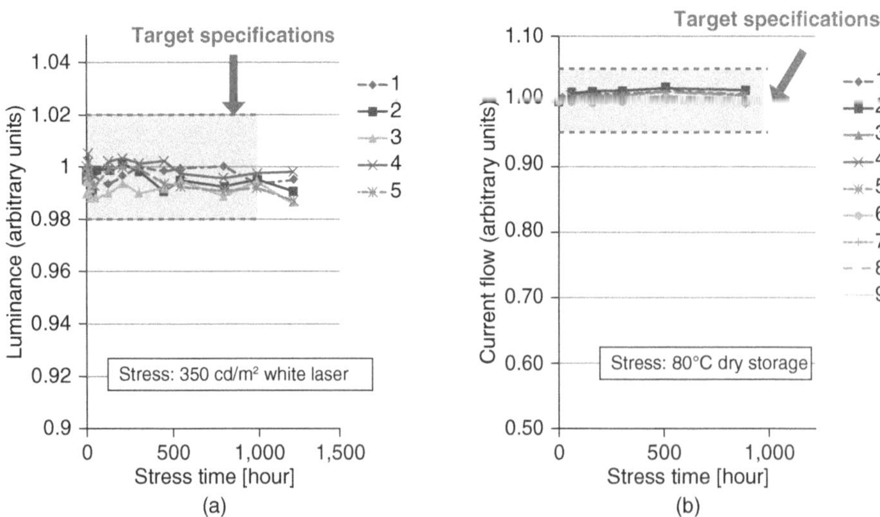

Figure 18.24 Luminance stability in accelerated panel reliability tests. (a) Drive stress test. (b) Storage stress test.

Figure 18.25 2.8-inch oxide-TFT-driven 350 ppi flexible printed OLED display prototype.

Table 18.8 Specifications of the 2.8-inch-diagonal 352-ppi flexible OLED display prototype.

Panel size	2.8 inches diagonal
Number of pixels	960(H)×152(V)
Resolution	352 ppi
Brightness	Peak: >350 cd/m^2
Contrast ratio	>1,000,000:1 (Dark)
Gate driver	Integrated
Substrate	Poly-imide

18.4 Summary

Oxide TFTs are expected for use in OLED displays because oxide TFTs have high mobility over 10 cm^2/Vs and can be uniformly formed on a large substrate. Although the stability of the oxide TFT has been an issue for driving OLED displays, this has been improved to the same level as that of the conventional LTPS-TFT by AlO_x passivation technology, the top-gate TFT technology, and the like. In the meantime, a printing technology has been applied for patterning the RGB photoluminescent materials of OLEDs. This technology also enables the scalable manufacturing of OLED devices for information display with oxide TFTs. By combining these technologies, highly productive OLED display manufacturing and low-price OLED displays are expected.

References

1 Fortunato, E., Barquinha, P., and Martins, R. (2012). Oxide semiconductor thin-film transistors: a review of recent advances. *Advanced Materials* 24-22: 2945–2986.

2 Park, J.S., Maeng, W.-J., Kim, H.-S., and Park, J.-S. (2012). Review of recent developments in amorphous oxide semiconductor thin-film transistor devices. *Thin Solid Films* 520-6: 1679–1693.

3 Arai, T. (2012). Oxide-TFT technologies for next-generation AMOLED displays. *Journal of the SID* 20-3: 156–161.

4 Tang, C.W. and Van Slyke, S.A. (1987). Organic electroluminescent diodes. *Applied Physics Letters* 51: 913–915.

5 Sasaoka, T., Sekiya, M., Yumoto, A. et al. (2001). A 13.0-inch AM-OLED display with top emitting structure and adaptive current mode programmed pixel circuit (TAC). *SID Symposium Digest of Technical Papers* 32-1: 384–387.

6 Tsujimura, T., Kobayashi, Y., Murayama, K. et al. (2003). A 20-inch OLED display driven by super-amorphous-silicon technology. *SID Symposium Digest of Technical Papers* 34-1: 6–9.

7 Lee, M.H., Seop, S.M., Kim, J.S. et al. (2009). Development of 31-inch full-HD AMOLED TV using LTPS-TFT and RGB FMM. *SID Symposium Digest of Technical Papers* 40-1: 802–804.

8 Han, C.-W., Kim, K.-M., Bae, S.-J. et al. (2012). 55-inch FHD OLED TV employing new tandem WOLEDs. *SID Symposium Digest of Technical Papers* 43-1: 279–281.

9 Arai, T. (2018). Innovative technologies for OLED display manufacturing. In: *2018 25th International Workshop on Active-Matrix Flatpanel Displays and Devices (AM-FPD)*, SP2-1.

10 Arai, T., Morosawa, N., Tokunaga, K. et al. (2011). Highly reliable oxide-semiconductor TFT for AMOLED displays. *Journal of the SID* 19-2: 205–211.

11 Dawson, R., Shen, Z., Furst, D.A. et al. (1998). Design of an improved pixel for a polysilicon active-matrix organic LED display. *SID Symposium Digest of Technical Papers* 29-1: 11–14.

12 Onoyama, Y., Yamashita, J., Kitagawa, H. et al. (2012). 0.5-inch XGA micro-OLED display on a silicon backplane with high-definition technologies. *SID Symposium Digest of Technical Papers* 43-1: 950–953.

13 Sanford, J.L. and Libsch, F.R. (2003). TFT AMOLED pixel circuits and driving methods. *SID Symposium Digest of Technical Papers* 34-1: 10–13.

14 Matsueda, Y., Kakkad, R., Park, Y.S. et al. (2004). 2.5-in. AMOLED with integrated 6-bit gamma compensated digital data driver. *SID Symposium Digest of Technical Papers* 35-1: 1116–1119.

15 Choi, A.M., Kwon, O.K., Komiya, N., and Chung, H.K. (2003). A self-compensated voltage programming pixel structure for active-matrix organic light emitting diodes. *In Proceedings of the International Display Workshops* 2003: 535–538.

16 Iida, Y., and Uchino, K. (2011). Pixel circuit, display, and method for driving pixel circuit. U.S. Patent 7,898,509 B2. Filed November 21, 2007; issued March 1, 2011.

17 Nomura, K., Ohta, H., Takagi, A. et al. (2004). Room-temperature fabrication of transparent flexible thin film transistors using amorphous oxide semiconductors. *Nature* 432: 488–492.

18 Powell, M.J. (1983). Charge trapping instabilities in amorphous silicon-silicon nitride thin-film transistors. *Applied Physics Letters* 43-6: 597–599.

19 Gelatos, A.V. and Kanicki, J. (1990). Bias stress-induced instabilities in amorphous silicon nitride/hydrogenated amorphous silicon structures: is the “carrier-induced defect creation” model correct? *Applied Physics Letters* 57-12: 1197–1199.

20 Libsch, F.R. and Kanicki, J. (1993). Bias-stress-induced stretched-exponential time dependence of charge injection and trapping in amorphous thin-film transistors. *Applied Physics Letters* 62-11: 1286–1288.

21 Cross, R.B.M. and De Souza, M.M. (2006). Investigating the stability of zinc oxide thin film transistors. *Applied Physics Letters* 89-26: 263513–263515.

22 Vygranenko, Y., Wang, K., and Nathan, A. (2007). Stable indium oxide thin-film transistors with fast threshold voltage recovery. *Applied Physics Letters* 91-26: 263508–263510.

23 Suresh, A. and Muth, J.F. (2008). Bias stress stability of indium gallium zinc oxide channel based transparent thin film transistors. *Applied Physics Letters* 92-3: 033502–033504.

24 Lee, J.-M., Cho, I.-T., Lee, J.-H., and Kwon, H.-I. (2008). Bias-stress-induced stretched-exponential time dependence of threshold voltage shift in InGaZnO thin film transistors. *Applied Physics Letters* 93-3: 093504–093506.

25 Park, J.-S., Jeong, J.K., Chung, H.-J. et al. (2008). Electronic transport properties of amorphous indium-gallium-zinc oxide semiconductor upon exposure to water. *Applied Physics Letters* 92-7: 072104–072106.

26 Kang, D., Lim, H., Kim, C. et al. (2007). Amorphous gallium indium zinc oxide thin film transistors: sensitive to oxygen molecules. *Applied Physics Letters* 90-19: 192101–192103.

27 Nomura, K., Kamiya, T., Hirano, M., and Hosono, H. (2009). Origins of threshold voltage shifts in room-temperature deposited and annealed a-In–Ga–Zn–O thin-film transistors. *Applied Physics Letters* 95-1: 013502–013504.

28 Sato, A., Abe, K., Hayashi, R. et al. (2009). Amorphous In–Ga–Zn–O coplanar homojunction thin-film transistor. *Applied Physics Letters* 94-13: 133502–133504.

29 Gosain, D.P. and Tanaka, T. (2009). Instability of amorphous indium gallium zinc oxide thin film transistors under light illumination. *Japanese Journal of Applied Physics* 48: 03B018.

30 Lee, K.-H., Jung, J.S., Son, K.S. et al. (2009). The effect of moisture on the photon-enhanced negative bias thermal instability in Ga–In–Zn–O thin film transistors. *Applied Physics Letters* 95-23: 232106–232108.

31 Morosawa, N., Ohshima, Y., Morooka, M. et al. (2012). Novel self-aligned top-gate oxide TFT for AMOLED displays. *Journal of the SID* 20-1: 47–52.

32 Park, J.-S., Jeong, J.K., Mo, Y.G., and Kim, H.D. (2008). Amorphous indium–gallium–zinc oxide TFTs and their application for large size AMOLED. In: *2008 15th International Workshop on Active-Matrix Flatpanel Displays and Devices (AM-FPD)*. 275.

33 Jeong, J.K., Yang, H.W., Jeong, J.H. et al. (2008). Origin of threshold voltage instability in indium-gallium-zinc oxide thin film transistors. *Applied Physics Letters* 93-12: 123508–123510.

34 Park, S.-H.K., Hwang, C.-S., Cho, D.-H. et al. (2009). Effect of channel/insulator interface formation process on the oxide TFT performance. *SID Symposium Digest of Technical Papers* 40-1: 276–279.

35 Hayashi, R., Sato, A., Ofuji, M. et al. (2008). Improved amorphous In–Ga–Zn–O TFTs. *SID Symposium Digest of Technical Papers* 39-1: 621–624.

36 Arai, T. (2015). The advantages of the self-aligned top gate oxide TFT technology for AM-OLED displays. *SID Symposium Digest of Technical Papers* 46-1: 1016–1019.

37 Hayashi, H., Murai, A., Miura, M. et al. (2017). AlO sputtered self-aligned source/drain formation technology for highly reliable oxide thin film transistor backplane. *In Proceedings of the International Display Workshops 2017* 324–327.

38 Hayashi, H., Yamada, T., Koshiishi, R. et al. (2019). Channel-dimension-scalable oxide thin-film transistor for high-resolution pixel and integrated gate driver. *SID Symposium Digest of Technical Papers* 50-1: 322–325.

39 Arai, T. (2019). High performance oxide TFT technology for med.-large size OLED displays. *Proceedings of the International Display Workshops* 2019: 493–496.

19

Mechanically Flexible Nonvolatile Memory Thin-Film Transistors Using Oxide Semiconductor Active Channels on Ultrathin Polyimide Films

Sung-Min Yoon, Hyeong-Rae Kim, Hye-Won Jang, Ji-Hee Yang, Hyo-Eun Kim, and Sol-Mi Kwak

Department of Advanced Materials Engineering for Information and Electronics, Kyung Hee University, Seoul, Republic of Korea

19.1 Introduction

In a new paradigm in consumer electronics, mechanically flexible and stretchable electronic devices prepared on flexible substrates find their ways to provide us with fascinating opportunities for a future Internet-of-things-based culture of living [1–3]. Actually, there are already diverse applications, including wearable sensors, foldable and rollable displays, and flexible electronic systems [4, 5]. Alternatively, system-embeddable nonvolatile memory (NVM) elements enhance the functions and performance of flexible and stretchable electronic device applications for saving power consumption as well as for storing information. The system power can be effectively reduced when the power supply required for given parts of the circuit can be shut off by temporarily transferring the logic data to the NVM elements integrated into the systems [6, 7]. A memory-in-pixel technology, developed to reduce the power consumption of flat-panel displays, is a good example [8]. For these purposes, the introduction of a nonvolatile-type memory device is quite demanding. Thus, many kinds of mechanically flexible NVMs have been fabricated on various flexible substrates, which can be classified into several groups according to the device structures and operating origins, such as floating-gate transistors inserting metal nanoparticles into the gate insulators [9], charge-trap memory transistors employing any charge-trapping layers [10], resistive-change two-terminal devices using oxide and organic thin films [11], and ferroelectric field-effect transistors [12]. Among them, the charge-trap-assisted memory thin-film transistor (CT-MTFT) has been suggested to be one of the most promising candidates for system-embeddable NVM elements. However, there have rarely been significant improvements for the charge-trap-type flexible memory devices, because the previously reported conventional devices showed critical problems such as long program time, asymmetric program operations, and unstable memory characteristics [13, 14]. From these viewpoints, flexible and stretchable memory devices should be designed and implemented to have the functions of both lower power operation and more reliable NVM characteristics under harsh mechanical bending and stretching conditions. In addition, the fabrication process temperature should be suppressed to be as low as possible.

To realize the system-embedded flexible memory transistors, we have proposed CT-MTFTs employing oxide semiconductor thin films as active channels and charge-trap layers (CTLs) for the first time, in which In-Ga-Zn-O (IGZO) and ZnO thin films have been uniquely utilized, respectively [15]. While the conventional charge-trap memory devices were developed to have dielectric CTLs such as SiON, SiN_x, and multilayered oxide insulators, in our work, the electronic nature of the semiconducting ZnO thin films was strategically controlled for the employment of CTL [16, 17]. As results, excellent NVM operations and superior device feasibility have been verified for our proposed CT-MTFTs fabricated on rigid and flexible substrates [18, 19]. In other words, we can say the memory device is one of the most promising technologies using oxide TFTs for various practical applications, including flexible and stretchable electronics.

Amorphous Oxide Semiconductors: IGZO and Related Materials for Display and Memory, First Edition.
Edited by Hideo Hosono and Hideya Kumomi.

In this chapter, we review the device structure, fabrication process details, NVM operations, and technical issues of our proposed CT-MTFTs using oxide semiconductor thin films from the viewpoint of developing mechanically flexible memory devices. First, in Section 19.2, the preparation method of flexible polyimide (PI) film substrates, device fabrication issues, and characterization methodologies for flexible memory transistors will be described. Then, we demonstrate the memory operations of flexible CT-MTFTs prepared on ultrathin PI films, with discussions on the basic device characteristics of flexible IGZO-TFTs, in Section 19.3. To enhance the mechanical flexibility of CT-MTFTs, some parts, such as source/drain (S/D) electrodes and gate insulators, can be replaced by organic thin films. The device operation and fabrication issues of prototype devices will be demonstrated and discussed in Section 19.4. Furthermore, the introduction of a vertical-channel structure can provide useful ways to scale the device and to improve the robustness against cyclic bending events. In Section 19.5, the device feasibility of CT-MTFTs with vertical channels will be described. For further advances of our proposed memory transistors, the remaining technical issues and future perspectives will also be briefly presented with closing remarks. This chapter, representing the flexible memory applications supported by oxide TFT technologies, corresponds to an extended version of our previous overview [3]. We hope that this overview can be a useful guide for understanding the NVM elements using oxide semiconductor thin films to realize highly functional, flexible, and stretchable future consumer electronic systems.

19.2 Fabrication of Memory TFTs

19.2.1 Substrate Preparation

The suitable choice of flexible substrate is most important to guarantee both requirements of good device performance and robust mechanical flexibility for the fabrication of flexible CT-MTFTs. Thus, in this work, the

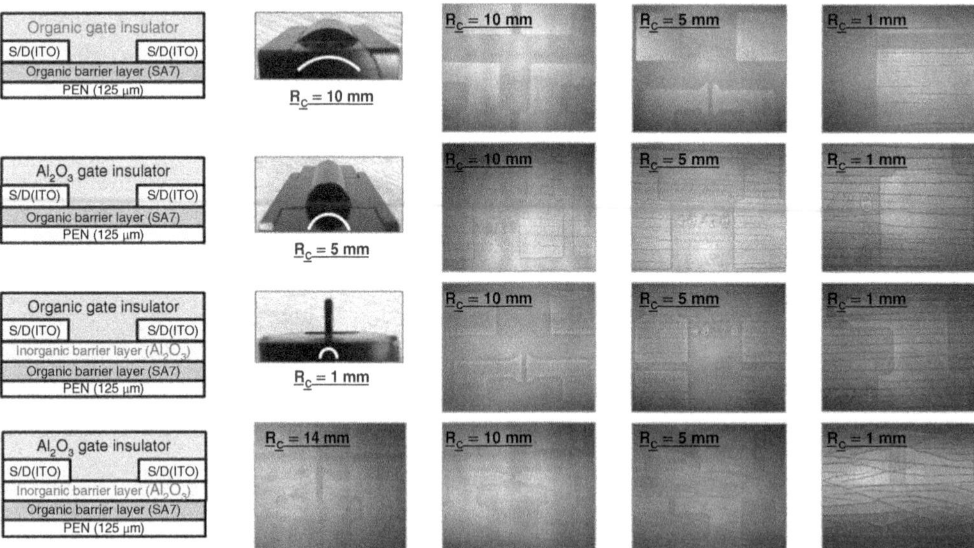

Figure 19.1 Cross-sectional diagrams of various multilayered film structures prepared on 125-μm-thick PEN substrates and photo images of the device planes bent with given values of R_C, in which typical bending curvatures corresponding to the R_C values of 10, 5, and 1 mm were also shown with custom-made bending jigs.

flexible substrate was chosen from the viewpoint of obtaining ultrathin film thickness, visible transparency, and solution-process compatibility. Even though we have demonstrated the device characteristics of flexible CT-MTFTs prepared on plastic poly(ethylene naphthalate) (PEN) [20], the PEN substrate suffered from larger mechanical strain under evaluations of mechanical flexibility. It is interesting to compare the situations of crack propagation into the device planes with various device structures. Figure 19.1 shows cross-sectional diagrams of various multilayered film structures prepared on 125-μm-thick PEN, and photo images of the device planes bent with given values of curvature radius (R_C), in which typical bending degrees corresponding to the R_C values of 10, 5, and 1 mm were also shown with custom-made bending jigs. Actually, for the reference structure composed of inorganic Al_2O_3 barrier and gate insulators, microcracks started to be developed into both inorganic layers prepared on the whole substrate from an R_C of 14 mm. Even when one of the layers of buffer and gate insulator was replaced by an organic layer, cracks were also developed at an R_C of 10 mm. Interestingly, even though all layers were replaced by organic layers, brittle ITO patterns were terribly broken at an R_C of 1 mm.

From these observations, in this work, the flexible substrate was replaced from 125-μm-thick PEN to 1.2-μm-thick colorless PI (CPI) film. In other words, the CPI has a very low CTE and a high glass transition temperature for the device fabrication. However, the main motivation was to prepare the flexible film substrate with a very thin film thickness using a simple spin-coating process. For the preparation of flexible substrates, the CPI film (synthesized by Kolon Industries) was spin-coated on carrier glass using PI varnish. For thermal curing, coated films were treated at 80 and 290 °C for 30 min on a hot plate and in a furnace, respectively. In preparing the flexible CPI film substrates for the device fabrication, the surfaces should be properly treated to suppress and control the physiochemical adsorption and/or absorption of water molecules from ambient. Figures 19.2a and 19.2b show a photo image of fully transparent IGZO-TFTs on CPI films laminated on carrier glass substrate and the Raman spectroscopy of the coated CPI film, respectively. The highest peak, observed at 1616 cm^{-1}, corresponds to the stretching COO^- anions in the carboxylic acid and ether. Thus, hydrogen and hydronium ions can be readily adsorbed at the surface of PI, and hence, positive charges can be trapped at the interfaces between the active layer and PI film, causing the backchannel effects on the IGZO channel. It was noteworthy that the introduction of an Al_2O_3 barrier layer was found to provide excellent backchannel and interface properties for enhancing the device performance of the IGZO-channel devices prepared on CPI film substrates. The effects of surface treatments and the improvements in device stabilities of the IGZO-TFTs will be described in detail in Section 19.3.1.

(a)

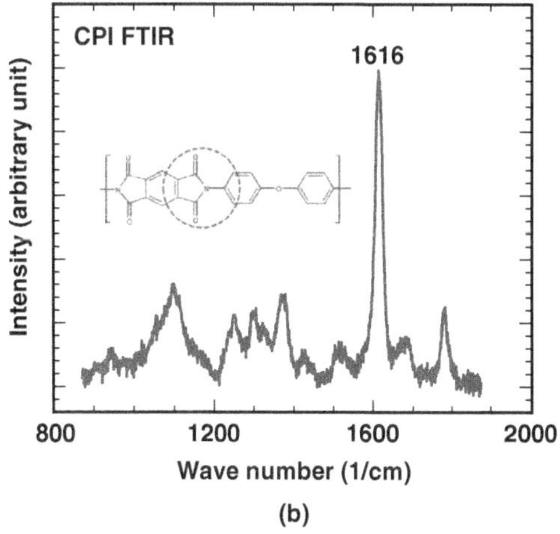

(b)

Figure 19.2 (a) Photo image of fully transparent IGZO-TFT on PI film laminated on carrier glass. (b) Raman spectroscopy of CPI film.

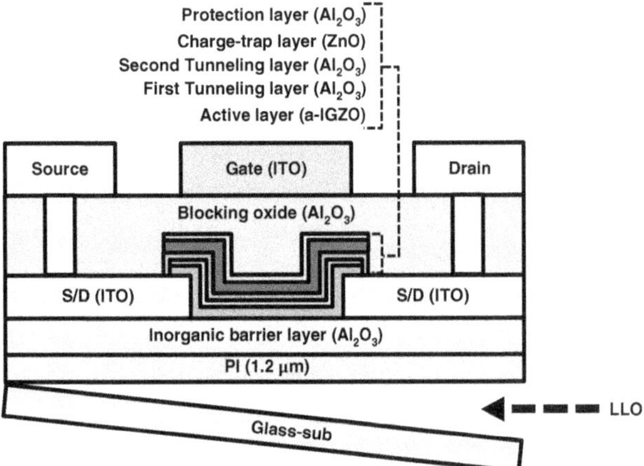

Figure 19.3 Schematic cross-sectional diagrams of the fabricated flexible CT-MTFTs fabricated on 1.2-μm-thick CPI film substrate.

19.2.2 Device Fabrication Procedures

Figure 19.3 shows the typical schematic cross-sectional diagram of the flexible CT-MTFTs fabricated on 1.2-μm-thick CPI film substrate. Details of the device fabrication procedures are summarized in Figure 19.4. First, a 50-nm-thick Al_2O_3 barrier layer was prepared on the coated PI film by atomic layer deposition (ALD) at 150 °C. Indium tin oxide (ITO, 150 nm) was deposited by direct-current (DC) magnetron sputtering and patterned by wet chemical etching into S/D electrodes. A 20-nm-thick IGZO active-channel layer was prepared by radiofrequency (RF) magnetron sputtering at room temperature (RT). Then, an Al_2O_3 (5 nm) thin film was formed by ALD at 180 °C as a first tunneling oxide. After patterning the IGZO channel and the Al_2O_3 first tunneling layers, Al_2O_3 (5 nm), ZnO (30 nm), and Al_2O_3 (3 nm) layers were successively deposited by ALD while keeping the vacuum at 180, 100, and 150 °C as second tunneling, charge-trap, and top-protection layers, respectively. Here, we have

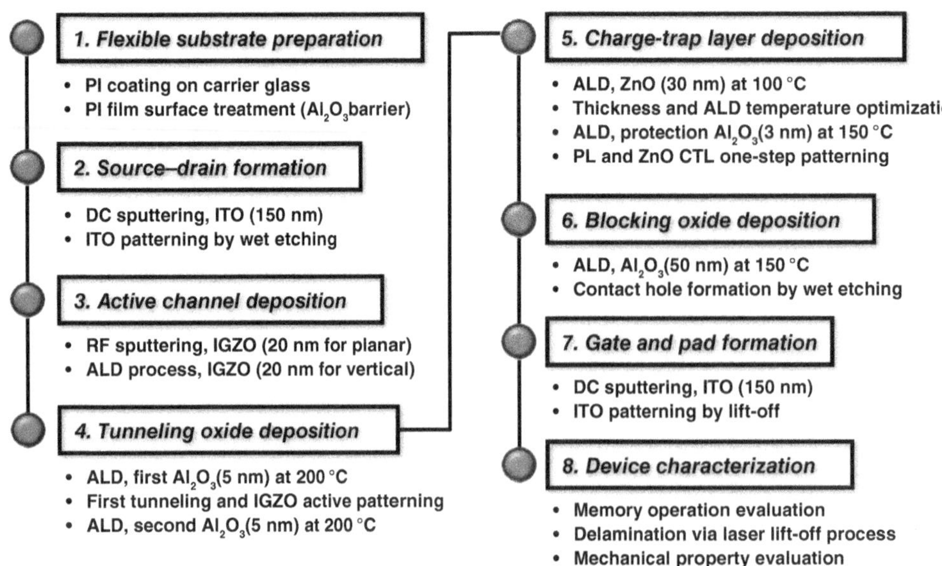

Figure 19.4 Flowchart of the full fabrication procedure and process conditions for the flexible CT-MTFTs.

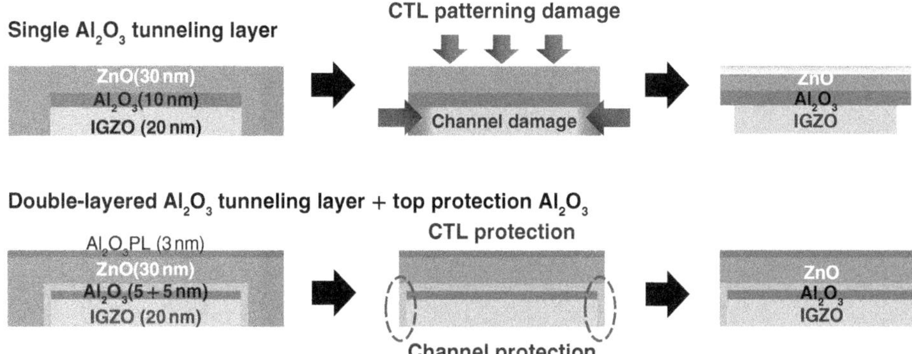

Figure 19.5 Schematic illustrations of the gate-stack structures during the patterning of CTL by means of wet etching with (a) a single tunneling oxide layer and (b) double-layered tunneling and top-protection layers.

to pick up three important fabrication issues for guaranteeing the sound device operations of our proposed CT-MTFTs: (i) The electronic nature, including the electrical conductivity and defect structures, of the ZnO CTLs was so modulated as to be available for the NVM operations of the flexible CT-MTFTs by properly controlling the deposition temperature during the ALD process (the effects of engineered ZnO CTLs on the device characteristics of the flexible CT-MTFTs will be discussed in Section 19.3.2); (ii) the double-layered tunneling oxide structure was designed to prevent process damages to the IGZO channel during the wet etching for patterning the CTL; and (iii) a top-protection Al_2O_3 layer was prepared to protect the surface of the ZnO CTL during the lithography process. Figures 19.5a and 19.5b illustrate schematic gate-stack structures during the wet-etching process for patterning the CTL when only a single tunneling oxide layer and double-layered tunneling oxide and top-protection layers were used, respectively. The strategically designed gate-stack structure was verified to improve the device characteristics by effectively reducing process damages into the active channel and CTL [18]. For the next step, the tri-layered gate stack was patterned by a one-step chemical etching. A 50-nm-thick Al_2O_3 blocking layer was formed by ALD at 150 °C. Finally, ITO films were deposited and patterned by a liftoff process as gate electrodes and S/D pads.

19.2.3 Characterization Methodologies

The device characteristics of the fabricated flexible CT-MTFTs were evaluated using a semiconductor parameter analyzer (Keithley 4200SCS) and a pulse generator (HP 8110A) in a dark box at RT. The mechanical flexibility of the fabricated devices was also examined with a bending jig after the delamination process of CPI film from the carrier glass. Here, we have to check the technical issues in evaluating the mechanical properties of the CT-MTFTs fabricated on flexible substrates and provide suitable methodologies to accurately examine the device characteristics under mechanically strained conditions. The first issue is related to the technique for adjusting the R_C for the devices fabricated on 1.2-μm-thick PI films. It was not practically easy to accurately set the values of R_C for the PI film substrate because of too-thin film thickness and mechanical vulnerability with a conventional bending-jig configuration. To solve this problem, a flexible printed circuit (FPC) for tab bonding and a printed circuit board (PCB) were designed, as shown in Figures 19.6a and 19.6b, respectively. The routing pads prepared on the edge of the substrate were connected to one side of the pads printed on the FPC, and the pins on the PCB for the measurements of devices were connected to another side of the pads. The number of pins was designed to be 120 with a distance of 70 μm. The laser liftoff (LLO) process was carried out for the delamination of PI film from the carrier glass, in which the dose and scan rate of laser irradiation were suitable adjusted, as shown in Figure 19.6c. Figures 19.6d and 19.6e show photo images of the delaminated PI film substrate, on which the device

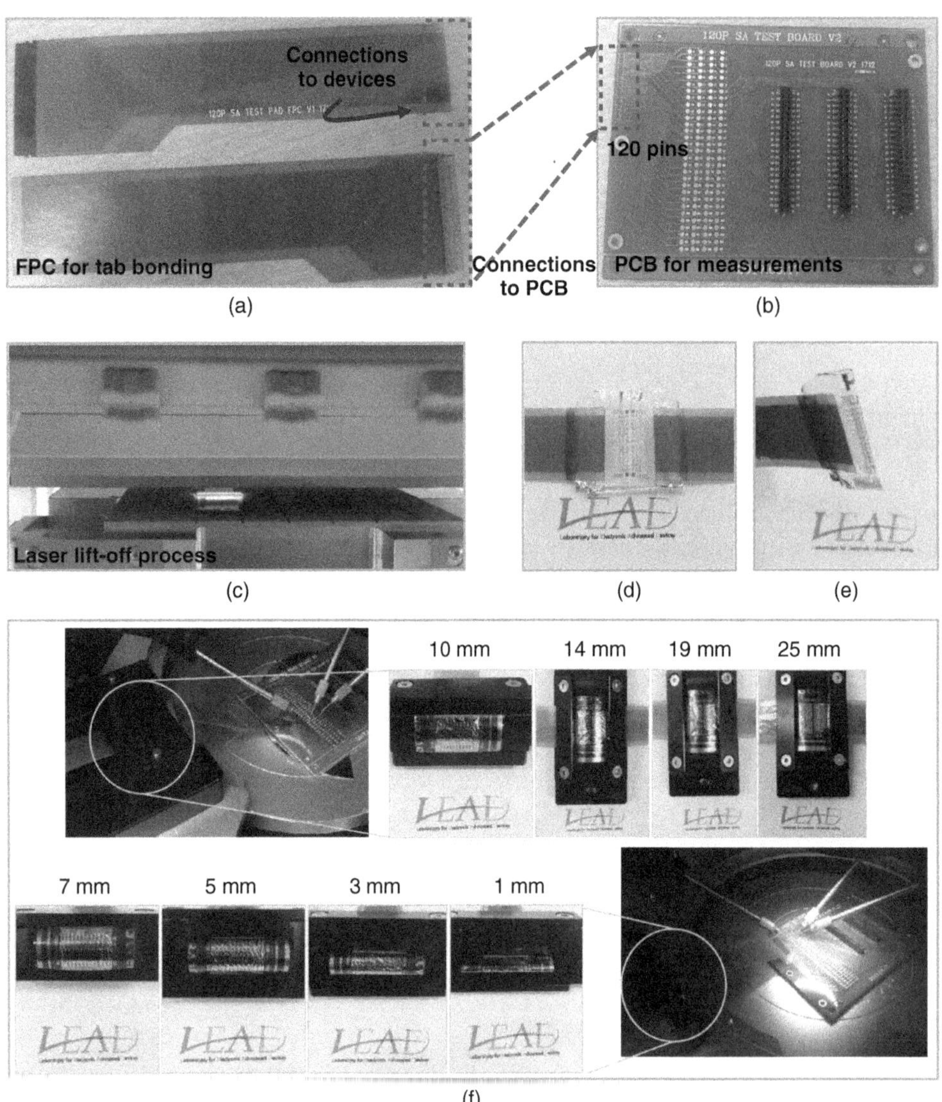

Figure 19.6 Photo images of (a) FPC cables for tab bonding between the (b) PCB and device pads. 120 pins were designed into the FPC and PCB for the connections of each device pad. (c) Photo image of the LLO process for the CPI film coated on the carrier glass. Laser was irradiated and scanned on the backside of the carrier glass for the delamination of the CPI film. Photo images of (d) the delaminated CPI film loaded with flexible devices bonded to FPC cables and (e) the mechanically bending situation of the CPI film substrate. (f) Evaluations of the device characteristics under mechanically strained conditions at various curvature radii using laboratory-made bending jigs with R_C's of 1 to 25 mm.

patterns were fabricated, with tab-bonded FPC cables in flat and mechanically bending states, respectively. For the accurate setting of R_C values for the PI film substrate, a set of laboratory-made bending jigs was manufactured with various R_C's from 1 to 25 mm, as shown in Figure 19.6f. The device characteristics of the fabricated flexible CT-MTFTs could be examined with accurate values of R_C by positioning the evaluated devices on center fields of bending areas.

19.3 Device Operations of Flexible Memory TFTs

19.3.1 Optimization of Flexible IGZO-TFTs on PI Films

As mentioned in this chapter, for the use of CPI film substrate, the introduction of Al_2O_3 barrier treatment was examined to significantly influence device characteristics, including bias–temperature stress stabilities for the IGZO-TFTs fabricated on CPI films [21]. Figure 19.7a shows the drain current–gate voltage (I_{DS}-V_{GS}) characteristics and gate leakage currents of the IGZO-TFTs fabricated with and without an Al_2O_3 buffer, after a postannealing process at 150 °C. The field-effect mobility at the saturation region (μ_{sat}) and the subthreshold swing (SS) of the TFT fabricated with the buffer layer were estimated to be 8.6 $cm^2\ V^{-1}\ s^{-1}$ and 0.16 V dec^{-1}, respectively. Alternatively, for the TFT fabricated without a buffer layer, the μ_{sat} and SS were obtained to be 5.1 $cm^2\ V^{-1}\ s^{-1}$ and 0.68 V dec^{-1}, respectively. Furthermore, a clockwise hysteresis was slightly detected in the transfer curve of the TFT fabricated without a buffer layer, even after the annealing at 150 °C. Thus, it was found that the TFT operations could be markedly improved by treating an Al_2O_3 barrier layer on the CPI film substrate, with which the adsorption and absorption of ions and/or water molecules could be effectively suppressed on the backchannel region of the PI surface.

It was more impressive to investigate the effects of the Al_2O_3 barrier layer on the device reliabilities of the flexible IGZO-TFTs fabricated on CPI films. Figure 19.7b shows the variations in the shifts of turn-on voltage (V_{ON}) of the IGZO-TFTs fabricated with and without an Al_2O_3 barrier treated on the CPI with a lapse of stress time under various bias and temperature stress conditions. A bias stress of −20 or +20 V was continuously applied to the gate terminal for 10^4 s for the tests of negative-bias stress (NBS) and positive-bias stress (PBS), and the drain bias was fixed to 10.5 V. The temperature stress was set as 40 °C. For the IGZO-TFT fabricated on a barrier layer, the V_{ON} was not shifted under an NBS, and the shift of V_{ON} (ΔV_{ON}) was measured to be as small as −1.0 V under a negative-bias temperature stress (NBTS). Interestingly, the device fabricated without a barrier layer did not exhibit

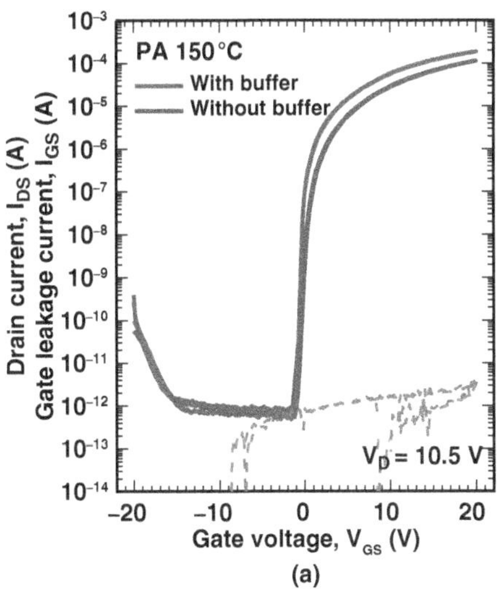

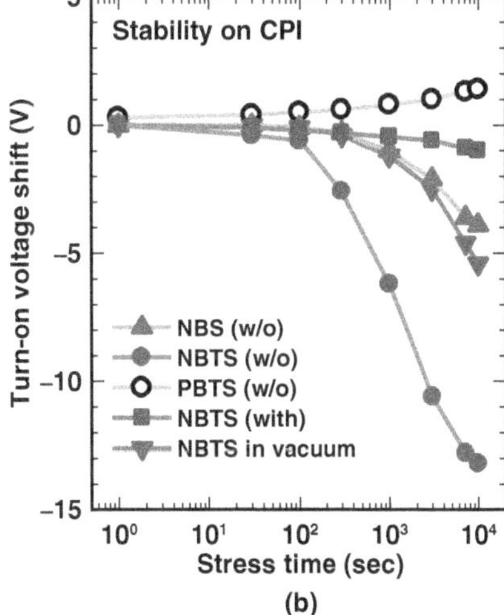

Figure 19.7 (a) I_{DS}-V_{GS} characteristics of the flexible IGZO-TFTs fabricated without and with an Al_2O_3 barrier layer on CPI film after annealing at 150 °C. (b) Variations in V_{on} shifts as a function of stress time for the flexible IGZO-TFTs with an Al_2O_3 barrier or without any barrier under NBS, NBTS, and PBTS in various conditions. Bias stress of −20 and +20 V was continuously applied for 10^4 s during the NBTS (and NBS) and PBTS tests, respectively. The drain bias was fixed at 10.5 V. The measurement temperatures were set as RT and 40 °C.

any ΔV_{ON} under a PBS, and the ΔV_{ON} was only 1.3 V even under a positive-bias temperature stress (PBTS), which suggests that the gate-stack structures including the interfaces were well formed during the device fabrication procedures. Contrarily, there was a considerable ΔV_{ON} of −3.9 V under the NBS condition when the PI film was not treated with an Al_2O_3 barrier layer. Since the well-designed IGZO-channel TFTs typically show good NBS stability due to its *n*-type nature, the main origin of degradations in device operation under the NBS could be speculated to be physiochemical absorption of water molecules on the surface of PI films owing to the absence of an inorganic barrier layer [22]. As a result, the ΔV_{ON} was measured to be −13.2 V under the NBTS at 40 °C. At higher temperature, the activation energy for absorption of water molecules might be lowered, and hence, the NBTS instabilities are accelerated. This explanation could also be verified by an NBTS test performed in a vacuum. The value of ΔV_{ON} was much smaller than that estimated in ambient air for the same evaluated device, because there was a smaller number of water molecules within a vacuum. A considerable amount of ΔV_{ON} detected, even in a vacuum, might result from positively charged ions that had already been adsorbed on the PI surface. Consequently, it can be evidently suggested from these results that the introduction of a suitable barrier layer, ALD-grown Al_2O_3 in this work, can be one of the most important strategies for guaranteeing the device performance of flexible TFTs employing oxide semiconductor channels on PI film substrates.

19.3.2 Nonvolatile Memory Operations of Flexible Memory TFTs

There has been much preliminarily work, as discussed in Section 19.3.1, to optimize the fabrication process conditions, including the surface treatments of CPI film substrate, for realizing high-performance mechanically flexible memory TFTs. Alternatively, irrespective of any limitation to bending characteristics, the CT-MTFTs fabricated on PEN substrates demonstrated excellent NVM device operations, including long-term stability [23]. However, the choice of a CPI film as flexible substrate with a film thickness as thin as 1.2 μm can extend application items demanding flexible memory devices in flexible and stretchable electronics. A higher glass transition temperature of PI enhances the process temperature available for the device fabrication. The thermal and chemical stabilities of the coated CPI film used in this work were preliminarily verified [21].

To optimize the memory device performance of the flexible CT-MTFTs prepared on PI films, the control of ALD temperatures for the formation of ZnO CTLs was chosen as the first technical strategy [24]. The electronic properties of the ALD-grown ZnO thin films were examined to be sensitively affected by the deposition temperature due to the changes in band structures [25]. In this work, the ALD temperature for the formation of ZnO CTLs was varied to 75, 100, and 125 °C to investigate the role of semiconducting ZnO thin film as the CTL for the flexible CT-MTFTs. Figure 19.8a shows the I_{DS}-V_{GS} characteristics of the fabricated devices employing ZnO CTLs deposited at different ALD temperatures. For all the fabricated devices, charge-trap-assisted clockwise hysteresis was clearly obtained, in which the memory window (MW) was defined as the width of the ΔV_{ON} during the sweep of V_{GS} in forward and reverse directions. The MW values were estimated to be 8.4, 13.0, and 16.8 V for the CT-MTFTs using ZnO CTLs prepared at 75, 100, and 125 °C, respectively, at a V_{GS} sweep range of ±20 V. When the ZnO thin films were deposited at higher ALD temperatures, a larger MW was obtained for the NVM operations owing to higher trap densities within the ZnO CTLs. Therefore, the NVM operation of our proposed CT-MTFTs could be modulated by easy controls in ALD temperature conditions for the formation of ZnO CTLs. Figure 19.8b shows the variations in the MW obtained for the flexible CT-MTFTs when the V_{GS} sweep range increased from ±10 to ±20 V. While the obtained MW was 7.2 V at a program V_{GS} of ±15 V for the device using CTL prepared at 125 °C, for the device using CTL deposited at 75 °C, only a small MW of 2.4 V was obtained. This suggests that the program voltage can be effectively reduced by appropriately controlling the ALD temperature for the formation of ZnO CTLs as a powerful solution to modulating the trap sites within the CTL. The program speed was also examined for all the fabricated devices using ZnO CTLs prepared at different ALD temperatures, as shown in Figure 19.8c. In these evaluations, the variations in the on- and off-programmed currents were estimated as a function of program pulse duration from 1 μs to 1 s at a fixed program voltage of ±20 V and at a given readout voltage (V_R). It is noteworthy

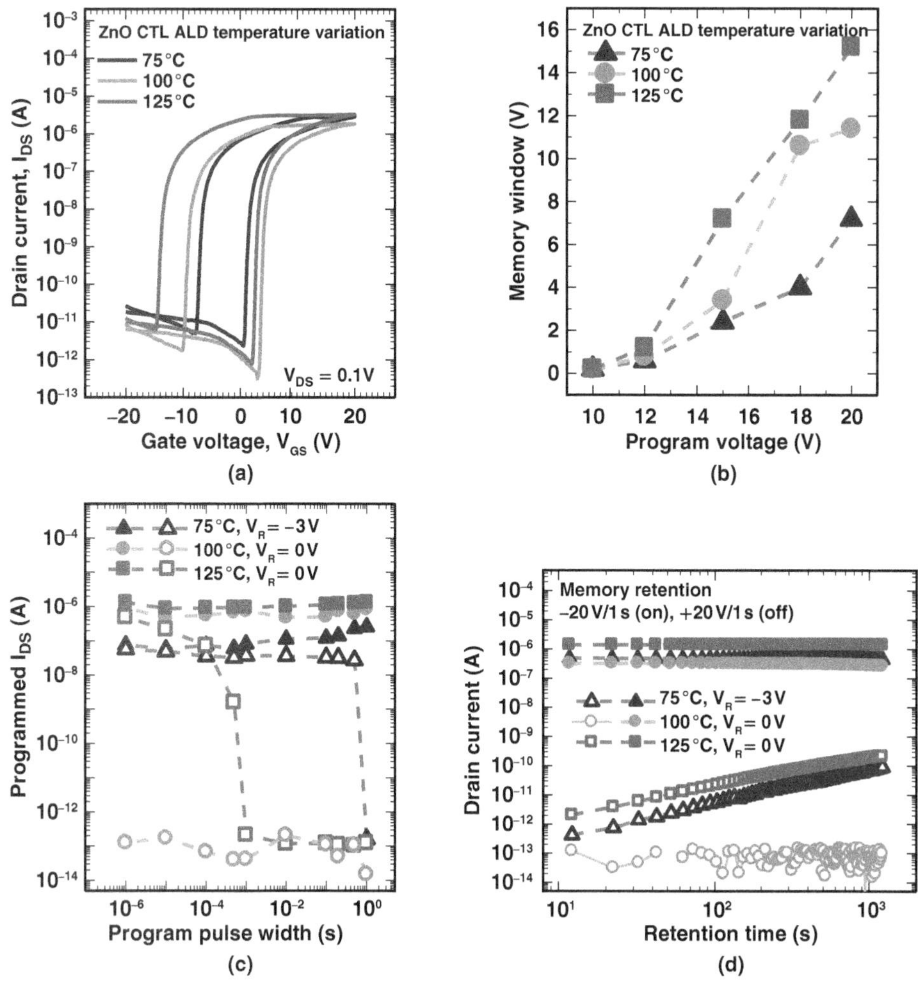

Figure 19.8 (a) Variations in I_{DS}-V_{GS} characteristics of the flexible CT-MTFTs fabricated on CPI films, employing ZnO CTLs prepared at various deposition temperatures of 75, 100, and 125 °C. (b) Variations in MW width for each fabricated flexible CT-MTFT at a V_{GS} sweep range of ±10 to ±20 V. Variations in on- (closed symbols) and off-program (open symbols) currents (c) as a function of program pulse durations and (d) with a lapse of memory retention time for 10^4 s when the CTL deposition temperature was varied to 75, 100, and 125 °C. The amplitude of program voltage pulses was set to ±15 V for the off- and on-program operations. The readout conditions were determined to be 0 or −3 V with respect to the position of the obtained MW of each device.

that the time duration required to completely program the memory off-states markedly varied with the deposition temperatures for the CTLs. When the CTL temperatures were varied to 75, 100, and 125 °C for the fabricated CT-MTFTs, the ratio between the on- and off-programmed drain currents (I_{ON}/I_{OFF}), which corresponds to NVM margins for the CT-MTFTs, was estimated to be 1.9, 1.7×10^7, and 6.0×10^2, respectively, at a program pulse duration of 500 μs. It is interesting to note that the device using CTLs deposited at 100 °C showed the most stable and fastest program operation, which originated from both factors of the number and position of charge-trap sites within the ZnO CTLs [24]. Additionally, the relative proportion between the number of trap sites at shallow and deep levels, which can be modulated by the deposition temperature of the ZnO CTLs, may be a critical feature to influence the program speed as well as the memory retention characteristic for the fabricated flexible CT-MTFTs. Figure 19.8d shows the variations in the on- and off-programmed currents for all the devices with a lapse of memory retention

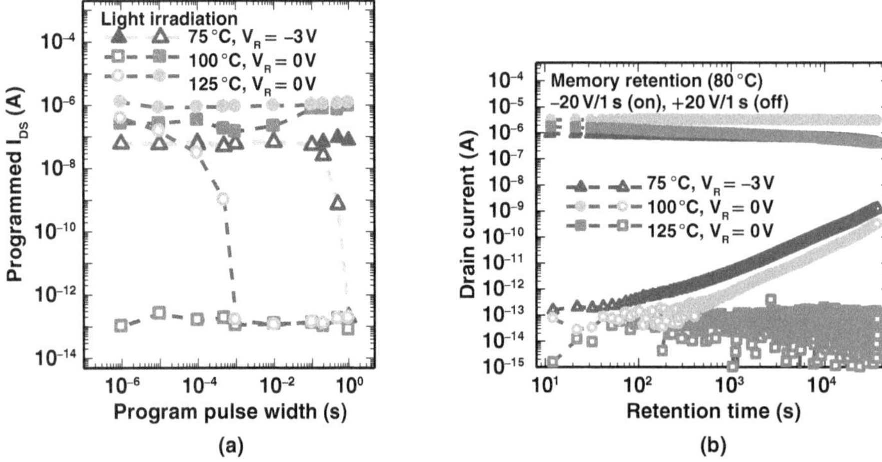

Figure 19.9 Variations in on- (closed symbols) and off-program (open symbols) currents of the fabricated flexible CT-MTFTs on CPI films (a) as a function of program pulse duration under white-light-irradiated conditions with an intensity of 0.1 mW and (b) with a lapse of memory retention time for 10^4 s at 80 °C when the CTL deposition temperature was varied to 75, 100, and 125 °C.

time for 10^4 s at RT. The device using a CTL deposited at 100 °C also showed the best long-term stability among the fabricated devices. Therefore, properly controlled charge-trap sites located in deep levels of the band gap were suggested to be absolutely desirable for guaranteeing fast program speed and long retention time for the flexible CT-MTFTs, which was successfully determined for the ZnO CTLs prepared at an ALD temperature of 100 °C.

The device stabilities of the flexible CT-MTFTs fabricated on 1.2-μm-thick CPI film were also examined under light-irradiated conditions or at a working temperature of 80 °C. As shown in Figure 19.9a, the on- and off-program operations did not exhibit any degradations for all the devices at an irradiation of a white wavelength with an intensity of 0.1 mW. The memory retention characteristics were evaluated at a temperature as high as 80 °C for 10^5 s, as shown in Figure 19.9b. The CT-MTFTs using CTLs prepared at 100 °C also showed excellent memory retention characteristics, even at a working temperature of 80 °C. Consequently, the NVM operations of the fabricated flexible CT-MTFTs were successfully confirmed on the barrier-treated ultrathin CPI films by means of properly designing the process conditions, including the ALD temperature conditions for the formation of ZnO CTLs. The obtained device characteristics were sufficiently comparable to those obtained for the devices fabricated on glass and PEN substrates [16, 19, 23].

The next concern for the flexible CT-MTFTs fabricated on 1.2-μm-thick CPI is their mechanical flexibility. To evaluate the device characteristics under mechanically bending states at the given R_C, the CPI films were first delaminated from carrier glass by an LLO process, as described in Section 19.2.3. Figures 19.10a and 19.10b show photo images of the detaching process of the CPI film and the delaminated film, respectively. Figure 19.10c compares the I_{DS}-V_{GS} characteristics of the flexible CT-MTFT fabricated on CPI film, which corresponds to the best device among the fabricated devices discussed here, before and after delamination via the LLO process. The charge-trap-assisted MW was not deteriorated after the LLO process and was normally obtained even when the R_C was set at 10 mm. Figure 19.10d shows the variations in the on- and off-program currents as a function of program pulse duration at an R_C of 10 mm. When the pulse duration was varied from 1 s to 1 μs, the I_{ON}/I_{OFF} did not exhibit any degradation even under the mechanically strained situations. To investigate the effects of continuous mechanical bending at an R_C of 10 mm on the memory device stabilities during the retention period, the on- and off-programmed currents were also examined with the evolution of retention time, as shown in Figure 19.10e. The memory I_{ON}/I_{OFF} showed a slight decrease from 9.3×10^6 to 2.2×10^6 after 10^4 s. However, the memory retention time required for securing three-orders-of-magnitude I_{ON}/I_{OFF} could be estimated to be three years by linear

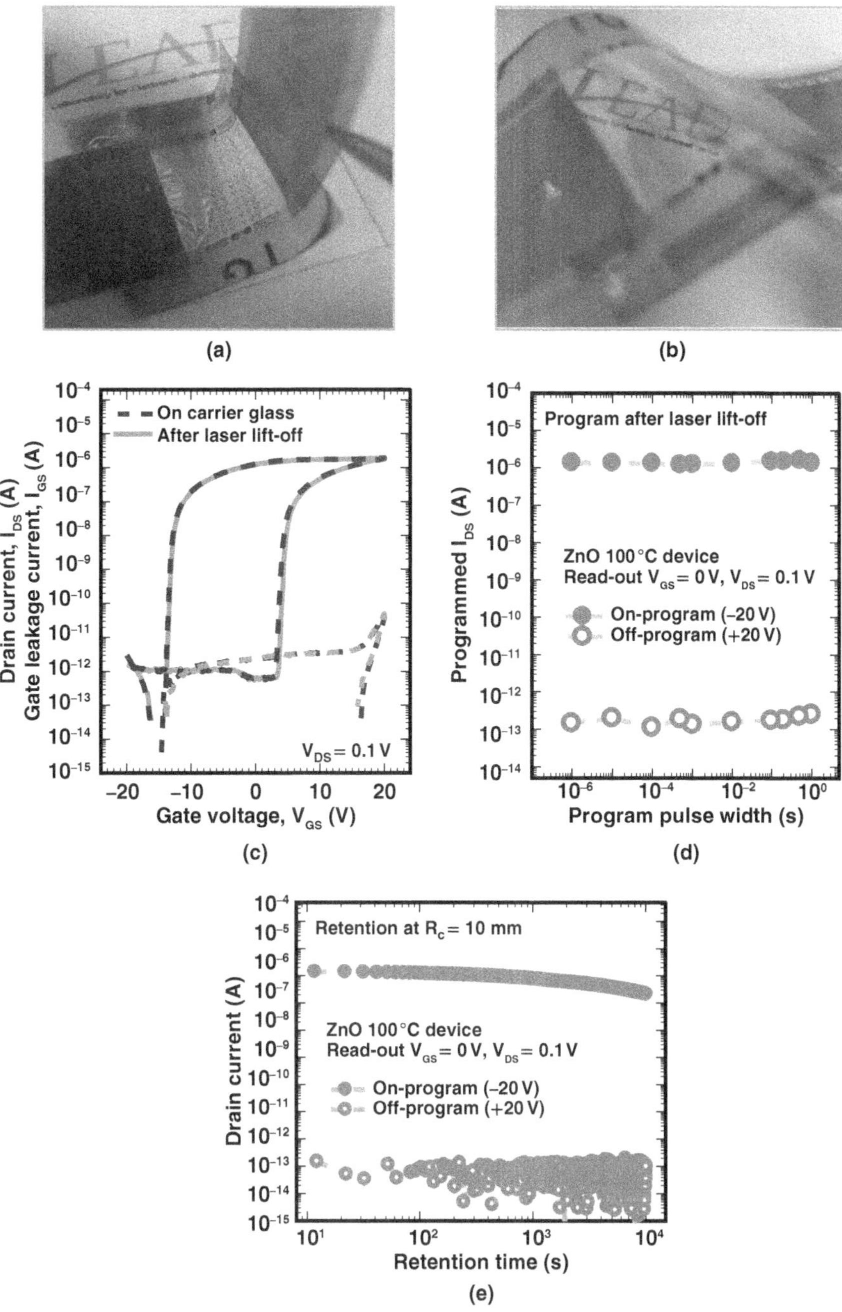

Figure 19.10 Photo images of (a) the delamination process of the CPI film from the carrier glass after the LLO process and (b) the delaminated CPI film substrate. (c) Comparison of I_{DS}-V_{GS} characteristics and (d) variations in program operations as a function of program pulse duration before and after the LLO process. (e) Variations in on- and off-program currents with a lapse of memory retention time for 10^4 s after the LLO process. The amplitude and duration of program voltage pulses were set, respectively, to ±20 V and 1 s for the off- and on-program operations.

extrapolation. Consequently, robust NVM operations, including a high-speed program and long-time retention, could be successfully examined for the flexible CT-MTFTs fabricated on ultrathin (1.2 μm) CPI film. However, mechanical durability at an R_C smaller than 1 mm and cyclic endurance of the fabricated flexible CT-MTFTs should be carefully investigated for practical applications. Thus, we have to optimize the reliable testing technique and reproducible delamination process, especially under harsher bending conditions, in our future works.

19.3.3 Operation Mechanisms and Device Physics

It was impressive that our proposed charge-trap-assisted flexible memory transistors employing oxide semiconductor active channels and CTLs exhibited high-speed program operations and long-term memory retention even under mechanically strained situations. These operational features of the fabricated devices could be totally different from those reported for conventional flexible memory devices in previous literature [26, 27]. We believe that these excellent memory device characteristics can be ascribed to the band structure of the ZnO thin film [28], which was selected as a CTL for the first time in our work. Although there have been many works that systematically investigated the physical origins of the device operations of our proposed CT-MTFTs, all results such as in-depth material analysis and device simulation have not yet been secured. However, the origins for the NVM operations can be explained with a band diagram of the gate-stack structure. Figure 19.11a illustrates the energy level and band offset among each layer composing the gate stack of the CT-MTFT when no external field is applied to the gate terminal. Alternatively, when a positive voltage pulse is applied, the conduction electrons accumulated in the IGZO channel can be transported and trapped in the ZnO CTLs, as shown by the arrows in Figure 19.11b. The threshold voltage (V_{TH}) of the MTFT is shifted to the positive direction, which corresponds to the memory off-program. As a result, a low-level programmed drain current can be determined at a subsequently carried-out readout operation at a given V_R. Contrarily, when a negative voltage pulse is applied, as shown in Figure 19.11c, the electrons located in the shallow levels of ZnO first come back to the IGZO channel, and the electrons trapped in the deep levels within the band gap are subsequently detrapped, as shown by the arrows. As a result, the V_{TH} of the MTFT is shifted to the negative direction, which corresponds to the memory on-program, and hence, a high level of programmed drain current can be obtained at a readout operation. These operational features efficiently support both benefits in the NVM operations of our proposed CT-MTFTs: high-speed program operation and long-time retention behavior. The robust retention of trapped charges with time evolution can be suggested to result from the hypothesis that most parts of electrons can be trapped in deep trap sites located in the midgap region of the ZnO CTL. In other words, a considerable amount of donor electrons occupying the shallow levels of ZnO can repel the additional electrons transported from the IGZO channel owing to self-limiting components. Thus, during the on-program events, the electrons can be readily positioned in deep trap sites, which is responsible for the long retention time of the CT-MTFTs. Furthermore, marked variations in program speeds between on- and off-program operations, as discussed in Figure 19.8c, can also result from this band diagram model. When the defect structure of the ZnO thin film was differently designed by controlling the ALD temperature, the pulse duration required for the off-program operation was examined and found to be markedly varied among the devices employing the CTLs prepared at different ALD temperatures. The off-program speed, which is associated with charge-trap events in the deep levels of ZnO, might be slower than the on-program speed related to the detrap events of trapped electrons from shallow-level states. Additionally, considering that Fowler–Nordheim tunneling phenomena can be regarded as one of the feasible origins of charge-trap/detrap events in the gate stack of MTFTs, the charge-trap process is more difficult to activate than the detrap process owing to a higher band offset between the IGZO and Al_2O_3 (3.35 eV) compared with that between the ZnO and Al_2O_3 (3.19 eV), as shown in Figure 19.11a. Consequently, the position and number of trap sites within the band gap of ZnO CTLs should be properly designed for guaranteeing excellent NVM performance, as mentioned in Section 19.3.2. Nevertheless, since there remain unclear issues to elucidate the operation mechanism and further improvements in device performance, we plan to perform additional investigations on our proposed CT-MTFTs.

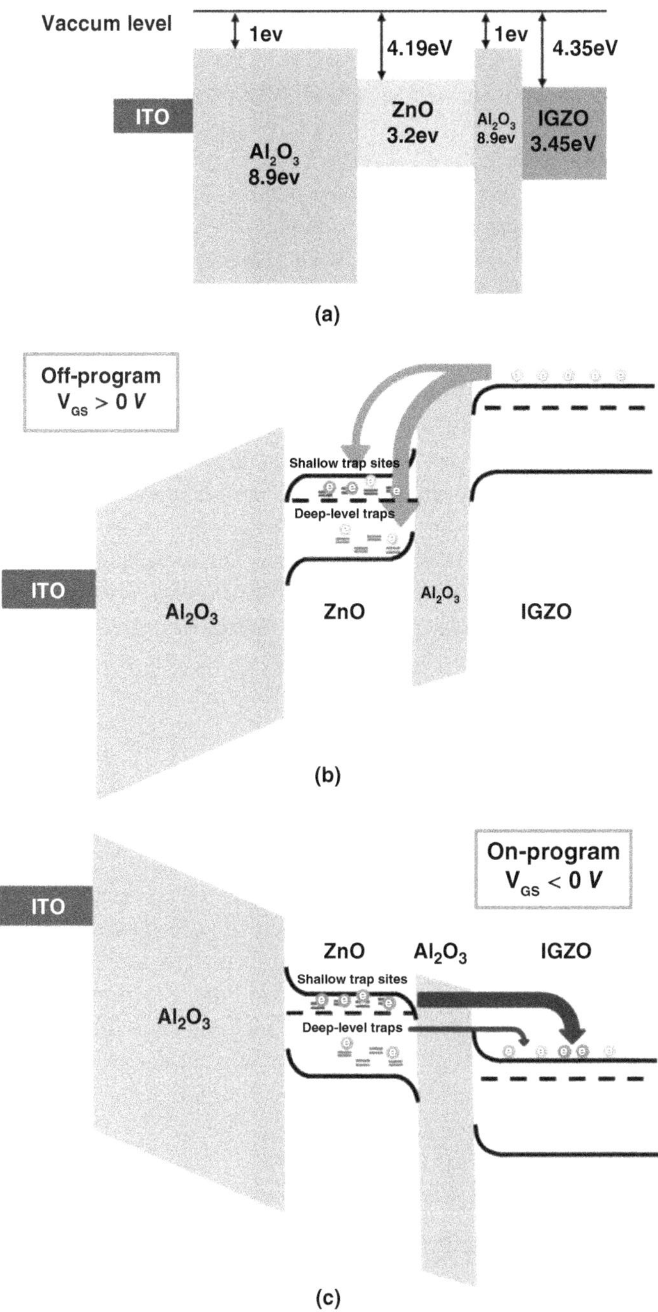

Figure 19.11 (a) Band offset and energy level of each layer composing the gate stack of the fabricated flexible CT-MTFTs without application of an external field. Schematic illustrations of the band structures during (b) off- and (c) on-program operations, respectively, in which (b) positive and (c) negative program voltages are applied to the gate terminal of the CT-MTFTs.

19.4 Choice of Alternative Materials

The mechanical flexibility of our proposed flexible CT-MTFTs has been supposed to be fatally damaged by the intrinsic brittle nature of oxide-channel materials. However, since the gate-stack structures composed of oxide thin films are individually patterned into small active areas on the substrate, the mechanical brittleness of oxide materials will not be a critical problem anymore. Furthermore, the mechanical strain induced on the device plane is determined by the flexible substrate thickness and the relative ratio of Young's modulus between the thin films prepared for the devices and the types of flexible substrates. Therefore, the substrate thickness is one of the most important design parameters to secure the mechanical durability of fabricated flexible devices. Especially for integrating large-scale memory arrays for flexible electronics, the mechanical strain induced on the device plane of a whole substrate can effectively be reduced by choosing the ultrathin film–type substrate, as discussed in Sections 19.3.1 and 19.3.2. Thus, from these viewpoints, we have to pay attention to the electrode materials composing the global interconnections and the oxide dielectric thin films, including the gate insulator and passivation layers. The electrically insulating oxides are typically prepared on the whole area of the substrate, and conducting interconnections are patterned as long lines running on every part of the substrate. Therefore, these two components are weakest to the strain applied during mechanical deformation situations. In this section, to enhance the mechanical flexibility of a single device as well as fully integrated flexible circuit systems fabricated on flexible substrates, two promising approaches are briefly introduced, such as an introduction of flexible-type conducting polymer electrodes and flexible organic gate insulators, as future perspectives of our proposed flexible CT-MTFTs, even though these techniques are anything but simple.

19.4.1 Introduction to Conducting Polymer Electrodes

The first issue is to introduce conducting polymer electrodes in place of conventional metal and/or ITO electrodes. We employed highly conductive and lithography-compatible poly(3,4-ethylenedioxythiophene) polystyrene sulfonate (PEDOT:PSS) films as S/D electrodes [29, 30]. The prototype devices using PEDOT:PSS S/D electrodes were fabricated on CPI film substrate and characterized to exhibit basic memory operations, including the charge-trap-assisted MW and program operations at an R_C of 1 mm, by carefully optimizing the patterning and transfer processes of PEDOT:PSS layers. Figure 19.12a schematically illustrates the formation of PEDOT:PSS electrode patterns, which were preliminarily developed for the first prototype devices. First, the organic layer, SA7 (TR-8857-SA7, manufactured by Dongjin Semichem, Republic of Korea), was chosen for facilitating the interactions between the polymer chains of SA7 and PEDOT:PSS (Step 1), resulting in enhanced adhesion of the PEDOT:PSS films transferred on the Al_2O_3 barrier layer treated on PI films during a photolithography process

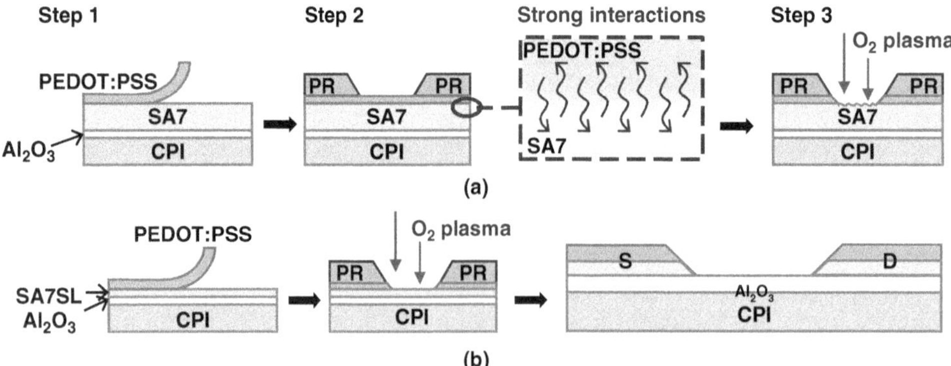

Figure 19.12 Schematic illustrations of the process flow for the PEDOT:PSS S/D patterns using (a) the previously developed conventional method and (b) the newly introduced sacrificial-layer method.

(Step 2). However, in this process, the surface of SA7, which corresponds to the backchannel for the IGZO active layer, was severely damaged by oxygen plasma employed during a dry-etching process of PEDOT:PSS. As results, the surface roughness of the backchannel region specifically degraded the device performance of the flexible CT-MTFTs. Based on this technical background, a novel process technique was developed to employ a sacrificial layer (SL) to improve the surface properties of the backchannel region and to successfully pattern the PEDOT:PSS. Figure 19.12b describes a newly developed process flow for the formation of S/D electrode patterns. First, the SL (SA7) was coated on the Al_2O_3 barrier layer, and then the PEDOT:PSS film was transferred onto the SL. Since stacked structures of PEDOT:PSS/SLs could be simultaneously patterned by a one-step dry-etching process, the Al_2O_3 layer was formed as a backchannel with a smooth surface owing to a self-regulating reaction. This can be a simple and efficient way to improve the lithography compatibility of PEDOT:PSS thin film as well as to suppress the process damage on the surface of backchannel regions. Considering that a too-thick total thickness of stacked patterns composed of SLs and PEDOT:PSS can cause the problems of higher contact resistance between the IGZO active layer and S/D electrode, the thickness of the SL should be designed to be less than 100 nm. Detailed procedures for optimization of this process can be found in our previous publication [30].

The flexible CT-MTFTs using PEDOT:PSS S/D electrodes were characterized to evaluate their memory device performance and mechanical flexibility. As discussed, various values of R_C's could be determined, as shown in Figure 19.13a. Figures 19.13b and 19.13c show photo images of the mechanically strained conditions at R_C's of 50 and 1 mm, respectively. As can be seen in these figures, application of an R_C of 1 mm corresponds to an almost folded situation of 1.2-μm-thick PI film substrate. The fabricated flexible CT-MTFTs did not exhibit any terrible degradation in device characteristics even at an R_C of 1 mm, as shown in Figure 19.13d. Figure 19.13e shows the variations in MW, SS, and I_{ON}/I_{OFF} of the flexible CT-MTFTs employing the PEDOT:PSS S/D electrodes when R_C was varied from 50 to 1 mm. The total fluctuations in MW and I_{ON}/I_{OFF} were estimated to be less than 7.5%, and the SS slightly increased from 1.11 to 1.87 V dec^{-1} due to the generation of additional defects within the

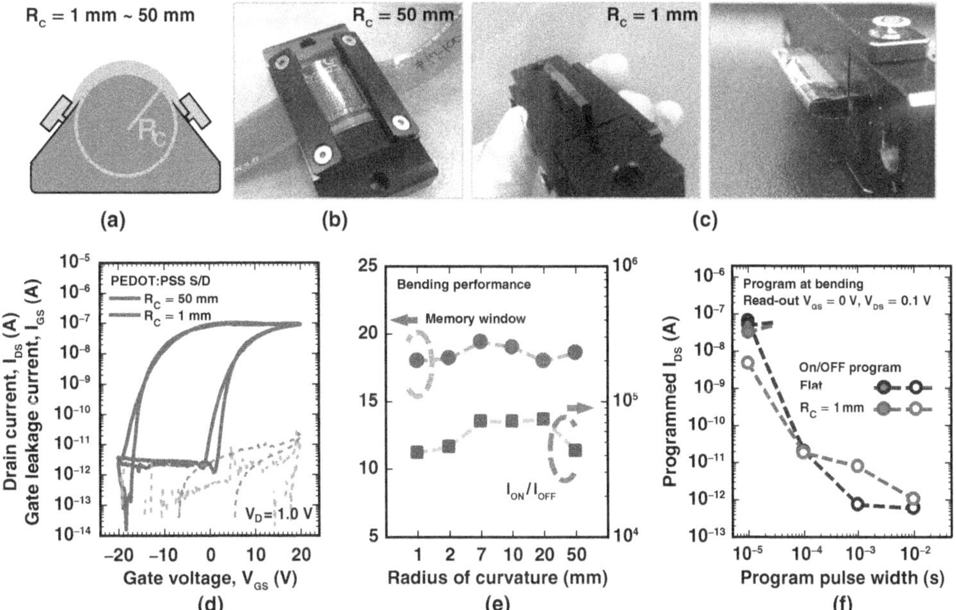

Figure 19.13 (a) Schematic illustrations of the custom-made bending jigs with various R_C values. Photo images of the mechanically strained situations at R_C's of (b) 50 and (c) 1 mm. (d) I_{DS}-V_{GS} characteristics of the fabricated flexible CT-MTFTs using PEDOT:PSS S/D electrodes at R_C's of 50 and 1 mm. (e) Variations in MW and I_{ON}/I_{OFF} when the R_C was varied from 50 to 1 mm. (f) Variations in on- and off-programmed I_{DS}'s when the program pulse duration was varied from 10 μs to 10 ms at an R_C of 1 mm.

active channel layer under the mechanically strained conditions. Stable program operations were verified with continuously applied mechanical strain at an R_C of 1 mm, as shown in Figure 19.13f. An I_{ON}/I_{OFF} higher than 10^4 was obtained at a program voltage pulse width of 100 μs. These obtained results suggest that the fabricated flexible CT-MTFTs exhibited excellent durability against harsh mechanical deformation, which can be attributed to the effective reduction of mechanical strain thanks to the use of conducting polymer electrodes and ultrathin PI film substrate. Alternatively, the thickness of flexible substrate may need to be thicker than several tens of micrometers (μm) for easy handling of thin-film substrates. Consequently, even though it is most important to control the thicknesses of both the device planes and the film substrate as a structural approach, an alternative approach to hybridize organic and inorganic materials can also be a powerful solution to enhance the mechanical flexibility of our proposed flexible memory devices for various practical applications.

19.4.2 Introduction of Polymeric Gate Insulators

The second issue is to introduce an organic thin film as an alternative flexible gate insulator for the flexible CT-MTFTs. In this work, one of the promising candidates, a coatable organic thin film, Zeocoat^{TM} (ES2110, produced by ZEON Corp., Japan), which is a thermally cross-linkable cyclo-olefin polymer, was employed as a flexible gate insulator [31]. There have been many publications on coatable polymeric dielectric thin films for use as gate insulators by means of a simple solution process for mechanically flexible TFT applications (see, e.g., Refs. [32, 33]). However, these organic thin films showed considerable gate leakage components and terribly suffered from bias–stress instability owing to lots of structural and electrical defects, such as pinholes and impurities in the polymeric gate insulators. The long-term operational stabilities of the devices using polymeric gate insulators have rarely been investigated. Alternatively, the IGZO-TFTs fabricated with the Zeocoat gate insulator exhibited superior performance in the SS, gate leakage, the ON/OFF current ratio, and the bias–stress stabilities. We have extensively investigated the physical properties of coated Zeocoat thin films, key process parameters for the device fabrication, and their impacts on the device characteristics of the IGZO channel devices in our previous publication [31].

Figures 19.14a and 19.14b show the sets of transfer characteristics of the fabricated IGZO-TFTs using Zeocoat gate insulator under NBS and PBS conditions for 3×10^4 s, respectively. The Zeocoat gate insulator was prepared to

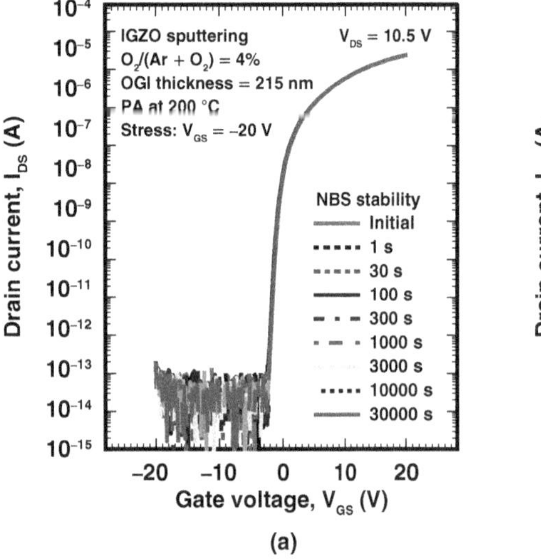

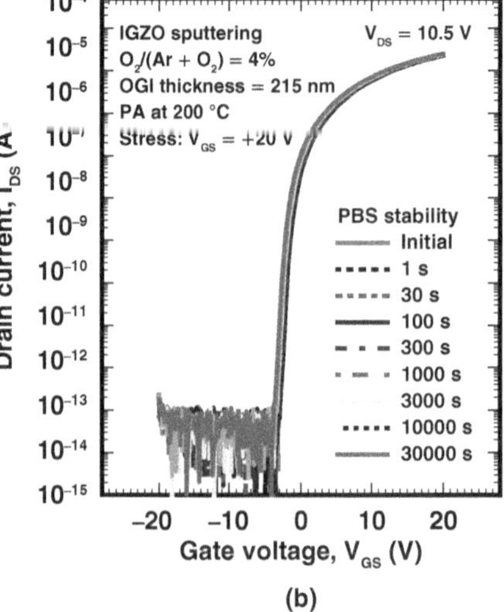

Figure 19.14 Sets of transfer characteristics of the fabricated IGZO-TFTs using a flexible organic GI of Zeocoat under (a) NBS and (b) PBS tests for 3×10^4 s. Gate bias stresses for the NBS and PBS were fixed at −20 and +20 V, respectively. The V_{DS} was set as 10.5 V.

have a film thickness of 215 nm and was finally annealed at 200 °C before the deposition of the IGZO active channel. The V_{GS} bias stresses for the NBS and PBS tests were set to −20 and +20 V, respectively. The excellent NBS and PBS stabilities could be successfully verified for the top-gate IGZO-TFTs, which suggests that the use of polymeric gate insulator can be feasible for improving the mechanical flexibility of our proposed flexible CT-MTFTs, even though total device performance should be properly designed. Furthermore, the Zeocoat thin films also can be employed for blocking layers and/or interlayer dielectrics for the flexible CT-MTFTs. Investigations on the mechanical and electrical characteristics of the flexible devices, using the Zeocoat thin films under harsh bending and stretching environments, are ongoing now.

19.5 Device Scaling to Vertical-Channel Structures

For specified future applications, the proposed flexible TFTs should also be scaled down to submicrometer channel size without any degradation in device characteristics. Additionally, short-channel effects (SCEs) should be systematically examined to verify the device feasibility, even for the scale of the submicrometer regime [34, 35]. Although oxide-channel TFTs have been mainly developed as the backplane devices for display applications, the SCE properties of oxide TFTs designed with submicrometer-long channel lengths have hardly been examined. However, for future applications of ultrahigh-resolution display panels on glass and/or plastic substrate as well as alternative-channel transistors for Si electronics, the introduction of vertical-shaped channels can be one of the most promising solutions for scaling the oxide TFTs as well as our proposed flexible CT-MTFTs [36, 37]. In this section, we introduce the device feasibility of the mechanically flexible vertical-channel IGZO-TFTs prepared on ultrathin CPI films. The memory operations and mechanical flexibility for the prototype device of vertical-channel CT-MTFTs are also presented as good examples to explore new approaches for securing the characteristics of aggressively scaling the device and further enhancing the mechanical flexibility.

19.5.1 Vertical-Channel IGZO-TFTs on PI Films

First, to verify the process and device feasibility of IGZO-TFTs with channel lengths shorter than 500 nm, vertical-channel devices were designed and characterized on CPI film substrates with a thickness of 1.2 μm [38]. Figure 19.15a shows the schematic diagram of the fabricated vertical TFTs (VTFTs), in which a spacer pattern was positioned between the source and drain electrodes. The sidewall of the spacer pattern corresponds to the channel area of the VTFT. The detailed fabrication flows and preliminarily optimized process conditions were described in our previous publication [38]. Figures 19.15b and 19.15c show a microscopic image and a cross-sectional focused-ion beam/scanning electron microscopy (FIB-SEM) image of a fabricated flexible VTFT fabricated on a PI film, which was featured to be implemented with the IGZO channel deposited by an ALD process and the coatable organic spacer of Zeocoat thin film. We have demonstrated excellent device characteristics, including bias–stress stability for the TFTs using IGZO active channels prepared by the ALD [39, 40]. The vertical profile of the Zeocoat spacer pattern was well formed, with smooth sidewalls with a vertical slope close to 90°, even though the drain ITO pattern was rather extruded from the spacer pattern. The choice of spacer material and the patterning process of the spacer pattern have continuously been explored for the realization of high-performance IGZO-VTFTs. As can be seen in Figure 19.15c, all the layers of IGZO-active, Al_2O_3 gate insulator, and Al-doped ZnO gate electrodes were conformally prepared by ALD processes. Furthermore, there were not any critical problems for obtaining sound device operations of the fabricated flexible IGZO-VTFTs. Figure 19.15d shows the I_{DS}-V_{GS} characteristics of the VTFTs fabricated by using ALD-grown IGZO active channels. The gate leakage current was estimated to be as low as 10^{-13} A. These obtained transfer characteristics suggested that the introduction of Zeocoat spacer and the optimization of a patterning process effectively improve the backchannel properties of the IGZO-VTFTs. The μ_{sat} of the fabricated VTFT and its standard deviation among the evaluated

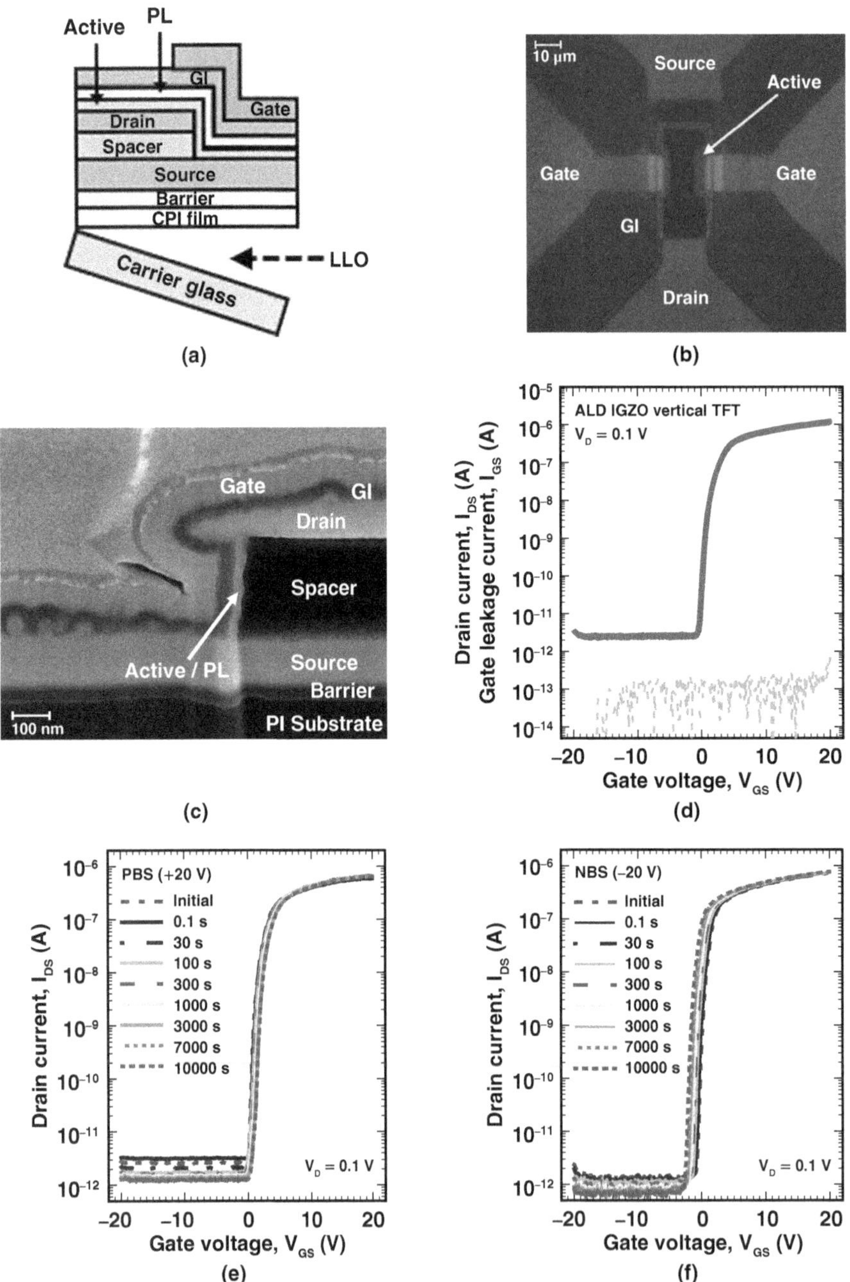

Figure 19.15 (a) Cross-sectional schematic diagram, (b) microscopic photo image, and (c) FIB-SEM image of the flexible vertical IGZO-TFTs prepared on CPI film, in which SA7 and Al_2O_3 thin films were employed as the spacer and protection layer (PL), respectively. (d) Transfer characteristics and gate leakage currents of the fabricated IGZO-VTFTs after the annealing process at 150 °C. Variations in transfer curves of the flexible IGZO-VTFTs with time evolution during (e) PBS and (f) NBS tests. A V_{GS} of +20 or −20 V was continuously applied for 10^4 s at a V_{DS} of 0.1 V.

devices were estimated to be 6.57 $cm^2\ V^{-1}\ s^{-1}$ and 0.08 $cm^2\ V^{-1}\ s^{-1}$, respectively. The SS and I_{ON}/I_{OFF} values were also examined to be 0.34 V dec^{-1} and 5.16×10^5, respectively. Figures 19.15e and 19.15f show the variations in the transfer characteristics for the IGZO-VTFTs with a lapse of stress time during the PBS and NBS tests, respectively. The gate bias stress of +20 or −20 V was continuously applied for 10^4 s during the PBS and NBS tests, respectively, and V_{DS} was fixed at 0.1 V, in which the ΔV_{ON}'s were estimated to be +1.0 and −1.8 V after a lapse of 10^4 s, respectively. These operational stabilities against the bias stresses for the flexible VTFTs were sufficiently comparable to those obtained from the well-fabricated conventional planar-channel IGZO-TFTs. It was clearly suggested from these results that the fabricated IGZO-VTFTs exhibited excellent device characteristics owing to careful designs of process conditions.

The PI films were delaminated by the LLO process to demonstrate the mechanical bending performance of the fabricated flexible VTFTs during mechanical deformations, which can be one of the structural merits of the VTFTs. Figures 19.16a and 19.16b show photo images of delamination of CPI film from carrier glass after LLO and the probing configuration for the bending test at an R_C of 1 mm, respectively. Figure 19.16c shows the variations in transfer characteristics of the flexible IGZO-VTFTs when the R_C was varied from infinity (flat state) to 1 mm. The flexible VTFTs did not exhibit any marked degradation in device operations, even under harsh mechanical bending situations at a R_C of 1 mm. The obtained excellent bending performance could originate from the factors of the structural advantages of vertical channels and the thin-film thickness of PI substrate. As discussed in this chapter, the reduction of flexible substrate thickness would be more desirable for suppressing the mechanical strain applied to the device plane and for securing the mechanical durability against cyclic bending events for flexible devices. However, the proposed VTFTs can also guarantee stable bending performance regardless of substrate thickness because in-plane propagation of microcracks does not occur along the vertical channel. Finally, the bias–stress stabilities were also examined under mechanically strained conditions to verify the superior durability of the fabricated flexible IGZO-VTFTs. Figures 19.16d and 19.16e show the variations in transfer characteristics at a R_C of 1 mm with evolution of stress time during the PBS and NBS, respectively. It was very impressive that the ΔV_{ON} values were estimated to be as small as +1.2 and −2.1 V after a lapse of 10^4 s during the PBS and NBS tests, respectively, even at a R_C of 1 mm. Considering that the excellent operational reliabilities at mechanically strained conditions have rarely been reported for the TFTs with vertical-channel structures, this is the first report on superior bending stability obtained for well-fabricated vertical-channel IGZO-TFTs on flexible PI film substrate. Now, further process optimization is ongoing to realize high-performance oxide-channel VTFTs with better device-to-device uniformity and run-to-run reproducibility.

19.5.2 Vertical-Channel Memory TFTs Using IGZO Channel and ZnO Trap Layers

Based on the experiences and insights obtained during the fabrication and characterization of flexible IGZO-VTFTs, vertical-channel CT-MTFTs were also fabricated on 1.2-μm-thick CPI film substrates [41]. The vertically formed gate stack of the fabricated CT-MTFT was verified to be well implemented on the vertical sidewall of the spacer patterns owing to the excellent conformality and self-limiting deposition mechanism of the ALD process. Figures 19.17a and 19.17b show an optical microscopic image and a magnified view of vertical-channel region of the fabricated flexible vertical CT-MTFTs, respectively. The gate-stack structures and material combination were designed to be identical with those for the conventional planar-channel device. The fabrication procedure details were described in our previous publication [41]. Figure 19.17c shows a cross-sectional FIB-SEM image of the fabricated device. The spacer pattern, which was a coated SA7 thin film with a thickness of 200 nm, was clearly formed with an angle close to 90° by means of oxygen plasma etching. All the layers, including IGZO active, ZnO CTL, the Al_2O_3 blocking layer, and the AZO gate electrode, were conformally prepared by ALD processes, even on the vertical-channel region.

Figure 19.17d shows the I_{DS}-V_{GS} characteristics and gate leakage current of the fabricated flexible vertical CT-MTFTs. The charge-trap-assisted clockwise hysteresis of I_{DS} was well confirmed. The MW and on-current of

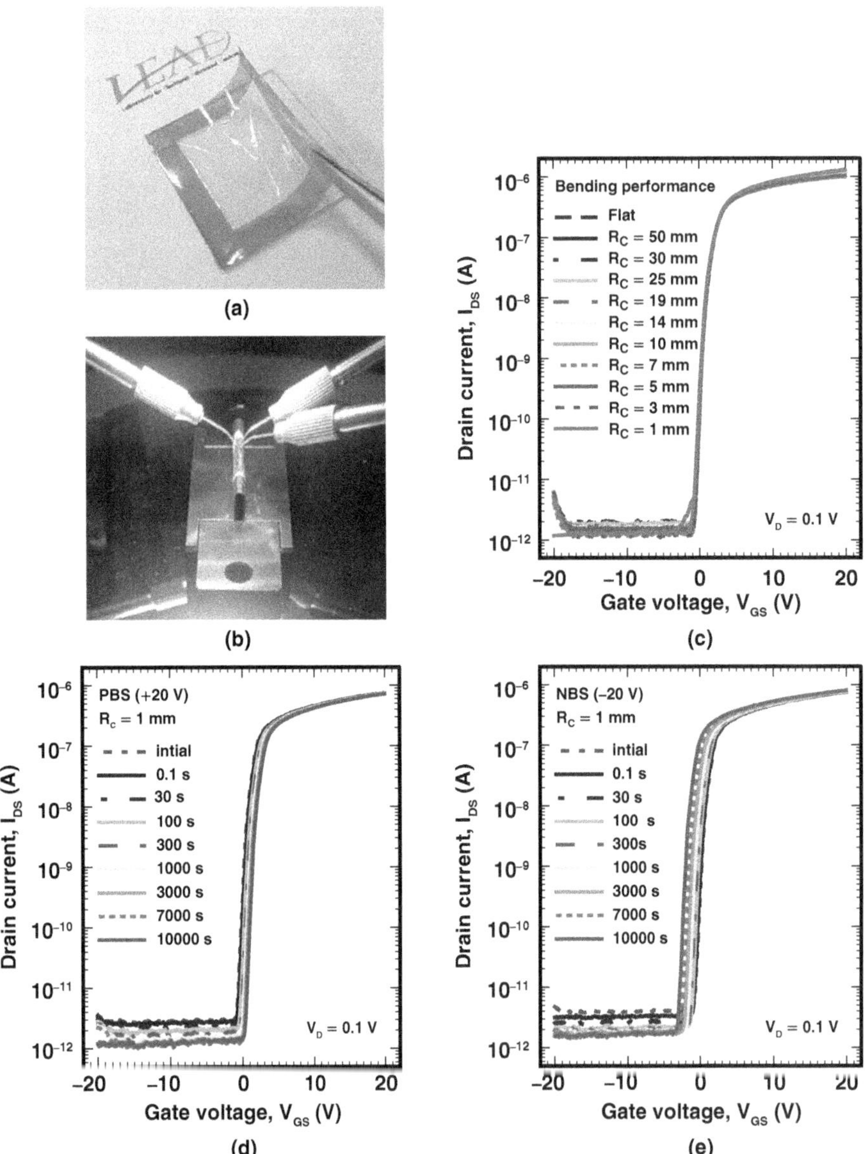

Figure 19.16 Photo images of (a) the delamination process for the flexible IGZO-VTFTs fabricated on 1.2-μm-thick CPI film from carrier glass after the LLO process and (b) the probing configuration for the mechanical bending test. (c) Variations in the transfer characteristics of the flexible IGZO-VTFTs under the bending conditions at various R_C's from infinity (flat state) to 1 mm. (a) Variations in transfer curves for the flexible IGZO-VTFTs with time evolution during (a) PBS and (b) NBS tests at an R_C of 1 mm. A V_{GS} of +20 or −20 V was continuously applied for 10^4 s at a V_{DS} of 0.1 V.

the fabricated device were estimated to be 20.8 V and 4.5×10^{-7} A, respectively, at a V_{GS} sweep range of ±20 V. The memory operations of the vertical CT-MTFTs were examined to be sensitively dependent on the deposition method of the IGZO channels. The introduction of the ALD process for the formation of the IGZO channel layer was suggested to be an imperative methodology in this work. In other words, uniform formation of an active channel can effectively reduce the structural defects within the IGZO layer, and hence, front- and backchannel interfaces could be well qualified for sound NVM operations of the fabricated device. The program operations and

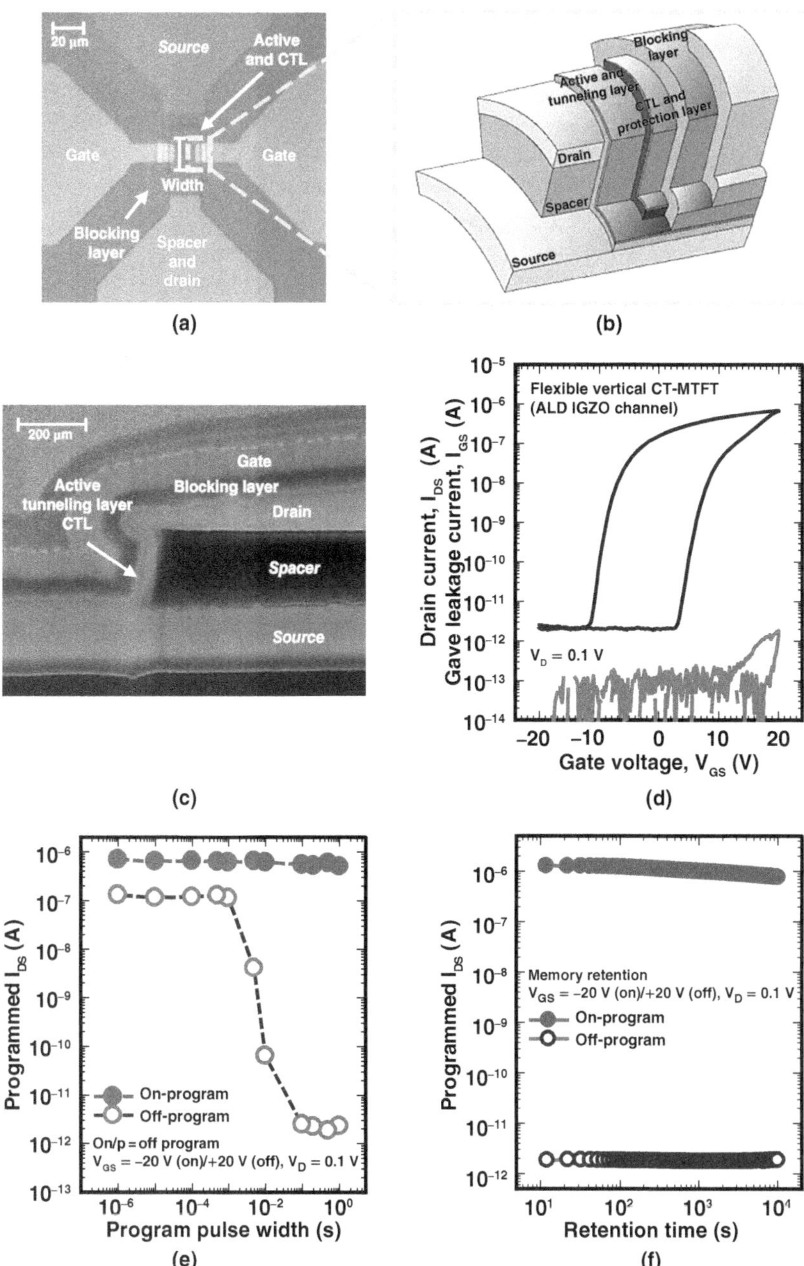

Figure 19.17 A (a) microscopic photo image and (b) magnified view of the vertical channel area of the fabricated flexible vertical CT-MTFTs. (c) A FIB-SEM cross-section of the fabricated flexible vertical CT-MTFTs. The active channel and CTL were prepared by ALD-IGZO and ZnO thin films. ITO S/D electrodes were separated by the coated organic spacer of SA7 with a thickness of 200 nm. (d) I_{DS}-V_{GS} characteristics of the fabricated vertical CT-MTFTs. Variations in the on- and off-programmed I_{DS}'s of the fabricated vertical CT-MTFTs (e) as a function of program pulse width at a V_{DS} of 0.1 V, in which the time duration of the program pulse was varied from 1 μs to 1 s; and (f) with a lapse of memory retention time for 10^4 s.

memory retention characteristics of the flexible vertical CT-MTFTs were also examined to evaluate the practical feasibility of the fabricated device. Figure 19.17e showed the variations in on- and off-programmed currents when the program voltage pulse width was varied from 1 s to 1 μs at a fixed program voltage of ±20 V. Even though it was noteworthy that the nonvolatile program operations were successfully verified for the fabricated flexible vertical CT-MTFTs, a time duration of program pulse was required to be longer than 100 ms to secure a full-range memory margin, which was much longer than that obtained from the CT-MTFTs with planar gate-stack structures. The deterioration of program speed might be caused by the surface quality of the backchannel formed on the vertical sidewall of an SA7 spacer pattern. Thus, material exploration and process optimization should be performed to improve the device performance of flexible vertical CT-MTFTs. Alternatively, Figure 19.17f shows the variations in the on- and off-programmed currents with a lapse of retention time for 10^4 s. The amplitude and durations of program pulses were set as ±20 V and 1 s, respectively. The initial memory margin of the fabricated device was obtained to be 6.9×10^5 and did not exhibit any marked degradation, even after a lapse of 10^4 s. The memory retention time of the flexible vertical CT-MTFTs using the ALD-IGZO channel could be extrapolated to be approximately 10 years with a memory I_{ON}/I_{OFF} as high as 4.1×10^5, which is much superior to those of previously reported vertical-channel memory transistors [42, 43]. The obtained memory retention characteristics were also found to be closely related to the conformal and uniform deposition of IGZO channel layers.

The mechanical flexibility of the fabricated device was finally examined by the delamination process described in this chapter. Figure 19.18a shows the variations in the I_{DS}-V_{GS} characteristics of the flexible vertical CT-MTFT when the R_C was varied from infinity (flat state) to 1 mm. It is noticeable that there were no marked changes in SS and MW values, even at an R_C of 1 mm, which corresponds to the mechanical strain of 0.17%. Then, to confirm the endurance of memory program operations during a mechanical deformation situation, the variations in programmed I_{ON}/I_{OFF} were evaluated during repetitive program operations of 10^4 cycles at an R_C of 1 mm, as shown in Figure 19.18b. There was no significant degradation in I_{ON}/I_{OFF} after the application of 10^4 program cycles at a R_C of 1 mm, which indicates that the proposed flexible vertical CT-MTFTs can guarantee robust bending durability even under severe mechanical bending conditions. This is the first demonstration of the sound NVM operations obtained from the flexible charge-trap-assisted NVM transistors with vertical-channel gate-stack structures. Consequently, these novel approaches to new designs for flexible memory devices and the obtained results can provide

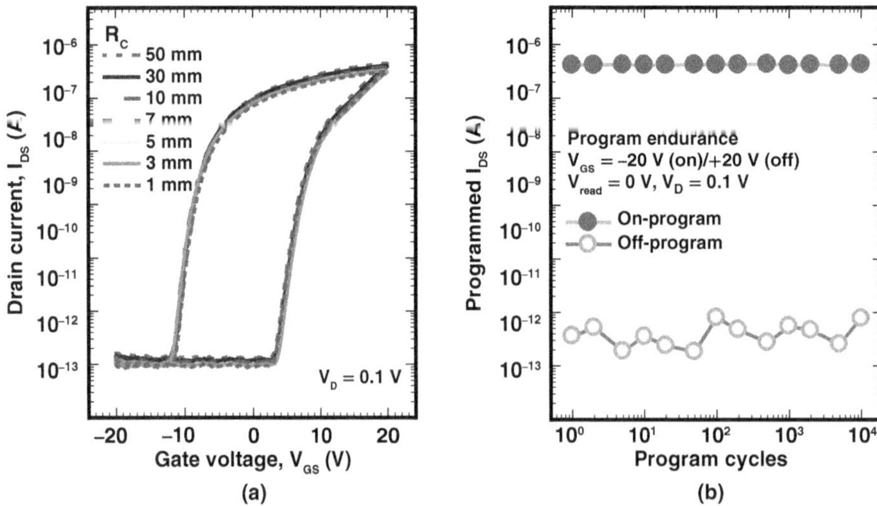

Figure 19.18 (a) Variations in the I_{DS}-V_{GS} characteristics for the flexible vertical CT-MTFTs under the bending conditions at various R_C's from infinity (flat state) to 1 mm. (b) Variations in the programmed I_{DS}'s for the vertical CT-MTFTs with repeated program cycles at an R_C of 1 mm. The amplitude and width of the program pulses were set as ±20 V and 1 s, respectively.

useful insight and guidelines in designing 3D architectures for post-DRAM (dynamic random access memory) and NAND flash memory as well as for memory-embedded monolithic-3D circuitries.

19.6 Summary

In this chapter, technical trends and recent advances for mechanically flexible charge-trap-assisted NVM TFTs using oxide semiconductor active channel and charge-trap layers were extensively overviewed from the device structure and fabrication process viewpoints. The remaining issues for further improving device performance and future perspectives of the proposed devices are finally described in this section.

19.6.1 Remaining Technical Issues

We have so far discussed the device characteristics and NVM operations of the flexible CT-MTFTs fabricated on ultrathin PI film substrate. Excellent memory device performance was successfully demonstrated, even under mechanically strained conditions, by means of strategic designs of the device fabrication process, including the preparation of flexible substrate. From this viewpoint, the specified methodologies for enhancing the mechanical flexibility and for scaling down the device size have already been stated in Sections 19.4 and 19.5, respectively. However, to provide suitable approaches for improving the device performance and to expand the application fields of our proposed flexible CT-MTFTs, several technical issues, especially in NVM operations, need to be pointed out. The NVM operations of the proposed device have been examined to be substantially feasible for practical flexible applications. However, the unified memory functions, such as low-power operation (low program voltage), high program speed, robust program endurance, and long memory retention time, will also be potentially required for future flexible and stretchable electronic applications. The suitable choice of materials and specified designs of structure of the CTLs can be the most important strategies. Engineered IGZO compositions [44] and stacked film structures of HfO_2/ZnO [45] have been explored as promising alternative CTLs. ZnO or Ag nanoparticles have also been employed for the discretely embedded shapes of CTLs for extending memory retention time [46, 47]. Film thickness variations of the Al_2O_3 tunneling oxide and ZnO CTLs have been examined to elucidate the structural effects of the gate stack on memory device performance [18]. Furthermore, the double-gate configuration, in which the second bottom gate is installed for the top-gate device, was newly introduced to lower the program voltage by appropriately controlling the fixed bias of the bottom gate and the capacitance coupling ratio between the top and bottom gates [48]. Although these approaches have exhibited meaningful improvements in parts of characteristics required for the unified memory functions, it was very hard to totally optimize the NVM performance of CT-MTFTs. Thus, the gate-stack structure proposed in Figure 19.3 has been explicitly proved to be one of the best solutions so far. Notwithstanding these achievements, the gate-stack structure should be additionally refined with bandgap engineering to meet the device specifications required for given applications demanding different degrees of memory operations and mechanical deformations.

19.6.2 Conclusions and Outlooks

As thoroughly overviewed on the proposed flexible CT-MTFTs, it would be quite demanding to integrate flexible memory elements with high performance and robust reliability into specified flexible and stretchable electronic systems. This embedded type of NVM device can be suggested to be highly desirable to enhance system performance and cost efficiency, considering that the externally connected memory chips should be individually developed for flexible electronic applications. Furthermore, the system-embedded memory elements can provide great benefits by significantly reducing power consumption owing to their functions of nonvolatile data storage. For example, the integration of NVM devices into the driving and pixel circuitries of flexible display panels achieved compact circuit size and low supply voltage by reducing the number of scan and data lines [7]. Thus, if highly functional memory devices featured to have mechanical flexibility, low-power operation, low process temperature compatibility, and robust device reliability can be secured, we can design numerous fascinating applications in flexible electronics.

As one of the future perspectives, the techniques developed for CT-MTFTs would be very useful for flexible electronics as well as for cutting-edge Si-based electronics. Actually, oxide-channel TFTs have been aggressively researched and developed as BEOL transistors with ultralow power consumption [49, 50]. Nanoscale 3D devices such as vertical NAND flash memory can also be included in the most promising application field for oxide-based CT-MTFTs. Thus, we have plans to further improve the device performance of CT-MTFTs by systematically exploring material properties and properly designing the device structures based on the operation mechanisms for various application fields in future work.

References

1 Sheng, J., Jeong, H.J., Han, K.L. et al. (2017). Review of recent advances in flexible oxide semiconductor thin-film transistors. *Journal of Information Display* 18 (4): 159–172.

2 Lee, N.E., and Trung, T.Q. (2017). Recent progress on stretchable electronic devices with intrinsically stretchable components. *Advanced Materials* 27 (3): 1603167.

3 Yoon, S.M., Yang, J.H., Kim, H.R. et al. (2018). Charge-trap-assisted flexible nonvolatile memory applications using oxide semiconductor thin-film transistors. *Japanese Journal of Applied Physics* 58 (9): 090601.

4 Fukuda, K., Sekitani, T., Zschieschang, U. et al. (2011). A 4 V operation, flexible braille display using organic transistors, carbon nanotube actuators, and organic static random-access memory. *Advanced Functional Materials* 8 (21): 4019–4027.

5 Myny, K., Rockelé, M., Chasin, A. et al. (2014). Bidirectional communication in an HF hybrid organic/solution-processed metal-oxide RFID tag. *IEEE Transactions on Electron Devices* 61 (7): 2387–2393.

6 Sakimura, N., Sugibayashi, T., Nebashi, R., and Kasai, N. (2009). Nonvolatile magnetic flip-flop for standby-power-free SoCs. *IEEE Journal of Solid-State Circuits* 44 (8): 2244–2250.

7 Kim, K.A., Byun, C.W., Yang, J.H. et al. (2016). Read-out modulation scheme for the display driving circuits composed of nonvolatile ferroelectric memory and oxide–semiconductor thin-film transistors for low-power consumption. *IEEE Transactions on Electron Devices* 63 (1): 394–401.

8 Lee, S.H., Yu, B.C., Chung, H.J., and Lee, S.W. (2017). Memory-in-pixel circuit for low-power liquid crystal displays comprising oxide thin-film transistors. *IEEE Electron Device Letters* 38 (11): 1551–1554.

9 Han, S.T., Zhou, Y., Xu, Z.X. et al. (2012). Microcontact printing of ultrahigh density gold nanoparticle monolayer for flexible flash memories. *Advanced Materials* 24 (26): 3556–3561.

10 Kim, S.M., Song, E.B., Lee, S. et al. (2012). Transparent and flexible graphene charge-trap memory. *ACS Nano* 6 (9): 7879–7884.

11 Ji, Y., Cho, B., Song, S. et al. (2010). Stable switching characteristics of organic nonvolatile memory on a bent flexible substrate. *Advanced Materials* 22 (28): 3071–3075.

12 Hwang, S.K., Bae, I., Kim, R.H., and Park, C. (2012). Flexible non-volatile ferroelectric polymer memory with gate-controlled multilevel operation. *Advanced Materials* 22 (28): 3071–3075.

13 Suresh, A., Novak, S., Wellenius, P. et al. (2009). Transparent indium gallium zinc oxide transistor based floating gate memory with platinum nanoparticles in the gate dielectric. *Applied Physics Letters* 94: 123501.

14 Yin, H., Kim, S., Kim, C.J. et al. (2008). Fully transparent nonvolatile memory employing amorphous oxides as charge trap and transistor's channel layer. *Applied Physics Letters* 93: 172109.

15 Park, J.Y., Ryu, M.K., Park, S.H.K. et al. (2014). Nonvolatile charge-trap memory transistors with top-gate structure using In–Ga–Zn-O active channel and ZnO charge-trap layer. *IEEE Electron Device Letters* 35 (3): 357–359.

16 Park, J.Y., Ryu, M.K., Park, S.H.K. et al. (2014). Impact of charge-trap layer conductivity control on device performances of top-gate memory thin-film transistors using IGZO channel and ZnO charge-trap layer. *IEEE Transactions on Electron Devices* 61 (7): 2404–2411.

17 Park, J.Y. and Yoon, S.M. (2014). High-performance transparent, all-oxide nonvolatile charge trap memory transistor using In-Ga-Zn-O channel and ZnO trap layer. *Journal of Vacuum Science & Technology B* 32 (6): 060604.

18 Park, J.Y., Kim, S.J., Byun, C.W. et al. (2015). Effects of thickness and geometric variations in the oxide gate stack on the nonvolatile memory behaviors of charge-trap memory thin-film transistors. *Solid-State Electronics* 111: 153–160.

19 Kim, S.J., Lee, W.H., Byun, C.W. et al. (2015). Photo-stable transparent nonvolatile memory thin-film transistors using In–Ga–Zn–O channel and ZnO charge-trap layers. *IEEE Electron Device Letters* 36 (11): 1153–1156.

20 Park, M.-J., Yun, D.-J., Ryu, M.-K. et al. (2015). Improvements in the bending performance and bias stability of flexible InGaZnO thin film transistors and optimum barrier structures for plastic poly(ethylene naphthalate) substrates. *Journal of Materials Chemistry C* 3: 4779–4786.

21 Jang, H.W., Kim, H.R., Yang, J.H. et al. (2018). Stability improvements of InGaZnO thin-film transistors on polyimide substrates with Al_2O_3 buffer layer. *Japanese Journal of Applied Physics* 57 (9): 090313.

22 Chien, Y.C., Chang, T.C., Chiang, H.C. et al. (2017). Role of H_2O molecules in passivation layer of a-InGaZnO thin film transistors. *IEEE Electron Device Letters* 38 (4): 469–472.

23 Kim, S.J., Park, M.J., Yun, D.J. et al. (2016). High performance and stable flexible memory thin-film transistors using In–Ga–Zn–O channel and ZnO charge-trap layers on poly(ethylene naphthalate) substrate. *IEEE Transactions on Electron Devices* 63 (4): 1557–1564.

24 Kim, H.R., Kang, C.S., Kim, S.K. et al. (2019). Characterization on the operation stability of mechanically flexible memory thin-film transistors using engineered ZnO charge-trap layers. *Journal of Physics D: Applied Physics* 52 (32): 325106.

25 Oh, B.Y., Kim, Y.H., Lee, H.J. et al. (2011). High-performance ZnO thin-film transistor fabricated by atomic layer deposition. *Semiconductor Science and Technology* 26 (8): 085007.

26 Chen, W.T. and Zan, H.W. (2012). High-performance light-erasable memory and real-time ultraviolet detector based on unannealed indium–gallium–zinc–oxide thin-film transistor. *IEEE Electron Device Letters* 33 (1): 77–79.

27 Jung, J.S., Rha, S.H., Kim, U.K. et al. (2012). The charge trapping characteristics of Si_3N_4 and Al_2O_3 layers on amorphous-indium-gallium-zinc oxide thin films for memory application. *Applied Physics Letters* 100: 183503.

28 Kamiya, T. and Hosono, H. (2005). Electronic structures and device applications of transparent oxide semiconductors: what is the real merit of oxide semiconductors? *International Journal of Applied Ceramic Technology* 2 (4): 285–304.

29 Yang, J.H., Yun, D.J., Kim, S.M. et al. (2018). Introduction of lithography-compatible conducting polymer as flexible electrode for oxide-based charge-trap memory transistors on plastic poly(ethylene naphthalate) substrates. *Solid State Electronics* 150: 35–40.

30 Yang, J.H., Kim, D.K., Yoon, M.H. et al. (2019). Mechanically robust and highly flexible nonvolatile charge-trap memory transistors using conducting-polymer electrodes and oxide semiconductors on ultrathin polyimide film substrates. *Advanced Materials Technologies* 4: 1900348.

31 Kawk, S.M., Kim, H.R., Jang, H.W. et al. (2019). Improvement in bias-stress and long-term stabilities for in-Ga-Zn-O thin-film transistors using solution-process-compatible polymeric gate insulator. *Organic Electronics* 71: 7–13.

32 Schroeder, R., Majewski, L.A., and Grell, M. (2005). High-performance organic transistors using solution-processed nanoparticle-filled high-k polymer gate insulators. *Advanced Materials* 17 (12): 1535–1539.

33 Gong, Y., Zhao, K., Yan, L. et al. (2018). Room temperature fabrication of high quality ZrO_2 dielectric films for high performance flexible organic transistor applications. *IEEE Electron Device Letters* 39 (2): 280–283.

34 Song, I., Kim, S., Yin, H. et al. (2008). Short channel characteristics of gallium–indium–zinc–oxide thin film transistors for three-dimensional stacking memory. *IEEE Electron Device Letters* 29 (6): 549–552.

35 Kim, H.W., Kim, E.S., Park, J.S. et al. (2018). Influence of effective channel length in self-aligned coplanar amorphous-indium-gallium-zinc-oxide thin-film transistors with different annealing temperatures. *Applied Physics Letters* 113: 022104.

36 Kim, Y.M., Kang, H.B., Kim, G.H. et al. (2017). Improvement in device performance of vertical thin-film transistors using atomic layer deposited IGZO channel and polyimide spacer. *IEEE Electron Device Letters* 38 (10): 1387–1389.

37 Kim, H.R., Yang, J.H., Kim, G.H., and Yoon, S.M. (2018). Flexible vertical-channel thin-film transistors using In-Ga-Zn-O active channel and polyimide spacer on poly(ethylene naphthalate) substrate. *Journal of Vacuum Science & Technology B* 37 (1): 010602.

38 Kim, H.R., Furuta, M., and Yoon, S.M. (2019). Highly robust flexible vertical-channel thin-film transistors using atomic-layer-deposited oxide channels and Zeocoat spacers on ultrathin polyimide substrates. *ACS Applied Electronic Materials* 1 (11): 2363–2370.

39 Yoon, S.M., Seong, N.J., Choi, K.J. et al. (2017). Effects of deposition temperature on the device characteristics of oxide thin-film transistors using In–Ga–Zn–O active channels prepared by atomic-layer deposition. *ACS Applied Materials & Interfaces* 9 (27): 22676–22684.

40 Yoon, S.J., Seong, N.J., Choi, K.J. et al. (2018). Investigations on the bias temperature stabilities of oxide thin film transistors using In–Ga–Zn–O channels prepared by atomic layer deposition. *RSC Advances* 8: 25014–25020.

41 Kim, H.R., Kim, G.H., Seong, N.J. et al. (2020). Comparative studies on vertical-channel charge-trap memory thin-film transistors using In-Ga-Zn-O active channels deposited by sputtering and atomic layer depositions. *Nanotechnology* 31 (43): 435702.

42 Bosch, G., Kar, G.S., Blomme, P. et al. (2011). Highly scaled vertical cylindrical SONOS cell with bilayer polysilicon channel for 3-D nand flash memory. *IEEE Electron Device Letters* 32 (11): 1501–1503.

43 She, X.J., Gustafsson, D., and Sirringhaus, H. (2017). A vertical organic transistor architecture for fast nonvolatile memory. *Advanced Materials* 27: 1703541.

44 Yun, D.J., Kang, H.B., and Yoon, S.M. (2016). Process optimization and device characterization of nonvolatile charge trap memory transistors using In–Ga–ZnO thin films as both charge trap and active channel layers. *IEEE Transactions on Electron Devices* 63 (8): 3128–3134.

45 Na, S.Y. and Yoon, S.M. (2019). Impacts of HfO_2/ZnO stack-structured charge-trap layers controlled by atomic layer deposition on nonvolatile memory characteristics of In-Ga-Zn-O channel charge-trap memory thin-film transistors. *IEEE Journal of the Electron Devices Society* 7: 453–461.

46 Seo, G.H., Yun, D.J., Lee, W.H., and Yoon, S.M. (2017). Atomic-layer-deposition-assisted ZnO nanoparticles for oxide charge-trap memory thin-film transistors. *Nanotechnology* 28 (7): 075202.

47 Park, M.J., Jeong, S.H., Hong, G.R. et al. (2016). Preparation of mono-layered Ag nanoparticles for charge-trap sites of memory thin-film transistors using In-Ga-Zn-O channel. *ECS Journal of Solid State Science and Technology* 6 (1): Q18–Q22.

48 Son, M.T., Kim, S.J., and Yoon, S.M. (2019). Impacts of bottom-gate bias control for low-voltage memory operations of charge-trap memory thin film transistors using oxide semiconductors. *Journal of Semiconductor Technology and Science* 19 (1): 69–78.

49 Saito, N., Sawabe, T., Kataoka, J. et al. (2019). High mobility (>30 cm^2 V^{-1} S^{-1}) and low source/drain parasitic resistance In–Zn–O BEOL transistor with ultralow <10^{-20} A μm^{-1}off-state leakage current. *Japanese Journal of Applied Physics* 58 (4): SBBJ07.

50 Ishizu, T., Yakubo, Y., Furutani, K. et al. (2019). A 48 MHz 880-nW standby power normally off MCU with 1 Clock full backup and 4.69-featuring 60-nm crystalline In-Ga-Zn oxide BEOL-FETs. In: *2019 Symposium on VLSI Circuits*, C48.

20

Amorphous Oxide Semiconductor TFTs for BEOL Transistor Applications

Nobuyoshi Saito and Keiji Ikeda

KIOXIA, Tokyo, Japan

20.1 Introduction

Continuous scaling of Si complementary metal-oxide semiconductor (CMOS) transistors has driven progress in our society by providing denser, higher performance, and cheaper CMOS large-scale integrations (LSIs) [1]; however, simple scaling of Si CMOS transistors reached its limit around the 130-nm technology node due to issues such as lithography, random dopant fluctuation, and increase of leakage current in metal-oxide semiconductor field-effect transistors (MOSFETs). Although transistor scaling has been pushed forward by introducing innovative technologies such as strained Si, high-k metal gates, and fin field-effect transistors (FinFETs), scaling of Si CMOS transistors is becoming more and more difficult. And the pace of scaling in recent technology nodes, such as 14 nm and 10 nm, has been slowed down, caused by increased process steps and their complexity [2]. Moreover, the scaling of interconnects in CMOS LSIs also causes increases of parasitic resistance and capacitance, which impact the deterioration of both the signal bandwidth and power consumption. Therefore, a new technology independent of 2D scaling has been strongly demanded. In this situation, 3D integration technology has attracted attention as an alternative to conventional 2D scaling. 3D integration technology is based on stacking layers with transistors and connecting them by vertical interconnects [3]. This integration scheme enables one to expand the functionality of LSIs continuously by increasing the number of transistors per chip while keeping the chip size. Moreover, this 3D integration can improve the power consumption and signal bandwidth by minimizing the interconnect length. There are several approaches in 3D integration, such as chip stacking with through-silicon vias (TSVs) [4], a stacking thin Si layer on prefabricated wafers [5, 6], and TFT fabrication in the back end of line (BEOL) [7]. Among these approaches, TFT fabrication in the BEOL has the advantages of enabling transistor formation in multiple layers with a low-cost process and the possibility to form ultrahigh-density ($>10^7/mm^2$) connecting-vias [8]. Previously, amorphous Si or polycrystalline Si has been used as channel material for BEOL transistors; however, the Si channel has a trade-off between its mobility and process temperature via its crystalline grain size. Consequently, it has been difficult to obtain a high-performance Si-channel BEOL transistor with a low thermal budget ($<400\ °C$) that is available in the BEOL process of a Cu interconnection layer.

From these viewpoints, amorphous oxide semiconductors (AOSs) such as InGaZnO are promising candidates for *n*-channel material for BEOL transistors owing to their low-temperature ($<400\ °C$) process and high-mobility ($>10\ cm^2/Vs$) characteristics [9, 10]. Moreover, thanks to their wide bandgap energy (>3 eV), InGaZnO TFTs have unique features such as high immunity to source/drain (S/D) punch-through [11] and extremely low off-state leakage current characteristics [12]. In addition to that, compared to channel materials with polycrystalline states, channel materials with amorphous states have the advantage of suppression of variations of TFT characteristics such as mobility and threshold voltage, because they are free from the influence of trap states at grain boundaries. Therefore, BEOL transistors using AOS channels are expected to not only serve *n*-channel transistors in

Amorphous Oxide Semiconductors: IGZO and Related Materials for Display and Memory, First Edition.
Edited by Hideo Hosono and Hideya Kumomi.

3D integration but also provide new functions to Si CMOS LSIs. Several groups have demonstrated novel devices by co-integrating InGaZnO BEOL transistors with Si CMOSs: high-voltage input/outputs (I/Os), image sensors, embedded memories, and normally-off CPUs [13–16].

However, several challenges must be solved to put these AOS-channel BEOL transistors into practical use. One is low thermal stability against H_2 annealing. The most practical AOS material, InGaZnO, has been reported to generate excess carrier by H_2 annealing and degrade to be metallic [17, 18]. This poor immunity to H_2 annealing is an obstacle to integration with Si CMOS LSIs, since the Si CMOS process often uses H_2 annealing for terminating defects at the Si/SiO_2 interface. Another challenge is the on-current of InGaZnO TFTs. InGaZnO TFTs show extremely low off-state leakage current characteristics, while on-current is insufficient for cooperation with Si CMOSs. To expand AOS-channel BEOL transistor applications, on-current improvement while maintaining extremely low off-state leakage current characteristics is essential. This chapter introduces our approaches to overcome these issues for BEOL transistor applications [19, 20].

20.2 Improvement of Immunity to H_2 Annealing

Figure 20.1 shows a concept for the improvement of H_2-anneal immunity by replacing Zn in InGaZnO with Si. This idea is based on a comparison of the dissociation energies of In-O, Ga-O, Zn-O, and Si-O bonds [21, 22]. Since the Si-O bond exhibits much larger dissociation energy than the Zn-O bond does, replacing Zn atoms with Si atoms is expected to suppress oxygen vacancy (V_o) formation during H_2 annealing.

Figure 20.2 shows the annealing temperature dependence of sheet resistance of deposited InGaSiO and InGaZnO films after annealing in N_2 (Figure 20.2a) and $H_2 + N_2$ (Figure 20.2b) (both 2%) atmosphere, respectively. Compared to N_2 annealing, the sheet resistance of AOS films decreased at lower temperature in H_2 annealing. It is supposed that hydrogen accelerates V_o formation in AOS films due to its reducing property. As shown in Figure 20.2b, the sheet resistance of InGaSiO (Si/In ratio > 0.47) maintained higher resistivity of more than five orders of magnitudes (>1×10^{10} Ω/sq.) after H_2 annealing at 320 °C and higher temperatures, while that of InGaZnO film decreased severely after annealing at 320 °C. This suggests that Si incorporation suppresses V_o formation in AOS films during H_2 annealing. Figure 20.3 shows the Si/In ratio dependence of sheet resistance of InGaSiO film after H_2 annealing at 370 °C. The InGaSiO films with a high Si/In ratio (>0.32) are expected to maintain semiconductor properties and show higher H_2-anneal immunity than that of conventional InGaZnO.

Figure 20.4 shows a comparison of I_d-V_g characteristics between an (a) InGaZnO TFT and (b) InGaSiO TFT (Si/In ratio = 0.47) after the H_2 annealing. Although InGaZnO film became metallic after H_2 annealing at 320 °C, as shown in Figure 20.4a, the InGaZnO TFT kept an enhancement-mode operation up to 360 °C. This indicates that

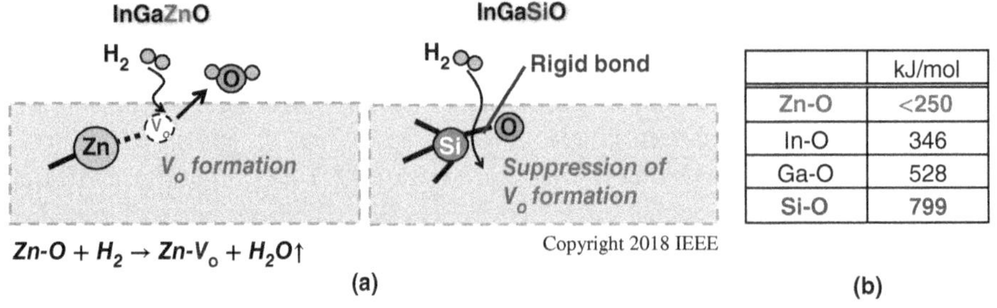

	kJ/mol
Zn-O	<250
In-O	346
Ga-O	528
Si-O	799

Figure 20.1 (a) Concept for the improvement of H_2-anneal immunity of AOS-channel TFTs, and (b) list of metal–oxygen bond dissociation energy of InGaZnO-related metals and Si [19, 21]. *Source*: Saito, N., et al. (2018). High-mobility and H2-anneal tolerant InGaSiO / InGaZnO / InGaSiO double hetero channel thin film transistor for Si-LSI compatible process. *IEEE Journal of the Electron Devices Society* **6**: 500–505.

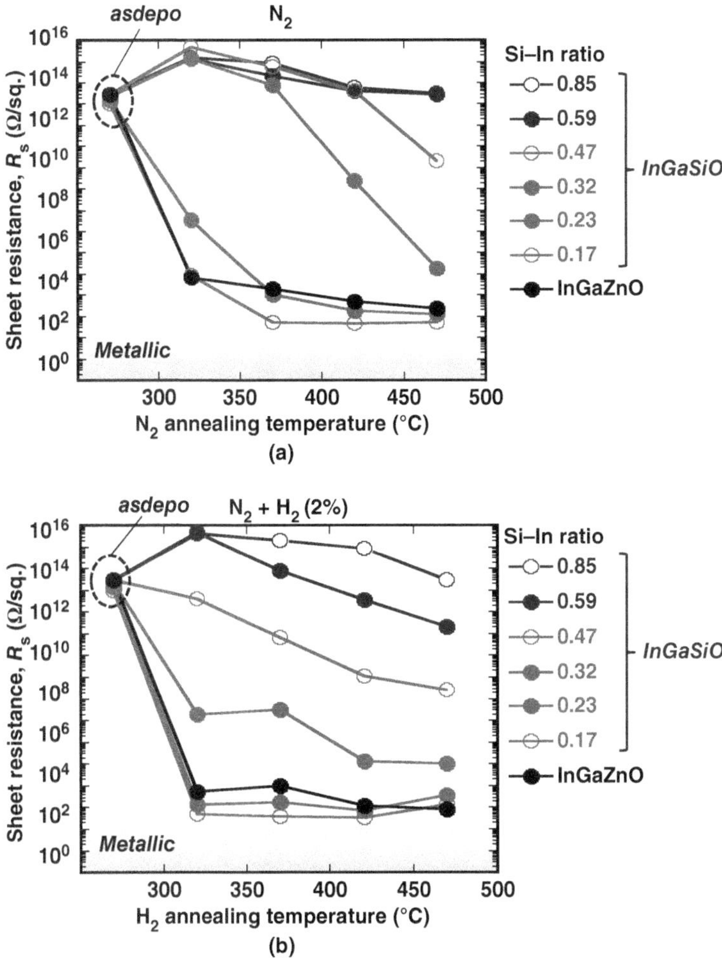

Figure 20.2 Annealing temperature dependence of sheet resistance of InGaSiO and InGaZnO films after annealing in (a) N_2 and (b) $H_2 + N_2$ atmosphere [19]. *Source*: Saito, N., et al. (2018). High-mobility and H2-anneal tolerant InGaSiO / InGaZnO / InGaSiO double hetero channel thin film transistor for Si-LSI compatible process. *IEEE Journal of the Electron Devices Society* **6**: 500–505.

an interlayer SiO_2 in the TFT structure may suppress V_0 formation in the InGaZnO channel by blocking oxygen out-diffusion from the InGaZnO channel, and it improves the H_2-anneal immunity to some degree; however, after the annealing at 380 °C, only the InGaSiO TFT remained in enhancement mode, while the InGaZnO TFT changed to depletion mode. This higher immunity to H_2 annealing of the InGaSiO TFT is consistent with the results in Figure 20.2.

Figure 20.5 shows a clear trade-off between the V_{th} stability and high I_{on} as a function of the Si/In ratio. Although a higher Si fraction improves the immunity to H_2 annealing, I_{on} is more degraded, and the electron mobility for the InGaSiO TFT was estimated to be 0.14 cm^2/Vs for the Si/In ratio = 0.47, which is only about 1% of the mobility for the InGaZnO TFT.

For the improvement of both mobility and H_2-anneal immunity, we examined a double-hetero (DH)-channel TFT composed of an InGaZnO core layer and thermally stable InGaSiO top and bottom barrier layers, as shown in Figure 20.6. Highly H_2-anneal immune InGaSiO encapsulates the high-mobility oxide semiconductor layer and

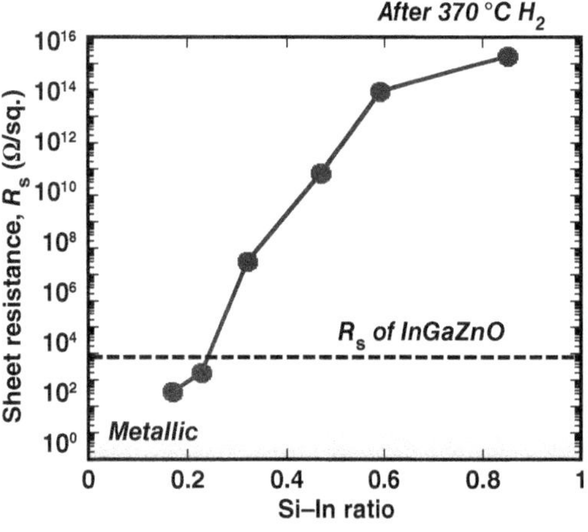

Figure 20.3 Si/In composition ratio dependence of sheet resistance of InGaSiO films after H_2 annealing at 370 °C [19]. *Source*: Saito, N., et al. (2018). High-mobility and H2-anneal tolerant InGaSiO / InGaZnO / InGaSiO double hetero channel thin film transistor for Si-LSI compatible process. *IEEE Journal of the Electron Devices Society* **6**: 500–505.

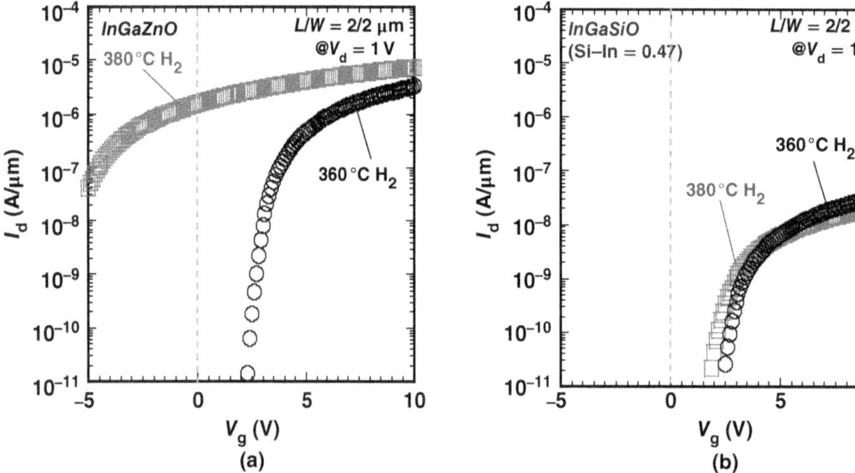

Figure 20.4 Comparison of I_d-V_g characteristics between (a) InGaZnO and (b) InGaSiO TFTs after H_2 annealing at 360 °C and 380 °C [19]. *Source*: Saito, N., et al. (2018). High-mobility and H2-anneal tolerant InGaSiO / InGaZnO / InGaSiO double hetero channel thin film transistor for Si LSI compatible process. *IEEE Journal of the Electron Devices Society* **6**: 500–505.

can suppress V_o formation in the high-mobility core layer by blocking oxygen diffusion. Additionally, X-ray photoemission spectroscopy and ultraviolet photoemission spectroscopy revealed that the bandgap energy of InGaSiO increases with the increase of the Si/In ratio. The conduction band offset of the proposed DH-channel structure was 0.7 eV, and the band structure of the DH-channel structure could be a type-I quantum well. Since electrons are confined in the InGaZnO layer with higher mobility than in the other, mobility improvement is expected by carrier separation from the SiO_2 interface via the InGaSiO barrier layers.

Figure 20.7 shows I_d-V_g characteristics of the InGaSiO–InGaZnO–InGaSiO (5 nm/10 nm/2 nm) DH-channel TFT after H_2 annealing at 380 °C. About two orders of magnitude higher I_{on} was observed compared to that of the InGaSiO TFT. Furthermore, the enhancement-mode operation was maintained, suggesting that the InGaSiO barrier layers suppress dissociation of Zn-O bonds in the InGaZnO core layer via blocking oxygen diffusion out from the core layer. Figure 20.8 shows effective electron mobility in the InGaZnO TFT and the DH-channel TFT extracted by a split C-V method (L/W = 50 μm/50 μm). Almost three times higher mobility (~30 cm^2/Vs) than that of the InGaZnO TFT was realized at $N_s = 2 \times 10^{12}$ cm^{-2} in the DH-channel TFT.

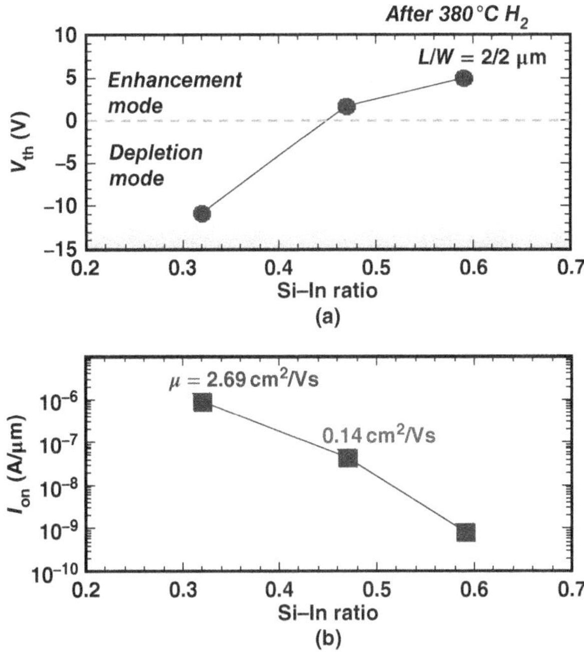

Figure 20.5 Si/In composition ratio dependence of (a) V_{th} and (b) I_{on} after H_2 annealing [19]. *Source*: Saito, N., et al. (2018). High-mobility and H2-anneal tolerant InGaSiO / InGaZnO / InGaSiO double hetero channel thin film transistor for Si-LSI compatible process. *IEEE Journal of the Electron Devices Society* **6**: 500–505.

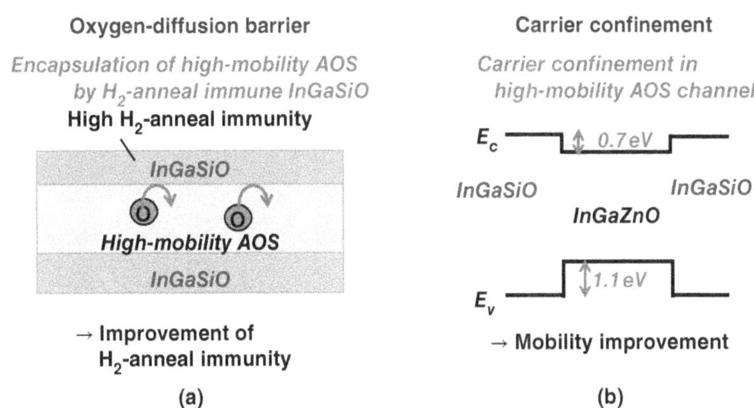

Figure 20.6 Concept for the improvement of both H_2-anneal immunity and mobility via a double-hetero-channel TFT composed of an InGaZnO core layer and H_2-anneal-immune InGaSiO top and bottom barrier layers [19]. *Source*: Saito, N., et al. (2018). High-mobility and H2-anneal tolerant InGaSiO / InGaZnO / InGaSiO double hetero channel thin film transistor for Si-LSI compatible process. *IEEE Journal of the Electron Devices Society* **6**: 500–505.

Figure 20.9 shows a relationship between mobility (at $N_s = 2 \times 10^{12}$ cm^{-2}) and H_2-anneal immunity (V_{th} after H_2 annealing at 380 °C, $L/W = 2$ µm/2 µm). It is shown that the proposed InGaSiO–InGaZnO DH-channel TFT achieved both better immunity (stable enhancement-mode operation) and higher mobility than InGaZnO and InGaSiO single-layer TFTs.

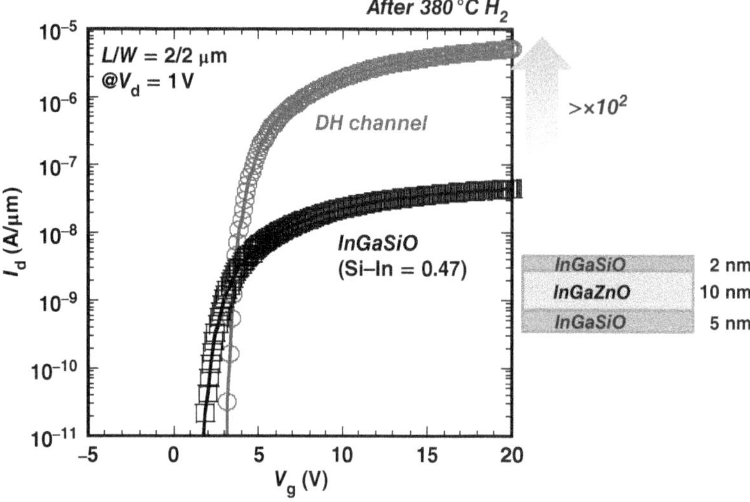

Figure 20.7 I_d-V_g characteristics of the double-hetero-channel TFT after H_2 annealing at 380 °C [19]. *Source:* Saito, N., et al. (2018). High-mobility and H2-anneal tolerant InGaSiO / InGaZnO / InGaSiO double hetero channel thin film transistor for Si-LSI compatible process. *IEEE Journal of the Electron Devices Society* **6**: 500–505.

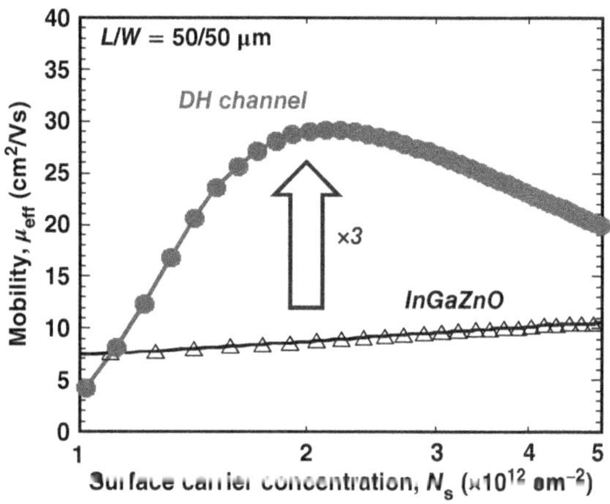

Figure 20.8 Comparison of mobility characteristics between InGaSiO–InGaZnO–InGaSiO double-hetero-channel and InGaZnO TFTs extracted by the split *C-V* method [19]. *Source:* Saito, N., et al. (2018). High-mobility and H2-anneal tolerant InGaSiO / InGaZnO / InGaSiO double hetero channel thin film transistor for Si-LSI compatible process. *IEEE Journal of the Electron Devices Society* **6**: 500–505.

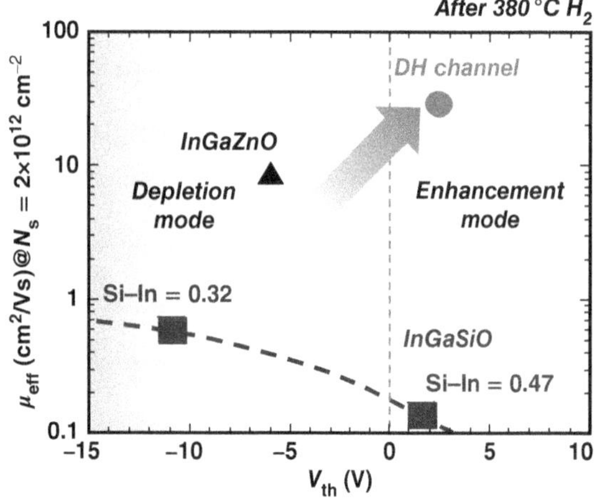

Figure 20.9 Relationship between mobility at $N_s = 2 \times 10^{12}$ cm^{-2} and H_2-anneal immunity (V_{th} after H_2 annealing at 380 °C, $L/W = 2$ μm/2 μm) [19]. *Source:* Saito, N., et al. (2018). High-mobility and H2-anneal tolerant InGaSiO / InGaZnO / InGaSiO double hetero channel thin film transistor for Si-LSI compatible process. *IEEE Journal of the Electron Devices Society* **6**: 500–505.

20.3 Increase of Mobility and Reduction of S/D Parasitic Resistance

For expanding BEOL transistor applications, it is key to not only increase the mobility but also reduce the S/D parasitic resistance (R_{para}) by maintaining extremely low off-state leakage current characteristics. Figure 20.10 shows our concept of a high-performance AOS-channel BEOL transistor with both high on-current (I_{on}) and extremely low off-state current (I_{off}). Here, we experimentally verify the advantages of the InZnO channel compared with the InGaZnO channel in both mobility and S/D parasitic resistance.

Figure 20.11a shows the mobility characteristics of InZnO TFTs and InGaZnO TFTs. The mobility of InZnO TFTs showed 33 cm^2/Vs at the surface carrier concentration (N_s) of 10^{13} cm^{-2}, which was about twice higher than that of typical InGaZnO TFTs. Figure 20.11b shows In composition ratio dependence of mobility of InGaZnO TFTs at N_s of 8×10^{12} cm^{-2}. Increase of the In composition ratio is effective to improve the mobility of InGaZnO TFTs [23, 24]. It was found that, when compared at the same In composition ratio of 0.56, InZnO TFTs showed higher mobility than InGaZnO TFTs. This suggests that Ga acts as a carrier-scattering source in InGaZnO [25, 26].

Figure 20.12a shows a R_{on}-L plot with gate overdrive voltages from 4 V to 16 V. The value of S/D parasitic resistance at each gate overdrive voltage was estimated from the vertical intersection point in the R_{on}-L plot. As shown in Figure 20.12b, S/D parasitic resistance decreased as the gate voltage increased. This gate voltage dependence of parasitic resistance can be explained by the potential modulation of the oxide semiconductor at S/D contact by gate voltage due to the gate–S/D overlapped structure. As illustrated in Figure 20.12c, the Schottky barrier becomes narrower as the gate voltage increases, which results in the increase of tunneling in addition to thermionic field emission. Figure 20.13 shows a comparison of gate overdrive voltage dependence of S/D parasitic resistance between InZnO TFTs and InGaZnO TFTs. It was found that the S/D parasitic resistance of InZnO TFTs was reduced to 3.6 kΩ μm at a gate overdrive voltage of 16 V, which was about 75% smaller than that of InGaZnO TFTs.

To investigate an origin of the S/D parasitic resistance reduction by InZnO TFTs, Schottky barrier height (SBH) was extracted. Assuming that the S/D parasitic resistance at the small gate overdrive voltage is mainly dominated by thermionic field emission at S/D contact, as shown in Figure 20.12b, we fitted S/D parasitic resistance behavior at the small gate overdrive voltage region (V_g-V_{th} < 1 V) with the following extended equation of the thermionic emission model [27]:

$$R_{para} = \frac{k}{qA * T} \exp\left(-\frac{q\left(\emptyset_B - \alpha V_g\right)}{kT}\right)$$

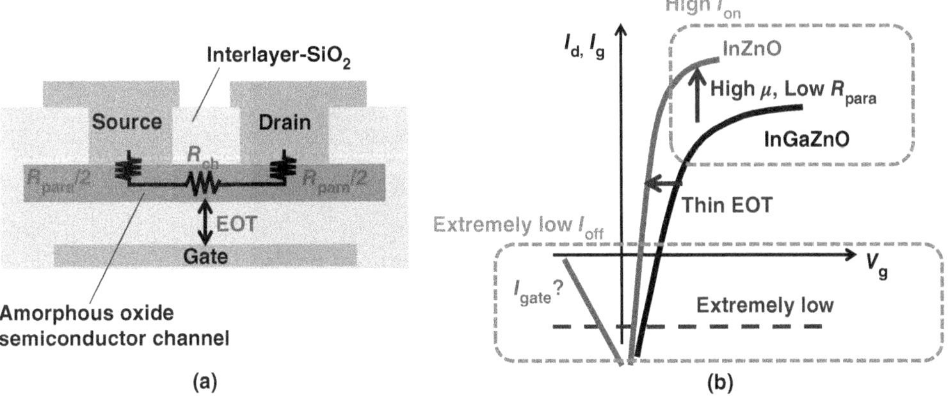

Figure 20.10 (a) Concept for I_{on} improvement, and (b) a proposed high-performance AOS-channel TFT with high I_{on} and extremely low I_{off} by using an InZnO channel with thin equivalent oxide thickness (EOT) [20]. *Source*: Saito, N., et al. (2019). High mobility (>30cm²/Vs) and low source/drain parasitic resistance In-Zn-O BEOL transistor with ultralow (<10^{-20}A/μm) off-state leakage current. *Japanese Journal of Applied Physics* **58**: SBBJ07.

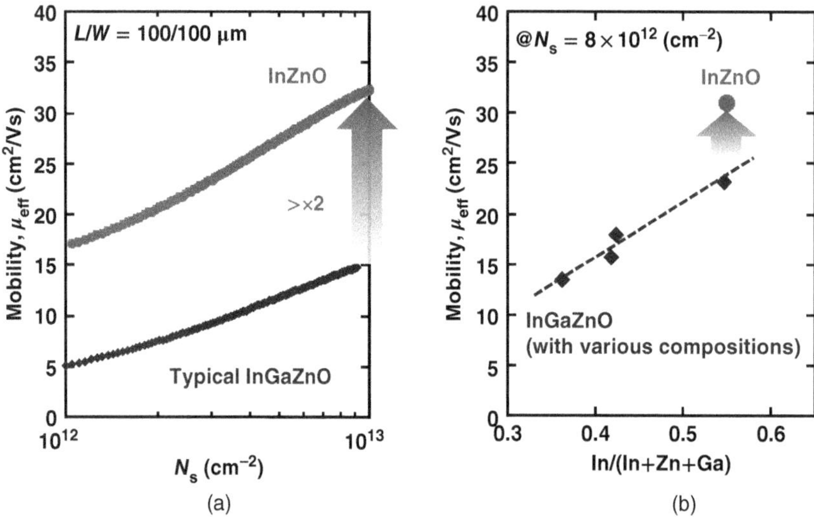

Figure 20.11 (a) Comparison of mobility characteristics between InZnO and InGaZnO TFTs, and (b) In composition ratio dependence of mobility [20]. *Source*: Saito, N., et al. (2019). High mobility (>30cm^2/Vs) and low source/drain parasitic resistance In-Zn-O BEOL transistor with ultralow (<10^{-20}A/μm) off-state leakage current. *Japanese Journal of Applied Physics* **58**: SBBJ07.

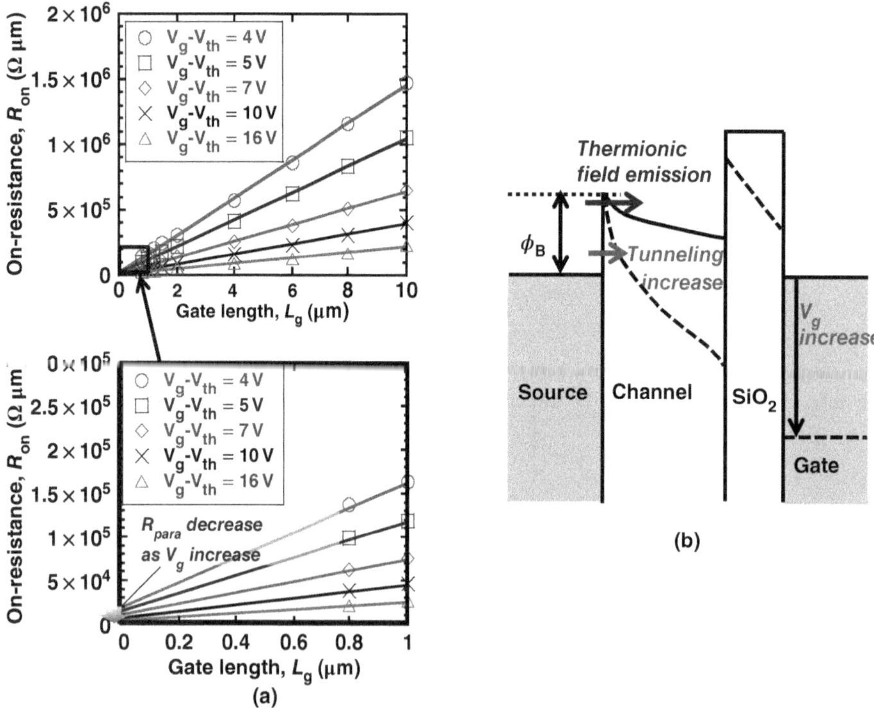

Figure 20.12 (a) Extraction method of S/D parasitic resistance from R_{on}-L plot, and (b) schematic illustration of the gate voltage dependence of S/D parasitic resistance [20]. *Source*: Saito, N., et al. (2019). High mobility (>30cm^2/Vs) and low source/drain parasitic resistance In-Zn-O BEOL transistor with ultralow (<10^{-20}A/μm) off-state leakage current. *Japanese Journal of Applied Physics* **58**: SBBJ07.

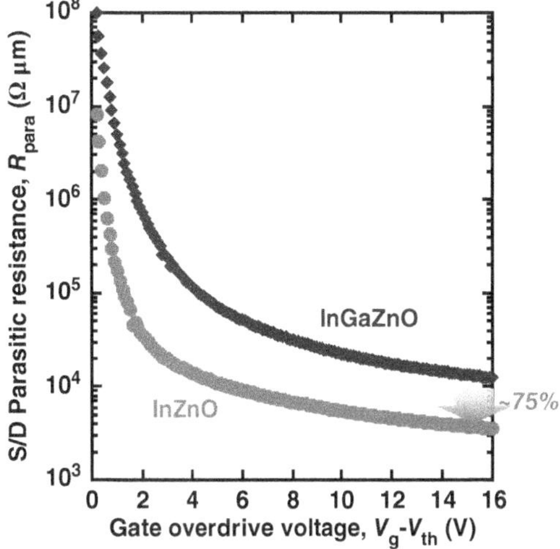

Figure 20.13 Comparison of S/D parasitic resistance between InZnO and InGaZnO TFTs [20]. *Source*: Saito, N., et al. (2019). High mobility (>30cm²/Vs) and low source/drain parasitic resistance In-Zn-O BEOL transistor with ultralow (<10⁻²⁰A/μm) off-state leakage current. *Japanese Journal of Applied Physics* **58**: SBBJ07.

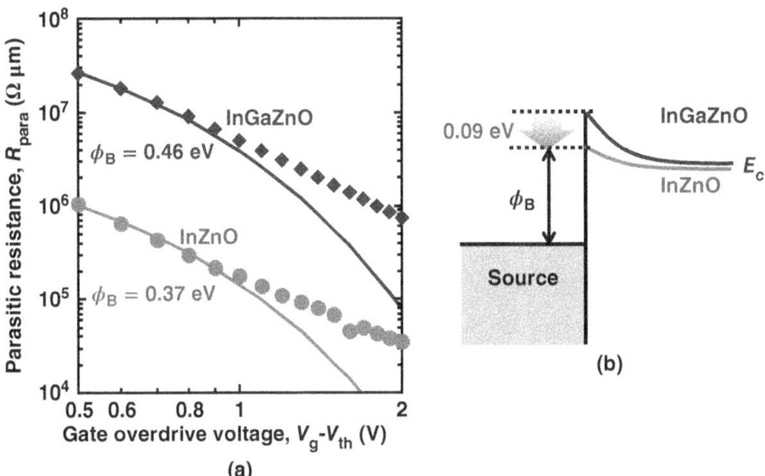

Figure 20.14 Extraction of Schottky barrier height from S/D parasitic resistance at small V_g-V_{th} region [20]. *Source*: Saito, N., et al. (2019). High mobility (>30cm²/Vs) and low source/drain parasitic resistance In-Zn-O BEOL transistor with ultralow (<10⁻²⁰A/μm) off-state leakage current. *Japanese Journal of Applied Physics* **58**: SBBJ07.

where q, k, T, and ϕ_B are electronic charge, Boltzmann's constant, temperature, and SBH, respectively. We introduced a gate bias voltage modulation factor as α in this model to represent the gate voltage dependence of S/D parasitic resistance by thermionic field emission, and we used ~0.1 as α and 41 A/cm^{2}K^2 as Richardson's constant (A^*) [28]. As shown in Figure 20.14, the SBH of InZnO TFTs was estimated to be 0.37 eV, while that of InGaZnO TFTs was 0.46 eV. The SBH of InZnO was found to be smaller by 0.09 eV than that of InGaZnO. This indicates that a reduction of S/D parasitic resistance by the InZnO channel results from a lowering of the conduction band minimum (E_c).

Since InZnO TFTs showed a lower parasitic resistance in addition to a higher mobility than InGaZnO TFTs, as shown in Figure 20.11 and Figure 20.13, InZnO TFTs can be expected to show higher I_{on} with proper gate length scaling. Figure 20.15 shows I_{on} improvement with gate length scaling. Solid circles and squares were experimental data, and dashed lines were calculated from each mobility and parasitic resistance. InZnO TFTs can be expected to show three times higher I_{on} at the gate length of 10 nm than InGaZnO TFTs.

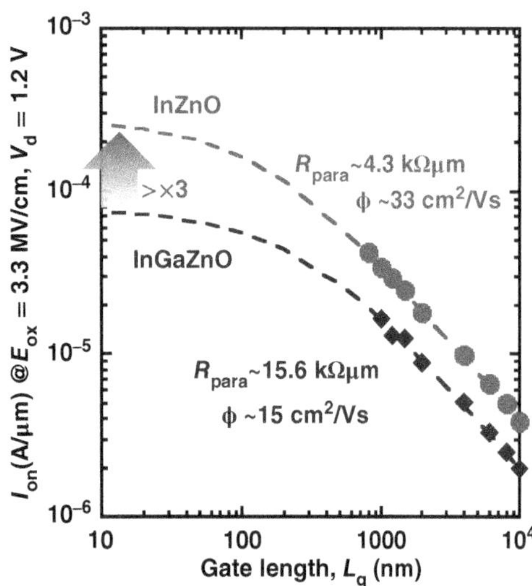

Figure 20.15 I_{on} improvement by gate length scaling calculated from extracted mobility and S/D parasitic resistance [20]. *Source*: Saito, N., et al. (2019). High mobility (>30cm²/Vs) and low source/drain parasitic resistance In-Zn-O BEOL transistor with ultralow (<10^{-20}A/μm) off-state leakage current. *Japanese Journal of Applied Physics* **58**: SBBJ07.

Figure 20.16a shows a comparison of J_g-V_g characteristics at the on-state between InZnO TFTs and InGaZnO TFTs with an EOT of 6.2 nm. Compared with InGaZnO TFTs, the J_g-V_g curve of InZnO TFTs shifted positively by 0.2 V, and the on-state gate leakage current was found to be suppressed by the InZnO channel. Figure 20.16b shows a Fowler–Nordheim plot of obtained data. On-state gate leakage current characteristics were confirmed to agree well with the Fowler–Nordheim tunneling model. This indicates that the on-state gate leakage current in InZnO TFTs and InGaZnO TFTs with an EOT of 6.2 nm is dominated by tunneling electron injection from the conduction band of the oxide semiconductor channel to the conduction band of the SiO_2 gate insulator. Conduction band offset between the oxide semiconductor channel and SiO_2 gate insulator was estimated using the following equations of

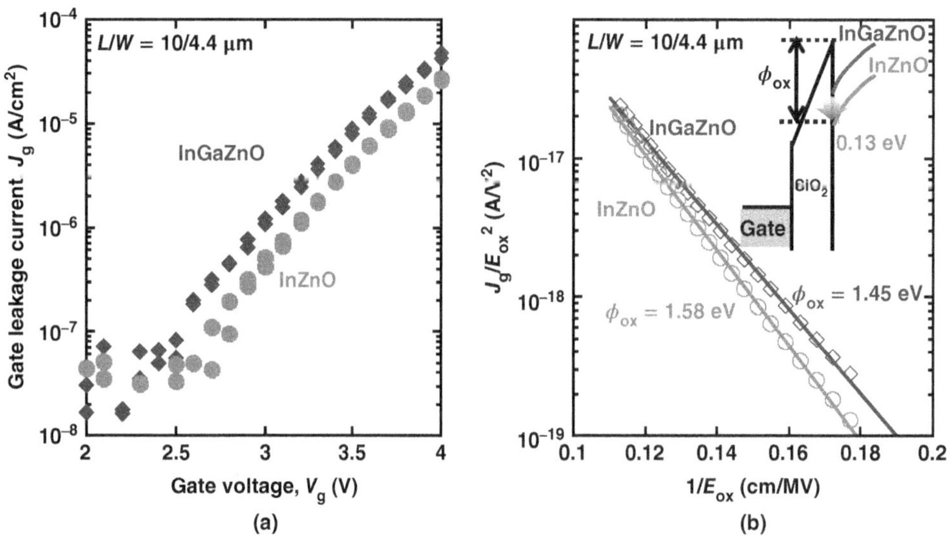

Figure 20.16 (a) J_g-V_g characteristics of InZnO and InGaZnO TFTs with an EOT of 6.2 nm at on-state, and (b) a Fowler–Nordheim plot [20]. *Source*: Saito, N., et al. (2019). High mobility (>30cm²/Vs) and low source/drain parasitic resistance In-Zn-O BEOL transistor with ultralow (<10^{-20}A/μm) off-state leakage current. *Japanese Journal of Applied Physics* **58**: SBBJ07.

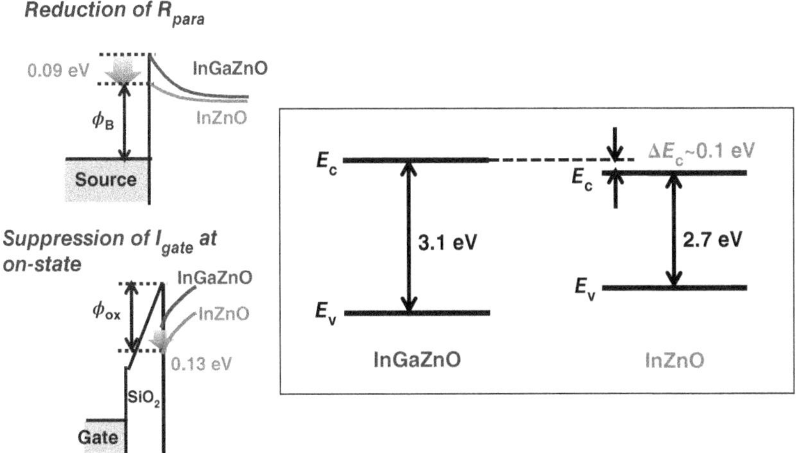

Figure 20.17 Schematic band diagram of InGaZnO and InZnO [20]. *Source*: Saito, N., et al. (2019). High mobility (>30cm^2/Vs) and low source/drain parasitic resistance In-Zn-O BEOL transistor with ultralow (<10^{-20}A/μm) off-state leakage current. *Japanese Journal of Applied Physics* **58**: SBBJ07.

the Fowler–Nordheim tunneling current (J_{FN}) [29]:

$$J_{FN} = A\frac{{E_{ox}}^2}{\emptyset_{ox}}\exp\left(-\frac{B{\emptyset_{ox}}^{3/2}}{E_{ox}}\right)$$

$$B = \frac{4\sqrt{2m^*q}}{3h}$$

where A, E_{ox}, ϕ_{ox}, m^*, q, and h are the constant, the oxide electric field, the band offset, the effective mass, electronic charge, and Plank's constant, respectively. Assuming that effective mass (m^*) is 0.34 [30], conduction band offsets between the oxide semiconductor channel and SiO_2 gate insulator were estimated to be 1.58 eV and 1.45 eV for the InZnO–SiO_2 and InGaZnO–SiO_2 interfaces, respectively, by using the above equation. The conduction band offset of InZnO–SiO_2 was found to be 0.13 eV larger than that of InGaZnO–SiO_2.

Figure 20.17 shows a comparison of extracted band diagrams of InGaZnO and InZnO. Here, bandgap energy was estimated by using reflection electron energy loss spectroscopy (REELS). As shown in Figure 20.13 and Figure 20.16, InZnO TFTs showed a reduction of S/D parasitic resistance and suppression of on-state gate leakage current. The difference between extracted SBH values of InZnO and InGaZnO at S/D contact was 0.1 eV, which is consistent with the difference in extracted band offset values between the oxide semiconductor channel and SiO_2 gate insulator. These results suggest that the advantages of reduction of parasitic resistance and on-state gate leakage current in InZnO originate from a lowering of E_c by the InZnO channel compared with the InGaZnO channel.

20.4 Demonstration of Extremely Low Off-State Leakage Current Characteristics

As the bandgap energy of InZnO is smaller than that of InGaZnO, there is concern about a degradation of extremely low off-state current characteristics. Figure 20.18a shows a cross-sectional transmission electron microscopy (TEM) image of fabricated InZnO TFTs with a thin EOT. The EOT was estimated to be 6.2 nm by *C-V* measurement (Figure 20.18b). Figure 20.18c shows typical I_d-V_g and I_g-V_g characteristics of an InZnO TFT with an EOT of 6.2 nm. The InZnO TFT showed enhancement-mode operation, and off-state gate leakage current was not observed in conventional direct-current (DC) measurement since the detection limit was about 0.1 pA.

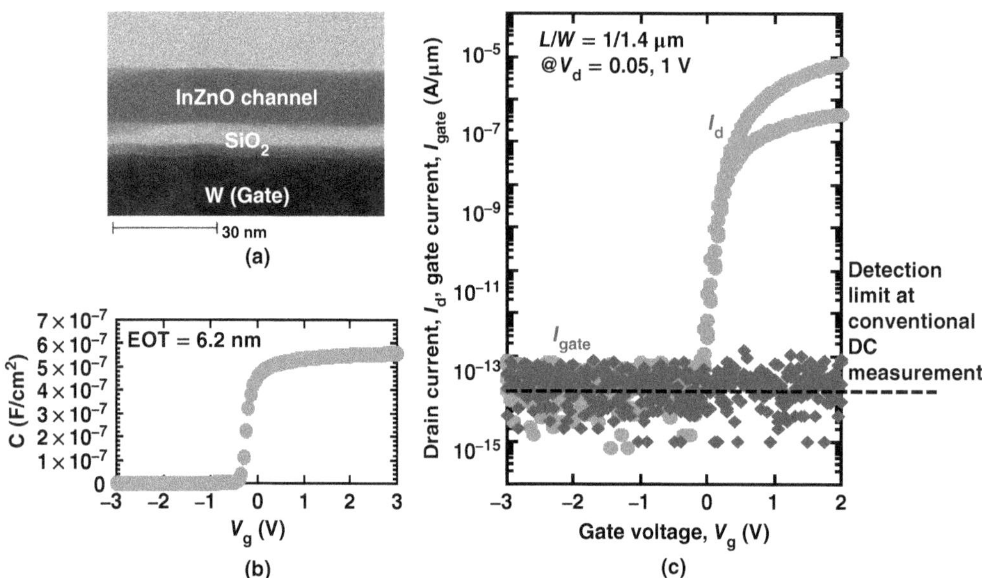

Figure 20.18 (a) Cross-sectional TEM image, (b) *C-V* characteristics, and (c) I_d-V_g and I_g-V_g characteristics of an InZnO TFT with an EOT of 6.2 nm [20]. *Source*: Saito, N., et al. (2019). High mobility (>30cm^2/Vs) and low source/drain parasitic resistance In-Zn-O BEOL transistor with ultralow (<10^{-20}A/μm) off-state leakage current. *Japanese Journal of Applied Physics* **58**: SBBJ07. With permission of IOP Publishing.

Figure 20.19 shows an evaluation method of extremely low off-state leakage current characteristics [31]. As shown in Figure 20.19a, we fabricated a simple circuit with the device under test (DUT), a source follower amplifier, and a capacitor. A thick SiO_2 film with a thickness of 150 nm was used for an interlayer of capacitor so that leakage current through the capacitor was negligible. Figure 20.19b shows a timing chart for evaluating extremely low off-state leakage current characteristics. Time dependence of the source follower amplifier's output voltage (V_{detect}) was monitored, and time dependence of the floating node voltage (V_{FN}) can be estimated by using the input–output characteristics of the source follower amplifier. Assuming that the voltage drop of the floating node is

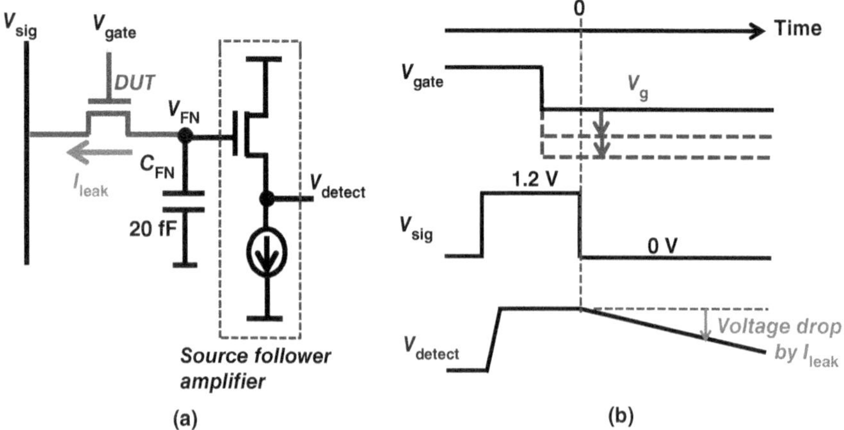

Figure 20.19 (a) Circuit diagram and (b) timing chart for evaluating extremely low off-state leakage current characteristics [20]. *Source*: Saito, N., et al. (2019). High mobility (>30cm^2/Vs) and low source/drain parasitic resistance In-Zn-O BEOL transistor with ultralow (<10^{-20}A/μm) off-state leakage current. *Japanese Journal of Applied Physics* **58**: SBBJ07.

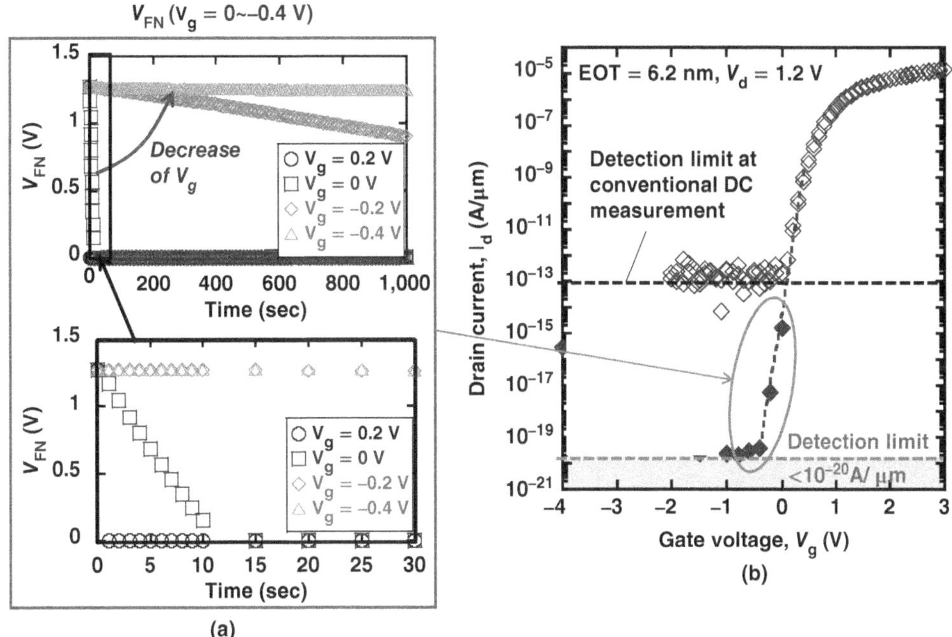

Figure 20.20 (a) Time dependence of V_{FN} at the gate voltage from 0 V to −0.4 V, and (b) I_d-V_g characteristics with extracted leakage current values [20]. *Source*: Saito, N., et al. (2019). High mobility (>30cm²/Vs) and low source/drain parasitic resistance In-Zn-O BEOL transistor with ultralow (<10^{-20}A/μm) off-state leakage current. *Japanese Journal of Applied Physics* **58**: SBBJ07.

caused by leakage current via the DUT, the off-state leakage current (I_{leak}) can be estimated by using the following equation:

$$I_{leak} = C_{FN} \times \Delta V_{FN} / \Delta t$$

where capacitance (C_{FN}) of the floating node was 20 fF, and each measurement time was 1,000 sec. This measurement was repeated with different V_g values in the off-state to obtain a gate voltage dependence of extremely low off-state leakage current.

Figure 20.20a shows time dependences of V_{FN} at a V_g from 0 to −0.4 V. V_{FN} at each time was calculated from measured V_{detect} by using the input–output characteristics of the source follower amplifier measured in advance. In this region, it was observed that the falling speed of V_{FN} decreased as the V_g was set more negatively, and V_{FN} no longer decreased at a V_g of −0.4 V. Moreover, as shown in the time dependences of V_{FN} at V_gs of 0 V and −0.2 V in Figure 20.20a, V_{FN} dropped to 0 V at constant speed. This means that the magnitude of the leakage current observed in this region does not depend on the potential difference between the source and drain of DUT. Since the subthreshold leakage current of TFTs is determined by electron diffusion with thermal energy, not by the potential difference between source and drain, the constant leakage current observed in Figure 20.20a is consistent with the characteristics of subthreshold leakage current. Figure 20.20b shows I_d-V_g characteristics with the extracted leakage current values. Extracted extremely low off-state leakage current values showed almost the same trend of the subthreshold characteristics obtained by conventional DC measurements. It was also confirmed that subthreshold characteristics were maintained below 10^{-20} A/μm (the detection limit of this method).

Figure 20.21a shows the time dependences of V_{FN} at a V_g less than −2 V. In this region, as V_g decreases negatively, V_{FN} drastically dropped. Unlike the constant leakage current observed in Figure 20.21a, the leakage current showed time dependence and the time derivatives were decreased as the drain voltage of the DUT was reduced due to the leakage current from the floating node. These behaviors of V_{FN} drop are consistent with the characteristics

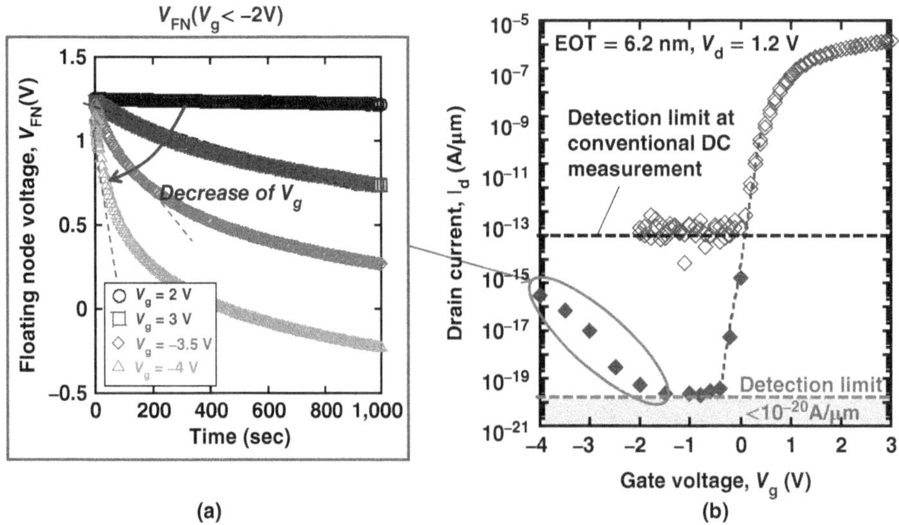

Figure 20.21 (a) Time dependence of V_{FN} at a gate voltage less than −2 V, and (b) I_d-V_g characteristics with extracted leakage current values [20]. *Source*: Saito, N., et al. (2019). High mobility (>30cm²/Vs) and low source/drain parasitic resistance In-Zn-O BEOL transistor with ultralow (<10^{-20}A/μm) off-state leakage current. *Japanese Journal of Applied Physics* **58**: SBBJ07.

of gate leakage current between the drain and gate of the DUT. Figure 20.21b shows I_d-V_g characteristics with the extracted leakage current values. Since the gate leakage current was reduced as the drain voltage of the DUT was decreased, the initial time derivative of V_{FN} was used to estimate a leakage current value at each gate voltage. It was confirmed that a gate leakage current over 10^{-20} A/μm was not observed in the range of $|V_g| < 1.5$ V, even at a thin EOT of 6.2 nm.

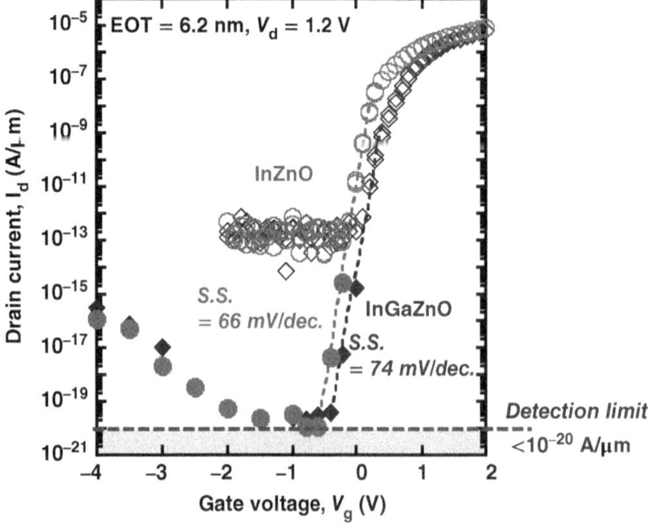

Figure 20.22 Comparison of extremely low off-state leakage current characteristics between InZnO and InGaZnO TFTs with an EOT of 6.2 nm [20]. *Source*: Saito, N., et al. (2019). High mobility (>30cm²/Vs) and low source/drain parasitic resistance In-Zn-O BEOL transistor with ultralow (<10^{-20}A/μm) off-state leakage current. *Japanese Journal of Applied Physics* **58**: SBBJ07.

Finally, we compared extremely low off-state leakage current characteristics between InZnO TFTs and InGaZnO TFTs extracted from the above evaluation method (Figure 20.22). The extremely low off-state leakage current values of InZnO TFTs and InGaZnO TFTs agree well with the trends of the subthreshold characteristics of InZnO TFTs and InGaZnO TFTs by conventional DC measurements, respectively. Subthreshold slope (*SS*) maintained below 10^{-20} A/μm was 66 mV/dec. and 74 mV/dec. for InZnO TFTs and InGaZnO TFTs, respectively. It was found that off-state gate leakage current characteristics were comparable between InZnO and InGaZnO TFTs. InZnO TFTs with an EOT of 6.2 nm exhibited extremely low off-state leakage current characteristics ($<10^{-20}$ A/μm), as well as InGaZnO TFTs in the range of $|V_g| < 1.5$ V.

References

1 Moore, G. (1965). Cramming more components onto integrated circuits. *Electronics* 38 (8): 114–117.

2 Bohr, M.T. and Young, I.A. (2017). CMOS scaling trends and beyond. *IEEE Micro* 37 (6): 20–29.

3 Baliga, J. (2004). Chips go vertical. *IEEE Spectrum* 41 (3): 43–47.

4 Van Olmen, J., Mercha, A., Katti, G. et al. (2008). 3D stacked IC demonstration using a through Silicon Via First approach. Paper presented at the 2008 IEEE International Electron Devices Meeting (IEDM), San Francisco, CA, December.

5 Batude, P., Fenouillet-Beranger, C., Pasini, L. et al. (2015). 3DVLSI with CoolCube process: an alternative path to scaling. Paper presented at the Symposium on VLSI Technology, Hsinchu, Taiwan, April.

6 Brunet, L., Fenouillet-Beranger, C., Batude, P. et al. (2018). Breakthroughs in 3D sequential technology. Paper presented at the 2018 IEEE International Electron Devices Meeting (IEDM), San Francisco, CA, December.

7 Naito, T., Ishida, T., Onoduka, T. et al. (2010). World's first monolithic 3D-FPGA with TFT SRAM over 90nm 9 layer Cu CMOS. Paper presented at the Symposium on VLSI Technology, Honolulu, Hawaii, June.

8 Batude, P., Vinet, M., Previtali, B. et al. (2011). Advances, challenges and opportunities in 3D CMOS sequential integration. Paper presented at the 2011 IEEE International Electron Devices Meeting (IEDM), Washington, DC, December.

9 Kljucar, L., Mitard, J., Rassoul, N. et al. (2019). IGZO integration scheme for enabling IGZO nFETs. Paper presented at the 2019 International Conference on Solid State Devices and Materials, Nagaya, Japan.

10 Chakraborty, W., Grisafe, B., Ye, H. et al. (2020). BEOL compatible dual-gate ultra thin-body w-doped indium-oxide transistor with Ion = 370 μA/μm, SS = 73 mV/dec and Ion/Ioff ratio >4x109. Paper presented at the Symposium on VLSI Technology, Honolulu, Hawaii, June.

11 Kaneko, K., Inoue, N., Saito, S. et al. (2011). A novel BEOL transistor (BETr) with InGaZnO embedded in Cu-interconnects for on-chip high voltage I/Os in standard CMOS LSIs. Paper presented at the Symposium on VLSI Technology, Kyoto, Japan, June.

12 Sekine, Y., Furutani, K., Shionoiri, Y. et al. (2011). Success in measurement the lowest off-state current of transistor in the world. *ECS Transactions* 37: 77–88.

13 Sunamura, H., Inoue, N., Furutake, N. et al. (2014). Enhanced drivability of high-Vbd dual-oxide-based complementary BEOL-FETs for compact on-chip pre-driver applications. Paper presented at the Symposium on VLSI Technology, Honolulu, Hawaii, June.

14 Jeon, S., Park, S., Song, I. et al. (2010). 180nm gate length amorphous InGaZnO thin film transistor for high density image sensor applications. Paper presented at the 2010 IEEE International Electron Devices Meeting (IEDM), San Francisco, CA, December.

15 Kobayashi, Y., Matsubayashi, D., Nagatsuka, S. et al. (2014). Scaling to 50-nm c-axis aligned crystalline In-Ga-Zn oxide FET with surrounded channel structure and its application for less-than-5-nsec writing speed memory. Paper presented at the Symposium on VLSI Technology, Honolulu, Hawaii, June.

16 Wu, T.H., Liu, M.Y., Wu, J.Y. et al. (2017). Performance boost of crystalline In-Ga-Zn-O material and transistor with extremely low leakage for IoT normally-off CPU application. Paper presented at the Symposium on VLSI Technology, Kyoto, Japan, June.

17 Hanyu, Y., Domen, K., Nomura, K. et al. (2013). Hydrogen passivation of electron trap in amorphous In-Ga-Zn-O thin-film transistors. *Applied Physics Letters* 103: 202114.

18 Nakashima, M., Oota, M., Ishihara, N. et al. (2014). Origin of major donor states in In–Ga–Zn oxide. *Journal of Applied Physics* 116: 213703.

19 Saito, N., Miura, K., Ueda, T. et al. (2018). High-mobility and H2-anneal tolerant InGaSiO / InGaZnO / InGaSiO double hetero channel thin film transistor for Si-LSI compatible process. *IEEE Journal of the Electron Devices Society* 6: 500–505.

20 Saito, N., Sawabe, T., Kataoka, J. et al. (2019). High mobility (>30cm^2/Vs) and low source/drain parasitic resistance In-Zn-O BEOL transistor with ultralow (<10^{-20}A/μm) off-state leakage current. *Japanese Journal of Applied Physics* 58: SBBJ07.

21 Luo, Y.R. (2009). Bond dissociation energies. In. In: *CRC Handbook of Chemistry and Physics*, 90the (ed. D.R. Lide) pp., 9–65. Boca Raton, FL: CRC Press.

22 Mitoma, N., Aikaw, S., Gao, X. et al. (2014). Stable amorphous In2O3-based thin-film transistors by incorporating SiO2 to suppress oxygen vacancies. *Applied Physics Letters* 104: 102103.

23 Itagaki, N., Iwasaki, T., Kumomi, H. et al. (2008). Zn-In-O based thin-film transistors: compositional dependence. *Physica Status Solidi (A): Applications and Materials Science* 90: 1915–1919.

24 Tsutsui, K., Matsubayashi, D., Ishihara, N. et al. (2015). Mobility enhancement in crystalline In-Ga-Zn-oxide with In-rich compositions. *Applied Physics Letters* 107: 262104.

25 Kang, Y., Cho, Y., and Han, S. (2013). Cation disorder as the major electron scattering source in crystalline InGaZnO. *Applied Physics Letters* 102: 152104.

26 Wang, W., Li, L., Lu, C. et al. (2015). Analysis of the contact resistance in amorphous InGaZnO thin film transistors. *Applied Physics Letters* 107: 063504.

27 Sze, S.M. (1981). *Physics of Semiconductor Devices*, 2nde. New York: Wiley.

28 Lee, D.H., Nomura, K., Kamiya, T., and Hosono, H. (2011). Diffusion-limited a-IGZO/Pt Schottky junction fabricated at 200°C on a flexible substrate. *IEEE Electron Device Letters* 32: 1695–1697.

29 Schroder, D. (2008). *Semiconductor Material and Device Characterization*, 3rde. Hoboken, NJ: Wiley.

30 Takagi, A., Nomura, K., Ohta, H. et al. (2005). Carrier transport and electronic structure in amorphous oxide semiconductor, a-InGaZnO4. *Thin Solid Films* 486: 38–41.

31 Kato, K., Shinori, Y., Sekine, Y. et al. (2012). Evaluation of off-state current characteristics of transistor using oxide semiconductor material, indium–gallium–zinc oxide. *Japanese Journal of Applied Physics* 51: 021201.

21

Ferroelectric-HfO_2 Transistor Memory with IGZO Channels

Masaharu Kobayashi

d.lab, School of Engineering, The University of Tokyo, Tokyo, Japan

21.1 Introduction

In the current highly information-oriented society, artificial intelligent (AI) and the internet of things (IoT) are driving new innovations, which need to load, store, and process big data. Data-intensive algorithms such as machine learning are to be used for edge computing. Therefore, high-density and low-power-consumption nonvolatile memories (NVMs) play key roles for developing IoT and AI devices using limited battery capacity and an energy-harvesting power supply. Ferroelectric memory is, in principle, a low-power memory because of its field-driven operation. However, there is no commercial high-density ferroelectric memory. Although ferroelectric RAM (FeRAM) is in the market, it consists of 1T-1C or 2T-2C, and the size of the capacitor and thus total cell size are quite large.

Recently, ferroelectric field-effect transistors (FeFETs) have attracted much attention [1–3] because ferroelectricity has been discovered in doped HfO_2 [4], which is a complementary metal-oxide semiconductor (CMOS)-compatible and highly scalable material. A FeFET is a 1T memory device with a ferroelectric gate insulator. Threshold voltage (V_{th}) of a FeFET can be shifted by changing the direction of ferroelectric polarization under positive or negative gate voltage (V_g), and thus a FeFET can work as a memory device. A FeFET with ferroelectric-HfO_2 (FE-HfO_2) has several advantages, such as nondestructive readout, high program and erase speeds, and low power consumption. 3D vertical structure [5–9] has achieved high density in 3D NAND flash memories. Inspired by 3D NAND flash memories, a 3D vertical-stack FeFET has been recently proposed and demonstrated [10]. FeFETs, especially 3D vertical-stack FeFETs, are a promising memory that target a unique segment of the market for high-speed, low-power, and high-density memory (Figure 21.1).

For 3D vertical-stack FeFETs, poly-Si is a natural choice as channel material for the vertical structure, just as with 3D NAND flash memory. However, there are several potential challenges with poly-Si channels, such as the low mobility of very thin poly-Si and the high thermal budget. Moreover, between poly-Si and gate oxide, a low-k interfacial layer is inevitably formed, which causes voltage loss and subthreshold swing (SS) degradation by charge trapping.

InGaZnO (IGZO) is an innovative material for transistor channels that was first reported in 2004 and became a commercial device as a driver transistor for flat-panel displays [11]. Due to its high mobility and high reliability [12, 13], IGZO is also suitable for large-scale-integration (LSI) applications. We think IGZO is a suitable material for 3D vertical FeFETs, as illustrated in Figure 21.2a. As shown in Figure 21.2b, thanks to the oxide–oxide interface between FE-HfO_2 and IGZO (unlike FE-HfO_2 and poly-Si) and junctionless transistor operation, interface charge

Amorphous Oxide Semiconductors: IGZO and Related Materials for Display and Memory, First Edition.
Edited by Hideo Hosono and Hideya Kumomi.

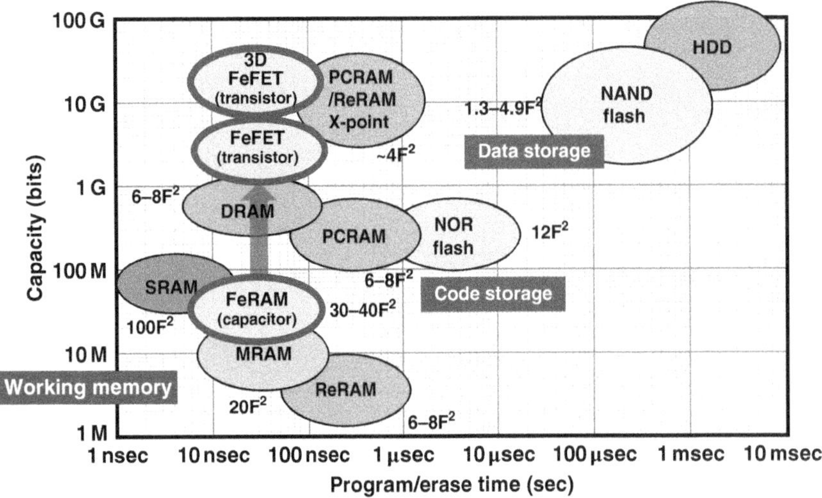

Figure 21.1 Positioning of FeFET memory devices among commercial memory devices.

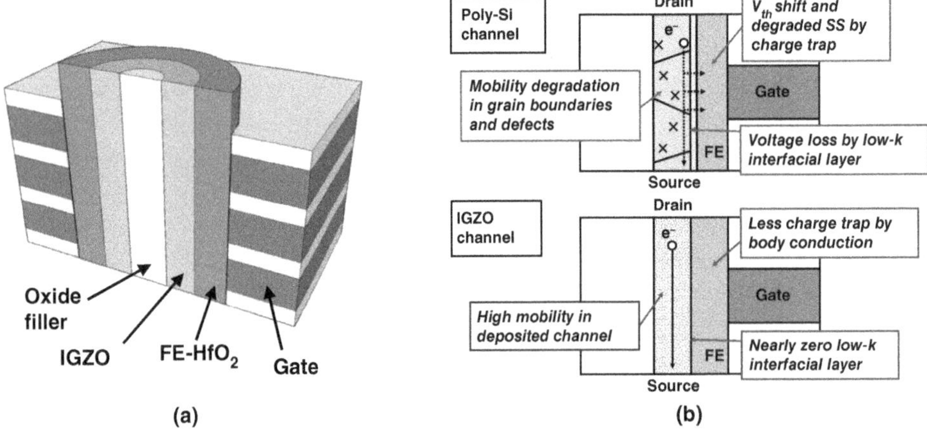

Figure 21.2 (a) Schematic of a 3D vertical-stack FeFET with an IGZO channel. (b) Schematic illustration of the challenges with a poly-Si channel and a possible solution by an IGZO channel.

trapping will not be severely involved in comparison to inversion-mode transistor operation. Moreover, there will be much less voltage loss due to the nearly zero low-k interfacial layer between FE-HfO_2 and IGZO channels. Therefore, the IGZO channel FeFET [14–17] is a promising candidate for high-density, low-power memory applications. However, the properties of the metal–FE-HfO_2–IGZO–metal structure and the device design of the IGZO channel–based FeFET with FE-HfO_2 thin film have not been fully investigated yet.

In this work, we design an ultrathin-body IGZO channel FE-HfO_2 field-effect transistor (FET). The device design is proposed to obtain a large memory window (MW) by discussing the dependence on IGZO, FE-HfO_2, and SiO_2 thickness. Then we develop the process of FE-HfO_2 capacitors with IGZO and investigate the property of the metal–IGZO–FE-HfO_2–metal structure, including its reliability. Finally, we fabricate and demonstrate IGZO FeFET memory functionality as a high-density and low-power memory device.

21.2 Device Operation and Design

The device operation principle of IGZO FeFETs is illustrated in Figure 21.3. Since IGZO is naturally an *n*-type semiconductor, IGZO FeFETs operate as junctionless transistors. In erase mode, negative V_g induces a high density of depletion charge, and V_{th} is shifted to a positive value. In program mode, positive V_g induces an accumulation charge, and V_{th} is shifted to a negative value. Spontaneous polarization can be flipped by the applied electric field and sustained by these induced charges.

For high-density memories, especially for 3D vertical-stack structure, channel thickness should be as thin as possible but without performance degradation. Moreover, V_{th} should be near 0 V for memory application. To find out an appropriate channel thickness, bottom-gate (front-gate) IGZO FETs with a 15 nm SiO_2 gate insulator were fabricated for investigating the IGZO thickness dependence. Please note that, in this work, "front/back" gate means gate/substrate electrode in the conventional definition of a metal-oxide semiconductor FET (MOSFET), while "top/bottom" gate means the structural definition in our fabricated device. Figure 21.4a shows the drain current (I_d) versus V_g curves of IGZO FETs with a 5~40 nm IGZO channel. Extracted V_{th} and SS of IGZO FETs are shown in Figure 21.4b. As the IGZO thickness decreases, V_{th} increases. This is because thinner IGZO has a thinner depletion layer, a smaller negative V_g is needed to turn off the channel, and the SS becomes steeper due to the smaller substrate capacitance [12]. Note that the 5 nm channel IGZO FET has significantly low drain current due to the nonuniform formation of IGZO film caused by a short deposition time in the sputter process, and thus the mobility degradation. 8 nm IGZO is chosen for device design with good SS, no drain current degradation, and a V_{th} near 0 V. We also extracted the defect density from the measured SS with different IGZO thicknesses. The interface state density is $\sim 1 \times 10^{11} cm^{-2}$, and the bulk defect is $\sim 6 \times 10^{17} cm^{-3}$, which indicates the high-quality interface at the oxide–oxide interface of gate oxide and IGZO.

To check the concept and design the device, we simulate IGZO FeFETs by technology computer-aided design (TCAD). The device structure is a bottom-gate structure, which consists of a bottom (front) gate, a FE-HfO_2 gate insulator, an IGZO channel, a SiO_2 layer, a top (back) electrode, and source/drain electrodes, as shown in Figure 21.9. IGZO thickness is 8 nm, as determined in the last paragraph. We assume FE-HfO_2 is $HfZrO_2$ (HZO), and the ferroelectric parameters are extracted from our previous work [18]. The dielectric

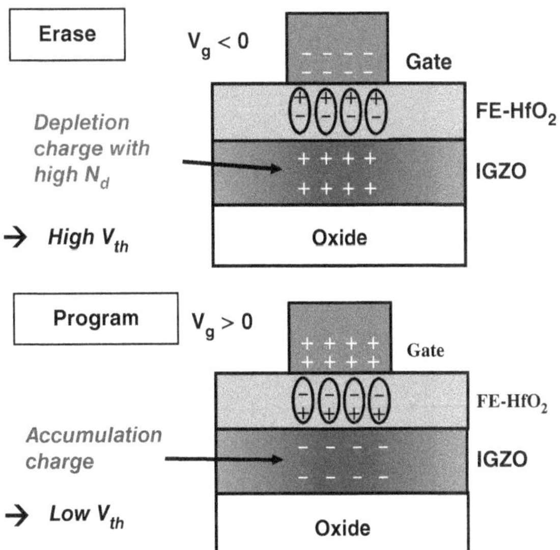

Figure 21.3 Operation principle of a FeFET illustrated in the form of a 2D transistor for simplicity.

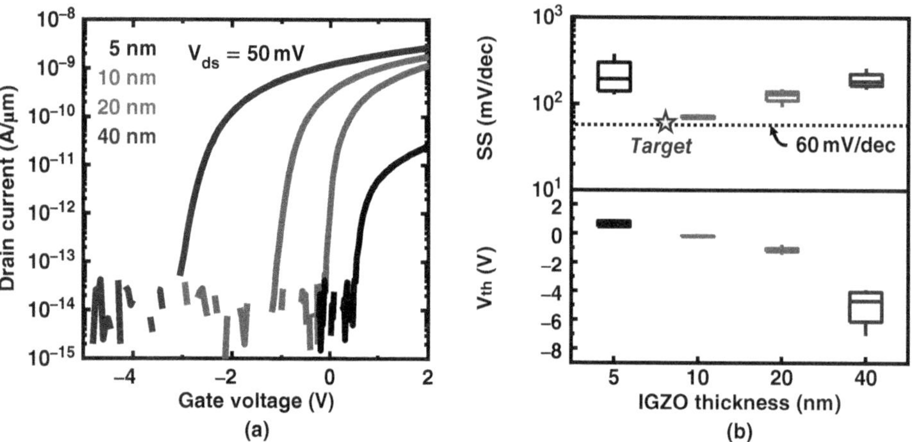

Figure 21.4 (a) Measured I_d-V_g curves of IGZO FETs with different IGZO thicknesses. (b) Extracted SS and V_{th} of IGZO FETs from (a).

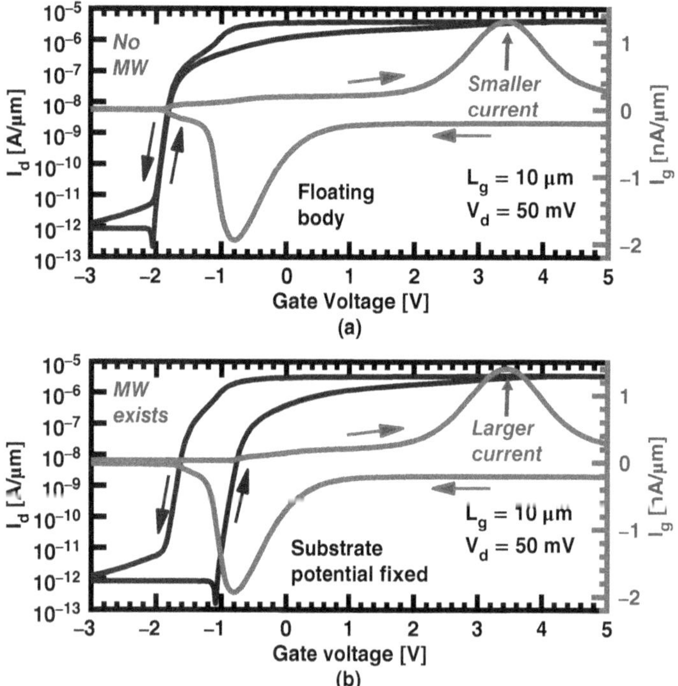

Figure 21.5 (a) Simulated I_d-V_g and I_g-V_g curves of the IGZO channel FET without a top gate. (b) Simulated I_d-V_g and I_g-V_g curves of the IGZO channel FET with a top gate.

constant (ε), remanent polarization (P_r), and coercive field (E_c) of HZO are 35, 20 $\mu C/cm^2$, and 1.16 MV/cm, respectively. We investigate the impact of the top (back) gate and the thickness dependence of HZO and SiO_2 as critical design parameters. MW is defined as the V_{th} difference between the program state and erase state.

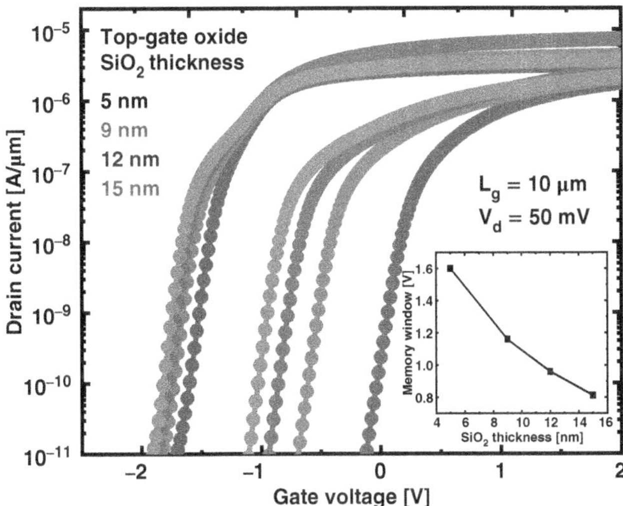

Figure 21.6 Simulated I_d-V_g curves of IGZO channel FeFETs with different SiO_2 thicknesses.

Firstly, we study the impact of the back gate. Figure 21.5a shows the simulated I_d-V_g and I_g-V_g curves with IGZO channel floating. The HZO thickness is 15 nm. The MW is not observed. This is because the IGZO body is floating and the voltage is not effectively applied to HZO for switching ferroelectric polarization, especially in erase mode where only depletion charges are available to balance with ferroelectric polarization charges. Note that the simulated I_g-V_g curve shows a polarization-switching current, which is caused under source and drain pad regions because the top source and drain pad overlap with the bottom gate. Then, we aim to fix the body potential by the back gate through the SiO_2 layer. Figure 21.5b shows the simulated I_d-V_g and I_g-V_g curves of the IGZO FeFET with the back gate fixed at 0 V. SiO_2 and HZO thicknesses are 12 nm and 15 nm, respectively. This time, the MW appears, and the gate current increases, which indicates polarization switching of FE-HfO_2 under the channel. This is because the IGZO body potential is fixed and the large electric field is effectively applied to the HZO layer, even in erase mode.

Secondly, we study the SiO_2 thickness dependence. SiO_2 thickness changes the electrical coupling between the back gate and IGZO body. Figure 21.6 shows the simulated I_d-V_g curves with a 5~15 nm SiO_2 layer. The HZO thickness is 15 nm. The back gate voltage is 0 V. As SiO_2 thickness decreases, the V_{th} of the erase state increases and the MW increases because the top gate is close to the channel and strongly fixes the body potential to zero. Thus, large voltage is applied to HZO for ferroelectric polarization switching. We decide to use a 12 nm SiO_2 layer to obtain a large MW but prevent leakage current from the top gate.

Thirdly, we study the HZO thickness dependence. Figure 21.7 shows the simulated I_d-V_g curves with different HZO thicknesses. The SiO_2 thickness is 12 nm. The back gate voltage is 0 V. The property of the HZO thin film may depend on its thickness [19]. For simplicity, however, we assume ε, P_r, and E_c of HZO are constant for all thicknesses. This assumption can be valid, as our simulation is limited to the relatively thick HZO region above 10 nm, in which the ferroelectric property of HZO does not vary much [20]. As HZO thickness increases, MW increases, as shown in the inset of Figure 21.7. This is because the maximum MW is to be twice the coercive voltage (V_c) and the thicker HZO has larger V_c, which is proportional to HZO thickness. Note that, as HZO thickness increases, the capacitance of the bottom gate oxide decreases, and thus higher (lower) voltage is applied to the HZO (IGZO) layer. Thus, with a thicker HZO layer, more negative V_g is required to turn off the device. A 15 nm HZO is chosen due to its V_{th} near 0 V and relatively large MW without causing leakage current between the bottom gate and source/drain pads.

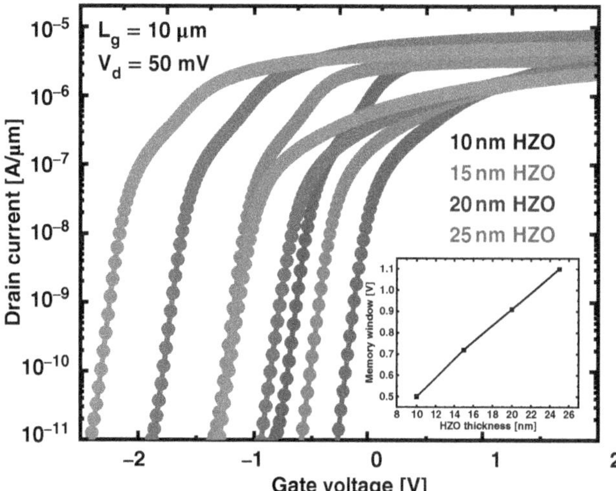

Figure 21.7 Simulated I_d-V_g curves of IGZO channel FeFETs with different HZO thicknesses.

21.3 Device Fabrication

To study the ferroelectricity in HZO with an IGZO layer, a metal–HZO–IGZO–metal stack was fabricated on an N^+ Si substrate. Figure 21.8 shows the capacitor structure and the process flow. The thickness of HZO and IGZO is determined in Section 21.2. The bottom TiN is deposited by radiofrequency (RF) sputtering on the RCA-cleaned Si substrate. 15 nm 50% Zr-doped $HfZrO_2$ is deposited by atomic layer deposition (ALD) at 250 °C. 8 nm IGZO is deposited by RF sputtering in ambient Ar using an IGZO target, followed by IGZO patterning by HCl solution. Al/Ti is deposited by EB evaporation. Lastly, the stack is annealed by rapid thermal annealing (RTA) to crystalize HZO in ambient N_2/O_2 (O_2:3%) at 500 °C for 10 sec.

We also fabricated an IGZO FeFET. A bottom-gate device structure is adopted for proof-of-concept, as shown in Figure 21.9. V_g is applied from the bottom gate. The fabrication process flow is shown in Figure 21.9. First, 20 nm TiN is deposited as a bottom gate on an RCA-cleaned SiO_2/Si substrate by RF sputtering. 15 nm 50% Zr-doped HfO_2 is deposited as a gate insulator by an ALD system at 250 °C. 8 nm IGZO is deposited as a channel by RF sputtering in ambient Ar using an IGZO target. The carrier concentration and Hall mobility of the IGZO layer are estimated to be $\sim 10^{19}$ cm^{-3} and 10 cm^2/Vs by Hall measurement, respectively. Then, IGZO is patterned by a diluted HCl solution. 12 nm SiO_2 is deposited as a passivation layer by RF sputtering. RTA is conducted to crystalize HZO

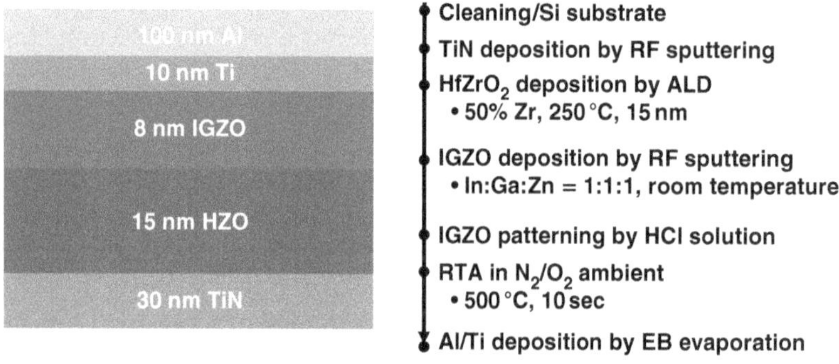

Figure 21.8 Schematic of the fabricated capacitor and the fabrication process flow.

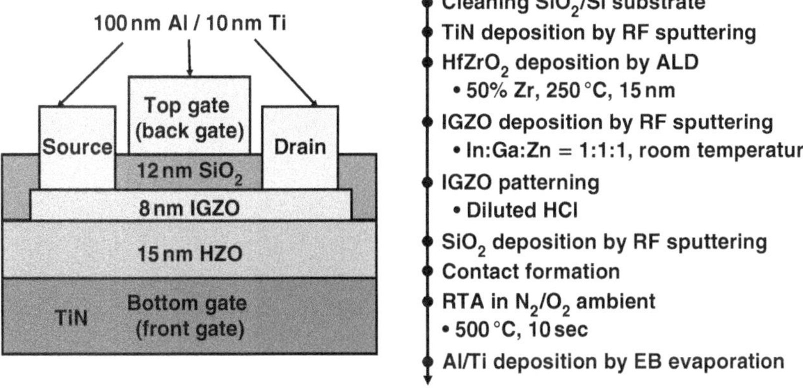

Figure 21.9 Schematic of the fabricated IGZO channel FeFET and the fabrication process flow.

in ambient N_2/O_2 (O_2:3%) at 500 °C for 10 sec. Al/Ti is deposited as an optional top gate to fix the IGZO channel potential by EB evaporation.

21.4 Experimental Results and Discussions

21.4.1 FE-HfO_2 Capacitors with an IGZO Layer

First, we characterize the physical properties of the capacitor. Figure 21.10 shows the measured GI-XRD spectra of the HZO film with and without the IGZO layer after annealing. The IGZO is a capping material on HZO that provides process-induced strain in the HZO film to preferentially induce the ferroelectric phase. The peak near 30.4° shows the orthorhombic phase is formed with the IGZO cap after annealing (shaded line). However, there is no similar peak (black line) without the IGZO cap. Thus, the ferroelectricity of HZO can emerge with IGZO capping. Figure 21.11 shows cross-sectional transmission electron microscopy (TEM) images of the fabricated HZO capacitor with IGZO capping. Each layer is uniformly formed. HZO is fully crystallized, and IGZO remains amorphous.

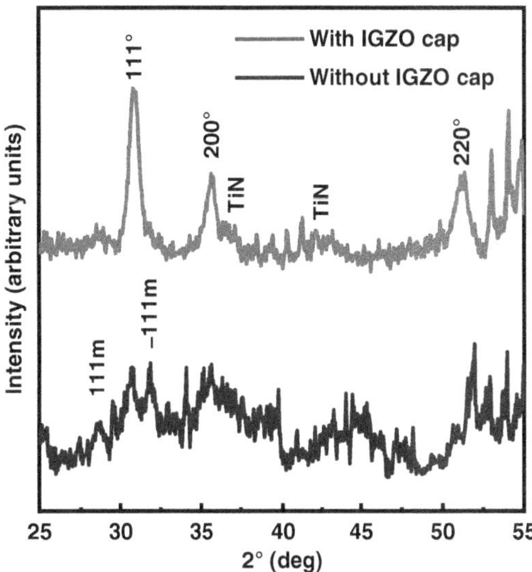

Figure 21.10 Measured grazing incidence X-ray diffraction (GI-XRD) spectra of the stack of HZO capacitors with and without IGZO layers after RTA.

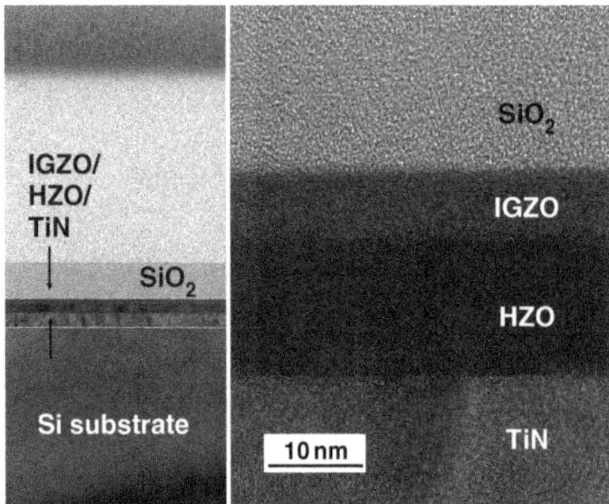

Figure 21.11 Cross-sectional TEM images of the fabricated HZO capacitor with IGZO capping. The left and right images are taken by low and high magnification of TEM.

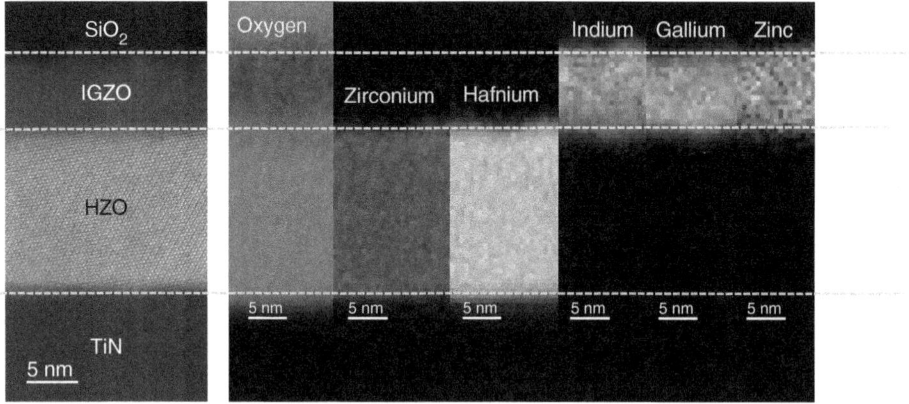

Figure 21.12 (Left) Cross-sectional STEM image, and (right) EDX analysis of the fabricated HZO capacitor with IGZO capping.

The interface of HZO and IGZO is free of the low-k interfacial layer. Figure 21.12 shows the cross-sectional STEM image and EDX analysis of the fabricated capacitor. There is no significant interdiffusion of the elements based on the given resolution of EDX analysis, and the sharp interface is realized between HZO and IGZO.

Next, we characterize the electrical properties of the capacitor. Figure 21.13a shows P-V and I-V curves of the HZO capacitor with IGZO capping. The capacitor shows clear ferroelectricity. P_r is about 22 $\mu C/cm^2$ with a 5 V maximum sweep voltage, which is as large as HZO with TiN top capping [21]. The large P_r is attributed to the similar thermal expansions coefficient of IGZO ($\approx 4.31 \times 10^{-6}$/K) to TiN [22], because the top electrode with lower thermal expansion coefficient induces thermal strain in HZO film during a rapid thermal process, which results in large P_r as reported [23]. The capacitor has a $V_c = 2$ V with a 15 nm HZO and 8 nm IGZO. Note that the positive and negative V_cs are almost symmetric. For the positive V_g, IGZO is in accumulation, while, for the negative V_g, IGZO is in depletion. However, unlike Si, IGZO has a dielectric constant as large as 15. Even with thin IGZO, the depletion layer capacitance of IGZO is large, and thus the large voltage can be applied on the HZO layer. The impact of the IGZO layer in series connected to the HZO layer on the capacitance is seen in the C-V curve in Figure 21.13b. In addition, the work function difference between IGZO and TiN preferentially shifts the P-V curve. These are the reasons for the symmetric V_c of this capacitor.

We also studied endurance and retention characteristics. Endurance characteristics of the capacitor are shown in Figure 21.14. A ±4 V/1 μs pulse is applied to program/erase the capacitor. Thanks to the nearly zero interfacial layer between HZO and IGZO, the capacitor can be programmed and erased more than 10^8 cycles without

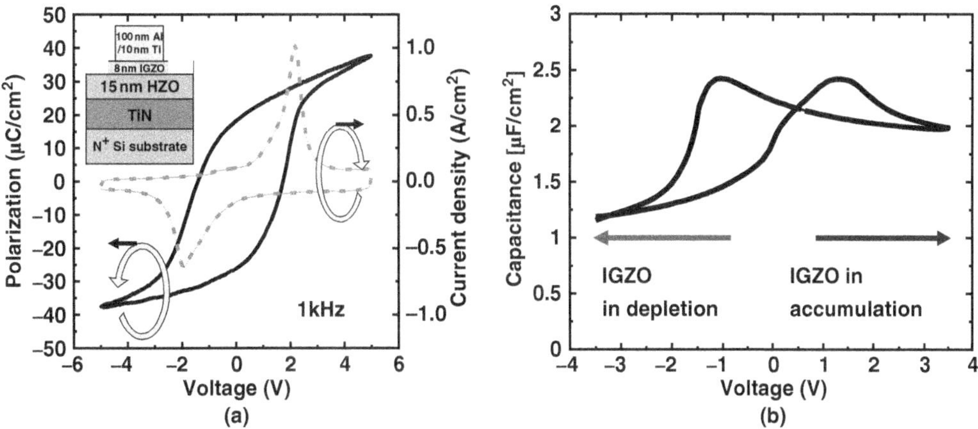

Figure 21.13 (a) Measured P-V and I-V curve, and (b) C-V curve of the fabricated capacitor.

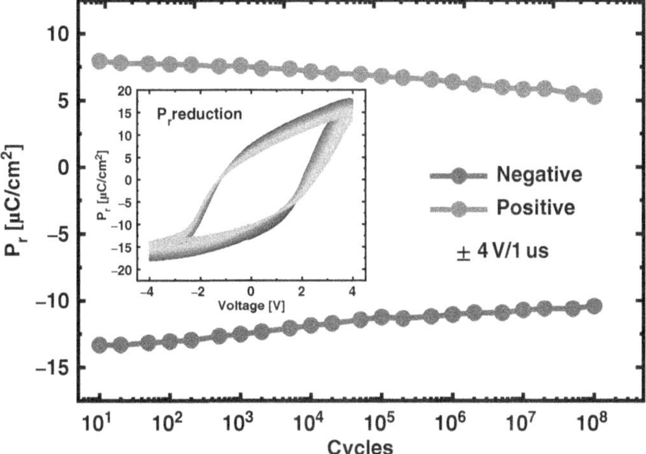

Figure 21.14 Measured endurance characteristics of the fabricated HZO capacitor with IGZO capping.

significant degradation. Moreover, there is no wakeup phenomenon. The possible reason is that the oxide–oxide interface without a low-k interfacial layer prevents charge trapping and the redistribution of charged defects. The wakeup-free operation is highly beneficial for manufacturing. Figure 21.15 shows the room temperature retention characteristics. The criterion of the failure is defined as the time when P_r becomes half of the initial value. The program/erase voltage is again ±4 V. The IGZO capacitor has at least one-year retention at both the program and erase states by extrapolating the result of Figure 21.15. It should be noted that FeFET reliability characteristics are largely affected by the depolarization field on HZO induced by the IGZO layer, which requires further study in our future work.

21.4.2 IGZO Channel FeFETs

Based on the understanding of the capacitor in Section 21.4.1, we characterize the fabricated IGZO channel FeFET. First, we check the channel mobility. We sweep V_g less than switching voltage to prevent polarization switching, leave the top gate floating, and measure I_d-V_g curves with drain voltage (V_{ds}) = 50 mV and 1 V, as shown in Figure 21.16a. I_d-V_g curves show nearly ideal junctionless FET characteristics for both linear and saturation

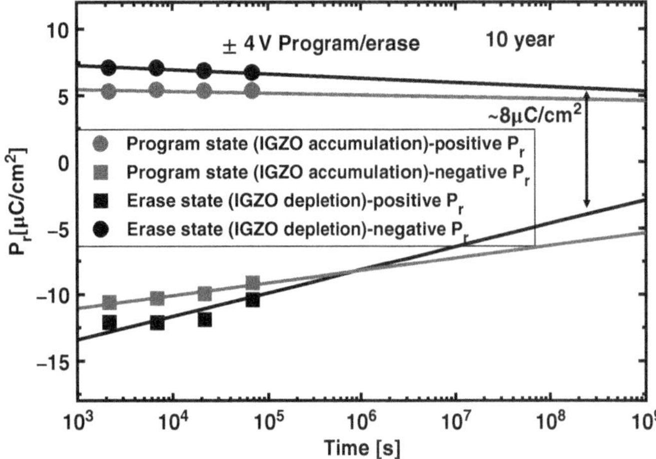

Figure 21.15 Measured retention characteristics of the fabricated HZO capacitor with IGZO capping.

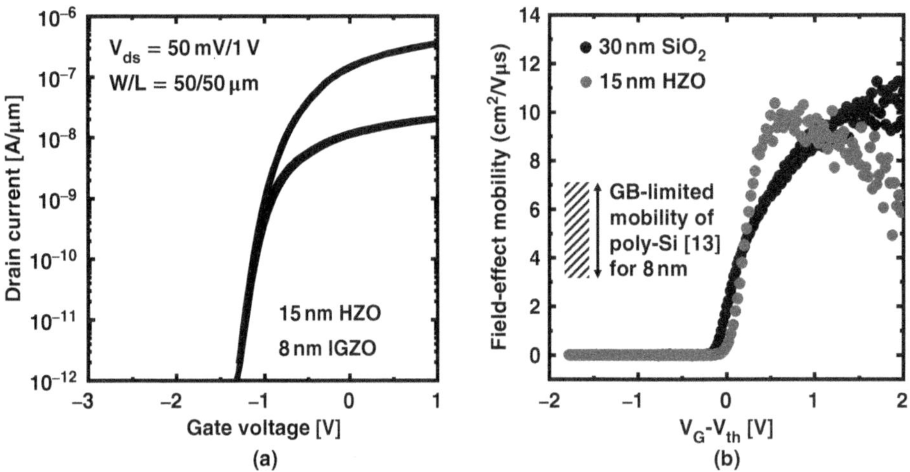

Figure 21.16 (a) Measured I_d-V_g curves of the fabricated IGZO channel FeFET by the narrow-range V_g sweep where polarization switching and hysteresis do not occur. (b) Extracted field-effect mobility of the IGZO channel FET with HZO and a SiO_2 gate insulator.

regions. Figure 21.16b compares the field-effect mobility with HZO and with a SiO_2 gate insulator extracted from the measured I_d-V_g curves. The field-effect mobility is consistent with Hall mobility ~10 $cm^2/V \cdot s$ and not significantly degraded by the HZO gate insulator compared with the SiO_2 gate insulator, thanks to the bulk conduction in junctionless FET operation and the interfacial-layer-free oxide–oxide interface with low interface state density.

Next, we examine the memory characteristics of the IGZO channel FeFET. Figure 21.17 shows the measured read I_d-V_g curves after applying erase and program pulse voltages from the bottom gate while the top-gate potential is fixed. The MW is ~0.5 V, as expected from the simulation results. The erase and program states occur thanks to the fixed body potential by the top gate. The extracted SS is shown in the inset of Figure 21.17. Nearly ideal SS is obtained for both erase and program states due to the low interface state and bulk defect densities. At the program state, the SS is slightly worse than at the erase state, which can be due to the residual interface states. The off current is mainly due to the leakage between the drain pad and the bottom gate.

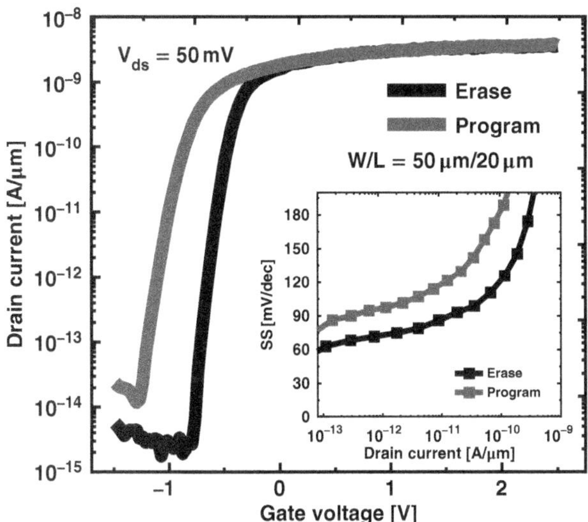

Figure 21.17 Measured read I_d-V_g curves of the IGZO channel FeFET after applying erase and program pulse voltages. The inset shows the extracted SS as a function of I_d.

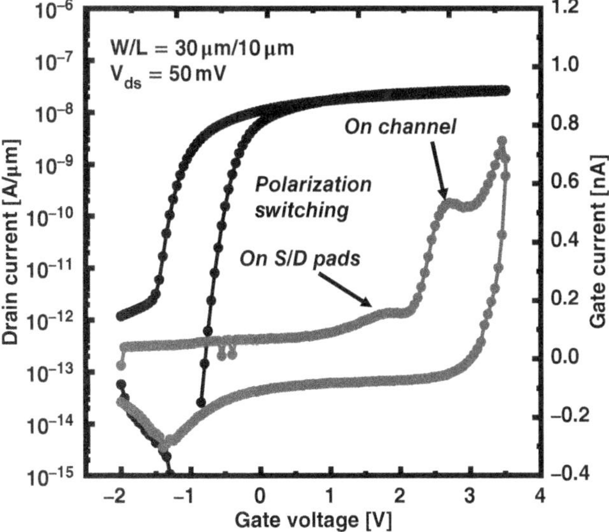

Figure 21.18 Measured I_d-V_g and I_g-V_g curves of the IGZO channel FeFET by bidirectional V_g sweep.

Polarization switching can be observed in quasi-static I_g measurement for a large V_g swing, as shown in Figure 21.18, similar to the simulation in Section 21.4.1. Particularly, in a positive V_g sweep after erase, two I_g peaks are observed. The first peak at lower V_g corresponds to the polarization-switching current between the bottom gate and drain/source pads. The voltage is almost the same as the polarization-switching voltage of the metal–HZO–IGZO–metal capacitor, while the second peak at higher V_g corresponds to the polarization-switching current between the bottom gate and the channel. Because the applied voltage partly drops on SiO_2, the polarization-switching voltage of HZO on the IGZO channel is higher than on the source/drain pad. Note that the peak in the negative V_g sweep after program is mainly due to the polarization switching between the bottom

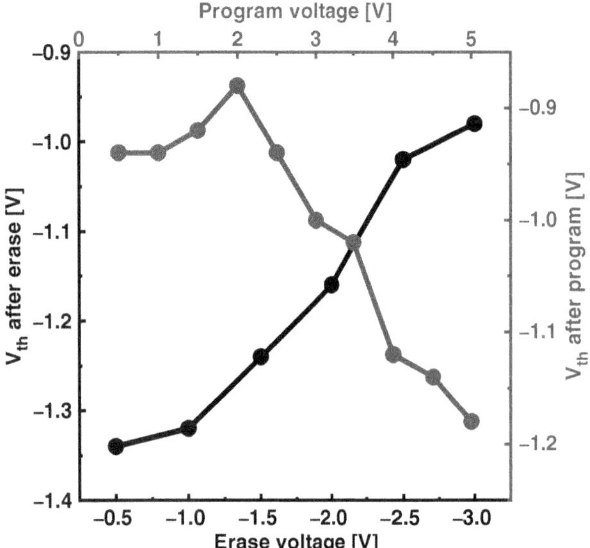

Figure 21.19 Measured voltage dependence of V_{th} in erase and program operations by gradually increasing the pulse amplitude.

gate and pads. During erase, both the depletion layer and SiO_2 capacitances are added to the HZO capacitance in series, the total capacitance becomes small, and thus the displacement current is much smaller than that under the pads.

Lastly, Figure 21.19 shows V_{th} versus the program and erase pulse voltages. The program/erase pulse width is 500 μs. The gradual V_{th} shift is observed. V_{th} shifts more when the pulse voltage becomes larger, which indicates that memory operation is controllable by the pulse write operation, and multibit operation is possible by using the intermediate states.

21.5 Summary

We proposed, designed, and fabricated a FE-HfO_2 FET with an ultrathin-body IGZO channel. The device shows the ideal SS and high channel mobility, thanks to the junctionless operation and the nearly zero interfacial layer at the oxide–oxide interface with low interface state density and bulk trap density. IGZO is an effective capping material that helps to form the ferroelectric phase in HZO. The metal–IGZO–HZO–metal stack capacitor shows clear ferroelectricity. Both endurance and retention characteristics show high reliability. The controllable memory device operation is demonstrated in the IGZO FeFET by fixing the body potential using the top gate. A FeFET with an ultrathin IGZO body is a promising candidate for high-density memory application.

Acknowledgments

This work was supported by JST PRESTO Grant No. JPMJPR1525, JSPS KAKENHI Grant No. JP18H01489, and Tokyo Electron Ltd.

References

1 Müller, J., Yurchuk, E., Schlösser, T. et al. (2012). Ferroelectricity in HfO_2 enables nonvolatile data storage in 28 nm HKMG. In: *2012 Symposium on VLSI Technology*, 25–26. HI: Honolulu.

2 Dünkel, S., Trentzsch, M., Richter, R. et al. (2017). A FeFET based super-low-power ultra-fast embedded NVM technology for 22 nm FDSOI and beyond. In: *2017 IEEE International Electron Devices Meeting* (IEDM). San Francisco, CA: 19.7.1–19.7.4.

3 Ni, K., Jerry, M., Smith, J.A., and Datta, S. (2018). A circuit compatible accurate compact model for ferroelectric-FETs. In: *2018 IEEE Symposium on VLSI Technology*, 131–132. HI: Honolulu.

4 Böscke, T.S., Müller, J., Bräuhaus, D. et al. (2011). Ferroelectricity in hafnium oxide thin films. *Applied Physics Letters* 99: 102903.

5 Tanaka, H., Kido, M., Yahashi, K. et al. (2007). Bit cost scalable technology with punch and plug process for ultra high density flash memory. In: *2007 Symposium on VLSI Technology*, 14–15.

6 Lue, H.-T., Hsu, T.-H., Hsiao, Y.-H. et al. (2010). A highly scalable 8-layer 3D vertical-gate (VG) TFT NAND flash using junction-free buried channel BE-SONOS device. In: *2010 Symposium on VLSI Technology*, 131–132. HI: Honolulu.

7 Choi, J. and Seol, K.S. (2011). 3D approaches for non-volatile memory. In: *2011 Symposium on VLSI Technology*, 178–179.

8 Goda, A. and Parat, K. (2012). Scaling directions for 2D and 3D NAND cells. In: *2017 IEEE International Electron Devices Meeting* (IEDM), 13–16.

9 Polakowski, P., Riedel, S., Weinreich, W. et al. (2014). Ferroelectric deep trench capacitors based on Al:HfO_2 for 3D nonvolatile memory applications. In: *2014 IEEE 6th International Memory Workshop* (IMW), Taipei, 1–4.

10 Florent, K., Pesic, M., Subirats, A. et al. (2018). Vertical ferroelectric HfO_2 FET based on 3-D NAND architecture: towards dense low-power memory. In: *2018 IEEE International Electron Devices Meeting* (IEDM). San Francisco, CA: 2.5.1–2.5.4.

11 Nomura, K., Ohta, H., Takagi, A. et al. (2004). Room-temperature fabrication of transparent flexible thin-film transistors using amorphous oxide semiconductors. *Nature* 432: 488–492.

12 Kawamura, T., Uchiyama, H., Saito, S. et al. (2008). 1.5-V Operating fully-depleted amorphous oxide thin film transistors achieved by 63-mV/dec subthreshold slope. In: *2008 IEEE International Electron Devices Meeting*, 1–4. San Francisco, CA.

13 Matsuda, S., Hiramatsu, T., Honda, R. et al. (2015). 30-nm-channel-length c-axis aligned crystalline In-Ga-Zn-O transistors with low off-state leakage current and steep subthreshold characteristics. In: *2015 Symposium on VLSI Technology*, T216–T217. Japan: Kyoto.

14 Lee, G.-G., Tokumitsu, E., Yoon, S.-M. et al. (2011). The flexible non-volatile memory devices using oxide semiconductors and ferroelectric polymer poly(vinylidene fluoride-trifluoroethylene). *Applied Physics Letters* 99: 012901.

15 Li, Y., Liang, R., Wang, J. et al. (2017). A ferroelectric thin film transistor based on annealing-free HfZrO film. *IEEE Journal of the Electron Devices Society* 5 (5): 378–383.

16 Besleaga, C., Radu, R., Balescu, L.-M. et al. (2019). Ferroelectric field effect transistors based on PZT and IGZO. *IEEE Journal of the Electron Devices Society* 7: 268–275.

17 Mo, F., Tagawa, Y., Jin, C. et al. (2019). Experimental demonstration of ferroelectric HfO_2 FET with ultrathin-body IGZO for high-density and low-power memory application. In: *2019 Symposium on VLSI Technology*, T42–T43. Japan: Kyoto.

18 Kobayashi, M., Ueyama, N., Jang, K., and Hiramoto, T. (2016). Experimental study on polarization-limited operation speed of negative capacitance FET with ferroelectric HfO_2. In: *2016 IEEE International Electron Devices Meeting* (IEDM). San Francisco, CA: 12.3.1–12.3.4.

19 Min, D., Kang, S.Y., Moon, S., and Yoon, S. (2019). Impact of thickness control of $Hf_{0.5}Zr_{0.5}O_2$ films for the metal–ferroelectric–insulator–semiconductor capacitors. *IEEE Electron Device Letters* 40 (7): 1032–1035.

20 Park, M.-H., Kim, H.-J., Kim, Y.-J. et al. (2013). Evolution of phases and ferroelectric properties of thin $Hf_{0.5}Zr_{0.5}O_2$ films according to the thickness and annealing temperature. *Applied Physics Letters* 102: 242905.

21 Kim, S.-J., Mohan, J., Summerfelt, S.R., and Kim, J. (2019). Ferroelectric $Hf_{0.5}Zr_{0.5}O_2$ thin films: a review of recent advances. *JOM: The Journal of The Minerals. Metals & Materials Society* 71: 246–255.

22 Kim, H.-J. and Kim, Y.-J. (2014). Influence of external forces on the mechanical characteristics of the a-IGZO and graphene based flexible display. In: *Global Conference on Polymer and Composite Materials (PCM 2014)*, 27–29. China: Ningbo.

23 Cao, R., Wang, Y., Zhao, S. et al. (2018). Effects of capping electrode on ferroelectric properties of $Hf_{0.5}Zr_{0.5}O_2$ thin films. *IEEE Electron Device Letters* 39 (8): 1207–1210.

22

Neuromorphic Chips Using AOS Thin-Film Devices

Mutsumi Kimura

Ryukoku University, Kyoto, Japan
Nara Institute of Science and Technology, Ikoma, Japan

22.1 Introduction

Artificial intelligence [1, 2] has been used in indispensable manners for various applications, such as letter and image recognition, information retrieval and provision, language translation and caption composition, expert systems, autonomous driving, artificial brains, and so on, and it is also promising in future societies. Neural networks [3–5] are representative technologies of artificial intelligence that mimic the operating principle of biological brains; their advantages are their self-organization, self-learning, parallel distributed computing, fault tolerance, and the like. However, because the conventional ones are long and complicated software executed on high-specification Neumann-type computer hardware, which is not optimized for neural networks, the machine size is unbelievably bulky, and the power consumption is incredibly huge. For example, one of the most famous cognitive systems [6, 7] occupies the size of 10 refrigerators and consumes approximately 100 kW in power. Moreover, some of the advantages are not acquired, because they are executed on Neumann-type computers. Neuromorphic systems or neuroinspired systems [8–10] are practical solutions and biomimetic systems that compose neural networks solely of optimized devices and hardware, and they have the same advantages as living brains: The machine size can be compact, the power consumption can be low, and the operation can be robust. Some neuromorphic systems [11, 12] are well known because they are hybrid systems—namely, multiple neuron elements are virtually emulated using time sharing of one neuron element, analog values are stored using multiple bits of digital memories, learning function is not implemented in itself, and so on—and the abovementioned advantages are only partially obtained. A great deal of effort is continuously being made to solve these problems [13].

On the other hand, thin-film semiconductor electronic devices are widely used, and the advantages are that they can be fabricated on large areas and three-dimensional layered structure can be acquired [14], whereas the unavoidable disadvantages are low performance and low yield. Neuromorphic systems are interesting applications for thin-film devices, because the advantages are available, and the disadvantages are acceptable. Although some other devices are used for neuromorphic systems [15], we are investigating neuromorphic systems with amorphous metal-oxide semiconductor (AOS) thin-film devices as synapse elements, because the AOS devices [16, 17] can possibly be layered using a printing process with low cost [18, 19]; it is expected that the neuromorphic systems will be three-dimensional integrated systems [20] in the future, the machine size can be excellently compact, and the power consumption can be remarkably low. Additionally, we are proposing modified Hebbian learning [21] as a learning rule done locally and automatically without extra control circuits. As a result, the conductance deterioration of the AOS thin-film devices can be ingeniously utilized as the strength plasticity of synaptic connections.

Amorphous Oxide Semiconductors: IGZO and Related Materials for Display and Memory, First Edition.
Edited by Hideo Hosono and Hideya Kumomi.

Although some other learning rules, such as spike timing–dependent plasticity (STDP), are used for neuromorphic systems [22], the modified Hebbian learning rule is one of the more promising ones. Moreover, because biological brains are substantially realized using the neuromorphic systems, it is also expected that all inherent functions are also realized, such as self-organization, self-teaching, parallel distributed computing, fault tolerance, damage robustness, and so on. In this study, a neuromorphic system with crosspoint-type amorphous Ga-Sn-O (α-GTO) thin-film devices as self-plastic synapse elements [23] and a neuromorphic system using a large-scale integration (LSI) chip and amorphous In-Ga-Zn-O (α-IGZO) thin-film devices [24] have been developed.

22.2 Neuromorphic Systems with Crosspoint-Type α-GTO Thin-Film Devices

We have developed neuromorphic systems with crosspoint-type α-GTO thin-film devices as self-plastic synapse elements [23]. The α-GTO thin-film device is a promising device using α-GTO semiconductor that has been presented recently by the authors [25–28]. Fundamental operations will be confirmed for neuromorphic systems with crosspoint-type α-GTO thin-film devices as self-plastic synapse elements. First, crosspoint-type α-GTO thin-film devices will be fabricated, and it will be found whether the electric current gradually decreases along the bias time. Next, a neuromorphic system will be actually implemented using a field-programmable gate array (FPGA) chip and crosspoint-type α-GTO thin-film devices, and it will be confirmed whether a necessary function for letter recognition is obtained after the learning process. Once the fundamental operations are confirmed, more advanced and useful functions will be obtained by increasing the device and circuit scales in the future, according to the technological history of neural networks proved using software prior to hardware [29, 30]. In comparison with our prior publication, which proposed double-layer α-GTO thin-film devices that were actually obtained later and the same system architecture as the neuromorphic system [31], it is meaningful that α-GTO thin-film devices have an extremely simple structure in this study.

22.2.1 Neuromorphic Systems

22.2.1.1 α-GTO Thin-Film Devices

The structure, conductance deterioration, and deterioration mechanism of the crosspoint-type α-GTO thin-film devices are shown in Figure 22.1. As shown in Figure 22.1a, the crosspoint-type α-GTO thin-film devices were fabricated as follows. First, a quartz glass substrate was used, whose thickness was 1 mm and size was 3 × 3 cm. Next, a Ti thin film was deposited as bottom electrodes using vacuum evaporation through a metal mask, whose thickness was 50 nm, line and space widths were 150 and 150 μm, and number of lines was 80. Sequentially, an α-GTO thin film was deposited as active layers using radiofrequency (RF) magnetron sputtering with a ceramic target of Ga:Sn = 1:3 and sputtering gas of Ar:O_2 = 20:1, whose thickness was 70 nm. Next, another Ti thin film was again deposited as top electrodes using the same fabrication process, whose thickness, line and space widths, and number of lines were the same as those of the top electrodes. The top and bottom electrodes are perpendicular to each other. After that, they were densified using furnace annealing at 350 °C for 1 hour. Finally, crosspoint-type α-GTO thin-film devices were completed, where α-GTO thin films were sandwiched between 80 top and 80 bottom electrodes, and 80×80 = 6,400 crosspoint-type α-GTO thin-film devices were fabricated.

As shown in Figure 22.1b, the conductance deterioration was measured as follows. First, the bias voltage was continuously applied between the top and bottom electrodes, whose voltage is 3.3 V. Next, the electric current was continuously measured through the α-GTO thin-film device. It is found that the electric current gradually decreases along the bias time, which is achieved by the abovementioned optimized process condition. As shown in Figure 22.1c, it is suggested that the conductance deterioration occurs because the trap states are generated in the α-GTO thin films [32]. Free electrons are accelerated by the electric field and collide with the GTO atoms, and trap states are generated in the α-GTO thin films. Some electrons are trapped, free electrons decrease, the movements

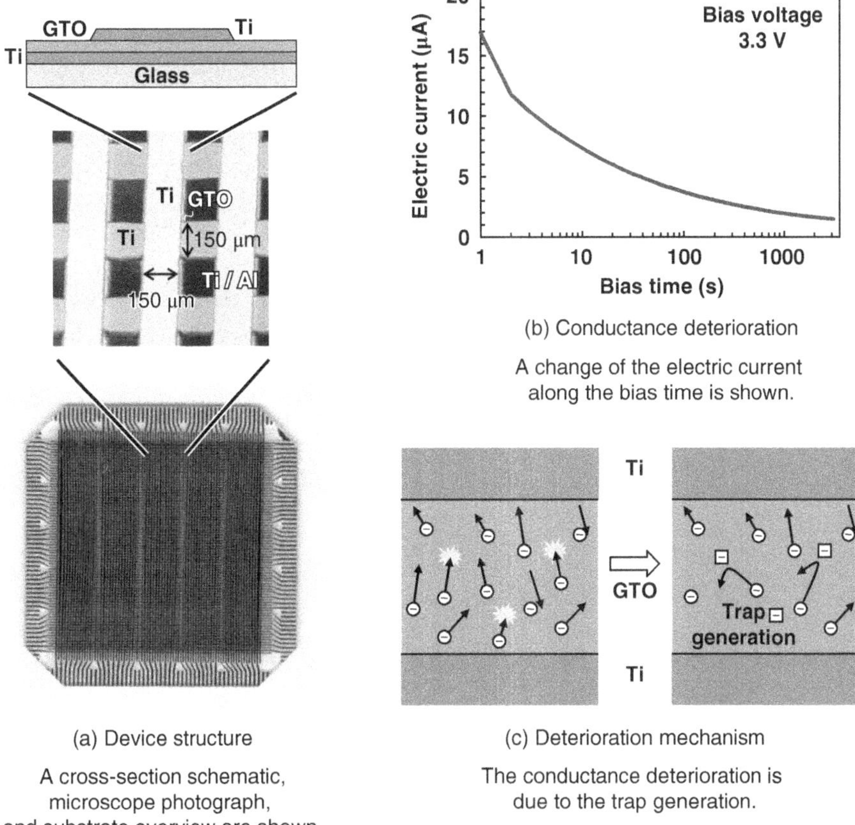

(a) Device structure
A cross-section schematic, microscope photograph, and substrate overview are shown.

(b) Conductance deterioration
A change of the electric current along the bias time is shown.

(c) Deterioration mechanism
The conductance deterioration is due to the trap generation.

Figure 22.1 Crosspoint-type α-GTO thin-film devices. (a) Device structure. A cross-sectional schematic, microscope photograph, and substrate overview are shown. (b) Conductance deterioration. A change of the electric current along the bias time is shown. (c) Deterioration mechanism. The conductance deterioration is due to the trap generation.

are disturbed by Coulomb scattering, and conductance deterioration occurs. With the modified Hebbian learning, the conductance deterioration of the α-GTO thin-film devices can be ingeniously utilized as strength plasticity of synaptic connections.

22.2.1.2 System Architecture

A system architecture, system implementation, system behavior, learning process, pixel pattern mapping, and majority-rule handling of the neuromorphic system are shown in Figure 22.2. As shown in Figure 22.2a, a Hopfield neural network was actually assembled by combining 80 neuron elements and 80×80 = 6,400 synapse elements. A Hopfield neural network is a neural network where input terminals of all neuron elements are connected from output terminals of all neuron elements through synapse elements. As shown in Figure 22.2b, the Hopfield neural network was actually implemented using neuron elements composed in an FPGA chip and synapse elements made by the crosspoint-type α-GTO thin-film devices. The neuron elements composed in the FPGA chip are just buffer circuits that offer a step function. First, as shown in the left figure of Figure 22.2c, during the learning process, a signal pattern was inputted to the neuron elements, and the same signal pattern was outputted and applied to the top electrodes of the crosspoint-type α-GTO thin-film devices and simultaneously to the bottom electrodes through the changeover switches. Next, as shown in the right figure of Figure 22.2c, during the recognizing process, a signal pattern was inputted, the same signal pattern was applied to the bottom electrode of the crosspoint-type α-GTO thin-film devices, the top electrode was floating, and the signal pattern was immediately released. Some analog signals were transmitted from the top electrodes of the crosspoint-type α-GTO thin-film devices and inputted to the neuron elements, and some signal pattern was outputted after dynamic behavior of the neuromorphic system.

As shown in Figure 22.2d, during the learning process, when the voltage difference existed between the top and bottom electrodes and the electric current flowed through the α-GTO thin-film device, the conductance deterioration occurred. The pattern generation of the conductance deterioration pattern corresponds to the learning achievement. As shown in Figure 22.2e, a two-dimensional pixel pattern of 9×9 was transformed to a one-dimensional signal pattern of 80 to use the neuromorphic system for letter recognition. Moreover, because our fabrication instruments are not very satisfying, some top or bottom electrodes, where the signal pattern is transmitted, are partially injured, as shown in the lower photograph of Figure 22.1a. Therefore, as shown in Figure 22.2f, majority-rule handling was introduced, where one pixel was assigned to nine signals and a state that the pixel was in on-state corresponded to a state that more than five signals were in on-state. The majority-rule handling not only is a compensation scheme in the neuromorphic system when some signals are partially injured but also may be simultaneously a representing scheme in biological brains, because all neurons do not work very well there and some image information processing is done on the retinal surface [33].

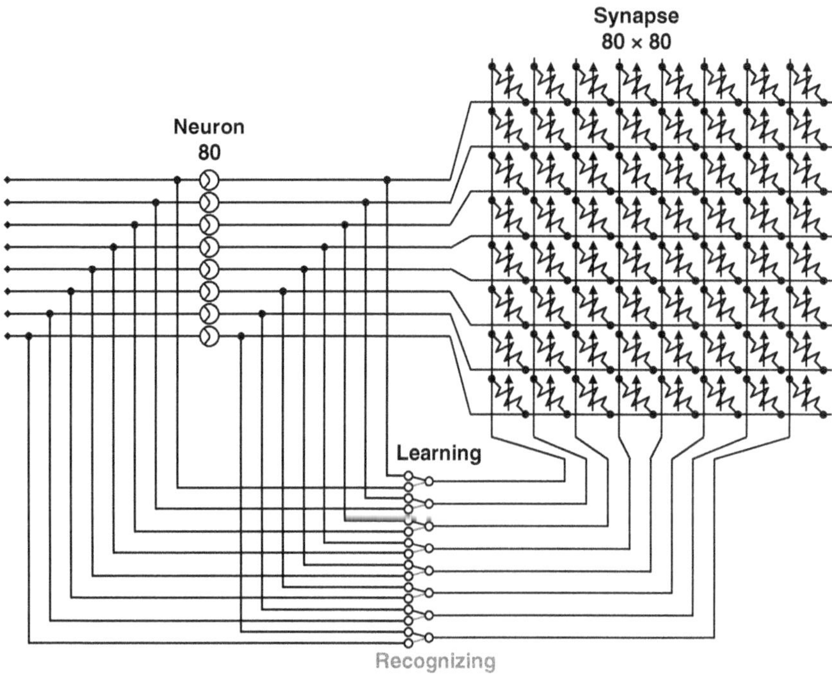
Synapse
80 × 80
Neuron
80
Learning
Recognizing
(a) System architecture
A Hopfield neural network is actually assembled by combining
80 neuron elements and 80 × 80 = 6400 synapse elements.

Figure 22.2 Neuromorphic system. (a) System architecture. A Hopfield neural network is actually assembled by combining 80 neuron elements, and 80×80 = 6,400 synapse elements. (b) System implementation. Neuron elements are composed in an FPGA chip, synapse elements are made by the crosspoint-type α-GTO thin-film devices, and they are connected with the changeover switches. (c) System behavior. (d) Learning process. Voltage differences are indicated by light and dark shaded lines, and electric currents are indicated by shaded pillars. (e) Pixel pattern mapping. A two-dimensional pixel pattern is transformed to a one-dimensional signal pattern. (f) Majority-rule handling. One pixel is assigned to multiple signals, and a state that the pixel is in the on-state corresponds to a state that more than half of the signals are in the on-state.

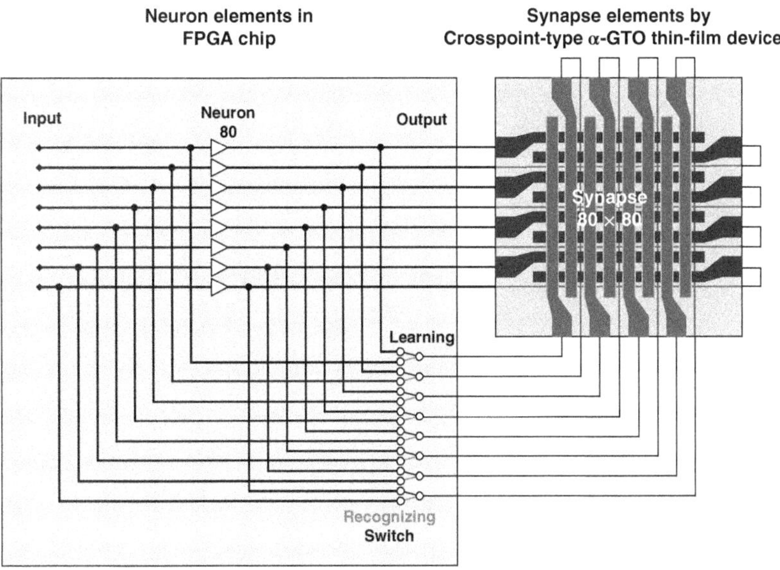

Neuron elements are composed in an FPGA chip, synapse elements are made by the crosspoint-type α-GTO thin-film devices, and they are connected with the changeover switches.

(b) System implementation

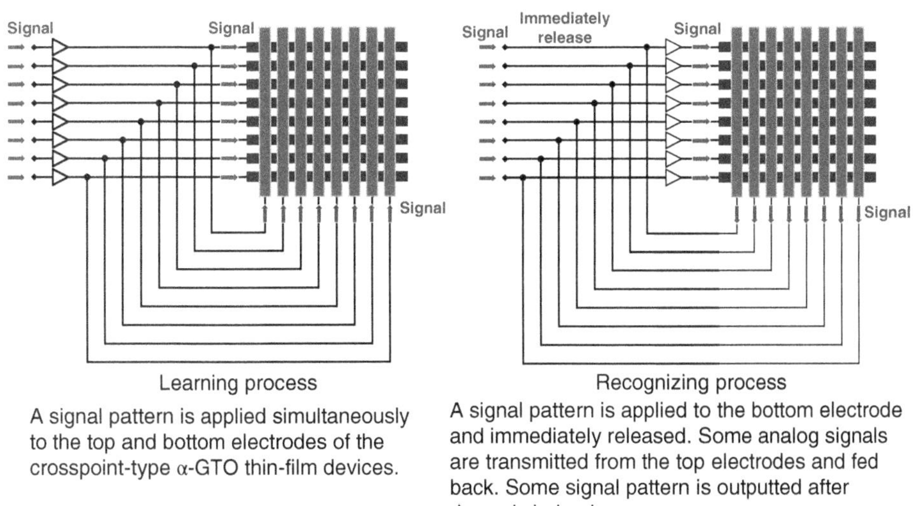

Learning process

A signal pattern is applied simultaneously to the top and bottom electrodes of the crosspoint-type α-GTO thin-film devices.

Recognizing process

A signal pattern is applied to the bottom electrode and immediately released. Some analog signals are transmitted from the top electrodes and fed back. Some signal pattern is outputted after dynamic behavior.

(c) System behavior

Figure 22.2 (*Continued*)

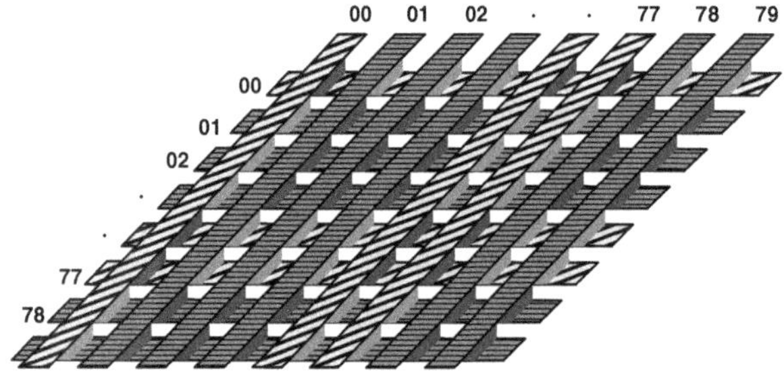

(d) Learning process

Voltage differences are indicated by striped and cross-hatched lines, and electric currents are indicated by gray pillars.

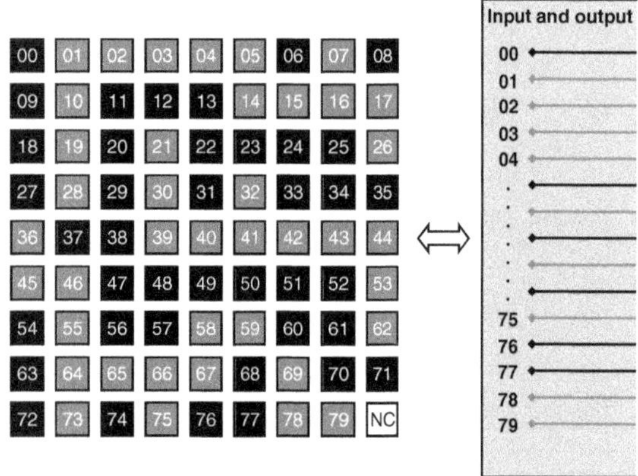

(e) Pixel pattern mapping

A two-dimensional pixel pattern is transformed to a one-dimensional signal pattern.

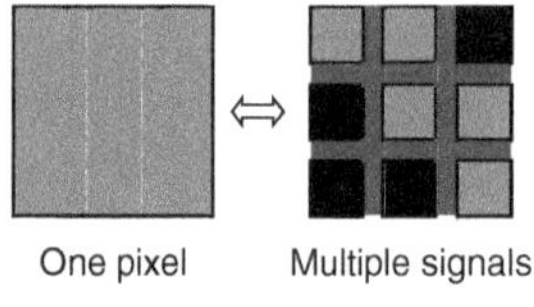

(f) Majority-rule handling

One pixel is assigned to multiple signals, and a state that the pixel is in on-state corresponds to a state that more than half of the signals are in on-state.

Figure 22.2 *(Continued)*

22.2.2 Experimental Results

Experimental results of letter recognition are shown in Figure 22.3. First, during the learning process, the alphabet letters of "T," "L," and "X" were learned. Initially, a pixel pattern of "T" was transformed to a signal pattern, and the signal pattern was inputted to the neuron elements for 1 second. The same signal pattern was outputted and applied to the top electrodes of the crosspoint-type α-GTO thin-film devices and simultaneously to the bottom electrodes, and the conductance deterioration occurred. Successively, a signal pattern of "L" was inputted, and a signal pattern of "X" was inputted by turns. Then these operations were repeated several hundred times. Next, during the recognition process, alphabet letters of "T," "L," and "X" were reformed. Initially, a slightly distorted pixel pattern (i.e., a one-pixel flipped pattern of "T") was transformed to a signal pattern with the majority-rule

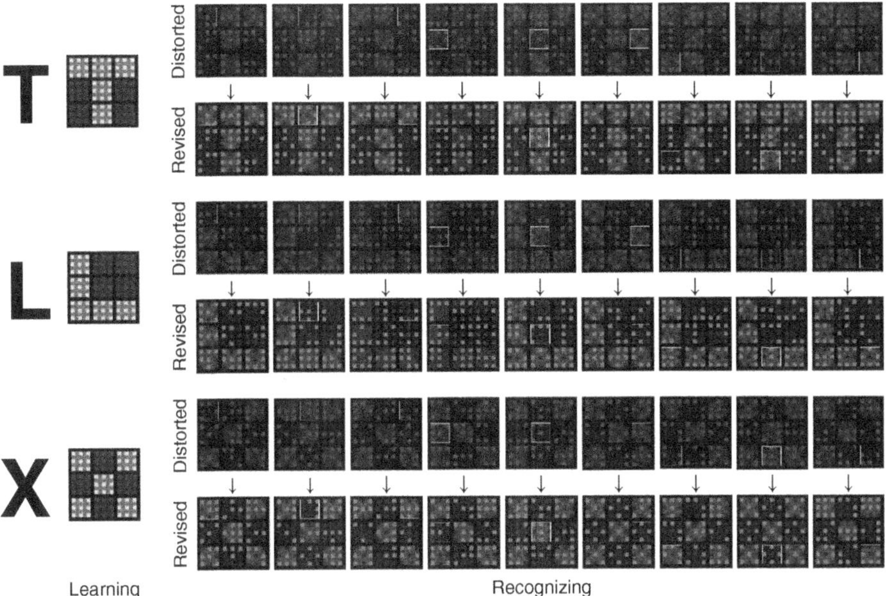
T
L
X
Distorted
Revised
Learning
Recognizing
Pixel patterns during learning process, slightly distorted pixel patterns inputted during recognizing process, and revised pixel patterns outputted during recognizing process are shown. The one-pixel flipped patterns are indicated by white squares.

Figure 22.3 Letter recognition. Pixel patterns during the learning process, slightly distorted pixel patterns inputted during the recognizing process, and revised pixel patterns outputted during the recognizing process are shown. The one-pixel flipped patterns are indicated by white squares.

handling (i.e., five signals were set to on-state when the pixel was in on-state, and vice versa), and the signal pattern was inputted. The same signal pattern was applied to the bottom electrodes of the crosspoint-type α-GTO thin-film devices, the top electrode was floating, and the signal pattern was immediately released. Subsequently, some revised signal pattern was outputted from the neuron elements and transformed to some revised pixel pattern with the majority-rule handling, namely, the pixel was set to on-state when more than five signals are in on-state, and vice versa. Then these operations were repeated for different distorted pixel patterns. Successively, slightly distorted signal patterns of "L" were inputted, and some signal patterns were outputted; and slightly distorted signal patterns of "X" were inputted, and some signal patterns were outputted. Although the same voltage of 3.3 V was applied during both the learning and recognizing processes, whereas the learning process was as long as 1 second and repeated several hundred times, because the recognizing process was as short as less than 0.1 second and not repeated, the conductance deterioration was negligible during the recognizing process. Finally, it is confirmed that the revised pixel pattern is the same as the pixel pattern of "T," "L," and "X." Because simple pattern matching is available once the pixel patterns are reformed, it can be concluded that a function of letter recognition is confirmed.

In this study, three alphabet letters were learned, whereas based on the basic theory [34], five alphabet letters can be learned in a neuromorphic system with 80 units. Although this Hopfield neural network is not a simple Hopfield neural network, it is believed that the results are sufficient from the viewpoint of the basic theory.

22.3 Neuromorphic System Using an LSI Chip and α-IGZO Thin-Film Devices [24]

A neuromorphic system is developed by using an LSI chip and depositing *a*-IGZO thin-film devices on it. Neuron elements are equipped in the LSI chip, and synapse functions are realized by the *a*-IGZO thin-film devices with

Figure 22.4 Neuron element equipped in the LSI chip.

modified Hebbian learning without any additional circuits to control synaptic weights. We will confirm whether the neuromorphic system can work as an associative memory by reproducing memorized letters.

22.3.1 Neuromorphic System

22.3.1.1 Neuron Elements

The neuron element equipped in the LSI chip is shown in Figure 22.4. The neuron element is a two-inverter circuit where two inverters are connected in series and composed of four transistors. The inverters generate a binary state, and the binary state is alternated whenever some input signal is received. Three terminals are unidirectional—namely, they constantly work as either input or output terminals—and three-ring electrodes are connected to the three terminals. A theoretical model for a neuron element is just a buffer block, whose function is completely the same as that of the two-inverter circuit. It should be noted that the two-inverter circuit is at least necessary to get the buffer block.

There are 25 × 25 neuron elements in the LSI chip, wherein 12 × 12 neuron elements on every other column and row line are accessible, which are employed as input/output (I/O) neurons to confirm whether the neuromorphic system can work as an associative memory, and hidden neuron elements are between the I/O neurons. The LSI chip is fabricated using a conventional silicon (Si) complementary metal-oxide semiconductor (CMOS) field-effect transistor (FET) process [35], and it is easy to fabricate the neuron circuit using the Si CMOS FET process because it is a general digital circuit.

22.3.1.2 Synapse Elements

The synapse element formed using the *a*-IGZO thin-film device is shown in Figure 22.5. A planar-type *a*-IGZO device is formed as follows. An *a*-IGZO thin film is deposited on the metal electrodes using RF magnetron sputtering. The sputtering target is an IGZO ceramic whose composition is In:Ga:Zn = 1:1:1, the sputtering gas flow rate is Ar:O_2 = 20:0.2 sccm, the deposition pressure 56 is 0.66 Pa, the plasma power is 60 W, and the deposition time is 5 min. These deposition conditions are roughly the same as those of the thin-film transistors (TFTs) with the best performance [36]. It is important that because the RF magnetron sputtering is done at room temperature and no additional annealing is executed, the *a*-IGZO thin film can be deposited on the LSI chip without any damage to the neuron elements.

An electrical current through the planar-type *a*-IGZO device is also shown. Here, we apply a bias voltage V = 1.8 V between the right and left metal electrodes and measure the electrical current through the planar-type

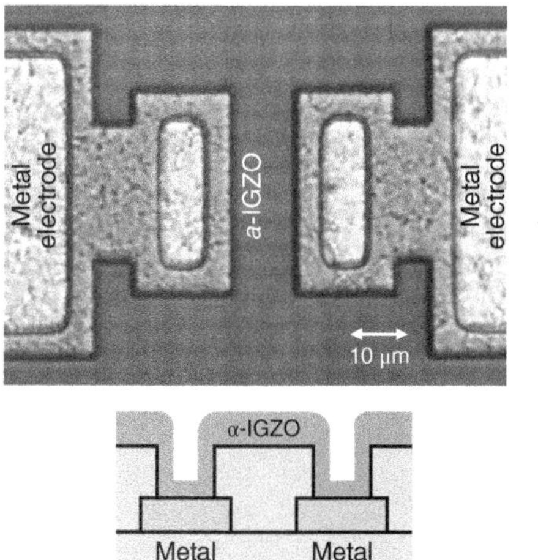

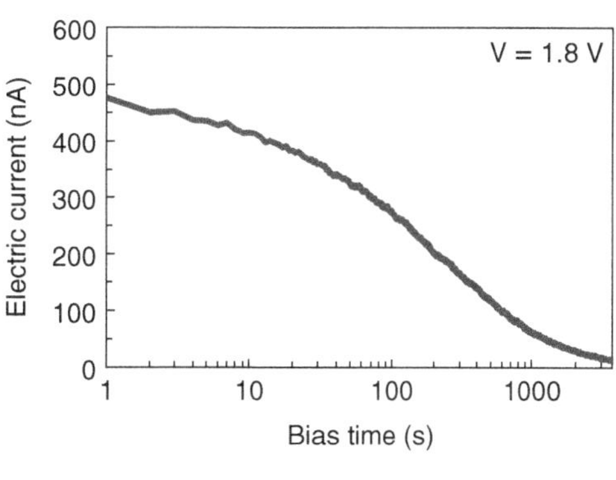

Figure 22.5 Synapse element formed using the α-IGZO thin-film device.

a-IGZO device. It is found that the electrical conductance continuously decreases while the electric current flows, and the conductance decrease is irreversible. Modified Hebbian learning, which is explained later in this chapter, can be employed using this phenomenon [37]. As shown in the graph, the conductance decrease is not so fast. However, because operation of the neural network is based on a delicate balance in majority rule, which is also explained later in the chapter, the slow change is sometimes convenient to avoid excess learning.

Working mechanisms for the conductance decrease are as follows. It is supposed that this phenomenon is caused by either generation of trap states in the a-IGZO thin film or injection of electric charges into the interface between the a-IGZO thin film and underlayer insulator film. In the case of the generation of the trap states, first, free electrons are accelerated and collide with the a-IGZO crystal. Next, the trap states are generated and capture the free electrons. Finally, the free electrons decrease because of the fixed charges in the trap states and are simultaneously scattered by them, which induces the conductance decrease. On the other hand, in the case of the injection of the electric charges, first, free electrons are again accelerated near the interface and are casually injected into the interface between the IGZO thin film and underlayer insulator film. Next, the free electrons are captured at the interface. Finally, the free electrons decrease because of the fixed charges of the injected electrons, which also induces the conductance decrease.

22.3.1.3 System Architecture

The neuromorphic system using an LSI chip and a-IGZO thin-film devices is shown in Figure 22.6. The aforementioned a-IGZO thin film is deposited on the aforementioned LSI chip. As a result, the neuron elements equipped in the LSI chip are jointed to the synapse elements formed using the a-IGZO thin-film devices through the metal electrode at the top surface of the LSI chip, and the layered structure for the neuromorphic system is realized. Here, a cellular neural network is built, where a neuron element is connected to the eight surrounding neuron elements, namely, four orthogonal front, back, right, and left ones, and four diagonal front-right, front-left, back-right, and back-left ones.

22.3.2 Working Principle

22.3.2.1 Cellular Neural Network

The cellular neural network for the neuromorphic system is shown in Figure 22.7. The cellular neural network is a neural network where a neuron element is connected to only the surrounding neuron elements, which

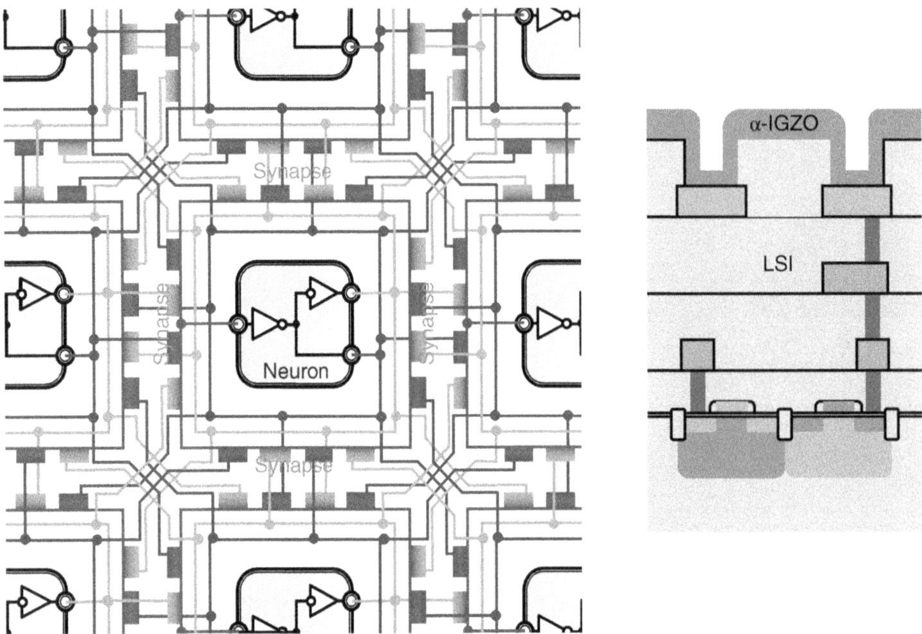

Figure 22.6 Neuromorphic system using an LSI chip α-IGZO and thin-film devices.

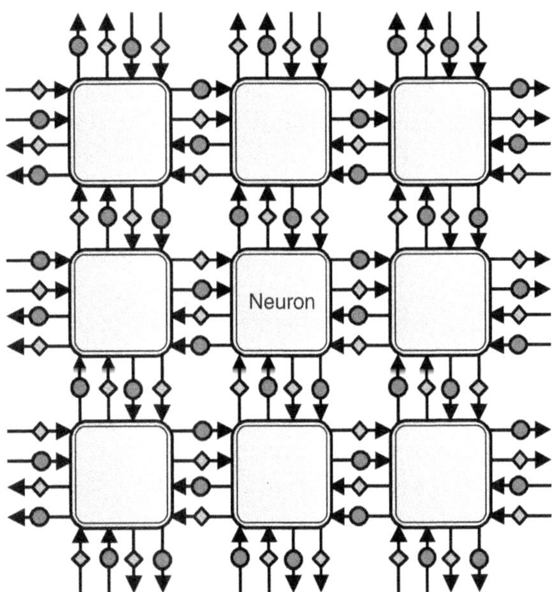

Figure 22.7 Cellular neural network for the neuromorphic system.

are remarkably suitable for integration of electron devices. They are promising for image processing, pattern recognition, associative memory, and so on. As a result, in comparison with other neural networks, a relatively larger number of the neuron elements exist, and a large number of the synapse elements still exist, but connection wiring occupies little areas or volumes.

Forward and backward synapse connections exist because the three terminals of the neuron elements are unidirectional, as written here, and the direction of signal transfer is defined for each synapse connection. The direction of the signal transfer is indicated by the arrows in Figure 22.7.

22.3.2.2 Tug-of-War Method

A tug-of-war method is proposed for the synaptic connections. Two-type synapse connections, a concordant connection and discordant connection, are prepared. The concordant connection connects the same logics of two neuron elements, namely, positive and positive logics or negative and negative logics, and tends to make the states of the two neuron elements the same. On the other hand, the discordant connection connects different logics of two neuron elements, namely, positive and negative logics, and tends to make the states of the two neuron elements different. The reason why two-type synapse connections are prepared is to obtain the same effect that the synaptic connection strength becomes both stronger and weaker even if the actual strength becomes either one. The concordant connections are indicated by shaded circles, and the discordant connections are indicated by shaded diamonds in Figure 22.7.

As a result, neighboring neuron elements are connected through four synapse elements corresponding to the combination of concordant, discordant, forward, and backward connections. Four diagonal connections are omitted to avoid overcomplexity in Figure 22.7.

22.3.2.3 Modified Hebbian Learning

Whereas Hebbian learning is a typical learning rule in biological and artificial neural networks [38], modified Hebbian learning is introduced into this neuromorphic system. Although the modified Hebbian learning is explained in detail elsewhere [37], it is explained again here. The modified Hebbian learning as a local learning rule is shown in Figure 22.8. Here, learning of NOT logic is tried. Initially, in the initial recognizing stage, some state is applied to the input element, and some state arises from the output element. Next, in the first learning stage, a stable state is applied to the input element, and a fire state is applied to the output element. Since the concordant connection connects the two terminals for the same logic in the two neuron elements and the binary states at both terminals are different, some voltage occurs and electric current flows through the concordant connection, and the synaptic connection strength gradually weakens. In the second learning stage, a fire state is applied to the input element, and a stable state is applied to the output element. Similarly, only the synaptic connection strength of the concordant connection gradually weakens. Finally, in the final recognizing stage, a stable state is applied to the input element, and a fire state arises from the output element, and vice versa, because the synaptic connection strength of the concordant connection becomes slightly weaker than that of the discordant connection. As a result, the learning of NOT logic succeeds. It is important that the learning is

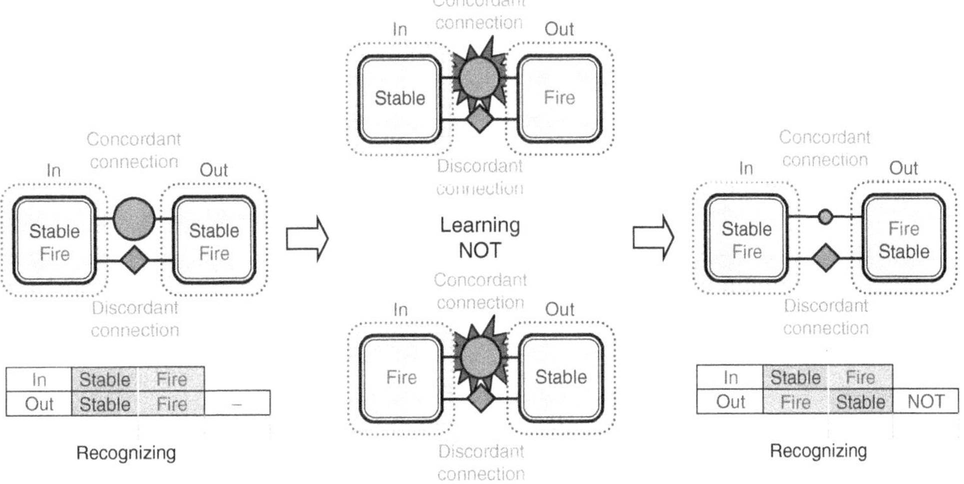

Figure 22.8 Modified Hebbian learning as a local learning rule.

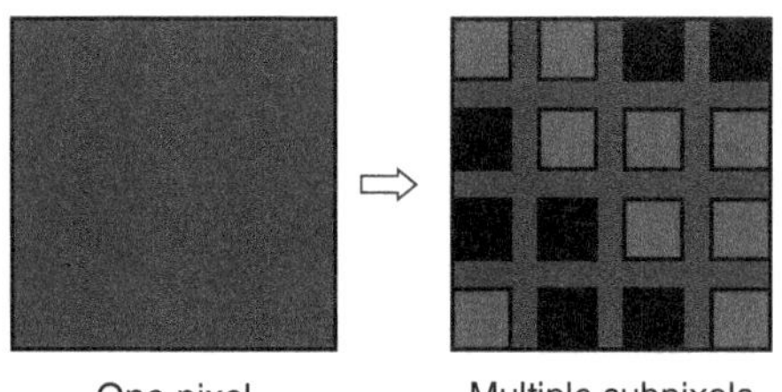

Figure 22.9 Majority-rule handling against partial malfunction.

completed without any additional circuits to control synaptic weights, namely, the modified Hebbian learning is a local learning rule done using only the local conditions.

22.3.2.4 Majority-Rule Handling

Majority-rule handling was invented to improve the robustness of this neuromorphic system. Majority-rule handling against partial malfunction is shown in Figure 22.9. One pixel corresponds to multiple subpixels, for example, 4×4 subpixels. When some letter pattern is inputted into the neuromorphic system, if a pixel is in the on-state, more than half of the subpixels—for example, nine subpixels—are randomly selected and set to be in the on-state, whereas the other subpixels are set to be in off-states. If a pixel is in an off-state, nine subpixels are set to be in the off-state, and so on. This subpixel pattern is inputted into the I/O neuron elements as fire and stable states. On the other hand, when some letter pattern is outputted, if the subpixels in the on-state are more than eight, the corresponding pixel is regarded as in the on-state. If the subpixels in the on-state are fewer than eight, the pixel is regarded as in the off-state. Majority-rule handling seems suitable for neural networks because it seems robust against the partial malfunction of neuron and synapse elements.

22.3.3 Experimental Results

22.3.3.1 Raw Data

We will confirm whether this neuromorphic system can work as an associative memory by reproducing memorized letters. An example of raw data of the letter reproduction is shown in Figure 22.10. The learning sequence is as follows. Initially, in the learning stage, standard patterns of the alphabet letters "T" and "L" are inputted into the I/O neuron elements. A steady pattern of the binary states is generated in the hidden neuron elements based on the normal theory of the dynamics of the neural network. After that, the synaptic connection strengths are changed, obeying modified Hebbian learning. Next, in the recognizing stage, slightly distorted patterns of the alphabet letters "T" and "L" are inputted into the I/O neuron elements using the aforementioned majority rule handling and immediately released. The pixel different from the standard pattern is indicated by a white square in the figure. Next, revised patterns are automatically outputted from the I/O neuron elements and transformed using the aforementioned majority-rule handling again. Finally, it is checked that the standard patterns of the alphabet letters "T" and "L" are reproduced in the revised patterns.

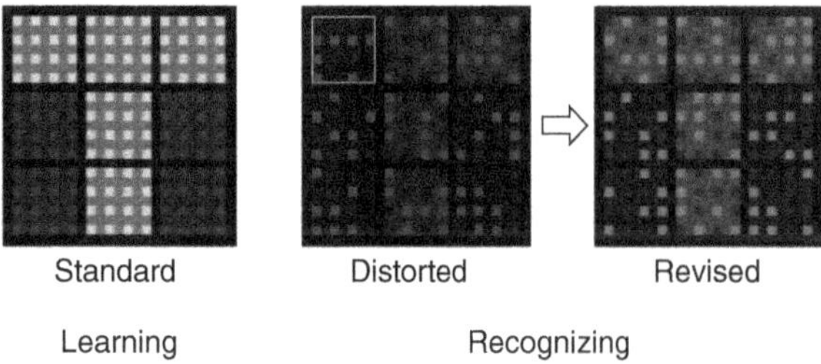

Figure 22.10 Raw data of the letter reproduction.

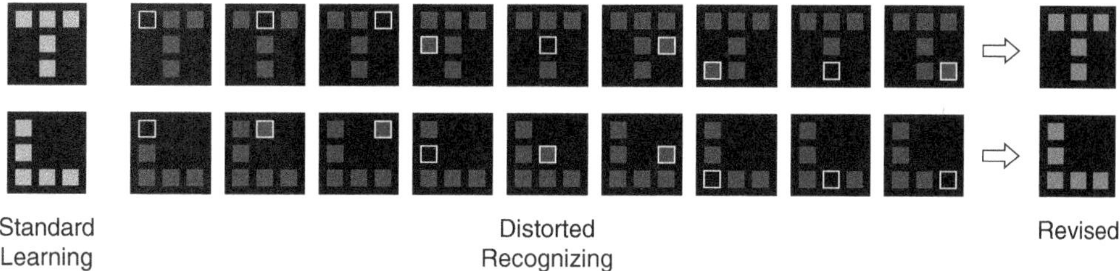

Figure 22.11 Experimental results of the letter recognition.

22.3.3.2 Associative Memory

The experimental results of the letter recognition are shown in Figure 22.11. It is found that the standard patterns can be reproduced in the revised patterns for the distorted patterns where only one pixel is flipped, which is indicated using a white square. It is concluded that this real neuromorphic system can work for the letter reproduction and, namely, as an associative memory.

22.4 Conclusion

First, we have developed neuromorphic systems with crosspoint-type α-GTO thin-film devices as self-plastic synapse elements, where the α-GTO device is a promising device that was presented recently by the authors, and fundamental operations were confirmed for neuromorphic systems. First, crosspoint-type α-GTO thin-film devices were fabricated, and it was found that the electric current gradually decreases along the bias time. Next, a neuromorphic system was actually implemented using an FPGA chip and crosspoint-type α-GTO thin-film devices, and it was confirmed that a function of letter recognition is obtained after the learning process. Once the fundamental operations are confirmed, more advanced and useful functions will be obtained by increasing the device and circuit scales in the future according to the technological history of neural networks.

In this study, although the device size is not so fine and they are not layered, it should be noted that the recognizing process finishes by only several operation steps, which leads to future high-speed and low-power computing. Moreover, it should be noted that the neuromorphic systems and α-GTO thin-film devices are scalable and adaptive. If it is assumed that the device size is 200×200 nm, which is feasible using current semiconductor technologies, and they are layered 1,000 times, which will be feasible by means of further efforts, the system size can be less than a cubic of 10^3 cm^3 for synapse elements of 200 trillion, which is the same number as those in a human brain. Moreover, if it is assumed that the operation voltage is reduced to 1/10—namely, 0.33 V, which has been attained in the newest semiconductor devices [39]—the power consumption can be roughly 30 W, which is similar to that in a human brain. In any case, it is expected that the machine size can be excellently compact and power consumption can be remarkably low.

Second, a neuromorphic system was developed by using an LSI chip and depositing *a*-IGZO thin-film devices on it. Neuron elements were equipped in the LSI chip, and synapse functions were realized by the *a*-IGZO thin-film devices with modified Hebbian learning without any additional circuits to control synaptic weights. We have confirmed that the neuromorphic system can work as an associative memory by reproducing memorized letters. It should be noted that this neuromorphic system has a basic architecture more similar to that of living brains than any other neuromorphic systems. Although only elementary results were obtained in this study, it is expected that more sophisticated functions will be acquired by increasing the number of neuron and synapse elements, and such neuromorphic systems will have advantages, such as compact size, low power, and robust operation in the future. For example, if the most advanced technology is used, the volume of a synapse element can be 1 μm^3, and even if the same number as a living brain is integrated, the volume of a neuromorphic system can be less than 1 ℓ. Because the power consumption of a synapse element is less than 1 μA, if the working ratio is 1%, the power consumption of a neuromorphic system can be less than 100 kW. The volume and power consumption are much smaller than those expected from a mere extension of the conventional ones.

Acknowledgments

This work is partially supported by KAKENHI (C) 16K06733, the Yazaki Memorial Foundation for Science and Technology, Support Center for Advanced Telecommunications Technology Research, Research Grants in the Natural Sciences from the Mitsubishi Foundation, the Telecommunications Advancement Foundation, the Laboratory for Materials and Structures in Tokyo Institute of Technology, RIEC Nation-wide Cooperative Research Projects, collaborative research with ROHM Semiconductor, collaborative research with KOA Corporation, and the Innovative Materials and Processing Research Center.

References

1 McCarthy, J., Minsky, M. L., Rochester, N., and Shannon, C. E. (1955). A proposal for the Dartmouth summer research project on artificial intelligence. Dartmouth Conference, August 31. http://www-formal.stanford.edu/jmc/history/dartmouth/dartmouth.html.

2 Russell, S. and Norvig, P. (2009). *Artificial Intelligence: A Modern Approach*. Prentice Hall, Upper Saddle River, NJ: Pearson Education.

3 McCulloch, W.S. and Pitts, W. (1943). A logical calculus of the ideas immanent in nervous activity. *Bulletin of Mathematical Biophysics* 5: 115–133.

4 Wasserman, P.D. (1989). *Neural Computing: Theory and Practice*. Scottsdale, AZ: Coriolis Group.

5 Dayhoff, J.E. (1990). *Neural Network Architectures: An Introduction*. New York: Van Nostrand Reinhold.

6 Ferrucci, D., Brown, E., Chu-Carroll, J. et al. (2010). Building Watson: an overview of the DeepQA project. *AI Magazine* 31: 59–79.

7 https://www.ibm.com/watson/index.html/.

8 Mead, C. (1989). *Analog VLSI and Neural Systems*. Reading, MA: Addison-Wesley.

9 Lande, T.S. (2013). *Neuromorphic Systems Engineering*. Springer, Boston: *Neural Networks in Silicon*.

10 Suri, M. (2017). *Advances in Neuromorphic Hardware Exploiting Emerging Nanoscale Devices*. New Delhi: Springer.

11 Merolla, P.A., Arthur, J.V., Alvarez-Icaza, R. et al. (2014). A million spiking-neuron integrated circuit with a scalable communication network and interface. *Science* 345: 668–673.

12 http://www.ibm.com/smarterplanet/jp/ja/brainpower/.

13 Neckar, A., Fok, S., Benjamin, B.V. et al. (2019). Braindrop: a mixed-signal neuromorphic architecture with a dynamical systems-based programming model. *Proceedings of the IEEE* 107: 144–164.

14 Kimura, M. (2017). Novel application of thin-film devices: sensing devices, electronics devices, etc. Paper presented at IDMC'17, Fri-S20-01, Los Angeles, CA.

15 Prezioso, M., Merrikh-Bayat, F., Hoskins, B.D. et al. (2015). Training and operation of an integrated neuromorphic network based on metal-oxide memristors. *Nature* 521: 61–64.

16 Nomura, K., Ohta, H., Takagi, A. et al. (2004). Room-temperature fabrication of transparent flexible thin-film transistors using amorphous oxide semiconductors. *Nature* 432: 488–492.

17 Wager, J.F. (2003). Transparent electronics. *Science* 300: 1245–1246.

18 Kamiya, T., Nomura, K., and Hosono, H. (2010). Present status of amorphous In-Ga-Zn-O thin-film transistors. *Science and Technology of Advanced Materials* 11: 044305.

19 Kim, S.J., Yoon, S., and Kim, H.J. (2014). Review of solution-processed oxide thin-film transistors. *Japanese Journal of Applied Physics* 53: 02BA02.

20 Nomura, K., Aoki, T., Nakamura, K. et al. (2010). Three-dimensionally stacked flexible integrated circuit: amorphous oxide / polymer hybrid complementary inverter using n-type a-In-Ga-Zn-O and p-type poly-(9,9-dioctylfluorene-co-bithiophene) thin-film transistors. *Applied Physics Letters* 96: 263509.

21 Kimura, M., Morita, R., Sugisaki, S. et al. (2017). Cellular neural network formed by simplified processing elements composed of thin-film transistors. *Neurocomputing* 248: 112–119.

22 Serrano-Gotarredona, T., Masquelier, T., Prodromakis, T. et al. (2013). STDP and STDP variations with memristors for spiking neuromorphic learning systems. *Frontiers in Neuroscience* 7: art. 2.

23 Kimura, M., Umeda, K., Ikushima, K. et al. (2019). Neuromorphic system with crosspoint-type amorphous Ga-Sn-O thin-film devices as self-plastic synapse elements. *ECS Transactions* 90: 157–166.

24 Kimura, M., Ikushima, K., Yamakawa, D. et al. (2019). Real neuromorphic system using LSI chip and thin-film devices. Paper presented at ICONS 2019: The Fourteenth International Conference on Systems, Valencia, Spain, March.

25 Matsuda, T., Umeda, K., Kato, Y. et al. (2017). Rare-metal-free high-performance Ga-Sn-O thin film transistor. *Scientific Reports srep44326.* .

26 Matsuda, T., Uenuma, M., and Kimura, M. (2017). Thermoelectric effect of amorphous Ga-Sn-O thin film. *Japanese Journal of Applied Physics* 56: 070309.

27 Okamoto, R., Fukushima, H., Kimura, M., and Matsuda, T. (2017). Characteristic evaluation of Ga-Sn-O films deposited using mist chemical vapor deposition. In: *International Meeting for Future of Electron Devices, Kansai*, 74. Japan, June: Kyoto.

28 Sugisaki, S., Matsuda, T., Uenuma, M. et al. (2019). Memristive characteristic of an amorphous Ga-Sn-O thin-film device. *Scientific Reports* 9: 2757.

29 Dayhoff, J.E. (1989). *Neural Network Architectures: An Introduction*. New York: Van Nostrand Reinhold.

30 Wasserman, P.D. (1989). *Neural Computing: Theory and Practice*. Scottsdale, AZ: Coriolis Group.

31 Kimura, M., Umeda, K., Ikushima, K. et al. (2018). Hopfield neural network with double-layer amorphous metal-oxide semiconductor thin-film devices as crosspoint-type synapse elements and working confirmation of letter recognition. In: *In Neural Information Processing: 25th International Conference, ICONIP 2018, Siem Reap, Cambodia, December 13–16, 2018, Proceedings, Part VII*, 637–646.

32 Kimura, M. and Imai, S. (2010). Degradation evaluation of α-IGZO TFTs for application to AM-OLEDs. *IEEE Electron Device Letters* 31: 963–965.

33 Vision Society of Japan (2017). *Visual Information Processing Handbook*. Tokyo: Asakura Publishing.

34 McEliece, R., Posner, E., Rodemich, E., and Venkatesh, S. (1987). The capacity of the Hopfield associative memory. *IEEE Transactions on Information Theory* 33: 461–482.

35 http://www.vdec.u-tokyo.ac.jp/.

36 Matsuda, T., Umeda, K., Kato, Y. et al. (2017). Rare-metal-free high-performance Ga-Sn-O thin film transistor. *Scientific Reports srep44326.* .

37 Kasakawa, T., Tabata, H., Onodera, R. et al. (2010). An artificial neural network at device level using simplified architecture and thin-film transistors. *IEEE Transactions on Electron Devices* 57: 2744–2750.

38 Hebb, D.O. (1949). *The Organization of Behavior*. New York: Wiley.

39 https://web.archive.org/web/20110825080659/http://public.itrs.net/home.html.

23

Oxide TFTs and Their Application to X-Ray Imaging

Robert A. Street

Palo Alto Research Center, Palo Alto, California, USA

23.1 Introduction

Metal-oxide thin-film transistors (TFTs) have developed into a new technology for flat-panel display backplanes [1–3], and their high mobility enables backplanes to drive organic light-emitting diode (OLED) displays [4]. The metal oxides have the advantage of a simpler TFT fabrication process compared to low-temperature polysilicon (LTPS), which is the primary commercial technology for high-mobility TFT backplanes. Digital X-ray detectors also rely on TFT backplane technology and currently use amorphous silicon for the pixel TFT and for the photodiode [5]. Oxide TFTs present opportunities to improve various aspects of detector performance, including enabling pixel amplifiers to reduce electronic noise and allowing high-frame-rate imaging for specialized medical procedures. This chapter describes the digital X-ray detector technology; recent examples of devices fabricated with oxide TFTs, particularly InGaZnO (IGZO); and possible future developments to improve digital X-ray imaging.

Numerous metal-oxide compounds have been shown to give TFTs with good performance [6–9], but IGZO is the most widely studied [1–3, 10–12] and is the dominant oxide in TFT production. IGZO is amorphous and provides higher device uniformity than polycrystalline materials in which a random number of grain boundaries can give variation in device characteristics. *n*-Channel, accumulation-mode TFTs can be fabricated over a wide temperature range from room temperature to above 300 °C, and in general the TFT properties improve with either deposition or annealing temperature. Details of the oxide TFTs and their fabrication are described in other chapters in this book.

The origin of the high mobility of oxide TFTs is believed to be that the metal *s*-orbitals are relatively insensitive to bond-angle disorder compared to a covalent semiconductor such as amorphous silicon. As a result, the electronic states of the amorphous oxides have reduced energy disorder. There are fewer localized band-tail states, and hence electronic transport is dominated by band conduction rather than trapping in band-tail states. Band tails typically have an exponential density of states proportional to $\exp(-E/E_0)$, where E is the energy from the band edge, and when $E_0 < kT$, then the occupation statistics are such that most carriers are in the band, and hence trapping in band-tail states does not limit the mobility significantly.

Oxide TFTs are usually deposited by sputtering in an oxygen environment at a temperature of 300–400 °C. However, lower temperature deposition at <200 °C also gives TFTs with good performance, and consequently deposition on many flexible plastic substrates is enabled. Solution deposition is also possible using a sol-gel process with various precursor chemicals [13, 14], giving further flexibility in the fabrication process. Early work found that oxide TFTs had a significant gate bias stress instability and also negative gate bias illumination stress (NBIS), but optimization of the deposition has largely eliminated these effects [15, 16].

Section 23.2 discusses digital X-ray imaging and the requirements for the TFT backplane. Section 23.3 describes some prototype oxide TFT detectors, and Section 23.4 discusses how oxide TFTs can improve X-ray imaging quality.

Amorphous Oxide Semiconductors: IGZO and Related Materials for Display and Memory, First Edition.
Edited by Hideo Hosono and Hideya Kumomi.

Sections 23.5 and 23.6 discuss radiation hardness and the possible use of oxides as the X-ray converter layer in a digital X-ray detector.

23.2 Digital X-Ray Detection and Imaging Modalities

The digital X-ray detection process is illustrated in Figure 23.1a. X-rays are incident on a converter material, which is either a phosphor or an X-ray photoconducting semiconductor. An incident X-ray excites a high-energy photoelectron that loses energy by the excitation of many electron–hole pairs as it moves through the converter. The ionization energy loss is characterized by a sensitivity parameter W, which is the average energy lost by the photoelectron to create one electron–hole pair. The maximum theoretical sensitivity (i.e., smallest W) is $W \sim 3E_G$, where E_G is the band gap of the converter material, and so the minimum value of W is typically 5–10 eV. Hence one 100 keV X-ray can theoretically generate 10,000–20,000 charges. A low value of W is important to give a detector with high gain, defined by the amount of measured charge per incident X-ray. High gain enables a high signal-to-noise (S/N) ratio, which leads to better image quality. The resulting ionization charge is stored by the pixel capacitance of the TFT backplane, and is then read out to external charge-sensitive amplifiers by active-matrix addressing of the backplane.

There are two techniques in which the X-ray energy is converted to an image-wise stored charge.

23.2.1 Indirect Detection Imaging

With indirect imaging (Figure 23.1b), X-rays interact with a scintillator, usually CsI or GdO_2S_2:Tb. The ionized electron–hole pairs from the photoelectron recombine to emit light, and a photodiode (typically, a-Si p-i-n) on each pixel of the TFT backplane absorbs the light from the scintillator and creates charge that is stored by the photodiode capacitance. The sensitivity of the detector is limited by the efficiency of the scintillator light absorption and emission, as well as the absorption efficiency of the photodiode. These losses typically result in a sensitivity parameter $W = 50$ eV, which corresponds to about 10–20% of the theoretical maximum gain. The scintillator is designed to scatter light to keep the luminescence emission near the incident X-ray, but the diffusion of the light causes the image to spread out by roughly the thickness of the scintillator, reducing spatial resolution in images with fine features. The CsI scintillator reduces the light diffusion because its internal structure has a light-piping effect.

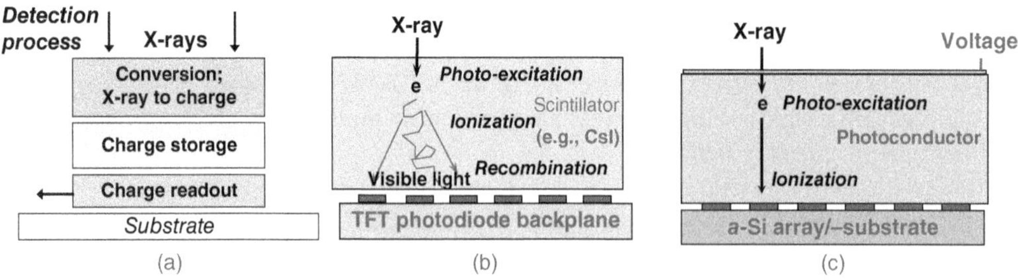

Figure 23.1 Schematic of the digital X-ray imaging process. (a) The main elements of the detector are the X-ray converter, which absorbs the X-ray and develops a corresponding charge; the charge storage, which is done through capacitance on the backplane; and the charge readout using the matrix-addressed TFT array. (b) Indirect detection showing an X-ray exciting a photoelectron and the subsequent creation of ionized electron–hole pairs that recombine to give light emission. (c) Direct detection in which the ionized electron–hole pairs are collected by an electric field.

23.2.2 Direct Detection Imaging

With the direct detection technique (Figure 23.1c), X-rays interact with an X-ray photoconductor. The ionization charge is collected by the electric field of an applied voltage across the photoconductor. In principle, direct detection can achieve the theoretical sensitivity without image spreading because the charge does not diffuse laterally. However, a material with the appropriate set of properties has not yet been found. Amorphous selenium deposited up to 1 mm thick is the only direct detection material in production and requires an applied voltage of about 10 kV to collect charge efficiently. There is minimal image spreading, but the sensitivity parameter $W{\sim}50\,\text{eV}$ is larger than the theoretical optimum. HgI_2 has demonstrated high sensitivity but so far has proved to have difficult fabrication problems. PbO has been shown to work but with low sensitivity. It is interesting to consider whether other metal oxides might provide a high-sensitivity alternative, as discussed in Section 23.6.

The technology of digital X-ray imaging prioritizes achieving the best possible image quality at the lowest X-ray dose, because the huge medical benefits of X-ray imaging come at a cost of exposing patients to possibly damaging X-rays. The paramount image-quality metric is the detective quantum efficiency (DQE), which is discussed further in Section 23.4.1, and a large component of the DQE arises from electronic noise in the various components, including the TFTs.

23.2.3 X-Ray Imaging Modalities

There are a number of medical imaging modalities that each have specific requirements for the TFT, the backplane, and other general properties.

- *Radiography*: Radiography is the capture of a single-exposure X-ray image. Approximately 1 billion radiographic medical images are taken each year with detectors that have a range of sizes and formats. A typical chest X-ray is exposed at 80–120 kVp X-rays, with most of the intensity centered at about half the maximum energy. (kVp refers to the generator accelerating voltage and is the upper limit on the X-ray energy spectrum.) The dose is 3–5 mR (30–50 μGy), which corresponds to an X-ray flux of 1–1.5 10^7 X-ray photons/cm^2. Approximately 10–20% of the X-rays pass through the person to the detector, and the image contrast (i.e., bone versus flesh) in the transmitted beam is 20–50%. The dynamic range of a digital X-ray detector needs to be >16,000 (i.e., 14–16-bit digitization), so that the low-contrast regions can be viewed without being significantly affected by digitization noise.
- *Fluoroscopy*: Fluoroscopy is video rate imaging intended to monitor a medical procedure while it is in progress. The emphasis is on extremely low doses, typically 1–10 μR per frame, which is about three orders of magnitude less than the radiographic dose, because the patient is exposed for an extended time. Sensitivity is gained by increasing the pixel size to 200 μm or larger, and the X-ray dose is arranged to provide the lowest contrast consistent with the requirement to visualize the procedure. An exposure of 1 μR corresponds to only about 10 X-ray photons/pixel/frame, and the photon-limited S/N ratio is just barely high enough to discern an image. In this regime, the DQE tends to be dominated by the electronic noise, and a good fluoroscopy imager should have high sensitivity and very low additive electronic noise.
- *Mammography*: Mammography imaging requires a significantly higher spatial resolution than most other radiographic procedures, in order to detect microcalcifications, which are very small features with low contrast that are precursors of breast cancer. Imaging requires a high X-ray dose and a pixel size of 50–100 μm.
- *Computed tomography (CT)*: Computed tomography gives a three-dimensional (3D) X-ray image by measuring the X-ray absorption along many projections through the object being imaged, usually with a rotating source and detector, from which the 3D image is computed. Conventional CT is performed with a linear array of solid-state detectors, but a two-dimensional (2D) detector TFT-addressed array has the potential to increase the data acquisition rate.

- *Nonmedical applications*: X-ray imaging is used for nondestructive testing (NDT) of electronic parts, pipelines, and many other structures. X-ray energy, array and pixel size, and other characteristics vary greatly depending on the application. Flexible detectors are valuable for imaging curved surfaces, inserting into tight spaces, and general robustness.

23.3 Oxide-TFT X-Ray Detectors

23.3.1 TFT Backplane Requirements for Digital X-Rays

The usual backplane circuit for digital X-ray imaging is shown in Figure 23.2 and comprises a pixel TFT along with a diode that can represent the photodiode in the case of indirect detection or the X-ray photoconductor in the case of direct detection. Radiographic images typically have exposures of up to 1 s, and a low off-current is therefore necessary for both the TFT and the photodiode/photoconductor to hold the charge on the pixel without loss. Some nonmedical applications need exposures of 10 s or even longer. Precise measurement of the signal charge requires a minimal trapped charge in the TFT channel because it is slow to release and may contribute to later image frames, an effect known as image lag. Sufficiently high mobility is needed for fast image readout for radiography. In addition, TFTs must be stable with low noise to achieve a 16-bit (64,000) dynamic range, and they should be radiation hard up to about 1 MRad exposure.

Figure 23.3 shows examples of IGZO-TFT characteristics; oxide-TFT data are discussed in much more detail in other chapters of this book. Figure 23.3a shows data from a well-optimized Gen 6 commercial process. The mobility is 7.5 cm^2/Vs, the subthreshold swing is 0.32 V/dec, the threshold voltage is 1.1 ± 0.25 V, and the on/off ratio is about 10^8, with a sub-pA leakage current. Figure 23.3b shows data from the author's laboratory, with slightly poorer turn-on characteristics and a mobility of 10–15 cm^2/Vs [17]. Oxide-TFT characteristics can vary widely, which is perhaps not surprising considering the multicomponent nature of the compound. Reported carrier mobility ranges up to about 40 cm^2/Vs [2]. Optimized IGZO-TFTs are also stable and exhibit only a small positive gate bias stress and NBIS.

23.3.2 An IGZO Detector Fabrication and Characterization

A prototype X-ray detector with an IGZO-TFT backplane was developed in the author's laboratory [17]. The flexible X-ray detector array was fabricated on a polyethylene naphthalate (PEN) substrate supported on a holder. The indirect detection array was designed with a bottom-gate IGZO-TFT and with a continuous *a*-Si p-i-n photodiode;

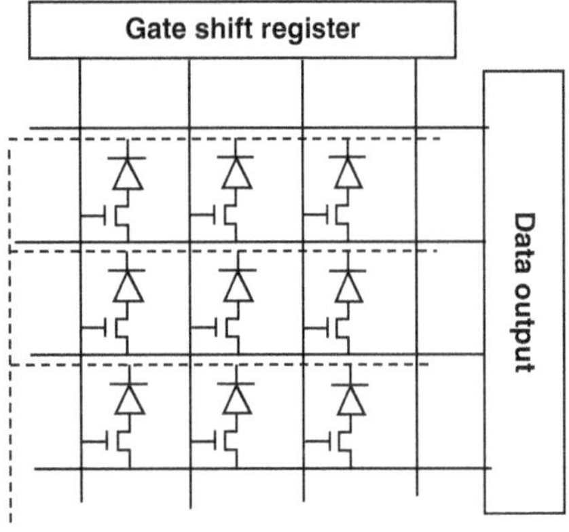

Figure 23.2 Schematic diagram of a TFT backplane showing a TFT and photodiode or photoconductor in each pixel, a shift register to switch the TFT gates, and data output to charge sensitive amplifiers.

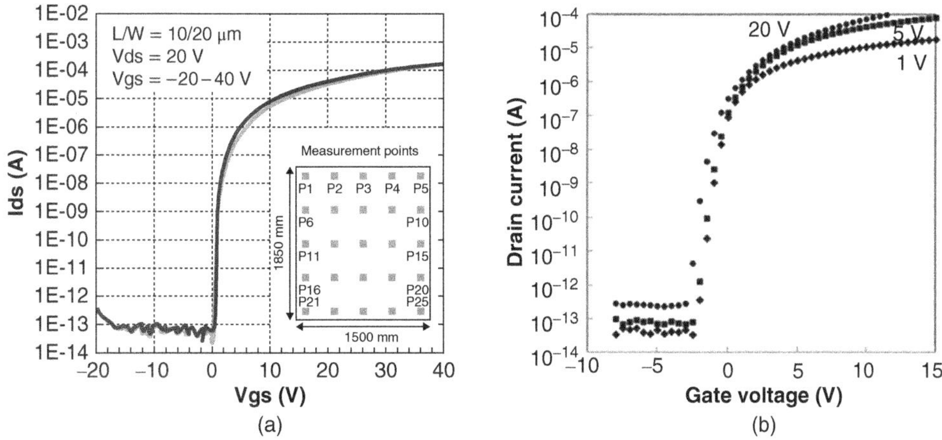

Figure 23.3 (a) TFT characteristics for a commercial IGZO process showing a sharp turn-on at $V_G = 0$ and excellent uniformity. (b) Example of a low-temperature IGZO-TFT on a flexible plastic substrate. *Source*: (b) Based on Lujan, R. A., and Street, R. A. (2012). Flexible X-ray detector array fabricated with oxide thin film transistors. *IEEE Electron Device Letters* **33**: 688.

Figure 23.4 Schematic side view of the detector pixel with a continuous photodiode sensor.

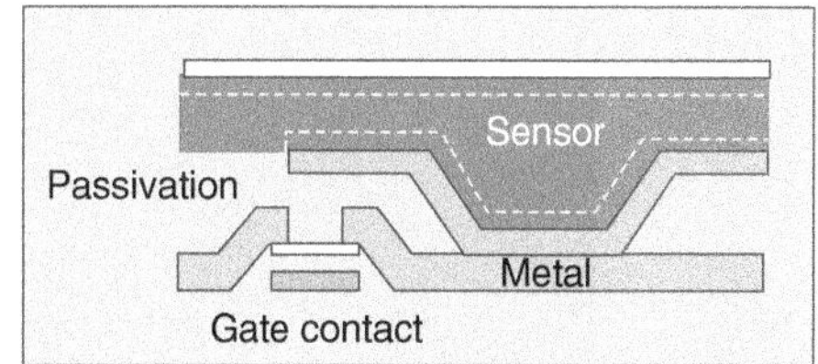

a side view of the pixel structure is shown in Figure 23.4. Conventional large-area photolithography was used to pattern the layers. Low-temperature processing was required with a temperature below 180 °C due to the PEN substrate. An oxide dielectric barrier layer was deposited on the substrate by plasma-enhanced chemical vapor deposition (PECVD), followed by the MoCr metal gate for the bottom-gate staggered TFT structure, and then a 160 nm PECVD SiN/SiO gate dielectric. Both the barrier and the gate dielectric were deposited at 160–170 °C, consistent with the requirements of the PEN substrate. The 15 nm IGZO channel material was sputtered at room temperature using a target with 1:1:1 composition of In:Ga:Zn, and it was followed by deposition of the MoCr source and drain contacts. The array was completed with a 1 μm parylene interlayer dielectric, then an Al/MoCr contact metal, an amorphous silicon p-i-n photodiode deposited by PECVD at 170 °C and with thickness of about 1.2 μm and an ITO top contact. The detector array has a 160×180 pixel format with 200 μm pixel size for an overall dimension of 3.2×3.6 cm. The array was annealed at 170 °C to improve the TFT properties.

A Gd_2O_2S:Tb (Lanex regular) scintillator was used for the indirect-detection detector mode, and it was placed in direct contact with the surface of the detector array. The Lanex emission at 550 nm is a good match for the *a*-Si photodiode. Figure 23.5 shows images of a standard X-ray resolution target and of an electronic circuit, both obtained with 80 kVp exposure and with 0.2 s integration time. The image is corrected for the dark background, but it is not normalized to the flood exposure field, since the response uniformity was found to be sufficiently good for the purpose of showing an image. The image is also shown without any spatial filter correction for line and pixel defects arising from the backplane processing. Aside from demonstrating good imaging properties, the image shows that the fabrication process is reasonably robust since there are relatively few defects, typical of a research process. Commercial oxide TFTs have very high yield.

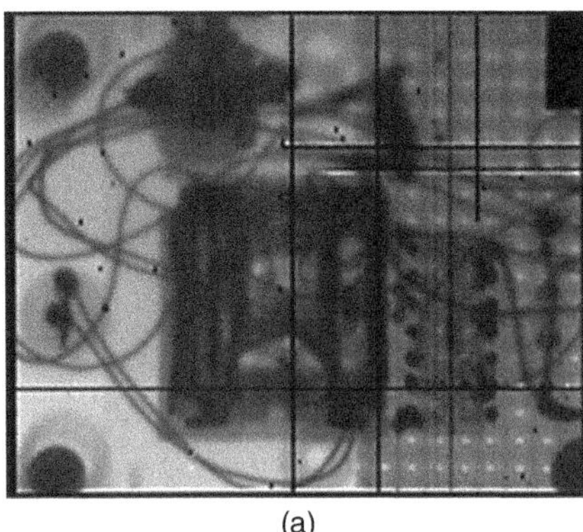
(a)

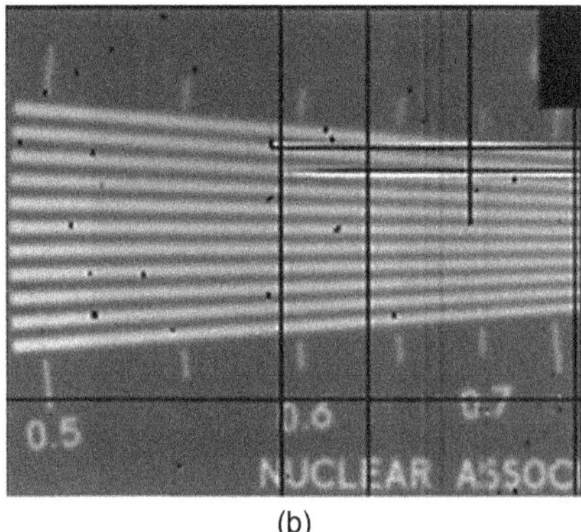

(b)

Figure 23.5 X-ray images of an electronic circuit and a resolution target obtained with 80 kVp exposure. Dark background subtraction is the only image correction.

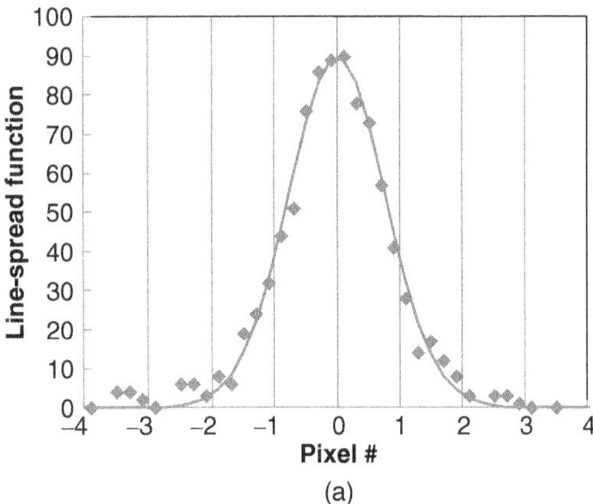

(a)

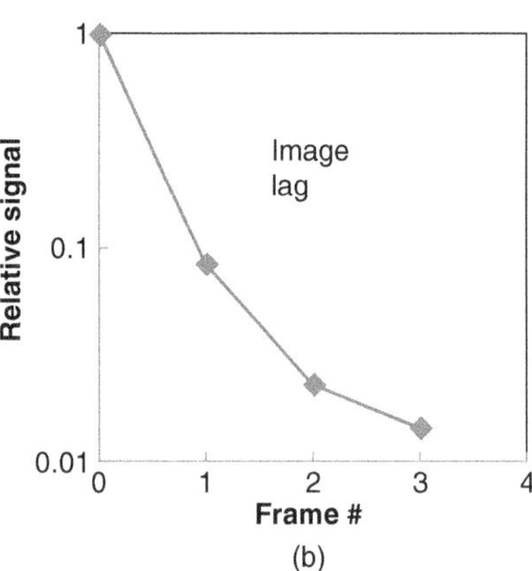

(b)

Figure 23.6 (a) Line spread function for the detector described in the chapter, showing spreading consistent with the scintillator. (b) Plot of residual signal for image frames measured after the X-ray source is turned off, quantifying the image lag.

The detector exhibited comparable performance to a similar detector with *a*-Si TFTs. The line-spread function (LSF) measures the spatial resolution and was obtained by the conventional technique of imaging a slit placed at a small angle with respect to the array pixel columns. The LSF shown in Figure 23.6a is described by a Gaussian with 1.8 pixels width, which is consistent with the known resolution of the Lanex screen. Measurement of the image lag, which is the fraction of the image charge that is not transferred to the external electronics within the frame time, found a 10% first-frame lag, decreasing to about 1% lag at the third frame, values that are typical of the a-Si photodiode. The data therefore show that the oxide TFT does not introduce any significant degradation of the imaging properties.

Flexible backplanes are of interest for X-ray detectors because the plastic substrate renders them more robust than a glass substrate, and also allows for curved detectors for applications such as computed tomography and NDT of items such as pipes. Furthermore, a sufficiently thin substrate can enable higher sensitivity detectors by placing phosphors next to the front and back surfaces, provided that both contacts to the photodiode are transparent.

23.3.3 Other Reported Oxide X-Ray Detectors

Metal-oxide TFT X-ray detectors have been reported in a variety of formats. Gelinck et al. made and characterized an X-ray detector on a 25-μm-thick plastic substrate [18]. It is an indirect conversion flat-panel detector, combining a standard scintillator with an organic photodetector (OPD) layer and metal-oxide TFT backplane. The solution-processed organic bulk heterojunction photodiode has the advantage of a low process temperature, compatible with plastic film substrates. In addition, some lithography steps are eliminated, giving the potential for lower production costs. The OPD met the requirements of low dark currents and high sensitivity. The proof-of-concept detector with 127 μm pixel size delivered high-resolution, dynamic images at 10 frames/s using X-ray doses as low as 3 μGy/frame.

A direct detection X-ray flat-panel detector with an HgI_2 photoconductor and metal-oxide pixel TFT was reported by Kim et al. [19]. This study used a HfInZnO TFT, which was shown to have a lower leakage current and higher mobility (10~20 cm^2/V.s) than *a*-Si TFTs, giving properties that enable an improved S/N ratio and image quality. The backplane format was 720×720 with 139 μm pixel size. The HgI_2 layer was fabricated by screen printing and had three times higher sensitivity than *a*-Se at the applied bias below 150 V, as well as better stability under a high accumulation dose and humidity.

IGZO-TFTs have been explored as a switching device for large-area flat-panel detectors in the context of an investigation of photon-counting capability for X-ray medical imaging [20]. Photon-counting detectors are of great interest because they can provide energy resolution and reduced noise. The study includes the design and fabrication of a charge-sensitive preamplifier and analog counter using IGZO-TFT processes and measurements of its performance. Photon counting requires a much more complex circuit than the single TFT pixel and can only be achieved with high-mobility TFTs.

Novel uses of IGZO are also reported in the context of digital X-ray detectors. Abbaszadeh et al. [21] investigated a thin (375 nm) IGZO film as a hole-blocking layer for an amorphous selenium (*a*-Se)-based detector to reduce the dark current of the photoconductor and improve the sensitivity of the detector. The dark current was measured below 1 pA/mm^2 and comparable to the thermal generation current in *a*-Se. The low dark current allowed an increase in the applied field, which resulted in a threefold improvement in external quantum efficiency compared to the conventional Se photoconductor. Liu et al. [22] also used a thin IGZO layer to improve sensitivity of an X-ray detector with an organic photodiode. Xu et al. [23] proposed a novel direct X-ray detector combining a vertical photoconductor with an IGZO-TFT readout designed as an active pixel sensor.

These studies show that oxide TFTs, and primarily IGZO, are well suited to form the backplane of digital X-ray detectors. The possible advantages of using oxides instead of the current *a*-Si technology are addressed in the remainder of this chapter.

23.4 How Oxide TFTs Can Improve Digital X-Ray Detectors

The prototype detectors demonstrate that oxide-TFT backplanes are successful for X-ray imaging applications on either glass or plastic substrates. The question remains whether oxides present any clear advantages for the backplane TFT over the current *a*-Si technology. The high mobility of the oxide-TFTs offers the possibility of an improved S/N ratio by a variety of strategies. This section discusses the various opportunities for oxide-TFTs to achieve improved detector performance, with most of the focus on an improved S/N ratio.

Aside from noise reduction, other ways that oxides could improve detector performance are:

- *Higher frame rates*: The time available for the TFT to address a column of pixels is the frame time divided by the number of columns. For large arrays with dense pixels and high frame rates, the short gate-on time can be challenging for *a*-Si TFTs because the discharge time must be several multiples of the RC time constant of the photodiode capacitance and the TFT resistance. The higher mobility of oxide-TFTs gives faster discharge rates and hence can enable higher frame rates.

- *Long integration times*: At the other extreme, radiographic imaging of a weak X-ray signal requires very long integration times. The integration time is limited by the photodiode and TFT leakage currents. Some oxide-TFTs have very low off-currents and therefore could enable longer integration times.
- *Reduced gate voltage swing*: As the gate voltage is switched on and off during the read cycle, there is capacitive transfer of charge to and from the photodiode and the readout amplifiers, arising from the channel capacitance and the parasitic capacitance of the TFT electrodes. Normally this charge is accurately cancelled by reading the signal before and after the gate pulse. However, this feedthrough charge can reduce the detector dynamic range, because the combined signal and feedthrough charge must not exceed the linear response region of the amplifier, otherwise a signal distortion will occur. The high mobility of oxides and sharp turn-on allow for a smaller voltage swing and smaller dimension TFTs with lower capacitance and hence a smaller feedthrough charge.

23.4.1 Noise and Image Quality in X-Ray Detectors

For medical X-ray imaging, it is critical to obtain the best possible image quality at the lowest X-ray dose because of the negative health consequences of X-ray exposure. Detector performance is described by the DQE, which is the ratio of the measured to the input S/N ratio [5]:

$$DQE = \frac{(S/N)^2_{OUT}}{(S/N)^2_{IN}}$$

The S/N ratio and DQE are also a function of spatial frequency (typically measured in units of line pairs/mm), meaning that the image quality depends on the charge spreading in the X-ray converter and on the pixel size of the detector array. $(S/N)_{IN}$ is governed by the Poisson statistics of the incident X-ray photons, and hence the photon noise is the square root of the number of incident photons. $(S/N)_{OUT}$ is governed by the fraction of X-rays absorbed; the system gain (the number of electrons detected per incident photon); the additive electronic noise introduced by the various components of the detection system; and, as a function of spatial frequency, the modulation transfer function (MTF). One incident X-ray can create 1,000–5,000 electrons in the backplane for a system with $W \sim 50$ eV, and about 10 times more for a system with the theoretical sensitivity. If the additive electronic noise has roughly the same magnitude, then the photon noise dominates at moderately high doses and the DQE at zero spatial frequency approaches unity, usually limited by the X-ray absorption. The goal is to achieve photon-limited noise at low doses and high spatial frequency, where the additive electronic noise typically dominates.

23.4.2 Minimizing Additive Electronic Noise with Oxides

The readout circuit for a conventional digital X-ray is shown in Figure 23.7. The charge developed at the pixel is transferred along the data lines without amplification to an external charge-sensitive amplifier in the form of a silicon integrated circuit. The input noise of the amplifier is the dominant additive noise source of the image for typical detectors. This source of noise is proportional to the input data line capacitance to the amplifier, which is substantial and on the order of 100 pF for a large-area detector typical of medical imaging. A large fraction of the data line capacitance arises from the pixel TFTs, particularly the overlap between the source/drain electrode and the gate, which is significant as there is not an established self-aligned technology for either *a*-Si or IGZO-TFTs. The gate and data line crossovers are another significant contribution to the data line capacitance. There are other sources of noise, including kTC noise arising from the pixel, but this is generally smaller than the amplifier noise.

Oxide-TFTs can reduce the amplifier noise in the conventional detector circuit. Freestone et al. [24] point out that since IGZO-TFTs have an electron mobility that is about 10 times higher than that of *a*-Si, the size of the TFT

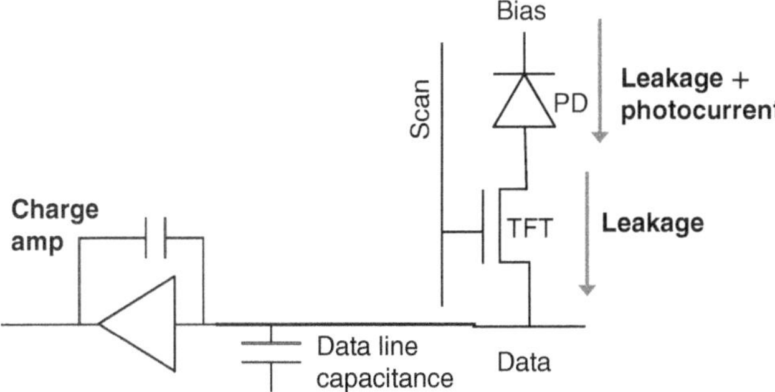

Figure 23.7 Equivalent circuit for the pixel, data line, and charge amplifier.

can be reduced for the same readout performance. The size reduction gives a corresponding decrease in the TFT parasitic capacitance and hence reduces the overall data line capacitance, and consequently also the amplifier noise. Alternatively, the high mobility of IGZO allows the pixel discharge time to be reduced, also improving the S/N ratio and increasing the detector readout rate. The study by Freestone et al. showed that the reduction in TFT size led to significant improvement in DQE at low doses in fluoroscopy detectors, based on data from a 31×31 cm IGZO detector, providing strong motivation to use IGZO in future detector backplanes.

23.4.3 Pixel Amplifier Backplanes

A widely studied approach to reducing the amplifier noise is a backplane containing a pixel amplifier, often referred to as an active pixel sensor (APS). A pixel amplifier made from high-mobility polysilicon (LTPS) TFTs has demonstrated electronic noise reduction in an X-ray detector [5]. Oxide TFTs offer the same opportunity, but with a technology that is expected to have lower fabrication cost than LTPS. An operational amplifier is another possible pixel amplifier circuit but requires more components that therefore occupy more space in the pixel. The APS pixel has three TFTs: the amplifier TFT, the read TFT, and a reset TFT. The gain of the pixel amplifier gives a proportional reduction in the effective readout amplifier noise, so that a gain of about 10 is sufficient to reduce its contribution sufficiently that other noise sources dominate, in particular the intrinsic noise of the pixel amplifier TFT.

Zhao and Kanicki explored the performance of an IGZO-TFT APS backplane in the context of digital breast tomosynthesis, where breast cancer detection is limited by the X-ray image quality [25]. The pixel circuit is shown in Figure 23.8a. The charge signal generated by the photodiode biases the gate of the amplifier TFT, which passes the signal through the read TFT to the external amplifier. After the signal is recorded, the reset TFT returns the gate voltage to the reference bias. Zhao and Kanicki investigated the electrical performance of the *a*-IGZO APS pixel circuit by SPICE simulation, with a careful analysis of the various noise sources. Figure 23.8b shows the resulting input-referred noise as a function of an X-ray dose comparing IGZO with *a*-Si and polysilicon in the APS circuit. IGZO is shown to have the lowest noise by a factor of 2–3 compared to *a*-Si and also has significantly lower noise than LTPS.

23.4.4 IGZO-TFT Noise

Since the TFT is the primary source of additive noise for a pixel amplifier and a secondary source for the simple TFT circuit, measurements of TFT noise are important to provide the information needed to evaluate the benefits of oxide-TFT detectors. There are a growing number of studies of IGZO-TFT noise, which find that 1/f noise tends to dominate, in common with other TFTs, and the IGZO 1/f noise evidently depends on the composition and structure of the TFT. Theodorou et al. [26] reported the low-frequency noise of bottom-gate IGZO-TFTs. The noise

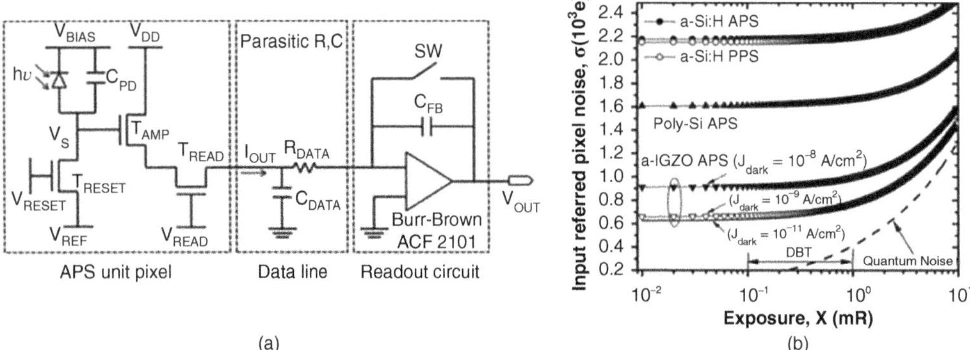

Figure 23.8 (a) Active pixel sensor circuit schematic with parasitic elements and external readout circuit is shown. The APS pixel consists of three transistors (T_{RESET}, T_{READ}, and T_{AMP}) and a photodiode with a capacitance of C_{PD}. (b) Input-referred pixel noise of 75 μm *a*-Si:H, poly-Si, and *a*-IGZO APS detectors under exposures ranging from 10 μR to 10 mR. The dashed line is the photon quantum noise. The pixel noise of the *a*-IGZO APS imager can be further reduced by reducing the dark current of OPD from 10^{-8} to below 10^{-9} A/cm². *Source*: (a,b) Zhao, C., and Kanicki, J. (2014). Amorphous In–Ga–Zn–O thin-film transistor active pixel sensor x-ray imager for digital breast tomosynthesis. *Medical Physics* **41**: 091902.

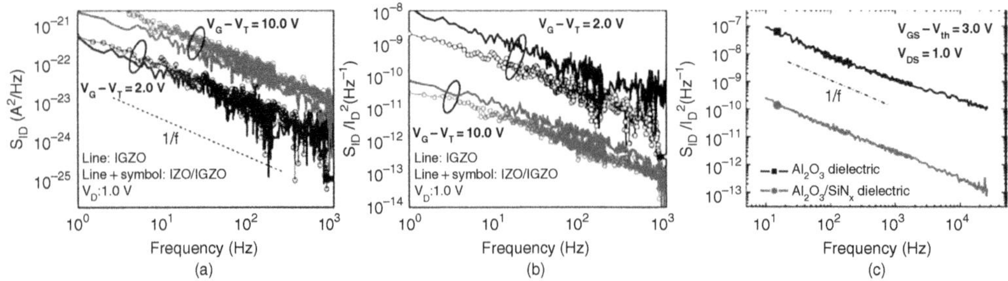

Figure 23.9 (a) Drain–current noise spectral densities (S_{ID}) and (b) normalized drain–current noise spectral densities (S_{ID}/I^2D) for the single-layer IGZO-TFT and bilayer IZO–IGZO-TFT that are measured at different gate–source voltages (V_G) of 2.0 and 10.0 V and a constant drain–source voltage (V_D) of 1.0 V. (c) Normalized noise power spectral density (SiD/I^2D) for devices with Al_2O_3 and Al_2O_3/SiN_x gate dielectrics, measured at the same gate overdrive voltage of 3.0 V in the linear region of device operation. *Source*: (a,b) Jeon, S., et al. (2010). Low-frequency noise performance of a bilayer InZnO–InGaZnO thin-film transistor for analog device applications. *IEEE Electron Device Letters* **31**: 1128. (c) Cho, I.-T., et al. (2009). Comparative study of the low-frequency-noise behaviors in a-IGZO thin-film transistors with Al_2O_3 and Al_2O_3/SiN_X gate dielectrics. *IEEE Electron Device Letters* **30**: 828.

spectra exhibit generation–recombination noise at drain currents <5 nA, which is attributed to bulk traps located in a thin layer of the IGZO close to the conducting channel. The 1/f noise observed at higher drain currents is attributed to carrier number fluctuations due to trapping/detrapping of carriers in slow oxide traps, located near the interface with uniform spatial distribution.

Jeon [27] reports that the normalized noise for a bilayer IZO–IGZO-TFT is three times lower than that for the single-layer IGZO-TFT (Figure 23.9a and 23.9b), attributed to the higher mobility of the thin interfacial InZnO layer. The carrier number fluctuation is the dominant low-frequency noise mechanism in both devices. Cho et al. [28] compared the low-frequency noise in IGZO-TFTs with Al_2O_3 and Al_2O_3/SiN_X gate dielectrics and found that the normalized noise for the Al_2O_3/SiNx device is two to three orders of magnitude lower than that for the Al_2O_3 device (Figure 23.9c). The SiN_X interfacial layer was thought to be very effective by suppressing the remote phonon scattering from the Al_2O_3 dielectric. The results show that material choices for the gate dielectric and the channel structures can have a substantial impact on the noise performance of oxide TFTs.

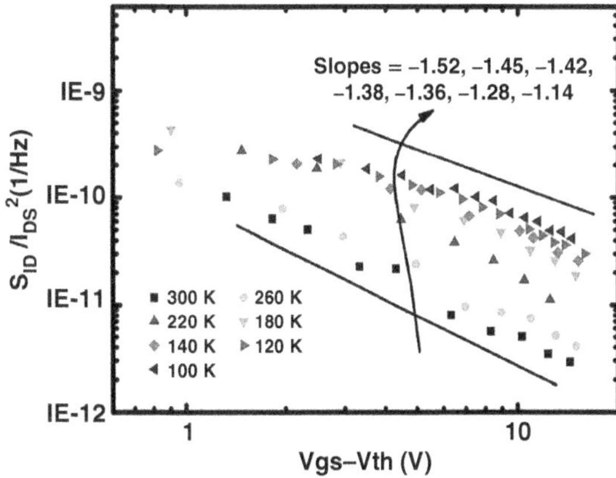

Figure 23.10 Normalized 10 Hz current noise versus effective gate voltage for IZO measured at various temperatures as shown, with $V_{DS} = 0.5$ V. *Source*: Liu, Y., et al. (2019). Temperature-dependent low-frequency noise in indium–zinc–oxide thin-film transistors down to 10 K. *IEEE Transactions on Electron Devices* **66**: 2192–2197.

The noise also depends on the electronic structure of the semiconductor. Kim et al. [29] investigated the relation between the low-frequency 1/f noise and subgap density of states of IGZO-TFTs by changing the annealing temperature from 150 °C to 300 °C. The density of tail states in TFTs annealed at 300 °C is lower than those in the TFTs annealed at 250 °C and 150 °C and has a 2× smaller 1/f noise. In addition, a TFT subjected to an AC gate voltage stress showed an increased density of states and a higher noise compared with one without electrical stress. Therefore, the work provides further evidence that the 1/f noise originates from carrier number fluctuations originating from trapping centers in the band-tail states and can be minimized by optimization of the deposition process.

Tai et al. report that X-ray exposure does not increase the noise [30]. The authors conclude that positive charge trapping in the oxide layer under X-ray exposure that gives a threshold voltage shift of the TFT does not affect its noise behavior. The data suggest that IGZO has less noise than polycrystalline-silicon TFTs, and hence the authors conclude that IGZO is a more suitable device for active pixel-sensing circuits, in agreement with Ref. [25]. The temperature dependence of the 1/f noise has been studied in IZO in the range from 10 to 300 K [31]. The TFT characteristics show an increasing effect of contact resistance at lower temperatures, which changes the slope of the normalized noise versus the effective gate voltage, as shown in Figure 23.10. The noise increases with decreasing temperature in the range of 200–300 K but decreases again below 200 K. The TFT measurements find thermally activated conduction in the range of 80–200 K and variable range hopping below 80 K.

23.5 Radiation Hardness of Oxide TFTs

Radiation hardness is obviously important for the materials and devices in an X-ray detector. Although the exposure for an individual medical X-ray is only 3–5 mR, frequent imaging over the lifetime of the detector results in exposures that can approach 1 MR. Limited studies indicate that oxide TFTs exhibit good radiation hardness. Cramer et al. [32] report that flexible IGZO-TFTs maintain a constant mobility of 10 $cm^2 V^{-1} s^{-1}$ even after exposure to X-ray doses of 410 krad, and they attributed the resistance to ionization damage to their intrinsic properties, such as independence of transport on the long-range structural order and the large heat of formation of the materials.

However, a dosimetry detection system has been developed based on flexible IGZO-TFTs with a negative threshold voltage shift of up to 3.4 V/gray upon exposure to ionizing radiation, as shown in Figure 23.11a [33]. The transistor is fabricated with a multilayer dielectric of silicon oxide and tantalum oxide. The threshold voltage shift is attributed to the accumulation of positive ionization charge in the dielectric layer resulting from high-energy photon absorption in the high-atomic-number dielectric. The high mobility combined with a steep subthreshold slope of the TFT allow for sensitive dosimetry. The report showed that it is possible to use low-cost, passive

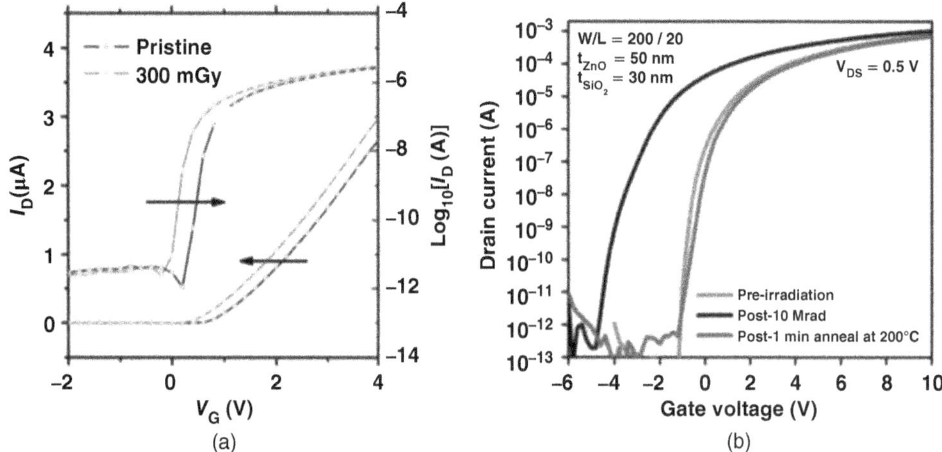

Figure 23.11 (a) IGZO-TFT dosimeter showing TFT transfer curves acquired before and after exposure to a 300 mGy radiation dose. (b) TFT characteristics for a pulsed-laser-deposited ZnO-TFT before irradiation, after 10 Mrad (SiO_2) ^{60}Co gamma ray irradiation, and after irradiation and a 200 °C, 1 min anneal in air. *Source*: (a) Cramer, T., et al. (2018). Passive radiofrequency x-ray dosimeter tag based on flexible radiation-sensitive oxide field-effect transistor. *Science Advances* **4**: eaat1825. (b) Ramirez, J. I., et al. (2015). Radiation-hard ZnO thin film transistors. *IEEE Transactions on Nuclear Science* **62**: 1399–1404.

radiofrequency identification sensor tags for the readout, because there is a large variation in transistor channel impedance upon exposure to radiation.

Ramirez et al. report effects of up to 100 Mrad ^{60}Co gamma-ray (~1 MeV) exposure on polycrystalline ZnO-TFTs with devices made by two different deposition techniques and with Al_2O_3 and SiO_2 gate dielectric [34]. The radiation induces a negative V_{ON} shift and a smaller V_T shift irrespective of whether a bias voltage is applied, as shown in Figure 23.11b, while the field-effect mobility remains nearly unchanged. Both V_{ON} and V_T shifts are almost completely reversed by annealing at 200 °C for 1 minute, and some recovery occurs even at room temperature. The authors conclude that these are the most radiation-hard TFTs reported to date.

Some radiation hardness studies have been done with electrons and ion beams, which are not directly relevant to X-ray detectors but give a general indication of radiation hardness. Costa et al. [35] studied the impact of highly energetic electron irradiation with fluence up to 10^{12} e^-/cm^2 and low operating temperatures down to 78 K on flexible IGZO transistors, to simulate 278 hours in low-earth orbit. The threshold voltage and the mobility of transistors that were exposed to electron irradiation are found to shift by only +0.09±0.05 V and −0.6±0.5 $cm^2V^{-1}s^{-1}$. Subsequent low-temperature exposure resulted in additional shifts of +0.38 V and −5.95 $cm^2V^{-1}s^{-1}$ for the same parameters.

The radiation effects of C-axis-aligned crystalline IGZO-TFTs in $^{12}C^{6+}$ beam irradiation showed a threshold voltage shift of −1 V and a subthreshold swing degradation of around 250 mV dec^{-1} after 130 krad dose irradiation [36]. Proton-beam irradiation on SnO-based *p*-type oxide-TFTs was performed using a 5 MeV proton beam with doses ranging from 10^{12} to 10^{14} p/cm^2 [37], and the results showed that the transfer curves of the TFT hardly changed. In common with other studies, the insensitivity of the current conduction path to SnO lattice disorder was considered a possible mechanism for the observed radiation tolerance of the *p*-type SnO-TFT. While the results in these studies vary for the different devices and exposure, it seems that oxide-TFTs have sufficient radiation hardness for operation as typical medical imaging detectors.

23.6 Oxide Direct Detector Materials

Commercial direct detection digital X-ray systems presently use selenium as the converter material. As noted in this chapter, these systems work well with a similar sensitivity as indirect systems, but are about 10 times less than the theoretical sensitivity. A direct detection material needs a high atomic number to give high X-ray absorption, deposition at a thickness of 0.1–1 mm, high charge collection, very low leakage current, and good uniformity and stability. The most widely studied alternative to selenium is HgI_2, which has demonstrated close to theoretical sensitivity but has had difficulty with meeting all the other requirements [38].

Metal oxides seem a generally promising class of materials for direct detection because there are a great number of possible compounds, many with high-atomic-number metals. A ZnO direct detection X-ray detector was demonstrated several years ago [39]. Samples were deposited by thermal evaporation to a thickness of about 300 μm. Charge collection was modest and required a field of 3 V/μm to obtain a good signal, which was possible because of the low dark current. An 18×20 cm^2 detector with 184 μm pixel size was reported and gave a DQE of 45% at zero spatial frequency. The charge collection was the main limitation to the detector, but an increase by a factor of perhaps 2–5 would result in excellent properties.

Liang et al. [40] developed a flexible X-ray detector using an amorphous Ga_2O_3 film deposited on a PEN substrate with careful control of the oxygen flux during the radiofrequency magnetron-sputtering process. Photodetectors with coplanar interdigitated metal electrodes were fabricated on the Ga_2O_3 film. Measurements under X-ray exposure found that a larger photocurrent occurs on the film deposited with smaller oxygen flux. A model combined with theoretical calculation is proposed to explain the enhancement of the X-ray photoresponse, which involves the slowing down of the recombination rate caused by the neutralization of more ionized oxygen vacancy states. No significant degradation of the device performance under UV and X-ray radiation is observed.

These publications suggest that there is an opportunity for radiation detectors based on amorphous oxide materials, and further studies of a wider range of materials would be very valuable.

23.7 Summary

Metal-oxide TFTs have potential advantages for flat-panel X-ray detectors, and IGZO-TFTs have been shown to give good performance in prototype detectors. The high mobility of oxides, compared to *a*-Si, can improve the detector readout speed and can also improve image quality. In particular, oxide-TFT pixel amplifiers have high potential to improve detector performance. Electronic noise in oxide-TFTs has been studied, but more definitive studies are needed to determine the optimum materials and device configuration for low noise. Exploration of oxide X-ray photoconductors could reveal new materials capable of high sensitivity in direct detection detectors.

References

1 Nomura, K., Ohta, H., Takagi, A. et al. (2004). Room-temperature fabrication of transparent flexible thin-film transistors using amorphous oxide semiconductors. *Nature* 432: 488–492.

2 Fortunato, E., Barquinha, P., and Martins, R. (2012). Oxide semiconductor thin-film transistors: a review of recent advances. *Advanced Materials* 24: 2945–2986.

3 Wager, J.F., Yeh, B., Hoffman, R.L., and Keszler, D.A. (2014). An amorphous oxide semiconductor thin-film transistor route to oxide electronics. *Current Opinion in Solid State and Materials Science* 18: 53–61.

4 Park, J.-S., Kim, T.-W., Stryakhilev, D. et al. (2009). Flexible full color organic light-emitting diode display on polyimide plastic substrate driven by amorphous indium gallium zinc oxide thin-film transistors. *Applied Physics Letters* 95: 013503.

5 Street, R.A. (2000). Large area image sensor arrays. In. In: *Technology and Applications of Amorphous Silicon* (ed. R.A. Street), 147–221. Springer, Cham, Switzerland.

6 Hosono, H. (2006). Ionic amorphous oxide semiconductors: material design, carrier transport, and device applications. *Journal of Non-Crystalline Solids* 352: 851–858.

7 Chiang, H.Q., Wager, J.F., Hoffman, R.L. et al. (2005). High mobility transparent thin-film transistors with amorphous zinc tin oxide channel layer. *Applied Physics Letters* 86: 013503.

8 Yaglioglu, B., Yeom, H.Y., Beresford, R., and Paine, D.C. (2006). High-mobility amorphous In_2O_3–10 wt% ZnO thin film transistors. *Applied Physics Letters* 89: 062103.

9 Ogo, Y., Hiramatsu, H., Nomura, K. et al. (2008). P-channel thin-film transistor using p-type oxide semiconductor. *SnO. Applied Physics Letters* 2008 (93): 032113.

10 Lee, S., Striakhilev, D., Jeon, S., and Nathan, A. (2014). Unified analytic model for current–voltage behavior in amorphous oxide semiconductor TFTs. *IEEE Electron Device Letters* 35: 84–86.

11 Nathan, A., Lee, S., Jeon, S., and Robertson, J. (2014). Amorphous oxide semiconductor TFTs for displays and imaging. *Journal of Display Technology* 10: 917.

12 Barquinha, P., Pereira, L., Gonçalves, G. et al. (2009). Towards high-performance amorphous GIZO TFTs. *Journal of the Electrochemical Society* 156: H161–H168.

13 Kim, M.-G., Kanatzidis, M.G., Facchetti, A., and Marks, T.J. (2011). Low-temperature fabrication of high-performance metal oxide thin-film electronics via combustion processing. *Nature Materials* 10: 382–388.

14 Street, R.A., Ng, T.N., Lujan, R.A. et al. (2014). Sol-gel solution-deposited InGaZnO thin film transistors. *ACS Applied Materials & Interfaces* 6: 4428.

15 Lee, S., Mativenga, M., and Jang, J. (2014). Removal of negative-bias-illumination-stress instability in amorphous-InGaZnO thin-film transistors by top-gate offset structure. *IEEE Electron Device Letters* 35: 930–932.

16 Ryu, B., Noh, H.-K., Choi, E.-A., and Chang, K.J. (2010). O-vacancy as the origin of negative bias illumination stress instability in amorphous In–Ga–Zn–O thin film transistors. *Applied Physics Letters* 97: 022108.

17 Lujan, R.A. and Street, R.A. (2012). Flexible X-ray detector array fabricated with oxide thin film transistors. *IEEE Electron Device Letters* 33: 688.

18 Gelinck, G.H., Kumar, A., Moet, D. et al. (2015). X-ray detector-on-plastic with high sensitivity using low cost, solution-processed organic photodiodes. *IEEE Transactions on Electron Devices* 63: 197.

19 Kim, S.I., Kim, S.W., Park, J.C. et al. (2011). Highly sensitive and reliable X-ray detector with HgI_2 photoconductor and oxide drive TFT. In: *2011 International Electron Devices Meeting*. (IEDM), 14.2.1–14.2.4.

20 Shimazoe, K., Koyama, A., Takahashia, H. et al. (2017). Prototype of IGZO-TFT preamplifier and analog counter for pixel detector. *Journal of Instrumentation* 12: C02045.

21 Abbaszadeh, S., Tari, A., Wong, W.S., and Karim, K.S. (2014). Enhanced dark current suppression of amorphous selenium detector with use of IGZO hole blocking layer. *IEEE Transactions on Electron Devices* 61: 3355–3357.

22 Liu, H., Hussain, S., and Kang, J. (2020). Improvement in sensitivity of an indirect-type organic X-ray detector using an amorphous IGZO interfacial layer. *Journal of Instrumentation* 15: P02002.

23 Xu, Y., Chen, J., Wang, K. et al. (2018). A direct-conversion x-ray detector based on a vertical x-ray photoconductor-gated a-IGZO TFT. In: *9th International Conference on Computer Aided Design for Thin-Film Transistors*, (CAD-TFT), Shenzhen, 1–1.

24 Freestone, S., Weisfield, R., Tognina, C. et al. (2020). Analysis of a new indium gallium zinc oxide (IGZO) detector. In: *Proceedings of SPIE 11312, Medical Imaging 2020: Physics of Medical Imaging*. 113123W.

25 Zhao, C. and Kanicki, J. (2014). Amorphous In–Ga–Zn–O thin-film transistor active pixel sensor x-ray imager for digital breast tomosynthesis. *Medical Physics* 41: 091902.

26 Theodorou, C.G., Tsormpatzoglou, A., Dimitriadis, C.A. et al. (2011). Origin of low-frequency noise in the low drain current range of bottom-gate amorphous IGZO thin-film transistors. *IEEE Electron Device Letters* 32: 898–900.

27 Jeon, S., Kim, S.I., Park, S. et al. (2010). Low-frequency noise performance of a bilayer InZnO–InGaZnO thin-film transistor for analog device applications. *IEEE Electron Device Letters* 31: 1128.

28 Cho, I.-T., Cheong, W.-S., Hwang, C.-S. et al. (2009). Comparative study of the low-frequency-noise behaviors in a-IGZO thin-film transistors with Al_2O_3 and Al_2O_3/SiN_X gate dielectrics. *IEEE Electron Device Letters* 30: 828.

29 Kim, S., Jeon, Y., Lee, J.-H. et al. (2010). Relation between low-frequency noise and subgap density of states in amorphous InGaZnO thin-film transistors. *IEEE Electron Device Letters* 31: 1236–1238.

30 Tai, Y.-H., Yeh, S., Chen, Z.-C., and Chang, T.-C. (2020). Effects of X-ray irradiation on the noise behavior of amorphous indium-gallium-zinc-oxide TFTs. *Journal of the Electrochemical Society* 167: 027512.

31 Liu, Y., He, H., Chen, Y.-Y. et al. (2019). Temperature-dependent low-frequency noise in indium–zinc–oxide thin-film transistors down to 10 K. *IEEE Transactions on Electron Devices* 66: 2192–2197.

32 Cramer, T., Sacchetti, A., Lobato, M.T. et al. (2016). Radiation-tolerant flexible large-area electronics based on oxide semiconductors. *Advanced Electronic Materials* 7: 1500489.

33 Cramer, T., Fratelli, I., Barquinha, P. et al. (2018). Passive radiofrequency x-ray dosimeter tag based on flexible radiation-sensitive oxide field-effect transistor. *Science. Advances* 4: eaat1825.

34 Ramirez, J.I., Li, Y.V., Basantani, H. et al. (2015). Radiation-hard ZnO thin film transistors. *IEEE Transactions on Nuclear Science* 62: 1399–1404.

35 Costa, J.C., Pouryazdan, A., Panidi, J. et al. (2018). Low temperature and radiation stability of flexible IGZO TFTs and their suitability for space applications. In. In: *48th European Solid-State Device Research Conference (ESSDERC)*, Dresden, 98–101.

36 Koyama, A., Miyoshi, H., Shimazoe, K. et al. (2017). Radiation stability of an InGaZnO thin-film transistor in heavy ion radiotherapy. *Biomedical Physics & Engineering Express* 3: 045009.

37 Jeong, H.-Y., Kwon, S.-H., Joo, H.-J. et al. (2019). Radiation-tolerant p-type SnO thin-film transistors. *IEEE Electron Device Letters* 40: 1124–1127.

38 Street, R.A., Ready, S.E., Van Schuylenbergh, K. et al. (2002). Comparison of Pb_{l2} and Hgl_2 for direct detection active matrix x-ray image sensors. *Journal of Applied Physics* 91: 3345.

39 Simon, M., Ford, R.A., Franklin, A.R. et al. (2004). Analysis of lead oxide (PbO) layers for direct conversion x-ray detection. *In IEEE Symposium Conference Record Nuclear Science Rome* 4268–4272.

40 Liang, H. (2019). Flexible x-ray detectors based on amorphous Ga_2O_3 thin films. *ACS Photonics* 6 (2): 351–359.

Part VII

New Materials

24

Toward the Development of High-Performance *p*-Channel Oxide-TFTs and All-Oxide Complementary Circuits

Kenji Nomura

Department of Electrical and Computer Engineering, University of California, San Diego, California, USA

24.1 Introduction

Since the invention of amorphous indium–gallium–zinc oxide (*a*-IGZO)-based *n*-channel oxide thin-film transistors (TFTs), oxide semiconductor device technology has been widely accepted as a promising technology to advance "giant microelectronics" from solid glass to soft flexible substrates [1]. In addition to superior device characteristics such as a high TFT mobility over 10 cm^2/Vs, a low operation voltage (threshold voltage (V_{th}) <3 V,) a small subthreshold slope (*s*-value, <200 mV/decade), and a low off-current characteristics, oxide-TFTs can take an advantage over conventional hydrogenated amorphous silicon-TFT in terms of optical transparency in the visible region, low-cost processing, excellent device stability, and reliability [2, 3]. So far, several state-of-the-art active-matrix flat-panel displays (AMFPDs), such as large-sized and high-resolution liquid crystal displays (AMLCDs) and active-matrix organic light-emitting diodes (AMOLEDs); flexible displays; and micro-LED displays have been successfully developed by adapting *a*-IGZO-TFT backplanes [4]. Beyond their use in display applications, oxide-TFTs have continuously extended their applications to broader areas, including low-power internet-of-thing (IoT) devices, energy harvesting, and biomedical sensing technology [5–7]. However, the device application of oxide-TFTs is currently limited to unipolar devices due to the absence of high-performance *p*-channel oxide-TFTs, which is a critical drawback of current oxide device technology [8, 9]. Since a low-power-enabled complementary metal-oxide semiconductor (CMOS) technology, which consists of *n*- and *p*-channel transistors, is an important building component in current digital logic circuit and analog circuit applications, developing oxide-TFT-based CMOS inverter circuits is the largest challenge faced in oxide device technology. Many efforts to develop high-performance *p*-channel oxide-TFTs have been made so far, but the device performance is still largely behind that of the *n*-channel TFT's performance. Clearly, it strongly requires significant breakthroughs at the levels of material and device developments in *p*-channel oxide-TFTs.

24.2 Why Is High-Performance *p*-Channel Oxide Difficult?

The poor device characteristics of a *p*-channel oxide-TFT mainly links to the absence of good *p*-type oxide materials. The main reasons are attributed to (i) the nature of the localized valence band maximum (VBM) mainly composed of oxygen 2*p* orbitals, (ii) the difficulty of acceptor doping due to a relatively deep VBM energy level from the vacuum level, (iii) the existence of a high-density subgap (i.e., in-gap) defect near the VBM, and (iv) unintentional electron doping by hydrogen impurity, among others. Moreover, (v) substitutional doping is currently not established yet for ionic amorphous oxides. All these reasons make hole doping and hole transport difficult in

Amorphous Oxide Semiconductors: IGZO and Related Materials for Display and Memory, First Edition.
Edited by Hideo Hosono and Hideya Kumomi.

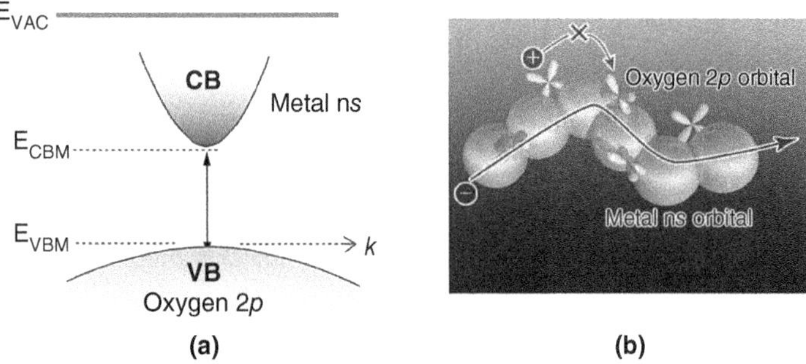

Figure 24.1 Schematic illustration of the (a) electronic structure and (b) carrier pathway for representative ionic oxide semiconductors, such as ZnO, In_2O_3, and *a*-IGZO.

oxide semiconductors and stand as a large technology barrier in the development of high-performance *p*-channel oxide TFTs. Figure 24.1 shows schematic illustration of the electronic structure of oxide semiconductors, which have an electronic structure in which the spherical metal *ns* orbital (*n*: quantum number) contributes to forming the conduction band minimum (CBM), while the VBM is mainly composed of the localized and anisotropic oxygen 2*p* orbital [10]. Owing to the spherical nature of metal ns orbitals such as Zn 4*s*, In 5*s*, and Sn 6*s* orbitals, the CBM possesses relatively high dispersion in In_2O_3, ZnO, and SnO_2. The feature explains the reason why many oxide semiconductors are good *n*-type semiconductors and offer high electron mobility, even in a disordered structure. In contrast, oxygen 2*p* orbitals do not form a dissipative valence band structure due to the strong directivity nature of 2*p* orbitals, and they form a localized hole-transport pathway. Thus, the hole effective mass is relatively large due to the nondispersive VBM structure in oxide. Moreover, hole transport is largely affected by bond angle distortions and structural defects, and is easily degraded.

Figure 24.2 shows the band lineups showing the band edge energy (i.e., the CBM and VBM energy) for several semiconductor materials, including Group IV, III–IV compounds, nitride, sulfide, and oxide semiconductors. For conventional covalent semiconductors such as Si and GaAs, the CBM and VBM energy levels lie within ~4–6 eV, which corresponds to the region where the *pn* polarity is relatively easily controlled by doping. In general, it is known that there is a tendency that a deep VBM energy level makes hole doping difficult, while electron doping is getting harder for semiconductors with shallow CBM levels. For several representative oxide semiconductors such as ZnO, SnO_2, and *a*-IGZO, the CBM energy levels are ~4–5 eV, and electron doping is relatively easy by doping. This also provides the reason why many oxide semiconductors exhibit intrinsically *n*-type behavior. On the other hand, the VBM levels are beyond ~7 eV and are very deep from the vacuum level in these oxides. This indicates that hole doping is difficult in many oxides.

Besides these issues related to the intrinsic electronic structure in ionic oxide semiconductors, the existence of a subgap defect state is considered to be an obstacle to the development of high-performance *p*-type oxides (Figure 24.3). A high-density subgap defect just above the VBM, called the "near-VB state," was observed in many oxide semiconductors [11]. The near-VB state broadly distributes from the VBM to the midgap region, and the density is estimated as over $\sim 10^{19}$ cm^{-3}. Such a high-density subgap defect not only makes hole doping difficult but also causes the pinning of the Fermi level (E_F) in a field-effect device. The density of the near-VB state is more than two orders of magnitude larger than the hole density inducible by a gate bias (V_g) in a conventional oxide-TFT structure (simply estimated to be $\sim 4 \times 10^{18}$ cm^{-3} for $V_g = 50$ V with a 100-nm-thick SiO_2 gate insulator). In such cases, the E_F in the channel is pinned by the subgap near-VB state, and free holes are not induced in the extended states in the valence band. The origin of the defect is not clear yet, but oxygen vacancy and low valence states of the metal cations are likely to be responsible for the defect formation in most oxides.

Another critical issue is the presence of hydrogen impurity, which is considered to act as an electron donor in many oxides [12]. Therefore, impurity hydrogen must be removed from the films to archive hole doping, but it is

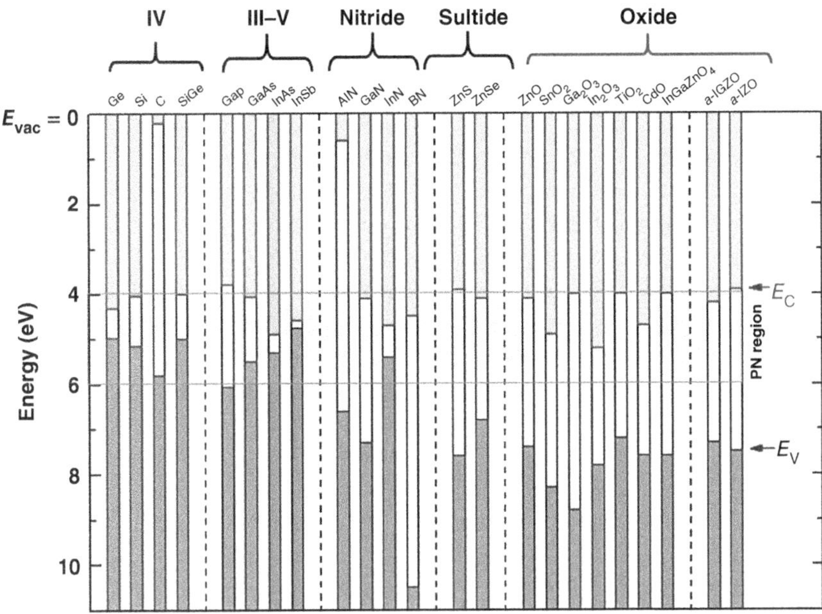

Figure 24.2 Band lineups for several semiconductor materials, including Group IV, III–IV compounds, nitride, sulfide, and oxide semiconductors.

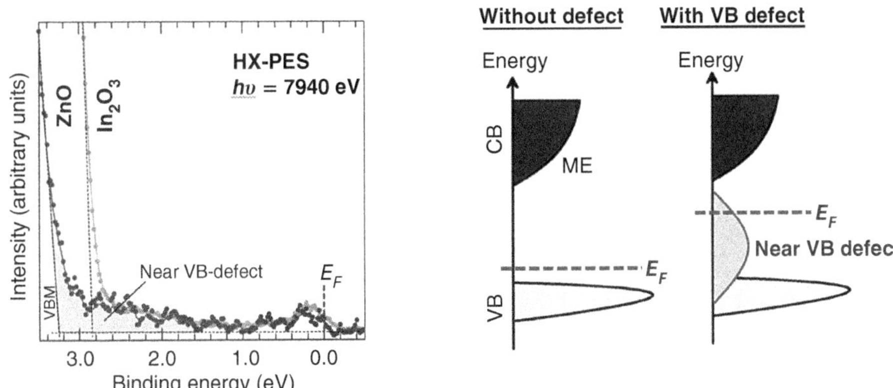

Figure 24.3 (a) Hard X-ray photoemission spectroscopy (HX-PES) spectra for the bandgap region for ZnO and In_2O_3. (b) Schematics of Fermi level pinning by a near-VB defect state.

very difficult because hydrogen exists everywhere (even in the vacuum chamber) and is always unintentionally introduced into oxides [13]. Moreover, substitutional doping is not confirmed in ionic amorphous oxide semiconductors due to the structural flexibility originating from strong ionicity, and it is also recolonized as an issue in the difficulty of polarity control [14]. (Substitutional doping, acceptor doping, in crystalline oxide semiconductors such as ZnO is available.) Because of these reasons, developing high-performance *p*-type oxide and *p*-channel oxide-TFTs is still challenging, particularly developing *p*-type amorphous IGZO, which has not succeeded yet. Therefore, the use of oxide semiconductors intrinsically exhibiting a *p*-type nature such as Cu_2O, SnO, NiO, and so on is the only way in the current development of *p*-channel oxide-TFTs.

24.3 The Current Development of *p*-Channel Oxide-TFTs

Figure 24.4 summarizes current oxide-TFT development based on the published literature (As of 2020). To my best knowledge, over 15,000 papers have been published on topics related to oxide-TFT technology, including *n* and *p*-channel TFTs and crystalline/amorphous channels, so far since the first report of *a*-IGZO-TFTs in 2004. Among them, 98% of the total publications relate to *n*-channel oxide-TFTs; 75% of them are amorphous oxide channels such as *a*-IGZO and *a*-IZO, and the remaining is for crystalline oxide devices such as ZnO-TFTs. Although the importance of *p*-channel oxide-TFT development is widely recognized to advance oxide-TFT technology, only ~1.4% of the total publications are found in the topic for *p*-channel oxide-TFT research. The low publication number implies the difficulty of developing *p*-type oxide and *p*-channel TFTs. Figure 24.4b also shows the trends of *p*-type oxide materials for oxide-TFT development from a total of ~150 publications. Many types of *p*-type oxide materials such as delafossite $CuMO_2$ (M = Al, Ga, In, Sc, Y, Cr, Fe, etc.), spinel oxide (ZnM_2O_4 [M = Co, Rh, Ir]), tetragonal oxide ($SrCu_2O_2$), foordite ($SnNb_2O_6$), and perovskite ($LaSrCrO_3$, $LaRhO_3$, $SrVO_3$, $CaVO_3$, $CuNbO_3$, etc.) systems have been discovered so far [15–18]. However, only three kinds of materials (i.e., SnO, Cu_2O, and NiO) are studied as an active channel material for oxide-TFTs. Among them, the SnO channel is intensively investigated due to its relatively easy and a low-temperature device fabrication, and it comprises ~60% of total publications for *p*-channel oxide-TFTs.

Figure 24.5 shows the progress of TFT mobility for oxide-TFTs, plotted by year. For comparison, the same data for *n*-channel oxides are also shown [19]. Since the first report of *a*-IGZO-TFTs with mobility of about 6–8 cm^2/Vs in 2004 [1], numerous efforts to improve TFT mobility have been made, mainly by the development of new channel chemical compositions. *a*-In-Sn-Zn-O (*a*-ITZO), *a*-In-Al-Zn-O (*a*-IAZO), and *a*-In-Zn-O (*a*-IZO) are well-known channel materials to exhibit higher TFT mobility than *a*-IGZO-TFTs [20–22]. Very high TFT mobility comparable to the mobility of low-temperature poly-Si (LTPS)-TFTs has often been reported in In-rich *a*-IZO and *a*-IGZO channels-TFTs. In contrast, although over 10 years have already passed since the first report (in 2008) of *p*-channel oxide-TFTs with mobility of about ~1 cm^2/Vs for SnO-TFTs [23], TFT mobility has barely improved in the last 10-plus years for *p*-channel oxide-TFTs.

Figure 24.6 also summarizes the progress of TFT mobility for *p*-channels such as Cu_2O, SnO, and NiO-TFTs, by the reported year [23–25]. Although Cu_2O film exhibits high hole mobility, the TFT mobilities for most Cu_2O-TFTs are less than 1 cm^2/Vs. The best Cu_2O-TFT was reported in 2011 [26]. The Cu_2O channel was grown by high-temperature pulsed laser deposition (PLD) at 500 °C, and the TFT showed excellent device performance such as the TFT mobility of ~ 2.7 cm^2/Vs, *s*-value of 137 mV/decade, and large on/off current ratio over 10^6. The best TFT mobility of 5.64 cm^2/Vs with a small *s*-value of 0.75 V/decade was also archived with the atomic layer deposition (ALD)-deposited Cu_2O channels [27]. On the other hand, TFT mobility of about 1–2 cm^2/Vs, which is

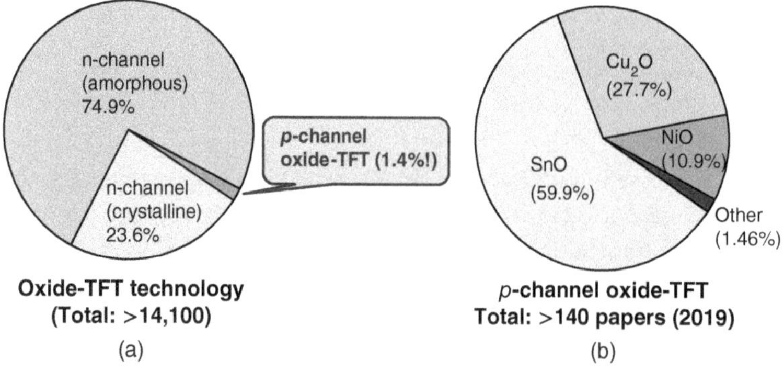

Figure 24.4 (a) The research trend of oxide-TFT technology. (b) The research trend of *p*-type oxide materials for *p*-channel oxide-TFTs.

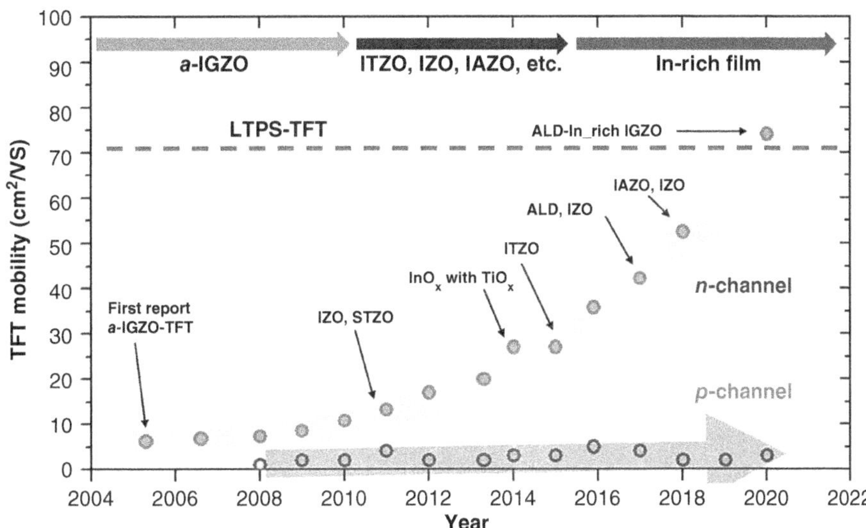

Figure 24.5 The progress of TFT mobility for oxide-TFTs, including both n-channel and p-channel devices.

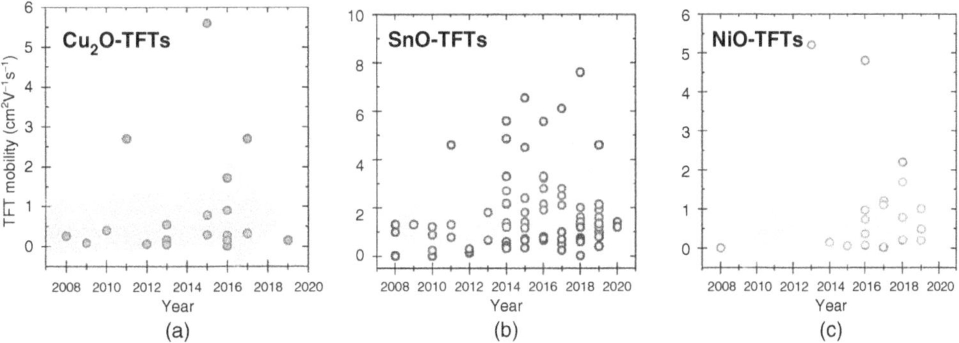

Figure 24.6 Progress of TFT mobility for p-channel oxide-TFTs by year. (a) Cu_2O, (b) SnO, and (c) NiO-TFTs.

comparable to the hole mobility of film, can be easily obtained in SnO-TFTs. Recently, very high-TFT-mobility devices have been demonstrated in several groups. High-performance SnO-TFTs with a high TFT mobility of 7.6 cm^2/Vs and an excellent s-value of ~140 mV/decade were reported by using a high-k HfO_2 gate with a 200 °C post annealing process [28]. The device performances are almost comparable to those of the n-channel a-IGZO-TFTs. It was also reported that the dual-gated SnO-TFTs significantly improved device performance [29]. The TFT mobility is 2–3 cm^2/Vs for the single-gate operation, while by operating the devices with dual gates, remarkably high TFT mobility over ~6 cm^2/Vs was achieved. Many high-mobility NiO-TFTs also have been reported very recently.

Figure 24.7 is the trend of fabrication methods for SnO and Cu_2O channels. Sputtering deposition, including the direct current (DC) and radiofrequency (RF) modes, is a major deposition technique for both the materials, and PLD is also widely used to grow them in academic research. Many cost-effective solution processes also have been reported for p-channel Cu_2O-TFTs [30]. High-performance solution-based Cu_2O-TFTs with TFT mobility of ~3 cm^2/Vs were demonstrated by using high-k Al_2O_3 gates [31]. The Cu_2O channel was prepared by a simple solution-based oxidation process of CuI precursor film. On the other hand, only one paper has been published for

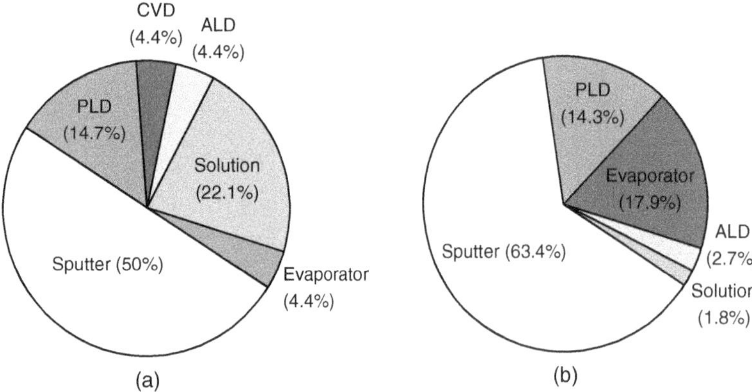

Figure 24.7 Trend of fabrication methods for (a) Cu_2O and (b) SnO channels for *p*-channel oxide-TFTs.

solution-processed SnO-TFTs so far [32]. The solution-based SnO channel was synthesized from $SnCl_2{\cdot}2H_2O$ precursor reacted with NH_4OH with post-thermal annealing at 450 °C. The TFT device with a staggered bottom-gate structure showed a TFT mobility of 0.13 cm^2/Vs.

24.4 Comparisons of *p*-Type Cu_2O and SnO Channels

p-Type oxides such as Cu_2O are SnO are characterized by the specific electronic structure with the different VBM nature, which are composed of a closed-shell Cu $3d^{10}$ configuration ($(n-1)\ d^{10}ns^0$) and Sn $5s^2$ configuration ($(n-1)\,d^{10}ns^2$) for Cu_2O and SnO, respectively. Another is a quasi-closed-shell nd^6configuration seen in spinel-type ZnM_2O_4 (M = Co, Rh, and Ir). Figure 24.8 illustrates the crystal structures and the electronic structure of Cu_2O and SnO. Cu_2O, which is called cuprous oxide, has a cubic crystal structure belonging to the space group (*pnm3*), which is built up of two interpenetrating diamond-like oxygen and copper networks [33]. The VBM is mainly derived from the hybridization of Cu 3*d* and O 2*p* orbitals, since the energy levels of the Cu 3*d* orbitals are comparable to those of the oxygen 2*p* levels (Figure 24.8b). Therefore, the Cu(I)-based material group creates the less localized hole-transport pathway and exhibits good hole-transport properties. On the other hand, SnO with PbO-type structure (tetragonal litharge structure), which belongs to the space group (*P4/nmm*), has a layered structure composed of a pyramid structure with four O atoms and one Sn atom [34]. The VBM of SnO is formed by the hybridization between oxygen 2*p* and spherical Sn 5*s* orbitals, and spherical *s* orbitals mainly contribute to the formation of the VBM, which also forms dispersed VBM structure (Figure 24.8d).

Figure 24.9a shows the band lineups for several intrinsically *p*-type oxide semiconductors. The VBM energy levels for these materials are 5–6 eV and are obviously shallower than typical *n*-type oxides (>7 eV). The CBM energy levels are relatively shallow, but electron doping for Cu_2O and SnO is possible. Figure 24.9b plots the reported hole mobility as a function of hole density for Cu_2O and SnO films. For comparison, the reported hole mobilities for *p*-type ZnO and NiO are also shown. The reported hole mobility of *p*-type ZnO film is still largely distributed, and the fabrication of *p*-type ZnO is still channeling. Although the data include different crystal quality (i.e., polycrystalline and epitaxial films) and film thickness, clear trends of hole mobility and carrier density are clearly recognized for Cu_2O and SnO films. The hole mobility of Cu_2O increases with the decrease of carrier density, and high hole mobility over 100 cm^2/Vs can be achieved if the carrier density is less than 10^{14} cm^{-3}. Very high hole mobility of ~256 cm^2/Vs was achieved in polycrystalline Cu_2O film deposited by DC sputtering at 600 °C [35]. The reported carrier density is distributed in the range of 10^{14}–$10^{17}cm^{-3}$, and typical Cu_2O film has a hole density of ~10^{15} cm^{-3}. On the other hand, the reported hole densities for SnO films are around ~10^{17}–$10^{18}cm^{-3}$,

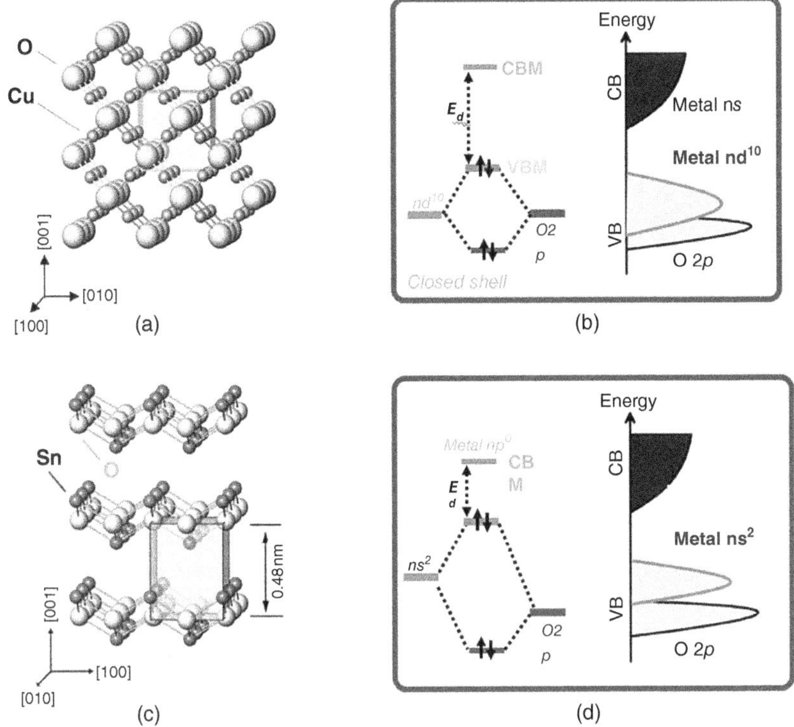

Figure 24.8 (a) Crystal structure and (b) schematic energy diagram for formation of bandgap for Cu_2O. (c) Crystal structure and (d) schematic energy diagram for SnO.

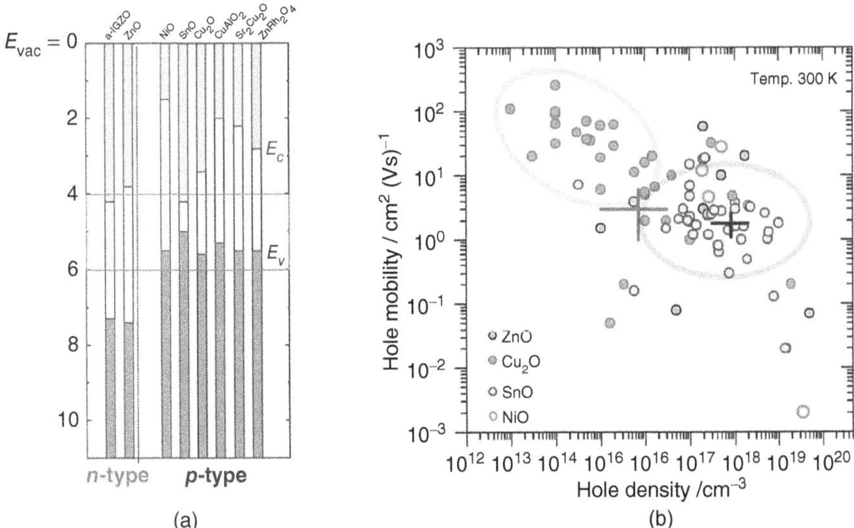

Figure 24.9 (a) Band lineups for several intrinsically *p*-type oxide semiconductors. (b) The reported film hole mobility versus carrier density for *p*-type oxide semiconductors, Cu_2O, SnO, and NiO films.

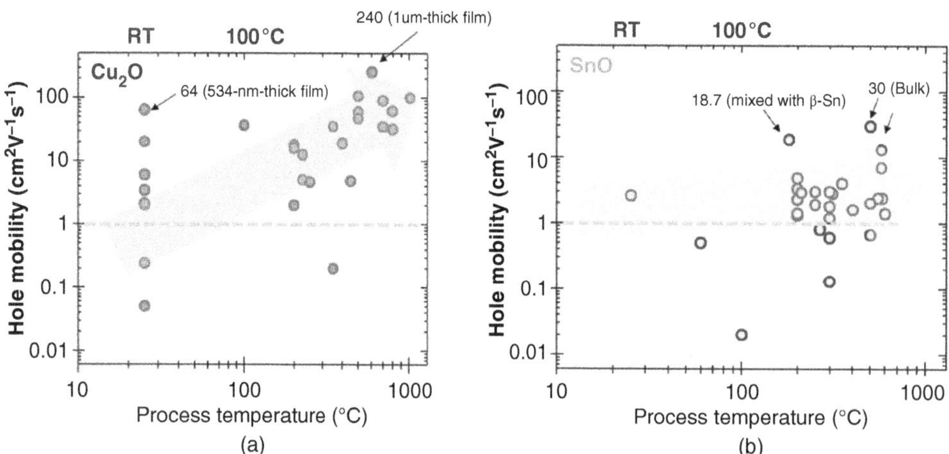

Figure 24.10 The variation of hole mobility by the process temperatures for (a) Cu_2O and (b) SnO films. The temperature includes deposition and a post-annealing process.

indicating that most SnO films are in a highly hole-doped state. The typical reported hole mobility is concentrated to ~1–3 cm^2/Vs in most SnO films, but very recently a remarkably high hole mobility of >10 cm^2/Vs was reported in the epitaxial SnO film grown on YSZ single-crystal substrate by high-temperature PLD depositions [36]. The theoretical material calculation also predicted that the intrinsic mobilities of SnO are ~60 cm^2/Vs [37]. For NiO, much higher hole mobility than a single-crystal sample (typically ~0.1 cm^2/Vs for single crystal) can be found in a few published works.

Hole mobility is also plotted by the process temperatures and the film thickness. Figure 24.10 shows the variations of hole mobility by the process temperatures for Cu_2O and SnO films. The process temperature includes the film deposition and post-thermal annealing temperatures. In Cu_2O films, the hole mobility strongly depends on the process temperatures and increases with increase of the process temperatures. Although there are a few reports of high hole mobility in low-temperature processes, these are very thick films (>0.5–1 um). A high process temperature of over 500 °C is generally required to attain higher mobility over 10 cm^2/Vs for Cu_2O films. On the other hand, SnO films are prepared at less than 300 °C and show very weak process temperature dependency for hole mobility. The hole mobility is around 1–3 cm^2/Vs in most cases. A remarkably high hole mobility of 18.7 cm^2/Vs was obtained, but the film is composed of a mixture of SnO and Sn metal [38].

Figure 24.11 also shows the variations of hole mobility by film thickness for Cu_2O and SnO films. Cu_2O films clearly exhibit the film thickness dependence on the hole mobility, and the hole mobility is monotony increased by increasing the film thickness. To obtain high mobility over 10 cm^2/Vs, a relatively thick film thickness of >100 nm is required in Cu_2O. On the other hand, the mobility of SnO films is almost independent of the film thickness, and even with very thin films (less than 10 nm), the film still exhibits hole mobility around 1 cm^2/Vs, which is obviously higher than the mobility in thin Cu_2O films with the same film thickness. These discussions are mainly attributed to the differences in polycrystalline grain structure between SnO and Cu_2O films.

Figure 24.12 shows atomic force microcopy (AFM) images of the film surface of epitaxial and polycrystalline films of Cu_2O and SnO. An X-ray diffraction (XRD) pattern confirms that both the polycrystalline films of Cu_2O and SnO do not have any preferred crystal orientations. For the polycrystalline film of Cu_2O, a relatively large polycrystalline grain structure with ~100 nm was observed, while the epitaxial film exhibits a very smooth surface without an apparent grain boundary. In this case, a large difference of hole mobility between polycrystalline and epitaxial films was confirmed, and the epitaxial film exhibited obviously higher hole mobility (~20 cm^2/Vs) than the polycrystalline films (~5 cm^2/Vs). Therefore, the crystal-grain structure of Cu_2O significantly affects the transport property and degrades the hole mobility. Although the hole mobility obtained from the epitaxial film

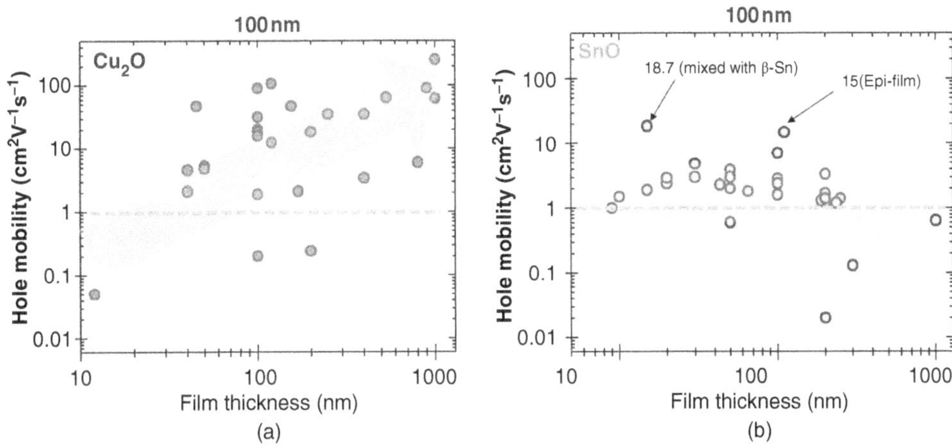

Figure 24.11 The variation of hole mobility by the film thickness for (a) Cu_2O and (b) SnO films.

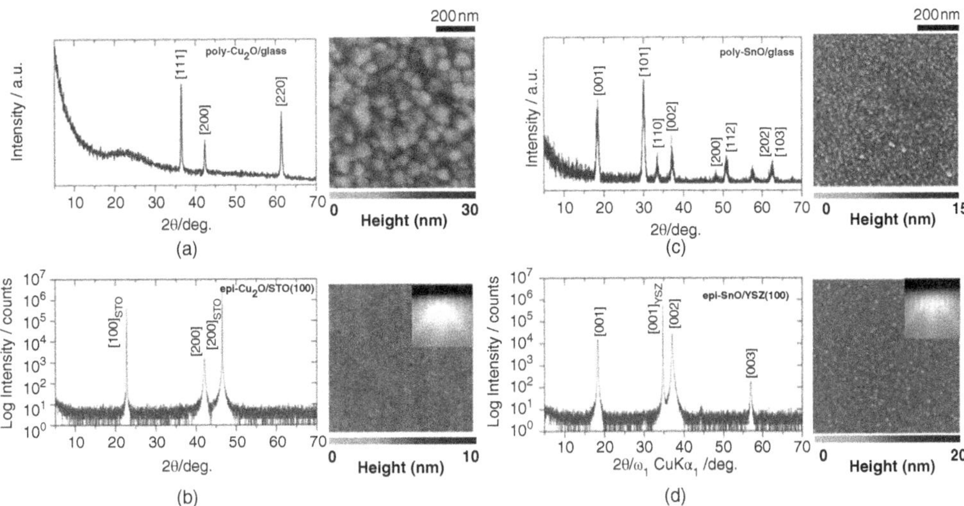

Figure 24.12 The XRD and AFM image of polycrystalline and epitaxial films (a,b) of Cu_2O and (c,d) for SnO films.

is slightly high, on the other hand, no significant degradation in the polycrystalline is observed in SnO films (~2 cm^2/Vs). The AFM image reveals that the film surface structures are very similar between polycrystalline and epitaxial films and are composed of small crystal grains with grain sizes of 10–20 nm. Unlike the Cu_2O films, the hole mobility for SnO does not strongly depend on the film quality. Therefore, this observation can explain that SnO doesn't show a strong dependency of the process temperature and film thickness on the hole mobility, while Cu_2O requires a high-temperature fabrication process and thick film to grow large crystal grains in order to achieve high hole mobility.

24.5 Comparisons of the TFT Characteristics of Cu_2O and SnO-TFTs

Figure 24.13 compares the TFT characteristics (24.13a: TFT mobility vs. *s*-value; and 24.13b: on/off current ratio vs. off-current) of the reported Cu_2O and SnO-TFTs. The most reported SnO-TFTs show TFT mobility

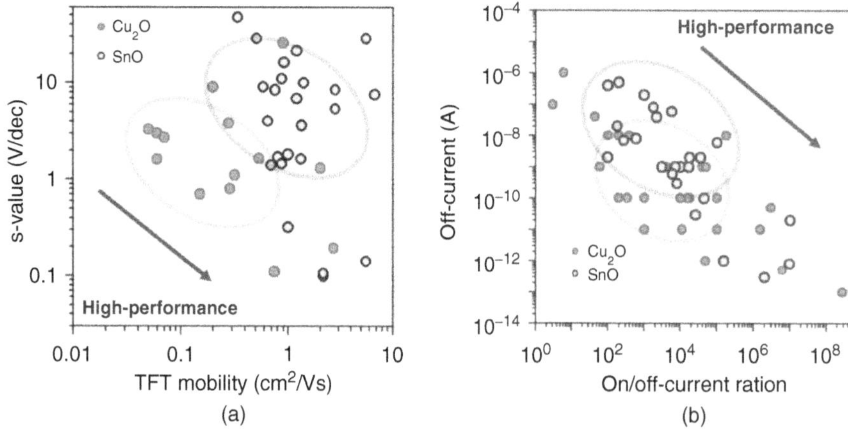

Figure 24.13 Comparison of TFT characteristics of Cu_2O and SnO-TFTs: (a) TFT mobility versus *s*-value, and (b) on/off current ratio versus off-current. The arrow indicates the direction for high performance.

over 1–3 cm^2/Vs, which is compatible with the hole mobility of film and is higher than the TFT mobility for Cu_2O-TFTs. For the *s*-values, the Cu_2O-TFT values are smaller than those of the SnO-TFTs. These observations strongly relate to the different TFT operation mechanisms mainly associated with different acceptor densities for Cu_2O and SnO channels. The Cu_2O-TFTs have slightly better off-state characteristics and can suppress the off-current at the level of <0.1 nA, but the on/off current ratio is limited to $\sim 10^4$ for both TFTs. Reducing the off-current is an important issue for *p*-channel oxide-TFTs.

These TFT parameters (i.e., TFT mobility, *s*-value, and on/off current ratio) are also plotted by the channel thickness. For the SnO-TFTs (Figure 24.14), the TFT mobility is almost independent of the channel thickness, which is in good agreement with the film thickness independent of hole mobility, as shown in Figure 24.11. In contrast, the *s*-values and on/off ratios strongly depend on the channel thickness and are largely improved by decreasing it. The trend can be comprehended by the channel depletion operation mechanism for TFT operations. Since the acceptor density of SnO is around $\sim 10^{17}$–10^{18} cm^{-3}, the SnO-TFT is expected to be operated by the carrier depletion modes, in which holes in the SnO channel are depleted by positive gate bias. The estimated width of the depletion layer for typical SnO-TFTs is ~15–20 nm, which corresponds to a maximum channel thickness that can control the carrier depletion to attain field effect modulation. If the channel thinness is thicker than ~20 nm, a source/drain bulk punch-through current around the backchannel appears, which causes the degradation of the *s*-values and on/off ratios. On the other hand, all TFT parameters are nearly unchanged by the channel thickness for the Cu_2O-TFTs (Figure 24.14b). Since the typical acceptor density of the Cu_2O channel is around $\sim 10^{15}$ cm^{-3}, the TFT operation is considered as carrier accumulation (enhancement-type) mode. Therefore, the effect of channel thickness on the TFT parameters may be small. Since the *s*-value strongly relates the subgap defect density around E_F, however, the *s*-value should be improved by decreasing the channel thickness irrespective of the TFT operation modes. The observation of no improvement of *s*-value by the channel thickness suggests that the TFT operation/performance is controlled by not only the channel bulk defect but also the front/back-channel defect.

Table 24.1 summarizes the material and device properties for SnO and Cu_2O-TFTs. These materials are crystalline; the amorphous films do not work as a *p*-channel layer due to very high-density subgap defects. Polycrystalline SnO is a relatively highly doped *p*-type semiconductor with a hole density of $\sim 10^{17}$–10^{18} cm^{-3}, while the hole density is $\sim 10^{15}$–10^{16} cm^{-3} for Cu_2O. The hole source for these materials is currently believed to be tin vacancy (V_{sn}) and copper vacancy (V_{cu}) with the low formation energy for SnO and Cu_2O, respectively, from density functional theory (DFT)-based computational studies [39, 40]. Unlike *n*-type oxide, in which the electron density can

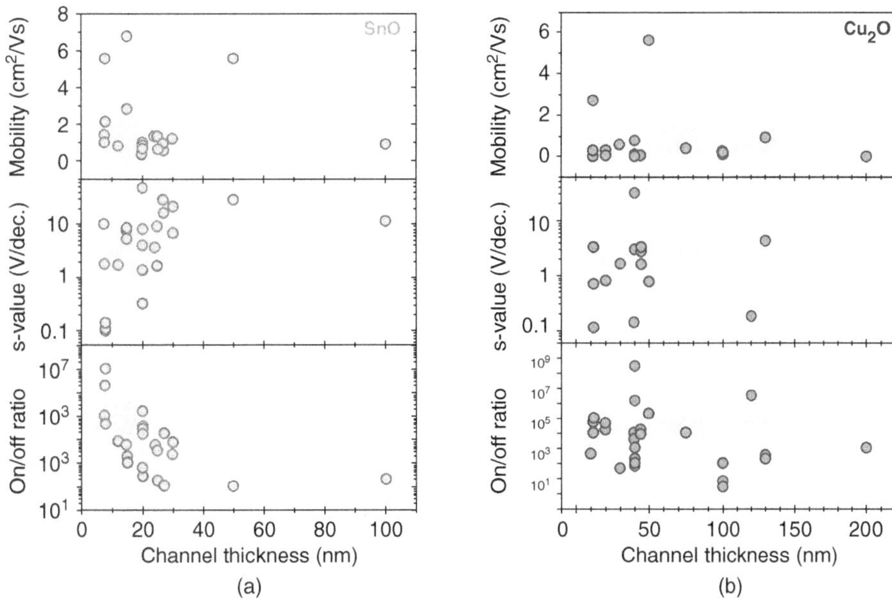

Figure 24.14 Comparison of channel thickness dependence of TFT parameters (i.e., TFT mobility, the s-value, and the on/off current ratio) for (a) SnO-TFTs and (b) Cu_2O-TFTs.

Table 24.1 Comparison of SnO and Cu_2O-TFTs.

	SnO (Crystalline)	Cu_2O (Crystalline)
Band gap (eV)	~0.8 (indirect)	~2.2 (direct)
Hole density (cm^{-3})	$\sim10^{17}$–10^{18}	$\sim10^{14}$–10^{15}
Hole mobility (cm^2/Vs)	1–2 (poly)	~5–10 (poly, t~100 nm)
TFT mobility (cm^2/Vs)	1–2	<1
TFT operation mechanism	Depletion-type	Enhancement-type
s-value (V/decade)	5–10	<5
Subgap defect density (cm^{-3})	$\sim1\times10^{19}$	$\sim8\times10^{18}$
Hole source	Tin vacancy	Cu vacancy
Hole trap source	Oxygen vacancy	Interstitial Cu

be controlled by oxygen partial pressure during the deposition, unfortunately, a technique to control the acceptor density of these materials has not been established yet. The hole density results in different TFT operation mechanisms for SnO- and Cu_2O-TFTs. The SnO-TFT operates in channel depletion-type, so the channel thickness is very important to achieve a small s-value and large on/off current ratio. On the other hand, most Cu_2O-TFTs operate by enhancement-type due to their low hole density. The largest issue in these p-channel oxide-TFTs is that both channel materials involve very high-density subgap defects to close to 10^{19} cm^{-3}, and except for a few papers achieve very good TFT performances. The defect density is obviously higher than that for n-channel IGZO with a defect density level of $\sim10^{17}$ cm^{-3}, and such a high-density subgap defect is mainly attributed as the root cause for the poor device performance of p-channel oxide-TFTs.

The origin of subgap defects can be predicated from computational theoretical study. For SnO, V_{Sn}, oxygen vacancy (V_o), and interstitial oxygen (O_i) are known as energetically favorable native point defects. Intraretinal

O_i is an electrically inactive defect, while V_o is expected to form a deep donor-like state near the VBM. The V_o would work as a hole trap in SnO. For Cu_2O, on the other hand, V_{cu}, V_o, O_i, and copper interstitial copper (Cu_i) are energetically favorable native point defects. Interestingly, several DFT studies conclude that V_o, unlike many oxide semiconductors, is an electrically inactive defect and does not contribute to subgap defect state formation in Cu_2O. The formation energy of Cu_i is slightly high, but Cu_i is considered to work as amphoteric defects in the bandgap and expected to act as a midgap donor-like defect state, which is attributed to the origin of high off-current in Cu_2O-TFTs. Therefore, the reducing V_o for SnO and Cu_i for Cu_2O are the key to improve TFT characteristics, and the developing effective defect termination methods are imperative to achieve high-performance *p*-channel oxide-TFTs.

24.6 Subgap Defect Termination for *p*-Channel Oxides

The reduction of subgap defects is a key to realize high-performance *p*-channel oxide-TFTs. So far, several attempts to reduce subgap defects of *p*-type oxides have been proposed to improve device performance, such as film growth optimization, formation of the passivation layer, and post-thermal annealing. Post-thermal annealing is the most simple and prevalent technique to improve film quality, and it is widely used for oxide-TFT technology. However, conventional thermal annealing with uncontrolled air atmosphere is not effective for these *p*-type materials, because higher oxidation metal cations such as Cu^{2+} in copper oxide and Sn^{4+} in tin oxide are more stable than Cu^{+} and Sn^{2+}, leading to the formation of undesirable high-oxidation states. The formation of these higher oxidation states results in severe degradation in *p*-channel TFT device performance. Therefore, it is imperative to develop effective defect termination techniques for *p*-type oxides to avoid degradation and improve the device performance of *p*-channel oxide-TFTs.

Figure 24.15 shows the typical transfer curves for the unannealed and SnO-TFTs annealed in several atmospheres at 250 °C for 0.5 hours. The unannealed SnO channel is amorphous, and the TFTs do not show any field-effect current modulation due to the high-density subgap defects. After post-annealing treatment over 250 °C, single crystalline phase SnO is formed, and all the devices show clear TFT action with *p*-channel mode, but the device characteristics are strongly varied by the annealing atmospheres. The vacuum-annealed TFT exhibits ambipolar characteristics (i.e., both *n*- and *p*-channel operations), while the *n*-channel mode is

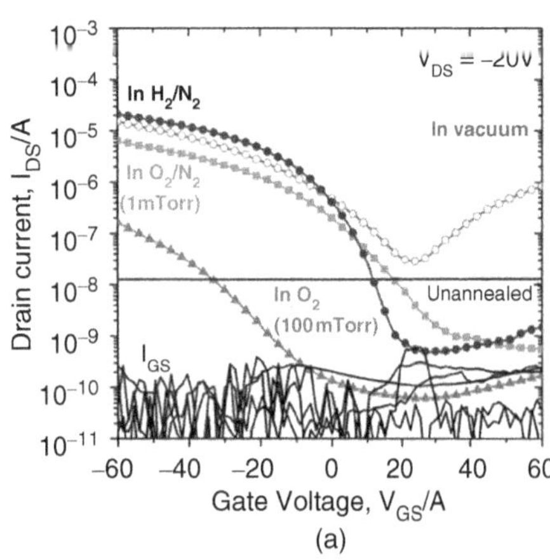

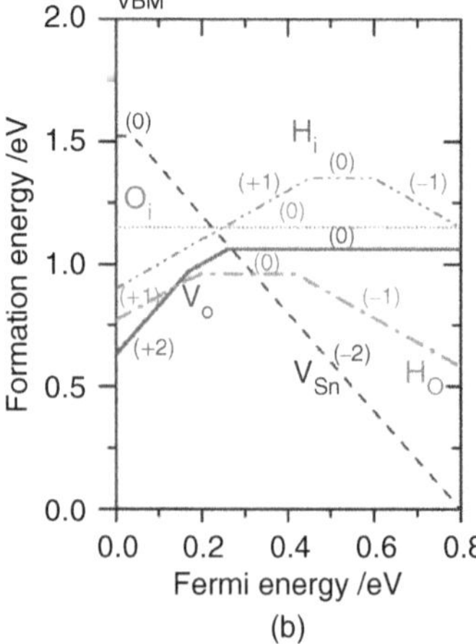

Figure 24.15 (a) The variation of transfer curves for the unannealed and TFTs annealed in several annealing atmosphere at 250 °C for 0.5 hour. (b) The formation energies of native defects as a function of the E_F under a Sn-rich limit.

well-suppressed in the oxygen-annealed device. The off-current was well suppressed to less than <1 nA, and a relatively large on/off ratio of $>10^3$ was obtained in the oxygen-annealed TFTs. The saturation mobility (μ_{sat}) and *s*-value were ~0.7 cm^2/Vs and ~15 V/decade, respectively, when the device was annealed at PO_2 = 1 mTorr. The off-current further lowers by increasing PO_2 in the annealing, but the mobility and *s*-values are also degraded due to the formation of SnO_2. Therefore, oxygen annealing is limited for SnO-TFTs. Hydrogen-containing annealing significantly improves TFT characteristics and offers the best TFT performance, with a high on/off current ratio of about $>5\times10^4$ and relatively high mobility of >1.6 cm^2/Vs. The *s*-value is ~5 V/decade, which corresponds to D_{it} ~1.2×10^{13}/cm^{2}eV^{-1} [41].

Figure 24.15b shows the formation energies of native defects as a function of the E_F under a Sn-rich limit obtained by DFT material calculation. The VBM sets to $E_F = 0$ eV. V_o exhibits the defect transition levels of ~0.2 eV for ε (0/2+) above the VBM, and forms a deep double donor state from the valence band. The energy level of the donor state is deep enough to work as an electron donor because it corresponds to E_c-0.6 eV, and it is not likely to work as an electron donor. This would mainly work as a hole trap state for *p*-channel oxide-TFTs. Hydrogen impurity is energetically stable in SnO. The formation energy of the H-V_o complex defect is low enough, suggesting that hydrogen can terminate oxygen vacancy by forming a Sn-H bond in a Sn-rich condition. Therefore, the hydrogen containing thermal annealing reduces a hole trap by terminating V_o defects. Rutherford backscattering spectrometry (RBS) and hydrogen forward scattering (HFS) analysis suggests that hydrogen can easily diffuse into the SnO film by hydrogen annealing and can produce a hydrogenated SnO channel with a high H concentration of ~10^{21} cm^{-3}.

On the other hand, poor TFT characteristics for conventional Cu_2O-TFTs originate from the backchannel donor-like defect, which is assigned to Cu_i defect. Therefore, eliminating backchannel-defect termination is critical to improving TFT characteristics. Figure 24.16a shows the improvement of TFT characteristics by a backchannel sulfur treatment using Thiourea solution [42]. The nontreated TFT showed poor device characteristics, such as a low μ_{sat} ~0.17 cm^2/Vs, a large *s*-value of >9.1 V/decade, a very large off-current of ~10^{-9} A, and a small on/off current ratio of ~10^4, which are nearly comparable with the device performances of the most-reported conventional Cu_2O-TFTs. Significant improvements with the μ_{sat} of 1.4 cm^2/Vs, the *s*-value of 2.4 V/decade, and the on/off current ratio of ~ 4.1×10^6 were observed after the sulfur backchannel-defect termination. The improvement of device characteristics was mainly attributed to the reduction of backchannel

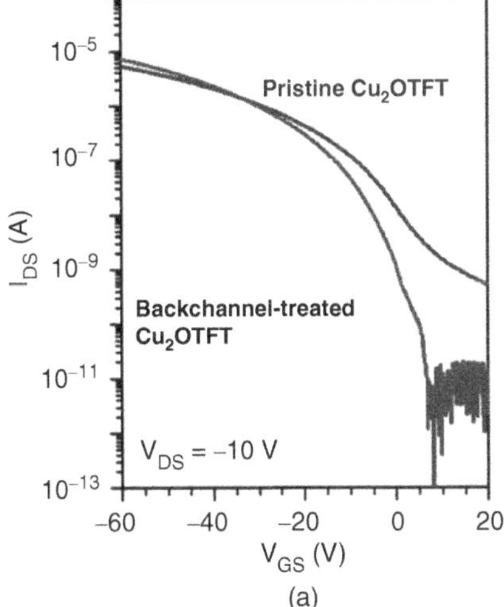

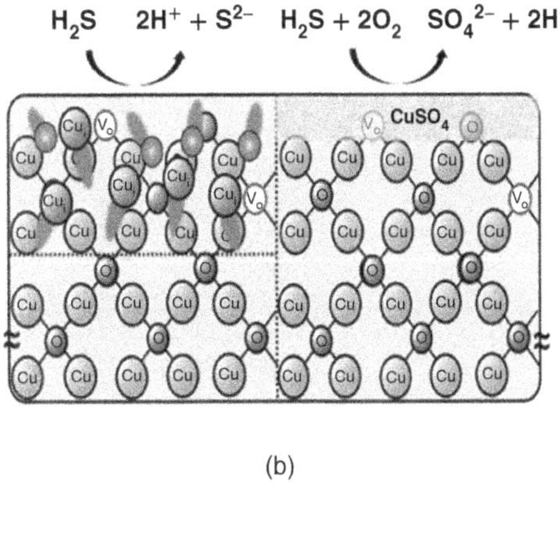

Figure 24.16 (a) Improvement of TFT characteristics by sulfur treatment for Cu_2O-TFTs. (b) Schematics of a backchannel defect termination mechanism by sulfur ions. The sulfate backchannel layer is formed with the assistance of 100 °C baking.

donor-like defects by the formation of a thin $CuSO_4$ backchannel passivation layer, which formed by chemical reaction of interstitial Cu_i defects with sulfur ions (Figure 24.16b).

24.7 All-Oxide Complementary Circuits

Developing oxide-TFT CMOS technology is critical for oxide-TFT technology to advance to the next stage, because CMOS technology offers many merits such as low-power dissipation, low waste-heat generation, high noise margins, high packing density, and simple circuit architecture for both analog and digital circuit applications. By taking the features of oxide-TFTs such as optical transparency and low-temperature processability, unique CMOS device applications for flexible, wearable, transparent, cost-effective electronics, which conventional semiconductors cannot meet, can be developed [43].

Figure 24.17 shows the channel material trend for oxide-TFT-based CMOS inverters and the progress of inverter voltage gain. *a*-IGZO-, SnO_x-, and In_2O_3-TFTs are used as *n*-channel TFTs, while *p*-channel TFTs are SnO- and Cu_2O-TFTs. The *n*-IGZO/*p*-SnO-TFTs are the current main combination for oxide-TFT CMOS devices due to the low-temperature device processability. Since the first demonstration of oxide-TFT-based CMOS inverters made up of *n*-In_2O_3/*p*-SnO_x with a gain of 11 [44], several efforts have been made to improve inverter performance accompanied by *p*-channel oxide-TFT development. A vertical geometric CMOS inverter composed of the stacked *n*-IGZO/*p*-Cu_2O-TFTs was demonstrated with a large output gain value of ~120 at V_{dd} = 20 V and high and low noise margin (NMH and NML) values of 11.68 and 6.01 V, respectively [45]. Several oxide-TFT CMOS inverters with *n*-IGZO/*p*-SnO-TFTs also have been reported, and a high-performance oxide CMOS with a large gain of ~226, NMH of 1.17 V, and NML of 1.79 V was achieved with an Al_2O_3 gate [46].

Figure 24.18b shows the typical voltage transfer characteristic (VTC) for oxide-TFT-based CMOS inverters that consist of the *p*-SnO/*n*-*a*-IGZO-TFTs. The transfer characteristics for *n*-*a*-IGZO and *p*-SnO-TFTs are also shown in Figure 24.18a. The μ_{sat} and the V_{th} were ~1.2 $cm^2V^{-1}s^{-1}$ and −3.5 V, respectively, for SnO-TFTs and 12.9 $cm^2V^{-1}s^{-1}$ and +2 V, respectively, for *a*-IGZO-TFTs. From the VTC, clear inverter action with the full swing close to the supply voltage was observed. The voltage gain is defined as dV_{OUT}/dV_{IN} of ~50 at maximum with V_{dd} = 10 V. The inverter characteristic is highly symmetric with respect to the switching threshold voltage (V_{inv}), and the V_{inv} values were ~−0.4 V at V_{dd} = 2 V, ~0.3 V at V_{dd} = 4 V, ~0.9 V at V_{dd} = 6 V, ~1.3 V at V_{dd} = 8 V, and ~1.7 V at V_{dd} = 10 V.

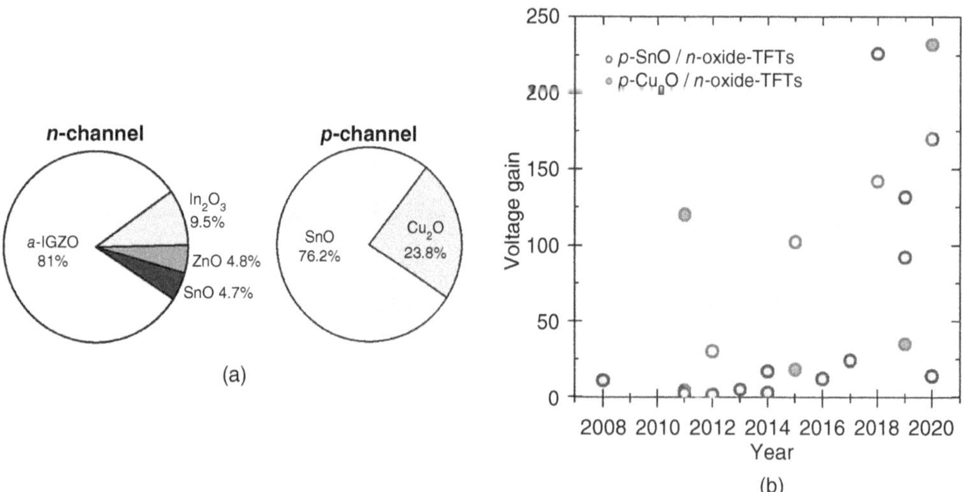

Figure 24.17 (a) The *n*/*p*-channel material for all-oxide-TFT-based CMOS inverters reported. (b) The progress of voltage gain in *p*-SnO/*n*-channel oxide-TFTs and *p*-Cu2O/*n*-channel oxide-TFTs.

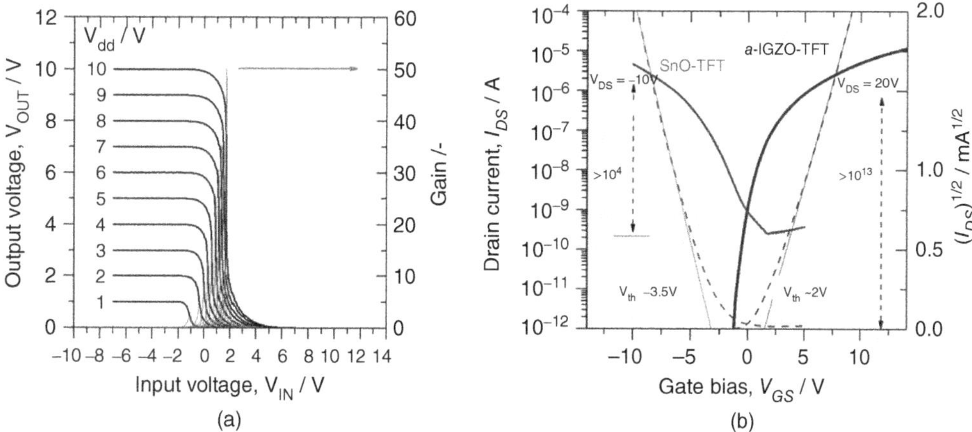

Figure 24.18 (a) A typical voltage transfer characteristic (VTC) for an oxide-TFT CMOS inverter consists of *n*-*a*-IGZO-TFTs and *p*-SnO-TFTs. (b) The transfer curves for the *n*-*a*-IGZO-TFTs and *p*-SnO-TFTs in the CMOS circuit.

The observed V_{inv} is largely deviated from the ideal V_{inv}, which provides ~3.8 V at V_{dd} = 2 V, ~4.9 V at V_{dd} = 4 V, ~5.9 V at V_{dd} = 6 V, ~6.9 V at V_{dd} = 8 V, and ~7.9 V at V_{dd} = 10 V. This would be related to the V_{th} shift of *p*-channel SnO-TFTs. The inverter performance is controlled by the *p*-channel TFT performance, and improving the *p*-channel device performance is vital to developing high-performance oxide-TFT-based CMOS inverters.

For SnO-TFTs, two types of operation modes, p-channel and ambipolar modes, are identified [47]. Ambipolar SnO-TFT-based CMOS-like inverters, in which unique inverter actions appear at both the first and third quadrants, have been demonstrated. The voltage gain of the first ambipolar SnO-TFT-based CMOS-like inverters was only 2.7 due to the unbalanced device performance for *n*- and *p*-channel modes [48], specifically, the low mobility of the *n*-channel mode. Significant improvement of gain by achieving a balanced performance of *n*- and *p*-modes was made by passivation, and a remarkably high gain of 100 was attained [49].

24.8 Conclusions

A review of the current status of *p*-channel oxide-TFT technology was presented. The reasons for the difficulties in *p*-type oxide development, comprehensive comparisons of materials and devices for *p*-type oxides, Cu_2O and SnO, and all-oxide-TFT-based complementary inverter circuit development have been discussed. The absence of high-performance *p*-channel oxide-TFTs and the CMOS circuit obviously hinders the high potential of oxide-TFTs for a broad field in next-generation electronic device applications and should be addressed. Developing *p*-type amorphous oxides with comparable device performance to *n*-channel TFTs such as *a*-IGZO-TFTs is extraordinarily difficult at the present time. Thus, the development of *p*-channel oxide-TFTs is progressing with crystalline Cu_2O and SnO channels, and several significant innovations to improve the device performance have been made, but these device' performance is still largely behind that of *n*-channel oxide-TFTs and not satisfactory yet. In particular, the presence of high-density subgap channel and backchannel defects should be eliminated to improve the device characteristics, and the developing effective defect termination technique is strongly demanded. In addition, a better understanding of material properties, particularly carrier transport and electronic/defect structure, for *p*-type oxide semiconductors is required to furthermore improve the device characteristics and the channel material quality. A significant technological breakthrough is strongly needed at the stage of material and device process developments in this field, in order to not end as just a dream that flexible and cost-effective high-level integrated circuits can be made by oxide-TFTs.

References

1 Nomura, K., Ohta, H., Takagi, A. et al. (2004). Room-temperature fabrication of transparent flexible thin-film transistors using amorphous oxide semiconductors. *Nature* 432 (7016): 488–492.

2 Ginley, D.S., Hosono, H., and Paine, D. (ed.) (2011). *Handbook of Transparent Conductors*. New York: Springer.

3 Kamiya, T. and Hosono, H. (2016). *Handbook of Visual Display Technology*, 2nde. Berlin: Springer Verlag.

4 Souk, J., Morozumi, S., Luo, F.-C., and Bita, I. (2018). *Flat-Panel Display Manufacturing*. Hoboken, NJ: Wiley.

5 Hung, M.H., Chen, C.H., Lai, Y.C. et al. (2017). Ultra-low voltage 1-V RFID tag implement in a-IGZO TFT technology on plastic. In: *Proceedings of IEEE International Conference on RFID (RFID)*. Phoenix, AZ, May 9–11, 193–197.

6 Wu, S.H., Jia, X., Kui, M. et al. (2016). Extremely low power C-axis aligned crystalline In-Ga-Zn-O 60 nm transistor integrated with industry 65 nm Si MOSFET for IoT normally-off CPU application. In: *Proceedings of IEEE Symposium on VLSI Technology*. Honolulu, HI, June 13–17, 1–2.

7 Wu, S.H., Jia, X.Y., Li, X. et al. (2017). Performance boost of crystalline In-Ga-Zn-O material and transistor with extremely low leakage for IoT normally-off CPU application. In: *Proceedings of Symposium on VLSI Technology*. Kyoto, Japan: June 5–8, T166–T167.

8 Shang, Z.-W., Hsu, H.-H., Zheng, Z.-W. et al. (2019). Progress and challenges in p-type oxide-based thin film transistors. *Nanotechnology Reviews* 8 (1): 422–443.

9 Wang, Z., Nayak, P.K., Caraveo-Frescas, J.A. et al. (2016). Recent developments in p-type oxide semiconductor materials and devices. *Advanced Materials* 28 (20): 3831–3892.

10 Hosono, H., Kikuchi, N., Ueda, N. et al. (1996). Working hypothesis to explore novel wide band gap electrically conducting amorphous oxides and examples. *Journal of Non-Crystalline Solids* 198–200 (1): 165–169.

11 Nomura, K., Kamiya, T., Ikenaga, E. et al. (2011). Depth analysis of subgap electronic states in amorphous oxide semiconductor, a-In-Ga-Zn-O, studied by hard X-ray photoelectron spectroscopy. *Journal of Applied Physics* 109 (7): 073726–073728.

12 Van de Walle, C.G. and Neugebauer, J. (2003). Universal alignment of hydrogen levels in semiconductors insulators and solutions. *Nature* 423 (6940): 626–628.

13 Nomura, K., Kamiya, T., and Hosono, H. (2013). Effects of diffusion of hydrogen and oxygen on electrical properties of amorphous oxide semiconductor, In-Ga-Zn-O. *ECS Journal of Solid-State Science and Technology* 2 (1): 5–8.

14 Funabiki, F., Kamiya, T., and Hosono, H. (2012). Doping effects in amorphous oxides. *Journal of the Ceramic Society of Japan* 120 (1407): 447–457.

15 Kawazoe, H., Yasukawa, M., Hyodo, H. et al. (1997). P-type electrical conduction in transparent thin films of $CuAlO_2$. *Nature* 389 (6654): 939–942.

16 Amini, M.N., Dixit, H., Saniz, R. et al. (2014). The origin of p-type conductivity in ZnM_2O_4 (M = Co, Rh, Ir) spinels. *Physical Chemistry Chemical Physics* 16 (6): 2588–2596.

17 Samizo, A., Kikuchi, N., Aiura, Y. et al. (2018). Carrier generation in p-type wide-gap oxide: $SnNb_2O_6$ foordite. *Chemistry of Materials* 30 (22): 8221–8225.

18 Zhang, K.H.L., Xi, K., Blamire, M.G. et al. (2016). P-type transparent conducting oxides. *Journal of Physics: Condensed Matter* 28 (38): 383002.

19 Lee, S.Y. (2020). Comprehensive review on amorphous oxide semiconductor thin film transistor. *Transactions on Electrical and Electronic Materials* 21 (3): 235–248.

20 Arai, T. and Sasaoka, T. (2011). Emergent oxide TFT technologies for next-generation AM-OLED displays. *SID Symposium Digest of Technical Papers* 42: 710–713.

21 Sheng, J., Hong, T.H., Lee, H.-M. et al. (2019). Amorphous IGZO TFT with high mobility of ~70 cm^2/Vs via vertical dimension control using PEALD. *ACS Applied Materials & Interfaces* 11 (43): 40300–40309.

22 Yang, J.-H., Choi, J.H., Cho, S.H. et al. (2018). Highly stable AlInZnSnO and InZnO double-layer oxide thin-film transistors with mobility over 50 $cm^2/V{\cdot}s$ for high-speed operation. *IEEE Electron Device Letters* 39 (4): 508–511.

23 Ogo, Y., Hiramatsu, H., Nomura, K. et al. (2008). p-channel thin-film transistor using p-type oxide semiconductor. *SnO. Applied Physics Letters* 93: –3. 032113-3.

24 Matsuzaki, K., Nomura, K., Yanagi, H. et al. (2008). Epitaxial growth of high mobility Cu_2O thin films and application to p-channel thin film transistor. *Applied Physics Letters* 93 (20): 202107–202103.

25 Shimotani, H., Suzuki, H., Ueno, K. et al. (2008). p-type field-effect transistor of NiO with electric double-layer gating. *Applied Physics Letters* 92 (24): 242107–242103.

26 Zou, X., Fang, G., Wan, J. et al. (2011). Improved subthreshold swing and gate-bias stressing stability of p-type Cu_2O thin-film transistors using a HfO_2 high-k gate dielectric grown on a SiO_2/Si substrate by pulsed laser ablation. *IEEE Transactions on Electron Devices* 58 (7): 2003–2007.

27 Maeng, W., Lee, S.-H., Kwon, J.-D. et al. (2016). Atomic layer deposited p-type copper oxide thin films and the associated thin film transistor properties. *Ceramics International* 42 (4): 5517–5522.

28 Shih, C.W., Chin, A., Lu, C.F. et al. (2018). Remarkably high hole mobility metal-oxide thin-film transistors. *Scientific Reports* 8 (1): 889.

29 Zhong, C., Lin, H., Liu, K. et al. (2015). Improving electrical performances of p-type SnO thin-film transistors using double-gated structure. *IEEE Electron Device Letters* 36 (10): 1053–1055.

30 Park, J.W., Kang, B.H., and Kim, H.J. (2020). Review of low-temperature solution-processed metal oxide thin-film transistors for flexible electronics. *Advanced Functional Materials* 30 (20): 1904632–1904640.

31 Liu, A., Nie, S., Liu, G. et al. (2017). In situ one-step synthesis of p-type copper oxide for low-temperature, solution-processed thin-film transistors. *Journal of Materials Chemistry C* 5 (10): 2524–2530.

32 Okamura, K., Nasr, B., Brand, R.A. et al. (2012). Solution-processed oxide semiconductor SnO in p-channel thin-film transistors. *Journal of Materials Chemistry* 22 (11): 4607–4610.

33 Ruiz, E., Alvarez, S., Alemany, P. et al. (1997). Electronic structure and properties of Cu_2O. *Physical Review B* 56 (12): 7189–7196.

34 Watson, G.W. (2001). The origin of the electron distribution in SnO. *The Journal of Chemical Physics* 114 (2): 758–763.

35 Li, B.S., Akimoto, K., and Shen, A. (2009). Growth of Cu_2O thin films with high hole mobility by introducing a low-temperature buffer layer. *Journal of Crystal Growth* 311 (4): 1102–1105.

36 Minohara, M., Kikuchi, N., Yoshida, Y. et al. (2019). Improvement of the hole mobility of SnO epitaxial films grown by pulsed laser deposition. *Journal of Materials Chemistry C* 7 (21): 6332–6336.

37 Hu, Y., Hwang, J., Le, Y. et al. (2019). First principles calculations of intrinsic mobilities in tin-based oxide semiconductors. SnO, SnO_2, and Ta_2SnO_6. *Journal of Applied Physics* 126 (18): 185701.

38 Caraveo-Frescas, J.A., Nayak, P.K., Al-Jawhari, H.A. et al. (2013). Record mobility in transparent p-type tin monoxide films and devices by phase engineering. *ACS Nano* 7 (6): 5160–5167.

39 Togo, A., Oba, F., and Tanaka, I. (2006). First-principles calculations of native defects in tin monoxide. *Physical Review B* 74 (19): 195128–195128.

40 Raebiger, H., Lany, S., and Zunger, A. (2007). Origins of the p-type nature and cation deficiency in Cu_2O and related materials. *Physical Review B* 76 (4): 045209.

41 Lee, A.W., Le, D., Matsuzaki, K. et al. (2020). Hydrogen-defect termination in SnO for *p*-channel TFTs. *ACS Applied Electronic Materials* 2 (4): 1162–1168.

42 Chang, H., Huang, C.-H., Matsuzaki, K. et al. (2020). Back-channel defect termination by sulfur for p-channel Cu_2O thin-film transistors. *ACS Applied Materials and Interfaces* 12 (46): 51581–51588.

43 Nomura, K. (2021). Recent progress of oxide-TFT-based inverter technology. *Journal of Information Display* 22 (4): 211–229.

44 Dhananjay, C.,.C.-W., Ou, C.-W., Wu, M.-C. et al. (2008). Complementary inverter circuits based on *p*-SnO_2 and n-In_2O_3 thin film transistors. *Applied Physics Letters* 92 (23): 232103–232103.

45 Dindar, A., Kim, J.B., Fuentes-Hernandez, C., and Kippelen, B. (2011). Metal-oxide complementary inverters with a vertical geometry fabricated on flexible substrates. *Applied Physics Letters* 99 (17): 172104–172103.

46 Yuan, Y., Yang, J., Hu, Z. et al. (2018). Oxide-based complementary inverters with high gain and nanowatt power consumption. *IEEE Electron Device Letters* 39 (11): 1676–1679.

47 Lee, A.W., Zhang, Y., Huang, C.-H. et al. (2020). Switching mechanism behind the device operation mode in SnO-TFT. *Advanced Electronic Materials* 6 (12): 200742–200747.

48 Nomura, K., Kamiya, T., and Hosono, H. (2011). Ambipolar oxide thin-film transistor. *Advanced Materials* 23 (30): 3431–3434.

49 Luo, H., Liang, L., Cao, H. et al. (2015). Control of ambipolar transport in SnO thin-film transistors by back-channel surface passivation for high performance complementary-like inverters. *ACS Applied Materials & Interfaces* 7 (31): 17023–17031.

25

Solution-Synthesized Metal Oxides and Halides for Transparent *p*-Channel TFTs

Ao Liu, Huihui Zhu, and Yong-Young Noh

Department of Chemical Engineering, Pohang University of Science and Technology, Pohang, Republic of Korea

25.1 Introduction

Displays nowadays play an indispensable role in our daily life as one of the most important visual human–computer interfaces. Over the past century, display technology appeared in forms from cathode ray tubes (CRTs) to plasma display panels (PDPs) to recent active-matrix (AM) liquid crystal displays (LCDs) and organic light-emitting diodes (OLEDs). The up-to-date flat-panel display backplane options are focused on hydrogenated amorphous silicon (*a*-Si:H), low-temperature polysilicon (LTPS), and metal-oxide thin-film transistor (TFT) technologies [1]. Since the commercialization of AMLCDs in around 1990, *a*-Si:H has dominated the backplane TFT market because of its low cost, process simplicity, and large-area meter-sized scaling with great uniformity. Unfortunately, its low mobility (μ_{FE}) and poor stability make *a*-Si:H inadequate for high-resolution AMLCDs and AMOLEDs with a superior viewing experience. In view of the intrinsic limitation of *a*-Si:H, LTPS is regarded as the replacement, because it possesses a high μ_{FE} of >50 $cm^2\ V^{-1}\ s^{-1}$ and stable electrical performance. The high channel-layer mobility is attractive for display application because the TFT size can be downscaled while keeping sufficient output driving current, and the TFT response time can be faster, enabling increased display refresh rates with lower power consumption. In addition, the high on-state current of LTPS allows peripheral circuit integration around the display edge for row and column driver function and overcomes the need for external silicon integrated circuits. However, LTPS scaling to large areas is difficult and expensive, which is caused by the polycrystalline microstructure and the requirements of ion implantation and excimer-laser recrystallization. Thus, the main products for LTPS are limited to small- or medium-area high-end displays, such as smartphones.

Besides μ_{FE}, the off-state drain current (I_{off}) is pivotal when evaluating a TFT's suitability for a display application. A low leakage current is desirable because low power is dissipated in the off-state and the TFT switch/drive can retain an internal pixel charge for a long time. Unfortunately, *a*-Si:H and LTPS have narrow band gaps (E_g)—1.7 eV for *a*-Si:H and 1.1 eV for LTPS—and their bipolar characters make the channel inversion at a sufficiently large reverse gate bias. This means the I_{off} in Si-based TFTs cannot keep at a very low level. In this case, the strengths of metal oxides (one typical example is *a*-InGaZnO) come out, which possess much lower I_{off} compared with Si-based semiconductors. This is beneficial due to not only the large E_g of over 3 eV but also the unipolar charge carrier transport property. Like *a*-Si:H, its amorphous microstructure allows low-cost deposition over a large area. In addition, metal oxide and *a*-Si:H manufacturing procedures are similar, so that upgrading an existing production line from *a*-Si:H to oxides is a feasible and sometimes promising option. Nowadays, the combination of LTPS and oxide, named LTPO, attracts great interest for pixel switching and peripheral circuit integration.

Amorphous Oxide Semiconductors: IGZO and Related Materials for Display and Memory, First Edition.
Edited by Hideo Hosono and Hideya Kumomi.

It is worth noting that all the metal-oxide semiconductors with high electrical performance are *n*-type (electron transporting), and very few *p*-type materials (hole transporting), such as Cu_xO, SnO, and NiO_x, were reported [2, 3]. The main reasons are (i) the naturally low formation energy of oxygen vacancies, which act as electron donors; and (ii) a highly localized valence band maximum (VBM) dominated by oxygen 2*p* orbitals, resulting in the large hole effective mass and thus low mobility. To explore the possible transparent *p*-type candidates with high electrical performance, several new compounds were designed and proposed, such as delafossite $CuMO_2$ (M = Ga, Gr, Y, etc.), spinel ZnM_2O_4 (M = Co, Rh, and Ir), and oxychalcogenides LnCuOCh (Ln: lanthanide). Unfortunately, their suitability as channel layers for TFT applications remains a concern. Attention was then paid toward non-oxide materials, and typical examples are bimetallic chalcogenide $CuInCh_2$ (Ch = Se and Te), (pseudo)halide CuSCN, and CuI, delivering decent *p*-channel transistor behaviors. The development roadmap is shown in Figure 25.1, and a detailed device description can be found in Ref. [4]. Despite the difficulties in the development of high-performance transparent *p*-channel TFTs, they are indispensable for the integration of high-speed and low-power-consumption complementary metal-oxide semiconductor (CMOS) circuits, which are the fundamental units in electronic devices. In addition, *p*-channel TFTs are superior to their *n*-channel counterparts for driving AMOLED displays because of the bottom hole-injection electrode structure [5]. The current input terminal of an OLED (an anode) is generally deposited before the cathode and is connected to the drain port of a *p*-channel driving TFTs or the source port of an *n*-channel driving TFTs. The latter connection, however, can result in nonuniform luminance or image sticking because the voltage change across the OLED leads to the gate bias variation on TFTs.

Nowadays, physical-vapor film growth techniques, such as sputtering and chemical vapor deposition, are widely used in the display industry. Nevertheless, owing to the strict and complex deposition conditions and thus the narrow processing window, the fabricated *p*-channel TFTs generally exhibited large lab-to-lab variations in device performance. By contrast, the solution-based film deposition approach has advantages in terms of simplicity and cost-effectiveness, and it is regarded as a promising alternative route for device manufacturing [6]. In our recent review paper, we summarized the progress of solution-processed inorganic *p*-channel TFTs before 2019 [4]. In this book chapter, we aim to update the newest achievements, especially the most recent works reported by our group. The first two parts focus on the conventional *p*-type metal oxides, emerging metal halides, and a novel doping approach for realizing high-performance transparent TFTs. In the end, present challenges and future opportunities of inorganic *p*-type semiconductors and related devices are discussed.

25.2 Solution-Processed *p*-Channel Metal-Oxide TFTs

The milestone of *p*-channel metal-oxide TFTs was built between 2008 and 2010 by H. Hosono, R. Martins, and their colleagues through reporting a series of Cu_2O and SnO TFTs using vacuum vapor deposition techniques [10–13]. Compared with Cu-based oxides, SnO TFTs exhibited a higher μ_{FE} of >1 $cm^2\ V^{-1}\ s^{-1}$ because of the more dispersed VBM with the hybridization of spherical Sn 5*s* and O 2*p* orbitals. However, SnO is metastable in air, and Sn^{2+} can be easily oxidized into Sn^{4+} with the formation of *n*-type SnO_2. To date, only a single study reported the *p*-channel SnO TFT using a solution process in stringent N_2/H_2 forming gas atmosphere [14]. Table 25.1 lists detailed parameters of recent achievements on solution-processed *p*-channel metal-oxide and (pseudo)halide TFTs. Nearly half of the studies focused on Cu-based oxide semiconductors, while the others included SnO, NiO, and Cu-based (pseudo)halides. As for the film deposition methods, spin-coating was preferentially used in laboratories due to its easy operation and low cost. For Cu-based *p*-type oxides, the holes mainly originate from negatively charged Cu vacancy (V_{Cu}) and possibly interstitial oxygen as acceptor. Despite the high Hall mobility (~90 $cm^2\ V^{-1}\ s^{-1}$) in the high-quality Cu_2O thin film, the fabricated TFT always delivered poor electrical performance with a low μ_{FE} of <1 $cm^2\ V^{-1}\ s^{-1}$. The reason might be related to the high defect density (>10^{18} cm^{-3}), arising mainly from oxygen

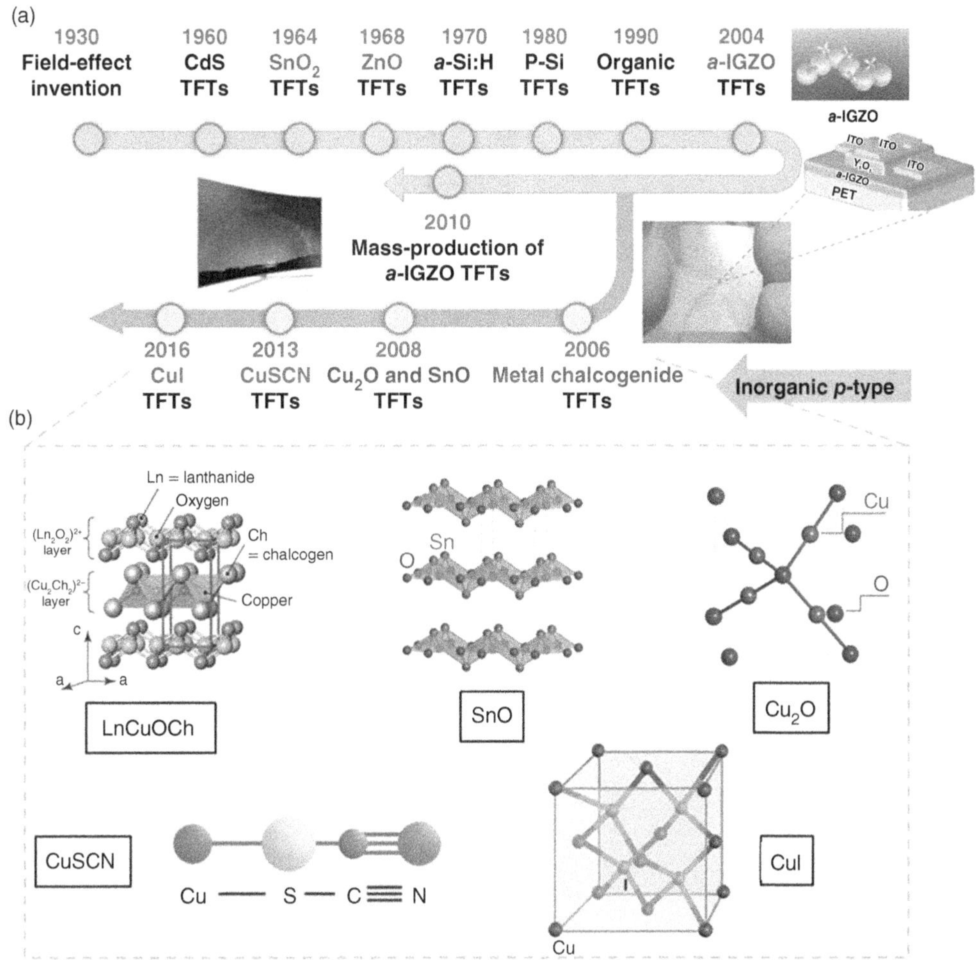

Figure 25.1 (a) Main landmarks achieved with TFTs. *Source*: Reproduced from Ref. [4], Liu, A., et al. (2019). Solution-processed inorganic p-channel transistors: recent advances and perspectives, *Materials Science & Engineering R: Reports* **135**: Figure 01, p. 86, with permission of ELSEVIER. https://doi.org/10.1016/j.mser.2018.11.001. (b) Crystal structures of *p*-type oxychalcogenides LnCuOCh, Cu2O, SnO, CuSCN, and CuI semiconductors. *Source*: LnCuOCh structure was reproduced with permission from Ref. [7], Hidenori, H., et al. (2004). Fabrication of heteroepitaxial thin films of layered oxychalcogenides LnCuOCh (Ln = La-Nd; Ch = S-Te) by reactive solid-phase epitaxy. *Journal of Materials Research* **19**: 2137. Copyright 2004, Materials Research Society. SnO and Cu2O were reproduced with permission from Ref. [3], Wang, Z., et al. (2016). Recent developments in p-type oxide semiconductor materials and devices. *Advanced Materials* **28**: 3831. Copyright 2016 Wiley-VCH. CuI was reproduced with permission from Ref. [8], Yan, W., et al. (2016). Hole-transporting materials in inverted planar perovskite solar cells. *Advanced Energy Materials* **6**: 1600474. Copyright 2016 Wiley-VCH. And CuSCN was reproduced with permission from Ref. [9], Pattanasattayavong, P., et al. (2015). Study of the hole transport processes in solution-processed layers of the wide bandgap semiconductor copper(I) thiocyanate (CuSCN). *Advanced Functional Materials* **25**: 6802. Copyright 2015 Wiley-VCH.

vacancies, and the CuO phase formation on the top surface. Due to the frequent hole interaction with magnons and phonons, CuO owns a low Hall mobility and is not a desired component for TFT application [4].

In Table 25.1, all the TFT parameters were collected on robust SiO_2 dielectric. Although the TFT performance with a high μ_{FE} of 5 to 10 $cm^2\ V^{-1}\ s^{-1}$ was reported based on solution-processed high-permittivity metal-oxide dielectrics, the μ_{FE} might be overestimated due to the existence of electric double-layer and pseudocapacitance phenomena. This was caused by the nonstoichiometric feature and certain amounts of impurity residues

Table 25.1 Recent achievements of solution processed *p*-channel metal-oxide and (pseudo)halide TFTs on SiO_2 gate dielectric.

Method	Channel (thickness, nm)	Annealing Temperature (°C)	μ_{FE} ($cm^2 V^{-1} s^{-1}$)	I_{on}/I_{off}	Inverter Gain (V/V)	Year	Reference
Spin coating	SnO (14.7)	450	0.13	85	×	2012	[14]
Spin coating	Cu_2O (100)	700	0.16	~10^2	×	2013	[15]
Spray pyrolysis	Cu_2O (40)	275	10^{-4}–10^{-2}	4×10^3	×	2013	[16]
Spin coating	CuSCN (20)	80	10^{-3}–10^{-2}	~10^4	2	2013	[17]
Spin coating	NiO (40)	500	0.14	—	×	2014	[18]
Spin coating	Cu_2O (17.5)	600	0.29	~10^4	×	2015	[19]
Spin coating	Cu_xO (−−)	300	0.26	~10^5	×	2015	[20]
Spin coating	CuO (−−)	500	0.01	~10^3	×	2016	[21]
Spin coating	NiO (−−)	250	0.07	~10^4	×	2016	[22]
Inkjet printing	CuI (100)	60	—	10~10^2	×	2016	[23]
Spin coating	CuO (12.2)	250	0.30	~10^4	×	2017	[24]
Spin coating	NiO (30)	350	0.01	2	×	2017	[25]
Spin coating	CuO (−−)	500	4×10^{-4}	~10^2	×	2018	[26]
Inkjet printing	NiO (15)	280	0.01	4×10^2	×	2018	[27]
Spin coating	CuSeCN (15)	140	0.002	~10^3	×	2018	[28]
Spin coating	$CuAlO_2$ (40)	1000	~0.1	~10^3	×	2018	[29]
Spin coating	$CuGrO_2$ (21)	800	0.59	~10^5	×	2018	[30]
Spin coating	CuO_x (20)	500	~10^{-3}	~10^4	×	2018	[31]
Spin coating	CuI (5)	RT	0.40	5×10^2	4	2018	[32]
Spin coating	CuO (−−)	600	~10^{-3}	~10^3	×	2019	[33]
Spin coating	$CuAlO_2$ (20)	900	0.33	~10^5	×	2019	[34]
Spin coating	CuO (7)	220	0.15	~10^4	37	2019	[35]
Inkjet printing	NiO (20)	175 (laser)	0.9	~10^5	10	2019	[36]
Spin coating	NiO (53)	250	0.48	~10^3	×	2019	[37]
Spin coating	$NdAlO_3$ ( )	700	0.19	10^5	×	2020	[38]
Spin coating	CuO (40)	550	10^{-4}	10^2	×	2020	[39]
Spin coating	CuI:$Zn_{5mol\%}$ (9)	80	5.3	7×10^6	56	2020	[40]

in the dielectric layers, especially those annealed at low temperatures. Also note that the frequencies used for the dielectric capacitance extraction were generally 20 Hz or even higher. This undoubtedly underestimated the capacitance value because the transistor measurement was carried out in the quasi-static status. Even the capacitance measurement at a very low frequency (0.1 Hz) is possible; the exact extraction of capacitance remains a concern. To investigate the effect of gate dielectric constant/capacitance on the TFT performance, the reliable insulating layers (e.g., ALD HfO_2 and Al_2O_3) should be used. In addition, the contact resistance between source/drain electrodes and oxide channel layers is neglected in many reports. If the Schottky contact or obvious current suppression was observed at low drain voltages of output curves, the μ_{FE} and other device parameters could be misestimated. Another serious problem is the absence of widely applicable and reproducible solution-based methods for *p*-channel oxide TFT fabrication. The experimental details in many reports are harsh, such as a long annealing time, high processing temperature, or specific annealing atmosphere. This will cause poor reproducibility and impede further optimization.

The commonly used commercial Cu salts for preparing precursor solutions are copper(II) nitrate hydrate [$Cu(NO_3)_2 \cdot xH_2O$] and copper(II) acetate hydrate [$Cu(COOCH_3)_2 \cdot xH_2O$]. Therefore, an additional reduction process is required to obtain Cu^+ from Cu^{2+}. The effective approaches include high-temperature annealing in inert/reduction atmosphere and the addition of reduction agents. The accurate control of phase reduction might be difficult because metallic Cu might be formed during the reduction process. By contrast, the gradual oxidization of thin metallic Cu film under controllable annealing temperature and time should be an easier approach with high reproducibility. The first attempt was carried out in 2013 by Kim et al. [15] using a two-step annealing process for Cu_2O TFTs. The metallic Cu-dominated phase was obtained after 400 °C annealing of Cu(II) acetate hydrate gel film in N_2 atmosphere. Subsequent annealing at 700 °C under different oxygen partial pressures oxidized Cu into Cu_xO. The optimized TFT annealed at 0.04 Torr oxygen partial pressure displayed a *p*-channel depletion mode with a μ_{FE} of 0.16 cm^2 V^{-1} s^{-1} and an on/off current ratio (I_{on}/I_{off}) of 10^2.

Two years later, an interesting polyol reduction method was adopted by Garlapati et al. to deposit the Cu layer from its nitrate precursor using glycerol at a low temperature of 150 °C in air [41]. The subsequent ambient thermal annealing at 400 °C achieved Cu_xO with composited phases of CuO (90%) and Cu_2O (10%). By employing a high-capacitance polymer electrolyte dielectric, the Cu_xO TFT was operated at a low voltage of 1.5 V and exhibited a μ_{FE} of 0.22 cm^2 V^{-1} s^{-1} and an I_{on}/I_{off} of 10^3. Following this strategy, recently, we conducted more detailed studies and aimed to deliver systematic understanding of solution-processed *p*-channel Cu_xO TFTs (Figure 25.2) [35]. Firstly, the annealing temperature, channel thickness, and solvent dependences on the TFT performance were investigated, and the results showed that 260 °C annealing, 5 nm channel thickness, and H_2O solvent achieved decent device performance and stable operation. We then studied the alternative polyols with lower boiling points and viscosities, which were expected to achieve smoother thin films at lower processing temperatures. The low annealing temperature is preferred during the Cu oxidization because of less of a phase transition from Cu_2O to CuO. The propylene glycol–derived TFTs exhibited the optimization electrical performance, including a μ_{FE} of 0.15 cm^2 V^{-1} s^{-1}, an I_{on}/I_{off} of ~10^4, and a threshold voltage (V_{TH}) of −7 V at 220 °C (Figure 25.2c). The low temperature analysis revealed the thermal activation dominated hole transportation with an activation energy of 0.16 eV (Figure 25.2d). The TFT performance under storage in dry (relative humidity, RH <10%) and humid (RH >60%) atmospheres was also investigated (Figure 25.2e). The moisture was found detrimental to the electrical performance, which could be attributed to the strong interaction of holes with polar H_2O at grain boundaries and the increased hole transport energy barrier. By contrast, the dry O_2 showed a positive impact on the device's performance improvement. This is contrary to previous observations on the sputtering-deposited *p*-channel oxide TFTs, for which O_2 was demonstrated as having negligible effect on the device behavior [42]. Using vacuum-vapor growth routes, the metal-oxide framework was formed completely after the finalization of film fabrication. However, for the as-deposited solution-processed Cu_xO thin films, the components were not stabilized after short-term thermal annealing, and many dissociated copper ions and oxygen vacancies existed. The gradually absorbed oxygen in dry air could react with these species, filling vacancy defects with enhanced Cu-O bond formation and facilitating hole transport (Figure 25.2f). The explanation was confirmed by X-ray photoelectron spectroscopy (XPS) analysis. Finally, a CMOS inverter was assembled with Cu_xO and *n*-channel InGaZnO TFTs, exhibiting a high gain of 37 at a supply voltage of 60 V (Figure 25.2g and 25.2h).

Considering the atmospheric solution process during the film coating/annealing, the environmental humidity may have a significant impact on the film component and device performance. To verify this, we fabricated Cu_xO TFTs under different environmental humidity (20–50%), as shown in Figure 25.3 [43]. The results showed that a dry air (RH ≤30%) condition was necessary to achieve good TFT performance, while those fabricated under humidity of >50% were inactive. The reason can be attributed to less Cu-OH in the "dry" Cu_xO because hydroxide groups act as trap defects and deteriorate the charge transport. The long-term negative bias stress (NBS) ($V_{GS} = V_{DS} = -30$ V) measurement indicated the highly stable operating character of the optimized Cu_xO TFTs,

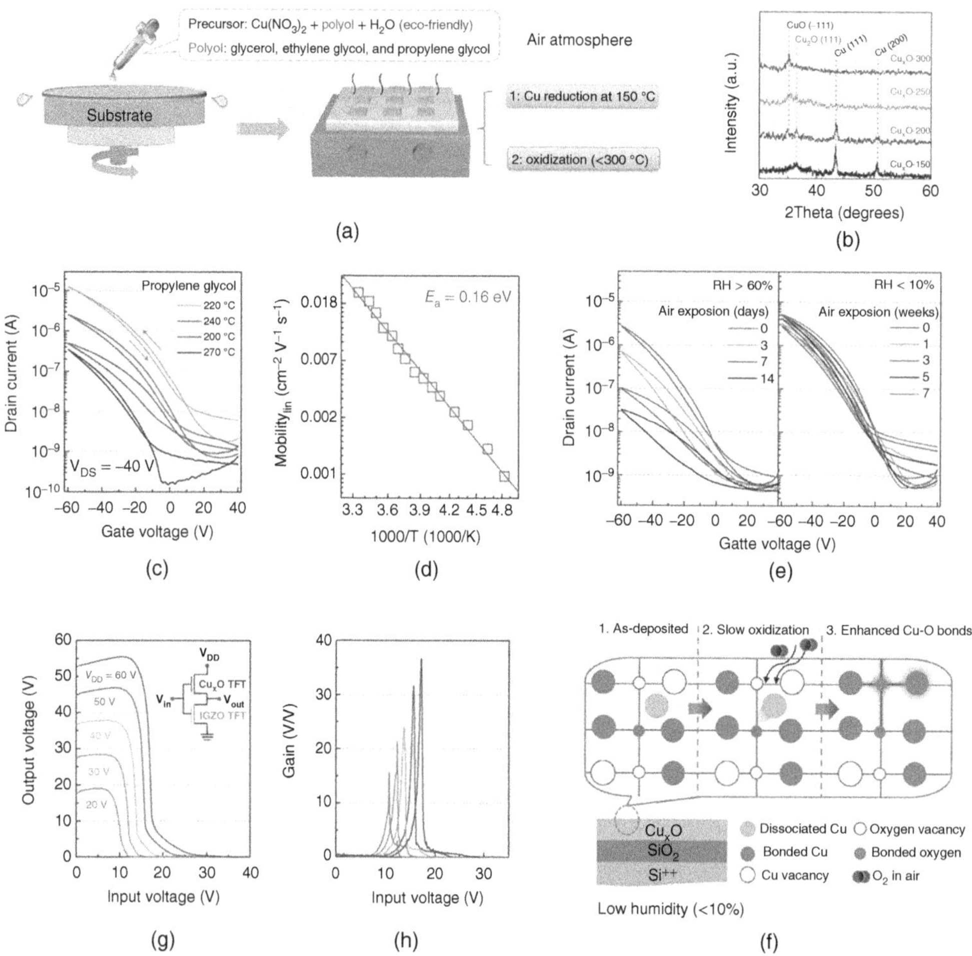

Figure 25.2 (a) Schematic illustration of a polyol-reduction process for the deposition of Cu_xO thin films. (b) X-ray diffraction patterns of the gel films annealed at indicated temperatures. (c) Transfer characteristics of the Cu_xO TFTs fabricated using propylene glycol as a function of annealing temperature. (d) Mobility variation for the propylene glycol–derived Cu_xO-220 TFTs measured at different temperatures. (e) Air stability of the Cu_xO TFTs stored in humid and dry environments. (f) Schematic of Cu-O bond formation stored in dry air. (g,h) Voltage transfer curves and gain voltage of the CMOS inverter based on *n*-channel IGZO and *p*-channel PG Cu_xO-220 TFTs. *Source*: (a–h) Reproduced with permission from Ref. [35], Liu, A., Zhu, H., Noh, Y.-Y. (2019). Polyol teduction: a low-temperature eco-friendly solution process for p-channel copper oxide-based transistors and inverter circuits, *ACS Applied Materials & Interfaces* **11**: 33157. Copyright 2019 American Chemical Society.

and a small ΔV_{TH} of 2 V was observed after a 5.5 h NBS test (Figure 25.3c). The stability is far beyond the current reports on *p*-channel oxide TFTs and even comparable to that of the solution-processed *n*-channel oxide counterparts.

The above achievements demonstrated the feasibility of fabricating reliable Cu_xO TFTs with high operational and environmental stabilities. However, there is still the absence of an effective and universal strategy to improve the electrical performance. Conventional substitution doping greatly speeds up the development of the semiconductor industry; however, this process is difficult to employ on *p*-type oxides. Firstly, the substitutional atoms are restricted to monovalent cations (e.g., alkaline metals) and trivalent nitrogen anion (N^{3-}), but the air-sensitive alkaline dopants and weak Cu-N bonds make this doping approach difficult and unstable. In addition, the

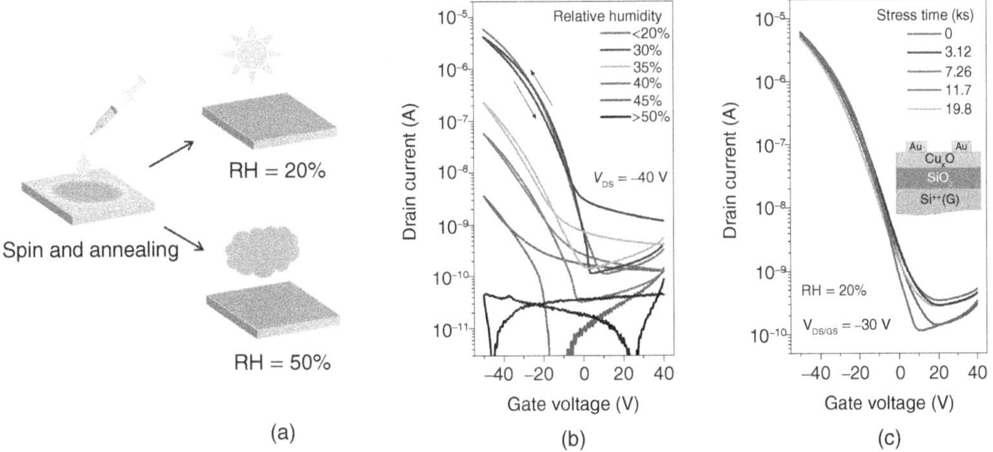

Figure 25.3 (a,b) Schematic process of the film coating and annealing and corresponding transfer curves of Cu_xO/SiO_2 TFTs under various humidity conditions. (c) Transfer curves of Cu_xO TFTs deposited at 20% RH under a long-term (5.5 h) NBS test. *Source*: (a–c) Reproduced with permission from Ref. [43], Zhu, H., et al. (2020). Impact of humidity on the performance and stability of solution-processed copper oxide transistors. *IEEE Electron Device Letters* **41**: 1033.

substitutional doping can inevitably perturb the host's oxide lattice thermodynamic equilibrium, resulting in the oxygen vacancy generation and, thus, counteracting the *p*-doping effect. Another problem is the possible dopant segregation and lattice distortion, which can reduce the doping efficiency and adversely degrade the charge transport.

Most recently, our group proposed an efficient molecule charge transfer doping (MCTD) method to enhance the electrical performance of *p*-channel Cu_xO TFTs [44]. The spin-coated tetrafluoro-tetracyanoquinodimethane (F_4TCNQ) was inserted between the Cu_xO channel and Au source/drain electrodes, and the electron transfer from Cu_xO to F_4TCNQ (*p*-doping) occurred owing to their work-function differences. This doping approach does not induce defects or impurities in the Cu_xO host lattice, thereby eliminating undesired carrier scattering while retaining the fundamental charge transport. In this study, we systematically investigated dopant configurations/types and found the insertion of a thin F_4TCNQ layer between Au and Cu_xO films can achieve efficient *p*-doping with μ_{FE} improvement and desired V_{TH} shift while maintaining high I_{on}/I_{off} (Figure 25.4a and 25.4b). The 7-nm-thick F_4TCNQ decoration achieved a 20-fold μ_{FE} enhancement to 0.25 $cm^2\ V^{-1}\ s^{-1}$. The electron transfer was also confirmed by ultraviolet photoemission spectroscopy (UPS) analysis, and the corresponding energy bending diagrams were shown in Figure 25.4c. The high-resolution transmission electron microscopy (HR-TEM) images revealed the sharp interface of laminated layers and the ultrathin Cu_xO channel thickness of ~5 nm (Figure 25.4d). The latter enabled high doping efficiency using MCTD, and the electrical performance could be improved dramatically if even very poor Cu_xO TFTs were deposited. After the device performance optimization, a CMOS inverter was assembled with IGZO TFTs and exhibited a high gain voltage of 50 and a noise margin of 67% at a supply voltage of 40 V. Besides the TFT application, the MCTD method can be applied to hole-transporting layers in various diode-based devices (e.g., an organic photovoltaic cell and organic light-emitting diode) to boost performance.

The CMOS technology offers several advantages, including low power consumption, fast switching speed, high noise margin, low heat dissipation, and high packing density. Even the CMOS inverter with high static-state gain voltage can be achieved through weakening *n*-channel oxide TFT performance; the overall low mobility will limit practical dynamic application. In addition, the *p*-channel oxide TFTs generally displayed a relatively high I_{off} of >0.1 nA, which can be attributed to the easy formation of oxygen vacancies, providing the minority carrier (electron) accumulation under the positive bias voltage. The high I_{off} will negate the key advantage of *n*-channel oxide TFTs (very low I_{off}) and will be a nightmare for oxide TFT CMOS integration due to the large

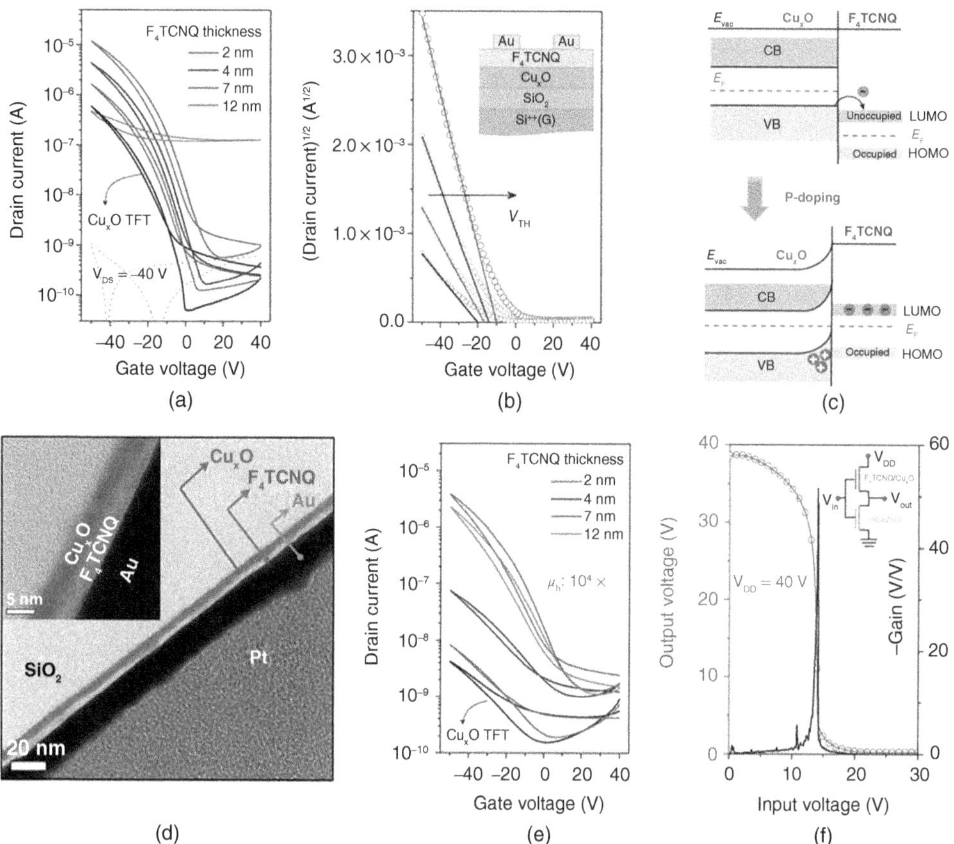

Figure 25.4 (a) Transfer and (b) V_{TH} curves for Cu_xO TFTs with different F_4TCNQ thickness; inset in (b) is the TFT structure. (c) Energy diagram of the electron transfer from Cu_xO to F_4TCNQ. (d) Corresponding device TEM images. *Source*: Reproduced from Ref. [44], Liu, A., et al. (2020). Molecule charge transfer doping for p-channel solution-processed copper oxide transistors. *Advanced Functional Materials* **30**: 2002625, Figure 04, p. 04, with permission of JOHN WILEY & SONS, INC. https://doi.org/10.1002/adfm.202002625. (e) Transfer curves of low-performance Cu_xO TFTs as a function of F_4TCNQ thickness. (f) Voltage transition and gain voltage curves of a CMOS inverter with *n*-channel InGaZnO and *p*-channel F_4TCNQ (7 nm)/Cu_xO TFTs. *Source*: (a–f) Reproduced with permission from Ref. [44], Liu, A., et al. (2020). Molecule charge transfer doping for p-channel solution-processed copper oxide transistors. *Advanced Functional Materials* **30**: 2002625. Copyright 2020 Wiley-VCH.

power consumption. The recent studies showed encouraging results that the backchannel post treatment can greatly lower the I_{off} to ~10^{-12} A [45, 46]. Overall, it is still extremely difficult to develop high-performance *p*-type oxide semiconductors, and as a result, many researchers no longer believe in the possibility of achieving a viable oxide TFT CMOS technology.

25.3 Transparent Copper(I) Iodide (CuI)-Based TFTs

Searching for *p*-type oxide alternatives with both high mobility and transparency has attracted great interest over the past years. In 2013, Anthopoulos's group reported transparent *p*-channel CuSCN TFTs using spin coating at a low annealing temperature of 80 °C [17]. The CuSCN thin film possessed a large E_g of >3.5 eV and a polycrystalline structure with surface roughness of 1–2 nm. The CuSCN VBM is composed of Cu 3*d* and S 3*p* hybridization orbitals, and the conduction band minimum (CBM) is dominated by C and N 2*p* states. The resultant TFTs combining high permittivity polymer P(VDF-TrFE-CFE) dielectric exhibited a μ_{FE} of ~0.1 cm^2 V^{-1} s^{-1} and an I_{on}/I_{off} of ~10^4. Great efforts have been paid later by the authors on the optimization of solution processability and electrical characteristics with the $\mu_{FE} \approx 0.18$ cm^2 V^{-1} s^{-1} and $I_{on}/I_{off} \approx 10^4$.

Another promising inorganic transparent *p*-type material is CuI ($E_g = 3.1$ eV), which possesses a high intrinsic hole mobility of over 40 cm^2 V^{-1} s^{-1} due to the small hole effective mass of 0.3 m_e [47]. CuI is composed of naturally abundant and environmentally friendly elements, making it appropriate for large-scale transparent electronics manufacturing. The first study on CuI was reported by Bädeker in 1907 [48]. Over the past century, various deposition techniques have been developed to fabricate CuI, and its versatility has extended to diverse research areas. The detailed summarization and the application progress can be found in our latest review paper [49]. In 2016, CuI transparent electrodes were achieved by Yang et al. [50] using radiofrequency (RF) sputtering with a record-high conductivity of 238 S cm^{-1}. CuI thin films have also been employed in thermoelectric devices, diodes, and solar cells. The above applications mainly relied on the relatively high hole concentration and conductivity of pristine CuI. Due to the low formation energy of V_{Cu}, the hole concentrations in the deposited CuI thin films were generally higher than 10^{19} cm^{-3}, which hindered their employment as semiconductor channel layers in transistors.

The first attempt at CuI TFTs was reported by Choi et al. [23] in 2016 using inkjet printing; and recently, our group carried out systematic investigations on solution-processed CuI thin films to clarify their feasibility as transistor channel components, and we reported improved TFT performance [32]. The polycrystalline phase was observed for the as-coated CuI thin film, and the crystallinity enhanced with annealing temperatures. The accelerated lattice iodine decomposition and surface aggregation occurred above 100 °C. The TFT performance was strongly related to the CuI channel thickness and annealing temperature. The TFT drain current (I_{DS}) exhibited positive correlation with CuI thickness while reduced at higher temperatures (Figure 25.5a). The latter can be ascribed to V_{Cu} compensation by donor type iodine vacancies, hence decreasing hole concentrations. The optimized TFTs with a 5-nm-thick CuI channel layer exhibited a μ_{FE} of 0.4 cm^2 V^{-1} s^{-1}, I_{on}/I_{off} of 10^2–10^3, and great reproducibility even without any posttreatment (Figure 25.5b). A CMOS inverter was then integrated by assembling *n*-channel IGZO TFTs, delivering clear rail-to-rail voltage inversion with a voltage gain of ~4 (Figure 25.5c). Due to the polycrystalline texture of pristine CuI, its application over a large area is limited. To address this issue, Hosono and coworkers proposed an amorphous design concept by incorporating Sn^{4+} into CuI (Figure 25.5d) [51]. Similar to amorphous ternary and quaternary oxide semiconductors, the multicomponent amorphization is also suitable to CuI. The crystallized CuI became amorphous from adding 5 mol% Sn^{4+} (Figure 25.5e) and kept the wide E_g of ~3 eV. A positive correlation between film Hall mobility and carrier concentration was observed for the amorphous CuSnI (Figure 25.5f), indicating that the percolation dominated the charge transport property.

Despite the initial success with CuI for transistor applications, the devices always delivered poor current modulation ($I_{on}/I_{off} \sim 10^2$) with high I_{off} of >1 nA. The main reason was caused by the excessive hole concentration ($n > 10^{19}$ cm^{-3}) in CuI owing to the easy formation of V_{Cu}. Although increasing the annealing temperature and reducing the channel thickness can achieve relatively low I_{off}, the I_{on} also sacrificed accordingly. Therefore, the development of an easy and controllable doping approach, while maintaining high optical transparency and hole mobility, is highly required. Most recently, our group proposed a substitutional doping method to reduce the hole concentration in CuI and revealed the key role of iodine vacancy passivation for the realization of high-performance *p*-channel TFTs [40, 52]. We conducted a set of theoretical and experimental trials and finally demonstrated that Zn^{2+} was the optimal *n*-type dopant for CuI with a stable doping process (Figure 25.6a). The suitable amounts of Zn^{2+} doping (≤10 mol%) kept the CuI host lattice structure with high optical transparencies ($E_g \approx 3$ eV). Meanwhile, the grain distribution became uniform with the enhanced film crystallinity, and the reasons can be attributed to the retardation of CuI fast crystallization and V_{Cu} passivation. As shown in Figure 25.6c, the Zn^{2+} doping reduced the μ_{FE} and shifted the V_{TH} toward the negative direction, representing typical *n*-doping behavior. The optimized TFTs with 5 mol% Zn^{2+} annealed at 80 °C exhibited a high μ_{FE} of over 5 cm^2 V^{-1} s^{-1} and high I_{on}/I_{off} of ~10^7 with good operational stability and excellent reproducibility. The output curves indicated the Ohmic contact between the CuZnI channel layer and Au electrodes (Figure 25.6d). Finally,

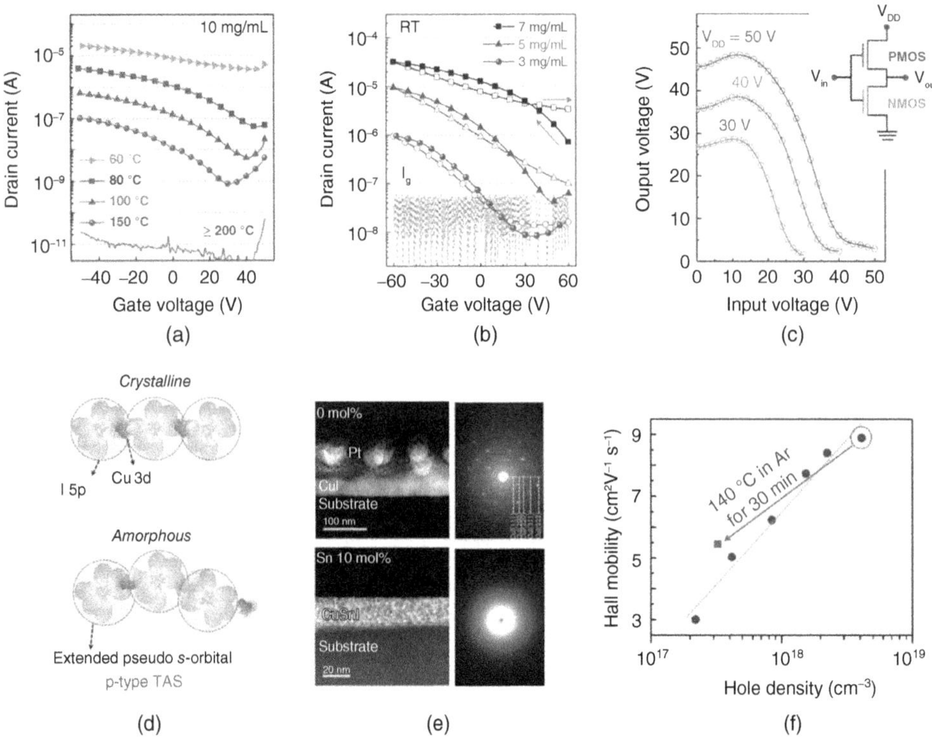

Figure 25.5 (a) Transfer curves of CuI TFTs as a function of annealing temperature. (b) Transfer curves of RT solution-processed CuI TFTs with different CuI thicknesses. (c) Voltage transfer characteristics of a CuI/IGZO CMOS inverter. *Source*: Reproduced with permission from Ref. [32], Liu, A., et al. (2018). Room-temperature solution-synthesized p-type copper(I) iodide semiconductors for transparent thin-film transistors and complementary electronics. *Advanced Materials* **30**: 1802379. (d) Schematic orbital drawing of CuI VBM. (e) TEM images and selected area electron diffraction patterns of CuI and 10 mol% Sn-added CuI. *Source*: From Ref. [51], Jun, T., et al. (2018). Material design of p-type transparent amorphous semiconductor, Cu–Sn–I. *Advanced Materials* **30**: 1706573, Figure 02, p. 03, with permission of JOHN WILEY & SONS, INC. https://doi.org/10.1002/adma.201706573. (f) The dependence of amorphous CuSnI Hall mobility on hole concentration. *Source*: Reproduced with permission from Ref. [51], Jun, T., et al. (2018). Material design of p-type transparent amorphous semiconductor, Cu–Sn–I. *Advanced Materials* **30**: 1706573.

a CMOS inverter was assembled with *n*-channel IGZO TFTs, exhibiting full rail-to-rail swings and rapid voltage transitions with a high peak gain of 56 and low power consumption of 0.25 μW (Figure 25.6e and 25.6f).

25.4 Conclusions and Perspectives

Recent decades have witnessed slow progress in *p*-type metal-oxide TFTs, and there is still no instructional guideline for performance improvement. It is still difficult to achieve phase-pure Cu_2O thin films using a solution process. The reason is the necessity to decompose the anion groups (e.g., NO^{3-}, $CH3COO^-$, and Cl^-) in oxygen-containing atmosphere, resulting in Cu^+'s oxidization into Cu^{2+}. As for the other oxide candidates, the enablement is more difficult using a solution process. By contrast, CuI shows more promising potential for transparent *p*-channel TFT application. Despite the excessive hole concentration, the doping approach has been demonstrated effectively for high-performance TFT fabrication. This component represents an up-and-coming *p*-type semiconductor, showing great compatibility with IGZO technology in the field of transparent electronics. In the near future, more attention should be paid to the enhancement of device performance and air/operational

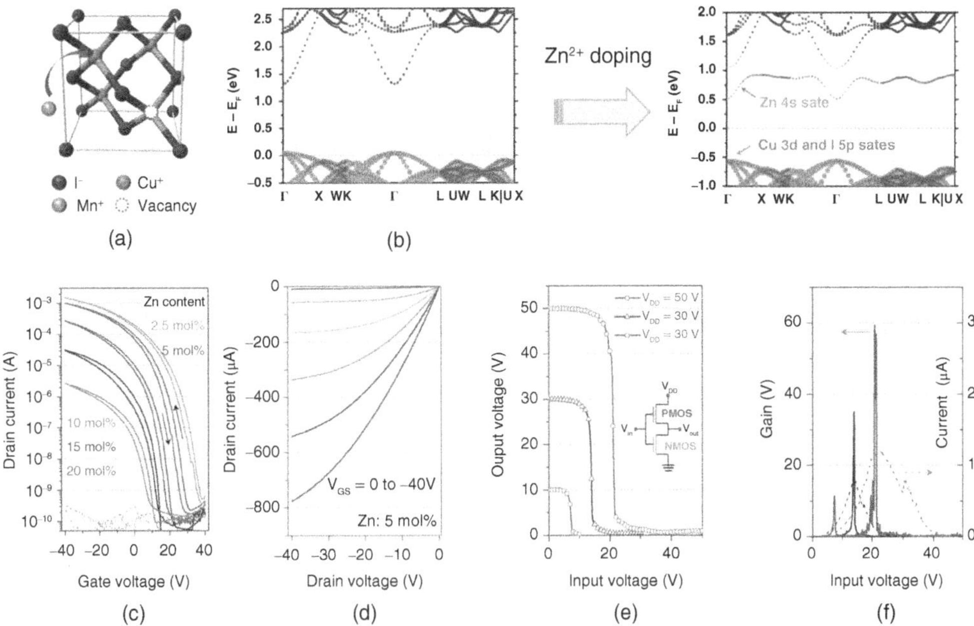

Figure 25.6 (a) Schematic illustrations of a unit cell of CuI, Cu vacancy, and the metal cation doping at a Cu^+ site. (b) (Left panel) Band structure of CuI; the Cu 3*d* (orange), Cu 4*s* (blue), and I 5*p* (purple) states are shown. (Right panel) Band structure of CuI with Zn^{2+} doping. (c) Transfer curves of CuI:Zn/SiO_2 TFTs as a function of Zn^{2+} content ($V_{DS} = -40$ V). (d) Output curves of the optimized CuI:$Zn_{5mol\%}$/SiO_2 TFTs. (e,f) Voltage transfer, gain, and current curves of the CMOS inverter. *Source*: Reproduced with permission from Ref. [40], Liu, A., et al. (2020). High- performance p-channel transistors with transparent Zn doped-CuI. *Nature Communications* **11**: 4309. Licensed under CCBY 4.0.

stabilities through film quality improvement (e.g., crystallinity and surface smoothness), interface modification, and device engineering. The use of computer-based materials design simulation can help screen new dopants and *p*-type semiconductors more effectively and well establish a fundamental understanding of materials and property relationships. Metal halides represent the up-and-coming *p*-type semiconductors, and their studies in transistors are still in the initial stage. Considering their superior charge transport property and easy processability, we expect greater progress in the near future.

Acknowledgments

This study was supported by the Ministry of Science and ICT through the National Research Foundation, funded by the Korean government (NRF-2021R1A2C3005401 and 2020M3F3A2A01085792) and Samsung Display Corporation.

References

1 Wager, J.F. (2020). TFT technology: advancements and opportunities for improvement. *Information Display* 36: 9.

2 Liu, A., Zhu, H., Guo, Z. et al. (2017). Solution combustion synthesis: low-temperature processing for p-type Cu:NiO thin films for transparent electronics. *Advanced Materials* 29: 1701599.

3 Wang, Z., Nayak, P.K., Caraveo-Frescas, J.A., and Alshareef, H. (2016). Recent developments in p-type oxide semiconductor materials and devices. *Advanced Materials* 28: 3831.

4 Liu, A., Zhu, H., and Noh, Y.Y. (2019). Solution-processed inorganic p-channel transistors: recent advances and perspectives. *Materials Science & Engineering R: Reports* 135: 85.

5 Zhu, H., Shin, E., Liu, A. et al. (2020). Printable semiconductors for backplane TFTs of flexible OLED displays. *Advanced Functional Materials* 30: 1904588.

6 Liu, A., Zhu, H., Sun, H. et al. (2018). Solution processed metal oxide high-κ dielectrics for emerging transistors and circuits. *Advanced Materials* 30: 1706364.

7 Hidenori, H., Kazushige, U., Kouhei, T. et al. (2004). Fabrication of heteroepitaxial thin films of layered oxychalcogenides LnCuOCh (Ln = La-Nd; Ch = S-Te) by reactive solid-phase epitaxy. *Journal of Materials Research* 19: 2137.

8 Yan, W., Ye, S., Li, Y. et al. (2016). Hole-transporting materials in inverted planar perovskite solar cells. *Advanced Energy Materials* 6: 1600474.

9 Pattanasattayavong, P., Mottram, A., Yan, F., and Anthopoulos, T.D. (2015). Study of the hole transport processes in solution-processed layers of the wide bandgap semiconductor copper(I) thiocyanate (CuSCN). *Advanced Functional Materials* 25: 6802.

10 Matsuzaki, K., Nomura, K., Yanagi, H. et al. (2008). Epitaxial growth of high mobility Cu_2O thin films and application to p-channel thin film transistor. *Applied Physics Letters* 93: 202107.

11 Fortunato, E., Barros, R., Barquinha, P. et al. (2010). Transparent p-type SnO_x thin film transistors produced by reactive rf magnetron sputtering followed by low temperature annealing. *Applied Physics Letters* 97: 052105.

12 Fortunato, E., Figueiredo, V., Barquinha, P. et al. (2010). Thin-film transistors based on p-type Cu_2O thin films produced at room temperature. *Applied Physics Letters* 96: 192102.

13 Ogo, Y., Hiramatsu, H., Nomura, K. et al. (2008). p-channel thin-film transistor using p-type oxide semiconductor. *SnO. Applied Physics Letters* 93: 032113.

14 Okamura, K., Nasr, B., Brand, R.A., and Hahn, H. (2012). Solution-processed oxide semiconductor SnO in p-channel thin-film transistors. *Journal of Materials Chemistry* 22: 4607.

15 Kim, S., Ahn, C., Lee, J. et al. (2013). P-channel oxide thin film transistors using solution-processed copper oxide. *ACS Applied Materials & Interfaces* 5: 2417.

16 Pattanasattayavong, P., Thomas, S., Adamopoulos, G. et al. (2013). p-channel thin-film transistors based on spray-coated Cu_2O films. *Applied Physics Letters* 102: 163505.

17 Pattanasattayavong, P., Yaacobi-Gross, N., Zhao, K. et al. (2013). Hole-transporting transistors and circuits based on the transparent inorganic semiconductor copper (I) thiocyanate (CuSCN) processed from solution at room temperature. *Advanced Materials* 25: 1504.

18 Liu, S., Liu, R., Chen, Y. et al. (2014). Nickel oxide hole injection/transport layers for efficient solution-processed organic light-emitting diodes. *Chemistry of Materials* 26: 4528.

19 Yu, J., Liu, G., Liu, A. et al. (2015). Solution-processed p-type copper oxide thin-film transistors fabricated by using a one-step vacuum annealing technique. *Journal of Materials Chemistry C* 3: 9509.

20 Liu, A., Liu, G., Zhu, H. et al. (2015). Water-induced scandium oxide dielectric for low-operating voltage n- and p-type metal-oxide thin-film transistors. *Advanced Functional Materials* 25: 7180.

21 Jang, J., Chung, S., Kang, H., and Subramanian, V. (2016). P-type CuO and Cu_2O transistors derived from a sol–gel copper (II) acetate monohydrate precursor. *Thin Solid Films* 600: 157.

22 Liu, A., Liu, G., Zhu, H. et al. (2016). Hole mobility modulation of solution-processed nickel oxide thin-film transistor based on high-k dielectric. *Applied Physics Letters* 108: 233506.

23 Choi, C., Gorecki, J.Y., Fang, Z. et al. (2016). Low-temperature, inkjet printed p-type copper(i) iodide thin film transistors. *Journal of Materials Chemistry C* 4: 10309.

24 Liu, A., Nie, S., Liu, G. et al. (2017). In situ one-step synthesis of p-type copper oxide for low-temperature, solution-processed thin-film transistors. *Journal of Materials Chemistry C* 5: 2524.

25 Li, Y., Liu, C., Wang, G., and Pei, Y. (2017). Investigation of solution combustion-processed nickel oxide p-channel thin film transistors. *Semiconductor Science and Technology* 32: 085004.

26 Lee, S., Lee, W., Jang, B. et al. (2018). Sol-gel processed p-type CuO phototransistor for a near-infrared sensor. *IEEE Electron Device Letters* 39: 47.

27 Hu, H., Zhu, J., Chen, M. et al. (2018). Inkjet-printed p-type nickel oxide thin-film transistor. *Applied Surface Science* 441: 295.

28 Wijeyasinghe, N., Tsetseris, L., Regoutz, A. et al. (2018). Copper (I) selenocyanate (CuSeCN) as a novel hole-transport layer for transistors, organic solar cells, and light-emitting diodes. *Advanced Functional Materials* 28: 1707319.

29 Li, S., Zhang, X., Zhang, P. et al. (2018). Preparation and characterization of solution-processed nanocrystalline p-type $CuAlO_2$ thin-film transistors. *Nanoscale Research Letters* 13: 259.

30 Nie, S.B., Liu, A., Meng, Y. et al. (2018). Solution-processed ternary p-type $CuCrO_2$ semiconductor thin films and their application in transistors. *Journal of Materials Chemistry C* 6: 1393.

31 Jung, T., Lee, H., Park, S. et al. (2018). Enhancement of switching characteristic for p-type oxide semiconductors using hypochlorous acid. *ACS Applied Materials & Interfaces* 10: 32337.

32 Liu, A., Zhu, H., Park, W. et al. (2018). Room-temperature solution-synthesized p-type copper(I) iodide semiconductors for transparent thin-film transistors and complementary electronics. *Advanced Materials* 30: 1802379.

33 Lee, H., Zhang, X., Kim, E., and Park, J. (2019). Structural and electrical characteristics of solution-processed copper oxide films for application in thin-film transistors. *Sensors and Materials* 31: 501.

34 Wang, C., Zhu, H., Meng, Y. et al. (2019). Sol–gel processed p-type $CuAlO_2$ semiconductor thin films and the integration in transistors. *IEEE Transactions on Electron Devices* 66: 1458.

35 Liu, A., Zhu, H., and Noh, Y.-Y. (2019). Polyol teduction: a low-temperature eco-friendly solution process for p-channel copper oxide-based transistors and inverter circuits. *ACS Applied Materials & Interfaces* 11: 33157.

36 Chen, C., Yang, Q., Chen, G. et al. (2019). Solution-processed oxide complementary inverter via laser annealing and inkjet printing. *IEEE Transactions on Electron Devices* 66: 4888.

37 Xu, W., Zhang, J., Li, Y. et al. (2019). p-type transparent amorphous oxide thin-film transistors using low-temperature solution-processed nickel oxide. *Journal of Alloys and Compounds* 806: 40.

38 Xin, Z., Ding, Y., Zhu, Y. et al. (2020). Solution-processed high-performance p-type perovskite $NdAlO_3$ thin films for transparent electronics. *Advanced Electronic Materials* 6: 1901110.

39 Trinh, B., Nguyen, H., Nguyen, Q. et al. (2020). Solution-processed cupric oxide p-type channel thin-film transistors. *Thin Solid Films* 704: 137991.

40 Liu, A., Zhu, H., Park, W. et al. (2020). High-performance p-channel transistors with transparent Zn doped-CuI. *Nature Communications* 11: 4309.

41 Garlapati, S.K., Baby, T.T., Dehm, S. et al. (2015). Ink-jet printed CMOS electronics from oxide semiconductors. *Small* 11: 3591.

42 Park, I., Jeong, C., Myeonghun, U. et al. (2013). Bias-stress-induced instabilities in p-type Cu_2O thin-film transistors. *IEEE Electron Device Letters* 34: 647.

43 Zhu, H., Liu, A., and Noh, Y. (2020). Impact of humidity on the performance and stability of solution-processed copper oxide transistors. *IEEE Electron Device Letters* 41: 1033.

44 Liu, A., Zhu, H., and Noh, Y.-Y. (2020). Molecule charge transfer doping for p-channel solution-processed copper oxide transistors. *Advanced Functional Materials* 30: 2002625.

45 Chang, H., Huang, C., Matsuzaki, K., and Nomura, K. (2020). Back-channel defect termination by sulfur for p-channel Cu_2O thin-film transistors. *ACS Applied Materials & Interfaces* 12: 51581.

46 Min, W., Park, S., Kim, H. et al. (2020). Switching enhancement via a back-channel phase-controlling layer for p-type copper oxide thin-film transistors. *ACS Applied Materials & Interfaces* 12: 24929.

47 Grundmann, M., Schein, F., Lorenz, M. et al. (2013). Cuprous iodide—a p-type transparent semiconductor: history and novel applications. *Physica Status Solidi (A): Applications and Materials. Science* 210: 1671.

48 Baedeker, K. (1907). Über die elektrische Leitfähigkeit und die thermoelektrische Kraft einiger Schwermetallverbindungen. *Annalen der Physik (Berlin)* 327: 749.

49 Liu, A., Zhu, H., Kim, M. et al. (2021). Engineering copper iodide (CuI) for multifunctional p-type transparent semiconductors and conductors. *Advanced Science (Weinheim)* 8: 2100546.

50 Yang, C., Kneiβ, M., Lorenz, M., and Grundmann, M. (2016). Room-temperature synthesized copper iodide thin film as degenerate p-type transparent conductor with a boosted figure of merit. *Proceedings of the National Academy of Sciences* 113: 12929.

51 Jun, T., Kim, J., Sasase, M., and Hosono, H. (2018). Material design of p-type transparent amorphous semiconductor, Cu–Sn–I. *Advanced Materials* 30: 1706573.

52 Liu, A., Zhu, H., Shim, K. et al. (2021). Key roles of trace oxygen treatment for high-performance Zn-doped CuI p-channel transistors. *Advanced Electronic Materials* 7: 2000933.

26

Tungsten-Doped Active Layers for High-Mobility AOS-TFTs

Zhang Qun

Department of Materials Science, Fudan University, Shanghai, China

26.1 Introduction

High-mobility amorphous oxide semiconductor thin-film transistors (AOS-TFTs) have attracted a great deal of attention not only in academic world but also in industry [1–3]. Since the high-performance demonstration of single-crystal IGZO-TFTs in science in 2003 [4] and flexible amorphous IGZO-TFTs (*a*-IGZO-TFTs) in technology in 2004 [5], considerable efforts have been made for *a*-IGZO-TFTs for a wider range of application possibilities, especially for emerging novel display devices. After more than 10 years, *a*-IGZO-TFT-based backplanes with high carrier mobility, low leakage current, large area uniformity, high optical transparency, low process temperature, and low production cost have been applied in large-sized, high-definition flat-panel displays [2]. Early in 1996, a working hypothesis was proposed by Hosono [6, 7] as a starting point for choosing multicomponent combinations of cations for the design of AOSs, though these AOS guidelines were originally formulated for the design of transparent conductive oxides (TCOs). The hypothesis predicts that amorphous oxides composed of heavy metal cations with an electronic configuration of $(n{-}1)d^{10}ns^0$ are promising candidates for a novel class of AOSs. Since then, conventional considerations for AOS materials have been constrained mostly to choose cations with a large ionic radius and spherically symmetric 4*s*, 5*s*, or 6*s* electron orbitals, that is, cations with a high degree of wave function overlap, electron delocalization, and relatively high electron mobility. Finally, of the 15 elements specified in the hypothesis, *a*-IGZO with an $InGaZnO_4$ composition is the current AOS commercial material of choice and provides a framework for undertaking future AOS material selection and design [8].

The upcoming display technologies with ultrahigh resolution, a high frame rate, large area displays, as well as vivid images and natural motion pictures demand even more high-mobility AOS-TFT backplanes to decrease the resistor–capacitor (RC) delay in signal lines and the charging time for each pixel [9]. Figure 26.1 shows the interrelationships between the electrical performance and stability of AOS-TFTs and the essential factors. Obviously, the electrical performance and stability of the devices are dependent on the active-layer materials, gate insulator materials, electrode materials, passivation materials, quality of these thin-film materials, device structures, and interface properties between the active layer and gate insulator layer and between the source–drain (S/D) electrodes and active layer. All these factors, especially active-layer compositions and qualities, as well as interface properties are extremely sensitive to and strongly related to the synthesis method, process conditions, and relevant theories and modeling. In contrast to the crystalline TCOs, where the electron mobility is governed primarily by the scattering of ionized impurities, phonons, and grain boundaries, the charge transport in AOSs is more complex [3]. Although there are no grain boundaries in amorphous materials, the structural long-range disorder and strong local distortions in the metal–oxygen (M-O) polyhedra give rise to several new terms

Amorphous Oxide Semiconductors: IGZO and Related Materials for Display and Memory, First Edition.
Edited by Hideo Hosono and Hideya Kumomi.

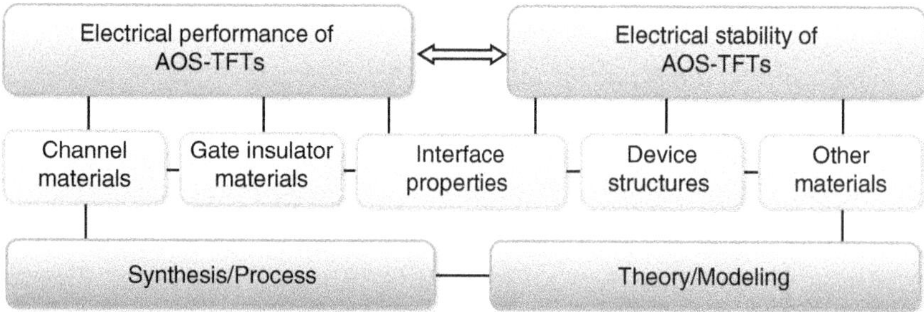

Figure 26.1 Schematic diagram of the interrelationships between the electrical performance and stability of AOS-TFTs and the essential factors.

in electron transport. Adhering to the Drude model, the electron mobility in an AOS can be represented by Equation (26.1):

$$\frac{1}{\mu} = \frac{m^*}{e}\left(\frac{1}{\tau_{\text{crystallinity}}} + \frac{1}{\tau_{\text{composition}}} + \frac{1}{\tau_{\text{defects}}} + \frac{1}{\tau_{\text{vibrons}}} + \frac{1}{\tau_{\text{strain}}}\right) \tag{26.1}$$

where the contributions to the overall relaxation time are due to (i) the size and density of nanocrystalline inclusions, (ii) the spatial distribution and clustering of incorporated cations, (iii) local point defects, (iv) thermal vibrations, and (v) piezoelectric effects associated with internal lattice strain. These five terms are likely to be intermixed for most AOS materials. For example, cation composition has a strong effect on crystallization processes, defect formation, and thermal properties, and can introduce significant lattice strain in amorphous structures. In AOS-TFTs, the contact effects, including interfacial trapping and work–function misalignment, influence severely the electrical performance and stability, especially the field-effect mobility (μ_{FE}) [10]. Therefore, much research to enhance the high mobility of AOS-TFTs has focused on active-layer composition and high-quality active layers for electron mobility improvement, or interface engineering for contact effect improvement in TFT devices.

Most active-layer materials of AOS-TFTs are derived from TCOs, a class of unusual materials with simultaneous high electrical conductivity and high optical transparency in the visible-light range. This peculiar combination of physical properties is achievable for materials with an energy band gap larger than 3.1 eV, carrier concentration higher than 10^{19} cm^{-3}, and mobility larger than 1 cm^{2}V^{-1}s^{-1} [11]. In fact, TCOs are a class of semiconductors in nature but are used as passive conductors or electrodes. The three best known, *n*-type, large-scale commercialized TCOs are tin-doped indium oxide (ITO), aluminum-doped zinc oxide (AZO), and fluorine-doped tin oxide (FTO) crystalline oxide semiconductors. The mobility of the *n*-type TCO thin films is normally in the range of ~10–40 cm^{2}V^{-1}s^{-1}. To obtain the conductivity of the TCOs as large as possible for optimal performance, intentional doping to extremely high-level carrier concentrations in the range of 10^{20}–10^{21} cm^{-3} is usually carried out. However, there is always a trade-off consideration between the high conductivity and high transparency for the TCOs.

Historically, In_2O_3, ZnO, and SnO_2 are the three primary oxide semiconductors serving as active layers for oxide TFTs. Despite the large band gaps, these oxide semiconductors can be processed with doping intrinsic or extrinsic defects to create free carriers. The electrical properties of these materials are dominated by the 4*s* orbital of Zn and the 5*s* orbitals of In and Sn. However, the larger radial extension of the 5*s* orbital relative to the 4*s* orbital might be expected to lead to higher mobility and conductivity for In_2O_3 and SnO_2. Compared to their crystalline counterparts, the structure of corresponding AOSs is extremely sensitive to cation composition, oxygen stoichiometry, lattice strain, and process conditions, giving rise to a wide range of tunable optical and electrical properties [3]. On the one hand, the large parameter space and the resulting complex deposition–structure–property relationships in AOSs make the currently available theoretical and experimental research data rather scattered and the design of new materials difficult. On the other hand, unlike Si-based semiconductors, AOSs were shown to exhibit optical, electrical, thermal, and mechanical properties that are comparable or even superior to those possessed by their crystalline counterparts.

The first report of an oxide TFT with an evaporated SnO_2 TCO as an active layer was in 1964 [12], and the first ZnO-based TFT was announced in 1968 [13]. In 2003, ZnO-based transparent TFTs were reported by three independent groups [14–16]. A few months later, Nishii et al. reported the fabrication of a ZnO-TFT with an improved μ_{FE} of $7\ cm^2V^{-1}s^{-1}$ at a maximum process temperature of 300 °C [17]. Individual In_2O_3 nanowire transistors as chemical sensors [18] and SnO_2 transparent TFTs were demonstrated in 2004 [19]. Although oxide TFT devices based on binary semiconductor oxides showed relatively good performance, there were limitations in controlling of the carrier concentration, threshold voltage, and device stability. These problems are attributed to the high defect concentration within the oxide thin films, resulting in an excess of free electrons. Another problem is the presence of grain boundaries in the polycrystalline binary oxide materials [20]. The uniformity in large areas is poor due to defects on grain boundaries, and the grain boundary–inhibited transport in polycrystalline materials has important influence on the performance uniformity and stability of the oxide TFTs, especially under various stress conditions such as bias stress, light irradiation, and thermal and environmental influences. Sparked by the *a*-IGZO-TFTs, the research work on AOS-TFTs has been growing exponentially since then, and the development of novel AOS active-layer materials for high-mobility AOS-TFTs is mainly reported in the doping of metal ions and a combination of two or more binary metal oxides into the host TCO semiconductors of In_2O_3, ZnO, and SnO_2.

Doping is an important technique for developing functional materials. It improves the properties of, or produces various functions in, the host materials. Properties of oxide semiconductor materials are drastically modified by the addition of a small amount of dopants [21]. The dopant, which supplies the carrier electrons, is the important factor to control the electrical properties of a semiconductor. The effective mass of mobile electrons depends largely on the size and the spatial directivity of atomic orbitals and their connectivity with the dopant and the host. If the conduction band is formed by largely spread *ns* orbitals ($n \geq 5$), the electron may have a small effective mass even in an amorphous state because the magnitude of the overlap between these neighboring *s*-orbitals is insensitive to the structure disorder. Off-stoichiometry/interstitial doping is much more effective in general for carrier generation. In amorphous oxides, electron doping is possible, like in the crystalline oxides when the stabilization energy of dopant is large. A cation that has large electronegativity and large Lewis acid strength and that forms a stronger bond with constituent oxygen may work as an effective dopant. It is noteworthy that the amorphous indium oxide–based TFTs with tungsten dopant, one of many doping metal ions, showed excellent device performance and stability. The role of tungsten dopant is discussed here, and the progress of tungsten-doped indium oxide–based TFTs is thoroughly reviewed and summarized.

26.2 Advances in Tungsten-Doped High-Mobility AOS-TFTs

In this section, tungsten dopant is used as the example to reflect the research progress on high-mobility active-layer materials and the corresponding high-mobility TFT devices. The single layers of amorphous tungsten-doped indium oxide (*a*-IWO) and amorphous tungsten-doped indium zinc oxide (*a*-IZWO) TFTs are reviewed and discussed. Recent progress on the dual active-layer structure, passivation layer, and post treatment on the backchannel surface, for the improvement of the electrical performance and stability of the *a*-IWO- and *a*-IZWO-TFTs, is analyzed and summarized. The main performance parameters of the tungsten-doped indium oxide–based TFTs are tabulated in Table 26.1.

26.2.1 *a*-IWO-TFTs

In 2005, Li et al. developed reproducible IWO TCO thin films with optimal resistivity of $2.7\times10^{-4}\ \Omega\cdot cm$, high carrier mobility larger than $57\ cm^2V^{-1}s^{-1}$, and high transmission in the visible-light region exceeding 80% by reactive

Table 26.1 Main performance parameters of tungsten-doped indium oxide–based thin-film transistors

Process	T_{SUB} (°C)	Post treatment (°C)	Active layer	Mobility ($cm^2V^{-1}s^{-1}$)	$V_{TH/ON}$ (V)	SS (V/decade)	I_{ON}/I_{OFF}	Reference
DC sputtering	RT	100	*a*-IWO	17.1	3.07	0.77	10^6	27
DC sputtering	RT	100	*a*-IWO	21.7	−0.5	0.47	8.9×10^9	28
DC sputtering	RT	150	*a*-IWO	39	−3.6	0.24	1.7×10^{10}	29
RF sputtering	RT	100	*a*-IWO	36.7	−3.4	0.39	—	30
RF sputtering	RT	100	*a*-IWO	27.55	0.5	0.5	1.7×10^8	31
Spin coating	RT	300	*a*-IWO	15.3	0.37	0.068	5×10^7	32
Spray pyrolysis	RT	120	*c*-IWO	~31	2.0	0.35	$\sim7\times10^5$	34
RF sputtering	RT	270	*a*-IWO	25.86	−1.5	0.3	5.6×10^5	35
RF sputtering	RT	—	*a*-IWO	5.9	0.17	0.107	5×10^7	36
RF sputtering	RT	100	*a*-IWO	26.5	−2.5	0.5	10^7	37
RF sputtering	RT	—	*a*-IWO	25.3	~0.1	0.063	10^9	38
RF sputtering	RT	None	*a*-IZWO	11.1	4.0	0.31	10^7	39
DC sputtering	RT	150	*a*-IZWO	40	−1.6 (V_{on})	—	1.8×10^{11}	40
RF sputtering	RT	250	IZWO	18.09	−0.32	0.30	—	41
RF sputtering	RT	300	*a*-IZWO	16.2	−7.98	0.2	1.6×10^8	42
RF sputtering	RT	300	*a*-IZWO	22.3	−1.48	0.36	—	43
RF sputtering	RT	250	IZWO	26.1	−3.35	0.38	7.39×10^6	44
RF sputtering	RT	250	IZWO	19.57	~ −0.3	0.14	—	45
RF sputtering	RT	250	IZWO	31.22	−1.13	0.31	$\sim10^7$	46
RF sputtering	RT	—	*a*-IZWO	23.8	−0.072	0.0726	9.4×10^8	47
Spin coating	RT	—	*a*-IZWO	30.5	~0.4	0.4	10^7	48
RF sputtering	RT	250	IZWO	23.6	−1.03	0.3	5.52×10^8	49
RF sputtering	RT	400	*a*-IWO/IWO	20.4	0.52	0.58	—	50
RF sputtering	RT	200	IZWO/IZWO	20.2	—	—	—	51
RF sputtering	RT	100	*a*-IWO/IWO:N	27.2	0.37	0.56	10^7	52
RF sputtering	RT	—	IWO/IZWO	21.1	−0.092	0.15	4.88×10^8	53
RF sputtering	RT	150	IZWO/IWO/IZWO	27.9	−0.106	0.079	6.05×10^7	54
RF sputtering	RT	200	*a*-IZWO	31.2	−0.01	0.07	4.73×10^6	55
Sputtering	RT	200	*a*-IWO/IGZO	27.92	−1.95	0.58	—	56
RF sputtering	RT	300	*a*-IZWO	52.5	−1.43	0.1077	10^7	57
RF sputtering	RT	350	IZWO	33.7	−0.1	0.09	—	58
RF sputtering	RT	—	*a*-IWO	98.91	0.209	0.082	5.11×10^7	59

"—": Not mentioned in the literature.

direct current (DC) magnetron sputtering at 380 °C [22, 23]. X-ray photoelectron spectroscopy (XPS) and X-ray diffraction (XRD) analysis revealed that hexavalent tungsten ions W^{6+} substituted for trivalent host indium ions In^{3+} without changing the crystalline structure of In_2O_3. The W^{6+} ions substituting for In^{3+} ions can contribute many more additional electrons. For the same amount of carrier concentration, fewer W dopant ions are needed, and thus the number of ion impurity-scattering centers will be decreased; or the number of oxygen vacancies that contribute two electrons can be decreased, and the trap densities will be decreased. Both factors may lead to the high carrier mobility at the different metal–oxygen ratio of the IWO thin film. The high conductivity is thought to be attributed to the improvement of carrier mobility rather than the conventional increase of carrier concentration. And high mobility can also avoid sacrificing the optical transparency simultaneously. To confirm this, polycrystalline IWO thin films with high transparency in the near-infrared region were prepared [24]. The optimized average transmittance of the films is about 94% in the visible-light region (400–700 nm) and approximately 90% in the near-infrared region (NIR; 700–2,500 nm), while the highest carrier mobility reaches 67 $cm^2V^{-1}s^{-1}$. The high transparency is ascribed to the low carrier concentration of 2.8×10^{20} cm^{-3} of the IWO films. For comparison, ITO thin film prepared under the same sputtering condition shows a similar resistivity of 3.2×10^{-4} Ω·cm, but a much lower mobility of 21 $cm^2V^{-1}s^{-1}$ and higher carrier concentration of 9.4×10^{20} cm^{-3}, with an average transmittance of about 92% in the visible-light region and 48% in the NIR. According to the Drude model, the plasma wavelength λ_p shifts to the longer wavelength with the decrease of carrier concentration. And the expanding of the transparency window to the NIR without sacrificing the conductivity can only be obtained by increasing the carrier mobility and decreasing the carrier density. In the case of almost the same resistivity, the carrier mobility of IWO is about three times larger than that of ITO, while the carrier density is one-third of that of ITO, which resulted in IWO thin films with high conductivity and high transmission but not high reflectivity in the NIR.

Feng et al. developed *a*-IWO TCO thin films prepared by reactive magnetron sputtering at room temperature [25]. The as-deposited *a*-IWO films with an optimum resistivity of 5.8×10^{-4} Ω·cm, and average transmittance of 92.3% from 400 to 700 nm were obtained at W content of 1 wt.%. The average transmittance in the NIR is 84.6–92.8% for *a*-IWO prepared under varied oxygen partial pressure. The mobility of the *a*-IWO films reaches its highest value of 30.3 $cm^2V^{-1}s^{-1}$ at the carrier concentration of 1.6×10^{20} cm^{-3}, representing their potential application as TCO thin films in various flexible devices. In 2011, Zhang et al. reported the *p-n* junction behavior of the *p*-type nickel copper oxide and the *n*-type indium tungsten oxide [26]. The *n*-type IWO thin film with a thickness of ~200 nm was deposited on an IWO TCO electrode coated on the glass substrate, then a 150 nm *p*-type $Ni_{0.9}Cu_{0.1}O$ layer was grown on the *n*-IWO layer. Through adjusting the oxygen partial pressure during the sputtering process, the conductivity of the *n*-type IWO layer is controlled to the semiconductor scope to match that of the *p*-type layer. Afterward, the total transparent *p-n* junction devices were prepared by replacing Al with IWO as the top electrode. Figure 26.2a shows the I-V characteristic curve of the *p*-$Ni_{0.9}Cu_{0.1}O$/*n*-IWO heterojunction diode. It exhibits nonlinear and rectifying characteristics, indicating the semiconductor nature of *n*-IWO and *p*-NiCuO. The ohmic contact between the IWO TCO electrode and *n*-type IWO semiconductor, as well as between the *p*-type *p*-$Ni_{0.9}Cu_{0.1}O$ and IWO electrode, was confirmed by the linearity of the I-V characteristics (Figure 26.2b).

After confirming the *n*-type semiconductor behavior of *a*-IWO thin films, in 2012 the same group tried to prepare AOS-TFTs with *a*-IWO as active layers on the polyvinylphenol (PVP) dielectric layer for flexible applications [27]. The as-deposited *a*-IWO-TFT shows a μ_{FE} of 7.1 $cm^2V^{-1}s^{-1}$, a subthreshold swing value (SS) of 1.57 V/dec, a threshold voltage V_{TH} of −0.13 V, and an on-off current ratio I_{ON}/I_{OFF} of 10^5, while that of an annealed device at 100 °C in air for 20 min shows a μ_{FE} of 17.1 $cm^2V^{-1}s^{-1}$, SS of 0.77 V/dec, V_{TH} of 3.07 V, and I_{ON}/I_{OFF} larger than 10^6. The results demonstrated that *a*-IWO is a promising active material and annealing played an important role in reducing the oxygen vacancies in *a*-IWO films and trap states between the *a*-IWO channel layer and PVP gate insulator layer. The primary work also confirmed that tungsten doping can modify the V_{TH} shift and improve the stability of the devices.

In 2013, Aikawa et al. demonstrated *a*-IWO-TFTs and discussed the role of W dopant in the In_2O_3-based semiconductor [28]. A patterned 10 nm thick IWO film was deposited at room temperature with an $O_2/(Ar+O_2)$

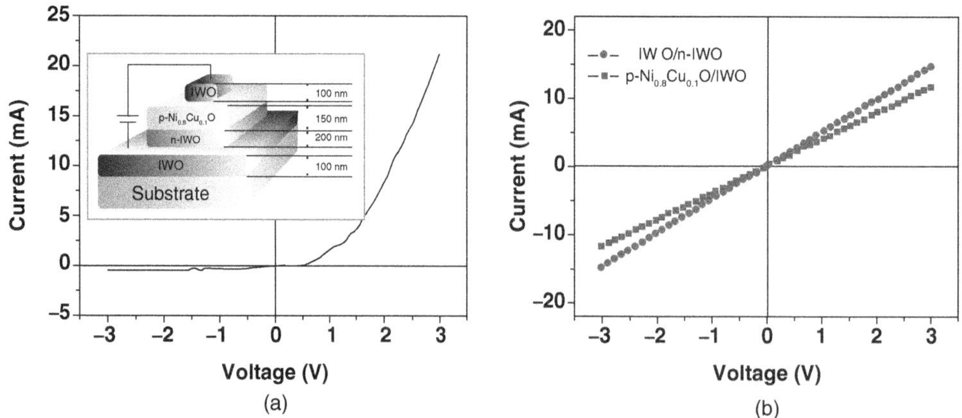

Figure 26.2 (a) I-V curve of the IWO/*n*-IWO/*p*-$Ni_{0.9}Cu_{0.1}O$/IWO *p-n* junction diode. Inset shows the structure of the diode. (b) Linear I-V characteristics of IWO/*n*-IWO and *p*-$Ni_{0.9}Cu_{0.1}O$/IWO contacts, respectively.

ratio of 0.11 by means of DC magnetron sputtering, from the IWO target with a WO_3 content of 1 wt.%. The IWO active layer was annealed three times at 100 °C for 5 min in N_2 using a rapid thermal-annealing method. Figure 26.3a–26.3d shows an optical micrograph of fabricated *a*-IWO-TFTs, a high-magnification image of a device, a schematic cross-sectional diagram of the device structure, and an atomic force microscopy (AFM) image of a postannealed IWO channel. The typical transfer characteristics (I_D-V_{GS}) curve of *a*-IWO-TFTs is shown in Figure 26.3e. Clearly, the TFT is in enhancement mode with a small hysteresis (~0.7 V) and a low I_{OFF} current (~10^{-14} A). At V_{DS} = 40 V, the estimated μ_{FE}, I_{ON}/I_{OFF}, V_{TH}, and SS are 21.7 $cm^2V^{-1}s^{-1}$, 8.9×10^9, −0.5 V, and 0.47 V/dec, respectively. It is thought that the presence of a very small amount of WO_3 prevents the crystallization of In_2O_3 film and leads to the formation of a very flat and smooth surface. It would reduce the degradation of carrier mobility caused by the rough surfaces. Figure 26.3f shows the I_D-V_{GS} curves of the *a*-IWO-TFT for different V_{DS} values in the sequence of 40 V to 0.1 V. No shifts or humps in the curves were observed, even though a high bias V_{DS} of 40 V was applied for 1 hour during the first measurement, implying its quite good electrical stability. Furthermore, Ga_2O_3 and ZnO are acid-soluble and very sensitive to wet-etching processes, which will damage the backchannel surface of the *a*-IGZO active layer when etching the source and drain electrodes. Whereas the *a*-IWO

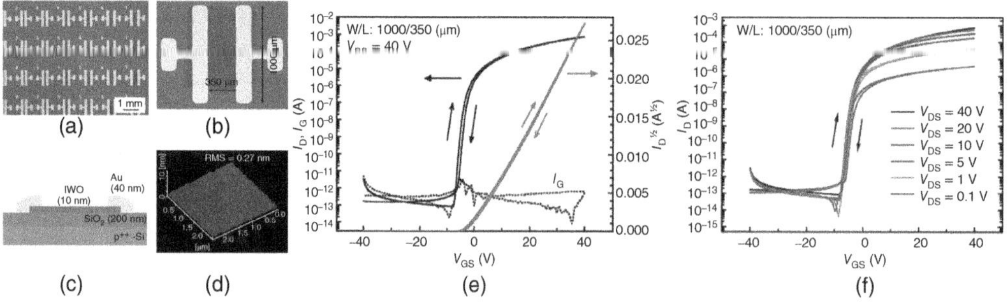

Figure 26.3 (a) Optical micrograph of fabricated *a*-IWO-TFTs. (b) High-magnification image of a device. (c) Schematic cross-sectional diagram of the device structure. (d) AFM image of a postannealed IWO channel. (e) Typical transfer characteristic curve of an *a*-IWO-TFT annealed three times at 100 °C for 5 min in N_2. (f) Transfer characteristics at various drain–source voltages. *Source*: (a) From [28], Aikawa, S., et al. (2013). Thin-film transistors fabricated by low-temperature process based on Ga- and Zn-free amorphous oxide semiconductor. *Applied Physics Letters* **102**: 102101, figure 1 (p. 102101-2) / with permission of AIP Publishing LLC. https://doi.org/10.1063/1.4794903. (a–f) Aikawa, S., et al. (2013). Thin-film transistors fabricated by low-temperature process based on Ga- and Zn-free amorphous oxide semiconductor. *Applied Physics Letters* **102**: 102101.

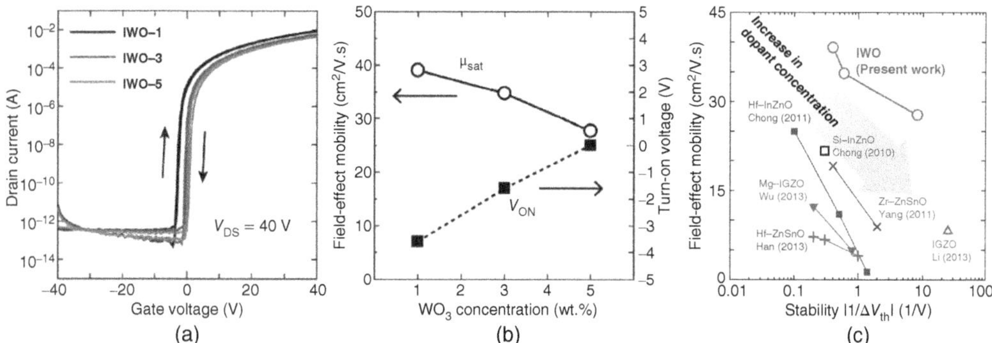

Figure 26.4 (a) Typical transfer characteristics of IWO-TFTs with different W concentrations. (b) The field-effect mobility in the saturation region (LSAT) and the turn-on voltage (VON) as functions of the WO3 concentration. (c) Relationship between the stability and the field-effect mobility of the *a*-IWO-TFTs and some other AOS-TFTs. The stability was estimated from the V_{TH} shift after a gate-bias stress for ~5,000 s. *Source*: Reprinted with permission from Kizu, T., et al. (2014). Low-temperature processable amorphous In-W-O thin-film transistors with high mobility and stability. *Applied Physics Letters* **104**: 152103.

active layer is free from Ga_2O_3 and ZnO, the thickness can be reduced to several nanometers, resulting in material saving and high throughput in the TFT backplane fabrication process. In 2014, the same group investigated the effects of W dopant content on the performance and stability of *a*-IWO-TFTs [29]. TFTs without any passivation layer on the backchannel surface were annealed in air at 150 °C for 5 min to improve the contact property between the S/D electrode and IWO active layer. It was found that the μ_{SAT} decreased from 39 to 28 $cm^2V^{-1}s^{-1}$, and V_{ON} shifted positively from −3.6 to 0 V with increasing of the WO_3 concentration from 1 to 5 wt.% (as shown in Figure 26.4a and 26.4b). The authors investigated the negative bias stress (NBS) behavior of the devices and confirmed that the change in V_{ON} under NBS for 5,000 seconds was only 0.1 V, showing an excellent electrical stability. The result is attributed to the high oxygen bond dissociation energy of W (720 kJ/mol), which suppressed the formation of excess oxygen vacancies and thus the trap density. Figure 26.4c shows the relationship between the stability and field-effect mobility of *a*-IWO-TFTs and some other AOS-TFTs. The oxygen vacancies provide conductive carriers, whereas the vacancies behave as carrier traps and cause instability in the electrical transport. When a dopant with high oxygen bond dissociation energy is introduced into the InO_x film, the oxygen vacancies decrease, which enhances the electrical stability. Decreased carrier density causes degradation in the mobility. Fortunately, the mobility of *a*-IWO-TFTs is still high even in high operation stability. The possible reason for this result is that W ions have six ionic charges W^{6+} and generate extra electrons. The comparison of their mobility–stability relationship with that of other AOS-TFTs indicated that the W atoms have an advantage as a dopant at achieving high mobility and high stability simultaneously.

Liu et al. investigated the electrical performance and reliability of *a*-IWO-TFT devices by modulating oxygen partial pressure (P_{O2}) during the sputtering deposition process, as well as the effect of P_{O2} on the evolution of interface trap states between SiO_2 and *a*-IWO active layers [30]. As P_{O2} decreased from 13% to 7%, the μ_{FE} increased from 22.4 to 36.7 $cm^2V^{-1}s^{-1}$, and the SS and hysteresis decreased from 0.55 to 0.39 V/dec and from 0.29 to 0.06 V, respectively. It is known that the SS is proportional to the volume density of the traps N_t in the *a*-IWO active layer and the area density of the interface traps D_{it} at the interface of the *a*-IWO–gate insulator layer. Analysis confirmed that the N_t and D_{it} values for the sample prepared at a P_{O2} of 13% were smaller than those of the samples prepared at a P_{O2} of 7%. The decrease in the trap density is consistent with the reduction in the SS and hysteresis. The fine fitting of the V_{TH} shift indicated that the improvement mechanism of bias stability was caused by the reduction of charge trapping at the interface of the active layer and gate insulator layer. Qu et al. studied the influence of annealing atmosphere on the electrical performance and stability of the *a*-IWO-TFTs [31]. The annealed devices in different atmosphere exhibit much better hysteresis characteristics. The dependencies of the I-V curves and ΔV_{TH} values on the NBS and positive bias stress (PBS) times present apparent improvement. The optimized performance of the O_2-annealed TFT with a μ_{SAT} of 27.55 $cm^2V^{-1}s^{-1}$, an I_{ON}/I_{OFF} larger than 10^8, a SS of 0.2 V, a V_{TH} of 0.5 V, and better stability can be obtained. The result can be explained by the reduction of oxygen vacancies and structural defects.

Shan et al. utilized an eco-friendly solution process with ethanol and water as solvents to form an *a*-IWO active layer and water-induced alumina gate insulator layer to prepare *a*-IWO-TFTs [32]. The electrical performance of the devices was systematically investigated as a function of tungsten concentration and annealing temperature. The solution-processed *a*-IWO/AlO_x-TFT presented high performance and stability with a μ_{FE} of 15.3 $cm^2V^{-1}s^{-1}$, an I_{ON}/I_{OFF} of 5×10^7, a small SS slope of 68 mV/dec, and a V_{TH} shift of 0.15 V under PBS for 2 hours. Solution-processed TFTs with crystalline IWO as active layers and Y_2O_3 as gate insulators were reported [33]. It was found that tungsten doping shifts up the Urbach tail energy of IWO films. NBS and PBS under dark ambient conditions of TFTs employing IWO (0.1 at.%) showed remarkable improvement in their stability characteristics compared to the undoped ones. For flexible AOS-TFTs, the carrier concentration and channel conductivity may be limited by the thermal budget of the flexible substrate. Ruan et al. reported flexible *a*-IWO-TFTs on a transparent polyimide substrate [34]. With a precise control of the tungsten content and active-layer thickness, high mobility and small SS value TFTs can be achieved via a 150 °C low-temperature process. Mathews's group demonstrated the usage of IWO active layers for flexible TFTs and investigated the role of its physical and optical characteristics on the performance and stability of the devices [35]. The 7-nm-thick IWO layers were formed onto the thermal atomic-layer-deposited Al_2O_3 at room temperature using RF magnetron sputtering. The IWO film, annealed at 270 °C for 30 min in air, showed a carrier concentration of 7.19×10^{18} cm^{-3} and a smaller contact angle of about 35.2°, possibly due to lowered surface roughness ($R_{rms} \approx 0.42$ nm) and the lower concentration of adsorbed organic molecules. The higher surface energy of the film implies better adhesion between the semiconductor and the ITO S/D contacts for TFTs. The device performance of μ_{FE} of 25.86 $cm^2V^{-1}s^{-1}$ and SS of 0.30 V/dec was reached. The transfer characteristics measured during the convex bending radius of 20 mm showed excellent flexibility and mechanical stability, with the $\mu_{FE} \approx 25$ $cm^2V^{-1}s^{-1}$ and almost unchanged SS and V_{TH} values. Moreover, synaptic behaviors can be extracted from IWO-TFTs, indicating TFTs' possible usage as building blocks of next-generation neural circuits and great potential applications in various flexible devices. Schmitt triggers with adjustable hysteresis windows based on ion–electron hybrid IWO-based TFTs were reported [36]. Such ion–electron hybrid devices are interesting for noise filtering, electrochemical sensors, and neuromorphic systems.

Switching TFTs in display applications are almost inevitably exposed to light during operation. To investigate the stability of *a*-IWO-TFTs under various light illumination conditions, the bottom-gate top-contact *a*-IWO-TFTs were prepared by RF magnetron sputtering at room temperature and annealed at 100 °C in air for 30 min [37]. The typical I_D V_{GS} curve shows good performance, with a μ_{FE} of 26.5 $cm^2V^{-1}s^{-1}$, SS value <0.5 V, I_{ON}/I_{OFF} >10^7, V_{TH} of −2.5 V, and hysteresis smaller than 0.3 V. Three wavelength lights of red light ($\lambda = 650$ nm, $E_{ph} = 1.91$ eV), green light ($\lambda = 520$ nm, $E_{ph} = 2.39$ eV), and blue light ($\lambda = 450$ nm, $E_{ph} = 2.76$ eV) at a power density of 0.1 mW/cm^2 were employed in the illumination stability measurements from 100 to 1,000 s. Figure 26.5a–26.5c shows that the shift of the transfer curves was wavelength and illumination time dependent. The changes in V_{TH}, μ_{FE}, SS, and the I_{ON}/I_{OFF} ratio become more obvious with shorter wavelengths and increasing illumination time. As is widely known, the relationship between the variation of SS and total trap density N_t within the active layer can be analyzed by using the following equation: $\Delta SS = \Delta N_t \ln(10)kT / C_i$, where k is the Boltzmann's constant, T the absolute temperature, and C_i the capacitance per unit area of the gate insulator layer. Clearly, the increase in ΔSS is related to the increase in ΔN_t. In other words, the N_t becomes larger when λ is shorter or illumination time is longer. Generally, the defect-induced subgap states are frequently related to oxygen vacancies (V_O), and neutral oxygen vacancies (fully occupied states) locate above the valence band maximum in high density. These neutral oxygen vacancies can be ionized by different-wavelength light illumination. Meanwhile, the band gap of In_2O_3 is so wide (~3.75 eV) that the electron excited from the deep trap state to the conduction band under illumination is difficult. The proposed mechanism for the creation of $V^{+}{}_{O}$ and $V^{2+}{}_{O}$ states under short-wavelength light illumination is

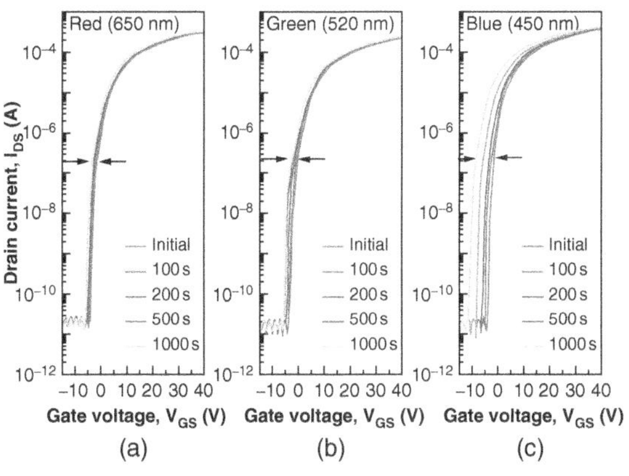

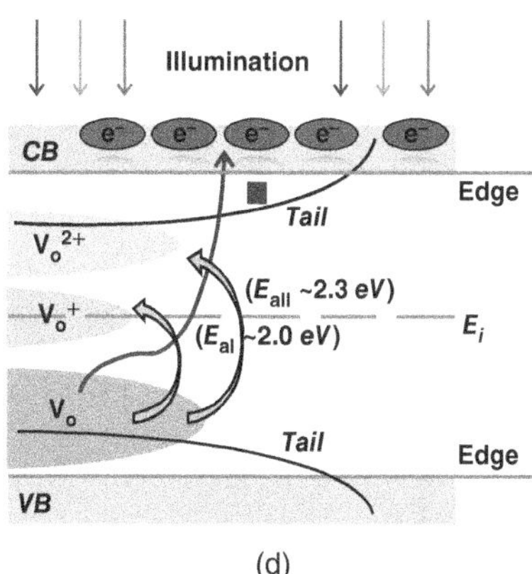

Figure 26.5 The measured transfer curves of *a*-IWO-TFTs before and during light illumination with different illumination times: (a) $\lambda = 650$ nm, (b) $\lambda = 520$ nm, and (c) $\lambda = 450$ nm. (d) Schematic of the proposed creation of $V^+{}_O$ and $V^{2+}{}_O$ states under short-wavelength light illumination. *Source*: Reprinted with permission from Yang, Z., et al. (2016). Stability of amorphous indium-tungsten oxide thin-film transistors under various wavelength light illumination. *IEEE Electron Device Letters* **37** (4): 437–440.

shown in Figure 26.5d. There are two steps: (i) the transition from V_O to $V^+{}_O$, and (ii) the transition from V_O to $V^{2+}{}_O$. And the activation energies needed for the transitions were reported to be ~2.0 eV (E_{aI}) and ~2.3 eV (E_{aII}), respectively. When irradiated by the red light of $E_{ph} = 1.91$ eV, which is slightly below the E_{aI} of ~2.0 eV for the transition from V_O to $V^+{}_O$, the device performance remains almost unchanged. When E_{ph} (λ <550 nm) is higher than E_{aII} (~2.3 eV), neutral oxygen vacancies excited to single-ionized ($V^+{}_O$) and double-ionized ($V^{2+}{}_O$) oxygen vacancies increased and resulted in the obvious SS degradation. Therefore, it can be concluded that short-wavelength light provides enough energy to ionize oxygen vacancies, which results in the increasing of N_t and SS degradation of the illuminated devices. In spite of this, *a*-IWO-TFTs still show relatively good light illumination stability.

Junctionless transistors with nanosheets of *a*-IWO as an active channel layer were demonstrated [38]. Figure 26.6a shows the structural diagram of *a*-IWO nanosheet junctionless transistors (NS-JLTs). HfO_2 gate insulator (GI) layers with thickness of 10 nm, 20 nm, or 30 nm were deposited by atomic layer deposition (ALD). An active layer of 4-nm-thick or 10-nm-thick *a*-IWO was deposited by RF magnetron sputtering at room temperature. To investigate the influences of different GI materials on the devices, the *a*-IWO NS-JLTs in bottom Si-sub-gate (BSG) configurations with an *a*-IWO NS channel = 4 nm were also fabricated on a heavily doped *n*-type Si wafer with a 30-nm-thick, high-quality thermal SiO_2 GI, as shown in Figure 26.6b. Figure 26.6c displays cross-sectional transmission electron microscope (TEM) images of the *a*-IWO NS-JLTs. The thickness of the *a*-IWO NS channel and HfO_2 GI are 4 nm and 10 nm, respectively. Obviously, the interfacial layer between the HfO_2 GI and the *a*-IWO NS channel is negligible, resulting in near-ideal SS and improved hysteresis characteristics. Figure 26.6d displays transfer characteristics of *a*-IWO NS-JLTs with different channel lengths. The extracted V_{TH} roll-off of *a*-IWO NS-JLTs with different channel lengths is plotted in the inset of Figure 26.6. As the channel length scaling down, the V_{TH} roll-off is a key parameter to verify the gate controllability over the channel region. The *a*-IWO NS-JLTs with different channel lengths have almost identical SS's of ~63 mV/dec and similar I_{OFF} characteristics, whereas the I_{ON} is nearly proportional to channel length. Thus, values of I_{ON}/I_{OFF} larger than 1×10^9 can be obtained for a device with 5 μm channel length at operating conditions of $V_{GS} - V_{TH} = 3$ V and $V_{DS} = 0.1$ V. The *a*-IWO NS-JLTs with very small V_{TH} roll-off exhibit high gate controllability and good short-channel-effects immunity due to the combined use of an *a*-IWO NS channel and thinner HfO_2 GI in the devices. The remarkable device characteristics also make the *a*-IWO NS-JLTs promising for system-on-panel and vertically stacked hybrid complementary metal-oxide semiconductor (CMOS) applications.

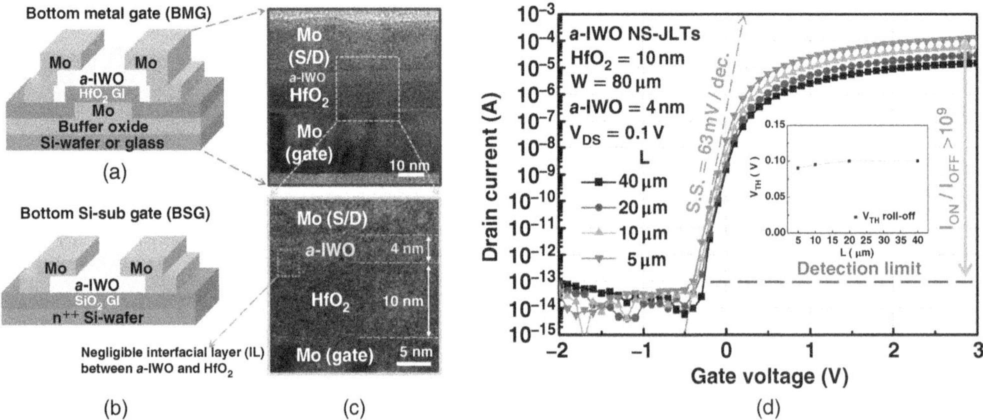

Figure 26.6 Structural diagrams of (a) proposed BMG *a*-IWO NS-JLTs with HfO_2 GI and (b) BSG *a*-IWO NSJLTs with SiO_2 GI; (c) cross-sectional TEM images of proposed BMG *a*-IWO NS-JLTs. (d) The transfer characteristics of *a*-IWO NS-JLTs with different channel lengths. The inset shows the extracted V_{TH} roll-off. *Source*: Adapted from Kuo, P. Y., et al. (2019). Two-dimensional-like amorphous indium tungsten oxide nano-sheet junctionless transistors with low operation voltage. *Scientific Reports* **9**: 7579.

26.2.2 *a*-IZWO-TFTs

Li et al. prepared *a*-IZWO-TFTs by RF magnetron sputtering at room temperature [39]. The XRD patterns of IZWO thin films with different W doping content show no diffraction peak until the sample annealed at 800 °C, indicating the amorphous structure of the films. XPS analysis indicates that when the W molar ratio increases from 0% to 11%, the ratio of the O 1*s* peaks related to the oxygen deficiency decreases from 50.6% to 38.9%, implying that the W doping actually suppresses the oxygen vacancies because of the stronger bond dissociation energy of W-O than that of In-O and Zn-O. This also can be explained by the fact that the electronegativity difference between W (1.47) and O (3.61) is larger than those of In (1.66) and Zn (1.59). In addition, W atoms in the films are fully oxidized to the W^{6+} ions state. Figure 26.7a shows the transfer characteristic curves as a function of tungsten doping contents. Apparently, the *a*-IZWO-TFTs have a lower off-current compared with that of the *a*-IZO-TFTs. Although the on-current also decreased, the total I_{ON}/I_{OFF} ratio is increased. At a W molar ratio of 6.2%, the μ_{FE} reaches 11.1 $cm^2V^{-1}s^{-1}$, the I_{ON}/I_{OFF} ratio increased to the maximum value of 10^7, and the SS decreased to the smallest value of 0.31 V/decade. The V_{TH} also exhibits a positive shift. But excessive doping content will degrade the performance obviously. These results reveal that the tungsten incorporation in the films actually serves as a carrier suppressor. The output curves shown in Figure 26.7b exhibit a typical *n*-channel field-effect character with a good pinch-off and hard saturation behaviors.

Kizu et al. demonstrated the *a*-IWZO-TFTs of a high mobility of 40 $cm^2V^{-1}s^{-1}$ by using a practical high-density sputtering target with a tungsten composition of 1 at.% and zinc of 1 at.% [40]. The oxygen vacancies from In-O and Zn-O were suppressed by the W incorporation because of the high oxygen bond dissociation energy. Figure 26.8 shows the transfer characteristic curves and typical output curves of the *a*-IZWO-TFTs with different active-layer compositions. All the devices exhibited good electrical performance with a low off-current of less than 10^{-12} A and a high I_{ON}/I_{OFF} ratio of more than 10^{10}. The I_D-V_G curves shift positively with increasing W concentration, indicating the carrier density reduction caused by the suppression of oxygen vacancies. The stability under NBS tests is particularly important because the duration of the off-states in the TFT operation is longer than the on-state in active-matrix liquid crystal display (AMLCD) operations. Under $V_G = -20$ V for 5,000 s, the V_{ON} shift can be decreased to 0.1 V when the W concentration increased. To clarify the role of tungsten dopant, Park et al. evaluated the device performance and stability of IZWO-TFTs as a function of the W doping concentration [41]. They investigated the electrical performance in terms of the evolution of the band alignment, chemical-bonding

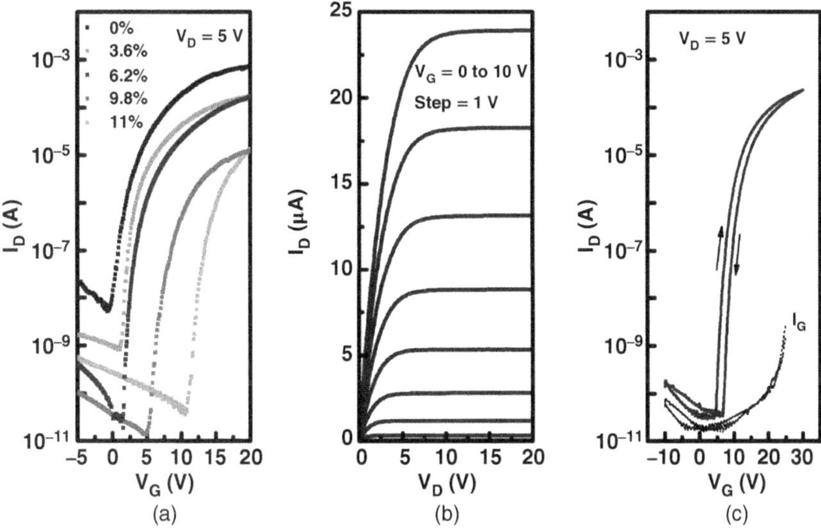

Figure 26.7 (a) Comparison of the transfer characteristics of *a*-IZWO-TFTs with different W doping content. (b) Output characteristics and (c) hysteresis curves of TFTs with 6.2% W doping content. *Source*: Reprinted with permission from Li, H. L., et al. (2013). Influence of tungsten doping on the performance of indium-zinc-oxide thin-film transistors. *IEEE Electron Device Letters* **34** (10): 1268–1270.

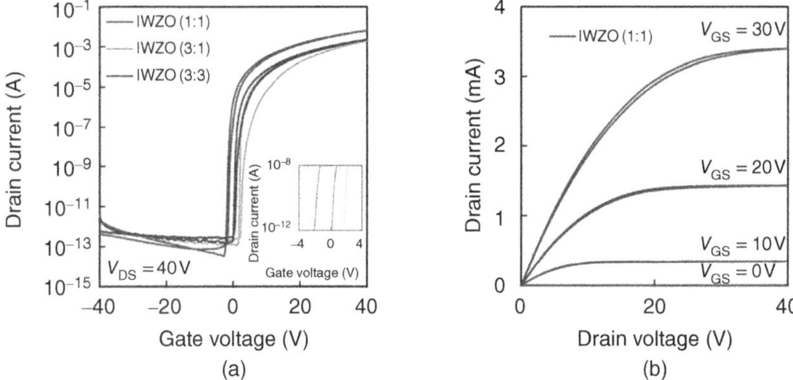

Figure 26.8 (a) The transfer characteristics of IWZO-TFTs with different film compositions. The inset shows the transfer characteristics in the gate–source voltage (V_{GS}) range of −4 V to 4 V. (b) Typical output characteristics of an IWZO (1:1) TFT. *Source*: Reprinted with permission from Kizu, T., et al. (2015). Codoping of zinc and tungsten for practical high-performance amorphous indium-based oxide thin film transistors. *Journal of Applied Physics* **118** (12): 125702.

states, and band-edge states, and found that tungsten doping concentration is a critical factor in changing the charge-trapped defects in the IZWO active layer and/or interface between the active layer and SiO_2 gate insulator layer. With increasing W doping content, the difference between the conduction band minimum and Fermi level decreased, whereas the relative concentration of W^{4+} and W^{5+} ions decreased, though this phenomenon is somewhat difficult to understand. At W doping content of 1.1%, the TFT showed optimized electrical performance of 18.09 $cm^2V^{-1}s^{-1}$ and remarkable improvement of the V_{TH} shift under NBS. The influence of annealing temperature on *a*-IZWO-TFTs has been investigated [42]. The μ_{FE} of the device increases as a function of annealing temperature and reaches 16.2 $cm^2V^{-1}s^{-1}$ at 300 °C, along with an I_{ON}/I_{OFF} ratio of $1.6{\times}10^8$. Meanwhile, there is a better electrical stability under PBS, benefiting from fewer oxygen-vacancy-related deep defects. It was reported that tungsten incorporation (0.08 at.%) in IZO improved the film quality by improving the surface roughness value (0.16 nm) and reducing the tail states ($\Delta E \sim 530$ meV) within the band gap of the semiconductor [43]. The IZO and IWO co-sputtered IZWO-TFT prepared even at 300 °C exhibits high electrical performance with a μ_{FE} of 22.30 $cm^2V^{-1}s^{-1}$.

The interfacial state between the oxide active layer and the adjacent layers plays an important role due to the stacked structure of the devices [44]. It is difficult to form sound electrical contacts between an AOS layer and S/D electrodes, and a highly conductive S/D electrode material is normally used. The contact resistance between the semiconductor and the S/D electrode is another factor for selecting a material and a process for electrodes. High-contact resistance may induce the current crowding at the output characteristics, and the signal delay will be increased even though the resistance is sufficiently low. Thermal annealing, UV irradiation, and plasma treatment were used to improve the contact resistance between the active layer and the S/D electrode. But some treatment like plasma treatment will be carefully applied; otherwise, performance degradation may occur. Park et al. fabricated homojunction- and heterojunction-structured *a*-IZWO-TFTs with two types of S/D electrode material, the conducting IZWO and ITO electrode. The homojunction-structured IZWO-TFT exhibited a superior device performance with a μ_{SAT} of 26.10 $cm^2V^{-1}s^{-1}$, a SS of 0.38 V/dec, and a hysteresis-based ΔV_{TH} within 0.07 V. It is thought to originate from the flat interfacial roughness and a small number of electron-trap sites (Figure 26.9). The electron barrier of the homojunction structure is 0.09 eV, which is lower than 0.21 eV of the heterojunction structure. The results suggest that control of the interfacial state is significant in the fabrication of high-performance TFT devices.

The dependency of the device performance of *a*-IZWO-TFTs on the film density, surface/interface roughness, band-edge state below the conduction band, refractive index, and composition along the depth direction was investigated [45]. At the optimized thickness of 10 nm, a high μ_{FE} of 19.57 $cm^2V^{-1}s^{-1}$ and small SS of 0.14 V/dec were obtained. The degradations of the device performance with increasing of active-layer thickness are attributed to the low film density, and a rough interface between the *a*-IZWO–SiO_2 gate insulator layer. A sequential ambient annealing process as an effective posttreatment method to enhance the device performance and stability of IZWO-TFTs was proposed [46]. Half of the time annealing in air and half of the time annealing in vacuum at 250 °C significantly enhanced the device performance and stability. The dominant mechanism responsible for the enhanced device performance and stability is considered to be a change in the shallow-level and deep-level band-edge states below the conduction band.

It is necessary to minimize the size of TFTs for high-resolution display technology. With shortening the channel length, short-channel effects will appear. One of the approaches to suppress this effect is to prepare a thinner active layer. The TFTs with an ultrathin active layer of *a*-IZWO and high-k gate insulator layer of HfO_2 were fabricated and investigated [47]. The SS and on currents of the devices are improved by an ultrathin active layer and scaling down the thickness of the HfO_2 gate insulator layer to 10 nm thick, owing to the good electrostatic

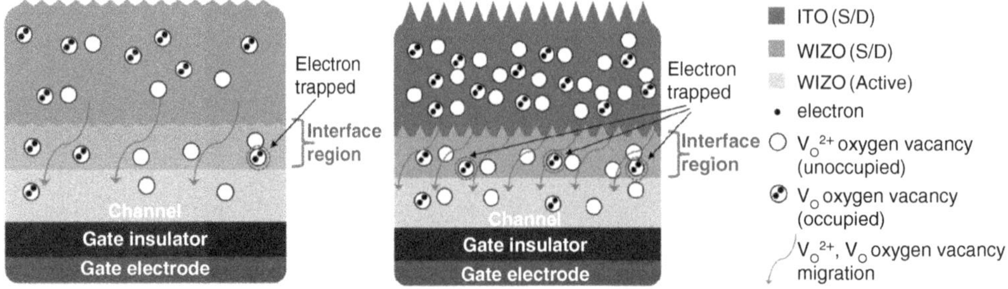

Figure 26.9 Schematic illustration of an electron-transport mechanism according to the S/D electrode materials. *Source*: Adapted from Park, H. W., Song, A. R., and Choi, D. H. (2017). Enhancement of the device performance and the stability with a homojunction-structured tungsten indium zinc oxide thin film transistor. *Scientific Reports* **7**: 11634.

controllability in devices. Outstanding characteristics of a low V_{TH} of ~−0.072 V, high μ_{FE} of ~23.8 $cm^2V^{-1}s^{-1}$, excellent SS of ~72.6 mV/dec, and low-voltage operation were obtained. Liao et al. investigated IZO-TFTs with different W doping levels via a solution process [48]. A maximum μ_{FE} of 30.5 $cm^2V^{-1}s^{-1}$ is obtained in the 0.2-wt.% W-doped IZO-TFTs, and the 0.5-wt.% W-doped IZO-TFT exhibits the best bias stress stability. The pulsed I-V, hysteresis characterizations, and electrical stability investigation reveal that the W-doping has significantly improved the charge trapping that induced the hysteresis and the device bias stress instability by effectively suppressing the formation of oxygen vacancies in the W-doped IZO-TFTs. The instability mechanism of IZWO-TFTs induced by a gate-bias stress was investigated from the dynamic electronic structure such as the unoccupied states in the conduction band, and the band-edge states below the conduction band, through applying a real-time gate-bias stress [49]. The NBS did not significantly change the V_{TH} shift of the IZWO-TFTs. However, the PBS dramatically changed it. The dominant mechanism responsible for the degraded device stability is strongly correlated to the change in the unoccupied states in the conduction band and the band-edge states below the conduction band.

26.2.3 Dual Tungsten-Doped Active-Layer TFTs

Bias stress–induced instability of AOS-TFTs has always been observed during their operation, and it is particularly deteriorated by negative bias illumination stress (NBIS) rather than positive bias illumination stress (PBIS). In addition, the device characteristics of AOS-TFTs are largely influenced by the environmental effects (H_2O, O_2, etc.) that occur during the operation. The suitable backchannel passivation layers (BPLs) are thereby needed to ensure their long-term stability. However, defects generation may also appear during the plasma-enhanced chemical vapor deposition (PECVD) process due to the ion bombardment on the backchannel surface. Therefore, a dual-layer *a*-IWO-TFT with a top oxygen-rich active layer is proposed for studying the backchannel layer (BCL) effect on TFT performance [50]. XPS analysis revealed that the oxygen deficiencies formed at the backside of an IWO active layer were observed after the formation of SiO_2-BPL by a plasma-based deposition process. However, with an oxygen-rich IWO layer back-stacked on the IWO channel, the oxygen vacancies at the backside of the IWO channel can be passivated by the oxygen atoms. As a result, after NBIS testing at 60 °C for 2000 s, a superior device performance and the immunity to NBIS can be effectively enhanced for the dual-layer *a*-IWO-TFT. Tsuji et al. used the oxide semiconducting material IZWO to develop backchannel-etched (BCE)-TFTs [51]. The proposed dual-layer structure was shown to be effective for realizing both high resistance to backchannel etching damage and high TFT mobility of 20.2 $cm^2V^{-1}s^{-1}$. In the proposed dual-layer structure, each oxide semiconductor layer is formed using different sputtering targets, so that the sputtering conditions could be individually adjusted for each layer to ensure high device reliability.

To enhance the electrical stability of the *a*-IWO-TFTs, Lin et al. fabricated the double-stacked amorphous IWO–IWO:N channel-layer TFTs and investigated the electrical performance and reliability of the devices [52]. XPS analysis showed that nitrogen doping could effectively suppress the formation of oxygen vacancies. With a double-stacked channel configuration, the TFTs exhibited smaller hysteresis and higher stability than the *a*-IWO and *a*-IWO:N single-layer structured TFTs. The nitrogen doping reduced the trap density of the channel layer. And the *a*-IWO:N channel layer can also serve as the passivation layer, preventing the moisture or oxygen absorbed on the backchannel surface. As a result, device performance with a μ_{FE} of 27.2 $cm^2V^{-1}s^{-1}$ can be reached, and immunity to NBS and NBIS can be effectively enhanced for the dual-layer IWO/IWO:N-TFTs compared to the counterparts of single-channel-layer devices. The BCE-type TFTs enable the channel length to be narrow and reduce parasitic capacitances owing to the shorter overlaps between the gate and S/D electrodes. Li et al. proposed a high-performance BCE-type oxide TFT with a stacked dual-layer IWO–IWZO as the active layer [53]. Respectively, IZWO exhibits a high resistance to BCE damage, and the IWO channel achieves a high mobility. The dual active-layer IWO–IWZO TFTs exhibit high-performance electrical characteristics, with a μ_{FE} of ~21.1 $cm^2V^{-1}s^{-1}$, SS of ~0.15 V/dec, V_{TH} of ~−0.092 V, and I_{ON}/I_{OFF} ratio of ~4.88×10^8.

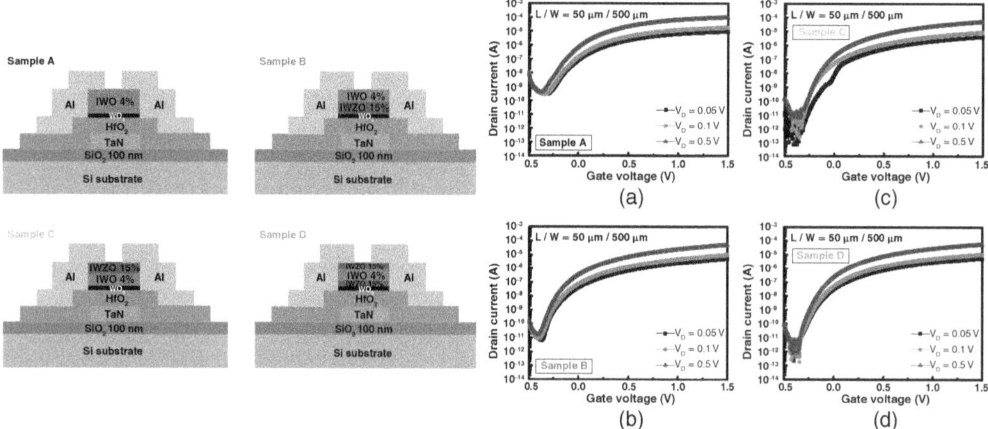

Figure 26.10 (Left) Schematic cross-sectional views of the bottom-gate staggered structures for the tungsten-doped multiactive-layer TFTs. (Right) The I_{DS}-V_{GS} curves of tungsten-doped TFTs with different multistacked channel structures under various V_{DS} for (a) sample A, (b) sample B, (c) sample C, and (d) sample D, respectively. *Source*: Reprinted with permission from Ruan, D. B., et al. (2018). The influence on electrical characteristics of amorphous indium tungsten oxide thin film transistors with multi-stacked active layer structure. *Thin Solid Films* **666**: 94–99.

To effectively enhance the carrier mobility and device stability simultaneously, a multistacked *a*-IWO active layer of TFTs is proposed [54]. A top-capping oxygen-rich IZWO thin film is used for suppressing the influence of the molecular adsorption or desorption from the ambient at the backchannel in the gate-bias stress test. A bottom-buffering oxygen-rich indium–tungsten–zinc oxide thin film is deposited to avoid oxygen vacancy generation during the following thermal process. On the other hand, a 1-nm-thick WO_3 film is inserted between the high-k gate insulator and active layer, which play important roles as the interfacial layer for improving the interface quality and keeping the stability of front-channel film. Besides, an HfO_2 film is chosen as the gate insulator layer for realizing the low-voltage operation (Figure 26.10). As a result, the tungsten-doped channel material with a both-sided oxygen-rich active layer exhibits a high I_{ON}/I_{OFF} ratio of $\sim 6\times10^7$ for low gate leakage current, attributed to the top-capping oxygen-rich thin film. Then, a high μ_{FE} of $\sim$27.9 $cm^2\ V^{-1}s^{-1}$ and a low SS of 0.079 V/dec are achieved by the good interface quality.

26.2.4 Treatment on the Backchannel Surface

Oxygen plasma treatment can effectively passivate the shallow oxygen vacancies, lower the channel conductivity, and achieve a desirable positive V_{TH} shift, but this is accompanied by a serious and inevitable degradation of carrier mobility. Fluorine plasma, with high electronegativity and a similar radius as oxygen, can be used to terminate donor-like oxygen vacancies by replacing the weakly bonded oxygen. It may introduce a positive V_{TH} shift while maintaining the carrier concentration and field-effect mobility, but may also introduce plasma damage and fluorine chemical etching. To simultaneously achieve good device stability and a desirable positive V_{TH} shift, as well as minimize the influence of the ion bombardment and etching damage without sacrificing the μ_{FE}, different types of remote plasma treatment processes with suitable power and gas flow were carried out for the *a*-IZWO-TFTs [55]. As a result, the device treated with CF_4/N_2+O_2 plasma exhibits a higher I_{ON}/I_{OFF} ratio of $\sim 4.73\times10^6$, a lower SS of 0.070 V/dec, a lower interfacial trap density value of $5.21\times10^{11}\ eV^{-1}cm^{-2}$, and a desirable positive V_{TH} shift and acceptable μ_{FE} of 31.2 $cm^2V^{-1}s^{-1}$. The experimental results confirmed that CF_4/N_2+O_2 plasma is an effective approach to improve the reliability and adjust the inevitable negative V_{TH} shift without sacrificing the carrier mobility of the devices. To prevent the diffusion of ambient hydrogen and oxygen into the AOS active layer, a good passivation layer is required to protect the devices from external contamination. Liu et al. investigated the effects of the backchannel passivation layer of SiO_2 (Device A) or Al_2O_3 (Device B) on the *a*-IWO–IGZO dual-active-layer TFTs under positive gate-bias stress (PGBS) [56]. The transfer characteristic curves of (a) Device A and (b) Device B under 25 V gate-bias stresses as functions of stress time show that the longer stress time causes large negative V_{TH} shifts in Device A but small positive V_{TH} shifts in Device B. After humidity stress (exposed to environment with 80% relative humidity for 4 hours), the V_{TH} shifts remarkably for Device A but is negligible for Device B. It is clear that TFTs with PECVD SiO_2 and ALD Al_2O_3 passivation layers exhibit different phenomena of electrical degradation, attributed to the difficulty levels of oxygen deficiency formation during passivation layer deposition as well as to the resistance of the passivation layer to moisture penetration. XPS analysis verified that the content of oxygen deficiency is obviously reduced by ALD Al_2O_3. The ALD Al_2O_3 film also exhibits superior resistance to moisture penetration compared to the PECVD SiO_2 film. This confirmed that the quality of the backchannel passivation layer plays an important role in the electrical degradation behavior of AOS-TFTs.

Liu et al. studied the mobility enhancement effect of passivation layers on the *a*-IZWO channel-layer TFTs [57]. Two types of passivation materials of HfO_2 and Al_2O_3 were selected to be deposited on the backchannel surface to resist gas and H_2O absorption. The *a*-IZWO channel layer was deposited by RF magnetron sputtering at room temperature and then annealed at 300 °C in oxygen atmosphere for 20 min. After patterned source and drain electrodes formed, a 30-nm-thick Al_2O_3 or HfO_2 BPL was deposited by e-gun evaporation to avoid plasma-induced ion bombardment effects, which may break oxygen bonds and generate excess oxygen vacancies, consequently leading to the carrier concentration increase and unexpected surface roughness. Figure 26.11a shows the transfer characteristic curves and μ_{FE} of the *a*-IZWO-TFTs with an Al_2O_3 BPL, a HfO_2 BPL, and the monitor TFT without a BPL, respectively. Compared to the monitor TFT with a peak μ_{FE} of 5.67 $cm^2V^{-1}s^{-1}$, the TFT with Al_2O_3 BPL presents a high peak μ_{FE} of 52.5 $cm^2V^{-1}s^{-1}$, whereas the TFT with HfO_2 BPL shows a highly conductive channel that cannot be depleted by a reasonable gate bias in the process conditions. According to the XPS analysis, the TFT with a HfO_2 BPL shows a higher proportion of donor-like oxygen vacancy bonds than the TFT with a Al_2O_3 BPL does. Meanwhile, the TFTs with a BPL show a larger decrease in oxygen–hydroxide signals than a monitor TFT without a passivation layer does, indicating that the BPL layer can block hydrogen diffusion and H_2O absorption effectively. Figure 26.11b shows the schematic band diagrams of the *a*-IZWO-TFT with and without a BPL. The difference in channel conductivity between the two BPL samples is thought to be due to three reasons: (i) The difference in Gibbs free energy between HfO_2 (−260.1 kcal/mol) and Al_2O_3 (−377.9 kcal/mol) may influence their ability to react with oxygen in the channel. (ii) The oxygen bond dissociation energies of Al-O of 791 kJ/mol and Hf-O of 512 kJ/mol are higher than that of In-O of 360 kJ/mol. This means that the oxygen in the *a*-IZWO channel can be desorbed by the BPL and cause an increase in oxygen vacancies with donor-like state behavior in the channel. (iii) There is a slight

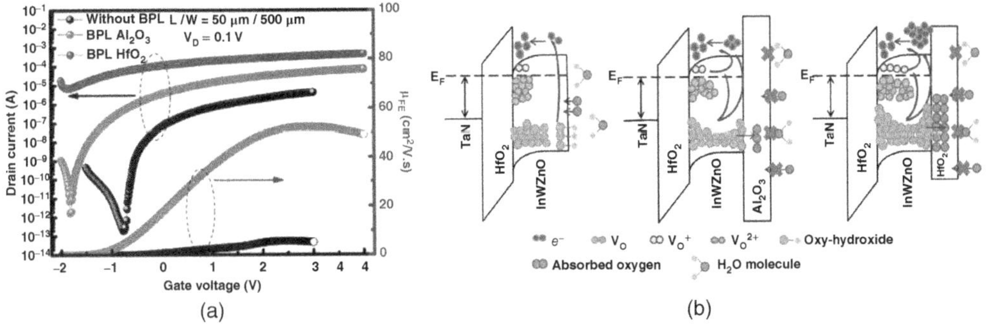

Figure 26.11 (a) The transfer characteristics and field-effect mobility of two types of backchannel passivation TFT device structures. (b) Schematic band diagrams of *a*-IZWO-TFTs with or without a BPL, which conceptually depict the mechanism of mobility enhancement caused by the increase in carrier concentration. *Source*: Reproduced from Ruan, D. B., et al. (2018). Mobility enhancement for high stability tungsten doped indium-zinc oxide thin film transistors with a channel passivation layer. *RSC Advances* **8** (13): 6925–6930.

increase in oxygen vacancy bonds at both the BPL–a-InWZnO and a-InWZnO–HfO_2 gate insulator interfaces. This can be attributed to more oxygen precipitation occurring at the back side and front side of the channel layer.

TFT properties are strongly affected by the annealing treatment, which may eliminate defects in bulk and at the interface of the channel layer. However, most of the reports on annealing effects were on nonpassivated TFTs. Furuta et al. investigated the influence of a PECVD-deposited SiO_2 passivation layer on the electrical properties and reliability of an IZWO-TFT under various postannealing temperatures T_A [58]. Although the TFT without passivation showed good transfer characteristics when the T_A is 150 °C, it has large hysteresis and poor reliability. Furthermore, the TFTs without a passivation layer changed from transistor to conductor when the T_A was 200 °C or higher. In contrast, the TFTs with passivation exhibited a switching property even at a T_A of 350 °C, and exhibited their best performance with a SS of 0.09 V, a V_{TH} of −0.1 V, and a μ_{FE} of 33.7 $cm^2V^{-1}s^{-1}$. The PBTS reliability of the TFTs with a passivation layer was significantly improved by increasing the T_A. After PBTS was applied at 60 °C for 10,000 s, the ΔV_{TH} gradually improved by increasing T_A and was extremely suppressed to +0.5 V at T_A = 350 °C (Figure 26.12). It can be concluded that a passivation layer is essential to increase the T_A, resulting in improvement of the electrical properties and reliability of IZWO-TFTs.

Supercritical-phase fluid (SCF) treatment was generally used to improve the performance of dielectric properties in resistive random-access memory, or to passivate defect states in hydrogenated amorphous silicon TFTs without any process damage. By means of a special high-pressure system, the supercritical-phase carbon dioxide ($SCCO_2$) can even be achieved around room temperature with the unique characteristics of high liquid-like solubility and a strong gas-like penetration ability. Liu et al. reported research work on high-mobility a-IWO-TFTs by introducing oxygen-rich hydrogen peroxide (H_2O_2) into a room-temperature $SCCO_2$ system as a special cosolvent and forceful oxidant for a-IWO active-layer modification and interface engineering [59]. The basic electrical characteristics of IWO-TFT devices without any treatment or with only vapor cosolvent treatment are relatively poor, especially for the on-off current ratio due to the high carrier density of the active layer (Figure 26.13a). Notably, the device with $SCCO_2 + H_2O_2$ treated exhibits the lowest SS of 82 mV/dec, the highest I_{ON}/I_{OFF} ratio of 5.11×10^7, the lowest D_{IT} of 8.76×10^{11} $eV^{-1}cm^{-2}$, and a sufficiently high μ_{FE} of 98.91 $cm^2V^{-1}s^{-1}$. Compared with the samples treated only with $SCCO_2$ fluid or with H_2O cosolvent in the SCF system, the remarkable enhancement of electrical performance can be attributed to the more forceful oxidation ability of the oxygen-rich hydrogen peroxide cosolvent. It can more effectively passivate the oxygen vacancy and repair the dangling bonds for the high-mobility channel material, even at room temperature. As a result, the smallest V_{TH} shift under both PGBS and NGBS is obtained (Figure 26.13b–26.13c).

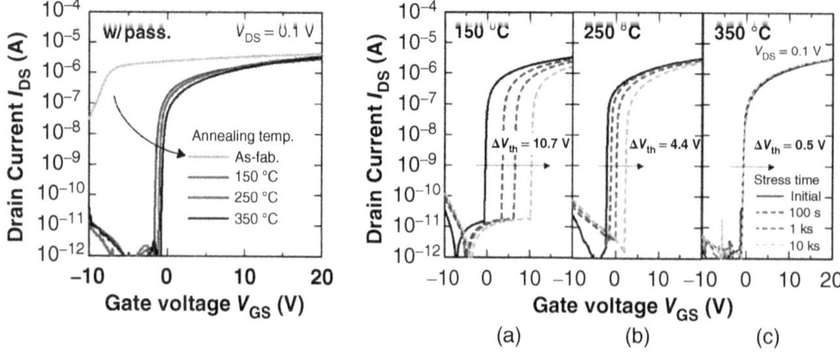

Figure 26.12 The dependency of the transfer characteristics of IZWO-TFTs with the passivation layer on the postannealing temperature T_A. The changes in transfer characteristics of the IZWO-TFTs with passivation at T_A of (a) 150 °C, (b) 250 °C, and (c) 350 °C under the PBTS at 60 °C (V_{DS} = 0.1 V, W/L = 1000/690 μm). *Source*: Reproduced with permission from Koretomo, D., Hashimoto, Y., and Hamada, S. (2019). Influence of a SiO_2 passivation on electrical properties and reliability of In-W-Zn-O thin-film transistor. *Japanese Journal of Applied Physics* **58** (1): 018003.

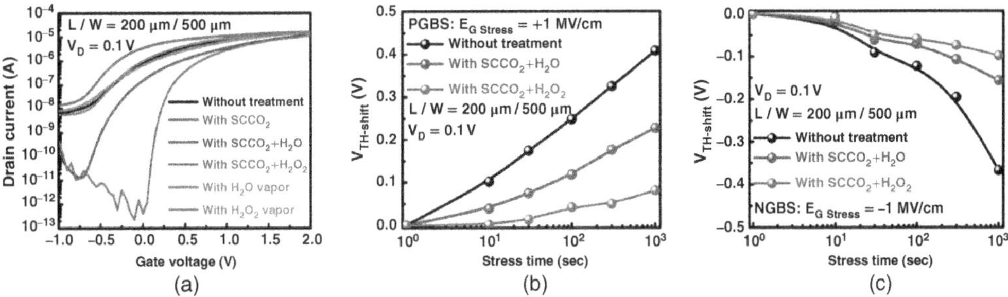

Figure 26.13 (a) Transfer characteristics of *a*-IWO-TFT devices with or without different cosolvent treatments in the SCF or vapor system. The V_{TH} shift as a function of (b) 1 MV/cm PBS and (c) −1 MV/cm NBS stress time for *a*-IWO-TFTs with or without different cosolvent treatments in the SCF system. *Source*: Reprinted with permission from Ruan, D. B., et al. (2019). Performance enhancement for tungsten-doped indium oxide thin film transistor by hydrogen peroxide as cosolvent in room-temperature supercritical fluid systems. *ACS Applied Materials & Interfaces* **11** (25): 22521–22530.

Schematic diagrams that conceptually depict the mechanism of performance enhancement of the *a*-IWO active layers with or without different cosolvent treatments in the SCF system are shown in Figure 26.14. It is known that the oxygen deficiency of different trap states can be significantly suppressed by an oxygen vacancy passivation process or an additional carrier suppressor dopant like Ga and W. In this study, the deep traps generated by dangling bonds or deep defects in the small or nanocrystal particle boundaries can be passivated by oxygen–hydroxide chemical groups. However, those bonds or defects passivated by oxygen–hydroxide groups are not perfect enough due to internal unreleased stress, which may induce lots of deep traps transforming into shallow traps and might cause an undesirable reliability issue in the device. In contrast, the forceful H_2O_2 oxidant can repair the dangling bonds without leaving stress in the film by an additional oxygen atom, instead of the weak strain bonds for the oxygen–hydroxide chemical group. Therefore, the SS, μ_{FE}, I_{ON}, and D_{IT} of the *a*-IWO-TFTs with $SCCO_2 + H_2O_2$ treatment can be improved significantly.

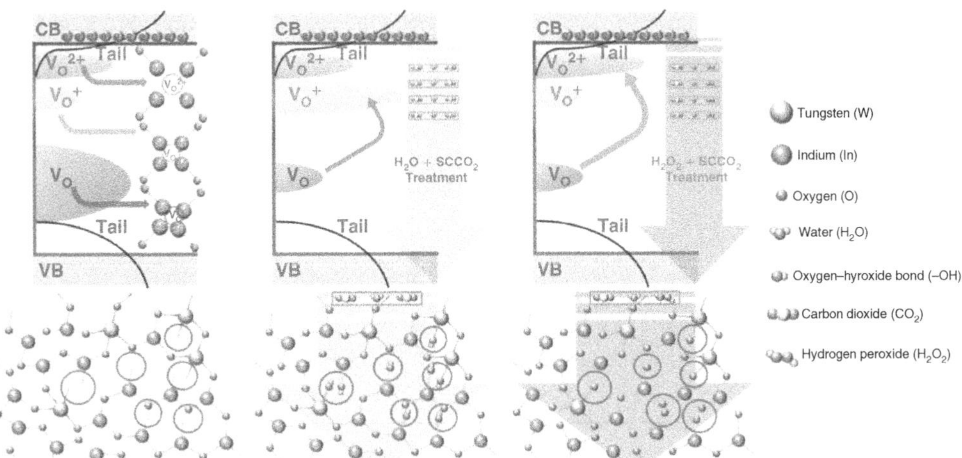

Figure 26.14 Schematic diagrams to conceptually depict the mechanism of performance enhancement of *a*-IWO active layers with or without different cosolvent treatments in the SCF system. *Source*: Reprinted with permission from Ruan, D. B., et al. (2019). Performance enhancement for tungsten-doped indium oxide thin film transistor by hydrogen peroxide as cosolvent in room-temperature supercritical fluid systems. *ACS Applied Materials & Interfaces* **11** (25): 22521–22530.

26.3 Perspectives for High-Mobility AOS Active Layers

Oxygen stoichiometry proves to be the decisive mechanism to generate free electrons in AOSs. Dopant and cation composition play important roles to control the oxygen–metal ratio, the degree of electron concentration, and carrier transport properties governed by composition-dependent characteristics of variable range hopping and percolation conduction [3]. The review and discussion of Section 26.2 indicate that tungsten-doped indium oxide–based TFTs show enhanced performance and improved stability. Tungsten dopant possesses the unique features of high oxygen bond dissociation energy, stronger electronegativity and Lewis acid strength compared to those of indium ions, almost the same cation radius as indium ions, and a high valence difference from host indium ions. In selecting dopant and developing a high-mobility active layer for AOS-TFTs, the following perspectives should be taken into consideration:

1. *Bond dissociation energy*: The bond dissociation energy (enthalpy change) for a bond A-B that is broken through the reaction $AB \rightarrow A + B$ is defined as the standard-state enthalpy change for the reaction at a specified temperature. The larger the oxygen bond dissociation energy, the stronger the binding force between the metal dopant and oxygen. The substitutional doping level and the type of incorporated metal species have a strong effect on the robustness of the metal–oxygen (M-O) chains and the AOS's nanostructure, and thus determine the oxygen vacancy concentrations and the limits of the carrier mobility in the amorphous regime [3]. Furthermore, the strength of the M-O bonding determines the clarity of the local polyhedral structure. The local distortions may affect the medium-range structure (e.g., edge- or corner-sharing between the M-O polyhedra and their integration into an extended network) and facilitate the formation of structural defects that govern the degree of electron localization near the valence and conduction band edges and deep inside the band gap.
2. *Electronegativity and Lewis acid strength*: Pauling originally defined electronegativity as "the power of an atom in a molecule to attract electrons to itself." Actually, when an element possesses different valence states, the electronegativity of the element is a function of their oxidation numbers. The higher the charge number of an element in a compound, the more strongly its atom attracts electrons [60]. Hence, the electron-attracting power of an element is entirely different when it is in different valence states. Therefore, the electronegativity of the element in valence states can be more physically defined as "the electrostatic force exerted by the effective nuclear charges on the valence electrons." Accordingly, the stability of a metal complex (the strength of the metal–ligand bond) should be the function of the electron-attracting power of the metal. Lewis acid is an acid that is an electron pair acceptor. Lewis acid strength could be composed of electrostatic force and covalent properties. An empirical equation of Lewis acid strength was proposed by Zhang [61]; it is $L = Z/r^2 - 7.7\chi + 8.0$. The first term Z/r^2, where Z is the charge number of the atomic core and r the ionic radius, is related to electrostatic force. The second term, the electronegativity of elements in valence states χ, is related to covalent bond strength. Obviously, Lewis acid strength is related to the effective charge, ionic radius, and electronegativity of the dopant. The high Lewis acid dopants polarize the electronic charge away from the oxygen 2*p* valence band more strongly than the weaker Lewis acids; this results in screening of the charge and weakens its activity as a scattering center, hence increasing the mobility. Kwon et al. discussed and reviewed the correlation of electrical performance and stability of the metal ion–doped AOS-TFTs and the Lewis acid strength and bonding strength of the dopants [62].
3. *Ionic radius and valence difference*: The ionic radius of the dopant has also a strong effect on the ionized carrier scattering, the strength of the M-O bond, and the M-O polyhedral. A smaller or larger dopant ion radius may change the extent of corner-sharing and edge-sharing M-O polyhedra or the number of voids in the active layer, leading to a great degree of local disorder, which may influence the carrier mobility in an extended range.

A high valence difference between the dopant ions with host ions may contribute more electrons if the substitutional doping effect existed in the nanocrystals or clusters in the amorphous structure. In this case, to keep the same amount of carrier concentration in the active layer, a smaller amount of high-valence-difference dopant ions is needed, and thus the number of ion impurity scattering centers will be decreased. Otherwise, the number of oxygen vacancies can be decreased, and the lost portion of carriers can be compensated by the high-valence-difference dopant. In addition, dopant may play a role in the quality or density of the active layer, the smoothness of the surface, or the interface.

4. *Quasi-2D free-electron layer*: Two-dimensional electron gas (2DEG) formation in ZnO–ZnMgO heterointerface by radical source molecular-beam epitaxy was reported [63]. It was shown that 2DEG could also form in ZnO–ZnMgO heterointerfaces grown on glass substrates via RF sputtering. Under certain circumstances, electron transfer and confinement at the heterointerface can occur because of conduction band offset in a process similar to that observed in conventional high-electron-mobility transistors made of AlGaAs–GaAs heterointerface channels. In these devices, the formation of a 2DEG at the critical heterointerface enables the realization of transistors with electron mobilities close to the theoretical limit set by phonon scattering in the absence of impurity scattering in the semiconductor [64]. By replacing the single-layer semiconductor channel with a low-dimensional, solution-grown In_2O_3–ZnO heterojunction, it was found that In_2O_3–ZnO transistors exhibit band-like electron transport, with mobility values significantly higher than those of single-layer In_2O_3 and ZnO devices by a factor of 2 to 100 [65]. This marked improvement is shown to originate from the presence of free electrons confined on the plane of the atomically sharp heterointerface induced by the large conduction band offset between In_2O_3 and ZnO. The existence of IWO/IWO-TFTs, IZWO/IZWO-TFTs, and IWO/IWO:N-TFTs with over $20\,cm^2V^{-1}s^{-1}$ mobility and good stability indicates that the dual-active-layer method is effective for enhancing the electrical performance and stability of AOS-TFTs.
5. *Defect modulation doping*: The classical doping approaches have in many cases reached their limits in regard to both achievable charge carrier density as well as mobility. Modulation doping, a mechanism that exploits the energy band alignment at an interface between two materials to induce free charge carriers in one of them, is shown to circumvent the mobility restriction. Due to an alignment of doping limits by intrinsic defects, however, the carrier density limit cannot be lifted using this approach. Defect modulation doping, a novel doping strategy using defects in a wide bandgap material to dope the surface of a second semiconductor layer of dissimilar nature, is proposed for TCOs [66]. It is shown that by depositing an insulator on a semiconductor material, the conductivity of the layer stack can be increased by seven orders of magnitude without needing high-temperature processes or epitaxial growth. This approach has the potential to circumvent limits to both carrier mobility and density. The research on *a*-IZWO-TFTs with an Al_2O_3 backchannel passivation layer presents a high peak μ_{FE} of $52.5\,cm^2V^{-1}s^{-1}$, and the *a*-IZWO-TFTs with a SiO_2 passivation layer exhibited the best performance with a μ_{FE} of $33.7\,cm^2V^{-1}s^{-1}$ imply the existence of a defect modulation doping mechanism. Therefore, though defect modulation doping is reported in enhancing the TCO property, it may be used in AOS-TFTs to improve the high mobility as well.
6. *Treatment on the backchannel surface*: The effectiveness of introducing oxygen-rich H_2O_2 into a room-temperature $SCCO_2$ system as a special cosolvent and forceful oxidant to modify the backchannel surface of the *a*-IWO active layer indicates the importance of selecting a temperate and appropriate posttreatment method for improving the electrical performance and stability of tungsten-doped indium oxide TFTs [59]. The deep traps generated by dangling bonds or deep defects in the small or nanocrystal particle boundaries can be passivated and repaired by forceful H_2O_2 oxidant without leaving stress in the active layers.

References

1 Kamiya, T. and Hosono, H. (2010). Material characteristics and applications of transparent amorphous oxide semiconductors. *NPG Asia Materials* 2 (1): 15–22.

2 Fortunato, E., Barquinha, P., and Martins, R. (2012). Oxide semiconductor thin-film transistors: a review of recent advances. *Advanced Materials* 24 (22): 2945–2986.

3 Medvedeva, J.E., Buchholz, D.B., and Chang, R.P.H. (2017). Recent advances in understanding the structure and properties of amorphous oxide semiconductors. *Advanced Electronic Materials* 3 (9): 1700082.

4 Nomura, K., Ohta, H., Ueda, K. et al. (2003). Thin-film transistor fabricated in single-crystalline transparent oxide semiconductor. *Science* 300 (5623): 1269–1272.

5 Nomura, K., Ohta, H., Takagi, A. et al. (2004). Room-temperature fabrication of transparent flexible thin-film transistors using amorphous oxide semiconductors. *Nature* 432: 488–492.

6 Hosono, H., Kikuchi, N., Ueda, N., and Kawazoe, H. (1996). Working hypothesis to explore novel wide band gap electrically conducting amorphous oxides and examples. *Journal of Non-Crystalline Solids* 198–200 (1): 165–169.

7 Hosono, H., Yasukawa, M., and Kawazoe, H. (1996). Novel oxide amorphous semiconductors: transparent conducting amorphous oxides. *Journal of Non-Crystalline Solids* 203: 334–344.

8 Wager, J.F., Yeh, B., Hoffman, R.L., and Keszler, D.A. (2014). An amorphous oxide semiconductor thin-film transistor route to oxide electronics. *Current Opinion in Solid State and Materials Science* 18 (2): 53–61.

9 Jang, Y.K. and Jae, K.J. (2015). Recent progress in high performance and reliable n-type transition metal oxide-based thin film transistors. *Semiconductor Science and Technology* 30 (2): 024002.

10 Liu, C., Chen, C.D., Li, X.J. et al. (2019). A general approach to probe dynamic operation and carrier mobility in field-effect transistors with nonuniform accumulation. *Advanced Functional Materials* 29 (29): 1901700.

11 Wager, J.F., Keszler, D.A., and Presley, R.E. (2008). *Transparent Electronics*. Boston: Springer.

12 Klasens, H.A. and Koelmans, H. (1964). A tin oxide field-effect transistor. *Solid-State Electronics* 7 (9): 701–702.

13 Boesen, G.F. and Jacobs, J.E. (1968). ZnO field-effect transistor. *Proceedings of the IEEE* 56 (11): 2094–2095.

14 Masuda, S., Kitamura, K., Okumura, Y., and Miyatake, S. (2003). Transparent thin film transistors using ZnO as an active channel layer and their electrical properties. *Journal of Applied Physics* 93 (3): 1624–1630.

15 Hoffman, R.L., Norris, B.J., and Wager, J.F. (2003). ZnO-based transparent thin-film transistors. *Applied Physics Letters* 82 (5): 733–735.

16 Carcia, P.F., McLean, R.S., Reilly, M.H., and Nunes, G. (2003). Transparent ZnO thin-film transistor fabricated by rf magnetron sputtering. *Applied Physics Letters* 82 (7): 1117–1119.

17 Nishii, J., Hossain, F.M., Takagi, S. et al. (2003). High mobility thin film transistors with transparent ZnO channels. *Japanese Journal of Applied Physics* 42 (4A): L347–L349.

18 Chao, L., Zhang, D.H., Liu, X.L. et al. (2003). In_2O_3 nanowires as chemical sensors. *Applied Physics Letters* 82: 1613–1615.

19 Presley, R.E., Munsee, C.L., Park, C.H. et al. (2004). Tin oxide transparent thin-film transistors. *Journal of Physics D: Applied Physics* 37 (20): 2810–2813.

20 Troughton, J. and Atkinson, D. (2019). Amorphous InGaZnO and metal oxide semiconductor devices: an overview and current status. *Journal of Materials Chemistry C* 7 (40): 12388–12414.

21 Funabiki, F., Kamiya, T., and Hosono, H. (2012). Doping effects in amorphous oxides. *Journal of the Ceramic Society of Japan* 120 (11): 447–457.

22 Li, X.F., Zhang, Q., Miao, W.N. et al. (2006). Transparent conductive oxide thin films of tungsten-doped indium oxide. *Thin Solid Films* 515 (4): 2471–2474.

23 Li, X.F., Zhang, Q., Miao, W.N. et al. (2006). Development of novel tungsten-doped high mobility transparent conductive In_2O_3 thin films. *Journal of Vacuum Science & Technology A* 24 (5): 1866–1869.

24 Yang, M., Feng, J.H., Li, G.F., and Zhang, Q. (2008). Tungsten-doped In_2O_3 transparent conductive films with high transmittance in near-infrared region. *Journal of Crystal Growth* 310 (15): 3474–3477.

25 Feng, J.H., Yang, M., Li, G.F., and Zhang, Q. (2009). Amorphous tungsten-doped In_2O_3 transparent conductive films deposited at room temperature from metallic target. *Journal of Non-Crystalline Solids* 355 (14–25): 821–825.

26 Yang, M., Shi, Z., Feng, J.H. et al. (2011). Copper doped nickel oxide transparent p-type conductive thin films deposited by pulsed plasma deposition. *Thin Solid Films* 519 (10): 3021–3025.

27 Zhang, Q. (2012). Investigation of the tungsten doped indium oxide thin film based diodes and TFTs. Speech presented at the 4th International Symposium on Transparent Conductive Materials, Crete, Greece, October 24.

28 Aikawa, S., Darmawan, P., and Yanagisawa, K. (2013). Thin-film transistors fabricated by low-temperature process based on Ga- and Zn-free amorphous oxide semiconductor. *Applied Physics Letters* 102: 102101.

29 Kizu, T., Aikawa, S., and Mitoma, N. (2014). Low-temperature processable amorphous In-W-O thin-film transistors with high mobility and stability. *Applied Physics Letters* 104: 152103.

30 Liu, P.T., Chang, C.H., and Chang, C.J. (2015). Reliability enhancement of high-mobility amorphous indium-tungsten oxide thin film transistor. *ECS Transactions* 67 (1): 9–16.

31 Qu, M.Y., Chang, C.H., Meng, T. et al. (2017). Stability study of indium tungsten oxide thin-film transistors annealed under various ambient conditions. *Physica Status Solidi A: Applications and Materials Science* 214 (2): 1600465.

32 Liu, A., Liu, G.X., Zhu, H.H. et al. (2016). Eco-friendly, solution-processed In-W-O thin films and their applications in low-voltage, high-performance transistors. *Journal of Materials Chemistry C* 4 (20): 4478–4484.

33 Paxinos, K., Antoniou, G., and Afouxenidis, D. (2020). Low voltage thin film transistors based on solution-processed In_2O_3:W. A remarkably stable semiconductor under negative and positive bias stress. *Applied Physics Letters* 116: 163505.

34 Ruan, D.B., Liu, P.T., Chiu, Y.C. et al. (2018). High mobility tungsten-doped thin-film transistor on polyimide substrate with low temperature process. In: *Proceedings: 2018 7th International Symposium on Next-Generation Electronics, ISNE 2018*, 1–2. Piscataway, NJ: IEEE.

35 Tiwari, N., Rajput, M., and John, R.A. (2018). Indium tungsten oxide thin films for flexible high-performance transistors and neuromorphic electronics. *ACS Applied Materials & Interfaces* 10 (36): 30506–30513.

36 Gao, Y., Wang, X.Y., and Luo, J. (2019). Schmitt triggers with adjustable hysteresis window based on indium–tungsten-oxide electric-double-layer TFTs. *IEEE Electron Device Letters* 40 (7): 1205–1208.

37 Yang, Z., Meng, T., Zhang, Q., and Shieh, H.-P.D. (2016). Stability of amorphous indium-tungsten oxide thin-film transistors under various wavelength light illumination. *IEEE Electron Device Letters* 37 (4): 437–440.

38 Kuo, P.Y., Chang, C.M., Liu, I.H., and Liu, P.-T. (2019). Two-dimensional-like amorphous indium tungsten oxide nano-sheet junctionless transistors with low operation voltage. *Scientific Reports* 9: 7579.

39 Li, H.L., Qu, M.Y., and Zhang, Q. (2013). Influence of tungsten doping on the performance of indium-zinc-oxide thin-film transistors. *IEEE Electron Device Letters* 34 (10): 1268–1270.

40 Kizu, T., Mitoma, N., Miyanaga, M. et al. (2015). Codoping of zinc and tungsten for practical high-performance amorphous indium-based oxide thin film transistors. *Journal of Applied Physics* 118 (12): 125702.

41 Park, H.W., Song, A.R., Kwon, S.R. et al. (2016). Improvement of device performance and instability of tungsten-doped InZnO thin-film transistor with respect to doping concentration. *Applied Physics Express* 9: 111101.

42 Fu, R.F., Yang, J.W., Chang, W.C. et al. (2018). The influence of annealing temperature on amorphous indium-zinc-tungsten oxide thin-film transistors. *Physica Status Solidi A: Applications and Materials Science* 215 (6): 1700785.

43 Chauhan, R.N., Tiwari, N., and Shieh, H.-P.D. (2018). Electrical performance and stability of tungsten indium zinc oxide thin-film transistors. *Materials Letters* 214 (1): 293–296.

44 Park, H.W., Song, A.R., and Choi, D.H. (2017). Enhancement of the device performance and the stability with a homojunction-structured tungsten indium zinc oxide thin film transistor. *Scientific Reports* 7: 11634.

45 Park, H.W., Park, K., and Kwon, J.Y. (2017). Effect of active layer thickness on device performance of tungsten-doped InZnO thin-film transistor. *IEEE Transactions on Electron Devices* 64 (1): 159–163.

46 Park, H.W., Song, A.R., and Kwon, S.R. (2018). Enhancing the performance of tungsten doped InZnO thin film transistors via sequential ambient annealing. *Applied Physics Letters* 112: 123501.

47 Liu, P.T., Kuo, P.Y., and Hsu, S.M. (2018). High performance amorphous In-W-Zn-O thin film transistor with ultra-thin active channel for low voltage operation. *ECS Transactions* 86 (11): 91–93.

48 Wan, D., Liu, X.Q., and Abliz, A. (2018). Design of highly stable tungsten-doped IZO thin-film transistors with enhanced performance. *IEEE Transactions on Electron Devices* 65 (3): 1018–1022.

49 Park, H.W., Kwon, S.R., and Song, A.R. (2019). Dynamics of bias instability in the tungsten-indium-zinc oxide thin film transistor. *Journal of Materials Chemistry C* 7 (4): 1006–1013.

50 Liu, P.T., Chang, C.H., and Chang, C.J. (2016). Suppression of photo-bias induced instability for amorphous indium tungsten oxide thin film transistors with bi-layer structure. *Applied Physics Letters* 108: 261603.

51 Tsuji, H., Nakata, M., and Nakajima, Y. (2016). Development of back-channel etched In-W-Zn-O thin-film transistors. *Journal of Display Technology* 12 (3): 228–231.

52 Lin, D., Pi, S.B., Yang, J.W. et al. (2018). Enhanced stability of thin film transistors with double-stacked amorphous IWO/IWO:N channel layer. *Semiconductor Science and Technology* 33 (6): 065001.

53 Li, Z.H., Kuo, P.Y., and Chen, W.T. (2018). Back-channel etched double layer In-W-O/In-W-Zn-O thin-film transistors. *ECS Transactions* 86 (11): 111–114.

54 Ruan, D.B., Liu, P.T., and Gan, K.J. (2018). The influence on electrical characteristics of amorphous indium tungsten oxide thin film transistors with multi-stacked active layer structure. *Thin Solid Films* 666: 94–99.

55 Ruan, D.B., Liu, P.T., and Chiu, Y.C. (2018). Performance improvements of tungsten and zinc doped indium oxide thin film transistor by fluorine based double plasma treatment with a high-K gate dielectric. *Thin Solid Films* 665: 117–122.

56 Liu, P.T., Chang, C.H., and Kuo, P.Y. (2018). Effects of backchannel passivation on electrical behavior of hetero-stacked a-IWO/IGZO thin film transistors. *ECS Journal of Solid State Science and Technology* 7 (3): Q17–Q20.

57 Ruan, D.B., Liu, P.T., and Chiu, Y.C. (2018). Mobility enhancement for high stability tungsten doped indium-zinc oxide thin film transistors with a channel passivation layer. *RSC Advances* 8 (13): 6925–6930.

58 Koretomo, D., Hashimoto, Y., and Hamada, S. (2019). Influence of a SiO_2 passivation on electrical properties and reliability of In-W-Zn-O thin-film transistor. *Japanese Journal of Applied Physics* 58 (1): 018003.

59 Ruan, D.B., Liu, P.T., and Yu, M.C. (2019). Performance enhancement for tungsten doped indium oxide thin film transistor by hydrogen peroxide as cosolvent in room-temperature supercritical fluid systems. *ACS Applied Materials & Interfaces* 11 (25): 22521–22530.

60 Zhang, Y.H. (1982). Electronegativities of elements in valence states and their applications. 1. Electronegativities of elements in valence states. *Inorganic Chemistry* 21 (11): 3886–3889.

61 Zhang, Y.H. (1982). Electronegativities of elements in valence states and their applications. 2. A scale for strengths of Lewis acids. *Inorganic Chemistry* 21 (11): 3889–3893.

62 Parthiban, S. and Kwon, J.Y. (2014). Role of dopants as a carrier suppressor and strong oxygen binder in amorphous indium-oxide-based field effect transistor. *Journal of Materials Research* 29 (15): 1585–1596.

63 Tampo, H., Shibata, H., Matsubara, K. et al. (2006). Two-dimensional electron gas in Zn polar ZnMgO/ZnO heterostructures grown by radical source molecular beam epitaxy. *Applied Physics Letters* 89 (13): 132113.

64 Labram, J.G., Lin, Y.H., and Anthopoulos, T. (2015). Exploring two-dimensional transport phenomena in metal oxide heterointerfaces for next-generation, high-performance, thin film transistor technologies. *Small* 11 (41): 5472–5482.

65 Faber, H., Das, S., Lin, Y.H. et al. (2017). Heterojunction oxide thin-film transistors with unprecedented electron mobility grown from solution. *Science Advances* 3 (3): e1602640.

66 Weidner, M., Fuchs, A., and Bayer, T.J.M. (2019). Defect modulation doping. *Advanced Functional Materials* 29 (14): 1807906.

27

Rare Earth- and Transition Metal-Doped Amorphous Oxide Semiconductor Phosphors for Novel Light-Emitting Diode Displays

Keisuke Ide[1], Junghwan Kim[2], Hideo Hosono[1,2], and Toshio Kamiya[1,2]

[1] *Laboratory for Materials and Structures, Tokyo Institute of Technology, Yokohama, Japan*
[2] *Materials Research Center for Element Strategy, Tokyo Institute of Technology, Yokohama, Japan*

27.1 Introduction

Organic light-emitting diodes (OLEDs) are widely used in state-of-the-art flat-panel displays because of their many advantages over conventional liquid crystal displays, such as wide color gamut, high contrast, and ultra-thinness [1]. However, OLEDs have serious issues for next-generation displays, such as their short lifetime due to chemical instability. Besides, it is difficult to use the photolithography process because the organic light-emitting materials are highly sensitive to air, water, and chemicals. These issues arise from the nature of organic materials. Such instability issues would be absent in inorganic light-emitting material. However, conventional inorganic light-emitting diodes (ILEDs) such as GaN are not suitable for pixels of flat-panel displays because they require crystal growth at high temperatures and expensive single-crystal substrates.

Thus, the amorphous oxide semiconductor (AOS) phosphor was proposed as a candidate for next-generation display. It is known that AOSs can be obtained by a low-temperature process on a large glass substrate, as seen in practical IGZO thin-film transistors (TFTs) [2, 3]. By doping (with an emission center) ions to the AOS material, several kinds of AOS phosphor, even via a room-temperature (RT) process, have been realized so far. In this chapter, rare earth (RE)-doped and transition metal (TM)-doped AOS phosphors are introduced and compared regarding luminance efficiency and electronic structures for several AOS hosts of amorphous (*a*-)In-Ga-Zn-O (*a*-IGZO), *a*-In-Mg-O (*a*-IMO), and *a*-Ga-O (*a*-GO) [4–7]. Importantly, it was revealed that the key to obtaining AOS phosphor is the control of defects similar to the TFT fabrication process [8].

27.2 Eu-Doped Amorphous Oxide Semiconductor Phosphor

RT fabrication of AOS:RE phosphor was firstly achieved using *a*-IGZO doped with Eu^{3+} ions as an emission center [4–6]. Figure 27.1a shows photoluminescence (PL) spectra for Eu-doped In-Ga-Zn-O (a-IGZO:Eu) deposited by pulsed laser deposition with optimized deposition conditions at RT. All of the emission peaks can be assigned to *f-f* transition in Eu^{3+} ions, while photoluminescence excitation (PLE) peaks were located at 270 nm attributed to the charge transfer process. Clear red emission excited by ultraviolet (UV) light of 270 nm can be observed, as shown in the photograph of Figure 27.1b.

To obtain strong luminescence from IGZO:Eu thin-film phosphor fabricated at RT, it is important to finely optimize the deposition condition, especially for the oxygen amount, similar to the IGZO-TFT fabrication process. Figure 27.2a shows PL intensity with oxygen partial pressure (P_{O2}) varied from 1 to 10 Pa during the deposition

Amorphous Oxide Semiconductors: IGZO and Related Materials for Display and Memory, First Edition.
Edited by Hideo Hosono and Hideya Kumomi.

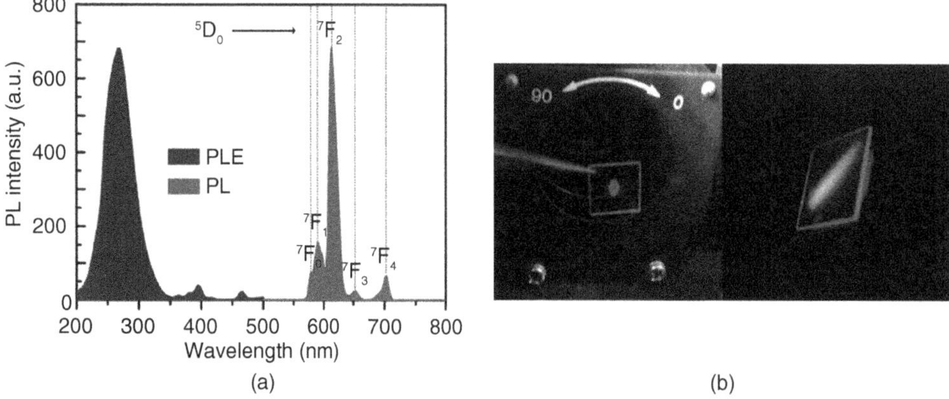

Figure 27.1 (a) PL and PLE spectrum for Eu-doped *a*-IGZO thin-film phosphors fabricated on a glass substrate at room temperature. (b) Photograph of Eu-doped *a*-IGZO thin-film phosphors excited by UV light at room temperature. *Source*: From Ref. [4], Kim, J., et al. (2016). Room-temperature fabrication of light-emitting thin films based on amorphous oxide semiconductor. *AIP Advances* **6** (1): 015106, Figure 01 [p. 015106-3] / CC BY 4.0 / with permission of AIP Publishing LLC. https://doi.org/10.1063/1.4939939.

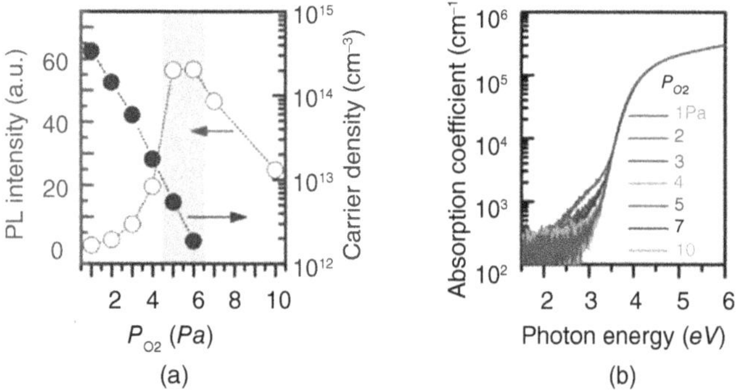

Figure 27.2 (a) PL intensity, carrier density, and (b) absorption coefficient of Eu-doped *a*-IGZO thin film deposited at various P_{O2} conditions. *Source*: [4] Kim, J., et al. (2016). Room-temperature fabrication of light-emitting thin films based on amorphous oxide semiconductor. *AIP Advances* **6** (1): 015106. Licensed under CC BY 4.0.

process. When the *a*-IGZO:Eu was deposited at P_{O2} = 1 Pa, almost no PL intensity was observed. On the other hand, PL intensity increased as increasing P_{O2}, and the optimum P_{O2} condition was found at 6 Pa. This result can be explained by the change of carrier density decreasing from 10^{14} to 10^{12} cm^{-3}, as shown in the right axis of Figure 27.2a, because free carriers cause an energy-back transfer process and suppress the luminance as known in many conventional phosphor materials. Besides, subgap states attributed to the oxygen deficiency (as shown in the optical measurement in Figure 27.2b) possibly caused deterioration of the luminance efficiency via energy transfer to the subgap states similar to Auger recombination as well. On the other hand, too high P_{O2} (larger than 6 Pa) deteriorates the luminance efficiency, although the free carriers were already reduced. It is known that too high P_{O2} conditions introduce excess oxygen in *a*-IGZO film and the excess oxygen works as an electron trap [9]. Therefore, it is considered that the radiative recombination was suppressed by the excess oxygen, resulting in the PL intensity degradation at $P_{O2} > 6$ Pa.

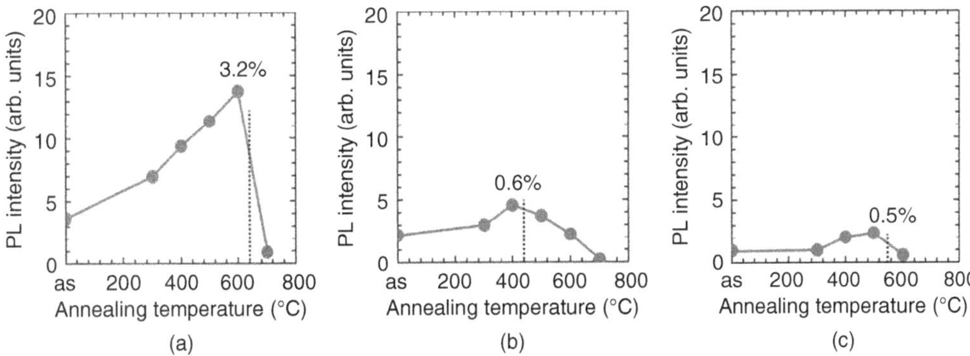

Figure 27.3 Relative PL intensity of Eu-doped (a) *a*-GO, (b) *a*-IGZO, and (c) *a*-IMO thin film phosphors fabricated on a glass substrate at room temperature. Crystallization temperatures are indicated by dashed lines and maximum internal quantum efficiencies are shown in the figure. *Source*: [6] Watanabe, N., et al. (2017). Amorphous gallium oxide as an improved host for inorganic light-emitting thin film semiconductor fabricated at room temperature on glass. *ECS Journal of Solid State Science and Technology* **6** (7): 410–415. Copyright 2017 IOP Publishing.

Figure 27.3 compares relative PL intensity and internal quantum efficiency of *a*-GO:Eu, *a*-IGZO:Eu, and *a*-IMO:Eu. Although deposition condition and doping concentration were optimized for each host material, the resulting internal quantum efficiencies were different and *a*-GO showed the highest luminance efficiencies. This would be due to the relatively low carrier density of *a*-GO compared with *a*-IGZO and *a*-IMO because free carriers enhance nonradiative recombination, as mentioned in this chapter. Besides, as seen in the dashed line in Figure 27.3, the optimum annealing temperature is different for different AOS host materials. Ref. [6] investigated the reason for the threshold annealing temperature by TDS and Hall, and X-ray diffraction (XRD) measurement, and concluded that crystallization largely suppresses the PL intensity in the case of AOS:RE material.

To achieve ILED using an emitting layer of AOS:RE, the electronic structure is important because efficient hole injection from the valence band maximum (VBM) to the 4*f* orbital of RE ions would be necessary. To investigate the energy level of the Eu 4*f* orbital, resonance photoemission spectroscopy (RPES) measurement was performed in Refs. [4, 6]. Figure 27.4a shows the X-ray absorption spectrum (XAS) of a Eu 3*d*-4*f* transition in *a*-IGZO, and Figure 27.4b shows RPES spectra for excitation energies corresponding to the XAS peaks. A Eu^{3+} 4*f* state was observed in the wide range of 3–13 eV from the Fermi level of *a*-IGZO when the excitation energies were corresponding to the XAS peak of Eu^{3+}. On the other hand, Eu^{2+} 4*f* states appeared at ~2.5 eV during resonances with the absorption energy of Eu^{2+}. From these results combined with hard X-ray photoemission spectroscopy (PES) and optical bandgap, Figure 27.4c and 27.4d illustrate the electronic structures of *a*-IGZO:Eu and *a*-GO:Eu. It can be seen that Eu^{3+} 4*f*-level states are buried in the valence band for both host materials of *a*-IGZO and *a*-GO. It means that other emission centers or host AOSs are required to realize an AOS-based ILED device, where hole injection is possible from the host valence band.

27.3 Multiple-Color Emissions from Various Rare Earth-Doped AOS Phosphors

The candidate of the RE emission center is not limited to Eu^{3+} [7]. Figure 27.5a–27.5e shows the PL spectra of the *a*-GO:RE thin films, where REs are Pr, Sm, Eu, Tb, and Dy ions. All the films show the detectable PL spectrum after optimization of deposition conditions and excitation energies. Figure 27.5f shows the photograph of *a*-GO:Pr, *a*-GO:Eu, and *a*-GO:Tb deposited at RT excited by 254 nm UV light, showing the bright visible-light emission, while the *a*-GO:Sm and *a*-GO:Dy films showed too weak light to see by our eyes.

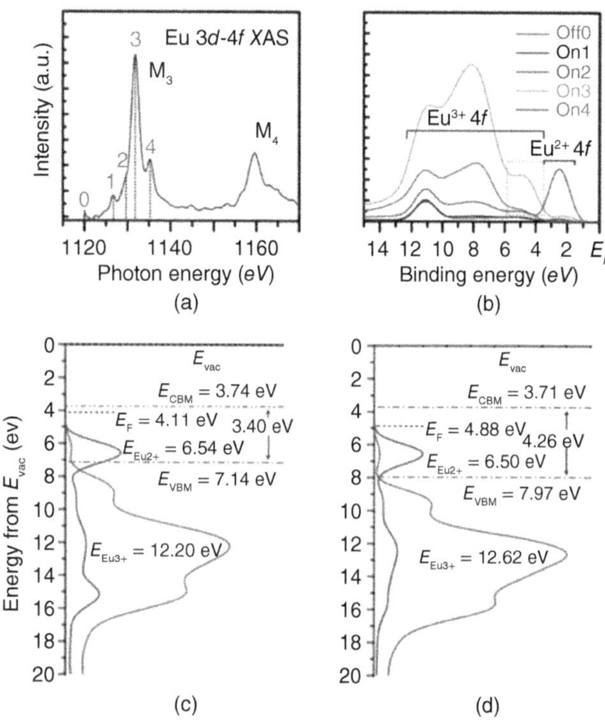

Figure 27.4 (a) XAS spectrum for Eu 3*d*-4*f* absorption and (b) RPES spectrum excited by the energies corresponding to the XAS result for Eu-doped *a*-IGZO thin film. RPES spectrum of (c) Eu-doped *a*-IGZO and (d) Eu-doped *a*-GO for Eu^{3+} and Eu^{2+}. The energy level of the VBM, CBM, and Fermi energy are indicated in the figure. *Source*: [4, 6] Kim, J., et al. (2016). Room-temperature fabrication of light-emitting thin films based on amorphous oxide semiconductor. *AIP Advances* **6** (1): 015106. Copyright 2016 AIP Publishing Watanabe, N., et al. (2017). Amorphous gallium oxide as an improved host for inorganic light-emitting thin film semiconductor fabricated at room temperature on glass. *ECS Journal of Solid State Science and Technology* **6** (7): 410–415. Copyright 2017 IOP Publishing.

Optimum deposition conditions and quantum efficiencies are different for each rare earth ion. Figure 27.6a–c shows P_{O2} dependence of PL intensity for *a*-GO:Eu, *a*-GO:Pr, and *a*-GO:Tb. Although the PL signals were not detected at P_{O2} <4 Pa for all cases, the PL intensity increased with increasing P_{O2} up to 10 Pa for Eu and Pr, while the optimum P_{O2} for Tb doping was 20 Pa. The PL intensity decreases with increasing P_{O2} when the P_{O2} increased further. Similar to the *a*-IGZO:Eu, both oxygen deficiencies at low P_{O2} and excess oxygen at high P_{O2} cause the nonradiative recombination, deteriorating PL intensity also in *a*-GO:RE thin-fim phosphor. The internal quantum efficiencies (IQEs) are 0.9%, 0.2%, and 1.8% for unannealed *a*-GO:Eu, *a*-GOPr, and *a*-GO:Tb, respectively, after optimizing deposition condition and doping concentration. As well as *a*-IGZO:Eu, the thermal annealing below crystallization temperature improved the IQE. It was reported that the maximum IQE was 3.2% for *a*-GO:Eu, 2.4% for *a*-GO:Pr, and 3.8% for *a*-GO:Tb.

Although the RPES measurement for the *a*-GO:Tb and *a*-GO:Pr films has not been reported, the energy level of the RE 4*f* orbital relative to the host of *a*-GO would be different for each RE dopant. It is well studied that the energies of trivalent RE ions are relative to the host valence band, such as for rare earth-doped $Y_3Al_5O_{12}$ in Ref. [10] and $Lu_2Si_2O_7$ in Ref. [11]. Those papers reported that the energy levels of Tb^{3+} and Pr^{3+} 4*f* orbitals are located 2–4 eV higher than those of Eu^{3+} 4*f* orbitals. Therefore, it is expected that the electronic structure of *a*-GO:Tb and *a*-GO:Pr would be preferable to inject holes from the VBM to the RE^{3+} 4*f*. To realize AOS-based self-emitting ILED devices, further studies are needed.

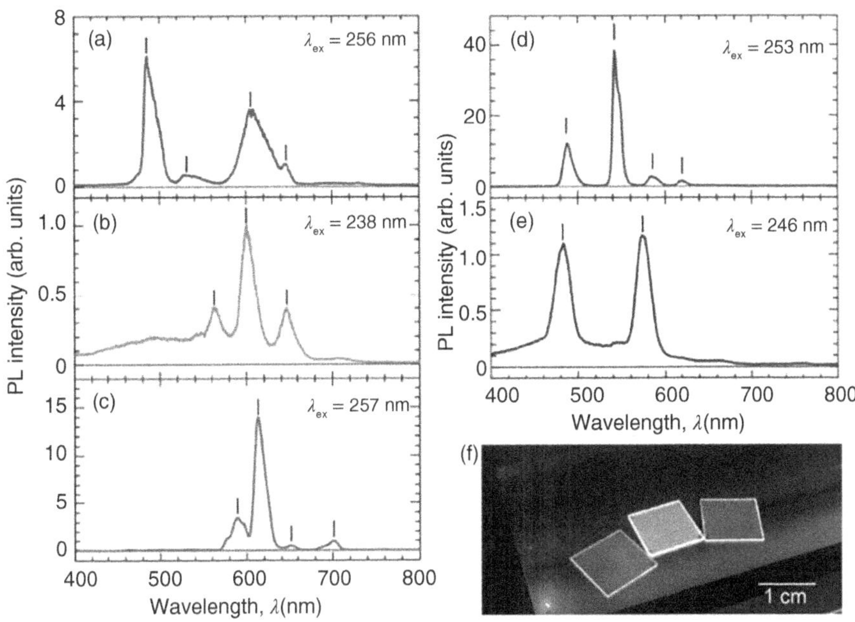

Figure 27.5 (a–e) PL spectrum for a-GO doped with Pr, Sm, Eu, Tb, Dy. Excitation energies were indicated in the figure. (f) Photograph of (left) *a*-GO:Eu, (center) *a*-GO:Tb, and (right) *a*-GO:Pr thin-film phosphors. All of the films were deposited at room temperature without heat treatment. *Source*: From [7], Watanabe, N et al (2018), Figure 01[p.02] / with permission of JOHN WILEY & SONS, INC. DOI-https://doi.org/10.1002/pssa.201700833.

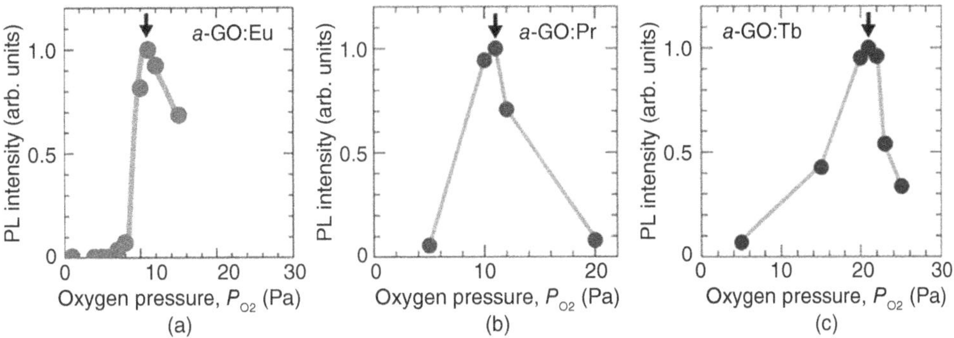

Figure 27.6 PL intensity of (a) Eu-doped, (b) Pr-doped, and (c) Tb-doped *a*-GO thin film deposited at various P_{O2} conditions. *Source*: (b,c) [7] Watanabe, N., Ide, K., Kim, J. et al. (2018). Multiple color inorganic thin-film phosphor, RE-doped amorphous gallium oxide (RE=rare earth: Pr, Sm, Tb, and Dy), deposited at room temperature. *Physica Status Solidi (A): Applications and Materials Science* **216** (5): 1700833.: Copyright 2018 John Wiley and Sons.

27.4 Transition Metal-Doped AOS Phosphors

Transition metals (TMs) were also explored as possible emission centers to replace RE for environmentally friendly, low-cost AOS-based thin-film phosphors [12]. Figure 27.7 shows the screening result of doping elements for *a*-GO-based phosphor using a polycrystalline bulk β-Ga_2O_3 sample. It is known that TMs such as Cu, Mn, and Cr ions in bulk oxide are known as a luminescent center of blue, green, or red colors. Figure 27.7a shows XRD patterns of the polycrystalline bulks of Cu-, Mn-, and Cr-doped β-Ga_2O_3. For Cu-doped Ga_2O_3, the diffraction peak from the spinel-type $CuGa_2O_4$ was observed at 35.9°, suggesting that the solid solution was not formed by sintering at 1200 °C. On the other hand, for Mn- and Cr-doped Ga_2O_3 bulks, there are no detectable impurity phases, suggesting that the solid solution was succesfuly obtained. For the PL spectrum in Figure 27.6b, broad PL around 400 nm originating from donor–acceptor recombination can be observed from undoped Ga_2O_3, while the Cu, Mn, and Cr doping suppressed the blue emission. As a result, only Cr-doped Ga_2O_3 exhibited clear PL originating from

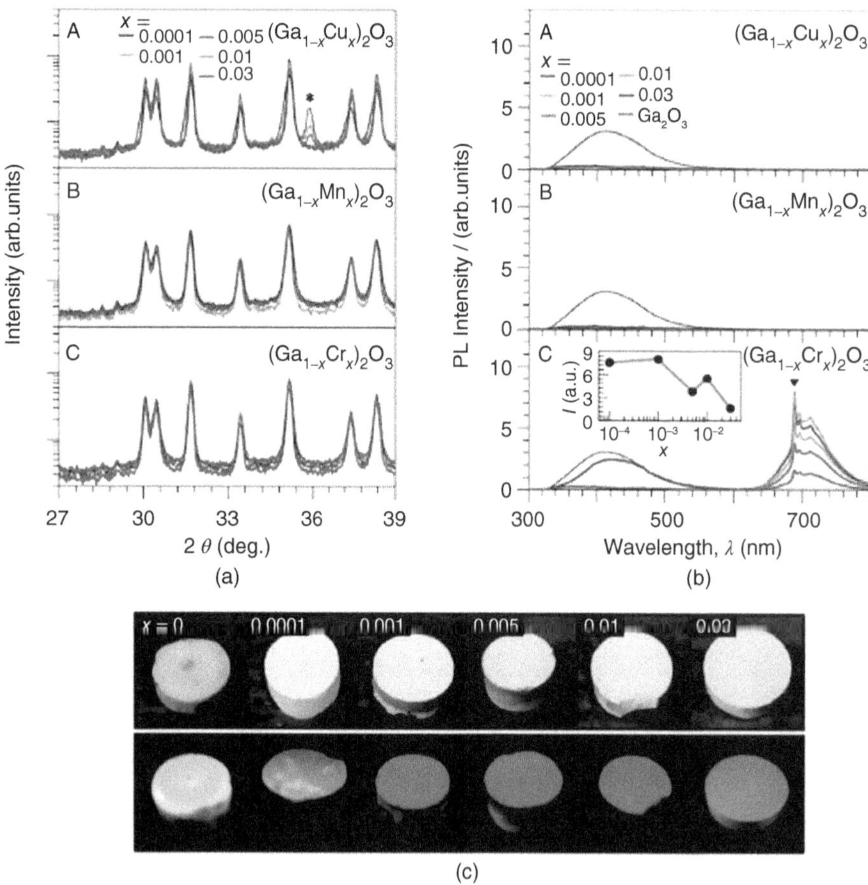

Figure 27.7 Screening test for choosing an appropriate TM emission center for AOS-based phosphor. (a,b) XRD pattern and PL spectrum of a TM-doped (TM = Cu, Mn, and Cr) β-Ga_2O_3 bulk sample. (c) Photograph of bulk sample of β-$(Ga_{2-x}Cr_x)O_3$ excited by UV light. *Source*: [12] Ide, K., et al. (2019). Transition metal-doped amorphous oxide semiconductor thin-film phosphor, chromium-doped amorphous gallium oxide. *Physica Status Solidi (A): Applications and Materials Science* **216** (5): 1800372. Copyright 2018 John Wiley and Sons.

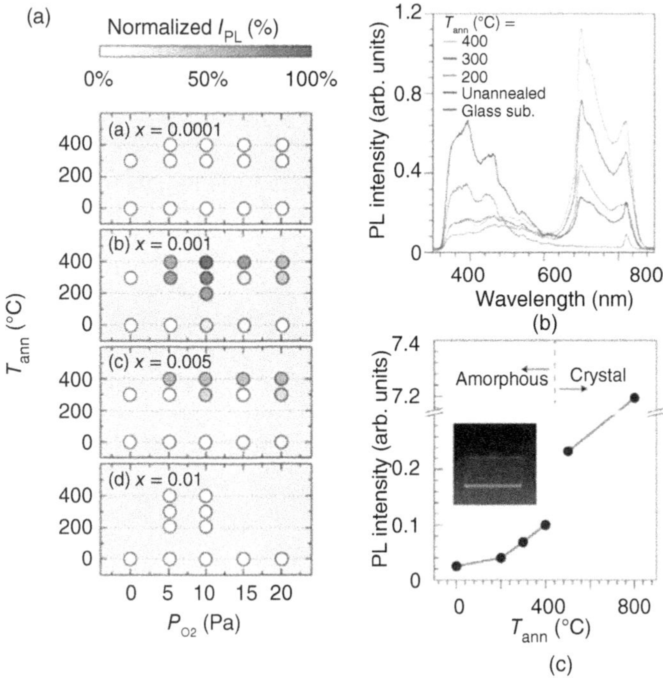

Figure 27.8 (a) Mapping for PL intensity of *a*-GO:Cr films, where P_{O2} and annealing temperature varied in the ranges of 0–20 Pa and 0–400 °C, respectively. Each panel is summarized by each doping concentration. (b) PL spectrum for *a*-GO:Cr with various annealing temperatures. (c) PL intensity was summarized as a function of annealing temperature. Inset photograph shows the *a*-GO:Cr fabricated at the optimized condition. *Source*: [12] Ide, K., et al. (2019). Transition metal-doped amorphous oxide semiconductor thin-film phosphor, chromium-doped amorphous gallium oxide. *Physica Status Solidi (A): Applications and Materials Science* **216** (5): 1800372. Copyright 2018 John Wiley and Sons.

TM ions in the range of 650–800 nm. Figure 27.6c summarizes the photographs of β-($Ga_{2-x}TM_x$)O_3 doped with Cr at the concentration of $x = 0.0001$–0.03, showing the blue and red emission excited by UV light. Thus, Cr was selected as the emission center for *a*-GO-based thin-film phosphor.

Figure 27.8a summarizes normalized PL intensities of *a*-GO:Cr, with various deposition conditions indicated by the colors inside the circles. For the as-deposited sample, weak red PL was observed at very limited conditions such as $P_{O2} = 10$–15 Pa for $x = 0.001$. On the other hand, 400 °C–annealed samples show clear emission in wide ranges of P_{O2} for $x \sim 0.001$ and 0.005. It is concluded that the optimum condition is $x = 0.001$, $P_{O2} = 10$ Pa, with an annealing temperature of 400 °C. Figure 27.8b and 27.8c compare the PL spectra and intensity for *a*-GO:Cr annealed at different temperatures. PL intensity increases with increasing annealing temperature. The maximum IQE of the amorphous film was 0.3% obtained for the 400 °C–annealed *a*-GO:Cr film. Interestingly, the PL intensity improved and the IQE reached 20% by 800 °C annealing opposite the AOS:RE phosphor. This difference is explained by the difference in their crystalline phases. Cr-doped β-Ga_2O_3 forms solid solutions, as confirmed in Figure 27.7, while RE-doped β-Ga_2O_3 doesn't form solid solutions due to large differences in the ion radii between the Ga and RE ions. It means crystallization causes segregated impurity crystalline phases such as $Eu_3Ga_5O_{12}$ (reported in Ref. [7]), which have too high RE concentrations resulting in concentration quenching. This highlights another advantage of AOS-based thin-film phosphors, because AOSs are free from degradation due to the solubility limits of emission centers.

References

1 Jin, D., Lee, J.-S., Kim, T.-W. et al. (2009). 65.2: Distinguished Paper: World-largest (6.5”) flexible full color top emission AMOLED display on plastic film and its bending properties. *SID Symposium Digest of Technical Papers* 40 (1): 983.

2 Nomura, K., Ohta, H., Takagi, A. et al. (2004). Room-temperature fabrication of transparent flexible thin-film transistors using amorphous oxide semiconductors. *Nature* 432 (7016): 488–492.

3 Kamiya, T., Nomura, K., and Hosono, H. (2010). Present status of amorphous In–Ga–Zn–O thin-film transistors. *Science and Technology of Advanced Materials* 11 (4): 044305.

4 Kim, J., Miyokawa, N., Ide, K. et al. (2016). Room-temperature fabrication of light-emitting thin films based on amorphous oxide semiconductor. *AIP Advances* 6 (1): 015106.

5 Kim, J., Miyokawa, N., Ide, K. et al. (2016). Transparent amorphous oxide semiconductor thin film phosphor, In–Mg–O:Eu. *Journal of the Ceramic Society of Japan* 124 (5): 532–535.

6 Watanabe, N., Kim, J., Ide, K. et al. (2017). Amorphous gallium oxide as an improved host for inorganic light-emitting thin film semiconductor fabricated at room temperature on glass. *ECS Journal of Solid State Science and Technology* 6 (7): 410–415.

7 Watanabe, N., Ide, K., Kim, J. et al. (2018). Multiple color inorganic thin-film phosphor, RE-doped amorphous gallium oxide (RE=rare earth: Pr, Sm, Tb, and Dy), deposited at room temperature. *Physica Status Solidi (A): Applications and Materials Science* 216 (5): 1700833.

8 Ide, K., Nomura, K., Hosono, H., and Kamiya, T. (2019). Electronic defects in amorphous oxide semiconductors: a review. *Physica Status Solidi (A): Applications and Materials Science* 1800372: 1–28.

9 Ide, K., Kikuchi, Y., Nomura, K. et al. (2011). Effects of excess oxygen on operation characteristics of amorphous In-Ga-Zn-O thin-film transistors. *Applied Physics Letters* 99 (9): 093507.

10 Thiel, C.W., Cruguel, H., Wu, H. et al. (2001). Systematics of electron energies relative to host bands by resonant photoemission of rare-earth ions in aluminum garnets. *Physical Review B: Condensed Matter and Materials Physics* 64 (8): 1–13.

11 Pidol, L., Viana, B., Galtayries, A., and Dorenbos, P. (2005). Energy levels of lanthanide ions in a $Lu_2Si_2O_7$ host. *Physical Review B: Condensed Matter and Materials Physics* 72 (12): 1–9.

12 Ide, K., Futakado, Y., Watanabe, N. et al. (2019). Transition metal-doped amorphous oxide semiconductor thin-film phosphor, chromium-doped amorphous gallium oxide. *Physica Status Solidi (A): Applications and Materials Science* 216 (5): 1800372.

28

Application of AOSs to Charge Transport Layers in Electroluminescent Devices

Junghwan Kim[1] *and Hideo Hosono*[1,2]

[1] *Materials Research Center for Element Strategy, Tokyo Institute of Technology, Yokohama, Japan*
[2] *Laboratory for Materials and Structures, Tokyo Institute of Technology, Yokohama, Japan*
[3] *National Institute for Materials Science, Tsukuba, Japan*

28.1 Electronic Structure and Electrical Properties of Amorphous Oxide Semiconductors (AOSs)

In a 2004 breakthrough, high-mobility thin-film transistors (TFTs) were achieved using AOSs such as *a*-IGZO; *a*-IGZO TFTs exhibit a quite high mobility of ~10 cm^2/Vs, which is larger by an order of magnitude than that of conventional *a*-Si:H TFTs [1–4]. Recently, much higher mobility TFTs exceeding 40 cm^2/Vs were already achieved by optimizing procedures [5, 6]. Here, we need to understand how transparent AOSs (TAOSs) can possess such a good electrical property. For AOSs, the conduction band minimum (CBM) is mainly composed of vacant *s*-orbitals of post–transition metal cations, whereas the valence band maximum (VBM) is formed by oxygen 2*p*-orbitals. The spatial overlap between the vacant *s*-orbitals of *p*-block metal cations is insensitive to structural randomness, explaining why high mobility is retained even in amorphous structures. Moreover, this is the major reason why conventional AOSs consisting of post–transition metal cations exhibit higher mobility via the small effective mass of the electrons. However, large dispersion of the conduction band, in general, leads to a deep CBM level (E_{CBM}) and a narrow E_g, as illustrated in Figure 28.1b. In contrast, AOSs based on light-metal cations such as Al and Si exhibit a wider E_g, but the higher E_{CBM} makes it difficult to dope the electrons, and the low dispersion results in little mobility (Figure 28.1c). As seen in Figure 28.1d, E_{CBM} deepens with the increase of heavy metal cation content [7].

28.2 Criteria for Charge Transport Layers in Electroluminescent (EL) Devices

As discussed in Section 28.1, E_{CBM} is widely tunable by combining cation elements in AOSs. In EL devices (also called "Els"), energy-level matching is a critical factor because carrier injection from electrodes to luminous layers is determined by the energy difference between the active layer and the contact layers. In this respect, the wide tunability of AOSs in energy levels works as a significant advantage that enables the design of charge transport layers for ELs. However, there are more requirements for efficient ELs. First, the optical band gap should be larger than the emission layer to suppress the reabsorption of photons. In this respect, the E_g of charge transport layers should be larger than 3 eV, which guarantees transparency in the visible-light region. Next, electrical properties such as mobility and carrier density seriously affect EL performance and productivity. Thus, we need to understand what "charge balance" is. Basically, charge transport layers are required to possess high mobility. However, the

Amorphous Oxide Semiconductors: IGZO and Related Materials for Display and Memory, First Edition.
Edited by Hideo Hosono and Hideya Kumomi.

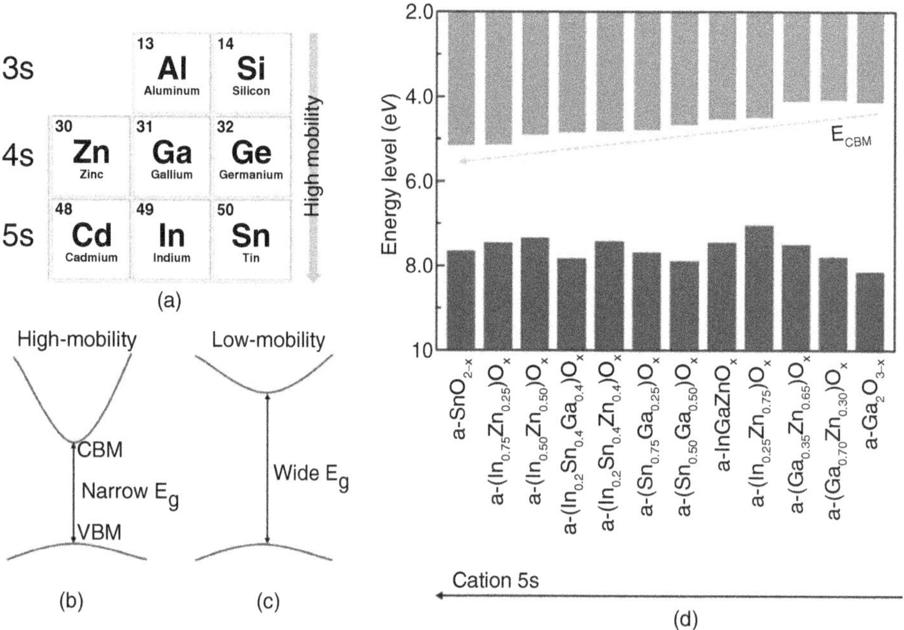

Figure 28.1 (a) Cation candidates for TAOS with vacant *s*-orbitals. Expected electronic structures and electrical properties of TAOSs with (b) high electron mobility and a narrow E_g, (c) low electron mobility and a wide E_g, and (d) experimentally obtained energy levels by UPS. *Source*: Kim, J., et al. (2019). Ultra-wide bandgap amorphous oxide semiconductors for NBIS-free thin-film transistors. *APL Materials* **7**: 022501.

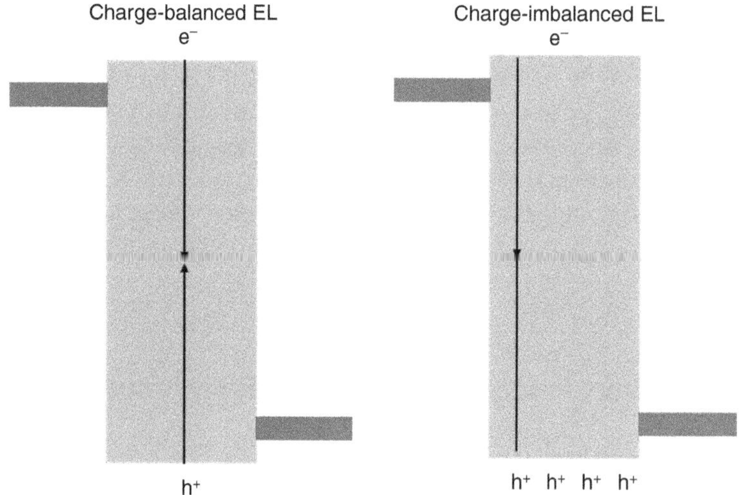

Figure 28.2 Schematic diagram of charge-balanced and charge-imbalanced ELs.

controllability of the electrical property is more important because of the charge balance. Figure 28.2 compares two ELs.

As is well known, when one electron and one hole recombine with each other, one photon is generated, and the quantum efficiency corresponds to 100%. In other words, the number of electrons and holes should be equal. If

Table 28.1 Physical properties of representative organic charge transport layers.

Material	Charge Transport Layer (CTL)	HOMO (eV)	LUMO (eV)	Mobility ($cm^2 V^{-1} s^{-1}$)
TPD	**HTL**	**5.4**	**2.4**	**1.0×10^{-3}**
NPD	HTL	5.4	2.3	8.8×10^{-4}
TCTA	HTL	5.7	2.4	2.0×10^{-5}
TAPC	HTL	5.5	2	1.0×10^{-2}
PVK	HTL	5.5	2.4	2.5×10^{-6}
Alq3	ETL	5.8	3	7.2×10^{-6}
BCP	ETL	6.7	3.2	5.6×10^{-6}
TPBI	ETL	6.2	2.7	3.3×10^{-6}
PBD	ETL	6.2	2.4	2×10^{-5}
Tm3PyPB	ETL	6.7	2.9	2×10^{-4}

ETL: Electron transport layer; HOMO: highest occupied molecular orbital; HTL: hole transport layer; LUMO: lowest unoccupied molecular orbital.

either electrons or holes are in excess, quantum efficiency decreases. This charge balance can be modified by controlling the electrical properties of the charge transport layer. However, for conventional charge transport layers, thickness variation is the only method to tune the electrical property, since organic charge transport layers do not possess free carriers. In this respect, AOSs are quite attractive since the electrical properties can be modified by altering the carrier concentration. Moreover, conventional organic charge transport layers possess little mobility, as seen in Table 28.1. Such a poor electrical property gives rise to an increase in the operating voltage of ELs.

28.3 Amorphous Zn-Si-O Electron Transport Layers for Perovskite Light-Emitting Diodes (PeLEDs)

Here, we introduce how AOSs work well as electron transport layers (ETLs) in ELs. Recently, lead halide perovskites such as $CsPbX_3$ (X: I, Br, and Cl), have attracted attention as promising candidates for emission layers (EMLs) due to their excellent photophysical properties: a very narrow full width at half maximum (FWHM) of emission, solution processability, and low-cost fabrication processes [8, 14, 19, 21]. Proper combination with ETLs should be a practical issue to realize efficient ELs. We reported a significant phenomenon that EL performance with 3D materials, such as $CsPbX_3$, is governed by adjacent charge transport layers, which is possibly due to non-radiative recombination resulting from the small exciton binding energy in 3D lead halide perovskites. To confirm how the energy alignment of $CsPbBr_3$ with neighboring layers affects photoluminescent (PL) or EL devices, we prepared several AOSs with different E_{CBM} values. Here, ZnO was chosen and modified to an amorphous Zn-Si-O (*a*-ZSO) system [9, 10]. *a*-ZSO thin films with different Zn/(Zn+Si) ratios were deposited on glass substrates using radiofrequency (RF) magnetron sputtering at room temperature. Figure 28.3 compares their optical band gap, electrical conductivities, and energy levels. It is evident that the optical band gap of *a*-ZSO increases when the Si content is increased, while electrical conductivity decreases and E_{CBM} becomes shallow. Note that the E_{CBM} of *a*-ZSO becomes close to that of $CsPbBr_3$ when Si content is ~15%, where the exciton confinement effect is expected to emerge. To confirm the exciton confinement effect, we investigated the correlation of E_{CBM} with PL lifetime, as seen in Figure 28.4.

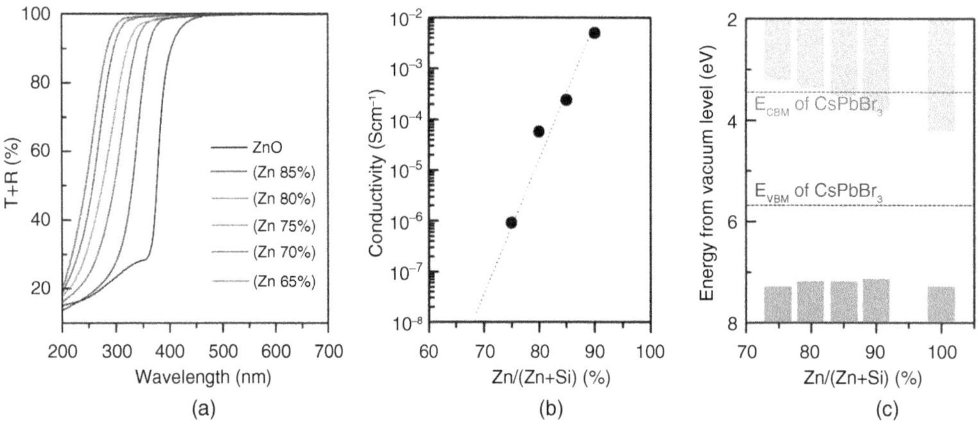

Figure 28.3 (a) Optical properties, (b) electrical conductivities at RT, and (c) energy levels are compared for *a*-ZSO thin films as a function of the Zn/(Zn+Si) ratio. *Source*: Sim, K., et al. (2019). Performance boosting strategy for perovskite light-emitting diodes. *Applied Physics Reviews* **6**: 031402.

Consequently, a remarkable correlation was observed between the energy alignments and PL lifetime, as seen in Figure 28.4b. The $CsPbBr_3$ on ITO exhibited the shortest PL lifetime arising from the Auger quenching process by a high concentration of free electron carriers, which agrees well with the photograph. On the other hand, the PL lifetime increased and saturated at ~17 ns from increasing the Si content of *a*-ZSO. Nearly the same value (~17 ns) was confirmed for the sample deposited on a bare glass substrate. These results imply that the *a*-ZSO (Si ≥20%) possesses an adequate energy level to confine excitons in $CsPbBr_3$, as shown in Figure 28.3, and no serious quenching occurs owing to the lack of a high free carrier concentration.

To confirm the ability of *a*-ZSO to be an ETL, we fabricated 1 mm × 1 mm PeLEDs composed of ITO (150 nm), 80ZSO (120 nm), $CsPbBr_3$ (60 nm), NPD (40 nm), MoO_x (5 nm), and Ag (100 nm), as shown in Figure 28.5. The $CsPbBr_3$ thin film was fabricated by spin-coating precursor solution on an 80ZSO ETL (Figure 28.5b) in a globe box. The precursor solutions were prepared by dissolving CsBr and $PbBr_2$ in dimethyl sulfoxide (DMSO; molar ratio of 1.05:1). To improve the film morphology, polyethylene oxide (Mv: 100,000; 2.5 mg/mL) was added to the precursor solutions. The precursor solution was stirred overnight at 70 °C.

As shown in Figure 28.5c, it was confirmed that the fabricated $CsPbBr_3$ thin films are quite uniform. Subsequently, NPD and MoO_n were thermally evaporated as a hole transport layer (HTL) and hole injection layer (HIL), respectively. The PeLED with the 80ZSO ETL exhibits an EL peak at ~523 nm with a FWHM of 16 nm (Figure 28.5d). On the other hand, significantly bright EL was confirmed at low operating voltage, that is, 10,000 cd/m^2 at 2.9 V and 120,000 cd/m^2 at 4 V (Figure 28.5e). Due to the low operating voltage, a high power efficiency of ~25 lm/W was achieved (Figure 28.5f), while significantly high brightness (180,000 cd/m^2) was confirmed at 5 V, as shown in Figure 28.5d. The L_{max} further increased to 500,000 cd/m^2 using a bilayered HTL of CBP 20 nm/NPD 20 nm, and current efficiency and power efficiency increased to 37 cd/A and 33 lm/W, respectively. Such high brightness substantiates efficient charge injection into $CsPbBr_3$ and good charge balance. This result demonstrates the critical importance of maintaining a balance between the luminescent and electronic transport properties to achieve high EL performance. As shown in Table 28.2, the obtained performance is noticeable among the recently reported PeLEDs. In addition, *a*-ZSO has another unique advantage: It makes a quasi-Ohmic contact with conventional cathode materials, such as Al and ITO, regardless of its shallow E_{CBM}. This feature enables the transport of electrons from the cathode to the EML with no injection barrier height. We also confirmed the charge balance of PeLEDs with respect to each different ZSO ETL. As shown in Figure 28.6, EL performance is significantly affected by the electrical properties of ETLs. It should be noted that EL efficiency deteriorates when an 85ZSO ETL with

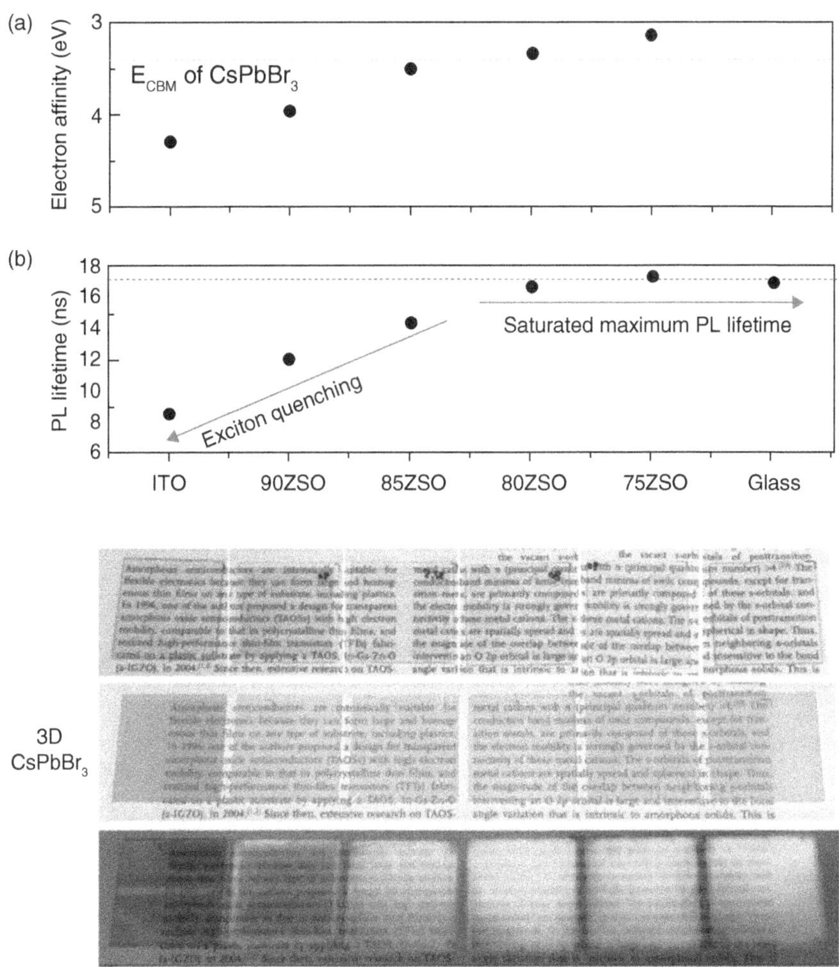

Figure 28.4 (a) Electron affinity values of ITO, 90ZSO, 85ZSO, and 75ZSO. (b) PL lifetimes and photo of PL from $CsPbBr_3$ thin films fabricated on ITO, 90ZSO, 85ZSO, 75ZSO, and bare glass substrates. *Source*: Sim, K., et al. (2019). Performance boosting strategy for perovskite light-emitting diodes. *Applied Physics Reviews* **6**: 031402.

the highest conductivity is used. This result demonstrates that usage of a TAOS ETL has a distinct advantage in optimizing the charge balance.

28.4 Amorphous In-Mo-O Hole Injection Layers for OLEDs

Transition metal oxides (TMOs) such as MoO_3 are often used as HILs owing to their large electron affinities (i.e., the deep position of E_{CBM}) [22–24]. However, the low mobility arising from open d-shell electrons, leading to the formation of the in-gap state of TMOs, restricts device performance and fabrication processes. Therefore, it will be a great advantage if we could design HIL materials based on AOSs. However, as described in Section 28.3, conventional AOSs possess relatively shallow E_{VBM} compared to TMOs. Here we describe a novel method to tune the electron affinity of conventional AOSs and an example material design of HIL AOSs. Recently, we found that the electronic structures of AOSs can be predicted from their constituent cation species (i.e., the electronic structure of the ternary AOS of a-ABO_x can be roughly expressed by a combination of a-AO_x and a-BO_x). Based on this

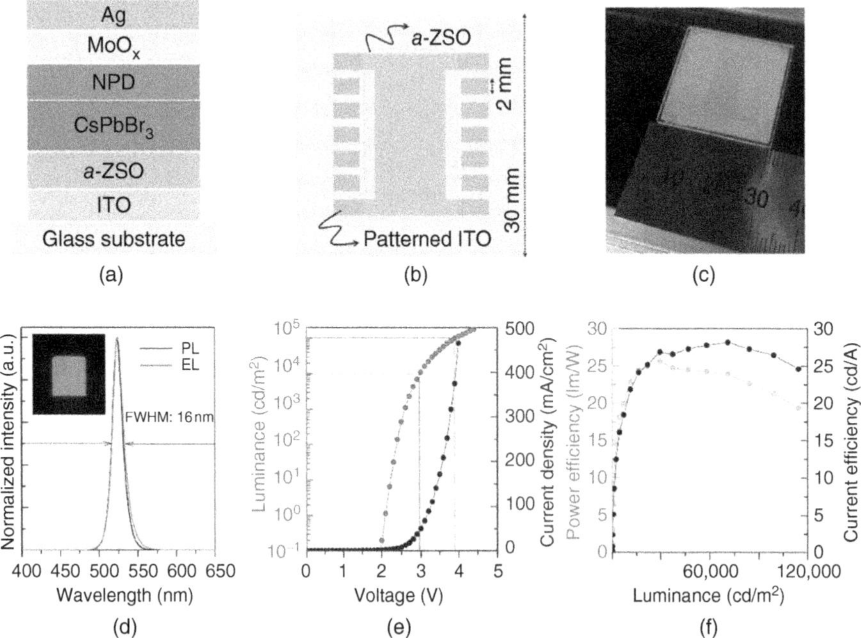

Figure 28.5 (a) Device structure of a PeLED and (b) ITO-patterned glass substrate onto which an 80ZSO thin film was deposited as an ETL. (c) Photograph of a spin-coated $CsPbBr_3$ thin film on an 80ZSO ETL (UV excitation: 365 nm). (d) Comparison of photoluminescence and electroluminescence spectra. (e) Luminance–current density–voltage characteristics. (f) Power efficiency (lm/W) and current efficiency (cd/A) of the PeLED with an 80ZSO ETL. *Source*: Sim, K., et al. (2019). Performance boosting strategy for perovskite light-emitting diodes. *Applied Physics Reviews* **6**: 031402.

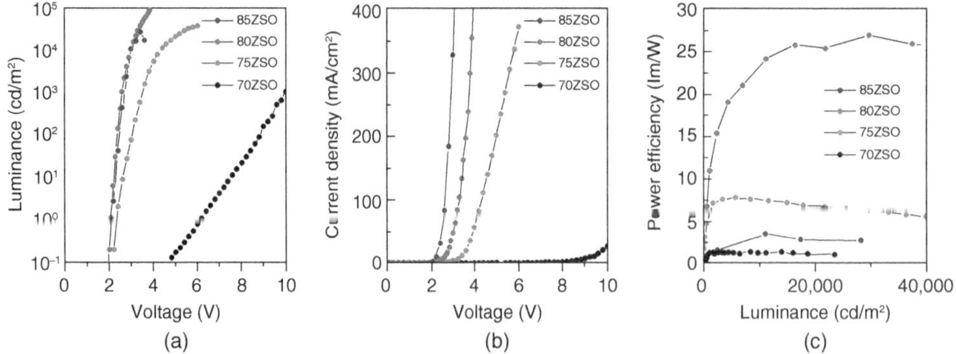

Figure 28.6 EL performance of PeLEDs with *a*-ZSO of different Zn content (85ZSO, 80ZSO, 75ZSO, and 70ZSO). Comparison of (a) luminance–voltage characteristics, (b) current density–voltage characteristics, and (c) power efficiency. *Source*: Sim, K., et al. (2019). Performance boosting strategy for perovskite light-emitting diodes. *Applied Physics Reviews* **6**: 031402.

simple idea, we tried to synthesize the AOS by mixing In_2O_x and MoO_3 [24, 25]. As a result, a conspicuous electron affinity deepening phenomenon was observed for the amorphous In-O system via admixing with Mo ions (In:Mo = 7:3). Hard X-ray photoemission spectroscopy (HAXPES) and in situ ultraviolet spectroscopy (UPS) measurements demonstrated that the electron affinities are primarily controlled by the oxidation states of the incorporated Mo ions. Furthermore, *a*-IMO exhibited high chemical stability against a variety of solvents, even though the MoO_x thin film was easily dissolved in water. OLEDs using *a*-IMO as an HIL exhibited a lower operating

Table 28.2 EL performance of recent green PeLEDs: Operating voltage, maximum luminance (L_{max}), maximum current efficiency ($C.E._{max}$), power efficiency (P.E.), and maximum external quantum efficiency (EQE_{max}).

EML Materials	Voltage (V) at 10^3, 10^4 cd/m^2	L_{max} (cd/m^2)	$C.E._{max}$ (cd/A)	P.E. (lm/W)	EQE_{max} (%)	Ref.
$CsPbBr_3$	2.5, 2.9	496,320	37	33	**9.3**	**This chapter**
$CsPbBr_3$	4.15, 5.15	51,890	21.4	14.9	4.76	[8]
$CsPbBr_3$	3.75, 5.2	23,828	9.5	5.4	2.94	[9]
$CsPbBr_3$	3.1, 4.1	53,525	15.7	12.3	4.26	[10]
$FA_xCs_{1-x}PbBr_3$	9.3, 17.6	9,834	14.5	3.4*	3.1	[11]
$MAPbBr_3$	4.6, 5.75	—	18	8.7*	3.8	[12]
$MAPbBr_3$	5.3, 8.5	20,000	42	14.7*	8.53	[13]
$MA_xCs_{1-x}PbBr_3$	3.2, 4.3	14,000	80	69	20	[17]
$MA_xCs_{1-x}PbBr_3$	4.4, 6.1	91,000	33.9	17.8*	10.4	[12]
Quasi-2D BABr:$CsPbBr_3$	3.85, 4.8	33,532	25.1	18.3*	8.42	[18]
$CsPbBr_3$ QDs	4.1, —	1,660	26.2	31.7	8.73	[15]
$CsPbBr_3$ QDs	6.2, 10.5	10,206	8.73	2.5*	4.63	[16]

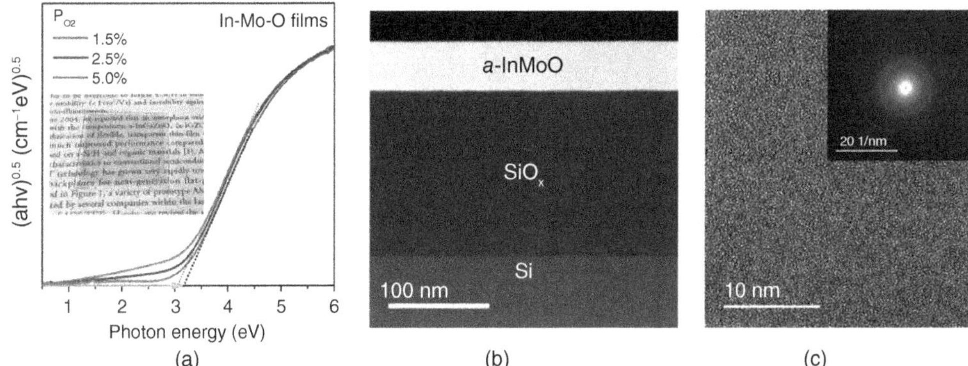

Figure 28.7 (a) A Tauc plot of *a*-IMO films deposited in various P_{O2}, and (inset) a photo of an *a*-IMO thin film deposited on a glass substrate; (b) a high-angle annular dark-field scanning transmission electron microscopy (HAADF-STEM) image; and (c) an enlarged high-resolution transmission electron microscopy (HR-TEM) image. The inset in (c) shows the electron diffraction pattern of a 50-nm-thick *a*-IMO film deposited on an Si substrate with a 150-nm-thick SiO_x layer. *Source*: Kim, J., et al. (2018). Electron affinity control of amorphous oxide semiconductors and its applicability to organic electronics. *Advanced Materials Interfaces* **5**: 1801307.

voltage and higher power efficiency than those using an MoO_x thin film as an HIL. The *a*-IMO thin films were fabricated on glass substrates by using RF magnetron sputtering without the intentional heating of the substrates. The *a*-IMO films exhibited high visible transparency that is attributed to a wide E_g of ~3 eV, which was determined from the Tauc plot shown in Figure 28.7a. It is observed that lower partial oxygen pressure (P_{O2}) values induce a higher density of the defect states attributable to the oxygen-deficiency defects or gap states due to Mo^{4+} or Mo^{5+}. A high-angle annular dark-field scanning transmission electron microscopy (HAADF-STEM) image was taken for a 50-nm-thick *a*-IMO film deposited on an Si/SiO_x substrate (Figure 28.7b). A dense bulk structure and a smooth surface were confirmed, thus implying that *a*-IMO suppresses the leakage current and short-circuit issues

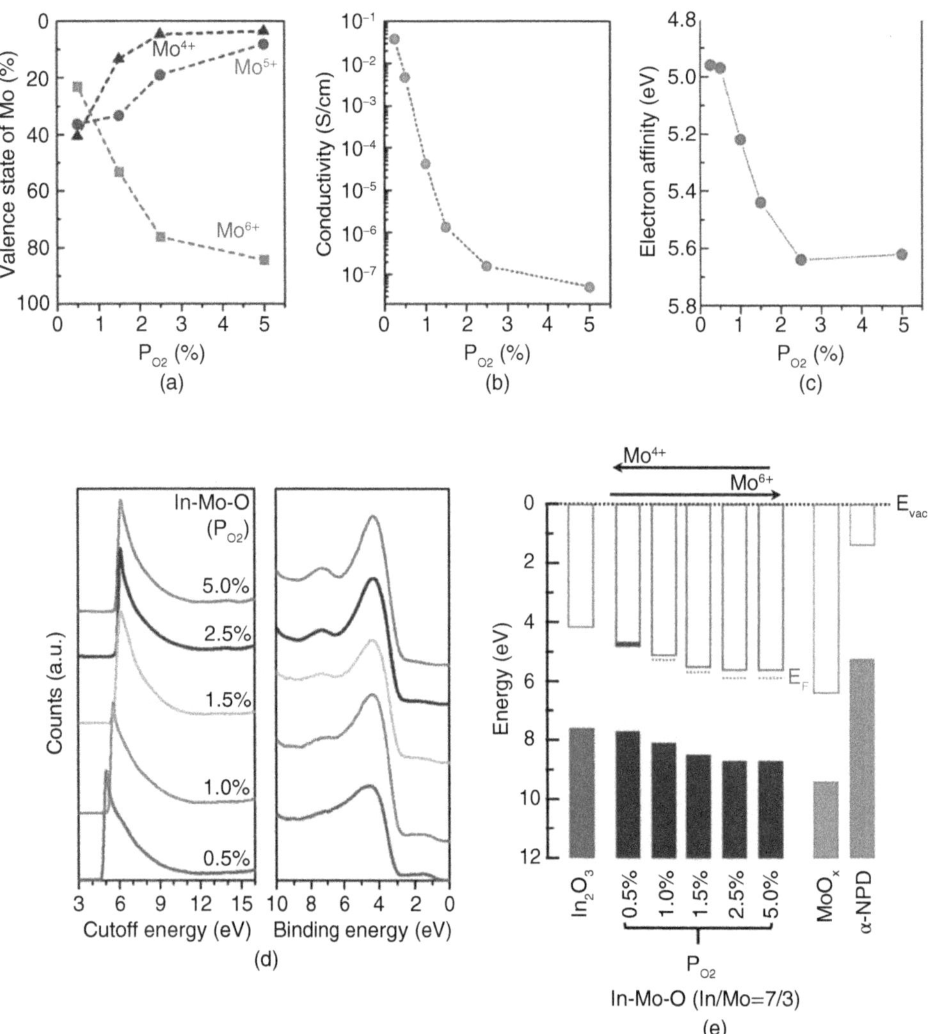

Figure 28.8 Comparison of (a) the ratio of Mo^{4+}, Mo^{5+}, and Mo^{6+} ions, (b) the conductivity; (c) the electron affinities; and (d) the UPS spectra as a function of the partial oxygen pressure (P_{O2}). (e) Energy lineup of *a*-IMO compared with In_2O_3, MoO_x, and *a*-NPD. *Source*: Kim, J., et al. (2018). Electron affinity control of amorphous oxide semiconductors and its applicability to organic electronics. *Advanced Materials Interfaces* **5**: 1801307.

effectively. The high-resolution transmission electron microscopy (HR-TEM) image confirmed the amorphous structure of *a*-IMO (Figure 28.7c).

X-ray photoelectron spectroscopy (XPS) results revealed that the oxidation states of the Mo ions varied greatly depending on the P_{O2} during sputtering. Figure 28.8a shows that the fraction of Mo^{6+} ions significantly decreased with decreasing P_{O2}, whereas the Mo^{4+} and Mo^{5+} ions increased. It is thought that the oxidation states of Mo easily change to maintain charge electroneutrality depending on the P_{O2}. Meanwhile, the *a*-IMO films exhibited a high conductivity dependence on the P_{O2}. The conductivities increased with decreasing P_{O2} (Figure 28.8b). Therefore, it is thought that the carrier generation in *a*-IMO is governed primarily by the oxygen vacancies, similar to conventional AOSs. An electron density of $\sim 10^{14}$ cm^{-3} and an electron mobility of ~ 1 cm^2/Vs were confirmed

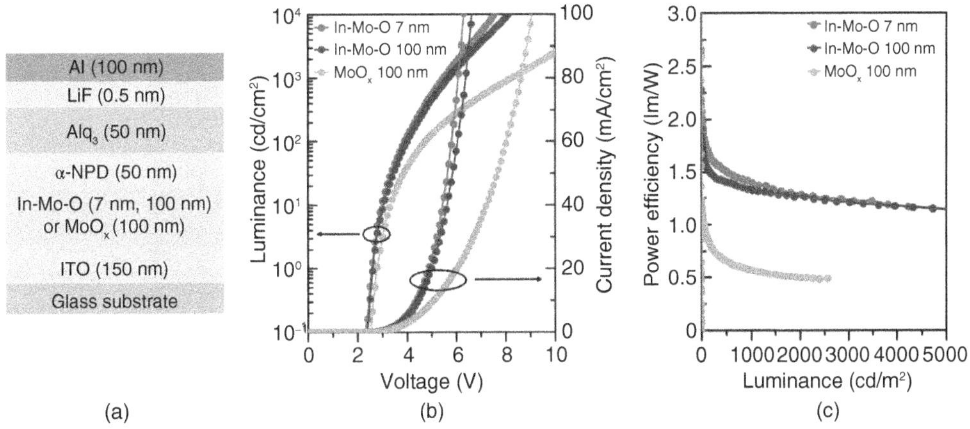

Figure 28.9 Comparison of OLEDs using 7- and 100-nm-thick *a*-IMO layers and 100-nm-thick MoO_x: (a) device structure, (b) current density–voltage–luminance (J–V–L) characteristics, and (c) current efficiency and power efficiency. *Source*: Kim, J., et al. (2018). Electron affinity control of amorphous oxide semiconductors and its applicability to organic electronics. *Advanced Materials Interfaces* **5**: 1801307.

from the Hall effect measurement for *a*-IMO (P_{O2}: 1.5%), thus revealing that *a*-IMO is an *n*-type semiconductor and has a larger mobility than that of MoO_x. We could not obtain the electrical properties for MoO_x because of its quite high resistivity.

However, a remarkable relationship between the oxidation states of Mo ions and the energy levels was clarified. As shown in Figure 28.8c, the electron affinity values ($E_{VAC} - E_{CBM}$) increased with increasing Mo^{6+} content (Figure 28.8e). This result supports our expectation that the band lineup of *a*-IMO is close to the weighted average of In_2O_3 and MoO_3. As a result, we could obtain the promising *a*-IMO HIL with a large electron mobility of $1\,cm^2/Vs$ and a deep E_{CBM} of 5.6 eV. To confirm the hole injection ability of *a*-IMO, two OLEDs were fabricated and compared with (i) a thin *a*-IMO layer of ~5 nm and (ii) a thick *a*-IMO layer of ~100 nm as HILs. For comparison, 100-nm-thick MoO_x was also employed. The device structures of the fabricated OLEDs are illustrated in Figure 28.9a. Both OLEDs using 5- and 100-nm-thick *a*-IMO layers exhibited nearly the same luminous characteristics, implying that the resistance of a 100-nm-thick *a*-IMO is still negligible from the viewpoint of total series resistance (Figure 28.9b). Meanwhile, a significant increase in operating voltage was observed for the OLED using 100-nm-thick MoO_x. This result means that the 100-nm-thick MoO_x contributed significantly to the total series resistance owing to its high resistivity. Figure 28.9c compares the power efficiency (lm/W) and current efficiency (cd/A). For both OLEDs, similar current efficiencies of ~3.5 cd/A were obtained, which is comparable to that for well-charge-balanced OLEDs using Alq_3 as an emission layer. The power efficiency of the OLED using 100-nm-thick *a*-IMO was higher than that of the OLED using 100-nm-thick MoO_x because of the large difference in the operating voltages. This result demonstrates why a low operating voltage is important to achieve efficient OLEDs.

Finally, the chemical stability of *a*-IMO was confirmed, which is an important factor for the application of *a*-IMO to electronics based on solution processes. The solubility of *a*-IMO and MoO_x in a variety of solvents, such as toluene, DMSO, and water, was studied. The film thickness was checked via X-ray reflectometry (XRR). As shown in Figure 28.10, *a*-IMO had a strong chemical stability against all solvents, whereas MoO_x was stable against only toluene, which is a nonpolar solvent. Therefore, it was concluded that various solvents are available on *a*-IMO. This chemical stability might result from a shielding effect of the majority In ions that surround Mo ions via oxygen ions. Thus, it is thought that the chemical stability of *a*-IMO deteriorates with the decrease in the In content.

In-Mo-O 1.5% film

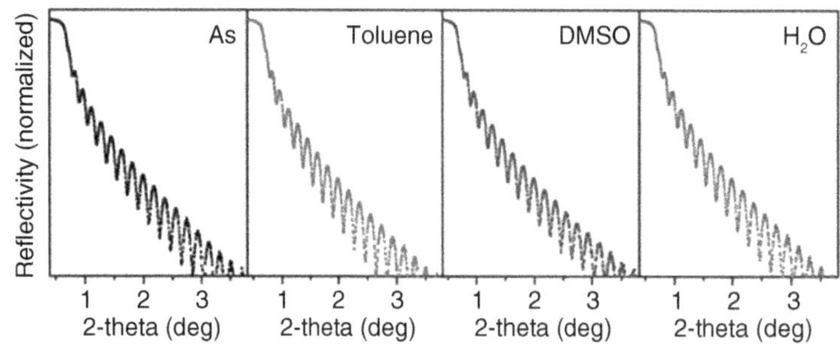

MoO_x film

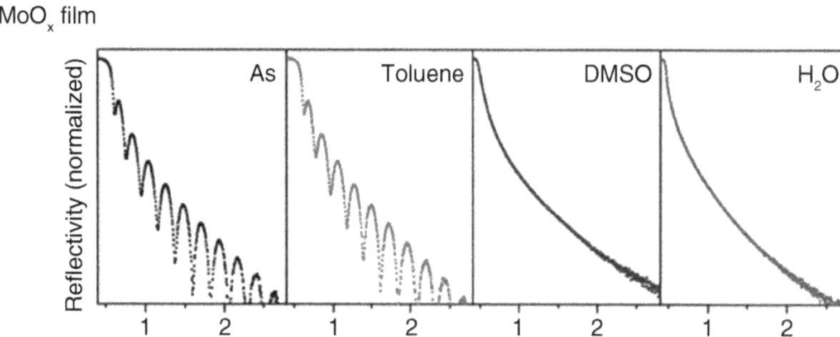

Figure 28.10 XRR results of *a*-IMO and MoO_x thin films after exposure to a variety of solvents. *Source*: Kim, J., et al. (2018). Electron affinity control of amorphous oxide semiconductors and its applicability to organic electronics. *Advanced Materials Interfaces* **5**: 1801307.

28.5 Perspective

Amorphous semiconductors ideally have the advantages of both amorphous materials and semiconductors. The clearest advantages are excellent homogeneity and easy tuning of physical properties based on the additive rule of each component. The essence of semiconductors is easy control of the Fermi level. The AOS materials described in this chapter, *a*-ZSO and *a*-IMO, used the tunability of energy of the CBM for *a*-ZSO as an ETL in perovskite LEDs, and of the VBM for *a*-IMO as an HIL in OLEDs. Such easy tunability is almost impossible for crystalline semiconductors. We think AOSs have a variety of applications in electronic devices by remembering that the contact of active layer to electrode is an almost universal issue in electronics, and confinement of carriers in luminous layers by electron and hole transport layers is key to realizing EL devices. In addition, the unique properties of *a*-ZSO are low work function (~3.5 eV), which is lower by ~1 eV than that of bulk ZnO, and easy formation of quasi-Ohmic contact with various metals regardless of work function. The latter property comes from the metallic nature of nanosized ZnO particles. *a*-IMO solved the issues of MoO_x as an HIL (poor chemical durability and carrier mobility) by combining with In_2O_3, which has high mobility and chemical durability. A combination of *a*-IMO and *a*-ZSO in a dual layer was applied to a charge generation layer in tandem-type OLEDs, and higher performance was obtained [20]. Rational materials design of AOSs is possible in many cases based on the additivity property of each component. It is essential work to choose the appropriate component for targeted properties. A further frontier in AOSs is to find cases in which additivity does not satisfy totally, and a distinct peak or valley is seen. A mixed-alkali effect in glass science is a typical example (i.e., a clear maximum in resistivity is commonly observed when two different alkali ions are mixed). This effect is often applied to improve the chemical durability and so on. No such phenomenon has been reported to date. The successful use of additivity in materials design and finding a drastic breaking of additivity are current challenges in AOS research.

As described in this chapter, AOSs enable one to design ETLs while focusing on the required functionality. In particular, the widely tunable energy levels and electrical properties are quite attractive points in using AOSs. In this respect, AOSs can be extended to optoelectronics such as ELs and solar cells. However, more effort will be made in terms of processability. As is well known, AOSs are currently fabricated mainly by a sputtering process. However, such a physical deposition method is rarely applicable to organic materials. Consequently, only a limited device structure is allowed together with AOSs. Fortunately, chemical deposition methods such as atomic layer deposition (ALD) have been extensively studied for AOSs. Besides, there is another unique method of inkjet printing for AOSs. Thus, we believe that a variety of functional materials based on AOSs will emerge in the future.

References

1 Park, J., Maeng, W., Kim, H., and Park, J. (2012). Review of recent developments in amorphous oxide semiconductor thin-film transistor devices. *Thin Solid Films* 520: 1679–1693.

2 Ide, K., Nomura, K., Hosono, H., and Kamiya, T. (2019). Electronic defects in amorphous oxide semiconductors: a review. *Physica Status Solidi (A): Applications and Materials. Science* 216: 1800372.

3 Kamiya, T. and Hosono, H. (2010). Material characteristics and applications of transparent amorphous oxide semiconductors. *NPG Asia Materials* 2: 15–22.

4 Nomura, K., Ohta, H., Takagi, A. et al. (2004). Room-temperature fabrication of transparent flexible thin-film transistors using amorphous oxide semiconductors. *Nature* 432: 488–492.

5 Sheng, J., Hong, T., Lee, H. et al. (2019). Amorphous IGZO TFT with high mobility of$\sim$ 70 $cm^2/(V\ s)$ via vertical dimension control using PEALD. *ACS Applied Materials & Interfaces* 11: 40300–40309.

6 On, N., Kim, B., Kim, Y. et al. (2020). Boosting carrier mobility and stability in indium–zinc–tin oxide thin-film transistors through controlled crystallization. *Scientific Reports* 10: 1–16.

7 Kim, J., Bang, J., Nakamura, N., and Hosono, H. (2019). Ultra-wide bandgap amorphous oxide semiconductors for NBIS-free thin-film transistors. *APL Materials* 7: 022501.

8 Sim, K., Jun, T., Bang, J. et al. (2019). Performance boosting strategy for perovskite light-emitting diodes. *Applied Physics Reviews* 6: 031402.

9 Hosono, H., Kim, J., Toda, Y. et al. (2017). Transparent amorphous oxide semiconductors for organic electronics: application to inverted OLEDs. *Proceedings of the National Academy of Sciences* 114: 233–238.

10 Nakamura, N., Kim, J., and Hosono, H. (2018). Material design of transparent oxide semiconductors for organic electronics: why do zinc silicate thin films have exceptional properties? *Advanced Electronic Materials* 4: 1700352.

11 Wu, C., Zou, Y., Wu, T. et al. (2017). Improved performance and stability of all-inorganic perovskite light-emitting diodes by antisolvent vapor treatment. *Advanced Functional Materials* 27: 1700338.

12 Wang, Z., Luo, Z., Zhao, C. et al. (2017). Efficient and stable pure green all-inorganic perovskite CsPbBr3 light-emitting diodes with a solution-processed NiOx interlayer. *Journal of Physical Chemistry C* 121: 28132–28138.

13 Ling, Y., Yuan, Z., Tian, Y. et al. (2016). Enhanced optical and electrical properties of polymer-assisted all-inorganic perovskites for light-emitting diodes. *Advanced Materials* 28: 8983–8989.

14 Cho, H., Kim, J.S., Wolf, C. et al. (2018). High-efficiency polycrystalline perovskite light-emitting diodes based on mixed cations. *ACS Nano* 12: 2883–2892.

15 Seo, H., Kim, H., Lee, J. et al. (2017). Efficient flexible organic/inorganic hybrid perovskite light-emitting diodes based on graphene anode. *Advanced Materials* 29: 1605587.

16 Cho, H., Jeong, S., Park, M. et al. (2015). Overcoming the electroluminescence efficiency limitations of perovskite light-emitting diodes. *Science* 350: 1222.

17 Wang, Z., Wang, F., Sun, W. et al. (2018). Manipulating the trade-off between quantum yield and electrical conductivity for high-brightness quasi-2D perovskite light-emitting diodes. *Advanced Functional Materials* 28: 1804187.

18 Chiba, T., Hoshi, K., Pu, Y.J. et al. (2017). High-efficiency perovskite quantum-dot light-emitting devices by effective washing process and interfacial energy level alignment. *ACS Applied Materials & Interfaces* 9: 18054–18060.

19 Shi, Z., Li, Y., Li, S. et al. (2018). Localized surface plasmon enhanced all-inorganic perovskite quantum dot light-emitting diodes based on coaxial core/shell heterojunction architecture. *Advanced Functional Materials* 28: 1707031.

20 Zhang, J., Bai, D., Jin, Z. et al. (2018). 3D–2D–0D interface profiling for record efficiency all-inorganic CsPbBrI2 perovskite solar cells with superior stability. *Advanced Energy Materials* 8: 1703246.

21 Zhang, L.Q., Yang, X.L., Jiang, Q. et al. (2017). Ultra-bright and highly efficient inorganic based perovskite light-emitting diodes. *Nature Communications* 8: 15640.

22 Yang, H., Kim, J., Yamamoto, K. et al. (2018). Surface tailoring of newly developed amorphous ZnSiO thin films as electron injection/transport layer by plasma treatment: application to inverted OLEDs and hybrid solar cells. *Applied Surface Science* 434: 995–1000.

23 Yang, H., Kim, J., Yamamoto, K., and Hosono, H. (2017). Efficient charge generation layer for tandem OLEDs: bi-layered MoO3/ZnO-based oxide semiconductor. *Organic Electronics* 46: 133–138.

24 Nakamura, N., Kim, J., Yamamoto, K. et al. (2017). Organic light-emitting diode lighting with high out-coupling and reliability: application of transparent amorphous ZnO–SiO_2 semiconductor thick film. *Organic Electronics* 51: 103–110.

25 Kim, J., Yamamoto, K., Iimura, S. et al. (2018). Electron affinity control of amorphous oxide semiconductors and its applicability to organic electronics. *Advanced Materials Interfaces* 5: 1801307.

29

Displays and Vertical-Cavity Surface-Emitting Lasers

Kenichi Iga

Laboratory for Future Interdisciplinary Research of Science and Technology (FIRST), Tokyo Institute of Technology, Yokohama, Japan

29.1 Introduction to Displays

"Display" comes from the Latin *dis* (reversal or negation) and *plicare* (to fold). In other words, the meaning of "display" is to expand something, by showing it wide; the Japanese word *hyouji* corresponds [1, 2].

Kenjiro Takayanagi's electronic display was selected as the IEEE Milestone in 2008 as the world's first such technology. After graduating from Tokyo Higher Technical School (later Tokyo Institute of Technology), Takayanagi became a professor at Hamamatsu Higher Technical School (later Shizuoka University, Faculty of Engineering) and studied electronic display devices. Figure 29.1 shows the milestone plaque submitted by Shizuoka University; with the permission of Shizuoka University, one of the nameplates is on display at the Tokyo Tech Museum.

Takayanagi's first electronic projection of images was the famous "イ" in Figure 29.2 [3]. It was the first attempt in 1926 using a cathode ray tube (CRT), which was invented in 1897 by Karl Ferdinand Braun of Germany. A Nipkow disk was used as the imaging device.

The display is seen by the human eye, and it is important to look gently. There are various ways of thinking, even when considering the types. These types are summarized in Table 29.1. In addition, there are the non-emissive type and self-emissive type; flat type, cylindrical type, and flexible type; and the like. Image types include static type, moving type, and fluid type.

29.2 Liquid Crystal Displays (LCDs)

29.2.1 History of LCDs

Table 29.2 outlines the research history of liquid crystal displays (LCDs) [1]. In 1888 the Austrian botanist Friedrich Reinitzer discovered a liquid–solid intermediate layer, a cholesterol derivative. In 1889, the German physicist Otto Lehmann, who researched at the request of Reinitzer, discovered that this intermediate layer had the optical anisotropy seen in solid crystals and the liquid crystal (in German, "fliessende Krystalle") [1].

The dynamic scattering mode (DSM) was used for calculators, but many LCDs thereafter used the TN (twisted nematic) mode, and further improved operating modes were developed. This TN mode was invented by Wolfgang Helfrich in 1969. Helfrich was in RCA's George Heilmeier group at this time, but the following year he moved to Hofmann-La Roche's Martin Schadt group. At the same time, James Lee Fergason of Kent State University's Liquid Crystal Research Center also discovered a TN mode that operates in an electric field in 1969. Both applied for patents, and there has been a patent dispute between Schadt's group and Fergason. The creators of the TN mode are supposed to be Helfrich, Schadt, and Fergason.

Amorphous Oxide Semiconductors: IGZO and Related Materials for Display and Memory, First Edition.
Edited by Hideo Hosono and Hideya Kumomi.

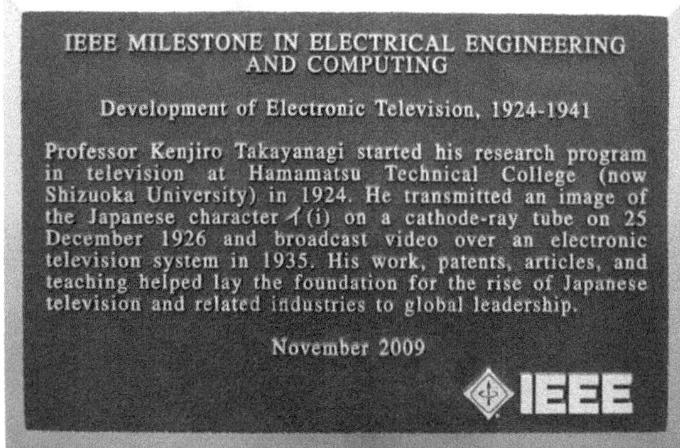

Figure 29.1 Kenjiro Takayanagi's electronic display IEEE Milestone, Tokyo Institute of Technology Museum. *Source*: Kenichi Iga.

Figure 29.2 The letter イ written on a mica plate as the subject used for sending and receiving by Kenjiro Takayanagi [3]. The basis of the display is the integration of its principle, basic mechanism, device, actual technology, and data transfer method [1, 2]. *Source*: Kenjiro Takayanagi Foundation. (2020). https://takayanagi.or.jp/index.html.

29.2.2 Principle of LCD: The TN Mode

"Liquid crystal" refers to a liquid crystal state that has a periodic structure such as a crystal. A liquid crystal called a nematic liquid crystal is used for LCDs, and it has the property (orientation) that elongated rod-shaped molecules are oriented in a substantially constant direction. The long axis of the molecule is called the liquid crystal director [1]. The director is a unit vector in the long-axis direction. Generally, the dielectric constant differs between the direction parallel to the director and the direction orthogonal thereto. If the permittivity of each is $\varepsilon_{\parallel}$ and $\varepsilon_{\perp}$, the difference is

$$\Delta\varepsilon = \varepsilon_{\parallel} - \varepsilon_{\perp} \tag{29.1}$$

The case of $\Delta\varepsilon>0$ is called a *p*-type liquid crystal, and the case of $\Delta\varepsilon<0$ is called an *n*-type liquid crystal. When an electric field is applied to the liquid crystal molecules, the liquid crystal molecules are oriented so that the axis with the large dielectric constant is parallel to the electric field. When the orientation of liquid crystal molecules changes, the refractive index anisotropy also changes, and the plane of polarization of light passing through the liquid crystal can be changed. This is the basic operation of LCDs. Note that $\Delta\varepsilon$ with respect to the frequency of light (visible light) is positive regardless of whether it is a *p*-type liquid crystal or an *n*-type liquid crystal. Therefore, the refractive index anisotropy is always positive:

$$\Delta n = n_{\parallel} - n_{\perp} = n_e - n_o > 0 \tag{29.2}$$

where n_o is the refractive index for ordinary rays, and n_e is the refractive index for extraordinary rays.

Table 29.1 Major displays and materials.

	Method		Head mount type	
Materials	**Panel**	**Projection**	**Half mirror**	**Retina projection type**
CRT	TV, measuring instrument	X	X	x
LCD	TV, PC, smartphone	TV, projector	Epson, etc.	x
OLED	TV, smartphone		Sony, etc.	x
LED	Advertisement (digital signage)	LED projector	x	x
Laser	TV	Movie theater		Quantum dot laser, etc.
Others		DLP		

CRT: Cathode ray tube; DLP: digital light processing; LCD: liquid crystal display; LED: light-emitting diode; OLED: organic light-emitting diode.

Table 29.2 History of LCDs and applied systems.

Date	Discovery or Invention	Inventor(s)
1888	Discovery of liquid crystals	Friedrich Reinitzer (Austria)
1964	Invention of dynamic scattering mode (DSM) LCDs	George Heilmeier (RCA)
1969	Invention of twisted nematic (TN) liquid crystals	Wolfgang Helfrich, Martin Schadt, and James Lee Fergason
1973	Calculator development	Sharp

Source: Iga, K., and Hatakoshi, G. (2020). *Treasure Microbox of Optoelectronics*, Vols. I–IV, e-version. Advanced Communication Media, Tokyo. https://www.adcom-media.co.jp/info-oe/2020/04/24/33770/; and Suematsu, Y., and Kobayashi, K. (2007). *Photonics-Optics and Its Progress*. Ohmsha, Tokyo.

As shown in Figure 29.3, let us explain the principle using the commonly used TN mode. When an electric field is applied as shown in Figure 29.3b, the liquid crystal is oriented in that direction, and the twist is eliminated. The refractive index for passing light is only $\varepsilon_{\perp}$, and the refractive index anisotropy disappears, and the polarization plane of light does not rotate. Therefore, this light does not pass through the analyzer rotated by 90°. Since the alignment direction of the liquid crystal molecules adjacent to the alignment film is fixed, when the electric field is turned off again, the liquid crystal in between aligns with the adjacent liquid crystal and returns to the state of Figure 29.3a. In other words, a spatial switch of light can be created. An image can be displayed by performing on-off for each small pixel.

The TN mode has the advantages of a simple structure and low cost, but it has problems such as large viewing-angle dependence and slow response speed. The viewing-angle dependence was greatly improved by using a retardation film that compensates for it. Regarding the response speed, high speed is achieved by an active-matrix drive that drives each pixel with a thin-film transistor (TFT).

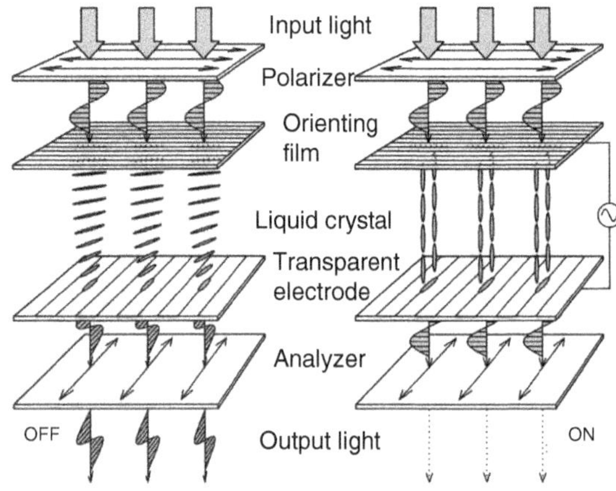

Figure 29.3 Principle of LCD (TN mode). *Source*: Iga, K., and Hatakoshi, G. (2020). *Treasure Microbox of Optoelectronics*, Vols. I–IV, e-version. Advanced Communication Media, Tokyo. https://www.adcom-media.co.jp/info-oe/2020/04/24/33770/.

29.2.3 Other LC Modes

There are other advanced modes in LCDs [1]:

STN: Super-twisted nematic
IPS: In-plane switching
MVA: Multidomain vertical alignment
PVA: Patterned-ITO vertical alignment
CPA: Continuous pinwheel alignment
OCB: Optically compensated bend

29.2.4 Light Sources

Since a normal light source, such as a fluorescent lamp or a light-emitting diode (LED) [1], has a random polarization, only the linearly polarized light that has been transmitted through a polarizing filter is used to obtain the linearly polarized light. Therefore, more than half of the light is wasted. When a laser light source is used, since it is originally linearly polarized, the loss is reduced.

Thin fluorescent tubes and high-intensity halogen lamps have been used for the backlight of liquid crystal panels for PCs. These had low power efficiency, the life of the tube was relatively short, and there was concern about when the lamp might run out.

With the advent of blue LEDs, white LEDs were quickly introduced into the backlight of the panel and the light source of the projector. In the case of LEDs, even if it is white, the light emission of the fluorescent substance is used from the excitation in the blue-violet range, so a measure to protect the eye was necessary. Also, the range covered by the chromaticity diagram is not necessarily large. However, from the viewpoint of power efficiency, most LCD light sources use LEDs.

Meanwhile, there are also attempts to use semiconductor lasers in the three primary colors for the backlight. In this case, if the color development range of the semiconductor laser is appropriately selected, it is possible to gain the advantage that a wider spectrum range can be reproduced. For the time being, TVs and high-fidelity medical displays will be applied.

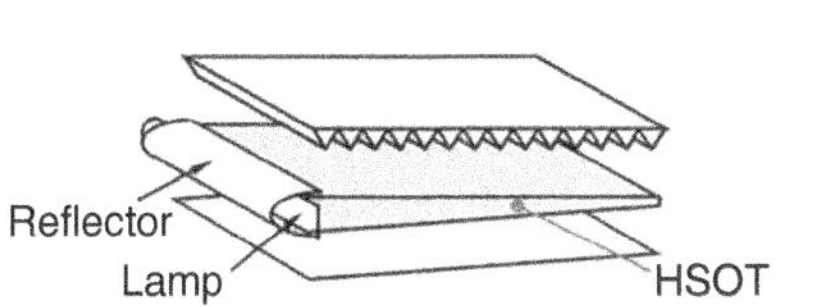

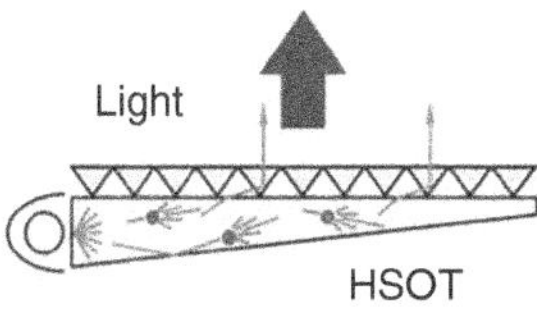

Figure 29.4 Diffusion light guide plate for LCD, by Yasuhiro Koike et al. *Source*: Horibe, A., et al. (1998). High-efficiency and high-quality LCD backlight using highly scattering optical transmission polymer. *IEICE Transactions on Electronics* **81-C** (11): 1697–1702.

29.2.5 Diffusion Plate and Light Guiding Layer

Light from fluorescent or LED lamp light sources must illuminate the liquid crystal panel evenly. Professor Yasuhiro Koike invented a transparent light guide plate that utilizes scattering by particles with different refractive indices introduced into plastics [1]. This principle has been proved and is used as a backlight for PCs The principle is shown in Figure 29.4. It is a transparent light guide plate named a HSOT (highly scattered optical transmission) polymer, and it was found to scatter light efficiently and uniformly. It has been applied to actual systems by contributing to lower power consumption, higher brightness, and uniform radiation of LCD panels.

29.2.6 Microlens Arrays

The flat microlens array also contributed to the luminous efficiency of the liquid crystal projector. In 1979, the author proposed a flat-plate microlens with a refractive index distributed in both the axial and radial directions [4]. A lens array is manufactured by the ion exchange diffusion method. After attaching a mask to the glass substrate, a circular window with a diameter of around 100 μm is provided by the photolithography method, and the refractive index distribution is formed by immersion in molten salt and ion exchange.

Nippon Sheet Glass Co. has developed a microlens array for use by pasting it on a liquid crystal panel, and mass production of uniform lens arrays has started [1]. A photograph of the microlens array is shown in Figure 29.5. Figure 29.6 shows the principle of improving the light transmittance by this method.

In addition, Sharp Corporation, in collaboration with Nippon Sheet Glass Co., applied it to a liquid crystal projector, as shown in Figure 29.7. With this technology, bright projected images can be obtained with liquid crystal projectors; LCD projectors play a major role in projectors used for lecture presentations and large displays. Figure 29.8 is a photograph of a successor model that has been made compact; it won the Good Design Award in 1997.

Figure 29.5 Planar microlens array for LCD projectors, Nippon Sheet Glass Co. *Source*: Kenichi Iga.

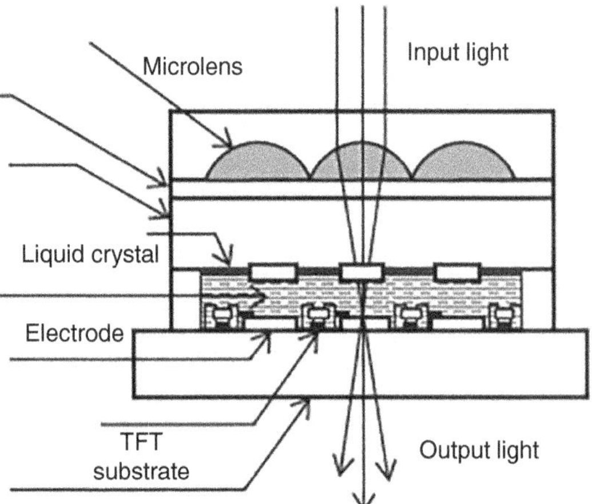

Figure 29.6 Microlens array improving transmission efficiency, Nippon Sheet Glass Co. *Source*: Iga, K., and Hatakoshi, G. (2020). *Treasure Microbox of Optoelectronics*, Vols. I–IV, e-version. Advanced Communication Media, Tokyo. https://www.adcom-media.co.jp/info-oe/2020/04/24/33770/.

Figure 29.7 LCD projector with planar microlens, 1997 Good Design Award model. *Source*: Sharp Corporation and Nippon Sheet Glass Co., https://www.g-mark.org/award/describe/23595.

After 2000, a microlens array that can be manufactured at low cost was developed, and it replaced the planar microlens array that the author had invented. This kind of thing should be a "karma" or "destiny" of the industry.

29.2.7 Short-Focal-Length Projection

Models that use a projection lens with an extremely short focal length, for use in conference rooms where it is difficult to secure the distance from the projector to the screen, have also been developed by Ricoh, Panasonic, and others.

29.3 Organic EL Display

Figure 29.8 shows the principle of luminescence transitions (fluorescence and phosphorescence) in organic electroluminescence (EL). In conventional EL materials, only 25% of excitations emit light through fluorescence. On the other hand, with TADF (thermally assisted delayed fluorescence) proposed by Chihaya Adachi, another 75% can be transferred to the singlet excited state by IISC (inverse intersystem crossing) and utilized for radiative transitions [1].

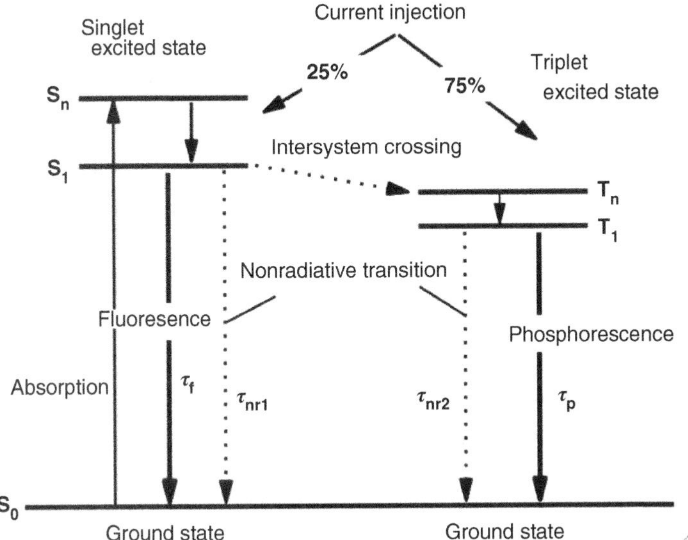

Figure 29.8 Typical organic EL principle. *Source*: Iga, K., and Hatakoshi, G. (2020). *Treasure Microbox of Optoelectronics*, Vols. I–IV, e-version. Advanced Communication Media, Tokyo. https://www.adcom-media.co.jp/info-oe/2020/04/24/33770/.

Unlike LCDs, organic EL displays are characterized as being self-luminous [1, 2]. Light-emitting materials used for organic EL include low-molecular-weight materials and high-molecular-weight materials.

RGB light sources are required for full-color display with organic EL. Subsections 29.1–29.3 describe three typical methods.

29.3.1 Method (a): Color-Coding Method

This is a method in which the R, G, and B organic EL elements are separately painted and manufactured, and the light from each element is mixed to obtain a full color. In principle, this method is highly efficient and can reduce power consumption. However, RGB element life is not always the same, so there is the problem of poor color balance [1].

29.3.2 Method (b): Filter Method

This is a method of extracting R, G, and B light with a color filter using a white organic EL element with a wide emission spectrum as a light source. The color filter technology established in LCD can be utilized. Due to the loss of color conversion and filters, the light utilization efficiency is lower than in method (a).

29.3.3 Method (c): Blue Conversion Method

This is a method of exciting a dye using a blue-emitting organic EL element as a light source to obtain a full color by color conversion. The blue color is used as it is, and the green color and the red color are excited by exciting the dyes for color conversion. Since there is only one type of organic EL element, there is no problem that the color balance changes with time as in method (a).

The application of organic EL to large industrial fields will be panel displays. Hideo Hosono of Tokyo Institute of Technology realized an oxide semiconductor composed of InGaZnO (IGZO) and applied it to TFTs. An organic EL panel using this as a driver has been put to practical use for TVs and the like.

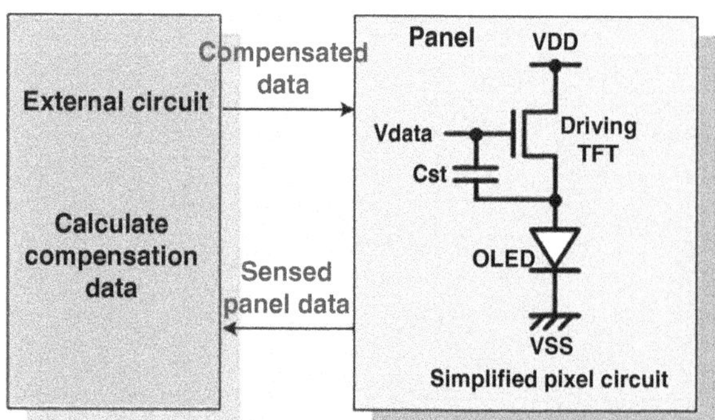

Figure 29.9 Circuits of display panels using organic EL. *Source*: Nomura, K., et al. (2003). Thin-film transistor fabricated in single-crystalline transparent oxide semiconductor. *Science* **30** (5623): 1269–1272.

Figure 29.10 Example of smartphone with organic EL panel [1], Ueno Park Peony Garden. *Source*: Kenichi Iga.

An example of a driver circuit using an oxide TFT as a driver is shown in Figure 29.9. OLED displays are used as display panels for medium-sized TV receivers and smartphones. It features self-luminous power efficiency. Figure 29.10 shows an example of a smartphone equipped with an organic EL panel.

29.4 Vertical-Cavity Surface-Emitting Lasers

29.4.1 Motivation of Invention

I conceived of a surface-emitting laser on March 22, 1977 [6]. The sketch of the idea is shown in Figure 29.11. In contrast to the conventional Fabry–Perot edge-emitting semiconductor lasers, this invention consists of a laser cavity vertical to the wafer surface on which many layers are monolithically grown, including an active layer. Soon later, it began to be called the vertical-cavity surface-emitting laser (VCSEL).

The contents include the features of the VCSEL concept, the motivation behind the invention, a breakthrough for the realization, and several technologies that became essential for later devices such as quantum wells, semiconductor Bragg reflectors, and the AlAs oxidation technique.

The motivations of its invention were to realize a laser [7] that

1) was manufactured by a monolithic process,
2) had single-mode operation, and
3) had wavelength reproducibility.

The motivations are illustrated in Figure 29.12. The items inside of the circle are those that were conducted in industries in the 1970s. We took a different way of fulfilling the aforementioned three conditions.

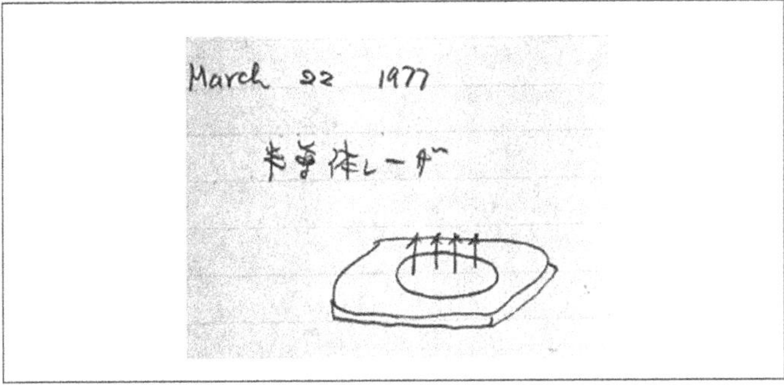

Figure 29.11 A sketch of a VCSEL idea in 1977. *Source*: Iga, K. (2018). Forty years of VCSEL: invention and innovation. *Japanese Journal of Applied Physics* **57** (8S2): 1–7; and Iga, K. (1977). Laboratory notebook, March 22, Figure 1, p. 2 / IOP Publishing / CC BY.

Figure 29.12 The motivation of research leading to VCSEL invention. *Source*: Iga, K. (2018). Forty years of VCSEL: invention and innovation. *Japanese Journal of Applied Physics* **57** (8S2): 1–7; and Iga, K. (1977). Laboratory notebook, March 22.

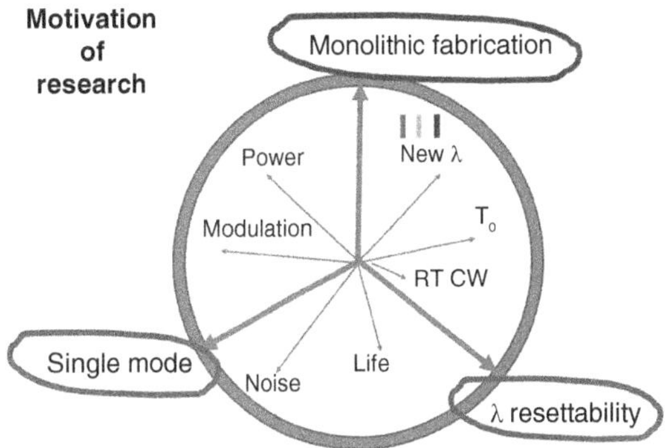

After critical consideration to solve the aforementioned issues associated with conventional edge-emitting lasers, the author reached a conclusion that the laser cavity should be made vertical, not to the transverse to the semiconductor wafer surface. The author thought that the cavity can be made by some semiconductor layers and/or dielectric layers, which can be fabricated by semiconductor processes, not by manual cleaving processes.

29.4.2 What Is the Difference?

The difference between stripe lasers and VCSELs for obtaining single-mode operation is shown in Figure 29.13. The upper figure shows the behavior of edge-emitting stripe lasers. The horizontal axis is the stripe width, and the vertical one shows the resonant frequency. f_0 means the central nominal frequency. Even in the case of a small stripe width on the order of microns, the lateral mode can be single, as indicated in the figure by an ellipse. On the other hand, resonance occurs at multiple frequencies. To realize single-frequency operation, some filtering structure is required such as distributed Bragg reflectors.

On the other hand, we consider the case of VCSELs shown in the lower figure. For small-diameter VCSELs, the lateral mode is single and the longitudinal mode also can be single, since the free spectral range (FSR) is large owing to the small cavity length.

29.4.3 Device Realization

The first room temperature continuous-wave (CW) operation was achieved in 1988 by Koyama and I using a GaAs system exhibiting a possibly engineered semiconductor laser, as shown in Figure 29.14 [8, 9]. We employed metal organic chemical vapor deposition (MOCVD) crystal growth and multilayer mirror formation techniques to lower the threshold current density. Also, we achieved a near-room temperature CW in a 1,300 nm VCSEL in 1993 [1].

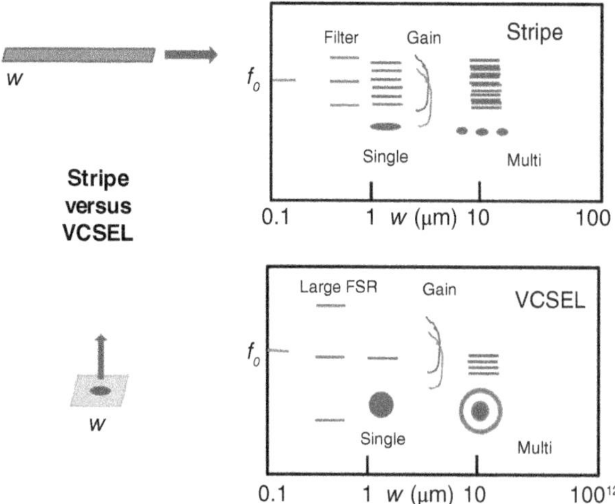

Figure 29.13 The difference between an edge-emitting stripe laser and a VCSEL. *Source*: Iga, K. (2018). Forty years of VCSEL: invention and innovation. *Japanese Journal of Applied Physics* **57** (8S2): 1–7.

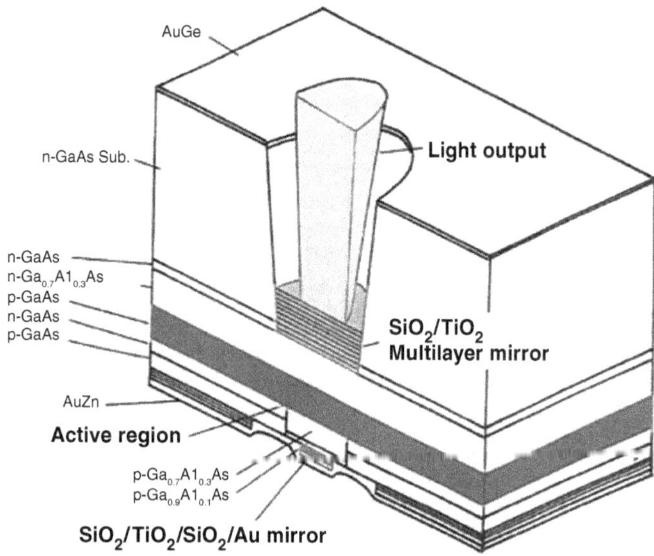

Figure 29.14 VCSEL that achieved the first room temperature CW operation. *Source*: Iga, K., and Hatakoshi, G. (2020). *Treasure Microbox of Optoelectronics*, Vols. I–IV, e-version. Advanced Communication Media, Tokyo. https://www.adcom-media.co.jp/info-oe/2020/04/24/33770/; Koyama, F., et al. (1988). Room temperature CW operation of GaAs vertical cavity surface emitting laser. *IEICE Transactions* **E71** (11): 1089–1090; Koyama, F., et al. (1989). Room-temperature continuous wave lasing characteristics of GaAs vertical cavity surface-emitting laser. *Applied Physics Letters* **55** (3): 221–222.

A strong research and development competition ensued in a worldwide scale from the end of the 1980s to the 1990s, where I was almost always in the forefront. A pulsed threshold current as small as 6 mA was reported at room temperature in a AlGaAs/GaAs VCSEL with a cavity length of 7 μm. This is the first time that a threshold current lower than 10 mA was realized in VCSELs at that time. In 1988, we achieved room temperature continuous operation for the first time [8, 9].

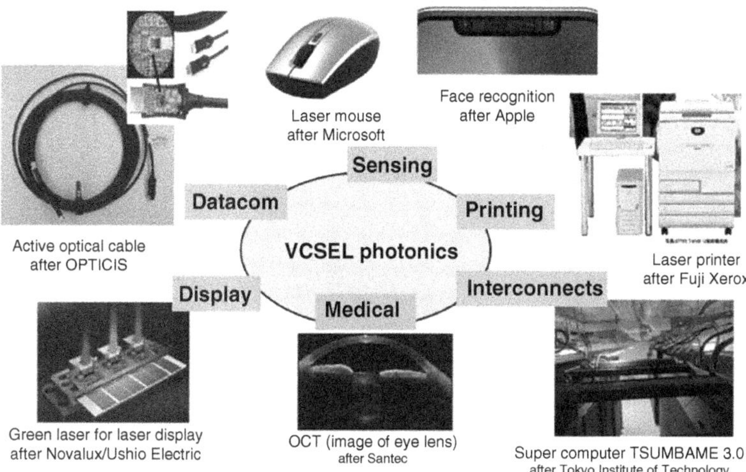

Figure 29.15 Some applications of VCSELs. *Source*: Iga, K. (2018). *VCSEL Odyssey*. Optronics Pub. Co., Tokyo, Figure 8, p. 11 / with permission of SPIE.

29.4.4 Applications

From 1999, the VCSEL was applied to high-speed local area networks (LANs) such as Gigabit Ethernet, and its mass production started. To date, VCSELs have been applied to various internet-of-things (IOT) fields, as shown in Figure 29.15. The market size of production was forecast to reach $2.5 billion (U.S.) in 2020 [10].

Today's smartphones are installed with many VCSELs, for example, for laser autofocus and proximity sensing by time-of-flight range detection and face recognition. One of the recently released smartphones, iPhone X, is said to have >500 pixels of VCSELs emitting 3 W peak power for face recognition [9]. The face recognition systems may be applied in security systems and for gesture recognition. A low-noise optical microphone is another application of VCSELs.

Application to new optoelectronic fields such as sensing makes use of the continuous sweep characteristics of wavelength advances. Features of VCSELs such as optical coherent tomography (OCT), hair removal treatment, laser surgery, laser heating, robot vision, and so on are being utilized.

The use of wavelength-swept VCSELs in medical OCT has a large impact. This is because the drive unit for wavelength tuning is microelectromechanical systems (MEMS), so that the size can be reduced. OCT is used to measure tomograms of the fundus.

29.5 Laser Displays including VCSELs

29.5.1 Laser Displays

There are some displays; one is to use a laser beam by the projection on a screen directly, and another is a liquid crystal backlit by RGB lasers. In both cases, the goodness of the chromaticity possessed by the laser is a major feature [1].

In particular, a method is adopted in which a nonlinear crystal for generating the second harmonic is inserted inside a resonator of a surface-emitting laser to cause green oscillation, and it is used for a large screen projector for a movie. The excellent color balance makes it an interesting form of the future. A projector equipped with a Novalux green laser manufactured by A. Mooradian has been introduced to a movie theater at the Osaka Expo site. He was a researcher at the MIT Lincoln Laboratory, and the author visited there several times.

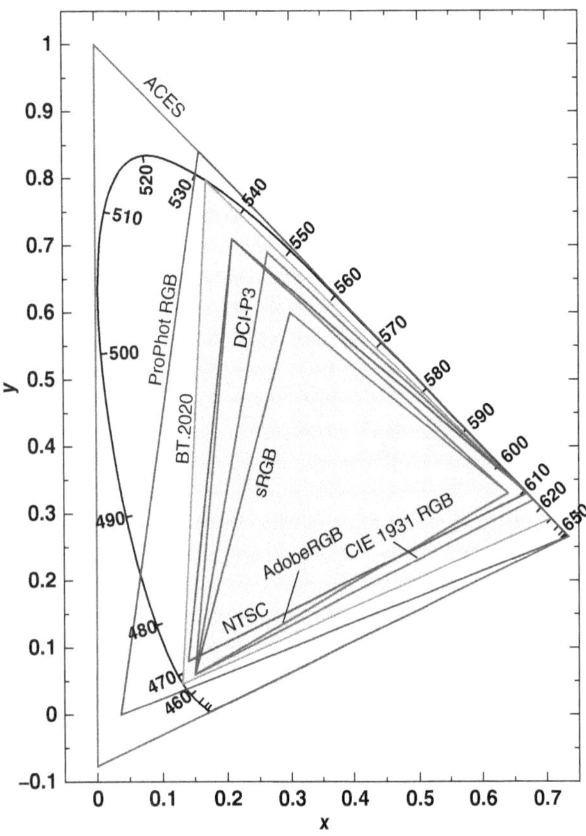

Figure 29.16 Various color gamuts. *Source*: Iga, K., and Hatakoshi, G. (2020). *Treasure Microbox of Optoelectronics*, Vols. I–IV, e-version. Advanced Communication Media, Tokyo. https://www.adcom-media.co.jp/info-oe/2020/04/24/33770/.

29.5.2 Color Gamut

The range of colors that can be displayed on a display or printing is called the color gamut, and there are various standards and specifications. Figure 29.16 shows the color gamut represented on the xy chromaticity diagram for each of these standards and specifications. Among them, ACES and ProPhoto RGB have colors that are not actually included in the color gamut, as you can see from the figure. ACES has R, G, and B chromaticity coordinates set so that it can represent all visible light.

As can be seen from Figure 29.16, BT.2020 (area covered by the figure), which is expected to be introduced in 8K/SHV broadcasting, has a significantly wider color gamut than the conventional standard. The vertices of the color gamut triangle are almost on the spectrum locus, so the stimulus purity is almost 1, and it can be called monochromatic light. The corresponding monochromatic light wavelengths are:

R (0.708,0.292): 630 nm
G (0.170,0.797): 532 nm
B (0.131,0.046): 467 nm

As can be seen in Figure 29.17, a semiconductor laser set for three primary colors is employed in the laser display for movie theaters. A laser with a wavelength of 445 to 460 nm is used for blue, while the blue in BT.2020 corresponds to 467 nm, which has a rather long wavelength. The semiconductor laser is an InGaN/GaN quantum

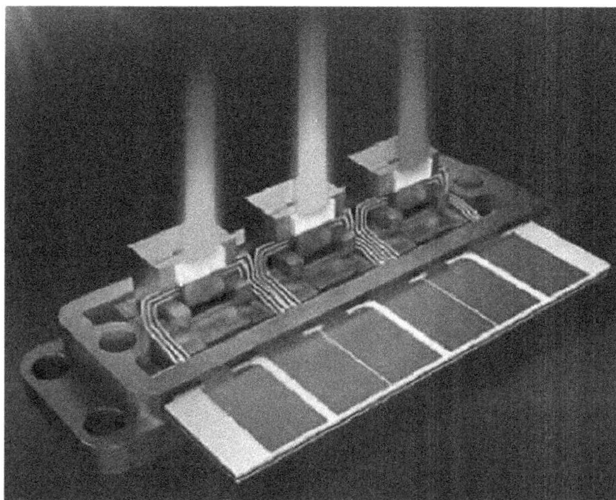

Figure 29.17 Image of semiconductor laser set for three-primary-color display by Novalux/USHIO INC. *Source*: Courtesy of USHIO Hidekazu Hatanaka.

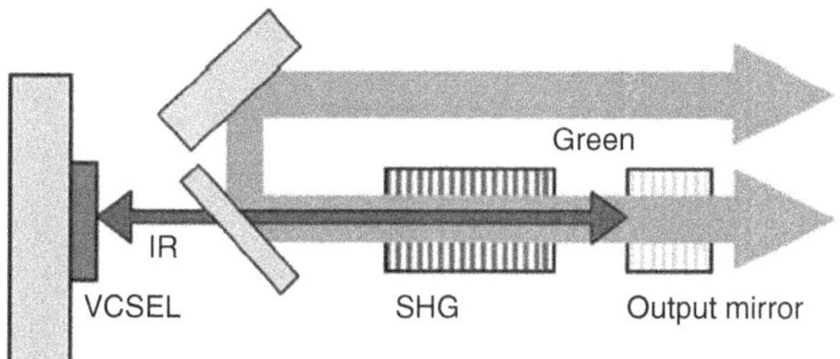

Figure 29.18 Method of extracting green color by a nonlinear crystal placed in the VCSEL resonator. PPRN: Periodically poled lithium–niobate; VBG: volume Bragg grating. Novalux/Ushio Electric Co., Ltd.

well semiconductor laser. For BD optical discs, the wavelength is 405 nm, which is shorter than the smaller focused spot, but for displays, the luminosity and color appearance are criteria for wavelength selection.

Green is the wavelength band in which the human eye has the highest visibility. 530 nm and 550 nm are used. The green wavelength in BT.2020 corresponds to 532 nm. InGaN, which has a large In content, is used, but when the In content is large, crystal growth tends to be difficult, and the increase in output is slightly behind that of blue.

The author went to see the screening in December 2016. At present, as a laser, Ushio Electric Co. handles a nonlinear crystal in a resonator to obtain a green beam of 532 nm from infrared light of 1064 nm, as shown in Figures 29.18 and 29.19. The 24 surface-emitting laser emitters are stacked in two stages, and 10 sets of them are used. The 4K movie was projected on a large screen, and the power and beauty were remarkable.

A single semiconductor's laser output exceeds 10 W class by stripe lasers, and a surface-emitting laser of more than several mW is being made with RGB. The array will be several times more. In addition, a μW-class low-power surface-emitting laser is requested for small head-mounted displays.

29.5.3 Laser Backlight Method

This method replaces the light source of a LCD with a RGB semiconductor laser, and its principle is shown in Figure 29.20. The three color beams are combined and guided to the light guide plate. There is a report on

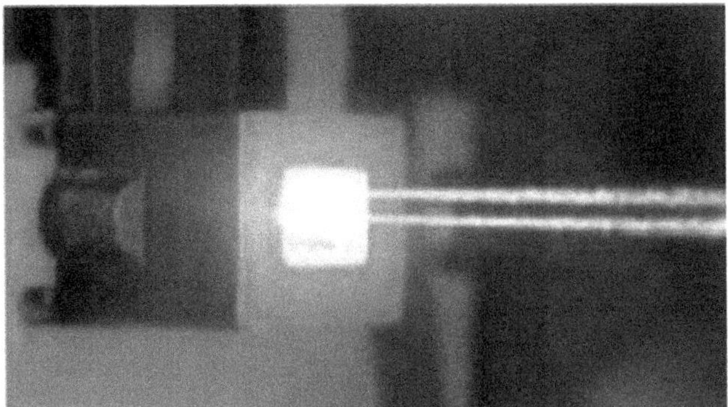

Figure 29.19 Second-harmonic-generation VCSEL [1] that generates green color in the resonator, Novalux/HUSIO ELECTRIC CO., LTD. *Source*: Courtesy of USHIO Hidekazu Hatanaka.

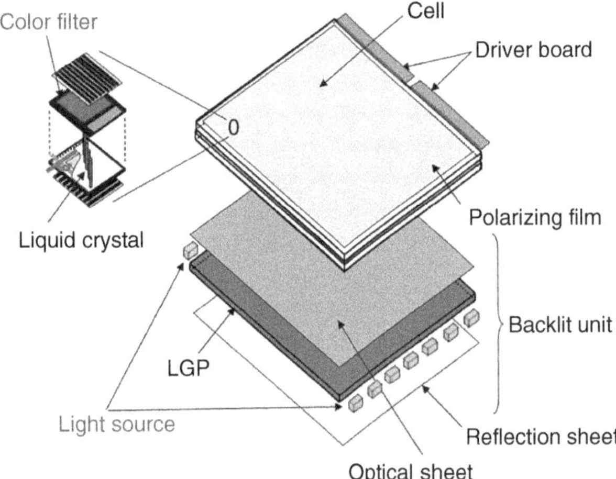

Figure 29.20 LCD panel engine for laser backlight display, Mitsubishi Electric Corp. *Source*: Kojima, K. (2016). Development of RGB laser backlit LCD. *Microoptics News* **34** (4): 25–30.

laser-backlit displays; see Ref. [11]. The laser light of three primary colors (RGB) is used for the backlight of the LCD.

Acknowledgments

I would like to express my deepest appreciation to Honorary Professor Yasuharu Suematsu for continuing advice. I thank Dr. Genichi Hatakoshi for coauthoring the book *Treasure Microbox of Optoelectronics*. Also, I thank Professor Emeritus Kohroh Kobayashi for accumulating the data on VCSELs as well as the information on VCSEL history. I thank Professors Fumio Koyama, Tomoyuki Miyamoto, Hiroyuki Uenohara, and Susumu Kinoshita, and former laboratory members and students for collaborations.

References

1 Iga, K. and Hatakoshi, G. (2020). *Treasure Microbox of Optoelectronics*, Vols, I–IV, e-version. Advanced Communication Media, Tokyo. https://www.adcom-media.co.jp/info-oe/2020/04/24/33770/.
2 Suematsu, Y. and Kobayashi, K. (2007). *Photonics-Optics and Its Progress*. Ohmsha, Tokyo.
3 Kenjiro Takayanagi Foundation. (2020). https://takayanagi.or.jp/index.html.
4 Iga, K., Oikawa, M., Misawa, S. et al. (1982). Stacked planar optics an application of the planar lenses. *Applied Optics* 21 (19): 3456–3460.
5 Nomura, K., Ohta, H., Ueda, K. et al. (2003). Thin-film transistor fabricated in single-crystalline transparent oxide semiconductor. *Science* 30 (5623): 1269–1272.
6 Iga, K. (1977). Laboratory notebook, March 22.
7 Iga, K. (2018). Forty years of VCSEL: invention and innovation. *Japanese Journal of Applied Physics* 57 (8S2): 1–7.
8 Koyama, F., Kinoshita, S., and Iga, K. (1988). Room temperature CW operation of GaAs vertical cavity surface emitting laser. *IEICE Transactions* E71 (11): 1089–1090.
9 Koyama, F., Kinoshita, S., and Iga, K. (1989). Room-temperature continuous wave lasing characteristics of GaAs vertical cavity surface-emitting laser. *Applied Physics Letters* 55 (3): 221–222.
10 Iga, K. (2018). *VCSEL Odyssey*. Optronics Pub. Co., Tokyo.
11 Kojima, K. (2016). Development of RGB laser backlit LCD. *Microoptics News* 34 (4): 25–30.

Index

Note: Figure page numbers styled in italics.

Amorphous Oxide Semiconductors: IGZO and Related Materials for Display and Memory, First Edition.
Edited by Hideo Hosono and Hideya Kumomi.